ST. MARY'S COLLEGE OF MARYLAND
ST. MARY'S CITY, MARYLAND
W9-AFC-613

THE PICTURE OF THE TAOIST GENII PRINTED ON THE COVER of this book is part of a painted temple scroll, recent but traditional, given to Mr Brian Harland in Szechuan province (1946). Concerning these four divinities, of respectable rank in the Taoist bureaucracy, the following particulars have been handed down. The title of the first of the four signifies 'Heavenly Prince', that of the other three 'Mysterious Commander'.

At the top, on the left, is Liu *Thien Chün*, Comptroller-General of Crops and Weather. Before his deification (so it was said) he was a rain-making magician and weather forecaster named Liu Chün, born in the Chin dynasty about +340. Among his attributes may be seen the sun and moon, and a measuring-rod or carpenter's square. The two great luminaries imply the making of the calendar, so important for a primarily agricultural society, the efforts, ever renewed, to reconcile celestial periodicities. The carpenter's square is no ordinary tool, but the gnomon for measuring the lengths of the sun's solstitial shadows. The Comptroller-General also carries a bell because in ancient and medieval times there was thought to be a close connection between calendrical calculations and the arithmetical acoustics of bells and pitch-pipes.

At the top, on the right, is Wên *Yuan Shuai*, Intendant of the Spiritual Officials of the Sacred Mountain, Thai Shan. He was taken to be an incarnation of one of the Hour-Presidents (*Chia Shen*), i.e. tutelary deities of the twelve cyclical characters (see p. 396). During his earthly pilgrimage his name was Huan Tzu-Yü and he was a scholar and astronomer in the Later Han (b. +142). He is seen holding an armillary ring.

Below, on the left, is Kou *Yuan Shuai*, Assistant Secretary of State in the Ministry of Thunder. He is therefore a late emanation of a very ancient god, Lei Kung. Before he became deified he was Hsin Hsing, a poor woodcutter, but no doubt an incarnation of the spirit of the constellation Kou-Chhen (the Angular Arranger), part of the group of stars which we know as Ursa Minor. He is equipped with hammer and chisel.

Below, on the right, is Pi *Yuan Shuai*, Commander of the Lightning, with his flashing sword, a deity with distinct alchemical and cosmological interests. According to tradition, in his earthly life he was a countryman whose name was Thien Hua. Together with the colleague on his right, he controlled the Spirits of the Five Directions.

Such is the legendary folklore of common men canonised by popular acclamation. An interesting scroll, of no great artistic merit, destined to decorate a temple wall, to be looked upon by humble people, it symbolises something which this book has to say. Chinese art and literature have been so profuse, Chinese mythological imagery so fertile, that the West has often missed other aspects, perhaps more important, of Chinese civilisation. Here the graduated scale of Liu Chün, at first sight unexpected in this setting, reminds us of the ever-present theme of quantitative measurement in Chinese culture; there were rain-gauges already in the Sung (+12th century) and sliding calipers in the Han (+1st). The armillary ring of Huan Tzu-Yü bears witness that Naburiannu and Hipparchus, al-Naqqāsh and Tycho, had worthy counterparts in China. The tools of Hsin Hsing symbolise that great empirical tradition which informed the work of Chinese artisans and technicians all through the ages.

SCIENCE AND CIVILISATION IN CHINA

PROBABLY another reason why many Europeans consider the Chinese such barbarians is on account of the support they give to their Astronomers—people regarded by our cultivated Western mortals as completely useless. Yet there they rank with Heads of Departments and Secretaries of State. What frightful barbarism!

FRANZ KÜHNERT (Vienna, 1888).

李約瑟著

中國科學技術史

冀朝鼎

39079

SCIENCE AND CIVILISATION IN CHINA

BY

JOSEPH NEEDHAM, F.R.S.

SIR WILLIAM DUNN READER IN BIOCHEMISTRY IN THE UNIVERSITY OF CAMBRIDGE, FELLOW OF CAIUS COLLEGE
FOREIGN MEMBER OF ACADEMIA SINICA

With the collaboration of

WANG LING, PH.D.

TRINITY COLLEGE, CAMBRIDGE
ASSOCIATE RESEARCH FELLOW OF ACADEMIA SINICA

VOLUME 3

MATHEMATICS
AND THE SCIENCES OF THE HEAVENS AND THE EARTH

CAMBRIDGE
AT THE UNIVERSITY PRESS
1970

PUBLISHED BY
THE SYNDICS OF THE CAMBRIDGE UNIVERSITY PRESS
Bentley House, 200 Euston Road, London, N.W.1
American Branch: 32 East 57th Street, New York, N.Y. 10022

© CAMBRIDGE UNIVERSITY PRESS 1959

Standard Book Number: 521 05801 5

First published 1959
Reprinted 1970

First printed in Great Britain at the University Press, Cambridge
Reprinted by photolithography in Great Britain
by Bookprint Limited, Crawley, Sussex

To

CHU KHO-CHEN

Vice-President of Academia Sinica, formerly Director of the Meteorological Institute of the Academy, and President of Chekiang University,
Hangchow, Tsun-i, Mei-than

deeply learned in the history of science among the Chinese people, and a constant encourager of the project of this book

and to

LI SSU-KUANG

Minister of National Resources and Director of the Chinese Geological Survey, Vice-President of Academia Sinica and formerly Director of its Geological Institute,
Nanking, Kweilin, Liang-fêng

the 土宿眞君 of our age

with gratitude and affection
this third volume
is dedicated

The Syndics of the Cambridge University Press desire to acknowledge with gratitude certain financial aid towards the production of this book, afforded by the Bollingen Foundation

CONTENTS

THE SCIENCES OF THE HEAVENS

2

LIST OF ILLUSTRATIONS

LIST OF TABLES

LIST OF ABBREVIATIONS

The following abbreviations are used in the text. For abbreviations used for journals and similar publications in the bibliographies, see pp. 686 ff.

B Bretschneider, E., *Botanicon Sinicum.*

B & M Brunet, P. & Mieli, A., *Histoire des Sciences* (*Antiquité*).

CCSS *Chiu Chang Suan Shu* (Nine Chapters on the Mathematical Art), completed +1st century.

CSHK Yen Kho-Chün (ed.), *Chhüan Shang-Ku San-Tai Chhin Han San-Kuo Liu Chhao Wên* (complete collection of prose literature (including fragments) from remote antiquity through the Chhin and Han Dynasties, the Three Kingdoms, and the Six Dynasties), 1836.

CTCS Li Kuang-Ti (ed.), *Chu Tzu Chhüan Shu* (collected works of the philosopher Chu Hsi).

G Giles, H. A., *Chinese Biographical Dictionary.*

HY Harvard-Yenching (Institute and Publications).

K Karlgren, B., *Grammata Serica* (dictionary giving the ancient forms and phonetic values of Chinese characters).

KCCY Chhen Yuan-Lung, *Ko Chih Ching Yuan* (Mirror of Scientific and Technological Origins), an encyclopaedia of +1735.

KCKW Wang Jen-Chün, *Ko Chih Ku Wei* (Scientific Traces in Olden Times), 1896.

MCPT Shen Kua, *Mêng Chhi Pi Than* (Dream Pool Essays), +1086.

N Nanjio, B., *A Catalogue of the Chinese Translations of the Buddhist Tripiṭaka*, with index by Ross (3).

NCNA New China News Agency.

P Pelliot numbers of the Chhien-fo-tung cave temples.

PTKM Li Shih-Chen, *Pên Tshao Kang Mu* (The Great Pharmacopoeia), +1596.

R Read, Bernard E., Indexes, translations and précis of certain chapters of the *Pên Tshao Kang Mu* of Li Shih-Chen. If the reference is to a plant, see Read (1); if to a mammal see Read (2); if to a bird see Read (3); if to a reptile see Read (4); if to a mollusc see Read (5); if to a fish see Read (6); if to an insect see Read (7).

RP Read & Pak, Index, translation and précis of the mineralogical chapters in the *Pên Tshao Kang Mu.*

S Schlegel, G., *Uranographie Chinoise*; number-references are to the list of asterisms.

SCTS *Chhin-Ting Shu Ching Thu Shuo* (imperial illustrated edition of the *Historical Classic*), 1905.

T Tunhuang Archaeological Research Institute numbers of the Chhien-fo-tung cave temples. In the present work we follow as far as possible the numbering of Hsieh Chih-Liu in his *Tunhuang I Shu Hsü Lu* (Shanghai, 1955) but give the other numbers also.

TH Wieger, L., *Textes Historiques.*

TKKW Sung Ying-Hsing, *Thien Kung Khai Wu* (The Exploitation of the Works of Nature), +1637.

TPYL Li Fang (ed.), *Thai-Phing Yü Lan* (the Thai-Phing reign-period (Sung) Imperial Encyclopaedia), +983.

TSCC *Thu Shu Chi Chhêng* (the Imperial Encyclopaedia of +1726). Index by Giles, L. (2).

TT Wieger, L. (6), *Tao Tsang* (catalogue of the works contained in the Taoist Patrology).

TW Takakusu, J. & Watanabe, K., *Tables du Taishō Issaikyō* (*nouvelle édition* (*Japonaise*) *du Canon bouddhique chinoise*), Index-catalogue of the Tripiṭaka.

YHSF Ma Kuo-Han (ed.), *Yü Han Shan Fang Chi I Shu* (Jade-Box Mountain Studio Collection of (reconstituted and sometimes fragmentary) Lost Books), 1853.

ACKNOWLEDGMENTS

List of Those who have kindly read through Sections in Draft

The following list, which applies only to this volume, brings up to date the list printed in Vol. 1, on pp. 15–16.

Prof. A. L. Basham (London)	Mathematics (Notations).
Prof. L. Bazin (Paris)	Astronomy (Calendar).
Dr A. Beer (Cambridge)	Mathematics, Astronomy and Seismology.
Prof. J. D. Bernal, F.R.S. (London)	All sections.
Mrs Margaret Braithwaite (Cambridge)	Mathematics.
Mr Robert Brittain (New York)	Geography.
The late Dr Herbert Chatley (Bath)	Astronomy.
Dr A. Christie (London)	Mathematics (Notations).
Prof. R. Cohen (Middletown, Conn.)	Mathematics (Concluding sub-section).
Rear-Admiral A. Day, Hydrographer of the Navy (London)	Seismology.
Dr D. W. Dewhirst (Cambridge)	Astronomy.
The late Dr W. N. Edwards (London)	Geology and Palaeontology.
Prof. V. Elisséeff (Paris)	All sections.
Sir Ronald Fisher, F.R.S. (Cambridge)	Mathematics.
Prof. W. Fuchs (München)	Geography and Cartography.
Dr A. R. Hall (Cambridge)	Astronomy.
Prof. D. G. E. Hall (London)	Mathematics (Notations).
Mr Brian Harland (Cambridge)	Geology and Mineralogy.
Dr K. P. Harrison (Cambridge)	Astronomy (Equatorial Mounting)
Prof. W. Hartner (Frankfurt a.M.)	Astronomy.
Dr Hsü Li-Chih (Cambridge)	Mathematics.
Mr P. A. Jehl (Paris)	Astronomy (Jesuit period).
Mr David H. Kelley (Jaffrey, N.H.)	Astronomy.
Dr Arnold P. Koslow (New York)	Mathematics.
Mr D. Leslie (Haifa)	Mathematics.
Dr Lu Gwei-Djen (Cambridge)	All sections.
Mr Scott McKenzie (Washington, D.C.)	Mineralogy.
Prof. K. Mahler, F.R.S. (Manchester)	Mathematics.
Prof. Gordon Manley (London)	Meteorology, Geography, Cartography.

Dr Stephen Mason (London)	Astronomy.
Mr Raymond Mercier (Cambridge)	Mathematics and Astronomy.
Mr Henri Michel (Brussels)	Astronomy and Meteorology.
Mr J. V. Mills (Richmond)	Geography and Cartography.
Mr Nakayama Shigeru (Tokyo)	Astronomy.
Dr Dorothy M. Needham, F.R.S. (Cambridge)	All sections.
Dr K. P. Oakley (London)	Geology, Palaeontology, Seismology.
Dr F. Parker-Rhodes (Cambridge)	Mathematics.
Prof. J. R. Partington (Cambridge)	Mineralogy.
Prof. Luciano Petech (Rome)	All sections.
Dr Derek Price (Washington, D.C.)	Astronomy.
Prof. R. A. Rankin (Glasgow)	Mathematics.
Dr Jerome Ravetz (Leeds)	Mathematics.
Prof. Keith Runcorn (Newcastle)	Mathematics.
Dr R. W. Sloley (Amersham)	Astronomy (Clepsydras).
Prof. E. G. R. Taylor (London)	Geography and Cartography.
Dr D. Twitchett (Cambridge)	Geography, Geology and Mineralogy.
Dr F. P. White (Cambridge)	Mathematics.
Dr W. A. Wooster (Cambridge)	Mineralogy.
Dr Wu Shih-Chhang (Oxford)	Mathematics (Notations).
Prof. A. P. Yushkevitch (Moscow)	Mathematics.

AUTHOR'S NOTE

With this volume we reach the ore of the work as a whole, and leave behind all tunnels and adits, all introductory explanation and interpreting. The purpose of this volume is to elucidate the contributions of traditional Chinese civilisation to mathematics and to the sciences of the heavens and the earth—astronomy and meteorology above, geography and geology below. The facts which have been here assembled may at first sight seem a little bewildering, but one must remember that they concern the culture of more than one-fifth of the human race, a people inhabiting for three millennia a land at least as large as Europe, and certainly no less gifted than others. Those who are best acquainted with the history sketched in this volume will feel its inadequacy rather than its wealth.

Yet as on a previous occasion we have in mind the needs of those whose time, often perhaps limited by active laboratory work, will not permit of extended study. Pointers are needed to guide their curiosity. A modern scientist may approach the history of science in China for at least four reasons. First, one may be interested in the nodal points of discovery and invention, the acts which have left permanent mark on the edifice of human knowledge. Thus this volume has something to say of the development of place-value in computing (Section 19*b*), the formulation of the triangle of binomial coefficients (19*i*, 9), the charting of the positions of stars (20*f*), the invention of the equatorial mounting and the clock-drive of telescopes (20*g*, 6), the setting up of the first seismograph (24*b*), and the beginnings of bio-geochemical prospecting (25*g*). Secondly, one may follow a more ethnological trail, curious to find out how science could grow up in a civilisation so deeply different from that of Western Europe. Hence there is interest in strange forms of algebraic notation (Section 19*i*, 8), in the polar-equatorial system of mansions of the moon (20*e*), quite unlike Graeco-Egyptian ecliptic astronomy, and in an oriental geographic tradition (22*d*, 5) far advanced beyond that of the Latin West. Thirdly, one may wish to explore contacts and transmissions, seeking to cast up a balance-sheet of indebtedness among the cultures of the Old World. Here something at least can be said regarding the travels of mathematical problems and methods (Section 19*j*), of the lunar mansions (20*e*, 3), of astronomical instruments (20*g*, 6, iv), and of cartographic techniques (22*h*). As aids to thought on these subjects we have prepared certain comparative charts (Tables 37 and 40). Finally there are many who realise that the ancient and medieval Chinese records of celestial and terrestrial phenomena, covering many centuries for which we have little other information, can still be of great value in current research, e.g. in radio-astronomy or meteorology. These we would refer to the Sections most relevant (20*i*, 21*b*, *d*, *h*, etc.).

One subject will be of common interest to all readers, namely the relation of mathematics to science in East and West. Can the study of Li Shih-Chen's China throw any light upon the birth of modern science in Galileo's Italy? This question is discussed in Section 19*k*, where we define the highest style of indigenous Chinese scientific and technical achievement as Vincean rather than Galilean, showing also that before 1600 there were in China no less than in Europe two groups which possessed portions of Galileo's method, the higher artisanate and the scholastic philosophers. Further precision on the social processes among which science developed in East and West we must leave to the concluding volume.

The interest of humanists may well take diverse forms analogous to those which we have sketched above. A particular disadvantage is theirs, however; unfamiliarity with the technical terms commonly used in science and its applications. The spectrum of specialisation being what it is, we could not hope to please everyone, and have had to be content with a rather arbitrary choice between terms which we would explain and terms which we would assume comprehensible. The difficulty is to decide what constitutes 'matter of general knowledge'. Thus we pass over such words as 'protein', 'crankshaft', 'anticline' and 'Vernier scale', but we devote some space to the definitions of the main features of spherical astronomy and we explain such phrases as 'the establishment of the port' or the 'Goldschmidt enrichment principle'. Inevitably some readers who mean to look into this subject will need to have a scientific dictionary handy, and its size will simply have to be in proportion to the purity of their humanistic learning.

In spite of this, we deeply hope that humanists and all of general culture will be attracted to a hitherto unopened page in the history of man's natural knowledge. Such studies are the only means of seeing the scientific activities of our own time in true perspective, one of the most useful ways of humanising technical education, and an indispensable part of the history of civilisation as a whole. The facts presented in these volumes simply go to show that in the history of science, as in other things, Europe cannot be thought of in isolation from the rest of the Old World. In an age of continually narrowing dimensions, sympathetic appreciation of the achievements and modes of life of cultures other than one's own is our *Quicunque Vult*.

Mention of technical terms raises a point of much significance. The first question which will occur to anyone who turns over the leaves of our book is this; how can the essential technical terms be recognised in their Chinese form and understood? One of our scholarly correspondents, writing to us on iron and steel technology in ancient and medieval China, asked what evidence there was that the terms for cast iron, wrought iron, or steel, could really be identified in ancient texts. Were we not reading meanings back into antiquity; interpreting ancient words too much in the light of modern knowledge? The answer is important. One must realise that there is an absolutely continuous tradition between the Chinese written language as found on the oracle-bones of the −14th century and the language as written and spoken today. Analogy with Sumerian or ancient Egyptian is therefore not valid; Hebrew doubtfully competes. The ancestry of many of the simpler technical terms begins with their bone forms. Again, the ancient pictograms which were used before the script became stylised and standardised often betray a technical characteristic. Thus the ancient form of the word *chou*,[1] boat, depicts the perennial Chinese transom-and-bulkhead

[1] 舟

construction, without sign of stem-post, stern-post or keel.[a] The ancient form of the word *kung*,[1] the bow, duly depicts the reflex composite bow.[b] And all this is applicable, *mutatis mutandis*, to the purely scientific words with which the present volume is concerned—for example the array of technical terms for the phenomena associated with mock suns and solar haloes (Section 21*e*).

In Chinese too we have to deal with a continuous lexicographical tradition going back at least to the −3rd century. The scholars of the Chi-Hsia Academy (founded in −318, just after the death of Aristotle),[c] or those who wrote the *Lü Shih Chhun Chhiu* (−239), or the editors of the *Artificers' Record* of the State of Chhi (*c.* −260),[d] frequently defined their terms, or used them in unmistakable contexts. Hsü Shen's dictionary, the *Shuo Wên Chieh Tzu* of +121, is as usable today as then. We know the technical terms for all the parts of the very complicated bronze crossbow triggers of the Han[e] partly because Liu Hsi described and named them with perfect clarity in his *Shih Ming* of +100. Indeed we find from time to time whole clusters of technical terms which mutually illuminate each other. Thus the description of the water-driven mechanical clock set up at Khaifêng by Su Sung and his collaborators in +1090 (the *Hsin I Hsiang Fa Yao*) has yielded more than 140 technical terms which go together in a veritable blue-print of Sung engineering.[f]

Daunting difficulties of course remain. A discovery or an invention may appear under various terminological disguises. Worse still, the term may continue though the thing changes. *Thung*[2] meant copper before it meant bronze.[g] The word *tho*[3] was assuredly used first of all to designate the steering-oar, but in medieval times it certainly meant the hinged rudder[h]—when did the invention occur? To cite a case from the present volume, the expression *hun hsiang*[4] in the Han meant a demonstrational armillary sphere with a model earth at the centre, but by the middle of the +5th century it certainly meant a solid celestial globe[i]—when did the change come about? Such problems can only be solved by the comparison of as many texts as possible. A probable conclusion, if not certainty, usually emerges. Mistakes which have been made in the past about technical terms in Chinese texts have generally been due to the fact that scholars had neither the desire nor the time nor the necessary knowledge of the natural sciences to pin down the terms in this way. But the present generation of sinologists is rapidly repairing this position.

In some fields, such as alchemy and pharmacy, Chinese writings contain the same bewildering abundance of synonyms as in the West, and no doubt for the same reason, precisely in order to bewilder the uninitiated. But the Confucian spirit of practical organisation could never let Taoist mysticism prevail, and so we find (what Anglo-Saxon England could hardly boast of) an admirable synonymic dictionary of minerals

[a] See Section 29.
[b] See Section 30*e*.
[c] Cf. Vol. 1, p. 95. It is likely that the *Kuan Tzu* book stems from them.
[d] The *Khao Kung Chi* incorporated in the *Chou Li*, cf. Vol. 1, p. 111.
[e] See Section 30*e* below.
[f] See Section 27*h* below.
[g] Sect. 36 will treat of this.
[h] Sect. 29*h*.
[i] Cf. pp. 382 ff. below.

[1] 弓 [2] 銅 [3] 柁 [4] 渾象

and drugs, the *Shih Yao Erh Ya*, compiled by Mei Piao as early as +818 and still useful today.[a] Japanese medicine followed suit, and Fukane no Sukehito's *Honzō Wamyō* of +918 also remains to guide us. Here again the juxtaposition of many texts is the only way—so far as we are aware, the expression *huo yao*[1] never once occurs in such a connection that it cannot mean that mixture of various proportions of sulphur, saltpetre and charcoal which in the medieval time it undoubtedly does.[b] Similarly, the *hou fêng ti tung i*[2] means the seismograph and never any other instrument.[c] At the end of the period of traditional science, analyses of Chinese drugs and minerals by 18th-century and modern chemists fixed the meanings of terms in such a way that we can now trace them far back into the past.[d] *Tshêng chhing*,[3] for instance, means malachite or copper carbonate, and as the line of pharmacopoeias fully parallels the tradition of the lexicographers, a term which was already fixed in the Han will not easily admit of other interpretations later. If *sêng thieh*[4] does not mean cast iron, if *shu thieh*[5] does not mean wrought iron, if *kang*[6] does not mean steel (as indeed they all still do in China today), none of the texts makes sense; conversely all make sense.[e]

'Interpreting the text in such a way', went on our correspondent, 'makes it sensible, but *was* the text sensible on a modern view?' We answer that ancient and medieval Chinese secular texts are always sensible, rational and comprehensible, if not too corrupted by copyists. A very distinguished critic complained of our second volume that it made the sayings of ancient Chinese philosophers sensible. For philosophers of course we would hardly dare stand guarantee, but where the practical men were concerned, the reckoners and star-clerks, the leeches, the miners and the ironmasters, there can be no possible doubt. If one cannot be sure of the details of the Mohist specifications for crossbow-artillery,[f] if some of the mathematical and astronomical methods of the Chhin and Han escape our penetration, it is because age has almost irretrievably jumbled the words. Even then, if one can once be quite sure what the ancient author was talking about, the whole pattern becomes clear and emendations may follow of themselves; such was the case, for example, with the root extraction methods of the *Chiu Chang Suan Shu*.[g] If after the Han there is any barrier it is because time's tooth has eaten the bamboo tablets and the paper scrolls, so that of such a man as Phei Hsiu, China's geographical Ptolemy (+224 to +271), we have but fragments.[h] Chinese scholars have lovingly edited many such, and have accomplished a mighty work in the establishment of correct texts. Moreover, many books survive in full from the Han onwards, and when we read such a work as the agricultural treatise of Chia Ssu-Hsieh, the *Chhi Min Yao Shu*, written about +450, we are astonished at the clarity of his exposition.[i]

[a] Cf. p. 644 below.
[b] See Section 30.
[c] Cf. p. 627 below.
[d] See Section 33.
[e] See Section 30*d*.
[f] See Section 30*h*.
[g] Cf. p. 66 below.
[h] Cf. p. 538 below.
[i] This book has recently been newly edited by my old friend Professor Shih Shêng-Han. It figures prominently in Sections 40 and 41 below.

[1] 火藥 [2] 候風地動儀 [3] 曾青 [4] 生鐵 [5] 熱鐵 [6] 鋼

All translations, someone said, are like stewed strawberries. Though there is no substitute for original freshness, we have done our best to use the deep-freeze technique, so that a quotation may come to life as much as possible when warmed by the friendly imagination of the reader. If many brackets are interspersed among our translated sentences, this is not because of excessive vagueness in the text itself, but because Indo-Europeans have to supply inflections and to add particles and other words, with which the laconic Chinese dispenses, to make the passage readable. Grammar and sentence-construction differ. Here considerable judgment, based on the knowledge of many parallel and relevant texts, may sometimes be required. Yet far from being vague, many of the Chinese statements are marvels of crystalline compression (for outstanding examples cf. Vol. 2, p. 482, and here, p. 432). Although so much is said of the ambiguity of classical Chinese we can remember surprisingly few passages where indelible doubt remained concerning the nature of the scientific proposition which the original intended to convey, or the technological process which it was discussing; parallel passages always being taken into account. Of course the information given is sometimes not as adequate as one would like, and the epigrammatic brevity of Confucian scholars has its faults where practical matters are concerned.

Though all translators may be traitors, the duty of translation has at least the merit that it forces decisions about meanings, provisional though they may have to be. A historian writing in his own language may quote an ancient text as general evidence for his argument without really explaining it or elucidating its technical terms. But when such a passage has to be translated into another language it is no longer possible to proceed in this way. When Chang Hêng is being englished—or Aristotle sinified—meanings and implications can no longer be left open; all linguistic interpretation is in fact necessarily the exposition of content. Indeed one might say that many technical texts of great interest first found their voice only in having to put on a different dress. Chinese scholars, newly conscious of this, are now beginning to make admirable versions in modern Chinese of some of their own ancient and medieval works.

One minor matter may also be mentioned. Everyone who translates Chinese texts into other languages meets with difficulty in rendering the numerous official titles therein. As yet there is no system of translation generally accepted for any dynastic period, though the current labours of sinologists are leading to this desirable end. In the meantime we assume that the titles of officials in the bureaucratic administration through the ages were meaningful, and as rational as any such human systems ever are. For this reason we choose to err on the side of modernisation, hoping thus to present the life of ancient and medieval China to the Western reader with a minimum of the foreign, the archaic, and the quaint. Thus, to take a pertinent example,[a] we use 'Astronomer-Royal' for *Thai Shih Ling*[1] because star-clerks did have a high official position in the bureaucratic hierarchy in China from a very early date, and from the beginning they carried out much good scientific astronomical work. This was doubtless because the calendrical needs of a great agricultural civilisation with an uncertain rain-

[a] Cf. pp. 190ff. below.

[1] 太史令

fall were at least as important as the political prognostications which rulers expected of State astrologers. So much for technicalities and translations.

Although every attempt has been made to take into account the most recent research in the fields here covered, we regret that it has generally not been possible to mention work appearing after December 1956.

This opportunity must not be lost of discharging (as far as is possible) certain accumulated debts. We are deeply grateful to a surrounding circle of experts upon whom we constantly rely for advice—Mr D. M. Dunlop for Arabic, Prof. Honda Minobu for Japanese, Dr Shackleton Bailey for Sanskrit, Mr R. L. Loewe for Hebrew, and Professor Reuben Levy for Persian questions. Their unstinting assistance and never-failing kindness cannot too highly be valued. Moreover, certain readers of the draft Sections have carried their help beyond what might be thought reasonable, and in particular therefore we should like to thank Professor K. Mahler, F.R.S., who not only raised many queries in the mathematical Section, but even undertook special researches where obscurities needed to be illuminated. Standing in the same relation to the astronomical Section is Dr Arthur Beer of the Cambridge Observatory, the genial editor of *Vistas in Astronomy*. To hear his familiar step on the threshold bringing new references or the solutions of puzzling problems has been inspiration as well as assistance. Moreover, in many aspects of the history of astronomical instruments the close collaboration of Dr Derek Price has been particularly precious.

Still other debts remain to be mentioned. Mr Derek Bryan, O.B.E., has shouldered a great burden of press work for us during the printing of the present volume. Mr Charles Curwen of Collet's Chinese Bookshop, formerly of the Baillie School at Shantan, has done us a valuable service in watching for the latest Chinese publications in the history of science and ensuring that they reach us. To Miss Muriel Moyle as before we are deeply indebted for the preparation of the elaborate index. Fr. Kenelm Foster, O.P., kindly assisted us with the tracing of quotations from the works of the European scholastic philosophers. And others, too numerous to be recited, have most kindly called attention, whether as reviewers or correspondents, to desirable emendations in the two preceding volumes. While the more elaborate of these must necessarily await a second edition, we are publishing a list of errata at the end of the present volume.

Acknowledgment of another kind is owing to Dr and Mrs Charles Singer and to Dr and Mrs Ludwik Rajchman. At Kilmarth near St Austell in Cornwall numberless ideas in the generation of this book took shape in the long library overlooking the sea, by the inspiration and kindness of its ever-hospitable inhabitants. For a Chinese, too, from Chiangsu on the shore of the Pacific, it was a reception at the court of Hsi Wang Mu, a palace of learning in Ultima Thule. The correction of proofs may be a weariness of the flesh, but not if it is done in circumstances and company so charming as those which are found at La Fosse Beauregard near Chenu in the Sarthe.

One person also there is who has read every word of this as well as of the preceding volumes—and even the legend of every illustration—Dr Dorothy Needham, F.R.S.—indeed, without her benign encouragement and moral support this volume would never have existed at all. And once again our deepest gratitude is due to the Syndics and Staff of the Cambridge University Press for all that they are contributing to the successful achievement of our enterprise. It is customary for them to observe an austere anonymity, but from Mr Peter Burbidge we have long received friendly help and co-operation in such generous measure that no convention could justly prevent an acknowledgment of warmest appreciation. And once again I am happy to thank the members of my own Faculty and Department, represented in particular by Professor F. G. Young, F.R.S., for that consistent sympathy and understanding without which the present work would have been quite impossible.

Last, but far from least, comes finance. Besides acknowledgment to the Bollingen Foundation elsewhere made, our grateful thanks for grants-in-aid are due to the Universities China Committee, to Gonville and Caius College, and to the Managers of the Ocean Steam-ship Company acting as Trustees of funds bequeathed by members of the Holt family. These means have been generously enlarged by the Wellcome Trust in grants which are enabling research to proceed in the fields which will be covered by the biological and medical volume.

19. MATHEMATICS

(*a*) INTRODUCTION

WITH this Section we enter upon the second half of the present work. Since mathematics and the mathematisation of hypotheses has been the backbone of modern science, it seems proper that this subject should precede all others in our attempt to evaluate Chinese contributions in the many specific sciences and technologies. Our immediate task will be an assessment of the Chinese achievement in this field. Opinions of historians of science hitherto have often oscillated between two extremes—Sédillot,[a] for instance, in 1868, reacting against traditional 18th-century sinophilism and the impossibly early datings of Chinese mathematical and astronomical works by J. B. Biot, was prepared to say (though without the authority which any acquaintance with texts might have given him) that the Chinese never did anything worth while in mathematics, and that such knowledge as they had was transmitted from the Greeks.[b] Later on, writers such as van Hée, in whom sinological competence wrestled with missionary disapproval, again insisted that the principal mathematical books of the Chinese were all inspired by foreign influence. But this point of view has gone by no means unopposed. How far away from the truth it always was will be apparent to anyone who reads through the present Section.

There is a very large literature on the history of mathematics in East Asia, though unfortunately (for most Westerners) by far the greater part of it is in the Chinese and Japanese languages. Those who are debarred from this original material inevitably have recourse to the well-known histories of mathematics in Western languages, such as those of Cantor (1), Loria (1), Cajori (2), D. E. Smith (1), and Karpinski (2). Cantor's famous work is now old (1880); he had to rely on the still earlier translations of E. Biot, while his other main source was a paper by Biernatzki of 1856. This, however, was a mere translation of the work of Wylie (4) of 1852—an excellent account which can still be read with profit today.[c] The most succinct modern description of Chinese mathematics in its historical context is that of Cajori (2), while the fullest is that in Smith (1), who arranged his two volumes chronologically in the first and according to subjects in the second.[d]

The work of all these scholars was vitiated by the fact that none of them possessed sufficient Chinese to permit any first-hand contact with the texts themselves.[e] This

[a] (2), vol. 2, p. xiii; (4).

[b] Some well-known historians of mathematics such as Rouse Ball (1) took Sédillot *au pied de la lettre* and are therefore of no use to us.

[c] Had the translation been accurate, Cantor would have been spared at least one serious mistake (see below, p. 121). Wylie (13) also prepared a glossary of mathematical terms.

[d] To these publications may with advantage be added the recent outlines of the history of mathematics by Archibald (1) and van der Waerden (3), though they exclude the Chinese contribution. Shorter articles, worth reading if nothing else is available, are those of D. E. Smith (2, 3) and Vacca (3). Those of Loria (2, 3) are simply reviews of the papers of van Hée.

[e] They were, in fact, sinologically at sea. They would cheerfully put the *Chou Pei* 1000 years too early, and at the same time cast doubt on the best-authenticated Sung texts. Loria's book (1) is, indeed, so misleading as to be almost useless. He suffered from an invincible suspicion that the Chinese *must* have borrowed all their ancient mathematical techniques from the West. The title of his relevant section 'l'Enigma Cinese' suited him well enough, but it will not do for us. See on him, Mikami (*4*).

criticism bears least forcibly upon D. E. Smith,[a] who himself spent some time in China and Japan, made collections of mathematical books there, and had the advantage of intimate collaboration with Asian mathematicians, notably Mikami Yoshio. Such collaboration is also evident in the distinguished recent contribution of A. P. Yushkevitch in Russian.

It is to the Japanese scholar that we owe a peculiarly important piece of work, *The Development of Mathematics in China and Japan* (1913), indispensable for any study of the subject. Mikami was the only historian of mathematics who, while saturated in the Chinese and Japanese texts themselves, yet had sufficient command of an occidental language to express himself in it with comparative ease and comprehensibility.[b] Whatever criticisms may have been levelled, therefore, against Mikami's judgment, the fact remains that he occupied a position or vantage-point quite unique in the field, the only possible comparison being Alexander Wylie in an earlier generation.[c] The best parallel to Mikami's book is the even less known monograph of Hayashi Tsuruichi which appeared in English in a Dutch journal, but this confines itself to Japanese mathematics.

Mention of Mikami's other books brings us into the realm of the history of mathematics in Chinese and Japanese. Originally written in the latter language, his *Chung-Kuo Suan-Hsüeh Thê-Sê* (Special Characteristics of Chinese Mathematics) is valuable.[d] He contributed the mathematical section (*Kagaku; Sūgaku*) to the large Japanese collective work, *The Trends of Oriental Thought*.[e] More recently Yabuuchi (*3*) has given us a valuable review.

Among Chinese historians of mathematics, two have been particularly outstanding, Li Nien and Chhien Pao-Tsung. The work of the latter, though less in bulk than the former's, is of equally high quality. Like D. E. Smith, Li Nien found it convenient to adopt a chronological and a classificatory treatment in different works. The first is found in his *Chung-Kuo Shu-Hsüeh Ta Kang* (Outline of Chinese Mathematics in History).[f] More complete treatment is found in his *Chung-Kuo Suan-Hsüeh Shih* (A History of Chinese Mathematics), abridged in *Chung-Kuo Suan-Hsüeh Hsiao Shih* (Brief History of Chinese Mathematics). The second method, that of choosing a number of topics for discussion, was adopted by him in his four-volume *Chung Suan Shih Lun Tshung* (Gesammelte Abhandlungen ü.d. chinesische Mathematik), now continued in a new five-volume series.

[a] But even Smith accepted impossibly early dates for the oldest Chinese mathematical books.

[b] Unfortunately, Mikami, like all the other historians of mathematics mentioned, staged a veritable ballet of fantastic romanisations of Chinese names, disguising them almost unrecognisably. I must also say that in spite of the presumed precision of the mathematical art, I have met in no other group of books with such an abundance of misprints and mistakes. This applies equally to works in Chinese and in Western languages.

[c] In mentioning the contributions of Wylie, the considerable bibliographical material in his *Notes on Chinese Literature*, pp. 90ff., should not be forgotten.

[d] Cf. Mikami (*12*); Fujiwara (*1*).

[e] Mikami died in 1950; life and bibliography by Ogura & Ōya (*1*) and Yajima (1). Autobiographical study, Mikami (*3*). Yajima (1) informs us that an unpublished manuscript of Mikami's on the history of mathematics in China, amounting to more than 1000 pages, still exists. Great responsibility rests with the Japanese National Academy to ensure that this work of scholarship is given to the world without delay.

[f] Only the first volume was ever published. It brings the story down to the Yuan (+14th century).

The principal work of Chhien Pao-Tsung is his *Chung-Kuo Suan-Hsüeh Shih* (A History of Chinese Mathematics), but there is also a shorter *Ku Suan Khao Yuan* (Über den Ursprung der chinesischen Mathematik). The treatment in the former is mainly chronological, coming down to the middle of the Ming; that in the latter topical, ending with the algebraists of the Sung, especially Chu Shih-Chieh. The publications of Hsü Chhun-Fang (*1–4*) are also of interest, and we shall often have occasion to quote the papers of Yen Tun-Chieh.

Some idea of the wealth of material in Chinese on this subject can be gained from recent bibliographies.[a] A list of articles on the history of Chinese mathematics[b] gives thirty-three important studies in the decade 1918–28. From the bibliography of Li Nien & Yen Tun-Chieh the number must have been about the same in the following decade, but between 1938 and 1944 it rose to sixty, and the most recent lists of Li Nien (*19*, *5*) give 104 between 1938 and 1949. Unfortunately, the great majority of these papers are in journals which have never been available in Western Europe, and they would not be very easy to collect even in China, unless one were to devote a great deal of time and effort to assembling this literature, as Li Nien did.[c]

While the history of mathematics in Japan is somewhat outside our province and in any case did not really begin until the end of the +16th century, it should be mentioned that a useful book by Smith & Mikami (in English) exists on the subject.[d] In Japanese may be mentioned the older work of Endō and the more recent résumé by Hosoi.[e] The lecture by Harzer (in German) is also worth consulting.

Mention has already been made[f] of the great collection of biographies of mathematicians published by Juan Yuan[1] in +1799, the *Chhou Jen Chuan*.[2] The analysis of this by van Hée (10) is not entirely without value, though the comments on it by Mikami (4, *5*) which were prompted by its mistakes, should also be studied.[g] The word *chhou* (K1090*l*) came to mean surveyor,[h] and then later on was applied to all computers (*chhou jen*), especially those surveyors of the heavens, the astronomers. Accordingly, the content of these biographies is concerned more with calendrical science than with mathematics as such.

In the previous volume the ideographic etymology of the chief word for 'calculation' was considered.[i] Of this character, *suan*,[3] no bone or bronze forms are known,[j] so

[a] Cf. Li Nien (*14*) for a list of the Chinese historians of mathematics, and (*15*) for an account of the progress of the last thirty years. Yen Tun-Chieh (*4*) describes collections of mathematical books in Shanghai.

[b] Prepared by one of us (Wang Ling), June 1945.

[c] Probably the greatest eighteenth-century Chinese mathematical bibliography is the *Li Suan Shu Mu*[4] of Mei Wên-Ting[5] (see below, p. 48). See now also that of Ting Fu-Pao & Chou Yün-Chhing (*2*).

[d] The authors seem to have assumed that no further work would ever be done on the subject, for they gave no characters (except haphazardly in the illustrations). A summary of his own studies is given in Mikami (*21*).

[e] Cf. Fujiwara (*2*). A bibliography of books and papers both in Japanese and European languages has been prepared by Mr Donald Leslie.

[f] Sect. 3*c*, Vol. 1, p. 50.

[g] On van Hée see also Li Nien (*12*).

[h] Like the *harpedonaptae* (rope-stretchers) of ancient Egypt (Gandz, 3). See the New Kingdom fresco in Klebs (3), p. 7, fig. 5. On the traces of land-mensuration in some technical terms of Chinese mathematics, see Wang Ling (2), vol. 1, pp. 132ff. Cf. p. 95 below.

[i] Vol. 2, p. 230, in Table 11. Other relevant words are also dealt with there.

[j] Or at least none has yet been recognised.

[1] 阮元 [2] 疇人傳 [3] 算 [4] 曆算書目 [5] 梅文鼎

it may not be older than the time of Li Ssu (−3rd century). While the suggestion that the later written form (see diagram) embodies an ancient (pre-Han) graph of an abacus[a] must certainly be rejected, it may well represent the counting-board[b] with its horizontal lines. There are two less usual ways of writing the word (*suan*[1,2]). The first of these derives from the older written form, the central part of which, though thought by Hsü Shen[c] to represent jade, is more probably a tally or a few counting-rods.[d] Hsü Shen connected it with the word *lung*,[3] 'to play with something', saying that what it signified was indicated by the bamboo radical in it (Rad. 118), namely, counting-rods 'six inches long, for calculating calendar and numbers'.[e] The second *suan* (K175), apparently late, is simply a duplication of Rad. 113, *shih* (K553), meaning 'to show, demonstrate, inform, reveal, as in a numinous manifestation of the divine'. This usage must surely derive from the association of calculation with the prognostication of fates.

K 173

Such an association, which is rather old, is evidenced by the constant use of the two words *suan*[4] and *shu*[5], in ancient and medieval texts, with the sense of foretelling the future. To give a single example, when the *Hsi Ching Tsa Chi* (Miscellaneous Records of the Western Capital)[f] says of the Later Han scholars Huangfu Sung,[6] Chen Hsüan-Thu[7] and Tshao Yuan-Li,[8] that they were expert in 'arithmetic art' (*suan shu*[9]), it is clear from the context that what they were able to do was to predict the length of their own lives and those of others. They do not, therefore, belong to the history of mathematics. It by no means follows, however, that further investigation of ancient Chinese divination procedures would be fruitless for the history of mathematics. What Shen Kua in the +11th century called *nei suan*[10] is still an unexplored field. We ourselves have watched blind soothsayers rapidly reckoning on their own finger-joints while divining the birth-year of their client. Some use the abacus. The arts of these men may in the past have involved a certain empirical knowledge of permutations and combinations, especially for use with the interlocking sexagesimal cogwheels of the Chinese calendar system.[g] Perhaps it was no coincidence that the monk I-Hsing in the Thang (late +7th century) won a great reputation both for combinatory calculations and for fate prognostication. Here is another proto-science which invites historical study.

[a] See below, pp. 74 ff.
[b] See below, pp. 62 ff. Or perhaps a field.
[c] The father of Chinese lexicography, *d.* +121 after finishing the *Shuo Wên*.
[d] See below, pp. 70 ff.
[e] This definition was taken from the *Chhien Han Shu*, ch. 21 A, p. 2*a*. It therefore probably goes back to Liu Hsin.
[f] Probably +6th century. Ch. 4, pp. 1*b* ff.
[g] See below, pp. 396 ff.

[1] 筭 [2] 祘 [3] 弄 [4] 算 [5] 數 [6] 皇甫嵩 [7] 眞玄菟
[8] 曹元理 [9] 算術 [10] 內算

(b) NUMERAL NOTATION, PLACE-VALUE AND ZERO

The basic information which we need[a] is summarised in Table 22. Though column A is described as containing the 'modern' forms, these are really ancient and medieval too; they are the numerals in written character as stabilised since the Chhin and Han[b] (−3rd century onwards). Their romanised pronunciations accompany them, and these all equally apply to the so-called 'accountants' forms' set forth in column C (*ta hsieh shu mu tzu*[1]). These more complex characters, which came gradually into use during and after the Han (−1st century),[c] were considered more elegant, and also less liable to falsification, but naturally they are little found in mathematical texts. The numbers in columns B and D give references to the etymological dictionary of Karlgren (1) for the standard and accountants' forms respectively.[d] For the lower digits, the characters in column A are unquestionably pictograms, but from the number 4 onwards there seems to have been a borrowing of homophones from primitive botanical and zoological nomenclature.[e] Next, columns E, F and G give the numerals as seen in the oracle-bone inscriptions (−14th to −11th centuries), and in the inscriptions on bronze vessels and coins of the Chou period (−10th to −3rd centuries).[f] Some of the forms seen are closely related to the 'counting-rod' characters or numerals of columns H and I, supposed to originate (and surely they did) from arrangements of actual calculating rods laid out on a flat board.[g] This trend is already evident for numbers between 11 and 14 on the oracle-bones (see Table 23), and for numbers less than 10 in the forms found on coins from the Warring States period.[h] All the later notations follow the counting-rod system.[i]

[a] There is a special monograph on the numeral notations by Glathe (1), but it is not very inspired.

[b] With the important exception of the zero, which we shall discuss shortly (pp. 10 ff.).

[c] Some of them indeed had originated long before, as was shown by a +14th-century scholar, Pai Thing,[2] whose interest was attracted to the question (*Chan Yuan Ching Yü*, ch. 1, p. 2*b*). For example, the full character for 1 appears already in the preface to the Book of Odes by Mao Hêng (−3rd century), and that for 2 is in Mencius (−4th).

[d] On etymologies see also Hopkins (14, 35), with the usual reservations.

[e] Four possibly from rhinoceros, six perhaps from mushrooms, one hundred perhaps from cypress-cone, ten thousand certainly from scorpion, and symbolising the multitudinous nature of insects (cf. Vol. 1, p. 30). The late graphs for 1000, depicting a man with a stroke across one leg, perpetuate the most ancient, oracle-bone, form (see Table 23 below).

[f] We are greatly indebted to Dr Wu Shih-Chhang for checking these lists according to the investigations of Kuo Mo-Jo (*3*), Sun Hai-Po (*1*), and other specialists. For the philology of the oracle-bone forms see further the compilation of Chu Fang-Pu (*1*). The lists given by Li Nien (*2*), p. 2, are wrong for the figure 9 in both cases.

[g] We here define counting-rod numerals as those sets in which the numbers 6–9 are indicated by the addition of one or more strokes at right angles to the original strokes. At present the earliest epigraphic evidence for these is −4th century (Wang Ling (2), vol. 1, pp. 83 ff.).

[h] Cf. the descriptions by de Lacouperie (2; 4, pp. 19, 122, 302, 311, 321, 329, 368); Wang Yü-Chhüan (1); Schjöth (1).

[i] Column J gives the forms which the visitor to China today will probably find on his restaurant bill. They are known by the name of *ma tzu*[3] or *an ma tzu*[4] (confidential weight numerals), and do not appear in print before the *Suan Fa Thung Tsung* of +1593 (see below, p. 51). They were associated with the formerly great commercial city of Suchow in Chiangsu. The curious form for 10,000 has been said to date only from the Thang, but one may see it on Chou knife-money (*Ku Chhüan Hui*, pt. 2, ch. 2, p. 8*b*).

[1] 大寫數目字 [2] 白珽 [3] 碼子 [4] 暗碼字

Table 22. *Ancient and medieval Chinese numeral signs.*

	A		B	C	D	E	F
	Standard modern forms			Accountants' forms		Shang oracle-bone forms (–14th to –11th centuries)	Bronze and coin forms (–10th to –3rd centuries)
1	一	*i*	395	弌 or 壹	395		
2	二	*erh*	564	弍 or 貳	564		
3	三	*san*	647	叁	647		
4	四	*ssu*	518	肆	509*h*		
5	五	*wu*	58	伍	58		
6	六	*liu*	1032	陸	1032*f*		
7	七	*chhi*	409	柒	—		
8	八	*pa*	281	捌	281		
9	九	*chiu*	992	玖	—		
10	十	*shih*	686	拾	—		
100	百	*pai*	781	佰	781	See Table 23	See Table 23
1,000	千	*chhien*	365	仟	365		
10,000	萬	*wan*	267	萬	267		
0	零	*ling*	—	零	—		

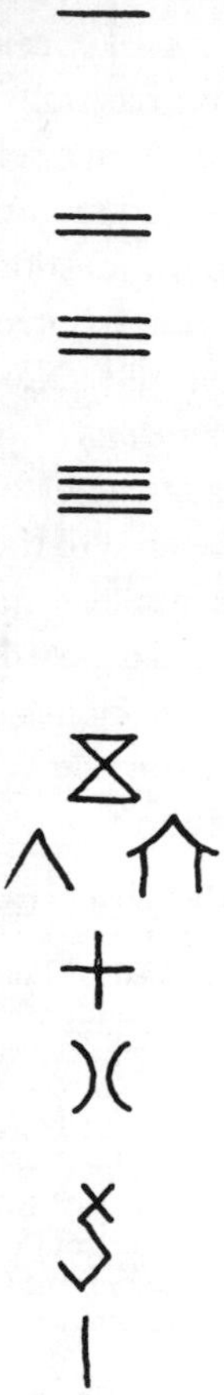

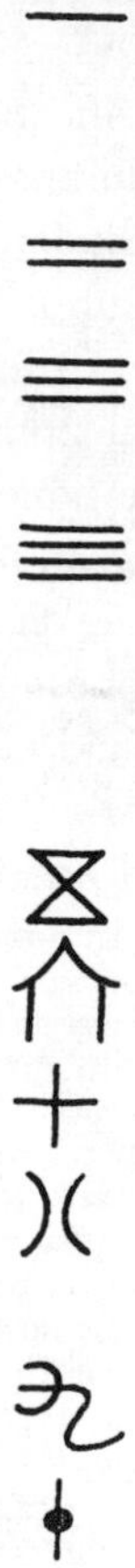

Table 22 (*continued*)

G	H		I		J
Other forms found on coins of Chou period (−6th to −3rd centuries)	Counting-rod forms (−2nd to +4th centuries)		Late counting-rod forms (+13th century)		Commercial forms (from +16th century)
	units	tens	units	tens	
	indicated by place		indicated by place		

blank space until +8th century ○ ○

It has been stated[a] that the earliest mathematical book in which the rod-numerals appear is the *Wu Tshao Suan Ching*,[b] written in the +5th (or perhaps the +4th) century. The editions of this work which we have seen do not in fact show rod-numerals, the calculations being simply written out in the standard way. But the question has little or no significance, for since the printing of mathematical works started from the +11th century onwards, and since, according to the epigraphic and numismatic evidence, the rod-numerals had been in use already more than a thousand years earlier, it must have depended upon the individual editors whether the rod-numerals were printed or not.[c] Moreover, the Han mathematical texts constantly use expressions such as *chih*,[1] 'set it up', and *lieh*,[2] 'spread it out', implying the use of counting-rods.

A famous rebus in the *Tso Chuan*, under date −542, has often been cited[d] to show that the rod-numerals go back to the middle of the Chou. The passage[e] concerns the determination of the age of an old man, and the cyclical character *hai*[3] is analysed to yield a two and three sixes, giving him 2666 decades of days. This shows an understanding of place-value. But in view of the subsequent remodelling of the text of the *Tso Chuan* it would perhaps be unsafe to accept this as evidence for a period earlier than that of the Warring States, for which in any case the coins bear witness. Of course if the character *suan*[4] is an *ancient* graph of arranged counting-rods, the numerals (as well as the instrument) may go back far into the −1st millennium. Some of the oracle-bone numerals, especially 5, 6, 7 and 10, certainly look like arranged counting-rods (see column E in Table 22).

During the Chhin and Han, the functions of the two kinds of numerals such as 𝍦 and 𝍯 were stabilised. The former were used for units, the latter for tens, the former for hundreds, the latter for thousands, and so on. By the +3rd century at least, they were termed respectively *tsung*[5] and *hêng*[6] numerals.[f] The *Sun Tzu Suan Ching*[g] of this period says:

> In making calculations we must first know the positions (and structure) (*wei*[7]) (of numerals). The units are vertical and the tens horizontal, the hundreds stand while the thousands lie down; thousands and tens therefore look the same, as also the ten thousands and the hundreds....When we come to 6 we no longer pile up (strokes), and the 5 has not got a one (a ligature)[h]

[a] Smith (1), vol. 2, p. 40, for instance; he dates the book much too early, as usual.

[b] See below, p. 34.

[c] The vertical lines did not encourage them, and for a long time they were regarded as insufficiently 'literary' to be printed. They may be seen in the Tunhuang MS. of the *Sun Tzu Suan Ching* (Bib. Nat. no. 3349; cf. Li Nien (*20*), p. 28), which we consider to be of Thang date.

[d] As by Wylie (4), p. 169; cf. *Lü Chai Shih-erh Pien*, ch. 23, p. 1*b*.

[e] Duke Hsiang, 30th year, tr. Couvreur (1), vol. 2, p. 544. Cf. *Chhien Han Shu*, ch. 21B, p. 28*a*; *Thang Chhüeh Shih*, ch. 2, p. 25*b*; *Kung Chhi Shih Hua*, ch. 9, 4*b*; *Hsiao Hsüeh Kan Chu*, ch. 1, p. 37*a*.

[f] This was another technical use of terms which had been applied to the diplomatic alliance of the Warring States period (cf. Vol. 1, p. 97 above).

[g] See below, p. 33.

[h] Ch. 1, p. 2*b*, tr. auct.

1 置 2 列 3 亥 4 筭 5 縱 6 橫 7 位

The system was thus stabilised[a] as follows:

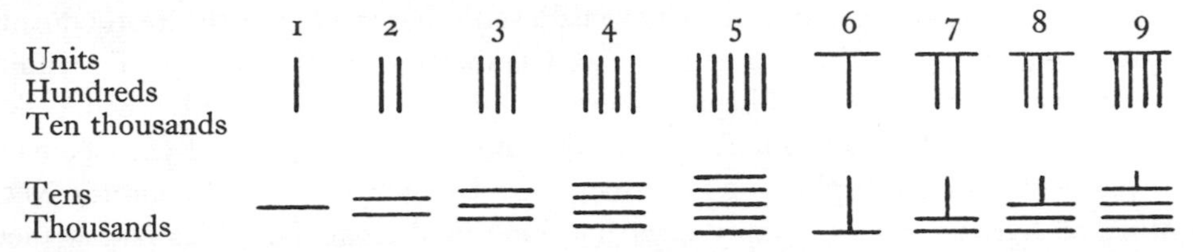

	1	2	3	4	5	6	7	8	9
Units Hundreds Ten thousands	𝍠	𝍡	𝍢	𝍣	𝍤	𝍥	𝍦	𝍧	𝍨
Tens Thousands	𝍩	𝍪	𝍫	𝍬	𝍭	𝍮	𝍯	𝍰	𝍱

Thus, for example, the number 4716 appeared as 𝍬 𝍦 𝍩 𝍥. The demarcation of neighbouring powers of ten in this way enabled the computers to use counting-boards without marked vertical columns.[b] In the Sung, figures tended to condense into monogrammatic forms,[c] such as in this case 𝍬𝍦𝍩𝍥. It is said[d] that sometimes forms such as 𝍦 were replaced, when occurring in the hundreds, by 𝍯, but this must be very rare.[e]

The word *wei*,[1] as used above, referred essentially to the positions of the rods in the columns on the counting-board, in other words, to local or place-value. Another word was 'rank' (*têng*[2]).[f] Before the +8th century, the place where a zero was required was always left vacant, as in the example from the *Sun Tzu Suan Ching* given by Li Nien.[g] This is well shown on some of the Thang MSS from the Tunhuang cave-temples. One of the rolls,[h] entitled *Li-Chhêng Suan Ching*,[3] contains multiplication tables in which the results appear both in written form and as rod-numerals. Here we find 405 shown as 𝍣 𝍤. From the Han counting-board to the 'matrix' notation of the Sung algebraists,[i] the *wei* was fundamental.[j]

[a] Perhaps some part in this stabilisation was played by Wang Mang. As Li Nien (*1*), p. 58; (*3*), p. 19, points out, one of the ways in which he attempted to mark the difference between his short-lived Hsin dynasty and the Han, at the beginning of the +1st century, was by placing on his coins the numeral 6 written with the cross-bar or ligature above the vertical line ┬ instead of ⊥ below it. This may have helped to suggest the convention later described by Sun Tzu. For an example of Wang Mang's practice see de Lacouperie's catalogue (4), p. 302, no. 106. Examples of the earlier practice occur on p. 160, no. 632, p. 162, no. 642, and p. 283, no. 1376. But the *tsung* six seems to have been used occasionally long before, in the late Warring States period; see p. 190, no. 779.

[b] It is a very striking fact that even as far back as the oracle-bone numerals of the −13th century, the symbols for 1 and for 10 were both straight lines, the former horizontal, the latter vertical (Table 22, col. E). As can be seen, this was just the opposite of the convention recorded by Sun Tzu, but the principle was identical.

[c] Doubtless for ease of printing in the vertical columns of books.

[d] Smith (1), vol. 2, p. 42.

[e] We ourselves have never seen this in any book, but it seems to be a coin form (Table 22, col. G). It would have been a fairly logical derivative from a very ancient practice (see Table 23 below).

[f] *Chang Chhiu-Chien Suan Ching*, ch. 1, p. 6*b*.

[g] (*1*), p. 59. The evidence for this is implicit in all the early descriptions of calculations which direct the movement of figures from one column to another. It is especially clear in root-extraction procedures (cf. Wang & Needham, 1).

[h] See below, pp. 36, 107; cf. Biot (7).

[i] See below, p. 129.

[j] It will be clear that the counting-board system, with its columns of figures arranged according to place-value, which must have been fully developed well before the end of the Chou period, was highly advanced for its time (e.g. the −4th century). But sometimes such pioneer developments have to be paid for by later fixations. It may be that the domination of the Sung algebraists by the chequer-board notation was, as it were, an arithmetical carry-over, hindering the free flight of symbolism. We may find further illustrations of this principle. Thus the very success of the Han mathematicians in finding a general method of solving numerical equations might explain the later absence of any theory of equations (cf. pp. 126, 112, 104 below). Place-value, so great a merit in arithmetic, handicapped algebraic symbolism. In another field, we may find that the advanced equatorial character of ancient Chinese astronomy delayed the discovery of the precession of the equinoxes (pp. 200, 270 below).

1 位　　2 等　　3 立成算經

The circular symbol for zero is first found in print in the *Su Shu Chiu Chang*[a] of Chhin Chiu-Shao (+1247), but many have believed[b] that it was in use already during the preceding century at least. The usual view is that it derived directly from India, where it first appears on the Bhojadeva inscriptions at Gwalior dated +870.[c] But there is no positive evidence for this transmission, and the form could perhaps have been borrowed from the philosophical diagrams of which the +12th-century Neo-Confucians were so fond.[d] In any case, the Sung mathematicians thus had at their disposal a fully developed notation, as in the example which Wylie chose a century ago from the work of Chhin Chiu-Shao, where the subtraction

$$1{,}470{,}000 - 64{,}464 = 1{,}405{,}536$$

appears as follows:

| ≡ O ≣ ||||| ≡ T | ≡ ╥ O O O O
T X |||| ⊥ X

From the tables in which ancient Indian numerals have been assembled[e] it can be seen that from the time of King Aśoka onwards (−3rd century) there was a steady development of forms akin to the 'Hindu-Arabic' numerals of today. It is worth noting that in all these systems the first three integers are written in just the same way as the Chinese; some of the ancient systems[f] also have a X for 4 (cf. Table 22, cols. I, J).[g] But in nearly all of them[h] there were separate symbols (containing no 'place-value component')[i] for 10 and multiples of 10, 20, 30, 40, 100, etc., and so long as any such practice prevailed place-value arithmetic could not exist.

While the first epigraphic evidence for the zero in India is, as has just been mentioned, of the late +9th century,[j] it has been discovered about two hundred years earlier in Indo-China and other parts of south-east Asia. This fact may be of much significance. The literary and epigraphic evidence for the antiquity of place-value in India has been conflicting. While the former pointed to a date well before +500 for the development of place-value and the zero concept,[k] it was never entirely convincing

a Brit. Mus. S/930. This MS. was first noted by Dr Hsiang Ta, and has been printed (with some inaccuracies) by Li Nien (7). It is on the verso of part of a Taoist book.

b Mikami (1), p. 73; (5); van Hée (15); Yen Tun-Chieh (6).

c Smith & Karpinski, p. 52; Datta & Singh (1), vol. 1, pp. 42, 118; Renou & Filliozat (1), vol. 2, p. 703. Also +876.

d On the *Thai Chi Thu* see Vol. 2, pp. 460ff. The suggestion is due to Wang Ling (2), vol. 1, pp. 97ff.

e One number on the Ragholi plates of King Jaivardhaṇa II would take it back to the +8th (Datta & Singh, vol. 1, pp. 40, 82).

f Smith & Karpinski, p. 25; Smith (1), vol. 2, p. 67; Datta & Singh (1), vol. 1, pp. 105ff.; Renou & Filliozat (1), vol. 2, pp. 705ff.

g E.g. the −1st-century Śaka inscriptions in Kharoshthi (Indo-Aramaic) script.

h Attempts of some early students, such as Kleinwachter (1), to derive the Hindu-Arabic numerals from the Chinese have long been abandoned.

i For example, in the Nāgarī numerals of the famous Nānā Ghāt inscriptions (*c.* −150), or the very complete series found on the Kshatrapa coins (*c.* +200). These are forms of Brāhmī script.

j See immediately below, p. 12.

k W. E. Clark (1); Datta & Singh (1), vol. 1, pp. 75ff. Two different things are involved here, though so closely connected. Place-value could and did exist without any symbol for zero, as in China from the late Chou onwards. But the zero symbol, as part of the numeral system, never existed, and because of the uncertain chronology of Indian history and the difficulty of dating documents. Kaye (1), in his critical examination of the dated inscriptions, could not take place-value back before the +8th century at the earliest. But Coedès (2) has shown that Indo-Chinese inscriptions use place-values much earlier (Cambodia +604, Champa +609, Java +732). They employed a system of 'symbolic words', i.e. names of things which were universally known to be associated with a given numerical value.[a] Thus in an inscription at Phnoṃ Bàyàṅ in Cambodia for +604, the 526th year of the Śaka era was expressed as 'the year (designated by) the (five) arrows, the (two) Aśvins, and the (six) tastes'.[b] Now it is soon after this period that the first inscriptions appear showing the zero (simultaneously in Cambodia and Sumatra, +683; and on Banka Island, +686). The 605th year of the Śaka era is represented as @•ξ, using a dot (*bindu*), and the 608th as @O∀, showing the modern zero itself.[c] The Indian numerals, without the zero, and encumbered with separate signs for the multiples of ten, were no improvement at all on the Greek and Hebrew alphabetical scripts. Yet Indo-China would seem at first sight a rather unlikely place for such a revolutionary discovery as that of the essential liberating element.

Coedès does not believe that the south-east Asian inscriptions indicate an east Asian origin for the symbolic word system (as Kaye had hinted might be possible), but rather that the Hinduising settlers of south-east Asia already had symbolic words and the old numerals when they first went there, or at any rate were soon followed by them.[d] So far so good, but we are free to consider the possibility (or even probability) that the written zero symbol, and the more reliable calculations which it permitted, really originated in the eastern zone of Hindu culture where it met the southern zone of the culture of the Chinese.[e] What ideographic stimulus could it have received at

could not have come into being, without place-value. It seems to be established that place-value was known to, and used by, the author of the *Pauliśa Siddhānta* in the early years of the +5th century, and certainly by the time of Āryabhaṭa and Varāha-Mihira (*c.* +500). And this was the decimal place-value of earlier China, not the sexagesimal place-value of earlier Babylonia. It may be very significant that the older literary Indian references simply use the word *śūnya*, 'emptiness', just as if they were describing the empty spaces on Chinese counting-boards. The earliest zero symbol used in computation, the dot (*bindu*), occurs in the Bakhshālī MS., but this cannot now be dated earlier than the +10th century (Renou & Filliozat (1), vol. 2, pp. 175, 679; Cajori (2), pp. 85, 89; (3), p. 77). Better evidence for the use of the dot comes from the +6th-century poem *Vasavadattā* of Subandhu (Renou & Filliozat (1), p. 703). This remains the earliest reference. The dot is still used in the Śāradā script of Kashmir.

[a] According to the rules of the 'symbolic correlations' discussed above in Sect. 13, Vol. 2, pp. 261, 271. It is rather surprising that the Chinese never adopted such a system, in view of the dominance of the symbolic correlations in their thought. Perhaps there were too many 'fives'. Smith & Karpinski, p. 38, discuss the sets of words used, and note that al-Bīrūnī, in his work on India, gave a list of them. See also Datta & Singh (1), vol. 1, p. 54; Renou & Filliozat (1), vol. 2, pp. 182, 708.

[b] In the original, the order of enumeration of the digits is inverted as was usual when words were employed instead of numerals.

[c] The exact date to which this corresponds depends upon present uncertainties as to the true date when the Śaka era began; see Tarn (1), p. 352; van Lohuizen de Leeuw (1); Thomas (1). If, as assumed in the other dates above, the commonest reckoning of +78 is adopted, the date of the first zero becomes +686; if +128 it is +736. This latter would yield precedence to the *Khai-Yuan Chang Ching* (see below). Of course both would probably have had a common source.

[d] They would doubtless be numerals related to the Gupta or Valabhī types (+4th to +7th centuries) which both included signs for all multiples of ten. On the great movement of intercourse between India and South-east Asia see Grousset (1) and Wales (1).

[e] It is interesting that Lattin (2), summarising Bubnov, contends that the Greek abacists also used blank spaces for zeros, but we have not found the evidence very convincing.

that interface? Could it have adopted an encircled vacancy from the empty blanks left for zeros on the Chinese counting-boards? The essential point is that the Chinese had possessed, long before the time of the *Sun Tzu Suan Ching* (late +3rd century) a fundamentally decimal place-value system.[a] It may be, then, that the 'emptiness' of Taoist mysticism,[b] no less than the 'void' of Indian philosophy, contributed to the invention of a symbol for *śūnya*,[c] i.e. the zero.[d] It would seem, indeed, that the finding of the first appearance of the zero in dated inscriptions on the borderline of the Indian and Chinese culture-areas can hardly be a coincidence.

How quickly it spread towards the latter has been insufficiently appreciated.[e] A zero symbol is mentioned in the *Khai-Yuan Chan Ching*,[1] that great compendium of astronomy and astrology edited by Chhüthan Hsi-Ta[2] between +718 and +729. The part of this work[f] which deals with the Chiu Chih[3] calendar[g] of +718 contains a section on Indian methods of calculation. After saying that the numerals are all written cursively with only one stroke each, the writer goes on to say:

> When one or other of the 9 numbers (is to be used to express a multiple of) 10 (lit. reaches 10), then it is entered in a column in front of (the unit digit) (*chhien wei*[4]). Whenever there is an empty space in a column (i.e. a zero), a dot is always placed (to signify it) (*mei khung wei chhu hêng an i tien*[5]).[h]

Here is the same dot which we met with less than half a century before in Cambodia.

The question then arises, exactly when had the Chinese adopted a decimal place-value system in writing numbers? Had they ever had symbols containing no place-value component for multiples of ten, hundreds and thousands in their written language? Apparently not. In the Shang period, as far back as we can go, on the

[a] The practice of writing out the place-value terms continued throughout Chinese history. The earliest example we have seen in which they were dropped occurs in the +14th-century *Ting Chü Suan Fa*, p. 19*a*. If this is not accidental, Arabic influence might be surmised. Of course, the *Khai-Yuan Chan Ching* (+718) had described (ch. 104) the similar Indian method of expressing numbers, but it had little or no influence. On the other hand, it had been a very ancient practice to leave out the place-value terms on coins, e.g. Chou knife-money (*Ku Chhüan Hui*, pt. 1, ch. 7, pp. 2*b* ff.; pt. 2, ch. 8, pp. 6*b* ff.).

[b] I shall mention later (p. 47) a hint, long subsequent, that Taoist 'emptiness' was indeed connected in the Chinese mind with abstract mathematical operations. The question, more philosophical than mathematical, as to whether 'one' should be regarded as a number or not (cf. Smith (1), vol. 2, pp. 26 ff.), presented itself in China as in the West, and the various answers to it have been considered by Solomon (1). Cf. the ideas of Leibniz on the construction of the world from unity and zero alone (by the binary arithmetic), Vol. 2, p. 340 above.

[c] Hence Arabic *al-ṣifr*; Byzantine *tziphra*; French *chiffre*; English 'cipher'. The word cypher, meaning a code, comes from an Aramaic word, *sifr*, a book.

[d] One remembers the Irishman's way of explaining how to make a pot—'take a hole and pour molten iron around it'.

[e] In the West. See Chhien Pao-Tsung (*1*), p. 95; Li Nien (*1*), p. 96; Mikami (1), p. 59; Yabuuchi (1), (*1*).

[f] Ch. 104, p. 1*b*.

[g] This was an adaptation of the *Navagrāha* (Nine Planet) calendar of Varāha-Mihira (*fl. c.* +505); Chhien Pao-Tsung (*1*), p. 94; Yabuuchi (1), (*1*).

[h] Tr. auct.

1 開元占經 2 瞿曇悉達 3 九執 4 前位
5 每空位處恒安一點

oracle-bones, we find numbers such as '547 days'[a] written *wu pai ssu hsün chhi jih*,[1] i.e. five hundreds (plus) four decades (plus) seven (of) days.[b] This was in the −13th century.[c] The actual symbols used in this period are assembled in Table 23.[d] The significance of the place-value components can be seen at once from the examples there written, 162 and 656, taken direct from bone inscriptions. Our previous number also appeared as [oracle-bone graphs: 5 over pine-cone, ≣, tens-of-days symbol, +, 日], the place-value being indicated by the component [pine-cone] for hundreds, and the separate symbols [tens-of-days symbol] for tens of days, and 日 for units of days.[e] From an inspection of Table 23 it can be seen that already at this most ancient period there were (above 49) no symbols which did not manifestly indicate their decimal position. For example, in contrast with Roman L or Greek ν′, the symbol for 50 showed a 5 surmounted by the place-value component for 10. This was a very short vertical stroke, just as the hundreds component was a 'pine-cone', and the thousands component was a man. Note also how well 50 was distinguished from 15 by position within the graph. Any number could therefore immediately find its place on a chequered counting-board where a blank space would be left for the zero. And instead of the Roman cumulative CCC for 300, the Shang Chinese (a thousand years earlier) wrote what was almost the equivalent of 3 C, i.e. a term ready for place-value calculations. Indeed it was better, for the 'C' could never stand as a numeral by itself. Only the signs for the lower multiples of 10 (20, 30, 40), consisting of upright strokes connected by a ligature, followed the cumulative system. These continued in modified form (*nien*,[2] *sa*,[3] *shu*[4]) throughout the centuries until now, but they were not used at all in mathematical work, nor often (except for poetry and pagination) in general literature.[f] The striking thought presents itself that where (as in India, Israel, Greece and Rome) an alphabet was available for the construction of numerals, there would be a strong temptation to run through all its letters, not stopping at 9, while this was not the case with ideographic script.[g]

In general, therefore, it will be seen that the Shang numeral system was more advanced and scientific than the contemporary scripts of Old Babylonia and Egypt.

[a] The significance of this number will appear later, Sect. 20, p. 293.

[b] See Tung Tso-Pin (*1*), pt. 2, ch. 4, pp. 4*b*, 5*a*, and for the dating, pt. 1, ch. 1, pp. 2*b* ff.

[c] Still earlier, the 'plus' had been explicit, written as *yu*.[5]

[d] We are again most grateful to Dr Wu Shih-Chhang for providing this list from the specialised works of Kuo Mo-Jo and others on Shang inscriptions. Table 23 also gives parallel forms from inscriptions on bronzes.

[e] This was a special calendrical usage. It is instructive to see that when additional symbols for tens or units were employed, the place-value component was omitted from the figure itself. The particular case given is especially striking because 40 is indicated by a digit 4 and not by the piling-up system.

[f] Lao Kan (*1*) gives examples of their use on Han dynasty bamboo-slip army records. It will be observed that these forms were and are pronounced as monosyllables, analysis of which shows them to be fused not from repeated tens, e.g. *shih-shih*, but from a digit and a ten (Karlgren, 16). For instance (in ancient pronunciation), *ńźi* (two) plus *źi̯ep* (ten) gives *ńźi̯ep* (twenty), and this last sound persists in Cantonese. Hence, though graphically cumulative, these forms were evidently thought of as combinations of a digit and a common multiplier, if not a place-value component. Furthermore, since the ancient pronunciation of the number one was *i̯ĕt*, the final consonant *p* was perhaps associated with the place-value component for ten. In that case twenty (*ńźi̯ep*) would have been composed of *ńźi* plus *p* and thirty (*sâp*) of *sâm* plus *p*. At all events the digit was always built in.

[g] This point was first suggested to us by Dr Derek Price.

[1] 五百四旬七日 [2] 廿 [3] 卅 [4] 卌 [5] 有

Table 23. *Notations for numerals higher than 10 on the Shang oracle-bones and on Chou bronze inscriptions*

	Bone forms	Coin forms	Bronze forms
11	11th month  perhaps also		
12			
13			
14	presumably analogous but no example known		
15			
20			
30			
40			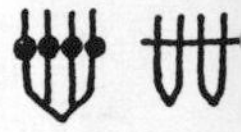
50			
56	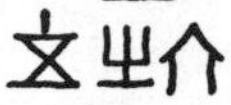 (i.e. *wu shih yu liu* 五 十 又 六; five tens plus six)		
60			
88	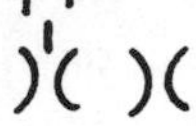		
90			
100			
162			
200			
209	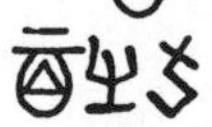 (i.e. *erh pai yu chiu* 二 百 又 九; two hundreds and nine)		
300			
500			
600			
656	 (i.e. *liu pai wu shih liu* 六 百 五 十 六; six hundreds, five tens, six.) This is the form which continued unchanged through the next three thousand years)		
1000			
3000			
4000			
5000			

All three systems agreed in that a new cycle of signs began at 10 and at each of its powers. With one exception already noted, the Chinese repeated all the original 9 numerals with the addition of a place-value component, *which was not itself a numeral.*[a] The Old Babylonian system, however, was mainly additive or cumulative[b] below 200, like the later Roman; and both employed subtractive devices, writing 19 as 20 − 1 and 40 as 50 − 10. But the multiplication process was also introduced, e.g. 10 × 100 representing 1000. Only in the sexagesimal notation of the astronomers,[c] where the principle of place-value applied, was there better consistency, though even then special signs were used for such numbers as 3600, and the subtractive element was not excluded. Moreover, numbers less than 60 were expressed by 'piled-up' signs. The ancient Egyptians followed a cumulative system, with some multiplicative usages.[d] It seems therefore that the Shang Chinese were the first to be able to express any desired number, however large, with no more than 9 numerals. The subtractive principle of forming numerals was never used by them.[e]

It will thus be seen that behind the 'Hindu' numerals, as the West subsequently knew them, there lay two thousand years of place-value in China.

There is of course a large literature on the transmission of the Hindu-Arabic numerals from India through Islam to Europe, but discussion of this lies outside our present plan. It was appropriate to refer at a much earlier stage to the comments of Severus Sebokht on them in the +7th century.[f] The transmission may be followed in the books of Smith & Karpinski (1), and of Cajori (2).[g] The most recent consideration of its broad significance is perhaps that of Kroeber,[h] who refers also to the parallel

[a] In the case of 1000, for instance, the pictogram of the man alone (see Table 23), without his crossbar, was no numeral. There seems no parallel to this in any other ancient civilisation. For example, Greek *pente-deka* ͲΔ (pente-deka sign) is a sign for 50, made by the multiplication of two symbols which themselves have substantive value. On the other hand, the dash in ,ϵ for 5000 did correspond to the Chinese system because the dash alone did not signify 1000. But this was only because the Greeks had run out of alphabetical letters, and since the principle was not consistently applied for all powers of ten it could not be very helpful. It is true that in Chinese literary usage after the Shang the unit digit required as a prefix to *pai* and *chhien* was often omitted, so that especially if heading further, smaller, numbers they stood alone for 100 and 1000, but its presence was always understood. In mathematical texts it was hardly ever omitted.

[b] Cajori (3), vol. 1, pp. 2ff.; van der Waerden (1).

[c] Neugebauer (9), p. 15.

[d] See Cajori (3); Menninger (1); van der Waerden (1).

[e] Nor the multiplicative system either—for the 'pine-cone' was not a number.

[f] Vol. 1, p. 219.

[g] It has been contested by Lattin (2) and Boyer (2), but unconvincingly.

[h] (1), p. 469.

NOTES to Table 23

(1) With regard to the deviation of the symbols for 20, 30, and 40 from the place-value principle, it is interesting to find that alternative forms which followed it were used on coins, e.g. Chou money (*Ku Chhüan Hui*, pt. 1, ch. 8, p. 17*b*; pt. 2, ch. 8, p. 6*a*, ch. 9, pp. 9*b*, 10*a*), as seen above. This point was remarked upon long ago in relation to a bronze vessel inscription by Ku Yen-Wu (*Jih Chih Lu*, ch. 21, p. 75). Possibly the deviant piling-up system originated because the adding of a single horizontal stroke to the single vertical stroke representing the tens place-value component, would have been likely to cause great confusion with the Shang numeral for 7. The component therefore became a numeral itself, like Roman X.

(2) When the symbol for 60 came to be made with the short vertical stroke at the top as the decimal digit component, it was found necessary to add a short horizontal stroke for the sake of distinctness.

Babylonian and Mayan inventions of the zero. These two, however, were very different. The former occurs in Seleucid Babylonian cuneiform tablets from about −300 onwards,[a] in numerous forms such as ०₀०, and its use continued in Hellenistic and Byzantine mathematical texts,[b] sometimes even appearing as the familiar empty circle.[c] But it indicated only blank spaces in tabulations, and was never used in computing.[d] The Maya zero, on the other hand, was a true one, associated with place-value.[e] But its place-value was inconstant, and neither decimal nor sexagesimal.[f]

The only question remaining concerns the origin and history of the Chinese written form for zero, *ling*.[1] The ancient meaning of this word was the last small raindrops of a storm, or drops of rain remaining afterwards on objects. This was its sense in the *Shih Ching* (Book of Odes). Afterwards it came to be applied to any 'remainder'[g] (especially when coupled with another word, as *chi ling*[2]), either non-integral, or as, for example, 'five over the hundred'. From this the transition to the use of the word for expressing the zero in a number such as 105 may readily be understood. Nevertheless, this use seems to have arisen late.[h] To determine exactly when would require a special investigation, but I can say that we have never met with it used in the sense of zero in any mathematical text earlier than the Ming, though of course the Sung algebraists used the symbol o extensively, and it is easy to find examples of numbers written out in which the term *ling* could well have been employed. The earliest book in which it is prominent is the *Suan Fa Thung Tsung* of the end of the +16th century[i] (just before the coming of the Jesuits), and afterwards it is widely found, for example, in such a book as the *Suan Hsüeh Pi Li Hui Thung*[3] (Rules of Proportion and Exchange) by Lo Shih-Lin[4] (+1818) and all contemporary works. Since written forms of numbers could be well understood without a written form of the zero, it is a little difficult to explain why the use of the *ling* came in during the Ming dynasty. Perhaps

[a] Perhaps even from the −6th or −7th century. See Neugebauer (9), pp. 13, 16, 20, 26ff.; Cajori (3), p. 7.

[b] Perhaps Indian also, if Datta & Singh (1), vol. 1, p. 76, are right in their discussion of the *Chandaḥ-sūtra* of Piṅgala, a work on prosody, said to date from about −200. See Renou & Filliozat (1), vol. 2, p. 104.

[c] Neugebauer (9), pp. 11, 14; Cajori (3), p. 28. Some (e.g. van der Waerden (3), p. 56) seek to derive the Indian zero symbol from this, but the Indo-Chinese evidence mentioned above does not encourage the idea of such a transmission.

[d] De Lacouperie (4), p. xl, noticed that there are occasional 'empty circles' on Chou dynasty coins. See, for example, the piece of knife-money figured in *Ku Chhüan Hui*, pt. 2, ch. 5, p. 7*b*; ch. 7, p. 8*a*; on which the number 50 seems to be written. But this may be a simplification of the 'pine-cone', in which case it would mean 500. If these symbols are real zeros, the connection with contemporary Babylonia becomes extremely interesting.

[e] Cajori (3), p. 43. There are also similarities between the Mayan and Chinese numerals.

[f] It began as vigesimal after the units 1 to 19, but then went on to three hundred and sixties, and eventually (in the fourth place) to seven thousand two hundreds.

[g] It still occurs in this sense as late as the +14th century (cf. *Ting Chü Suan Fa*, pp. 18*b*, 19*a*; *Ming I Thien Wên Shu*, ch. 2, pp. 33*b*, 42*b*).

[h] A number of other words had previously been used to signify zero, at least as far back as the Thang; see Yen Tun-chieh (*14*), p. 10.

[i] See below, p. 52, for the description of this work. As it contains terms of the type *i chhien ling ling i* for 1001, the identification of *ling* with zero is very clear. But this usage was not systematically adhered to, and in later writings *ling* is often not repeated when several zeros come in succession.

[1] 零 [2] 奇零 [3] 算學比例滙通 [4] 羅士琳

the zero symbol had been pronounced 'ling' from the time of its first general use in the Sung, and it may well be that the application of the old character arose not only because it had long meant 'remainder' (i.e. that there were some more digits to come after the zero), but also because the o symbol was shaped like a spherical raindrop.[a]

References to numbers in old Chinese literature can often only be clearly understood in the light of the rod-numeral system. For example, in the *Hsi Yu Chi* (the +16th-century novel so well known in its English translation as *Monkey*), one of the Thang emperors visits the underworld, where the officials, looking into their dossiers, kindly add 20 years to his life by marking two extra strokes in the figure for the length of his reign as previously allotted.[b] This was a change from 13 to 33, i.e. —III to ≡III, and it must have been in counting-rod numerals, since the change in written numerals would normally have required the addition of three strokes.[c] Another example where we can see the manipulation of decimal places occurs in a very lively satire on minor government officials written by the astronomer Shu Hsi[1] about +280. In his *Chhüan Nung Fu*[2] we find:[d]

> The established officials of a district are various and enjoy different functions, but if we study these lowly positions in the administration, we shall find that none is more fair than that of the Official who Encourages the Farmers. His power is absolute over the whole village and everybody in it. When the green banners[e] (are flown to) restrict wanderers and idlers, and when taxes on land are being assigned by acreage, the size of the assessment is determined by him alone, and full say regarding the quality of the land is his. The winning of his favour depends upon (gifts of) rich meat, and the securing of his support upon good wine. When the harvest work is finished and the levies are to be made, and he gathers the head men of the village and summons the chiefs of the hamlets to register holdings and names—then chickens and pigs fight their way to him and bottles and containers of wine arrive from all directions. Then it is that a 'one' can become a 'ten', and a 'five' a 'two'.[f] I suppose this is because hot food is twisting his belly, and the wine-god obstructing his stomach.[g]

Whatever might be the reasons for his arithmetical inexactitudes, it is clear that in the first case I had simply become —, while in the second case the crossed rods perhaps disappeared from the inside of ⊠ leaving =.[h] Or more probably the middle three strokes were removed from IIIII.

[a] This suggestion is due to Dr Lu Gwei-Djen.

[b] Ch. 11.

[c] Unless the abbreviated form of *san shih* was used. Not perhaps appreciating this point, Waley (17), p. 106, gave the change as three strokes in his translation, but the story was rectified by Duyvendak (20), p. 12.

[d] *CSHK* (Chin section), ch. 87, pp. 2*a*, *b*.

[e] Sympathetic fertility magic.

[f] Literally, 'one fixed (in its place-value) so as to become ten, and five turned into two by hooking (away some of its strokes)'.

[g] Tr. Yang Lien-Shêng (5), p. 134.

[h] It is of course true that this would mean 20, not 2, in the system as finally stabilised after the *Sun Tzu Suan Ching*, but Shu Hsi was just contemporary with the probable *floruit* of Master Sun himself. If Shu Hsi was following older variants, which worked just the other way round, the one and the ten, as we have shown them, would require inversion. But the general principle is not affected.

[1] 束皙 [2] 勸農賦

(c) SURVEY OF THE PRINCIPAL LANDMARKS IN CHINESE MATHEMATICAL LITERATURE

In accordance with the system which other writers have found convenient, it will be desirable first to survey the most important books which the Chinese produced through the centuries in this field. But if the description mentions less than twenty major works, it must not be thought that the Chinese mathematical literature was so restricted. Generation after generation commented upon the mathematical 'classics', as they came to be called, and in addition there were in every century new books added to the list. The *Ku Chin Suan Hsüeh Tshung Shu*[1] collection of 1898 reproduces seventy-three in its third section alone.[a] Li Nien's own library, of which he has published catalogues (*6*), contained some 450. In 1936 Têng Yen-Lin & Li Nien published a union catalogue of Chinese mathematical books in the libraries of Peking[b] which contains somewhat over 1000 titles, though some of them are European books translated either by the 17th-century Jesuits or by later foreign scholars such as Wylie, Edkins and Mateer.[c] The bulk of the material, however, is prior to the European influence, or by Chinese mathematicians working largely independently during the Chhing dynasty. Even if we assume that this vast literature has nothing to contribute to post-Renaissance and modern mathematics, it remains a sobering thought that the history of mathematics has so far been written almost without access to so large a body of material. It must also be remembered that a high proportion of the earlier books on mathematics, before the Sung (+13th century), are irretrievably lost—we know them by the titles which appear in the bibliographies of the official histories, and by references in other writings. We shall have occasion to mention a few of such lost works in what follows. It must also be pointed out that some even of the books we still possess were lost for centuries until reprinted from rare copies, or inserted in imperial collections designed to preserve rare texts.

Probably the *editio princeps* of the mathematical classics is the *Suan Ching Shih Shu* (The Ten Mathematical Manuals). Edited by officials in +656 for use as a text-book and first printed in +1084, most of it was copied into the *Yung-Lo Ta Tien* encyclopaedia in the 15th century, but subsequently became rare. The corpus was recovered by Tai Chen and printed, with his editing, in the Palace collection *Wu Ying Tien Chü Chen Pan Tshung-Shu* in the years preceding +1794.[d]

[a] Edited by Liu To. This year saw also the appearance of the important Chinese mathematical bibliography of Liu To (*2*).

[b] *Peiping Ko Thu-Shu-Kuan so Tshang Chung-Kuo Suan-Hsüeh Shu Lien-Ho Mu-Lu.*

[c] The index excludes translated modern text-books, i.e. since 1900.

[d] See above, Vol. 2, pp. 513ff.; Hummel (2), pp. 160, 697. It was printed with movable type.

[1] 古今算學叢書

(1) From Antiquity to the San Kuo Period (+3rd Century)

Though the *Chou Pei Suan Ching*[1] is generally considered the oldest of the mathematical classics, the first firm dates which we can connect with it are nearly two hundred years later than those associated with the *Chiu Chang Suan Shu*[2] (Nine Chapters on the Mathematical Art) which will be described next in succession. But there are reasons for retaining the traditional order, as will be seen (p. 257). For the title of the *Chou Pei Suan Ching* we adopt the translation 'The Arithmetical Classic of the Gnomon and the Circular Paths of Heaven', though there are several alternatives. The first word has often been taken to refer to the Chou dynasty, at which period the dialogues in the book are supposed to have taken place, but since the word also means 'circumference' and since the book contains much on the declination of the sun at different times of the year, together with a diagram of the concentric declination-circles[a] through which its movements take it, we may accept the view of Li Chi[3] of the Sung, in his *Chou Pei Suan Ching Yin I*,[4] that *chou* refers to these 'circular paths of the heavens'.[b] The word *pei*[c] originally meant a femur, and some have therefore seen a reference in it to counting-rods made of bone, but the text itself definitely says that here it means a *piao*[5] or gnomon. While largely concerned with primitive astronomical calculations, the book opens with a discourse on the properties of right-angled triangles as used for the measurement by proportions of heights and distances,[d] both terrestrial and celestial.

The question of its date is a difficult one. First, there is no mention of it in the bibliography of the *Chhien Han Shu*, though it purports to be a text of the Chou period; this must mean either that about +100 it was thought so unimportant as not to be worthy of inclusion (which is hardly believable), or it did not exist at all, or perhaps it was not then known under its present title. This last explanation seems quite plausible, since practically all the eighteen calendrical and twenty-two astronomical books mentioned in the bibliography have long been lost. There was, for example, a *Hsia Yin Chou Lu Li*[6] (Treatise on the Calendars of the Hsia, Shang, and Chou Dynasties), and another on the positions of the sun and moon (*Jih Yüeh Hsiu Li*[7]), which may well have contained some of the material now in the *Chou Pei*.

Since there is a mention of Lü Pu-Wei in the text (though the quotation given is not an exact one from the *Lü Shih Chhun Chhiu*, cf. p. 195, as we know it) it has been argued that the *Chou Pei* must be later than the −3rd century. But this quotation may well have been inserted by an early Han editor. On the other hand, the book cannot be later than its first commentator, Chao Chün-Chhing,[8] whose generally accepted

[a] Cf. below, p. 257.

[b] He says *suan jih yüeh chou thien hsing tu*.[9] The word 'path' or 'road' has a special significance which will be appreciated later (p. 256).

[c] According to Li Chi the correct pronunciation should be *pi*.

[d] I.e. *kou ku tshê wang*[10]—observations and measurements by means of the right angle and the line, or the gnomon and the shadow.

[1] 周髀算經 [2] 九章算術 [3] 李藉 [4] 周髀算經音義 [5] 表
[6] 夏殷周魯曆 [7] 日月宿曆 [8] 趙君卿 [9] 算日月周天行度
[10] 句股測望

date, though not exactly known, is around the end of the Later Han (probably +3rd century).[a] He cannot have written on it earlier than +180, since he refers to the Chhien-Hsiang[1] calendar, formulated by Liu Hung[2] between +178 and +183. Then Tshai Yung,[3] in one of his lost writings such as the *Piao Chih*[4] (Memorandum on the Gnomon) written between +133 and +192, referred to the *Chou Pei*.[b] The *Chou Pei* undoubtedly took a place in the cosmological discussions characteristic of the Han between the supporters of the Kai Thien[5] and the Hun Thien[6] theories,[c] for Chao Chün-Chhing says in his preface that while Chang Hêng's[7] *Ling Hsien*[8] (The Spiritual Constitution of the Universe)[d] was the chief book of the latter school, the *Chou Pei* was the chief book of the former. It has been noticed that some of the arguments of Wang Chhung in the Shuo Jih[9] chapter (on the sun)[e] in his *Lun Hêng*[10] closely resemble some of those in the *Chou Pei*. Moreover, there are close parallels with the Ssu Fên Li[11] calendar of just the same time (+85).[f] Chhien Pao-Tsung (1) has noted resemblances of detail between the *Chou Pei* and the San Thung Li[12] (Three Sequences Calendar) of Liu Hsin[13] (formulated in −26), as also with the Thai Chhu Li[14] calendar of −104. Thus, in sum, there has been a tendency to regard the *Chou Pei Suan Ching* as essentially a Han book.

Such a conclusion may be accepted unreservedly for the date of final composition, but most of the book is so archaic, and especially so much more so than the *Chiu Chang Suan Shu*, that it is difficult not to believe that it goes back to the Warring States period (late −4th century), or even earlier.[g] Li Nien accepts this view, as do Nōda (1) and Liu Chao-Yang (*3*). But no one now holds the traditional estimate, represented for instance by Pao Kan-Chih[15] of the Sung or Chu Tsai-Yü,[16] prince of the Ming, taken over by European sinologists such as Biot, and perpetuated in histories of mathematics[h] until today, that the *Chou Pei* gives us 'a very good record of the mathematics of about −1100'.

Subsequently famous commentators on the *Chou Pei* included Chen Luan[17] and Hsintu Fang[18] about +565, Li Shun-Fêng[19] (late +7th century), and Li Chi[20] of the Sung. It has been translated in full by E. Biot (4),[i] and in part by Vacca (4).

[a] Wang Ling (2), vol. 2, p. 161. Mikami (1) consistently refers to Chao as Chang.

[b] *CSHK* (Hou Han sect.), ch. 70, pp. 8*b*, 9*a*. Cf. p. 210 below.

[c] See pp. 210 ff. below.

[d] A few pages, all that time has left of this work, are to be found in Ma Kuo-Han's collection of fragments, *YHSF*, ch. 76, p. 61*a*.

[e] Ch. 32.

[f] Nōda (1).

[g] Astronomical evidence given below (pp. 256 ff.) indeed suggests rather strongly that the oldest parts of the *Chou Pei* may go back to the time of Confucius (late −6th century) or the generations immediately preceding him. The astronomical content of the book includes many features which closely parallel Old Babylonian astronomy as we know it from the cuneiform tablets of the Ea-Anu-Enlil and the Mul-Apin series (−14th to −10th and −9th to −8th respectively). And we shall see that the *Chou Pei* was regarded as representing the most archaic of the cosmological schools.

[h] Such as Smith (1) and, I regret to say, Mikami (1). The latest histories of mathematics published in 1953 are no better informed; cf. Becker & Hofmann, p. 132; Struik (2), p. 34.

[i] This translation is not always very sure of itself, having been made just over a century ago.

1 乾象　2 劉洪　3 蔡邕　4 表志　5 蓋天　6 渾天
7 張衡　8 靈憲　9 說日　10 論衡　11 四分曆　12 三統曆
13 劉歆　14 太初曆　15 鮑澣之　16 朱載堉　17 甄鸞
18 信都芳　19 李淳風　20 李籍

A brief description of this ancient work would run somewhat as follows. The first portion (which we may call I*a*) consists of a dialogue between Chou Kung[1] (the duke of Chou) and a personage[a] named Shang Kao[2] on the properties of the right-angled triangle, in which the Pythagorean theorem is stated, though not proved in the Euclidean way.[b] A second portion of this dialogue (I*b*) refers to the use of the gnomon, the circle and the square, and the measurement of heights and distances. Part II*a* introduces two new personages in a dialogue, Chhen Tzu[3] and Jung Fang,[4] who continue to talk about the sun's shadow, now estimating its length difference at different latitudes, now describing the measurement of the sun's diameter by means of a sighting-tube method.[c] Part II*b* commences with a diagram of right-angled triangles with different bases, called Jih Kao Thu[5] (diagram of the sun's altitude). From here onwards there seems to have been an intercalation of texts, Chhen Tzu and Jung Fang gradually fading out, as Biot well says, while paragraphs more and more frequently begin with the words *Fa yüeh*[6] (According to the method...), or *Shu yüeh*[7] (According to the art...). Part II*c* again begins with diagrams, especially the one already referred to, the Chhi Hêng Thu[8] (diagram of seven declination-circles,[d] the pole star occupying the centre). This is where appears the reference to Lü Pu-Wei. After some pages comes the point at which the text traditionally divides into the first and second chapters, and though we may term what follows Parts III*a*, III*b*[e] and III*c*, it does not differ in style from II*c*. It contains calculations concerning the annual movements of the sun, mentions the use of the water-level in obtaining a horizontal surface for the sun-shadow and gives a list of sun-shadow lengths for each of the fortnightly periods of the year. It also describes the determination of the meridian from sunrise and sunset observations, the culminations of stars, the twenty-eight lunar mansion constellations,[f] the nineteen-year period, and other astronomical matters.[g]

What of the mathematical content of the *Chou Pei Suan Ching*, the point which mainly interests us here? Apart from the consideration of the right-angled triangle, which will be referred to in a moment, there is first some use of fractions, their multiplication and division, and the finding of common denominators. Though the process of working out square roots is not given, they were certainly used, as in the passage *kou ku ko tzu chhêng, ping erh khai fang chhu chih*,[9] i.e. multiply both the height

[a] Traditionally supposed to be a worthy from the Shang dynasty.

[b] We cannot now regard it, with Biot, as certainly 'five or six centuries earlier than Pythagoras' (*fl.* −530), but there is not much reason for putting it any later, and it may well be older.

[c] Cf. Maspero (4), p. 273.

[d] See definitions, p. 179 below, and p. 256.

[e] This is the place where there is mention of the interesting *hsüan chi*[10] sighting-tube instrument (see below, p. 334). The passage comes at the beginning of Pt. III*b* (ch. 2, p. 2*a*).

[f] See below, pp. 233 ff.

[g] E. Biot (4), pp. 198, 623, from the data in the text, calculated that the declination of the pole-star given, assuming α Ursae Minoris, would have held good for the year +247, which is reasonable in view of other evidence. In his later emendatory note, however, he calculated for an adjacent passage a date about −1100; this was based on the positions of the nearest star to the pole, observed by the sighting-tube (see below, pp. 261 ff.), assuming β Ursae Minoris. The data are probably not reliable enough for these calculations, however.

1 周公 2 商高 3 陳子 4 榮方 5 日高圖 6 法曰
7 術曰 8 七衡圖 9 句股各自乘，并而開方除之 10 璿璣

of the post (*ku*) and the shadow length (*kou*) by their own values, add (*ping*) the squares, and take the square root (*khai fang chhu chih*) of the sum.[a] This is

$$c=\sqrt{(a^2+b^2)}.$$

Elsewhere expressions such as the square root of 5 are given to the nearest integer 'and a bit' (*yu chi*[1]). There is an idea of arithmetical series because the declination circles are said to be each 19,833 *li* apart, and in each fortnightly period (the 24 *chhi*[2]) (see p. 405 below) the augmentation and diminution of the shadow is 9 *tshun* $9\frac{1}{6}$ *fên*.

The discussion of the right-angled triangle occurs at the beginning, in the oldest part of the text. It is worth reproducing in full (see Fig. 50), on account of its rude antiquity.[b]

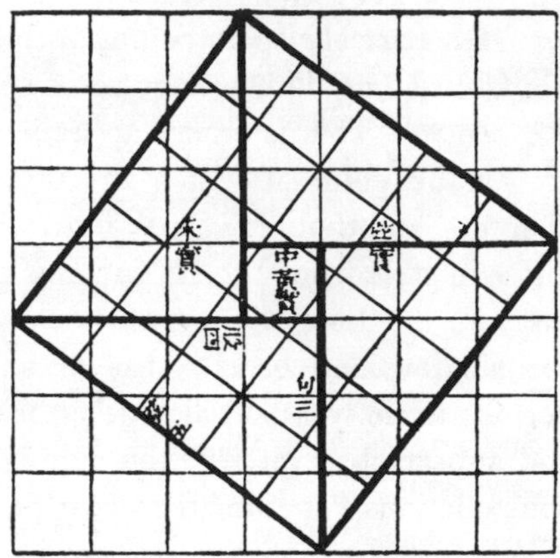

Fig. 50. The proof of the Pythagoras Theorem in the *Chou Pei Suan Ching*.

(1) Of old, Chou Kung addressed Shang Kao, saying, 'I have heard that the Grand Prefect (Shang Kao) is versed in the art of numbering. May I venture to enquire how Fu-Hsi anciently established the degrees of the celestial sphere? There are no steps by which one may ascend the heavens, and the earth is not measurable with a foot-rule. I should like to ask you what was the origin of these numbers?'

(2) Shang Kao replied, 'The art of numbering proceeds from the circle (*yuan*[3]) and the square (*fang*[4]). The circle is derived from the square[c] and the square from the rectangle (lit. T-square or carpenter's square; *chü*[5]).

(3) The rectangle originates from (the fact that) 9×9=81 (i.e. the multiplication table or the properties of numbers as such).[d]

(4) Thus, let us cut a rectangle (diagonally), and make the width (*kou*[6]) 3 (units) wide, and the length (*ku*[7]) 4 (units) long. The diagonal (*ching*[8]) between the (two) corners will then be 5 (units) long. Now after drawing a square on this diagonal, circumscribe it by half-rectangles like that which has been left outside, so as to form a (square) plate. Thus the (four) outer half-rectangles of width 3, length 4, and diagonal 5, together make (*tê chhêng*[9]) two rectangles (of area 24); then (when this is subtracted from the square plate of area 49)

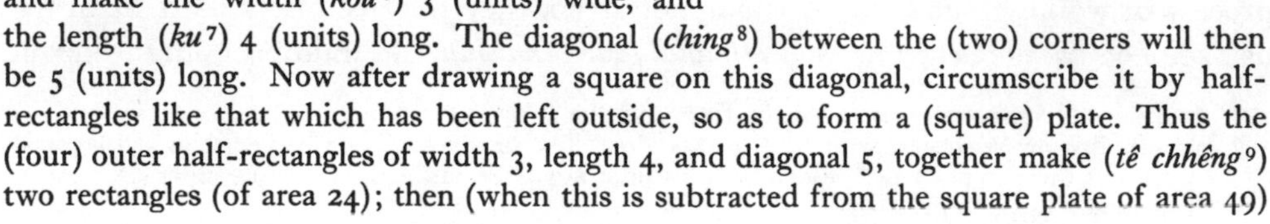

[a] Ch. 1, p. 9*a* (Biot (4), p. 606).

[b] The numbering of the paragraphs was introduced by Wylie (4). Cf. Mikami (16).

[c] Presumably the writer was thinking of the diameter of a circle as equal to the diagonal of its inscribed square; perhaps also of exhaustion methods for getting π.

[d] Chao Chün-Chhing explains that it is necessary to know the properties of numbers before one can work with geometrical figures. Note how radically different this is from the Euclidean method, in which actual numerical values are irrelevant provided the basic axioms and postulates are accepted. Here an arithmetical square is significantly given.

[1] 有奇 [2] 氣 [3] 圓 [4] 方 [5] 矩 [6] 句 [7] 股
[8] 徑 [9] 得成

the remainder (*chang*[1]) is of area 25. This (process) is called "piling up the rectangles" (*chi chü*[2]).[a]

(5) The methods used by Yü the Great in governing the world were derived from these numbers.'

> It will be remembered that the legendary Yü was the patron saint of hydraulic engineers and all those concerned with water-control, irrigation and conservancy. Epigraphic evidence from the Later Han, when the *Chou Pei* had taken its present form, shows us, in reliefs on the walls of the Wu Liang tomb-shrines (*c.* +140), the legendary culture-heroes Fu-Hsi and Nü-Kua holding squares and compasses (see Fig. 28 in Vol. 1, p. 164). The reference to Yü here undoubtedly indicates the ancient need for mensuration and applied mathematics.

(6) Chou Kung exclaimed, 'Great indeed is the art of numbering. I would like to ask about the Tao of the use of the right-angled triangle (lit. T-square).'[b]

(7) Shang Kao replied, 'The plane right-angled triangle (laid on the ground) serves to lay out (works) straight and square (by the aid of) cords. The recumbent right-angled triangle serves to observe heights. The reversed right-angled triangle serves to fathom depths. The flat right-angled triangle is used for ascertaining distances.

(8) By the revolution of a right-angled triangle (compasses) a circle may be formed. By uniting right-angled triangles squares (and oblongs) are formed.

(9) The square pertains to earth, the circle belongs to heaven, heaven being round and the earth square.[c] The numbers of the square being the standard, the (dimensions of the) circle are (deduced) from those of the square.[d]

(10) Heaven is like a conical sun-hat (*li*[3]). Heaven's colours are blue and black, earth's colours are yellow and red. A circular plate is employed to represent heaven, formed according to the celestial numbers; above, like an outer garment, it is blue and black, beneath, like an inner one, it is red and yellow. Thus is represented the figure of heaven and earth.[e]

(11) He who understands the earth is a wise man, and he who understands the heavens is a sage. Knowledge is derived from the straight line.[f] The straight line is derived from the right angle.[g] And the combination of the right angle with numbers is what guides and rules the ten thousand things.'

(12) Chou Kung exclaimed 'Excellent indeed!'[h]

No further commentary is needed on this classical passage, except by way of emphasis on what seems a deeply significant point, namely, the statement in sentence (3) that geometry arises from mensuration. As has already been indicated, this seems to show the Chinese arithmetical-algebraic mind at work from the earliest times, apparently

[a] Note the use of the same word, *chi*, found elsewhere in ancient writings meaning condensation or agglomeration (Vol. 2, pp. 41 etc.). In the translation of this vital passage on the Pythagoras theorem we have adopted the interpretation of Mr Arnold Koslow. We believe that the numbers given in the text were intended only as typical examples of the lengths of the three sides, each of which has its special technical term.

[b] Biot (4), following commentators, translated this *chü*[4] as gnomon; this emphasises the astronomical connection. The sense of the text makes it necessary to render the word in more than one way.

[c] There was a good deal of speculation behind this, for 3, the number of heaven and the nearest approximation to π, was a male or Yang number, while 4, the number of earth, was female or Yin.

[d] Cf. paragraph (2) above.

[e] Although *li*[3] is here in question (whatever it was), I cannot help suspecting a reference to the diviner's board (*shih*[5]), with its two plates, one round and one square; cf. Sect. 26*i* below.

[f] The shadow.

[g] Or gnomon.

[h] Tr. auct.; adjuv. Wylie (4), Biot (4), Mikami (1).

1 長 2 積矩 3 笠 4 矩 5 式

not concerned with abstract geometry independent of concrete numbers, and consisting of theorems and propositions capable of proof, given only certain fundamental postulates at the outset. Numbers might be unknown, or they might not be any particular numbers, but numbers there had to be. In the Chinese approach, geometrical figures acted as a means of transmutation whereby numerical relations were generalised into algebraic form.

Without anticipating what will have to be said in the section on algebra below, it is striking that the first commentator, Chao Chün-Chhing, embarked at once on an algebraic treatment of the Pythagoras theorem, involving quadratic equations.[a] This was not expressed, of course, in anything analogous to modern notation, but in written words and sentences. The technical phrases of greatest interest will be given later (p. 96).

As was pointed out by Chhen Chieh[1] of the Chhing, in his *Suan Fa Ta Chhêng*[2] (Complete Survey of Mathematics, 1843) the greatness of the *Chou Pei* is that at a time when astrology and divination were universally dominant, it speaks of the phenomena of the heavens and the earth without the slightest admixture of superstition.

We come now to the *Chiu Chang Suan Ching* (or, as it is generally called, the *Chiu Chang Suan Shu*,[3] Nine Chapters on the Mathematical Art).[b] It is paradoxical that although the content of the *Chiu Chang Suan Shu* is much fuller and more advanced than that of the *Chou Pei Suan Ching*, the first definite dates which we can connect with it are rather earlier than in the latter case. When Liu Hui[4] wrote the present introduction to the *Chiu Chang* just before +260, he stated what may have been the tradition in his time, that the book had first been arranged and commented on by Chang Tshang[5] (*fl.* −165, *d.* −142) and Kêng Shou-Chhang[6] (*fl.* −75 to −49),[c] both of the Former Han dynasty. Unfortunately, there is no mention of it in the biographies of either of these scholars. Moreover, like the *Chou Pei*, the *Chiu Chang* does not appear in the bibliography of the official history of the Former Han (*Chhien Han Shu*, *i wên chih*) completed about +100. But the same argument applies again, that its contents may have been included in some book which then had quite a different title, rendering it unrecognisable to us now. And hence there is little force, as Chang Yin-Lin (*3*) points out, in the fact that no mention of the *Chiu Chang* occurs in the *Chhi Lüeh*[7] (Bibliography in Seven Sections) of Liu Hsin (−50 to +20).[d] The full title first appears in the inscriptions on two bronze standard measures[e] dated +179.

[a] See Li Nien (*1*), pp. 24ff.; Chhien Pao-Tsung (*1*), pp. 28ff.; Wang Ling (2), vol. 2, pp. 133*j* ff.

[b] We are informed by Prof. Yushkevitch that a complete Russian translation of this work has been made by E. I. Berezkina. One of us (W.L.) is engaged upon an English translation of it.

[c] Hardly recognisable under the romanisation used by Mikami, and subsequently copied by all historians of mathematics—Ching Ch'ou-Ch'ang.

[d] This was of course the basis of the official bibliography of the *Chhien Han Shu*. A reconstruction of it from quotations in later works is contained in the Ma Kuo-Han collection (*YHSF*, ch. 64, p. 29*a*), and in Hung I-Hsüan's[8] *Ching Tien Chi Lin*.[9]

[e] Jung Kêng (*2*), ch. 3 (rubbings), pp. 12*a*, 13*a* (modern readings), p. 1*b*. We are indebted for this reference to Prof. A. F. P. Hulsewé.

[1] 陳杰 [2] 算法大成 [3] 九章算術 [4] 劉徽 [5] 張蒼
[6] 耿壽昌 [7] 七略 [8] 洪頤煊 [9] 經典集林

Surer ground is afforded by the commentaries on the *Chou Li* (Record of the Rites of Chou). In the entry for the Pao Shih,[1] the official charged with the education of the imperial princes, it is said[a] that among the matters in which they are to be instructed is the 'Nine Calculations' (*chiu shu*[2]). This may perhaps have meant, as some[b] have thought, the multiplication table, but the first commentator, Chêng Chung[3] (*fl.* +89, *d.* +114), is quoted by his great successor, Chêng Hsüan[4] (+127 to +200), as defining the Nine Calculations in such terms as to form a list almost identical with the chapter headings of the *Chiu Chang* as we have them today.[c] Thus it would seem that something very similar to the present text must have existed in the latter half of the +1st century. What connection it had with the texts known to Chang Tshang and the early Han scholars it is impossible to say. As for Chêng Hsüan himself, his biography tells us that he was a master of the *Chiu Chang*,[d] and adds that Liu Hung,[5] about +180, made a commentary upon it. So also did Tsu Chhung-Chih[6] (late +5th century),[e] but Liu Hui's is the one still printed.

One thing alone is certain, namely, that the *Chiu Chang* represents a much more advanced state of mathematical knowledge than the *Chou Pei*. If the latter is put back into the Warring States period, then the *Chiu Chang* is plausibly Former Han. If the *Chou Pei* is considered Former Han, then the *Chiu Chang* must be of the +1st century. But of course neither text can have sprung into existence suddenly. Perhaps the safest view is to regard the *Chou Pei* as a Chou nucleus with Han accretions, and the *Chiu Chang* as a Chhin and Former Han book with Later Han accretions.[f]

The *Chiu Chang Suan Shu*, in its influence perhaps the most important of all Chinese mathematical books, contains nine chapters with a total of 246 problems. The contents may be briefly described as follows:

(1) *Fang thien*[7] (Surveying of land). In this correct rules are given for the area of rectangles, trapezoid figures, triangles, circles ($\frac{3}{4}d^2$ and $\frac{1}{12}c^2$ with π taken as 3),[g] arcs and annuli. Rules are given for addition, subtraction, multiplication and division of fractions, as also for their reduction. The area of the segment of a circle is given[h] as $\frac{1}{2}(c+s)s$.

(2) *Su mi*[8] (Millet and rice); percentages[i] and proportions. The last nine problems

[a] Ch. 4, p. 8*b*, (ch. 13) tr. Biot (1), vol. 1, pp. 298, 299.

[b] E.g. Liu Tshao-Nan (*1*).

[c] The slight differences are analysed by Chang Yin-Lin (*3*) and need not be enumerated here. Sun Wên-Ching (*1*) has a special study of the chapter headings.

[d] *Hou Han Shu*, ch. 65, p. 12*a*.

[e] See Li Nien (*20*), p. 66.

[f] Chhien Pao-Tsung (*1*) and Chang Yin-Lin (*3*) have noted that the titles of certain officials mentioned in the statement of some of the problems are those current in Chhin and early Han times (−3rd and early −2nd centuries). There are also passing references which must indicate a taxation-system of −203. For a fuller analysis of the dating of these books, see Wang Ling (2), vol. 1, pp. 46ff.

[g] Where *d* is the diameter and *c* the circumference.

[h] Where *c* is the chord and *s* the sagitta of the segment. This expression is later found in the Indian *Gaṇitasārasaṃgraha* of Mahāvīra (*c.* +850). On this see Renou & Filliozat (1), vol. 2, p. 174. An analogous but more complex formula giving a closer approximation occurs in Heron (+1st century; *Opera*, vol. 3, pp. 73ff., vol. 4, p. 357, vol. 5, p. 187) and in the *Mishnāh ha Middot*, a Hebrew work which may be of the +2nd century (Gandz, 5).

[i] These are also in the *Arthaśāstra* (cf. Datta, 1). This work is now placed in the +3rd century (Kalyanov, 1).

[1] 保氏 [2] 九數 [3] 鄭衆 [4] 鄭玄 [5] 劉洪 [6] 祖沖之 [7] 方田 [8] 粟米

of this section were suited for the use of indeterminate equations, but these were avoided by reasoning on proportions.[a]

(3) *Tshui fên*[1] (Distribution by progressions) discusses partnership problems and rule of three. It contains questions involving ratios which seem to have strayed from the previous chapter; conversely the last nine problems of the previous chapter would seem to have belonged to this one. *Tshui fên* includes problems in taxation of goods of different qualities, and also others in arithmetical and geometric progression, all solved by proportion.[b]

(4) *Shao kuang*[2] (Diminishing breadth) deals with finding the sides of figures when areas and single sides are known. In this chapter there is much on square and cube roots. The former process led naturally to the quadratic equation of chapter 9.

(5) *Shang kung*[3] (Consultations on engineering works) is devoted to the mensuration and determination of volumes of solid figures (prism, cylinder, pyramid, circular cone, frustum of a cone, tetrahedron, wedge, etc.).[c] Walls, city-walls, dykes, canals and rivers are considered. Mikami (*1*) suggests that volumes may have been determined empirically in the first instance by the construction of models.

(6) *Chün shu*[4] (lit. Impartial taxation) deals with problems of pursuit[d] and alligation, especially in connection with the time required for the people to carry their grain contributions from their native towns to the capital. There are also problems of ratios in connection with the allocation of taxation burdens according to population.

(7) *Ying pu tsu*,[5] or *Ying nu*[6] (Excess and deficiency). Both these phrases apply to the full moon and to the new moon, indicating a condition of 'too much or too little'. The chapter is devoted to a Chinese algebraic invention, the 'Rule of False Position' (see p. 117 below), used mainly for solving equations of the type $ax = b$.

(8) *Fang chhêng*[7] (The way of calculating by tabulation). In all later usage, this expression came to mean any equation. It presumably grew up because of the way of writing equations used in the Han, and afterwards, in which the various quantities were placed in a table of rectangular columns. The chapter deals with simultaneous linear equations, using both positive (*chêng*[8]) and negative (*fu*[9]) numbers. This is the earliest appearance of negative quantities in any civilisation.[e] The last problem in the chapter, involving four equations and five unknowns, foreshadows indeterminate equations.

[a] Cf. Mikami (*1*), p. 24; Wang Ling (*2*), vol. 1, pp. 187ff.; on these equations see below, p. 119.

[b] See Wang Ling (*2*).

[c] Coolidge particularly admires the treatment of the truncated triangular right prism, where the formula is equivalent to $\frac{(b^1 + b^2 + b^3)\, dl}{6}$, b^1, b^2, b^3 being the three breadths of the wedge, d the depth and l the length. This is usually attributed to Legendre (Prop. 20, Bk. 6) and is not in Euclid.

[d] These are not found till much later in the West (Smith (*1*), vol. 2, p. 546). Alcuin of York about +780 gives some in his collection of puzzle problems.

[e] Their use was extended to quadratic equations probably by the time of Tsu Chhung-Chih (+5th century) and certainly by the time of Liu I (+11th century). Cf. pp. 101, 41 below.

1 衰分 2 少廣 3 商功 4 均輸 5 盈不足 6 盈朒
7 方程 8 正 9 負

(9) *Kou ku*[1] (Right angles), an elaboration, in algebraic terms, of the properties of the right-angled triangle already described in the *Chou Pei Suan Ching*. The twentieth of the twenty-four problems in this section involves the equation

$$x^2+(20+14)\,x-2\times 20\times 1775=0,$$

and though we cannot grant the text the antiquity which Smith & Mikami accepted, this example still remains a very ancient one. In this chapter appears the problem of the reed growing 1 ft above the surface at the centre of the water in a pond 10 ft square, which just reaches the surface when drawn to the edge, it being required to find the depth of the pond. There is also the problem of the broken bamboo which forms a natural right-angled triangle (Fig. 51). These are found also in later Indian mathematical books[a] and made their way to medieval Europe. Of the importance of similar right-angled triangles in measurement of heights and distances mention has already been made.

All the diagrams of the *Chiu Chang*'s first commentary were lost centuries ago, and modern editions[b] have them as added by Chhing scholars (see Fig. 52).[c] Each problem usually begins with the words *Chin yu*[2] ('now we have here such and such a situation...'). The solution starts with the words *Ta yüeh...*[3] ('the answer is...'). Further explanations are given under the heading *Shu yüeh...*[4] ('according to the art...'). All this is the *Ching* itself. Commentaries in small characters by Liu Hui (and, where so mentioned, by Li Shun-Fêng of the Thang) follow. The working out of the problems is described in written form by the modern editors, such as Li Huang, under the opening words *Tshao yüeh...*[5] ('the explanation is...'). If there is a diagram to be explained, he adds *Shuo yüeh...*[6] ('the caption says...').

Throughout later Chinese history the *Chiu Chang* was studied generation after generation. All kinds of scholars participated. If, for example, we read the biography of the mathematician Yin Shao,[7] who flourished *c.* +430 to +460 under the Northern Wei,[d] we find that his teachers included the hermit Chhêngkung Hsing,[8] the Buddhist monk Than-Ying[9] and a Taoist with a Buddhist name, Fa-Mu.[10] The appearance of Buddhists in the Chinese mathematical tradition may well indicate intellectual contacts with India.

Apart from the works so far described, it is certain that there were many other mathematical books current during the Han dynasty. Unfortunately, all were afterwards lost. But we know the titles of some of them, such as the *Lü Li Suan Fa*[11]

[a] E.g. Mahāvīra in the +9th century. Cf. Smith & Mikami (1), p. 14.

[b] Such as the *Chiu Chang Suan Shu Hsi Tshao Thu Shuo*[12] (The *CCSS* Carefully Explained and Illustrated with Diagrams) by Li Huang[13] (*fl.* +1790).

[c] Those of Tai Chen himself are found at the conclusion of each chapter of the *Chiu Chang* in the Palace edition, under the heading *CCSS Ting Ê Pu Thu*.[14]

[d] *Pei Shih*, ch. 89, p. 5*b*.

1 句股 2 今有 3 答曰 4 術曰 5 草曰 6 說曰
7 殷紹 8 成公興 9 曇影 10 法穆 11 律曆算法
12 九章算術細草圖說 13 李潢 14 訂訛補圖

(Mathematical Methods concerned with the Pitchpipes and Calendar), though the author's name has not survived. There were also *Suan Shu* books by two mathematicians of the Former Han dynasty (−1st century), Tu Chung[1] and Hsü Shang.[2]

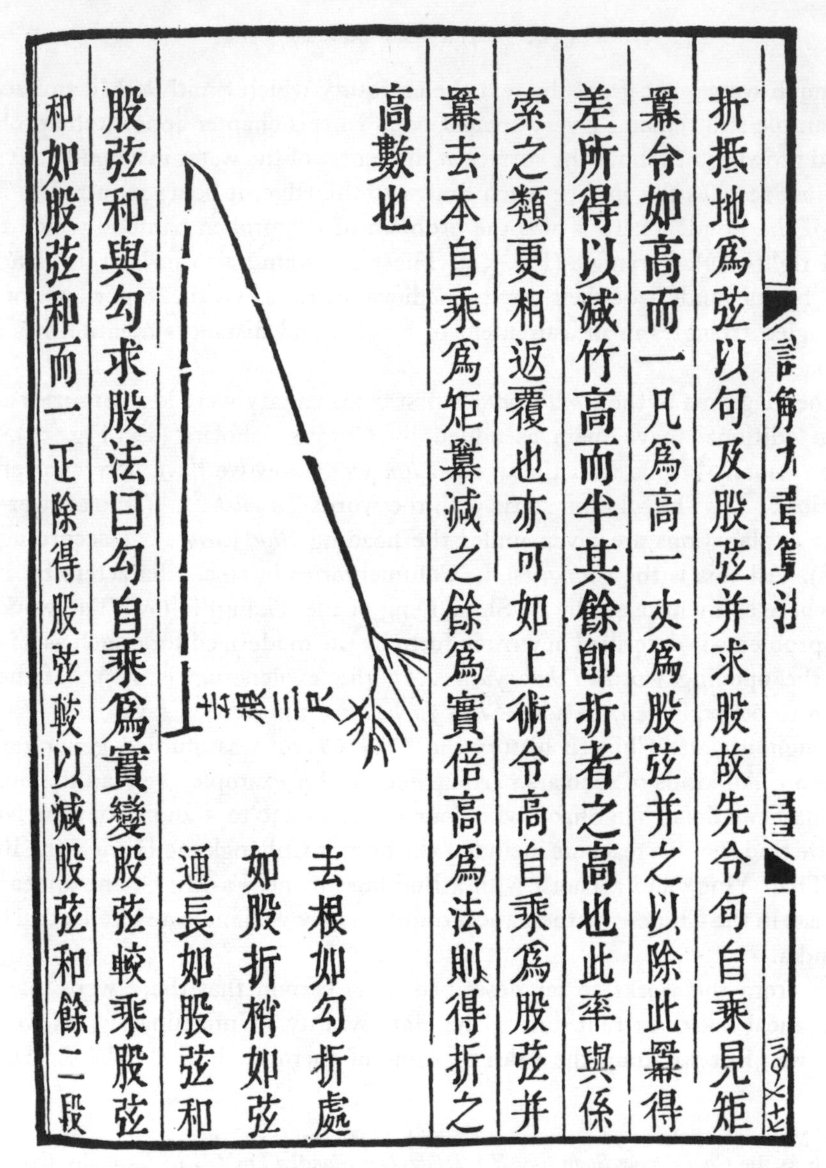

詳解九章算法

折抵地爲弦以句及股弦并求股故先令句自乘見矩冪令如高而一凡爲高一丈爲股弦并之以除此冪得差所得以減竹高而半其餘卽折者之高也此率與係索之類更相返覆也亦可如上術令高自乘爲股弦并冪去本自乘爲矩冪減之餘爲實倍高爲法則得折之高數也

去根如勾折處
如股折梢如弦
通長如股弦和

股弦和與勾求股法曰勾自乘爲實變股弦較乘股弦和如股弦和而一正除得股弦較以減股弦和餘二段

Fig. 51. The Problem of the Broken Bamboo (from Yang Hui's *Hsiang Chieh Chiu Chang Suan Fa*, +1261).

It would be very interesting to know what relation these had to the material at present found in the *Chiu Chang Suan Shu*. The traces of these old works remain in the bibliographical chapter of the *Chhien Han Shu* and the *Chhi Lüeh* fragments.

[1] 杜忠 [2] 許商

A book of considerable importance,[a] and quite different in style from any of those yet mentioned, is the *Shu Shu Chi I*[1] (Memoir on some Traditions of Mathematical Art) by Hsü Yo[2] of the Han. He appears in the *Chin Shu*,[b] discussing calendrical

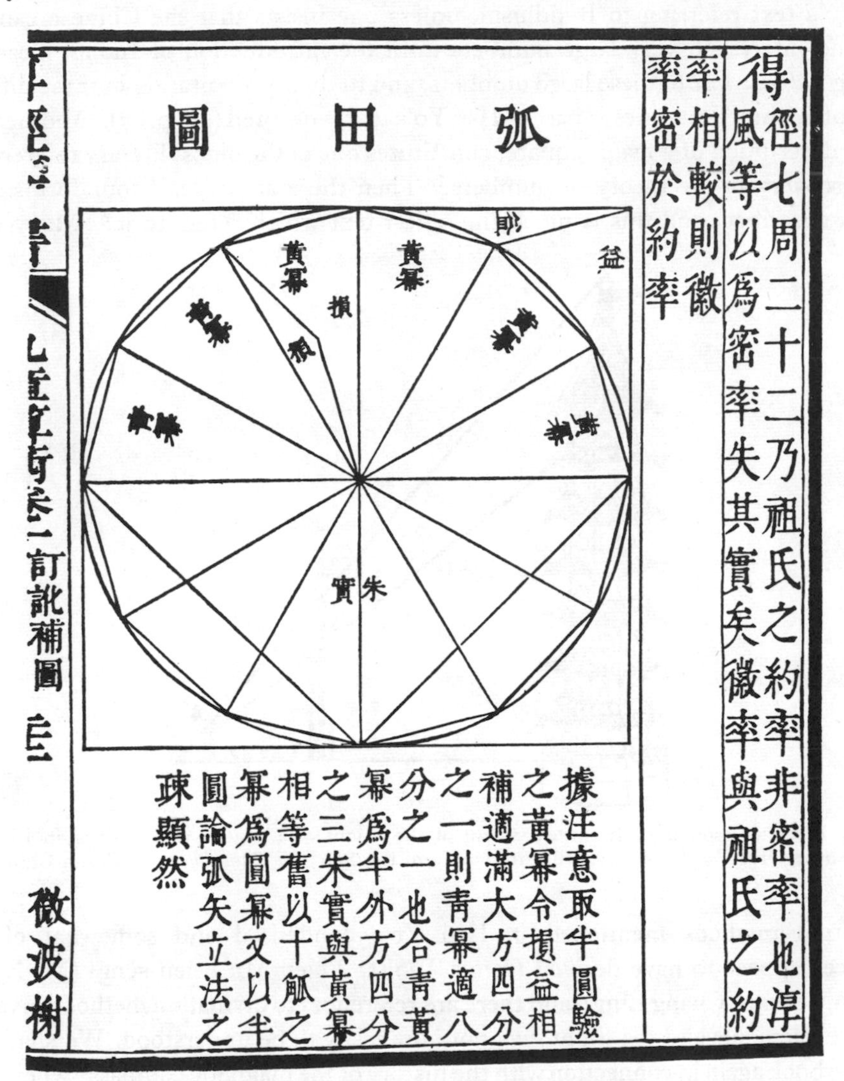

Fig. 52. One of Tai Chen's illustrations of the *Chiu Chang Suan Shu*, explaining Liu Hui's exhaustion method (+264) of finding the approximate value of π.

astronomy with Liu Hung,[c] Kaothang Lung[d] and Han I,[e] and must therefore have flourished about +190. The commentary on the book was written by Chen Luan (*fl.* +570). The *Shu Shu Chi I* is much nearer to Taoism and divination than the

[a] Not so much from the viewpoint of the history of mathematics as such, but for other reasons.
[b] Ch. 17, pp. 2*a*, 3*a*.
[c] *Fl.* +178 to +206. He was apparently the teacher of Hsü Yo.
[d] *Fl.* +213 to +235. We shall meet with him again in Sect. 27*c*.
[e] *Fl.* +223.

[1] 數術記遺 [2] 徐岳

other books described in this Section, and there are Buddhist references in Chen Luan's commentary, facts which have led many scholars[a] to regard the whole text as a later fabrication, perhaps by the commentator himself. Nevertheless, there is nothing in Hsü Yo's text referring to Buddhism, unless one insists that the Chinese can have taken no interest in very large numbers until the introduction of Indian ideas concerning *kalpas*. For to these large numbers, and to their presentation in three different forms of arithmetical series, a part of Hsü Yo's text is devoted (see p. 87). Another part, a clear description of a magic square, constitutes one of the oldest literary references to this discovery in the theory of numbers.[b] Then there are at least four forms of the abacus described, and this is much the oldest text which refers to it.[c] Many of the

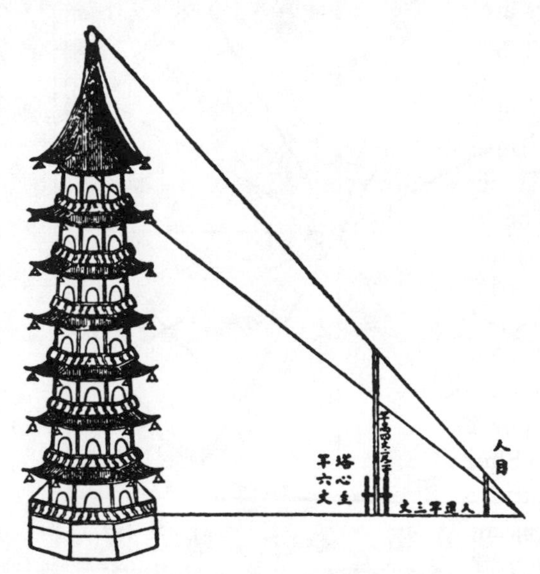

Fig. 53. Practical geometry; the measurement of the height of a pagoda as explained in Liu Hui's +3rd-century *Hai Tao Suan Ching* (illustration from the *Shu Shu Chiu Chang* of Chhin Chiu-Shao).

calculation methods mentioned in Hsü Yo's condensed and somewhat obscure sentences appear to have derived from a Taoist, Thien Mu hsien-sêng[1] (Mr Eye-of-Heaven, perhaps Chang Ling), and there are references to divination methods involving the Five Elements and the Eight *kua*, not now easily to be understood. We shall meet with the book again in connection with the history of the magnetic compass,[d] where it has an importance not dependent on whether Hsü Yo's text really goes back to the end of the +2nd century or not. Lastly, several systems of counting-rods are referred to in it.

The name of Liu Hui, whose redaction of the 'Nine Chapters' we have just spoken of, remains attached to another important book, the *Hai Tao Suan Ching*[2] (Sea Island Mathematical Manual), which appeared in +263 in the State of Wei in the Three

[a] Followed by Wylie (1), p. 92.
[b] See p. 58.
[c] See p. 77.
[d] See below, Sect. 26*i*.

[1] 天目先生 [2] 海島算經

Kingdoms period. It is no longer than one of the chapters of the *Chiu Chang*, at the end of which it is often printed, and it seems indeed to have been intended as an extension of the last chapter of the *Chiu Chang*. Another name for this last chapter was 'Chhung Chha',[1] which may be translated 'The Method of Double Differences', i.e. the properties of similar right-angled triangles; hence there is a clear relation to the contents of the 'Sea Island Manual'.[a] This is entirely concerned with the measurements of heights and distances using, when necessary, tall surveyor's poles and horizontal bars fixed at right angles to them. The following cases are considered: (*a*) measurement of the height of an island seen from the sea; (*b*) of the height of a tree on a hill; (*c*) of the size of a distant walled city; (*d*) of the depth of a ravine (henceforward cross-bars are needed); (*e*) of the height of a tower on a plain (cf. Fig. 53) seen from a hill; (*f*) of the breadth of a river-mouth seen from a distance on land; (*g*) of the depth of a transparent pool;[b] (*h*) of the width of a river seen from a hill; (*i*) of the size of a city seen from a mountain. The military, as well as civil, significance of such measurements will be obvious. The style of the book is very similar to that of the *Chiu Chang*. It has been translated by van Hée (7, 8).

Wylie (1) described the book as 'nine problems in practical trigonometry', but this is somewhat misleading,[c] since although it is wholly concerned with similar triangles, it never considers angular properties such as the sine, cosine, etc.

In the late Sung or Yuan dynasties, after having received a commentary by Li Shun-Fêng of the Thang, the book became rare and no editions of those times have survived. Fortunately it was copied into the Yung-Lo encyclopaedia (+1403 to +1407), and it was from this manuscript collection that it was recovered by Tai Chen in the 18th century. It so happens that the very volume of the *Yung-Lo Ta Tien* which contains quotations from the *Hai Tao Suan Ching* is preserved in the Cambridge University Library, and the first page of it is here reproduced (Fig. 54).

Mikami (1) observes that the method of working out the calculations of the unknowns required from the similar right-angled triangles set up, as described in the written text, was essentially algebraic. For example, the problem of the size of a city seen from a hill (see diagram) is worked out in 'rhetorical' forms equivalent to:

$$x=\frac{(d-b)c}{\dfrac{cd}{a}-b}, \quad y=\frac{\left(d-\dfrac{cd}{a}\right)b}{\dfrac{cd}{a}-b},$$

where CB in the accompanying diagram $=x$ and $BD=y$, $DE=a$, $DF=b$, $DG=c$,

[a] Before the Thang, the *Hai Tao Suan Ching* was known as the *Chiu Chang Chhung Chha Thu*.

[b] By viewing a white stone on the bottom, but no allowance is made for refraction.

[c] Many historians have followed Wylie here, e.g. in 1953, Struik (2), p. 35, who has to add that 'Western influence may not be excluded'.

[1] 重差

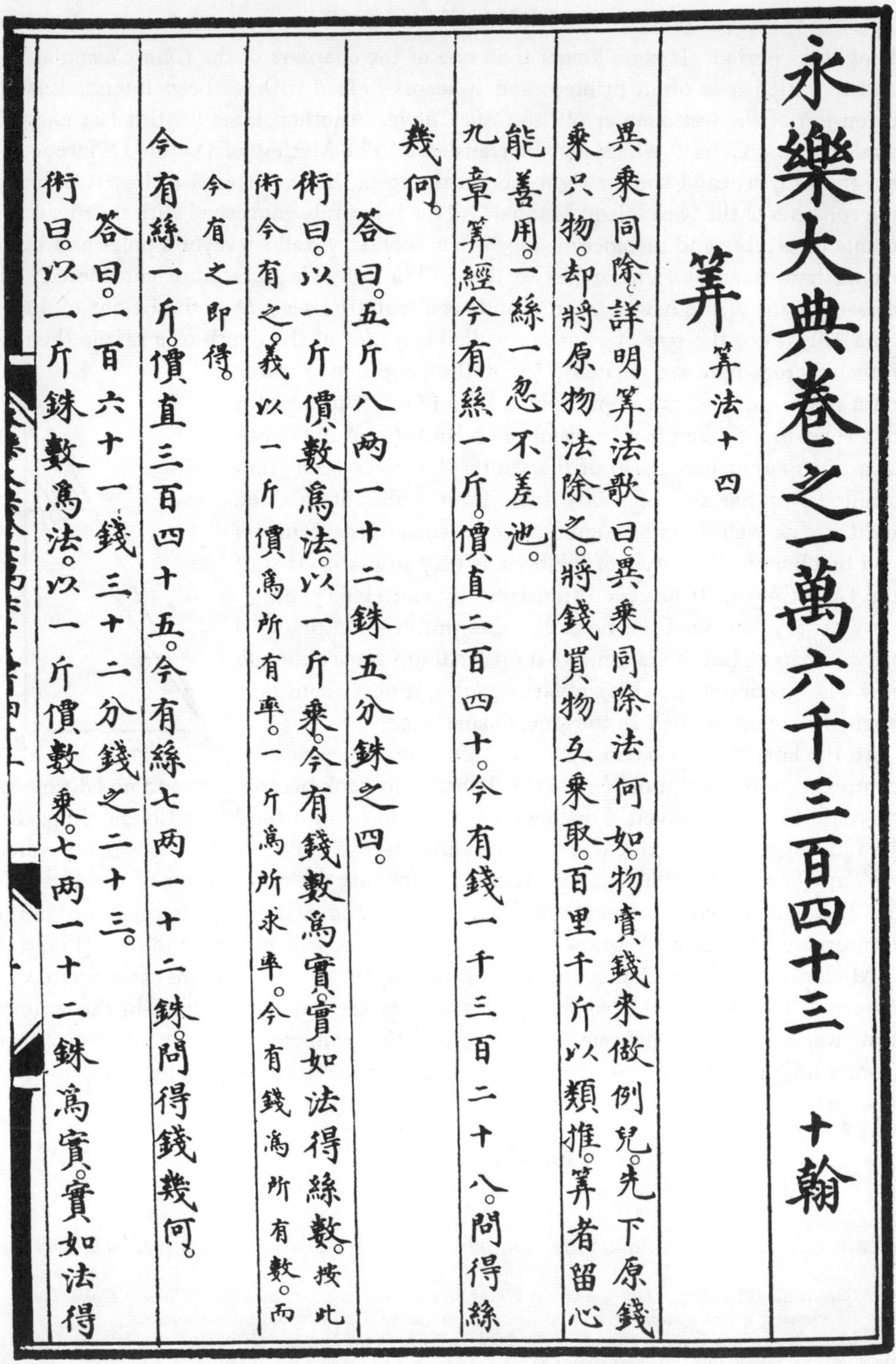

永樂大典卷之一萬六千三百四十三　十翰

筭　筭法十四

異乘同除。詳明筭法歌曰。異乘同除法何如。物賣錢來做例兒。先下原錢乘只物。却將原物法除之。將錢買物互乘取。百里千斤以類推。筭者留心能善用。一絲一忽不差池。

九章筭經今有絲一斤。價直二百四十。今有錢一千三百二十八。問得絲幾何。

答曰。五斤八兩一十二銖五分銖之四。

術曰。以一斤價數爲法。以一斤乘。今有錢數爲實。實如法得絲數。按此術今有之義。以一斤價爲所有率。一斤爲所求率。今有錢爲所有數。而今有之即得。

今有絲一斤。價直三百四十五。今有絲七兩一十二銖。問得錢幾何。

答曰。一百六十一錢三十二分錢之二十三。

術曰。以一斤銖數爲法。以一斤價數乘。七兩一十二銖爲實。實如法得

Fig. 54. The opening page of one of the chapters (ch. 16,343) of the MS. *Yung-Lo Ta Tien* (+1407). This chapter contains excerpts from the *Hsiang Ming Suan Fa*, a +14th-century work probably by Chia Hêng, the *Chiu Chang Suan Ching*, and, later on, the *Hai Tao Suan Ching*. Preserved in the Cambridge University Library.

DA=*d*. *A* and *F* are observer's positions, *E* and *D* are two poles connected with a cord at eye-level, and *C* and *B* the two ends of the city wall. But of course no such algebraic generalisations were ever thought necessary by the old Chinese calculators, who dealt only with specific practical problems.

(2) From the San Kuo to the Beginning of the Sung (+10th century)

During the following centuries there appeared half a dozen books which achieved renown, but which modern scholars have not found easy to date. The earliest of these is doubtless the *Sun Tzu Suan Ching*[1] (Master Sun's Arithmetical Manual). It has nothing to do with Sun Wu (cf. Sect. 30), the −6th-century general and author of the military treatise; but seems to be of the San Kuo, Chin, or (Liu) Sung periods, i.e. between +280 and +473. Its mention of 'Buddhist books', and its references to the distance between Chhang-an and Loyang, preclude it from being appreciably earlier.[a] There is next the *Wu Tshao Suan Ching*[2] (Mathematical Manual of the Five Government Departments); this is perhaps late Chin (+4th century). There follow the *Hsiahou Yang Suan Ching*[3] (Hsiahou Yang's Mathematical Manual) and the *Chang Chhiu-Chien Suan Ching*[4] (Chang Chhiu-Chien's Mathematical Manual). Unfortunately the dates of both these authors have been lost, and all one can say is that they were before Chen Luan,[5] who in a sense closes the period, and whose *floruit* was definitely Northern Chou (+560 to +580). He made commentaries on the books of both Hsiahou Yang and Chang Chhiu-Chien. Moreover, the latter must have been later than the former, since he mentions Hsiahou. Internal evidence shows that Chang must have written between +468 and +486.[b] Hsiahou Yang mentions a change in volume standard which took place in +425. It must be, therefore, that both Hsiahou and Chang lived under the Northern Wei.

The *Sun Tzu Suan Ching*, to the importance of which in relation to the rod-numerals we have already alluded, is a straightforward book concerned with the operations of multiplication and division, mensuration of areas and volumes, the handling of fractions, and square and cube roots. It begins with a description of the weights and measures in use at the time, and adds a short list of the densities of metals (gold, silver, copper, lead and iron), jade and 'stone'. Besides the other subjects discussed, the *Sun Tzu* gives the earliest example of a worked-out problem in indeterminate analysis (linear congruences);[c] this is the chief advance of the book

[a] A reference to a taxation method characteristic of the beginning of the Chin dynasty gives the earlier of these dates, a change in mensuration scale the latter (cf. Wang Ling (2), vol. 2, p. 43).

[b] See Wang Ling (2), vol. 2, p. 66.

[c] See below, p. 119.

[1] 孫子算經 [2] 五曹算經 [3] 夏侯陽算經 [4] 張邱建算經
[5] 甄鸞

over the 'Nine Chapters'.[a] But the solutions of the problems are not more clearly expressed.[b]

The *Wu Tshao* book,[c] the name of which suggests that it was an elementary handbook for government officials, is somewhat retrograde; its first section (*Thien tshao*[1]) is mostly concerned with the measurement of areas, and many of the formulae which it gives, and which require only multiplication and division, are either very rough approximations or positively wrong. The other sections (e.g. *Ping tshao*[2], military; *Chi tshao*[3] and *Tshang tshao*[4], granary storage; *Chin tshao*[5], exchange, and so on) all involve nothing more than multiplication and division.

Hsiahou Yang's book, as we have it now, is not much more interesting; it contains calculations of percentages and roots, as well as the ordinary logistic operations. Some of the mistakes of the *Wu Tshao* are repeated, e.g. the rule for determining the area of a figure consisting of two congruent trapezoids and a quadrilateral. However, the book seems to have been wholly re-written by Han Yen[6] (+780 to +804) in the Thang,[d] and the original may have been better. Van Hée (11), who devoted a special article to the *Hsiahou Yang Suan Ching*, suspected an Indian origin for the names of certain fractions used therein, such as *chung pan*[7] ('dead on half') for $\frac{1}{2}$, *shao pan*[8] ('lesser half')

[a] We call it the 'earliest example' advisedly. It is agreed that the Indian mathematicians who treated of this subject all lived much later than Sun Tzu (e.g. Āryabhaṭa the elder, *fl.* +476 to +510; Brahmagupta, *fl.* +598 to +628). But Dickson (1), vol. 2, p. 58, says that Nicomachus of Gerasa (*fl. c.* +90, probably a younger contemporary of Wang Chhung) gave the same problem as Sun Tzu, and its correct solution, 23. Here Dickson made a mistake. The problem occurs, not in either of the two parts of the *Introduction to Arithmetic* (*Eisagoge Arithmetike*, *εἰσαγωγὴ ἀριθμητική*) of Nicomachus, but in Problem no. 5 of the supplementary problems printed by Hoche in his edition of the text (pp. 148 ff.), but not translated by d'Ooge, Robbins & Karpinski (1) in their English version. This supplement is found only in the Codex Cizensis and its copies, i.e. in but two or three of the nearly fifty extant MSS. of Nicomachus. The codex itself dates from the late +14th or early +15th century. Still more, three of these five problems are signed by (or attributed by the writer to) one Isaac, an individual who was identified by Hoche himself with Isaac the Monk, i.e. Isaac Argyros. Isaac was a Byzantine mathematician and astronomer of merit who flourished about the middle of the +14th century (+1318 to +1372 approximately; Sarton (1), vol. 3, p. 1511; Fabricius (1), vol. 11, pp. 126 ff.). It seems clear, therefore, that the problem in indeterminate analysis is here of late medieval origin, and does not go back to Nicomachus himself. How Isaac got hold of it is another matter, but as the example given by Brahmagupta (Colebrooke (2), p. 326) is quite different numerically, it may be legitimate to think of a direct overland transmission. A more truly ancient example of an indeterminate equation is embodied in the famous 'Cattle problem' ascribed, somewhat doubtfully, to Archimedes (−3rd century); see Archibald (2); Smith (1), vol. 2, pp. 453, 584; Heath (5), pp. 142 ff. One of the lost books of Diophantus may have discussed this Pell equation. The exact date of Diophantus himself is not certain, but as most probably of the second half of the +3rd century he may well have been a contemporary of Sun Tzu. Diophantus describes many indeterminate quadratic equations which he solves by nearly as many different methods (cf. Heath (5), Gow (1), pp. 114 ff.). This branch of algebra was much cultivated later among the Arabs, e.g. Abū Bakr al-Farajī in the +11th century (cf. Woepcke, 3). One might derive Isaac Argyros wholly from the Arabs if he had not given the identical problem first stated and solved by Sun Tzu.

[b] It is interesting that among the last problems in *Sun Tzu* is one which relates to fortune-telling (the sex of an expected child). It may have slipped in to the text from an extraneous source, but calculation and fate-calculation were never far apart in those times. On the other hand, the appearance of beings or animals with many heads and limbs looks like an importation from the Indian pantheon.

[c] Smith (1), vol. 1, p. 141, confuses the *Sun Tzu* and the *Wu Tshao*, treating them as a single book. Becker & Hofmann, as recently as 1953, repeat this mistake (p. 134).

[d] This was long suspected by Chhien Pao-Tsung (*1*), p. 51, and Li Nien (*2*), 2nd edition, now agrees with him.

[1] 田曹 [2] 兵曹 [3] 集曹 [4] 倉曹 [5] 金曹 [6] 韓延
[7] 中半 [8] 少半

for $\frac{1}{3}$, *thai pan*[1] ('greater half') for $\frac{2}{3}$, and *jo pan*[2] ('weak half') for $\frac{1}{4}$—but the arguments are not convincing. The terms were clepsydra graduations, and similar expressions were used in astronomical measurements (cf. p. 268 below). *Thai pan* and *shao pan* occur in the *Chiu Chang* and others in the *Hou Han Shu*.

These books can only be placed in a proper perspective by comparing them with those of Europe and other parts of the world at the same time. From surveys such as that of Smith (1) it can be seen that since so much of Greek mathematics had been forgotten in Europe, India alone could compete with Chinese mathematical knowledge from the +3rd to the +8th centuries.

The book of Chang Chhiu-Chien (*c.* +500) is considerably longer than either of the two last named. His preface says that multiplication and division are not difficult, but that fractions give rise to many troubles; much of the book is therefore given over to them. Chang gave examples of the modern rule of division (already given in the *Chiu Chang*);[a] multiplying by the reciprocal of the divisor. This was known to the Indians, such as Mahāvīra in the +9th century, but then dropped out until re-discovered by Stifel in +1544.[b] Chang Chhiu-Chien's book also takes up surveying problems using similar right-angled triangles like the *Hai Tao*, percentages, rule of false position, simultaneous equations, rule of three, and indeterminate analysis. There are new problems in arithmetical progression connected with the weaving of textiles, correctly solved. Problems involving geometrical progression enter, too, with the right answers, obtained by proportion (as in the *Chiu Chang*) and not by the general formulae now known. There is extraction of square and cube roots.

Chen Luan (*c.* +570), who wrote so many commentaries on earlier mathematical works, also wrote a *Wu Ching Suan Shu*[3] (Arithmetic in the Five Classics), but it has not survived in full.[c] A famous problem in indeterminate analysis occurs in his commentary on Hsü Yo's book considered above, together with several surveyor's problems of classical type. He made a calendar for the Northern Chou dynasty. We have met with him before in connection with the Taoist religion,[d] which he attacked after his conversion to Buddhism.

Probably the most important lost book of this period is the *Chui Shu*[4] of Tsu Chhung-Chih[5] (+430 to +501). It had a very great reputation, and was supposed to require far more study than any of the other mathematical books; few could understand it. The probability is that it was connected with his remarkably accurate determination of the value of π which will be mentioned in its place (p. 99), and with the Method of Finite Differences in astronomy and calendrical theory. Chhien Pao-

[a] Ch. 1, p. 7*a*.

[b] Smith (1), vol. 2, p. 226.

[c] Or so it is usually said, but editions of the *Suan Ching Shih Shu* contain a book of some length with this title. The biography of Chen Luan's elder contemporary Hsintu Fang[6] in the *Pei Shih*, ch. 89, p. 13*b*, describes him as copying a *Wu Ching Suan Shih*.[7] If this was the same book it was probably a good deal older than either of them.

[d] Vol. 2, p. 150.

1 太半　2 弱半　3 五經算術　4 綴術　5 祖沖之
6 信都芳　7 五經算事

Tsung[a] brings forward some rather impressive arguments for thinking that Tsu expounded the approximation of fractions, and introduced the two new equations:

$$x^2 - ax = k,$$
$$x^3 - ax^2 - bx = k.$$

Our information about the book is derived mainly from the *Sui Shu* (History of the Sui Dynasty).[b] In the Sung, Shen Kua was interested in it and discussed it in his *Mêng Chhi Pi Than.*[c]

Tsu Chhung-Chih's book seems to be the chief exception to the rather static character of the Chinese mathematical literature after the *Chiu Chang Suan Shu.* Most of the books between the +3rd and the +6th centuries repeated the mistakes of the *Chiu Chang* while adding little that was new to its numerous solid achievements. It is interesting in connection with the governmental associations of the *Wu Tshao* that it was the worst of all. Apart from that, each writer made some special contribution. Sun Tzu gave details of the fundamental logistic processes which in the *Chiu Chang* had been implicit, and opened the new field of linear congruences in a single problem. Hsiahou stated more clearly than his predecessors the properties of powers of ten, and in one problem came extremely close to the developed conception of decimals by abandoning separate names for the successive places. Lastly, Chang Chhiu-Chien was the first to deal with true indeterminate equations, and to give the two formulae for arithmetical progressions; moreover, he cleverly found new applications for the quadratic equations of the *Chiu Chang.*

In the Thang dynasty (+618 to 906) mathematics began to make progress again. A corpus of the most important earlier works already mentioned was brought together and used as the official text-book for the imperial examinations. Fragments of arithmetical books of this period have been discovered among the Tunhuang manuscript library remains[d] and published by Li Nien;[e] the longest piece includes a complete multiplication table, and a series of sums of products.[f] The most important book of the early +7th century was the *Chhi Ku Suan Ching*[1] (Continuation of Ancient Mathematics) by Wang Hsiao-Thung.[2]

The 'Nine Chapters' in the Han had already employed a numerical quadratic equation for solving one of its problems.[g] The process is revealed by the technical term used to express it.[h] Then in Chang Chhiu-Chien's book the equation is expressed

[a] (*1*), p. 58.

[b] *Lü li chih*, ch. 16. There are short biographies of Tsu Chhung-Chih by Li Nien (2) and Chou Chhing-Chu (*1*).

[c] Ch. 8, para. 6.

[d] In Paris: Bib. Nat. Pelliot, nos. 2490, 2667, 3349; in London: Brit. Mus. Stein, nos. 19,930, 5779; O/N, nos. 42,518, 39,813, 39,760.

[e] (*7*); (*4*), vol. 1, pp. 123 ff.

[f] Its title is *Li-Chhêng Suan Ching*,[3] cf. above, p. 9, and below, p. 107.

[g] Mikami (1), p. 23, expressed the formulae correctly in modern notation, but the literal quadratic equation which he thought was embodied in the problem does not represent what the writer of the *Chiu Chang* did.

[h] I.e. *Khai fang chhu chih*,[4] meaning 'extract the square root', implying that the standard procedures for doing that should be followed. It is also said 'take the number in question as the *tshung fa*[5]', indicating what part of the procedures should be chosen. See Wang & Needham (1).

[1] 緝古算經 [2] 王孝通 [3] 立成算經 [4] 開方除之 [5] 從法

in algebraic form, using words (connected with a mensuration formula) instead of specific numbers. Quadratic equations and their solutions were therefore certainly known from the Later Han onwards. Equations of the third degree, however, are found for the first time in the *Chhi Ku Suan Ching*, which must have been produced in the near neighbourhood of +625. They arose, as usual, from the practical needs of engineers, architects and surveyors, as, for instance, in the problem: 'There is a right-angled triangle, the product of two sides of which is $706\frac{1}{50}$ and whose hypotenuse is greater than the first side by $30\frac{9}{60}$. It is required to know the lengths of the three sides.' The answer obtained involved the use of the equation $x^3+\frac{S}{2}x^2=\frac{P^2}{2S}$, where P is the product and S the surplus. Other problems concern the volumes of granaries. Wang Hsiao-Thung did not discuss equations of higher degrees.

It will be remembered from the historical introduction[a] that by the time of the Sui there had been a considerable introduction of Indian knowledge doubtless through Buddhist channels. Books with the prefix 'Brahmin' dealt with astronomy, calendrical science and mathematics.[b] Unfortunately, since all were subsequently lost, one cannot now estimate what they contributed. It is certain, however, that during the +7th and +8th centuries Indian scholars were employed in the Astronomical Bureau at the Chinese capital. Wang Hsiao-Thung may well have known some of the first of these men, such as Chiayeh Hsiao-Wei,[1] who was there shortly after +650, occupied with the improvement of the calendar, as were most of his later Indian successors.[c] The greatest of them was Chhüthan Hsi-Ta,[2] who became President of the Board, and produced the *Khai-Yuan Chan Ching*[3] about +729, a work of much importance, often elsewhere mentioned.[d] It seems that these Indians brought an early form of trigonometry, a technique which was then developing in their country.[e] All this indicates again the close association of mathematics with calendrical science in medieval China.

The same complex is clearly seen in the life of the most famous Thang mathematician, the monk I-Hsing,[4] who receives no less than three whole chapters in the *Chhou Jen Chuan*, more than any other scholar in any age, largely on account of his contributions to calendrical science. Between +721 and +727 he prepared by imperial order a calendar which, though it did not have much to do with indeterminate analysis, was known as the Ta Yen Li[5] (cf. below, p. 120). It is regrettable that so few records have come down to us from which the exact nature of I-Hsing's mathematical discoveries could be appreciated.[f] All his books are lost.

[a] Sect. 6*f* (Vol. 1, p. 128).
[b] One must bear in mind that the title of a book such as *Po-lo-mên Suan Fa* (see Vol. 1, p. 128) does not necessarily mean that it concerned mathematics; the subject discoursed of by the Indian author or the Chinese transmitter of Indian learning could have been fate-calculation and divination. See Li Nien (*2*), p. 60, and Sarton (1), vol. 1, p. 450.
[c] See *Chiu Thang Shu*, ch. 33, p. 17*a*.
[d] See above, p. 12, and below, p. 203.
[e] Yabuuchi (*1*), p. 154; (1), p. 588. See below, pp. 148, 202. The *Khai-Yuan Chan Ching* includes a table of sines.
[f] See pp. 119, 139, 202 below.

[1] 迦葉孝威 [2] 瞿曇悉達 [3] 開元占經 [4] 一行 [5] 大衍曆

In the *Ming Huang Tsa Lu*[1] of Chêng Chhu-Hui[2] written in +855 there is an account of I-Hsing's life, from which a quotation may be permitted at this point by way of enlivening an otherwise austere Section:

(Before he was introduced to the emperor) I-Hsing had studied under Phu-Chi at Sung Shan. There was an entertainment of monks and *śramaṇas* from a hundred *li* around. Among them was a very learned one named Lu Hung. This hermit was asked to write an essay commemorating the meeting, and did so, using very difficult words, and saying that he would take as his pupil anyone who could read and understand it. I-Hsing came forward smiling, looked at it, and laid it down. Hung despised his offhand manner, but when I-Hsing repeated it without one mistake, he was quite overcome and said to Phu-Chi, 'This student is not a person you can teach, you had better let him travel.'

So I-Hsing, wishing to study the *thai yen* (indeterminate analysis), travelled to look for a good teacher, even as far as a thousand *li*. Thus he came to the astronomical observatory at the Kuo Chhing Ssu temple, where he saw a courtyard with an old pine-tree and a spring flowing in front of it. I-Hsing stood still and held his breath for he heard from inside the sound of a monk making mathematical calculations (with the counting-rods). Meanwhile the old monk inside was saying, 'Today there must be someone coming to learn my mathematical arts. He should be at the door by now. Why is no one bringing him in?' Then he went back to his work, but after a while he said again, 'Before the gate the waters are meeting and flowing to the west—my student should be arriving.' So I-Hsing entered and knelt down before him. And the old monk began to teach him his computing methods and systems, and immediately the water before the gate changed its direction, and flowed eastwards.[a]

The story, charming in itself, suggests the difficulties which mathematicians had in those days in communicating with each other, and shows how easily discoveries and improvements might die with their authors.[b]

This was also the period of Li Shun-Fêng,[3] who has already been mentioned, probably the greatest commentator of mathematical books in all Chinese history. His work is described by Li Nien.[c] As an original discoverer in mathematics, however, he can hardly be compared with Wang Hsiao-Thung.

(3) The Sung, Yuan and Ming Periods

Between these men and the great figures of the Sung and Yuan algebraic period (+13th and +14th centuries) there stand few mathematicians of importance. The most interesting is Shen Kua,[4] whose many-sided genius has already been mentioned in the historical Section.[d] His book, the *Mêng Chhi Pi Than*[5] (Dream Pool Essays) of +1086, is not a formal mathematical treatise, for it contains notes on almost every science known in his time, but there is much of algebraic and geometric interest to be

[a] App. p. 2*a*, *b*; tr. auct.

[b] Another story in the same text illustrates I-Hsing's friendly relations with the Taoists, one of whom lent and discussed with him the *Thai Hsüan Ching* of Yang Hsiung (cf. Vol. 2, p. 350).

[c] (*2*), pp. 40 ff. Cf. pp. 27, 31 above.

[d] Vol. 1, p. 135. His name appears in histories of mathematics as 'Ch'ön Huo' and other disguises, hardly to be penetrated even by the elect.

[1] 明皇雜錄 [2] 鄭處誨 [3] 李淳風 [4] 沈括 [5] 夢溪筆談

found in it. In particular, Shen Kua, whose duties as a high official had placed him in charge of considerable engineering and survey works, made progress in plane geometry. His method for the determination of the lengths of arcs of circles formed the basis, as Juan Yuan and Gauchet (7) have shown, of the spherical trigonometry evolved by Kuo Shou-Ching[1] in the +13th century.[a]

In the 18th chapter of the *Mêng Chhi Pi Than* (para. 4) Shen Kua says:

For effecting the division of a circumference I have another way. Take the diameter (*ching*[2]) of a circular area (*yuan thien*[3]) and halve it, then let this (radius) be taken as the hypotenuse (*hsüan*[4]) of a right-angled triangle. Let the difference which arises when the radius is diminished by the divided part (the sagitta) be the first side of the triangle (*ku*[5]). Subtract the square of this side from that of the hypotenuse, extract the square root (*khai fang*[6]) of the remainder; this will give the second side (*kou*[7]). Twice this is the chord (*hsüan*[4]) of the segmental area (*hu thien*[8]). Square the divided part (the sagitta, *shih*[9]),[b] and double the result. Divide the result by the diameter, and add the chord, which will give the corresponding arc.

This is equivalent to the expression

$$a = c + 2\frac{s^2}{d},$$

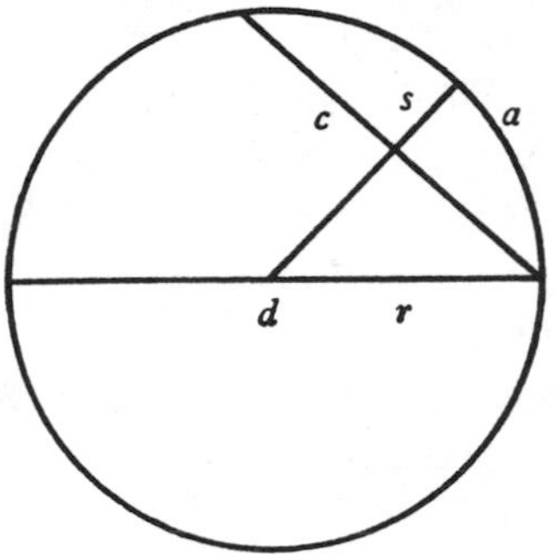

where a is the arc, c the chord, s the sagitta and d the diameter. This method was known as the *hu shih ko yuan chih fa*,[10] the arc-sagitta method for the determination of lengths of arcs of circle segments. This may be compared with the formula of Kuo Shou-Ching:

$$d^2\left(\frac{a}{2}\right)^2 - d^3s - (d^2 - ad)s^2 + s^4 = 0,$$

which is much more accurate, but two centuries later.[c]

Shen Kua also gives in his book (chapter 18) the first instance of the summation of a series in Chinese mathematics; it relates to the number of kegs which can be piled up in layers in a space shaped like the frustum of a rectangular pyramid. This was important from several points of view (see below, pp. 138 and 142).

Printing had begun in the +8th century, and had been applied to the dissemination of the classics in the +9th. While it seems possible that mathematical books were printed then also,[d] the first edition of which we have knowledge was that of the *Hai Tao*

[a] See p. 109 below. Hu Tao-Ching (*1*) has recently produced a critical edition of the *Mêng Chhi Pi Than*, and a brief biography of Shen Kua has been written by Chhien Chün-Hua (*1*).

[b] There are textual difficulties at this point, on which we have followed in the main Mikami (1) rather than Gauchet (7). Most of the technical terms inserted, though not *shih*, occur in Shen Kua's own words. Cf. Hu Tao-Ching (*1*), vol. 2, pp. 575 ff.

[c] See Chhien Pao-Tsung (*1*), p. 149; van Hée (13). Shen Kua developed his method because he was dissatisfied with the roughly approximative and indeed incorrect formula for circle segment areas given in the *Chiu Chang*, $\frac{1}{2}(sc+s^2)$ (cf. p. 98 below).

[d] Cf. Carter (1), 2nd ed., p. 60.

1 郭守敬 2 徑 3 圓田 4 弦 5 股 6 開方 7 句
8 弧田 9 矢 10 弧矢割圓之法

Suan Ching and nine other books in +1084, and that of Chang Chhiu-Chien's book in the following year. These were distributed to all government libraries. In +1115 there was an important edition of the *Chiu Chang Suan Ching*. It will be remembered that this was the period of Wang An-Shih and the reformist movement in the Sung dynasty. It is of much interest that his successor Tshai Ching[1] is specifically stated[a] to have encouraged the study of mathematics for the imperial examinations (+1108). Yen Tun-Chieh (*9*) has traced a connection between the Sung mathematicians and the use of paper money at that time.

In what follows only the greatest figures of the Sung and Yuan will be mentioned, but it must not be thought that during the +12th to 14th centuries they were isolated. A wealth of books is noted for this period by Li Nien (*1*, *2*). Some of them were contained in bibliographies of the Sung which still survive, such as the library catalogue of Yu Mou[2] (*Sui Chhu Thang Shu Mu*[3]),[b] where ninety-five titles are listed in the mathematical section. Among the works of this period may be mentioned Tshai Yuan-Ting's[4] *Ta Yen Hsiang Shuo*[5] (Explanation of Indeterminate Analysis) by a student of the *I Ching* who was interested in the permutations and combinations of the *kua*.[c]

Whether or not the fact is significant, it was in the later phase of Sung science (i.e. after +1200), that the principal contributions were made. In fact the most brilliant work was done by several contemporaries in a period covering almost exactly the last half of the +13th century. It opens with the *Shu Shu Chiu Chang*[6] (Mathematical Treatise in Nine Sections) of Chhin Chiu-Shao,[7] which appeared in +1247. The nine sections of this important work have nothing to do with the divisions of material in the 'Nine Chapters' of the Han. Though we do not know the dates of Chhin Chiu-Shao's birth and death,[d] we know enough about him[e] to say that he must have been rather an intriguing character; in his youth he was an army officer[f] and famous both for athletic and literary achievements. In love affairs he had a reputation similar to that of Avicenna. He rose to the rank of Governor of two cities.

In +1248, only a year later than Chhin Chiu-Shao's book, there appeared an equally important work, the *Tshê Yuan Hai Ching*[8] (Sea Mirror of Circle Measurements) by Li Yeh,[9] and in +1259 it was followed by the same mathematician's *I Ku Yen Tuan*[10] (New Steps in Computation). Li Yeh[g] (+1178 to +1265) was a northerner and lived

[a] *Sung Shih*, ch. 20, pp. 8*a*, 9*b*; *TH*, p. 1608. Cf. Li Nien (*21*), vol. 4, pp. 238ff., 260.

[b] Contained in ch. 28 of the *Shuo Fu* collection.

[c] Forke (9), p. 204. Cf. p. 599 below.

[d] In Sarton's judgment 'one of the greatest mathematicians of his race, of his time, and indeed of all times'.

[e] From the memoir on him contained in the *Kuei-Hsin Tsa Chih Hsü Chi*[11] of Chou Mi,[12] ch. 2, p. 5*b*.

[f] In this he is reminiscent of Descartes, but we do not know whether, like the French philosopher, he joined the army in order to have leisure for mathematical studies.

[g] There has been some doubt whether the reading Li Yeh was not a misprint for Li Chih,[13] which is written in a very similar way. But Miu Yüeh has shown (*1*) that his original name was Li Chih and, finding that it was the same as one of the Thang emperors, he changed it in middle life to Li Yeh.

1 蔡京　2 尤袤　3 遂初堂書目　4 蔡元定　5 大衍詳說
6 數書九章　7 秦九韶　8 測圓海鏡　9 李冶　10 益古演段
11 癸辛雜識續集　12 周密　13 李治

in the state of Chin; he was Governor of a city when the province fell into the hands of the Sung, so after his escape he found retirement to pursue mathematical studies. After the overthrow of the Chin by the Mongols in +1234 he was consulted by the Yuan government, and later shown many honours, but did not again enter the civil service. Although both Chhin Chiu-Shao and Li Yeh used the same symbols, and in many ways their works complement each other, it is improbable that they ever met, since the former lived in southern Sung and the latter in northern Chin and Yuan, states which were almost continuously at war during the lifetimes of the two men.[a]

The algebraic works of Chhin and Li were soon supplemented by those of another gifted mathematician, Yang Hui,[1] whose *Hsiang Chieh Chiu Chang Suan Fa Tsuan Lei*[2] (Detailed Analysis of the Mathematical Rules in the 'Nine Chapters' and their Reclassification) appeared in +1261. This was extended by other books later collected in the *Yang Hui Suan Fa*[3] (Yang Hui's Computing Methods), one of which is dated +1275. Practically nothing is known of his life, save that he was a southerner and lived under the Sung. Yang Hui does not mention Chhin or Li, but refers to Liu I[4] as his predecessor; unfortunately no other information has come down to us concerning Liu, but he must have been an algebraist. Moreover, Yang Hui does not use the expression *thien yuan shu*[5] ('array of the coefficients of unknowns'), characteristic of the other writers, for algebra.

Last to appear was the greatest of the group,[b] Chu Shih-Chieh,[6] whose *Suan Hsüeh Chhi Mêng*[7] (Introduction to Mathematical Studies) appeared in +1299, followed by the famous *Ssu Yuan Yü Chien*[8] (Precious Mirror of the Four Elements) in +1303. Very little is known of the circumstances of Chu Shih-Chieh's life, or of the dates of his birth and death, but he seems to have been a wandering scholar, earning his living by teaching the mathematical art. In his preface he hints that his tradition had descended through Yuan Hao-Wên,[9] an official of the Chin dynasty (therefore prior to +1230) who had been a friend of Li Yeh and had commented on an (otherwise unknown) earlier algebraic work by Liu Ju-Hsieh,[10] the *Ju Chi Shih Hsiao*[11] (Full Explanation of Tabulated Equations). This would take the celestial element algebra well back into the +12th century, the time of the great Neo-Confucian philosophers.

It is strange, as Sarton has well said,[c] that the relations between these four men, Chhin, Li, Yang and Chu, following one another in half a century, were so remote. Chhin perfected one terminology for the solution of equations. Li perfected another, though the general system was the same. Yang quoted neither of them, but stemmed from other mathematicians, Liu I and Chia Hsien, about whom we know nothing else. Chu cited none of his three predecessors (though there is internal evidence that he knew of Yang), and indicates that his affiliations were with other men again unknown,

[a] Cf. Vol. 1, p. 134. On the possible effects of wars on the history of mathematics in China and Japan, see Mikami (*18*).

[b] And perhaps, Sarton thinks, the greatest of all medieval mathematicians.

[c] (1), vol. 3, p. 138.

1 楊輝　2 詳解九章算法纂類　3 楊輝算法　4 劉益　5 天元術
6 朱世傑　7 算學啓蒙　8 四元玉鑑　9 元好問　10 劉汝諧
11 如積釋鎖

Yuan Hao-Wên and Liu Ju-Hsieh.[a] 'All this suggests that our knowledge of the medieval Chinese school is still very fragmentary. We know only a few of the treatises, and even these have been insufficiently investigated.'[b] One's impression that there are illuminating discoveries still to be made in the mass of Chinese mathematical literature is strongly reinforced.[c]

Another interesting point is the difference in social position between Sung and pre-Sung mathematicians. While Li Shun-Fêng and Wang Hsiao-Thung in the Thang had occupied high official positions, like Tsu Chhung-Chih in Liu Sung, the greatest mathematical minds were now (with the exception of Shen Kua) mostly wandering plebeians or minor officials. Moreover, their attention was devoted less to calendrical calculations, and more to practical problems in which the common people and technicians were likely to be interested. In fact one might almost point to the freedom from bureaucratic control in the Jurchen Chin and Mongol Yuan empires, together with the obstacles to official careers which Chinese scholars then encountered, as among the major liberating factors of the upsurge of Chinese mathematics in this period.

The first section of Chhin Chiu-Shao's *Shu Shu Chiu Chang* is concerned with indeterminate analysis (linear congruences), called by him the *Ta Yen Chhiu I*.[1] A number is sought such that it will give when repeatedly divided by $m_1, m_2, m_3, \ldots, m_n$, the remainders $r_1, r_2, r_3, \ldots, r_n$. It is here that a famous phrase appears for the first time in the term *thien yuan i*[2] (lit. 'heaven origin unit'), i.e. the unity symbol (1) placed at the left-hand top corner of the counting board before the beginning of one of the most important parts of the operation. It was probably so called to distinguish it from the unity symbol (1) of the last remainder. Chhin refers to the use of such methods in calendrical computations.

In later sections of the book there are calculations of complex areas and volumes. Treatment of equations of higher degrees, involving powers up to 10, and any combination of signs of terms, appeared for the first time in Chhin's work, arising in connection with such problems as the determination of the diameter and circumference of a circular walled city from a distance (Fig. 55). It is interesting to find problems of allocation of irrigation water, the construction of dykes (Fig. 56), and financial affairs involving

[a] And they were only part of an extended school, to judge from one of the prefaces to Chu Shih-Chieh's book. There had been Chiang Chou,[3] who had written an *I Ku Chi*[4] as early as +1080 if he is the same person as Chiang Shun-Yuan;[5] Li Wên-I[6] who wrote the *Chao Tan*[7] (Bright Courage—in solving equations?), and Shih Hsin-Tao,[8] who wrote a *Chhien Ching*[9] (Mathematical Key). We also learn that Li Tê-Tsai[10] had introduced the 'earth' element into the Thien Yuan algebra and written a *Liang I Chhün Ying Chi*;[11] while the 'man' element had been introduced by Liu Ta-Chien,[12] author of a *Chhien Khun Kua Nang*.[13] All, all were lost.

[b] Sarton adds that not one of the Sung algebraic texts is yet available in a critical edition, nor is there a complete translation of any one of them properly annotated by sinologists and mathematicians.

[c] On the background of the Sung algebraists, including the appearance of the zero, there is an important recent paper by Yen Tun-Chieh (*6*).

[1] 大衍求一 [2] 天元一 [3] 蔣周 [4] 益古集 [5] 蔣舜元
[6] 李文一 [7] 照膽 [8] 石信道 [9] 鈐經 [10] 李德載
[11] 兩儀羣英集 [12] 劉大鑑 [13] 乾坤括囊

arithmetical progressions and simultaneous linear equations.[a] As has already been mentioned, the use of the zero occurs freely in Chhin Chiu-Shao's and all later books. Negative and positive numbers were distinguished by black type for the former and red for the latter.[b] In the solution of numerical higher equations Chhin developed an ancient method substantially the same as that rediscovered by Horner in +1819.[c] In Europe the advantage of always equating to zero in the study of general equations was not recognised until the beginning of the +17th century; the merit belongs either to Napier (1594), Bürgi (1619) or Harriot (1621).[d] But a similar achievement was made

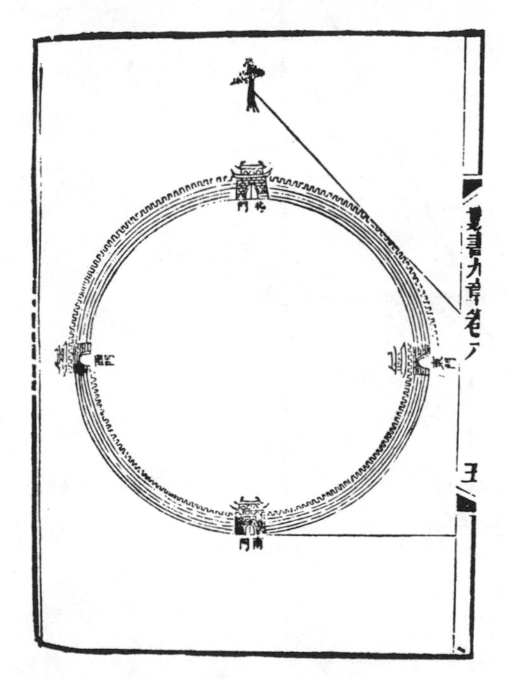

Fig. 55. Determination of the diameter and circumference of a circular walled city from distant observations (*Shu Shu Chiu Chang*, +1247).

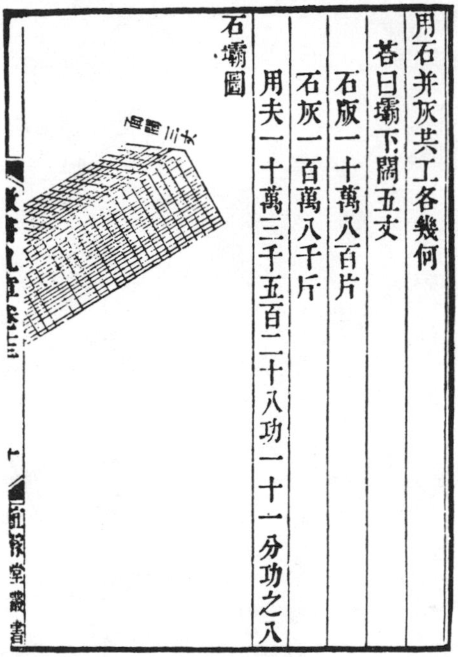

用石并灰共工各幾何
答曰壩下闊五丈
石版一十萬八百片
石灰一百萬八千斤
用夫一十萬三千五百二十八功一十一分功之八
石壩圖

Fig. 56. A problem in dyke construction from the *Shu Shu Chiu Chang* (+1247).

by Chhin Chiu-Shao and all the Sung and Yuan algebraists after him, in always arranging that the absolute term should be negative.

Naturally enough, it is not difficult to find traces of Neo-Confucian ideas in the work of Chhin Chiu-Shao. In his first chapter there is mention of the *Thai Yin*,[1] *Hsiao Yang*[2] and so on, in connection with the five elements, and one of the central

[a] It is hard to understand the comment of Smith (1), vol. 1, p. 270, that Chhin 'showed little interest in applying his knowledge of algebra to the solution of practical problems, preferring to look upon it as a pure science'.

[b] In Chhin Chiu-Shao and all later mathematical books there is much use of the monogrammatic way of writing numbers, which fits into the vertical columns of Chinese type.

[c] Wang & Needham (1); Smith (1), vol. 2, p. 471; see below, p. 66. [d] Smith (1), vol. 2, p. 431.

[1] 太陰 [2] 小陽

features of the algebraic notation was, as we shall see (p. 129), the use of the word *Thai* (standing for *Thai Chi*[1]) as the symbol for the known quantity.[a]

None of Chhin Chiu-Shao's work has been translated into a Western language. In later centuries there were many explanations of it after its rediscovery in the Chhing, notably the *Shu Shu Chiu Chang Cha Chi*[2] of Sung Ching-Chhang[3] (1842).

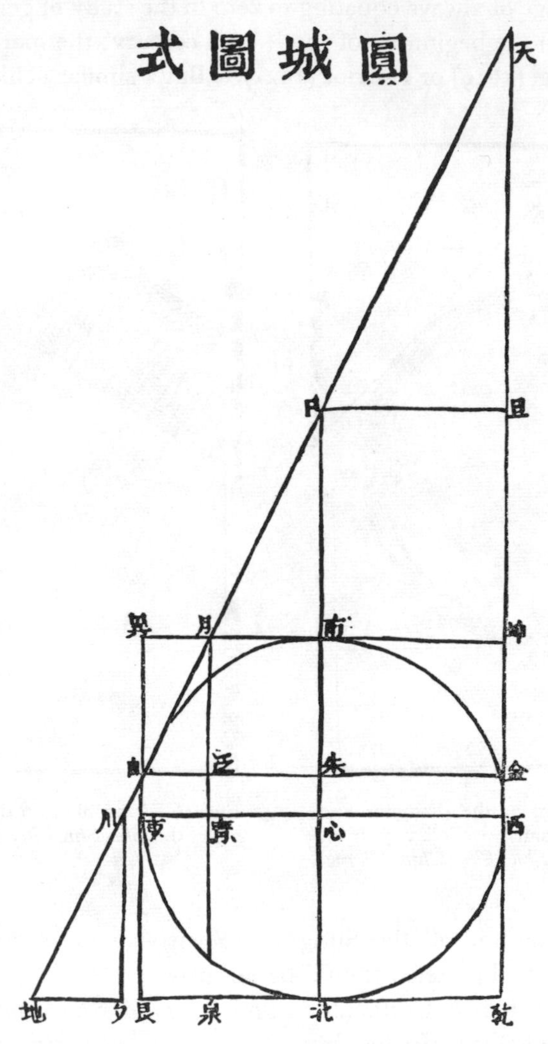

Fig. 57. The diagram prefixed to Li Yeh's *Tshê Yuan Hai Ching* (+1248) illustrating the properties of circles inscribed in right-angled triangles (see p. 129).

Li Yeh's 'Sea Mirror of Circle Measurements' deals partly with the properties of circles inscribed in triangles (Fig. 57 shows the illustration prefixed to his book), but is principally concerned with the solution of equations. The treatment is entirely

[a] Cf. Vol. 2, pp. 460ff.

[1] 太極 [2] 數書九章札記 [3] 宋景昌

algebraical.[a] His *I Ku Yen Tuan* was also concerned with algebra, though the notation in the two books is somewhat different. Li Yeh uses the term 'coefficient array' (*thien yuan shu*[1]), and applies it not to indeterminate analysis, as Chhin Chiu-Shao had done, but to the notation of numerical equations. In Li Yeh's method of writing equations the usual vertical columns and horizontal rows were used, the unknown being labelled *yuan* or element, and the absolute term coming under the linear term and being labelled *thai*.[b] The square and the cube term, if present, come above the unknown. Thus the following notation

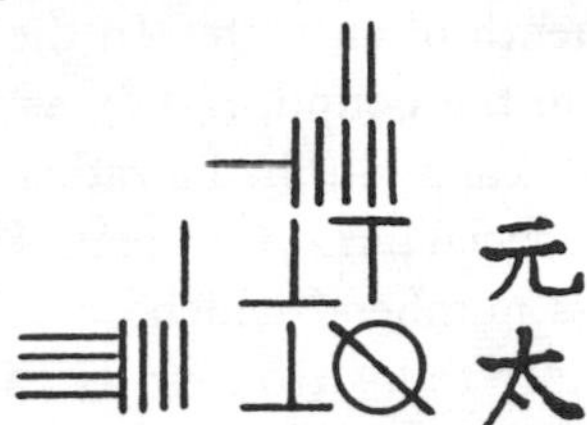

represents the equation $2x^3+15x^2+166x-4460=0$. It will be seen that there is a diagonal stroke through one of the digits of one of the numbers; this indicates that it is negative. Such a notation was an improvement on the earlier red and black colours, and was generally adopted after Chhin and Li. Later writers returned to Chhin's custom, however, of placing the absolute term uppermost and all the others below it, and Li himself does this in the *I Ku Yen Tuan*. The column system in any case doubtless represented the actual arrangement of counting-rods on the counting-board.

There is no translation of the *Tshê Yuan Hai Ching* of Li Yeh,[c] but the sixty-four algebraical problems of the *I Ku Yen Tuan* have been translated by van Hée (4). Among articles in the recent Chinese literature there is one on this book by Liu Ping-Hsüan (*1*).

Yang Hui stands a little apart from the rest of the group. He was interested in arithmetical progressions. In his *Suan Fa* he gives rules for summing the series

$$1+(1+2)+(1+2+3)+\ \dots\ +(1+2+3+\ \dots\ +n)$$

and

$$1^2+2^2+3^2+\ \dots\ +n^2,$$

but no explanation or analysis of them. He was also interested in alligation problems. He considered questions in compound proportion (*chhung hu huan*,[2] lit. 'doubled alternate exchanges') and in simultaneous linear equations dealt with up to five unknowns. Yang Hui was well able to manipulate decimal fractions. He has a problem of a rectangular field of breadth 24 paces $3\frac{4}{10}$ ft. and length 36 paces $2\frac{8}{10}$ ft.; wishing to multiply these numbers, he expresses them in decimal parts of the pace, as $24{\cdot}68\times36{\cdot}56=902{\cdot}3008$. This was equivalent to our use of the decimal point.

[a] See below, p. 129.

[b] In practice he uses only one of these symbols to fix the meaning of the lines in the column.

[c] Since Biernatzki his name has appeared in histories of mathematics under the brilliant disguise of Le-Yay-Jin-King, owing to a confused use of his literary name as well as his *ming-tzu*.

[1] 天元術 [2] 重互換

This needs a word of explanation. In the Sung, the first three places of decimals were named by Yang Hui *fên*,[1] *li*[2] and *hao*[3] respectively, even when, as in the foregoing example,[a] he was dealing with a problem involving lengths. Now, apart from the fact that in his time these terms, or at least the second and third of them, had become measures of weight rather than of length,[b] he does not use, in the above problem, the terms which would be expected. The real foot (*chhih*[4]) and the real inch (*tshun*[5]) do not appear in his expression, and the first place of decimals in his 0·68, although more than half a pace (*pu*[6]), he calls *fên*.[1] Yet this word, if correctly applied as a length measure, meant a tenth of an inch. In the same way, another example[c] expresses 6 *liang*[7] (ounces, 16 to the pound, *chin*[8]), as 3 *fên*,[1] 7 *li*,[2] 5 *hao*,[3] i.e. 0·375 *chin*.[8] All this goes to show that Yang Hui had a rather highly developed conception of decimal places, and adopted special terms for them. His decimals were by no means restricted to the 'little nameless numbers' which occurred in root extraction, as had been the case in Liu Hui's time. In the +3rd century Liu Hui came upon his decimals naturally while pursuing the root extraction process, and with him they remained 'subconscious' since he expressed all his results in decimal fractions. But in the +13th Yang Hui deliberately avoided common fractions by using decimal expressions. Although he had no symbol for the point like Stevin, he certainly had in his mind an imaginary vertical counting-board line during operations. In the following century Ting Chü again uses *fên*,[1] *li*[2] and *hao*[3] as the first decimal places with a great roll of cloth (*phi*[9]) as the unit,[d] thus showing that they had a strictly technical mathematical sense. Ting interposes the character *yü*[10] (remainder) between the integer and the first decimal place—this was the functional equivalent of the decimal point.

Yang Hui's writings are also the first in which quadratic equations with negative coefficients of x appear, though he says that Liu I had considered them before him. None of his work has been translated.

With Chu Shih-Chieh the high-water mark of Chinese algebra is reached. In the first of his books, the *Suan Hsüeh Chhi Mêng*, the rule of signs for algebraic addition and multiplication is given, and it forms, as a whole, a general introduction to algebra. There is a division table (expressed in words) suitable for use with the abacus.[e] This book contains nothing that was not in the work of Chu's predecessors, but it exercised an immense influence on the development of mathematics in Japan, though lost in China until a Korean reprint of 1660 was discovered in 1839 (Yen Tun-Chieh, *1*).

It was in the *Ssu Yuan Yü Chien* that Chu published his really important discoveries.

[a] Cf. Mikami (1), p. 86. Note how these terms resembled place-value components in function (see p. 13 above), yet deliberately borrowed from small measures (see p. 85 below) to indicate values less than integral.

[b] Especially due to the imperial edict of +992.

[c] Cf. Chhien Pao-Tsung (*1*), p. 78.

[d] *Ting Chü Suan Fa*, p. 10*a*.

[e] See below, p. 75. Yang Hui, Ting Chü and Chia Hêng[11] (Chia Hong in Mikami, 1), a younger contemporary of Chu Shih-Chieh's, all gave one of these. They afford strong evidence for the habitual use of the abacus at that time. See Mikami (*10*).

[1] 分 [2] 釐 [3] 毫 [4] 尺 [5] 寸 [6] 步 [7] 兩
[8] 斤 [9] 匹 [10] 餘 [11] 賈亨

A special paper has been devoted to it by van Hée (12),[a] and there is a description of the book in English by Konantz (1), who collaborated on a complete translation with the late Chhen Tsai-Hsin,[1] unfortunately not yet published.[b]

In the preface to Chu's book, Tsu I-Chi[2] wrote as follows:

> People come like clouds from the four quarters to meet at his gate in order to learn from him.... By the aid of geometrical figures he explains the relations of heaven, earth, men and things (technical terms for the algebraic notation). Heaven corresponds to the base of the right-angled triangle, earth to the height, man to the hypotenuse, and things to the diameter of a circle inscribed in the triangle (*huang fang*[3]), as may be seen from his diagrams. By moving the expressions upwards and downwards, and from side to side, by advancing and retiring, alternating and connecting, by changing, dividing and multiplying, by assuming the unreal for the real and using the imaginary for the true, by employing different signs for positive and negative, by keeping some and eliminating others and then changing the positions of the counting-rods, by attacking from the front or from one side, as shown in the four examples—he finally succeeds in working out the equations and roots in a profound yet natural manner.... By not using (a thing) yet it is used; by not using a number the number required is obtained. Mathematicians aforetime could not attain to the mysterious principles contained in the present profound book.[c]

The interest of this quotation lies in the Taoist paradoxes which it contains, suggesting that they may have afforded some inspiration to the Chinese algebraists.

The work opens with a diagram identical with that which later on in the West came to be known as Pascal's triangle[d] (Fig. 80). It is a device for finding the coefficients in the binomial theorem to develop $(a+b)^n$ for any integral value of n. In Europe it first appeared in print in a book of Peter Apianus in +1527, but was most fully investigated in Pascal's *Traité du Triangle Arithmétique* of +1654. It is curious, however, that in Chu Shih-Chieh's book of +1303 he entitles it 'Diagram of the Old Method for finding Eighth and Lower Powers', suggesting that it had been known at least for some time. We shall return to the subject later (p. 134), when giving a brief description of the new algebraical methods employed by Chu Shih-Chieh.[e] The 'four-element process' which he described was essentially to take accessory unknowns beside that one which is sought for, and then from the known relations given by the data of the problem to get rid of the accessory unknowns. Hence the remark of Tsu I-Chi quoted above. Chu's procedure for thus solving simultaneous equations of less than five variables was practically identical with the method of elimination and substitution. In some points his reasoning is very similar to Sylvester's dialytic elimination, except

[a] This is very difficult for non-mathematicians and almost incomprehensible without the help of Mikami (1) and Smith & Mikami (1).

[b] Cf. Sarton (1), vol. 3, p. 703. According to private information from Miss E. M. Hancock, the manuscript of this has now been lost.

[c] Tr. Konantz (1); add. et mod. auct.

[d] Smith (1), vol. 2, p. 508.

[e] They were new rather as regards the notation than as regards the essential processes which had all been known to the Han and San Kuo mathematicians.

1 陳在新 2 祖頤季 3 黃方

that he did not use determinants.[a] His methods were elucidated by many Chinese mathematicians of the nineteenth century, notably Ting Chhü-Chung,[1] who edited[b] the famous collection of ancient mathematical texts *Pai Fu Thang Suan-Hsüeh Tshung-Shu*[2] (1875).

Hitherto, as we have seen, Chinese mathematicians had applied their methods to all kinds of practical problems, and no doubt demands for solutions arising from taxation, irrigation and fortification had often stimulated their work. Already in the Thang the higher mathematics of the time had been applied to calendar-making in the work of I-Hsing, and now under the Yuan (Mongol) dynasty this need led to the work of a very notable mathematician and astronomer, Kuo Shou-Ching.[3] Though none of his original writings has survived, his methods can be studied in the long calendrical sections of the *Yuan Shih* (History of the Yuan Dynasty)[c] and the *Ming Shih* (History of the Ming Dynasty).[d] Moreover, on the basis of documents not now available, Hsing Yün-Lu[4] (*fl.* +1573 to +1620) in his *Ku Chin Lü Li Khao*[5],[e] and Mei Wên-Ting[6] (+1633 to +1721) in his *Li Suan Chhüan Shu*[7],[f] were able to elucidate Kuo's work. All this has been considered by Gauchet (7) who regards him as the founder of spherical trigonometry in China, deriving his inspiration from the key work of Shen Kua on circle segments in the +11th century.

Kuo Shou-Ching (+1231 to +1316) was at first (+1262) engaged under Khubilai Khan in hydraulic engineering works,[g] and began his astronomical and calendrical investigations about fourteen years later. Some of the instruments constructed by him still survive, and will be described below in the Section on astronomy (p. 369). Here we are concerned only with his mathematics. His spherical trigonometry (if such it can be called, since it did not, so far as we know, employ sines, cosines, etc.)[h] was concerned with the spherical figures made by the intersections of the equator, the ecliptic, and the moon's path on the celestial sphere. We shall consider it more fully in its proper place. Kuo employed biquadratic equations, and a method called *chao chha*[8] ('calling the differences') for the summation of power progressions.[i] This latter was a way of finding the values of the constants A, B, C, so that the values of an observed quantity x could be predicted by the formula $Ax+Bx^2+Cx^3$ for all values of another

[a] The determinant is a Japanese development of Chinese mathematics (Smith & Mikami (1), p. 124). Determinants were first explained by the great Seki Kōwa in +1683, ten years before Leibniz communicated his independent discovery in a letter not published till after his death. The first connected exposition was that of Vandermonde in +1771, and the term 'determinant' was first used by Gauss in +1801. See below, p. 117.

[b] In collaboration with the Jesuits; see van Hée (5).

[c] Chs. 52–7.

[d] Chs. 25, 31 and following.

[e] 'Investigation of the Chinese Calendars New and Old', chs. 67 ff.

[f] 'Complete Works on Calendar and Mathematics', section *Chhien Tu Tshê Liang*.[9]

[g] See Sect. 28*f* below.

[h] Unless these are concealed under the ancient terms *kou*,[10] *ku*[11] and *shih*,[12] which he continued to use. Chhien Pao-Tsung (*1*), p. 150, has shown the equivalence of these, in Kuo Shou-Ching's methods, to sine, cosine and secant respectively.

[i] Cf. pp. 123 ff. below.

[1] 丁取忠 [2] 白芙堂算學叢書 [3] 郭守敬 [4] 邢雲路
[5] 古今律曆考 [6] 梅文鼎 [7] 曆算全書 [8] 招差 [9] 塹堵測量
[10] 句 [11] 股 [12] 矢

quantity *n*. This is equivalent to the Method of Finite Differences. The principle was not new, since Li Shun-Fêng in the Thang had, as we shall see, already used the term *phing*[1] (lit. 'floating' difference) for the empirically observed varying observations, and *ting*[2] (lit. 'fixed' difference) for what we should now call an arbitrary constant. The addition of a term *li*[3] improved the approximation. The whole method was used for the calculation of the angular speed of the sun's apparent motion. It was equivalent, in a way, to the fitting of equations to curves in post-Cartesian science.

Exactly what was the Arab influence on Kuo Shou-Ching and his contemporaries, and how far it extended, is still an unsolved question. In the Yuan time the Arabs (they must, in fact, have been largely Persians and Central Asians) played a role in Chinese science and technology quite similar to the Indians in the Thang. The 203rd chapter of the *Yuan Shih* informs us about the Muslim gunners[a] A-Lao-Wa-Ting[4] and I-Ssu-Ma-Yin,[5] who were in the Mongol service in +1271; the former died about +1295, and the latter in +1274. They could certainly have transmitted mathematical information.[b] There was also the Syrian Nestorian Ai-Hsüeh,[6] 'Īsa Tarjaman ('Īsa the Interpreter), a much more educated man,[c] who worked for the Mongol Khans from +1250 till his death in +1308, during which time he rose to be a Hanlin Academician and minister of State. Then the Persian astronomer, Cha-Ma-Lu-Ting[7] (Jamāl al-Dīn),[d] devised for Khubilai Khan in +1267 a new calendar, the Wan Nien Li,[8] which was afterwards lost, and in any case failed in competition with the Shou Shih Li[9] of Kuo Shou-Ching (+1281), of which the Ming calendar Ta Thung Li[10] (started +1364) was but a modification. Nevertheless, the 'Arab' influence in Chinese learned circles was so great at the end of the Yuan dynasty that in +1368, the year in which the Ming Dynasty was established, a Muslim (*Hui-hui*[11]) Astronomical Bureau was set up[e] in parallel with the ordinary Astronomical Bureau, to which, however, after a couple of years it became subordinate. The president of it was a Muslim, Hai-Ta-Erh[12],[f] and we still possess a work, the *Ming I Thien Wên Shu*[13] (Astronomy, Officially Translated (from Arabic?) by order of the Ming),[g] which he presented to the emperor in +1382. This contains, however, no mathematics, but consists of abundant data on

[a] Perhaps 'Alā' al-Dīn and 'Ismā'īl.

[b] We shall meet them again in connection with military technology, Sect. 30 below.

[c] He was skilled as a mathematician and astronomer as well as in medicine and pharmacy. His biography says that he came from Fu-Lin, i.e. Byzantium, but he was certainly a Syrian Arab.

[d] See p. 372 below.

[e] The history of this really goes back to +1271 when Jamāl al-Dīn himself had been appointed to direct a new Bureau. There is an interesting recent book on the Muslim calendar and its specialists by Ma Chien (*1*).

[f] Alias Hei-Ti-Erh.[14]

[g] Not so far noticed by historians of science, such as Sarton. It is contained in the *Han Fên Lou Pi Chi*[15] collection. It was translated by the Hanlin Academicians Li Chhung[16] and Wu Po-Tsung[17] in collaboration with the Muslims A-Ta-Wu-Ting,[18] Ma-Sha-I-Hei,[19] Ma-Ha-Ma[20] and others. The imperial commission is in *Ming Shih*, ch. 37, p. 1*a*.

[1] 泙 [2] 定 [3] 立 [4] 阿老瓦丁 [5] 亦思馬因 [6] 愛薛
[7] 札馬魯丁 [8] 萬年曆 [9] 授時曆 [10] 大統曆 [11] 回回
[12] 海達兒 [13] 明譯天文書 [14] 黑的兒 [15] 涵芬樓祕笈
[16] 李翀 [17] 吳伯宗 [18] 阿答兀丁 [19] 馬沙亦黑 [20] 馬哈麻

positional astronomy, together with much astrological material. A great deal of computation must have been needed, however, for the planetary ephemerides according to the Muslim methods, re-issued as the *Chhi Chêng Thui Pu*[1] (On the Motions of the Seven Governors)[a] by Pei Lin[2] in +1482. When the Jesuits reached Peking at the end of the 16th century, the descendants of these 'Arabian' astronomers were still working in their observatory. There can thus be no doubt that there was every opportunity for Arabic and Persian mathematical influences (as from the observatories of Marāghah and Samarqand) to enter Chinese traditions. But whether any important effect was actually exerted will not be known until scholars who can combine sound knowledge of the history of mathematics and astronomy with first-hand acquaintance with both Arabic and Chinese texts arise to subject the problem to a thorough examination—at present we can only speculate.[b] One of the difficulties is that there occurred,[c] after the period of culture contact, a marked decline in Chinese mathematics, so that there is hardly any work of value to notice between +1400 and +1500.

Before passing on to the 16th century mention should be made of a number of mathematicians who were working in the half-century immediately following the death of Kuo Shou-Ching. Thus there was Ting Chü[3] whose dates we can only guess by his *Suan Fa*[4] published in +1355. Like the anonymous *Thou Lien Hsi Tshao*[5] (The Mathematical Curtain Pulled Aside) of about the same date, it is a collection of fairly simple arithmetical problems. In +1372 appeared Yen Kung's[6] *Thung Yuan Suan Fa*[7] (Origins of Mathematics) containing indeterminate equations with very incomplete solutions. It so happens that these treatises, together with a few others,[d] are contained in chapters 16,343–4 of the *Yung-Lo Ta Tien* encyclopaedia of +1406, which are preserved in the Cambridge University Library (Figs. 54, 65 show two pages from this manuscript).[e] They have been discussed by Li Nien.[f] The diagram, which gives the geometrical figure illustrating the process of calculating square roots, comes from Yang Hui's *Hsiang Chieh Chiu Chang Suan Fa Tsuan Lei* of +1261.

During the first century and a half of the Ming dynasty there was little of any interest occurring in mathematics, but after +1500 mathematicians begin to reappear.

[a] Yabuuchi (1) has made a special study of this work, which shows that the planetary part was based on the Ilkhanic Tables of the Marāghah Observatory. But the book also contains an interesting catalogue of 277 star positions in and around the zodiacal zone; this must have been made in one of the Muslim countries about +1365. An abridged form of the *Chhi Chêng Thui Pu* is contained in chs. 37–9 of the *Ming Shih* under the title Hui Hui Li Fa.[8]

[b] Cf. Li Nien (*18*); Chhien Pao-Tsung (7).

[c] Contrary to the impression created by certain writers such as van Hée.

[d] Such as the *Suan Fa Chhüan Nêng Chi*[9] (Record of Do-Everything Mathematical Methods) by Chia Hêng[10] and the *Hsiang Ming Suan Fa*[11] (Explanations of Arithmetic) by An Chih-Chai[12] and Ho Phing-Tzu[13] (but Chia Hêng may have written this too, for what remains is almost identical: Li Nien (*4*), vol. 3, p. 42). Cf. Sarton (1), p. 1536. There is also the *Chin Nang Chhi Mêng*[14] (Brocaded Bag of Books for the Relief of Ignorance), which is anonymous.

[e] It will be remembered that only a few volumes of this once immense collection remain, scattered in libraries all over the world. Cf. Vol. 1, p. 145.

[f] (*4*), vol. 2, pp. 83 ff.

1 七政推步 2 貝琳 3 丁巨 4 丁巨算法 5 透簾細草
6 嚴恭 7 通原算法 8 回回曆法 9 算法全能集 10 賈亨
11 詳明算法 12 安止齋 13 何平子 14 錦囊啓蒙

Thang Shun-Chih[1] (+1507 to +1560), a military engineer as well as a mathematician, achieved renown for his work on circle measurements. He published five works, of which the *Hu Shih Lun*[2] (Discussion on Arcs and Sagittae) is probably the most important. A contemporary was Ku Ying-Hsiang,[3] Governor of Yünnan, whose *Tshê Yuan Hai Ching Fên Lei Shih Shu*[4] (Classified Methods of the Sea Mirror of Circle Measurement), which appeared in +1550, distinguished equations of the various degrees according to the signs of the coefficients, and gave fuller explanations for their solution. His *Hu Shih Suan Shu*[5] of two years later systematised the formulae developed up to that time for dealing with arcs and circle segments, for example:

$$\frac{c}{2}=\sqrt{\left[\left(\frac{d}{2}\right)^2-\left(\frac{d}{2}-s\right)^2\right]} \quad \text{or} \quad s=\frac{d}{2}-\sqrt{\left[\left(\frac{d}{2}\right)^2-\left(\frac{c}{2}\right)^2\right]},$$

$$a=\frac{2s^2}{d}+c,$$

$$d=s+\frac{(\frac{1}{2}c)^2}{s},$$

$$A=\tfrac{1}{2}(s+c)s,$$

$$s=\frac{d^2(\frac{1}{2}a)^2}{(d^3-d^2s')-(ad-s')s'},$$

where d = diameter, c = chord, s = sagitta, A = area of segment of circle, a = arc, and s' = a rough value for the sagitta. Yet another contemporary of Thang Shun-Chih's was Chou Shu-Hsüeh;[6] he wrote mostly on calendrical calculations, one of his books[a] appearing in +1558.

None of these mathematicians of the Ming, however, was a master of the Sung and Yuan algebra. It fell completely out of use, and was not revived until long after the introduction of algebra from Europe by the Jesuits and others, when men such as Mei Ku-Chhêng recognised medieval Chinese algebra under its cloak of unfamiliar notation, and re-investigated it.

One of the most interesting of the Ming mathematicians was Chhêng Ta-Wei,[7] whose *Suan Fa Thung Tsung*[8] (Systematic Treatise on Arithmetic)[b] appeared in +1593. It is important historically because, though so late, it was the first to give an illustration of the Chinese abacus, with instructions for its use.[c] We shall shortly

[a] Titles of them are given by Chhien Pao-Tsung and Li Nien in their books already quoted, to which the reader is referred also for the names and works of the minor mathematicians in this and other periods. Mikami (9) discusses the work of Chhing mathematicians on circle segments.

[b] Occupying no less than 13 chapters (113–25) it shares with the *Chou Pei* (chs. 109–11) the honour of quotation *in extenso* in the mathematical section of the *Li fa tien* of the *Thu Shu Chi Chhêng* encyclopaedia. The rest is made up of a mixed bag of quotations, including one fragment from Hsieh Chha-Wei of the Sung (cf. p. 79). The *Suan Fa Thung Tsung* has been closely studied by Takeda (*2*).

[c] Though it seems that there was some discussion, if not an illustration, of it in the earlier (+1578) work of Ko Shang-Chhien,[9] *Shu Hsüeh Thung Kuei*[10] (Rules of Mathematics).

[1] 唐順之 [2] 弧矢論 [3] 顧應祥 [4] 測圓海鏡分類釋術
[5] 弧矢算術 [6] 周述學 [7] 程大位 [8] 算法統宗 [9] 柯尙遷
[10] 數學通軌

summarise what is known about the history of this important aid to computation. A translation of the section headings of the *Suan Fa Thung Tsung* was given a little over a century ago by Biot (5), who reported (6) the fact that it contains[a] an illustration of the Pascal Triangle, closely similar to that of Chu Shih-Chieh. Although it is a very practical book, with much attention given to mensuration and the determination of plane areas of peculiar shapes, and to problems such as the mixing of constituents of alloys, it also includes a considerable number of magic squares. Chhêng Ta-Wei gives the titles of many mathematical books which existed in his time, but most have since disappeared, with the exception of those of Ku Ying-Hsiang already mentioned. Special studies of the literature of the period have been made by Mikami (7) and Takeda (*1*); they give details of a dozen works of minor importance in the +16th century.[b]

With the coming of the Jesuits to Peking at the beginning of the 17th century, the period of what may be called 'indigenous mathematics' for the purpose of the present book comes to an end. Those who wish to study the period of collaboration between Chinese scholars and the Jesuit mathematicians will find a careful chapter in Mikami (1) on the subject.[c] The extent of the appreciation with which Matteo Ricci and his companions were received may be gauged when we remember that they were almost the only persons of foreign birth who ever attained the distinction of having their biographies admitted to the Chinese official histories. The translation of the first six books of Euclid into Chinese was undertaken by Ricci (Li Ma-Tou[1]) and Hsü Kuang-Chhi;[2] it was completed in +1607.[d] The *Thung Wên Suan Chih*[3] (Treatise on European Arithmetic) was dictated by Ricci[e] and recorded by Li Chih-Tsao;[4] it was printed in +1614. These were followed by books on more advanced geometry and surveying. Later in the century there came (+1669) the *Hsin Fa Suan Shu*[5] (Mathematical Methods of the New Calendrical System) which had been compiled by Adam Schall von Bell (Thang Jo-Wang[6]) and other Jesuits before +1635.[f] Logarithms first appear in the *Thien Pu Chen Yuan*[7] (True Course of Celestial Motions), a treatise on eclipses by Nicholas Smogułęcki (Mu Ni-Ko[8]);[g] and his pupil Hsüeh Fêng-Tsu[9] produced in +1653 the first Chinese logarithmic tables together with a discussion of

[a] Ch. 6, p. 2*b*. The Triangle is ascribed to a Mr Wu. This is Wu Hsin-Min.[10] See p. 79.

[b] Cf. also Forke (9), p. 424 on Lo Hung-Hsien.[11]

[c] Most people imagine that the Jesuits were bringing 'age-old' European mathematical knowledge to the Chinese. Only for Euclid's *Elements* is this true, however; we shall emphasise below, p. 114, that the non-geometrical mathematical discoveries and techniques which the Jesuits transmitted were extremely recent in Europe. Moreover, the Chinese were so impressed by the new knowledge partly because of the decay into which their own mathematics had fallen.

[d] *Chi Ho Yuan Pên*.[12] The edition they used was that of Christopher Clavius (+1537 to +1612), *Euclidis Elementorum libri XV*, who naturally appeared as Ting hsien-sêng[13] to puzzle posterity. The copy still exists in the Pei Thang library at Peking, where I have had the pleasure of seeing it. Cf. d'Elia (2), vol. 2, pp. 356ff., (6); Trigault (Gallagher ed.), p. 476.

[e] From Clavius' *Epitome Arithmeticae Practicae*.

[f] Much of it is contained in *TSCC*, *Li fa tien*, chs. 51–72 and 85–8. Bernard-Maître (7) has closely studied this work, on which see further below, pp. 447ff.

[g] Biography by Kosibowicz (1).

[1] 利瑪竇 [2] 徐光啓 [3] 同文算指 [4] 李之藻 [5] 新法算書
[6] 湯若望 [7] 天步眞原 [8] 穆尼閣 [9] 薛鳳祚 [10] 吳信民
[11] 羅洪先 [12] 幾何原本 [13] 丁先生

them.[a] After the beginning of the 18th century, compendia of mathematics were compiled and issued by imperial order in the Khang-Hsi reign period; there was the *Lü Li Yuan Yuan*[1] (Ocean of Calendar Calculations) of +1713 (in which van Vlacq's logarithm tables of +1628 were reprinted in Chinese form),[b] and the *Shu Li Ching Yün*[2] (Collected Basic Principles of Mathematics) of +1722. From this time onwards the Chinese mathematical literature becomes voluminous, but, though still somewhat isolated, it is part of the world literature.

It was now that the Chinese began to discover that they also had had a science of algebra. Owing to the decay of indigenous science, and the first flush of enthusiasm for the *a-erh-jo-pa-la*[3] which the Jesuits had brought with them, this had been overlooked. It was Mei Ku-Chhêng[4] (+1681 to +1763), the grandson of Mei Wên-Ting already mentioned, who first realised that autochthonous Chinese mathemetics had gone very far before the 17th century under unfamiliar notations, and put forward this discovery in his *Chhih Shui I Chen*[5] (Pearls recovered from the Red River).[c] By some obscure play upon words it was thought that the Europeans admitted that their algebra had come from the East,[d] and although the intense nationalism of some later Chinese writers in this respect[e] has been much castigated by 19th-century Europeans, the fact remains that whatever future research may reveal about transmission, algebra was just as essentially Indian and Chinese as geometry was Greek. Actually, there is some evidence of transmission from the Arabs to the Chinese in the +13th and +14th centuries, and much more from the Chinese earlier to India and Europe.[f] A great deal of careful historical work will be needed before any final conclusions can be drawn.

With this we reach the end of what had to be said about the main lines of ancient and medieval Chinese mathematics and the landmarks in the literature. What follows will treat of the various branches of mathematics specifically, in an endeavour to elucidate some of the methods and technical terms employed, and to set the accomplishments of the Chinese in better perspective with those of other peoples. For an account of late 18th- and 19th-century mathematics in China, the reader is referred to the historical works mentioned at the beginning of this Section.[g]

[a] Van Hée (6) has given a paraphrased translation of Juan Yuan's biography of Hsüeh Fêng-Tsu. See also Li Nien (*13*). Cf. p. 454 below.

[b] Feldhaus (23) has recalled how Charles Babbage (1) identified their origin in 1827 by a comparison of the errors in a copy which Antoine Gaubil had sent to the Royal Society in 1750. Cf. p. 448 below.

[c] He was followed by many men in the nineteenth century, such as Li Shan-Lan.[6]

[d] The very name was of course Arabic.

[e] As in the *Ko Chih Ku Wei* and the *Ying Hai Lun*, cf. Vol. 1, p. 48 above.

[f] It is, however, surprising to read in Smith (1), vol. 1, p. 269, that our knowledge of Western travellers (de Plano Carpini, William Rubruck, King Haython of Armenia, Marco Polo, and so on) 'clears up such questions as whether the algebra of China could have found its way to Italy in the +13th century. We repeat that it would be a cause for wonder if it had failed to do so.' But surely the methods and the notation were too different from those of Europe. Would not the Thien Yuan Shu have left some trace of itself in the West if it had been carried there? However, see p. 128 below.

[g] Though a considerable number of names of Chinese mathematicians from these later periods will, for convenience, be found in the biographical glossary in the final volume.

[1] 律曆淵源 [2] 數理精蘊 [3] 阿爾熱巴拉 [4] 梅瑴成
[5] 赤水遺珍 [6] 李善蘭

(d) ARITHMETICA AND COMBINATORIAL ANALYSIS

(1) Elementary Theory of Numbers

In ancient times the term *arithmetica* did not mean the simple computations which go by the name of arithmetic today, but concerned rather the elementary aspects of the theory of numbers.[a] From the 'Pythagorean' atmosphere of number-mysticism and numerology (cf. Vol. 2, pp. 268ff.), common to both Greek and Chinese beginnings, there emerged an appreciation of the existence of prime and composite numbers, figurate numbers, amicable numbers, and the like. Certain books in the *Elements* of Euclid[b] systematise such knowledge, and the −3rd-century 'sieve of Eratosthenes', a method of finding prime numbers by sifting out the composite numbers in the natural series, is well known.[c] In the +2nd century, contemporary with Liu Hung (cf. p. 29), the Greeks Nicomachus of Gerasa and Theon of Smyrna added many new propositions to number theory; as did Diophantus of Alexandria, Liu Hui's contemporary, in the +3rd.[d]

The distinction between odd and even numbers must have been the first to arouse interest. In the West odd numbers were known as gnomonic numbers,[e] since the gnomon with its shadow forms a figure of the type $2n+1$, and hence must be an odd number. This necessarily involves the counting of rows of squares, a practice which

[a] The elaborate work of Dickson (1) treats mainly of the more advanced aspects of the theory of numbers. See also P. G. H. Bachmann.

[b] Parts of Bks. 2, 5, 7, 8, 9 and 10.

[c] Heiberg (1), p. 22; Sarton (1), vol. 1, p. 172; R. A. Fisher (1).

[d] One of the fundamental theorems in number-theory is that of Fermat (+1640), which states that if p is any prime and x any integer not divisible by p, then $x^{p-1}-1$ is divisible by p. Dickson (1), vol. 1, pp. v, 59, says that this was known to the Chinese for the case $x=2$ from ancient times. We have not been able to trace this with certainty to its origin, but the source of the statement, which other mathematicians copy (e.g. Rouse Ball (2), p. 63; Smith (1), vol. 2, p. 29, with misprint), is a mysterious remark by J. H. Jeans in a note of 1897 (written while he was still an undergraduate) that 'a paper found among those of Sir Thomas Wade and dating from the time of Confucius contains this theorem and also states [wrongly] that it does not hold if p is not a prime'.

Possibly this arose from a misunderstanding by Jeans of what was only a statement of the fact that even numbers can be divided by 2 and that odd ones cannot. The reference to Confucius would be readily understandable if the early sinologists were being misled by something in the *Chiu Chang*, since its true date was not then understood. The *Chiu Chang* (ch. 1, p. 3a) says:

'*kho pan chê pan chih*:[1] If (both numerator and denominator) are divisible by 2, then halve them both.'

'*pu kho pan chê, fu chih fên mu tzu chih shu, i shao chien to, kêng hsiang chien sun*:[2] If they are not divisible by 2, then set up the numbers for numerator and denominator respectively (continually and) alternately subtracting the smaller from the larger.'

'*chhiu chhi têng yeh*:[3] And seek their equality (i.e. continue until the last minuend is equal to the last subtrahend).'

Somebody realised that a rule was being described here but failed to understand that it was in fact the finding of a Greatest Common Divisor by a process of continued division. Possibly Sir Thomas Wade or Prof. H. A. Giles, thinking of $\frac{x^{2-1}-1}{2}$, saw that the statement began by having 2 as denominator, and went on to talk about subtracting the smaller from the larger. *Kêng* could be taken as 'again', to suit the second −1. The rest they could not translate. They thus read into this text (if our suggestion is acceptable) what was not there at all. The Han mathematicians certainly never thought of x in terms of x^{2-1}.

[e] Smith (1), vol. 2, p. 16.

[1] 可半者半之 [2] 不可半者副置分母子之數以少減多更相減損 [3] 求其等也

was also used in contemporary China, though for other purposes. Theon of Smyrna was well aware of the corresponding fact that the sum of the first *n* odd numbers, including 1, is a square. It was natural that the odd and even numbers should be associated with the two sexes, and this is found in ancient Chinese discussions just as among the Pythagoreans.[a] The Chinese also shared the widespread superstition that odd numbers were lucky and even ones unlucky.[b]

The aliquot parts of an integral number are defined as the integral and exact divisors of the number, including unity but not including the number itself. A number is said to be deficient, perfect, or abundant, according as it is greater than, equal to, or less than the sum of its aliquot parts. These definitions were known to the Greeks, who by the end of the +2nd century had given the first four perfect numbers. The fifth was not found until about +1460, though the rule which still holds good for finding them was first enunciated by Fibonacci in +1202. Two integral numbers are said to be amicable if each, as in the case of 220 and 284, is equal to the sum of the aliquot parts of the other. This interested Arabic mathematicians in the +9th century, but the second pair were not discovered till eight hundred years later (Fermat, +1636). All this was rather foreign to Chinese mathematics, where the preference was for concrete number, not numbers as such. The Greeks were also much interested in figurate numbers, i.e. those the units of which will form geometrical figures. Thus 3, 6 and 10 are triangular numbers, 4 and 9 are square, 5 and 12 are pentagonal, and so on. The Chinese were aware of this as well.[c] Biot (5) found some reference to triangular numbers in the late (+16th-century) treatise, *Suan Fa Thung Tsung*, but the idea was far older.[d] In +1247 Chhin Chiu-Shao had discussed 'square and round arrays' (*fang chen*,[1] *yuan chen*[2]), found by the 'arrow-bundle method' (*shu chien fa*[3]). The origin of this goes back to the *Chiu Chang*'s expression of series, 'progressive rows in a pyramidal array' (*chui hang tshui*[4]).

(2) Magic Squares

Another Chinese interest was in combinatorial analysis, the construction of magic figures, i.e. arrangements of numbers in tables of various geometrical forms in such a way that when simple logistic operations are performed on them, such as addition, the sum or product is the same in whichever way the addition is made.[e] Owing to the acceptance by Western historians of quite untenable datings of Chinese classics, this branch of combinatorial analysis has been given a much higher antiquity than it deserves, but even when we take a justifiably conservative view of the sources the priority still seems to be Chinese. Since what has been well called 'the undatable

[a] Cf. p. 57 below. Chhien Pao-Tsung (*1*), p. 80, gives an interesting exceptional case.

[b] See, for example, Granet (5), p. 293.

[c] Both in China and Greece it led to the study of series and progressions (Heath (5), p. 247; Gow (1), p. 103; Smith (1), vol. 2, p. 499).

[d] Cf. Li Nien (*11*).

[e] See the excellent articles by Dudeney (1) and Frost & Fennell (1). Among the books on the subject may be mentioned that of Andrews (1) and the old treatise of Violle (1). Cf. the relevant chapter in Rouse Ball (2), pp. 193 ff.

[1] 方陣 [2] 圓陣 [3] 束箭法 [4] 錐行衰

Chinese tradition' goes back into the legendary period of history, the facts are rather difficult to ascertain, but they are approximately as follows.[a]

One of the components of the corpus of legend was the story that to help him in governing the empire the engineer-emperor Yü the Great was presented with two charts or diagrams by miraculous animals which emerged from the waters which he alone had been able to control. Thus the Ho Thu[1] was the gift of a dragon-horse which came out of the Yellow River, and the Lo Shu[2] the gift of a turtle from the River Lo. The former (the 'River Diagram') was generally described as green, or in green writing, and the latter (the 'Lo River Writing') was traditionally red. There can be no doubt that this story is of great antiquity. It can hardly be later than the −5th century, since there are mentions of it in the *Lun Yü* (Conversations and Discourses of Confucius)[b] and the *Shu Ching*[c] (Historical Classic).[d] At the beginning of the −4th century there is a reference in *Mo Tzu*,[e] and at the end another in *Chuang Tzu*.[f] This latter is the first which associates numbers with the diagrams, speaking of the nine elements of the Lo Shu.[g] In the −2nd century the picture begins to develop. The reference in *Huai Nan Tzu*[h] does not throw any further light on what these diagrams were, but in the Hsi Tzhu portion of the *I Ching* (Book of Changes), which, as we have already seen, was probably put together about this time (cf. Vol. 2, p. 307), there are two significant passages. In chapter 9 of the first part of this 'Great Appendix' it is said that 'Heaven is one, earth is two, heaven three, earth four, heaven five, earth six, heaven seven, earth eight, heaven nine, and earth ten';[i] the tradition of the Sung commentators[j] on this was that it refers to the two diagrams as ways of arranging the first series of numbers to the base 10—the two diagrams themselves are then mentioned in chapter 11 in the usual way as having emerged from the rivers as patterns for the sages.[k]

In the *Shih Chi* Ssuma Chhien records one curious story[l] which is probably connected with the Ho Thu and Lo Shu, namely, that Master Lu,[3] one of Chhin Shih Huang Ti's magicians, presented him with a book by a certain Lu Thu,[4] containing

[a] An essay of Li Nien's is specially devoted to the history of magic squares, (*4*), vol. 3, p. 59, (*21*), vol. 1, p. 175.

[b] IX, viii (Legge (2), p. 83). Confucius, complaining of the decadence of the age, sighs that 'the River gives forth no more diagrams'. A minority of scholars, e.g. Creel (4), p. 218, suspect that this passage is a later interpolation.

[c] Ch. 42 (Ku Ming) in the Chou Shu (Legge (1), p. 239; Medhurst (1), p. 299; Karlgren (12), p. 71), where the '*thu*' is displayed among the royal treasures at the death of a king. This chapter is one of the few which everyone regards as genuinely pre-Confucian. Legge's note (p. 138) that the *Shu Ching* says nothing of the Ho Thu is misleading.

[d] Khung An-Kuo's commentary to this, written about −100, says that the diagrams each had the numbers up to nine (cf. Medhurst (1), pp. 198, 199).

[e] Ch. 19 (Mei (1), p. 113).

[f] Ch. 14 (Legge (5), vol. 1, p. 346).

[g] The text says only 'Lo' but later commentators have always assumed that the Lo Shu was meant.

[h] Ch. 2 (Morgan (1), p. 54) and ch. 18.

[i] R. Wilhelm (2), vol. 1, p. 234.

[j] E.g. the relevant passage in the *Chou I Pên I*[5] of Chu Hsi[6] in +1177.

[k] R. Wilhelm (2), vol. 1, p. 244.

[l] Ch. 6, p. 21*b* (Chavannes (1), vol. 2, p. 167).

[1] 河圖 [2] 洛書 [3] 盧 [4] 錄圖 [5] 周易本義 [6] 朱熹

prophecies.[a] This account, which would date from about −100, reporting events of about −230, refers in all probability to the beginning of the process whereby the ancient diagrams became a kind of nucleus of crystallisation for the magical-divinatory material incorporated during the Han in the Chhan-Wei apocryphal treatises, as already described (Vol. 2, p. 380). Tsêng Chu-Sên[b] and Chhen Phan[c] have critically examined several books with Ho Thu and Lo Shu in their titles, occurring in this corpus. Nearly all the host of references to the Ho Thu which may be found in the *Thai-Phing Yü Lan* encyclopaedia of the +10th century are quotations from these books.

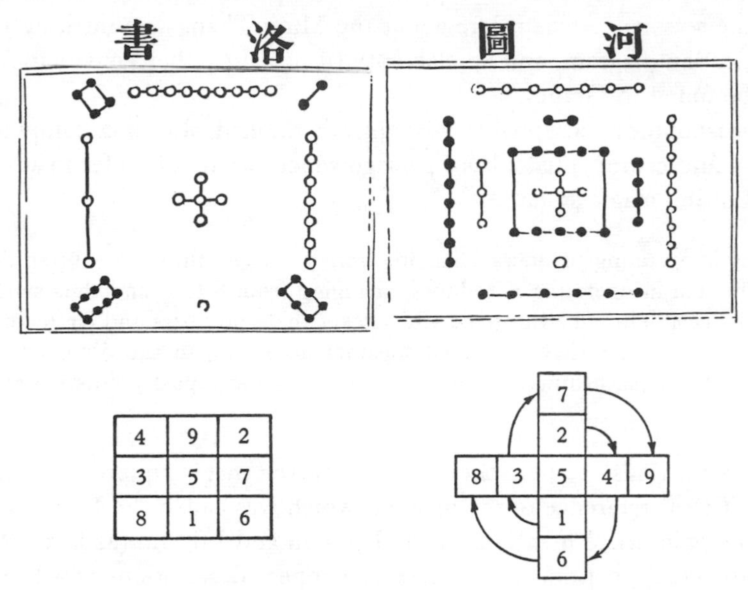

Fig. 58. The Lo Shu diagram.

Fig. 59. The Ho Thu diagram.

Up to this point one can find only hints that numbers were concerned. Before going further we should see what, according to the universal interpretation of later scholars,[d] the diagrams of the Ho Thu and Lo Shu were. They represented, indeed, a simple magic square and a cruciform array of the numbers from 1 to 10. Even or Yin numbers are represented in black and odd or Yang ones in white.[e]

Figs. 58 and 59 show the form in which they are generally found, for example, in modern editions of the *I Ching* and elsewhere. The Lo Shu is a straightforward magic square in which the figures, added up along any diagonal, line or column, make 15,

[a] Lu Thu was taken as the name of a person, later identified with Lao Tzu, but Chhen Phan's view is that the name is the title of the book, which concerned the Ho Thu.

[b] (1), pp. 103 ff.

[c] (*1–4*).

[d] It is known that Chu Hsi in the +12th century interverted the names of the two diagrams (*Shih Chia Chai Yang Hsin Lu*, ch. 1, pp. 6, 7). This usage is followed here.

[e] I follow the description of Chhen Wên-Thao as well as those of Li Nien. The most complete discussion will be found in Granet (5), pp. 176 ff. Gaubil (2) was probably the first (1732) to recognise the identity with the magic squares of Europe.

and from which can be developed a figure of swastika form.[a] The Ho Thu is so arranged that disregarding the central 5 and 10, both odd and even number sets add up to 20.[b] The earliest text which gives a clue as to these simple arrangements of numbers seems to be the *Ta Tai Li Chi* (Record of Rites compiled by Tai the Elder) of about +80, but including much earlier material. This gives the Lo Shu figures in the order 'two, nine, four; seven, five, three; six, one, eight'[c] though it does so while speaking of the *chiu kung*[1] or nine halls of the Ming Thang[2] (Bright Palace), the mystical temple-dwelling which the emperor was supposed to frequent, carrying out the rites appropriate to the seasons. Ancient lore about the Ming Thang was intricately connected with the Lo Shu diagram, and an elaborate (if not altogether convincing) exposition of it will be found in Granet.[d]

One at least of the apocryphal treatises just mentioned, almost certainly of Late Han (+1st- or +2nd-century) date, has a passage which can hardly refer to anything other than the Lo Shu magic square.[e]

The Yang in operating advances, changing from 7 to 9 and thus symbolising the waxing of its *chhi*. The Yin in operating withdraws, changing from 8 to 6 and thus symbolising the waning of its *chhi*. Thus the Supreme Unity takes these numbers and circulates among the Nine Halls.[f] (Whether they be added together according to the direction of) the four compass-points, or (according to) the four intermediary compass-points,[g] they always add up to 15.[h]

Thus the Nine Halls are here the nine cells of the magic square. Another of these works has a clear reference to the diagram which was called the Ho Thu.[i]

After this point mere mentions of the Ho Thu and Lo Shu (as in the *Chhien Han Shu*, about +100)[j] continue, but there also appear descriptions which indicate that magic square diagrams existed. If Hsü Yo's *Shu Shu Chi I* (Memoir on Some Traditions of Mathematical Art) is really of about +190 its reference to a 'Nine Hall' method of calculating is significant; it says that this method is like that of the Five Elements, and that it worked 'as if following round a circle', which may well be a reference to the Ho Thu. Its +6th-century commentator, Chen Luan, at once explains that 'two and four are the shoulders, six and eight the feet, there is three on the left and seven on the right; it wears nine on its head and is shod with one, while

[a] On the swastika and China, see Loewenstein (1).

[b] The real significance of this was that it embodied the numbers appropriated to the four seasons, five elements, etc., in their proper symbolic correlations (cf. Vol. 2, pp. 261 ff.).

[c] Ch. 67.

[d] (5), pp. 177 ff.; (1), pp. 116, 478. Cf. Soothill (5).

[e] *I Wei Chhien Tso Tu* (Apocryphal Treatise on the *Book of Changes*; A Penetration of the Regularities of the *kua* Chhien), ch. 2, p. 3*a*. Quoted in the commentary of the *I Wei Ho Thu Shu* (Apocryphal Treatise on the *Book of Changes*; The Numbers of the Ho Thu), in *Ku Wei Shu*, ch. 16, p. 2*a*.

[f] Like the emperor.

[g] Three additions longitudinally, three vertically, and two diagonal.

[h] Tr. auct. adjuv. Bodde, in Fêng Yu-Lan (1), vol. 2, p. 101.

[i] *I Wei Ho Thu Shu*, in *Ku Wei Shu*, ch. 16, p. 1*a*.

[j] Ch. 27A, p. 1*a*.

[1] 九宮 [2] 明堂

five is in the middle'. This plainly describes the diagram which was afterwards identified with the Lo Shu. Somewhat more than a century earlier, Kuan Lang,[1] a geomancer who wrote an *I Chuan*,[2] had given an equally plain description of the Ho Thu, saying that 'seven is in front, six at the back, eight to the left and nine to the right'.[a] In the +6th century these matters must have been generally understood, for Lu Pien,[3] a contemporary of Chen Luan's, who commented on the *Ta Tai Li Chi*, interpreted the mysterious numbers in the same way.[b]

Though doubt remains concerning Hsü Yo and Kuan Lang, it is already clear that the true inventor of magic squares lived long before the identification of the Ho Thu and Lo Shu as diagrams of this type by Chhen Thuan,[4] the famous Taoist who flourished between the Thang and the Sung dynasties (*c.* +940).[c] His responsibility was first suggested by Hu Wei in the 18th century.[d] Chhen paid much attention to the subject in his *I Lung Thu*[5] (The Dragon Diagrams of the *Book of Changes*). His ideas descended through a series of intervening scholars to Liu Mu[6] of the early Sung (+10th century) and his *I Shu Kou Yin Thu*[7] (The Hidden Number-Diagrams of the *Book of Changes* Hooked Out).[e] A great mass of quotations from these Sung and later speculators is preserved in the *Thu Shu Chi Chhêng* encyclopaedia.[f] Needless to say, most of their concern was either with divination (as in the case of Hsing Khai's[8] *Than Chai Thung Pien*[9] (Miscellaneous Records of the Candid Studio)[g] of *c.* +1220), or with the setting of the Ho Thu and Lo Shu diagrams in the framework of the symbolic correlations (cf. Vol. 2, pp. 261 ff.), e.g. in connection with music, as in Chiang Yung's[10] *Lü Lü Hsin Lun*[11] (New Discourse on Acoustics and Music) of about +1740.

Before the +13th century the development of magic squares was conspicuously absent from the main current of mathematical thought. Knowledge of the *tsung hêng thu*[12] (vertical-horizontal diagrams), as they were called, was first studied as a mathematical problem by Yang Hui in his *Hsü Ku Chai Chhi Suan Fa*[13] (Continuation of Ancient Mathematical Methods for Elucidating the Strange (Properties of Numbers)), of +1275. Some of the magic squares he made are very complicated (Fig. 60). He also gave some simple rules for their construction. For example, if the numbers from 1 to 16 are

[a] This was accepted and quoted by Chu Hsi in his *I Hsüeh Chhi Mêng*[14] (Introduction to Knowledge of the *Book of Changes*) of +1186, but later scholars have cast doubt upon its authenticity, suggesting that it dates rather from the first half of the +11th century. Kuan Lang (or the later writer, Juan I[15]) also described the Lo Shu. Quotations from his book, now lost, are to be found in *TSCC, Ching chi tien*, ch. 51.

[b] They were also well understood in the Thang, as in the book *Thai I Chin Ching Shih Ching*[16] (Golden Mirror of the Great Unity Classic of the Divining Board), by Wang Hsi-Ming.[17] I should like to mention in this connection another statement of the number arrangement in the Lo Shu, by Su Chia-Chhing.[18] It appears in *TSCC, Ching chi tien*, ch. 51, but we cannot date him.

[c] We have already met with him, Sect. 16*d*, Vol. 2, p. 467, in connection with the *Thai Chi Thu*.

[d] In his *I Thu Ming Pien* of +1706.

[e] Liu Mu transposed the Ho Thu and Lo Shu, but Tshai Yuan-Ting corrected his mistake before Chu Hsi inserted them in the *Book of Changes*.

[f] *Ching chi tien*, chs. 51–8.

[g] P. 7*b*.

[1] 關朗 [2] 易傳 [3] 盧辯 [4] 陳摶 [5] 易龍圖 [6] 劉牧
[7] 易數鈎隱圖 [8] 邢凱 [9] 坦齋通編 [10] 江永 [11] 律呂新論
[12] 縱橫圖 [13] 續古摘奇算法 [14] 易學啓蒙 [15] 阮逸
[16] 太乙金鏡式經 [17] 王希明 [18] 蘇嘉慶

placed in an array of four columns and four horizontal rows, and the numbers at the corners of both inner and outer square transposed, a magic square will be produced in which all columns, lines and diagonals add up to 34. Yang Hui's work was continued by Chhêng Ta-Wei in the *Suan Fa Thung Tsung* of +1593, who gave fourteen diagrams (Fig. 61). A few more were added by Fang Chung-Thung[1] in his *Shu Tu Yen*[2] (Generalisations on Numbers), +1661; and many by Chang Chhao[3] in the Suan Fa

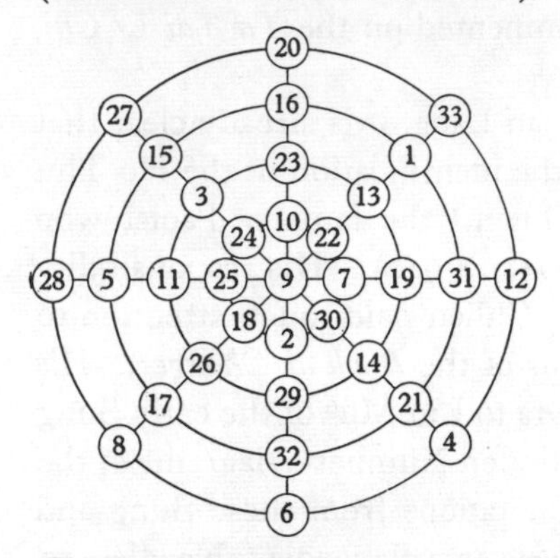

Fig. 60. A magic square from Yang Hui's *Hsü Ku Chai Chhi Suan Fa* (+1275), after Li Nien.

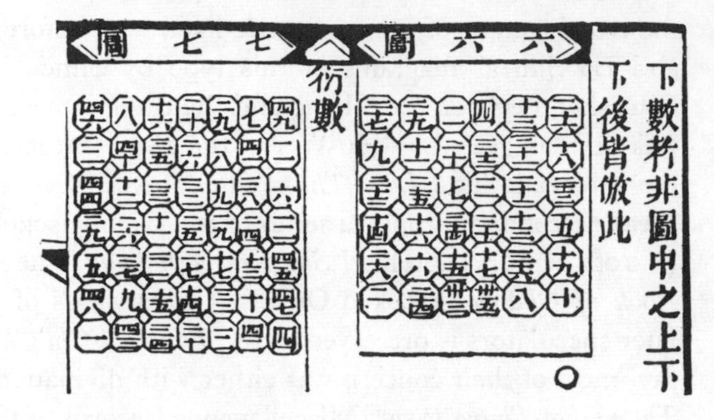

27	29	2	4	13	36
9	11	20	22	31	18
32	25	7	3	21	23
14	16	34	30	12	5
28	6	15	17	26	19
1	24	33	35	8	10

Fig. 61. Two magic squares from Chhêng Ta-Wei's *Suan Fa Thung Tsung* (+1593). The one on the right is shown below in Arabic numerals corrected by Li Nien.

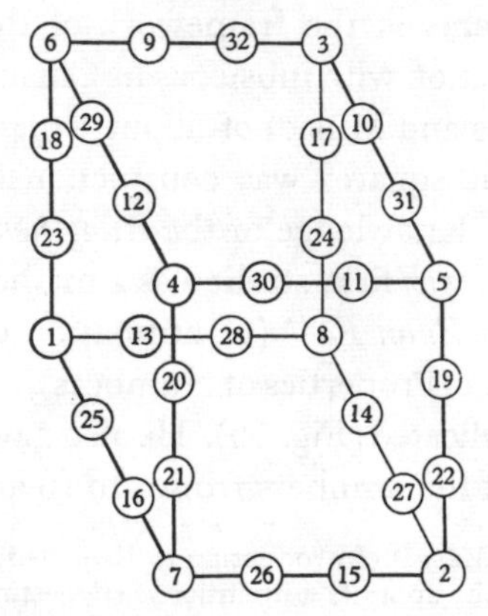

Fig. 62. A three-dimensional magic square from Pao Chhi-Shou's *Pi Nai Shan Fang Chi*.

Thu Pu[4] (mathematical diagrams chapter) of his contemporary *Hsin Chai Tsa Tsu.*[5] At this time and during the eighteenth century Japanese mathematicians were also extremely interested in these aspects of combinatorial analysis.[a] The interest still continued in China, and a gifted amateur such as Pao Chhi-Shou[6] could publish, in the latter part of the 19th century, a book such as the *Pi Nai Shan Fang Chi*[7] (Pi Nai Mountain Hut Records) containing three-dimensional magic squares (see Fig. 62).

[a] Smith & Mikami (1), pp. 57, 69, 116, 177.

[1] 方中通 [2] 數度衍 [3] 張潮 [4] 算法圖補 [5] 心齋雜俎
[6] 保其壽 [7] 碧奈山房集

The preceding account may perhaps have seemed more sinological than mathematical, but it was necessitated by the approach still current among Western historians of science:[a] 'Magic squares were invented by Huang Ti, who, as is well known, ruled over China in the 27th century B.C.' No comparison with the European story is possible unless the Chinese dates are clarified. We can now make such a comparison. The earliest discussion of a magic square in the West occurs in the works of Theon of Smyrna, already mentioned, about +130.[b] It is thus distinctly later than the *Ta Tai Li Chi*, and if we accept the correctness of the Chinese tradition that the Ho Thu and Lo Shu were indeed magic squares, it may be several centuries later than their first appearance in China. Since there must remain a considerable element of doubt about this, the Chinese origins may at least be said to antedate the Greek by perhaps a couple of centuries. There is certainly the possibility of transmission at such a period, but it seems more likely that the origins were independent.

After Theon there was little development in the West until the Arabs began to take an interest in magic squares, and then, just around the time of Yang Hui, three works concern them. Those of the Arabic writers were primarily magical—the *Kitāb al-Khawāṣṣ* (Book of (Magic) Properties) by Abū-l-'Abbas al-Būnī (*d.* +1225),[c] and the essay on magic squares dedicated to al-Manṣūr by Najm al-Dīn al-Lubūdī (+1211 to +1267).[d] But the real corresponding figure to Yang Hui was the Byzantine Greek Manuel Moschopoulos, probably his younger contemporary (*fl.* +1295 to +1316), who wrote, at the request of Nicholas Rhabdas, a work on 'square numbers' (*tetragonon arithmon*, τετραγώνων ἀριθμῶν), describing means of arranging the numbers 1 to n^2 in such a way that the sum of them in each row, column or diagonal equals $\frac{1}{2}n(n^2+1)$.[e] Like the work of Yang Hui, this was purely mathematical, not magical.[f] It is interesting that so many centuries after the first construction of magic squares, a thoroughly scientific treatment of them should have been undertaken almost simultaneously in China and in the Byzantine empire at the end of the +13th century.[g] Later, they played an important part in the mystical philosophy of such figures as Agrippa of Nettesheim (+16th century).[h]

Somewhat allied to magic square questions is the so-called Josephus problem.[i] A given number of persons are arranged in a circle. The persons are of two kinds (Turks and Christians, sacrificial victims and others, heirs and disinherited, etc.), and the problem consists in so arranging them that in counting round, every person in a given periodicity (e.g. every fifteenth) shall be of one kind. It has been thought that this may go back to the Roman military practice of decimation, but it is not mentioned in Europe much before the +10th century. Its origin is obscure, for it was probably one of the permutation problems dealt with by Fujiwara Michinori[1] in the +12th

[a] Cf. in 1953, Struik (2), p. 34.
[b] Sarton (1), vol. 1, p. 272 (tr. Dupuis).
[c] Sarton (1), vol. 2, p. 596.
[d] Sarton (1), vol. 2, p. 624.
[e] Sarton (1), vol. 3, pp. 119, 679.
[f] Tr. Tannery (2).
[g] The exact Lo Shu itself has been found in +13th-century European MSS. (Smith (1), vol. 2, p. 597).
[h] Nowotny (1); Calder (1).
[i] Cf. Ahrens (1).

[1] 藤原通憲

century, and certainly attracted great interest among later Japanese mathematicians (cf. Fig. 63).[a] We are not yet able to point to an instance of the occurrence of this *chi tzu li*[1] (*mamako-date*) in a Chinese mathematical work.

Fig. 63. The Josephus problem, from a Japanese book, the *Jingōki* of Yoshida Mitsuyoshi (+1634).

(*e*) LOGISTIC OF NATURAL NUMBERS

The number of fundamental operations in arithmetical computation has not always been recognised as four (addition, subtraction, multiplication and division). At various times in history[b] others were included too, e.g. duplation and mediation (doubling and halving), and the extraction of roots. It would seem, however, that the Chinese always recognised only the modern four.[c]

(1) The Four Basic Processes

About addition (*ping*,[2] or for fractions *ho*;[3] giving the sum, sometimes called *tu shu*[4])[d] there is little to say. Texts earlier than the +3rd century always write out the numbers in full, but it is clear that from the Warring States time the additions must have been made with counting-rod numerals on a counting-board, using a place-value system in

[a] Smith (1), vol. 2, p. 541; Smith & Mikami (1), pp. 80ff.; Papinot (1), p. 99.

[b] As Smith (1), vol. 2, pp. 32 ff.; Menninger (1); and Tropfke (1) have shown.

[c] It was common in late medieval times down to the 18th century to derive the fundamental operations from the Ho Thu and Lo Shu; these fanciful analyses have been briefly described by Chêng Chin-Tê (1).

[d] Since the technical terms in ancient Chinese mathematical texts differ considerably from those in modern use, they will be included in this account. Those who wish to go further into the subject may be helped thereby.

[1] 繼子立 [2] 并 [3] 合 [4] 都數

which blanks were left where we should put zeros. Though Chinese writing has always run downwards from top to bottom of columns, numerals seem to have been written horizontally from left to right, the sum of the addends being placed separately below. The carrying-over process must have been very ancient, and would have been a natural outcome of counting-board operations.

Subtraction was carried out with rod-numerals in the same way. It has been called *chien*[1] from the *Chiu Chang* until now. The remainder, *yü*,[2] or *chha*[3] (difference), was placed, like the sum, separately but above.

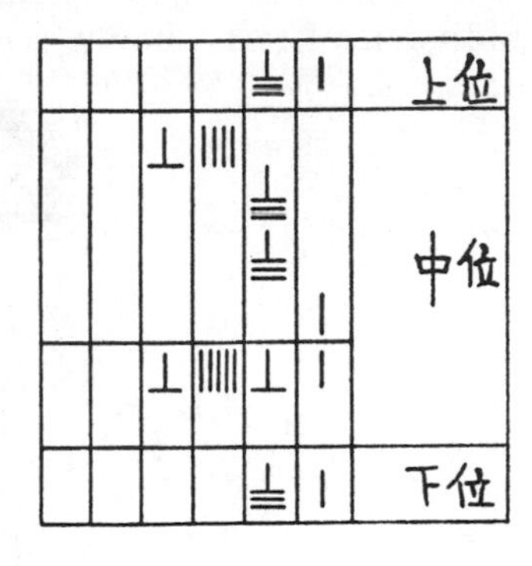

Just as addition was a device for obtaining results that could have been more laboriously obtained by counting, so multiplication was an 'abridgement of addition', a 'folding together' of many addends. The word *chhêng*[4] seems to have been applied by a similar process of thought; its common meaning is to ride in or upon something, and here the addends were thought of as a team of horses controlled by a chariot-driver. The multiplier was indeed placed in the 'uppermost position' (*shang wei*[5]). This was the converse of division, and the schemes were as follows:

		multiplication	*division*
shang wei	上位	multiplier	quotient
chung wei	中位	product	dividend
hsia wei	下位	multiplicand	divisor

The bottom line was multiplied by the top one (*i shang ming hsia*[6]) and for every figure occupying the tens column the product figure was moved over one column to the left (*yen shih chi kuo*[7]). The process can be followed in the accompanying diagram, set up to show the multiplication of 81 by 81. First they occupy their places in the top and bottom positions. Then 'the upper 8 calls the lower 8' (*shang pa hu hsia pa*[8]), and 64 is written two spaces to the left representing 6400. Then the upper 8 calls the lower 1, and 8 is written in the middle position. The upper 8 is then withdrawn, the upper 1 is 'called' by the lower 8, the lower 8 withdrawn, and '1 calling 1' is set down as 1 in the units column. On addition (*chia*[9]) of the products the final product is obtained.[a] This process resembles one which is unusual for Europe but found in the Treviso arithmetic of +1478.[b] Probably our common modern system, known in medieval Europe as the chessboard (*scacchero*) method, originated from such calculations. Forms intermediate between them and the old Chinese counting-board might

[a] Cf. Lagrange's advocacy of 'inverted multiplication'. The particular example chosen was given because it is explained in detail in *Sun Tzu Suan Ching*, ch. 1, pp. 4*a*, *b*.

[b] Smith (1), vol. 2, p. 109.

1 減 2 餘 3 差 4 乘 5 上位 6 以上命下
7 言十卽過 8 上八呼下八 9 加

be found in the works of Bhāskara[a] and other Indian mathematicians. The *gelosia*, or grating, method of multiplication, which was Indian or Arabic in origin,[b] is not found in Chinese books until the *Suan Fa Thung Tsung* of +1593 (see Fig. 64)—it was known as the *yin chhêng thu*,[1] or *phu ti chin*.[2]

The multiplication table (*chiu chiu ko chüeh*[3]) is naturally ancient in China. Unlike the Babylonian practice of arranging the numbers in columns, the table appears in the old Chinese texts simply written out in words. Li Nien (*8*) and Yen Tun-Chieh (*2*)

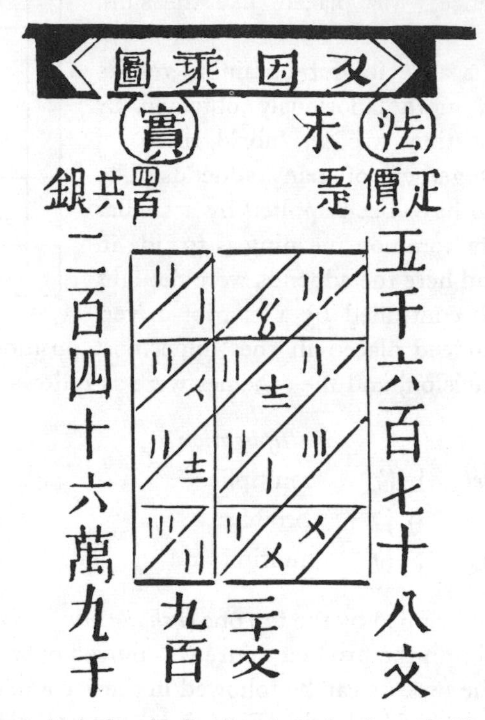

Fig. 64. The *gelosia* method of multiplication, from the *Suan Fa Thung Tsung* (+1593).

have collected notes of the texts in which it appears; among the earliest may be the fragments embodied in the *Kuan Tzu* book,[c] perhaps before the −4th century. Yen Tun-Chieh has also described the multiplication tables on the bamboo tablets[d] recovered from the sand-buried city of Chü-yen at Edsin Gol—these are of Han date, *c.* −100. Portions of such tables continue to occur in texts down to the Sung, as, for example, in the *Sou Tshai I Wên Lu*[4] (In search of Strange Things),[e] by Yung Hêng.[5] The so-called Pythagorean form of the table, in which the numbers are arranged upon

[a] *C.* +1150; Taylor (1) tr. p. 10. And in the +10th-century Bakshālī MS. See Renou & Filliozat (1), vol. 2, p. 175.

[b] Smith (1), vol. 2, p. 115.

[c] Chs. 58, 80.

[d] These bamboo slips also had written on them a number of problems similar to those in the *Chiu Chang Suan Shu*.

[e] Ch. 3, p. 4*a*.

[1] 因乘圖 [2] 鋪地錦 [3] 九九歌訣 [4] 搜採異聞錄 [5] 永亨

two coordinates (like a London bus-fare table), seems to appear in China, in triangular array,[a] about the +8th century.[b]

The square of a number was naturally called *fang*[1] by the Han and later mathematicians, *chhêng fang*[2] in the Sung, and in modern times *tzu chhêng*,[3] the number multiplied by itself. Another modern term is *phing fang*[4] (the flat square), which contrasts with the name of the cube both ancient and modern, *li fang*[5] (the solid upstanding square thing).

Division was carried out using the rod-numerals in columns as sketched in the case of multiplication. The word *chhu*[6] was invariably used for this operation. The divisor was called *fa*[7] and the dividend *shih*[8] (real). The first of these words is of course the same as that used for law, especially positive law.[c] If division first arose in connection with land measurement, the divisor would naturally have been the unit measure fixed by law. The dividend would have been the real length of the field. Sun Tzu (+3rd century) says[d] that if the dividend (*shih*[8]) has a remainder (*yü*[9]), then the divisor (*fa*[7]) must be taken as 'mother' (*mu*[10]), i.e. denominator, and the remainder as 'child' (*tzu*[11]), i.e. numerator. Thus $6562 \div 9 = 729\frac{1}{9}$. Gradually the terms *fa*[7] and *shih*[8] acquired specific technical meanings in a variety of other operations, including root extraction, and quadratic, cubic and higher numerical equations.[e] This shows that many complex processes originated in the handling of numbers in division. In Hsiahou Yang's time the quotient was called *shang*.[12] The galley method[f] was not used in China until after the time of the Jesuits.[g] Division tables (using words) were common from the Sung onwards in connection with abacus computations.

(2) Roots

The ancient processes for the extraction of square roots (*khai fang*,[13] *khai fang chhu chih*,[14] etc.) and of cube roots (*khai li fang*[15]) were essentially the same as those of today. Arabic mathematicians conceived a square number to grow out of a root, while Latin writers thought of it as the side of a geometric square; the term *radix* is thus derived from the Arabs[h] and *latus* from the Romans. In ancient China also the procedure arose from geometrical considerations, and the Han mathematicians refined their techniques so far as to lay a firm foundation for the solution of numerical equations by the great

[a] Smith (1), vol. 2, pp. 125, 127.
[b] Li Nien (7), editing the Tunhuang MSS. Cf. pp. 9, 36 above.
[c] See Sections 12 and 18*f* above.
[d] Ch. 2, p. 1*b*.
[e] See below, pp. 66, 126. Cf. Wang Ling (2), vol. 1, pp. 132ff., 217ff.
[f] Smith (1), vol. 2, p. 136.
[g] Li Nien (2), p. 227.
[h] Discussion took place between Gandz and Ma on the origin of the modern Chinese mathematical term *kên*[16] (root), and the demonstration by Ma that the word is an ancient one did nothing to settle the matter. Neither scholar seemed to be aware that the old Chinese terms for square and cube root were quite different. The use of the term *kên* does not go back beyond the +14th-century Arab influence, and probably not beyond the Jesuits.

1 方 2 乘方 3 自乘 4 平方 5 立方 6 除
7 法 8 實 9 餘 10 母 11 子 12 商
13 開方 14 開方除之 15 開立方 16 根

algebraists of the Sung and their anticipation of Horner's method.[a] The extent to which this method may be considered to have been already implicit in the text of the *Chiu Chang Suan Shu* has been discussed at length by Wang & Needham (1), who explain the successive stages of operations on the Han counting-board. Reconstruction of the counting-board positions used in computing *Chiu Chang* root problems by the Han mathematicians shows how the Thien Yuan notation of the Sung[b] would have been a natural development from the processes used.

The geometrical basis of the square-root method was almost certainly pictured by Liu Hui. Modern editions of the *Chiu Chang Suan Shu* carry a diagram supplied by Tai Chen, showing that if a straight line is divided into two parts the square on the whole line is equal to the sum of the squares on the two parts, plus twice the rectangle contained by them, i.e. algebraically, $(a+b)^2 = a^2 + b^2 + 2ab$. These quantities were named as shown in the diagram, *fang*[1] for the larger square, *yü*[2] for the smaller square, and two *lien*[3] for the rectangles. The oldest surviving diagram is that of Yang Hui's *Hsiang Chieh Chiu Chang Suan Fa Tsuan Lei* of +1261 preserved in MS. form in chapter 16,344 of the *Yung-Lo Ta Tien* encyclopaedia (Fig. 65). Actually these terms appear first in the *Sun Tzu*, the areas being represented by Liu Hui in different colours. In the tabulated calculations, *shang*,[4] which occupies the top line, is the root (to be obtained), *shih*[5] is the number of which the root is wanted (occupying the second line), while *fang fa*,[6] *lien fa*,[7] *yü fa*[8] and *hsia fa*[9] (occupying lower lines) are factors temporarily appearing in the calculation.

The figure used by the Han people was similar to that given by Euclid (*c.* −300).[c] He demonstrated geometrically the above proposition, though this was probably known much earlier.[d] But at this early stage, a and b could be any numbers, and the idea of identifying them with figures of differing place-value (so that, for example, $a = 10b$) was not present. It occurs, however, quite consciously, in the *Chiu Chang* (perhaps about −100), and Liu Hui (*fl.* +260) knew that the process could be extended indefinitely. In Europe it seems to come first in the work of Theon of Alexandria,[e] about +390 (i.e. nearer to the time of Hsiahou Yang than Liu Hui), who extracted square roots in terms of sexagesimal fractions, showing clearly the idea of having a and b in definite numerical progression. The modern processes of extraction of square and cube roots arose in Europe between the time of Maximus Planudes (*c.* +1340), who still used the methods of Theon's time, and that of Pacioli, who in +1494 gave a method similar to that of the *Chiu Chang*, in which the digits of the root were worked out one by one, and place-valuc was isolated.[f] We do not know what stimulus reached Europe between these dates, but certain it is that about +1430 Ghiyāth al-Dīn Jamshīd

[a] See below, p. 126.
[b] See below, p. 129.
[c] *Elements*, bk. II, prop. iv.
[d] Smith (1), vol. 2, p. 145.
[e] Not to be confused with the +2nd-century Theon of Smyrna, who was the first European to study magic squares; cf. Sarton (1), vol. 1, pp. 272, 367. See pp. 54, 61 above.
[f] Smith (1), vol. 2, p. 146.

1 方 2 隅 3 廉 4 商 5 實 6 方法 7 廉法
8 隅法 9 下法

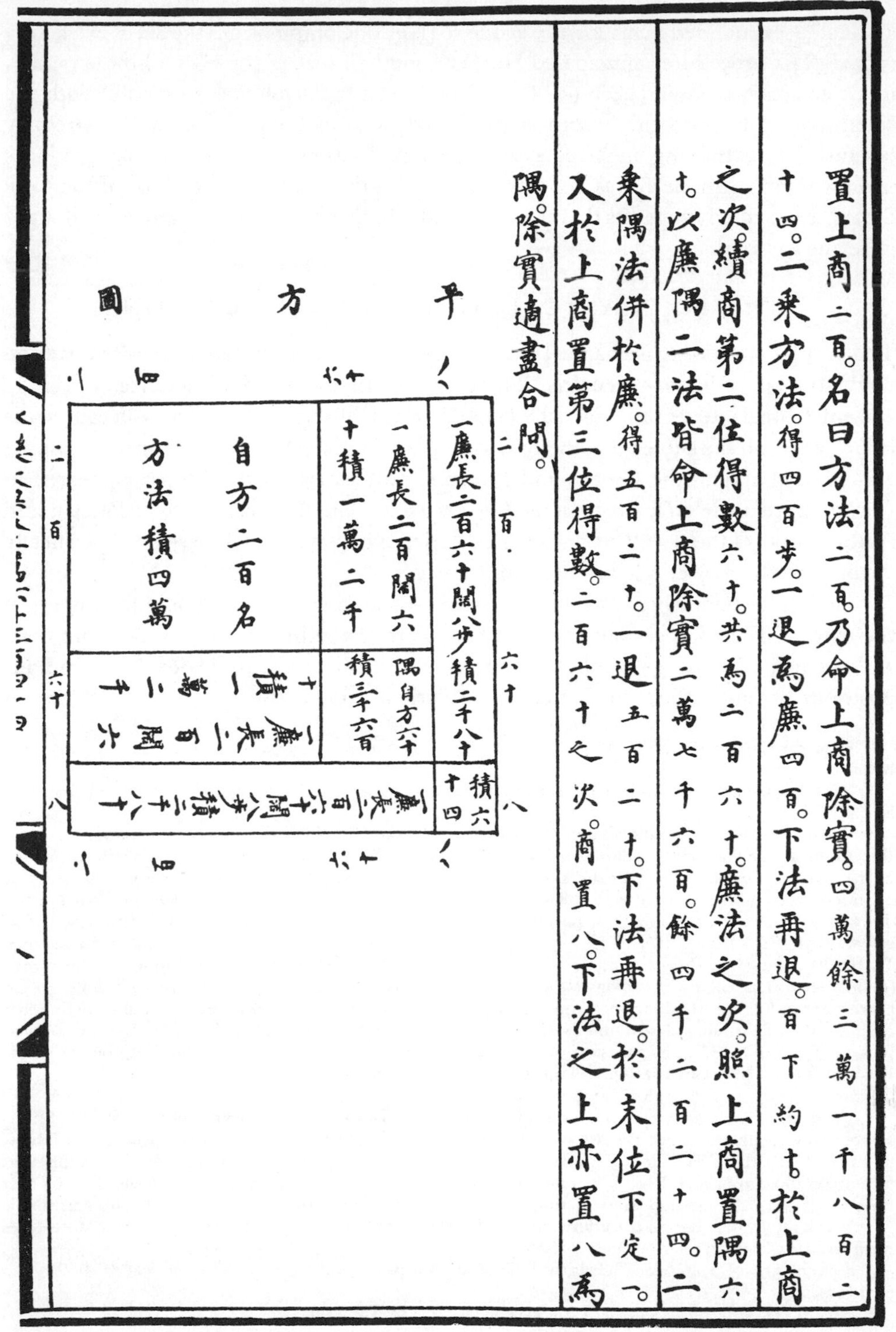
置上商二百。名曰方法二百。乃命上商除實。四萬餘三萬一千八百二
十四。二乘方法。得四百步。一退爲廉四百。下法再退。百下約六十於上商
之次。續商第二位得數六十。共爲二百六十。廉法之次。照上商置隅六
十。以廉隅二法皆命上商除實二萬七千六百。餘四千二百二十四。二
乘隅法併於廉。得五百二十。一退五百二十。下法再退。於末位下定一。
入於上商置第三位得數。二百六十之次。商置八。下法之上亦置八爲
隅。除實適盡合問。

Fig. 65. Diagram to illustrate the process of square-root extraction, from Yang Hui's *Hsiang Chieh Chiu Chang Suan Fa Tsuan Lei* (+1261), as preserved in the MS. *Yung-Lo Ta Tien* (+1407), ch. 16,344 (Cambridge University Library).

al-Kāshī,[a] expounded in his *Miftāḥ al-Ḥisāb* (Key of Computation) the art of extracting roots.[b] His procedures anticipated Horner's method just as those of Chinese mathematicians such as Chia Hsien (*c.* +1100) had done.[c] Pacioli was successful with the square root. Even in the +16th century European mathematicians were extremely backward in extracting cube roots, and the visualisation of the process using blocks (solid geometric models) was not introduced until the +17th.[d] The tenor of the *Chiu Chang* text strongly suggests that the Han mathematicians used such models or thought in terms of them.[e]

(*f*) MECHANICAL AIDS TO CALCULATION

The earliest form of a mechanical aid to calculation, if such it could be called, was no doubt the art of finger-reckoning. There are not many obvious references to this in ancient Chinese texts, but Eastlake (1) and Wei Chü-Hsien (*2*) have collected some information on the subject.[f] Although, as L. J. Richardson has shown, finger-reckoning was general among the Greeks and Romans, no treatise on the subject appeared in Europe until Bede's *De Loquela per Gestum Digitorum* of about +710.[g] Persian and Arabic works of the +13th and +14th centuries exist,[h] but no Chinese text specifically devoted to the matter has yet been found.

Lemoine distinguishes three types of finger-reckoning, one using only the fingers of the extended hand, a second in which the joints of the fingers are allotted numbers, and a third, the most complicated, in which numbers are indicated by numerous positions of fingers flexed or extended in various combinations. The first is used in

[a] First director of the observatory set up by Ulūgh Beg at Samarqand in +1420. See further, p. 89 below.

[b] This book has been studied by Luckey (1) and translated into Russian by Rosenfeld & Yushkevitch (1).

[c] See p. 137 below. A transmission of the technique from China is visualised by Yushkevitch (1) and Rosenfeld & Yushkevitch, p. 386. Relations between Chinese and Islamic scholars had at this time long been close. As far back as +1221 the Taoist Chhiu Chhang-Chhun had recorded meeting at Samarqand a Chinese astronomer, Mr Li, who was superintendent of an observatory (*Chhang-Chhun Chen Jen Hsi Yu Chi*, ch. 1, p. 13*b*; Waley (10), p. 97). At Marāghah about +1260 there was at least one Chinese astronomer, Fu Mêng-Chi, and in +1267 Jamāl al-Dīn took his astronomical delegation to Peking. During the +14th century a number of Muslim mathematicians resided there, and much of their work (mainly astronomical and astrological) was translated into Chinese (see p. 49 above). Then in +1420 there went to Peking an important Timurid embassy from Samarqand, sent by Shāh Rukh (Ulūgh Beg's predecessor), with Ghiyāth al-Dīn-i Naqqāsh as recorder, and a retinue of no fewer than seventy-seven persons (cf. *Ming Shih*, ch. 332, p. 23*b*). It is interesting that the *Tables* of Ulūgh Beg, begun by al-Kāshī, have a chapter on Chinese calendrical science (cf. Sédillot, 3).

[d] Smith (1), vol. 2, p. 148.

[e] From the *Chiu Chang* commentaries one can see that Liu Hui in the +3rd century and Li Shun-Fêng in the +7th envisaged the use of blocks for geometrical demonstration of the procedures (ch. 4, pp. 14*a*, 18*a*); this was called *tieh chhi*.[1] In view of all the foregoing facts it is hardly believable that Cronin could suppose (1), p. 197, that the works of Clavius, relayed by Ricci, 'revealed for the first time in Chinese the method of extracting square and cubic roots from whole numbers and fractions'.

[f] Wei's attempt to derive the etymology of the number characters from presumed ancient gestures need not be taken very seriously.

[g] Sarton (1), vol. 1, p. 510. The classical illustration is perhaps that in Pacioli's *Summa de Arithmetica* (+1494).

[h] Sarton (1), vol. 3, p. 1533; and the excellent study by Lemoine.

[1] 叠棊

bargaining, and must be connected with the famous 'finger-game' (*huo chhüan*[1] or *tshai chhüan*[2]), so familiar to everyone who has been present at convivial entertainments in China.[a] This seems to be identical with the *micatio digitis* of the Romans (Ar. *mukhāraja*; It. *morra*; Fr. *mourre*).[b] The second method,[c] widely used in China, is connected with early calendrical calculations, since the number of joints on the hands may be made, according to the convention adopted, 19, 28, etc. The third method, which is found over the greater part of Asia, seems to be Babylonian in origin, and Lemoine has shown a detailed correspondence between its description in Bede and in a number of Persian and Arabic writings. But there were Chinese systems which differed from these. One late form, given in the *Suan Fa Thung Tsung*,[d] represented decimals and powers of ten by the different fingers.[e]

A very simple device, used rather for recording numbers than for calculating, was the system of knotted strings best known in the form of the Peruvian *quipu*, which is described in detail by Locke.[f] Ancient Chinese literature contains a number of distinct references to the use of the *quipu*. The *locus classicus* is perhaps the *I Ching* (Book of Changes),[g] where the reference may date from the −3rd century: 'In the most ancient times the people were governed by the aid of the *quipu* (*chieh shêng*[3]).' But there is also a mention in *Chuang Tzu*, and in a famous chapter of the *Tao Tê Ching*.[h] Li Nien (*8*) gives several later references.[i] Of particular interest is the description by Simon (1) of the use of the *quipu* by the aboriginal inhabitants of the Liu-Chhiu Islands;[j] quite possibly it might still be found in China among the tribesfolk such as the Miao or the Yi. Here is another of those strange similarities between East Asian and Amerindian culture.

During the historical period Chinese mathematicians have had at their disposal, apart from the universal counting-board itself (Fig. 66), three main types of mechanical calculating aids: (*a*) plain counting-rods; (*b*) counting-rods marked with numbers, analogous to Napier's bones; and (*c*) the abacus. There has been much discussion concerning the historical origin of these, but we are not completely in the dark, as the following account will show. It will be best to establish first the most probable course of events for China and to postpone till the end the difficult question of parallel developments in the West.

Naturally these questions are fully discussed by the Chinese historians of mathematics. Li Nien (*9*) has devoted a special paper to the history of the abacus. Perhaps

[a] The players extend one or more fingers simultaneously, at the same time guessing the total number extended; he who guesses correctly wins.

[b] The Arabs use it in divination in connection with discoidal planispheres (see below, Sect. 26*i*, where the relation of these with Byzantine star-chess and the Chinese diviner's boards of antiquity is discussed).

[c] Cf. Bayley (1).

[d] Ch. 12, p. 9*b*.

[e] See also *TSCC, Li fa tien*, ch. 125, p. 10*b*. Cf. Leupold (2).

[f] Some specimens show clear traces of use in calendrical calculations (Nordenskiöld, 1).

[g] Hsi Tzhu (Great Appendix), 2nd part, ch. 2 (R. Wilhelm (2), vol. 2, p. 256). Cf. p. 95 below.

[h] Ch. 80.

[i] There is no evidence, however, for the *use* of the *quipu* in China in historical times.

[j] Cf. Mikami (*14*).

1 豁拳 2 猜拳 3 結繩

the most valuable contributions in Western languages are those of de Lacouperie (2) and Vissière (1); the latter gave a partial translation of the *Ku Suan Chhi Khao*[1] (Enquiry into the History of Mechanical Computing Aids) by Mei Wên-Ting of about +1700.[a] The best summary of what is known about the counting-rods is by Li Nien.[b]

(1) Counting-Rods

Epigraphic evidence as to the antiquity of the plain counting-rods (*suan*,[2] *chhou*[3] or *tshê*[4]) exists in the appearance of the counting-rod numerals on the coins of the

Fig. 66. 'Discussions on difficult problems between Master and Pupil', a view of a counting-board (frontispiece of the *Suan Fa Thung Tsung*, +1593).

Warring States period (−4th and −3rd centuries). But there are literary evidences from the same time. Perhaps the most famous is that in the *Tao Tê Ching*,[c] where Lao Tzu characteristically says 'Good mathematicians do not use counting-rods'

[a] At the time when Mei wrote, some of the most important Chinese mathematical books had not yet been rediscovered, but Vissière supplemented his argument where necessary. Mei's monograph forms part of his *Li Suan Chhüan Shu*[5] (Complete Works on Calendar and Mathematics) published in +1723.

[b] (*4*), vol. 3, pp. 29 ff.

[c] Ch. 27.

[1] 古算器考 [2] 筭 [3] 籌 [4] 筴 [5] 曆算全書

(*Shan shu pu yung chhou tshê*[1]).[a] After the beginning of the Han, however, mentions of the rods are much more frequent.[b] One occurs in a poem by Mei Chhêng[2] (d. −140) similar to those in the *Chhu Tzhu* (Elegies of Chhu).[c] The *Chhien Han Shu*[d] says that they were bamboo sticks $\frac{1}{10}$ inch in diameter and 6 inches long. 271 of them fitted together into a hexagonal bundle (*liu ku*[3]) conveniently held in the hand.[e] 'They permitted the determination of lengths to an accuracy of a hundredth or thousandth of an inch; of volumes and of weights without the loss of a single grain of millet.'[f] In the *Shih Chi*,[g] Ssuma Chhien describes a conversation between the first Han emperor and Wang Ling[4] in which he says that he was not so good as his three great generals and advisers in various respects, but he alone knew how to use them all; one of these talents was 'planning campaigns with counting-rods in the headquarters tent' (*yün chhou tshê wei chang chih chung*[5]). This refers to −202. From the biography of Ma Jung[6] in the *Hou Han Shu*[h] we know that Chhen Phing,[7] another minister of State in the early Han (*d.* −178), was famous for his use of the counting-rods. There was, moreover, a tradition that Chao Tho,[8] a minister of Chhin Shih Huang Ti, who afterwards ruled over the south as an independent king, had several different kinds of counting-rods made before he led his army there.[i] These were afterwards preserved in the museum of the emperor An (+397 to +419) of the Chin; they were each 1 foot long, some white, made of bone, others black, made of horn. We also know from the *Chhien Han Shu*[j] that Sang Hung-Yang[9] (−152 to −80), the famous minister of State who appears as the protagonist of the nationalisation of salt and iron in the *Yen Thieh Lun* (cf. Sect. 48), was renowned as a computer because he could do so much mentally without the aid of the counting-rods.

It would be tedious to add further references, some of which may be found in Li Nien (*4*). We may just note a few more points. Wang Jung[10] (+235 to +306), minister of Chin and patron of water-mill engineers, 'taking his ivory counting-rods in his hands, used to calculate all through the night, as if he could hardly stop'.[k] Hence the proverbial expression *ya chhou chi*.[11] In the +9th century the rods were

[a] A misprint in the *Thai-Phing Yü Lan* encyclopaedia caused de Lacouperie to make nonsense of this reference. To show the confusion which has existed in this field, one may mention that his mistake was copied even by a Chinese mathematician writing in English (Chêng Chin-Tê, 2).

[b] *Huai Nan Tzu*, ch. 14, pp. 5*b*, 10*a*.

[c] Edkins (10), p. 222.

[d] *Lü li chih*, ch. 21 A, p. 2*a*.

[e] An early example of a figurate number.

[f] The *Shuo Wên* (+121) repeats the statement about the size of the rods. It corresponds to about four of our inches.

[g] Ch. 8, p. 30*a* (tr. Chavannes (1), vol. 2, p. 383, where, however, the significance of the text is not brought out).

[h] Ch. 90 A, p. 5*b*.

[i] This would be about −215. The information comes from the pre-Sui book *I Yuan*[12] by Liu Ching-Shu,[13] quoted in *TPYL*, ch. 750, p. 3*b*. It may well be a reference to special rods for negative numbers.

[j] Ch. 24 B, pp. 11*a*, 13*a*; Swann (1), pp. 272, 285. Cf. *Chi Chiu Phien* (*c.* −40), ch. 4, p. 34*a*.

[k] *Shih Shuo Hsin Yü*[14] and *Chin Shu*,[15] quoted in *TPYL*, ch. 750, p. 2*a*.

1 善數不用籌策 2 枚乘 3 六觚 4 王陵 5 運籌策帷帳之中
6 馬融 7 陳平 8 趙佗 9 桑宏羊 10 王戎
11 牙籌計 12 異苑 13 劉敬叔 14 世說新語 15 晉書

made of cast iron.[a] Thang administrators and engineers used to carry a bag of counting-rods (*suan nang*[1]) at their girdle, as appears from the biography[b] of Li Ching.[2] Another name for this was *suan tai*,[3] and there was a legend that a certain kind of fish was derived from the bag of counting-rods which Chhin Shih Huang Ti once threw away in the eastern ocean.[c] In the +11th century Shen Kua, describing one of his contemporaries, the astronomer Wei Pho,[4] says that 'he could move his counting-rods as if they were flying, so quickly that the eye could not follow their movements before the result was obtained'.[d] This description makes one visualise rather the speed with which the abacus can be manipulated. After the late Ming, less is heard of the counting-rods, no doubt because they were ousted by the abacus.[e] In all the above cases it is supposed that calculations were effected by the actual formation of the rod-numerals by means of the rods on a counting-board; this had the advantage over writing that it was easier to cancel numbers which were no longer wanted. In the Thang, at any rate, the transverse bars of the numerals on the board were called 'reclining' (*wo suan*[5]) and the longitudinal ones 'standing up' (*li suan*[6]).

The counting-rods have left their traces in orthography since most of the terms for calculation (*suan*,[7,8,9] *chhou*,[10] *tshê*[11]) have the 'bamboo' radical (no. 118), and many expressions such as *thui suan*,[12] 'to push the rods about', *chhih chhou*,[13] 'to grasp the rods', etc., have survived as expressions for computing.

(2) Graduated Counting-Rods

Counting-rods with numbers marked on them were probably a late development in Chinese mathematics. They seem to be practically identical with Napier's bones. Napier, in his *Rhabdologiae, seu Numerationis per Virgulas Libri Duo* of +1617, described a system of rods arranged on the basis of the *gelosia* method of multiplication (see above, p. 64), as if each column had been separated and made to slide on the others independently. These slide rods, which continued in use throughout the 17th century (cf. Leybourn's *The Art of Numbring by Speaking-Rods; Vulgarly termed Nepeir's Bones*, +1667), were quickly introduced into China and Japan, where they attracted considerable interest. The best-known treatise on them is the *Tshê Suan*[14] of +1744 by the famous scholar and mathematician Tai Chen[15] (cf. Vol. 2, p. 513). Fig. 67 shows them in their East Asian form. The set as used in China comprised also a zero rod, and square and cube rods; the method of employing them has been described in detail by Chêng Chin-Tê (2). They retained the same name as the ancient plain computing rods, a fact which has sometimes given rise to confusion. In the 19th century Lao

[a] *KCCY*, ch. 49, p. 7*b*, quoting the *Chhing I Lu*[16] of Thao Ku[17] of the Five Dynasties period (+10th century).

[b] *Hsin Thang Shu*, ch. 93, p. 4*b*.

[c] *Yu-Yang Tsa Tsu*, ch. 17, p. 1*b*.

[d] *MCPT*, ch. 8, para. 11.

[e] Their use is described, however, in the *Suan Hsüeh Chhi Mêng* of +1299.

1 筭囊 2 李靖 3 筭袋 4 衛朴 5 臥算 6 立算
7 算 8 筭 9 笇 10 籌 11 策 12 推算 13 持籌
14 策算 15 戴震 16 清異錄 17 陶穀

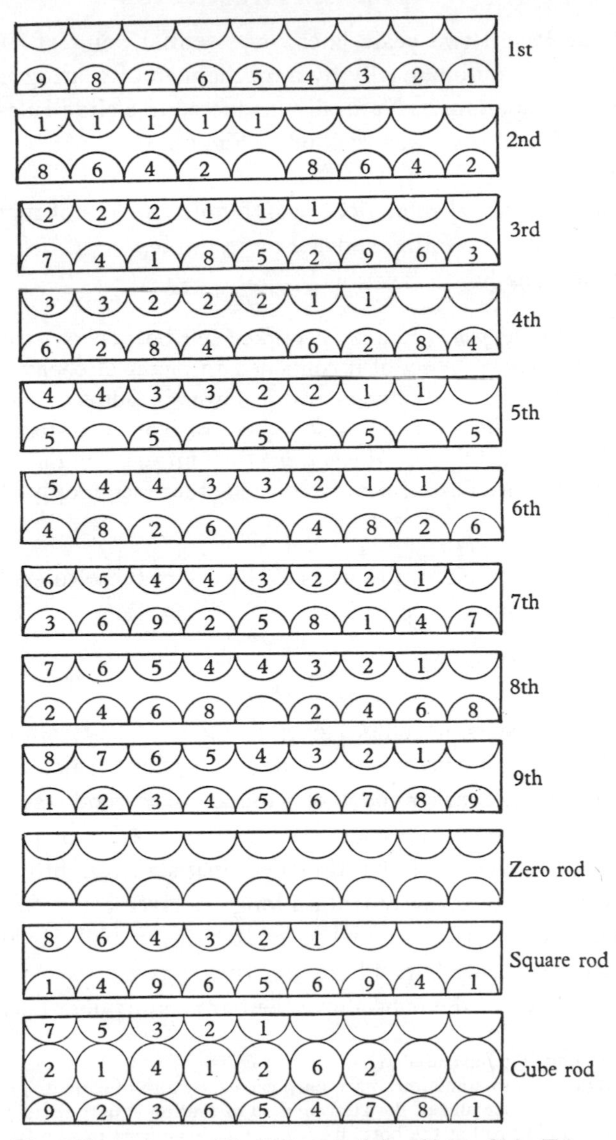

Fig. 67. The Chinese form of Napier's rods (after Chêng Chin-Tê).

Nai-Hsüan[1] wrote much on Napier's bones and the earlier rods, e.g. his *Ku Chhou Suan Khao Shih*[2] of 1886, and its supplement *Ku Chhou Suan Khao Shih Hsü Pien*.[3] The system would perhaps have been more useful if the two inventions of the logarithmic slide-rule and the adding-machine had not rapidly followed upon Napier's device. The former was due to the work of Edmund Gunter in +1620 and William Oughtred in +1632,[a] the latter, suggested by Johann Ciermans in +1640, was per-

[a] Smith (1), vol. 2, p. 205; Cajori (4).

[1] 勞乃宣 [2] 古籌算考釋 [3] 古籌算考釋續編

fected by Blaise Pascal two years later; its essential point of superiority was that it carried over the tens automatically by a mechanism of gearing instead of this having to be done by the operator.[a] Naturally in due course slide-rules[b] and simple calculating machines found their way to China. Fig. 68, from Michel (5), shows a Chinese slide-rule of +1660.

One can find occasional references to unfamiliar pieces of mathematical apparatus resembling graduated counting-rods. For example, the *Chieh An Man Pi*[1] (Notes from the Fasting Pavilion)[c] says that

> there was a *chhien hsing pan*[2] (lit. a hand-worked star board)[d] constructed by Ma Huai-Tê[3] (*fl. c.* +1064) at Suchow. A set of it contained 12 pieces of ebony ranging in length from 7 inches downwards. They were graduated in finger-breadths up to 12, and each had also smaller divisions as well. There was also a piece of ivory with its four corners left blank, and 2 inches long. It had the words *pan chih*[4] (half finger), *pan chio*[5] (half angle) and *i chio*[6] (full angle) carved on it. These rulers faced each other reciprocally. They were called *Chou Pei Suan*[7] rulers.[e]

No explanation has come down to us as to the use of these rulers but they would seem to have been for geometrical purposes.

(3) The Abacus

As to the Chinese abacus, which has given rise to a large literature,[f] it will be advisable to begin by describing it briefly.[g] It is called *suan phan*[8] (calculating plate) or *chu suan phan*[9] (ball-plate). It consists today of a rectangular wood frame, the longer sides of which are connected by means of wires forming a series of parallel columns (*wei*,[10] *hang*[11] or *tang*[12]). Each of these has threaded on it seven slightly flattened balls (*chu*[13]), which can be moved to or away from a long transverse bar (*liang*[14] or *chi liang*[15]), dividing the abacus into two unequal parts, so that two balls always remain above, and five below, the separating bar. There are usually twelve wires, but there may be as many as thirty. Each of the balls above the bar is equivalent to the five balls below in

[a] Lilley (1); Taton (1); Baxendall (1).

[b] Misunderstanding may sometimes have been caused by an unfortunate phrase used by Mikami (1), pp. 13, 14. In giving an account of the Han method of finding square roots, he referred to the placing of one ordinary counting-rod at the bottom of the counting-board to signify 1, 10 or its powers. This he called 'borrowing a unit calculator'. Needless to say, this has nothing remotely to do with borrowing a slide-rule from a friend.

[c] Probably part of the *Chieh An Chi*[16] by Chin Kuei[17] (*fl. c.* +1500).

[d] The first of these characters is a rare one. The 'stars' may have been pins hinging the rulers together.

[e] Tr. auct.

[f] From the mass of papers, the following may be selected as valuable for consultation: Goschkevitch (1), Rodet (1), Westphal (1), van Name (1), Knott (1), Kuo Mai-Ying (1), Yen Tun-Chieh (3), Leavens (1), Rohrberg (1), Yoshino (1).

[g] This instrument cannot be judged by the dwindled and degenerate form of it which till recently persisted in European infant schools. It is used in China and Japan in a particularly scientific way, and still holds its own outside the narrow circle of those concerned with higher mathematics.

1 戒菴漫筆 2 捀星板 3 馬懷德 4 半指 5 半角
6 一角 7 周髀算 8 算盤 9 珠算盤 10 位 11 行
12 檔 13 珠 14 梁 15 脊梁 16 戒菴集 17 靳貴

PLATE XXI

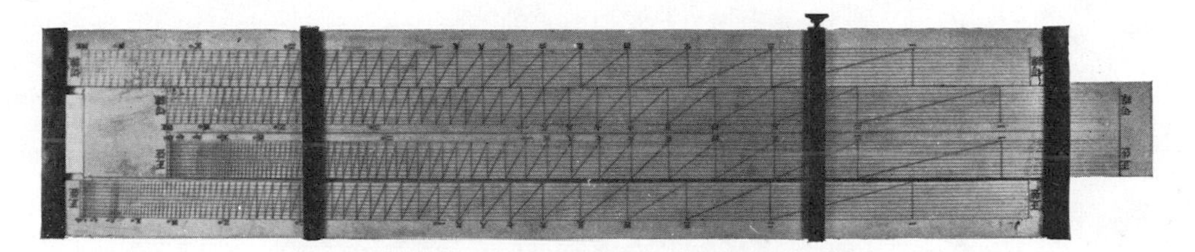

Fig. 68. A Chinese slide-rule of +1660 (photo. Michel).

Fig. 69. Chinese abacus (orig. photo.).

the same vertical column, and each of the columns differs by ten, so that each ball in any given column is equivalent to ten similarly placed balls of the column immediately on its right. Exactly what units each column stands for varies at the will of the operator. The use of the abacus may be seen from Fig. 69 where the number 123,456.789 has been set up.[a] It is possible to carry out the first three fundamental operations using only one of the balls in the upper register, but for division it is convenient to be able to indicate on any given column a number higher than 10, so that two upper balls as well as five lower ones are provided, giving a total of 15. The left side of the abacus is considered the front (*chhien*[1]) and the right the back (*hou*[2]); 'advancing' a number (*chin*[3]) means raising it by powers of ten, while 'retiring' it (*thui*[4]) means the opposite. Setting up a number is to '*shang*[5]' it; for cancelling a number the verbs *chhi*,[6] *chhu*[7] or *chhü*[8] are used. The left-hand column of a calculation is the 'head' (*shou*[9] or *thou*[10]), the last or right-hand column is the 'tail' (*wei*[11] or *mo*[12]).[b] The column which is being operated upon is the '*body*' (*shen*[13]) or principal position (*pên wei*[14]). The detailed operations in the use of the abacus, including the extraction of square and cube roots, are described by Knott (1) and by Smith & Mikami (1). Its special advantages, for example, in accounting, even at the present day, are alluded to by Leavens. In order to illustrate the astonishing speed which Chinese and Japanese computers, trained to the use of the instrument from childhood, can attain, it may be mentioned that when in 1946 a competition was staged in Tokyo between an abacus clerk and an American sergeant using an electric calculating machine, the abacus won for speed in all operations except multiplication, and also in making fewer mistakes (Kojima, 1). Of course, with all simple mechanical aids to computation, the intermediate stages leave no trace, and checking is therefore difficult.

We have now to examine the history of the instrument. The fact that there is no complete description of the abacus in its modern form before the *Suan Fa Thung Tsung* of Chhêng Ta-Wei (+1593)[c] (see Fig. 70) has led many, including Mei Wên-Ting, to conclude that it did not become known in China until the end of the +15th century. Yet the earliest illustration of it is in the *Hsin Pien Tui Hsiang Ssu Yen*[15] of +1436 (Goodrich, 5).[d] Besides, no one has noticed that the *Lu Thang Shih Hua*[16] (Foothill Hall Essays) of +1513, written by Li Tung-Yang,[17] describes the abacus clearly as the

[a] It will be seen that only those balls count in the calculation which have been placed in contact with the transverse bar.

[b] These terms occur as early as the +3rd century in the biography (*San Kuo Chih*, ch. 63, pp. 4*a* ff.) of a famous mathematician and diviner of the Wu State, Chao Ta[18] (*fl.* +225 to +245). He had a method of computation called *thou chhêng wei chhu*[19] (multiplying at the head and dividing at the tail), which was vainly sought after by contemporary officials such as Kungsun Thêng.[20] However, the interpretation of this remains uncertain and we shall return to the matter in Sect. 26*i*.

[c] There were many descriptions after that time; references to books especially devoted to abacus calculations in the 17th and 18th centuries will be found in Wieger (3), p. 264, and Wylie (1), p. 103.

[d] This is of much interest in itself, as the oldest illustrated children's primer in the world, more than two centuries earlier than the *Orbis Sensualium Pictus* of Comenius (+1658).

1 前 2 後 3 進 4 退 5 上 6 起 7 除
8 去 9 首 10 頭 11 尾 12 末 13 身 14 本位
15 新篇對象四言 16 麓堂詩話 17 李東陽 18 趙達
19 昷乘尾除 20 公孫滕

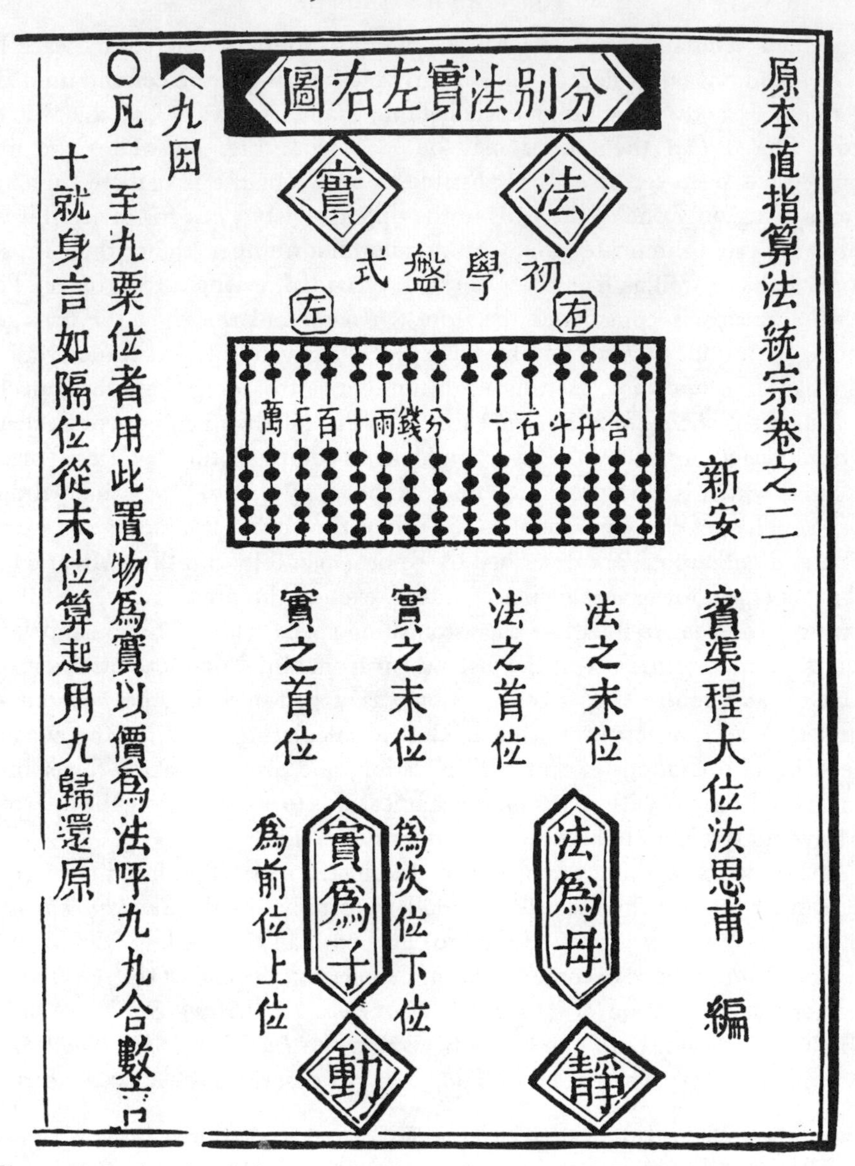

Fig. 70. An early printed picture of the abacus, from the *Suan Fa Thung Tsung*, +1593.

'moving-ball plate' (*chu chih tsou phan*[1]) which was operated according to rules and standard methods.[a] But here enters in the difficult problem of the *Shu Shu Chi I* (Memoir on some Traditions of Mathematical Art), which is attributed to Hsü Yo at the end of the Later Han (about +190), but which may possibly have been actually written by its commentator, Chen Luan (*c.* +570). In either case, it is much the earliest work which speaks of 'ball arithmetic'. According to Hsü Yo, his teacher

[a] Pp. 3*a*, 4*b*.

[1] 珠之走盤

Liu Hui-Chi[1] visited a Taoist adept, Thien Mu hsien-sêng,[2] who explained to him fourteen old methods of calculation, one of which was actually called ball-arithmetic (*chu suan*[3]). As this has not before been englished, it is given here:

(Text) The ball-arithmetic (method) holds and threads together (*khung tai*[4]) the Four Seasons, and fixes the Three Powers (heaven, earth and man) like the warp and weft of a fabric.

(Commentary) A board is carved with three horizontal divisions, the upper one and the lower one for suspending the travelling balls (*yu chu*[5]), and the middle one for fixing the digit[a] (*ting suan wei*[6]). Each digit (column) has five balls. The colour of the ball in the upper division is different from the colour of the four in the lower ones. The upper one corresponds to five units, and each of the four lower balls corresponds to one unit. Because of the way in which the four balls are led (to and fro) it is called 'holding and threading together the Four Seasons'. Because there are three divisions among which the balls travel, so it is called 'fixing the Three Powers like the warp and weft of a fabric'.[b]

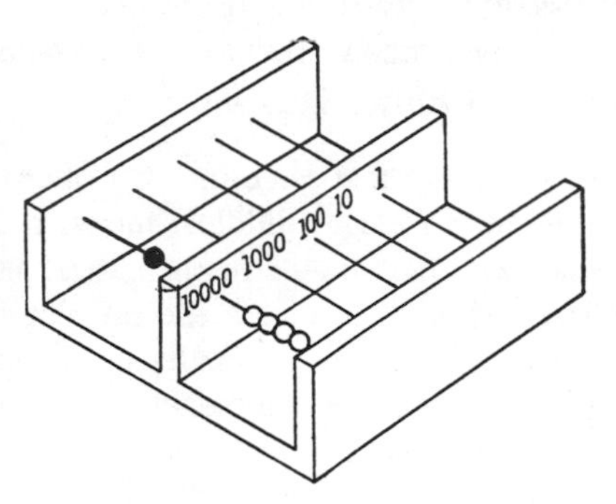

It must be admitted that this is a remarkably clear description of some kind of abacus, obviously one in which the total available number of units in each column was 9. It could be pictured as a trough-and-ball instrument if it were not for the word *tai*[7] (belt, ribbon) which has unmistakably the sense of a wire.

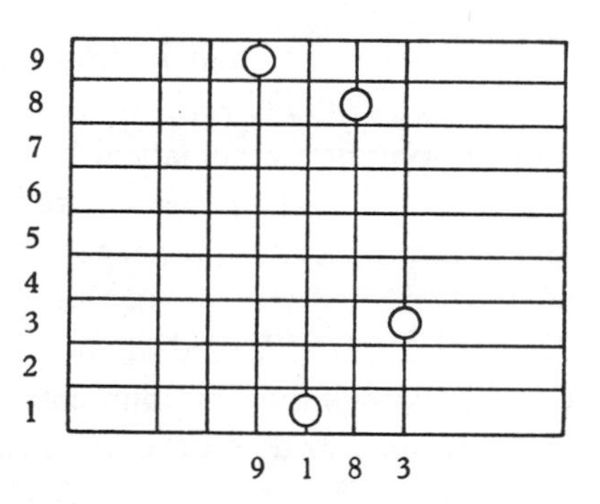

The commentary on three of the other methods also mentions balls. Mikami (*1*) has made an effort to elucidate what they were from their obscure descriptions. In the *Thai I*[8] (Great Unity) method, the text speaks of something coming and going along nine *tao*,[9] ways or lines (perhaps troughs?). The commentary says that there was one ball (hence the name of the method) for each vertical column, which was divided into nine horizontal divisions; hence by moving the balls up and down, any number which it was desired to retain could be set up. This method brings out clearly the way in which coordinate geometry was latent in the abacus system, the graduation into powers of ten forming the x-axis, and the graduation of numbers less than ten forming the y-axis. If the balls could ever have been persuaded, even in thought, to move along continuous curves, what a Cartesian world of graphs would have opened out!

[a] I.e. whether tens, hundreds, etc. This was a good idea. The modern abacus is always unmarked.
[b] Tr. auct.

[1] 劉會稽 [2] 天目先生 [3] 珠算 [4] 控帶 [5] 游珠
[6] 定算位 [7] 帶 [8] 太一 [9] 九道

Another method, the *Liang I*[1] (Heaven and Earth method), used balls of two different colours, yellow and blue, one relating to a *y*-axis on the left and the other to a *y*-axis on the right. Numbers were set up in the same manner as the *Thai I* method. This bears an astonishing resemblance to the practice in modern graphic representations of curves where points of different types relate to different superimposed axis graduations. In the third of these methods, balls of three different colours were used, and there were only three horizontal positions. In this way again any desired number could be set up. This was the *San Tshai*[2] (Three Powers)[a] method. On the whole, these systems, even if as late as the +6th century, show an interesting appreciation of coordinate relationships.[b]

A reference to the abacus may perhaps be contained in the biography of the metallurgist Chhiwu Huai-Wên[3] (*fl.* +550 to +570) of the Northern Chhi dynasty.[c]

> Many say that once in the College at Chinyang, when a Jun-Jun (i.e. Hun) guest was present, a foreign Buddhist monk, pointing to (Chhiwu) Huai-Wên, said to him, 'Here's a man who has strange mathematical arts!' And indicating a jujube tree in the courtyard he asked Huai-Wên to use the calculator (*suan tzu*[4]) and say how many dates there were on the tree. The trial was made, and Huai-Wên stated not only the total number but how many were ripe, unripe and partly ripe. When the dates were counted and tested they were only one short, but the mathematician said, 'It can't be wrong, just shake the tree again.' And sure enough, one more fell down.[d]

Here counting-rods may be implied, though the 'calculator' sounds like an abacus.

After the *Shu Shu Chi I* there is a long silence about the instrument. However, the bibliography of mathematical books which Chhêng Ta-Wei annexed to his work of +1593 states that between +1078 and +1162 there existed four books which, to judge from their titles, concerned the abacus. These were the *Phan Chu Chi*[5] (Record of the Plate and the Balls), the *Tsou Phan Chi*[6] (Record of the Moving Plate), the *Thung Wei Chi*[7] (Record of the Communicating Small Objects), and the *Thung Chi Chi*[8] (Record of the Communicating Machine).[e] None of them has come down to us, and it is doubtful whether Chhêng Ta-Wei himself had seen them. Vissière was

[a] Not mathematical powers, but Heaven, Earth and Man.

[b] Cf. my remark above (Vol. 1, p. 34) concerning rhyme-tables. Of course the coordinate concepts in Apollonius' conic sections (−3rd century), involving continuous variables, more fundamentally foreshadow analytic geometry.

[c] *Pei Shih*, ch. 89, p. 20*a*.

[d] Tr. auct. For an Indian parallel to this story see the legend of Nala and Rituparna in the *Mahābhārata* (Ray tr., vol. 3 (Vana Parva), ch. 72, p. 215). Chhiwu Huai-Wên's great importance in the development of iron and steel technology will be seen in Sect. 30*d*.

[e] All these books were said by Chhêng Ta-Wei to have been mentioned by Yang Hui in his +1275 book on magic squares, already referred to, but they are not in the text of that as we now have it. None of them is mentioned in the HY Index of titles in the official bibliographies, with the exception of a *Thung Wei Tzu*[9] (if it is the same book) of the Thang, which included a *Shih Wu Chih*,[10] 'Record of the Ten Things' (Balls?). The Sung bibliography also has a *Thung Wei Miao Chüeh*[11] (Mysterious Secret of the Communicating Small Objects), anonymous but possibly relevant. See *Suan Fa Thung Tsung*, ch. 12, p. 20*b*.

[1] 兩儀 [2] 三才 [3] 綦毋懷文 [4] 筭子 [5] 盤珠集
[6] 走盤集 [7] 通微集 [8] 通機集 [9] 通微子 [10] 十物志
[11] 通微妙訣

sceptical about their relevance, but the impression which their titles give of having concerned the abacus is really rather strong.

Mei Wên-Ting was of the opinion that the use of the abacus had first become general about the time of the mathematician Wu Hsin-Min[1] (*fl. c.* +1450).[a] He also considered that the computations for the Ta Thung Li calendar of +1384, which had been prepared under the leadership of Yuan Thung[2] with the help of Kuo Po-Yü,[3] a descendant of the great Kuo Shou-Ching, had been made with the abacus.[b] But, as Li Nien has pointed out,[c] Mei overlooked the fact that Chhêng Ta-Wei quoted fragments from the now lost book of Hsieh Chha-Wei,[4] who was probably a contemporary of Shen Kua and would have flourished in the last quarter of the +11th century, which clearly refer to the abacus and even use the expression *chi-liang*[5] (backbone) for the dividing bar. Yen Tun-Chieh (*3*) has added a poem of +1279 by Liu Yin[6] which mentions the term *suan phan*.

A few words must now be said about the comparative history of the abacus in other civilisations. The etymology of the Latin word is obscure, but it probably originated from a Semitic word *abq* (dust), recalling the fact that the earliest precursor of the abacus was probably a dust or sand tray. The next step was to have a surface ruled in lines on which pebbles (*calculi*) or counters could be placed, and there is some evidence that this was first used in ancient India,[d] though Herodotus says that the Egyptians had been accustomed to reckon with pebbles.[e] Concrete evidence is afforded by the famous Salamis abacus, with its parallel lines ruled in marble;[f] unfortunately it cannot be dated. In Latin literature there are numerous references to abaci having ruled lines and used with pebbles,[g] and in modern times several examples of metal plates with knobs running in slots have found their way into museums; these however, like the Salamis marble, are of quite uncertain date. The position can be summed up by saying that if these specimens are placed as late as the +3rd or +4th century (which would be quite reasonable), and if the dating of Hsü Yo's text is accepted as of the end of the +2nd century, then the Chinese practice would slightly antedate the European.[h] Since so

[a] His *Chiu Chang Pi Lei Suan Fa*[7] (Comparative Study of the 'Nine Chapters') has long been lost, but it is known to have mentioned the abacus.

[b] Chhien Ta-Hsin[8] (*Shih Chia Chai Yang Hsin Lu*,[9] ch. 17, p. 3*b*) was able to push this back a few decades by noting certain references in the *Cho Kêng Lu*[10] of Thao Tsung-I[11] of +1366.

[c] (*2*), 1st ed. p. 171, 2nd ed. p. 162, (*4*), vol. 3, p. 37, (*21*), vol. 4, p. 21. The reference is to *Suan Fa Thung Tsung*, ch. 1, p. 2*b*. Cf. Mikami (1), p. 61.

[d] Kaye (2).

[e] Smith (1), vol. 2, p. 160.

[f] Rangabé (1); Kubitschek (1).

[g] Smith (1), vol. 2, p. 165.

[h] What is the relation of the abacus to the rosary used for religious or magical purposes? There seems to be general agreement that India was its original home, and that the *japa-mala* (muttering chaplet) was used by Buddhists there in the +1st century. Islam had it by the +9th and Christendom by the +11th; one of the first Chinese references concerns a court eunuch of the +8th. The development and spread of the rosary seems to go parallel to that of the abacus; may not the chaplet have given rise to the instrument? Cf. Kirfel (1).

[1] 吳信民 [2] 元統 [3] 郭伯玉 [4] 謝察微 [5] 脊梁 [6] 劉因
[7] 九章比類算法 [8] 錢大昕 [9] 十駕齋養新錄 [10] 輟耕錄
[11] 陶宗儀

much uncertainty is involved, however, it is not possible to consider the question as in any way settled.[a]

The abacus did not become common in Europe until the +11th and +12th centuries.[b] There were several treatises on calculation *super lineas et per projectiles*, such as that of Hermann the Lame about +1050, which would correspond to such books as the *Phan Chu Chi*. A persistent European medieval tradition derived the abacus from the Arabs, but such a transmission is not established, and the origin of the Arabic abacus itself, which had ten balls on each wire, and no centre bar, is much in doubt.[c] The boards with lines were known as counting-tables, and the pebbles used were called jettons or casting-counters.[d] The counting-table was also, because of its chessboard appearance, an 'exchequer'—hence the modern term Chancellor of the Exchequer. It has been suggested that one form of it in Europe, that of the Banque des Argentiers, had been a direct importation of the Chinese *suan phan*,[1] and this may well be true, irrespective of other origins or transmissions, since it would have been something in which the +13th-century commercial travellers in Asia, such as Marco Polo, would naturally have taken an interest. The Russians, who maintained the use of the abacus down to very recent times, have sometimes attributed a Chinese origin to it.[e] In its specifically Chinese form it attracted much interest in 17th- and 18th-century Europe, and was described by Martini (+1658), Spizel (+1660), de la Loubère (+1691) and by two authors in the *Philosophical Transactions of the Royal Society*, Gamaliel Smethurst in +1749 and Robert Hooke himself in +1686.

In disposing of this subject we may notice a few final points. It seems unlikely that the character *suan*[2] originated from a graphic representation of an abacus (cf. p. 4 above). The argument of de Lacouperie that if the *suan phan*[1] were really pre-Han it would have been represented by a single character instead of a compound expression is still not without weight. Knott's suggestion that it had a foreign origin because the numbers are set up from left to right on it, not from right to left, loses all plausibility when we remember not only that each individual Chinese character is written from left to right, but also that the ancient counting-board was worked in that way. Perhaps the best provisional conclusion is that of Sarton,[f] that independent inventions took place. It will not be the last time that we shall have to fall back on this suspended judgment.

[a] It is clear that the whole treatment of this subject by D. E. Smith is out of date and requires reconsideration. He repeats (vol. 2, p. 169) the error of de Lacouperie (2) which arose from the misprint in the *Thai-Phing Yü Lan* (see p. 71 above). One of the latest histories of mathematics, Becker & Hofmann (1953) places the Chinese abacus not before the +12th century, thus neglecting Chen Luan, to say nothing of Hsü Yo (p. 133). What may perhaps be a representation of an abacus in the reliefs of the Han tombs at I-nan (*c.* +193), contemporary with Hsü Yo himself, has just been discovered.

[b] Sarton (1), vol. 1, p. 756. But Chen Luan's contemporary, Boethius, knew and used it.

[c] Gandz (2).

[d] Barnard (1).

[e] Not, however, Spassky (1), in the most recent study of the *schioty*.

[f] (1), vol. 1, p. 757.

[1] 算盤 [2] 算

(g) ARTIFICIAL NUMBERS

(1) Fractions

Of the treatment of fractions something has already been said in the remarks on the principal ancient Chinese mathematical books (pp. 18ff.). In earliest times in all parts of the world there was a tendency to avoid fractions by creating a great number of ever smaller units of weight and measure, but the relations of these were in some cases more judicious than others; thus the Romans had multiples of 12 and 16, and the Babylonians of 60, while, as we shall see, the Chinese often chose powers of 10. The most ancient effort to deal satisfactorily with fractions seems to be that of the Ahmes Papyrus of about −1500, but the Egyptians were unable to represent fractions with numerators other than unity. This method survived remarkably late, down to the +17th century.[a] But it was never characteristic of China.[b]

Chinese mathematics, from the time when we can first observe it,[c] was accustomed to deal with fractions (*fên*[1]).[d] The *Chou Pei* has problems which involve such numbers as $247\frac{933}{1460}$, though they are given in words, not symbols. The expression in this and other Han works is *a fên chih*[2] *b* for *b/a*. In the *Chiu Chang* the numerator and denominator were *tzu*[3] and *mu*[4] before an operation,[e] and *shih*[5] and *fa*[6] afterwards. Addition (*ho fên*[7]), subtraction (*chien fên*[8]), multiplication (*chhêng fên*[9]) and division (*ching fên*[10]) were all familiar from the early Han onwards. In a division such as that of 119,000 by $182\frac{5}{8}$ both were first multiplied by 8, and indeed all the modern rules were used. The method of *yo fên*[11] meant the reduction of the fraction to its simplest form by finding the greatest common divisor (*têng shu*[12]). This was done by continued division (*kêng hsiang chien sun*[13]). In adding fractions, each numerator was multiplied by the denominators of the other fractions (*mu hu chhêng tzu*[14]), and then all the denominators multiplied by one another. Each fraction was thus 'equalised' (*chhi*[15]) above and below, and given a common denominator (*thung*[16]); thus they were made to 'communicate' with one another (*thung*[17]). The numerators were then added. Subtraction of fractions (*chien fên*[18]) was carried out by multiplying the numerators by the alternate

[a] Smith (1), vol. 2, p. 213.

[b] Mikami (1), p. 12, gives an erroneous impression here. The unit fractions which he correctly transcribes from the *Chiu Chang* were very exceptional, intended only as a simplified exercise in finding the lowest common denominator.

[c] This time goes back to a period contemporary with the Egyptian 20th Dynasty (*c.* −12th century), since, according to Tung Tso-Pin's investigations on the Shang calendar (*ku ssu fên li*[19]), the men of that culture knew the length of the year as $365\frac{1}{4}$ days. In the −2nd century *Huai Nan Tzu*, ch. 3, p. 8*a*, gives the length of the month as $29\frac{499}{940}$ days.

[d] See especially Mikami (3).

[e] The choice of 'son' for the numerator and 'mother' for the denominator is quite revealing. It shows that the ancients were thinking of proper fractions, where the lower figure would be the larger (i.e. pregnant as it were). The difference of 'sex' (Yang and Yin) reminded them that the multiplication of the one was equivalent to the division of the other.

1 分 2 分之 3 子 4 母 5 實 6 法 7 合分
8 減分 9 乘分 10 經分 11 約分 12 等數 13 更相減損
14 母互乘子 15 齊 16 同 17 通 18 減分 19 古四分曆

denominators, subtracting the smaller of the two products from the greater, and dividing the remainder by the product of the denominators. The lowest common denominator was known as *tsui hsia fên mu*.[1]

The use of the bar for fractions, which seems to have been an Arab development, was not current in China until the 17th century.[a] On the other hand, the facility of the Han mathematicians in using lowest common denominators and greatest common divisors seems remarkably advanced when we remember that these were not employed in Europe until the +15th and +16th centuries.[b] One of the effects of the *Chiu Chang* on Chinese mathematics probably was that its elaborate treatment of fractions prevented the popularisation of decimals, though, as we shall see, these, in various forms, developed early.

Sumerian and Babylonian arithmetic was essentially sexagesimal,[c] and there can be little doubt that the sexagesimal fractions of the Greeks and Alexandrians, together with the division of the circle into 360°, were derived from them. The speculation has often been made that the *chia tzu*[2] system (the recurrent 60-day cycle) of the Chinese, which is certainly very ancient (cf. pp. 396ff.), had the same origin, but there is no real evidence for it. Moreover, the number of degrees of the old Chinese circle was $365\frac{1}{4}$, not 360. Sexagesimal fractions, therefore, never played any part in Chinese calculations. An argument against strong Babylonian influence may be drawn from the fact that Chinese never contained a philologically unitary symbol for $\frac{2}{3}$, that fraction so important in Mesopotamia.[d]

(2) Decimals, Metrology, and the Handling of Large Numbers

When we come to discuss the Chinese development of decimal fractions we find ourselves involved in the development of Chinese metrology, for the system of measures of length had steps by powers of 10 from a remarkably early period. The standard introduction to Chinese metrology, though little known in the West, is the book of Wu Chhêng-Lo.[e] Besides this, a number of sinologists have concerned themselves with such matters as the changing length of the foot-measure during the different dynasties; the reader may be referred to papers by Wang Kuo-Wei (2), Daudin (1), Ma Hêng (1), Marakuev (1), and Ferguson (3), through which the older literature may be reached. There is general agreement that the length of the standard foot showed a continuous tendency to increase through the three millennia from the Chou to the Chhing.[f] But this is not what here interests us; we must rather examine the parts into which it was divided, and the multiples of it which were recognised.

[a] Unless indeed it was originally derived from the horizontal lines on the Chinese counting-board.

[b] Smith (1), vol. 2, pp. 222, 223.

[c] Archibald (1); Thureau-Dangin (1).

[d] But certain words such as *lê*[3,4] could stand for one-third as well as one-tenth (cf. *Chou Li*, Kao Kung Chi, ch. 11, p. 6*b*).

[e] (2), based on the important investigations of such men as Wu Ta-Chhêng (*1*) and Lo Fu-I (*1*). See also Yang Khuan (*4*).

[f] From 0·195 to 0·308 m.

[1] 最下分母 [2] 甲子 [3] 阞 [4] 仂

Perhaps the earliest text which we can cite as showing an understanding of decimal place-value is one of the propositions in the *Mo Ching* (Mohist Canon), dating from about −330. It runs as follows:

Ch 59/—/37/51.[a] *Decimal Notation*

C 'One' is less than two and (yet) more than five. The explanation is given under 'establishing a position' (*chien wei*[1]).

CS Within 'five' there is 'one' (i.e. there are 'ones', because the rod-numeral for five is ǀǀǀǀǀ). (Yet) within the 'one' there is 'five' (because — in the rod-numeral for six ⊤ represents five). (And one horizontal stroke in the tens position means) 'ten', (that is to say, the equivalent of) 'two' (of these symbols for five). (auct.)

From this, and from evidence previously given (pp. 9, 13), it is clear that although the idea may have been temporarily lost or not always widespread, there was an apprecia-tion of place-value[b] in China already fifteen hundred years before it is found (as we saw above, p. 8) in the *Sun Tzu Suan Ching*.[c]

As we have already explained,[d] one finds that on the Shang oracle-bones and in the Chou bronze inscriptions numbers from 50 upwards were expressed just as in modern times by combining a figure with the word defining its place-value, e.g. *wu pai*[2] for 500. The following special symbols were used:

10^2	*pai*	百
10^3	*chhien*	千
10^4	*wan*	萬

These separate symbols were not a hindrance to computation because they differed essentially from those for 18, 19, 30, 40, 500 and so on, which were used in other early civilisations, such as Egypt, Greece and India.[e] For they were not themselves numerals; they were simply terms denoting particular place-values. We shall return in a moment to the expression of large numbers.

During the Chou period (first millennium down to −221) measures of length were variable and not always advanced by powers of 10. Wu Chhêng-Lo[f] has collected together an approximate statement of the case[g] from which it is clear that these earliest

[a] For the explanation of these conventions see Vol. 2, p. 172.

[b] Cf. Biot (7) and Edkins (6); but purely for their interest for the history of sinology.

[c] Occasionally, in later times, other philosophers referred to decimal notation—for example, in the Sung, *c.* +1200, Tshai Chhen,[3] one of the Neo-Confucian philosophers (cf. Vol. 2, p. 273 above), even though given to 'Pythagorean' numerology and mutationism (Forke (9), p. 279; Sarton (1), vol. 2, p. 625).

[d] Cf. Kuo Mo-Jo (*3*) and Sun Hai-Po (*1*). On this subject Li Nien (*21*), vol. 5, p. 1, supersedes (*8*), but should still be read in conjunction with the works of the philological archaeologists.

[e] Cf. Smith (1), vol. 2, pp. 40 ff.

[f] (*2*), pp. 89, 90.

[g] From texts such as *Shuo Wên*, *Huai Nan Tzu*, *Khung Tzu Chia Yü*, *Khung Tshung Tzu*, etc.

[1] 建位 [2] 五百 [3] 蔡沉

measures were based on parts of the human body, the finger, the woman's hand, the man's hand, the forearm, the foot, and so on.[a] We get the following:

8 *tshun*	寸 = 1	*chih*	咫
10 *tshun*	寸 = 1	*chhih*	尺
8 *chhih*	尺 = 1	*hsün*	尋
2 *hsün*	尋 = 1	*chhang*	常
10 *chhih*	尺 = 1	*chang*	丈
4, 7 or 8 *chhih*	尺 = 1	*jen*	仞
5 *chhih*	尺 = 1	*mo*	墨
2 *mo*	墨 = 1	*chang*	丈
2 *chang*	丈 = 1	*tuan*	端
2 *tuan*	端 = 1	*liang*	兩
2 *liang*	兩 = 1	*phi*	疋

This table includes several independent systems but no one of them used powers of 10 exclusively.[b] When Chhin Shih Huang Ti first unified the empire he chose the number 6 (correlative with the colour black and the element Water) as his emblem,[c] and carried out his famous standardisation of weights and measures.[d] Although the double-pace (*pu*[1]) was fixed at 6 *chhih*[2],[e] the emperor's Fa-Chia advisers took advantage of Mohist decimal notation for the principal measures of length below the *chhih*, which were henceforward arranged in powers of 10:

1 *chhih*	尺 = 10	*tshun*	寸
1 *tshun*	寸 = 10	*fên*	分
1 *fên*	分 = 10	*li*	釐
1 *li*	釐 = 10	*fa*	髮
1 *fa*	髮 = 10	*hao*	毫

As Chia I[3] put it in his *Hsin Shu*[4] of about −170,[f] these made six measures, in accordance with the system of the 'former Ruler', but in fact there was still the *chang*[5] of 10 *chhih* and the

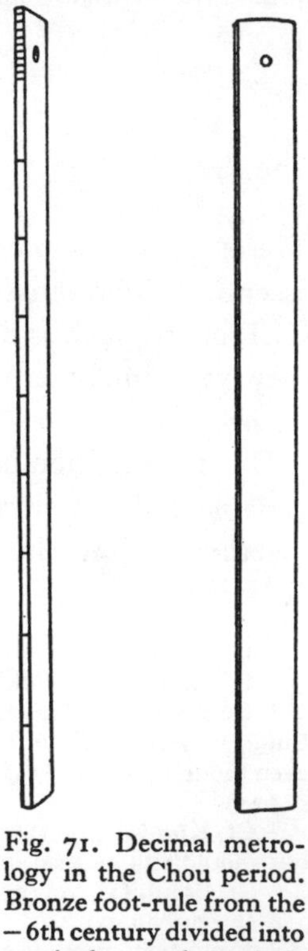

Fig. 71. Decimal metrology in the Chou period. Bronze foot-rule from the −6th century divided into ten inches, and one inch into tenths (after Ferguson).

[a] Cf. Vol. 2, p. 229, Table 11, no. 73, the ideographic etymology of *tu*,[6] degree.

[b] We do not know how far this idea goes back. Ferguson (3) has described a bronze foot-rule, apparently of the Chou period, in his possession (Fig. 71). It is divided into ten inches and one of these inches into tenths; its total length is 0·231 m. An accompanying object, a bell, from the same tomb, bore a date which can only be interpreted as −550 or −404, and Ferguson preferred the former. Cf. the Chou bronze foot-rules described by Weinberger (1). There is some evidence for decimal metrology from Mohenjo-daro (Sarton, 5).

[c] *Shih Chi*, ch. 6, p. 12*a*; Chavannes (1), vol. 2, p. 130.

[d] *Shih Chi*, ch. 6, p. 13*b*; Chavannes (1), vol. 2, p. 135.

[e] *Shih Chi*, ch. 6, p. 12*a*.

[f] Ch. 48.

1 步 2 尺 3 賈誼 4 新書 5 丈 6 度

yin[1] of 10 *chang*. This metrological system was current throughout the Han,[a] and (with slight modifications of nomenclature) in later ages. A famous inscription by Liu Hsin on a standard measure of volume dated +5 (the *chia liang hu*[2]) uses the terms in a decimal sense, speaking of a length correct to 9 *li*, 5 *hao*. Several standard measures of length, in copper or bronze, have come down to us from this time, one of date +12 and another of +81 (Daudin).

In the San Kuo period this trend continued. About +300 Sun Tzu says that the diameter of freshly spun silk threads gives the standard for the *hu*,[3] and that

10 *hu* 忽 = 1 *miao* 秒
10 *miao* 秒 = 1 *hao* 毫
10 *hao* 毫 = 1 *li* 釐
10 *li* 釐 = 1 *fên* 分

This is the system for small lengths which persisted till very late times, with the exception that the *miao*[4] was replaced in Sung by the *ssu*.[5] It is clear also that by Sun Tzu's time measures of volume and weight were following suit to some extent since he gives:

For volume: 1 *ho* 合 = 10 *shao* 勺
1 *shao* 勺 = 10 *tsho* 撮
1 *tsho* 撮 = 10 *miao* 秒
1 *miao* 秒 = 10 *kuei* 圭
1 *kuei* 圭 = 6 *su* 粟 (an exception)
For weight: 1 *shu* 銖 = 10 *lei* 絫
1 *lei* 絫 = 10 *shu* 黍

Nevertheless this was not universally accepted, and in the Liang and the Thang there are many references to units of 2, 4, 6 and 12 parts for volume and weight. In +992, however, the decimal system for weight was officially fixed, using the order *liang*[6]—*chhien*[7]—*fên*[8]—*li*[9]—*hao*[10]—*ssu*[5]—*hu*[3], descending by tens. Calculations with the older systems persisted all the same as late as the Ming.

The expression of decimal fractions by the use of the metrological units runs through the whole of Chinese mathematics. Liu Hui, in his +3rd-century commentary on the *Chiu Chang*, expresses a diameter of 1·355 feet as 1 *chhih*,[11] 3 *tshun*,[12] 5 *fên*,[8] 5 *li*.[9] In extracting square roots the *Chiu Chang* itself had spoken of cases where the root would be non-integral (*pu chin*[13]), in which case the remainder would be left as such (*i mien ming chih*[14]).[b] Liu Hui, however, was concerned about these 'little nameless

[a] *Chhien Han Shu*, ch. 21A, pp. 9*b*ff. Cf. Swann (1), pp. 360ff.; Dubs (2), vol. 1, pp. 276ff.

[b] The work of Wang Ling (2), vol. 1, pp. 251ff., (3), suggests that this phrase, correctly interpreted, means the continuation of the root extraction to one digit of the decimal portion, by the operation of which a common fraction was obtained; e.g. $\frac{C}{2R+r}$, C being the remainder, R the integer portion of the root, and r the first digit of the decimal portion.

1 引 2 嘉量斛 3 忽 4 秒 5 絲 6 兩 7 錢
8 分 9 釐 10 毫 11 尺 12 寸 13 不盡
14 以面命之

numbers' (*wei shu wu ming*[1]), and said that first 10 should be taken as denominator (*mu*[2]), then 100, and so on, giving rise to a series of decimal places expressed as above,[a] which could go to any number of places recordable in figures, though after the fifth (*hu*[3]), nameless. These decimal roots were undoubtedly computed with the counting-rods, and the results expressed in decimal fractions.

After this time there was little change in the methods used. In the early +5th century Hsiahou Yang, mentioning a type of division which he calls *pu chhu*[4] ('by steps'), says that if the divisor is 10 or a power of 10, it is unnecessary to divide. He quotes[b] a book, the *Shih Wu Lun*,[c] of the +3rd century, which gives the rule *shih chhêng chia i têng, pai chhêng chia erh têng*[5]—'multiplying by ten means one step up, multiplying by a hundred means two steps up'—a clear use of the word *têng* (which sometimes meant the greatest common divisor in adding fractions) in the sense of powers. The converse rule is also given. Thus implicitly in the *Chiu Chang*[d] and explicitly in the *Shih Wu Lun* and Hsiahou Yang, familiarity is shown with what we should now express as 10^{-1}, 10^{-2}.

As Li Nien (*10*) has pointed out, the *Sui Shu* (+635) expresses a decimal 3·1415927 as 3 *chang*,[6] 1 *chhih*,[7] 4 *tshun*,[8] 1 *fên*,[9] 5 *li*,[10] 9 *hao*,[11] 2 *miao*,[12] 7 *hu*;[13] and the word *wei*[14] was coming into use for yet a further place. Han Yen[15] in the Thang (*fl.* +780 to +804) seems to have been the first to drop these expressions sometimes and simply write down the numbers as in modern decimal notation,[e] using mensuration words such as *tuan*[16] or *wên*[17] to mark the last integer.[f] But the introduction of a unified system and terminology and its application in general operations did not occur until the +13th century, when, as we have seen above (p. 46), Yang Hui made much use of them. Chhin Chiu-Shao (+1247) employed the following terminology. The position immediately before the decimal point (the integral number) he called *yuan shu*;[18] it is, he said, the number the 'tail positions' (*wei wei*[19]) of which are zero. Actual decimal numbers he called *shou shu*;[20] this term included all the places which other mathematicians knew as *fên*,[9] *li*,[10] etc. The modern term for decimals is *hsiao shu*,[21] and it is interesting that this goes back to the Sung, for it was used by Chu Shih-Chieh.

[a] Chhien Pao-Tsung (*1*), p. 77; Li Nien (*1*), p. 70.

[b] *Hsiahou Yang Suan Ching*, p. 2*b*.

[c] Perhaps the *Shih Wu Lun*[22] by Yang Wei[23] (*fl.* +237) of which other fragments remain in the Ma Kuo-Han collection (*YHSF*, ch. 74, p. 32*a*).

[d] Cf. its expressions *chung thui i*,[24] *hsia thui erh*,[25] 'retire one column, retire two columns', in root extraction.

[e] Chhien Pao-Tsung (*1*), p. 78; but it may have been Hsiahou Yang himself.

[f] In the calendar calculations of +660, +665 and +705, Tshao Shih-Wei,[26] Li Shun-Fêng[27] and Nankung Yüeh[28] respectively used a decimal system in which one word represented two places, so that 365·2448 was expressed as 365, *yü*[29] 24, *chi*[30] 48. See *Chiu Thang Shu*, ch. 33, pp. 18*a*, 24*a*, 25*a*.

1 微數無名 2 母 3 忽 4 步除 5 十乘加一等百乘加二等
6 丈 7 尺 8 寸 9 分 10 釐 11 毫 12 秒
13 忽 14 微 15 韓延 16 端 17 文 18 元數
19 尾位 20 收數 21 小數 22 時務論 23 楊偉
24 中退一 25 下退二 26 曹士蔿 27 李淳風 28 南宮說
29 餘 30 奇

Li Yeh and Chu Shih-Chieh, as well as Chhin Chiu-Shao, continued their root extractions to several places of decimals.

Before comparing this whole picture with developments elsewhere in the world it is necessary to glance at the interest which the Chinese took in expressing very large numbers. Several terms for very large numbers had appeared already in texts of Chou date, such as the *Shih Ching* (Book of Odes), but later commentators interpreted them in different ways, and it is likely that at first they did not have fixed meanings (cf. 'myriads'). The *Shu Shu Chi I* (about +190) has a very interesting passage on the matter. Hsü Yo says that there are three methods of interpreting these large numbers, the 'upper, middle, and lower', and his statements may be represented as follows:[a]

	Upper (*shang* 上)	Middle (*chung* 中)	Lower (*hsia* 下)
wan 萬	10^4	10^4	10^4
i 億	10^8	10^8	10^5
chao 兆	10^{16}	10^{12}	10^6
ching 京	10^{32}	10^{16}	10^7
kai 垓	—	10^{20}	10^8
tzu 秭	—	10^{24}	10^9
jang 壤	—	10^{28}	—
kou 溝	—	10^{32}	—
chien 澗	—	10^{36}	—
chêng 正	—	10^{40}	—
tsai 載	—	10^{44}	—

The interpretations of all the commentators on ancient texts show that they were following one or other of these nomenclatures for the powers of 10.[b]

The particular interest which the Indians, especially the Buddhists, took in the expression of very large numbers, is well known, though its significance for the progress of mathematics is somewhat doubtful.[c] The ancient Chinese system just described cannot, however, be of Indian origin, for even if the *Shu Shu Chi I* itself is of +6th century rather than +2nd, the classical commentators (such as Chêng Hsüan[1] or Mao Hêng[2]) are much too early to have been affected by Buddhist influence. The first two terms of the 'upper' and 'middle' series are to be found already in the *Chiu Chang*.[d]

[a] Taking into account emendations proposed by Chhien Pao-Tsung (*1*), p. 76. The list of Hsü Yo does not go beyond *ching*; the higher values are inserted following a traditional list given by Knott.

[b] Shen Kua (*MCPT*, ch. 18, para. 7) says that the 'lower' method was the older. Vacca (2) has noted a similarity between the 'upper' method in which the index number is doubled each time, and a progression adopted by Archimedes in his *arenarius* or sand-grain reckoner (Heiberg (1), p. 26), but the coincidence seems likely to be accidental. Sir Ronald Fisher has pointed out to us that in modern scientific usage there is still confusion between the three systems. The most frequent current American usage corresponds with the 'lower' method, that advocated by Eddington corresponds with the 'middle', while Fisher and Yates and other statisticians use the 'upper'. The confusion between the American and European usages of the words 'million' and 'billion' is well known. Modern China uses the 'lower' system; modern Japan more commonly the 'middle'.

[c] Cf. Renou & Filliozat (1), vol. 2, p. 171. The Tibetans had words for powers up to 10^{60} (Edgar, 1).

[d] Ch. 4, p. 15*b*. Cf. *Lü Shih Chhun Chhiu*, ch. 58 (vol. 1, p. 113).

[1] 鄭玄 [2] 毛亨

The *Fêng Su Thung I*[1] of +175 gives two complete 'lower' series,[a] as long as the longest of the *Shu Shu Chi I*. Moreover, the Indian system, as it appears, for instance, in the *Ta Pao Chi Ching*[2] (*Mahāratnakūta Sūtra*)[b] translated by Yüeh-Pho-Shou-Na[3] (Upaśūnya) in +541, has quite different progressions and names.[c] All the Indian mathematicians from the +4th century onwards had words for powers of 10, e.g. Āryabhaṭa, *b*. +476 and so Hsiahou Yang's contemporary;[d] Mahāvīra, *c*. +830;[e] Bhāskara, +1114 to +1178;[f] Nārāyana, *fl*. +1356.[g] But in view of what has already been said about (*a*) the way of writing numbers as far back as the Shang period, (*b*) the clear appearance of the decimal idea in the Mohist Canon about −330, (*c*) the interpretations of high numbers in powers of 10 by Mao Hêng about −200, (*d*) the decimalisation of measures proposed in −170, and (*e*) the whole system of counting-board operations in Han mathematics, Sarton cannot possibly be right in suggesting[h] that the decimal notation was introduced to the Chinese by Chhüthan Hsi-Ta[i] or by the Po-lo-mên books from India in the +7th or +8th century.[j]

Indeed, the positive Chinese spirit reacted strongly against the Indian association of mathematics and mysticism. In an interesting passage, not hitherto noticed, Shen Tso-Chê writes (about the +12th century):

> Nowadays even children learn mathematics from printed Buddhist textbooks (*Phu-Sa Suan Fa*[4]) which deal with the counting of infinite numbers of sand-grains (*wu liang sha*[5]), so that their numbers can be known.[k] They also tell how to enumerate the differences between the Ten Worlds (*shih fang shih chieh*[6]). But how can the Buddhas know the answers where there are no definite numbers and no precise principles? That which is vague and obscure can have no place in matters connected with number and measure. Whether the numbers or dimensions be large or small, problems can all be solved and the answers distinctly stated. It is only things which are beyond shape and number (*hsiang shu*[7]) which cannot be investigated. How can there be mathematics beyond the reach of shape and number?[l]

Another aspect in which Chinese familiarity with the decimal idea showed itself was in their cartography. From the time of Phei Hsiu[8] in the +3rd century, as we shall see later (p. 538), maps with rectangular grids were made, each gradation corresponding to 100 *li*. Transmitted through Chia Tan[9] in the +8th century, this

[a] Preserved in ch. 750, p. 21*b* of *Thai-Phing Yü Lan*, and in the fifth *chih*[10] section of *Kuang Yün*. *Fêng Su Thung I* Chung-Fa critical edition, *i wên*, ch. 4, p. 111. Cf. *Lü Shih Chhun Chhiu*, ch. 62 (vol. 1, p. 121).

[b] N 23 (23).

[c] Cf. McGovern (2), pp. 39 ff.

[d] Sarton (1), vol. 1, p. 409.

[e] Sarton (1), vol. 1, p. 570.

[f] Sarton (1), vol. 2, p. 212.

[g] Sarton (1), vol. 3, p. 1535. On all of these men see Renou & Filliozat (1), vol. 2.

[h] Sarton (1), vol. 1, pp. 321, 444, 450, 513. One of the most recent books on the history of mathematics (Becker & Hofmann in 1953) continues to affirm that the 'decimal position system' was the greatest mathematical discovery of India, dating from the +7th century (p. 118).

[i] Cf. p. 37 above and p. 203 below.

[j] Sect. 6*f*; Vol. 1, p. 128, above.

[k] A distant echo of Archimedes' *arenarius*?

[l] *Yü Chien*, ch. 7, p. 14*a*, tr. auct.

1 風俗通義 2 大寶積經 3 月婆首那 4 菩薩算法
5 無量沙 6 十方世界 7 象數 8 裴秀 9 賈耽 10 旨

tradition culminated in the splendid stone-inscribed maps of +1137. Yet in Europe, apart from the quantitative geography of Eratosthenes and Ptolemy, which entirely died out, maps with decimal grids were not known until the portolan charts of the beginning of the +13th century.[a]

As for the handling of decimal fractions, there was an old rule for the extraction of the nth root of a by using an expression which would, in modern form, be $\frac{\sqrt[n]{(a \,.\, 10^{kn})}}{10^k}$. The root extraction was similar to the modern decimal process, but this device avoided awkward vulgar fractions. It was ascribed by Smith[b] to Indian sources, but Wang & Needham (1) have shown that it occurs in the 1st-century *Chiu Chang Suan Shu* for $n=2$ or 3. It was known in +12th-century Europe and gave rise to tables of square roots expressed in quasi-decimal notation common in the +16th century. Arabic mathematicians, doubtless inspired by Indian sources, were dealing with decimal fractions in the +11th century (e.g. Aḥmad al-Nasawī, a Persian, *c.* +1030),[c] and Jewish and Latin scholars in the +12th (e.g. Abraham ben Ezra, *c.* +1150;[d] John of Seville, *c.* +1140).[e] There was also the expression of π to 16 places of decimals by Ghiyāth al-Dīn Jamshīd al-Kāshī about +1427,[f] in which 3 was separated as the 'complete' or 'integral' from the other figures. The first appearance of the decimal point occurs in the arithmetic of Pellos (+1492), but its use was not fully clarified until Stevin's *La Disme* of +1585.[g]

To sum up, the use of decimal notation was extremely ancient among the Chinese, going back to the −14th century. In this they were unique among early civilisations.[h] In their application of the decimal system to metrology they were particularly

[a] Sarton (1), vol. 2, p. 1049. See p. 532 below.

[b] (1), vol. 2, p. 236.

[c] Mieli (1), p. 109; Hitti (1), p. 379; Sarton (1), vol. 1, p. 719, (2). Significantly, he entitled his book *al-Muqni' fīl-Ḥisāb al-Hindī* (The Convincer on Indian Calculations).

[d] Sarton (1), vol. 2, p. 187.

[e] Sarton (1), vol. 2, p. 169.

[f] In his *Risālat al-Moḥīṭīje* (Treatise on the Circumference), which has recently been studied by Luckey (2) and translated into Russian by Rosenfeld & Yushkevitch (1), pp. 364ff. Cf. Smith (1), vol. 2, pp. 240, 310; Cajori (2), p. 108. Decimal fractions appear also in his other works, see, for example, Luckey (1), Rosenfeld & Yushkevitch (1), p. 88. Smith made the remarkable mistake, however (vol. 1, p. 289), of confusing al-Kāshī with no less than two other mathematical astronomers of the same period, under the impression that only one man was concerned. They were in fact the three successive directors of the observatory founded by Ulūgh Beg at Samarqand in +1420. The first was our friend Ghiyāth al-Dīn Jamshīd ibn Mas'ūd ibn Maḥmūd al-Kāshī, who died in +1436 (Suter (1), p. 173, no. 429; Brockelmann (2), vol. 2, p. 211, suppl. vol., p. 295). He was followed by Ṣalāḥ al-Dīn Mūsa ibn Muḥammad ibn Maḥmūd Qāḍī Zāde al-Rūmī ('the son of the Turkish judge', *b.* +1357, *d.* between +1436 and +1446; Suter (1), p. 174, no. 430; Brockelmann (2), vol. 2, p. 212). The third director was another Turk, 'Alā' al-Dīn 'Alī ibn Muḥammad al-Qūshchī (Suter (1), p. 178, no. 438; Adnan Adivar (2), p. 33; Süheyl Ünver (3); Sarton (1), vol. 3, pp. 1120ff.). It was under Ibn al-Qūshchī that the famous planetary tables and star catalogue (the *Zīj Ulūgh Beg*) were finally completed, but after the fall of the Timurids he joined the Osmanli Turks and died in newly conquered Istanbul in +1474. There has been much confusion in the literature about these mathematicians, but the facts are clearly set forth in the *Eš-Šaqā'iq en-No'mānijje* (The Peonies of No'man) by Tašköprüzade (vol. 1, p. 78), translated by Rescher (1), pp. 7ff., 102ff., and in the Prolegomena to the *Tables*, translated by Sédillot (3), pp. 5, 225ff. See also the excellent general account of Ulūgh Beg's observatory, and the men who worked in it, by the Uzbek scholar, Kari-Nyazov (1). This case of three persons in one footnote shows at any rate how hard it is for historians of other cultures to master the names of Arabic, Persian and Turkish scholars.

[g] Smith (1), vol. 2, p. 242; Cajori (3), p. 314: Sarton (2).

[h] Because they used the same nine symbols at all power levels, the latter being indicated by place-value components.

advanced.[a] For that Europe had to wait, as Sarton says,[b] until the French Revolution. When they applied the system to cartography they ante-dated the Arabs and Europeans by nearly a thousand years. But the little symbol which alone permitted of its revolutionising all mathematical computations had to await the Renaissance in the West.

(3) Surds

This sub-section must conclude with a reference to surds. The Greek tradition (as given by Proclus) was that the Pythagoreans had discovered the incommensurability of the diagonal and the sides of a square, i.e. the geometric view of the irrationality of $\sqrt{2}$. The Greek idea of an irrational was something which could not be expressed as the ratio of two integers. It was not 'unreasonable' but 'un-ratio-able'. The Chinese mathematicians, with their early use of decimal fractions in expressing roots, seem to have been neither attracted nor perplexed by irrationals, if indeed they appreciated their separate existence.[c]

(4) Negative Numbers

The Chinese also found no difficulty in the idea of negative numbers. As has already been mentioned, positive coefficients (*chêng*[1]) were represented by red counting-rods, and negative ones (*fu*[2]) by black counting-rods, probably as early as the Former Han dynasty (−2nd century). Another method was to use counting-rods of triangular cross-section for the positive numbers, and square-sectioned rods for the negative ones.[d] The law of signs was familiar to the Sung algebraists,[e] and stated, for instance, in the *Suan Hsüeh Chhi Mêng* of +1299. Of course Diophantus (+275) spoke of equations which would result in a negative number as 'absurd', and the Chinese ignored them too. In India, Brahmagupta (*c.* +630) is the first to mention negative numbers. Their first satisfactory treatment in Europe was that of Jerome Cardan in his *Ars Magna* of +1545. The Sung algebraists had two methods of writing negative

[a] It could of course be argued that this metrological precocity acted as a retarding influence on the development of decimal symbolism. For it meant that each decimal place retained, as it were, a personal name, and this was hard to drop. This obscured the distinction between integers and decimal places. Moreover, there was no standardisation of the names until the time of Yang Hui (+13th century). This is not the first time that we shall meet with the paradoxical situation that early achievements tended to delay later ones. As we shall see below (p. 270), the early adoption in China of equatorial coordinates for charting the heavens, though entirely in line with modern practice, probably delayed the discovery of the precession of the equinoxes.

[b] Sarton (1), vol. 2, p. 5. In no other part of the world was there so early a decimalisation of weights and measures, and the Romans were especially backward in the matter. The subject of comparative metrology is too vast to enter upon here, but a glimpse of the Hellenistic situation can be had from Pernice (1) and a +10th-century treatise in Syriac has been translated by Sauvaire (1).

[c] As Prof. K. Mahler points out to us, the Greeks also had techniques for approximating roots, but these were probably not considered as gentlemanly as the rigorous theory of incommensurables. It is, moreover, rather paradoxical that in China, where developed atomic theories never had any success, the square should have been regarded as composed of multitudes of smaller squares of suitable sizes, decreasing according to the powers of 10, and permitting the approximation of square roots to any desired degree of accuracy. Cf. Neugebauer (9), p. 142.

[d] Li Nien (*1*), p. 36.

[e] Part of it had been given already in the *Chiu Chang*. On this see Wang Ling (2), vol. 2, p. 163.

[1] 正 [2] 負

numbers: one was to write or print them in black, distinguishing them thus from the red positive numbers; the other was to draw a diagonal stroke through the right-hand digit figure of the negative number.[a] Perhaps this practice originated from an old method, mentioned in Liu Hui's commentary, of representing negative numbers by rod-numerals placed in a slanting position. It seems that the Chinese paid no attention to the problem of the square roots of negative quantities, though the Indians, such as Mahāvīra and Bhāskara, were aware of it, and the significance of complex or imaginary numbers was not tackled until the Renaissance, indeed until the end of the 17th century in Europe.

(h) GEOMETRY

(1) The Mohist Definitions

It is often said that all early geometry, other than that of the Greeks, was intuitive in nature, not demonstrative, i.e. that it sought facts relating to mensuration without attempting to prove any geometrical theorems by deductive reasoning. It is questionless that demonstrative geometry was the central feature of Greek mathematics, culminating in Euclid (*c.* −300) and Apollonius (*fl.* −200); and it is equally certain, as may be seen from many pages of this Section, that the genius of Chinese mathematics lay rather in the direction of algebra. In China there never developed a theoretical geometry independent of quantitative magnitude and relying for its proofs purely on axioms and postulates accepted as the basis of discussion.[b] But just as in living organisms one type will show, upon dissection, vestigial or abortive organs characteristic of another type, even as between two mammalian sexes; so also we find that while Greek mathematics was not without its algebra (culminating in Diophantus, *fl. c.* +250 to +275), neither was Chinese mathematics devoid of certain sprouts of theoretical geometry. That these did not develop is one of the characteristics of Chinese civilisation. The propositions in which they are contained occur in the Mohist Canon,[c] and seem to have remained hitherto unknown, for some reason or other, to historians of mathematics.[d] This ancient work, which touches at various points almost all the branches of physical science, is now believed to date from the close neighbourhood of −330. Let us look at the statements in it which deal with geometrical matters.

Cs 61/—/24.54. *'Atomic' definition of the geometrical point*

C The definition of 'point' (*tuan*[1]) is as follows: The line is separated into parts, and that part which has no remaining part (i.e. cannot be divided into still smaller parts) and (thus) forms the extreme end (of a line) is a point.

[a] Cf. above, p. 45.

[b] Hence the significance of the title *Chi Ho Yuan Pên*[2] chosen for the +17th-century Jesuit translation of Euclid. Perhaps the first two words were a pun on the syllable 'geo'; at all events their invariable meaning since the Chou dynasty had been 'how much?', 'how many?'.

[c] *Mo Ching*; cf. Sect. 11*b*, Vol. 2, pp. 171 ff. above.

[d] The key to the conventions used in what follows will be found in Vol. 2, p. 172 above.

[1] 端 [2] 幾何原本

CS A point may stand at the end (of a line) or at its beginning like a head-presentation in childbirth. (As to its indivisibility) there is nothing similar to it. (auct.)

This seems to be identical with Euclid's first and third definitions (bk. 1). Compare Plato's ἀρχὴ γραμμῆς, 'the beginning of a line' (Heath (1), p. 155).

Ch 60/270/39.52. *The same*

C That which is not (able to be separated into) halves cannot be cut further and cannot be separated. The reason is given under 'point' (*tuan*[1]).

CS If you cut a length continually in half, you go on forward until you reach the position that the middle (of the fragment) is not big enough to be separated any more into halves; and then it is a point. Cutting away the front part (of a line) and cutting away the back part, there will (eventually remain an indivisible) point in the middle. Or if you keep on cutting into half, you will come to a stage in which there is an 'almost nothing', and since nothing cannot be halved, this can no more be cut. (auct.)

Compare the obvious reference to these same discussions in paradox PC/31 of the group of dialecticians associated with Hui Shih (Vol. 2, p. 191 above). It seems that they were concerned (like Zeno) to prove the impossibility of the physically discontinuous, while the Mohists, as here, took the concept of 'insecability' as valid. This is the interpretation of Fêng Yu-Lan, and we agree with Ku Pao-Ku (1), p. 81, in preferring it to that of Hu Shih, who thought that the Mohists were at one with the Logicians. Ssuma Piao[2] (+3rd century), commenting on the relevant passage in the 33rd (Thien Hsia) chapter of *Chuang Tzu*, said: 'If it can still be divided, there is always the "two"; if it comes to an undividable state, there is only the "one" existing.' Chang Chan[3] (*fl.* +320 to +400), commenting on the relevant passage in the 4th (Chung Ni) chapter of *Lieh Tzu*, said 'Within an area (*yü*[4]) of real size (*tshu yu*[5]) there is always something (*yu*[6]). At its margin there are always single elements in existence which can never be separated into anything smaller.' Such quotations show that later Chinese thinkers retained some appreciation of the 'atomic' geometrical point, though it never acquired a leading place in Chinese thought.

There is a parallel of much interest here with theories which the Greek atomists may have held about infinitesimals, according to Luria (1). Apparently mathematicians associated with the school of Democritus had the notion of a 'geometrical atom', large (but finite) numbers of which built up lines, areas and volumes. But occidental antiquity preferred the more rigorous method of exhaustions (see the discussion of Struik (2), pp. 54 ff.).

Cs 53/—/8.49. *Lines of equal length*

C (Two things having the) same length, means that two straight lines finish at the same place.

CS It is like a straight door-bolt which can be placed flush with the edge of the door. (auct.)

Equivalent to Euclid, def. 4 (bk. 1).

Cs 68/—/38.61. *Comparison of lengths*

C Comparison (*phi*[7]) means finding that sometimes (two lines) will coincide with each other and sometimes they will not.

1 端 2 司馬彪 3 張湛 4 域 5 麤有 6 有 7 仳

CS In both methods there is a fixed point so that the comparison may be carried out. (auct.)

The two methods here mentioned were presumably (*a*) superimposed or parallel measurement, and (*b*) placing one leg of a compass at the fixed point from which both lines radiated.

Cs 52/—/6.49. *Parallels*

C Level means (being supported by props) of the same height.

CS Like two persons carrying (a beam on their shoulders), who should be of the same height like brothers. (auct.)

Cf. Euclid, defs. 30 and 31 of bk. I, and def. I of bk. II.

Cs 40/—/82.38. *Space*

C Space (*yü*[1]) includes all the different places (*i so*[2]).

CS East, west, south and north, are all enclosed in space. (auct.)

Cf. Euclid, postulates 1 and 2.

Cs 41/—/84.39. *Bounded space*

C Outside a bounded space (*yü*[3]) no line can be included (because the edge of an area is a line, and beyond that line is outside the area).

CS A plane area cannot include every line since it has a limit. But there is no line which could not be included if the area were unbounded (*wu chhiung*[4]). (auct.)

Cf. Euclid, defs. 13 and 14 of bk. I, as for the next.

Cs 62/—/26.55. *The same*

C Having (two-dimensional) space (*chien*[5]) means that something is contained in it.

CS Like the 'door-ear', the space being the part between the door and the jamb. (auct.)

Cs 63/—/28.56. *The same*

C A plane space (*chien*[5]) does not reach its sides (*phang*[6]).

CS The plane space is what is inserted between lines. The line is in front of the plane (*chhü*[7]) but behind the point (*tuan*[8]), and yet it is not 'between' them. (auct.)

The lines themselves were not considered part of the space, so it was said to be inserted between the lines. The line was considered to be the piling up of points, and the space or area the piling up of lines. The line could unite two points or two planes, but not a point and a plane. So also Aristotle (cf. Heath, I). See below, pp. 142 ff.

Cs 59/—/20.52. *Rectangles*

C Rectangular shapes (*fang*[9]) have their four sides (*chu*[10]) all straight, and their four angles all right angles.

CS Rectangular means using the carpenter's square so that the four lines all just meet (each other). (auct.)

Cf. Euclid, defs. 30 and 31 of bk. I, and def. I of bk. II.

Cs 69/—/40.62. *Piling up*

C As for piling up (*tzhu*[11]), where there is no space between (i.e. as in planes which have no thickness), they cannot mutually touch (and therefore cannot be piled up).

1 宇 2 異所 3 域 4 無窮 5 間 6 旁 7 區
8 端 9 方 10 柱 11 次

CS Things which have no thickness (*wu hou*[1]) (exemplify this principle). (auct.)

Identical with Euclid, def. 7, bk. 1. Compare with Hui Shih's paradox HS/2 (Vol. 2, p. 190 above); 'That which has no thickness cannot be piled up, but it can cover a thousand *li* in area.' A plane, with no third dimension.

Cs 47/—/10 and 18?.51? *Circle*

C A circle (ring, *huan*[2]) can rest on any point of its circumference.

CS (missing). (auct.)

Cf. the *Khao Kung Chi* (Artificers' Record) in the *Chou Li* (Record of the Rites of Chou), where it is said (Chêng Hsüan's commentary) that a wheel must be as perfectly circular as possible in order to move at the slightest touch (Biot (1), vol. 2, p. 465).

Cs 58/—/18.51. *Centre and circumference (diameter)*

C A circle (*yuan*[3]) is a figure such that all lines drawn through the centre (and reaching the circumference) have the same length.

CS A circle is that line described by a carpenter's compass which ends at the same point at which it started. (auct.)

Cf. Euclid, def. 17 of bk. 1.

Cs 54/—/10.49. *Centre and circumference (radius)*

C (As for the circle, there is a) centre (from which the distance to any point on the circumference is) of the same length.

CS The centre is (like a) heart, from which (a point moving to any part of the circumference) travels in all cases the same distance. (auct.)

Cf. Euclid, defs. 15 and 16 of bk. 1.

Cs 55/—/12.50. *Volume*

C (Every volume) has a thickness dimension (*hou*[4]) which gives a size (to the body).

CS Without a thickness dimension, there is no bodily size. (auct.)

Cf. Euclid, def. 1 of bk. XI. But it is probable that the Mohists were thinking of some definite sizes, and not of dimensions in the abstract.

It seems evident from the above text that the Mohists were thinking along lines which, if continued, could have developed into a geometrical system of Euclidean type. We cannot indeed be certain that they did not go beyond the point represented in these propositions or definitions, for the *Mo Ching* has come down to us in a very corrupt and fragmentary state, but if they did, their deductive geometry remained the mystery of a particular school and had little or no influence on the main current of Chinese mathematics. In any case, what is left, together with the many other evidences of geometrical thought in ancient and medieval China, precludes any suggestion that this was wholly lacking—though it was a knowledge of facts rather than of the logical reasons for them, and the algebraic trend, with its own form of logical reasoning, always dominated. Yet the Mohists were clearly attempting to pass from the practical to the philosophical, and in a way which perhaps strengthens the impression one gets from the close correspondence in date, that they worked quite independently of any western influence.

[1] 無厚 [2] 環 [3] 圓 [4] 厚

It is a pity that the Mohist geometry, as we have it, contains no reference to the properties of the right-angled triangle which is so prominent in the *Chou Pei Suan Ching* (p. 22). That part of its text was almost certainly written before the Mohist school came into existence. There is no telling how far back the appreciation of the value of right-angled triangles in surveying and mensuration may have gone in China; it was probably understood contemporaneously with Thales' determination of the heights of the Egyptian pyramids (*c.* −600), but we have no evidence to prove it.[a] We are not in a position to observe much of the mathematics of the Chinese until the Warring States period. It must not be forgotten, however, that the compasses (*kuei*[1]) and carpenter's square (*chü*[2]) play a part in legends coming down from before this time. In the Warring States, Chhin and Han periods there are numerous references[b] to these instruments, as also to the plumb-line (*shêng*[3]). Fig. 28 shows[c] a relief from the Wu Liang tomb-shrines in Shantung (*c.* +129 to +147) depicting the divine culture-hero Fu-Hsi holding a carpenter's square, and his consort Nü-Kua with a *quipu* (personified) between them.[d] Chariot-axles and other objects discovered with the oracle-bones at Anyang, and dating from the −13th or −12th century, are ornamented with remarkably complex geometrical figures, pentagons, heptagons, octagons and nine-sided polygons in various combinations. Numerous examples of Chou pottery and Han bricks also show geometrical figures. And in later ages the ingenuity of Chinese designers produced an extraordinary wealth of geometrical complexity in those wooden lattices which, covered with paper, filled the windows of palaces, houses and temples.[e]

(2) The Theorem of Pythagoras

Reference has already been made (p. 22) to the discussion of the theorem of Pythagoras (Euclid, bk. 1, prop. 47) in the *Chou Pei*.[f] This has been analysed by Mikami (2) and Li Nien,[g] who discuss the diagrams traditionally annexed to it which probably date from the +3rd-century commentary of Chao Chün-Chhing. The principal diagram (called *Hsüan Thu*) (Fig. 50) shows the square on the hypotenuse (made by the *chi chü*[4] process) folded backwards, as it were, so as to rejoin the plane of the paper, and demonstrably containing three further identical triangles together with a square of the difference between the base and the altitude. Liu Hui called this figure 'the

[a] See however the discussion on gnomons in the astronomical Section, pp. 284 ff. below.
[b] Collected by Li Nien (*8*).
[c] Vol. 1, p. 164.
[d] Cf. Yen Tun-Chieh (*13*). The knotted cords used for reckoning and recording are best known by their Peruvian name, here used. See also p. 69 above.
[e] There is a rich collection of these in Dye's *Grammar of Chinese Lattice*.
[f] It will be remembered that *kou*[5] stood for the base, *ku*[6] for the altitude, and *ching*[7] or *hsüan*[8] for the hypotenuse. This last word, read *hsien*, means a bow-string or a lute-string; the idea of stretching of cords (by surveyors) invites a comparison with the 'rope-stretchers' of another irrigation-agriculture civilisation, that of ancient Egypt, the *harpedonaptae* (Gandz, 3). Cf. p. 3 above.
[g] (*4*), vol. 1, p. 1, (*21*), vol. 1, p. 44; also (*1*), p. 24.

1 規 2 矩 3 繩 4 積矩 5 句 6 股 7 徑
8 弦

diagram giving the relations between the hypotenuse and the sum and difference of the other two sides (*kou ku chha, kou ku ping, yü hsüan hu chhiu chih thu*[1]) whereby one can find the unknown from the known'. In the time of Liu and Chao, it was coloured, the small central square being yellow and the surrounding rectangles red. The same figure is given by the Indian Bhāskara in the +12th century.[a] The algebraic formulation, given (in words) in the text, is: $h^2=4\frac{ab}{2}+(a-b)^2=a^2+b^2$, where h is the hypotenuse, a the altitude, and b the base. This proof is quite different from Euclid's.[b]

It would be a mistake, however, to suppose that all Greek geometry was axiomatic and deductive. We have just seen that Chinese geometry contained the beginnings of this kind of thought not only in the propositions of the Mohists. And conversely, there was a form of Greek geometry (to which Neugebauer has drawn attention)[c] which stood very close to the empirical and algebraic variety dominant in China; this was the content of the *Metrica* and the *Geometrica* associated with the name of Heron of Alexandria (+1st century).[d] He thinks that this 'geometrical algebra' was derived from Babylonian origins. If so, they had already generated similar bodies of thought farther east, for, as we have seen, though the *Chou Pei Suan Ching* commentary does not go back before the +3rd century, the *Chiu Chang Suan Shu* was being put together from the −3rd to the −1st. It is now known that an algebraical formulation of the Pythagoras theorem was familiar to the Old Babylonian mathematicians contemporary with the Shang period (−14th to −11th centuries).[e]

In the course of time the Chinese developed algebraic expressions for finding any unknown side or angle, the length of one side or the differences between sides, or other data, being given. All such problems on the right-angled triangle were reduced to twenty-five types of equation by Li Jui[2] at the end of the 18th century in his *Li shih I Shu*[3] (Mathematical Remains of Mr Li). For example, the twenty-third is as follows: knowing $h+a$ and $h-a-b$, then

$$-b^2+sb=-\left[\left(\frac{s+d}{2}\right)^2-\left(\frac{s-d}{2}\right)^2\right],$$

where s is the known sum (*ho*[4]) and d the known difference (*chiao*[5]). The others were similar. This treatment, translated by van Hée (2), was criticised by him as showing

[a] But the relations between the sides of right-angled triangles must have been known early in India, as some typical dimensions, as well as a general statement, are given in the *Āpastamba Śulvasūtra* (see Datta, 2). This, like the other *Śulvasūtras*, is associated with the Vedic corpus, and concerns the construction of fire-altars, hence plane and solid geometry. Perhaps the most probable date is between the −5th and the −2nd century. There is, however, no proof of any kind in this early material, and it seems extremely probable that Bhāskara's treatment derives from the *Chou Pei* (see Wang Ling (2), vol. 2, p. 162). On the *Śulvasūtras* see Renou & Filliozat (1), vol. 2, p. 172.

[b] Coolidge, in his *History of Geometrical Methods*, says that it is perhaps the easiest proof of all. Heath (1), p. 355, speaks of it as being 'of quite a different colour' from Greek modes of geometrical thought.

[c] (9), pp. 140, 152, 172.

[d] And Euclid, bk. 2.

[e] Evidence in Neugebauer (9), pp. 35 ff.

[1] 句股差句股并與弦互求之圖 [2] 李銳 [3] 李氏遺書 [4] 合
[5] 較

poor geometrical sense on the part of the Chinese.[a] But Petrucci (1) pointed out that the Chinese method, while purely algebraical, was none the worse for that, and was simply a way of treating the matter unfamiliar to those whose minds had been formed in the Euclidean deductive conventions natural to Europeans. Moreover, the Chinese deserved credit for the mastery of the handling of negative quantities which had been necessary in following this path. It amounted to an exhaustion of all the problems which could arise concerning right-angled triangles.

Throughout Chinese history the interest in right-angled triangles was mainly practical, for survey purposes. In the Sung, Shen Kua contrasts[b] this art (*chuan shu*[1]),[c]

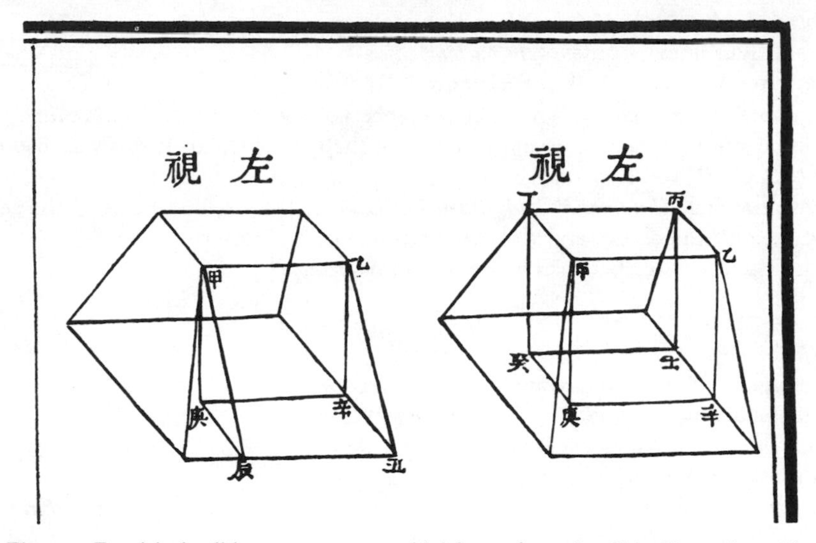

Fig. 72. Empirical solid geometry; pyramidal frusta from the *Chiu Chang Suan Shu*.

which concerns measurement of geometrical shapes which can be seen, with calendrical computations based on the geometry of the heavens (*chui shu*[2]), i.e. circles and curved arcs which have to be imagined.

(3) Treatment of Plane Areas and Solid Figures

By the end of the Former Han correct or approximately correct formulae had been worked out for ascertaining the area of many plane shapes, and the volume of many solid figures, though in all cases without deductive geometrical demonstrations. Use may have been made of models, and the more complex reduced experimentally to the simpler. This knowledge was embodied in the *Chiu Chang Suan Shu* (see Fig. 72);

[a] Just to make this subject more difficult than it need be, van Hée refers to the compiler of these equations only as 'Li Chang Tche'—the reader has to know (or to find out) that he is talking about Li Jui, whose *hao* was Shang-Chih.[3]

[b] *MCPT*, ch. 18, para. 3.

[c] The unusual word *chuan*[4] he regards as having been originally a drawing of the carpenter's ink box (*mo tou*[5]). See above, Vol. 2, p. 126, and below, Sect. 27*a*.

[1] 𡨥術 [2] 綴術 [3] 尙之 [4] 𡨥 [5] 墨㪷

the formulae are given in Li Nien (*1*). Here it may be of interest simply to list the figures with which the Chinese geometers were familiar.[a]

These were as follows:

Areas (*chi*[1])

Square (*fang*,[2] *fang thien*,[3] *phing fang*,[4] *shih*[5]).

Rectangle (*kuang thien*,[6] *chih thien*,[7] *mi*,[8] this last word meant originally a rectangular cloth food-cover).

Isosceles triangle (*kuei thien*[9]).

Trapezium (*chi thien*[10]). Its base was called *chung*[11] and its short upper side *shê*.[12] The altitude of any such figure was called *chêng tshung*.[13]

Rhomboid (*hsieh thien*,[14, 15] *hsiao thien*[16]).

Trapezoid (four unequal sides) (*ssu pu têng thien*[17]).

Double trapezium (*ku thien*,[18] *yao ku thien*,[19] *shê thien*[20]).

Circle (*yuan*,[21] *yuan thien*[22]). The circumference is *chou*,[23] the diameter is *ching*.[24] There was no special word for radius, except *pan ching*[25] (half the diameter); this was also used in the West (Smith (1), vol. 2, p. 278).

Segment of a circle (*hu thien*,[26] *kung thien*[27]). *Hu*[28] is the circumference of the segment, or the arc, *hsüan*[29] the chord, and *shih*[30] the altitude or sagitta.

Annulus (annular space between two circles) (*huan thien*[31]).

Volumes[b]

Cube (*li fang*[32]).

Parallelepiped with two square surfaces (*fang pao thao*[33]).

Parallelepiped with no square surfaces (*tshang*,[34] *fang chiao*[35]).

Pyramid (*yang ma*[36],[c] *fang chui*[37]).

Frustum of pyramid with square base (*fang thing*,[38] *chiao*[39]).

Frustum of pyramid with rectangular base of unequal sides (*chiao*,[39] *chhü chhih*,[40] *phan chhih*,[41] *chhu thung*,[42] *ming ku*[43]).

Prism (*chhien tu*[44]) (a wedge with the 'cutting edge' the same length as the base and one side at right angles to it).

Wedge with rectangular base and both sides sloping (*chhu mêng*[45]).

Wedge with trapezoid base and both sides sloping (*yen chhu*[46]).

[a] In the hope that this may be useful to those who may like to devote some study to old Chinese mathematical texts.

[b] Curiously, a specific word for volume as opposed to area seems to have lacked in the old mathematical books. The modern term is *thi chi*[47] 'piled body'. In the answers and calculations the word 'cubic' is understood. Of course there were also the standard measures of volume, used, for instance, in the Su Mi chapter of the *Chiu Chang*.

[c] This strange term seems to derive from the divergent rays of a beam of sunlight (*thai yang*) entering a dark building through a small square opening. Later it was adopted in architecture (cf. *Ying Tsao Fa Shih*, ch. 5, p. 6*a*).

1 積 2 方 3 方田 4 平方 5 實 6 廣田
7 直田 8 冪 9 圭田 10 箕田 11 踵 12 舌
13 正從 14 邪田 15 斜田 16 蕭田 17 四不等田
18 鼓田 19 腰鼓田 20 蛇田 21 圓 22 圓田 23 周
24 徑 25 半徑 26 弧田 27 弓田 28 弧 29 弦
30 矢 31 環田 32 立方 33 方堡壔 34 倉 35 方窖
36 陽馬 37 方錐 38 方亭 39 窖 40 曲池 41 盤池
42 芻童 43 冥谷 44 塹堵 45 芻甍 46 羨除 47 體積

Tetrahedral wedge (*pieh ju*[1]).

Frustum of a wedge of the second type (*chhêng*,[2] *yuan*,[3] *kou*,[4] *chhien*,[5] *thi*,[6] *chhü*,[7] *chhiang*[8]). This figure was perhaps the most important of all for the old engineers. The terms *chhêng*, *yuan*, and *chhiang* recall that this was the shape in which city-walls had to be built. The term *thi* refers to the building of dykes and embankments for flood control. The terms *kou*, *chhü* and *chhien* all mean excavated canals and waterways, including moats.

Cylinder (*yuan tshang*,[9] *yuan pao thao*,[10] *yuan chiao*,[11] *yuan chhün*[12]). Here the second expression recalls, as above, fortification towers, and the first and fourth, granaries.

Cone with circular base (*yuan chui*,[13] *wei su*,[14] *chü su*[15]).

Frustum of a cone (*yuan thing*,[16] *yuan thuan*[17, 18]).

Sphere (*li yuan*,[19] *wan*[20]). The area of a spherical segment was known as *chhiu thien*,[21, 22] *wan thien*,[23] and *yuan thien*.[24]

Liu Hui, the +3rd-century commentator on the *Chiu Chang*, was one of the greatest expounders of this 'empirical' solid geometry. He saw, for example, that a wedge of the first type could be broken down into a pyramid and a tetrahedral wedge; and that a wedge of the second type could be made to give two tetrahedral wedges separated by a pyramid. The frustum of a pyramid with a rectangular base of unequal sides had, he said, a rectangular parallelepiped in the centre, with a prism on each side and a pyramid at each corner. By these simple methods he was able to arrive at formulae for the volumes.

It is evident that when circles were involved the value of π was required, and it was only in such cases that the *Chiu Chang* formulae were approximations. To the Chinese evaluations of π we must therefore now turn.

(4) Evaluation of π

Historians have devoted much attention to the efforts of the old mathematicians to arrive at approximations to the value of the ratio between the diameter and the circumference of a circle (*yuan chou lü*[25]), possibly because the increasing accuracy of the results seems to offer a kind of measure of the mathematical skill of succeeding generations. There are two chapters on the subject in Mikami (1) and several sections in Li Nien (*1*), besides many articles, among which may be mentioned those of Mao I-Shêng (*1*), Mikami (19) and Chang Yung-Li (*1*).[a]

Although there is evidence[b] that the ancient Egyptians and Old Babylonians had values such as 3·1604 and 3·125, the commonest practice in ancient civilisations was to take the ratio simply as 3. This we find in the two great Han arithmetics (*Chou Pei* and *Chiu Chang*), as also in the *Khao Kung Chi* (Artificers' Record) in the *Chou Li* (Record of the Rites of Chou).[c] As a rough approximation this persisted for centuries.

[a] The special monograph of D. E. Smith (4) has not been accessible to us.

[b] Gow (1), p. 127; Smith (1), vol. 2, p. 270; Neugebauer (9), pp. 46, 53.

[c] Biot (1), vol. 2, p. 469.

1 鱉臑 2 城 3 垣 4 溝 5 塹 6 隄 7 渠
8 牆 9 圓倉 10 圓堡壔 11 圓窖 12 圓囷 13 圓錐
14 委粟 15 聚粟 16 圓亭 17 圓圖 18 圓篅 19 立圓
20 丸 21 邱田 22 丘田 23 丸田 24 宛田 25 圓周率

The first indication that a more exact value was sought for arises from the Chia Liang Hu standard measure, prepared for Wang Mang between +1 and +5 by Liu Hsin,[1] an object of great archaeological interest which is still preserved in Peking (Ferguson, 3). This is simply a cubical space cut out of a solid bronze cylinder. The inscription[a] reads:

The standardised *chia liang hu*[2] (has) a square with each side 1 *chhih*[3] (foot) long, and outside it a circle. The distance from each corner of the square to the circle (*thiao phang*[4]) is 9 *li*[5] 5 *hao*.[6] The area of the circle (*mi*[7]) is 162 (square) *tshun*[8] (inches), the depth 1 *chhih*[3] (foot), and the volume (of the whole) 1620 (cubic) *tshun*[8] (inches).

From this Chhien Pao-Tsung (*1*) saw that Liu Hsin must have used a value for π of 3·154, but there is no record of how he arrived at it.

The first explicit effort to obtain a more accurate figure was that of Chang Hêng[9] about +130.[b] According to his biography in the *Hou Han Shu* he 'threw a network about heaven and earth and calculated (the movements and dimensions)'.[c] Probably his value of π was contained in his lost book, the *Suan Wang Lun*[10] (Discussion of Calculations on the (Universal) Network). We know it only because of the reference to it in the commentary on the *Chiu Chang*;[d] it was 3·1622 (i.e. $\sqrt{10}$). A later mention in the *Khai-Yuan Chan Ching* (+718)[e] expresses it by the fraction $\frac{92}{29}$, a slightly higher value.[f] In the +3rd century Wang Fan,[11] mathematician and astronomer of the Wu State in the Three Kingdoms period, recalculated the value, getting $\frac{142}{45}$ or 3·1555; this would be in the neighbourhood of +255. His contemporary, Liu Hui,[12] however, who worked in the Wei State to the north, did much better. Liu Hui's method was to inscribe a polygon within a circle and to calculate the perimeter on the basis of the properties of the right-angled triangles formed by each half-segment. Starting from the simplest case of a hexagon, he obtained,[g] from the area of a polygon of 192 sides, a rough value of $\frac{157}{50}$ or 3·14. But he also gave[h] two values, a lower one $3{\cdot}14\frac{64}{625}$ and a higher one $3{\cdot}14\frac{169}{625}$, the right figure being somewhere between the two. The higher of these (3·142704) was a little better than the famous fraction $\frac{22}{7}$ (3·1428)

[a] See Jung Kêng (*2*), ch. 3 (modern readings), pp. 1*a*, *b*.
[b] Valuable biographies of him by Sun Wên-Chhing (*2*, *3*, *4*); Li Kuang-Pi & Lai Chia-Tu (*1*).
[c] *Wang lo thien ti erh suan chih.*[13] Ch. 89, p. 2*a*. Cf. p. 538 below.
[d] Ch. 4, p. 17*a*.
[e] Ch. 1, pp. 25*b*, 26*a*.
[f] This fraction is given only in the Thang compilation. What Chang Hêng actually did, according to the *Chiu Chang* commentary, was to follow the empirical formulae of that book, though conscious of their crudeness, in comparing the volumes of a cube and a cylinder and sphere inscribed within it. Somehow or other, perhaps even by weighing, he arrived at the expression $\frac{V_{inscribed\ sphere}}{V_{cube}} = \sqrt{\frac{25}{64}} = \frac{5}{8}$. He must therefore have considered $\pi = \sqrt{\frac{16 \times 5}{8}} = \sqrt{10}$.
[g] Commentary to *Chiu Chang*, ch. 1, pp. 11*a* ff., 15*a*.
[h] Commentary to *Chiu Chang*, ch. 1, pp. 12*a* ff.

1 劉歆 2 嘉量斛 3 尺 4 庣旁 5 釐 6 毫
7 冪 8 寸 9 張衡 10 算罔論 11 王蕃 12 劉徽
13 網絡天地而算之

which Archimedes had found about −250 by the use of a 96-sided polygon.[a] Elsewhere,[b] still striving for greater accuracy, Liu Hui continued his process to a polygon of 3072 sides, and got his best value, a fraction equivalent to 3·14159. He knew he could go further if necessary. This figure was better than that accepted by Ptolemy (*c.* +150).[c]

At this time, therefore, the Chinese had more than caught up with the Greeks, but they took a leap forward in the +5th-century calculations of Tsu Chhung-Chih[1] and his son[d] Tsu Kêng-Chih,[2] which set them ahead for a thousand years. Tsu Chhung-Chih (+430 to +501)[e] was the most distinguished mathematician, astronomer and engineer of his time (the Liu Sung and Chhi dynasties). For π he gave two values,[f] an 'inaccurate' one (*yo lü*[3])[g] which was the same as that of Archimedes, and an 'accurate' one (*mi lü*[4]) of $\frac{355}{113}$ or 3·1415929203. This latter was not equalled anywhere until the end of the +16th century (Anthoniszoon). Conscious, however, that his figure still lacked precision, Tsu found further approximations[h]—an 'excess value' (*ying shu*[5]) of 3·1415927, and a 'deficit value' (*nu shu*[6]) of 3·1415926, between which the true ratio must lie. The figures given in +1593 by Vieta (who certainly never knew of his predecessor) fell exactly half-way between these limits.[i]

The calculations were no doubt contained in Tsu's book, the *Chui Shu*,[7] but this has long been lost,[j] and most of what we know about him and his son (who may have been the joint author or editor of the book), comes from the calendrical and astronomical chapters of the *Sui Shu* (History of the Sui Dynasty), written by Chhangsun Wu-Chi[8] and others before +656, and the biographical sections of the *Nan Shih*

[a] Sarton (1), vol. 1, p. 169; B. & M. p. 407. A transmission from West to East is often assumed, but we doubt it. The Greek method had a polygon outside the circle as well as one within it, and did not include calculations of area. These involved a knowledge of approximation of fractions which was not available in Greece. What Archimedes proved was that the true value must lie between $\frac{223}{71}$ and $\frac{22}{7}$.

[b] Commentary to *Chiu Chang*, ch. 1, pp. 14*a* ff.

[c] 3·141666, Halma ed. VI, 7; Heath (6), vol. 1, p. 233; Smith (1), vol. 2, p. 308.

[d] Biography of him by Yen Tun-Chieh (*5*).

[e] Biography of him by Chou Chhing-Chu (*1*).

[f] *Sui Shu*, ch. 16 (*Lü li chih*), p. 3*b*. See Yen Tun-Chieh (*8*).

[g] This was generally used in his time, as by the astronomer Ho Chhêng-Thien[9] (*fl.* +460). Other mathematicians, e.g. Phi Yen-Tsung[10] (*fl.* +445), also worked on this subject, but the results have not survived.

[h] *Sui Shu*, ch. 16, p. 3*b*; cf. Chhien Pao-Tsung (*1*), pp. 57, 58. The 'accurate value' fraction was an extraordinary achievement for it is one of the continued-fraction convergents.

[i] Narrien in 1833 promoted Kuo Shou-Ching to the imperial yellow, but greatly misjudged the Chinese work on π. 'The pure sciences', he wrote, (1), p. 350, 'have always been in a low state among this ancient people. The missionaries found that before the time of Cocheou Kong, who reigned in the thirteenth century, they considered the proportion between the diameter of a circle and its circumference to be exactly as 1 to 3..., nor does it appear that they had advanced one step further until they were instructed by the Europeans...'. Narrien was misled by Gaubil (2), p. 115, but the egregious error was perpetuated by Sédillot (2), p. 642 and others. Such were the sources of the authoritative Whewell.

[j] We do not even know the exact significance of its title—'threading together'. Shen Kua in the +11th century (*Mêng Chhi Pi Than*) hints that it had something to do with calendrical computations (cf. p. 394). The *Chui Shu* was still in use in the Thang dynasty as a text for the imperial examinations, but it was considered by far the most difficult of the mathematical books, though elegantly written (*Hsin Thang Shu*, ch. 44, p. 2*a*). Cf. des Rotours (2), pp. 140, 154.

1 祖沖之 2 祖暅之 3 約率 4 密率 5 盈數 6 朒數
7 綴術 8 長孫無忌 9 何承天 10 皮延宗

(History of the South) and *Nan Chhi Shu* (History of the Southern Chhi Dynasty).[a] Li Shun-Fêng of the Thang referred to Tsu's value and praised it.[b] About +1300 Chao Yu-Chhin,[1] the eccentric author[c] of the *Ko Hsiang Hsin Shu*[2] (New Elucidation of the Heavenly Bodies), returned to the question, and by the continued use of inscribed polygons with up to 16,384 sides, confirmed that Tsu's value was very accurate. After that it was forgotten, and, as Goodrich (4) has pointed out, in the time of Khang-Hsi the Chinese relied entirely on the methods of the Jesuits such as Verbiest and Schall. Not until later was this 'pearl in the Red River' rediscovered. The $\sqrt{10}$ was still being used in the mid-18th century, as by Wang Yuan-Chhi[3] and Chhien Thang.[4]

As for parallel developments elsewhere, Āryabhaṭa, the contemporary of Tsu Chhung-Chih and Tsu Kêng-Chih, was content with 3·1416;[d] and Brahmagupta, a century later, used 3·162.[e] In Europe, Franco of Liège,[f] an +11th-century contemporary of Shen Kua, had a very poor value, 3·24. But in the middle of the +15th century, al-Kāshī worked out the ratio to sixteen places of decimals,[g] while about +1600 Adriaan Anthoniszoon obtained a result identical with that of Tsu Chhung-Chih.[h] Finally, in the 17th century van Ceulen took it to the 35th place, and in +1853 it was carried to the 707th by William Shanks.[i] The proof of the transcendency of π was first given by Lindemann in +1882, thus showing the impossibility of squaring the circle by finite constructions with ruler and compasses alone. Outstanding in the Chinese contribution is the +3rd- and +5th-century work of Liu Hui and Tsu Chhung-Chih.

(5) Conic Sections

Chinese solid geometry, non-demonstrative in character, developed from the practical needs of mensuration and never went much beyond them. There developed in China no counterpart to Apollonius of Pergamon (*fl.* −220) and his great work on conic sections.[j] The study of the ellipse, the parabola and the hyperbola had to await the 17th century.[k] There were, however, certain geometrical problems concerning curved figures which were then studied persistently in East Asia, notably those concerning tangent circles, where the problem was how many circles and of what size could be inscribed in given figures such as semicircles, fans (annulus sectors) and ellipses (Fig. 73).

[a] Cf. Li Nien (*10*).
[b] *Chiu Chang* commentary, ch. 1, p. 15*a*.
[c] Consistently mis-romanised by Mikami (1) as Chang Yu-Chin.
[d] Sarton (1), vol. 1, p. 409; Karpinski (3).
[e] Sarton (1), vol. 1, p. 474.
[f] Sarton (1), vol. 1, p. 757.
[g] As we have already seen, p. 89.
[h] Smith (1), vol. 2, pp. 255, 310.
[i] Chinese and Japanese mathematicians also participated in these 19th-century developments of series; the details will be found in Mikami (1) and Smith & Mikami (1). It is now known that the last hundred of Shanks' digits were erroneous. Some 10,000 places have now been obtained on electric computers.
[j] Cf. the views of Neugebauer (3) on the derivation of this branch of mathematics from observations of sun-dial gnomons (p. 307 below).
[k] Li Nien (*16*, *17*, *21*, vol. 3, pp. 519ff.).

1 趙友欽　2 革象新書　3 王元啓　4 錢塘

Smith[a] illustrates only Japanese examples, but says that these were inherited from the Chinese; Smith & Mikami (1) devote a whole chapter to them. This matter has some relation to the origin of calculus (see below, p. 141).

Greek interest in polyhedra was not represented in China. Nor did the Chinese come up against two of the three famous problems of Greek geometry, the trisection of any angle, and the duplication of a cube (the Delian problem).[b]

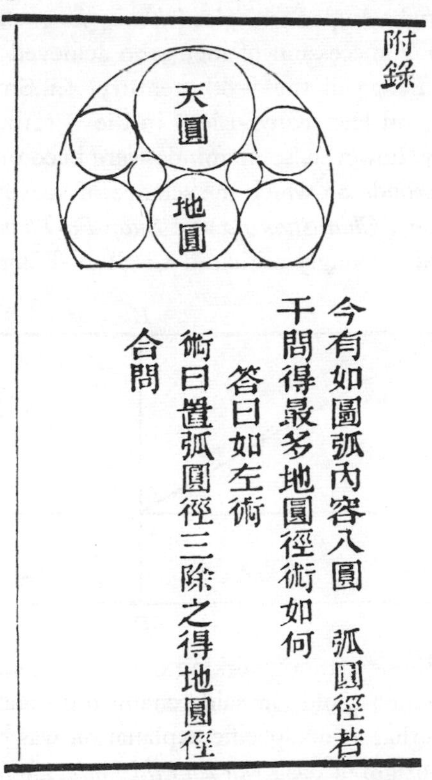

Fig. 73. A problem in the packing of circles, from the *Suan Fa Yuan Li Kua Nang* (*Sampō Enri Katsunō*) of Kaetsu Denichiro (1851).

(6) Yang Hui and the Coming of Euclid

That Chinese geometry was always entirely empirical and non-demonstrative is a statement which (as we have seen, p. 94) cannot be made in an unqualified way. The Chinese proof of the Pythagoras theorem was indeed a proof. Later commentators on the Han books, such as Liu Hui and Chao Chün-Chhing, were accustomed to speak in words which show that they used colours to distinguish the various areas and figures which they were comparing. For example, 'Replace the blue piece by the red piece; there will be no excess or deficit (*chhing chhu chu ju, hsiang pu* [1]).'[c] We have

[a] (1), vol. 2, p. 536.

[b] The quadrature of the circle (the third famous problem) was to some extent another way of stating the problem of the evaluation of π.

[c] *Chiu Chang*, ch. 9, p. 1*b* (comm.).

[1] 青出朱入相補

already spoken of the treatise on surveying by means of similar (right-angled or other) triangles by Liu Hui in +263, the *Hai Tao Suan Ching* (Sea Island Mathematical Manual) (p. 30). This 'double-difference' (*chhung chha*[1]) method had certainly been used before him, in the Han, for the *Chou Pei* has references, connected with the gnomon, for 'observing that which is distant' (*wang yuan*[2]); and about the end of the +1st century Chang Hêng, in his *Ling Hsien* (Spiritual Constitution of the Universe) speaks of using double right-angled triangles (*chhung yung kou ku*[3]). After the time of Liu Hui there was a long succession of men who achieved fame as surveyors—one might mention Hsintu Fang[4] in the +6th century, Li Shun-Fêng[5] in the +7th, Hsia Ao[6] in the +10th and Han Kung-Lien[7] in the +11th.[a]

In the +13th century, however, some minds were becoming very dissatisfied with the mainly empirical methods on which the science of surveying had been based. In his two books *Hsü Ku Chai Chhi Suan Fa* and *Suan Fa Thung Pien Pên Mo*,[b] both of about +1275, Yang Hui[8] strongly criticised Li Shun-Fêng[9] and Liu I,[10] who had

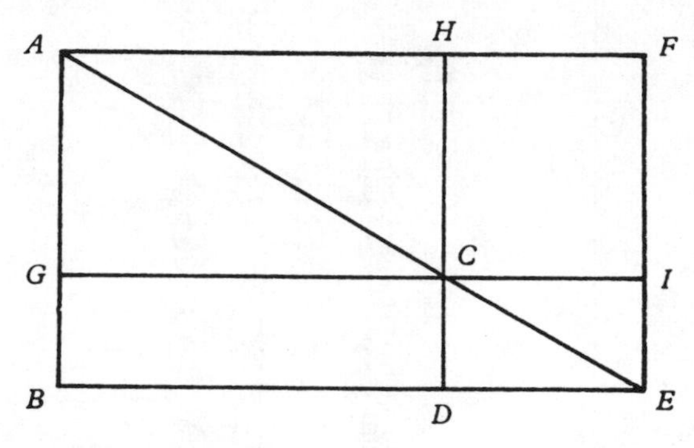

been content to use methods without working out their theoretical origin (*yuan*[11]) or principle (*chih*[12]).[c] 'The men of old', he said, 'changed the name of their methods from problem to problem, so that as no specific explanation was given, there is no way of telling their theoretical origin or basis (*ku jen i thi i ming, jo fei shih ming, tsê wu i chih chhi yuan*[13]).'[d] This was an extremely modern attitude. Yang Hui proceeded to give a theoretical proof of the proposition that the complements of the parallelograms which are about the diameter of any given parallelogram are equal to one another. This is the same as the forty-third proposition of the first book of Euclid, only that Yang Hui took the case of a rectangle and gnomon, as in bk. II, def. 2. In the accompanying diagram, *AB* was considered the *kou*,[14] *BE* the *ku*,[15] *AE* the *hsüan*,[16] *CD* the *yü kou*,[17]

[a] The last-named has a more important place in history as a horological engineer (see Sect. 27*h*).

[b] Respectively, 'Continuation of Ancient Mathematical Methods for Elucidating the Strange (Properties of Numbers)'; and, 'Alpha and Omega of the Mathematics of Mutually Varying Quantities.'

[c] Whether or not a coincidence, this, it will be remembered, is the same term which the School of Logicians in the −3rd century had used for 'universals'. I do not know whether the *Kungsun Lung Tzu* would have been available to Yang Hui. Cf. Vol. 2, p. 185.

[d] Li Nien (*4*), vol. 1, pp. 39, 54.

1 重差 2 望遠 3 重用句股 4 信都芳 5 李淳風
6 夏翱 7 韓公廉 8 楊輝 9 李淳風 10 劉益 11 源
12 旨 13 古人以題易名若非釋名則無以知其源 14 句 15 股
16 弦 17 餘句

and *CI* the *yü ku.*[1] It was shown that the rectangle *BC* (*kou chung yung hêng*[2]) and the rectangle *CF* (*ku chung yung chih*[3]) were of equal area (*erh chi chieh thung*[4]). The procedure was then extended to two similar right-angled triangles like those used in surveying. This necessitated speaking of a larger and smaller *yü kou.*[5]

Could such proofs have been extended the Chinese might have developed an independent deductive geometry. In any case, by the end of the +13th century, during the Mongol wars, some Chinese minds (if we may judge by Yang Hui) were prepared to appreciate the Euclidean system. It is therefore of great interest that there may have been a translation of Euclid into Chinese at that time, due to the Chinese-Arabic contact already referred to (p. 49). Yen Tun-Chieh (7) and others[a] have noticed that the chapter on Muslim books in the *Yuan Pi Shu Chien Chih*[6] mentions[b] that just about the time of Yang Hui the official astronomers studied some useful Western works, including the *Ssu-Pi Suan Fa Tuan Shu*,[7] in fifteen books, by a certain Wu-Hu-Lieh-Ti.[8] This was in the imperial library in +1273. Since the second character of the author's name was probably then read Khu, the probability is that this was a transliteration of Euclid.[c] It has been thought that the words Ssu-Pi may have been a transliteration of an Arabic term meaning 'original text'.[d] The number of books has also seemed suspicious.[e] Yen Tun-Chieh thinks that the transmitter here was Naṣīr al-Dīn al-Ṭūsī[f] (cf. Vol. 1, p. 217), the founder, under Hulagu Khan, of the famous observatory of Marāghah in Persia. The evidence does not absolutely indicate that the books in question were a translation of Euclid into Chinese, though it seems probable that they were.[g] In any case they had no perceptible effect, and during the following centuries the survey geometers, such as Thang Shun-Chih[9] and Chou Shu-Hsüeh[10] in

[a] Cf. Li Nien (*2*), p. 149, (*18*).

[b] Ch. 7. A collection of official records of the Yuan dynasty, made in the middle of the +14th century.

[c] The word *Tuan* in the title suggests 'propositions'.

[d] Other possibilities seem more likely, however, and for the following conjectures we are indebted to Mr D. M. Dunlop. For the *Elements* the *Fihrist* has *Usūl al-Handasah* (Steinschneider (3), p. 82), while al-Khwārizmī (van Vloten ed., p. 202) has *al-Usṭuqusāt*, this word being a typical malformation of the Greek *stoikheia*. But Euclid's geometry was also known in Arabic as the *Kitāb Uqlīdis fī al-Ḥisābī*, i.e. the 'Book of Euclid on Calculation' (Houtsma (1), vol. 1, p. 135). The last word of this title, then, might have given rise to *ssu-pi*. One notes that the final syllable of the Arabic genitive *-i* was often retained in Chinese transcriptions, as we shall see later on (p. 373 below) in the case of *ha-la-chi* for *ḥalaq-i*. Another possibility would derive *ssu-pi* from the name of one of the translators of Euclid from Greek into Arabic, namely, Thābit ibn-Qurrah (+826 to +901), sometimes called al-Harrānī, but sometimes also al-Ṣābī, the Sabian. See Sarton (1), vol. 1, p. 599.

[e] Yet, as d'Elia (4) points out, it is very doubtful whether the Elements had more than thirteen books in any Arabic version. Only at the Renaissance were the last two books added (the fourteenth by Hypsicles, *c.* −190, and the fifteenth by a pupil of Isidore in the +6th). See Sarton (1), vol. 1, p. 154.

[f] Mieli (1), pp. 150 ff.

[g] Such at least is our view. D'Elia (4) suggests, however, that since the title occurs in a list of 195 'Muslim books useful at the observatory', the text may have remained in Arabic while the title only was translated. A MS. of this kind is figured by Sarton (1), vol. 3, p. 1530, and others probably still exist. In any case the possibility is not favoured, as d'Elia thinks, by the 'otherwise inexplicable' lag in the spread of knowledge of deductive geometry in China. Running counter as it did to two millennia of mathematical tradition, a translation made in the imperial observatory might well, we think, have failed to evoke wide interest.

[1] 餘股 [2] 句中容橫 [3] 股中容直 [4] 二積皆同 [5] 餘句
[6] 元秘書監志 [7] 四擘算法段數 [8] 兀忽列的 [9] 唐順之
[10] 周述學

the +16th century, continued quite naturally to follow the ways of their predecessors in the Thang and before.

The story of the translation of the first six books of Euclid by Ricci & Hsü Kuang-Chhi in +1607 has already been told (p. 52). The remaining books were done into Chinese by A. Wylie & Li Shan-Lan in +1857, using the same title as in the 17th century, *Chi Ho Yuan Pên*.[1] In +1865 a complete corrected edition of the whole was issued by Tsêng Kuo-Fan, the famous official, and there are interesting translations of his preface, together with the original ones of Ricci and Hsü Kuang-Chhi, by G. E. Moule (1) and d'Elia (4). Chhen Yin-Kho (2) and van Hée (14) have described a Manchu translation.

(7) Co-ordinate Geometry

This sub-section must end with a few words on analytic geometry. Three principal steps were involved in its development: (*a*) the invention of a system of coordinates; (*b*) the recognition of a one-to-one correspondence between geometry and algebra; and (*c*) the graphic expression of a function such as $y=f(x)$. Both the first steps were pre-Renaissance, and both were taken in China.

The idea of coordinates in land utilisation must of course be very ancient. The Egyptian hieroglyph for a district was a grid, and in Chinese there was the character *ching*,[2] fields round a well, later simply a well. Hipparchus (*c.* −140) was one of the first, if not the first, to locate points both in the heavens and on the earth's surface by means of their *mekos* (μῆκος) and *platos* (πλάτος), longitude and latitude.[a] Eratosthenes before him (*c.* −284) had laid down meridians and parallels for his geography.[b] When quantitative cartography was totally lost in the West, after the time of Ptolemy, it began to flourish in China, probably from the +2nd century onwards, on the foundations laid by Chang Hêng and Phei Hsiu, as we shall see in Section 22*d* on geography; and in China the grid system was never lost.

Another aspect of coordinates, to which historians of mathematics have paid singularly little attention, is the growth of tabulation systems. Mention has already been made (Vol. 2, p. 279) of the strange tables which are to be found in the *Chhien Han Shu* (History of the Former Han Dynasty). About +120 Pan Ku and his sister Pan Chao supplied this famous work with eight chronological tables.[c] In one of the most remarkable of these, the names of nearly 2000 legendary and historical characters are arranged according to an arbitrary scale of nine ratings as to virtue, in a grid, still preserved as chapter 20 of this dynastic history, where time is, in a sense, one axis, and virtue the other. The points, if plotted, would form a descending curve, since the legendary culture-heroes received so many high ratings. Such a system of coordinates

[a] B. & M. pp. 534, 544.
[b] B. & M. p. 477.
[c] See the interesting analysis by Bodde (5). Actually, the pattern of having tables in a historical work had been set some two hundred years previously by Ssuma Chhien. Chs. 13–22 of the *Shih Chi* are occupied by chronological tables, but here only one axis was quantitative, namely, the time axis; the other axis consisted of the various feudal States the events in which had to be marked down.

[1] 幾何原本
[2] 井

long antedated the chessboard which has often been called the earliest set of coordinates.[a]

Elsewhere above (Vol. 1, p. 33) reference had to be made to the philological system of 'spelling' Chinese characters, by indicating their initial and final sounds, in a system which, though originating in the +3rd century, was brought to perfection in the *Chhieh Yün* (Dictionary of Sounds) of Lu Fa-Yen at the beginning of the +7th. We mentioned also that by the +11th or +12th century this dictionary was no longer usable owing to linguistic shifts, so that in the *Chhieh Yün Chih Chang Thu*,[1] ascribed to Ssuma Kuang,[b] the characters were rearranged in a coordinate tabulation system generally known as the 'rhyme-tables' (Fig. 1 in Vol. 1, p. 35). As already noted, numerous other compilations followed this system in later centuries.

Of course in China from early times there was tabulation of numerical quantities, just as there had been in Old Babylonia.[c] We have already seen[d] that the term *Li-Chhêng*[2] designates this in book titles; meaning 'immediate success' it is equivalent to 'ready reckoner'. An early example of the use of this is the *Chiu Kung Hsing Chhi Li-Chhêng*,[3] tables compiled by Wang Chhen[4] to summarise the astronomical data used by Li Yeh-Hsing[5] in his calendar of +548.[e] Similar tables frequently occur afterwards in the calendrical chapters of the dynastic histories.

Essentially the abacus was a coordinate system, a fact which emerges with particular clarity from some of the early attempts at developing it (p. 77 above). This was, of course, much less advanced than the use of rectangular axes by the Greeks in dealing with curves of various forms. Smith[f] thinks that Menaechmus (*c.* −350) may have used that property of the parabola expressed by the equation $y^2 = px$; and Apollonius (*c.* −250) was very conscious of coordinate axes.[g] In the Middle Ages of Europe there were a number of pre-Cartesian graphs; descriptions of them have been published.[h] In the +14th century Nicholas d'Oresme[i] took considerable steps forward. His *Tractatus de Latitudinibus Formarum* and *Tractatus de Uniformitate et Difformitate Intensionum*, of about +1370, used the words *longitudo* and *latitudo* in the Cartesian sense of ordinate and abscissa, and functions were represented by curves.

But we doubt whether Smith[j] is right in tracing the first appreciation of the identity of geometrical and algebraic relations to Arabic mathematicians such as the great al-Khwārizmī of the +9th century.[k] From the beginning of mathematics in China

[a] We must not here anticipate what will have to be said about the chessboard in Section 26*i* on physics and magnetism. Chess in its modern form appears to be a +7th-century Indian invention, but some of its roots are connected with the Chinese divining-board.

[b] Cf. Tung Thung-Ho (*1*).

[c] On the cuneiform 'table texts' see Neugebauer (9), pp. 29ff.

[d] See pp. 9 and 36 above.

[e] *Chiu Thang Shu*, ch. 47, p. 7*b*; cf. p. 542 below.

[f] (1), vol. 2, p. 317.

[g] Heath (2).

[h] Günther (1); Funkhouser (1); Lattin (1).

[i] Sarton (1), vol. 3, p. 1486.

[j] (1), vol. 2, p. 320.

[k] One may indeed legitimately ask whether Arabic mathematics may not have been influenced by the Chinese in this respect. It would be rewarding to study the original works of al-Khwārizmī from this point of view. He was sent as an ambassador to Khazaria by the Caliph between +842 and +847 (see Dunlop (1), p. 190), and Khazaria lay athwart some of the trade routes between China and the West (see on, p. 682).

1 切韻指掌圖 2 立成 3 九宮行碁立成 4 王琛 5 李業興

geometrical propositions were expressed in generalised algebraic form, and when geometrical figures were used, as in the +3rd-century *Hai Tao Suan Ching*, the treatment was uniquely algebraic. In Europe this realisation came relatively late; Fibonacci (*c.* +1220) was the first mathematician of prominence to recognise the value of relating algebra to geometry. The difficulty with which the Chinese contended was that they had no well-developed geometry of curves with which to relate it. The first text-book on algebraic geometry[a] was that of Marino Ghetaldi in +1630. About this time the basic conceptions of analytic geometry developed in the minds of Fermat and Descartes, and the latter's *La Géométrie*, from which all the graphical treatments so universal in modern science derive, appeared in +1637. Leibniz provided our names 'coordinate', 'abscissa' and 'ordinate'. This was reasoning from equations to geometrical figures; what the Chinese had always done was to transform geometrical figures into equations.

(8) TRIGONOMETRY

There is not a great deal to say about trigonometry[b] in old Chinese mathematics, since the modern theory of the trigonometric functions is a post-Renaissance development. As in all ancient civilisations the properties of the right-angled triangle were early studied in connection with the astronomical measurements for which the gnomon was required, and in a sense, therefore, the *Chou Pei* recognised the importance of the ratios of its sides. But the most important steps were taken by the Greeks. There is reason to think that Aristarchus of Samos (*c.* −260) used ratios similar to the tangent of an angle, and that Hipparchus (*c.* −140) carried out graphic solutions of spherical triangles (Braunmühl). The chief advances in spherical trigonometry,[c] however, were made by Menelaus of Alexandria about +100 in his *Spherics*;[d] he first formulated the famous 'rule of six quantities' in which sines were involved. Ptolemy (*c.* +150) extended the table of chords which had been begun by Hipparchus.

It was then the Indians who brought trigonometry into its modern form. The notion of sines and versed sines appears for the first time in the *Pauliśa Siddhānta* shortly after +400.[e] Āryabhaṭa (*c.* +510) was the first to give a special name to the function, and to draw up a table of sines for each degree. His contemporary Varāha-Mihira, in the *Pañca Siddhāntikā* (*c.* +505), gave formulae which in modern terms would include both sines and cosines. The Indian work was taken over by the Arabs and by them transmitted to Europe. Ibn Jābir ibn Sinān al-Battānī (+858 to +929),[f] for example, used sines regularly with a clear realisation of their superiority over the Greek chords, and introduced concepts from which the tangent and cotangent are derived. Abū'l-Wafā' al-Būzjānī (+940 to +998)[g] introduced the secant and cosecant. The first work in which plane trigonometry appears as a science by itself

[a] This term is here used of course not in its modern technical sense.
[b] *San chio mien shu.*[1]
[c] *San chio hu mien shu.*[2]
[d] Tr. Krause.
[e] Sarton (1), vol. 1, p. 387.
[f] Sarton (1), vol. 1, p. 602; Mieli (1), p. 83.
[g] Sarton (1), vol. 1, p. 666; Mieli (1), p. 108.

[1] 三角面術 [2] 三角弧面術

was the *Kitāb al-Shakl al-Qaṭṭā'* (Book of Sharp Sector Figures) by Naṣīr al-Dīn al-Ṭūsī (+1201 to +1274).[a] This was in Mongol Persia; there was nothing equivalent to it in Europe until Regiomontanus' *De Triangulis* of +1533. The transmission of Arabic trigonometry to Europe has been described in many histories of mathematics.[b]

In the meantime there had been some progress in China;[c] for instance, the work of Shen Kua in the +11th century on chords and arcs, already mentioned. With their terminology for the sides of the right-angled triangle, the Chinese seem not to have felt the need for special names for angle functions. In the practical use of plane trigonometry, *kou/hsüan* was sufficient for the sine, *kou/ku* for the tangent, and *hsüan/ku* for the secant. The *chhung chha*[1] method, already noticed, was a kind of empirical substitute for trigonometric functions.

By the time of Naṣīr al-Dīn al-Ṭusī, however, the Chinese were becoming anxious to improve their calendrical and astronomical calculations, as the work of Kuo Shou-Ching[2] shows. His spherical 'trigonometry'[d] has been carefully studied by Gauchet (7). Most unfortunately none of Kuo's own writings has survived,[e] but his methods can be reconstructed from the books mentioned above.[f] Fig. 74 shows the most important construction with which he dealt. It was a quadrangular spherical pyramid, the basal quadrilateral of which consisted of one equatorial and one ecliptic arc, together with two meridian arcs, one of which passed through the summer solstice point. In the accompanying figure, *AB* is the equator and *CD* the ecliptic; *CA* and *DB* are the two meridian arcs; *O* is the centre of the celestial sphere, and *D* the summer solstice point. *CMNK* was the rectangle which Kuo formed for his calculations. For the line *AP* would evidently be the sine of the arc *AB*, and the line *DR* would be the sine of the arc *DB*,[g] and these would be parallel to *KN* and *MN* respectively. By such methods he was able to obtain the *tu lü*[3] (degrees of equator corresponding to degrees of ecliptic), the *chi chha*[4] (values of chords for given ecliptic arcs) and the *chha lü*[5] (difference between chords of arcs differing by 1°).[h] The whole method was described as 'the consideration of right-angled triangles, chords and sagittae, squares and rectangles, all contained, inclined or straight, within circles'.[i] Fig. 75 shows the kind of figure which must have been used by Kuo Shou-Ching in his calculations; it is taken from the *Ku Chin Lü Li Khao* (Investigation of the Calendars, New and Old) by Hsing Yün-Lu (+1600).

[a] Sarton (1), vol. 2, p. 1003. *Mafātih al-'Ulūm*, p. 207.
[b] E.g. Smith (1), vol. 2, pp. 609 ff.
[c] Later on we shall mention the probability that an early phase of trigonometry reached China from India in the Sui and Thang, but failed to take root there (pp. 148, 202). The *Khai-Yuan Chan Ching* contains (ch. 104) a table of sines translated into Chinese in +718 (cf. Yabuuchi, 1).
[d] I place the word in inverted commas, since names for the angle functions were not used in it.
[e] Unless, as Gauchet says, they are still sleeping in some forgotten library, as the works of the Sung algebraists did for so long.
[f] In the description of the chief mathematical works, p. 48.
[g] The radius being taken as unity.
[h] Tables of these, taken from the *Yuan Shih*, are included in *TSCC*, *Li fa tien*, chs. 35–7.
[i] *Kou ku hu shih fang yuan hsieh chih so yung*.[6]

[1] 重差 [2] 郭守敬 [3] 度率 [4] 積差 [5] 差率
[6] 句股弧矢方圓斜直所容

To what extent, if any, Kuo was influenced by the Persian astronomers whom he almost certainly knew at the Chinese court is very difficult to say. In Li Nien's opinion[a] he had at his disposal a partial table of sines, but Gauchet doubts it. Gauchet, following Juan Yuan, claims, moreover, that the work of Shen Kua gave Kuo Shou-Ching all he needed for his astronomical determinations and computations, much more exact than anything previously done in China.[b] It may yet turn out, however, that trigonometrical tables and methods used by Kuo Shou-Ching have survived in

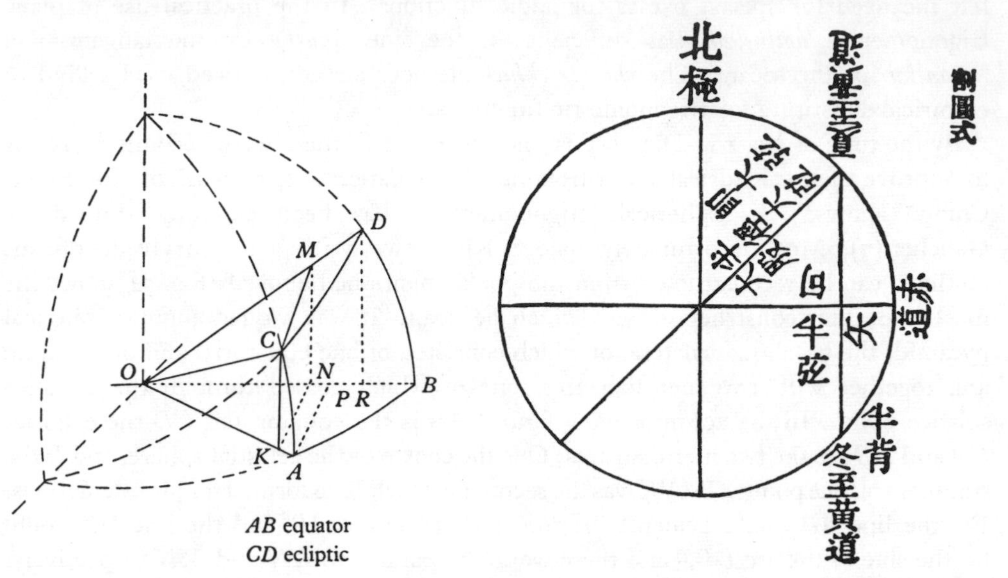

Fig. 74. Diagram to illustrate the spherical trigonometry of Kuo Shou-Ching (+1276).

Fig. 75. Diagram illustrating problems in spherical trigonometry, from Hsing Yün-Lu's *Ku Chin Lü Li Khao* (+1600), ch. 70.

Mongol form if not in Chinese. Such at any rate is the possibility which arises from the interesting recent study by Baranovskaia (1) of a Mongol MS. *Treatise on Coordinates* current at Karakoron *c.* +1712. This contains full tables of functions and calculations in spherical trigonometry for lat. 44° N.

After Kuo's time, there was nothing of importance until the day came when Hsü Kuang-Chhi & Matteo Ricci produced in +1607 the first modern trigonometry in Chinese, the *Tshê Liang Fa I*.[1] In +1631 Hsü added a book, the *Tshê Liang I Thung*,[2] in which he showed that the new terms for angle functions (side-ratios) had been implicit in the old *kou-ku-hsüan* treatments of triangle geometry.[c]

[a] (*4*), vol. 3, p. 381; (*21*), vol. 3, p. 237.

[b] Embodied in the Shou Shih calendar of +1281. The remark of Goodrich (1), p. 177, that Kuo 'discovered spherical trigonometry', is too strong, in view of the much greater contributions of the Indians and Arabs, to say nothing of Menelaus. Trigonometry was not one of the important Chinese gifts to the world. But Gauchet's view is borne out by the recent study of Chhien Pao-Tsung (7).

[c] On the introduction of trigonometry into China and its later development there cf. Li Nien (*4*), vol. 3, p. 323; (*21*), vol. 3, p. 191.

[1] 測量法義 [2] 測量異同

(9) PROBLEMS AND PUZZLES

For want of any better place, a word may here be said about other mathematical problems, and puzzles. In Europe there has long been a tendency to dub many of these 'Chinese', but it is not at all clear how many of them really did have an East Asian origin. Perhaps Europeans were inclined to ascribe to puzzles the name of what was, to them, a puzzling civilisation. This question invites a special investigation in the byways of the history of mathematics, which would probably not be unrewarding; it could take the book of Rouse Ball (2) as a starting point, and go back through Montucla and Ozanam. Many of these puzzles involve several branches of mathematics, and relate to material objects of various kinds.

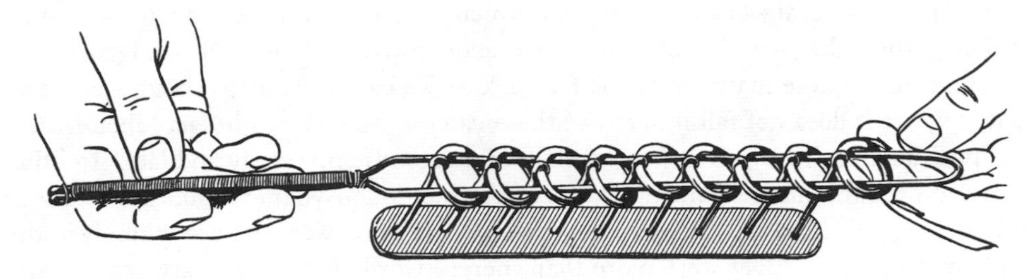

Fig. 76. The Puzzle of the Linked Rings (drawing of an example purchased by Mr Brian Harland at Lanchow).

For example, the topological 'Puzzle of the Chinese Rings'[a] (which may have been derived from the abacus) is first found in Cardan (+1550)[b] and was treated mathematically at some length by John Wallis (+1685).[c] Gros gave it its most elegant mathematical treatment in the 19th century using binary arithmetic notation.[d] The puzzle was commonly known in China at the beginning of this century as the 'Ring of Linked Rings' (*lien huan chhüan*[1]) but its origin is quite obscure[e] (see Fig. 76). Another, geometrical, puzzle, the rearrangements of a set of wooden pieces (a square, a rhombus and five triangles of different sizes),[f] said to be 'one of the oldest amusements of the East', known to the Chinese as the 'Seven Subtle Shapes' (*chhi chhiao thu*[2]) and to Europeans as the 'tangrams', is related to geometrical dissections, static games, anallagmatic tessellation, etc.[g] It is also related to the great wealth of geo-

[a] A number of rings hung on a bar in such a manner that the ring at one end can be taken off or put on the bar at pleasure, but any other ring can be taken off or put on only when neighbouring rings are in certain positions. Rouse Ball (2), p. 305; Ahrens (1).

[b] *De Subtilitate*, bk. XV, 2 (Sponius edn., III, 587).

[c] *De Algebra Tractatus* (*Opera*, II, p. 472).

[d] Cf. above, Sect. 13*g*, in Vol. 2, p. 340.

[e] Cf. Vol. 2, p. 191 above; the paradoxes of the Ming Chia.

[f] Rouse Ball (2), p. 113. Cf. what was said above (p. 96) concerning the methods of the San Kuo commentators.

[g] The Cambridge University Library possesses a number of small books on this subject, some containing prefaces by Sang Hsia-Kho, and dating from the early years of the 19th century.

1 聯環圈　　2 七巧圖

metrical forms employed by Chinese builders through the centuries in the lattice-work of windows.[a] Another widespread art, more related to topology like the Linked Rings, was that of the folding of paper (*chê chih shu*[1]), mentioned in a famous poem of Tu Fu.[b]

(*i*) ALGEBRA

As has been well said, the early history of algebra cannot be discussed without deciding what is meant by the term. If we thus designate the art which allows us to solve such an equation as $ax^2+bx+c=0$ expressed in these symbols, it is a 16th-century development. If we allow other less convenient symbols, it goes back to the +3rd century at least; if purely geometrical solutions are allowed, it begins in the −3rd century; and if we are to class as algebra any problem which would now be solved by algebraic methods, then the −2nd millennium was acquainted with it.[c] Now algebra was dominant in Chinese mathematics as far back as we can trace it (to about the −2nd century), yet it does not fall into any of these categories.[d] It was in fact 'rhetorical'[e] and positional, using symbols (as generally understood) only rarely and late. In other words, it brought into play an abundance of abstract monosyllabic technical ideograms indicating generalised quantities and operations.[f] If these were not yet symbols in the mathematical sense, they were more than merely words in the ordinary sense. And then in the course of the work the counting-board with its numbers was laid out in such a way that certain positions were occupied by specific kinds of quantities (unknowns, powers, etc.). Thus a permanent filing system of mathematical patterns was established.[g] But since the types of equation always retained their connection with concrete problems, no general theory of equations developed. However, the tendency to think in terms of patterns finally evolved from the counting-board a positional notation so complete (as far as it went) that it rendered unnecessary most of our fundamental symbols. Unfortunately, though the achievement was magnificent, it led to a position from which no further advance was possible.

Just as the Chinese from the earliest times mastered their geometrical problems by the use of algebra, the Greeks solved many comparatively difficult algebraic problems in a purely geometrical way. Thus Euclid solved the equivalent of $x^2+ax=b^2$, sub-

[a] See the collection of Dye (1). These lattice designs have aroused the interest of mathematicians such as Hermann Weyl (1), who has found in them examples of all the seventeen possible essentially different kinds of symmetry in two-dimensional frameworks (pp. 103 ff.).

[b] Vacca (7) has been the only Western mathematician to take notice of the scientific interest of this. Since it led to such intriguing questions as Peano-Schwarz surfaces, the theory of knots, and regular polygons considered as the two-dimensional forms of ideal cravats, modern topologists might well find interesting an excursion into the origin and development of the art in China.

[c] So Smith (1), vol. 2, p. 378. Apart from the standard histories of mathematics, there are valuable accounts of the history of algebra by Zeuthen (1) and Thureau-Dangin (2).

[d] Cf. Wang Ling (2), vol. 1, pp. 286 ff.

[e] This is the term of Nesselmann (1842) for algebra fully written out in words.

[f] For a good example see Wang & Needham (1), p. 377.

[g] On the philosophical relations of Chinese algebra cf. Vol. 2, pp. 292, 336.

[1] 摺紙術

stantially by completing the geometric square and neglecting negative roots.[a] But Greek algebra was late in developing, and it was not until the *Arithmetica* of Diophantus of Alexandria, who flourished in the quarter-century +250 to +275, which makes him the close contemporary of Liu Hui, that Western algebra acquired some sort of notation and an independent existence. The suggestion has been made[b] that Diophantus was of Sarmatian origin, and that his algebra derived from Iranian and Chinese influences. But recent investigations by Neugebauer and others[c] have demonstrated that the algebra of the Babylonians was more advanced than had been realised, including problems involving cubic and biquadratic equations. Even where the foundation of problems in cuneiform texts is apparently geometric, the essence is strongly algebraic, for operations are often performed which do not correspond to any possible geometrical interpretation.[d] In view of its great age, one cannot help wondering whether this Babylonian algebra could have been transmitted in seminal forms to lay the foundations for Indian and Chinese algebra on the one hand,[e] and for the Hellenistic developments on the other.[f]

During the decay of Western science in the early Middle Ages the algebra of the Diophantine period was forgotten, and when the great Arab scientific movement took place Arabic algebra was certainly indebted to that of India, and perhaps to a lesser extent to China. Its best known name is Muḥammad ibn Mūsa al-Khwārizmī (*fl.* +813 to +850),[g] from whose book, the *Ḥisāb al-Jabr w'al-Muqābalah*,[h] our present word, algebra, is derived. The two words of the title mean 'restoration'[i] and 'confrontation'. Thus, given

$$bx+2q=x^2+bx-q,$$

then by the process of *al-jabr*

$$bx+2q+q=x^2+bx,$$

and by that of *al-muqābalah*

$$3q=x^2.$$

Thus the former has the idea of the transposition of a negative quantity, and the latter the cancellation of positive quantities with the simplification of each side of the equation. In Chinese mathematics there were no technical words exactly corresponding to these, because the algebraic operations were carried on entirely without

[a] Bk. II, prop. 2. A century ago it was customary to append to the *Elements* notes giving the algebraic formulations of each proposition, as in the edition of Potts.

[b] E.g. by Mazaheri (2).

[c] References in Archibald (1).

[d] This is in contrast to Egyptian mathematics, which was superior in geometry (Gandz, 4).

[e] We have already seen an example, in relation to astrology, where there seems to be some detectable connection between Babylonian and Chinese systems (Vol. 2, p. 353).

[f] We have also seen (Sect. 14*a*) other results of the incorporation of Babylonia into the Hellenistic world.

[g] Sarton (1), vol. 1, p. 563. Besides this most compact and reliable source I should like to cite also a short but attractive article on him by Dumont.

[h] Translation into English by Rosen (1), and (from the Latin of Robert of Chester) Karpinski (1). Cf. Winter & Arafat (1).

[i] Hence the strange fact that the verb *jabara* in Moorish Spain gave rise to the Spanish word *algebrista*, meaning a bone-setter.

the use of the = sign, and the terms were arranged in tabulated columns.[a] Nevertheless, the casting out of bx on both sides of the equation might be said to correspond to the 'subtraction of terms of the same sign' (*thung ming hsiang chhu*[1]) mentioned in the *Chiu Chang*. Similarly, the transposition of $-q$ to $+q$ would be a case of 'addition of terms of different signs' (*i ming hsiang i*[2]).[b] Simplification by addition of terms was called by Li Yeh *ping ju*,[3] and cancellation *hsiang hsiao*.[4]

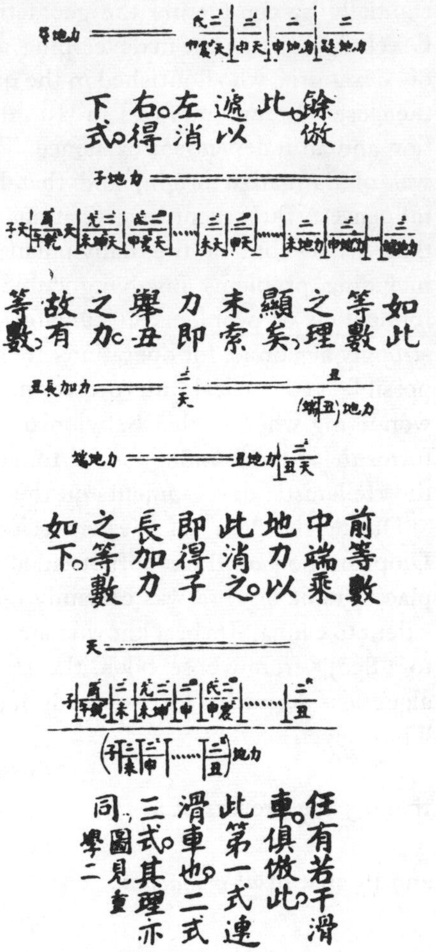

It is often forgotten that the symbolism of algebra in Europe was extremely slow in developing.[c] The words plus and minus were first used in connection with the 'Rule of False Position',[d] the latter in the +13th and the former not till the +15th century. The modern signs for + and − appeared first in an arithmetic of +1489. The multiplication symbol × was not developed until about +1600, and the division ÷ sign not until later in the 17th century. François Viète, the great French mathematician, working in the last three decades of the +16th century, was largely responsible for introducing the representation of numbers by letters.[e] Not until the Ming dynasty, therefore, had the European algebraic symbolism established itself. Now this process occurred just at the time when the old Chinese algebraic methods were at a low ebb in China; and when the Jesuits introduced European algebra there, they were not bringing something which had had a long tradition behind it, but something which in its techniques at least, was relatively new. Perhaps the most important of all symbols, that which made equations possible, was the sign = for equality. Various marks for this had been used in Babylonia and Egypt, but the first generally recognised symbols were those of Diophantine notation, *esti* (ἐστί) and *isos* (ἴσος), shortened to *is* (ἴσ) and *i*ˢ (ισ). In the Middle Ages there was

Fig. 77. Long 16th-century equality signs in a Chinese book on physics (from the *Chung Hsi Suan-Hsüeh Ta Chhêng* by Yeh Yao-Yuan *et al.*).

[a] Before the +13th century, all Chinese 'equations' ended tacitly with '$=+n$', n being the constant term. Then, in the Sung, Chhin Chiu-Shao introduced a practice equivalent to having '=0', the constant term joining the rest of the terms as $-n$.

[b] Ch. 8, p. 4*b*. These phrases actually refer to the treatment of simultaneous equations.

[c] There is a good account of it in Smith (1), vol. 2, pp. 395 ff.

[d] See below, p. 117.

[e] Smith (1), vol. 1, p. 311.

1 同名相除 2 異名相益 3 併入 4 相消

great confusion in the signs used for equality, and the modern = is not seen until Recorde's *Whetstone of Witte* (+1557); he clearly stated that he chose two parallel lines of equal length since nothing could be more equal than they. Yet it was not universally adopted until the 18th century. In Recorde's book, and other works of the period, the lines of the = sign were often prolonged to considerable lengths in the setting up of the calculation, and this practice persisted in China until a very late date (Fig. 77 shows a page of calculations from a work on physics of 1889). It is a matter for reflection how far Chinese algebra was inhibited from developments of post-Renaissance type by its failure to produce a sign which would permit the setting up of equations in modern form. Nor did Chinese algebra have any sign for exponents and powers; the value of a quantity, whether x^2 or x^3, depending entirely on its place in the 'matrix' tables which the Chinese used. Our present integral exponents also appeared very late, not until Descartes (+1637), many other forms, such as xxx for x^3, having previously been in use.

(1) Simultaneous Linear Equations

Simultaneous linear equations are much in evidence in the Han book, *Chiu Chang Suan Shu* (about −1st century), though the Greeks, even Diophantus, were not as successful with them as the Babylonians earlier. The counting-rods, in Chinese style, were placed in different squares in a table so as to represent coefficients of the different unknowns, and the constant terms. The *Chiu Chang* (chapter 8) has numerous problems requiring the solution of equations of the type

$$ax+by=c, \quad a'x+b'y=c'.$$

It was a question of arranging the coefficients, multiplying crosswise, and adding or subtracting. Such equations as:

$$\begin{aligned} x+2y+3z&=26\\ 2x+3y+\ z&=34\\ 3x+2y+\ z&=39 \end{aligned}$$

1	2	3	*shang ho ping shu*[1]
2	3	2	*chung ho ping shu*[2]
3	1	1	*hsia ho ping shu*[3]
26	34	39	

were arranged in tabular form as shown, the top line representing the x terms (no. of bushels of first-quality cereal), the second line representing the y terms (no. of bushels of second-quality cereal), the third line representing the z terms (no. of bushels of third-quality cereal), and the fourth line giving the constant terms. Generalising: in the case of the two equations given above, the first is multiplied by a' and the second by a, giving, after subtraction,

$$y=\frac{ca'-c'a}{ba'-b'a}.$$

[1] 上禾秉數 [2] 中禾秉數 [3] 下禾秉數

This was called the *chih chhu*[1] method. It runs through many later books.[a] In certain problems Liu Hui, in his *Chiu Chang* commentary, was able to avoid cross-multiplication of the coefficients of the first unknown term by rearranging and combining the equations so as to get rid of x. This would of course involve changes of sign, presumably indicated on the counting-board table by changing the colour of the counting-rods from black to red or vice versa.[b]

In the Han and the San Kuo periods the rules for solving simultaneous linear equations were never divorced from specific problems; it remained for Yang Hui in the Sung[c] to state them in a generalised way. He spoke of the consecutive equations as *chia hang*,[2] *i hang*,[3] etc., and of the coefficient of the first unknown as *shou wei*.[4] Equations with more than two unknowns were called *hang fan chê*.[5]

A story[d] in the *Thang Chhüeh Shih*[6] shows this branch of algebra being used about +855 as a test for minor functionaries.

Yang Sun[7] (a high official)[e] was famous for selecting and promoting members of the civil service not through private influence or personal preference but by taking general opinions on their merits and weighing all criticisms which might be brought forward. He applied this principle even to petty clerks and menial officers (*hsi hsü chien tsu*[8]).[f]

Once there were two clerks who held the same rank and had equal lengths of government service. They had even acquired the same commendations, and the criticisms in their personal dossiers were identical. The responsible official of middle rank was quite baffled by the problem of their promotion, and appealed to Yang Sun, who was his superior. Yang Sun thought the matter over, and then said, 'One of the best merits of minor clerks is to be quick at computations. Let both the candidates now listen to my question. Whoever first gets the right answer will obtain the advancement. The problem is this: Someone, when walking in the woods, overheard a number of robbers discussing how best to share the rolls of cloth which they had stolen. They said that if each had 6 rolls there would be 5 rolls left, but if each had 7 rolls they would be 8 rolls short. How many robbers were there, and what was the total number of the rolls of cloth?' This problem was taken down by another minor clerk, and Yang Sun then asked the two candidates to reckon it out with counting-rods on the stone steps of the hall. After a short time one of them in fact got the right answer. He was duly given the better position, and the officials dispersed having nothing to complain of or criticise in the decision.[g]

[a] E.g. the *Chang Chhiu-Chien Suan Ching*, of the +5th century, and the *Chiu Chang Hsi Tshao*[9] of Liu Hsiao-Sun[10] of the +7th.

[b] In two problems he first got rid of the constant terms and then used the substitution method to eliminate the unknowns.

[c] We have already seen his desire for precise geometrical proof.

[d] Which seems to have escaped the eagle eyes even of Mikami and Chhien Pao-Tsung. Li Nien (*1*), p. 85, mentions Yang Sun's question, but we noticed this only after finding and translating the passage.

[e] Biography in *Hsin Thang Shu*, ch. 174; *Chiu Thang Shu*, ch. 176.

[f] Here is more than a hint at the socially low valuation of mathematicians. Cf. pp. 152 ff. below.

[g] Ch. 2, p. 24*b*, tr. auct.

[1] 直除 [2] 甲行 [3] 乙行 [4] 首位 [5] 行繁者 [6] 唐闕史
[7] 楊損 [8] 細胥賤卒 [9] 九章細草 [10] 劉孝孫

(2) Matrices and Determinants

The Chinese method of representing the coefficients of the unknowns of simultaneous linear equations by means of rods on a counting-board led naturally to the discovery of simple methods of elimination. The arrangement of the rods was precisely that of the numbers in a matrix. Chinese mathematics therefore developed at an early date the idea of subtracting columns and rows as in the simplification of a determinant. Yet it was not until it had been taken over by Japanese scholars that the idea of determinants assumed independent form. Seki Kōwa,[a] the greatest of the 17th-century Japanese mathematicians, wrote in +1683 a work called *Kai Fukudai no Ho*,[b] in which the concept of determinants and their expansion was clearly described.[c] Since his work certainly ante-dated that year, and since the first European statement of determinants[d] was that of Leibniz in +1693, the credit for the advance must be ascribed to Seki Kōwa, whose thought arose more from Chinese than from Western influences.[e] The only surprising thing about the discovery is, as Smith says, that it had not been stated much earlier, for example by the Sung algebraists.

(3) The Rule of False Position

Simple linear equations were much more troublesome in ancient times than after the elaboration of a good symbolism. It seems impossible, says D. E. Smith,[f] that the world should ever have been troubled by an equation like $ax+b=0$, but in order to solve it the old mathematicians had recourse to a somewhat cumbrous method known in later Europe as the Rule of False Position. The chief form which this took was that called 'Double False'. In relation to the above equation it may be explained as follows: Let g_1 and g_2 be two guesses as to the value of x, and let f_1 and f_2 be the failures, that is, the values of ag_1+b and ag_2+b, which would be equal to 0 if the guesses were right. Then

$$ag_1+b=f_1 \qquad (1)$$

and

$$ag_2+b=f_2, \qquad (2)$$

whence

$$a(g_1-g_2)=f_1-f_2. \qquad (3)$$

From (1)

$$ag_1g_2+bg_2=f_1g_2,$$

and from (2)

$$ag_1g_2+bg_1=f_2g_1,$$

[a] Seki Takakusu[1] is a better reading.
[b] 'Methods of solving Problems with Determinants'.
[c] Smith (1), vol. 2, p. 475; Smith & Mikami (1), p. 124; Hayashi (1); Katō (*1*).
[d] Muir (1); Lecat (1).
[e] Dr Kobori Akira (private communication) strongly concurs. Smith & Mikami (1), pp. 132ff., made an interesting study of the possible channels of occidental influence, which remain tantalisingly obscure.
[f] (1), vol. 2, p. 437.

[1] 關孝和

whence

$$b(g_2-g_1)=f_1g_2-f_2g_1. \quad (4)$$

Dividing (4) by (3)

$$-\frac{b}{a}=\frac{f_1g_2-f_2g_1}{f_1-f_2}.$$

But, since

$$-\frac{b}{a}=x,$$

x may be found.

This 'Regula Falsae Positionis' was certainly transmitted to Europe by the Arabic mathematicians. It occurs in the works of al-Khwārizmī (*c.* +825), Qusṭā ibn Lūqa al-Ba'albakī (*d.* 922)[a] and several later writers. The Arabic name for the rule was *ḥisāb al-khaṭā'ain*, hence all kinds of forms such as *elchataym* (Fibonacci, +13th), *el cataym* (Pacioli, +15th), *Regola Helcataym* (Tartaglia, +16th), *Regole del Cattaino* (Pagnani, +16th), etc.[b] It may have had a Chinese origin, for, as Chhien Pao-Tsung,[c] followed by Chang Yin-Lin (*3*), pointed out, the rule is nothing else than the Chinese method *Ying pu tsu*[1] (Too Much and Not Enough) which is actually the title of the seventh chapter of the *Chiu Chang Suan Shu* of the −1st century.[d] Liu Hui called it the *thiao nu*[2] rule, both these words being taken from lunar movements, the first meaning the latest appearance of the waning moon and the second being the earliest appearance of the waxing moon. In the explanation of the solutions of the relevant problems,[e] the words *chia ling*,[3] 'let us now assume',[f] are used for the adoption of the first proposal g_1 (whether excess or deficiency), and the words *ling chih*[4] are used for the adoption of the second proposal g_2 (whether excess or deficiency).[g] Something like

[a] Sarton (1), vol. 1, p. 602; Hitti (1), p. 315; a Christian Arab of the Lebanon.

[b] The evocation by these names of the word Cathay (Khiṭāi), and the fact that there are certain examples of Arabic nomenclature originating just thus (see Sect. 30 on gunpowder), has been a seduction to some (cf. Chhien Pao-Tsung (*3*); Chang Yin-Lin (*3*), p. 306). Yet the similarity seems to be but superficial, for *Khaṭā'ain* means only 'double false' (cf. the description in *Mafātih al-'Ulum*, p. 201). We are much indebted to Mr D. M. Dunlop for help on this point. Besides, if there was a transmission it can hardly have occurred so late as the time of the Western Liao Kingdom (+1124 to +1211), as Chhien Pao-Tsung suggested, for al-Khwārizmī in the +9th century knew and used the method. Presently (p. 457 below), we shall find another Chinese historian pointing to the Western Liao (Qarā-Khiṭāi) as a means of transmission of knowledge from east to west. This is well worth bearing in mind, though it seems not to apply in the present case.

[c] (*3*), (*1*), p. 36.

[d] There has been confusion about this because Mikami (1), p. 16, only examined the first eight problems in the chapter. These all contain an excess and a deficiency *given in the problem*, and can be solved by ordinary simultaneous linear equations, as he explained. Smith (1), vol. 2, p. 433, simply copied from him. But problems 9–20 are all set in such a way that the excess and deficiency have to be *assumed*, as in the guesses of the Rule of False Position.

[e] The seventh chapter deals mostly with linear equations, but it also has three problems involving quadratic or higher powers. These the author treated in the same way, perhaps not realising that the solutions would only be approximate.

[f] This is an idiom, but as the sense of the character *chia* alone is actually 'false', foreigners translating word for word could easily have obtained that implication.

[g] Elsewhere in the *Chiu Chang* (chs. 3 and 5) the method is used to simplify problems of arithmetical and geometrical progression so that they became simple proportions.

1 盈不足 2 朓朒 3 假令 4 令之

the Rule of False Position seems to have been known in ancient Egypt[a] and in India.[b] The philosophical aspects of the idea of excess and deficiency are important,[c] and evoke Greek biology as well as all ancient mathematics.[d]

(4) Indeterminate Analysis and Indeterminate Equations

In the survey of Chinese mathematical literature mention was made several times of indeterminate analysis. When n equations are given, involving more than n variables, there may be an infinite number of sets of solutions. Of course, in some cases, the nature of the problem may be such that only those solutions are wanted which have positive integral values. Indeterminate analysis was always a marked mathematical interest of the Chinese, at least from the +4th century, when the *Sun Tzu Suan Ching* gave the following problem:[e]

> We have a number of things, but we do not know exactly how many. If we count them by threes we have two left over. If we count them by fives we have three left over. If we count them by sevens we have two left over. How many things are there?

Sun Tzu determined the 'use numbers' (*yung shu*[1]) 70, 21 and 15, multiples of 5×7, 3×7 and 3×5, and having the remainder 1 when divided by 3, 5 and 7 respectively. The sum $2 \times 70 + 3 \times 21 + 2 \times 15 = 233$ is one answer, and by casting out a multiple of $3 \times 5 \times 7$ as many times as possible the least answer, 23 is obtained.[f] This was the only one that Sun Tzu gave. Written in modern form, the expression would be:

$$N \equiv 2 \pmod 3, \quad \equiv 3 \pmod 5, \quad \equiv 2 \pmod 7.$$

Indeterminate analysis has attracted much attention from historians of mathematics; Wylie (4) devoted several pages to it, and the best modern discussions are those of Li Nien[g] and Chhien Pao-Tsung.[h]

In the course of time indeterminate analysis acquired the name *Ta yen shu*,[2] derived from an obscure statement in the *I Ching*[i] that the 'Great Extension Number' is 50.[j] In the first decade of the +8th century I-Hsing, in his *Ta Yen Li Shu*[3] (Book

[a] Cajori (2), p. 13. See p. 147 below.

[b] Datta & Singh (1), vol. 2, p. 37. This is the Bakhshālī MS., now placed not earlier than the +10th century. There are striking similarities between the actual problems in *Chiu Chang Suan Shu*, ch. 7 (esp. nos. 15, 16 and 18) and those in Brahmagupta (+628); see Colebrooke (1), p. 289; and in Mahāvīrā (+9th century), see Datta & Singh (1), vol. 1, p. 205.

[c] Cf. Sects. 13, 16, 18, Vol. 2, pp. 270, 286, 463, 566 above, in connection with the Neo-Confucian school.

[d] Cf. d'Arcy Thompson (1).

[e] Ch. 3, p. 10*b*.

[f] Cf. Dickson (1), vol. 2, p. 57.

[g] (*4*), vol. 1, p. 61; (*21*), vol. 1, p. 122.

[h] (*2*), pp. 45 ff.

[i] Great Appendix, 1, ch. 9. See above, Vol. 2, pp. 305 ff.

[j] The reason for the adoption of this technical term is clear enough. In the classical method of consulting the oracle, which the *I Ching* (Book of Changes) describes (cf. R. Wilhelm (2), vol. 1, pp. 236, 280; Baynes tr., pp. 334, 392), one of the fifty stalks or rods is set aside before the forty-nine are divided into two random heaps symbolising the Yin and the Yang. It was very natural, therefore,

1 用數　　2 大衍術　　3 大衍曆書

of the Ta Yen Calendar), must have made use of indeterminate analysis.[a] The details of his calculations are not given in the text as we now have it,[b] only the answer, 96,961,740 years. This, the period which he sought, was the time which had elapsed between the beginning of the Grand Cycle (*Shang Yuan*[1])[c] and the 12th year of the Khai-Yuan reign-period (+724). His problem, expressed in modern form, was:

$$1,110,343y \equiv 44,820 \pmod{60 \times 3040}, \quad \equiv 49,107 \pmod{89,773}.$$ [d]

Five centuries later Chhin Chiu-Shao embodied in his *Shu Shu Chiu Chang* a full explanation of the subject,[e] and this was the source which enabled Wylie to understand it. In the above Sun Tzu problem, the numbers 3, 5 and 7 were called *ting mu*[2] (fixed 'denominators'), their least common multiple, 105, was the *yen mu*[3] (multiple 'denominator'), the quotients after its division, 35, 21, 15, were the *yen shu*[4] (multiple numbers), and the figures obtained by analysis, 2, 1, 1, were the *chhêng lü*[5] (multiplying terms). The finding of these was the essential process.

The second problem of Chhin's first chapter deals with calendrical calculations, but those in his third chapter (especially the third problem) approach most nearly to the methods of I-Hsing, and though the terminology is different, give the clues whereby we can understand them.[f] The first problem of chapter 3 concerns the calculation of the *chhi*[6] (time from the winter solstice to the end of the sexagenary day-cycle) in any year; the second the *jun*[7] (corresponding time from the eleventh month). The fourth of ch. 3 and the first of ch. 4 deal with inequalities of annual and diurnal revolutions and planetary motions. The other problems of the first and second chapters concern public works, such as embankments, treasuries, granaries, the movement of troops, etc.[g]

that mathematicians, seeking for remainders of one by continued divisions, should have remembered this. Other 'remainders' follow, for the divining process continues by throwing out the residual rods when each successive heap is counted through by fours. In China, therefore, indeterminate analysis was connected with, if not actually derived from, an ancient method of divination using yarrow stalks. An association of these with the counting-rods at some time may perhaps be presumed. The classical exposition of the *I Ching* process in mathematical form was given by Chhin Chiu-Shao (*Shu Shu Chiu Chang*, ch. 1, pp. 1*a* ff.). Wylie (4) gives a translation of this problem and its explanation; cf. also Chao Jan-Ning (*1*).

[a] So Chang Tun-Jen (*2*), one of the deepest students of this subject.

[b] *Chiu Thang Shu*, ch. 34; *Hsin Thang Shu*, chs. 28 A and B.

[c] I.e. the last time when the winter solstice occurred exactly at midnight on a first day of an eleventh month which was also a *chia-tzu* day initiating a 60-day cycle. Cf. Chatley (16) and p. 408 below.

[d] The first of these figures (*tshê shih*[8]) is the numerator of the improper fraction representing the number of days in the tropic year, the third (*hsiao shu*[9]) is the usual cycle, the sixth (*shê fa*[10]) is the numerator of the improper fraction representing the number of days in the synodic month, and the fourth (*thung fa*[11]) is the denominator of both these improper fractions. The second figure is the numerator of the improper fraction representing the number of days between the winter solstice of +724 and midnight of the previous *chia-tzu* day; the fifth (*kuei yü chih kua*[12]) is the numerator of the fraction summing this with another improper fraction representing the number of days between that solstice and the first day of the eleventh month. Though the expression as given above is formally correct, it is very unlikely that I-Hsing worked on these large numbers without any simplification. In the analogous problem solved by Chhin Chiu-Shao later they were reduced to smaller numbers by a series of intermediate steps. But the method of indeterminate analysis was presumably the same.

[e] Chs. 1 and 2, on which see also Mikami (1), pp. 65 ff.

[f] See the special study of Wang Ling (5).

[g] Van Hée (3) gives a translation of the third problem of ch. 1 and the third of ch. 2, with their explanations.

[1] 上元 [2] 定母 [3] 衍母 [4] 衍數 [5] 乘率 [6] 氣
[7] 閏 [8] 策實 [9] 爻數 [10] 揲法 [11] 通法 [12] 歸餘之卦

Subsequent problems, concerned with rainfall and snowfall, do not come in the Ta Yen category.[a]

Wylie's account became known in Europe through Biernatzki's translation, but owing to certain inaccuracies in this, Cantor was led to doubt the validity of the Chinese rule. The rule, however, was defended by Matthiessen (1), who pointed out its identity with Gauss' formula. If $m = m_1 m_2 \ldots m_k$, where $m_1, m_2, \ldots, m_k$ are relatively prime in pairs, and if

$$a_i \equiv 0 \left(\mathrm{mod}\ \frac{m}{m_i}\right), \quad a_i \equiv 1 \ (\mathrm{mod}\ m_i) \quad (i = 1, 2, \ldots, k),$$

then $x = a_1 r_1 + a_2 r_2 + \ldots + a_k r_k$ is a solution of

$$x \equiv r_1 \ (\mathrm{mod}\ m_1) \quad (i = 1, 2, \ldots, k).$$

The contributions of Chhin Chiu-Shao dealt also with cases where the moduli m_i are not relatively prime.[b] The method is as follows. Choose positive integers $\mu_1, \mu_2, \ldots, \mu_k$ which are relatively prime in pairs such that each μ_i divides the corresponding m_i and that further the L.C.M. of $\mu_1, \mu_2, \ldots, \mu_k$ equals that of $m_1, m_2, \ldots, m_k$. Then every solution x of

$$(\mu): \quad x \equiv r_1 \ (\mathrm{mod}\ \mu_1), \quad \equiv r_2 \ (\mathrm{mod}\ \mu_2), \ \ldots, \quad \equiv r_k \ (\mathrm{mod}\ \mu_k)$$

also satisfies

$$(m): \quad x \equiv r_1 \ (\mathrm{mod}\ m_1), \quad \equiv r_2 \ (\mathrm{mod}\ m_2), \ \ldots, \quad \equiv r_k \ (\mathrm{mod}\ m_k).$$

Hence the system (m) may be solved by applying the earlier rule to (μ). However, this construction is valid only if each difference $r_i - r_j$ is divisible by the G.C.D. of the corresponding moduli m_i and m_j. No such necessary condition is mentioned in the Chinese text, but it is satisfied in the example which Chhin Chiu-Shao gives.

His first problem, in which (as in others) he used the kind of terminology mentioned above, amounted to

$$x \equiv 1 \ (\mathrm{mod}\ 1), \quad \equiv 1 \ (\mathrm{mod}\ 2), \quad \equiv 3 \ (\mathrm{mod}\ 3), \quad \equiv 1 \ (\mathrm{mod}\ 4).$$

Let 1, 1, 3, 4 be factors, relatively prime in pairs, of 1, 2, 3, 4, respectively, such that $1 \times 1 \times 3 \times 4$ (or 12) is the L.C.M. of the latter numbers. Then $k_1 = 1$, $k_2 = 1$, $k_3 = 1$ and $k_4 = 3$, satisfy the congruences $\frac{12}{1} k_1 \equiv 1 \ (\mathrm{mod}\ 1)$, $\frac{12}{1} k_2 \equiv 1 \ (\mathrm{mod}\ 1)$, $\frac{12}{3} k_3 \equiv 1 \ (\mathrm{mod}\ 3)$ and $\frac{12}{4} k_4 \equiv 1 \ (\mathrm{mod}\ 4)$. The smallest positive solution is therefore given by

$$x \equiv 1 \times 1 \times \tfrac{12}{1} + 1 \times 1 \times \tfrac{12}{1} + 3 \times 1 \times \tfrac{12}{3} + 1 \times 3 \times \tfrac{12}{4} - 9 \times 12 \ (\mathrm{mod}\ 12).$$

The commonest form taken by indeterminate problems in Chinese mathematics was that of the 'Hundred Fowls'. This first appears just about +475.[c] Chen Luan of the +6th century, Li Shun-Fêng of the +7th and Hsieh Chha-Wei of the +11th

[a] But they are very interesting in connection with the use of rain gauges at that time, see below, Sect. 21 (p. 471).

[b] Matthiessen (2, 4, 5) was very much confused, attributing Chhin Chiu-Shao's first problem to I-Hsing, and supposing that it was concerned with the numbers of workmen building dykes, instead of with the *I Ching* divination technique. These misunderstandings were faithfully reproduced by Dickson (1), vol. 2, p. 57. We are greatly indebted to Prof. K. Mahler for criticisms which led to the elucidation of this whole subject, and for his active participation therein.

[c] *Chang Chhiu-Chien Suan Ching*, ch. 3, pp. 37*a* ff.

have it thus: 'If a cock is worth 5 cash, a hen 3 cash, and 3 chickens together only 1 cash; how many cocks, hens and chickens, in all 100, may be bought for 100 cash?'[a] Van Hée (3) gives detailed accounts of the passages in which problems of this type are discussed.[b] Chang Chhiu-Chien solved his problem by indeterminate equations (incompletely expressed), but the others found that they could get an answer by simpler methods and did so. Analysis was not applied to this famous problem until the Chhing (Lo Thêng-Fêng, *1*).

Indeterminate analysis finally came to be called *Ta yen chhiu i shu*,[1] the 'searching for unity' referring to the final step in the process.[c] In the Sung and Yuan there were, however, other names—*Kuei Ku suan*[2] (presumably because of some attribution to the legendary philosopher Kuei Ku Tzu), *Ko chhiang suan*[3] ('Behind the arras' computations), or *Chien kuan shu*[4] ('Cutting lengths of tube'). Popular names connected it with military matters, e.g. 'The Prince of Chhin's Secret Method of Counting Soldiers'.[d]

The Chinese *Ta yen* procedure was similar to the *kuṭṭaka* (cuttaca, or pulveriser) method in Indian mathematics,[e] first found in the work of the elder Āryabhaṭa (*b.* +476). This derived its name from the 'pulverising' multiplier, p, such that if n_1, n_2, n_3 are given numbers, then pn_1+n_2 shall be divisible by n_3.

Curiously, the algebra of Diophantus, in so far as it touches this subject, deals with indeterminate quadratic equations almost solely; whereas in China these were not studied. Indeterminate problems were known in Europe as early as the +9th century, and afterwards became common puzzles. Their analysis was known as the 'Regula Coecis' or 'Regula Virginum' or 'Regula Potatorum', since the problems generally concerned the payment for drinks at parties of men and girls.[f] Chhin Chiu-Shao's older contemporary Leonardo Pisano dealt with remainder problems in his *Liber Abaci* of +1202.[g]

Problems of indeterminate analysis were often connected with questions concerning alligation, or the mixture of constituents in, for example, alloys. There is an interesting discussion in Smith[h] on this point at which mathematics and chemistry touch (as they did all too rarely before modern times). The chief Chinese work which concerned itself with this, though the problems were simple enough, was the *Suan Fa Thung Tsung*[i] of +1593. But the *Chiu Chang* had dealt with oil and lacquer mixtures long before.[j]

[a] Exactly the same problem reappears in the works of the Egyptian algebraist Abū Kāmil al-Miṣrī (*c.* +900). Cf. Suter (2); Mieli (1), p. 108.

[b] Mikami (1), p. 32, deals with another example.

[c] In *MCPT*, ch. 18, para. 9 for example. Cf. Hu Tao-Ching (*1*), p. 594.

[d] *Chhin Wang an tien ping*.[5] Or Han Hsin's Method (*Han Hsin tien ping*,[6] the famous Han general), *Suan Fa Thung Tsung*, ch. 5, pp. 21*b*ff.

[e] Datta & Singh (1), vol. 2, pp. 87ff.; Dickson (1), vol. 2, pp. 41ff. The argument of Matthiessen (3) that they were very different does not carry conviction.

[f] Smith (1), vol. 2, p. 586.

[g] Dickson (1), vol. 2, p. 59.

[h] (1), vol. 2, pp. 588ff.

[i] Ch. 2, p. 36*a*, *b* (E. Biot (5), p. 205).

[j] Ch. 7, p. 9*a*.

1 大衍求一術 2 鬼谷算 3 隔牆算 4 翦管術
5 秦王暗點兵 6 韓信點兵

(5) Quadratic Equations and the Method of Finite Differences

Quadratic equations were early used in Chinese mathematics. The *Chiu Chang* has[a] a problem which was solved by finding a positive root of a quadratic equation equivalent to $x^2+ax=b$. Quadratics were sometimes avoided by conversion to linear form and solution of the square root resulting,[b] a process which had already been employed by the Old Babylonian mathematicians.[c] Numerical higher equations of special kinds also occur, as far as the third degree, in this Han book.[d] Another equation (as always, written out in characters), found in the +5th-century *Chang Chhiu-Chien Suan Ching*,[e] is $x^2+cx=c^2-36\frac{a}{b}$. By the time of the Sung algebraists expressions equivalent to the following[f] had appeared: $x^2-ax=b$. Van Hée has devoted a special paper (1) to these equations of the second degree, basing his exposition, however, largely on the 18th-century mathematician Li Jui,[1] whose principal relevant work was the *Khai Fang Shuo*[2] (Theory of Equations of Higher Degrees). Gauchet (6) has given us a similar study of the *Chiu Chang* equations based on the work of Lo Shih-Lin.[3]

One of the most interesting methods in which quadratic expressions were involved was the process of finding arbitrary constants in formulae for celestial motions. It was the same as that now called the Method of Finite or Divided Differences. How far this 'Chao Chha Fa'[4] goes back is not quite clear, but it was certainly used by Li Shun-Fêng[5] in the preparation of the Lin Tê[6] calendar of +665.[g] Chhien Pao-Tsung[h] notes Sung and Chhing opinions that it was in fact the original *Chui Shu*[7] of Tsu Chhung-Chih[8] (late +5th century), a view which the meaning of the word *chui*, 'threading', might support—but there is no real evidence.[i] The method seems to have been introduced by Liu Chhuo[9] in his Huang Chi[10] calendar of +604. It depends on the equation[j] $Ax+Bx^2=C$, which is still used today, extended to as many higher powers beyond the square as may happen to be wanted. Newton was interested in it, and used it.

[a] Cf. p. 26 above.

[b] *Chiu Chang*, ch. 9, p. 5*b*, for example. The procedure, which involved the choosing of a different unknown, is described in Wang Ling (2), vol. 2, pp. 133*g*ff.

[c] Neugebauer (9), p. 40.

[d] If we may interpret the cube root extraction of a given number as a special case of a third-degree equation.

[e] Ch. 3, p. 9*a*. Cf. ch. 2, p. 21*b*.

[f] Cf. Mikami (1), pp. 87, 109, etc.; Smith (1), vol. 2, p. 448.

[g] My attention was drawn to this subject by a query (July 1949) from Dr Arthur Waley, and I wish to acknowledge with gratitude the help of Sir Ronald Fisher in arriving at an understanding of the work of Li Shun-Fêng. We now have a special study of the interpolation formulae of medieval Chinese mathematicians by Li Nien (1).

[h] (*1*), pp. 57, 63, 147.

[i] See above, p. 35, for the mystery which has surrounded this lost work.

[j] More correctly given by Sarton (1), vol. 1, p. 494, than by Mikami (1), p. 104. Mikami's whole treatment of the subject is unnecessarily difficult to follow, unlike that of Li Nien (*1*), p. 179, but the latter is of course in Chinese. There is some difference in the explanations.

1 李銳 2 開方說 3 羅士琳 4 招差法 5 李淳風
6 麟德 7 綴術 8 祖沖之 9 劉焯 10 皇極

Li Shun-Fêng's objective was to state (and therefore presumably to predict) the irregularities of the sun's angular motion. If he had lived after Descartes, instead of a thousand years before him, he would have plotted them on coordinates as a graph, making a curve in agreement with his equation. The x is an observed quantity—in this case the number of days or more exact time intervals between successive observations of the sun's position. C is also an observed quantity—in this case the number of degrees through which the sun has moved between two observations. The point then is to find A and B, two arbitrary constants. The procedure is very modern; all current natural science is full of equations with arbitrary constants designed to fit empirical curves. The course of the argument is as follows. For successive data gathered:

$$Ax + Bx^2 = C,$$

$$Ax_1 + Bx_1^2 = C_1,$$

$$Ax_2 + Bx_2^2 = C_2 \quad \text{and so on.}$$

Then

$$A(x_2 - x_1) + B(x_2^2 - x_1^2) = C_2 - C_1,$$

$$A + B(x_2 + x_1) = \frac{C_2 - C_1}{x_2 - x_1},$$

$$A + B(x_3 + x_2) = \frac{C_3 - C_2}{x_3 - x_2}.$$

Subtracting, to remove A,

$$B(x_3 - x_1) = \frac{C_3 - C_2}{x_3 - x_2} - \frac{C_2 - C_1}{x_2 - x_1}.$$

B = a numerical answer. By a similar procedure A can also be obtained.[a] The technical terms used in the method can now be understood. The 'floating difference' (*phing chha*[1, 2]) would refer to the empirically observed slight differences seen at each observation, while the 'fixed difference' (*ting chha*[3]) would be the arbitrary constants. In general, the greater the number of observations made, and the higher the powers introduced into the calculation, the more accurate the determination of the constants will be, and the better prediction will be possible.

Li Shun-Fêng's method is contained, neither in his numerous commentaries on older mathematical books, nor in the three or four quasi-astrological books attributed to him, but in the calendrical chapters of the *Sui Shu* (History of the Sui Dynasty),[b] where he discusses the previous work of Liu Chhuo. The compilers of the *Chiu Thang Shu* (History of the Thang Dynasty)[c] and the writer of the *Khai-Yuan Chan Ching* omitted his method 'because the numbers were too large'.[d]

[a] If more than two equations are used, several possible values for A and B are obtained. Presumably mean values were chosen—an early problem in *Ausgleichungsrechnung*. A thorough study of the method of finite differences as used by the medieval Chinese astronomers is much needed.

[b] E.g. ch. 18, p. 3*a*; partly reproduced in *TSCC*, *Li fa tien*, ch. 9, *hui khao* 9, p. 10*b*, and ch. 10.

[c] And the 'New History' (*Hsin Thang Shu*) also.

[d] Chhien Pao-Tsung (*2*), p. 47.

[1] 泙差 [2] 平差 [3] 定差

The greater accuracy obtained by the introduction of the higher power seems to have been realised by Kuo Shou-Ching[1] in his calculations for the Shou Shih[2] calendar of +1281. The third arbitrary constant was termed *li chha*,[3] suitably, since the word *li* stood for the third dimension of solid figures, and hence for the extraction of the cubic root, as in the expression *khai li fang*.[4] Gauchet (7) noted that in the *Ming Shih*[5] (History of the Ming Dynasty), where Kuo's calendrical work is described,[a] as the basis of later calendars, the writer complains (1739) that the Chao Chha method had been for a time no longer understood. In fact it would have been clear enough if the Chinese mathematicians of the 17th century had been able to recognise processes similar to those which the Jesuits had taught them, under the unfamiliar terminology of the +13th century.[b] One may say again that it was primarily due to the decay of the +15th and +16th centuries that the Jesuit contribution of the +17th seemed so strikingly new and advanced.

The Chao Chha method was related to the procedure used by Chu Shih-Chieh[6] in the *Ssu Yuan Yü Chien* of +1303 for summing up certain series. It seems to have been a remarkable anticipation on the part of the Chinese, for it was not until the 17th and 18th centuries in Europe that the method was taken up and thoroughly worked out. The names of François Nicole (+1717) and Brook Taylor (+1716) are particularly associated with it.

(6) Cubic and Higher Equations

Though Kuo Shou-Ching used cubic equations at the end of the +13th century, they had first been considered by the Chinese more than six hundred years earlier, by Wang Hsiao-Thung,[7] in the generation before Li Shun-Fêng. But his work was mostly restricted to numerical cubics,[c] such as those which will be our next subject of discussion. In the West, among Greek and later Arabic mathematicians, cubic equations were closely connected with the knowledge of conic sections,[d] so it is not surprising that they did not develop in China. Real progress was not made until the time of the famous controversy between Cardan and Tartaglia in the +16th century.[e] It was Hudde in +1658 who used Cartesian symbolism and brought the subject to its modern status.

Equations of the fourth to the ninth degree were handled in Sung China if numerical. Chia Hsien began about +1200 with biquadratics of special form. No one had much success anywhere with non-numerical biquadratic equations until the +16th-century Italian algebraists just mentioned, and non-numerical equations of higher degrees had to await the 19th century in Europe for satisfactory treatment.

[a] Ch. 33, p. 6*a*.
[b] Gauchet quotes a striking passage about this from ch. 30 of the *Chhou Jen Chuan* (his p. 155). Cf. Mikami (1), p. 120.
[c] Having positive terms only. Cf. p. 37 above.
[d] Smith (1), vol. 2, p. 456.
[e] Smith (1), vol. 2, p. 460.

[1] 郭守敬 [2] 授時 [3] 立差 [4] 開立方 [5] 明史
[6] 朱世傑 [7] 王孝通

(7) Numerical Higher Equations

The solution of numerical higher equations for approximate values of roots begins, as far as we know, in China. It has been called the most characteristic Chinese mathematical contribution. That it was well developed in the work of the Sung algebraists has long been known, but it is possible to show[a] that if the text of the Han *Chiu Chang* is very carefully followed, the essentials of the method are already there at a time which may be dated as of the −1st century. Mikami[b] was the first to see that even in its earliest form it was similar to a procedure of the early 19th century. In 1802 an Italian scientific society offered a gold medal for the improvement of the solution of numerical higher equations; it was won by Paolo Ruffini. Quite independently Horner in 1819 developed the same method, though neither of them ever knew that the Chinese had been familiar with it in the +13th century, and indeed (for a special case of it) far earlier.[c] Of course the treatment of the subject by Ruffini and Horner was recondite and intricate, involving higher analysis as well as elementary algebra, but their methods were widely adopted, and are still given in common algebraic textbooks.

Among the problems in the *Chiu Chang*, one,[d] the finding of a cube root, may be written $x^3 = 1,860,867$, and another,[e] involving the square and the simple term, $x^2 + 34x = 71,000$. Naturally these equations were not written as such, but set up with

1	8	6	0	8	6	7	*shih*[1]
							fa[2]
							chung[3]
						1	*chieh suan*[4]

counting-rods on a board.[f] The uppermost line in the table is to contain the radical solution sought, the second line (*shih*[1]) is the constant term before the operation, the third (*fa*[2] or, later in the operation, *ting fa*[5]) is for numbers obtained in its successive stages, the fourth (*chung*[3]) receives another number found at an intermediate stage,

a Wang & Needham (1).
b (1), p. 25; (6), p. 180.
c Cajori (2), p. 271, gives an account of the Ruffini-Horner discoveries. Cf. de Morgan (1).
d Ch. 4, p. 13*a*.
e Ch. 9, p. 11*a*.
f There is an element of conjecture in describing exactly what the Han people did on their counting-boards, but we can reconstruct it following the known practices of the Sung. In the accompanying diagram modern numerals are used, but it should be remembered that with the counting-rods a blank space would have been left instead of the zero.

1 實 2 法 3 中 4 借算 5 定法

and the fifth (*chieh suan*[1]) is the coefficient of x^3 before the operation. The process starts by raising the 1 in the *chieh suan* line to 10^6 (let $x = 100x_1$) so that

$$1{,}000{,}000x_1^3 - 1{,}860{,}867 = 0.$$

After this the text, not at all explicit, uses the word *i*,[2] 'discussion', which must mean the choosing of a first root figure a (e.g. 1). This is then inserted in its proper place on the top line. Then follows a set of operations by which the numbers on the counting-board are converted into a pattern corresponding to Horner's transformed equation:

$$1000x_2^3 + 30{,}000x_2^2 + 300{,}000x_2 - 860{,}867 = 0.$$

Here $x_2 = 10(x_1 - a)$. With the transformed equation, 'discussion' (*i*[2]) again takes place, and the next digit of the solution is inserted. The same process is repeated (with modifications)[a] to find the third digit, at which point the *Chiu Chang* procedure ends. In this particular case, the answer was 123. Liu Hui in the +3rd century, however, pointed out that the process could be continued past the decimal point (as we should say) as long as required.[b] The *Chiu Chang* author himself seems to have carried square-root extraction to one decimal place.[c]

The system can be found in all the later mathematical books[d] with terminological variations, and minor changes in the steps, generally not improvements.[e] Numerical equations of degrees higher than the third occur first in the work of Chhin Chiu-Shao around +1245.[f] He deals very clearly with equations such as

$$-x^4 + 763{,}200x^2 - 40{,}642{,}560{,}000 = 0.$$[g]

Li Nien[h] and Mikami (*11*) have given an exhaustive account of the treatment of numerical higher equations by the Sung algebraists, including Chu Shih-Chieh. The stabilised terminology, as found in Li Yeh's books, can be seen from the following table, based on the type-equations:

$$ax^6 + bx^5 + cx^4 + dx^3 + ex^2 + fx + e = 0$$

[a] I.e. the multiplication by three is dismantled into a series of additions. In this respect the method anticipated Horner's particularly closely.

[b] In his commentary on the *Chiu Chang*, ch. 4, p. 14*a*, he refers to the decimal part of the root as *wei shu*.[3] Li Shun-Fêng, commenting (p. 15*a*), added: *khai chih pu chin chê, chê hsia ju chhien*[4] (if it doesn't finish at the integral part, we can go on (progressively) cutting down (the place value of the terms) as before). See p. 85 above.

[c] See p. 66 above.

[d] *Sun Tzu*, *Hsiahou Yang*, *Chang Chhiu-Chien*, etc., and in Liu I's 'Tai Tsung Khai Fang[5]' method of the early Sung.

[e] Thus it was a pity that Brahmagupta (+7th century) took over the *Sun Tzu* treatment rather than that of the *Chiu Chang* (Colebrooke, p. 280). Cf. p. 146 below.

[f] Or perhaps Chia Hsien[6] if he really lived about +1100, in his *Huang Ti Chiu Chang Hsi Tshao*[7] (Detailed Explanations of the Yellow Emperor's Nine Chapters of Mathematics). In any case he was concerned only with the special case $x^4 = a$.

[g] Described, Mikami (1), pp. 74 ff.

[h] (*4*), vol. 3, p. 127; (*21*), vol. 1, p. 246.

[1] 借算 [2] 議 [3] 微數 [4] 開之不盡者折下如前 [5] 帶從開方
[6] 賈憲 [7] 黃帝九章細草

and $$-ax^6-bx^5-cx^4-dx^3-ex^2-fx-e=0:$$

$+ax^6$	*yü* 隅, *yü fa* 隅 法, *chhang fa* 常 法[a]	$-ax^6$	*i yü* 益 隅, *hsü yü* 虛 隅, *hsü fa* 虛 法, *hsü chhang fa* 虛 常 法
$+bx^5$	*ti ssu lien* 第 四 廉	$-bx^5$	*ti ssu i lien* 第 四 益 廉
$+cx^4$	*ti san lien* 第 三 廉	$-cx^4$	*ti san i lien* 第 三 益 廉
$+dx^3$	*ti erh lien* 第 二 廉	$-dx^3$	*ti erh i lien* 第 二 益 廉
$+ex^2$	*ti i lien* 第 一 廉	$-ex^2$	*i lien* 益 廉, *ti i i lien* 第 一 益 廉
$+fx$	*tshung* 從, *tshung fang* 從 方	$-fx$	*i tshung* 益 從, *i fang* 益 方, *hsü tshung* 虛 從
$+e$	*shih* 實 (the absolute term) or *wu chhêng fang shih* 五 乘 方 實	$-e$	*shih* 實, *wu chhêng fang shih* 五 乘 方 實

Early Greek and Indian mathematics seem to have contributed little or nothing to the solution of higher numerical equations.[b] The first noteworthy work on them in Europe was due to Fibonacci at the beginning of the +13th century. In +1225 he gave a solution for $x^3+2x^2+10x=20$, expressing the answer, which was very nearly correct, in sexagesimal fractions (degrees and minutes). 'How this result was obtained', says Smith,[c] 'no one knows, but the fact that numerical equations of this kind were being solved in China at the time, and that intercourse with the East was possible, leads to the belief that Fibonacci had learned of the solution in his travels....'[d] At any rate, one cannot help noticing that the equation is exactly characteristic of those which Wang Hsiao-Thung used to solve in Thang China (+7th century), i.e. the coefficient of the highest term is unity, and all the terms are positive (as thus written). Subsequent progress in Europe was due to Viète (+1600) and Newton (+1669). In considering the long lead of Chinese mathematics in this field, one may suggest that the counting-board, with its lines representing increasing powers, was particularly convenient for the purpose, though of course no attempt was made, even in the Sung, to produce a generalised theory of these equations.

[a] These terms refer, of course, only to the coefficient, as the power is indicated by the position of the term in the table.

[b] The old Indian rule, already mentioned (p. 89), for getting square roots in decimals would seem to be the converse of the first operation in the Chinese system.

[c] (1), vol. 2, p. 472.

[d] Later, as in the Sections on horology, textile technology, explosives and siderurgy, many examples will be adduced of marked Chinese influence on Europe in the Mongol period (Vols. 4 and 6).

(8) The *Thien Yuan* Notation

We come now to the general system of notation used by the Sung algebraists for the expression of numerical equations. It was of 'matrix' character.[a] Chu Shih-Chieh's classical introduction to it involved a problem concerned with a circle inscribed in a right-angled triangle (Fig. 57 shows such a diagram taken from Li Yeh's *Tshê Yuan Hai Ching*, at the front of which it is always found). The four unknowns of the problem were as follows:

The hypotenuse	*jen*[1] (man)	corresponding to z
The altitude	*ti*[2] (earth)	corresponding to y
The base	*thien*[3] (heaven)	corresponding to x
The diameter of the inscribed circle (*huang fang*;[4] the yellow magnitude)	*wu*[5] (thing)[b]	corresponding to u

The 'square' or matrix of the four directions[c] was set up as shown in the accompanying diagram. The centre compartment was occupied by the absolute term, if any; it was called *thai*[6] (an abbreviation of *thai chi*[7]).[d] *Jen*[1] (z and its powers) were written to the right of *thai*; *ti*[2] (y and its powers) were written to the left of it; *thien*[3] (x and its powers) were written below it, and *wu*[5] (u and its powers) were written above it. Proceeding outwards from *thai* in any straight direction, the first compartment was for the insertion of the simple term (e.g. $10x$); these were the four *yuan*.[8] The second compartment outwards was for the square, the third for the cube, the fourth for the fourth power and so on indefinitely. Proceeding outwards from *thai* diagonally the first compartments were for the insertion of multiplied terms, such as xy or xz. But since, with four unknowns, there could be six of these, it was necessary to insert a small number in the centre compartment

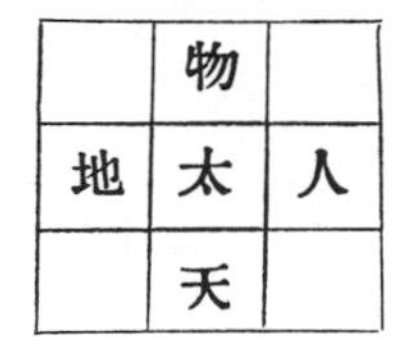

[a] Detailed explanations have been given by van Hée (9, 12), Mikami (1), ch. 14, and of course by all the Chinese historians of mathematics already quoted.

[b] Smith & Mikami (1), p. 51, supported by Smith (1), vol. 2, pp. 392, 393, drew attention to the parallel here between this and Latin *res* and Italian *cosa* for the unknown. The old name for algebra, the 'Cossic Art', was derived from the latter word. Sarton (1), vol. 3, p. 701, thinks it very unlikely that there could have been any connection. But the Arabs also called the unknown *shai'*, the thing (cf. Gow (1), p. 111). Further research will no doubt reveal where this usage originated and how it spread. Later on (p. 146 below) we shall find a rather striking instance of semantic equivalence between the Chinese and Sanskrit technical terms for the Rule of Three.

[c] *Ssu yuan*;[9] hence the title of Chu's book, p. 41 above. *Yuan* has sometimes been translated 'elements', but this invites confusion with the Five Elements (see Vol. 2, pp. 243ff., 253ff.) of physical theory, and is unwarranted. The translation 'monads' is even worse. The *yuan* are really the points of origin of the four radiating directions along which the coefficients of the various powers of the unknowns are arrayed in this positional algebra. See the diagram on the following page (top, right).

[d] Note especially the use of the same technical term as that which we have already found among the Neo-Confucian philosophers of the Sung, see above, Vol. 2, p. 460. The mathematicians presumably used it in the present sense because the absolute term was a starting-point ($+a = ax^0$), or separating-point between positive and negative powers of x.

1 人 2 地 3 天 4 黃方 5 物 6 太 7 太極
8 元 9 四元

alongside *thai*. The diagram on the left below illustrates the way in which was written $x+y+z+u$. The middle diagram illustrates the writing of the more complex expression

	元	
元	太	元
	元	

$$x^2+y^2+z^2+u^2+2xy+2xz+2xu+2yu+2yz+2zu.$$

It is necessary to imagine several of these counting-boards in use at once, as the equations solved were generally treated as simultaneous. It is not possible in the short space available, where so many other subjects have to be discussed, to give a full account of the working out of a problem; for this the reader is referred to van Hée (12) and other sources mentioned. It may, however, be helpful to take a glance at a fragment of one of the boards during the actual process of the solution of an equation with four unknowns. The right-hand diagram represents the expression

$$2y^3-8y^2-xy^2+28y+6yx-x^2-2x.$$

It will be remembered that the negative signs were represented by a diagonal line crossing out the rod-numerals set up in a given square. Zeros, as always, indicate that the terms for the compartments which they occupy do not occur in the expression.

	1	
1	太	1
	1	

		1		
	2	0	2	
1	0	2 太 2	0	1
	2	0	2	
		1		

2	−8	28	太
0	−1	6	−2
0	0	0	−1

In the books of the Sung algebraists, as they have come down to us, there are never any of these diagrams in the text, except in so far as they have been inserted by modern editors (Fig. 78 shows some), yet their nature is certain from the description given. We do find, however, calculations in the simpler form of the *Thien yuan shu*[1] algebra where only one unknown was considered. These are arranged in the vertical columns of the printed page, and usually only the word *thai*[2] or *yuan*[3] is given, since if one line only was fixed it could be seen at once what the others represented. The equation

$$x^3+15x^2+66x-360=0$$

looked, therefore, as in the small diagram on p. 133. Fig. 79, taken from the *Tshê Yuan Hai Ching*, shows how they appear in the text. This kind of array was, as it were, a single axis detached from the matrix with which four unknowns could be handled, but if it were continued downwards below the *thai*[2] (absolute term), it gave the possibility of dealing with negative powers (x^{-2}, x^{-3}, etc.),[a] and this was made use of by Li Yeh.[b]

[a] x^{-1} was called *yuan chhu thai*,[4] and x^{-2} *yuan tsai chhu thai*[5] and so on.
[b] See Li Nien (*1*), p. 142.

1 天元術　2 太　3 元　4 元除太　5 元再除太

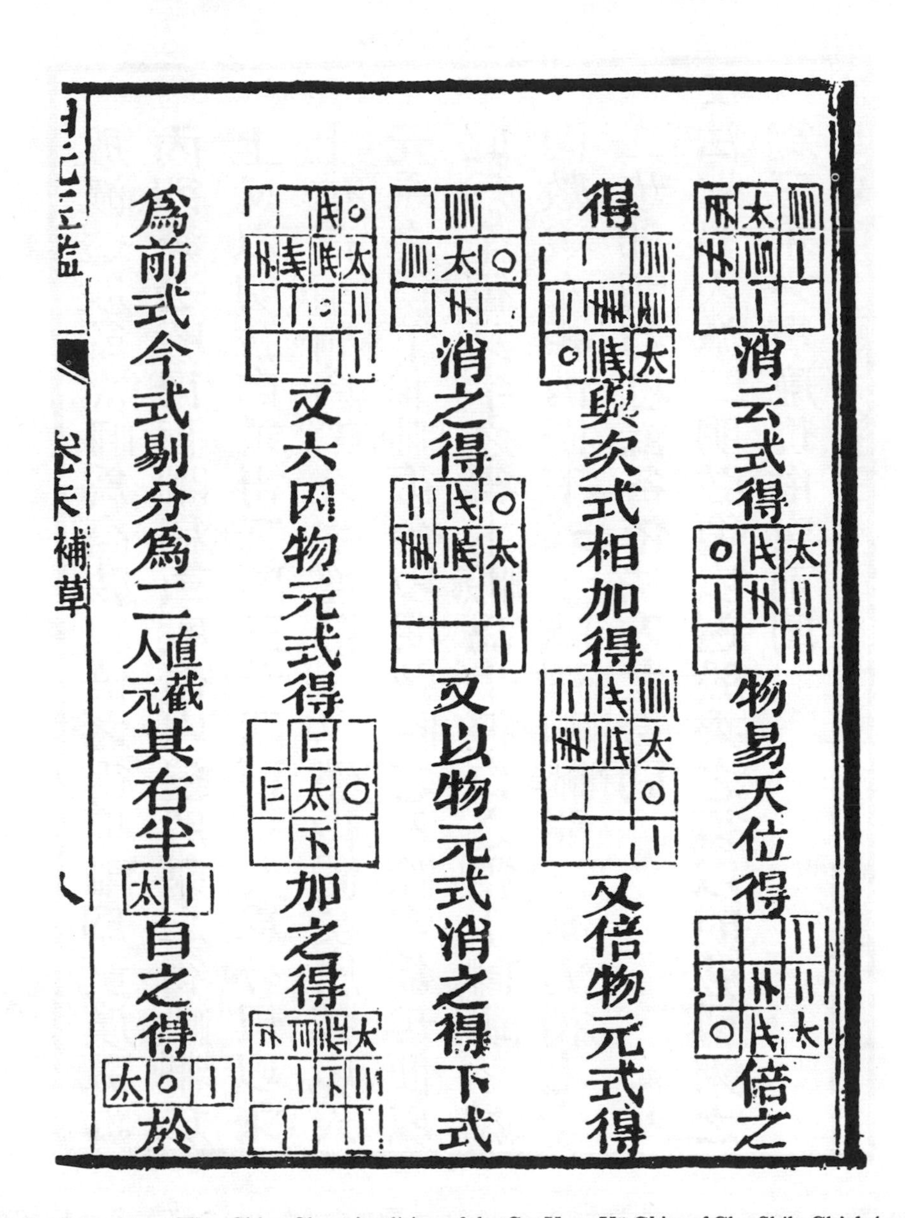

Fig. 78. A page from Ting Chhü-Chung's edition of the *Ssu Yuan Yü Chien* of Chu Shih-Chieh (+1303) showing the 'matrices' of the *Thien Yuan* algebraic notation. The middle frame in the first column on the right is analogous to the example given on the opposite page (right-hand diagram), showing $xy^2 - 120y - 2xy + 2x^2 + 2x$.

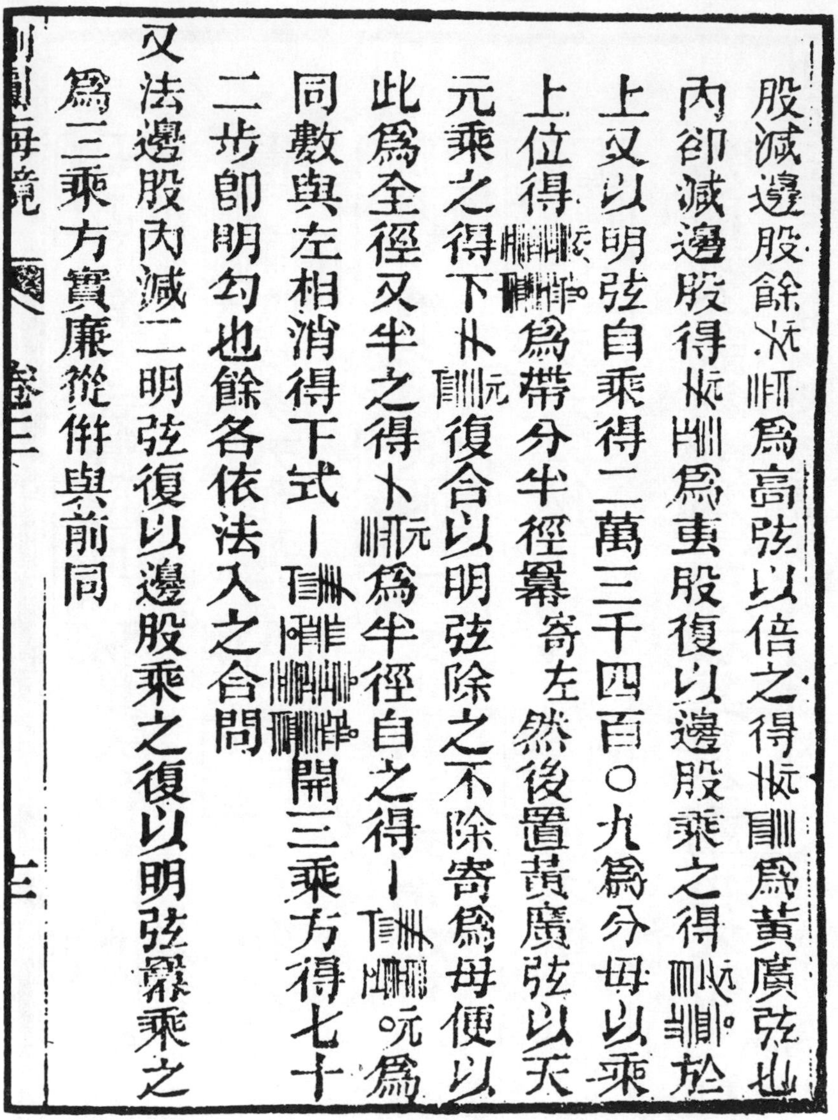

股減邊股餘[illegible]為高弦以倍之得[illegible]為黃廣弦也
內卻減邊股得[illegible]為重股復以邊股乘之得[illegible]於
上又以明弦自乘得二萬三千四百〇九為分母以乘
上位得[illegible]為帶分半徑冪寄左然後置黃廣弦以天
元乘之得下[illegible]復合以明弦除之不除寄為母便以
此為全徑又半之得[illegible]為半徑自之得[illegible]為
同數與左相消得下式[illegible]開三乘方得七十
二步即明勾也餘各依法入之合問
又法邊股內減二明弦復以邊股乘之復以明弦冪乘之
為三乘方實廉從併與前同

測圓海鏡 卷三 三

Fig. 79. Single axes of the 'matrix' notation of the *Thien Yuan* algebra, from the *Tshê Yuan Hai Ching* of +1248 by Li Yeh. Note the monogrammatic form of the counting-rod numerals, the zero and minus signs, and the character *yuan* fixing the position of the simple term. For example, in the fifth column from the right, the expression $-2x^2+654x$ will be seen.

As we have already seen in the description of books (pp. 41 ff.) there is plenty of evidence that an active movement in algebra preceded the generation of Li Yeh and Chhin Chiu-Shao. We may safely say that it was developing in the +12th century. The earliest method would quite naturally have involved a columnar arrangement such as that here shown, deriving from the ancient counting-board. The system of the *Tshê Yuan Hai Ching* just figured would have its logical ancestor in the *Chiu Chang*'s division process.[a] The alternative system used in the *I Ku Yen Tuan*, which has the constant term at the top,[b] would go back similarly to the root-extraction process of the *Chiu Chang*. Li Yeh, in his *Ching-Chai Ku Chin Chu*,[1] tells us[c] that (presumably in his youth, which must have been before +1200) he found a certain (unnamed) mathematical book dealing with this art; it explained nineteen technical terms for distinguishing the 'storeys of figures, whether above or below' (*shang hsia tshêng shu*[2]). Probably this dealt with only one variable, in which case the central 'storey', *jen*,[3] must have been the absolute term, and the ninth above it must have corresponded to x^9.[d] The ninth below it (*kuei*[4]) would have corresponded to x^{-9}. But not even the title of this book is now known.[e]

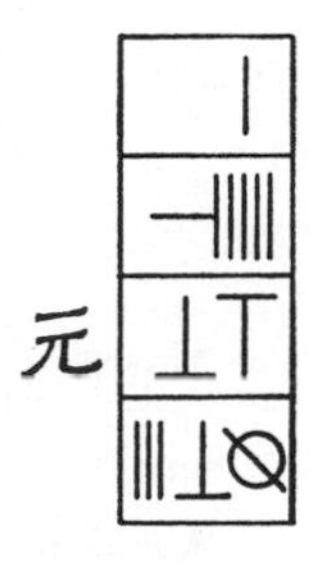

(9) Binomial Theorem and the 'Pascal' Triangle

It will be evident that in the solution of their higher numerical equations the Sung algebraists needed the Binomial Theorem. The development of the binomial $(x+a)^n$ for any integral value of n amounts to a device for finding the coefficients of the intermediate terms in the expansion. For example

$$(x+1)^3 = x^3 + 3x^2 + 3x + 1,$$

which was just the kind of expression with which they were dealing. In modern terminology the generalised form of this is

$$(x+a)^n = x^n + {}_nC_1 x^{n-1}a + {}_nC_2 x^{n-2}a^2 + \ldots + {}_nC_n a^n,$$

the last being the absolute term. From this it is easy to derive a table of binomial coefficients, from which the coefficients of the intermediate powers in the series can

[a] See p. 65 above.
[b] See p. 45 above.
[c] 'Commentary on Things Old and New' by (Li) Ching-Chai, ch. 3.
[d] The name of this storey, *hsien*,[5] suggests that the origins of this algebraic movement were (like those of the Han book *Shu Shu Chi I*) not unconnected with the Taoists.
[e] Also in the same book of Li Yeh's he says that Phêng Tsê,[6] one of his predecessors, departed from ancient practice by 'establishing the Thien Yuan underneath'; presumably underneath the *thai*, whereas positive powers of x ought to be written above it, according to Li Yeh.

[1] 敬齋古今注 [2] 上下層數 [3] 人 [4] 鬼 [5] 仙 [6] 彭澤

be read off for any given value of n. Here $_nC_0$ is the coefficient of the highest power, and the following ones are those of the descending powers down to the absolute term:

n	$_nC_0$	$_nC_1$	$_nC_2$	$_nC_3$ etc.
1	1	1		
2	1	2	1	
3	1	3	3	1
etc.				

This array has been known since the 17th century in Europe as Pascal's Triangle, since it was in +1665 that Blaise Pascal's *Traité du Triangle Arithmétique* was posthumously published. Actually it had already appeared in print for the first time more than a century earlier, on the title-page of Apianus' *Arithmetic* (+1527),[a] and in the +16th century had been fairly widely known.[b] But Apianus, Oughtred, Stifel and Pascal would have been rather surprised if they could ever have seen Chu Shih-Chieh's *Ssu Yuan Yü Chien* of +1303, from which the triangle of binomial coefficients is here reproduced in Fig. 80. It is called the *Ku fa chhi chhêng fang thu*[1] (Old method Chart of the Seven Multiplying Squares). The series $n-1$ *chhêng chi*[2] corresponds to x^n; the series $n-1$ *chhêng yü*[3] corresponds to a^n; and at the base of the triangle *lien*[4] marks off the successive terms.[c]

The fact that Chu speaks of the triangle as old or ancient suggests that the binomial theorem had already been understood at least by the beginning of the +12th century.[d] The only other contribution had been a statement by 'Umar al-Khayyāmī[e] about +1100 in Persia that he could find the fourth, fifth, sixth and higher roots of numbers by a law he had discovered which did not depend on geometrical figures. He says that this was explained in another book, but it has not come down to us.[f] Now although the earliest extant Chinese representation of the triangle is in Yang Hui's *Hsiang Chieh Chiu Chang Suan Fa* of +1261, it is known (from this book) to have existed long

[a] Reproduced by Smith (1), vol. 2, p. 509, who calls it 'as first printed', contradicting thus what he says about Chu Shih-Chieh on the previous page.

[b] It had of course been studied earlier by Arabic writers such as Jamshīd al-Kāshī about +1425; see Rosenfeld & Yushkevitch (1), pp. 59ff., 387ff. His work on this subject was made known to the West more than a century ago by Tytler (1).

[c] After Chu Shih-Chieh, the triangle was reproduced in several later Chinese books, such as that of Wu Hsin-Min (*c.* +1450) and the *Suan Fa Thung Tsung* of +1593.

[d] The case of $n=2$ had of course been appreciated by Euclid (*Elements*, bk. II, prop. 4), and the *Chiu Chang* covered $n=3$ in its treatment of cube roots. The extension of the process beyond the cube was a step of particular importance for it liberated algebra from the geometrical bonds of three-dimensional space (cf. Wang Ling (2), vol. 1, p. 246).

[e] Commonly known as Omar Khayyam (Sarton (1), vol. 1, p. 759).

[f] The passage has been translated by F. Woepcke (1), p. 13, and reproduced by Winter & Arafat (1), p. 34, and Luckey (4), p. 218. Cf. the comments of Smith (1), vol. 2, p. 508. Al-Khayyāmī makes it quite clear that he regarded his methods as arising from the coefficients described by earlier Indian mathematicians for squaring or cubing the sum of two terms (see e.g. Brahmagupta, +628, in Colebrooke (2), p. 279). Similar coefficients occur in *Sun Tzu*, ch. 2, pp. 7ff. (+3rd or +4th century), and even in the *Chiu Chang* (ch. 4, pp. 9*b*ff., 13*a*ff.); see Wang & Needham (1), pp. 350, 356, 390.

[1] 古法七乘方圖 [2] 乘積 [3] 乘隅 [4] 廉

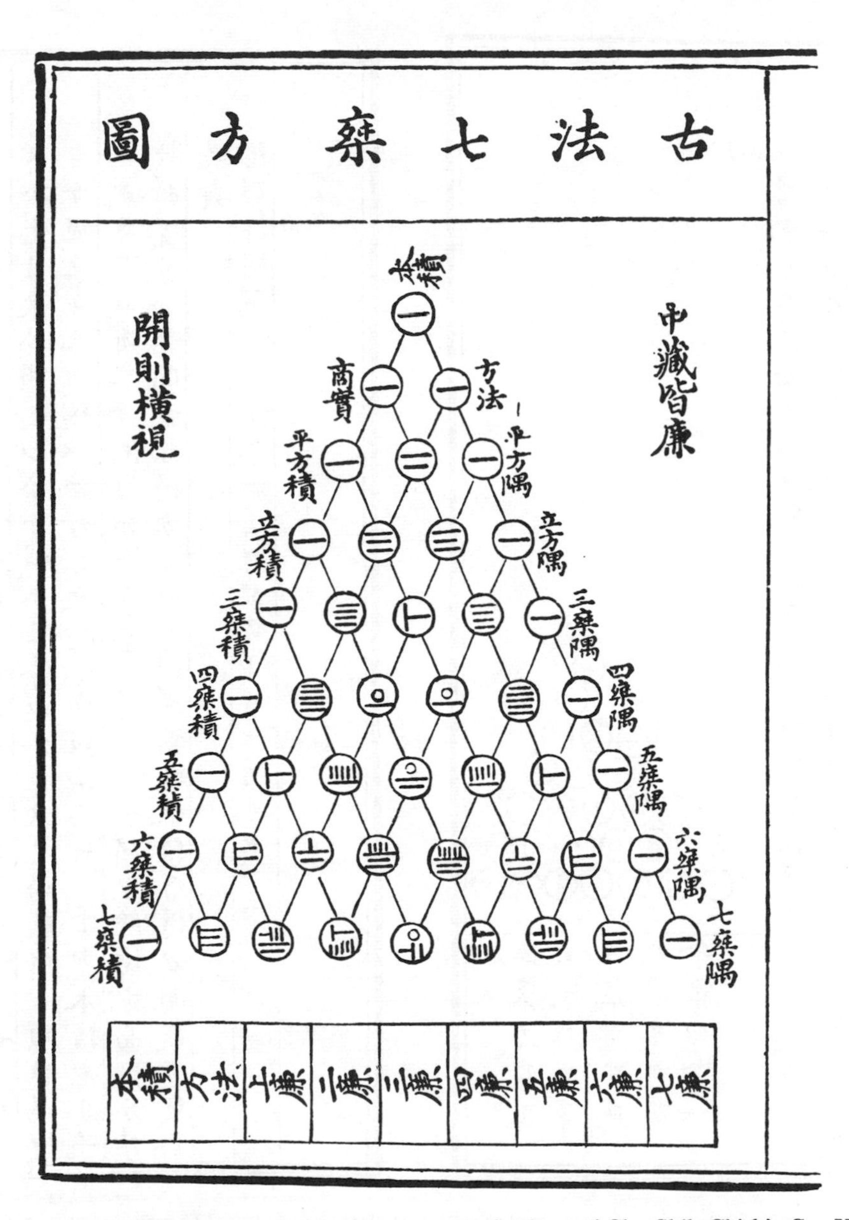

Fig. 80. The 'Pascal' Triangle as depicted in +1303 at the front of Chu Shih-Chieh's *Ssu Yuan Yü Chien*. It is entitled 'The Old Method Chart of the Seven Multiplying Squares' and tabulates the binomial coefficients up to the sixth power.

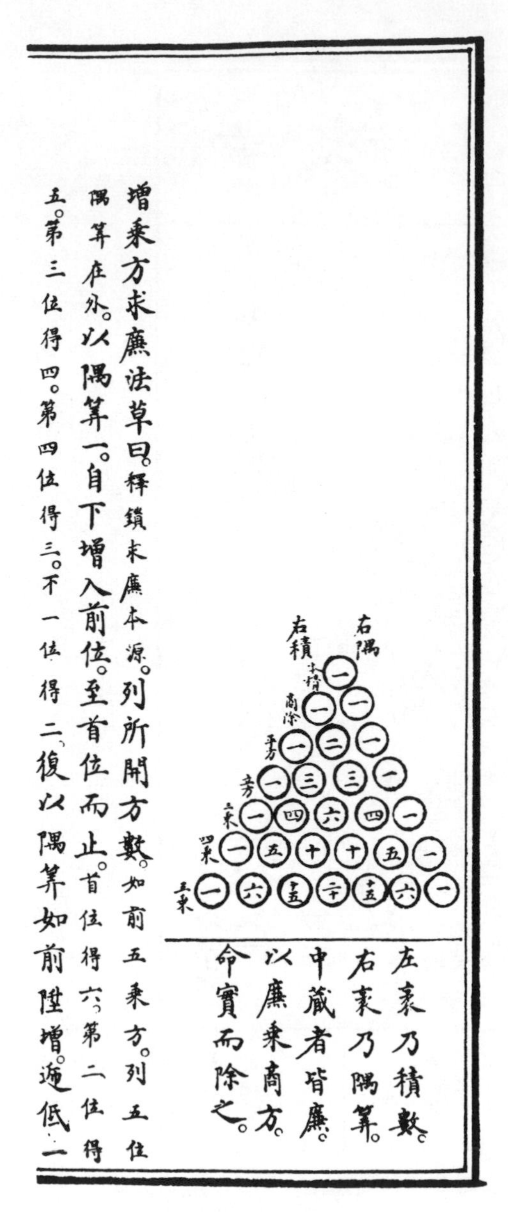

分母以命分子之數。 再求積數還源術曰置方面全步以分母通之併入分子。自乘於頭。又以分子減分母。餘以分子乘之得數。併入頭位爲實。商除還原。無此一段。以分母自乘爲法。實如法而一。 平方本積有分子。即是原方面有之分。術曰。分母乘全步。併入分子。開方除得方面散分積數。别置原分母。開方除得方面分母。以除前段散積。乃得方面幾步幾分之幾。

楊輝詳解開方作法本源。出釋鎖算書。賈憲用此術。

原本空

增乘方求廉法草曰。釋鎖求廉本源。列所開方數。如前五乘方。列五位隅筭在外。以隅筭一。自下增入前位。至首位而止。首位得六。第二位得五。第三位得四。第四位得三。不一位得二。復以隅筭如前陞增。遞低一

Fig. 81. The oldest extant representation of the 'Pascal' Triangle in China, from ch. 16,344 of the MS. *Yung-Lo Ta Tien* of +1407 (Cambridge University Library). As the seventh column of characters from the right shows, it was taken by Yang Hui in his *Hsiang Chieh* (*Chiu Chang Suan Fa Tsuan Lei*) of +1261 from an earlier book, *Shih So Suan Shu*. The text adds that the Triangle was made use of by Chia Hsien (*fl. c.* +1100).

before. Chia Hsien[1] expounded it about +1100. The method used by Chia was called '*Li-chhêng shih so*'[2] (the tabulation system for unlocking binomial coefficients). It was probably first described in the (now lost) book *Ju Chi Shih So*[3] (Piling-up Powers and Unlocking Coefficients) by another mathematician, Liu Ju-Hsieh,[4] who seems to have been contemporary with him (see Fig. 81).

Here a very interesting point arises. In Fig. 80 it will be noticed that the counting-rod numerals are so turned as to presuppose that the bottom of the triangle originally stood vertically on the left. Thus the powers of the unknown would have stood on successive horizontal rows of the counting-board, exactly as we know (from the *Chiu Chang*) was the ancient (Han) practice in extracting square or cubic roots. Here again we can see the logical continuity between the ancient counting-board and the notation of the Sung algebra. The network of compartments was a natural development from the ancient horizontal rows. It therefore seems rather likely that this triangle of coefficients originated in China. 'Umar al-Khayyāmī's work was derivative from Indian traditions which seem to have been themselves influenced much earlier by Chinese root-extraction methods.[a]

(10) Series and Progressions

The subject just discussed is not far from the general question of series problems. Greek mathematicians studied arithmetic, geometric and harmonic progressions, and these are also found in Indian and Arabic works. The first treatment of arithmetical series is probably that of the ancient Egyptians in the Ahmes papyrus. There was an interest in series from the beginning of Chinese mathematics,[b] indications of it occurring first in the *Chou Pei*. The *Chiu Chang*, in its third (*tshui fên*[5]) chapter, has many problems involving progressions, one, for example, about the distribution of venison among five ranks of officials. This is an arithmetical progression which we would write

$$a+(a+d)+(a+2d)+(a+3d)+(a+4d)=5,$$

where a is what the lowest rank received and d the difference between two successive ranks. Only one solution is given, that in which both a and $d=\frac{1}{3}$. In the Han and San Kuo periods there was evidently considerable interest in textile production, for several problems concern the output of women weavers. Both the *Chiu Chang* and the *Sun Tzu* have a question running thus: 'A girl skilful at weaving doubles each day the

[a] Arabic mathematicians before al-Khayyāmī may have been acquainted with some binomial coefficients. For instance Abū'l Wafā al-Būzjānī (+940 to +998) wrote a book the title of which implies a knowledge of the extraction of roots of the third, fourth and seventh powers: F. Woepcke (2), p. 253, Luckey (4). Also Abū Bakr al-Karajī (*fl.* +1019 to +1029) knew the binomial coefficients of the third and fourth degrees: Luckey (3, 5), Levi della Vida (1). Al-Khayyāmī systematised matters like his contemporary Chia Hsien. A claim for a great antiquity of the Pascal triangle in India has been made by Singh (2). But though a similar numerical pattern (the *Meru-prastāra*) was derived from a passage in Piṅgala's *Chandaḥ-sūtra*, VIII, 23 (*c.* −200) by the +10th-century commentator Halāyudha, it concerns prosodic combinations only and has nothing to do with binomial coefficients (cf. Luckey (4), p. 219).

[b] There is an elaborate discussion of this by Li Nien (*11*); (*4*), vol. 3, pp. 197ff.; (*21*), vol. 1, pp. 315ff.; and a short paper by van Hée (16).

[1] 賈憲 [2] 立成釋鎖 [3] 如積釋鎖 [4] 劉汝諧 [5] 衰分

output of the previous day. She produces 5 feet of cloth in 5 days. What is the result of the first day, and of the successive days respectively?' Here the progression is[a] $a+ar+ar^2+ar^3+ar^4=5$, with $r=2$, and the answer $a=\frac{1}{10}+\frac{19}{310}$, i.e. $\frac{5}{31}$. This would be a geometrical progression.[b]

Chang Chhiu-Chien, in his late +5th-century *Suan Ching*, to some extent generalised the procedures. He has a problem[c] of a weaver who increases her speed of weaving each day, so that $a+(a+d)+(a+2d)+\ldots n$ terms $=S$. According to the text (in words, of course) $d=\frac{(2S/n)-2a}{n-1}$. There is also a descending series: 'There is a girl who weaves 5 feet the first day, but then her output diminishes day after day, until on the last day she weaves 1 foot only. If she has worked 30 days, how much has she woven in all?' The answer is given as $S=\frac{n}{2}(a+z)$, where z is the last term. For these weaving problems Chang Chhiu-Chien uses the following terminology. The first term, a, of a series, is called *chhu jih chih shu*;[1] the last, z, is *mo jih chih shu*;[2] and the number of terms, n, is *chih chhi jih shu*.[3] The sum, S, is *chih shu*,[4] and the common difference, d, is *jih i*.[5] These are the first formulae in China for arithmetical progression.

It will be remembered that according to the *Chhien Han Shu* 271 counting-rods fitted together into a hexagonal bundle.[d] This was an example, not only of a figurate number, but also of an arithmetical progression, and it probably gave the inspiration for an elaborate analysis of series by Chu Shih-Chieh at the end of the +13th century.[e] His *Ssu Yuan Yü Chien* includes[f] a discussion of higher series at an advanced level. He deals with bundles of arrows (*chien*[6]) made into various cross-sections, such as circular (*yuan*[7]) or square (*fang*[8]); and with piles of balls, e.g. triangular (*chiao tshao to*[9]), pyramidal (*san chio to*[10]), conical (*yuan chui*[11]), etc. Assuming

$$r^{|p|}=r(r+1)\ldots(r+p-1),$$

r and p being positive integers, Chu obtained relations equivalent to

$$\sum_{r=1}^{n}\frac{r^{|p|}}{1^{|p|}}=\frac{n^{|p+1|}}{1^{|p+1|}}$$

[a] Li Nien (*4*), vol. 3, p. 215.

[b] It must be understood that the *Chiu Chang* mathematicians always solved their series and progression problems by using the rule of false position (cf. p. 117) and the rule of three, thus breaking them down into separate proportions. They developed no new formulae for obtaining d, r or a, respectively. The presence of these problems, therefore, does not justify the impression which might easily be gained from the discussions of Mikami, Li Nien, and other historians, that the Han people made a contribution to this branch of mathematics.

[c] Ch. 1, p. 19*b*.

[d] P. 71 above.

[e] It must of course be remembered that it was Shen Kua who began the Sung treatment of series by his Kua Chi Shu,[12] 'Finding volumes by what they will contain' (*Mêng Chhi Pi Than*, ch. 18), about +1078; in which he considers the piling up of spherical chessmen, altar-step bricks, and wine-kegs, allowing for the intervening spaces, in a solid figure equivalent to the frustum of a wedge. Shen Kua's work was extended by Chhin Chiu-Shao and Yang Hui. See p. 142 below.

[f] Ch. 2, pp. 6*a*ff. Cf. Li Nien (*21*), vol. 1, pp. 339ff.

[1] 初日織數 [2] 末日織數 [3] 織訖日數 [4] 織數 [5] 日益
[6] 箭 [7] 圓 [8] 方 [9] 茭草垛 [10] 三角垛
[11] 圓錐 [12] 括積術

and $$\sum_{r=1}^{n} \frac{r^{|p|}}{1^{|p|}} \frac{(n+1-r)^{|q|}}{1^{|q|}} = \sum_{r=1}^{n} \frac{r^{|p+q|}}{1^{|p+q|}},$$

and many others of a similar nature, but gave no theoretical proofs.

Beyond this point the Chinese treatment of series did not advance until after the coming of the Jesuits, though several Yuan and Ming mathematicians continued to solve problems in this field.

(11) Permutation and Combination

In view of the general matrix character of Chinese mathematical notation, one would expect that chessboard problems would have reached Europe from China, and also, of course, from India, in view of the invention of chess in its modern form in that country. Some of these involved series, others permutations and combinations, and probability. One of the best-known European medieval problems related to the number of grains which could, theoretically speaking, be placed on a chessboard, one being put on the first square, two on the second, four on the third, and so on in geometrical progression. The total amounts to $2^{64}-1$.

Another famous problem, which has been discussed by Vacca (2), is associated with the name of the +8th century Thang monk I-Hsing. In the *Mêng Chhi Pi Than*, Shen Kua says:[a]

The story-tellers say that I-Hsing once calculated the total number of possible situations (*chü*[1]) in chess,[b] and found he could exhaust them all. I thought about this a good deal, and came to the conclusion that it is quite easy. But the numbers involved cannot be expressed in the commonly used terms for numbers. I will only briefly mention the large numbers which have to be used. With two rows (*fang erh lu*[2]) and four pieces (*tzu*[3]) the number of probable situations will be of 81 different kinds. With three rows and nine pieces the number will be 19,683. Using four rows and sixteen pieces the number will be 43,046,721. For five rows and twenty-five pieces, the number will be 847,288,609,443.... Above seven rows we do not have any names for the large numbers involved. When the whole 361 places are used, the number will come to some figure (of the order of) $10{,}000^{52}$ (*lien shu wan tzu wu shih erh*[4])....

Shen Kua goes on to explain I-Hsing's methods, saying that they are capable of enumerating all possible changes and transformations occurring on the chessboard.[c]

[a] Ch. 18, para. 7, tr. auct.; cf. Chhien Pao-Tsung (*1*), p. 102; Hu Tao-Ching (*1*), vol. 2, p. 590.

[b] Presumably *wei chhi*.[5] This 'chess' is of course a different game from that known in the West, and must go back at least to the +3rd century (*Chhou Jen Chuan*, ch. 1). See Culin (1), p. 868, where it can be seen that the modern board has 19 rows. Shen Kua says that the oldest board had 17 rows, making 289 positions on the grid, and that there were 150 white and 150 black pieces. The 'situations' referred to mean therefore either 'white-occupied', 'black-occupied' or 'empty'. The board is empty at the beginning of the game. See also Volpicelli (1); H. A. Giles (6).

[c] I.e. 10^{208}. In Shen Kua's further explanation there occurs the expression 3^{361} (equivalent to 10^{172}), which is the correct answer. Perhaps his first figure ought to be textually emended to $10{,}000^{42}$. It is interesting to find in a text so early as the *Chhien Han Shu* (ch. 21A, p. 2*a*) a form of the expression 3^{12}.

1 局 2 方二路 3 子 4 連書萬字五十二 5 圍棊(棋)

It looks as if the name for these calculations was *Shang chhü*, *Ta yin*, *Chhung yin*[1] (driving upwards, overlapping factors, and repeating factors).[a]

In any discussion of the study of permutations and combinations in China one immediately thinks of the *I Ching* (Book of Changes) with its eight trigrams and sixty-four hexagrams, already discussed at length in Section 13*g*.[b] One would have expected that it would have given rise to some mathematical study of all its possible arrangements, and if the results of such a study are not obvious, there is reason for thinking that they were maintained in certain circles (perhaps the Taoists) as an esoteric doctrine. Here the *Shu Shu Chi I* of +190 (see above, p. 29), with its undoubted Taoist background, is interesting. In this book there is mention of several methods of calculation obviously connected with divination, e.g. the Pa Kua Suan[2] and the Kuei Suan,[3] where the trigrams are arranged in different ways at the eight compass-points,[c] and the Pa Thou Suan,[4] which may have been concerned with throws of dice.[d] While discussing the origin of the abacus above (p. 77), we had occasion to describe several calculation arrangements mentioned in the *Shu Shu Chi I*, in which balls, sometimes of different colours, were placed on numbered or graduated coordinates. If these were merely for the object of recording numbers, their value is not too obvious, since mathematicians were already very skilled in the use of the counting-rods. We may therefore suggest that what they were really concerned with was the study of permutations and combinations. When we look at the *Thai I*[5] system, for example, with the number 9183 marked on it, it seems plausible to suppose that its object was to help the answering of the question 'How many different numbers can be made out of 9183?' It would be assumed that the first, third, eighth and ninth row balls would be interchangeable. Similarly, the *San Tshai*[6] method may have been designed to answer some such question as 'In how many ways can nine letters be arranged of which three are *a*'s, three are *b*'s, and three are *c*'s?' Where the hexagrams come in, the question would be similar to 'In how many ways can eight people seat themselves at a round table?'

Here one must refer to the arrangement of the hexagrams of the Book of Changes by Shao Yung in the +11th century, about which we have already spoken (Sect. 13*g*, Vol. 2, p. 340). This was what Leibniz recognised as nothing other than the numbers

[a] Ch. 18, para. 9. Dr H. J. Winter tells us that in the same century, al-Bīrūnī discussed some similar problems. Cf. Rouse Ball (2), pp. 161 ff., and especially, on Arabic interest in this field, Wiedemann (10). It starts with al-Ya'qūbī about the end of the +9th century. Cf. Hu Tao-Ching (*1*), vol. 2, p. 594.

[b] More than one observer, such as Mikami (6), has drawn attention to the fact that the *I Ching* hexagrams were composed of long and short sticks and may thus reasonably be thought to have had some connection with the counting-rods. Cf. p. 120 above.

[c] Since the description here includes a mention of a 'needle', this book constitutes an important piece of evidence to which we shall have to return in Section 26*i* on magnetism.

[d] Mikami (*1*), p. 57; see Culin (2). The *Wên Wang kho*[7] system of divination, still used in China, consists in throwing coins in sixes, each marked with a different legend on the two sides (something like 'heads or tails'). The interpretations are looked up in books on divination. We shall see later how such practices were connected with the history of the magnetic compass.

[1] 上驅搭因重因 [2] 八卦算 [3] 龜算 [4] 把頭算 [5] 太乙
[6] 三才 [7] 文王課

1 to 64 written in binary notation (see Vol. 2, p. 340). The hexagrams also inspired a Japanese feudal lord, Fujiwara Michinori, to produce about +1157 what was one of the earliest works of Japanese mathematics, the *Keishi-zan*;[1] though now lost, it is known to have contained a mathematical treatment of the hexagram combinations.[a]

I believe that a search through the obscure and difficult Chinese medieval divination texts, if conducted by a sinologist who was also a good mathematician, would be rewarding for this field. In the introduction to his *Shu Shu Chiu Chang* of +1247 Chhin Chiu-Shao said that there were no less than thirty schools of mathematicians, some of which were based on books of the *san shih*[2] type and dealt with Thai I Jen Chia[3] (divination and fate-calculation), but that all this was *nei suan*[4] (esoteric mathematics). Whether or not any valuable early contributions to permutation-combination theory could be found in the bulky remains of their work is a matter for further research. The investigation would be interesting in view of the extreme paucity of discoveries in the subject in Europe before Abraham ben Ezra (+1140)[b] or in India before Bhāskara (*c.* +1150). The subject did not really progress till the time of Pacioli at the end of the +15th century, and the first book on it did not come until +1713 — the *Ars Conjectandi* of Bernoulli.

(12) Calculus

There have been, says D. E. Smith,[c] four chief steps in the development of what is generally called the calculus. The first, found already among the Greeks as early as the −5th century with Antiphon,[d] and among the Chinese about the +3rd century, consisted in passing from commensurable to incommensurable magnitudes by the method of exhaustion. When the early seekers after the true value of π inscribed polygons within circles, it was an attempt to 'exhaust' the residual area so formed. The second general step was the method of infinitesimals, which began to attract attention in the +17th century in the work of Kepler and Cavalieri, and was used by Newton and Leibniz. The third was the fluxions of Newton, and the fourth, that of limits, is also due to him.[e]

The nearest approach to integration among the Greeks was the work of Archimedes (*c.* −225) on the area of a parabolic segment, which he found by inscribing in each space between the parabola and the largest triangle insertable in it, a new triangle with the same base and the same height as the space. This operation was repeated indefinitely so that the last of these triangles could be extremely small and thin. His method was really the earliest example of the summation of an infinite series. One can readily see that here was a close intellectual connection with theories of atoms, and

[a] Smith & Mikami (1), p. 17. See p. 61 above.
[b] Sarton (1), vol. 2, p. 187.
[c] (1), vol. 2, pp. 676 ff.
[d] Only known to us from a mention in Aristotle (Heiberg (1), p. 5; Sarton (1), vol. 1, p. 93). Cf. van der Waerden (3), p. 131.
[e] Reference should be made, in connection with this subject, to the excellent book of Boyer (1). Cf. also Sergescu (1); Struik (1).

[1] 計子算 [2] 三式 [3] 太乙壬甲 [4] 內算

with the 'atomic' definition of the geometrical point (see above, p. 91). At the beginning of the −3rd century some unknown Peripatetic wrote a book, *De Lineis Insecabilibus*, which opposed these ideas. They long endured in dormant state, so that in the early +12th century the Provençal Judah ben Barzillai wrote, 'It has been said that there is no form in the world except the rectangle, for every triangle or rectangle is composed of rectangles too small to be perceived by the senses'. We should therefore expect that since atomism was so foreign to Chinese thought, the roots of the calculus would not be likely to have grown there.[a] As we shall see in a moment, this expectation is not quite fulfilled, but first a few more words must be said on the history of the subject in the West.

About +1609 Kepler, who was extremely interested in the focal sectors of ellipses, which he supposed that the planets must describe in equal travel times, considered that solids might be composed 'as it were' of infinitely small cones or thin discs, the summation of which would give the answer required. This led Cavalieri to develop his method of 'indivisibles'.[b] Lengths, areas and volumes were to be found by the summation of an infinite number of these indivisibles, or infinitesimals. Many mathematicians now began to push forward along this line—Fermat, Roberval, Wallis, Barrow and others. Up to this time, the approach had always been static; Newton and Leibniz broke fundamentally new ground by considering not the mass of small 'particles', but the dynamic movement of points. So Newton thought of a curve as described by a flowing point, calling the infinitely short path traced in an infinitely short time the moment of the flowing quantity.[c]

One may say at once that Chinese and Japanese thought never attained to this dynamic conception until after the arrival of the Jesuits and the consequent unification of world science. But there were certain roots of the idea of infinitesimals, exhaustion and integration which are worth looking at. Less than a century before the Provençal rabbi wrote the words already quoted, Shen Kua, in his *Mêng Chhi Pi Than*,[d] spoke of the *Tsao wei chih shu*,[1] the art of piling up very small things, by which he certainly had in mind something almost equivalent to the summation of indivisibles by Cavalieri 600 years later. For volumes he spoke of the *chhi chi*;[2] 'crack-volumes' or 'interstice-volumes', i.e. exactly the residual space which it was the object of the exhaustion methods to assess. For areas he spoke of a *Ko hui chih shu*,[3] an art of cutting and making to meet. With regard to the former he discussed the outer indentations (*kho chhüeh*[4]) of, and the empty interstices (*hsü chhi*[5]) between, piled *wei-chhi* pieces (i.e. balls), or altar-step bricks (*tshêng than*[6]), or wine-kegs (*ying*[7]).[e] He must have

[a] Atomism was of course at home in India, so it is not surprising that medieval Indian mathematicians had a tendency to regard 0 as an infinitesimal (Sengupta, 1).

[b] *Geometria Indivisibilibus*, +1635.

[c] Here, of course, we cannot speak of the famous controversy between the followers of Newton and Leibniz as to the invention of the calculus; on the whole the former seems to have had a certain priority, but the latter introduced substantially the notation used today.

[d] Ch. 18, para. 4. See Hu Tao-Ching (*1*), pp. 574ff.

[e] His work has already been mentioned in connection with series, above, p. 138.

[1] 造微之術 [2] 隙積 [3] 割會之術 [4] 刻缺 [5] 虛隙
[6] 層壇 [7] 罌

known that the smaller the units were the more fully it would be possible to exhaust any given volume or area. This had been Liu Hui's method in finding π by increasing the number of sides of a polygon inscribed within a circle. So also Shen Kua used the words *tsai ko*[1]—cutting again and again. It is an interesting coincidence that modern Chinese mathematicians adopted the expression *wei fên*[2] for differentiation in the calculus, and *chi fên*[3] for integration, perhaps not knowing that these words had been used for essentially the same ideas by a Chinese thinker in the +11th century. We have also seen, moreover, that the ideas of forming a line by piling up 'atomic' points, and

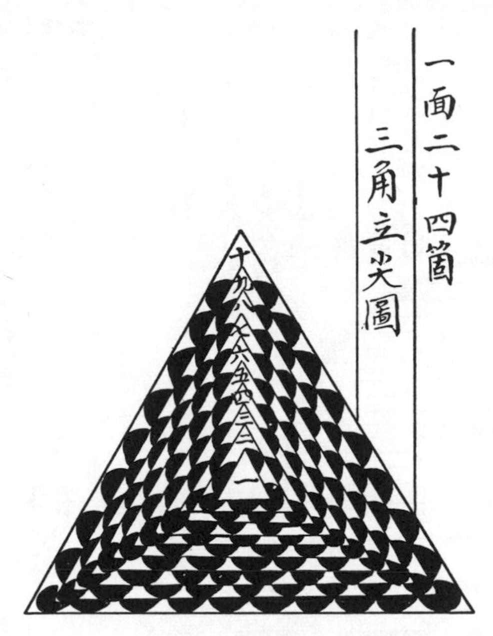

Fig. 82. Packing problems; a pyramid of spheres seen from above (from Chou Shu-Hsüeh's *Shen Tao Ta Pien Li Tsung Suan Hui* of +1558).

a plane by piling up lines, had been present from the dawn of Chinese philosophy in the Mohist definitions (p. 91 above). The idea of continuity, of infinite secability, had also been clearly expressed by the logical paradoxers, the friends of Hui Shih (early −4th century).[a] But of course these arguments were covered by the dust of centuries before being appreciated in our own time.

Questions concerning the piling up of small units within a given volume continued to attract the attention of Chinese mathematicians down to the +16th century. Chou Shu-Hsüeh,[4] in his *Shen Tao Ta Pien Li Tsung Suan Hui*[5] of +1558 (now a rare book), gave illustrations showing, for example, the piling up of spheres in ten layers within a pyramid. In Fig. 82 the pyramid can be seen looked at from the top. He made no particular advances in the mathematics of series, however.

[a] *Chuang Tzu*, ch. 33; see Sect. 11*c*, in Vol. 2, p. 191 above (PC/31).

[1] 再割 [2] 微分 [3] 積分 [4] 周述學 [5] 神道大編曆宗算會

In the +17th century the Japanese mathematicians produced a good deal of work quite similar to Cavalieri's. Muramatsu Kudayū Mosei (*d.* +1683) cut a sphere into parallel planes or segments of equal altitude, and obtained an approximation to the volume of the sphere by regarding each slice as the section of a cylinder. Nozawa Teichō (*c.* +1664) extended this by taking thinner laminae. Sawaguchi Kazuyuki (*c.* +1670) gave, in his book *Kokon Sampō-ki*[1] (Old and New Mathematical Methods), illustrations (cf. Fig. 83) showing approximation to circle area measurement by a crude

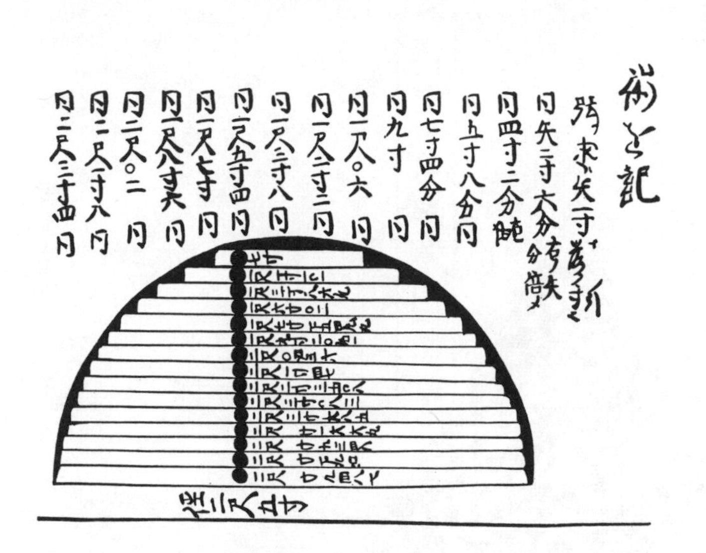

Fig. 83. Integration of thin rectangles in circle area measurement, from the *Kaisan-ki Kōmoku* (+1687) of Mochinaga Toyotsugu & Ōhashi Takusei, derived from Sawaguchi Kazuyuki's *Kokon Sampō-ki* (*c.* +1670).

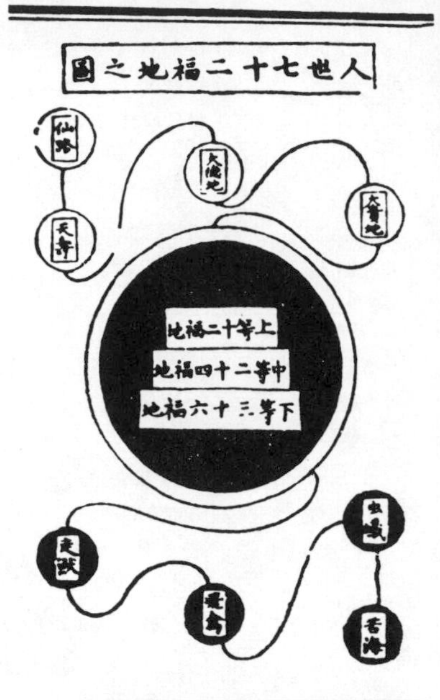

Fig. 84. Rectangles inscribed within a circle (from Hsiao Tao-Tshun's *Hsiu Chen ThaiChi Hun Yuan Thu*, *c.* +11th century).

kind of integration of thin rectangles. This approach is found in several contemporary works of the period.

While it has been suspected that this exhaustion method originated in China it has not so far been possible to point to any instance of it in a Chinese mathematical work. We are able to reproduce an illustration, however, from a Taoist work of the Sung (perhaps +11th century?) attributed to Hsiao Tao-Tshun,[2] the *Hsiu Chen Thai Chi Hun Yuan Thu*[3] (Restored True Chart of the Great-Ultimate Chaos-Origin).[a] This contains a set of obscure diagrams, with some Buddhist influence. Fig. 84 shows a circle containing three rectangles (ranks of paradises in multiples of twelve) within it. It is just possible that here, under Taoist and Neo-Confucian auspices, we may be in

[a] *TT* 146.

[1] 古今算法記 [2] 蕭道存 [3] 修眞太極混元圖

presence of one of the germs of the idea of the exhaustion of a circular area by the inscribing of rectangles.[a]

Shen Kua's problems were essentially concerned with what modern physicists and crystallographers would call 'packing'.[b] Robert Hooke, in his *Micrographia*, was perhaps the first (+1665) to apply packing theory to the study of crystals.[c] Shen Kua had similarly considered, in the +11th century, the packing of round objects in plane-bounded spaces. Although we do not see clearly the intermediate channels, there is no doubt that packing problems became one of the ruling passions of Japanese mathematicians from the seventeenth century onwards. The survey of Smith & Mikami (1) is full of illustrations taken from their works depicting the fitting of smaller circles into large ones and into annulus segments, of balls into spheres and cones,[d] etc. Seki Kōwa (+1642 to +1708) solved Sawaguchi's problem (+1670) of inscribing three circles in tangent contact within a larger one. Takebe Kenkō (+1664 to +1739) made use of infinite series in an Archimedean approximation to circle segment areas,[e] and thus duplicated, to some extent, the work which Cavalieri had done in Europe in the previous generation. Ajima Chokuyen (+1739 to +1798) solved the Problem of Malfatti, of inscribing three circles in a triangle, each tangent to the other two. All these studies went under the name of *yuan li* or *enri*,[1] circle-principles, and some have considered them as closely akin to the differential and integral calculus as developed in Europe.[f] They were, however, undoubtedly to some extent derivative, since Pierre Jartoux, one of the Jesuits (Tu Tê-Mei[2]), had brought to China in +1701 a number of formulae for infinite series, and this stimulated much Chinese work, for example, the *Ko Yuan Mi Lü Chieh Fa*[3] (Quick Method for Determining Segment Areas) by the Manchu mathematician Ming An-Thu,[4] which appeared posthumously in +1774. A similar work was the valuable *Ko Yuan Lien Pi Li Thu Chieh*[5] of Tung Yu-Chhêng[6] (+1819).[g] In the 17th and 18th centuries the Japanese were in some ways more cut off from European influence than the Chinese, and the precise extent to which their work was original constitutes a difficult question of historical transmission.[h] The essential point is that all the indigenous Chinese and Japanese work remained at the static level, while the dynamic approach was due to Newton and Leibniz alone.

[a] One can see here the possible mental connection between the idea of close-packing and the characteristic Buddhist preoccupation with innumerable worlds within a single drop of water, etc. (cf. Sect. 15*e* above, Vol. 2, pp. 419ff.).

[b] Cf. Rouse Ball (2), p. 148.

[c] Andrade (1).

[d] This branch of mathematics was called *yung thi*[7] (*yōdai*) and *yung shu*[8] (*yōjutsu*).

[e] The title of his work, *Fukyū Tetsujutsu*, embodied the words *Chui Shu*, the title of the lost +5th-century book of Tsu Chhung-Chih (see p. 35 above), which he may have read.

[f] See Mikami (*6, 7, 8*), (18, 21).

[g] Li Nien (*4*), vol. 2, p. 129, (*21*), vol. 3, p. 254, devotes a very long essay to the work of these men and their associates.

[h] For details the books of Mikami and of Smith & Mikami, and the papers of Mikami (17) and Harzer (1), may be consulted.

[1] 圓理 [2] 杜德美 [3] 割圓密率捷法 [4] 明安圖
[5] 割圓連比例圖解 [6] 董祐誠 [7] 容題 [8] 容術

(*j*) INFLUENCES AND TRANSMISSIONS

A pause may be made at this point to put together what little we have gathered concerning the contacts which seem to have taken place between Chinese mathematics and that of the other great culture-areas of the Old World. In the first place, reasons have been given[a] for thinking that here the detailed influence of prior Mesopotamian culture was limited. Apart from certain traditional cycles, there is little sign of sexagesimal arithmetic, and no special symbol for the $\frac{2}{3}$ fraction. Moreover, the Chinese treatment of fractions is radically different from that of the ancient Egyptians.

When, however, we ask what mathematical ideas seem to have radiated from China southwards and westwards, we find a quite considerable list.[b]

(1) A numeral notation of only nine figures combined with place-value components appears already as early as the Shang period in China (−14th century). Yet special symbols for multiples of 10 were not abandoned in India until the +6th;[c] and India was more advanced in this respect than Europe, for the oldest appearance of the 'Hindu numerals' there[d] is in a Spanish MS. of +976, and the use of the zero was not understood until the following century. The zero in its most primitive form, i.e. the blank space on the counting-board, goes back to the Warring States period in China (−4th century). Of written zero symbols we shall speak again in a moment.

(2) The extraction of square and cube roots is found highly developed in −1st-century China. The treatment of square roots by Sun Tzu[e] in the +4th century, and of cube roots by Chang Chhiu-Chien[f] in the +5th, closely resembles the rules given in Brahmagupta's work of about +630. The treatment of roots of higher degrees used in China from Chia Hsien (+11th century) onwards seems to have influenced al-Kāshī (+15th century), and traces of these advanced methods are shortly afterwards found in Europe.

(3) Rule of Three, though generally attributed to India,[g] is found in the Han *Chiu Chang*, earlier than in any Sanskrit text. Noteworthy is the fact that the technical term for the numerator is the same in both languages—*shih*[1] and *phala*, both meaning 'fruit'.[h] So also for the denominator, *fa*[2] and *pramāṇa*, both representing standard unit measures of length.[i]

(4) The way of expressing fractions in vertical columns, found in all the medieval Indian mathematicians,[j] was the same as that used on the Han counting-boards.

(5) Negative numbers, appearing first in −1st-century China,[k] were not handled in India until Brahmagupta's time (*c.* +630).

[a] P. 82 above.

[b] Kaye (3); Mikami (1), (*15*) have given some of the examples which follow, though by no means all.

[c] Datta & Singh (1), vol. 1, p. 40.

[d] See, for example, Smith (1), vol. 2, p. 75.

[e] *Sun Tzu Suan Ching*, ch. 2, pp. 8*a* ff.

[f] *Chang Chhiu-Chien Suan Ching*, ch. 3, pp. 31*a* ff. Actually the full description of the method here is in the Sui commentary of Liu Hsiao-Sun[3] (+576 to +625), which even so is before Brahmagupta. But he was only elaborating what was implicit in Chang's text, and must have been understood in Chang's time.

[g] E.g. Smith (1), vol. 2, pp. 483, 488.

[h] Cf. Colebrooke (1), p. 283; Wang Ling (2), vol. 1, pp. 213 ff.

[i] Even the third known term in the relationship can be identified in the two languages. For *icchā*, 'wish, or requisition', reflects *so chhiu lü*,[4] i.e. ratio number sought for.

[j] Cf. Colebrooke (1), p. 285.

[k] See p. 90 above.

[1] 實 [2] 法 [3] 劉孝孫 [4] 所求率

(6) The 'Hsüan Thu' proof of the Pythagoras theorem given in the +3rd-century commentary of Chao Chün-Chhing on the *Chou Pei*[a] is reproduced precisely by Bhāskara in the +12th. It does not occur anywhere else.

(7) Geometrical survey material appearing in the *Chiu Chang* and in Liu Hui's +3rd-century commentary is found later on in the +9th-century work of Mahāvīra. For example, the 'broken bamboo' problem,[b] and the travellers meeting on the hypotenuse.[c]

(8) The *Chiu Chang*'s rule[d] for the area of a circle segment is given again by Mahāvīra, while those for the volumes of a cone[e] and a truncated pyramid[f] reappear in many of the Indian books. The mistakes in the Chinese formulae for the segment and the cone are duly reproduced in the Indian statements.

(9) The fundamental identity of algebraic and geometrical relations, which had been consciously appreciated by Chinese mathematicians throughout the previous millennium, was expressed elsewhere for the first time by the Persian mathematician al-Khwārizmī (+9th century). On both logical and geographical grounds an influence may not unreasonably be suspected, though there is no positive evidence for it save the embassy of al-Khwārizmī to Khazaria.[g]

(10) The Rule of False Position, found in the Han *Chiu Chang*, appears in +13th-century Italy under such names as 'Regola Elchataym', indicating that the Arabs had transmitted it. They probably derived it from India but China may well have been its origin.[h]

(11) Indeterminate analysis, first undertaken in the *Sun Tzu*[i] (+4th century), next shows itself in Āryabhaṭa (late +5th) and especially Brahmagupta (+7th). Knowledge of this branch of algebra may have reached the +14th-century Byzantine monk Isaac Argyros as well overland as through Arabic and Indian intermediation. The problems and methods of Diophantus (late +3rd century) were rather different from those of Sun Tzu.

(12) An almost identical problem involving indeterminate equations (the 'Hundred Fowls') occurs first in the *Chang Chhiu-Chien Suan Ching*[j] (*c.* +500), and then in Mahāvīra (+9th) and Bhāskara (+12th).

(13) In the Thang period (+7th century), Wang Hsiao-Thung successfully solved cubic numerical equations. In the later Sung (+12th and +13th centuries) the Chinese algebraists were particularly skilled in the management of higher numerical equations.[k] In Europe, Fibonacci (+13th) was the first to give a solution to a problem of Wang's type. There is some reason for thinking that he may have been influenced by East Asian sources.

(14) The Pascal Triangle of binomial coefficients[l] was known in China by +1100. It seems to have originated about the same time in Persia, as a result of contact with Indian root-extraction methods which may themselves have owed much to earlier Chinese work. It reached Europe not long before the +16th century, where it was first published in +1527.

[a] See p. 22 above.
[b] *Chiu Chang Suan Shu*, ch. 9, p. 7*a*.
[c] *Chiu Chang Suan Shu*, ch. 9, p. 7*b*.
[d] *Chiu Chang Suan Shu*, ch. 1, p. 17*b*.
[e] *Chiu Chang Suan Shu*, ch. 5, p. 8*a*.
[f] *Chiu Chang Suan Shu*, ch. 5, p. 7*b*.
[g] See p. 107 above.
[h] *Chiu Chang Suan Shu*, ch. 7, pp. 2*a* ff. Strictly speaking, 'Double False'. The method used by the ancient Egyptians, which has sometimes been given the same name (Rule of False Position), as by Cajori (2), p. 13, was distinctly different.
[i] Ch. 3, pp. 10*a* ff.
[j] Ch. 3, p. 37*a*.
[k] See p. 126 above.
[l] See p. 133 above.

Thus in spite of all the 'isolation' of China, and all the inhibitory social factors which we shall presently examine, it seems probable that between −250 and +1250 a good deal more came out than went in.[a]

Only about this later date did influence from south or west begin to be noticeable, and even then it took little root.[b] We may note the following:

(1) In the Sui and Thang, there was an influx of knowledge about Indian mathematics and astronomy. The 'Po-lo-mên' (Brahmin) books[c] are witnesses of this. Indian astronomers, such as Chhüthan Hsi-Ta in his great *Khai-Yuan Chan Ching* of about +729, described Indian calendar calculations. These experts brought an early phase of trigonometry, but their influence seems on the whole to have been small (see pp. 37, 202).

(2) There is reason to think that Euclid's geometry first reached China through Arab intermediation about +1275. Yet few scholars showed any interest in it, and if there was a translation it was soon lost.[d] Not until the opening of the modern period in the +17th century did this fundamental work begin to exert an effect upon Chinese thinking.

(3) About the same time (+1270) Chinese mathematicians, especially Kuo Shou-Ching, seem to have begun to use trigonometrical methods. The Arabic influence of men such as Jamāl al-Dīn[e] would be the most likely channel, but direct Indian contacts cannot be ruled out.

(4) In the following century Ting Chü first began to write numbers without specifying terms for place-value. Again this was perhaps an Arab introduction.

(5) The *gelosia* method of multiplication[f] appears in Chinese mathematical books about +1590, possibly owing to contact with the Portuguese.

But if China conserved with completeness her own characteristic style in mathematics until the beginning of the modern period, if India was the more receptive of the two cultures, one illustrious invention there was which seems to have occurred in the very border marches of the two great civilisations, into both of which it quickly spread. This was nothing less than the development of a specific written symbol for nil value, emptiness, *śūnya*, i.e. the zero. Perhaps we may venture to see in it an Indian garland thrown around the nothingness of the vacant space on the Han counting-board.

This raises, however, the question of the origin and peregrinations of that fundamental manifestation of mathematical order, place-value of figures. Here the position of China is now fairly clear. In the Shang period (late −2nd millennium) the system of writing numbers was more advanced and scientific than in any other culture of high antiquity. The nine digits appeared in all units, characterised at higher levels by place-value components. These components were not themselves numerals, but in the course of time conducted, as it were, the nine digits to their proper places in the columns of the counting-boards on which calculations were made. The existence of such counting-boards is implicit in all that we know of Chhin and Han mathematics, and it was quite natural that the absence of a digit at any unit level should be represented by an empty

[a] Of course reserve is necessary, for we do not know whether there were earlier books in India (between the *Śulvasūtras* and Varāha-Mihira) which have not survived.

[b] We are thinking here particularly of mathematics as such. In the following Section several examples will be given of Indian work in the field of astronomy which did gain considerable acceptance in China.

[c] Cf. Vol. 1, p. 128; and p. 37 above.

[d] See p. 105 above.

[e] See p. 49 above and p. 372 below.

[f] See p. 64 above.

space. That this system was already complete about the −4th century, the *Mo Ching* and the coin inscriptions testify.

In view of what has already been said, the derivation of the place-value principle in China from that form of it which the Old Babylonian astronomers (early −2nd millennium) were the first to use, seems doubtful. Many principles of numeration (additive, subtractive, multiplicative) were current in ancient Mesopotamia, and even on the mathematical and astronomical cuneiform tablets, the sexagesimal place-value arrangement was combined with other principles for values lower than 60. Yet place-value in China from the Shang to the Han and onwards was decimal, not sexagesimal. And with one slight exception it was not combined with any other principle. Since therefore in other fields strong evidence of transmission of techniques and inventions from ancient Mesopotamia to China is noted,[a] we might perhaps regard the conception of place-value as a case of stimulus diffusion;[b] the travel of a bare idea and not its concrete implementation.

But when we consider Indian developments, the conviction grows that the appreciation and use of the place-value principle was much posterior to its appearance in China. It cannot be traced in India with certainty before the early years of the +5th century, the date of the *Pauliśa Siddhānta*, though by the time of Āryabhaṭa (+499) and his contemporaries such as Varāha-Mihira (author of the *Pañca-Siddhāntikā*) it was without doubt employed. Significantly, it was the decimal place-value of China, not the sexagesimal place-value of ancient Babylonia. Then, towards the end of the +7th century in Indo-China, and in the +8th and +9th in India itself, the inscriptions were made which attest the existence of the zero symbol, both in its dot and space-enclosing forms. These simply confirm that by then the place-value of figures was well understood. Now the period in question (+300 to +900) was approximately that to which we must ascribe many of the mathematical transmissions from China just mentioned, and it was also that which saw the great expansion of Buddhism in the Chinese culture-area. Could it be that the travelling monks exchanged mathematics for Indian metaphysics? An epic story of culture-contact may well await excavation from the monastic biographies of the *Kao Sêng Chuan*. We probably know only a few fragments of it, such as the fact, for instance, already referred to,[c] that about +440 a monk of the Northern Wei, Than-Ying, could be a distinguished teacher of the mathematics of the *Chiu Chang*.

The discovery of the alphabetic principle in ancient Western Asia gave convenience to the writing of languages all over the world. Yet its application to numeral notation was strongly inhibitory to arithmetical computing. No doubt the temptation was to use all the available letters, and not stop at nine. Paradoxically, it was in a non-alphabetic culture faithful to the ideographic principle that there developed the earliest form of that decimal place-value system which is now universally employed by mankind, and without which the unified modern world would hardly have been possible. Chinese reckoners and star-clerks paved the way for those Indian computing methods

[a] Cf. Vol. 2, p. 353; and p. 256 below.
[b] Cf. Sect. 7*l* (8) above (Vol. 1, p. 244).
[c] P. 27 above.

needing only nine signs which, said Severus Sebokht, the Syrian bishop, in +662, surpass description.[a] 'If those who believe, because they speak Greek', he added, 'that they have reached the limits of science, should know these things, they would be convinced that there are also others who know something.'

(k) MATHEMATICS AND SCIENCE IN CHINA AND THE WEST

If the foregoing pages have been numerous, we must reflect that from many points of view mathematics has always been a discipline of its own of equal rank with the whole of the natural sciences. The conclusion of the account of Chinese mathematics brings us to what might be described as a focal point in the plan of the present work. What exactly were the relations of mathematics to science in ancient and medieval China? What was it that happened in Renaissance Europe when mathematics and science joined in a combination qualitatively new and destined to transform the world? And why did this not happen in any other part of the world? Such are some of the questions which now arise.

The first thing to do is to get the perspective right. Few mathematical works before the Renaissance were at all comparable in achievement with the wealth and power of the developments which took place afterwards. It is pointless, therefore, to subject the old Chinese contributions to the yardstick of modern mathematics. We have to put ourselves in the position of those who had to take the earliest steps and try to realise how difficult it was *for them*. Measured in terms of human labour and intellectual attack, one can hardly say that the achievements of the writers of the *Chiu Chang Suan Shu* or of the initiators of the Thien Yuan algebra, were less arduous than those of the men who opened new mathematical fields in the 19th century. The only comparison which can be made is between the old Chinese mathematics and the mathematics of other ancient peoples, Babylonian[b] and Egyptian[c] in their day, Indian[d] and Arabic.[e] The description in the present Section shows that Chinese mathematics was also quite comparable with the pre-Renaissance achievements of the other medieval peoples of the Old World. Greek mathematics[f] was doubtless on a higher level, if only on account

[a] The full passage has been quoted in Vol. 1, p. 220 (Sect. 7*j*). In view of its date, the fact that the bishop spoke of nine signs only is interesting. He must have been in contact with men who worked with empty spaces, not with zero symbols.

[b] This subject has been greatly advanced by the work of Neugebauer (9, and bibliography in Archibald, 1). See also a brief and clear paper by Gordon Childe (8).

[c] Again the best recent review, with bibliography, is that of Archibald. I add a short paper by Dawson (1).

[d] The work of Datta & Singh, already mentioned, is, like nearly all books on Indian mathematics, unreliable as to dates, owing to the lack of an established Indian chronology. Indian historians of mathematics (Singh, Gurjar, and others) are often at variance with the chief European authorities on the subject; Kaye (3); Clark (2). But even for those who do not wish to go deeply into it, the translations of Bhāskara's *Līlāvatī* by Taylor and Colebrooke (though now old) are worth study. Āryabhaṭa's work has been translated by Clark (3). The most judicious accounts of the history of Indian mathematics are those of Thibaut (2) and Renou & Filliozat (1).

[e] Here the principal work is that of Suter (1), but a mass of information is found in the broader book of Mieli (1).

[f] Any attempt to cover the vast bibliography of this field would be outside the scope of this book, but reference might be made to an excellent summary by Heath (3).

of its more abstract and systematic character, seen in Euclid; but, as we nave noted, it was weak or tardy just where the mathematics of India and China (more faithfully based, perhaps, on those of the Babylonians) were strong, namely, in algebra.[a]

'On account of its more abstract and systematic character'—so came the words of themselves to the keys of the machine. Systematic, yes, there no doubt is possible, but abstract—was that wholly an advantage? Historians of science are beginning to question whether the predilection of Greek science and mathematics for 'the abstract, the deductive and the pure, over the concrete, the empirical and the applied' was wholly a gain.[b] According to Whitehead:

> It is a mistake to think that the Greeks discovered the elements of mathematics, and that we have added the advanced parts of the subject. The opposite is more nearly the case; they were interested in the higher parts of the subject and never discovered the elements.... Weierstrass' theory of limits and Cantor's theory of sets of points are much more allied to Greek modes of thought than our modern arithmetic, our modern theory of positive and negative numbers, our modern graphical representation of the functional relation, or our modern idea of the algebraic variable. Elementary mathematics is one of the most characteristic creations of modern thought—characteristic of it by virtue of the intimate way in which it correlates theory and practice.[c]

How far the mathematics of ancient and medieval China helped to lay the foundations for the more difficult, because most simple and elementary, techniques of handling the concrete universe will, we hope, have been made clear by the foregoing Section. In the flight from practice into the realms of the pure intellect, Chinese mathematics did not participate.[d]

Here, then, are some of the answers to the doubts and uncertainties of the writers who were mentioned at the beginning of the Section (p. 1). Better informed observers have drawn attention to certain particular weaknesses which we can now more fully appreciate. Mikami (*1*) considered that the greatest deficiency in old Chinese mathematical thought was the absence of the idea of rigorous proof, and correlated this (as do some modern Chinese scholars, such as the late Fu Ssu-Nien[e]) with the failure of formal logic to develop in China,[f] and with the dominance of associative (organic) thought.[g] Cajori (3), evaluating, in his history of mathematical notations, the Thien Yuan algebra, says that one is impressed both by its beautiful symmetry

[a] And indeed arithmetic. Cf. Becker & Hofmann (1), p. 119.

[b] See, for example, the interesting discussion of Fréchet (1).

[c] (7), pp. 132 ff. The essay of Farrington (8) which comments upon this passage is well worth reading.

[d] It is interesting that the converse process seems to have taken place with logic. While the Greeks and Indians paid early and detailed attention to formal logic, the Chinese (as we saw many times in Vol. 2) showed a constant tendency to develop dialectical logic. The corresponding Chinese philosophy of organism paralleled Greek and Indian mechanical atomism. In these fields the 'West' was 'elementary' and China 'advanced'. And the characteristic Chinese passion for the concrete fact would have been as much inimical to the abstractness of Greek geometry as it was to the metaphysical idealism of the Buddhists. Both were flights from the practical and the empirical, the concrete and the real.

[e] In a letter to the author, dated 13 Aug. 1944.

[f] See above, Sect. 11 on the School of Logicians; and below, Sect. 49.

[g] See above, Sect. 13 on the symbolic correlations and correlative thinking.

and by its extreme limitations. After an initial burst of advance, the Sung science of algebra did not experience rapid and extended growth. The standstill which it reached after the +13th century he attributes to an inelastic and cramping notation. Moreover, although so advanced in many respects (for example, the very early appreciation of decimal place-value and the zero 'blank') Chinese mathematicians never spontaneously invented any symbolic way of writing formulae, and until the time of the coming of the Jesuits, mathematical statements were mainly written out in characters. Strangely, in a people who carried algebra so far, the equational form remained implicit, and there was no indigenous development of an equality sign (=). How far the widespread use of counting-board and abacus acted as an inhibiting factor is a moot point; they certainly allowed calculations to vanish without trace, leaving no record of the intermediate stages by which the answer was reached.[a] But it seems hard to believe that what was essentially a mechanical aid to calculation would not have been helpful if more modern mathematical methods had developed.

Turning to social factors, it is striking that throughout Chinese history the main importance of mathematics was in relation to the calendar. It would be hard to find a mathematician in the *Chhou Jen Chuan* who was not called upon to remodel the calendar of his time, or to help in such work. For reasons connected with the ancient corpus of cosmological beliefs, the establishment of the calendar was the jealously guarded prerogative of the emperor, and its acceptance on the part of tributary states signified loyalty to him. When rebellions or famines occurred, it was often concluded that something was wrong with the calendar, and the mathematicians were asked to reconstruct it.[b] It has been thought that this preoccupation fixed them irretrievably to concrete number, and prevented the consideration of abstract ideas; but in any case the practical and empirical genius of the Chinese tended in that direction. In the calendrical field mathematics was socially orthodox and Confucian, but there are reasons for thinking that it also had unorthodox Taoist connections. In the +2nd century Hsü Yo was certainly under Taoist influence, and this can be detected in the mysterious +11th-century book which inspired Li Yeh; moreover, there was Hsiao Tao-Tshun's strange figure in the Sung.[c] What one misses, however, is any contact between such personalities as Ko Hung the great alchemist and Sun Tzu the mathematician, in all probability his contemporary; doubtless such thought-connections would have been impossible anywhere before the Renaissance.[d] Lastly, a factor of great importance must be sought in the Chinese attitude to 'laws of Nature'. This

[a] Cf. Mikami (20). This may partly explain the universal absence of explanations in the old books of the ways in which their rules were applied, an absence which left Loria (1) with the persistent, but entirely unjustified, suspicion that they were all copying from some half-understood Greek books which had found their way to China.

[b] For example, Han Kao Tsu believed that Chhin Shih Huang Ti's fall had been due to his erroneous calendar, so Chang Tshang was charged with making a new one (*Chhou Jen Chuan*, ch. 2, p. 1*a*).

[c] See p. 144 above.

[d] There was a distinctly romantic element about the Taoists, ensconced in their temples among the mountains and forests. Though they busied themselves with alchemical furnaces, they also inspired poets. But the mathematicians seem to have been very plain practical men, men in the retinues of provincial officials. Their style of writing was quite unliterary. Unlike Indian mathematical knowledge, that of China was rarely enshrined in verses. No doubt the Chinese mathematician also had his beautiful and intelligent Līlāvatī—but he kept her out of his books.

we studied in detail at the end of the last volume (Sect. 18); here it need only be repeated that the absence of the idea of a creator deity, and hence of a supreme law-giver,[a] together with the firm conviction (expressed by Taoist philosophers in high poetry of Lucretian vigour) that the whole universe was an organic, self-sufficient system, led to a concept of all-embracing Order in which there was no room for Laws of Nature, and hence few regularities to which it would be profitable to apply mathematics in the mundane sphere.

In taking our leave of the twenty centuries of autochthonous Chinese mathematics we may cast a brief backward glance over the successive periods and their qualities. The two dynasties which stand out for mathematical achievement are the Han and the Sung. For the −1st century, the time of Lohsia Hung and Liu Hsin, the *Chiu Chang Suan Shu* (Nine Chapters on the Mathematical Art) was a splendid body of knowledge. It dominated the practice of Chinese reckoning-clerks for more than a millennium. Yet in its social origins it was closely bound up with the bureaucratic government system, and devoted to the problems which the ruling officials had to solve (or persuade others to solve). Land mensuration and survey, granary dimensions, the making of dykes and canals, taxation, rates of exchange—these were the practical matters which seemed all-important.[b] Of mathematics 'for the sake of mathematics' there was extremely little. This does not mean that Chinese calculators were not interested in truth, but it was not that abstract systematised academic truth after which sought the Greeks. During all this time the masses of the people remained illiterate, having no access to the manuscript books which the government commissioned, copied and distributed to the various nodes of the administrative network. Artisans, no matter how greatly gifted, flourished upon the other side of an invisible wall which separated them from the scholars of literary training. It was only because Shen Kua (in the +11th century) was so exceptional a man that he took notice of the *Mu Ching*[1] (Timberwork Manual)[c] of a great architect Yü Hao,[2] who had probably had to dictate it to a scribe. But other artisans, some centuries before, under the significant inspiration of Taoists and Buddhists, had taken a decisive step to explode this situation—they had invented printing.[d] This without doubt fostered the second flowering of Chinese mathematics, in the Sung, when a group of truly great mathematicians, themselves either commoners or subordinate officials, broke out into fields much wider than the traditional bureaucratic preoccupations. Intellectual curiosity could now be abundantly satisfied. But the upsurge did not last. The Confucian scholars who had practised calligraphy on all the last copies of Tsu Chhung-Chih's *Chui Shu*[e] swept back into power in the nationalist reaction of the Ming,[f] and mathematics was again

[a] A very strong statement of the extent to which the Newtonian world-view depended on a belief in a supreme creator and upholder deity will be found in Koyré (1), p. 308.

[b] For Arabic parallels see Cahen (2) on such treatises as the *Kitāb al-Ḥāwī* of about +1035.

[c] See below, Sect. 28*d*.

[d] See below, Sect. 32.

[e] See above, p. 35.

[f] Significantly, the Pa Ku[3] essay examination system, with all its deadening influence, was introduced first in +1487.

1 木經 2 喻皓 3 八股

confined to the back rooms of provincial yamens. When the Jesuits entered upon the scene, there was no one even able to tell them of China's past mathematical glories.

What was it, then, that happened at the Renaissance in Europe whereby mathematised natural science came into being? And why did this not occur in China? If it is difficult enough to find out why modern science developed in one civilisation, it may be even more difficult to find out why it did not develop in another. Yet the study of an absence can throw bright light upon a presence. The problem of the fruitful union of mathematics with science is, indeed, only another way of stating the whole problem of why it was that modern science developed in Europe at all.

Pledge[a] hits the target when he contrasts Galileo[b] (who must be considered the central figure in the mathematisation of natural science) with Leonardo da Vinci,[c] saying that in spite of all the latter's deep insight into nature and brilliance in experimentation, no further development followed because of his lack of mathematics. Now Leonardo was not the isolated genius that many have been led to suppose him; he was, as Zilsel (2), Gille (3) and others have shown, the most outstanding of a long line of practical men in the +15th and +16th centuries — artist-engineers and architects, such as Brunelleschi;[d] artist-metallurgists such as Cellini;[e] gunners such as Tartaglia;[f] surgeons such as Ambroise Paré;[g] miners who found a voice in Agricola;[h] shipbuilders such as those of the Arsenal at Venice which was the setting for Galileo's *Discourse* of +1638; gunpowder-millers and other chemical technologists whom Biringuccio[i] represents; and instrument-makers such as Robert Norman,[j] whose *Newe Attractive* of +1581 greatly stimulated William Gilbert's work on the magnet.[k] All these men busied themselves with the investigation of natural phenomena, and most of them produced experimental data ready, as it were, for the magic touch of mathematical formulation. Roughly speaking, they had their Chinese counterparts,[l] such as Sung Ying-Hsing[m] (who may be called the Chinese Agricola); or an architect like Li Chieh;[n] or the prince of pharmacists, Li Shih-Chen;[o] or the horticulturist Chhen Hao-Tzu[p] or the gunner Chiao Yü.[q] Whatever field one chooses will yield parallels; thus in horological engineering (allowing for the difference of century) a de Dondi[r] can

[a] (1), p. 15.
[b] +1564 to +1642.
[c] +1452 to +1519.
[d] +1377 to +1446.
[e] +1500 to +1571.
[f] +1500 to +1557.
[g] +1510 to +1590.
[h] +1490 to +1555.
[i] *D.* +1538.
[j] *Fl.* +1590.
[k] See Zilsel (3).
[l] This is a very important point, for those who say, quite rightly, that China never produced a Galileo, a Vesalius or a Descartes, generally forget that China did produce men of a type similar to Agricola, Gesner, and Tartaglia.
[m] *Fl.* +1600 to +1650, author of the *Thien Kung Khai Wu*. Biographies by Lai Chia-Tu (*1*) and Yabuuchi Kiyoshi (*11*).
[n] *D.* +1110, author of *Ying Tsao Fa Shih*.
[o] +1518 to +1593, author of *Pên Tshao Kang Mu*. Biographies by Yen Yü (5); Li Thao (*1*); Chang Hui-Chien (*1*).
[p] *Fl.* +1688, author of the *Hua Ching*.
[q] *Fl.* +1412, author of the *Huo Lung Ching*.
[r] +1318 to +1389.

be matched by a Su Sung.[a] But in Europe, unlike China, there was some influence at work for which this stage was not enough. Something pushed forward beyond it to make the junction between practical knowledge, empirical even when quantitatively expressed, and mathematical formulations.

Part of the story undoubtedly concerns the social changes in Europe which made the association of gentlemen with the technicians respectable. As Gabriel Harvey wrote in +1593:

> He that remembereth Humphrey Cole a mathematicall mechanician, Matthew Baker a shipwright, John Shute an architect, Robert Norman a navigator, William Bourne a gunner, John Hester a chymist, or any like cunning and subtile empirique, is a proud man, if he contemn expert artisans, or any sensible industrious practitioners, howsoever unlectured in Schooles or unlettered in Bookes.[b]

William Gilbert's treatise on the magnet in +1600 was the first printed book composed by an academically trained scholar which was based entirely on personal manual laboratory experimentation and observation. Yet it neither employed mathematical formulations nor spoke in terms of laws of Nature. His contemporary, Francis Bacon, though the first writer in the history of mankind to realise fully the basic importance of modern scientific research for the advancement of human civilisation, understood no better the enormous part which mathematics was soon to play[c].

Not the mathematics of the Middle Ages, however. As Koyré (1) has put it in his brilliant essay on the origins of Newtonian-Cartesian science, mathematics itself had to be transformed, mathematical entities had to be brought nearer to physics, subjected to motion,[d] viewed not in their 'being' but in their 'becoming' or 'flux'. The calculus was the crowning achievement of this movement. In +1550 European mathematics had been hardly more advanced than the Arabic inheritance of Indian and Chinese discoveries. But there followed an astounding range of things basically new—the elaboration of a satisfactory algebraic notation at last by Viète (1580) and Recorde (1557), the full appreciation of what decimals were capable of by Stevin (1585), the invention of logarithms by Napier (1614) and the slide-rule by Gunter (1620), the establishment of coordinate and analytic geometry by Descartes (1637), the first adding machine (Pascal, 1642), and the achievement of the infinitesimal calculus by Newton (1665) and Leibniz (1684). No one has yet fully understood the

[a] +1020 to +1101, author of the *Hsin I Hsiang Fa Yao*. This book, finished in +1092, gives a detailed description of the great astronomical clock which had just then been erected. The story of this enterprise gives us one of the most outstanding examples of the application of mathematical methods to engineering construction in medieval China (see Sect. 27*h* below). For Su Sung's chief engineer, Han Kung-Lien, is recorded as having written a memorandum on the geometry of the clockwork. But although we have here a direct connection between mathematics and technology, it was still to serve a bureaucratic purpose, and the essential connection between mathematics and experimental science was still lacking. Cf. pp. 363ff. below.

[b] Quoted by Taylor (3), p. 161. Cf. especially Taylor (7).

[c] Both these excellent characterisations are due to Zilsel (2).

[d] How central the problem of motion was in Galileo's time can be seen from the penetrating accounts of Koyré (2, 3) and Olschki (2, 3).

inner mechanism of this development.[a] It has often been said that whereas previously algebra and geometry had evolved separately, the former among the Indians and the Chinese, and the latter among the Greeks and their successors, now the marriage of the two, the application of algebraic methods to the geometric field, was the greatest single step ever made in the progress of the exact sciences. It is important to note, however, that this geometry was not just geometry as such, but the logical deductive geometry of Greece. The Chinese had always considered geometrical problems algebraically, but that was not the same thing.

The birth of the experimental-mathematical method, which appeared in almost perfect form in Galileo, and which led to all the developments of modern science and technology, presents the history of science with one of its most important and complex questions. Though we cannot do it justice, a brief analysis here will not be out of place, for only in this way can we gain some idea of how it was exactly that mathematics and science came together at the Renaissance, and how far they had remained apart in earlier medieval, as in Chinese, society.[b] If we dissect the Galilean method we find that it comprised the following phases:

(*a*) Selection, from the phenomena under discussion, of specific aspects expressible in quantitative terms.

(*b*) Formulation of a hypothesis involving a mathematical relationship (or its equivalent) among the quantities observed.

(*c*) Deduction of certain consequences, from this hypothesis, which were within the range of practical verification.

(*d*) Observation, followed by change of conditions, followed by further observation—i.e. experimentation; embodying, in so far as possible, measurement in numerical magnitudes.

(*e*) Acceptance, or rejection, of the hypothesis framed in (*b*).

(*f*) An accepted hypothesis then served as the starting-point for fresh hypotheses and their submission to test.[c]

That the 'new, or experimental, philosophy' was characterised by the search for measurable elements in phenomena, and the application of mathematical methods to these quantitative regularities, has long been recognised.[d] A world of quantity was substituted for the world of quality.[e] But the advance into abstraction went further than this, for motion was considered apart from any particular moving bodies.[g] The

[a] The best description of it is that of Zeuthen (2).

[b] Apart from the classical studies which will be referred to in the following pages, it is now possible to profit from the brief but valuable expositions of Dingle (1) and Lilley (4). One Chinese thinker has reviewed the field—Lin Chi-Kai (1). He was a pupil of Abel Rey, and though he must often have had Chinese parallels and problems in mind, they were rigidly excluded (unfortunately) from his thesis. The same is true of most modern discussions of scientific method in Chinese (cf. Wang Ching-Hsi, *1*).

[c] Here gradually, with growing confidence, entered in the element of scientific 'prediction'.

[d] As by Whitehead (1), p. 66.

[e] Koyré (1), p. 296. Everywhere men had difficulty in accepting the uniform application of space-time coordinates to all regions of the universe, as witness the opposition of Ho Wên-Phao (*fl.* +1640 to +1670) to the Jesuits (*CJC*, 2nd. add., p. 79).

[f] And Galileo did not hesitate to use concepts of the unobserved and the unobservable—such as a perfectly frictionless plane, or the motion of a body in empty infinite space. Cf. Crombie (1), p. 305, and the discussions of Moody (1) and Wiener (2) on the ideas which stimulated Galileo.

motion of a body had no longer anything to do with its other characteristics or qualities, and could not be derived from them. Moreover, motion was recognised as being the same everywhere in the universe. This was indeed a fundamental change in outlook,[a] for the 'uniformisation' of the Cosmos was also, in a sense, its extinction and death.[b] The geometrisation of space, the substitution of homogeneous, abstract, dimensional, Euclidean space, for the concrete and differentiated place-continuum of pre-Galilean physics and astronomy, was the liquidation of what had been a morphological Cosmos.[c] In fact, the world was no longer to be conceived of as a finite and hierarchically ordered whole, qualitatively and ontologically differentiated; but as an open, indefinite, even infinite, universe, held together only by the identity and universal applicability of simple fundamental laws. There was nowhere in the universe, for example, where the writ of the law of gravitation would not run—once the concept of gravitation had been formulated.

It is evident that the denial of 'inherent tendencies' of bodies to move towards certain places was but one aspect of a general breaking up of the organic unity of the material object. It seemed, as Dingle says, such an obvious unity, with its compact properties of shape, weight, colour and movement, that only a mind of the highest originality, goaded by centuries of frustration, could take the revolutionary step of rejecting this unity, and maintaining that a wooden ball and a planet of unknown substance had more in common than the motion and the colour of the same ball. And, indeed, the Galilean revolution did destroy the organic world-view which medieval Europeans had possessed, to some extent in common with the Chinese, replacing it by a world-view essentially mechanistic, and fully ripe for fortuitous concourses of atoms. The sense of loss and disorientation experienced by minds steeped in the traditional world-view was expressed by John Donne:[d]

> And new Philosophy calls all in doubt,
> The Element of Fire is quite put out;
> The Sun is lost, and th' earth, and no man's wit
> Can well direct him where to looke for it.
> And freely men confesse that this world's spent,
> When in the Planets, and the Firmament
> They seeke so many new....
> 'Tis all in Peeces, all cohaerance gone;
> All just Supply, and all Relation....

But the dramatic irony of fate brought it about that by the time of the death of Newton himself (1727) the seeds of a new organic view of the world, destined ultimately to

[a] See especially on this Koyré (1, 2, 3, 5). [b] Koyré (1), p. 295.

[c] Time also became continuous, undifferentiated and homogeneous; in contrast with the separate and divided times of medieval thought, both Eastern and Western (cf. Vol. 2, p. 288). There is an echo of this in current discussions among historians concerning Chinese historiography. As van der Sprenkel (1) has said, most Chinese writers (though by no means all) worked with a compartmentalised chronology of dynastic periods, reigns and reign-periods; discrete units of time being inhabited by particular events. The time-continuum of modern history is surely derived from the world-outlook of modern science.

[d] 'Anatomie of the World', First Anniversary, ll. 205–14.

replace or to correct the mechanistic view, had been sown by Leibniz.[a] Perhaps some of these originated in China, but that is another argument which cannot be retraced here.[b]

That the hypothesis formed should be a mathematised hypothesis (phase (*b*) above) was of enormous importance. Mathematics was the largest and clearest body of connected logical thought then available. That the logic of experimentation was not absolutely bound to mathematical expression became, no doubt, apparent in the science of physiology from William Harvey and J. B. van Helmont to Claude Bernard, and in chemistry too. But it formed the model.[c] Much discussion has centred around the origins of the mathematisation of hypotheses, and the historical problem is still far from its solution. Burtt, in a well-known book (1), and Koyré, have emphasised the persistence of Pythagorean and Platonic influences, exemplified in the view of the universe as a mathematical design, and mediated through such men as Ficino and Novara. There was also the perennial importance of mathematics in astronomical science, and no doubt a certain stimulus from the rediscovery of Greek writers such as Archimedes. Galileo himself certainly said:[d]

> Philosophy is written in that great book which ever lies before our gaze—I mean the universe—but we cannot understand it if we do not first learn the language and grasp the symbols in which it is written. The book is written in the mathematical language, and the symbols are triangles, circles, and other geometrical figures, without the help of which it is impossible to conceive a single word of it, and without which one wanders in vain through a dark labyrinth.

Nevertheless, E. W. Strong has convincingly shown that before Galileo, and during his lifetime, mathematics was increasingly utilised by the practical technicians and artisans of whom we have already spoken. Some of these, such as Nicolò Tartaglia and Simon Stevin, were among the best mathematicians of their time. Their interests in gunnery, in shipbuilding, in hydraulic engineering and building technology, invited them at all points to apply to their problems the quantitative and the mathematical. They were men of measure and rule. The great Renaissance craftsmen would have found natural enough Galileo's isolation of particularly simple examples of motion in phase (*c*) above, and the measurement of numerical magnitudes in phase (*d*). In fact, as Whitehead has said,[e] the idea of functionality had been born. One had to see how much change of a single specific condition corresponded to how much variation in the effect produced. 'Mathematics supplied the background of imaginative thought with which men of science approached the observation of Nature. Galileo produced formulae, Descartes produced formulae, Huygens produced formulae,

[a] This is well appreciated by Koyré, who says, (1), p. 310, that the evolution of 19th-century scientific thought, proceeding under the aegis of the field concept, was an essentially anti-Newtonian development.

[b] See Vol. 2, in Sects. 13*f*, 16*f*, 18*e*.

[c] Mechanics occupied a special position as the starting-point of modern science because the direct physical experience of man is predominantly mechanical, and the application of mathematics to mechanical magnitudes was relatively simple. Cf. Sambursky (1), p. 234.

[d] *Opera*, vol. 4, p. 171.

[e] (1), p. 46.

Newton produced formulae.' Everyone began to draw curves showing the relations between natural phenomena, and to find equations to fit them.

Perhaps the Galilean innovation may best be described as the marriage of craft practice with scholarly theory.[a] Tremendous though the consequences were which followed from this conjunction, it is important to realise that it was not in every respect unique. Situations of a somewhat similar kind had occurred before on lower planes, each generating something new. As Eliade has pointed out,[b] Hellenistic alchemy, with all its technical import for the progress of chemistry, was a marriage of practical chemical craftsmanship with Orphic and Gnostic philosophy. And we are now familiar with an earlier example, the origin of proto-scientific Taoism in the alliance of the shamanist magician-technicians with the hermit philosophers of the Tao of Nature.[c]

In what way, then, had the instinctive experimentation of the technologists and craftsmen[d] differed from the conscious experimental test of precise hypotheses which formed the essence of the Galilean method? The question is of great importance, for the higher artisanate (as we might perhaps call it) included as many Chinese as Europeans. Dissected in the same way as before, it might give us something like this:

(*a*) Selection, from the phenomena under discussion, of specific aspects.

(*d*) Observation, followed by change of conditions, followed by further observation—i.e. experimentation, embodying, in so far as possible, measurement in numerical magnitudes.

(*b*) Formulation of a hypothesis of primitive type (e.g. involving the Aristotelian elements, the Tria Prima of the alchemists, or the Yin-Yang and Five-Element theories).

(*f*) Continued observation and experimentation, not too strongly influenced by the concurrent hypothetical considerations.

In such empirical ways it was possible to accumulate great stores of practical knowledge, though the lack of rationale necessitated a handing down of technical skill from one generation to the next, through personal contact and training. With due regard to different times and places there was not much to choose between China and Europe regarding the heights of mastery achieved; no westerners surpassed the bronze-founders of the Shang and Chou,[e] or equalled the ceramists of the Thang and Sung.[f] The preparations for Gilbert's definitive study of magnetism had all taken place at the other end of the Old World. And it could not be said that these technological operations were non-quantitative, for the ceramists could never have reproduced their effects in glaze and body and colour without some kind of temperature control, and the discovery of magnetic declination could not have occurred if the geomancers had not been attending with some care to their azimuth degrees.[g]

The first step in the series of procedures, both of the Galileans and of the artisans,

[a] This point has often been made by Bernal, cf. (1), pp. 865ff., 869.
[b] (5), p. 149; following Festugière (1); Sherwood Taylor (2, 3) and many other scholars.
[c] Cf. Vol. 2, pp. 34ff.
[d] Reference is again made to the admirable review of Zilsel (2) on this subject.
[e] Cf. Sects. 30*d*, 36 below.
[f] Cf. Sect. 35 below.
[g] Cf. Sect. 26*i* below.

has passed without remark. Yet as several writers have emphasised,[a] the isolation of specific phenomena from the flux of things for the purpose of systematic study confers upon all experimentation a highly artificial character. Hence the interest of the fact that the medieval artisans (unlike the Greeks) had shown themselves able to do this, in China as well as in the West, long before the scientists of the 17th century. Indeed, it arose naturally because each wright and craftsman was interested only in a limited group of techniques. These men had also done another thing, they had realised the importance of the repetition of experiments for the confirmation of results. Greek astronomy had been so successful because the repetitions were provided by Nature in the cyclical recurrences of celestial phenomena. It took the medieval artisans (Chinese as well as European) to organise those earthly repetitions, varying in known degrees, which paved the way for modern physics and all the terrestrial sciences.

But the inhibition lay in the realm of hypothesis-making, as one may see in the relative theoretical backwardness of Leonardo. Duhem,[b] after describing some of his achievements and inventions connected with matter in the gaseous state, points out that his ideas on air and fire, smoke and vapour, were so impregnated with medieval physics that what he did and suggested seems almost inexplicable. While sketching a hygrometer, a helicopter, or a centrifugal pump, he was capable of explaining that the moisture of a wet rag has an intrinsic tendency to move to the fire, and that its less material parts accompany the ascent of that pure element towards the empyrean, for fire has a quasi-spiritual power of carrying light things up with it. There is no need to illustrate this point more abundantly, but it is an important one, since it helps us to understand what remarkable technical achievements may be effected without adequate scientific theory. It therefore throws light on the Chinese situation, and defines the point reached by indigenous Chinese science and technology as Vincean, not Galilean. But what exactly happened in Europe remains difficult to unravel.

Historians have long recognised that the middle of the +12th century was a turning-point in the history of European thought. Whether or not because of the stimulus of new contacts with the Islamic world,[c] the +12th and +13th centuries saw a vast movement away from anthropocentric symbolism towards genuine interest in objective Nature.[d] This can be traced in every department of thought and art, from the growing naturalism of Gothic stone-carving[e] to the rise of new realism in theology, liturgy and drama. It is not possible to overlook this naturalistic movement in tracing the roots of modern science.

The higher artisanate was not the only group which possessed part of the Galilean method before Galileo. It has long been maintained[f] that within European scholastic

[a] Levy (1), pp. 118, 700, etc., and more recently Sambursky (1), pp. 233ff.

[b] (1), p. 329. Cf. Randall (2); Hart (3). And as Zilsel (5) points out, even in Copernicus, many medieval and animistic ideas persisted.

[c] There is, at any rate, an exact contemporaneity here with such important translators as Adelard of Bath, whose chief work ended about +1142.

[d] This has been illuminatingly expounded by Goetz (1) and Lynn White (2). See also the parallel account in Raven (1), pp. 40, 58 ff.

[e] Reference may be made to the standard work of Jalabert (1) on the capitals of columns in church architecture.

[f] Especially by Duhem (1) in his studies on the precursors of Leonardo da Vinci.

philosophy there was a trend towards experimentation which, starting from Aristotle,[a] led to Galileo through Leonardo. Aristotle distinguished between knowledge of facts and knowledge of the reasons or causes of facts,[b] but never gave any clear account of how these could be ascertained by experiment.[c] Early in the +13th century philosophers at Oxford began to interest themselves in the possibilities of a deeper understanding of natural phenomena, and much attention was given to the forming of hypotheses and the manner of their testing. These ideas came later to be associated with the university at Padua,[d] where Averroism was strong and logic was studied as a preliminary to medicine, not law or theology. Discussions there, between the +14th and the +16th centuries,[e] led to a methodological theory which, except for the important element of mathematisation, showed some similarity to the eventual practice of Galileo. The theory (called at Padua *regressus*),[f] dissected, looks something like this:

(*a'*) Selection, from the detailed phenomena under discussion, of features which seemed to be common to all of them (*analysis*,[g] *resolutio*), complete enumeration being recognised as unnecessary because of faith in the uniformity of Nature and the representativeness of samples.

(*b'*) Induction of a specific principle by reasoning on the essential content of these features (also *resolutio*).

(*c'*) Deduction (*synthesis*[g] in thought, *compositio*) of the detailed consequences of this hypothetical principle.

(*d'*) Observation of the same, and perhaps also similar, phenomena, leading to *verificatio* or *falsificatio* by experience, and in rare cases, by arranged experiment.

(*e'*) Acceptance, or rejection, of the hypothetical principle formulated in (*b'*).

Thus while the practice of the higher artisanate was akin to the second or experimental part of the Galilean method, the theorising of the scholastics foreshadowed the first or speculative part. But how widely they were aware that agreement with empirical fact was the ultimate test of hypotheses seems doubtful, nor is it clear that they always understood the importance of examining new phenomena in phase (*d'*) which had not already been used as the source of the hypothesis under test. Moreover, they rarely succeeded in advancing beyond the primitive style in their hypotheses. Robert Grosseteste of Lincoln (+1168 to +1253) has been selected as the key figure in this natural philosophy,[h] but the dual process of induction and deduction goes back to Galen[i] and

[a] On Aristotle's scientific method see McKeon (1); W. D. Ross (1); Peck (1, 2).

[b] *Post. Analyt.* I, 13; 78 a 22.

[c] To what extent the full experimental method was practised in ancient Greece has been the subject of much discussion. The systematic observations of the Pythagoreans on acoustics, of Archimedes on the lever, and of Straton of Lampsacus and Heron on pneumatic phenomena, certainly come near it, but may not go beyond the level of the higher artisanate. Cf. Heath (6, 8); Brunet & Mieli (1); Farrington (4, 15), and the recent excellent discussion of Sambursky (1), pp. 222ff., 237.

[d] It will be remembered that both Galileo and Harvey studied and taught at Padua. Galileo used the technical terms of the school, calling the experiment in his phase (*d*) a further *resolutio* (Crombie (1), p. 307).

[e] Of which a detailed account has been given by Randall (1).

[f] Crombie (1), p. 297.

[g] These were the earliest terms—used by the Greek geometers and Galen (cf. Crombie (1), p. 28).

[h] As in the interesting book of Crombie (1).

[i] Kühn ed. vol. 1, p. 305; vol. 8, p. 60; vol. 14, p. 583. Galen certainly made physiological experiments himself, as did Herophilus and Erasistratus.

the Greek geometers,[a] probably reaching Grosseteste through Arabic sources, such as the encyclopaedist Abū Yūsuf Ya'qūb ibn-Isḥāq al-Kindī (*d.* +873)[b] and the medical commentator 'Alī ibn Riḍwān (+998 to +1061).[c] Though Grosseteste may have believed that organised experimentation beyond mere further experience should be used to verify or disprove hypotheses, it is not claimed that he himself was an experimentalist. He does seem however to have influenced the +13th-century group of practical scientific workers which included the Englishmen Roger Bacon (1214 to 1292) and Thomas Bradwardine (1290 to 1349) in physics, the Frenchman Petrus Peregrinus (*fl.* 1260 to 1270) in magnetism, the Pole Witelo (*c.* 1230 to 1280) in optics, and the German Theodoric of Freiburg (*d.* 1311) with his admirable theory of the rainbow.[d] It is curious that during the period in which these men were working, China was the scene of a scientific movement quite comparable.[e] But after the early years of the 14th century there was a marked regression, and verbal argument again dominated in Europe until the time of Galileo himself. Or such at any rate was the case so far as theoretical science was concerned, for the latter part of the +14th century and the +15th saw the rise of the military engineers, largely German, whose practical achievements foreshadowed those of the higher artisanate in the century before Galileo. The originator of this new phase was Konrad Kyeser (1366 to after 1405) with his *Bellifortis*, begun in +1396, but earlier figures had shown the way, especially Guido da Vigevano (1280 to after 1345)[f] whose work on machines and war machines had been finished only a couple of decades after the death of Theodoric of Freiburg. Kyeser was followed by many other technologists who certainly owed part of their inspiration to the new techniques of gunnery and gunpowder,[g] such as Giovanni de' Fontana (*fl.* 1410 to 1420), the anonymous engineer of the Hussite Wars (*fl.* 1420 to 1433), Abraham of Memmingen (*fl.* 1422), and so on.[h] There was thus a continuous line of experimentalists in Europe from Roger Bacon to Galileo, but after about +1310 the contribution of scholastic philosophy ceased, and for three centuries practical technology was the order of the day.[i]

[a] Heath (1), pp. 137 ff.; Zeuthen (3), pp. 92 ff.

[b] Mieli (1), p. 80; Hitti (1), p. 370.

[c] Mieli (1), p. 121.

[d] In which he had been anticipated by Arab physicists (cf. Sects. 21*e*, 26*g* below), but only by a very short time (see Sarton (1), vol. 2, p. 23; vol. 3, p. 141). Since their work was not available in Latin, the simultaneity probably arose from the use of a common source, Ibn al-Haitham. See Winter (4).

[e] It included the outstanding mathematicians Chhin Chiu-Shao (*fl.* +1240 to +1260) and Yang Hui (*fl.* +1260 to +1275), the astronomer Kuo Shou-Ching (+1231 to +1316), the geographer Chu Ssu-Pên (*c.* +1270 to +1337), as well as Wang Chen (*fl.* +1280 to +1315), encyclopaedist of agriculture and engineering, and Sung Tzhu (*fl.* +1240 to +1250), the founder of forensic medicine—a particularly experimental discipline. The period also saw the travels of men such as Marco Polo (in Peking, *c.* +1280), but it is not likely that any of them were sufficiently versed in the sciences to transmit anything more than techniques or fragments of techniques. On the 'golden period' of science and technology in Sung China, see further Vol. 2, pp. 493 ff.

[f] Sarton (1), vol. 3, p. 846.

[g] Which had been a direct Chinese transmission to Europe, see Sects. 30, 34 below.

[h] These men will be referred to frequently in Sects. 27, 28 below. For a good summary of their work and books see Sarton (1), vol. 3, pp. 1550 ff.

[i] The late +14th and +15th centuries were not great periods of scientific or technological achievement in China, where the +11th had been relatively much more important than in Europe. But it is not difficult to find men of worth—such as the prince Chu Hsiao (*fl.* +1382 to +1425) who main-

Someone should raise the question whether the theorising of the Neo-Confucians of the +11th and +12th centuries[a] about the acquisition of natural knowledge was not as advanced in its way as that of the +13th-century European scholastics.[b] The induction of a specific principle from many observations (*b'*, the second part of *resolutio*) was represented in Neo-Confucian thought by the search for the underlying or intrinsic patterns (*li*[1]). Someone said to Hsü Hêng[2] (+1209 to +1281):

If we fully apprehend (lit. exhaust) the patterns of the things of the world, will it not be found that every thing must have a reason why it is as it is (*so i jan chih ku*[3])? And also a rule (of co-existence with all other things)[c] to which it cannot but conform (*so i tang jan chih tsê*[4])? Is not this just what is meant by Pattern (*li*[1])?[d]

Hsü Hêng agreed, saying that this brought out very well the meanings of the technical terms employed. All the spatio-temporal relations of all organisms and events in the universe were determined by the ubiquitous manifestations of *li*. 'Wherever there is *li*,' said Chhêng I-Chhuan[5] (+1033 to +1108), 'east is east and west is west.'[e] The key phrase of the Neo-Confucians for the process akin to induction was taken from the *Ta Hsüeh* (Great Learning), a text of about −260, 'the extension of knowledge consists in the investigation of things' (*chih chih tsai ko wu*[6]).[f] For them this meant a kind of sudden insight into the natures and relations of things, as if the components of a pattern were seen suddenly to 'fall into place'. The figures of Nature fitted themselves, as it were, into a meaningful array on the counting-board of the universe.

Apprehension of specific natural patterns present in a multiplicity of phenomena (*a'*) was reached by a process of 'relating' or 'threading together' (*kuan*[7]), or 'interrelating' (*kuan thung*[8]). The image was that of holed cash threaded on a string. As one of the Chhêng brothers said, 'Whenever men hear a saying or learn of an affair, and their knowledge is still confined to this one saying or affair, it is simply because they cannot interrelate.'[g] Another of the formulations of the Chhêngs was the following:[h]

In labouring to apprehend (lit. exhaust) patterns fully, we are not necessitated to attempt an exhaustive and complete research into the patterns of all the myriad phenomena in the

tained a botanical garden and wrote a valuable work on plants suitable for food in emergency, or the astronomer Huangfu Chung-Ho (*fl.* +1437) who made new instruments similar to those of Kuo Shou-Ching, or Ma Huan (*fl.* +1400 to +1430), Yunnanese Muslim interpreter, geographer on the staff of the admiral Chêng Ho.

[a] See Sects. 16*d* and 18*e* in Vol. 2.

[b] A convenient introduction to the problem is afforded by the study of Graham (1), esp. pp. 192ff.

[c] For a full discussion of the term *tsê* as 'rules applicable to parts of wholes' see Vol. 2, pp. 557ff., 565ff.

[d] *Sung Yuan Hsüeh An*, ch. 90, pp. 2*b*, 3*a* (*WYWK* ed. vol. 22, p. 128), tr. auct. A very similar definition is given by Wu Chhêng[9] (+1249 to +1333) in *Hsing Li Ching I*, ch. 9, p. 29*b*.

[e] *Honan Chhêng shih I Shu*, ch. 22A, p. 14*b*.

[f] Cf. Vol. 1, p. 48. See Legge (2), p. 222.

[g] *Honan Chhêng shih Wai Shu*, ch. 3, p. 2*b*.

[h] *Honan Chhêng shih I Shu*, ch. 22A, p. 22*b*; tr. auct. adjuv. Graham (1).

1 理 2 許衡 3 所以然之故 4 所以當然之則 5 程伊川
6 致知在格物 7 貫 8 貫通 9 吳澄

world. Nor can we attain our aim by fully apprehending only a single one of these patterns. It is simply necessary to accumulate (lit. pile up and tie together, *chi lei*[1]) a large number (of phenomena). Then (the patterns) will become visible spontaneously.

The conviction of the Chhêng brothers that concentration on one thing or on one small group of things was not the way to natural knowledge is particularly interesting in view of the failure of Chinese scholars in later times to appreciate the scientific method.[a]

Someone asked Chhêng I-Chhuan, 'Is it necessary to investigate all things, or can the innumerable patterns be known simply by the investigation of a single thing?' (I-Chhuan replied): 'No indeed, for in that case how could there be comprehensive inter-relation? Even a Yen Tzu[b] would not attempt to understand the patterns of all things by the investigation of only one thing. What is necessary is to investigate one thing after another day after day. Then after long accumulation of experience (the things) will suddenly reveal themselves in a state of inter-relatedness (*kuan thung*[2]).'[c]

Introspection was no substitute for the study of external nature.

Again someone asked Chhêng I-Chhuan, 'In observing outer things and in searching into the self, should one look back into oneself to seek what has already been seen in things?' (He replied): 'There is no need to put the matter in that way. The external world and the self have one single great Pattern in common; as soon as "that" is understood, "this" becomes clear. This is the Tao of the union of the internal and the external. The scholar should try to observe and understand all Nature—at one extreme the height of the heavens and the thickness of earth—at the other, why a single (tiny) thing is as it is.'

Somebody else said: 'In extending knowledge, what do you say to seeking (the patterns of the world) first in the "Four Beginnings" (*ssu tuan*[3])?'[d] (The philosopher answered): 'To seek them in our own nature and passions is of course simple and near at home, but every (blade of) grass and every tree have their own patterns, and these must be investigated.'[e]

'A wide knowledge of the names (and properties) of birds, animals, plants and trees, is one of the means of reaching an understanding of pattern.'[f]

If this is not yet the natural science of the Renaissance, it seems no further away from it than the ideas of the medieval European scholastics.

Besides, the converse process of deduction from principles (*c'*, *compositio*) seems to have its counterpart in 'extending the pattern-principle' (*thui li*[4]) or 'enlarging the

[a] We are, of course, thinking of the idealist school of Wang Yang-Ming in the early +16th century; see Vol. 2, p. 510.

[b] The favourite pupil of Confucius.

[c] *Honan Chhêng shih I Shu*, ch. 18, p. 5*b*.

[d] The allusion is to *Mêng Tzu*, II (1), vi, 5 (Legge (3), p. 79). The feeling of commiseration is the beginning of human-heartedness, that of shame and dislike the beginning of righteousness; modesty and complaisance lead to good customs, and deciding between 'is' and 'is not' to knowledge. Cf. *Hsiao Hsüeh Kan Chu*, ch. 3, p. 16*a*.

[e] *Honan Chhêng shih I Shu*, ch. 18, pp. 8*b*, 9*a*.

[f] *Honan Chhêng shih I Shu*, ch. 25, p. 6*b*; both passages tr. auct. adjuv. Graham (1). The reference to the value of natural history is a quotation from *Lun Yü*, XVII, ix, 7. Parallel passage in *Yang Kuei-Shan Chi*, ch. 3, p. 66 (§41).

[1] 積累 [2] 貫通 [3] 四端 [4] 推理

class (of things or processes with the same pattern)' (*thui lei*[1]).[a] Sometimes the latter phrase is used in the sense of 'inferring by analogy'. Chhêng Ming-Tao[2] (+1032 to +1085) wrote:

> The myriad things all have their opposites; there is alternation of the Yin and the Yang, the good and the bad. When the Yang waxes the Yin wanes, when good increases evil is diminished. Far and wide is the spread of this pattern-principle (*li*[3]).[b]

And elsewhere:

> In investigating things to apprehend fully their patterns, there is no question of completely exhausting all the phenomena in the world. If the pattern is fully apprehended in one matter only, inferences can be made about other matters of the same class (*kho i lei thui*[4]).[c]

Both processes (induction followed by deduction, *b′*, *c′*) seem to be mentioned in these words of Chhêng I-Chhuan:

> To learn (the patterns) from what is outside, and to grasp them within, may be called 'understanding' (*ming*[5]). To grasp them within and to connect them (*chien*[6]) with what is outside may be called 'integration' (*chhêng*[7]).[d] Now integration and understanding are one.[e]

When we come to verification or invalidation by experiment (*d′*), the Neo-Confucians were no more enlightened than the Scholastics. But the idea of submission to the test of experience was always vaguely present, and the form which it took in the Chinese milieu, so dominated by ethics and sociology, was the contrast of knowledge and practice. This subject was debated throughout the centuries. The famous tag 'Action is easy but knowledge difficult' (*hsing i chih nan*[8])[f] was affirmed, reversed or modified in every age.[g] 'Knowledge', said Wang Chhuan-Shan[9] in the +17th century,[h] 'is the beginning of practice, and practice is the completion of knowledge.' In fact the epistemological problem received answers which varied in accordance with the current trend (so far as Chinese thought allowed of it) towards metaphysical idealism or materialism. Many examples of these have already been given as, for example, in the criticism of Wang Chhung (+1st century) on the Mohists;[i] while in our own time

[a] The technical term *thui* is one which goes back to the Mohists (cf. Vol. 2, pp. 183ff.), but its sense had quite changed. It is constantly found in scientific contexts. For example, in +19 Wang Mang ordered his Astronomer-Royal to compute (*thui*) a calendar which would be valid for 36,000 years (*Chhien Han Shu*, ch. 99c, p. 4*b*; cf. Dubs (2), vol. 3, p. 379). And about +260 Wang Fan, in his calculations of cosmic distances, deduces (*thui*) certain figures from others (*Chin Shu*, ch. 11, p. 6*b*).

[b] *Honan Chhêng shih I Shu*, ch. 11, p. 5*a*.

[c] *Honan Chhêng shih I Shu*, ch. 15, p. 11*a*; both passages tr. auct. adjuv. Graham (1).

[d] This technical term was discussed at length in Vol. 2, pp. 468ff.

[e] *Honan Chhêng shih I Shu*, ch. 25, p. 2*a*; the passage is based on *Chung Yung*, XXI.

[f] This was the form made famous by Sun Yat-Sen in modern China, and used, for example, as the device of the National Peiping Academy. It is the opposite of the phrase as it made its first appearance—in one of the spurious (i.e. +4th century) chapters of the *Shu Ching* (Historical Classic), ch. 17B (Yüeh Ming), *fei chih chih chien, hsing chih wei chien*[10] (cf. Legge (1), p. 116; Medhurst (1), p. 173). The speaker was Fu Yüeh,[11] a semi-legendary minister of the Shang.

[g] There seems to be no complete historical study in a Western language, but an interesting beginning has been made by Nivison (1).

[h] Cf. Fêng Yu-Lan (1), vol. 2, p. 604.

[i] Cf. Vol. 2, p. 170.

[1] 推類 [2] 程明道 [3] 理 [4] 可以類推 [5] 明 [6] 兼
[7] 誠 [8] 行易知難 [9] 王船山 [10] 非知之艱行之惟艱
[11] 傅說

the solutions of the Chinese philosophical schools have received a Marxist critique.[a] In any case, there was no clear understanding, either among the Neo-Confucians or the Scholastics, that in the study of Nature precise hypotheses must be put to the test of agreement with further ranges of empirical fact. The important point is that just as in China there were representatives of the 'higher artisanate' in abundance, counterparts of Norman and Tartaglia, so also there were medieval thinkers who corresponded to Grosseteste and the Paduans.

Perhaps it would not be fruitful to compare too closely the European Scholastics with the Neo-Confucians. At all events the former do not always pre-empt our sympathy. Two strangely contrasting statements may indicate which of the two schools was really the more scientifically minded. Thomas Aquinas (+1226 to +1274) wrote that 'a little knowledge about the highest things is better than the most abundant knowledge about things low and small'.[b] But Chhêng Ming-Tao (+1032 to +1085) said of the Buddhists: 'When they strive only to "understand the high" without "studying the low",[c] how can their understanding of the high be right?'[d]

No question is more difficult than that of historical causation. Yet the development of modern science in Europe in the +16th and +17th centuries has either to be taken as miraculous or to be explained, even if but provisionally and tentatively. This development was not an isolated phenomenon; it occurred *pari passu* with the Renaissance, the Reformation, and the rise of mercantile capitalism followed by industrial manufacture. It may well be that concurrent social and economic changes supervening only in Europe formed the milieu in which natural science could rise at last above the level of the higher artisanate, the semi-mathematical technicians.[e] The reduction of all quality to quantities, the affirmation of a mathematical reality behind all appearances, the proclaiming of a space and time uniform throughout all the universe; was it not analogous to the merchant's standard of value? No goods or commodities, no jewels or monies there were, but such as could be computed and exchanged in number, quantity and measure.

Of this there are abundant traces among our mathematicians. The first literary exposition of the technique of double-entry book-keeping is contained in the best mathematical text-book available at the beginning of the +16th century, the *Summa de Arithmetica* (+1494) of Luca Pacioli. The first application of double-entry book-keeping to the problems of public finance and administration was made in the works of the engineer-mathematician Simon Stevin (+1608). Even Copernicus wrote on monetary reform (in his *Monetae Cudendae Ratio* of +1552). The book of Robert Recorde, in which the equality symbol was first used (*Whetstone of Witte*, +1557),

[a] As in Mao Tsê-Tung's *On Practice*, and the interesting commentary of Fêng Yu-Lan (6) upon it.

[b] *Summa Theologiae*, Ia, i, 5 ad 1. Cf. Aristotle, *De Partibus Animalium* I. 5.

[c] Phrases from *Lun Yü*, XIV, xxxvii, 2 (Legge (2), pp. 152, 153).

[d] *Honan Chhêng shih I Shu*, ch. 13, p. 1*b*.

[e] Well-known expositions of this point of view are to be found in the book of Borkenau (1), the works of R. K. Merton, and the celebrated essay of Hessen (1) on the factors in the intellectual climate of Newton's time which led him to prefer certain subjects for investigation rather than others. It has also had its critics, notably Grossmann (1), G. N. Clark (1), and A. R. Hall (1) who thoroughly studied the details in a key-subject, ballistics.

was dedicated to 'The Governors and the reste of the Companie of Venturers into Moscovia', with the wish for 'continuall increase of commodities by their travell'. Stevin's *Disme* opens with the words 'To all astronomers, surveyors, measurers of tapestry, barrels and other things, to all mintmasters and merchants, good luck!'[a] Even a relative of the great and selfless missionary, Francesco Ricci, published a work on accounting at Macerata in +1659. Such examples could be indefinitely multiplied. Commerce and industry were 'in the air' as never before.

The problem of the exact relations between modern science and technology and the socio-economic circumstances of its birth constitutes, perhaps, the Great Debate of the history of science in Europe. We shall have to return to it later on.[b] I believe that in due course the study of parallel civilisations such as that of agrarian-bureaucratic China will throw some light on the events which took place in the West. For example, Koyré,[c] in criticising the socio-economic theory of the causation of post-Renaissance science, urges,[d] with Cassirer (1), that there was a purely theoretical current stimulated by the rediscovery of Greek mathematics, and under recognisably Platonic and Pythagorean inspiration. This is doubtless part of the truth.[e] He also urges that the supporters of the socio-economic theory insufficiently allow for what he calls the autonomous evolution of astronomy. But here is the kind of point where it is profitable to compare parallel events in China. The Chinese should have been interested in mechanics for ships, in hydrostatics for their vast canal system (like the Dutch), in ballistics for guns (after all, they had possessed gunpowder[f] four centuries before Europe), and in pumps for mines. If they were not, could not the answer be sought in the fact that little or no private profit was to be gained from any of these things in Chinese society, dominated by its imperial bureaucracy? Their techniques and industries, described in the books of men akin to the writers of the 'semi-mathematical' stage mentioned above, were all essentially 'traditional', the product of many centuries of slow growth under bureaucratic oppression or at best tutelage, not the creations of enterprising merchant-venturers with big profits in sight.[g] As for astronomy, no organisation stood more in need of it than the Chinese imperial court, which by immemorial custom gave forth the calendar to be accepted by all under heaven.[h] And,

[a] These examples are due to Zilsel (2).

[b] Below, Sects. 48 and 49 on the social factors affecting science in China and the West.

[c] (1), (4), p. 294.

[d] Nevertheless, he sees very clearly (p. 310) how congruent the mechanistic conception of fortuitous concourses of atoms was with the social ideas of capitalism, in which harmony would prevail, it was thought, if every man devoted himself to the selfish pursuit of gain. A book such as Malyne's *Lex Mercatoria* shows how conscious this analogy was in the 17th century.

[e] Another strand derived from certain schools of medieval nominalism which denied the objective reality of relations and considered individual things as isolated. The relevance of this to the mathematical and mechanical view of the world has been elucidated in an unjustly neglected book by Conze (5).

[f] And used it in war; not only for fireworks, as a common but mistaken opinion maintains.

[g] In this connection it would be valuable to make a study of all the different types of problems contained in the old Chinese mathematical books. We have noted much practical area and volume mensuration, summation of series, indeterminate analysis, alligation, etc., but no statistical examination of the material exists.

[h] It is regrettable that no complete study of Chinese calendrical science has been made, either in Chinese or a Western language. The best perhaps is that of Chu Wên-Hsin (*1*). In Japanese there are the valuable works of Yabuuchi Kiyoshi.

as we shall see in the next Section, Chinese astronomy was far from negligible. If an 'autonomous evolution' of astronomy was ever going to give rise to the mathematisation of natural science, it is hard to see why this did not occur, or had not already occurred, in China. If the need had been sufficiently great, there would surely not have been wanting those who could have burst the bonds of the old mathematical notation, and made the discoveries which in fact were only made in Europe. But this was evidently not the force from which modern science could spring, and indigenous Chinese mathematics went down into a kind of tomb, from which the filial care of Mei Ku-Chhêng and his successors only later succeeded in resurrecting it.

Put in another way, there came no vivifying demand from the side of natural science. Interest in Nature was not enough, controlled experimentation was not enough, empirical induction was not enough, eclipse-prediction and calendar-calculation were not enough—all of these the Chinese had. Apparently a mercantile culture alone was able to do what agrarian bureaucratic civilisation could not—bring to fusion point the formerly separated disciplines of mathematics and nature-knowledge.

THE SCIENCES OF THE HEAVENS

20 ASTRONOMY

(a) INTRODUCTION

With this Section we enter upon the first of the natural sciences considered in this book. Astronomy was a science of cardinal importance for the Chinese[a] since it arose naturally out of that cosmic 'religion', that sense of the unity and even 'ethical solidarity' of the universe, which led the philosophers of the Sung to their great organic conceptions, about which much has already been said.[b] The establishment of the calendar by the emperor of an agricultural people, and its acceptance by all those who acknowledged allegiance to him, are threads which run continuously through Chinese history from the earliest times. Correspondingly, astronomy and calendrical science were always 'orthodox', Confucian, sciences, unlike alchemy, for example, which was typically Taoist and 'heterodox'.[c] As has been well said, 'While, among the Greeks, the astronomer was a private person, a philosopher, a lover of the truth (as Ptolemy said of Hipparchus), as often as not on uncertain terms with the priests of his city; in China, on the contrary, he was intimately connected with the sovereign pontificate of the Son of Heaven, part of an official government service, and ritually accommodated within the very walls of the imperial palace.'[d]

This is not to say that the Chinese astronomers of ancient and medieval times were not also lovers of the truth, but it did not appear necessary to them that it should be expressed in that highly theoretical and geometrical form which was characteristic of the Greeks. Apart from the Babylonian records, so many of which are presumably wholly lost, those of the Chinese show that they were the most persistent and accurate observers of celestial phenomena anywhere in the world before the Arabs. Even today, as will be shown below,[e] there is a long period (from about the −5th to the +10th centuries) for which Chinese records are almost the only ones available, and modern astronomers have in many cases (e.g. the recurrent apparitions of comets, especially Halley's)[f] had recourse to these records with valuable results. An outstanding instance is the appearance of novae and supernovae, important in current cosmological

[a] The essay of O. Franke (6) is perhaps the best brief statement of the place of astronomy in the ancient Chinese world-outlook.

[b] Especially in Sects. 16*d* and 18*f* above.

[c] Nevertheless, the Taoists had a good deal to do with astronomy, especially in the Han and earlier times. Ssuma Chhien learnt his astronomy from his father Ssuma Than, whose teacher had been the Taoist Thang Tu.[1] Lohsia Hung[2] also had Taoist affiliations. These men were younger contemporaries of the Taoist group around the Prince of Huai Nan (cf. below, pp. 199, 224, 248, 250).

[d] De Saussure (16*e*). The Austrian Kühnert is good on this: '...wahrscheinlich sind die Chinesen auch deshalb in den Augen manches Europäers Barbaren, weil sie sich unterfangen, die Astronomen—ein höchst unnützes Völkchen nach der Ansicht dieser Erdenpilger im hoch culturellen Westen—im Range gleichzuhalten den Sectionschefs und ersten Ministerialsecretären—o grässliche Barbarei!' (Kühnert, 2).

[e] Cf. pp. 409ff.

[f] Cf. pp. 431ff.

[1] 唐都 [2] 落下閎

theory, for which the Chinese records cover the whole of that period between Hipparchus and Tycho Brahe during which the rest of the world remained in almost complete ignorance of the fact that 'new stars' sometimes appear in the heavens.[a] In other cases (e.g. sun-spots) phenomena were regularly observed by the Chinese for centuries, which Europeans not only ignored, but would have found inadmissible upon their cosmic preconceptions.[b] All these form no small contribution to the history of man's knowledge of celestial events, nor is it vitiated by the fact that the observations of the earlier centuries were often due to the belief in the importance of prognostication for State affairs.[c] After all, astrology in Europe lasted on (and in a narrower, individual form) until the time of Kepler, and was not fully abandoned till the 18th century.[d] If, then, we may adopt the general conclusion that Chinese astronomy participated in the fundamental empiricism characteristic of all Chinese science, this had, at any rate, the effect of allowing a suspension of judgment on subjects such as celestial mechanics, and when, after Matthew Ricci's arrival in China in the late 16th century, he discussed astronomy with Chinese scholars, their ideas, which he preserved in his account of the colloquy,[e] sound in many ways more modern today than his own Ptolemaic-Aristotelian world-view.

The literature in Western languages on Chinese astronomy is much more voluminous than that on Chinese mathematics. Unfortunately, it is confused, controversial and repetitive. From the first, European understanding of Chinese astronomy was affected by the advantages which the Jesuit missionaries saw that they could gain by acquainting the Chinese with the scientific advances of the European Renaissance, and introducing themselves into official circles by their superior calendrical calculations and eclipse predictions. They then strove on the one hand to impress the Chinese and win them over to Christianity by belittling indigenous Chinese astronomical knowledge, while on the other hand they praised it in numerous European publications as part of their general campaign, undertaken within the framework of occidental ecclesiastical politics, to fortify their own position in the mission field.[f] Moreover, from the first, the Jesuit understanding of Chinese astronomy, very honestly undertaken in itself, was vitiated by fundamental misconceptions. These mostly stemmed from the fact, which will receive clarification later, that Chinese astronomy was essentially polar and equatorial, depending largely on observations of the circumpolar stars, while Greek and medieval European astronomy had been essentially ecliptic, depending largely on heliacal risings and settings of zodiacal constellations and their

[a] Only about nine were recorded during this period in the West, and some of these are doubtful: Clerke (1); Stratton (1).

[b] The heavens were thought to be perfect.

[c] See above, Sect. 14*a*, in Vol. 2, pp. 351 ff., where Chinese astrology was discussed. Some of the greatest Greek astronomers, such as Hipparchus, figured prominently in Hellenistic astrological literature (cf. Neugebauer (9), p. 178).

[d] And of course persists even today as a popular superstition in western 'scientific' civilisation.

[e] Cf. below, pp. 438 ff.

[f] 'To have propagated their Religion only in a barbarous and uncultivated Nation, would not have been so much for the Credit of the Mission, as to have been able to introduce it among a People civilised and polished by Arts and Literature' (Costard, +1747). The best single reference to all these questions is the work of Pinot (1, 2).

paranatellons.[a] The Jesuits were naturally quite unprepared for the possibility that another entire system of astronomy might have existed, equal in scope and value, but different in method from that of the Greeks, and their erroneous identifications led to a series of misapprehensions which were not cleared up until the end of the 19th century.

The enormous difficulties which confronted the Jesuit pioneers must not be underestimated. Ricci, Schall, Verbiest, and in a later generation, Gaubil, were in China at a period of spontaneous decline of indigenous science, the Ming dynasty and early Chhing, a decline which had nothing obviously to do with the forces which sent them there and permitted them to stay. Exceedingly few Chinese scholars were available who had enough learning to expound with clarity the traditional Chinese system and to find and translate the essential passages in old books. As we noted in the mathematical section, many important ancient books were wholly lost, and not recovered till the end of the 18th century. There was, of course, the almost insuperable difficulty of language at a time when sinology hardly existed and no good dictionaries had been made. In view of all these handicaps the surprising thing is that a man like Gaubil should have got as far as he did. Moreover, much of the Jesuit interest was given to chronology, since no simple means existed of identifying dates in Chinese history with corresponding dates in that of the rest of the world.[b] This was not strictly a matter concerning the natural sciences.

Though more satisfactory understanding of the Chinese system of astronomy was reached by J. B. Biot (1–7) in the middle of the last century, and by de Saussure (1–34) at the beginning of the present, controversies arose between sinologists, indianists and arabists. This was because a so-called 'lunar zodiac', or chain of constellations situated on or near the equator, was found to be common to the astronomical systems of all these peoples. Firm positions were taken, and categorical statements often made, by writers with little or no access to the original texts of these various cultures. It will be suggested later that perhaps this system did not originate in any of them but was derived by all from Babylonian sources.

We may distinguish two types of interest taken by Westerners in Chinese astronomy. One related to the history of Chinese astronomy as such, considered as part of the history of science in general. The other arose from the attempt to make use of Chinese observations in computations concerning saecular trends, such as the variation of the solstice points, the obliquity of the ecliptic, the precession of the equinoxes, etc.[c] The success of this has depended on the source from which the observations were taken; when the historical records were reliable, as from the Former Han period onwards, the results were valuable; but when weight was placed on texts of the semi-legendary

[a] I.e. constellations on the same hour-circle but away from the ecliptic.

[b] Again I would refer to the works of Pinot, who vividly describes the commotion brought about in Europe by the apparently secure datings of Chinese events in the −3rd millennium. A great part of 18th-century argumentation is now quite nugatory, since Europeans are no longer committed to the ecclesiastical chronology then orthodox, nor are the old Chinese datings, recognised as belonging to legendary rather than historical times, now acceptable.

[c] This was the motive which led Laplace to publish manuscripts of Gaubil nearly a century after he had sent them from Peking to Paris.

period, such as the *Shu Ching*, the dating of which is very difficult, or on alleged observations by Chou Kung about −1000, the results were not, and only served to discredit Chinese material.[a] No small amount of paper and ink has been wasted by writers and computers whose sinological basis was highly insecure.

European literature on Chinese astronomy has also been afflicted by the careless perpetuation of frank mistakes by many authors. Dubs (20) has gone to the trouble of correcting one of these. Yule, in his famous *Cathay and the Way Thither*, says[b] that 'about the same time (+164), and perhaps by means of this embassy,[c] the Chinese philosophers were made acquainted with a treatise on astronomy, which had been brought from Ta-Tsin;[d] we are told that they examined it and compared it with their own.' Such a statement is obviously of much importance, since the possession by the Chinese of Ptolemy's *Syntaxis Mathematica* or *Almagest* (as it was later called), which was finished about +144, would greatly affect any estimate of the development of astronomy in China.[e] Yule gave as his authority C. L. J. de Guignes, in a paper of +1784, and de Guignes refers back to Gaubil (2), i.e. the history of Chinese astronomy by Gaubil which appeared in the second of the three volumes edited by Souciet.[f] Here Gaubil wrote[g] that in the year +164 people arrived in China who said that they had been sent by An-Tun, the king of Ta-Tsin. 'The astronomy of this country has much similarity with that of the Chinese', he added, appending to his statement the following note: 'It is not the History of the Han itself which says that the astronomy of Ta-Tsin had some relation to that of the Chinese, (but the) *Wên Hsien Thung Khao*, the author of which (Ma Tuan-Lin) lived at the end of the Sung.[h] However that may be, it seems certain that there was some kind of comparison of the two astronomical systems, and that could not have been done without a knowledge of both of them.'[i] Though Dubs did not identify the passage in Ma Tuan-Lin's book to which Gaubil referred, it was almost certainly a sentence which in fact says something quite different.[j] There was thus no basis in ancient literature, either Western or Chinese, for the state-

[a] Astronomers such as von Zach were conscious of this difficulty as early as 1816.

[b] Vol. 1, p. 53.

[c] Yule confused the embassies of +120 and +166 (see above, Vol. 1, p. 192).

[d] I.e. Ta-Chhin; Roman Syria.

[e] Sédillot, whose avowed object it was to discredit Chinese astronomy as much as possible in favour of the Arabs, gave great prominence to it, (2), pp. 482, 608 ff., 616. At the same time, in his desire to show that every good thing in East Asia derived from the West, he antedated the Jewish community at Khaifêng by about a thousand years. Perhaps he was misled by Gaubil (2), p. 26 in this.

[f] See below, p. 183.

[g] P. 118; cf. p. 26.

[h] This great historical encyclopaedia was published in +1319, a very long time indeed after the period in question.

[i] Gaubil may have had Persian rather than Graeco-Roman astronomy in mind, since on the following page he identifies Ta-Tsin with the whole region between the Mediterranean and the Caspian. He recognised An-Tun as Marcus Aurelius Antoninus, however.

[j] In the entry on Ta-Chhin, in ch. 339 (p. 2659.3), after speaking of the embassy, Ma says: 'Although in that country the sun, moon and constellations appear no different from what we see in China, former historians have said that going 100 *li* west of Thiao-Chih one comes to the place where the sun sets—this is indeed far from the truth.' Thus the comparison was between constellations, not astronomies. And a hurried reading might have confused An-Tun with An-Hsi (Parthia), which is mentioned just above as being on the way to Ta-Chhin. Perhaps de Guignes read Gaubil no more hastily than Gaubil read Ma. The latter's remark was not taken from his chief basic texts (*Thung Tien*, ch. 193 and *Thung*

ment of de Guignes and Yule, and no reason whatever for thinking that Greek astronomical works reached the Chinese at the end of the +2nd century. The whole story rests on a misunderstanding of Ma by Gaubil. But as usual, the errors of orientalists were dutifully copied by historians of science. Cordier, who ought to have known better, repeated Yule's statement in his four-volume history;[a] and we meet with the same sinological legend in Delambre's monumental *Histoire de l'Astronomie Ancienne*,[b] in Houzeau & Lancaster,[c] and even in Zinner's standard *Geschichte d. Sternkunde* of 1931.[d] It is time it received its quietus.

As a pendant to this story, we offer the legend of 'Ku-Tan'. Gaubil,[e] speaking of the terms for the nodes of the moon's orbit, Rahu and Ketu,[f] says that these, with other words, were contained in an astronomical treatise called 'Kieou-tche', which came from the West. And he adds that this treatise was translated into Chinese by an astronomer named 'Ku-Tan' who was himself a foreigner from some occidental kingdom, possibly Syria.[g] The true facts are recognisable from this description. The astronomer was Chhüthan Hsi-Ta,[1] and the book which he compiled was the *Khai-Yuan Chan Ching*,[2] finished soon after +718. Far from being a translation, however, this indeed constitutes the greatest collection of Chinese astronomical fragments, from the −4th century onwards, which we possess.[h] Perhaps the only translation in it is from an Indian calendar, the *Navagrāha*, the Chiu Chih[3] (i.e. 'Kieou-tche'),[i] which had reached China not long before. Chhüthan Hsi-Ta himself[j] was an Indian Buddhist, the most eminent member of one of three clans of Buddhist astronomers and calendar experts originating from India and resident at the Chinese capital during Thang times. Gaubil's confusion of India with some occidental country was grist to the mill of certain 19th-century scholars, who were sure that this 'translated' astro-

Chih, ch. 196) but from the much earlier *Wei Shu*, ch. 102, p. 21*a* and *Pei Shih*, ch. 97, p. 21*a*; it must therefore derive from +6th-century contacts between Chinese, Persian and Byzantine travellers (cf. Vol. 1, p. 186). The whole texts were translated by Hirth (1), pp. 48, 77ff. We are indebted to Dr Lu Gwei-Djen for this elucidation.

[a] (1), vol. 1, p. 281.

[b] In vol. 1, p. 370, where the implication is that Chinese astronomy was a 'parody' of Greek. I find it difficult to share the very high estimate of Delambre's book entertained by Neugebauer (1), p. 130; he speaks of its unequalled contact with original sources, but this can only refer to Greek texts—on Chinese matters it is distinctly poor.

[c] (1), vol. 1, p. 135. So also Narrien (1), p. 346.

[d] (1), p. 199. I suspect that the same legend lies behind the similar presentation of R. Berthelot in 1949!

[e] (2), pp. 89, 124ff.

[f] Cf. p. 228 below.

[g] It is interesting that in an appended footnote Gaubil says that he had sought for a copy of this book in vain.

[h] According to the preface, a copy of it was found in the +16th century within the body of an old statue in a temple, after having been lost for many years, perhaps centuries.

[i] This occupies only one chapter (ch. 104) out of a total of 120. The calendar was never officially adopted in China, and had little influence. It is not prominently mentioned in the astronomical or calendrical chapters of either of the *Thang Shu*. Only the *Hsin Thang Shu*, ch. 28B, p. 14*a*, devotes half a page to it. The general view is that is was similar to the calendrical material in the +6th-century *Pañcha Siddhāntikā* of Varāha-Mihira. The meaning of the calendar's title, the 'Nine Upholders' or 'Nine Forces', refers to the sun, moon, five planets, Rahu and Ketu. See further Chu Wên-Hsin (*1*), pp. 153ff.

[j] The name certainly transliterates Gautama Siddhārtha. Cf. p. 203 below.

[1] 瞿曇悉達 [2] 開元占經 [3] 九執

nomical treatise had come from the West,[a] and must have conveyed to China the wisdom of Ptolemy[b] or the first-fruits of Arabic research.

It is not the purpose of this section to rake over old controversies.[c] The basic facts about the history of Chinese astronomy are not so difficult to disentangle and not now in dispute, though there are, of course, still many unsolved problems. The historical vicissitudes of the Chinese calendar will, indeed, remain a matter of some difficulty, as the definitive monograph on this subject has not yet been written, either in Chinese or a Western language. Fortunately, however, it is not of primary scientific importance. The various shifts to which the calendar experts were put by their inaccurate knowledge of precession, planetary cycles, etc., need not delay us too much. What seem really interesting in Chinese astronomy are such questions as the ancient and medieval cosmic theories, the mapping of the heavens and the coordinates used, the understanding of the great circles of the celestial sphere, the use of circumpolar stars as indicators of the meridian passages of invisible equatorial constellations, the study of eclipses, the gradual development of astronomical instruments (which by the +13th century had attained a level much higher than that of Europe), and the thorough recording of observations of important celestial phenomena.

One basic question may here be raised before proceeding further, namely, the antiquity of Chinese astronomy. As Maspero (15) and others have pointed out, European opinion has been much confused regarding this, on account of what may be an erroneous interpretation of a certain famous text. Maspero held that Chinese astronomy was the only ancient system of astronomy which arose at a period sufficiently recent for us to be able to gain a full view of its progress. While Babylonian inscriptions show that an advanced state had already been reached by the end of the −2nd millennium, Chinese astronomy, in Maspero's view, did not arise until the −6th or −5th century,[d] and its primitive form, without calculations, measurements or regular observations, was still perceptible. Maspero finally came to think that it had developed entirely independent of Babylonian influences, but there can be no doubt that he overstated the case for this. We may agree with him that Chinese astronomy was particu-

[a] Sédillot (2), pp. 609, 634.

[b] There were, of course, Ptolemaic traces in the Indian material, but this was a very different thing from an effective translation of the *Almagest*. Nearly a century ago Wylie (15, p. 42) acutely identified Greek words in the Indian material of the Chiu Chih calendar. Thus *li-to*,[1] a minute, was Skr. *liptā*, but that had once been λεπτή. So also Skr. *horā* (hour), from ὥρα, appears in a Chinese title as *huo-lo*.[2] The book is the *Fan Thien Huo-Lo Chiu Yao*[3] (The Horā of Brahma and the Seven Luminaries), attributed to I-Hsing but in reality not translated before +874 (Chavannes & Pelliot (1), p. 160). All this constitutes a concrete case of the travel of unimportant details and the filtration effect of distance and language on fundamental scientific ideas. Moreover, as Yabuuchi (1) has pointed out, the *Chiu Chih Li* contains no mention whatever of epicyclic theory; it was a selection of material strictly relevant to calendar computations.

[c] Cf. the discussion in 1775 between Voltaire, who defended the Indian and the Chinese achievements, and Bailly (2), who minimised them. Bailly made a good guess, however, in his suggestion that part at least of their sciences had been transmitted from a civilisation older than either.

[d] Maspero was not able to take into account, when he wrote, the facts regarding Chinese astronomy revealed by the oracle-bones of the −14th and −13th centuries.

[1] 立多 [2] 火羅 [3] 梵天火羅九曜

larly dependent upon the progress of invention of astronomical instruments, but again he overstated the view that this was because of a persistent backwardness of mathematics; it would be more convincing to say that Chinese astronomy suffered from the characteristically non-geometrical nature of Chinese mathematics.

The text to which reference has just been made is found in the Yao Tien[1] chapter of the *Shu Ching*,[2] and will be discussed in its proper place;[a] it cannot date from earlier than the −8th or −7th century. It deals with the meridian passages of certain stars which were placed in relation with the seasons. Since these stars no longer have this relation, it appeared to nearly all students of Chinese astronomy[b] that the date of the text could be ascertained by calculating the positions of these stars relative to the equinoctial and solstitial points according to the law of precession of the equinoxes. All such computations have led to dates as early as the −3rd millennium, for example the −24th century.[c] However, the difficulty is that the text fails to mention the exact day and hour of the observation, and one hour's difference will make a difference of many centuries. Hashimoto (*1, 4*), for instance, taking 7 p.m. instead of 6 p.m. as de Saussure did, could bring the date down to the −8th century or later still. Yet the question should not be regarded as closed, since there were certain reasons for adopting the hour as 6 p.m. (see below, p. 246), and no one has considered the possibility that the reference to the −3rd millennium might indeed be correct, but not refer to Chinese observations. In other words, this link between stars and seasons might possibly be a part of the traditional patrimony of knowledge about the heavens derived from Babylonian sources, and this particular connection might then really be Babylonian. In any case, it does not affect the general conclusion that all literary attributions[d] of observations to Chou Kung and other characters of the −9th or −10th century, or to Huang Ti in the −2nd or −3rd millennium, are to be considered purely legendary.[e] What the oracle-bones tell us about Chinese astronomy in those times we shall see in due course.

[a] Below, p. 188.

[b] For example, Gaubil (2) in the 18th century, then J. B. Biot (1), Chalmers (1), de Saussure (3).

[c] Biot, for example, firmly accepted −2357.

[d] The fabulous datings accepted by the early Jesuits seem to be ineradicable from Western literature, as witness (for instance) the standard work of Abetti (1), p. 24.

[e] This is, of course, something quite different from interpolation or conscious falsification. Entering as we now are upon the first of the Sections of this work in which concrete inventions and discoveries are to be discussed, it is worth while to remember a point which was made earlier (Vol. 1, p. 43). The overwhelmingly humanistic emphasis of ancient and traditional Chinese scholarship makes it quite unlikely that any ancient or medieval texts have been manipulated or tampered with in order to prove that a scientific or technical discovery was made earlier than it really was. Until very recent times it would never have occurred to any Chinese scholar that kudos was to be gained by such a claim. Neither science nor technology had social prestige. This situation is also found, to some extent, in other Asian cultures, as Sédillot saw, a hundred years ago, (2), p. 55, in connection with the discovery of the third lunar anomaly by Abū'l-Wafā'.

[1] 堯典 [2] 書經

(b) DEFINITIONS

Before proceeding further, it will be desirable to define the terms in spherical astronomy necessary for this Section, some of which indeed have already been used. The study of the history of astronomy naturally requires at least an elementary knowledge of astronomy itself, and while in this book it will not be the practice to furnish definitions of all the technical terms in the various natural sciences, for which the reader must have recourse to standard glossaries and text-books, those of astronomy, as being perhaps less familiar than is the case in some other sciences, must be briefly mentioned. The reader will find it convenient, in any consideration of the history of astronomy, to have at hand a manual of what is known as spherical astronomy.[a]

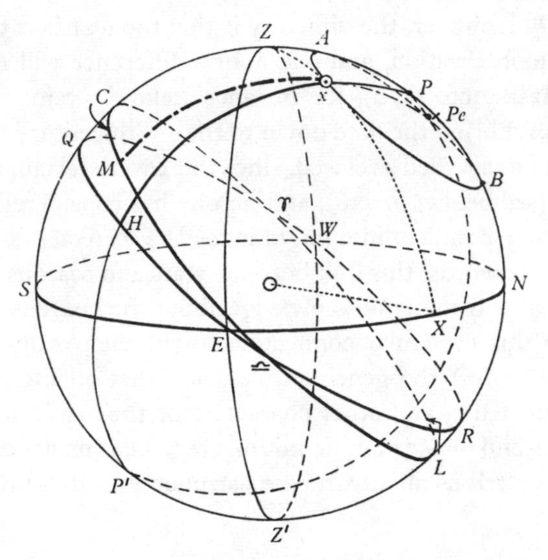

Fig. 85. Diagram of the circles of the celestial sphere (see text). For greater clarity of drawing, the obliquity of the ecliptic is made less than its correct value.

The earliest observers of the stars must have felt it easiest to imagine that they were bright points scattered about on the interior of a hollow spherical dome, the centre of which would be at the eye of the observer. Many centuries before any information became available as to the real distances of these points of light from the earth, it was found possible to measure their directions, and this is the subject-matter of spherical astronomy. In Fig. 85 a diagram is provided by the aid of which one may remind oneself of the chief arcs and angles necessary in talking about the positions of the celestial bodies. It should be needless to add that the poles and equator of which we here speak are those of the celestial, not the terrestrial, sphere.

The following are the essential definitions. The horizon of the observer at *O* is represented by the great circle *NESW* which passes through the four cardinal points

[a] Each country has of course its own; for English readers the obvious one is that of Smart (1), but a simpler treatment will be found in Barlow & Bryan (1).

(*ssu fang*[1]). *P* is the (celestial) north pole (*pei chhen*[2] or *pei chi*[3]), and *P′* the (celestial) south pole; the line joining them is the axis of the apparent diurnal rotation of the heavens. The equator (the 'red road', *chhih tao*[4]), the great circle perpendicular to this axis, is represented by *EQWR*; and the meridian, that is to say, the great circle passing through the north and south and the zenith of the observer, by *ZSZ′N*, *Z* being the zenith and *Z′* the nadir. The great circle *ZEZ′W* is known as the prime vertical. Then the great circle of the ecliptic (the 'yellow road', *huang tao*[5]) is seen crossing the equator at the two equinoctial points, (*a*) the First Point of Aries ♈; and (*b*) the First Point of Libra ♎, thus being represented by ♈*C*♎*L*. It is, of course, the mean path of the sun and planets. *C* and *L* are the summer and winter solstice points.

Let a given star occupy the point *x*. This point can obviously be defined by two coordinates: (1) on the great circle originating from the pole and passing through *x* and through two points, one on the ecliptic and one on the equator; (2) on a small circle which the star *x* will trace out during the course of the apparent diurnal revolution of the heavens. The great circle appears on the diagram as *PxHP′*, *H* being the point where it crosses the equator. The small circle appears as *xAB*, *A* and *B* indicating the transits which the star makes across the meridian. These have particular importance for Chinese astronomy, as we shall later see, and in the diagram it will be noticed that *x* has been chosen as a star quite near the pole, one of the 'circumpolar' stars, so that its upper transit *A* (or culmination) and its lower transit *B* will both be visible. At this point we find a certain fluidity of terminology. The best term for the great circle *PxHP′* is the hour circle, since the positions of stars may be defined by their hour angle, *QPx* (= *QH*), a regular succession of hour circles coming to coincide with any given meridian during the hours of night when the stars are visible. It is evident that the hour circle of a star will fix its times of rising and setting relative to the horizon. In Fig. 85, *x* on the diagram would have been called by the Greeks a 'paranatellon' of those stars or constellations near the zodiac (ecliptic), or equator, which share its hour circle. The hour circle is sometimes called the star's meridian. Some writers, however,[a] call it the declination circle, while others[b] apply the term declination circle to the small circle *xAB*; this is more usually called the parallel of declination. But the term 'declination circle' for *xAB* seems not illogical, since this small circle links all stars which have the same declination, and we shall so use it in the present discussion.[c]

A few further definitions remain. The altitude of a star, *Xx*, i.e. its height above the horizon, measured in degrees, minutes and seconds of arc, was the fundamental observation for which the medieval astrolabe was used. The complement of this measurement is the zenith distance, *Zx*. Similarly, the complement of the star's declination, *Hx*, is the north polar distance, *Px*. The declination is one of the two coordinates most universally used in modern positional astronomy, the other being the right ascension ♈*QH*. This point is of importance, since both these measurements

[a] E.g. Barlow & Bryan (1), p. 17. [b] E.g. Moreau (1), p. 16.

[c] When this paragraph was first written, no one seemed to have called attention to these discrepancies of definition, but Michel (8) has now done so.

1 四方 2 北辰 3 北極 4 赤道 5 黃道

depend on the equator, and are therefore in agreement with the old Chinese astronomical system. The celestial latitude, Mx, and the celestial longitude, ♈CM, which were more emphasised in Hellenistic and medieval European astronomy, are today hardly used at all. These are based on the ecliptic, Mx being an arc of the great circle through the pole of the ecliptic, Pe, which is different from the pole of the equator. The last type of measurement which may be mentioned is the azimuth, $SWNX$, corresponding to the angle SZx. One can see that the altitude and the azimuth were data based on the horizon, corresponding to the declination and the right ascension respectively based on the equator, and the celestial latitude and longitude respectively based on the ecliptic. An important angle seen on the diagram is C♎Q, the obliquity of the ecliptic. The observer's own latitude (terrestrial) is naturally given by ZQ or NP. Since the obliquity of the ecliptic is 23° 27′, the sun can only be vertically overhead at any time of the year within a zone north and south of the terrestrial equator by this amount; the two limits of this are the (terrestrial) tropics of Cancer in the north and Capricorn in the south. On the other hand, near the terrestrial poles, within the Arctic and Antarctic zones, the sun will during some periods of the year behave like a circumpolar star, never rising and never setting.

Summary of definitions

O	observer's position.
$NESW$	horizon.
Z	zenith.
Z'	nadir.
$ZSZ'N$	meridian.
P	celestial north pole.
P'	celestial south pole.
$EQWR$	celestial equator.
$ZEZ'W$	prime vertical.
$ZQ = NP$	observer's terrestrial latitude.
♈C♎L	ecliptic. (The zodiac consists of a series of constellations along the ecliptic.)
Pe	pole of the ecliptic.
♈ and ♎	equinoctial points, the First Point of Aries and the First Point of Libra.
C and L	solstitial points.
x	position of a given star.
$PxHP'$	great circle passing through the star, the hour circle.
H	hour circle's crossing of the equator.
M	point of intersection of the ecliptic with the great circle of celestial longitude through Pe and the star x.
xAB	parallel of declination (declination circle).
A	upper meridian transit of star (culmination).
B	lower meridian transit of star.
QPx	360° minus the hour angle (the hour angle being always expressed westwards from the meridian).
Xx	altitude of star.
Zx	zenith distance.
Mx	celestial latitude.

Hx	declination.
Px	north polar distance.
SWNX	azimuth (reckoned westwards from *S*).
♈*CM*	celestial longitude (reckoned eastwards from ♈, i.e. in the direction of the sun's annual motion, to *M*).
♈*QH*	right ascension (reckoned eastwards from ♈ to the star's hour-circle at *H*).
C≏*Q*	obliquity of ecliptic.

Sidereal time at a given moment is defined as the hour angle of the First Point of Aries. For any given star this is therefore equal to the sum of its right ascension and its hour angle.

The spherical astronomy of the ancient Greeks and the ancient Chinese may be said to be still in use in modern positional astronomy. This is why it is possible to write, as R. Wolf did (1), an excellent manual of general astronomy in which both the spherical geometry and the observations of the ancients find their place in a story continuous with the post-Renaissance discoveries which depended on new sciences such as optics and electricity, unknown to the ancients. Such a presentation would be difficult in biology and impossible in medicine. The historian of science, however, who has to soak himself in writings belonging to the ages before the establishment of the heliocentric view of the solar system, may have to make a certain effort to recall that the 'sun's path' is only its apparent path, and that the declination circles of the fixed stars are only those in which they appear to move.

For the purposes of this Section it has to be assumed that the reader is acquainted with the elements of modern cosmology, e.g. that the earth is revolving in an ellipse with the sun at one focus, at a speed of some 19 miles/sec., and faster at perihelion (the point nearest the sun) than at aphelion (the point farthest away). Upon such facts our measures of time depend.[a] Thus the sidereal year is one revolution of the earth in its orbit (365·2564 solar days). Observationally, it is the time taken by the sun to complete one apparent revolution returning to any given starting-point among the stars. It thus differs from the tropic year,[b] which is the time elapsing between two successive passages of the sun through the vernal equinox. The plane of the terrestrial equator is inclined to the earth's orbit at approximately $23\frac{1}{2}°$ (the angle *C*≏*Q*), the 'obliquity of the ecliptic', already mentioned. It is this inclination which accounts for the existence of the equinoxes and solstices. As the earth moves in its orbit, and rotates to give what we call day and night, its axis also performs a slow gyrating motion with a period of 26,000 years. It thus describes a cone about a line through the earth's centre perpendicular to the plane of the orbit. This motion causes the equinox points to perform a clockwise movement on the ecliptic (i.e. oppositely to the earth's orbital revolution), and this is their 'precession', amounting to approximately 50 seconds of arc each year. This small motion, however, is large enough to make a difference of 20 minutes between the sidereal year and the tropic year.

Parallel with these definitions are those of the days. The sidereal day is the time between two successive transits of the first point of Aries (♈). The true or apparent

[a] Cf. the elementary exposition of Bolton (1).

[b] 365·2422 mean solar days.

solar day is the time between two successive transits of the sun. It is 4 minutes longer than the sidereal day because of the apparent eastward movement of the sun in the heavens. The number of sidereal days in the year is one more than the number of true solar days, since the sun appears to revolve once relatively to the celestial sphere in the course of the year, in the opposite direction to its daily apparent rotation. The number of true solar days (the length of which varies) is therefore $365\frac{1}{4}$, as the Chinese and other ancient peoples very early knew from their gnomonic observations. The mean solar day is the length of the true solar days averaged over a tropic year. The difference between mean solar time and true solar time at any instant is known as the equation of time.

(c) BIBLIOGRAPHICAL NOTES

(1) The History of Chinese Astronomy

(i) *European literature*

For the benefit of any who wish to gain the most succinct information as to the best recent opinion on the problems of Chinese astronomy, few words will suffice to say that it may be sought with confidence in two papers, those of Chatley (9) and Maspero (15). To these should be added, perhaps, the first and the last of the long series of papers by de Saussure (2 and 30), together with his summarising papers (16 and 16*a*). At a more detailed level, examples of how the subject should be treated will be found in the two brilliant contributions of Maspero (3, 4) on Chinese astronomy down to the end of the Han. Certain reviews of Eberhard (10, 11), whose detailed papers will be referred to in their place, may also be consulted.[a] Those who desire to avoid too much detail need read no further, but to anyone with a concern to go more deeply into the subject, there is more to be said.

The Interpreter-General and Father-Superior of the history of Chinese astronomy is, as has already been mentioned, Antoine Gaubil. Born in 1689, he was a Jesuit missionary in China from 1723 until his death in 1759. He had had a considerable astronomical training under Cassini and Maraldi at the Paris Observatory, and after his departure from France carried out what may truly be called titanic and indefatigable labours in acquiring an almost perfect knowledge of Chinese, collecting all possible texts bearing on astronomy, conferring with the few Chinese scholars of his time who were proficient in astronomy and mathematics, and himself making astronomical observations. So perfect was his knowledge of Chinese and other Asian languages, that he was frequently called upon by the emperor to act as verbatim interpreter at State interviews. Just as Johann Schreck, a century before him, had been in correspondence with Kepler from Peking,[b] so Gaubil kept up an active relationship with many outstanding European scholars, such as Fréret,[c] and with academic bodies such

[a] For the shortest, best and most up-to-date general review of the history of ancient astronomy, particularly Babylonian and Greek, I would refer the reader to O. Neugebauer (1, 9).

[b] This was about 1637 (Gaubil (5), p. 285).

[c] Who defended his views on Chinese chronology. Cf. Cassini (1).

as the Royal Society.[a] An admirable account of his life has recently been given by Dehergne (2).[b]

The bulk of Gaubil's earlier work was published by Souciet in three volumes of a series entitled *Observations Mathématiques, Astronomiques, etc., tirées des anciens Livres Chinois ou faites nouvellement aux Indes et à la Chine par les Pères de la Compagnie de Jésus.*[c] The first volume, of 1729, includes many miscellaneous observations (1), but the main work is the *Brief History* of Chinese astronomy in the second (1732), and the *Treatise* on it in the third, which appeared in the same year (2 and 3). There was then a long silence until many years after Gaubil's death, when another 'History' was published in the *Lettres Edifiantes et Curieuses* for 1783; this (4) actually dates from 1749. In this same year Gaubil had sent back his *Traité de la Chronologie Chinoise* (5), but it was not printed till Laplace found it in 1814. An important manuscript (Gaubil, 6) still exists at the Paris Observatory, parts only of which were published by Laplace between 1809 and 1811 (7–9). All of Gaubil's writings are somewhat confused in presentation, and constitute, as has been well said, a mine in which one must know how to dig.[d] The material which concerned the legendary period has long lost its value, but there is a mass of useful information for the historical periods which has as yet been imperfectly exploited. For any thorough study of Chinese astronomy Gaubil is needed even today.

In 1782 C. L. J. de Guignes, who had been consul at Canton, published a Chinese celestial planisphere, with an alphabetical star catalogue (de Guignes, 2), now only of historical interest. Again in 1819 Reeves drew up a similar list of identifications, while Wylie (6) prepared a very complete one about 1850 using the star-maps of Chhien Wei-Yüeh[1] (1839) as his basis. This is still useful. Reeves' catalogue served as the basis for the *Uranographie Chinoise* of Schlegel (5), which remains the most important reference work on the positional astronomy of the stars and constellations, accompanied as it is by excellent maps.[e] Unfortunately, owing to misunderstandings and miscalculations, Schlegel put forward a quite absurd chronology, setting the beginnings of a knowledge of the heavens in China about the year −16,000; this naturally received little support and only served to discredit what real historical research might reveal. In the meantime ancient Asian astronomy had attracted the interest of J. B. Biot, the many-sided French astronomer and chemist,[f] who in 1840, taking advantage of a book on Chinese chronology by Ideler, sketched a general survey of Chinese astronomy in the *Journal des Savants* (4). In this he was aided by his son

[a] The corresponding secretaries were Cromwell Mortimer and Thomas Birch.

[b] This includes an indispensable index to Gaubil (1–5). The biography by Rémusat (3) may also be consulted.

[c] Unfortunately, Souciet was a very careless editor, and the work of Gaubil's is full of mistakes which Gaubil vainly tried to correct.

[d] The Chinese material in Montucla's famous *History of Mathematics* is a mere boiling down of Gaubil.

[e] References to stars and constellations discussed in Schlegel's book will hereinafter be given by their serial numbers, thus: S555. Page number references to it will be given in the usual way.

[f] It will be remembered that it was J. B. Biot who, as an old man, encouraged the young Louis Pasteur when he told him of his discovery of the two stereoisomeric forms of tartaric acid.

[1] 錢維樾

E. Biot, a young sinologist whose early death was a great blow to the history of science in China, by Abel Rémusat, Stanislas Julien and other orientalists. Similar studies of Indian astronomy followed (2, 3, 6) and were incorporated into a book still as important as those of Gaubil, the *Etudes* of 1862 (Biot, 1). Biot's treatment was much more systematic than any which had preceded it; he was perhaps the first to recognise clearly the equatorial character of the twenty-eight so-called 'lunar mansions' (*hsiu*[1]), and the importance which the Chinese had attached to the meridian passages of the circumpolar stars.

During the 19th century other luminaries arose on the orientalist horizon and engaged in extended discussions concerning the *hsiu* in relation with similar divisions of the sky in Hindu and Arabic astronomy, as will be told in its place. They came into conjunction, so to say, in 1907 in one of the earliest works of de Saussure—a conjunction very unfavourable to them. Leopold de Saussure[a] was a sailor and navigator, hence a more practical astronomer than Biot, and possessing considerable sinological knowledge, though certainly much less than Gaubil.[b] His long series of papers is still indispensable, though it must be remembered that they were interim research reports rather than a finished presentation; he made many mistakes, and afterwards corrected them, changing his views as he went along. The luminaries in question were, first, the great Legge himself, then Chalmers, who had written an astronomical introduction to his *Shu Ching* translation; Russell, a professor at the Thung Wên Kuan College at Peking; Schlegel and Ginzel. De Saussure (3) showed that none of these had understood the reference to times of meridian passages of stars in the *Shu Ching*, and that all had failed to appreciate the equatorial character of the *hsiu*,[c] and the way in which these were keyed to circumpolar stars. He also demolished the arguments of Whitney (1, 2), who had misunderstood Biot's work; of L. P. Sédillot (2, 5), who had claimed an Indian or an Arabic origin for the *hsiu*; and of Kühnert (1), who inclined to Schlegel's impossible chronology.

Among the more recent articles of value[d] on Chinese astronomy may be mentioned those of Chu Kho-Chen (7), Chatley (8) and Michel (1, 2, 4, 7). Long ago Wylie (13) essayed a glossary of terms, which should not be forgotten. As for the descriptions of Chinese astronomy contained in works on the history of astronomy or science in general, nearly all are unsatisfactory, from Delambre in 1817 to Zinner in 1931.[e] Of the general histories of astronomy, those of Grant and von Mädler (mainly topical), though old, are good, as also is that of Berry (mainly chronological).[f]

[a] Biography by R. de Saussure (1).

[b] Exactly how competent a Chinese scholar de Saussure was is hard to say; he curiously left untouched a mass of texts which would have been very useful to him.

[c] This mistake seems ineradicable; Yampolsky and Weinstock continue to confuse in 1950 what Sédillot (2, p. 477) confused in 1849.

[d] That of Fu (1) is misleading in many ways, and should be avoided. Those of Stuhr (1) and Narrien (1) were written before Biot's time, and are now of purely historical interest. The work of T. Ferguson may also be mentioned.

[e] Rey's account was written very much at second hand, and he cannot now be regarded as a trustworthy guide.

[f] It will be wise not to overlook the material scattered through the great bibliographical compendia of Houzeau (1) and Houzeau & Lancaster (1). Gundel's review (1) is also a mine of information.

[1] 宿

(ii) *Chinese and Japanese literature*

The bibliography of modern Chinese works on the history of astronomy in east Asia naturally blends into the general astronomical literature of the 18th and 19th centuries, about which something must later be said (p. 454). One of the first books on the history of Chinese astronomy must have been that of Yang Chhao-Ko[1] about 1790, the *Li Tai Lun Thien*[2] (Discussions on the Heavens in Different Ages). Shortly afterwards, Chhen Mou-Ling[3] produced his *Ching Shu Suan-Hsüeh Thien-Wên Khao*[4] (Investigations of the Astronomy and Mathematics of the Classics), which was modelled on Chen Luan's +6th-century *Wu Ching Suan Shu* (cf. p. 35).

Mention was made above of planispheres, star-maps and lists of constellations identifying Chinese and occidental star-names, published in Western languages. This work had of course been under way in China, after the coming of the Jesuits, a century before the Western-language works began to appear. Thus in 1723, Mei Wên-Mi[5] (the brother of Mei Wên-Ting) issued his *Chung Hsi Ching Hsing Thung I Khao*[6] (Investigation of the Similarities and Differences between Chinese and Western Star-Names). In the middle of the following century, three such works were prepared about the same time: the *Hêng Hsing Chhih-Tao Chhüan Thu*[7] (Complete Map of the Fixed Stars, based on the Equator) of Yeh Thang,[8] and two atlases with the same title, *Hêng Hsing Chhih-Tao Ching-Wei Tu Thu*[9] (Map of the Fixed Stars according to their Right Ascension and Declination), one by Liu Yen[10] and the other by Li Chao-Lo[11] (*2*). These were much more complete than that of Chhien Wei-Yüeh,[12] which had been the basis of Wylie's star-catalogue (6), and probably served among Schlegel's sources.[a]

In the present century we have a book of 1900 by Lei Hsüeh-Hung, the *Ku Ching Thien Hsiang Khao* (Investigation of Celestial Phenomena as recorded in the Ancient Classics), but most of the works in current use date from the second and third decades. Thus Chu Wên-Hsin has given us a useful history of Chinese calendrical science (1), a study of the eclipses recorded in the official histories (2), two short but excellent histories of Chinese astronomy (3, 4), and an indispensable commentary on the Thien Kuan[13] (Celestial Officials) chapter of the *Shih Chi* (Historical Record) of Ssuma Chhien.[b] Chhen Tsun-Kuei has produced the most up-to-date identification-list of Chinese and Western star-names (*1*, *3*). Short but valuable articles on the history of Chinese astronomy in general are also due to him (*2*), as also to Chang Yü-Chê (*1*); and Chu Kho-Chen (*5*).

[a] The titles of these atlases are significant. In adopting a coordinate-system based on the equator, Mei, Yeh, Liu and Li were in agreement with modern astronomical practice, and diverged from that of the Jesuits. Thus in 1746, Ignatius Kögler (Tai Chin-Hsien[14]) had published his *Huang-Tao Tsung Hsing Thu*,[15] in which the stars were arranged on an ecliptic basis with celestial latitudes and longitudes.
[b] The translation of this is of course available in Chavannes (1).

1 楊超格 2 歷代論天 3 陳懋齡 4 經書算學天文考
5 梅文鼐 6 中西經星同異考 7 恆星赤道全圖 8 葉棠
9 恆星赤道經緯度圖 10 六嚴 11 李兆洛 12 錢維樾
13 天官 14 戴進賢 15 黃道總星圖

In Japanese also there is a large literature on the history of Chinese astronomy.[a] Two schools formed, led by Shinjō Shinzō and Iijima Tadạo, the former considering Chinese astronomy fundamentally independent of any Western influences, and the latter striving to prove Greek (or at least Babylonian) derivations. The fullest summary of all their views is perhaps that of Nōda Chūryō (*3*), but Eberhard (10) (in German) may be consulted, and recently Yampolsky has translated some pages of conclusions of the two Japanese protagonists. Two of the works of Shinjō (*1* and *3*) have been translated into Chinese, and one summary of his views is available in English (Shinjō, 1). Liu Chao-Yang (*3*) has summarised and criticised the views of Iijima. Iijima's conviction of the Western origin of many Chinese astronomical ideas rested largely on the appearance of the cycles of Meton and Callippus, and on what he believed was the saros cycle. Shinjō saw no reason why the former should not have been independently developed, and pointed out many notable differences, for example, the use of the 60-day and 60-year cycles, the division of the month into 10-day periods instead of weeks, the graduation of the circle in $365\frac{1}{4}°$ instead of 360°, the totally different constellation-names, and even the selection of quite different groups of stars to form constellations (see below, pp. 271 ff.).

The books of Yabuuchi (*2*) and Chhen Tsun-Kuei (*5*, *6*) did not become available to us until after the present Section had been written. The review of Yabuuchi & Nōda on the calendrical parts of the *Chhien Han Shu* was also unavailable to us. Nor did the outstanding bibliography of old Chinese astronomical books by Ting Fu-Pao & Chou Yün-Chhing (*1*) appear until our work on this Section had been completed.

(2) The Principal Chinese Sources

(i) *The 'official' character of Chinese astronomy*

First a few words on the basic quality of Chinese astronomy already noted, namely, its official character and intimate connection with the government and the bureaucracy. This situation is fixed from the dawn of historical records in China, and its *locus classicus* is the famous commission of the legendary emperor Yao[1] to the 'astronomers' Hsi[2] and Ho[3] (Fig. 86). The story is found in the first chapter of the *Shu Ching* (Historical Classic), where two separate texts have been conflated by some later scholiast. One concerned the two (or rather six) astronomers; the other was an excerpt from an ancient calendar such as those which we shall shortly mention. The former is here relevant, the latter is the text already referred to which has been used by generations of chronologists making their calculations as to the date of the book, and we shall examine it in its proper place.[b] Only the vaguest guesses can be made about the dates

[a] See particularly Yabuuchi (*8*, *1*) and Kamada (*1*). A brief bibliography, with western-language titles only, is given in Anon. (28).

[b] Cf. p. 177 and p. 245. We shall find that it constitutes a riddle, for if certain assumptions are made, the observations which it contains may well be of the −3rd millennium, but it does not follow that they were made in China; in any case the calendar which included it was incorporated in the *Shu Ching* far later.

[1] 堯 [2] 羲 [3] 和

Fig. 86. A late Chhing representation of the legend of the Hsi and Ho brothers receiving the government commission from the emperor Yao to organise the calendar and pay respect to the celestial bodies. From *SCTS*, ch. 1, Yao Tien (Karlgren (12), p. 3).

of these two texts, but roughly it may be said that the 'government commission' text could be of the −7th or −8th century, while the 'stars and seasons' text could be of the −5th or −6th. Though much more recent than traditional views would have held, the combination remains a venerable document.

Here, then, is the 'government commission' portion:

(Yao) commanded the (brothers) Hsi and Ho, in reverent accordance with the august heavens, to compute and delineate the sun, moon and stars, and the celestial markers (*chhen* [1]),[a] and so to deliver respectfully the seasons to be observed by the people.

He particularly ordered the younger brother Hsi to reside among the Yü barbarians (at the place called) Yang-Ku and to receive as a guest the rising sun, in order to regulate the labours of the east (the spring).

He further ordered the youngest brother Hsi to go and live at Nan-Chiao in order to regulate the works of the south and pay respectful attention to the (summer) solstice.

He particularly ordered the younger brother Ho to reside in the west (at the place called) Mei-Ku, and to bid farewell respectfully to the setting sun, in order to regulate the western (autumnal) accomplishment.

He further ordered the youngest brother Ho to go and live in the region of the north (at the place called) Yu-Tu, in order to supervise the works of the north.[b]

For nearly three thousand years this remained, as it were, the foundation charter of Chinese official astronomy, but modern research has thrown quite new light on its legendary basis. Maspero (8) has shown that everywhere else in pre-Han literature Hsi-Ho is not the name of two or six persons, but a binome, the name, in fact, of the mythological being who is sometimes the mother and sometimes the chariot-driver of the sun. This name then in some way became split up and applied to four magicians or cult-masters who were charged by the mythological emperor to proceed to the four 'ends' of the world in order to stop the sun and turn it back to its course at each solstice, and to keep it going on its way at each equinox.[c] This legend must have derived quite naturally from fears which primitive minds might well entertain that winter might go on indefinitely getting colder, and summer might intensify to lethal heat.[d] At the same time, the mythological magicians were charged with the prevention of eclipses.

[a] There is no telling what the *chhen* meant in the time of the *Shu Ching*; Shinjō (*1*) has brought evidence to show that later the word meant the three main stars of Orion, the seven of the Great Bear, the pole star and Antares.

[b] Legge (1), p. 32; Medhurst (1), p. 3; tr. Maspero (8), p. 7, eng. auct., adjuv. Karlgren (12).

[c] The learned researches of Karlgren into the *Shu Ching* text, (2), p. 262, (11), p. 49, have confirmed the general interpretation of Maspero.

[d] This anxiety about the solar path is clearly found in other ancient civilisations. Many scholars, such as R. Berthelot (1), pp. 55 ff., connect it with the protean manifestations of fire-magic and fire-worship. In some cultures it took a terrible form, as, for instance, the mass human sacrifices which the Aztecs thought were necessary to prevent the standstill or death of the sun (Soustelle (1), pp. 19 ff.). In the +6th century, Procopius has an interesting passage about the Scandinavians. He relates how it is said that in Scandinavia at midsummer the sun never sets for forty days and nights, while at midwinter it never rises during a similar period, so the people send scouts up the mountains to watch for its return, 'because, I imagine,' he goes on, 'they always become afraid that it may fail them some time' (*De Bello Gothico*, II, 15, 6–15; cf. J. O. Thomson (1), p. 388).

[1] 辰

A later chapter in the *Shu Ching* (the Yin Chêng;[1] originally Yün Chêng[2]) deals with precisely this, and purports to record a punitive expedition sent out against the magicians by the emperor owing to their failure to prevent an eclipse.[a] What lies behind this is probably some ancient comminatory ritual directed against the semi-divine magicians at the four ends of the world who were thought to have neglected their duties.[b]

In any case, the astronomical, or rather astral, character of ancient Chinese State-religion comes out clearly from the above. As H. Wilhelm[c] has well said, astronomy was the secret science of priest-kings. An astronomical observatory (*ling thai*[3]) was from the beginning an integral part of the Ming Thang,[4] that cosmological temple which was also the emperor's ritual home.[d] For an agricultural economy, astronomical knowledge as regulator of the calendar was of prime importance. He who could give a calendar to the people would become their leader. More especially was this true, says Wittfogel,[e] for an agricultural economy which depended so largely upon artificial irrigation; it was necessary to be forewarned of the melting of the snows and the consequent rise and fall of the rivers and their derivative canals,[f] as well as of the beginning and end of the rainy monsoon season. In ancient and medieval China the promulgation of the calendar by the emperor was a right corresponding to the issue of minted coins, with image and superscription, by Western rulers.[g] The use of it signified recognition of imperial authority. There can be little doubt that very concrete social reasons may be found for the official and governmental character of Chinese, in contrast to Greek, astronomy.[h]

All this can, of course, be seen in the importance given to astronomy in the *Chou Li* (Record of the Rites of Chou), put together, no doubt, in the former Han dynasty, about the −2nd century. In the opening sentence of the whole book, the emperor is named as he who fixes the four cardinal points (by observations of the pole star and the sun).[i] Of the imperial astronomer (Fêng Hsiang Shih[5]) we read:

> He concerns himself with the twelve years (the sidereal revolution of Jupiter), the twelve months, the twelve (double-) hours, the ten days, and the positions of the twenty-eight stars

[a] Legge (1), p. 81; Medhurst (1), p. 125. The idea that it was because they had failed to *predict* an eclipse was the cause of some mirth among 19th-century historians of astronomy. So also, even in 1954, Abetti (1), p. 24. But the whole chapter is a +4th-century forgery anyway (Maspero (8), p. 46).

[b] There is no need to suppose, with Fotheringham (3), that the punishment of unskilled calendar-makers is what lies behind the legend.

[c] (1), p. 16.

[d] See the elaborate studies of Soothill (5) and Maspero (25) on this, as well as many discussions in the books of Granet. The chief work on the Ming Thang in Chinese is the *Ming Thang Ta Tao Lu*[6] of Hui Shih-Chhi,[7] *c.* +1736. Soothill draws an interesting parallel between the Ming Thang and the Roman Regia or official palace of the Pontifex Maximus; which also had calendrical connections.

[e] (2), p. 98.

[f] For example, at Chungking today the annual rise and fall of the Yangtze may be of the order of 100 ft.

[g] Cf. Huard (1); Soothill (5).

[h] Cf. below, in Section 48 on social and economic background. Wittfogel contrasts the comparative development of astronomy and mathematics in China and Japan; in the latter culture these sciences were far behind the Chinese level until the 17th century, when under mercantile influence, as in the West, they took great strides ahead.

[i] Ch. 1, p. 1*a* (Biot (1), vol. 1, p. 1).

[1] 胤征 [2] 允征 [3] 靈臺 [4] 明堂 [5] 馮相氏
[6] 明堂大道錄 [7] 惠士奇

(the determinative stars of the *hsiu*[1]). He distinguishes them and orders them so that he can make a general plan of the state of the heavens. He takes observations of the sun at the winter and summer solstices, and of the moon at the spring and autumn equinoxes, in order to determine the succession of the four seasons.[a]

The office was important and hereditary; the title implied that the holder had to keep watch at night upon the astronomical observatory tower or platform (*thien thai*[2]).[b] Of the imperial astrologer (Pao Chang Shih[3]) we likewise read:

He concerns himself with the stars in the heavens, keeping a record of the changes and movements of the planets (*chhen*[4]), the sun and the moon, in order to examine the movements of the terrestrial world,[c] with the object of distinguishing (prognosticating) good and bad fortune. He divides the territories of the nine regions of the empire in accordance with their dependence on particular celestial bodies. All the fiefs and principalities are connected with distinct stars, and from this their prosperity or misfortune can be ascertained. He makes prognostications, according to the twelve years (of the Jupiter cycle), of good and evil in the terrestrial world. From the colours of the five kinds of clouds,[d] he determines the coming of floods or drought, abundance or famine. From the twelve winds he draws conclusions about the state of harmony of heaven and earth, and takes note of the good or bad signs which result from their accord or disaccord. In general he concerns himself with the five kinds of phenomena, so as to warn the emperor to come to the aid of the government, and to allow for variations in the ceremonies according to the circumstances.[e]

This office was again important and hereditary; its title refers specifically to the keeping of records. Both these officials had (or were supposed to have) a considerable staff.[f]

A third official, the Shih Chin,[5] was charged with observations of a somewhat more meteorological character, which may, however, have included eclipses (cf. Section 21). Lastly, the official in charge of the clepsydras (water-clocks)[g] should not be forgotten, the Chhieh Hu Shih.[6] He and his staff may be deferred till the discussion on time-keeping (p. 319 below); here it need only be noted that the *Chou Li* did not forget him.

Throughout the course of Chinese history this official system, described in somewhat idealised form in the *Chou Li*, persisted, and the majority of the observers who thought and calculated and wrote about astronomical problems were in State service.[h] For more than two millennia they were organised in a special government department,

[a] Ch. 6, p. 44*b* (ch. 26, p. 15), tr. Biot (1), vol. 2, p. 112, eng. auct.

[b] The expression is used to this day to mean astronomical observatory.

[c] This phrase does not, of course, refer to terrestrial movements in the astronomical sense, but to events on earth which could be related to those seen occurring in the heavens.

[d] This certainly included the observation of auroras; see below, p. 482.

[e] Ch. 6, p. 45*a* (ch. 26, p. 18), tr. Biot (1), vol. 2, p. 113; eng. auct.

[f] Biot (1), vol. 1, pp. 413, 414 translating ch. 5, p. 8*b* (ch. 17, p. 29).

[g] Ch. 7, p. 27*a* (ch. 30, p. 28), Biot (1), vol. 2, pp. 146, 201. The title meant Official in Charge of the Raised Vessels. The office was hereditary.

[h] We noted above (p. 185) that the stars were known as 'celestial officials', and later (p. 273) we shall see that the whole system of Chinese star nomenclature was based on a parallel with the earthly emperor and his bureaucracy. See also Sect. 48.

1 宿 2 天臺 3 保章氏 4 辰 5 眡祲 6 挈壺氏

the Astronomical Bureau or Directorate, which went by various different names.[a] To the end it retained high prestige, even when this was no longer justified by the scientific competence of the scholars who held office, as in the time of the Jesuits. The official astronomers indeed enjoyed a special 'benefit of clerks', since, as late as the last century, exceptionally light punishments in case of offences by members of the Astronomical Bureau were provided for in the Chhing code.[b] Probably the most ancient title of their Director or President was Thai Shih Ling.[1] Though quite conscious of his astrological functions we feel that the true astronomical and calendrical element in the work of his department was always amply sufficient to warrant the translation 'Astronomer-Royal'.[c]

Under his control, of course, in all ages, was the imperial observatory and its equipment. More remarkable (and less well known) is the fact that in some periods it was customary to have two observatories at the capital, both furnished with clepsydras, armillary spheres and other apparatus. Phêng Chhêng,[2] one of Shen Kua's contemporaries, tells us[d] about these in the Northern Sung (+11th century). One, the Astronomical Department (Thien Wên Yuan[3]) of the Hanlin Academy, was located within the imperial palace itself,[e] the other, the Directorate of Astronomy and Calendar (Ssu Thien Chien[4]), presided over by the Thai Shih Ling[1] himself, was outside. The data from the two observatories, especially concerning unusual phenomena, were supposed to be compared each night and presented jointly so as to avoid false or mistaken reports. This is a remarkable example of the sceptical and indeed scientific temper of the Confucian bureaucracy in medieval China. But in the middle of the century astronomy, according to Phêng Chhêng, was in a bad way. When he himself became Astronomer-Royal about +1070, he found that the observers of the two institutions had been simply copying each others' reports for years past and saw nothing strange in it. Nor did they make any use of the observatory equipment, but

[a] For example, in the Thang (+7th and +8th centuries) the Bureau was sometimes a *chü*,[5] i.e. a department of the Imperial Library, and sometimes a *chien*,[6] i.e. an independent directorate. The name was always being changed (cf. *Thang Hui Yao*, ch. 44, pp. 16*a*ff., *Yü Hai*, ch. 121, pp. 33*a*ff. and des Rotours (1), vol. 1, pp. 208, 210). In +758 it became definitively independent as the Ssu Thien Thai.[7] Ssu Thien Chien[4] was a common form, but after +1370, for instance, it was for some time the Chhin Thien Chien.[8] A monograph on the history of the Astronomical Bureau and its organisation is urgently needed.

[b] Staunton (1), p. 21.

[c] It is true that 'Grand Lord Chronologer' would better represent the ancient association of his office with historical recording as well as with the computing of the calendar (cf. Schafer, 4).

[d] In his *Mo Kho Hui Hsi*[9] (Flywhisk Conversations of the Scholarly Guest), ch. 7, p. 8*a*. The passage is fully translated in Needham, Wang & Price (1).

[e] A similar arrangement had existed in the Thang (early +8th century). The College of All Sages Chi Hsien Shu Yuan[10]) was a kind of duplicate Academy; like the Hanlin it had the duty of drafting edicts and checking their texts, but entrance to it was by nomination not examination (cf. des Rotours (1) vol. 1, pp. 17, 19). Apparently it could make use of the services of scholars and experts who were not in the strict Confucian tradition, since most of the work of the great astronomer-monk I-Hsing was done in an observatory within its walls. This we know from Wei Shu's[11] *Chi Hsien Chu Chi*,[12] written about +750 and quoted at length in *Yü Hai*, ch. 4, pp. 24*a*ff. (full translation in Needham, Wang & Price).

1 太史令 2 彭乘 3 天文院 4 司天監 5 局 6 監
7 司天臺 8 欽天監 9 墨客揮犀 10 集賢書院 11 韋述
12 集賢注記

contented themselves with presenting ephemerides very roughly computed. Phêng Chhêng had six officials punished, but could not achieve full reform. Shen Kua,[1] who succeeded him as Astronomer-Royal, was more successful, and no less critical of the former neglect of astronomy. He wrote:[a]

In the Huang-Yu reign-period [between +1049 and +1053] the Ministry of Rites arranged for the examination-candidates to be asked to write essays on the instruments used for gaining knowledge of the heavens. But the scholars could only write confusedly about the celestial globe. However, as the examiners themselves knew nothing about the subject either, they passed them all with a high class.

But Shen Kua himself was the sign of a new period of brilliance, marked by such men as Su Sung and Han Kung-Lien. Lastly, we need only add that besides the two or more observatories at the capital, astronomical observations in many ages were certainly made in outlying parts of the country and the results forwarded to the Bureau of Astronomy.[b]

Much could also be written on the place of astronomy and mathematics in the Imperial University (Kuo Tzu Chien, cf. Vol. 1, p. 127). The *Thang Yü Lin* tells us[c] that to the two degrees in classics which the Sui dynasty had established, the Thang added four others, including one in mathematics, 'Ming Suan';[2] but nobody wanted to take it since it was not likely to lead to high advancement in the bureaucracy. We are also told[d] of a corresponding teaching department, the Suan Kuan,[3] which had its Professor (Po-shih[4]) and Lecturer (Chu-chiao[5]) like the others.

With this background in mind it is easy to see why this Section on astronomy cannot be prefaced by a 'survey of the chief landmarks in Chinese astronomical literature' similar to that which formed the beginning of the Section on mathematics. A great portion of the surviving Chinese astronomical literature is to be found in the chapters dealing with astronomy, calendrical science, and unusual natural phenomena, in the official dynastic histories, and the identity of the writers of these chapters is not always very clear. Apart from these, the great majority of the ancient books on astronomy have perished, and if today available in any sense, are known only in the form of brief fragments. The reason for this must surely be that whereas the mathematical texts were of very general use (for example, to officials in charge of public works, to merchants, and to military commanders), the astronomical writings were, within the framework of Chinese society, regarded as far more technical, and only of interest to the restricted circles connected with the Astronomical Directorate. They were therefore probably copied in much fewer numbers, and tended to accumulate in palace and governmental libraries, where they disappeared in one or other of the successive holo-

[a] *MCPT*, ch. 7, para. 11, tr. auct.; cf. Hu Tao-Ching (*1*), vol. 1, pp. 295 ff.

[b] We shall find a number of evidences of this as we go on (e.g. pp. 274, 293 and 420 below). Dubs (*2*), vol. 3, pp. 544 ff., 546 ff., has proved it for the solar eclipses recorded in the *Chhien Han Shu*. One of the most remarkable cases is that of −145, which though visible only at sunrise at the tip of the Shantung peninsula, was observed and reported to the capital (vol. 1, p. 338).

[c] Ch. 8, p. 16*b*.

[d] Ch. 5, p. 17*b*.

[1] 沈括 [2] 明算 [3] 算館 [4] 博士 [5] 助教

causts which accompanied the disturbances at changes of dynasty. After the development of printing this situation changed somewhat, but not entirely.

Furthermore, owing to the close association between the calendar and State power, any imperial bureaucracy was likely to view with alarm the activities of independent investigators of the stars, or writers about them, since they might secretly be engaged upon calendrical calculations which could be of use to rebels interested in setting up a new dynasty. New dynasties always overhauled the calendar and issued one with a new name, and this might happen even in successive reign-periods under the same emperor. No less grave was the political significance of celestial events which could be memorialised as portents according to the rules of the age-old State astrology.[a] Hence it is not at all surprising to find exhortations to security-mindedness addressed century after century to the astronomical officials. For example:

> In the twelfth month of the 5th year of the Khai-Chhêng reign-period (+840)[b] an imperial edict was issued ordering that the observers in the imperial observatory should keep their business secret. 'If we hear', it said, 'of any intercourse between the astronomical officials or their subordinates and officials of other government departments or miscellaneous common people, it will be regarded as a violation of security regulations which should be strictly adhered to. From now onwards, therefore, the astronomical officials are on no account to mix with civil servants and common people in general. Let the Censorate look to it.'[c]

As a sociological phenomenon, therefore, there was nothing new about Los Alamos or Harwell. But whether or not the best and greatest scientific achievements happen under such conditions is another question. Even a Galileo and a Priestley had their difficulties with the powers that were.

From early times Chinese astronomy had benefited from State support, but the semi-secrecy which it involved was to some extent a disadvantage.[d] Realisation of this was occasionally expressed by Chinese historians themselves—for example, in the *Chin Shu* we read:[e]

> Thus astronomical instruments have been in use from very ancient days, handed down from one dynasty to another, and closely guarded by official astronomers. Scholars have therefore had little opportunity to examine them, and this is the reason why unorthodox cosmological theories[f] were able to spread and flourish.

Nevertheless, it would be a mistake to push this too far. In the Sung dynasty, at any rate, we have clear indications that the study of astronomy was quite possible in scholarly families connected with the bureaucracy. Thus we learn that in his earlier years Su Sung had model armillary spheres of small size in his home, and so came

[a] On this see particularly the studies of Bielenstein (1, 2). The *Chin Shu*, ch. 12, pp. 12*a*ff. and ch. 13, pp. 1*a*ff. (tr. Ho Ping-Yü), constitutes a veritable treasury of portents and their interpretations between about +250 and +450.

[b] There had been some particularly distressing comets—four of them in +837 alone (one of which was Halley's).

[c] *Chiu Thang Shu*, ch. 36, p. 15*b*, tr. auct.

[d] The extent to which this prohibition of mathematical and astronomical studies by unauthorised persons acted as a seriously inhibiting factor on Chinese science will be considered later (Sect. 49).

[e] Ch. 11, p. 5*a*, tr. auct. adjuv. Ho Ping-Yü.

[f] Lit. 'the Hsüan (Yeh) and Kai (Thien) theories'.

gradually to understand astronomical principles, though it was not until he became a high official that he could think of receiving an imperial command to make a full-scale one.[a] About a century later the great philosopher Chu Hsi had an armillary sphere in his house, and tried hard to reconstruct the water-power drive of Su Sung, though unsuccessfully.[b] Still, the general tradition explains well enough why Matteo Ricci's mathematical books were confiscated when he was on his way to the capital in +1600. 'In China', he wrote, 'it is forbidden under pain of death to study mathematics without the king's authorisation, but this law is no longer observed....' Fortunately, they were returned to him, by mistake, before he went on to Peking in the following year.[c]

In sum, therefore, we do not have so clear a succession of conserved complete treatises in astronomy as in mathematics. This, however, does not mean that the Chinese astronomical literature is not very large. We must describe some of its chief features.

(ii) *Ancient calendars*

Astronomical data are contained in the two oldest calendars which have come down to us.[d] One of these is known as the *Hsia Hsiao Chêng*[1] (The Lesser Annuary of the Hsia Dynasty), the other as the *Yüeh Ling*[2] (Monthly Ordinances).

The *Hsia Hsiao Chêng* has nothing to do with the Hsia 'dynasty'. It is substantially a farmer's calendar, but it includes comments on the weather, the stars, and animal life, arranged under the twelve lunations of the year.[e] It was commented many times in the Chhing dynasty,[f] but the most accessible modern commentary is probably that of Hung Chen-Hsüan, the *Hsia Hsiao Chêng Su I*. In a careful study of the astronomical content Chatley (10) concluded that the most probable date of composition of the calendar is in the neighbourhood of −350, i.e. just about the time when, as we shall see in a moment, much astronomical activity was going on. However, the brevity of the details given would permit a dating as far back as the −7th century, though that would be unlikely. A fair estimate would perhaps be the −5th. The Lesser Annuary of the Hsia was incorporated in the +1st century in the *Ta Tai Li Chi* (Record of Rites of Tai the Elder).[g] The astronomy in the book does not go much beyond that of

[a] *Chhü Wei Chiu Wên*, ch. 8, p. 10*a* (translation in Needham, Wang & Price), by Chu Pien, *c.* +1140.

[b] *Sung Shih*, ch. 48, p. 19*b* (translation in Needham, Wang & Price).

[c] D'Elia (2), vol. 2, p. 122; Bernard-Maître (1, 5); Trigault (Gallagher ed.), p. 370. It also explains the impression which the Dominican missionary, Gaspar da Cruz, had received in +1556. He wrote, 'And though there were some Portugals who reported without any certainty that the Chinas did study natural philosophy, the truth is that there are no studies nor universities in it, nor private schools, but only the schools-royal of the laws of the kingdom. The truth is that some are found who have knowledge of the courses of Heaven, whereby they know the eclipses of the sun and of the moon. But these if they know it by some writings that are found among them, they teach it privately to some person or persons, but of this there are no schools.' See Boxer (1), p. 161.

[d] Research is of course proceeding on the calendar indications which may be derived from the Shang oracle-bones, but this is not yet a settled matter; cf. below, p. 391.

[e] Translations by Douglas (now outdated); Chatley (10), in part; R. Wilhelm (6); Soothill (5).

[f] As by Jen Chhao-Lin (*Hsia Hsiao Chêng Chu*); Chhêng Hung-Chao (*Hsia Hsiao Chêng Chi Shuo*) and Huang Mu (*Hsia Hsiao Chêng Fên Chien*).

[g] Ch. 47.

1 夏小正 2 月令

the monthly night-sky maps of modern newspapers, being confined to observation of the more prominent constellations.

The *Yüeh Ling*, which came to be incorporated (in the −1st century) in the *Hsiao Tai Li Chi* (Record of Rites of Tai the Younger),[a] is a much longer document.[b] It is practically identical with the first twelve chapters of the *Lü Shih Chhun Chhiu* (Master Lü's Spring and Autumn Annals),[c] in which, however, each chapter (each month) of the *Yüeh Ling* is followed by four chapters of other material supposed to be on matters arising out of it. All the *Yüeh Ling* chapters are built on a similar pattern; the first sentences give the astronomical characteristics of the months, then follow details about the associated musical notes, numbers, tastes, sacrifices, and so on,[d] the bulk of the chapter describing the imperial ceremonies to be performed. Prohibitions of various activities bring up the rear, and each chapter ends with a statement of what calamities will happen if the proper observances are not carried out. These Monthly Ordinances are often considered to date from the −3rd century, and they certainly cannot be later, since the *Lü Shih Chhun Chhiu* is an authentic compilation of −240 or −239, but there is no reason why the Ordinances may not be a good deal earlier. Nōda (2, *2*), for instance, has made calculations from the star data given which lead him to the conclusion that a better approximation to the time of their collection would be −620 with a range of a couple of centuries on either side. Again, therefore, the −5th century is a likely possibility. We shall have occasion to recur to these calendars in connection with the observation of meridian passages of the *hsiu*.

Meanwhile we may pause a moment to take a closer look at the work in which the *Yüeh Ling* was embodied. On previous occasions the 'Spring and Autumn Annals of Master Lü' have often been mentioned,[e] and there will hardly be a single subsequent section of this book in which they will not appear. The biography of Lü Pu-Wei in the *Shih Chi* gives an unexpectedly detailed account of the composition of this encyclopaedia of the knowledge of the age.

At that time (−3rd century) in the State of Wei[1] there was the Lord of Hsin-Ling;[2] in Chhu there was the Lord of Chhun-Shen;[3] in Chao there was the Lord of Phing-Yuan;[4] and in Chhi there was the Lord of Mêng-Chhang.[5] All of them had been members of the lesser gentry (*hsia shih*[6]), and they delighted in having (about them) visiting (scholars) by means of whom they could compete with each other (in argument). Lü Pu-Wei was ashamed that Chhin, with all its power, was still not equal (to these States in scholarship), so he summoned scholars to come, and entertained them lavishly, until he had three thousand visitors whom he supported.[f] At this time among (the entourage of) the feudal lords there were many disputing scholars, such as the followers of Hsün Chhing,[7] who wrote books

[a] This is the *Li Chi* as now known; ch. 6.
[b] Translations by Legge (7), vol. 1, p. 249; Couvreur (3).
[c] Translation by R. Wilhelm (3). One of the most recent studies of the *Yüeh Ling* is that of Hsiang Tsung-Lu (*1*).
[d] Cf. Sect. 13*d* above.
[e] Vol. 1, pp. 98, 150, 223; vol. 2, pp. 36, 55, 72, 131, 563; and p. 19 above.
[f] He was outdoing the Chi-Hsia Academy of the State of Chhi (see Vol. 1, p. 95).

1 魏 2 信陵君 3 春申君 4 平原君 5 孟嘗君 6 下士
7 荀卿

which were spread throughout the world.[a] Lü Pu-Wei now had all his guests record what they had learned, and he collected their discussions to form eight 'Observations' (*lan*[1]), six 'Discussions' (*lun*[2]), and twelve 'Records' (*chi*[3]), (totalling) more than 200,000 words.[b] He maintained that all matters pertaining to Heaven, earth, and the myriad things (in the universe) were contained (in this work). He entitled it 'Master Lü's Spring and Autumn Annals'.[c] It was displayed at the gate of the market-place at Hsien-yang,[d] where 1000 catties of gold were suspended above it, and notification was issued to the travelling scholars and visitors of the feudal lords that any one among them who could add or subtract a word, would win this treasure.[e]

The postface to the first part of the *Lü Shih Chhun Chhiu* itself[f] then tells us that it was completed in −239. This was not the last time in history when a politically powerful but (even in its own opinion) relatively uncultured society would make great efforts to attract to itself the best of the light and learning which existed elsewhere; examples could be found readily enough in Mongol Persia and later on the North American continent.

These are not the only ancient calendars known. There is, for example, a Later Han one, the *Ssu Min Yüeh Ling*[4] (Monthly Ordinances for the Four Sorts of People) attributed to Tshui Shih;[5] if it has astronomical references they have not been investigated.

(iii) *Astronomical writings from the Chou to the Liang* (+*6th century*)

There is in Mencius an interesting reference to astronomy. The passage is one of distinctly Taoist flavour, in which he is criticising the scholars of his time for forcing facts and going against Nature.

All who speak about the natures (of things) have only cause and effect (*ku*[6]) to reason from, and nothing else. The value of phenomena lies in their naturalness. What I hate in your learned men is the way they bore out their conclusions. If they would act as did Yü the Great when he conveyed away the waters, there would be nothing to dislike in their learning. He did it in such a way as to give himself no trouble (i.e. by realising that water will flow downhill and not trying to make it do the opposite). If your learned men would act thus, their knowledge would be great. Consider the heavens so high and the stars so distant. If we have investigated their phenomena (*ku*[6]) we may, while yet sitting in the same place, go back to the solstice of a thousand years ago.[g]

[a] See Vol. 2, p. 19.

[b] Early in the +3rd century, Kao Yu, the first commentator, reckoned 173,054 words, so this estimate was rather close. The tripartite division still exists in the book as we have it today; the calendar of the *Yüeh Ling* coming under the head of 'Records'.

[c] In challenging imitation, of course, of the *Chhun Chhiu* classic (see Vol. 1, p. 74), but with no resemblance.

[d] The capital of the State of Chhin.

[e] Ch. 85, p. 5*a*, tr. Bodde (15), p. 5.

[f] Now ch. 61 (vol. 1, p. 118).

[g] *Mêng Tzu*, IV (2), xxvi, tr. Legge (3), p. 207, mod.

[1] 覽 [2] 論 [3] 紀 [4] 四民月令 [5] 崔寔 [6] 故

Perhaps Mêng Kho had in mind some of his contemporaries, for there were then living two of the greatest, and earliest, astronomers in Chinese history, Shih Shen [1] of the State of Chhi and Kan Tê [2] of the State of Wei. Their work must be placed between about −370 and −270. It was they, together with a third astronomer, Wu Hsien [3],[a] who drew up the first star-catalogues, of which something will be said below (p. 263).[b] Their work, which was quite comparable to that of Hipparchus, was thus carried out two centuries earlier than his (−134).

The original titles of their books, or star-lists, were the *Thien Wên* [4] (Astronomy) of Shih Shen, and the *Thien Wên Hsing Chan* [5] (Astronomical Star Prognostication) of Kan Tê. The list of Wu Hsien simply bore his name. These books seem to have lasted until the Liang (+6th century), but after that they disappear from the bibliographies of the dynastic histories. In the Sui (end of the +6th) they were apparently largely incorporated in the *Ku Chin Thung Chan* [6] (Compendium of Astrology, New and Old) of Wu Mi,[7] but this in its turn was lost about the Mongol period. Portions of their work have thus reached us in four forms: (*a*) a book known as the *Hsing Ching* [8] (Star Manual); (*b*) the astronomical chapters of the *Chin Shu* [9] (History of the Chin Dynasty) [c] compiled in the +7th century, almost certainly by Li Shun-Fêng [10] the mathematician;

[a] This was not his real name, which is unknown; his work was attributed to a legendary minister of the Shang dynasty.

[b] These three astronomers are generally regarded as the first pioneers of positional astronomy in China, the first to give positions of stars in degrees in a coordinate system (see pp. 266 ff. below). But in the introduction to the astronomical chapters of the *Chin Shu* (ch. 11, p. 1*a*) a number of other names are listed with them. 'In the history of the feudal lords', it says, 'Tzu Shen[11] of the State of Lu, Pu Yen[12] of the State of Chin, Pei Tsao[13] of the State of Chêng, (Shih) Tzu-Wei[14] of the State of Sung, Kan Tê of the State of Chhi, Thang Mo[15] of the State of Chhu, Yin Kao[16] of the State of Chao, and Shih Shen of the State of Wei—all possessed deep knowledge of astronomy, and all discussed the charts (of the stars) and their verification.' Wu Hsien is mentioned in the preceding sentence as a personage of the Shang period, and Shih I[17] as an astrologer-astronomer of the early Chou. The oldest of the others was Pu Yen, who predicted the fall of a city which Duke Hsien of Chin was besieging (−675 to −650). Tzu Shen came in the following century, for he served Duke Hsiang of Lu (−570 to −540). Of the early −5th century, just about the time of the death of Confucius, were Pei Tsao, who predicted the downfall of Chhen State in −478, and Shih Tzu-Wei, who gave a famous answer to Duke Ching of Sung in −480 (*Shih Chi*, ch. 38, p. 15*b*; Chavannes (1), vol. 4, p. 245). A fragment said to be written by him survives as the *Sung Ssu-Hsing Tzu-Wei Shu*[18] (Book of the State Astronomer Tzu-Wei of Sung), *YHSF*, ch. 77, p. 12*a*. The date of Thang Mo is unknown, but Yin Kao must have been after −403 when his State originated, so he was probably a contemporary of Shih Shen, who served another of the succession States of Chin. That all these men were called astrologers, and known mainly as such to following generations, proves nothing; that their names were associated with those of Shih Shen, Kan Tê and Wu Hsien may mean that positional astronomy in China was a good deal older than the middle of the −4th century. Quite possibly Shih Tzu-Wei, for example, was the Timocharis to Kan Tê's Hipparchus.

The information in the *Chin Shu* draws great authority from a parallel passage in the *Shih Chi*, ch. 27, p. 38*a* (Chavannes (1), vol. 3, p. 402). Ssuma Chhien, however, omitted Tzu Shen and Pu Yen, and added the name of a Chou colleague of Shih I's, Chhang Hung,[19] about whom nothing else is known.

Some of the information in the foregoing note was assembled by Ho Ping-Yü (1).

[c] Ch. 11, pp. 7*a* ff.

[1] 石申 [2] 甘德 [3] 巫咸 [4] 天文 [5] 天文星占
[6] 古今通占 [7] 武密 [8] 星經 [9] 晉書 [10] 李淳風
[11] 梓愼 [12] 卜偃 [13] 裨竈 [14] 史子韋 [15] 唐眛 [16] 尹皐
[17] 史佚 [18] 宋司星子韋書 [19] 萇弘

(*c*) the compendium[a] called the *Khai-Yuan Chan Ching*[1] (The Khai-Yuan reign-period Treatise on Astrology);[b] and (*d*) an important manuscript astrological work of +621 discovered by Pelliot at Tunhuang.[c] Maspero (3) has analysed the contents of these four sources. The present *Hsing Ching* is incomplete, containing only the stars and constellations of the 'Central Palace' (circumpolar region) and of the Eastern and Northern 'Palaces'.[d] Its text is considered less good than that in the *Chin Shu*. All these contain measurements in degrees, and we shall discuss later whether these were in the original work or were additions of later observers. The *Khai-Yuan Chan Ching* has the fullest data.[e] There can be little doubt that all the texts derive from the conflation of the works of the Chou astronomers made by Chhen Cho[2] about +310, each being a different series of extracts from it. There would be great value in a systematic reconstruction of the whole remaining material, but this has not yet been done. The date of the present *Hsing Ching* must be at least as early as the Sui and very probably goes back to the +5th century. Strangely, Maspero (3) did not mention that it was incorporated into the *Tao Tsang*, where it appears[f] under the title *Thung Chan Ta Hsiang Li Hsing Ching*.[3] The Tunhuang manuscript has a section on planetary motion which has been lost from the other recensions.

After this period of preliminary observational work in the late Chou time, the Han period was notable particularly for the flourishing of cosmological theories. These we shall notice in the next sub-section. But the questions had been raised in the −4th century by the School of Naturalists[g] in the north and by the famous poet Chhü Yuan[4] in the south. One of his odes (written in the close neighbourhood of −300) takes the form of a series of semi-rhetorical questions about the universe; it is the *Thien Wên*[5] (Questions about the Heavens).[h] We met with this already in Section 18 in connection with the development of the concept of laws of Nature. Some have believed[i] that the verses were meant to accompany cosmological wall-paintings in a temple (as Wang I[6] had said in the +2nd century), but this is now considered improbable.[j] The notions on the structure of the universe which the poem contains have been the subject of an interesting commentary by Wên I-To (*2*). Among them there occurs the idea that heaven has nine storeys or layers (*chiu chhung*[7]), curiously reminiscent of Ptolemaic-Aristotelian spheres.[k] But this had almost no

[a] Put together just before +729 by the Indian Chhüthan Hsi-Ta[8] (Gautama Siddhārtha); cf. p. 37 above.

[b] Chs. 65–70.

[c] Bibliothèque Nationale, no. 2512.

[d] Cf. below, p. 240.

[e] This source has been carefully studied by Yabuuchi (*6*).

[f] *TT* 284.

[g] Cf. Sect. 13*c* in Vol. 2.

[h] Translation by Conrady & Erkes (1). Biography in *Shih Chi*, ch. 84.

[i] E.g. Erkes (8).

[j] See H. Wilhelm (7), who regards the questions as some kind of liturgical riddles.

[k] That this theory was actually derived from the spheres of the Greeks is the first thought which would occur to most historians of science. Chhen Wên-Thao, indeed, (*1*), p. 38, affirmed it with a diagram. But the dates do not fit, quite apart from the difficulty of any transmission before the time of Chang Chhien (cf. Sect. 7*e*, *g*, in Vol. 1). Eudoxus, who worked in the half-century immediately before

1 開元占經 2 陳卓 3 通占大象曆星經 4 屈原 5 天問
6 王逸 7 九重 8 瞿曇悉達

echoes in later Chinese astronomical thought, though occasionally mentioned during the Han.[a]

Here the *Chou Pei Suan Ching*[1] (The Arithmetical Classic of the Gnomon and the Circular Paths of Heaven) comes before us again,[b] for it stated the theory of one of the principal schools of thought;[c] as we saw, it is to be regarded as a Chou nucleus with Chhin and Former Han accretions. Another school of thought is represented by the latter part of the 3rd (or Thien Wên[2]) chapter of the *Huai Nan Tzu* book (about −150).[d] Contemporary with Liu An, the Prince of Huai Nan, was Lohsia Hung,[3] to whom was ascribed the earliest beginnings of astronomical instruments in China (though this can hardly be true); his book, the *I Pu Chhi Chiu Chuan*[4] (Old Discourse of an Experienced Elder for the benefit of the (Astronomical) Bureau),[e] now exists only in fragments in the *Thai-Phing Yü Lan* encyclopaedia.

Still pursuing the development of cosmological theory, one would expect to find something of it in the so-called 'false' or Weft writings, apocryphal treatises which grew up during the Han time.[f] This is indeed the case, for a considerable amount of two relevant texts, the *Shang Shu Wei Khao Ling Yao*[5] (Investigation of the Mysterious Brightnesses), and the *I Wei Thung Kua Yen*[6] (Verifications of the Powers of the Kua in the Book of Changes), both commented on by Sung Chün[7] (*d.* +76), has come down to us, forming parts of the *Ku Wei Shu*[8] collection (Ming). The first of these espouses the cosmological theory of the *Chou Pei*, the second that of another school, the chief representative of which was the great astronomer and geodesist Chang Hêng[9] (+78 to +139) and his book *Ling Hsien*[10] (The Spiritual Constitution of the Universe).[g]

Retracing our steps a little, mention must be made of the very important astronomical chapter of Ssuma Chhien's *Shih Chi* (Historical Record), finished in −90. This Thien Kuan[11] (Celestial Officials, or Governors of Heaven) chapter is written in a systematic way,[h] the author, who had himself filled the highest astronomical and astrological offices of State, passing first in review the stars and constellations of the

that of Chhü Yuan, only adumbrated the system of concentric spheres, and epicycles did not come in before Apollonius about −200 (Neugebauer (9), pp. 147 ff.; B. & M., pp. 426 ff.). Of course the vague ideas of −5th-century Pythagoreans might conceivably have been the source of the theory. Edkins (10) suggested that there might have been a transmission in the opposite direction, but the obstacles to travel hold good just as much, and it would be most unlikely in view of the rather characteristic Greek love of circles and other geometrical figures.

[a] E.g. *Huai Nan Tzu*, ch. 3, p. 16*a*. The 'Nine Heavens' occur also in other poems of Chhü Yuan, cf. Waley (23), p. 42; Yang & Yang (1), pp. 3, 28.

[b] Cf. above, p. 21.

[c] Cf. below, p. 210.

[d] Of this an unpublished English translation exists, by Chatley (1). I have been privileged to make use of it. Maspero (3) also translated part of this chapter. An important discussion of it in Chinese is that of Chhien Thang[12] (+1788): *Huai Nan Thien Wên Hsün Pu Chu.*[13]

[e] In later times there were several other books with the same engaging title.

[f] See above, Vol. 2, pp. 380, 391.

[g] Now available only in fragmentary form in the Ma Kuo-Han collection (*YHSF*, ch. 76). We have met with it already in the Section on the concept of laws of Nature (Vol. 2, pp. 556 ff.). The same chapter contains also another work of Chang Hêng's, the *Hun I Chu*[14] (Commentary on the Armillary Sphere), but not complete.

[h] Ch. 27. Translation by Chavannes (1), vol. 3, pt. 2, p. 339.

1 周髀算經 2 天文 3 落下閎 4 益部耆舊傳
5 尙書緯考靈曜 6 易緯通卦驗 7 宋均 8 古微書 9 張衡 10 靈憲
11 天官 12 錢塘 13 淮南天文訓補注 14 渾儀注

five 'Palaces' (Circumpolar, East, South, West and North). He then devotes an elaborate discussion to the planetary movements, including retrogradations. Then follow the astrological associations of the *hsiu* with specific terrestrial regions, the interpretation of unusual appearances of the sun and moon, comets and meteors, clouds and vapours (including auroras), earthquakes and harvest signs. The chapter ends by the reflections of the historian. He says that there had never been a time when the rulers had not carefully observed the heavens, and recalls the great advances which had been made, naming a long list of star-clerks which ends with Shih Shen. He then mentions the number of eclipses recorded at different periods, exceptional showers of meteors, and the corresponding earthly events which they had presaged or accompanied. The chapter is a text of the highest importance for ancient Chinese astronomy.[a]

The first of the official dynastic histories, the *Chhien Han Shu*, also contains astronomical chapters,[b] the author of which was probably Ma Hsü.[1] At this time (*c.* +100) the historical data of the Former Han dynasty were combined with the more advanced astronomical knowledge of the Later Han. The material, which includes detailed information concerning the calculation of synodical cycles and the attempted prediction of eclipses, has been greatly elucidated,[c] but much remains to be done. There is no translation of this important source as a whole.

The ecliptic had played very little part in Chinese astronomy before the +1st century, but about the time of the calendar reform of Chia Khuei[2] in +85 instruments were made with which to measure it, and this new knowledge, including its obliquity stated in degrees, appears in the *Lü Li Chih*[3] (Memoir on the Calendar) of Liu Hung[4] and Tshai Yung[5] in +178. The obliquity had already been measured by Eratosthenes in the latter part of the −3rd century. It was shortly after the time of Liu Hung and Tshai Yung, in the San Kuo period, that Chhen Cho[6] made his compilation of the Chou positional lists, but we are not even certain what title it bore. On the other hand, there has survived an important work of his older contemporary Wang Fan,[7] *Hun Thien Hsiang Shuo*[8] (Discourse on Uranographic Models), written in the State of Wu about +260; this has been translated and commented on by Eberhard & Müller (2). Another astronomer of the same time was Yao Hsin,[9] fragments of whose *Hsin Thien Lun*[10] (Discourse on the Diurnal Revolution) survive in encyclopaedias; he produced a variant of the theory of the celestial sphere whose chief protagonist had been Chang Hêng.[d] Among his +3rd-century contemporaries it was represented by the *Wu Li Lun*[11] (Discourse on the Principles of Things) of Yang Chhüan.[12]

In the next century came the discovery of the precession of the equinoxes by Yü Hsi[13] (*fl.* +307 to +338), whose book, the *An Thien Lun*[14] (Discussion on the Conforma-

[a] Doubts which had been raised as to its authenticity have been effectively set at rest by the thorough treatment of Liu Chao-Yang (*3*, *4*).

[b] Especially ch. 26.

[c] Eberhard & Henseling (1); Eberhard, Müller & Henseling (1); and Eberhard & Müller (1).

[d] Cf. below, p. 216.

1 馬續 2 賈逵 3 律曆志 4 劉洪 5 蔡邕 6 陳卓
7 王蕃 8 渾天象說 9 姚信 10 昕天論 11 物理論
12 楊泉 13 虞喜 14 安天論

tion of the Heavens)[a] still exists in the form of fragments. We also have the *Chhiung Thien Lun*[1] (Discourse on the Vastness of Heaven),[b] written by his grandfather Yü Sung.[2] A hundred years later came Chhien Lo-Chih[3] with his star-maps of unknown title, now lost, and Tsu Kêng-Chih[4] with his *Thien Wên Lu*[5] (Résumé of Astronomy), fragments of which are preserved in the *Khai-Yuan Chan Ching*.

(iv) *Astronomical writings from the Liang to the beginning of the Sung (+10th century)*

At the end of the +6th century, in the Sui, came the compiling work of Wu Mi already mentioned, and an interesting astronomical poem by Wang Hsi-Ming,[6] whose Taoist *nom de plume* was Tan Yuan Tzu,[7] called *Pu Thien Ko*[8] (The Song of the March of the Heavens).[c] He might perhaps be termed the Aratus[d] or the Manilius[e] of China, though so much later than they. This poem was held in much repute through the following centuries, and in the 18th (+1726), when the *Thu Shu Chi Chhêng* encyclopaedia was assembled, each section of the heavens described in chapters 44–54 of the first part (*chhien hsiang tien*), with its list of stars and their coordinates, was preceded by the appropriate section of the *Pu Thien Ko*. At the end there follows a *Hsi Pu Thien Ko*,[9] a 'Western revised version' produced by the Jesuits or their assistants. A contemporary of Wang Hsi-Ming's was Li Po,[10] the father of Li Shun-Fêng the mathematician. Li Po was a Taoist who wrote a good description of the great constellations in the heavens towards the end of the Sui or at the beginning of the Thang (*Thien Wên Ta Hsiang Fu*[11]).

In the Thang, in the +7th century, both the *Chin Shu*[12] (History of the Chin Dynasty) and the *Sui Shu*[13] (History of the Sui Dynasty) were in course of preparation around +630. Their astronomical chapters, especially those of the *Chin Shu*, are a mine of information,[f] and it seems probable that Li Shun-Fêng the mathematician had a large share, with Chhangsun Wu-Chi, in the writing of both of them. In the +8th came the *Khai-Yuan Chan Ching* already mentioned, and posterity owes a debt

[a] And whether there can be a movement such as precession. Cf. pp. 270, 356 below.

[b] A significant title. The Chinese astronomers were not, on the whole, disposed to conceive of solid spheres, crystalline or other.

[c] Tr. Soothill (5).

[d] −315 to −245; wrote the *Phaenomena*, a description of the stars and constellations in Greek verse; a contemporary of Hsün Chhing. Cf. Mair (1); Böker (1).

[e] *Fl.* about +20; wrote the *Astronomicon*, an astrological poem in Latin verse; a contemporary of Liu Hsin. In the twentieth century a lifetime of study was devoted to the work of Manilius by the English poet A. E. Housman.

[f] We know a little about the sources of the *Chin Shu* chapters though we should like to know more. For example, when Liu Piao[14] was governor of Chingchow about +200 he commissioned one of his officials, Liu Jui[15] to write a compendium of astrological astronomy, the *Ching-chou Chan*.[16] This was especially concerned with non-planetary phenomena, e.g. comets, novae, meteors, sun-spots, haloes, parhelia, etc. Its text has long been lost, but the +7th-century astronomers certainly used its material.

1 穹天論　2 虞聳　3 錢樂之　4 祖暅之　5 天文錄
6 王希明　7 丹元子　8 步天歌　9 西步天歌　10 李播
11 天文大象賦　12 晉書　13 隋書　14 劉表　15 劉叡
16 荊州占

of gratitude to its author for preserving so many passages from the ancient writings on astronomy, however astrological his interests may have been. This was also the period of the activity of I-Hsing[1], [a] and the Indian astronomers resident in China.

Though most of their writings failed to survive, something more should be said here of these Indian astronomers and calendar-experts of the Sui and Thang. The story begins with the books of 'Brahmin' astronomy such as the *Po-lo-mên Thien Wên Ching*,[2] mentioned in the *Sui Shu* bibliography, but now long lost.[b] These must have been circulating about +600. Then during the following two centuries we meet with the names of a number of Indian astronomers resident at the Chinese capital. In +759 a Buddhist astrological work, the *Hsiu Yao Ching*[3] (Hsiu and Planet Sūtra),[c] was translated by Pu-Khung[4] (Amoghavajra), and five years later a Chinese pupil of his, Yang Ching-Fêng,[5] also an astronomer, added a commentary. He wrote:

Those who wish to know the positions of the five planets adopt Indian calendrical methods. One can thus predict what *hsiu* (a planet will be traversing). So we have the three clans of Indian calendar experts, Chiayeh (Kāśyapa), Chhüthan (Gautama), and Chümolo (Kumāra), all of whom hold office at the Bureau of Astronomy. But now most use is made of the calendrical methods of Master Chhüthan, together with his 'Great Art',[d] in the work which is carried out for the government.[e]

And there indeed they were. Chiayeh Hsiao-Wei[6] assisted Li Shun-Fêng in the preparation of the Lin-Tê calendar of +665, and later kinsmen (Chiayeh Chih-Chung[7] around +708 and Chiayeh Chi[8] eighty years later) seem to have combined astrology with military service. The first of the Gautama clan was Chhüthan Lo,[9] who

[a] A special monograph on the contribution of this remarkable cleric is much needed. There is a little in Chu Kho-Chen (5). Very few of I-Hsing's individual works have survived, but the Tripiṭaka still contains his *Hsiu Yao I Kuei*[10] (The Tracks of the *Hsiu* and Planets) and his *Pei Tou Chhi Hsing Nien Sung I Kuei*[11] (Mnemonic Rhyme of the Seven Stars of the Great Bear and their Tracks). A Tantric Buddhist monk, he was one of the greatest astronomers and mathematicians in Chinese history (+682 to +727). Much influenced by Indian (and therefore indirectly also Hellenistic) astronomy, he made measurements in ecliptic coordinates and, with Liang Ling-Tsan, constructed armillary spheres with ecliptically mounted sighting-tubes. He was also, with Liang, the inventor of the first escapement for a mechanical clock. This was used in the water-powered celestial sphere and planetarium with jack-work which was completed for the emperor Hsüan Tsung in +725. We have already noted (pp. 119, 139 above) some of I-Hsing's contributions to mathematics. He was famous for his eclipse computations and prepared the Ta Yen calendar of +728.

[b] Cf. Sect. 6*f* (vol. 1, p. 128).

[c] See below, pp. 204, 258.

[d] This was presumably either Chhüthan Chuan or Chhüthan Yen. What could his 'Great Art' have been? One suggestion (cf. Yabuuchi, *1*) is that the main contribution of the *Po-lo-mên* books and the clans of Indian astronomers to Chinese mathematics in this period was an early phase of trigonometry. Indeed, the *Khai-Yuan Chan Ching* contains (ch. 104) a table of sines, which Yabuuchi (1) reproduces. As we saw (p. 108 above), the naming and tabulating of angle functions (sine, cosine, etc.) began in India about +420 with the *Pauliśa Siddhānta*, and the further developments by Āryabhaṭa and Varāha-Mihira (*c.* +500) would have prepared the way for the transmission of this novel approach to China by Buddhist travellers. There is no sign, however, that it had any permanent effect (cf. pp. 37, 109 and 148 above).

[e] TW 1299 (*Taishō*, p. 391.3), N 1356; tr. auct.

[1] 一行 [2] 婆羅門天文經 [3] 宿曜經 [4] 不空 [5] 楊景風
[6] 迦葉孝威 [7] 迦葉志忠 [8] 迦葉濟 [9] 瞿曇羅
[10] 宿曜儀軌 [11] 北斗七星念誦儀軌

produced two calendar systems[a] in +697 and +698, but the greatest was Chhüthan Hsi-Ta[1] (Gautama Siddhārtha, often mentioned elsewhere),[b] who compiled the *Khai-Yuan Chan Ching* (+729) in which a zero symbol and other innovations[c] appeared. This clan was opposed to the views of the Chinese Buddhist monk I-Hsing on calendrical matters. In the year after his death, +728, his treatise, the *Ta Yen Li Shu*, had been edited (under imperial commission) by Chang Yüeh[2] and Chhen Hsüan-Ching.[3] But in +733 the latter joined Chhüthan Chuan[4] in declaring[d] that I-Hsing's Ta Yen calendar was but a plagiarism of the Chiu Chih (Navagrāha) system which Chhüthan Hsi-Ta had translated,[e] with added mistakes. Though backed by an able Chinese astronomer, Nankung Yüeh,[5] they failed to shake its dominance,[f] and having been officially adopted in +729 it remained in use till +757. Then in the sixties of the century the clan was represented by Chhüthan Yen.[6] The Chümolo[7] family, on the other hand, were associated with I-Hsing, for one of them contributed a method of computation of solar eclipses to the Ta Yen calendar (+728) as well as producing an astrological manual.[g]

Were the Indian experts monks, one may ask, like so many other Indian Buddhists? They can hardly have been celibate since their clans can be traced for nearly two centuries, and one should perhaps picture them simply as lay technicians, wedded no doubt to Chinese women. It is remarkable that their activities seem to have exerted little influence on the course of Chinese astronomy. The equatorial mansions remained as before, the circle continued to have $365\frac{1}{4}°$, Indian trigonometry was not taken up, the zero symbol slumbered for another four centuries, and naturally enough the Greek zodiac remained buried in bizarre transliterations. Further research is needed to establish whether the Indian clans contributed anything of permanent value which was accepted from their time onwards, and if so, what it was. Contemporaries such as Sulaimān al-Tājir recorded[h] the impression that Indian astronomy was more advanced than Chinese, but this may well have been due to conversations with such men as Chiayeh Chi. Apart from the important Indian development of trigonometric methods there was no objective basis for it. In any case the paradox remains that we owe to an Indian, Chhüthan Hsi-Ta, the greatest collection of ancient and medieval Chinese astronomical fragments.

[a] *Thang Hui Yao*, ch. 42.

[b] Pp. 12, 37, 148.

[c] Such as the division of the circle into 360 degrees, and sexagesimal minutes and seconds.

[d] Sarton (1), vol. 1, p. 475, misled by Mikami's slip (1), p. 58, placed Chhüthan Chuan just over a century too early.

[e] In +718. See above, p. 175. The translation was not literal and all the computations were recast for the latitude of Chhang-an.

[f] *Hsin Thang Shu*, ch. 27A, p. 1*a*. Tests in the imperial observatory, under the supervision of Huan Chih-Kuei[8] and others, vindicated the accuracy of I-Hsing's calendar compared with those of Li Shun-Fêng and Chhüthan Hsi-Ta. The story was told long ago by Gaubil (2), p. 89.

[g] We are grateful to Prof. E. Pulleyblank for the communication of some investigations on these men. Cf. Yabuuchi (*1*), pp. 40 ff.

[h] Renaudot ed. (1), p. 46; Sauvaget (2), p. 26.

1 瞿曇悉達 2 張說 3 陳玄景 4 瞿曇譔 5 南宮說
6 瞿曇晏 7 俱摩羅 8 桓執圭

To complete the setting, one must remember that other influences were also at work. As we noted earlier,[a] Persian astronomers also came to China, as in +719 when a Ta-Mu-Shê[1] arrived[b] from Jaghānyān.[c] Persian astrological terms are to be found in Chinese Buddhist writings of this century,[d] and the *Hsiu Yao Ching* gave Sogdian planet names[e] in transliterated form (+764). A number of astronomical texts[f] were translated from Sogdian and Persian. The Manichaean word for Sunday finally came to rest as *mi*[2] in local usages in Fukien and Japan. Nearly a century ago this was correctly identified by Dudgeon as the first syllable of Mithras. But the Manichaeans were not the only Western foreigners involved, for we know that one at least of the Nestorians, Adam, the writer of the famous inscription (+781),[g] translated an astronomical-astrological text (presumably from Sogdian) the *Ssu Mên Ching*[3] (Manual of the Four Gates, i.e. the repartition of the *hsiu* among the four palaces).

Such activities helped the growth of quite a rich astronomical-astrological literature in the +8th and +9th centuries. Some books and fragments are contained in the Tripiṭaka, such as the *Chhi Yao Hsing Chhen Pieh Hsing Fa*[4] (The Different Influences of the Seven Luminaries and the Constellations),[h] which lists the *hsiu* and gives the number of the stars in each one.[i] A calendar with a title of this style, the Chhi Yao Li,[5] the work of Wu Po-Shan,[6] was officially adopted about +755, though it did not last more than a couple of years. Now it has long been suspected that books with titles mentioning the Seven Luminaries were in some way connected with influences from Persia and Sogdia,[j] the term being associated with the planetary seven-day week, certainly introduced to China, though not adopted there, from the Iranian culture-area.[k] This influence is important, and we must retrace our steps a little to consider it.

Chavannes & Pelliot[l] knew no use of the expression Chhi Yao Li before the beginning of the +6th century. But in +437 the King of the Hunnic kingdom of Pei Liang[7] (Northern Liang) in Kansu, about to be extinguished by the Liu Sung, presented to the Sung emperor certain books by his Astronomer-Royal[m] Chao Fei,[8]

[a] Sect. 7*h* (Vol. 1, p. 205).

[b] *Tshê Fu Yuan Kuei*, ch. 971, p. 3*b*. The name was a title: see Chavannes & Pelliot (1), p. 153, and Grousset (1), vol. 1, p. 352.

[c] Tokharestan, former Bactria.

[d] Huber (2); Eberhard (12); Chavannes & Pelliot (1).

[e] See p. 258 below, and Wylie (15).

[f] Such as the curious *Tu-Li-Yü-Ssu Ching*,[9] which Ishida (*1*) has studied. This was translated by Chhü-Kung[10] about +800.

[g] His Chinese name was Ching-Ching.[11] He must have been a catholic man in the best sense for he also translated a Buddhist treatise (see Chavannes & Pelliot (1), p. 134).

[h] TW 1309.

[i] Other works in the Chinese Buddhist patrology will be mentioned from time to time hereafter. See also Eberhard (12).

[j] A preliminary study of this subject has been made by Yeh Tê-Lu (*1*).

[k] See F. W. K. Müller (3); Chavannes & Pelliot (1), p. 174.

[l] (1), p. 170.

[m] The story is given in *Sung Shu*, ch. 98, p. 19*b*. Cf. Li Nien (*20*), p. 63. There is even a mention of the Chhi Yao in the *Nei Ching*, ch. 66, p. 4*a*, though no doubt a Thang or pre-Thang interpolation. The Thang commentator, Wang Ping, says that 'the foreigners use them nowadays for the prognostication of good and evil'.

1 大慕闍 2 密 3 四門經 4 七曜星辰別行法 5 七曜曆
6 吳伯善 7 北涼 8 趙𢾺 9 都利聿斯經 10 璩公
11 景淨

including one with the title *Chhi Yao Li Shu Suan Ching*[1] (Mathematical Treatise on the Seven Luminaries Calendar).[a] The significance of the north-western location of this star-clerk, who was a prolific scholar, is self-evident.[b] Then after +500 the trend intensifies. We find, for instance,[c] an official of the Liang[2] dynasty, Yü Man-Chhien[3] (*fl.* +520 to +570), writing a commentary[d] on a *Chhi Yao Li Shu*,[4] as well as on the mathematical classics.[e] The *Sui Shu* records[f] no less than twenty-two books with Chhi Yao in their titles.[g] About +660 one was written by Tshao Shih-Wei,[5] an astronomer who bore (perhaps significantly) one of the nine Sogdian surnames. In all the dynastic bibliographies there must be at least forty books of the kind.

What do these facts mean? They simply illuminate one of the ways by which Babylonian mathematics and astronomy, especially the computation of solar, lunar and planetary ephemerides, passed to China for further development. Above (p. 123) we discussed the interpolation formulae which seem to have started there with Liu Chhuo[6] towards the end of the +6th century. Below (p. 394) we shall refer to the algebraic methods used for these purposes by the Seleucid Babylonian astronomers.[h] These were being developed during and after the time of Alexander the Great, about −300 onwards. By +100 'Chaldea' was in Parthian hands, and in +226 embodied in Sassanian Persia; there was thus no bar to the spread of knowledge north-eastwards into Central Asia.[i] Then, in alliance with the Persians, the Turks took Samarqand from the Ephthalite Huns[j] about +560. This opened up the eastern roads to that commercial enterprise which made the Sogdians the greatest merchants of Turkestan and the Chinese marches for four or five centuries;[k] indeed, by the end of the century

[a] It is listed, with other books of his on astronomy and calendrical science, in *Sui Shu*, ch. 34, pp. 17*b*, 19*a*.

[b] Actually two *Chhi Yao* books are mentioned already in the *Hou Han Shu*, i.e. before +210; one by the well-known mathematician Liu Hung,[7] the other by an obscurer man, Liu Thao.[8] The expression itself occurs in Liu Hung's memorial on eclipses (see *CSHK* (Hou Han sect.), ch. 66, p. 8*b*); and in Liu Chih's *Lun Thien* of +274 (*CSHK* (Chin sect.), ch. 39, p. 5*a*). Another calendar similarly named was the *Chi Wang Chhi Yao*[9] prepared by Hsü Kuang[10] and mentioned in +425 by his brother-in-law Ho Chhêng-Thien, when presenting a calendar of his own.

[c] *Liang Shu*, ch. 51, p. 27*b*.

[d] This is listed under almost identical title in *Sui Shu*, ch. 34, p. 18*b*.

[e] Yü Man-Chhien belonged to a great family of astronomers and astrologers. His father, Yü Shen[11] (*d.* +532), a Taoist and a Buddhist as well as a noted mathematician, had written an 'Imperial Calendar', Ti Li,[12] in twenty chapters (*Liang Shu*, ch. 51, pp. 25*b*, 27*a*). His son, Yü Chi-Tshai[13] (*fl.* +560 to +600), was one of the most distinguished astronomers of the Sui dynasty (p. 27*b*). The Taoist connection is marked by Yü Chhêng-Hsien[14] (p. 28*a*), a contemporary of Yü Shen, and by Yü Hsin,[15] who served the Northern Chou and wrote on astrological chess. Finally the clan provided a Thang astronomer-royal, Yü Chien[16] (*fl.* +610 to +630).

[f] Ch. 34, pp. 18*a*ff.

[g] Two of these were due to the mathematician Chen Luan[17] (*fl.* +535 to +577), a convert to Buddhism from Taoism, whom we have already met several times (pp. 29, 33, 76 above). Among them there are ten named one for each of the reign-periods of the Chhen and Sui dynasties (+557 to +604).

[h] Cf. Neugebauer (9) and especially (10).

[i] Sassanian Persia in the +3rd century was the scene of serious astronomical activity; cf. Taqizadeh (1); Neugebauer (12).

[j] Cf. Hudson (1), pp. 79, 123.

[k] Cf. Cordier (1), vol. 1, pp. 392ff.; Pulleyblank (3, 4).

1 七曜曆數算經 2 梁 3 庾曼倩 4 七曜曆術 5 曹士蔿
6 劉焯 7 劉洪 8 劉陶 9 既往七曜曆 10 徐廣
11 庾詵 12 帝曆 13 庾季才 14 庾承先 15 庾信
16 庾儉 17 甄鸞

Sogdia[a] had become a vassal State of China. Thus we should expect to find just what we do, a great expansion of ephemerides computation with marked advances there after the middle of the +6th century, even though one can see the beginnings of it as early as the +3rd.

Another work of the Thang was a literary composition by the poet Liu Tsung-Yuan,[1] who set out to give answers, about +800, to the questions about the heavens posed in the poem *Thien Wên* by Chhü Yuan (*c.* −300). His replies were entitled[b] *Thien Tui.*[2]

(v) *Sung, Yuan and Ming*

In view of all that we know about the great flourishing of the natural sciences in the Sung dynasty (cf. Vol. 2, pp. 493 ff.) we should expect that astronomy would then have flourished also, and that this would be reflected in a proliferation of astronomical literature. One would expect monographs on specific subjects similar to those which were produced in the biological sciences at this time (+10th to +13th centuries). And there does, indeed, seem to have been a rich literature at this period, but unfortunately little of it has survived. The Second Sung emperor (+976 to +997) had an astronomical library (Thien Wên Ko[3]) containing books with a total of 2561 chapters.[c] Accordingly it is interesting to look at some of the library catalogues which have come down to us from the +12th century.

In Chêng Chhiao's[4] *Thung Chih Lüeh*[5] (Historical Collections) there is a large catalogue based on the imperial library about +1150. Here we find the titles of no less than 369 books on astronomy and related subjects. It seems worth while to glance at the sections into which it was divided, and to look at some of the titles.

ASTRONOMY, GENERAL (seventy-three books), including:

Ling Hsien Thu Chi[6] (Diagrams illustrating the Spiritual Constitution of the Universe). This may or may not have been an illustrated version of Chang Hêng's book of the early +2nd century. In any case we would very much like to see these diagrams today.

Hun Thien Thu Chi[7] (Diagrams illustrating the Celestial Sphere).

An Thien Lun[8] of Yü Hsi (Discussion of whether the Heavens are at Rest) doubtless complete (+4th century).

Hsin Thien Lun[9] of Yao Hsin (Discourse on the Diurnal Revolution) (+3rd century).

Erh-shih-pa Hsiu Erh-pai-pa-shih-san Kuan Thu[10] (Planisphere showing the 28 Hsiu and the 283 Constellations).

Lun Erh-shih-pa Hsiu Tu Shu[11] (Measurements of the 28 *Hsiu* in Degrees).

Thai Hsiang Hsüan Chi Ko[12] (Poem on the Mysterious Mechanism of the Celestial Bodies) by Lü Chhiu-Chhung.[13]

[a] Ancient Khang-Chü; cf. Vol. 1, p. 175.
[b] This work will be found in *TSCC, Chhien hsiang tien*, ch. 11. See p. 198 above.
[c] *Kuei Thien Lu* (+1067), ch. 14, p. 1*b*.

[1] 柳宗元 [2] 天對 [3] 天文閣 [4] 鄭樵 [5] 通志畧
[6] 靈憲圖記 [7] 渾天圖記 [8] 安天論 [9] 昕天論
[10] 二十八宿二百八十三官圖 [11] 論二十八宿度數 [12] 太象玄機歌
[13] 閭丘崇

Thai Hsiang Hsüan Wên[1] (Essay on the Mysteries of the Celestial Bodies) by Li Shun-Fêng[2] (+7th century).

Hsing Shu[3] (Description of the Stars) by Chhen Cho.[4] This may have been his +4th-century compilation of the Chou material, mentioned above.

Hun I Fa Yao[5] (Essentials of the Technique of Measurement with the Armillary Sphere). This was presented by Han Hsien-Fu[6] in or shortly after +995, and perhaps enlarged in +1010.[a]

Ssu Thien Chien Hsü Chih[7] (What Every Member of the Astronomical Bureau ought to Know).

ASTRONOMY, INDIAN (six books), including several of the works with titles beginning *Po-Lo-Mên*[8] (Brahmin), cf. Vol. 1, p. 128, and one by I-Hsing.[9]

ASTROLOGY, GENERAL (forty-three books), including the *Khai-Yuan Chan Ching*.

ASTROLOGY, PLANETARY (fifteen books).

ASTROLOGY, SOLAR AND LUNAR (eighteen books).

ASTROLOGY using miscellaneous celestial phenomena, including COMETS AND METEORS (ten books).

CALENDRICAL SCIENCE (one hundred and fifty-seven books).

PLANETARY CYCLES in relation to calendar (thirty-two books).

CLEPSYDRA (Water-Clock) (fifteen books).

Some of the titles in this section will be referred to below in the sub-section on the water-clock.

The list of books given by Chêng Chhiao may usefully be compared with another similar list drawn up by his contemporary Yu Mou[10] (+1127 to +1194), in this case for a private library, in a work called *Sui-Chhu Thang Shu Mu*[11] (Catalogue of the Books in the Sui-Chhu Hall). The relevant section has ninety-five books, but they are not all strictly astronomical; what is interesting is that, though the catalogues are of very similar date, very few of the titles seem to occur in both—an evidence for the bulk of literature at the time. Among the titles are:

Ssu Li Po Shih Khao[12] (Investigation of Eclipses in relation to the Four Kinds of Calendar). This looks as if it might have been a work of the monograph type.

Chi Yuan Li Ching[13] (Calendrical Manual based on the Jupiter Cycle). This was actually adopted in +1105.

Kao-Li Jih Li[14] (Korean Solar Calendar).

Thu Kuei Fa[15] (How to Use the Gnomon Shadow Template). Cf. p. 286 below.

Yin Ho Chü Pi Chüeh[16] (Confidential Remarks on the Appearances of the Milky Way).

Yang Shih Tsuan Wei[17] (Observation of Minute Details in the Heavens).

Sometimes an emperor himself would write a preface for an astronomical work. The *Ling Thai Pi Yao*[18] (Essential Knowledge for Official Astronomers), written by an Assistant Astronomer-Royal, Wang Hsi-Yuan,[19] in +1006, was honoured in this way.[b]

[a] Cf. *Sung Shih*, ch. 48, pp. 4*b*ff., ch. 206, p. 9*b*; *Yü Hai*, ch. 4, pp. 30*a*ff., 32*b*ff.

[b] *Sung Shih*, ch. 461, p. 3*b*.

1 太象玄文 2 李淳風 3 星述 4 陳卓 5 渾儀法要
6 韓顯符 7 司天監須知 8 婆羅門 9 一行 10 尤袤
11 遂初堂書目 12 四曆剝蝕考 13 紀元曆經 14 高麗日曆
15 土圭法 16 銀河局祕訣 17 仰視纂微 18 靈臺秘要
19 王熙元

In general, then, one may say that the wealth of astronomical literature in the Sung was considerable. Unfortunately, only a minimal part of it has been preserved.

We are lucky enough, however, to possess one astronomical work of capital importance from the Sung dynasty, namely, the *Hsin I Hsiang Fa Yao*[1] (New Description of an Armillary Clock) by Su Sung,[2] begun in +1088 and finished in +1094. The first chapter describes a complex and perfected armillary sphere instrument, illustrations being given not only of the whole apparatus but of each of its most important parts. The second chapter describes a celestial globe and includes star-maps in which the central palace (circumpolar region) and south polar region are planispheres, while the stars of the more equatorial-ecliptic regions are arranged on a cylindrical projection very similar to Mercator's.[a] The third chapter describes the mechanism which kept the globe and a number of jack-wheels continuously in motion while supplying the sphere with a clock-drive. Power was provided by a water-wheel equipped with a special kind of escapement. Here again illustrations of the separate parts are provided.[b] Su Sung was a friend of Shen Kua,[3] whose *Mêng Chhi Pi Than*[4] of +1086, already often mentioned, contains much of astronomical interest which urgently needs translation.

A few Sung books on the constellations, also still extant, have been carefully studied by Yabuuchi (*10*). The *Ling Thai Pi Yuan*[5] (Secret Garden of the Observatory) had originally been written by Yü Chi-Tshai[6] under the Northern Chou (*c.* +580). Sung material is also embodied in the *Kuan Khuei Chi Yao*.[7] A substantial work on astronomy and calendrical science, the *Liu Ching Thien Wên Pien*[8] (Treatise on Astronomy in the Six Classics), written by Wang Ying-Lin,[9] seems not to have been noticed by modern scholars.

The Yuan period, naturally enough, was one of close collaboration between Chinese and Muslim (Persian and Arab) astronomers. We have already seen something of this (p. 49 above) and shall return again to it later (pp. 372, 380).

Thenceforward the literature seems less voluminous until the time of the Jesuits. But even after their arrival, large compendia on the borderline of astrology and astronomy continued to be published, such as the *Thien Wên Ta Chhêng Kuan Khuei Chi Yao*[10] (Essentials of Observations of the Celestial Bodies through the Sighting Tube) by Huang Ting.[11] This was the latter end of a tradition which derived from the *Khai-Yuan Chan Ching* and all its predecessors, through the +11th-century *Hsing Ming Tsung Kua*[12] (General Description of Stars and their Portents) by Yehlü Shun[13] of the Liao; and the *Ko Hsiang Hsin Shu*[14] (New Elucidation of the Heavenly Bodies) by Chao Yu-Chhin[15] of the Yuan, revised by Wang Wei[16] of the Ming.[c] As has already been

[a] +1512 to +1594; the projection +1569. Cf. Struik (1).

[b] Though the sphere will be more fully described in this Section below, pp. 351 ff., the account of the clockwork must be deferred till the appropriate place in Vol. 4.

[c] He added the words *Chung Hsiu*[17] to the title.

[1] 新儀象法要 [2] 蘇頌 [3] 沈括 [4] 夢溪筆談 [5] 靈臺秘苑
[6] 庾季才 [7] 管窺輯要 [8] 六經天文編 [9] 王應麟
[10] 天文大成管窺輯要 [11] 黃鼎 [12] 星命總括 [13] 耶律純
[14] 革象新書 [15] 趙友欽 [16] 王禕 [17] 重修

pointed out, it is very unfortunate that none of the writings of the greatest astronomer of the Yuan dynasty, Kuo Shou-Ching,[1] has survived. One may note, however, that three years after his death, in +1319, there appeared the *Wên Hsien Thung Khao*[2] (Historical Investigation of Public Affairs) of Ma Tuan-Lin[3] which collected, among many other things, elaborate lists of the appearances of comets, novae, meteors, etc.; these formed to a large extent the basis of modern lists in Western languages which will be mentioned in their place.

Astronomy seems to have shared in the general decline of science during the Ming dynasty. Apart from the work of Wang Wei just mentioned (*fl.* +1445), there was that of Wang Kho-Ta[4] somewhat later, e.g. his *Hsiang Wei Hsin Phien*[5] (New Account of the Web of Stars).[a] After the coming of the Jesuits, astronomy re-awakened, and there was a burst of publications, but consideration of them is better postponed till a later sub-section (p. 454).

In the *Thu Shu Chi Chhêng* encyclopaedia of +1726 there is a great deal of astronomical material, particularly in the first three main divisions.[b] Though much of it was Jesuit, attention should be drawn to the copious collection of quotations from old writers (including lost books) on cosmology (chapters 4–14, 28 and 31 of the first), and on the stars in detail (chapters 55–63 of the first). There is a general list of all the calendars in Chinese history in chapter 79 of the second, historical excerpts about them in chapters 80–2, and a full history of Chinese astronomical instruments in chapters 83–4. The third (*Shu chêng tien*) is important for its extremely full lists of solar eclipses and sun-spots (chapters 18–24), lunar eclipses (chapters 25–6), and novae, comets, meteors and star and planet colours (chapters 27–59).

It will be seen from the foregoing that the Chinese astronomical literature, though much more confused, scattered and fragmentary than that on mathematics, is probably rather larger, even apart from all losses. It is only equalled by the botanical-zoological-pharmaceutical literature, and only surpassed by the medical writings. In view of the wide prospect which this brief review has opened out, one can readily see that everything which has been done in the West, from Gaubil to de Saussure, has only touched the fringe of the subject.[c] While excellent work is in progress on many specific problems,[d] it would be a challenging task to survey the whole field of Chinese astronomy taking into account the numerous sources which have not yet been drawn upon at all. Here in the present book nothing more than an orienting survey, keyed in scope to the general plan, can be attempted.

[a] Quoted in *TSCC*, *Chhien hsiang tien*, ch. 9.

[b] *Chhien hsiang tien*, *Li fa tien*, and *Shu chêng tien*.

[c] How extraordinary it seems that just over a century ago, a European scholar as eminent as Whewell, who was unable to read a word of Chinese, could write that 'we find no single observation, or fact, connected with astronomy, in the Chinese histories, and their astronomy has never advanced beyond a very rude and imperfect condition' (*History of the Inductive Sciences*, vol. 1, p. 166). But even in 1954 an outstanding modern astronomer tells us that 'a legend of advanced astronomical science, even in earliest times, is widespread among the Chinese—but there are very few documents which inform us with any reliability of their true advancement' (Abetti (1), p. 24).

[d] Such as that of Eberhard on Han and San Kuo astronomy, or Tung Tso-Pin on the Shang calendar, or Michel on Chou astronomical instruments.

[1] 郭守敬 [2] 文獻通考 [3] 馬端臨 [4] 王可大 [5] 象緯新篇

(*d*) ANCIENT AND MEDIEVAL COSMOLOGICAL IDEAS

As we have already seen, the period of the late Warring States and the earlier and later Han was one of intense speculation in cosmology and astronomy.[a] The chief groups which formed were thus described about +180 by Tshai Yung,[1] himself a skilled astronomer, in a memorial presented to the emperor:

> Those who discuss the heavens form three schools. The first is that known as the Chou Pei[2] school, the second is the Hsüan Yeh[3] school, and the third is the Hun Thien[4] school. The teaching of the Hsüan Yeh school has been interrupted and there is now no master in it. As for the Chou Pei theory, though its methods and computations still remain, it proved incorrect and lacking in many ways when tested in explaining the structure of the heavens. Therefore the official (astronomers) do not use it. Only the Hun Thien theory approximates to the truth.[b]

Tsu Kêng-Chih,[5] writing at the end of the +5th century, in his *Thien Wên Lu*[6] (Astronomical Résumé),[c] says the same thing, but gives to the Chou Pei school its alternative name, Kai Thien.[7]

The words Chou Pei are the same as those of the title of the oldest mathematical-astronomical work which was described in the Section on mathematics,[d] and may be taken to mean 'the gnomon and the circular paths of heaven'[e]—this work was regarded as the most important text of the 'Heavenly Cover' (Kai Thien) school, which figured the heavens as a hemisphere covering the earth. The words Hsüan Yeh were explained by Yü Hsi[8] in the +4th century as meaning 'brightness and darkness', though modern scholars[f] suggest 'all-pervading night'. As for Hun Thien, there is no doubt that the term means 'celestial sphere'.[g]

(1) The Kai Thien Theory (a Hemispherical Dome)

It seems convenient to begin with the Kai Thien theory, since on internal evidence it must surely be regarded as the most archaic of the three.[h] The heavens were imagined as a hemispherical cover, and the earth as a bowl turned upside down, the distance between them being 80,000 *li*, thus making two concentric domes. The Great Bear

[a] One of the best discussions on this is in the book of Nōda (*4*), in Japanese.

[b] *TPYL*, ch. 2, p. 4*a*, eng. auct., adjuv. Maspero (3). Also in *Chin Shu*, ch. 11, p. 1*b*. Its title was probably *Thien Wên Chih*.[9] Cf. *CSHK* (Hou Han sect.), ch. 70, p. 8*b*; and p. 20 above.

[c] The passage is preserved in Ma Kuo-Han's fragment collection, appended to the *Chhiung Thien Lun* (*YHSF*, ch. 77, p. 6*b*).

[d] P. 19 above.

[e] Further evidence that this was the original meaning of the words is contained in Yü Hsi's +4th-century *An Thien Lun*.

[f] Forke (6), p. 23, who, with Puini (1), assembled material for the field covered by this sub-section.

[g] Note that the first word is the same as that in the technical term *hun-tun*, 'chaos', about which much was said in the Section on Taoism (cf. Vol. 2, pp. 107, 114 above).

[h] In spite of the doubts of Maspero (3), p. 348. What is generally considered a reference to this theory occurs in ch. 13 of the −3rd-century *Lü Shih Chhun Chhiu* (see Chhien Pao-Tsung (*1*), p. 16).

1 蔡邕 2 周髀 3 宣夜 4 渾天 5 祖暅之 6 天文錄
7 蓋天 8 虞喜 9 天文志

was in the middle of the heavens and the *oikoumene* of man in the middle of the earth. Rain falling upon the earth flowed down to the four edges to form the rim-ocean.[a] At the earth's edge the sky was 20,000 *li* above it, and therefore lower than the highest parts of the earth.[b] The heavens were round but the earth was square.[c] The vault rotated like a mill from right to left, carrying with it the sun and moon, which nevertheless had a proper motion of their own from left to right, but much slower than that of the vast wheel to which they were attached. Risings and settings of the heavenly bodies were, however, only illusions, in fact they never passed below the base of the earth. Yü Sung,[1] a forebear, probably the grandfather, of Yü Hsi,[2] the discoverer of the precession of the equinoxes, wrote about +265, in his *Chhiung Thien Lun*[d] (Discourse on the Vastness of Heaven):[e]

The shape of the heavens is lofty, and concave like the membrane of a hen's egg. Their edges meet the surface of the four seas (the rim-ocean). They float on the *yuan chhi*[3] (primeval vapour). It is like a bowl upside down which swims on water without sinking because it is filled with air. The sun turns round the pole, disappearing at the west and returning from the east, but neither emerges from nor goes below (lit. enters) the earth. Heaven has a pole just as a cover has a domed top. The northern heaven is lower than the earth by 30°. (The axis of the celestial) pole inclines towards the north, also making an angle of 30° (as seen from) the due east-west line. Now men live over a hundred thousand *li* south of the pole's due east-west line. Hence the centre of the earth (i.e. the *oikoumene*) is not directly below the (celestial) pole. (This centre) just corresponds to the due east-west line (and the prime vertical) of heaven and earth. The sun, following the (path of the) ecliptic, encircles the (celestial) pole. (At the winter solstice) the position of the pole is 115° north of the ecliptic, and the ecliptic is 67° south of it.[f] These figures are determined from the positions of the two solstice (points).[g]

According to the *Chou Pei*,[h] the sun could illuminate an area only 167,000 *li* in diameter; people outside this would say it had not risen, while those inside would be enjoying daylight. The sun was thus regarded essentially as a circumpolar star, illuminating continually one or other part of the earth's surface[i] as if by a kind of searchlight-beam. But its distance from the pole varied according to the season as it

[a] An idea certainly connected with Tsou Yen's theories (cf. above, Vol. 2, p. 236). Cf. Fêng Yu-Lan (1), p. 160; J. O. Thomson, p. 43.

[b] *Chou Pei Suan Ching*, ch. 2, p. 1*b*.

[c] At the base.

[d] Preserved in *YHSF*, ch. 77, p. 5*a*, from *Chin Shu*, ch. 11, p. 2*b* and elsewhere.

[e] Maspero (3), p. 389, makes one of his rare mistakes in attributing this book to Yü Hsi instead of Yü Sung.

[f] I.e. measured round below the pole through the northern horizon. The pole referred to is of course the north celestial pole, not the pole of the ecliptic, and the figures are only approximate since 115 and 67 do not make $182\frac{5}{8}$ Chinese degrees, the half of $365\frac{1}{4}$°. These round numbers had been given first (so far as we can tell by extant documents) by Chia Khuei about +85 (see below, p. 287).

[g] Tr. auct., adjuv. Ho Ping-Yü (1).

[h] Ch. 2, p. 1*b*.

[i] Wang Chhung (*Lun Hêng*, ch. 32; tr. Forke (4), vol. 1, p. 262; repeated in *Chin Shu*, ch. 11, p. 3*a*) argued that the sun's setting was only apparent, like the disappearance of the light of a torch carried by a man away from an observer on a level plain. The *Chin Shu* goes on to give Ko Hung's refutation of this.

1 虞聳　2 虞喜　3 元氣

followed one or other of the roads (*chien*[1]) between seven parallel declination-circles (*hêng chou*[2]), the outermost being that of the winter solstice and the innermost that of the summer solstice—this is explained in the *Chou Pei* where the diagram of concentric circles is given.[a]

A detailed study of the Kai Thien universe has been made by Chatley (11) based on the measurements and calculations given in the *Chou Pei Suan Ching*.[b] His diagram is reproduced here as Fig. 87. As he says, there is just enough physical truth in the scheme to render it acceptable to very archaic geometers having little more than the Pythagoras theorem at their disposal. One's impression of its antiquity is strengthened

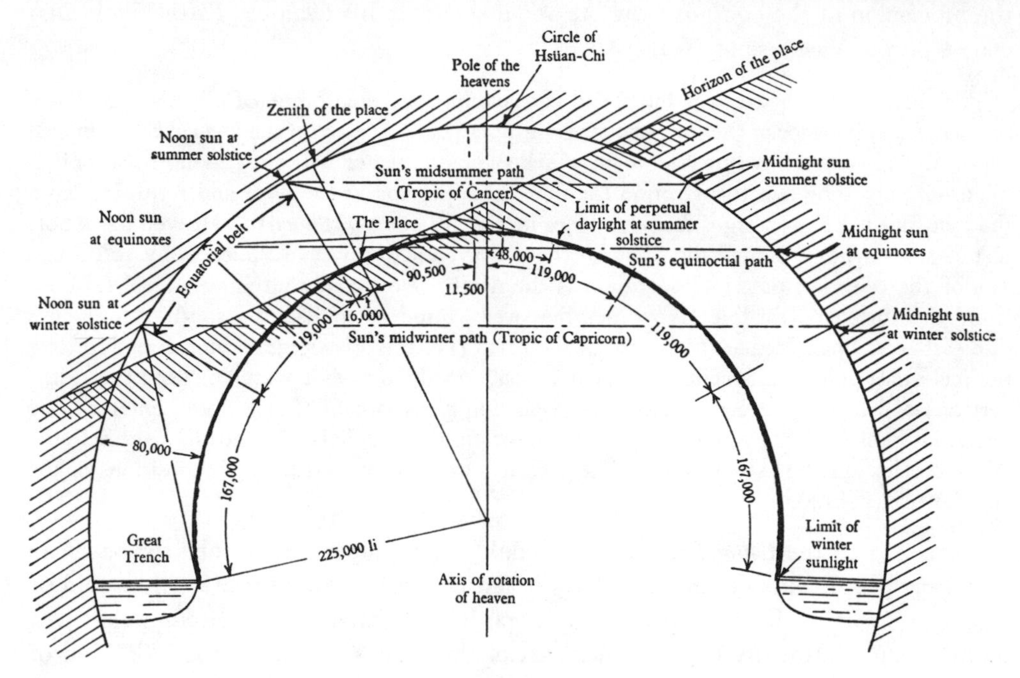

Fig. 87. Reconstruction of the Kai Thien cosmology (after Herbert Chatley).

by the circumstance that a similar double-vault theory of the world existed in Babylonia.[c] It would have been one of the culture-traits which passed both westward to the Greeks and eastward to the Chinese, to be developed later in both civilisations into the theory of the celestial sphere. Rather characteristically Chinese, however, was the insistence that the heavens were circular and that the earth was square, an idea

[a] Ch. 1, p. 17*a*. Already referred to in Section 19 on mathematics, p. 21 above. See also p. 256 below. Chatley and others have believed that for the Kai Thien cosmologists the sun itself moved in right ascension and declination, making sudden jumps from time to time; the existence of the ecliptic being ignored or denied. Whether this was really their view seems rather difficult to substantiate, but the ideas of some of the Old Babylonian astronomers before about −500 may have been quite similar.

[b] The paper of Mikami (*13*) on the same subject has unfortunately not been available to us.

[c] Bouché-Leclercq (1), p. 40; Jastrow (2), P. Jensen (1), and others. Hence also in ancient Israel (Schiaparelli (1), pp. 170, 184).

[1] 間 [2] 衡周

which would arise naturally enough from the circles of the celestial sphere on the one hand and the four cardinal points of earthly space on the other.[a]

That the Kai Thien cosmology was very ancient, Chinese tradition itself maintained, as we see from the following passage in the *Chin Shu*:

(The Kai Thien theory) originated from Pao-Hsi's[b] setting up of degrees for the circumference of the heavens and for the calendar. As for the handing down (of the theory), the Duke of Chou received it from the Yin and Shang people, and men of Chou recorded it. Hence it became known as the Chou Pei, because *pei* means the upright side of a right-angled triangle, and also the vertical gnomon (used to determine sun shadow lengths).

(The Kai Thien theory) says that the heavens seem to be a cover like a bamboo hat enclosing the earth like an inverted basin. Both the centre of the heavens and the centre of the earth are elevated, while the outer regions are low. Beneath the north (celestial) pole is the centre of both the heavens and the earth. There the earth is highest, sloping down in all four directions. The Three Lights,[c] sometimes hidden and sometimes shining, give day and night. The centre of the heavens is higher than the outermost barrier (declination-circle; *hêng*[1]) where the sun is at winter solstice, by a distance of 60,000 *li*.[d] The earth below the north (celestial) pole is higher than it is in the outer regions under the outer barriers (declination-circles) by 60,000 *li*. The outermost barrier (declination-circle) of the heavens is higher than the earth beneath the pole by 20,000 *li*. The heavens and the earth fit into each other in their contours (like concentric domes). The constant vertical distance of the sun from the earth is 80,000 *li*.

The sun is attached to the heavens. It shifts its position with even motion[e] between the seasons of winter and summer. Its movement (cuts across) the seven barriers (declination-circles) and the six roads (between them). The diameter and circumference of each barrier in *li* can be worked out mathematically by using the method of similar right-angled triangles and observing the (lengths of) the shadows (of the gnomon).[f] The measurements of distances of the pole, and of the motions, whether near or far, are all obtained from the use of the gnomon and the right-angled triangle which it forms. Hence the method is called that of Chou Pei.

The Chou Pei school also asserted that the heavens are round in shape like an open umbrella, while the earth is square like a chessboard. The heavens rotate sideways towards the left in the same manner as the turning of a mill-stone. The sun and moon both move towards the right, but at the same time they have to follow the heavens, which rotate towards the left. Hence, though in reality they are moving east, they are dragged along by the turning of the heavens, so that they appear to set in the west.[g]

The inclination of the polar axis must have been among the earliest of astronomical

[a] But this was often criticised, cf. p. 220 below. Thus the *Ta Tai Li Chi* says, about +80, 'if the heavens were really round and the earth really square, its four corners could not be covered properly' (ch. 56; cf. R. Wilhelm (6), p. 128).

[b] Synonym for Fu-Hsi, legendary deified culture-hero.

[c] Sun, moon, and stars.

[d] The numerical magnitudes given in this version differ very much from those in the *Chou Pei Suan Ching*; here the concentric domes are much flatter than in that work.

[e] Another very archaic trait.

[f] See the table in Maspero (3), p. 345.

[g] Ch. 11, p. 1*b*, tr. auct., adjuv. Ho Ping-Yü (1).

[1] 衡

observations. It is not surprising to find, therefore, among the Chinese myths, one which has reference to this. In the *Huai Nan Tzu* (*c.* −120) we find:

> In ancient times Kung Kung[1][a] strove with Chuan Hsü[2][b] for the Empire.
> Angered, he smote the Unrotating Mountain,
> Heaven's pillars broke, the bonds with Earth were ruptured,
> Heaven leaned over to the north-west (*thien chhing hsi pei*[3]);
> Hence the sun, moon, stars and planets were shifted,
> And earth became empty in the south-east.[c]

Writing in the +5th century, Tsu Kêng-Chih tells us that in the Kai Thien school there were several ideas about the polar axis:

> One is that heaven is like the cover of a carriage (*chhê kai*[4]) and that it travels between the eight limits (*chi*[5]). Another is that it is like a conical bamboo hat (*li*[6]) with its centre high and its edges bending downwards. A third says that it is like the cover of an inclined carriage (*i chhê kai*[7]),[d] higher at the south (i.e. zenith and equator) and lower at the north (i.e. the pole).[e]

There seems little difference between these representations, but it is at any rate sure that the celestial motion cannot have been thought of as occurring along the edges where the celestial vault met the rim-ocean (i.e. around the horizon).[f] The reclining umbrella, hat or carriage-roof must have been supposed to rotate round the polar axis. I would suggest that the real technological model may have been the edge-runner mill, only that the wheel would have to revolve in place instead of travelling round the circumference.[g]

Wang Chhung has an interesting analogy in the *Lun Hêng* (+83) about the apparent contrary motion of the sun and moon relative to the fixed stars:

> The sun and moon [he says] are attached to heaven, and follow its movements during the four seasons. Their movement may be compared to that of ants crawling on a rolling mill-stone (*jo i hsing yü wei*[8]). The movements of the sun and moon are slow, while heaven moves very fast. Heaven carries the sun and moon along with it, so that they really move eastward, but are turned westward.[h]

Exactly the same analogy is made by Vitruvius[i] about a century previously, but there can be no question of a transmission of the thought. Mo Tzu much earlier seems to

[a] One of the legendary rebels, see above, Vol. 2, p. 117.
[b] One of the legendary emperors.
[c] Ch. 3, p. 1*b*, tr. Chatley (1). Also *Lieh Tzu*, ch. 5 (Wieger (7), p. 131).
[d] The umbrella-like roof of a Han chariot with its shafts lowered to the ground would incline at an angle.
[e] *YHSF*, ch. 77, p. 6*b*, tr. auct.
[f] As Maspero (3), p. 339, appears to suggest.
[g] See on, in Section 27*c* on engineering. Perhaps this connection might serve as a suggestion of the antiquity of the device. Cf. Chatley (2), p. 126.
[h] Ch. 32, tr. Forke (4), vol. 1, pp. 266, 267. Repeated by Ko Hung in *Pao Phu Tzu* (cit. *TPYL*, ch. 762, p. 8*a*; ch. 947, p. 5*b*).
[i] IX, i, 15. And others also (v. Ueberweg-Heinze (1), vol. 1, p. 180).

1 共工 2 顓頊 3 天傾西北 4 車蓋 5 極 6 笠
7 欹車蓋 8 若蟻行於磑

refer to it.[a] Elsewhere Wang Chhung compares the millstone of heaven to a potter's wheel (*thao chün chih yün*[1]),[b] and this again is found in Nigidius Figulus, a Roman astrologer (*d.* −44),[c] who apparently derived his cognomen from this analogy.

On the Kai Thien theory the nature of the bearings of the polar axis must have caused much difficulty. This is evident from what remains of Yao Hsin's *Hsin Thien Lun* (Discourse on the Diurnal Revolution), which dates from about +250. He explains a theory of 'two earths' (*liang ti chih shuo*[2]), the lower one of which furnished supports for the bearings on which the celestial axis could revolve, and accounts for its inclination by a macrocosm-microcosm analogy, the chin of the human head pointing forwards and downwards but not backwards. The most curious part of his scheme was that the celestial vault not only revolved on the polar axis, but slid up and down along it, the pole being much farther away from the earth in summer than in winter.

> At the winter solstice the pole is low (on the polar axis), and the heavens in their rotation move near the south, so the sun is far from man and the Great Bear near him. Thus the *chhi* of the northern sky arrives and (the weather) becomes icy cold. At the summer solstice the pole rises (on the polar axis), and the heavens in their rotation move near the north, so that the Great Bear is far from man, and the sun near him. The *chhi* of the southern sky arrives, so (the weather) becomes steamy and hot. When the pole is lifted high (on its axis) the sun moves underneath (our position on) the earth but slightly, so nights are short and days are long. When the pole sinks low (on its axis), the sun goes deeply below (us), so that days are short and nights are long. Thus heaven is nearer to the *hun*[3] in winter, and to the *kai*[4] in summer.[d]

No doubt Yao Hsin's contemporaries asked him to explain why there were not more stars visible south of the equator in summer than in winter.[e] The rather obscure last sentence suggests that he may have been trying for a compromise between the two chief schools. The theory of the lifting and sinking of the pole had, however, been current already in the +1st century, for Wang Chhung mentions it.[f]

[a] *Mo Tzu*, ch. 35, p. 1*b*; cf. Mei Yi-Pao (1), p. 182, but we do not feel confident that his translation is correct.

[b] Ch. 32; Forke (4), vol. 1, p. 266. The expression *ta chün*,[5] the 'Great Potter's-Wheel', was common in Han times (Couvreur, 2). Cf. Yü Hsi's *Chih Lin Hsin Shu*[6] in *YHSF*, ch. 68, p. 40*a*.

[c] Sarton (1), vol. 1, p. 207; Bouché-Leclercq (1), p. 256.

[d] Tr. auct.

[e] Cf. the remarks of the *Chin Shu*, ch. 11, p. 3*a*: 'Yü Hsi, Yü Sung and Yao Hsin were all fond of strange ideas and willing to accept peculiar theories. Their descriptions of the heavens were not based on perfect calculations.'

[f] *Lun Hêng*, ch. 32 (Forke (4), vol. 1, p. 259). What is much more extraordinary is that the idea of the celestial vault moving up and down, opening and closing a gap between the rims of the sky and the earth, is a widely distributed motif of folklore cosmology not restricted to North Asia but found also among many Amerindian peoples from Mexico northwards, and in the Philippines: Erkes (17); Hatt (1). Migratory birds were sometimes supposed to fly in and out through it. Conceivably this idea might have been at the origin of Yao Hsin's speculations.

[1] 陶鈞之運 [2] 兩地之說 [3] 渾 [4] 蓋 [5] 大鈞
[6] 志林新書

If some interpreters are right, a theory very similar to this was expounded in the *Timaeus* of Plato.[a] The earth was supposed to slide up and down the polar axis, oscillating about a central point. But the passage is notoriously obscure.

(2) The Hun Thien School (the Celestial Sphere)

The school of Hun Thien (the Celestial Sphere) corresponded to the conception of spherical motions centred on the earth which had been developing slowly among the Greek pre-Socratics, and came to be associated particularly with Eudoxus of Cnidus (*c.* −409 to −356). While it must have been known in China at least as early as the −4th century when Shih Shen was making his star lists, the earliest exponent of it whose name has come down to us was Lohsia Hung[1] (*fl. c.* −140 to −104). In his *Fa Yen*,[2][b] Yang Hsiung[3] has been thought to say that Lohsia Hung invented it, Hsienyü Wang-Jen[4] (*fl.* −78) measured it, and Kêng Shou-Chhang[5] (*fl.* −75 to −49) represented it.[c] The oldest full description of it which we have comes from the pen of the great +1st-century astronomer Chang Hêng.[6] In his *Ling Hsien*[7] (Spiritual Constitution of the Universe), he says:

Formerly the sage-kings, wishing to trace (*pu*[8]) the ways of heaven (*thien lu*[9]), and to fix the sublime tracks (*ling kuei*[10]) (the paths of the heavenly bodies), and to ascertain the origins of things, first set up a celestial sphere (*hun thi*[11]), thus rectifying their instrument and establishing degrees (*chêng i li tu*[12]), so that the imperial pole[d] was fixed. All turned round the heavenly axis in a reliable way which could be studied. After this was set up and observed, it was seen that heaven had a normal regularity (*chhang*[13]). The sages had no preconceived theories and based their thinking on what the phenomena showed. To explain these I therefore write the *Ling Hsien*. [There follows a cosmogonic passage, here omitted.]

In the heavens there are phenomena, and on the earth there are formed shapes. Heaven has nine positions (*wei*[14]) as earth has nine continents (*yü*[15]).[e] In heaven there are the three *chhen*[16] (presumably sun, moon, and stars), and on earth the three *hsing*[17] (shapes; perhaps land, water, and air). Both the phenomena and the shapes can be observed and measured. There are thousands of different things communicating with each other, influencing and attacking each other. They follow a natural principle of spontaneity, mutually producing each other (*tzu-jan hsiang sêng*[18]).[f] The sage, as the essence of all mankind, traces the *chi-kang*[19] (nexus of connections in Nature),[g] fixing the celestial co-ordinates (*ching wei*[20]) and

[a] 40B; see Heath (4), p. xli.
[b] 'Admonitory Sayings' (about +5), ch. 7, p. 2*b*.
[c] This passage more probably concerns the invention of the armillary sphere, see p. 354 below.
[d] As will later be seen, the emperor and his bureaucracy served as the model on which the naming of the constellations was carried out, and the pole naturally corresponded to the emperor.
[e] Reference back to Tsou Yen (cf. Vol. 2, p. 236). The nine positions of heaven are probably the nine divisions of the twenty-eight *hsiu* as given in *Huai Nan Tzu*, ch. 3.
[f] Taoist influence here (cf. Vol. 2, pp. 50, 255 above).
[g] See, on this term, especially Vol. 2, pp. 554ff. above.

1 落下閎 2 法言 3 揚雄 4 鮮于妄人 5 耿壽昌
6 張衡 7 靈憲 8 步 9 天路 10 靈軌 11 渾體
12 正儀立度 13 常 14 位 15 域 16 辰 17 形
18 自然相生 19 紀綱 20 經緯

the eight limits (*chi*[1]).[a] The diameter of the 'bond' (*wei*[2]) (which holds the sphere together) is 2,032,300 *li*, but in the north-south direction it is 1000 *li* shorter than this and in the east-west 1000 *li* longer. The distance from earth to heaven is half that between the eight limits, and the depth below the earth is the same distance. The measurements are made with the graduated *hun* instrument (armillary sphere) (*thung erh tu chih tsê shih hun i*[3]). For calculations the method of two right-angled triangles is used.[b] The shadow of the gnomon faces the heavens and (explains the) meaning of the spheres celestial and terrestrial (*hun ti chih i*[4]). A difference of a thousand miles south or north in the gnomon's position means a difference of one inch in the shadow's length. These things can all be calculated, but what is beyond (the celestial sphere) no one knows, and it is called the 'cosmos' (*yü chou*[5]). This has no end (*wu chi*[6]) and no bounds (*wu chhiung*[7]). In heaven the two symbols (*i*[8]) (the sun and moon) dance round in the line of the ecliptic (*wu tao chung*[9]), encircling the pole star (*shu hsing*[10]) at the north pole. The pole in the south is not visible, so the sages gave no name to it. [There follows a passage on macrocosm and microcosm, here omitted.]

Thus heaven moves following a principle of normality (*thien i shun tung*[11]) and never loses its centre; the four seasons succeed each other regularly in warmth and cold, nourishing all living beings....[c]

In his *Hun I Chu*[12] (Commentary on the Armillary Sphere), Chang Hêng is more precise:

The heavens are like a hen's egg and as round as a crossbow bullet; the earth is like the yolk of the egg, and lies alone in the centre. Heaven is large and earth small. Inside the lower part of the heavens there is water. The heavens are supported by *chhi*[13] (vapour), the earth floats on the waters.

The circumference of the heavens is divided into $365\frac{1}{4}°$; hence half of it, $182\frac{5}{8}°$, is above the earth, and the other half is below. This is why, of the 28 *hsiu* (equatorial star groups) only half are visible at one time. The two extremities of the heavens are the north and south poles, the former, in the middle of the sky, is exactly 36° above the earth, and consequently a circle with a diameter of 72° encloses all the stars which are perpetually visible. A similar circle around the south pole encloses stars which we never see. The two poles are distant from one another 182° and a little more than half a degree. The rotation goes on like that around the axle of a chariot.[d]

These words of Chang Hêng are precious for several reasons. His natural philosophy is of interest in connection with the beginnings of the idea of laws of Nature, which we discussed in the preceding volume. He attributes the visualisation of the celestial

[a] These might be the 6 limiting lines of the star-palaces plus the 2 poles; or perhaps the 4 tropical points plus the 2 circumpolar circles of perpetual visibility and invisibility plus the 2 poles.

[b] Cf. above, in the mathematical Section, p. 31.

[c] *YHSF*, ch. 76; *CSHK* (Hou Han sect.), ch. 55, tr. auct. Our translation follows the former, though the latter is a little fuller.

[d] Tr. Maspero (3), p. 335; eng. auct. This passage does not occur in the fragments of the *Hun I Chu* in Ma Kuo-Han's collection (*YHSF*, ch. 76), where the armillary sphere itself is described. Only the first few sentences are in *TPYL*, ch. 2, p. 8*b*. The *Chin Shu*, ch. 11, p. 3*a*, *b* has a little more, and the fullest is the *Khai-Yuan Chan Ching*, ch. 1, p. 4*b*. The analogy with the yolk of the hen's egg is often found in later texts, e.g. *Sui Shu*, ch. 19, p. 6*b*.

1 極 2 維 3 通而度之則是渾已 4 渾地之義
5 宇宙 6 無極 7 無窮 8 儀 9 儛道中 10 樞星
11 天以順動 12 渾儀注 13 氣

sphere to a much earlier time than his own, and clearly shows how the conception of a spherical earth, with antipodes, would arise naturally out of it. So also would the possibility of the first astronomical instruments, armillary rings and armillary spheres. Moreover, Chang Hêng realises that space may be infinite, and he is able, so to say, to look through the immediate mechanism of sun and stars to the unknown lying beyond.

There are numerous later expositions of the Hun Thien theory, as by Yang Chhüan[1] in his *Wu Li Lun* (Discourse on the Principles of Things) and Wang Fan[2] in his *Hun Thien Hsiang Shuo*[3] (Discourse on Uranographic Models);[a] both of the +3rd century. Wang Fan again uses the analogy of the concentric spheres of the bird's egg,[b] and so do many other writers, such as Ko Hung a few years later.[c] In this connection Forke (6) alludes to the ancient and widespread myths about the cosmic egg,[d] but though this may well have been derived from the impression of a celestial hemisphere which must have been gained in remote antiquity, it is not quite to the point here, since the Chinese writers of the Han and later were using the egg simply as a homely analogy. The uniform and apparently circular rotation of the heavenly bodies is mentioned many times in texts which are probably much older than Chang Hêng or even Lohsia Hung, for example the *Chi Ni Tzu*[4] book[e] and the *Wên Tzu*[5] book. The former (perhaps of −3rd or −4th century) speaks of the sun's path as a turning ring (*hsün huan*[6]) with limits but no starting-point (*wei shih yu chi*[7]) ever rotating (*chou hui*[8]) and never still, the sun moving 1° each day.[f] The latter[g] speaks of endless rotation (*lun chuan wu chhiung*[9]). Other passages which may be early evidence of Hun Thien ideas occur in the poems of Chhü Yuan (late −4th century). In the *Chiu Ko*[10] (Nine Songs), the last line of the Tung Chün[11] (Song of the Eastern Lord) speaks of the sun travelling back to the east during the hours of darkness.[h] At the end of the +1st century, Wang Chhung, though advancing some arguments[i] against the Kai Thien hypothesis, found himself still more unable to accept the Hun Thien, as he felt it would imply that the sun, the fiery essence of Yang, would have to move through water. Ko Hung exerted himself to prove,[j] in favour of the Hun Thien

[a] The full text of this work was lost, but fragments contained in *Chin Shu*, *Sui Shu* and other places have been reassembled, as in *CSHK* (San Kuo sect.), ch. 72, pp. 1*a*ff. Forke (6), p. 20, misunderstanding *Sung Shu*, ch. 23, p. 1*b*, gave an erroneous title, but *Hun Thien Hsiang Chu*[12] is admissible as an alternative. The translation by Eberhard & Müller (2) is incomplete; cf. Maspero (4), p. 331. See Needham, Wang & Price; also p. 200 above, and p. 386 below.

[b] As quoted in *Chin Shu*, ch. 11, p. 5*b*.

[c] *Chin Shu*, ch. 11, pp. 3*a*, 7*a*.

[d] Cf. Needham (2), p. 9.

[e] On this, see especially above, Vol. 2, p. 554. Its other title is *Chi Jan*.[13] Ch. 2, p. 3*a*.

[f] The extremely early appearance of this graduation system, which led to 365¼° instead of 360° in the Chinese circle, should be noted.

[g] Ch. 2, p. 14*b*.

[h] Tr. Pfizmaier (83); Waley (23), p. 45; the point is missed by Yang & Yang (1), p. 29. I am indebted to my friend Professor Chhen Shih-Hsiang for this reference.

[i] *Lun Hêng*, ch. 32 (tr. Forke (4), vol. 1, pp. 258ff., 261), repeated in *Chin Shu*, ch. 11, p. 3*a*.

[j] *Chin Shu*, ch. 11, p. 3*b* (tr. Ho Ping-Yü, 1); cf. Forke (6), pp. 18, 22.

1 楊泉 2 王蕃 3 渾天象說 4 計倪子 5 文子
6 循環 7 未始有極 8 周迴 9 輪轉無窮 10 九歌
11 東君 12 渾天象注 13 計然

hypothesis, that this was not impossible, for dragons, which are very Yang, can live in water. But soon the archaic conception of the 'waters under the earth' fell into abeyance. A vivid account of another argument, between Yang Hsiung[1] (−53 to +18) and Huan Than[2] (−40 to +30), has been preserved in several places.[a] These two scholars were in attendance in the western gallery of the White Tiger Hall in the Imperial Palace. As the sun sank, and they could no longer warm their backs, Huan Than explained the theory of the celestial sphere to his friend. By the end of the Han, the Hun Thien view seems to have been generally accepted. Chêng Hsüan[3] (+127 to +200) and Lu Chi[4] (+3rd century) both supported it.

(3) The Hsüan Yeh Teaching (Infinite Empty Space)

The earliest name associated with the Hsüan Yeh system is a relatively late one, Chhi Mêng,[5] who flourished in the Later Han, and may have been a younger contemporary of Chang Hêng, though we do not know his exact dates or much else about him. Somewhat over a century later, Ko Hung wrote:

> The books of the Hsüan Yeh school were all lost, but Chhi Mêng, one of the librarians, remembered what its masters before his time had taught concerning it. They said that the heavens were empty and void of substance (*wu chih*[6]). When we look up at it we can see that it is immensely high and far away, having no bounds (*wu chi*[7]). The (human) eye is (as it were) colour-blind (*mou*[8]), and the pupil short-sighted; this is why the heavens appear deeply blue. It is like seeing yellow mountains sideways at a great distance, for then they all appear blue. Or when we gaze down into a valley a thousand fathoms deep, it seems sombre and black. But the blue (of the mountains) is not a true colour, nor is the dark colour (of the valley) really its own.[b]
>
> The sun, the moon, and the company of the stars float (freely) in the empty space (*fou khung chung*[9]), moving or standing still. All are condensed vapour (*chieh chi chhi*[10]).[c] Thus the seven luminaries sometimes[d] appear and sometimes disappear, sometimes move forward and sometimes retrograde, seeming to follow each a different series of regularities; their advances and recessions are not the same. It is because they are not rooted (to any basis) or tied together that their movements can vary so much. Among the heavenly bodies the pole star always keeps its place, and the Great Bear never sinks below the horizon in the west as do other stars. The seven luminaries all fall back (lit. move) eastwards, the sun making 1° a day and the moon 13°. Their speed depends on their individual natures, which shows that they are not attached to anything, for if they were fastened to the body of heaven, this could not be so.[e]

[a] *TPYL*, ch. 2, p. 7*a*; *CSHK* (Hou Han sect.), ch. 15, p. 2*b*; *Chin Shu*, ch. 11, p. 4*a*.

[b] Cf. the speculations in *Chuang Tzu* (ch. 1) on the blueness of the sky, in the −4th century; quoted above, Sect. 10*e* (Vol. 2, p. 81).

[c] Cf. *Lieh Tzu* (ch. 1), quoted in Vol. 2, p. 41 above. The *Chin Shu* text has: 'Whether they stop, or move about, *chhi* must be present.'

[d] Sun, moon and five planets (*Chhi Yao*[11]).

[e] Tr. auct., adjuv. Forke (6), p. 23; Maspero (3), p. 341; Ho Ping-Yü (1). The passage occurs in *TPYL*, ch. 2, p. 2*a*, and in *Chin Shu*, ch. 11, p. 2*a*. The former source opens with the words 'Pao Phu Tzu said...', but we have not been able to find it in either of the existing parts of *Pao Phu Tzu*, and apparently Maspero could not either. The latter quotes Tshai Yung as saying that all the books were lost (p. 1*b*).

[1] 揚雄 [2] 桓譚 [3] 鄭玄 [4] 陸績 [5] 郄萌 [6] 無質
[7] 無極 [8] 瞀 [9] 浮空中 [10] 皆積氣 [11] 七曜

These cosmological views are surely as enlightened as anything that ever came out of Hellas. The vision of infinite space, with celestial bodies at rare intervals floating in it, is far more advanced (and the point is worth emphasizing) than the rigid Aristotelian-Ptolemaic conception of concentric crystalline spheres which fettered European thought for more than a thousand years. Though sinologists have tended to regard the Hsüan Yeh school as uninfluential, it pervaded Chinese thought to a greater extent than would at first sight appear.[a] In the passage from Chang Hêng just quoted, there was an echo of Hsüan Yeh ideas when he said that even beyond the great circles of the celestial sphere there was an infinity of space. Chinese astronomy is often reproached for its overwhelmingly observational bias, but the lack of theory was an inevitable result of the lack of deductive geometry. It might well be argued that the Greeks had had too much, since the apparent mathematical beauty of 'cycle on epicycle, orb on orb' came eventually to constitute a strait-jacket posing unnecessary difficulties for a Tycho, a Copernicus and a Galileo. That Hsüan Yeh ideas persisted until the time of Matthew Ricci will appear below (p. 439). 'Native scholars suppose', said Wylie,[b] 'that there is a close resemblance between the Hsüan Yeh and the system introduced by (post-Jesuit) Europeans.' They were not far wrong.

Yü Hsi, the Chinese discoverer of the precession of the equinoxes, was inclined to Hsüan Yeh ideas. In his *An Thien Lun*[1] (Discussion of whether the Heavens are at Rest),[c] written in +336, he says:

> I think that the heavens are infinitely high, and that (the space below) the earth is unfathomably deep. Undoubtedly the form of the heavens above is in a permanent state of rest, and the body of the earth below also remains quiet and motionless. One envelops the other; if the one is square so is the other, if the one is round, so must the other be; they cannot differ as to squareness and roundness.[d] The luminaries are distributed, each pursuing its own course like the high and low tides of the sea and its rivers, and like the thousands of living creatures which sometimes come out and sometimes hide away.[e]

Yü Hsi goes on to attack the mythological legends about the sun's course and resting-places,[f] and to urge an allegorical interpretation of references to round and square in the rites.[g]

[a] It is significant that the above passage from Ko Hung mentions measurements in degrees, but says nothing about distances and dimensions of the universe in *li* (miles). The ancient texts are full of figures for these dimensions based on erroneous calculations; one of the best sources for them is the *Chin Shu*, ch. 11, p. 6*b*, reporting Wang Fan's summary (+3rd century). Maspero (3, p. 347) has acutely pointed out that most of them can be traced to the Kai Thien world picture, and not to the Hun Thien. It would seem, then, that the impossibility of establishing such figures was recognised rather early, and that the astronomers fell back on the degrees of the Hun Thien celestial sphere, with a background of Hsüan Yeh emptiness, akin to the infinite space so often discussed by the Taoists.

[b] (1), p. 86.

[c] In Ma Kuo-Han's collection; *YHSF*, ch. 77, p. 2*a*; also *Chin Shu*, ch. 11, p. 2*b*.

[d] A hit at the Kai Thien school.

[e] I.e. the stars are not attached to the Kai Thien hemisphere or the Hun Thien spheres. Tr. auct.

[f] A mass of myths about the sun-stations is discussed in the works of Granet, and summarised in *Huai Nan Tzu*, ch. 3, p. 10*a* (tr. Chatley, 1).

[g] He refers several times to a shadowy figure named Chhen Chi-Wei[2] as having taught the Hsüan Yeh doctrines before Chhi Mêng, and seems to connect him with the *Chou Pei*, so that he might be the Chhen Tzu[3] who is one of the interlocutors in that ancient work.

[1] 安天論 [2] 陳季胃 [3] 陳子

The Hsüan Yeh system has a distinctly Taoist flavour, which may account for the disappearance of the oldest writings concerning it. One senses a connection with the 'great emptiness' (*hsü wu*[1]) of Lao Tzu, and with the idea of heaven as 'piled-up *chhi*' (*chi chhi*[2]) in Lieh Tzu. Significantly, most of what we know about it comes from Ko Hung[a] and Li Shun-Fêng.

Buddhism, however, also contributed. At an earlier stage (Sect. 15*e* above) attention was drawn to its conceptions of infinite space and time and to a plurality of worlds.[b] From the Chin onwards the indigenous Hsüan Yeh theory must have received much support from this Indian source. A late statement may be drawn from the +13th-century book *Po Ya Chhin*[3] of Têng Mu.[4]

Heaven and Earth are large, yet in the whole of empty space (*hsü khung*[5]) they are but as a small grain of rice.... It is as if the whole of empty space were a tree and heaven and earth were one of its fruits. Empty space is like a kingdom and heaven and earth no more than a single individual person in that kingdom. Upon one tree there are many fruits, and in one kingdom many people. How unreasonable it would be to suppose that besides the heaven and earth which we can see there are no other heavens and no other earths![c]

To such minds the discovery of galaxies other than our own would have seemed full confirmation of their beliefs. Finally, Chu Hsi gave to these views his great philosophical authority—the Heavens, he said,[d] are bodiless and empty (*thien wu thi*[6]).

To substantiate the important conclusion that the Hsüan Yeh world-picture came to form, with the Hun Thien spherical motions, the background of Chinese astronomical thinking, it is necessary to look briefly at the subsequent history of theoretical cosmology in China. The Kai Thien system lasted on into the +6th century, and was even officially accepted at a conference called by the Liang emperor Wu Ti in the Chhang-Chhün Hall about +525.[e] In the +5th and +6th centuries a good deal of effort was put into reconciling the Kai Thien and Hun Thien, as by Tshui Ling-Ên[7] about +520 and Hsintu Fang;[8] it was urged that to see only half the sphere was to see the truth, but only half of it.[f] But from that time onwards the official histories consider the Hun Thien sphere as the only correct view. In later times, in the Ming, the Hun Thien conception was demonstrated in rather crude ways; for instance, Huang Jun-Yü,[9] in his *Hai Han Wan Hsiang Lu*[10] (The Multiplicity of Phenomena),

[a] The *Chin Shu*, after quoting the passage from Yü Hsi's *An Thien Lun* just given, goes on: 'Ko Hung heard of this and laughed, saying, "If the stars and constellations are not attached to the heavenly (vault), then it is quite useless. One might as well say that it doesn't exist at all. Why should one maintain that it exists if one admits that it doesn't move?" From this it is clear that Ko Hung knew how to argue' (ch. 11, p. 2*b*, tr. auct.). This story is repeated in the *Sui Shu*, ch. 19, p. 5*b*.

[b] That such beliefs reacted back on Europe in the early Jesuit period has been suggested above, Sect. 16*f*, Vol. 2, p. 499. We shall have more to say on this later on (pp. 440ff.).

[c] P. 21*b*, tr. auct.

[d] *Chu Tzu Yü Lei*, ch. 1, p. 7*b*. Perhaps one of the earliest statements of the Hsüan Yeh theory is that in the *Nei Ching*, ch. 67, pp. 11*b*ff.; it may well be of the -2nd century.

[e] *Sui Shu*, ch. 19, p. 5*b*.

[f] *Liang Shu*, ch. 48, p. 21*a*; *Pei Chhi Shu*, ch. 49, p. 3*b*. A Thang discussion on Hsüan Yeh theory (+676) occurs in Yang Chiung's *Hun Thien Fu*.

1 虛無 2 積氣 3 伯牙琴 4 鄧牧 5 虛空 6 天無體
7 崔靈恩 8 信都芳 9 黃潤玉 10 海涵萬象錄

describes how when young he made a model of the universe by floating a ball upon water inside a bladder.[a]

But running through the centuries was an additional conception, that of the 'hard wind' (*kang chhi*[1] or *kang fêng*[2]), which helped the Chinese (who could not know of the absence of atmosphere in the outer space of the celestial bodies) to imagine that stars and planets could be borne along without being attached to anything.[b] This is generally thought to have been of Taoist origin, and we have already met with it in the Section on Neo-Confucian philosophy.[c] I should like to suggest that it probably goes back to the early use of bellows in metallurgy, when technicians noticed the resistance of a powerful jet of air. Yü Sung may have been thinking of it when in the +3rd century he said that the rim of the Kai Thien hemisphere floats on the primeval vapour (*yuan chhi*[3]).[d] In the +11th and +12th centuries Shao Yung and Chu Hsi constantly refer to the 'hard wind' in the heavens which supports the luminaries and bears them on their way. Chu Hsi, as we saw,[e] suggested that there were nine layers of *chhi* differing in speed and therefore in hardness, and corresponding to the nine layers or storeys imagined by ancient authors such as Chhü Yuan.[f] Chu Hsi's older contemporary, Ma Yung-Chhing,[4] expressed the general point of view well when he wrote, in his *Lan Chen Tzu*[5] (Book of the Truth-through-Indolence Master) about +1115:

My friend Chêng Chêng,[6] talking about the heavens, said, 'Many people have made many words on this subject formerly, and still do, but they have no proof of anything. Only one thing is sure, that as Lieh Tzu said, we are all the day moving and resting in the midst of the heavens. On this Chang Chan[7] commented that everything above the earth is the heavens. For long I disbelieved this, but once I was sent as government examiner to Chinchow and had to climb over the highest passes; they seemed ten or twenty miles high, and I realised that it was as if I was in the heavens and looking down at the earth far below. I remembered the words of Ko Hung: 'Miles above the ground the *hsien*[8] move upon the support of the hard wind (*kang chhi*[9]).' So also above, far above, float the sun, moon, and stars. The heavens are nothing but a condensation of air[g] and are not like shaped substances which must have limits. The higher you go the farther away you will be. I do not believe that there is any definite distance from the ground to the heavens.[h]

Among the Neo-Confucian philosophers, Chang Tsai gave particular attention to cosmological theory.[i] When he spoke of *thai hsü wu thi*,[10] the great emptiness without

[a] Cit. in *TSCC, Chhien hsiang tien*, ch. 7, p. 12*a*; cf. Forke (6), p. 105.

[b] Cf. the quotation from Giordano Bruno below, p. 440. We shall meet with the idea again in relation to the prehistory of aviation; Sect. 27*j* below.

[c] Vol. 2, pp. 455, 483 above. In medieval times it was no doubt reinforced by similar ideas coming from India through the Buddhists; they are found, for instance, in the *Sūrya Siddhānta*, II, 1–5 (Burgess (1), p. 53).

[d] P. 211 above. [e] See Vol. 2, p. 483 above. [f] See p. 198 above.

[g] Mem. the concepts of rarefaction and condensation in the ancient Taoist philosophers.

[h] Ch. 1, p. 12*a*, tr. auct.

[i] *Chêng Mêng* (about +1076), esp. chs. 2, 3 and 5; in *Sung Ssu Tzu Chhao Shih*, ch. 4 (*Tshan Liang*), and *Chang Tzu Chhüan Shu*, ch. 2.

1 剛氣 2 剛風 3 元氣 4 馬永卿 5 嬾眞子 6 鄭正
7 張湛 8 仙 9 剛氣 10 太虛無體

substance, of the heavens, he was following the Hsüan Yeh tradition. Earth, he said, consisted of pure Yin, solidly condensed at the centre of the Universe; the heavens, of buoyant (*fou*[1]) Yang revolving to the left[a] at the periphery. The fixed stars are carried round endlessly with this floating rushing *chhi*.[b] Then in order to explain the opposite direction of the annual movements of the stars, sun, moon and five planets, Chang Tsai had recourse to an interesting conception of viscous drag: these bodies, he thought, were so much nearer the earth that the latter's *chhi* impeded their forward motions. 'The *chhi* of the earth, driven by some internal force, (also) revolves continually to the left',[c] but more slowly (because of the effect of the stationary earth), with the result that relatively, though not absolutely, the movement of the bodies of the solar system is opposite to that of the fixed stars.[d] The significance of so strong a statement of a principle of action at a distance, in the +11th century, will not be missed. The degree of retardation of these bodies depended on their constitution: the moon was Yin, like the earth, and therefore most affected; the Yang sun least, and the planets to an intermediate extent.[e]

Such ideas retained vitality until the time of fusion with modern science after the coming of the Jesuits. For the Ming one can adduce an interesting passage in the +14th-century *Liang Shan Mo Than* (Jottings from Two Mountains)[f] by Chhen Thing, in which the cosmic wind is also called *ching fêng*[2] and *kang chhi*,[3] this *kang* referring to the four stars of the 'box' or 'bowl' of the Great Bear.

In a word, the Chinese astronomers were practically free from the cramping orthodoxy of Hellenistic and medieval Europe which supposed that the heavenly bodies were fixed to a series of concentric material spheres with the earth as centre.[g] These spheres were essentially nothing but an unjustified concretisation of spherical geometry, and it is rather paradoxical that the Chinese, whose views are so often accused of being excessively materialistic and concrete, should have remained comparatively free from them. They may have had no deductive geometry, but they had no crystalline spheres either.[h]

[a] Looking at the sky above the pole-star, of course.

[b] Bodde, in Fêng Yu-Lan (1), vol. 2, p. 485, translated Chang Tsai's expression *hsi*[4] as 'attached' to the heavens, but this is unfortunate, since Chang undoubtedly thought of them as composed of gaseous *chhi* in violent rotation, and not as anything solid, like the crystalline spheres of Europe, to which objects could be fixed.

[c] *Ti chhi chhêng chi tso hsüan yü chung*.[5]

[d] 'The bodies which lie towards the centre follow Heaven's leftward motion, but with some delay, hence they (seem to) move to the right' (*Thien tso hsüan, chhu chhi chung chê shun chih shao chhih, tsê fan yu i*[6]). Thus is explained the annual westward motion of the sun, moon and planets.

[e] Each was composed of one of the five elements, and therefore of a particular mixture of Yin and Yang.

[f] Ch. 18, p. 10*a*.

[g] In this connection we shall remember the numerous places in the Section on Taoist thought where we found the Taoists emphasising the infinity of the universe (Vol. 2, pp. 38, 66, 81). They might be compared with the pre-Socratics, but the point is that China did not subsequently pass through a period of Aristotelian-Ptolemaic-Thomist rigidity.

[h] Nor an Inquisition to enforce belief in them.

1 浮 2 勁風 3 罡氣 4 繫 5 地氣乘機左旋於中
6 天左旋處其中者順之少遲則反右矣

If the celestial bodies moved on the hard wind, perhaps the earth moved also? Not a few ancient Chinese thought that it did, though the motion envisaged was not at first one of rotation, but rather of oscillation. This is that strange theory of the 'four displacements' (*ssu yu*[1]). Hints of it have already been seen in the Mohist Canon[a] and it must be of considerable antiquity.[b] The *Shang Shu Wei Khao Ling Yao*[c] (−1st century), says:

> The earth has four displacements. At the winter solstice, being high and northerly, it moves to the west 30,000 *li*. At the summer solstice, being low and southerly, it moves to the east 30,000 *li*. At the two equinoxes it occupies a middle position (neither high nor low). The earth is constantly in motion, never stopping, but men do not know it: they are like people sitting in a huge boat with the windows closed; the boat moves but those inside feel nothing.[d]

In a way, this was the inverse of the theory of Yao Hsin quoted above concerning the lifting and sinking of the heavens along the polar axis. Its implications, when followed up, were rather complex, as Chêng Hsüan found in the +2nd century when he came to comment upon it.[e] The main point of interest to us is that the highly anthropocentric conviction of a central and immobile earth, which so dominated European thinking, was not marked in Chinese thought. Nor was the terrestrial displacement theory purely ancient or short-lived; it is referred to by Sung scholars such as Chang Tsai,[f] Chu Hsi and Chhu Yung;[g] and contested in the Ming by Wang Kho-Ta[2] and Chang Huang.[3] Chang Tsai and other Neo-Confucians combined the idea of a periodical rise and fall of the earth with tides of Yang and Yin force within it, to explain the seasonal heat and cold. They also linked it with the phenomenon of sea tides.

(4) Other Systems

What Maspero (3) claims as the residue of a fourth theory, different from either Kai Thien, Hun Thien or Hsüan Yeh, is to be found in the third chapter of the *Huai Nan Tzu* book[h] (*c.* −120). He has himself translated it, and there is also the version of Chatley (1). The rather long passage, concerned wholly with gnomons[i] and their shadows, is distinctly obscure, and no satisfactory interpretation has yet been made. It seems inescapably to point, however, to a theory on which the sun at the meridian is five times further away from the earth than at its rising and setting, which would at least involve a very elliptical cover or shell.

[a] See above, Sect. 11*d* in Vol. 2, p. 193ff.

[b] If the book is genuine, which seems doubtful, one of the first references to it would be in the *Shih Tzu*[4] of Shih Chiao[5] (*TPYL*, ch. 37, p. 3*a*), *c.* −330.

[c] 'Apocryphal Treatise on the *Book of History*; Investigation of the Mysterious Brightnesses.'

[d] Ch. 1, p. 3*b*, tr. auct., adjuv. Maspero (3), p. 336; quoted in *Po Wu Chih*, ch. 1, p. 4*b*. The same idea, and even the same analogy, recur in Nicholas of Cusa (*De Docta Ignorantia*, II, 12; cf. Koyré (5), pp. 15, 17).

[e] Full discussion in Maspero (3), p. 337.

[f] In his *Chêng Mêng*, ch. 4 (Tshan Liang), in *Sung Ssu Tzu Chhao Shih*; Fêng Yu-Lan (1), vol. 2, p. 486.

[g] In his *Chhü I Shuo Tsuan* (Discussions on the Dispersion of Doubts), ch. 1, p. 22*a*.

[h] P. 18*b*.

[i] As many as four or six are involved.

[1] 四遊 [2] 王可大 [3] 章潢 [4] 尸子 [5] 尸佼

In any case it is quite clear that in the −2nd century the Chinese were busy applying the knowledge of the properties of similar right-angled triangles to celestial measurements. Part of the text runs:

To find the height of heaven (i.e. of the sun) we must set up two ten-foot gnomons and measure their shadows on the same day at two places situated exactly 1000 *li* apart on a north-south line. If the northern one casts a shadow of 2 feet in length, the southern one will cast a shadow 1 and 9/10 foot long. And for every thousand *li* southwards the shadow diminishes by one inch. At 20,000 *li* to the south there will be no shadow at all and that place must be directly beneath the sun. (Thus beginning with) a shadow of 2 ft. and a gnomon of 10 ft. (we find that southwards) for 1 ft. of shadow lost we gain 5 ft. in height (of gnomon). Multiplying therefore the number of *li* to the south by 5, we get 100,000 *li*, which is the height of heaven (i.e. of the sun).[a]

This is, of course, employing the principle which Thales used to measure the distance of terrestrial objects[b] for determining the distance of celestial ones.[c] The curvature of the earth's surface not being taken into account, it must fail. Aristarchus, with his ingenious use of the exact half-moon,[d] had not been much more successful. But by inverting the method in order to find the circumference of an assumed spherical earth, Eratosthenes, with his gnomons at Alexandria and Syene, had achieved, only half a century prior to the work of the group around Liu An, a very fair estimate.[e] Computations of cosmic distances are contained in some of the Han apocryphal books (−1st century),[f] and these are further elaborated in Wang Fan's monograph (+260).[g] They deal chiefly with the measurement of the 'circumference' or 'diameter' of heaven and the distance corresponding to one degree of arc.

The fact that the Chinese and the Greeks were engaged in such observations at about the same time is of interest in itself. But one must add a story which has come down to us, interesting because it illustrates the chilling lack of interest of Confucianism in scientific problems, a social factor against which the Greeks did not have to contend. The passage is in *Lieh Tzu* and may date from any time between the −4th and −1st centuries:

When Confucius was travelling in the east, he came upon two boys who were disputing, and he asked them why. One said 'I believe that the rising sun is nearer to us and that the midday sun is further away.'[h] The other said 'On the contrary, I believe that the rising and setting sun is further away from us, and that at midday it is nearest.'[i] The first replied

[a] Tr. Chatley (1).

[b] Singer (2), p. 10; B. & M., p. 188.

[c] On the history of our knowledge of the distances of the heavenly bodies see, for example, Schiaparelli (1), vol. 1, p. 329; Eichelberger (1).

[d] Singer (2), p. 59.

[e] Singer (2), p. 72.

[f] *Lo Shu Wei Chen Yao Tu*, in *Ku Wei Shu*, ch. 36, p. 1*a*; *Chhun Chhiu Wei Khao I Yu*, in *Ku Wei Shu*, ch. 10, p. 5*a*; and *Shang Shu Wei Khao Ling Yao*, in *Ku Wei Shu*, ch. 1, p. 2*a*.

[g] The *Hun Thien Hsiang Shuo*, in *CSHK* (San Kuo sect.) ch. 72, p. 1*a*, quoted in *Chin Shu*, ch. 11, pp. 6*a*ff. None of the figures are the same as those of the Kai Thien theory in the *Chou Pei Suan Ching* discussed by Chatley (11). Ho Ping-Yü (1) has recalculated all the derived ones given in the course of the argument and finds arithmetical mistakes in each; presumably the scribes were not interested in the figures which the ancient astronomers thought important, and copied them carelessly.

[h] He was evidently an adherent of the *Huai Nan Tzu* school.

[i] A view compatible with Kai Thien ideas.

'The rising sun is as big as a chariot-roof, while at midday the sun is no bigger than a plate. That which is large must be near us, while that which is small must be further away.' But the second said 'At dawn the sun is cool but at midday it burns, and the hotter it gets the nearer it must be to us.' Confucius was unable to solve their problem. So the two boys laughed him to scorn saying, 'Why do people pretend that you are so learned?'[a]

This story belongs to a whole corpus of legend and folk-tale centring on a quasi-Taoist *puer senex*, Hsiang Tho,[1] who together with other small boys always defeats Confucius in argument or riddle.[b] Sometimes he seems to be an incarnation of Lao Tzu himself. But though the Taoists and Naturalists might laugh at the Confucians, it was the latter who became more and more the dominant social group, and indifference to natural philosophy grew with their dominance.

Records of real discussions on these problems remain in the *Hsin Lun*[2] (New Discussions) of Huan Than[3] (*c.* +20), who describes how Kuan Tzu-Yang[4] a century earlier had supported the former of the two views.[c] He may have derived directly from the *Huai Nan Tzu* school. And Ko Hung used the larger size of the setting sun to combat Wang Chhung's analogy of the receding torch.[d] From the standpoint of modern science, the matter is not so simple as it looks.[e] Consideration will show that when the sun or moon is seen on the horizon the observer is further from the object by a distance equal to the earth's radius than when the object is at the zenith; it therefore has a very slightly smaller angular diameter. In addition, refraction by the earth's atmosphere causes a readily perceptible flattening of the sun and moon, due to a reduction in the vertical diameter only (Colton). The apparent gross enlargement of the sun or moon near the horizon is a subjective effect, which has been investigated by Dember & Uibe. The eye sees the cloudless sky not as a hemisphere, but as the cap of a sphere; the zenith appears nearer than the horizon. This apparent shape of the sky can be explained on a physical basis, and Dember & Uibe have shown that if the eye is assumed to imagine the sun or moon to be at the same distance as the apparent sky surface, its apparent size is in good agreement with that theoretically predicted by the geometry of the apparent sky. Though the angle subtended at the eye is the same, the 'screen' of the horizon sky *seems* further away, and the image on it correspondingly larger.[f] It is remarkable that the terrestrial nature of the phenomenon was recognised already in the Han, for Chang Hêng, in his *Ling Hsien*, explains the effect as an optical one. Shu Hsi[5] of the Chin (late +3rd or +4th century) also urged that the sun never

[a] Ch. 5, p. 14*b*, tr. auct., adjuv. Wieger (7), p. 139; Maspero (3), p. 354; R. Wilhelm (4), p. 55; often quoted afterwards, as in *Po Wu Chih*, ch. 8, p. 2*b*; *Chin Lou Tzu*, ch. 4, p. 10*a*.

[b] See the interesting paper of Soymié (1).

[c] *CSHK*, Hou Han sect., ch. 15, p. 4*a*; *Sui Shu*, ch. 19, p. 10*b*, cit. in *TSCC*, *Chhien hsiang tien*, ch. 1, p. 13*a*. Cf. also *Lun Hêng*, ch. 32 (tr. Forke (4), vol. 1, p. 263). Both Kuan Tzu-Yang and Wang Chhung thought that the sun was composed of fire or fiery Yang essence. Another argument of Huan Than's, in favour of the Hun Thien theory against the Kai Thien, will be found in *Chin Shu*, ch. 11, pp. 3*b*, 4*a*.

[d] *Chin Shu*, ch. 11, p. 4*b*.

[e] Cf. Pernter & Exner (1), pp. 5ff.

[f] Thanks are due to Dr Beer and Mr Dewhirst for unravelling this seemingly simple problem, and our paragraph is based on a note of theirs. Cf. O. Thomas (1), p. 238.

1 項橐 2 新論 3 桓譚 4 關子陽 5 束晳

changed in size, and that if it seemed to do so, this was due only to a deception of our senses (*jen mu chih huo*[1]), of which he gave several good examples.[a] Others, however, inclined to explanations which attributed the effect to the earth's atmosphere. The influence of the earth's atmosphere on the appearance of the sun was explained by Chiang Chi about +400, but he was concerned with its colour rather than its size (see below, p. 479).

(5) General Notions

This sub-section may end with a few remarks on general notions about the chief celestial bodies. That the sun was of a fiery Yang (male) nature and that the moon was Yin (female) and watery was a commonplace from the earliest phases of Chinese science.[b] The earth was also Yin.[c] Chatley (5) has drawn attention to the fact that from an early time the sun was termed Thai Yang[2] (the Greater Yang) and the fixed stars Hsiao Yang[3] (the lesser Yang), while to the moon (Thai Yin[4]) corresponded the planets (Hsiao Yin[5]); distinction thus being made quite correctly between those bodies which shine by their own light and those which shine by reflected light. But perhaps this was only a coincidence; we do not know of any early or medieval text which distinctly states that the fixed stars were of the nature of far-distant suns.

Parmenides of Elea (*fl. c.* −475, and therefore a younger contemporary of Confucius) was apparently the first of the Greeks to state definitely that the moon shines with a purely reflected light, and by the time of Aristotle (−4th century) this was accepted as a matter of course. Presumably the oldest statement of it in Chinese literature is that of the *Chou Pei*: 'The sun gives to the moon her appearance, so the moonlight shines brightly forth (*jih chao yüeh, yüeh kuang nai chhu, ku chhêng ming yüeh*[6])'.[d] This cannot be later than early Han, and may well be of the −4th century, if not as old as the −6th.[e] In the latter half of the −1st century Ching Fang[7] wrote:

> The moon and the planets are Yin; they have shape but no light. This they receive only when the sun illuminates them. The former masters regarded the sun as round like a cross-bow bullet, and they thought the moon had the nature of a mirror. Some of them recognised the moon as a ball too. Those parts of the moon which the sun illumines look bright, those parts which it does not, remain dark.[f]

[a] His arguments are reported in the *Sui Shu*, later in the same passage, ch. 19, p. 11*a*. Cf. Forke (6), p. 85.

[b] Cf. below, in Section 26*g* (physics), on burning-mirrors and dew-mirrors (Vol. 4). See also Forke (6), pp. 79, 83.

[c] For parallels in myths of other cultures, see Eliade (2).

[d] Ch. 2, p. 1*b*, tr. E. Biot (4), p. 620.

[e] Another ancient statement, of similar antiquity, would be that in the *Chi Jan* of Chi Ni Tzu: 'The moon is of the essence of water, reflecting like a concave mirror' (*TPYL*, ch. 4, p. 9*a*).

[f] Tr. auct., adjuv. Forke (6), p. 90; quoted by Kuo Pho about +300 in his commentary on the *Erh Ya* dictionary.

[1] 人目之惑 [2] 太陽 [3] 小陽 [4] 太陰 [5] 小陰
[6] 日兆月月光乃出故成明月 [7] 京房

Erroneous theories were introduced in the +6th century with the translation of the *Lokasthiti Abhidharma Śāstra* (*Li Shih A-Pi-Than Lun*[1]),[a] but they did not affect the general acceptance of the correct view, which is stated over and over again, as by Shao Yung[2b] and Shen Kua[3c] in the +11th century, Chu Hsi[4d] in the +12th, and Li Chhung[5e] at the end of the +14th. As in all other ancient civilisations there was a patrimony of myth about imaginary beings in the moon,[f] but by the Sung time even a poet could share in the sceptical tradition and attack such childlike ideas.[g]

Another erroneous Indian theory which travelled to China with Buddhism[h] was that of the two imaginary invisible planets Rahu (Lo-Hou[6]) and Ketu (Chi-Tu[7]) which 'personified' the ascending and descending nodes of the moon's path, and were doubtless devised to account for lunar eclipses.[i] These 'dark stars' were numbered among the planets. François Bernier[j] described this idea in the 17th century.[k] But as we shall see,[l] the Chinese had themselves imagined, from ancient times, the existence of a 'counter-Jupiter' which moved round diametrically opposite the planet itself. There was a Greek parallel to this in the strange Pythagorean theory of the 'counter-earth', apparently due to Philolaus of Tarentum (late −5th century), which was devised either to bring the number of planets up to a perfect number, 10, or to explain lunar eclipses.[m] Perhaps both originated from a more ancient Babylonian theory.

[a] Translation in Forke (6), p. 92.

[b] Quoted in *I Thu Ming Pien* by Hu Wei, ch. 3, p. 5*b*.

[c] *MCPT*, ch. 7, para. 14. See Hu Tao-Ching (*1*), vol. 1, p. 309. The passage is translated in full below, p. 415.

[d] *Tshan Thung Chhi Khao I*, p. 7*a*; *Chu Tzu Yü Lei*, ch. 1, p. 9*a*.

[e] *Jih Wên Lu*[8] (Daily Notes), p. 6*b*. He was one of the collaborators of the Muslim astronomers, cf. p. 49 above.

[f] Cf. Sect. 45 for the legend of the rabbit in the moon engaged in pounding drugs; a favourite Taoist motif. Cf. also Hentze (1).

[g] Su Tung-Pho[9] in his Chien Khung Ko Shih,[10] quoted by Yung Hêng[11] of the same period in his *Sou Tshai I Wên Lu*[12] (A Bunch of Strange Stories), ch. 3, p. 8*b*; cf. *I Chao Liao Tsa Chi*, ch. 1, p. 11*a*.

[h] See Soper (1).

[i] See pp. 416, 420 below. I-Hsing and the Chhüthans agreed at least on this.

[j] (1), vol. 2, p. 114.

[k] See also E. Burgess (1), pp. 56, 149; Berry (1), p. 48. One of the best accounts of Rahu and Ketu in India and Islam is that of Hartner (6).

[l] P. 402 below. This idea presents an odd parallel to the problem currently faced by astronomers, namely the existence of intense sources of radiation in parts of the heavens where no visible bodies exist (Lovell, 1). We shall return to this subject presently (p. 428).

[m] Berry (1), p. 25; Freeman (1), p. 227.

[1] 立世阿毗曇論 [2] 邵雍 [3] 沈括 [4] 朱熹 [5] 李翀
[6] 羅睺 [7] 計都 [8] 日聞錄 [9] 蘇東坡 [10] 鑒空閣詩
[11] 永亨 [12] 搜採異聞錄

(*e*) THE POLAR AND EQUATORIAL CHARACTER OF CHINESE ASTRONOMY

It is now established beyond question that ancient (and medieval) Chinese astronomy was based upon a system quite different from that of the Egyptians, Greeks and later Europeans, though in no way less logical or useful.[a] This was not understood by the first Jesuits, though Gaubil (2, 3, 4) acknowledged it, as did J. B. Biot (4). De Saussure (3) expressed the matter by saying that while Greek astronomy was ecliptic, angular, true and annual, Chinese astronomy was equatorial, horary, mean and diurnal.[b] This calls for some explanation.

De Saussure never dealt with the question more clearly than in the first of all his papers (2). Early astronomers faced the great difficulty that the star which determines the seasons (the sun) dims the other stars to invisibility by its brilliance, so that its position among them, unlike that of the moon, cannot easily be obtained. Simultaneity of observation being thus impossible,[c] there remained only the methods of contiguity and opposability. The method of contiguity was that adopted by the ancient Egyptians[d] and the Greeks; it involved the observation of heliacal risings and settings, i.e. the risings and settings of stars near the ecliptic just before sunrise and just after sunset. The heliacal rising or setting of a star indicates to within a few days the date in the annual day-cycle. One of the most famous of all ancient scientific observations was that of the heliacal rising of Sirius[e] which warned the ancient Egyptians of the imminent inundation by the Nile. The apparent annual course of the sun among the stars has the effect of varying, according to the season, the stars which set heliacally at a given hour. Constellations which, at a certain time in the year, appear at dusk in the southern part of the sky, advance progressively towards the west, so that the duration of their visibility constantly diminishes, and after three months they are so near the horizon that they set almost as soon as they have become visible. They then rise in the east just before dawn, again to be visible only a short while, and the cycle recommences. Such observations required no knowledge of pole, meridian or equator, nor any system of horary measurement; but naturally led to the recognition of ecliptic (zodiacal) constellations, and to stars appearing and disappearing simultaneously with them nearer or farther away from the ecliptic (their paranatellons). Attention was concentrated on the horizon and the ecliptic.

The method of opposability, on the contrary, was that adopted by the ancient Chinese. They concentrated attention, not on heliacal risings and settings, not on the

[a] Connections were rather, as we shall see, with Babylonia and India.

[b] Moreover Greek astronomy was geometrical, Chinese astronomy arithmetical-algebraical.

[c] To all intents and purposes. But throughout the history of astronomy there runs a suspicion that zenith stars could be seen in full daylight if observed from the bottom of a deep well. Whether there was ever any ground for this belief is still doubtful; we shall return to it later (p. 333).

[d] Neugebauer (1); Chatley (13).

[e] The Chinese knew this as Thien lang,[1] the Celestial wolf. Canis Major, to which Sirius belongs, is one of the few constellations which in China shares a symbolic significance with occidental nomenclature (see below, p. 272).

[1] 天狼

horizon, but on the pole star and on the circumpolar stars which never rise and never set. Their astronomical system was thus closely associated with the concept of the meridian (the great circle of the celestial sphere passing through the pole star and the observer's zenith), and they determined systematically the culminations and lower transits (meridian passages) of these circumpolar stars.[a] The circumpolars were of course familiar to the Greeks also, as is shown by the famous passage in Homer:

> The never-wearied sun, the moon exactly round,
> And all those stars with which the brows of ample Heaven are crowned,
> Orion, all the Pleiades, and those seven Atlas got,
> The close-beamed Hyades; the Bear, surnamed the Chariot,
> That turns about heaven's axle-tree, holds ope a constant eye
> Upon Orion, and of all the cressets in the sky
> His golden forehead never bows to the Ocean's empery.[b]

And there was a Greek story to the effect that the sentinels at the siege of Troy changed their guard according to the vertical or horizontal positions of the tail of the Great Bear. There are, conversely, indications that some heliacal risings and settings were noted by the ancient Chinese.[c] For example, Chhü Yuan says in his −4th-century *Thien Wên* (Questions about the Heavens), 'When Spica (α Virginis) is rising just before dawn, where is the great Luminary (the Sun) hiding himself? (*Chio hsiu wei tan, yao ling an tshang?*[1])'.[d] But the emphasis in China was quite different from that in Greece.

The pole was thus the fundamental basis of Chinese astronomy. It was connected therein with a background of microcosmic-macrocosmic thinking. The celestial pole corresponded to the position of the emperor on earth, around whom the vast system of the bureaucratic agrarian state naturally and spontaneously revolved. For this there was high Confucian authority:

> The Master said: 'He who exercises government by means of his virtue may be compared to the pole star, which keeps its place while all the stars turn around it.[e]

And we find a familiar echo in the speech which Shakespeare gave to Caesar:

> But I am constant as the northern star,
> Of whose true-fix'd and resting quality
> There is no fellow in the firmament.
> The skies are painted with unnumber'd sparks,

[a] The Chinese system had the advantage that the exact times of heliacal risings and settings are difficult to determine accurately, owing to mists on the horizon and other atmospheric effects (de Saussure, 15).

[b] *Iliad*, XVIII, 488, tr. Chapman. Cf. *Odyssey*, V, 273. And Virgil, *Georgics*, I, 246: 'Arctos Oceani metuentes aequore tingi'; Ovid, *Metam.* XIII, 725. Aratus of Soli (*c.* −280) speaks (l. 48) of 'the Bears that shun evermore the blue sea', and says (l. 37) that the Greeks steered by Ursa Major (Helice) and the Phoenicians by Ursa Minor (Cynosura).

[c] Especially of Antares acronycally; see Chatley (10), pp. 528, 529; Chu Kho-Chen (1), p. 7, emphasised in (8); Liu Chao-Yang (1).

[d] Cf. Edkins (10).

[e] *Lun Yü*, II, 1.

[1] 角宿未旦曜靈安藏

They are all fire and every one doth shine,
But there's but one in all doth hold his place;
So in the world, 'tis furnish'd well with men...
Yet in the number I do know but one
That unassailable, holds on his rank,
Unshaked of motion, and that I am he....[a]

The meridian was a very understandable derivative from what was probably the most ancient astronomical instrument of all, the gnomon, used for measuring the different lengths of the sun's shadow. Looking south, the observer measured the shadow at noon; looking north during the night, he measured the times at which the various circumpolars made their upper and lower transits across the meridian. This is said in so many words in the Khao Kung Chi (Artificers' Record) Section of the *Chou Li* (Record of the Rites of Chou). Speaking of certain artificers (the Chiang Jen[1]) it says:

By day they collected observations of the length of the sun's shadow, and by night they investigated the culminations of stars, so that they might set in order mornings and evenings (*chou tshan chu jih chung chih ching; yeh khao chih chi hsing; i chêng chao hsi*[2]).[b]

It follows from what has already been said that at different times of the year, but at a given hour, different stars will be culminating. There can be no doubt that such observations are those referred to in what may be the oldest surviving Chinese astronomical text, the Yao Tien chapter of the *Shu Ching*. Before alluding to this, however, some further explanations are required.

Just as the influence of the Son of Heaven on earth radiated in all directions, so the hour-circles radiated from the pole.[c] During the −1st millennium the Chinese built up a complete system of equatorial divisions,[d] defined by the points at which these hour-circles transected the equator—these were the *hsiu*.[3] One has to think of them as segments of the celestial sphere (like segments of an orange) bounded by hour-circles and named from constellations which provided determinative stars (*chü hsing*[4]),[e] i.e. stars lying upon these hour-circles, and from which the number of degrees in each *hsiu* could be counted.[f] Many European scholars have found it almost impossible to believe that a fully equatorial system of astronomy could have grown up without passing through an ecliptic (zodiacal) phase, yet that undoubtedly happened.[g] Now once having established the boundaries of the *hsiu*—and for this the declinations of the

[a] *Julius Caesar*, act III, sc. 1.
[b] Ch. 12, p. 15*b* (ch. 43, p. 21), tr. auct., adjuv. Biot (1), vol. 2, p. 555. Cf. p. 571 below.
[c] 'Like the spars of an umbrella', as Shen Kua said in +1086 (*MCPT*, ch. 7, para. 13).
[d] The celebrated lunar mansions.
[e] Or 'reference stars'. Two instances of the use of the term will be found in *Chin Shu*, ch. 13, p. 2*b*.
[f] It was not necessary that *all* the stars of a constellation should be actually in the hour-angle segment (*hsiu*) to which the constellation gave its name. In certain cases (Tou, no. 8, Khuei, no. 15, Pi, no. 19, and Hsing, no. 25) much of the constellation extended back into the preceding *hsiu*. This can be seen well in Fig. 94.
[g] It happened in Babylonia first (cf. Neugebauer, 9). So fixed in the Western mind was the zodiacal framework that when the Babylonian planispheres of *c.* −1200 were first discussed, as by Bosanquet & Sayce (1), it was assumed that the ecliptic was represented on them. See p. 256 below. To Sédillot (2), p. 592, the minor part played by the ecliptic in the Chinese system was very shocking.

[1] 匠人 [2] 晝參諸日中之景夜考之極星以正朝夕 [3] 宿 [4] 距星

hsiu-asterisms and the determinative stars, whether nearer or farther from the equator, did not matter at all—the Chinese were in a position to know their exact locations, even when invisible below the horizon, simply by observing the meridian passages of the circumpolars keyed to them. And this was the way in which the sidero-solar problem was resolved, for the sidereal position of the full moon is in opposition to the invisible position of the sun.[a] Here was the real essence of the Hun Thien theory discussed above; once having gained a clear understanding of the diurnal revolution of the heavens, then the culminations and lower transits of the circumpolar stars would fix the position of every point on the celestial equator.[b] Hence the position of the sun among the stars could be known; solar and stellar coordinates could be correlated.[c]

(1) Circumpolar Stars and Equatorial Mark-points

That transits of circumpolars were indeed employed as indicators of the positions of invisible *hsiu* may be illustrated most clearly from a passage in the Thien Kuan chapter of the *Shih Chi*. Ssuma Chhien says:

> Piao[1] is attached (*hsi*[2]) to the Dragon's Horn (Chio;[3] *hsiu* no. 1). Hêng[4] hits the Southern Dipper (Nan Tou;[5] *hsiu* no. 8) in the middle. Khuei[6] is pillowed (*chen*[7]) on the head of Orion (Shen;[8] *hsiu* no. 21).
>
> The dusk indicators (*hun chien chê*[9]) (those of which the transits are noted at dusk) are the Piao[1] stars. The midnight indicator (the star of which the midnight transit is noted) is Hêng.[4] The dawn indicators (those whose transits are noted at dawn) are the Khuei[6] stars....[d]

This can be understood immediately when the nomenclature of the stars of the Great Bear is known. This was as follows:

(*a*) the 'bowl' or 'box', Khuei[6] ('The Chiefs'):

α	Dubhe	Thien shu,[10] 'Celestial pivot'.
β	Merak	Thien hsüan,[11] 'Celestial template'.
γ	Phecda	Thien chi,[12] 'Celestial armillary'.
δ	Megrez	Thien chhüan,[13] 'Celestial balance'.

[a] Cf. Gaubil (2), p. 45. De Saussure (20) is very good on this. 'If you ask literary men', he wrote, 'when the full moon rises, they generally answer, "How should I know? I'm not an astronomer." But if you ask a poacher, or a smuggler, or indeed any honest countryman, whose profession has led him to observe Nature and the heavens, his reply will be, as it was in remote antiquity, "It rises as the sun goes down." It is therefore not necessary to call up Copernicus and Newton to confirm that the full moon, being in opposition to the sun, rises on the eastern horizon when the latter sets in the west.' And on this indeed depended the sacrifices with which many ancient peoples thought fit to signalise the passage of the lunar months. See also Hashimoto (2).

[b] De Saussure (30).

[c] An interesting example of this in practice may be seen in the *Chin Shu*, which mentions (ch. 13, pp. 1*b*, 23*b*) two daylight appearances of portents, in one case Venus, in the other a large meteor. Then the text says 'Reasoning from the graduations on the sundial' (*i kuei tu thui chih*[14]), their positions among the *hsiu* and stars were known, and hence what parts of the country were affected by the presage. On Venus as a daylight object in Chinese records see Dubs (2), vol. 3, pp. 349ff.

[d] Ch. 27, p. 2*b*, tr. auct., adjuv. Chavannes (1), vol. 3, p. 341. This passage did not escape de Saussure (1), p. 566. Three corresponding astrological indications are omitted.

[1] 杓 [2] 擕 [3] 角 [4] 衡 [5] 南斗 [6] 魁 [7] 枕
[8] 參 [9] 昏建者 [10] 天樞 [11] 天璿 [12] 天璣 [13] 天權
[14] 以晷度推之

(*b*) the 'handle', Piao[1] ('The Spoon'):

ϵ	Alioth	Yü hêng,[2] 'Jade sighting-tube'.
ζ	Mizar	Khai Yang,[3] 'Opener or introducer of heat, or of the Yang'.
η	Benetnash	Yao kuang,[4] 'Twinkling brilliance'.

Consequently the meaning of the passage is that the position of Chio[5] can be ascertained from the position of the last two stars of the handle. In effect, lines drawn through α Ursae Minoris (Thien huang ti[6] or Thien chi[7]) and Khai Yang,[3] and through β Ursae Minoris (Thien ti hsing[8]) and Yao kuang[4] will meet at Chio[5] (α Virginis; Spica). Similarly a line drawn from Yü hêng[2] parallel with that between Thien chi[7]

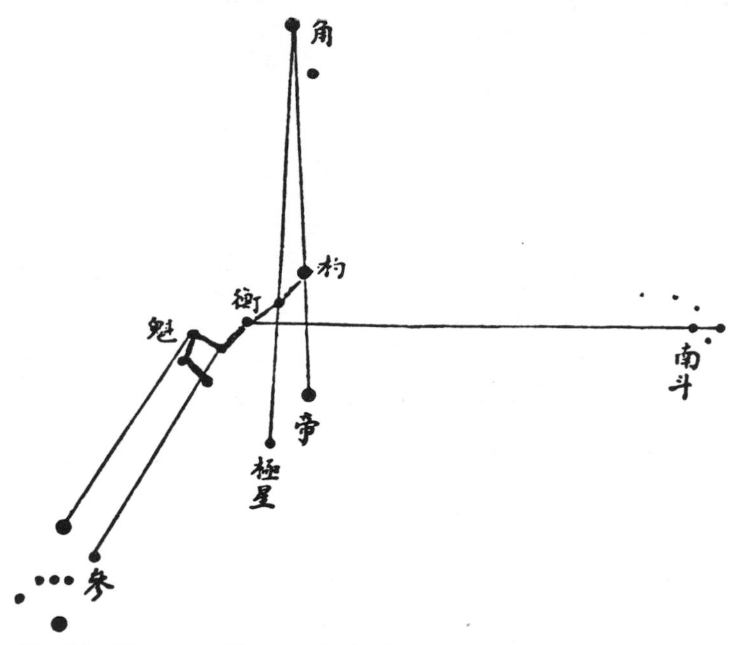

Fig. 88. Diagram to illustrate the keying of the circumpolar with other stars.

and Thien chhüan[9] will indicate the position of Nan tou[10] (ϕ Sagittarii). And the prolongation of the 'top' and 'bottom' of the box (i.e. Thien chhüan[9]/Thien shu[11] and Thien chi[7]/Thien hsüan[12]) will give lines meeting in Shen[13] (Orion). Cf. Fig. 88 from Chu Wên-Hsin (5).

Before going further, it will be convenient to examine the system of the *hsiu* in its full development. The data are given in Table 24. By comparing columns 5 and 8*a* it will be seen that there was practically no parallel between the ancient names of the relevant constellations in China and the West.[a] As is shown by col. 8*d* some of the

[a] Cf. below, pp. 271 ff.

[1] 杓 [2] 玉衡 [3] 開陽 [4] 搖光 [5] 角 [6] 天皇帝
[7] 天極 [8] 天帝星 [9] 天權 [10] 南斗 [11] 天樞 [12] 天璿
[13] 參

Explanation of Table 24

Col. 1. The 'Palace' (*kung*[1]). The central, or circumpolar, palace is of course not involved, since the constellations which give their name to the *hsiu* are all on one or other side of the equator, though they may be as much as 35° away from it.

Col. 2. The number of the *hsiu* according to the order in, for example, *Huai Nan Tzu*, ch. 3, p. 15*a*.

Col. 3. Name of the *hsiu* (romanised and Chinese).

Col. 4. References to numbers in the star-list of Schlegel (5), p. 805. His descriptions of the asterisms are to be consulted for much fuller information.

Col. 5. Probable ancient significance of the name (including a few references to Karlgren (1) on ancient forms of the characters).

Col. 6. Number of stars in the constellation which gives its name to the *hsiu* (Chu Kho-Chen, 1). Detailed identifications are of course in Schlegel (5) and Wylie (6). Other lists of all the stars in the *hsiu* asterisms are those of Hartner (2); Kingsmill (1); Chhen Tsun-Kuei (3).

Col. 7. (*a*) Equatorial extension of the *hsiu* in Chinese degrees (365¼), as given by *Huai Nan Tzu*, ch. 3, p. 15*a* (*c.* −120). Cf. Maspero (3), p. 282, who gives three other ancient lists, not greatly differing. The measurements made in the Thang period by I-Hsing (*c.* +700) are given by Gaubil (3), p. 108. Yuan measurements in *Yuan Shih*, ch. 54, p. 8*a*; ch. 56, p. 9*b*.

(*b*) Conversion of the *Huai Nan Tzu* figures to modern degrees (360) by Nōda (2), p. 19; (3), p. 64.
(*c*) True extensions along the equator calculated for −450 by Nōda (2, 3), same page references.

Col. 8. (*a*) Identification of the determinative star of the *hsiu*, i.e. that lying on the hour-circle with which it begins (Chu Kho-Chen, 1). There are slight differences between this list and that accepted by J. B. Biot (4).
(*b*) Magnitude of the determinative star (Delporte, 1).
(*c*) Right ascension of the star for +1900 (Chu Kho-Chen, 1).
(*d*) Declination of the star for +1900 (Chu Kho-Chen, 1).

Col. 9. Correlated circumpolar phenomena; from J. B. Biot (4), p. 246. This list gives the culminations and lower transits of circumpolar stars (mostly in Ursa Major, Ursa Minor and Draco) which would closely correspond to the invisible meridian transits of the *hsiu* below the horizon. It should be noted that this series was calculated by Biot for a region between the 34th and 40th parallels of terrestrial latitude and for the year −2357. There would be modifications if a similar list was to be constructed for the −4th century, for example, but the general principle would not be affected. For the Chinese characters of these circumpolar asterisms see Table 25 immediately following. Note that Biot's identifications sometimes differ from those now accepted, as in the lists of Chhen Tsun-Kuei (3), but generally agree with those of Schlegel (5).

[1] 宮

Notes will be found on p. 237.

1 Palace	2 No.	3 Name rom.	3 Name Ch.	4 S	5 Probable ancient significance	6 No. of stars	7 Equatorial extension a Ch.	7 b mod.	7 c calc.	8 Determinative star a ident.	8 b mag.	8 c R.A. (+1900) h. m. s.	8 d Decl. (+1900)	9 Correlated circumpolar phenomena
Eastern	1	CHIO	角	165	Horn	2	12°	11·83°	11·70°	α Virginis	1·2	13 19 55	−10° 38′ 22″	lower transits of α Ursae Minoris (Thien huang ta ti),[a] S451; and of a 3233 Ursae Minoris (Shu tzu), S42 culmination of i Draconis (Thien i), S611
	2	KHANG	亢	133	Neck	4	9°	8·87°	8·81°	κ Virginis	4·3	14 07 34	−09° 48′ 30″	transits of α and β Centauri (Nan mên), S244[b]
	3	TI	氐	423	Root	4	15°	14·78°	14·46°	$α^2$ Librae	2·9	14 45 21	−15° 37′ 35″	none
	4	FANG	房	55	Room	4	5°	4·93°	5·25°	π Scorpii	3·0	15 52 48	−25° 49′ 35″	culmination of α Draconis (Yu shu), S736
	5	HSIN	心	318	Heart	3	5°	4·93°	4·14°	σ Scorpii	3·1	16 15 07	−25° 21′ 10″	none
	6	WEI	尾	714	Tail	9	18°	17·74°	18·95°	$μ^1$ Scorpii	3·1	16 45 06	−37° 52′ 33″	none
	7	CHI	箕	151	Winnowing-basket	4	11¼°	11·0°	10·22°	γ Sagittarii	3·1	17 59 23	−30° 25′ 31″	lower transit of κ Draconis, and upward perpendicular position of the tail of the Great Bear (Pei tou), S280
Northern	8	NAN TOU	南斗	416	Southern Dipper	6	26°	25·8°	26·54°	φ Sagittarii	3·3	18 39 25	−27° 05′ 37″	culmination of ι Draconis (Tso shu), S678
	9	NIU or CHHIEN NIU[c]	牛 牽牛	252	Ox / Herd boy	6	8°	7·89°	7·90°	β Capricorni	3·3	20 15 24	−15° 05′ 50″	lower transits of α Ursae Majoris (Thien shu), S568; and of β Ursae Majoris (Thien hsüan), S543 culminations of α Lyrae (Chih nü) (Vega), S396; and of β Lyrae (Chien thai), S660
	10	NÜ or HSÜ NÜ[d]	女 須女	254	Girl / Serving-maid	4	12°	11·83°	11·82°	ε Aquarii	3·6	20 42 16	−09° 51′ 43″	none
	11	HSÜ	虛	101	Emptiness	2	10°	9·86°	9·56°	β Aquarii	3·1	21 26 18	−06° 00′ 40″	lower transits of γ Ursae Majoris (Thien chi), S465; and of δ Ursae Majoris (Thien chhüan), S488

Table 24 (*continued*)

1 Palace	2 No.	3 Name rom.	3 Name Ch.	4 S	5 Probable ancient significance	6 No. of stars	7 Equatorial extension *a* Ch.	7 *b* mod.	7 *c* calc.	8 Determinative star *a* ident.	8 *b* mag.	8 *c* R.A. (+1900) h. m. s.	8 *d* Decl. (+1900)	9 Correlated circumpolar phenomena
Northern	12	WEI	危	711	Rooftop	3	17°	16·76°	16·64°	α Aquarii	3·2	22 00 39	−00° 48′ 21″	lower transits of δ to ε Ursae Majoris (Yü hêng), S747; and of 42 and 184 Draconis (Thai i), S364[e]
	13	SHIH or YING SHIH	室 營室	32	House Encampment	2	16°	15·77°	16·52°	α Pegasi	2·6	22 59 47	+14° 40′ 02″	lower transits of ε Ursae Majoris (Yü hêng), S747; and of 42 and 184 Draconis (Thai i), S364
	14	PI or TUNG PI	壁 東壁	287	Wall Eastern wall	2	9°	8·87°	8·44°	γ Pegasi	2·9	00 08 05	+14° 37′ 39″	culmination of β Ursae Minoris (Thien ti hsing), S575; lower transit of ζ Ursae Majoris (Khai Yang), S130
Western	15	KHUEI	奎	197	Legs	16	16°	15·77°	15·66°	η Andromedæ	4·2	00 42 02	+23° 43′ 23″	culmination of a 3233 Ursae Minoris (Shu tzu), S42
	16	LOU	婁	206	Bond	3	12°	11·83°	10·83°	β Arietis	2·7	01 49 07	+20° 19′ 09″	lower transit of η Ursae Majoris (Yao kuang), S725; and of i Draconis (Thien i), S611
	17	WEI	胃	710	Stomach	3	14°	13·8°	15·2°	41 Arietis[f]	3·7	02 44 06	+26° 50′ 54″	none
	18	MAO	昴	232	graph of a group of stars (Pleiades) K1114*h*	7	11°	10·84°	10·44°	η Tauri[g]	3·0	03 41 32	+23° 47′ 45″	lower transit of α Draconis (Yu shu), S736
	19	PI	畢	297	Net (Hyades)	8	16°	15·77°	17·86°	ε Tauri	3·6	04 22 47	+18° 57′ 31″	none
	20	TSUI or TSUI CHUI	觜 觜觿	686	Turtle	3	2°	1·97°	1·47°	λ¹ Orionis[h]	3·4	05 29 38	+09° 52′ 02″	culmination of κ Draconis
	21	SHEN	參	647	graph of 3 stars; K647	10	9°	8·87°	6·93°	ζ Orionis	1·9	05 35 43	−01° 59′ 44″	downward perpendicular position of the tail of

	Tung Ching	東井		Eastern Well									S 508; and of ρ Ursae Majoris (Thien hsüan), S 543; these form the end of this very broad *hsiu*
23	Kuei or	鬼		Ghosts	4	4°	3·94°	4·46°	θ Cancri	5·8	08 25 54	+18° 25′ 57″	the same as the preceding
	Yü Kuei	輿鬼	198	Ghost-vehicle									
24	Liu	柳	217	Willow	8	15°	14·78°	15·16°	δ Hydrae	4·2	08 32 22	+06° 03′ 09″	culmination of β Ursae Majoris, as for the preceding culmination of α Puppis (Lao jen) (Canopus), S 205[i]
25	Hsing or	星	320	Star	7	7°	6·9°	6·86°	α Hydrae	2·1	09 22 40	−08° 13′ 30″	culminations of γ Ursae Majoris (Thien chi), S 465; and of δ Ursae Majoris (Thien chhüan), S 488
	Chhi Hsing	七星		Seven Stars									
26	Chang	張	378	Extended net	6	18°	17·74°	17·13°	μ Hydrae[j]	3·9	10 21 15	−16° 19′ 33″	none
27	I	翼	741	Wings	22	18°	17·74°	17·81°	α Crateris	4·2	10 54 54	−17° 45′ 59″	lower transit of γ Ursae Minoris (Thai tzu), S 361 culminations of 42 and 184 Draconis (Thai i), S 364 and of ϵ Ursae Majoris (Yü hêng), S 747
28	Chen	軫	399	Chariot platform	4	17°	16·76°	16·64°	γ Corvi	2·4	12 10 40	−16° 59′ 12″	lower transit of β Ursae Minoris (Thien ti hsing), S 575 culmination of ζ Ursae Majoris (Khai Yang), S 130

[a] Chhen Tsun-Kuei (*3*) makes α Ursae Minoris the first star of the Kou Chhen constellation (see p. 261 below).
[b] Stars of the southern hemisphere, not circumpolars; mentioned in *Hsia Hsiao Chêng*.
[c] Properly speaking, Chhien Niu is α Aquilae (Altair).
[d] Not to be confused with Chih nü, the Weaving Girl (see Table 25 below).
[e] There is some doubt about the identification of the small star or stars near the pole which corresponded to Thai i; cf. de Saussure (13) and Maspero (3), p. 323.
[f] Chhen Tsun-Kuei (*3*) gives 35 Arietis.
[g] Chhen Tsun-Kuei (*3*) gives 17 Tauri.
[h] Chhen Tsun-Kuei (*3*) gives 37 ϕ^1 Orionis.
[i] This star is supposed to have been just visible above the Chinese horizon, at any rate in the −3rd millennium (J. B. Biot (4), p. 250), but whether its transit was noted in connection with the position of Liu is very uncertain. Like Nan mên, it is, of course, a star of the southern hemisphere, and not a circumpolar.
[j] Chhen Tsun-Kuei (*3*) gives λ Hydrae; Chu Kho-Chen (8) ν_1 Hydrae.

Table 25. *Stars referred to in col. 9 of Table 24*

Chinese name		Translation	Identifications: Chhen Tsun-Kuei (*3*)
character	romanisation		
漸臺	Chien thai	Clepsydra terrace	10 β Lyrae
織女	Chih nü	Weaving girl	3 α Lyrae
開陽	Khai Yang	Introducer of heat, or of the Yang	16 ζ Ursae Majoris
老人	Lao jen	Old man	α Argus
南門	Nan mên	Southern gate	α^2 and ϵ Centauri
北斗	Pei tou	Northern dipper (Great Bear)	Ursa Major
庶子	Shu tzu	Son of (imperial) concubine	5 Ursae Minoris
太一	Thai i	Great unity, or the Great first one	—
太子	Thai tzu	Imperial prince	13 γ^2 Ursae Minoris
天權	Thien chhüan	Celestial balance	69 δ Ursae Majoris
天璣	Thien chi	Celestial armillary (see below, p. 334)	64 γ Ursae Majoris
天璿	Thien hsüan	Celestial template (see below, p. 286)	48 β Ursae Majoris
天皇大帝	Thien huang ta ti	Great emperor of august heaven	32 H Cephei
天乙	Thien i	Celestial unity, or the Heavenly first one	10 i Draconis
天樞	Thien shu	Pivot of heaven	50 α Ursae Majoris
天帝星	Thien ti hsing	Star of the heavenly emperor, or Sovereign star	7 β Ursae Minoris
左樞	Tso shu	Pivot of the left	12 ι Draconis
搖光	Yao kuang	Twinkling brilliance	21 η Ursae Majoris
玉衡	Yü hêng	Celestial sighting-tube (see below, p. 336)	77 ϵ Ursae Majoris
右樞	Yu shu	Pivot of the right	11 α Draconis

determinative stars were quite far from the equator, e.g. Wei[1] as much as 37° south, and Mao[2] as much as 23° north. The calculations of Nōda (2, *3*) (cols. 7*a b, c*) show that in the −2nd century there was a considerable variation of accuracy in the measurements of the equatorial extensions of the *hsiu*. Some, such as Niu,[3] were remarkably accurately observed, but in other cases, such as Shen[4] and Pi,[5] they were out by more than 1°; this would hardly be surprising, since the observers were confined to naked-eye measurements with armillary rings and alidades. An extremely important point which appears at once from col. 8*b*, is that the determinative stars were chosen largely irrespective of their magnitude. Only two were of the second magnitude, including

[1] 尾 [2] 昴 [3] 牛 [4] 參 [5] 畢

Spica (α Virginis, Chio[1]), and no less than four of the fifth, while one (θ Cancri, Kuei[2]) was as low as the sixth. This shows that what the ancient astronomers were interested in was a geometrical division of the heavens which would neglect bright stars if not useful for their purposes, and it was J. B. Biot (4) who was the first to realise that the choice of the *hsiu*-determinatives had depended on their having the same (or approximately the same) right ascensions as the constantly visible circumpolars.[a] In this connection it might be pointed out that the system of the *hsiu* distinctly foreshadowed the accurate division of the heavens into delimited constellation fields, such as is used today (Delporte, 1).

Why was the number of the *hsiu* (lunar mansions) just 28? The question is not quite so simple as it looks. The most ancient forms of the character indicate a shed made of matting.[b] These segments of the heavens must thus have been thought of as the temporary resting-places of the sun, moon, and planets, like the tea-houses scattered along the roads on earth; but especially of the moon, the greatest nightly luminary.[c] The line of the mansions was thus a graduated scale on which the motion of the moon could be measured, and probably their number was an ancient compromise between the time-spans of its fundamental periods.[d] For while the moon takes 29·53 days to complete its phasic cycle from full to full or new to new (the lunation or synodic month),[e] it takes only 27·33 days to return to the same place among the stars (the sidereal month).[f] These periods are always out of step but 28 was a very convenient average.[g]

[a] About +1080 the administrator of the Imperial Observatory asked Shen Kua why the equatorial extensions of the *hsiu* were so unequal (*MCPT*, ch. 7, para. 13). He replied that it was because of the convenience of having them all in whole numbers of degrees. The origin of the system had thus been forgotten.

[b] K 1029; Hopkins (11).

[c] Cf. Hashimoto (2).

[d] Other suggestions, however, have been made. Menon (1), for instance, has pointed out that the Indian astrological custom of dividing a square (not a circle) into twelve houses, will lead to twenty-eight, not twenty-four, houses if the number of spaces along each side is 'doubled', the corners being each divided twice. But no cuneiform tablets with such diagrams have been found and the *dodecatopos* system of Hellenistic astrology is a relatively late development.

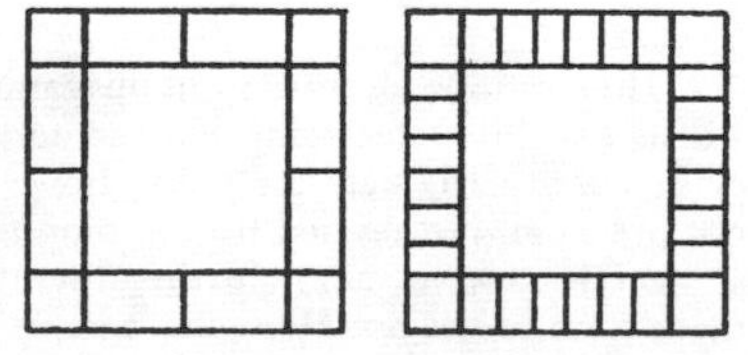

[e] The development of more precise approximations of this constant in ancient and medieval China will be discussed on p. 392 below.

[f] The 30 and 31 days of the calendar month, connected as they are with the length of the tropic year, represent of course the attempt to reconcile lunar and solar periodicities. Cf. O. Thomas (1), p. 310.

[g] Moran (1) has sought to show that the letters of the earliest alphabets were derived from the symbols of the 28 lunar mansions, and thus to explain the fact that most sets of alphabetic letters have from 25 to 30 components (Diringer, 1). A full phonetic alphabet needs 46. The thesis is original, the hypothesis even seductive, but the presentation marred by too much special pleading. The small schematic drawings of *hsiu* constellations according to the ball-and-link convention (cf. p. 276), commonly used in China, can hardly have affected, as Moran seems to think, the forms of ancient Semitic letters, since apart from all geographical considerations they are of the Warring States or Han period rather than Shang (cf. Fig. 93), and therefore much too late. If any set of lunar mansion symbols lies behind the oldest alphabetical letters, it can only be that which represented the constellations of the Road of Anu (see p. 256 below).

[1] 角　　[2] 鬼

In Table 24 the *hsiu* are grouped, seven each, in the four equatorial palaces. The symbolic names for these palaces, which corresponded to the seasons, are given below (p. 242).[a] Here the principle of opposability had a curious result, namely, the 'inter-version' of the spring and autumn palaces.[b] Hsin (no. 5, Antares, Scorpio) receives the visit of the sun in autumn, but is associated with the spring; Shen (no. 21, Orion) receives it in spring, but is associated with autumn. This is because the relations sought were of opposition, not of conjunction (as would be the case with heliacal risings and settings). Spring full moons do appear in the 'spring' *hsiu*, and autumn ones in the 'autumn' *hsiu*. Such difficulties were bound to arise when the cardinal point system of the terrestrial horizon (associated as it was by correlative thinking[c] with the seasons) was extended to the celestial equator (see Fig. 89).[d]

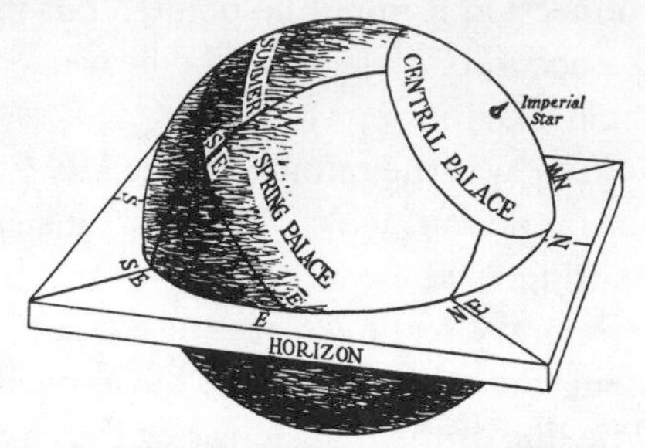

Fig. 89. Diagram of the ancient Chinese divisions of the celestial sphere and their relations with the horizon (after de Saussure, 16*b*).

The addition of a fifth (central, i.e. circum polar) palace, extremely characteristic of Chinese cosmology, and important for all questions of mutual influence with respect to other cultures (e.g. Iranian),[e] brought the fields of the heavens into line with all the other five-fold divisions which have already been discussed, in connection with 'correlative thinking'.[f] In view of the obvious analogy between the emperor and the pole star, and the peculiar respect enjoyed by the circumpolar stars in the Chinese system, it was entirely natural that the latter should be viewed as the residences or offices of the principal members of the (celestial) imperial bureaucracy. Fig. 90, taken from the Wu Liang tomb-shrine reliefs (later Han), shows one of them seated in the Great Bear.

[a] Illustrations of the symbolic animals associated with them are very common in Chinese art; Rufus (2) illustrates some Korean tomb-paintings of about +550 which are typical. Archaeological evidence of early date for them, however, is not at all abundant. Yetts (6) concluded that they were not pre-Han, but revised this opinion later (7); in view of what is now known about Shang ideas on the subject (below, p. 242), further investigation seems likely to bring to light more representations of them such as the pre-Han glass plaques which Yetts has described. Cf. Chhen Tsun-Kuei (5), pp. 97, 98; Cohn (1).

[b] De Saussure (7), p. 259; (6), pp. 151, 160; (1), pp. 121, 157, 282 and many other places, where the matter is discussed minutely in relation to the animal cycle, the five elements, etc.

[c] Cf. below, p. 257.

[d] Cf. Vol. 2, pp. 261 ff., 279 ff.

[e] Perhaps this is what led Michel (17) to say that the Chinese confused azimuth measurements, hour angles and celestial longitudes. In fact, as we shall see on p. 266 below, their use of equatorial coordinates was quite consistent from the time when their positional astronomy began, and celestial longitudes and latitudes never flourished among them. Cf. pp. 279, 398 below.

[f] In Section 13*d* (fundamental scientific ideas), Vol. 2, pp. 262, 264.

Fig. 90. The Great Bear carrying one of the celestial bureaucrats; a relief from the Wu Liang tomb shrines (*c.* +147). *Chin Shih So*, *Shih* sect., ch. 3. On the original relief, the spirit with the isolated star is in line with, and beyond the last of, the stars of the 'handle', which is represented in a straighter line. The isolated star must therefore be Chao yao (γ Bootis, see p. 250).

(2) The Development of the System of the *Hsiu*

The first question which arises is that of the antiquity of the *hsiu*. Two centuries of wrangling have been set to rest by the discoveries of the oracle-bones at Anyang which date from the Shang period (*c.* −1500 onwards). From these a great deal of astronomical and calendrical information has been gathered, notably by Kuo Mo-Jo (*3*), Liu Chao-Yang (*1*, *2*), and in a highly systematic way by Tung Tso-Pin in his monumental *Yin Li Phu.*[1] When this is fully worked out, we shall have a much firmer basis of fact than was available to any of the older scholars. Gaubil, Biot, and to some extent de Saussure, relied far too much on what we now recognise as legendary material, and used early datings no longer acceptable; while their opponents, such as Weber and Whitney, who put their faith in the (imaginary) holocaust of books during the Chhin, maintained that nothing could be known about Chinese astronomy before the −3rd century. As late as 1932 Maspero (15) emphasised what he believed to have been its very recent origins, suggesting that nothing certain could be ascertained before the −7th century. Yet today we can be sure that the system of the *hsiu* was growing up gradually from the middle of the Shang, since its nucleus can be found in the −14th century.

This was the time of the ruler Wu-Ting[2] (−1339 to −1281), and on inscribed bones of his reign (as, indeed, on others earlier and later) there are mentions of stars.[a] Of greatest importance were Niao hsing,[3] the Bird Star or Constellation, to be identified with Chu chhiao[4] (the Red bird), i.e. the 25th *hsiu*, Hsing[5] (α Hydrae), central to the southern palace (the Vermilion Bird)[b]—and the Huo hsing,[6] the Fire Star or Constellation, to be identified with Antares (α Scorpii) and with the 4th and 5th *hsiu*, Fang[7] and Hsin,[8] central to the eastern palace.[c] Chu Kho-Chen plausibly infers from these names that the scheme of dividing the heavens along the equatorial circle into four main palaces (the Blue Dragon (Tshang lung[9]) in the east, the Vermilion Bird (Chu niao[10]) in the south, the White Tiger (Pai hu[11]) in the west, and the Black Tortoise (Hsüan wu[12])[d] in the north) was growing up already at this time.[e] The arrangement is clearly seen in Fig. 91. Besides these two star-names already mentioned, the bones refer also to an important star presumably pronounced Shang,[13] which has not yet been identified,[f] and to another, the 'Great Star', Ta hsing.[14] Possibly these

[a] Tung Tso-Pin (*1*), pt. II, ch. 3, sect. 3, pp. 7 ff.
[b] Chu Kho-Chen (1), p. 12.
[c] Chu Kho-Chen (1).
[d] Often translated the 'Sombre Warrior', cf. Yetts (5), p. 145.
[e] Cf. *Huai Nan Tzu*, ch. 3, p. 3*a*. See the maps in Chhen Tsun-Kuei (*5*), pp. 97, 98.
[f] It may, however, be very significant that the *Shuo Wên* gives Shang hsing[15] as another name for Hsin[8] (no. 5, Antares), cf. Chu Wên-Hsin (*4*), p. 125. The *Tso Chuan* also has a legend (Duke Chao, 1st year; Couvreur (1), vol. 3, p. 31) concerning the equinoctial *hsiu* Shen and Hsin (see below, p. 249) in which the word *shang* occurs.

1 殷曆譜 2 武丁 3 鳥星 4 朱雀 5 星 6 火星
7 房 8 心 9 蒼龍 10 朱鳥 11 白虎 12 玄武
13 鷸 14 大星 15 商星

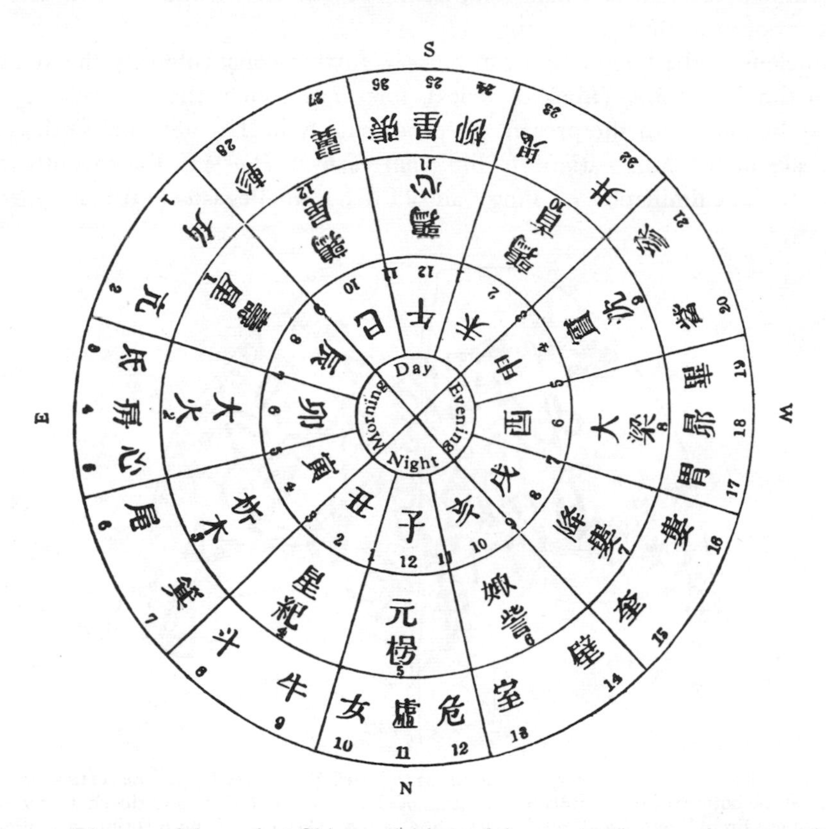

Fig. 91. Diagram of the ancient Chinese divisions of the equator (hour-angle segments).

Outer ring, mansions (*hsiu*): 1 Chio, 2 Khang, 3 Ti, 4 Fang, 5 Hsin, 6 Wei, 7 Chi, 8 Tou, 9 Niu, 10 Nü, 11 Hsü, 12 Wei, 13 Shih, 14 Pi, 15 Khuei, 16 Lou, 17 Wei, 18 Mao, 19 Pi, 20 Tshui, 21 Shen, 22 Ching, 23 Kuei, 24 Liu, 25 Hsing, 26 Chang, 27 I, 28 Chen.

Middle ring, Jupiter stations (*tzhu*), cf. Table 34, p. 403.
Eastern Palace (*Tshang lung*): 1 Shou Hsing, 2 Ta Huo, 3 Hsi Mu.
Northern Palace (*Hsüan wu*): 4 Hsing Chi, 5 Hsüan Hsiao, 6 Chhü Tzu.
Western Palace (*Pai hu*): 7 Chiang Lou, 8 Ta Liang, 9 Shih Chhen.
Southern Palace (*Chu niao*): 10 Shun Shou, 11 Shun Hsin, 12 Shun Wei.

Inner ring, cyclical signs (*chih*); numbered for the double-hours of day and night, starting from the noon double-hour, wu, wei, shen, yu, hsü, hai, tzu; then starting from the midnight double-hour tzu, chhou, yin, mao, chhen, ssu, wu.

The arrangement shown derives from Chou Shih-Chang's *Wu Ching Lei Pien* of +1673, copied by Schlegel (5), p. 39, de Saussure (16*a*) and others.

would complete the four cardinal-point *hsiu*.[a] Fig. 92 shows one of the oracle-bones which mentions the Bird-Star.

The nucleus of the *hsiu* system can be seen further constituted by the stars mentioned in the *Shih Ching* (Book of Odes), folksongs which, though collected much later, may be placed for the present purpose roughly in the −8th or −9th century. One[b] speaks of the culmination of Huo (*chhi yüeh liu Huo*[1]) in the seventh month; another[c] of the culmination of Ting,[2] an old name for Pegasus (13th and 14th *hsiu*)

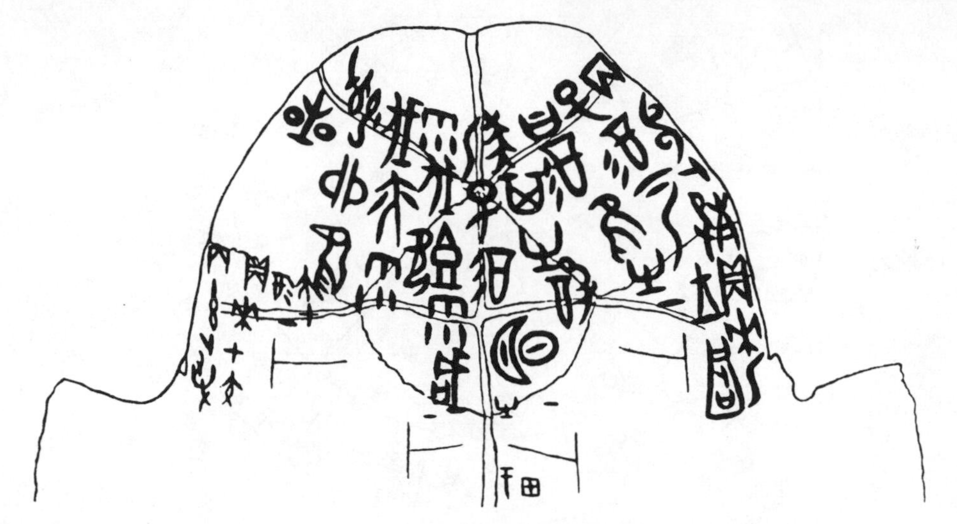

Fig. 92. An oracle-bone inscription which mentions the Bird-Star (α Hydrae). The pictogram of a bird can be seen at the bottom of the penultimate column of characters on the left, and the character for 'star' follows it alone. From Tung Tso-Pin (*1*), Pt. 2, ch. 3, p. 1*b*. The date of the inscription is the time of the King Wu-Ting, i.e. between −1339 and −1281.

(*Ting chih fang chung*[3]). Still another[d] mentions Mao[4] (the Pleiades, the 18th *hsiu*) under its old name of Liu;[5] and Shen[6] (Orion, 21st *hsiu*) as *san hsing*:[7]

> As often as a ewe has a ram's head,
> As often as Orion is in the Pleiades,
> Do people today, if they find food at all,
> Get a chance to eat their fill.[e]

[a] Of great interest is the fact that the bone-inscriptions make frequent references to eclipses, some of which Tung Tso-Pin believes he has identified (cf. p. 410 below), to meteors (cf. p. 433 below), and to a 'New Great Star' (*hsin ta hsing*[8]) which is interpreted as referring to a nova or supernova (cf. p. 424 below); Chou Kuang-Ti (1).

[b] Odes of Pin, 1 (Legge (1), XV, 1; Waley (1), p. 164; Karlgren (14), p. 97).

[c] Odes of Yung, 6 (Legge (1), IV, 6; Waley (1), p. 281; Karlgren (14), p. 33).

[d] Decade of Tu Jen Shih, 9 (Legge (1), VIII, 9 and Karlgren (14), p. 184 mistranslated it; Waley (1) p. 324 correctly).

[e] Tr. Waley (1).

1 七月流火　2 定　3 定之方中　4 昴　5 留 or 罶
6 參　7 三星　8 新大星

There are also mentions of Pi[1] (Hyades, in Taurus; the 19th *hsiu*), of Chi[2] (Sagittarius, the 7th *hsiu*), of Chhien niu[3] the herd-boy (here α Aquilae, but thought to be an ancient determinative for the 9th *hsiu*), and of Chih nü[4] (the Weaving Girl), which is a circumpolar rather than a *hsiu* star. At least eight out of the twenty-eight *hsiu* are therefore to be found in the *Shih Ching*.

Reference has already been made to the Yao Tien chapter (The Canon of Yao)[a] in the *Shu Ching* (Historical Classic) in connection with the astronomical legend about the brothers Hsi and Ho. That 'government commission' text is interwoven with a 'stars and seasons' text, and the latter is the one which now concerns us. It should be considered in relation to two other such texts, the *Hsia Hsiao Chêng* (Lesser Annuary of the Hsia), and the *Yüeh Ling* (Monthly Ordinances of the Chou), bibliographical details about which have already been given. The safest dating for this *Shu Ching* text on philological grounds is between the −8th and the −5th century.[b] What it says is as follows:

> The day of medium length and the (culmination of the) star Niao[5] (serve to) adjust the middle of spring....The day of greatest length and the (culmination of the) star Huo[6] (serve to) fix the middle of the summer....The night of medium length and the (culmination of the) star Hsü[7] (serve to) adjust the middle of autumn...The night of greatest length[c] and the (culmination of the) star Mao[8] (serve to) fix the middle of the winter....The year has 366 days.[d] The four seasons are regulated by means of intercalary months (*jun yüeh*[9]).[e]

Here then quite clearly are the 25th, 5th, 11th and 18th *hsiu*, each one central to its palace or equatorial quadrant. At first sight indeed they seem to be associated here with the wrong seasons. As can be seen in Fig. 94, in the early part of the −2nd millennium Hsing (Niao) and Hsü were solstitial *hsiu*, while Fang & Hsin (Huo) and Mao were equinoctial ones. But this applies of course to the moment of solar conjunction, when the stars would be invisible. One of the basic observations of the old Chinese astronomers was that the quarters of the diurnal rotation correspond every

[a] Ch. 1; it is one of the 'genuine' or Chin Wên chapters (cf. Vol. 2, p. 248).

[b] A minority opinion (Ku Chieh-Kang (*8*); Creel (4), p. 202) brings down the date of this chapter to the time between Confucius and Mencius, or even to the Han. But that makes any astronomical interpretation even more difficult. Besides, it does not seem to have been noticed that the expression of the number 366 is extremely archaic in this passage—*san pai yu liu hsün yu liu jih*,[10] i.e. 'three hundreds plus six tens plus six days'. This insertion of a particle between the powers of ten is typical of the way in which such numbers were written (as Tung Tso-Pin (*1*) has shown) on the oracle-bones of the −14th century. And moreover he finds this practice already dropped in the writings on the later oracle-bones. Cf. p. 13 above. Perhaps this text should be regarded as a very early Chou fragment, whatever its context may seem to make it.

[c] Some texts have 'the shortest day'.

[d] This was not a very inspired remark, since, as we now know, the Shang people in the −13th century were aware that the length of the sidereal year was 365·25 solar days approximately (Tung Tso-Pin, *1*). But perhaps the figure of 366 days for the *chi*[11] or year-length simply meant that it was over 365, as Tung Tso-Pin (*8*) suggests.

[e] Tr. de Saussure (1, 3, 4, 16*a*), eng. auct.; adjuv. Chavannes (1), vol. 1, pp. 44ff. Cf. Legge (1), p. 33; Medhurst, (1), p. 3; both of whose translations are erroneous. All four verbs in brackets could grammatically be rendered in the past tense, as in Karlgren (12).

1 畢 2 箕 3 牽牛 4 織女 5 鳥 6 火 7 虛
8 昴 9 閏月 10 三百有六旬有六日 11 朞

three months with the quadrants of the annual revolution.[a] Thus the *hsiu* which culminates at 6 p.m. at the winter solstice (in this case Mao) could be identified as that in which the sun would stand at noon of the following spring equinox, and so on successively all through the yearly round. This procedure was entirely in character for ancient Chinese astronomy, which solved its sidero-solar problems by deducing the positions of invisible bodies from those of visible ones, all being firmly held in a polar-equatorial coordinate network.

The apparent exactness of this passage has long offered to scholars an irresistible invitation to determine its date by the precession of the equinoxes. Thus J. B. Biot[b] was able to show that the four *hsiu* mentioned would have occupied the equinoctial and solstitial points (0°, 90°, 180° and 270°) about the year −2400. Indeed, there is not much escape from this conclusion.[c] Maspero (15), in criticising the assumptions involved (the chief of which relates to the hour of observation of culmination), and Hashimoto (*1, 4*), assuming 7 p.m. and so reducing the date to the −8th century or rather later, perhaps hardly gave enough weight to the venerable tradition preserved by Gaubil[d] that transit observations were made at 6 p.m. and checked, if daylight prevented them, by the clepsydra. But the great difficulty of any exact determination of the date was pointed out a century ago by Pratt. One of the most recent discussions is that of Chatley (10*a*), who while recognising the strength of the case of Biot and de Saussure adds further uncertainties. The term Niao (Bird Constellation), in its widest sense, can include no less than seven *hsiu* (the whole of the eastern palace), and Huo (Fire Constellation) can include three. On the other hand, Hsü and Mao, which are individual *hsiu* of small equatorial extension, agree very badly together, the former indicating a date about −350 and the latter one about −1400. The question is far from settled, and perhaps the oracle-bone inscriptions may throw further light on it. In view of all that we now know about ancient Chinese history,[e] it seems very unlikely that the data in our text could refer to a time earlier than about −1500 at the most generous estimate, and therefore Hashimoto's conclusion[f] is perhaps the most attractive. But the possibility remains open that the text is indeed the remnant of a very ancient observational tradition, not Chinese at all but Babylonian. In any case, for the present argument the greatest interest about the passage is not the attempts at exact dating, but the clear indication which it gives us of the systematic use of quadrantal *hsiu* for the determination of the seasons and the positions of the sun among the stars at solstices and equinoxes.

The sidereal equinoctial and solstitial points, and hence the sidereal limits of the seasons, become displaced, owing to precession, by 1° in 71·6 years, or 31° approximately in twenty-two centuries.[g] As de Saussure (30) pointed out, one must therefore either modify the limits of the four quadrantal palaces (as the Europeans did by

[a] The clearest discussions of this point are to be found in de Saussure (1), pp. 16ff., (3), pp. 316ff., (4), p. 141.

[b] (1), pp. 363 ff.

[c] See, for instance, Liu Chao-Yang (*10*); Chu Kho-Chen (*3*), and Nōda (2, *2*, *6*) as well as de Saussure (1, 3, 4).

[d] (3), p. 8. Emphasised by de Saussure (1), p. 35, (3), p. 335, (4), p. 139.

[e] Cf. Vol. 1, pp. 83ff.

[f] Now supported also by Chu Kho-Chen (*3*, *8*).

[g] Modern degrees.

attaching their zodiacal signs to the tropic rather than the sidereal year),[a] or keep them fixed, in which case the quarters of the firmament will cease to correspond with the seasons. In the −1st century Ssuma Chhien still retained the cosmological cardinal points of the Yao Tien (*hsiu* nos. 25, 4, 11, 18), but they did not correspond to the tropic seasons. The movement could not be indefinitely disregarded, and at the time of the Thai Chhu[1] calendar reform (by Têng Phing and Lohsia Hung), the winter solstice was found to be between Tou and Niu (nos. 8 and 9) instead of in Hsü (no. 11).[b] This was in −104. By +85, when Li Fan and Pien Hsin were working on the Ssu Fên[2] calendar, it was $9\frac{3}{4}°$ further on in Tou.[c] In the *Chhien Hsiang Li Shu*[3] (Calendrical Science based on the Celestial Appearances) of Tshai Yung and Liu Hung, written in +206,[d] the traditional system was modified to the extent of shifting all four quadrantal points backwards by one *hsiu* (i.e. replacing the ancient ones by nos. 24, 3, 10 and 17—Liu, Ti, Nü and Wei). This text, preserved in later compilations,[e] was brought by St. Julien to the attention of Biot,[f] who by calculation found that its arrangement would have been correct for −1100. Both he and de Saussure[g] believed that it represented a tradition of the time of Chou Kung, but more probably it was a late Han attempt at adjustment. All these discussions were among those which led to the full recognition of the precession of the equinoxes by Yü Hsi in the +4th century.

As for the *Hsia Hsiao Chêng*, six *hsiu* are mentioned (nos. 4, 5, 6, 18, 21 and 24), two of which, Liu[4] and Wei,[5] appear for the first time. The astronomical data given are found[h] to be consistent with a date about the middle of the −4th century contemporary with Shih Shen and Kan Tê.[i] The much fuller *Yüeh Ling* mentions all the *hsiu* except five[j] and might therefore date from about the same time. But since Nōda (2, *2, 4*) has shown in elaborate calculations, based on the data actually given in the text, that the epoch of the observations can only be −620 ±200 years, the lack of information in the former book (Lesser Annuary of the Hsia) may well be due to its brevity rather than to an earlier date. What the *Yüeh Ling* says is worth citing in view of the description of the general system so far given; we may take a single month as sufficient example:

> In the third month of the autumn (i.e. at the beginning of the third month), the sun is in Fang. Hsü culminates at dusk. Liu culminates at dawn (*Jih tsai Fang*; *hun Hsü chung*; *tan Liu chung*[6]).[k]

Here we see clearly the deduction of the invisible *hsiu* which the sun occupies.[l]

[a] Eisler (1), p. 112.
[b] *Hou Han Shu*, ch. 12, p. 3*b*. Actually $21\frac{1}{4}°$ in Tou.
[c] *Hou Han Shu*, ch. 12, p. 4*a*. Chinese degrees.
[d] Cf. Gaubil (2), p. 26.
[e] *Huang Chhing Ching Chieh*, ch. 289, p. 34*a*; *Wu Li Thung Khao*, ch. 182, p. 33*a*.
[f] (4), pp. 147, 150.
[g] (3), pp. 348, 352.
[h] By Nōda (*7, 4*), Hartner (1) and Chatley (10).
[i] Cf. Knobel (2).
[j] The missing ones are Hsin, Chi, Mao, Kuei and Chang (5, 7, 18, 23, and 26). Certain other star-groups occur instead, however, notably Hu,[7] a group in Canis Major and Puppis with an R.A. similar to Ching *hsiu*; and Chien,[8] a group in Sagittarius very close to Nan tou *hsiu*.
[k] Tr. Nōda (2).
[l] Note the difference between these calendars and the Greek *parapegmata* (Diels (1), p. 5), where only heliacal risings and settings are given. For a Roman parallel see Soothill (5), p. 64.

1 太初 2 四分 3 乾象曆術 4 柳 5 尾
6 日在房昏虛中旦柳中 7 弧 8 建

As already mentioned, these two books were incorporated in the *Li Chi* (Record of Rites) and the *Lü Shih Chhun Chhiu* (Master Lü's Spring and Autumn Annals), of Chhin and Han date (−3rd to −1st centuries), and if the other contents of these works is taken into account, the complete *hsiu* system is fully present.[a] The *hsiu* are also complete in other Former Han books, such as the *Huai Nan Tzu*[b] and the *Erh Ya*.[c] It seems almost certain that they were already complete in the mid −4th century, when Shih Shen and Kan Tê were working.[d] Ueta, indeed, by plotting the north polar distances given in the *Hsing Ching*[1] (Star Manual) and similar texts, has shown that the data handed down correspond, for six of the *hsiu*-determinatives, to observations made about −350. The rest must have been re-checked about +200, perhaps by Tshai Yung, as their north polar distances correspond with that epoch.[e] Thus we see that it is possible to trace a continuous development of the system of the *hsiu* from the −14th to the −5th and −4th centuries, after which no further variation occurs.

The completed round of hour-angle segments is seen in Fig. 94, where the determinative stars of each *hsiu* are shown as white circles, and the other stars of each constellation in black.[f]

The general distribution of the *hsiu*-determinatives, with their very varying declinations, seems at first sight strange. It has occurred to more than one observer, however,[g] that they would perhaps be expected to lie, not along the equator of the present day, but along that of the period when they were first chosen. At present, the rough line which they trace out crosses the ecliptic between Mao and Shen (18, 21), descends to Chang and I (26, 27) about 20° declination south, then returns to the ecliptic which it recrosses between Ti and Hsin (3, 5), to describe another curve through Pi and Khuei (13, 15) about 20° declination north. If now, taking into account 10° of precessional movement each 716 years, we draw the approximate position of the celestial equator in −1600 (dotted line in Fig. 94), it can be seen that many more of the *hsiu* fall upon it. This date corresponds to a time just before the Shang period. It is thus significant enough that Mao and Fang[h] lie very close to the equinoctial nodes,[i] while Hsing and

[a] Chavannes (7) describes (p. 104) and illustrates a mirror of the Thang period upon which all the *hsiu* are represented. I reproduce one here as Fig. 93.

[b] Where (ch. 3, p. 2*b*) they are divided into the Nine Spaces (*Chiu Yeh*[2]), as well as the usual Five Palaces.

[c] Ch. 9, pp. 1*a* ff.

[d] Strong evidence for this is afforded by the astronomical inscriptions on Chou bronze vessels which Shinjō (3) has analysed; there the names of *hsiu* occur.

[e] One or two other evidences for the date of the attainment of the complete system were adduced by de Saussure (14). The *Kuo Yü*[3] (Discourses on the States) has a passage mentioning twenty-eight *hsiu* in four palaces referring to −554; but this text cannot safely be taken as a pre-Chhin document. Also the *Shih Chi* quotes a passage from the *Shu Ching* which speaks of the seven directors and the twenty-eight *shê*[4] (an unusual term), but this passage is not in the *Shu Ching* as we have it today (Chavannes (1), vol. 3, p. 300).

[f] Cf. the diagram of de Saussure (6), p. 131; (1), p. 101.

[g] De Saussure (3*a*), Chu Kho-Chen (1).

[h] This opposition of the Pleiades and Antares can be clearly seen on the planisphere constructed by Jeremias (1) for Babylon at date −3200. There are curious parallels in other civilisations where a fire-god star marked the beginning of the year, e.g. Agni in India and Xiuhtecutli in Aztec Mexico.

[i] Where the equator and the ecliptic cross.

[1] 星經 [2] 九野 [3] 國語 [4] 舍

Fig. 93. Bronze mirror of the Thang period (between +620 and +900) showing constellation diagrams of the twenty-eight *hsiu* (second circle from the outside), the eight trigrams (next circle), the twelve animals of the animal cycle (next innermost circle), and the four symbolic animals of the Celestial Palaces (innermost disc). From the collection of the American Museum of Natural History, this mirror is of a type which has several times been illustrated (e.g. *Chin Shih So*, *Chin* sect., ch. 6; *Hsi Chhing Ku Chien*, ch. 40, p. 45; reproduced and discussed by Chavannes (7), pp. 104ff.). The outer circle is inscribed with a poem beginning at the floret on the right (at 3 p.m.).

'(This mirror) has the virtue of *Chhang-kêng* (the Evening Star, Hesperus, Venus)
And the essence of the White Tiger (symbol of the Western Palace),[a]
The mutual endowments of Yin and Yang (are present in it),
The mysterious spirituality of Mountains and Rivers (is fulfilled in it).
With due observance of the regularities of the Heavens,
And due regard to the tranquillity of Earth,
The Eight Trigrams are exhibited upon it,
And the Five Elements disposed in order on it.
Let none of the hundred spiritual beings hide their face from it;
Let none of the myriad things withhold their reflection from it.
Whoever possesses this mirror and treasures it,
Will meet with good fortune and achieve exalted rank.' (tr. auct. adjuv. Chavannes, 7.)

[a] Both planet and palace preside over metal in the system of symbolic correlations.

Hsü lie close to the solstitial points.[a] And it is precisely Hsing and Fang which we have found mentioned in the most ancient Chinese astronomical documents, the oracle-bones, of about −1300. As the Shang people certainly had two of the quadrantal points, they must surely have been aware of the other two also. The *Shu Ching* text mentions all four.[b] The significant thing is that these four occupy the quadrantal points of the equatorial circle which also satisfies the majority of the *hsiu*-determinatives.

Chu Kho-Chen (1) has made calculations to ascertain the date of the equatorial circle which satisfies the maximum number of *hsiu*-determinatives, with the following result:

Date	No. of *hsiu* lying partly or wholly within a belt of 10° north or south Declination
+1900	11
0	14
−2300 to −4300	18–20
−6600	15
−8800	6

He is tempted, therefore, to return once again to the −3rd millennium for the establishment of the *hsiu* system. The great difficulty about this is that all the archaeological and literary evidence is against so early a date.

Our question is complicated by the fact that, as already pointed out, the declination of a star was irrelevant to its selection as a *hsiu*-determinative. Several *hsiu* (e.g. Shen, no. 21; Wei, no. 6; Chi, no. 7; Tou, no. 8) will not fit into any likely position of the equatorial circle. What was important in the selection was whether the determinative star could be keyed to a circumpolar or not, and most of these exceptions could. The fact, therefore, that some of the other *hsiu* appear to fit well on to the equator circle of Shang times may be an illusion; their selection may have been much later and due to other motives.

The equinoctial opposition of Mao and Fang would have been shared, at a different time, by a similar opposition between Shen (no. 21, Orion) and Hsin (no. 5, Antares, Scorpio).[c] This raises the question of the meaning of the important word *chhen*[1] (K455), which constantly occurs in ancient texts.[d] In later usage, it acquired many meanings; besides being the fifth of the duodenary cyclical characters, it came to be used as one of the terms for the twelve standard double-hours, for auspicious or inauspicious conjunctions of celestial bodies, for lucky or unlucky stars, and any definite time or moment. Sometimes the Three Chhen were spoken of, and sometimes the Twelve Chhen. Hopkins (18), however, has suggested that the most ancient forms of the character represent the tail of a scorpion or dragon, and that the

[a] Where the equator and the ecliptic are farthest from one another.

[b] Owing to the relatively late discovery of equinoctial precession in China, this text, if a forgery, would have to be later than the +4th century, which is sinologically improbable in the highest degree. The doubts expressed by Rey (1), vol. 1, p. 350, need not be entertained.

[c] Which, as Fig. 94 illustrates, have positions adjacent to the other pair in right ascension.

[d] Cf. Chu Wên-Hsin (*4*), p. 122; de Saussure (1), p. 106.

1 辰

graph should be regarded as a drawing of part of the constellation Scorpio. Hence the significance of the explanation given[a] in one of the ancient commentaries on the *Chhun Chhiu* (Spring and Autumn Annals), that of Kungyang Kao.[1] There[b] the Great Chhen (Ta chhen[2]) are defined as Ta huo[3] (Antares, central star of Hsin), Fa[4] (the halberd of Orion, Shen), and the pole star (Pei chi[5]).[c] It will be remembered that Hsin *hsiu* is in Scorpio, in fact the heart of the dragon of the Eastern Palace. Here again we find the texts, and even the structure of a character, referring back to a celestial situation earlier than −2000. It is quite impossible to believe that these evidences were manufactured after the precession of the equinoxes was known. The ancient meaning of *chhen* would thus be 'celestial mark-point'.

Precessional change inevitably had a considerable effect upon the whole system of the *hsiu*. With the shifting of the celestial equator, certain stars which had formerly been circumpolar ceased to be so. We have already seen that the position of the tail of the Great Bear (whether pointing upwards, downwards, east or west) was taken as a seasonal indicator.[d] Now the *Hsing Ching* (Star Manual) mentions as an old tradition that the Great Bear originally consisted of nine stars, not seven, but that two of them had been lost sight of.[e] And in effect, if we prolong the handle of the Great Bear, we come to a number of stars in Bootes which could well have been considered as belonging to it. In the *Huai Nan Tzu* (*c.* −120) there is a whole chapter[f] which gives social-ceremonial directions for the months, enumerating them according to the point indicated by the star Chao yao[6] (Twinkling indicator). It is said that when Chao yao points to Yin[7] (N. 60° E.), it is the first month of spring, when it points to Mao[8] (due E.), it is the second month, etc.[g] As Chao yao is probably to be identified with γ Bootis,[h] a star which must have left the area of perpetual visibility about −1500, it can be seen that this text seems to report a very ancient tradition.[h] Here is another piece of evidence which should induce caution in accepting too late a dating for the origins of Chinese astronomy.

Moreover, with the advance of time, precession can bring about changes in the right ascensions of stars, and the magnitude of the effect varies according to the star's position with respect to the equinoctial nodes and the colures. Determinative stars of

[a] To a passage in *Tso Chuan*, Duke Chao, 17th year (Couvreur (1), vol. 3, p. 280).

[b] Ch. 23, p. 5*a* (Duke Chao, 17th year); the occasion was a comet of −524. If not of the −5th century, as traditionally, the commentary is certainly pre-Han. Cf. Wu Khang (1), p. 6; Schlegel (5), p. 146; de Saussure (6), p. 138, (19).

[c] This accounts for the Three Chhen; Shinjō (*1*), p. 13, suggests that the Twelve Chhen were made up by taking the three stars of Orion, plus Antares, plus the pole star, plus the seven stars of the Great Bear.

[d] *Hsia Hsiao Chêng*, *Yüeh Ling*, *Shih Chi*, and especially *Huai Nan Tzu*, ch. 3, p. 10*b*. Also in the *Ho Kuan Tzu*[9] (Book of the Pheasant-Cap Master), which purports to be of the −4th century, but much of which is late Han (see Forke (13), p. 528); ch. 5, p. 16*a*. Cf. Chu Kho-Chen (8).

[e] Chu Kho-Chen (1), p. 14.

[f] Ch. 5.

[g] For each month dusk and dawn culminations of *hsiu* are systematically given.

[h] Chu Kho-Chen (1), Chhen Tsun-Kuei (*3*). Its declination is 38° 43′. Schlegel identifies it, however, with β Bootis, more to the north, and Ueta with λ Bootis. The essence of the argument is not affected. Bulling (5) has identified it in a Han relief (see Fig. 90).

1 公羊高 2 大辰 3 大火 4 伐 5 北極 6 招搖
7 寅 8 卯 9 鶡冠子

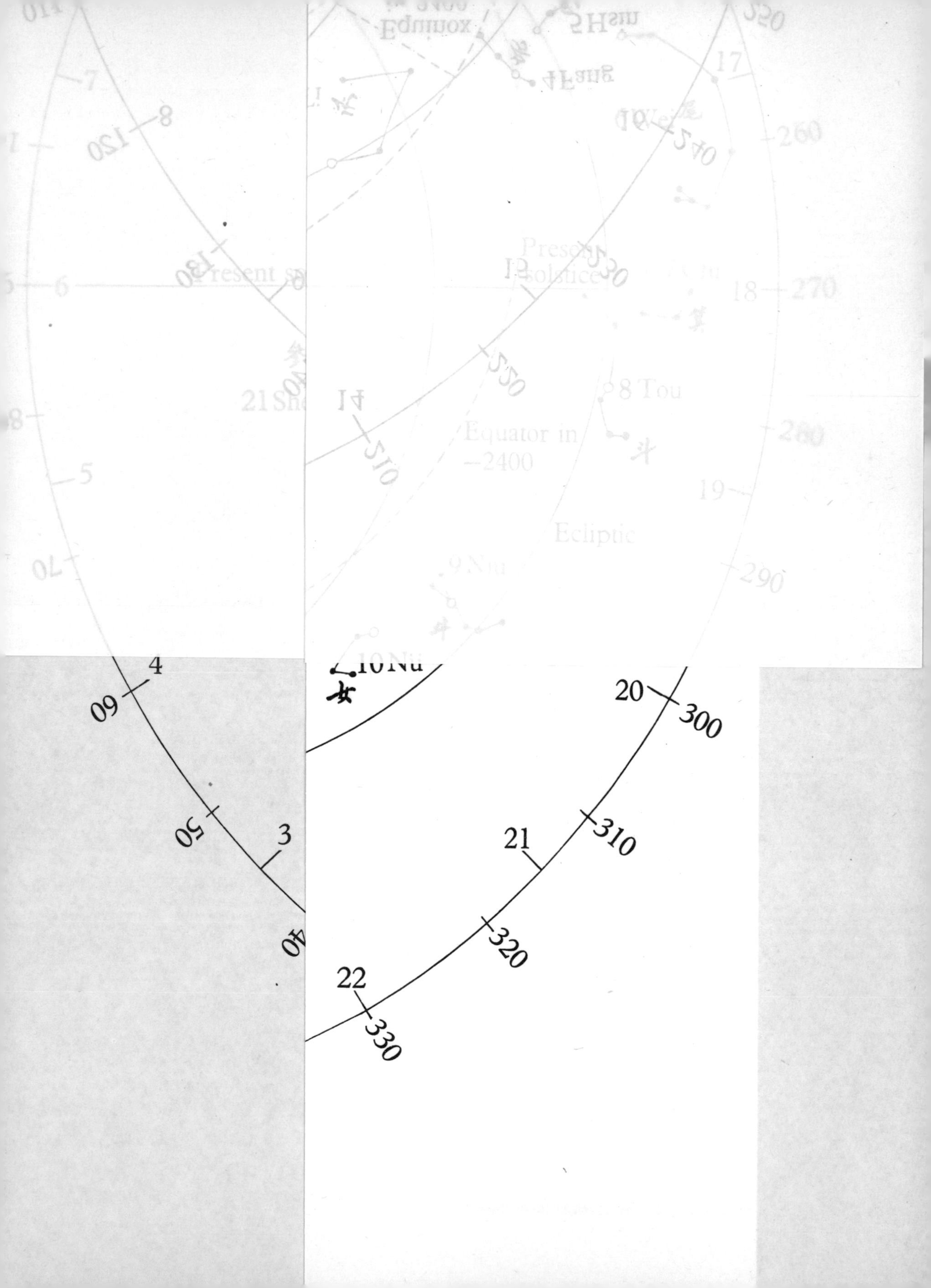

10 Nu
女
4
60
50
3
40
20
300
21
310
320
22
330

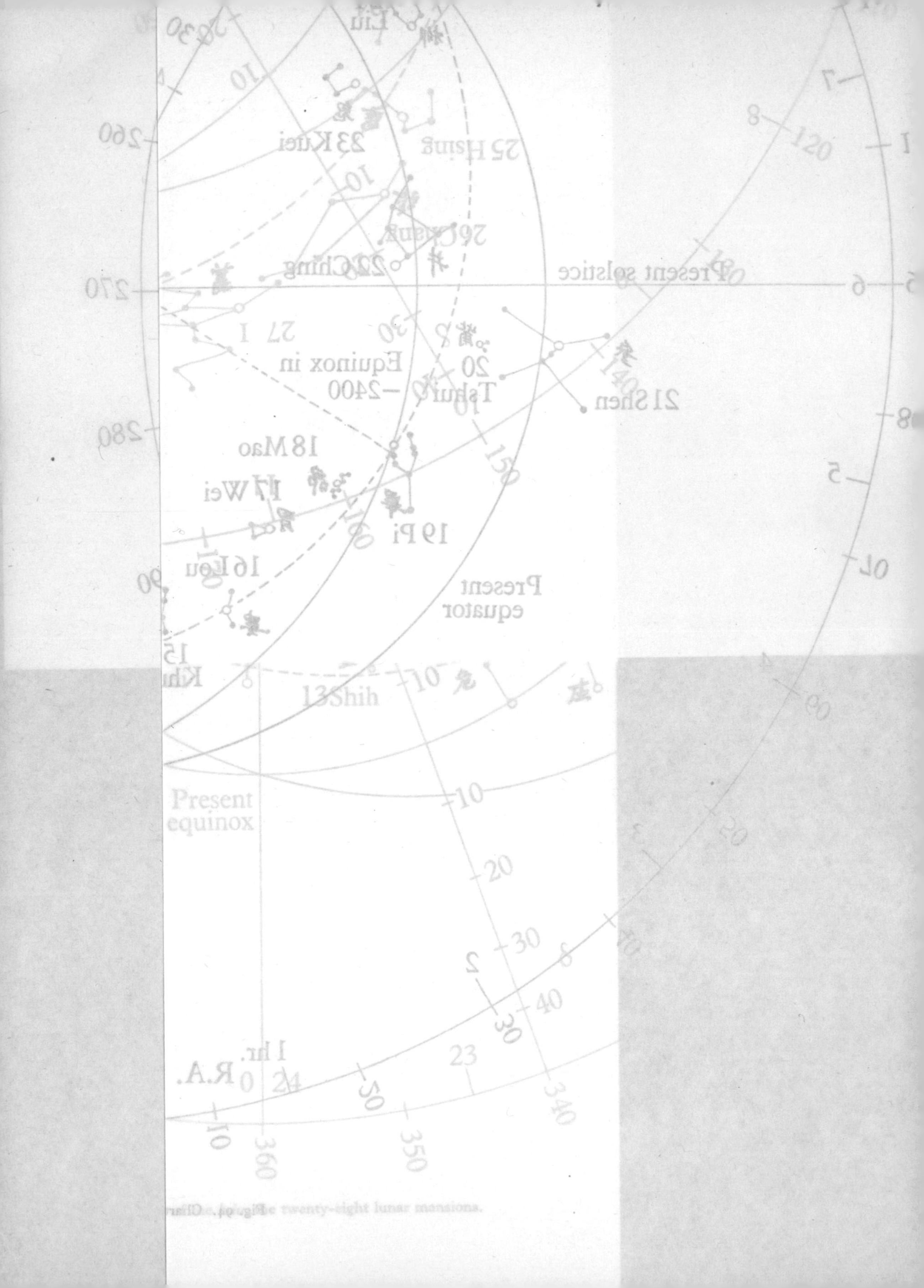

The twenty-eight lunar mansions.

neighbouring constellations with small differences in right ascension, but with large differences in declination, may in time change places in the enumeration order. The keying of the *hsiu* to the circumpolars will also be affected. Thus it is thought that the bright Chhien niu[1] (the Herd-boy; Altair; α Aquilae) was replaced at some time between the period of the *Shih Ching* (Book of Odes) and that of the Han writers, by the weak Nü[2] (ε Aquarii). Similarly, the bright Chih nü[3] (Weaving girl; Vega; α Lyrae) gave place to the weak Niu[4] (β Capricorni).[a] And Ta chio[5] (Arcturus; α Bootis), which may once have been one of the 'prolongation' stars of the Great Bear's handle, was superseded by Chio[6] (Spica; α Virginis).[b] The best-known case of an interversion of the *hsiu* occurred in the +13th century when, under the Yuan dynasty, the determinative star of Shen (δ Orionis) overtook that of the preceding *hsiu*, Tshui (λ Orionis).[c] Kuo Shou-Ching's observations around +1280 showed that Tshui was about to be extinguished.[d] The question was settled in the Ming by a change of the determinative of Shen from δ to ζ Orionis, thereby allowing Tshui to persist as a very narrow *hsiu*.[e]

The position of Ta chio (Arcturus) in Bootes needs a further remark. It has on each side, east and west, two small groups of three stars each, the 'left' and 'right' Shê-thi[7] or 'Assistant conductors'.[f] The fact that the first year in the Jupiter cycle (see below, p. 404) was called *shê-thi-ko*[8] led Chalmers and others[g] to the view that it was a transliteration of the Sanskrit *Bṛhaspati-cakra* (Jupiter cycle), but de Saussure (11) more cautiously translated it 'the rule of the Shê-thi'. Ta chio and Chio *hsiu* (the Greater Horn and the Lesser Horn of the Caerulean Dragon of the Spring (East) Palace) appeared necessarily in the spring, and Ta chio, by reason of its rather northern declination, participated in the function of the Great Bear's handle, since though not strictly circumpolar, its period of visibility was long. The 'rule of the Shê-thi' was thus, according to de Saussure,[h] one of the few cases where the Chinese noted a heliacal

[a] Chu Kho-Chen (1) thinks that there had also been an interversion between Vega and Altair about −3600, which the Babylonians may have noted.

[b] It will have been noticed that the usual order of enumeration of the *hsiu* begins with Chio. This may be because, in Shang times, the first full moon of the year appeared there (de Saussure, 15), or because the Bear's handle pointed to it (Shinjō, 1).

[c] De Saussure (6), p. 172; (1), p. 142.

[d] Gaubil (3), p. 107. One of the Indian *nakshatra* (see below, p. 253), Abhijit, was actually extinguished in this way; de Saussure (15); Renou & Filliozat (1), vol. 2, p. 721.

[e] It is obvious, from the chart in Fig. 94, that it had in any case always been anomalous to have Tshui as well as Shen. De Saussure suggests that Shen (with Hsin) had to be incorporated into the system because of their great antiquity and prestige as the asterisms occupying the equinoctial points at the time when systematic astronomy began, though Shen did not fit in at all well, clashing with Tshui and being a good deal south of the then equator. This may indicate the conflation of more than one ancient system. See de Saussure (28); (1), p. 547; (14), p. 271.

[f] A −1st-century text says: 'They lead by the hand Arcturus on one side and the Great Bear on the other, connecting them with what is below' (*Chhun Chhiu Wei Yuan Ming Pao* (Apocryphal Treatise on the *Spring and Autumn Annals*; the Mystical Diagrams of Cosmic Destiny), in *Ku Wei Shu*, ch. 7, p. 5*b*). This interesting astrological-astronomical work of the Han deserves closer study than it has yet received.

[g] Including recently Chu Kho-Chen (1) who suggests *Kṛttikā*, the assumed intermediate form being Cantonese *chip-thai-kak*.

[h] (1), p. 388.

1 牽牛 2 女 3 織女 4 牛 5 大角 6 角
7 攝提 8 攝提格

rising. But the effect of precession was to spoil the seasonal significance of the phenomenon as the ecliptic moved away from the neighbourhood of Arcturus, and hence we find Ssuma Chhien saying, 'The Shê-thi could no longer serve as indicators'.[a]

In this connection it has been pointed out that the word *lung*,[1] which means the rising of the moon, must have originated from the fact that at the beginning of the year in Han times it rose somewhere between the two horns of the Spring Dragon.[b] The character combines the radicals for moon and dragon. Hence the oft-repeated art motif of the dragons and the pearl—the moon (Fig. 95).[c]

Another list of instructions for ascertaining the season by the observation of circumpolar or near-circumpolar stars occurs in an astronomical MS. of about +800 in the Stein Tunhuang collection.[d] Here the position of Thien yü[2] (the Celestial Prison) is to be observed. Unfortunately, this term applies to four stars or star-groups, any of which might be meant—β Aurigae (S616), the Pleiades (S615), or Lou *hsiu* in Aries (S614), but the most probable is the Shê-thi in Bootes (S617).

(3) The Origin of the System of the *Hsiu*

The replacement of Vega and Altair and other bright stars as mark-points by weaker ones in more strategic positions did not occur in other civilisations, e.g. India. And this is where we come to the problem which has caused so much controversy, namely, the relation of the Indian *nakshatra* and Arabic *al-manāzil*, 'moon-stations', to the Chinese *hsiu*. First brought to the attention of Western scholars by Colebrooke in 1807, many lists of the *nakshatra* are available,[e] while Biot (6) analysed the data on them which al-Bīrūnī had collected in the +11th century. Lists of the Arabic *al-manāzil* will be found in Higgins and others. They were certainly pre-Koranic,[f] and the Hebrews knew them as the *mazzaloth*,[g] while they even got into Coptic (Chatley,

[a] *Shih Chi*, ch. 26, p. 2*b*; tr. Chavannes (1), vol. 3, pp. 325, 345; cf. Chu Wên-Hsin (5), p. 22.

[b] Chatley (5); de Saussure (1), pp. 168, 589.

[c] It is important to note that the dragon here is the symbol of a palace, i.e. a whole quarter of the heavens. The limits of the region happen to correspond approximately with those of the occidental constellation Draco, but this can hardly be more than coincidence. In any case the Chinese sky-dragon had originally nothing whatever to do with the idea of a dragon whose head and tail represented the moon's nodes and which accordingly 'ate the sun' at eclipses. From this no doubt we derive our term 'draconitic months' to designate the time between the moon's successive passages through these points Berry (1), p. 48). Noteworthy in this connection is the fact that the value of the period, 27·2122 days, was obtained by Tsu Chhung-Chih already in the +5th century correct to four places of decimals (cf. Li Nien, 2). But since the nodes are of course not fixed, rotating round the ecliptic once in 18·6 years approximately, the original symbolism cannot have applied to them, and if it was transferred to them later on this can only have been due to the importation of the Indian idea of the monster (or monsters) Rahu and Ketu (see p. 228 above). The ball was then no longer the 'moon-pearl' (*yüeh*[3]) but the sun, and is indeed sometimes represented as red in colour, with issuing flames. In a case like this multiple layers of symbolism are naturally to be expected; see, for example, de Visser (2), pp. 103ff.

[d] B.M., Stein collection, no. 2729.

[e] As in E. Burgess (1); J. Burgess (1); Brennand (1); Chu Kho-Chen (1); Whitney (1), etc. Brennand fell into the common error of regarding them as zodiacal or ecliptic. See also for Mongolia Baranovskaia (2); for Siam Bailly (1); and for Cambodia Faraut (1).

[f] Cf. Pellat (2, 3).

[g] Cf. 2 Kings xxiii. 5. The right interpretation of this text remains, however, somewhat uncertain. Cf. Schiaparelli (1), p. 215.

1 朧 2 天獄 3 玥

Fig. 95. The symbolism of the dragon and the moon; a portion of the Nine Dragon Screen wall in the Imperial Palace at Peking (photo. Nawrath).

12). Iranian versions of the moon-stations are also known.[a] The earliest reference outside Asia is in a Greek papyrus of the +4th century[b] analysed by Weinstock.

The common origin of the three chief systems (Chinese, Indian and Arab) can hardly be doubted,[c] but the problem of which was the oldest remains. That of the *manāzil* is not a competitor, but the other two have elicited from time to time remarkable displays of vicarious chauvinism on the part of indianists and sinologists.[d] Nine out of the twenty-eight *hsiu*-determinatives are identical with the corresponding *yogatārā* (junction-stars), as they are called, while a further eleven share the same constellation, though not the same determinative star. Only eight *hsiu*-determinatives and *yogatārā* are in quite different constellations, and of these, two are Vega and Altair (i.e. possible *hsiu*-determinatives in earlier times). Both in China and India the new year was reckoned from the mansion corresponding to α Virginis, and in both cultures the Pleiades was one of the four quadrantal asterisms.[e]

Most of the arguments currently put forward for the Chinese origin of the *hsiu* system are summarised by Chu Kho-Chen (1). In no other civilisation is it yet possible to trace (as has been done above) the gradual development of the system from the earliest mention of the four quadrantal asterisms. The *nakshatra* do not show so clearly the 'coupling' arrangement discovered by Biot (4), whereby *hsiu* of greater or lesser equatorial breadth stand opposite each other.[f] In Indian astronomy there was no keying of the *hsiu* with the circumpolars, though this was of the essence of the Chinese system.[g] Particularly interesting is the fact that the distribution of the *nakshatra* is much more scattered than that of the *hsiu*, following even less closely the position of the equator in the −3rd millennium.[h] Then Indian calendrical science, following Indian climatic conditions, divided the year into three or six seasons, not four,[i] though the *nakshatra* embody four palaces like the *hsiu*.

However, as regards documentary evidence, the Indians have not much to yield. Filliozat (7, 8) is quite justified in saying that the list of *hsiu* appears about as early in India as it does in China.[j] The facts, so far as we yet know them, are as follows. In the

[a] Anquetil-Duperron (2), vol. 2, p. 349. In the *Būndahišn*, a Persian encyclopaedia finished in +1178. But Henning (1) dates the coming of the *nakshatra* from India to Persia at about +500.

[b] B.M. Papyrus 121.

[c] Unless of course one should take the view that every civilisation using a primarily lunar calendar inevitably needed a system of lunar mansions, so that independent invention occurred. This may be tenable astronomically but hardly historically or ethnographically.

[d] Max Müller and A. Weber were in opposition to J. B. Biot, Gustav Schlegel and F. Kühnert. A purely Chinese origin is still upheld by Chhen Tsun-Kuei (5), p. 89.

[e] See Chu Kho-Chen (8).

[f] Biot noticed that some of the *hsiu* (e.g. Ching and Tou) had to be very wide because there were no circumpolars to which narrower divisions could be keyed. De Saussure (1), pp. 537, 538, 550; (14), believed, on the other hand, that the 'coupling' effect had been sought for first, and that the 'keying' was secondary.

[g] Most historians of science agree (e.g. Datta (3); Kaye (4), p. 77; Brennand, p. 38) that positional stellar astronomy was not the strong point of the ancient Indians who were more interested in the motions of the sun, moon and planets. Nothing has come down to us corresponding to the star catalogues of Shih Shen and his contemporaries.

[h] De Saussure (14) and (1), p. 533. Sometimes (as p. 541), de Saussure was tempted to think that the *nakshatra* might represent a more archaic and undeveloped system than the *hsiu*, not a degenerate or parallel one.

[i] See Sengupta (2, 3).

[j] Actually he sometimes says that it appears earlier in India than in China, but this is going too far at present.

Ṛg Veda hymns (about −14th-century, contemporary with the Shang oracle-bones), the word *nakshatra* seems to refer to any star (Keith, 4), and just two of the later lunar mansion *nakshatra*[a] make their appearance in a portion of the text[b] which is (by general consent) an insertion of several centuries later.[c] The lists, however, are found complete in the *Atharvaveda*[d] and in all three recensions of the Black *Yajurveda*.[e] Although some would accept this as evidence that the system was fully organised by about −1000, others[f] would regard −800 as a safer terminal date. If we now compare with this the Chinese evidence, we find that the four quadrantal *hsiu* at least were known in the −14th century, and that eight of them appear in the *Shih Ching* (−9th or −8th). Since the *Yüeh Ling* may come from as far back as −850, mentioning twenty-three out of the twenty-eight *hsiu*, there is not much to choose in date between the earliest Chinese and Indian textual evidence. There would be no point here in tracing further the dates of later lists in either civilisation, but it is worth noting that the *hsiu*-determinatives and the *yogatārā* cannot be compared until the time of the *Siddhāntas* (*c.* +5th century). One might conjecture therefore that in ancient times the two systems developed separately, but were brought into relation to some extent in later ages when Sino-Indian cultural contacts so much intensified.[g]

But must they not have had a common origin? An important paper by Oldenberg (2) nearly fifty years ago proposed that both were derived from a Babylonian prototype 'lunar zodiac' which was received by all Asian peoples.[h] It will be remembered that in a previous Section[i] we noticed some suggestive correspondences in ideas concerning planetary astrology, which Bezold (1) found in comparing cuneiform texts with the traditions reported by Ssuma Chhien.[j]

The chief attempt to identify the 'moon-stations' as such in cuneiform texts was that of F. Hommel, who believed that the Babylonians had twenty-four, of which he felt able to give sixteen determinative stars. Of these only three corresponded to those in the Chinese *hsiu*, while ten were identical with those in the Indian *nakshatra*. The connection was not, therefore, very striking.[k] But the identification of determinative stars in such ancient material may be too much to hope for. There are quite other grounds for thinking that it may be possible to find a common origin in Babylonian astronomy for all 'moon-station' systems. Assyriologists have long been familiar with a number of cuneiform tablets which were preserved in the library of King Assurbanipal

[a] Their names are different from the later standard forms.

[b] x, 85, 13; Geldner (1), vol. 3, p. 269.

[c] Keith (4), p. 140; (5), vol. 1, pp. 4, 25, 79.

[d] xix, 7; Whitney & Lanman, vol. 2, p. 906 (twenty-eight *nakshatra*).

[e] The *Taittirīya Saṃhitā*, iv, 4, 10; Keith (6), p. 349 (twenty-seven *nakshatra*). The *Maitrāyaṇi Saṃhitā*, ii, 13, 20 (twenty-eight *nakshatra*). The *Kāthaka Saṃhitā*, xxxix, 13 (twenty-seven *nakshatras*).

[f] Keith (4, 5). For the most recent views on Vedic dating see Renou & Filliozat (1), vol. 1, pp. 270ff., 310.

[g] This view was expressed to me in private correspondence by Dr J. Filliozat (1 Sept. 1951).

[h] This caused some modification in the views of de Saussure (14), p. 252; (1), p. 528.

[i] Section 14*a*, in Vol. 2, pp. 353ff. above.

[j] Edkins (8), Boll (3) and Keith (4) all supported the hypothesis of a Babylonian origin of the *hsiu*, while Weber (1) and Whitney (1, 2) inclined to it. So also now does Chu Kho-Chen (8).

[k] As Thibaut (1) hastened to point out.

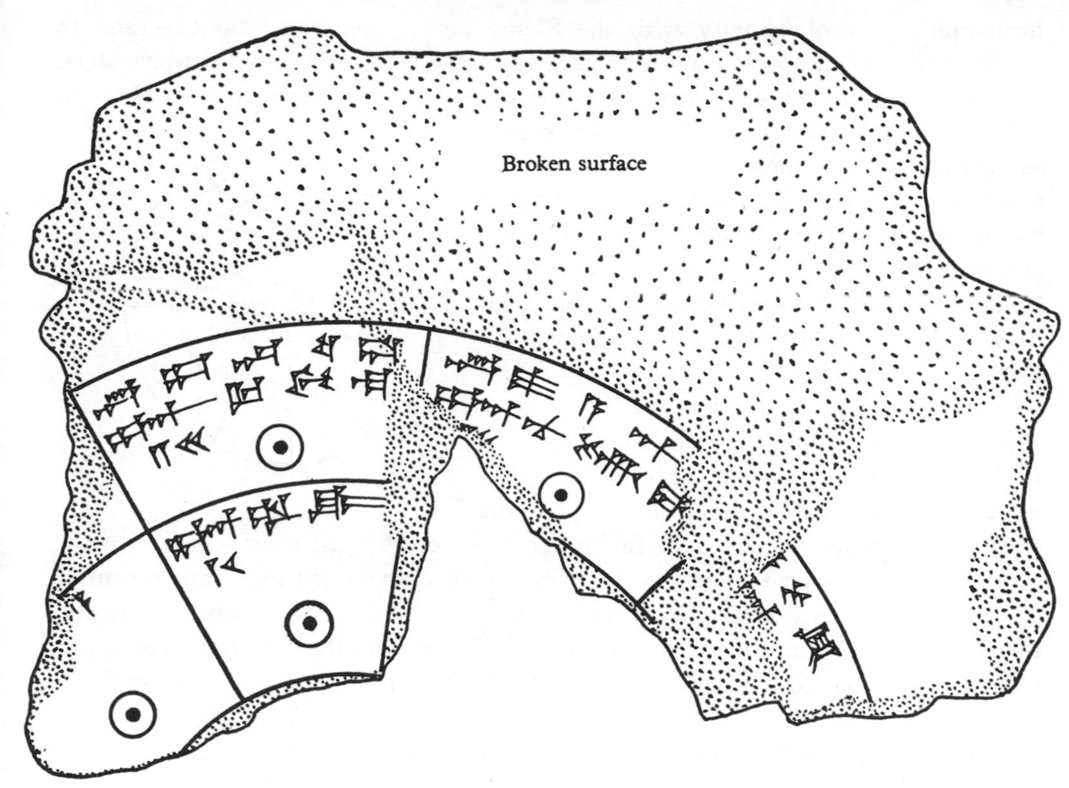

Fig. 96. Fragment of Babylonian planisphere (*c.* −1200), from Budge (3).

(Ashur-bāni-apli, −668 to −626) at Nineveh, but which date as to contents from the late −2nd millennium.[a] These show diagrams[b] of three concentric circles, divided into twelve sections by twelve radii. In each of the thirty-six fields thus obtained there are constellation-names and certain numbers, the exact significance of which has not yet been explained.[c] Fig. 96 illustrates one of these tablets.[d] Much light has been thrown upon the diagrams by modern attempts at reconstruction and interpretation.[e] Could they not be regarded as primitive planispheres, showing both the circumpolar stars and the equatorial 'moon-stations' corresponding to them?

Such a view seems to be supported by the most recent researches. The tablets with 'planispheres' belong to the class now known as that of the 'Enuma (or Ea) Anu Enlil'[f] series. The corpus contains some 7000 astrological omens, and the time of its

[a] Cf. Neugebauer (1); Boll (5), p. 55.

[b] Often called, most erroneously, 'astrolabes'.

[c] The numbers increase and decrease in arithmetic progression and are surely connected with a schematic twelve-month calendar.

[d] Budge (3) ed. *Cuneiform Texts from Babylonian Tablets*, 1912, vol. 33, plates 11 and 12, and p. 6. Further description by Bosanquet & Sayce.

[e] See especially van der Waerden (2); Weidner (1), pp. 62, 76, and A. Schott. Reproduction in Eisler (1), pp. 83, 84.

[f] These were the gods of Elam, Akkad and Amurru.

formation was contemporary with the Shang period, i.e. from about −1400 to −1000.[a] The planispheres consist of 'three roads',[b] each marked with twelve stars, one for each of the months according to the times of their heliacal risings. Those of the central road,[c] the equatorial belt, were known as the Stars of Anu; those of the outer road were asterisms south of the equator (Stars of Ea), while the inner road was travelled by the northern and circumpolar asterisms (Stars of Enlil). After this time the planispheres are no more seen, but their place is taken by lists of stars, modified and much improved, on the tablets of the 'Mul-Apin' series (*c.* −700).[d] Thus eighteen star-gods were supposed to be always visible at one time. Now in these texts there is never mention of any zodiac or of constellations lying along the ecliptic; the earliest documentary evidence of this conception occurs[e] just after −420. On the other hand, the Seleucid Babylonian cuneiform texts of the −3rd and −2nd centuries give great prominence to the zodiac, and use ecliptic coordinates exclusively. Finally, the thirty-six Old Babylonian asterisms were confused with the Egyptian decans and twelve of them ousted to make room for the zodiacal constellations. One might fairly surmise, therefore, that the equatorial moon-stations of East Asia originated from Old Babylonian astronomy before the middle of the −1st millennium and probably a long time before.

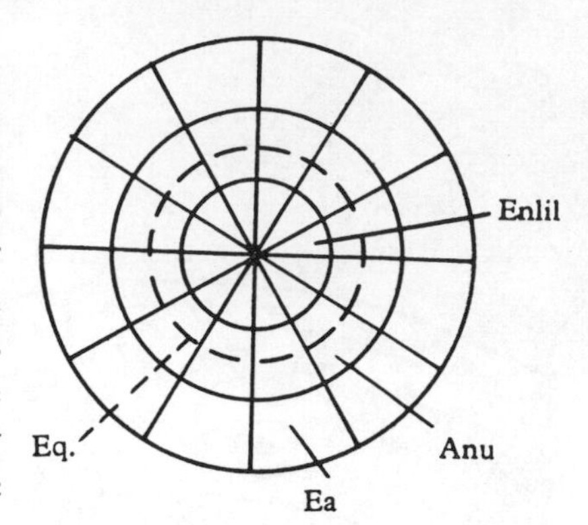

In this connection it is rather interesting (though hitherto unnoticed) that the *Chou Pei Suan Ching* contains[f] a diagram closely resembling the Babylonian planispheres. The pole (*pei chi*[1]) is surrounded by six concentric roads (*chien*[2])[g] separated by seven divisions (*hêng*[3]), as if the ancient Three Roads had each been double-tracked. The whole diagram is called the Chhi Hêng Thu,[4] and the commentators say that originally

[a] These were the tablets from which Bezold (1) chose excerpts for comparison with later Chinese State astrology. A comprehensive description of them is given by Weidner (2).

[b] Van der Waerden (2) could show that these 'three roads' were still known and mentioned, as late as +362, by the Emperor Julian, in his *Oratio ad Solem*. But by then they were 'Chaldean' mystical dogma, not Greek 'hypotheses'.

[c] In width about 30°.

[d] So called because one of the three lists begins with the 'star' (mul) Apin, i.e. a constellation composed of γ Andromedae and stars in Triangulum. The Chinese also recognised this group, as the 'Great general of the heavens' (Thien ta chiang-chün[5]) S 550; he set heliacally in autumn, the season for punishments. On the whole, however, there seems to be much closer correspondence between Babylonian and European constellation names (see van der Waerden, 2) than between those of Babylonia and China. The best descriptions are those of Bezold, Kopff & Boll (1), and Weidner (3).

[e] See Rehm (1). On the whole subject see also Neugebauer (9), pp. 14, 94 ff., 133.

[f] Ch. 1, p. 17*a*.

[g] This technical term was very appropriately chosen, since the original pictograph shows the moon and a gate (K 191), and the word meant 'space between', or 'between' in general. It doubtless derived from light shining in through a door. A later form substituted sun for moon (*chien*[6]).

[1] 北極 [2] 閒 [3] 衡 [4] 七衡圖 [5] 天大將軍 [6] 間

it was coloured blue, with the ecliptic marked on it in yellow.[a] The accompanying text sets forth the motions of the sun through these zones at different times of the year, and adds a number of somewhat naïve computations of the diameters and circumferences of the Roads in *li* (miles).[b] This is strangely reminiscent of the Old Babylonian (*c.* −1400) Hilprecht tablet.[c] The text ends by giving a table of sun-shadow lengths (computed except for the winter and summer solstices), again reminiscent of Old Babylonian tablets,[d] and problems of lunar motion. It is relevant to recall that the *Chou Pei* represents the most archaic of the ancient cosmological theories.[e] Another circumstance worth noting is that the Altar of Heaven and the Temple of Heaven at Peking both retain to this day three circular terraces as if to symbolise the Three Roads of Ea-Anu-Enlil. Here the officiating emperor naturally occupied the central polar position. If Chinese equatorial astronomy derived, at least in part, from Old Babylonian equatorial astronomy, such traces are just what we might expect to find.

Assuredly ancient Persia was one of the intermediate stages in the transfer of Babylonian ideas to China and India. Persian astrological terms have been detected in Chinese Buddhist writings by Huber (2). But how ancient, and how important, the lunar mansions were in Iran still remains doubtful. De Saussure tells us (19) that his presentations of the Chinese system found more response among iranists than among assyriologists, and indeed it is claimed[f] that there was in Persia (at least from the first centuries before our era) the same division of the heavens into four equatorial palaces and one central palace, the same number of equatorial divisions, and the four quadrantal asterisms. From the Persian system de Saussure was able (20, 21) to derive the Arab *manāzil*, and to explain certain aspects of Hebrew symbolism.[g]

That there were Coptic lists of the 'moon-stations' has already been mentioned. It is strange to find that rumours of them even reached the medieval Europeans. Steinschneider has reported (1, 2) the existence of a number of late Latin manuscripts, the contents of which have not yet been adequately studied, which deal with the 'mansions of the moon'.[h] But such transmissions (presumably through Arabic) had no effect upon the deeply rooted ecliptic preconceptions of the West. Conversely, we know[i] that from the Sui onwards (perhaps even as early as the +3rd century), accounts, or at least lists, of the Indian *nakshatra* were translated into Chinese in the

[a] It may well be that this diagram had some influence on the designs of the backs of circular mirrors. A Thang specimen, from the *Chin Shih So*, is represented in Fig. 93 (cf. Chavannes, 7). Cf. the zodiacally adapted planispheres, Graeco-Egyptian in Boll (1), pl. VI, reprod. Eisler (1), pl. IX; Japanese in Boll, Bezold & Gundel (1*b*), fig. 34, reprod. Eisler (1), pl. IX.

[b] Cf. Maspero (3), p. 345, who tabulates.

[c] Cf. Neugebauer (9), p. 94.

[d] Cf. Weidner (3).

[e] Cf. p. 210 above. The date of the older parts of the book may therefore be about the time of Confucius (−6th century). Dr W. Hartner concurs with this view.

[f] De Saussure (17, 19, 20, 21, 24); Chatley (12). Filliozat (7) disagrees.

[g] The colour symbolism, for example, in Ezechiel, Zekariah and Revelation (the four horses of the Apocalypse) can be clearly related to the colours associated with the sky palaces and cardinal points in Chinese and Iranian correlative thinking.

[h] E.g. *Capitulum cognitionis mansionis* (sic) *lunae*, Paris (Bib. Nat.) 9335, f. 140, 141.

[i] Eberhard (12), p. 211; Yampolsky (1).

form of Buddhist texts.[a] We also know[b] that the twelve Greek zodiacal signs found their way into Chinese Buddhist documents from the +6th century onwards, sometimes converted into Chinese equivalents,[c] sometimes appearing in the form of unintelligible transliterations from the Sanskrit,[d] which Confucian astronomers must have written off as yet one more item of Buddhist hocus-pocus.

Then, a thousand years later, when the Jesuits reached China, they entirely misconceived the nature of the Chinese system.[e] The Chinese had possessed,[f] from fairly early times, a twelve-fold division of the year and the equator, which, being based upon *hsiu* and palaces, consisted of unequal parts.[g] The Jesuits mistook these unequal divisions for distorted or degenerate Greek zodiacal signs, and aided by circumstances

[a] Particulars of these are worth attention. Apparently the earliest was the *Śārdūlakarṇā-vadāna Sūtra* translated as the *Mo-Têng-Chhieh Ching*[1] (*Mātangī Sūtra*) by Chu Lü-Yen[2] and Chih-Chhien[3] about +220 (TW 1300). Shinjō doubts whether this was a purely Indian work as it contains some observations which must have been made as far north as lat. 43°. Eberhard has even found reasons for placing the whole text as late as the +8th century. The same book was again translated about +300 by Chu Fa-Hu[4] (Dharmarakṣa) as the *Shê-Thou-Chien Thai-Tzu Erh-shih-pa Hsiu Ching*[5] (TW 1301) but the present text may also be Thang. Both versions have *nakshatra* lists. It is not necessary to believe that all these Sanskrit or partly-Sanskrit writings came from India. For example, the *Mahāvaipulya-mahā-saṃnipāta Sūtra* translated in, or just before, the Sui period (i.e. between +566 and +585) by Narendrayaśas (Na-Lien-Thi-Yeh-Shê[6]) under the title *Ta Fang Têng Ta Chi Ching*[7] (TW 397) was probably written entirely in Eastern Turkestan (modern Sinkiang) and embodies much Turkic calendrical (e.g. animal cycle) material. A fourth work in the Chinese *Tripiṭaka* enjoys the long title of the 'Sūtra of the Discourses of the Bodhisattva Mañjuśrī and the Immortals on Auspicious and Inauspicious Times and Days, and on the Good and Evil Hsiu and Planets' (*Wên-Shu-Shih-Li Phu-Sa Chi Chu Hsien so Shuo Chi-Hsiung Shih Jih Shan-O Hsiu Yao Ching*[8]). This is sometimes called simply the *Hsiu Yao Ching*. It was done into Chinese (as we have already seen, p. 202) by Pu-Khung[9] (Amoghavajra) in +759, and a commentary provided by Yang Ching-Fêng[10] in +764 (TW 1299). This gives twenty-seven *nakshatra* only, the Indian number at the time. It enumerates also the seven planets by their Sanskrit, Sogdian and Persian names, and associates them with the days of the week, telling the Chinese reader, if he does not remember the names of these days, to 'ask a Sogdian or a Persian, or the people of the Five Indies, who all know them' (cf. Bagchi (1), p. 171; Chavannes & Pelliot (1), p. 172). For Buddhist pictorial representations of the spirits of the planets, as on the Tunhuang frescoes, the paper of Meister (1) may be consulted. Whatever the place of origin of these various books they existed at some time or other in Sanskrit form, but it is probable that certain astronomical texts entered Chinese directly from Sogdian and not from Sanskrit (see p. 204 above). Perhaps the most important of these is the *Chhi Yao Jang Tsai Chüeh*[11] (Formulae for avoiding Calamities according to the Seven Luminaries), translated shortly after +806 by Chin Chü-Chha[12] (TW 1308). He seems to have added some Chinese material to his compilation, which in spite of its title is more astronomical than astrological. It contains planetary ephemerides (see p. 395 below), *nakshatra* extensions with ecliptic data, and of course the day-names of the Sogdian seven-day week. But all this, though of absorbing interest for medieval scientific culture contacts, has no bearing whatever on the problem of the ancient origin of the twenty-eight lunar mansions.

[b] Chavannes (7), pp. 37 ff.; Eberhard (12), p. 256.

[c] E.g. *Chhi Yao Jang Tsai Chüeh* (TW 1308), also TW 397, 1299, 1312.

[d] E.g. *Ta Fang Têng Ta Chi Ching* (TW 397).

[e] Cf. de Saussure (1), pp. 181, 513; and especially (16*d*), pp. 336, 339. Modern Jesuits, such as Bernard-Maître (1), fully admit this.

[f] Cf. below, p. 403.

[g] This is one reason why the correlations which have been attempted between *nakshatra* and occidental zodiac symbolism remain unconvincing (e.g. Weinstock (1); Gibson, 1).

1 摩登伽經 2 竺律炎 3 支謙 4 竺法護
5 舍頭諫太子二十八宿經 6 那連提耶舍 7 大方等大集經
8 文殊師利菩薩及諸仙所說吉凶時日善惡宿曜經 9 不空 10 楊景風
11 七曜攘災訣 12 金俱吒

which happened to be favourable to them,[a] proceeded to effect an unnecessary calendar-reform which struck a mortal blow at the ancient system.[b]

(4) The Pole and the Pole-Stars

The basic importance of the celestial pole for Chinese astronomy has already been made clear. But the effect of precession on the position of the pole is considerable, causing it to describe a large circle having the pole of the ecliptic for its centre. It is now, of course, extremely close to α Ursae Minoris, the pole-star of contemporary astronomy (Polaris) (Thien huang ta ti[1]),[c] but some 11,000 years hence it will be at the other extremity of its 'orbit', about 45° north declination, in Lyra not far from Vega. Hence in −3000 it was to be found at about 64° north declination and approximately 14 hrs. R.A. It is therefore a fact of the greatest interest that we find, along the whole length of the path which it has traversed since that date, stars which have preserved Chinese names indicating that they were at various times pole-stars, but later ceased to be so. Special attention has been given to this matter by de Saussure (13), Maspero[d] and Chu Kho-Chen (1). The facts seem hard to reconcile with Maspero's eventual low estimate of the antiquity of the Chinese astronomical tradition (15), since unless there had really been a succession of pole-stars in the Chinese sky, one would hardly expect to find a string of abandoned ones along the polar trajectory and nowhere else in the circumpolar area.[e] It does not, of course, follow that the Chinese observations themselves go back to −3000, since there might have been a direct taking-over of Babylonian star-names—but this has not been demonstrated.

The situation can be appreciated by the diagram in Fig. 97.[f] A zone of some 15° radius around the present pole was bounded, according to Chinese star-maps, by two 'barriers' of stars, enclosing the 'Purple forbidden enclosure' (Tzu wei yuan[2])—an analogy with the imperial court. The 'Eastern boundary' (Tung fan[3]) consisted of the stars ι, θ, η, ζ, ϕ Draconis, χ and γ Cephei,[g] and 21 Cassiopeiae.[h] The 'Western boundary' (Hsi fan[4]) consisted of α, χ and λ Draconis, *d*2106 Ursae Majoris, and 43, 9 and 1 H¹ Camelopardi.[i] It is then significant to find that on each side of the space

[a] Especially the desire of the Khang-Hsi emperor to escape (+1669) from the tutelage of his Regents (cf. du Halde (1), vol. 3, p. 285). He made use of the errors which the Jesuits thought they had discovered in the Chinese calendar to inaugurate a calendar-reform such as was customary enough at the beginnings of reign-periods. Cf. pp. 443, 446ff.

[b] 'Father Verbiest', wrote de Saussure (16*d*), 'reproached the Vice-President of the Astronomical Bureau for not having established, according to true ecliptic motion, what Chinese astronomy had always represented as mean equatorial motion.' But the Jesuits were successful, and Chinese astronomy (in its pure state) descended into a limbo from which only the last century of study has resurrected it.

[c] For the meanings of the Chinese names, refer back to Table 25 on p. 238.

[d] (3), p. 323.

[e] Cf. Zinner (1), p. 217.

[f] Cf. Nōda (*4*), p. 104.

[g] These are Schlegel's identifications; Chhen Tsun-Kuei (*3*) gives instead 52 υ Draconis for ϕ Draconis; 73 Draconis for χ Cephei, and 33 π Cephei for γ Cephei.

[h] Here Schlegel gave another star, the old *n* Rangiferi as the end-star.

[i] Instead of the three last-mentioned stars, which are given by Chhen Tsun-Kuei (*3*), Schlegel gave 924 C, 1316 L and 579 A. Chhen Tsun-Kuei prefers 5 κ to χ Draconis.

1 天皇大帝　2 紫微垣　3 東藩　4 西藩

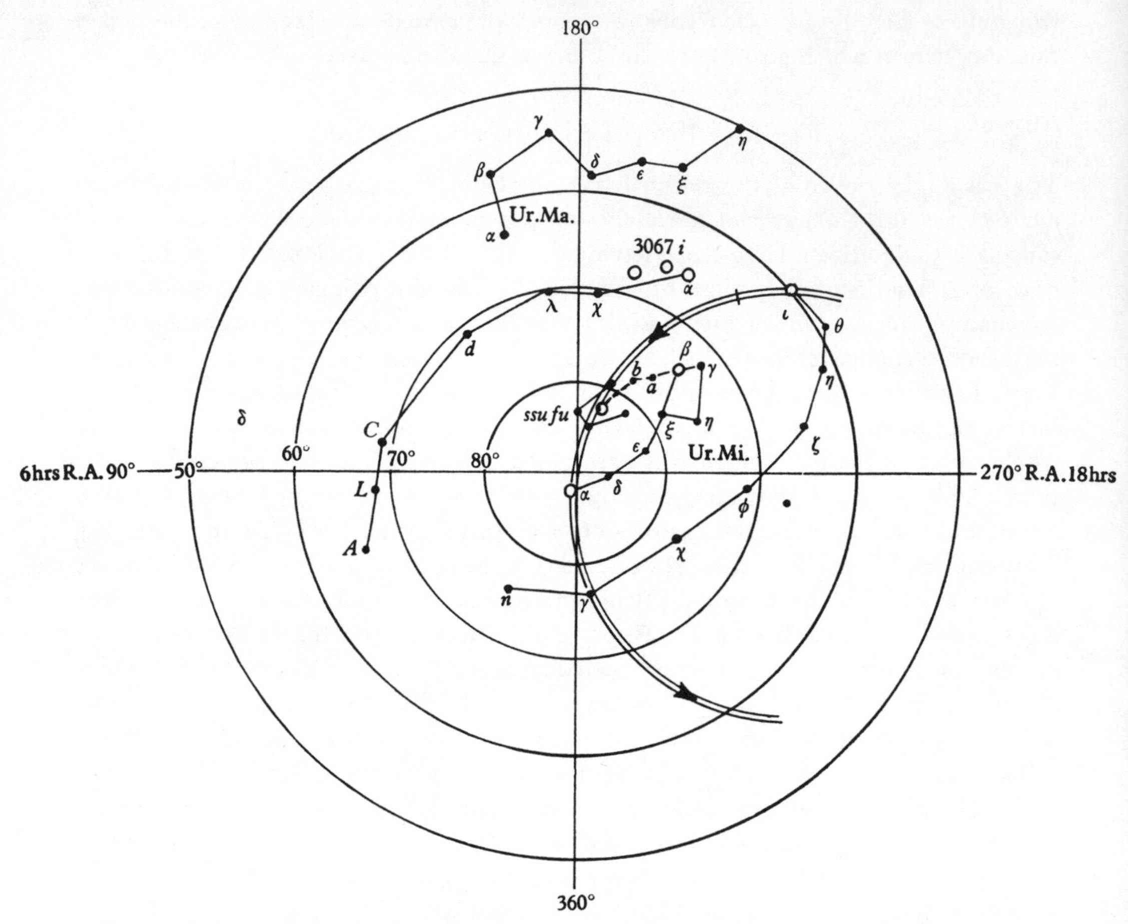

Fig. 97. Polar projection showing the trajectory of the equatorial pole, and the ancient pole-stars.

between the two 'northern' ends of the 'boundary-walls' (Tzu kung mên,[1] 'Gate of the Purple Palace'; or Chhang ho mên[2]), the end stars were called on one side the 'Left Pivot' (Tso shu[3]) and on the other, the 'Right Pivot' (Yu shu[4]). Precisely between them lies the point which the celestial pole would have occupied about −3000.[a]

Nearby, furthermore, are found two stars the names of which point very suspiciously to former duty as pole-stars. Just outside the western boundary are Thien i[5] (the 'Celestial Unique'), which has been identified as 3067 i Draconis; and Thai i[6] (the 'Great Unique'), which is probably either 42 or 184 Draconis.[b] Both these can be seen on the +13th-century planisphere reproduced in Fig. 106. They are small stars of the 5th magnitude, and might have been considered pole-stars in the earlier and

[a] It seems that these two stars were known to the Greeks as the 'pegs' of the pole (Zinner (1), p. 22).

[b] There has been some difficulty in the identification of these stars; cf. the chief references just given. Ueta gives χ Draconis and 4 Draconis respectively. On Thai i see Chhien Pao-Tsung (*4*). It was the residence of a deity important in Han times; see Waley (23), p. 24.

[1] 紫宮門 [2] 閶闔門 [3] 左樞 [4] 右樞 [5] 天乙 [6] 太乙

later part of the −2nd millennium respectively. Though not very near the polar trajectory they are not much further away than β Ursae Minoris (Thien ti hsing;[1] the 'Celestial Emperor'; Kochab), which it seems was used about −1000.

Ursa Minor was not seen by Chinese astronomers as a constellation similar to that which Western observers drew (see Fig. 97). Instead of this, Shih Shen and his contemporaries in the −4th century distinguished a string of stars, the Pei Chi[2] ('North Pole' constellation), which started with γ Ursae Minoris (Thai tzu,[3] the 'Crown prince', suitably next to his father, the former pole-star); then the imperial star itself just mentioned;[a] then Shu tzu[4] (the 'Son of the imperial concubine'), a small star, a3233 Ursae minoris;[b] then Chêng fei[5] or Hou kung[6] (the 'Empress' or 'Imperial concubine'), another small star b3162 Ursae Minoris;[c] and lastly a third small star almost exactly on the polar path, Thien shu[7] (the 'Celestial pivot') or Niu hsing[8] (the 'Knot-star'), 4339 Camelopardi.[d] This star was surrounded by what the Chinese saw as an enclosure on three sides formed by four small stars known as the Ssu fu[9] (the 'Four supports').[e] It was certainly the pole-star of the Han.[f] The rest of our Ursa Minor, together with three small stars[g] surrounding α Ursae Minoris (Thien huang ta ti;[10] Polaris) just as the Ssu-fu surrounded the Thien-shu, was known as Kou chhen,[11,12] which Maspero (3) translated as the 'Curved array', but which might perhaps be better rendered by Chatley's 'Angular arranger' (1), since, after all, this was the point of the sky from which all the *hsiu* segments radiated.[h]

Since none of these stars was exactly on the polar path, we should expect to find some efforts to determine the position of the pole more accurately. The earliest suggestion which we have of this is a passage in the *Chou Pei Suan Ching*[i] which speaks of the 'four excursions' (*ssu yu*[13]) of the pole-star. The 'great star in the middle of the North Pole constellation' was to be observed through the *hsüan-chi*[14] instrument (probably a sighting-tube fitted with a template),[j] and its displacements in the four directions measured.[k] With the passage of time, the Han pole-star moved more and

[a] Maspero (3), it should be said, doubted whether β Ursae Minoris was ever used as a pole-star, though admitting the high antiquity of its imperial name.

[b] Or A, or 5.

[c] Or b, or 4.

[d] Or 32² H.

[e] Identified by Schlegel (5) as 32 H, 207 B, 223 B; and the Piazzi star xiii h 133 Camelopardi. Chhen Tsun-Kuei (3) prefers 29 H and 30 H Camelopardi and 1H Draconis. They can be made out in the polar region of the planisphere in Fig. 106. A fragment from Wu Hsien confirms that they surrounded the pole-star of his time; *Khai-Yuan Chan Ching*, ch. 69, p. 1*b*.

[f] The exact identification of the pole-star of Eudoxus and Hipparchus is apparently not clear, but the latter said that it formed a triangle with two other small stars (Zinner (1), p. 86) so that it may quite probably have been identical with the Han pole-star.

[g] Identified as ζ, ε, δ and 6B Ursae Minoris, with the Piazzi star vi h 21 (46 Camelopardi ?) and 323 B Cephei.

[h] I am much indebted to Dr A. Beer and Dr J. Haas for help in all these star identifications.

[i] Ch. 2, p. 2*b* (tr. E. Biot (4), p. 623). This text has been greatly misunderstood.

[j] See below, p. 336.

[k] The author of the *Chou Pei* undoubtedly knew the pole-star of Shih Shen, but probably found Kochab easier to observe on account of its respectable magnitude.

[1] 天帝星 [2] 北極 [3] 太子 [4] 庶子 [5] 正妃 [6] 后宮
[7] 天樞 [8] 紐星 [9] 四輔 [10] 天皇大帝 [11] 鈎陳 [12] 勾陳
[13] 四游 [14] 璿璣

more; still accepted by Chhen Cho at the end of the +3rd century,[a] it was found to be distant from the true pole more than 1° by Tsu Kêng-Chih in the +5th.[b] By the time of Shao O (+12th century) the distance was $4\frac{1}{2}$°,[c] and his contemporary, the great Chu Hsi, was well aware that at the true polar point there was no star.

We have an interesting account of the use of the sighting-tube by Shen Kua a century earlier, in determining the true pole:

Before Han times it was believed that the pole star was in the centre of the sky, so it was called Chi hsing[1] ('Summit star'). Tsu Kêng (-Chih) found out with the help of the sighting-tube that the point in the sky which really does not move was a little more than 1° away from the summit-star. In the Hsi-Ning reign-period (+1068 to +1077) I accepted the order of the emperor to take charge of the Bureau of the Calendar. I then tried to find the true pole by means of the tube. On the very first night I noticed that the star which could be seen through the tube moved after a while outside the field of view. I realised, therefore, that the tube was too small, so I increased the size of the tube by stages. After three months' trials I adjusted it so that the star would go round and round within the field of view without disappearing. In this way I found that the pole-star was distant from the true pole somewhat more than 3°. We used to make diagrams of the field, plotting the positions of the star from the time when it entered the field of view, observing after nightfall, at midnight, and early in the morning before dawn. Two hundred of such diagrams showed that the 'pole-star' was really a circumpolar star. And this I stated in my detailed report to the emperor.[d]

One would like to know more about the measures taken by Shen Kua to ensure the stability of his instrument. It is interesting that not long afterwards similar observations were being made in Europe. U. T. Holmes has drawn attention to an old French book of scientific questions and answers dating from about +1250, the *Sydrac*, where we find:

Toutes (etoiles) tornent o lui (le firmament) sans une, qui est apellee Guioure, laquelle guie ceus de mer et de terre...mais au declinement dou firmament elle mue une foiz la montance d'une paume...elle torne en tel maniere com la cheville de la pierre soretaine dou molin.[e]

The analogy of the pin or set-screw of a millstone would have pleased Shen Kua. Eventually it became necessary to abandon the Han pole-star and to adopt α Ursae Minoris, but this does not seem to have been done until the end of the Ming.

To illustrate what has been said earlier about the circumpolar stars, and to give some idea of what Shen Kua would have seen through his sighting-tube if he had had a camera, I reproduce a photographic record of the motions of the stars nearest the pole (Fig. 98).[f] In +1687 Robert Hooke suggested a pole-finding telescope fitted with a glass plate in the eyepiece etched with concentric circles so as to permit the observation of the stars nearest the pole in their diurnal revolution.[g]

[a] *Chin Shu*, ch. 11, p. 7*b*.
[b] *Sui Shu*, ch. 19, p. 28*a*.
[c] *Sung Shih*, ch. 48, p. 20*a*.
[d] *MCPT*, ch. 7, para. 11, tr. auct., adjuv. Maspero (4), p. 295. Cf. Hu Tao-Ching (*1*), vol. 1, pp. 296ff.
[e] Question 210, ii.
[f] Cf. Berry (1).
[g] Derham; reprinted in Gunther (1), vol. 7, p. 704.

[1] 極星

(*f*) THE NAMING, CATALOGUING AND MAPPING OF STARS

(1) Star Catalogues and Star Co-ordinates

Something has already been said in the bibliographical introduction to this Section[a] about the work of the great −4th-century astronomers, Shih Shen,[1] Kan Tê[2] and Wu Hsien,[3] and the titles of their lost books were given. These were still listed in the bibliography of the Liang dynasty (+6th century), but after that they appear no more. The catalogues of stars which the books contained had, however, been

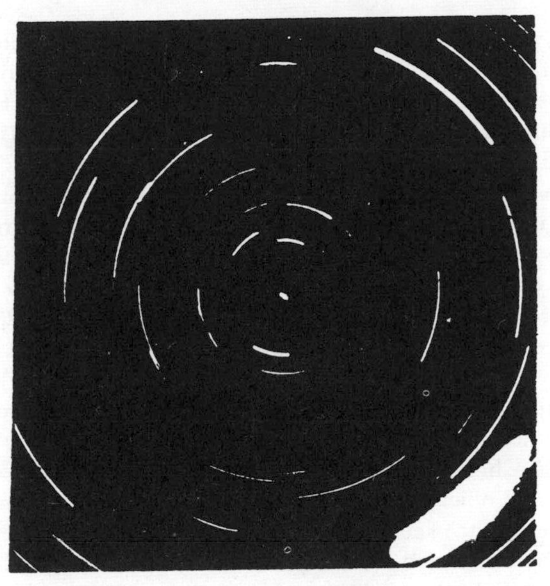

Fig. 98. A time-exposure photograph showing the rotation of the circumpolar stars, taken in 1925. Polaris is responsible for the over-exposed trail in the lower right-hand corner (Yerkes Observatory reproduced by Pruett, 1).

conflated and commented on by the astronomer Chhen Cho[4] at the beginning of the +4th century, who constructed a star-map from them.[b] Between +424 and +453 another astronomer, Chhien Lo-Chih,[5] made an improved planisphere, marking in different colours the stars which had been determined by the three ancient observers; those of Shih Shen were coloured red, those of Kan Tê black, and those of Wu Hsien white.[c] This was not due, as Maspero (3) points out, to any particular interest in the history of science, but to the belief that the three men had had different methods of astrological interpretation and that it was therefore necessary to know which system to apply.[d] These maps (or at least the second one) still existed at the time of Chhüthan

[a] P. 197 above. [b] *Chin Shu*, ch. 11, p. 7*b*.

[c] This marking was changed to yellow in the Thang.

[d] Although these colours had nothing to do with the observed colours of stars, the Chinese did record such characteristics in much detail; see, for instance, the elaborate description of the constellations in *Chin Shu*, ch. 11, pp. 7*b* ff., ch. 12, pp. 4*a* ff. (tr. Ho Ping-Yü). So also had the Babylonians (Boll & Bezold (1); Boll (5), pp. 45 ff.) before them. Arabic tables had special columns for star colour (private communication from Professor Juan Vernet).

[1] 石申 [2] 甘德 [3] 巫咸 [4] 陳卓 [5] 錢樂之

Hsi-Ta[1] (*c.* +715) when the *Khai-Yuan Chan Ching* (Khai-Yuan Treatise on Astrology) was being compiled. This is the work still extant, which, together with the *Hsing Ching* (Star Manual) in its various forms,[a] gives us the fullest data concerning observations from the −4th century. That star-maps showing the traditional colours still existed in +1220 we know from a story about an unfortunate examination candidate Hsü Tzu-I[2] which refers to approximately that date.[b] One MS. example of the coloured star-map of Chhien Lo-Chih, dating from about +940, has, moreover, been preserved to the present day[c] (see Figs. 99 and 100).

Here is the story as the *Sui Shu* tells it:[d]

> The (star-) maps which had been cast (*chu*[3]) by (Chang) Hêng[4] got lost in the disturbances (at the end of the Han), and the names and details of the stars and constellations which they showed were not preserved. But then Chhen Cho[5] (who had been) Astronomer-Royal of the Wu State (in the Three Kingdoms period) first constructed and made (in +310) a map of the stars and constellations according to the three schools of astronomers, Master Kan, Master Shih, and Wu Hsien, adding an explanation with an astrological commentary.[e] There were 254 constellations, 1283 stars, and 28 *hsiu*, with 182 additional stars, making in all 283 constellations[f] and 1565 stars.[g] Then in the (Liu) Sung dynasty, in the Yuan-Chia reign-period (+424 to +453), the Astronomer-Royal Chhien Lo-Chih[6] cast a bronze astronomical instrument (i.e. a celestial globe) using marks of three colours, red, black and white, to distinguish the three schools of astronomers, the total numbers (of each kind) agreeing with (the lists of) Chhen Cho. Then (at the beginning of the Sui dynasty), when Kao Tsu (the first emperor) conquered the Chhen (dynasty), he captured their astronomical expert Chou Fên[7] and the instruments which had been handed down from the (Liu) Sung time. Whereupon he ordered Yü Chi-Tshai[8] and others to check for size and accuracy the old (star-) maps, both private and official, dating from the (Northern) Chou, Chhi, Liang and Chhen dynasties, and formerly in the keeping of Tsu Kêng-Chih,[9] Sun Sêng-Hua[10] and others. The object of this was the construction of hemispherical maps (*kai thu*[11]) following the positions of the stars of the Three Schools.

[a] See p. 198 above.

[b] *Ssu Chhao Wên Chien Lu*, ch. 1, p. 23*b*.

[c] This is the MS. 3326 in the Stein (Tunhuang) collection in the British Museum. Though mentioned by L. Giles (5), its great astronomical interest has not hitherto been brought out. Each *hsiu* is depicted by itself, in a cylindrical orthomorphic projection like Mercator's centring on the equator, with columns of text in between, while at the end of the scroll there is a planisphere centred on the north celestial pole. I discovered this extremely interesting map in conjunction with my friend Prof. Chhen Shih-Hsiang. Its probable date makes it about contemporary with the maps in the 'Book of the Fixed Stars' (*Kitāb Suwar al-Kawakib*, cf. Winter, 6), written by 'Abd al-Rāhman ibn 'Umar al-Sufī (+903 to +986, cf. Suter (1), no. 138), but the oldest MS. of this dates only from +1010. The catalogue itself is for +964 (Destombes, 2).

[d] Ch. 19, pp. 2*a*, *b*, tr. auct.

[e] The titles of some of his works on this subject have been preserved (*Sui Shu*, ch. 34, pp. 15*b*, 16*a*); for instance, the *Ssu Fang Hsiu*[12] (Mansions of the Four Quarters) and *Thien Kuan Hsing Chan*[13] (Prognostications from Stars and Asterisms).

[f] Seemingly a mistake for 282, but *Chin Shu*, ch. 11, p. 7*b*, also gives 283.

[g] Seemingly a mistake for 1465.

[1] 瞿曇悉達 [2] 徐子儀 [3] 鑄 [4] 張衡 [5] 陳卓
[6] 錢樂之 [7] 周墳 [8] 庾季才 [9] 祖暅之 [10] 孫僧化
[11] 蓋圖 [12] 四方宿 [13] 天官星占

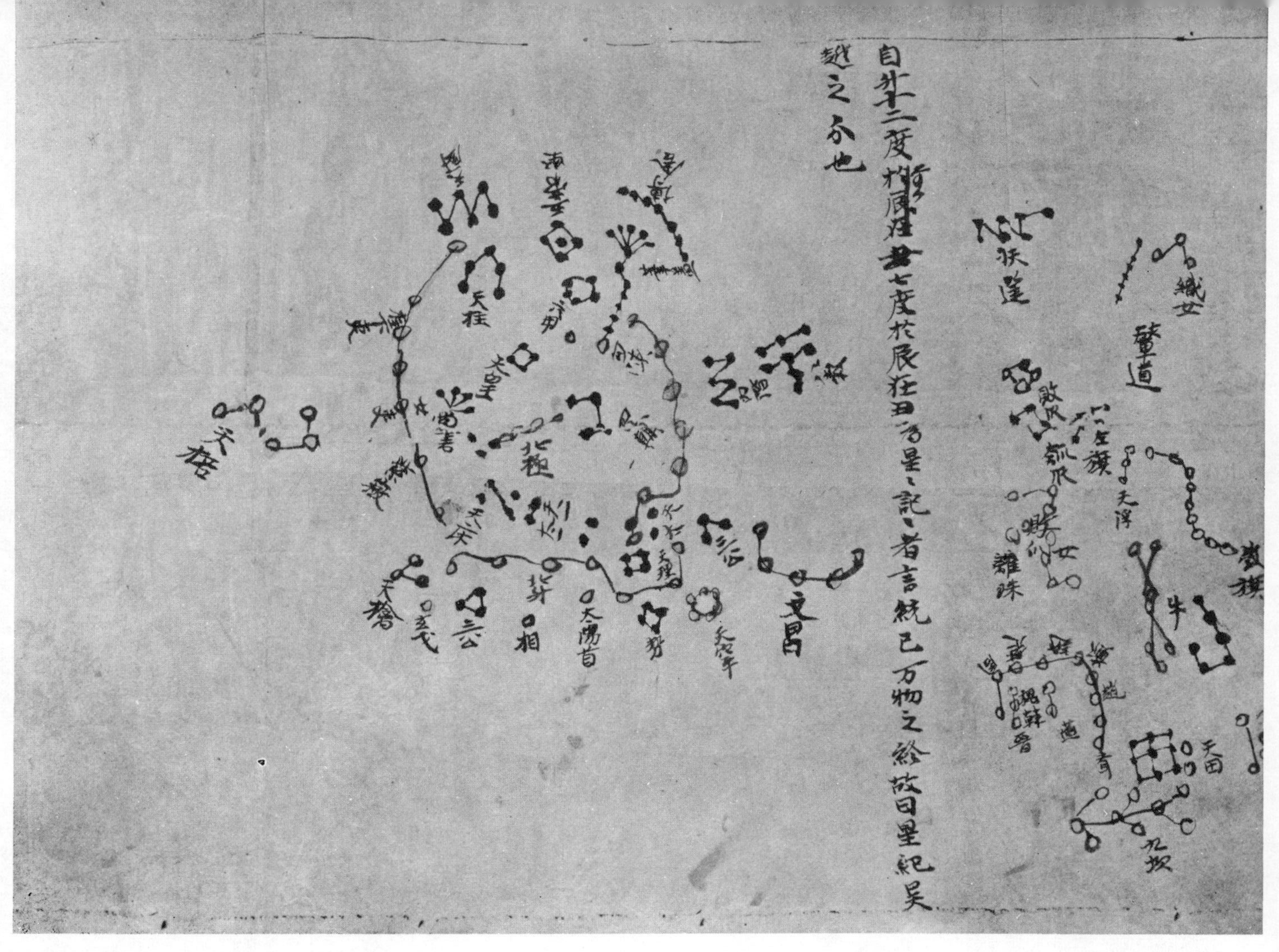

Fig. 99. The Tunhuang MS. star-map of *c.* +940 (Brit. Mus. Stein no. 3326). To the left, a polar projection showing the Purple Palace and the Great Bear. To the right, on 'Mercator's' projection, an hour-angle segment from 12° in Tou *hsiu* to 7° in Nü *hsiu*, including constellations in Sagittarius and Capricornus. The stars are drawn in three colours, white, black and yellow, to correspond with the three ancient schools of positional astronomers.

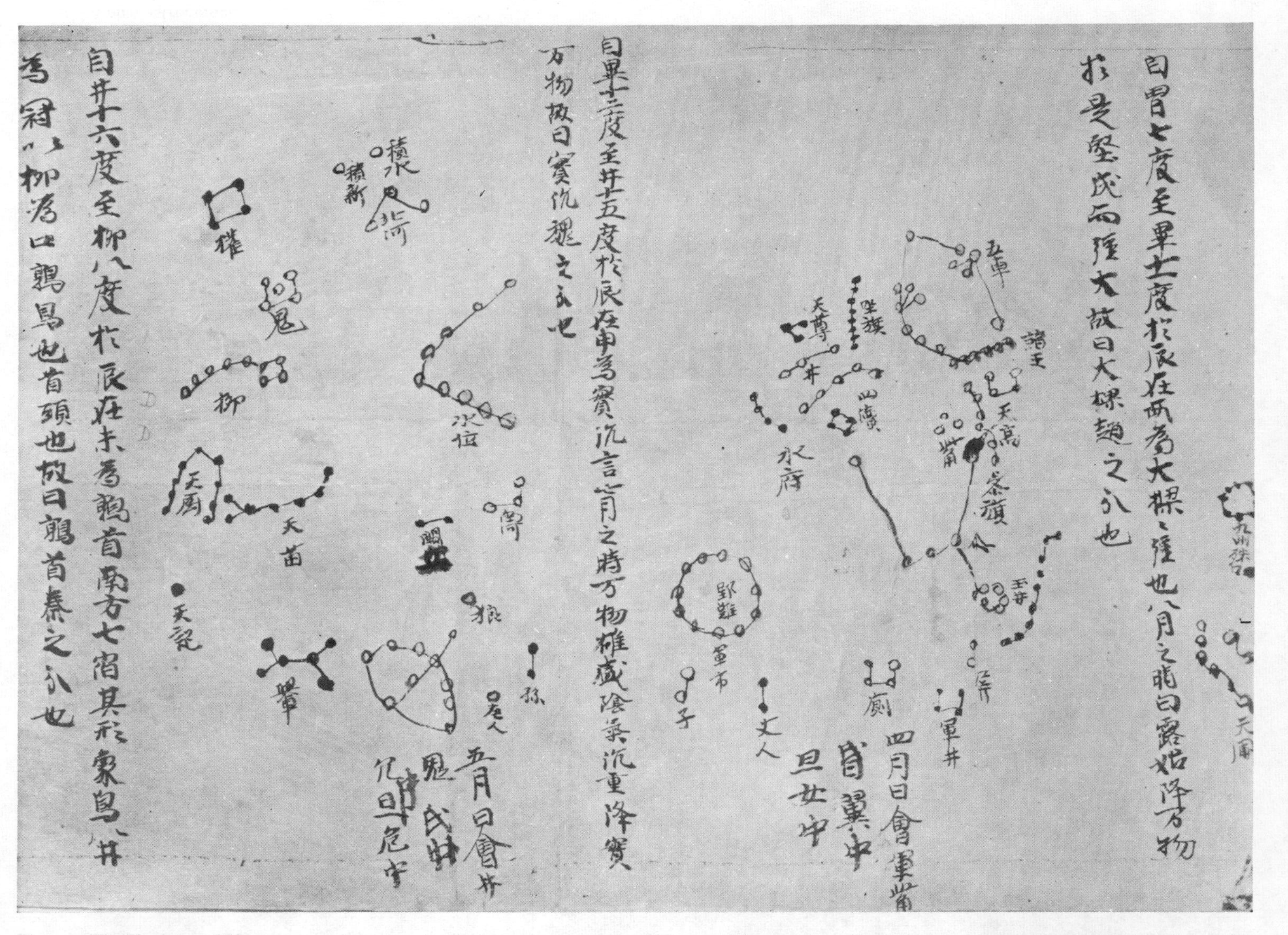

Fig. 100. The Tunhuang MS. star-map of *c.* +940; two hour-angle segments on 'Mercator's' projection. To the right, a segment from 12° in Pi *hsiu* to 15° in Ching *hsiu*, which includes Orion, Canis Major and Lepus; to the left, another from 16° in Ching *hsiu* to 8° in Liu *hsiu*, which includes Canis Minor, Cancer and Hydra.

Table 26. *Breakdown of star totals in ancient star-lists*

	'Chairs'	Stars	'Chairs'	Stars
SHIH SHEN				
Internal (*chung*[1]), i.e. north of the equator	64	270	—	—
External (*wai*[2]), i.e. south of the equator	30	257	—	—
The 28 *hsiu*[3]	28	282	—	—
Total of 'red' stars	—	—	122	809
KAN TÊ				
Internal	76	281	—	—
External	42	230	—	—
Total of 'black' stars	—	—	118	511
WU HSIEN				
Internal and External; total of 'white' or 'yellow' stars[a]	—	—	44	144
	—	—	284	1464

We are evidently in the presence of a very long and continuous tradition of celestial cartography. The total number of stars now in these ancient star-lists amounts to 1464, grouped into 284 constellations (*kuan*,[4] officials; or *tso*,[5] chairs of office).[b] The breakdown is shown in Table 26. Naturally some other figures are found in ancient texts. Thus about +130 Ma Hsü[6] said that there were some 118 named and charted constellations, containing 783 stars.[c] His contemporary, Chang Hêng,[7] wrote in the *Ling Hsien*:

North and south of the equator there are 124 groups which are always brightly shining. 320 stars can be named (individually). There are in all 2500, not including those which the sailors observe (*erh hai jen chih chan wei tshun*[8]).[d] Of the very small stars there are 11,520. All have their influences on fate.[e]

[a] The fact that the lists of Kan Tê and Wu Hsien show fewer stars than that of Shih Shen does not mean that their original catalogues contained fewer, but simply that his was selected as the core, and theirs contributed additional stars which were not in Shih Shen's.

[b] The identification of the stars in the three ancient catalogues is the subject of a special paper by Wu Chhi-Chhang (*1*).

[c] *Chhien Han Shu*, ch. 26, p. 1*a*; *Chin Shu*, ch. 11, p. 7*a*. Ma Hsü was the author of the astronomical chapter of the former history.

[d] This is at first sight a curious statement. Chinese civilisation of that time is not generally thought of as having much connection with sea-faring. Perhaps the reference is to the astrological experts from the region of the old coastal state of Chhi. But later on (Sect. 29) evidence will be presented that in Chang Hêng's own time Chinese envoys and merchants were habitually voyaging to many countries in south-east Asia by sea, and often in the ships of those foreign countries. Malayan and Indian sailors had no doubt their own astronomical traditions. In this connection it is interesting that the *Khai-Yuan Chan Ching* (*c.* +720) constantly quotes (e.g. ch. 30, p. 3*a*; ch. 71, pp. 9*a*, 9*b*) from a *Hai Chung Chan*[9] (long lost), i.e. an 'Astrology of the People in the Midst of the Sea'. This seems to have some Indian connection.

[e] Tr. auct. Cf. above, p. 216; below, p. 355. This passage is on p. 4*a* of the fragment in Ma Kuo-Han's collection, *YHSF*, ch. 76. Cf. *Chin Shu*, ch. 11, p. 7*b*.

[1] 中 [2] 外 [3] 宿 [4] 官 [5] 座 [6] 馬續 [7] 張衡
[8] 而海人之占未存 [9] 海中占

This fragment gives us some idea of what was on Chang Hêng's charts, but whether they were flat or spherical we cannot be sure.

The most thorough examination of these ancient star-lists by a competent astronomer is that of Ueta (1, *1*).[a] In all these texts the data are given in much the same way, thus: (1) the name of the asterism; (2) the number of stars which it contains; (3) its position with respect to neighbouring asterisms; and (4) measurements in degrees (on the $365\frac{1}{4}°$ basis, of course) for the determinative or principal star in the group. These measurements invariably include: (*a*) the hour angle of the principal star measured from the first point of the *hsiu* in which it lies; and (*b*) the north polar distance of the star. The first of these corresponds to right ascension. To take a concrete case, it is said[b] of the constellation Tung hsien[1] (Eastern harmony)[c] that its southernmost star is 2° forwards from the beginning of the *hsiu* Hsin (*ju Hsin erh tu*[2]). The second of these is the complement of declination. In the same case, it is said that from the star to the north pole the distance is 103° (*chhü chi* (or *pei chhen*) *i pai san tu*[3]).[d]

Alone of these texts, the *Khai-Yuan Chan Ching* gives, in addition, (*c*) the celestial latitude. For the same example, the text adds that the star is north of the ecliptic 2° (*tsai huang tao nei erh tu*[4]). This is of much interest, and necessitates a little explanation.

Let us amplify what has already been said in the sub-section on definitions.[e] In Fig. 101 the three great types of celestial coordinates are shown. Since the Renaissance, and especially since the work of Tycho Brahe,[f] modern astronomy has made universal use of equatorial coordinates, using right ascension and declination (R.A. and δ in Fig. 101 A). It will be seen that the ancient Chinese system did this too, the north polar distance (N.P.D.) being the complement of declination, and the degrees from the first point of each *hsiu* (marked as cross-lines on the equator in Fig. 101 A) being equivalent to the right ascension. Thus both ancient Chinese and all modern astronomy agree in using the star's hour-circle and the equator as the two fundamental great circles required. On the other hand, ancient Greek and medieval European astronomers made use of the ecliptic and the circle of celestial longitude as the two fundamental great circles, expressing the star's position in terms of celestial longitude (Fig. 101 C, λ) and latitude (Fig. 101 C, β). Hence what we find in the *Khai-Yuan Chan Ching* added to the classical Chinese coordinates may be considered a piece of typically Greek astronomy which found its way to China through the compilation of Chhüthan Hsi-Ta,

[a] Besides the chief sources already mentioned (pp. 197ff. above) Ueta used certain rare works now preserved only in Japan, though both in fragmentary state. One is the *Thien Wên Yao Lu*[5] (Record of the most important Astronomical Matters) by Li Fêng[6] about +664. The other is the *Thien Ti Jui Hsiang Chih*[7] (Record of Auspicious Phenomena in the Heavens and Earth) by the monk Shou-Chen[8] about +666. Ueta's investigations were later reviewed by Nōda (*5*).

[b] *Khai-Yuan Chan Ching*, ch. 65, p. 15*b*.

[c] χ, ψ, ω Ophiuchi, and 24 Scorpii (S634).

[d] Such N.P.D. measurements occur in many places other than the *Hsing Ching* star-lists themselves; see, for example, *Chou Pei Suan Ching*, ch. 2, p. 6*a*.

[e] Pp. 178 and 180 above. [f] Cf. below, pp. 341, 378, in connection with armillary spheres.

1 東咸 2 入心二度 3 去極(北辰)一百三度 4 在黃道內二度
5 天文要錄 6 李鳳 7 天地瑞祥志 8 守眞

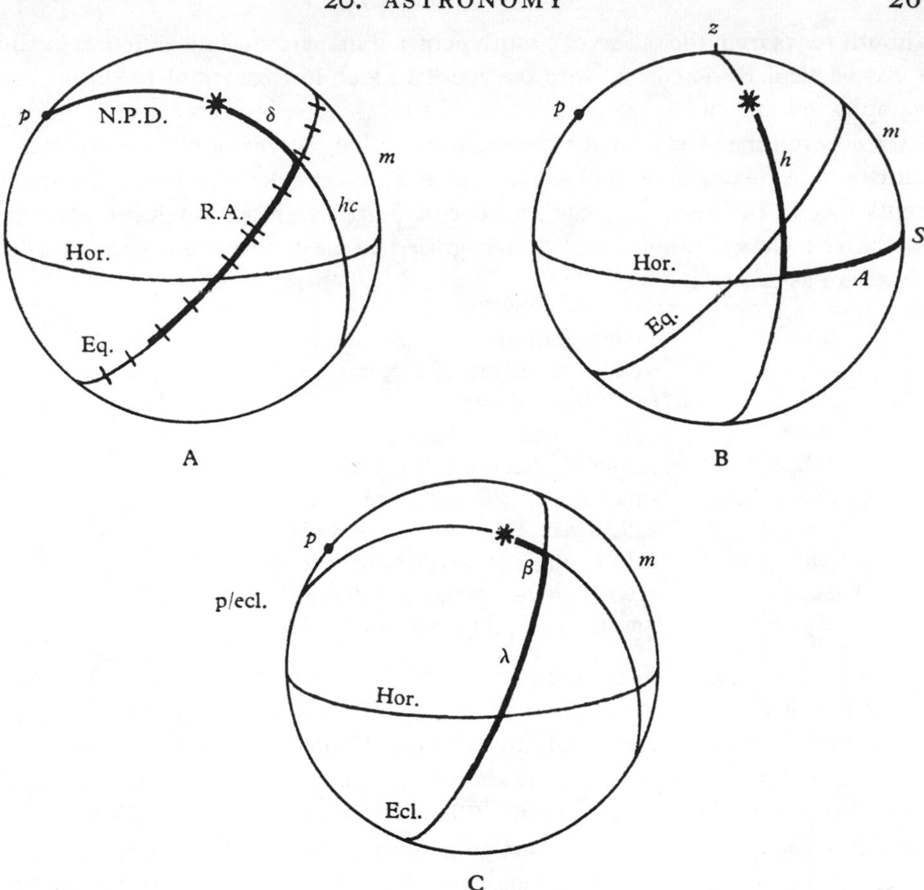

Fig. 101. Diagrams to illustrate the three main types of celestial co-ordinates (see text).

who, if not himself born in India, had close Indian connections.[a] It does not seem to have had any effect; the Chinese continued to use equatorial coordinates for their star-positions.[b] The third system, that of the azimuth (Fig. 101 B, *A*), and the altitude (Fig. 101 B, *h*), which was rather characteristic of Arabic astronomy, had the great disadvantage that it is applicable only to a specific point on the earth's surface (the

[a] Indian astronomers did not follow the Greek ecliptic coordinates (cf. E. Burgess (1), pp. 52, 202 ff.; Brennand (1), p. 42), but used a peculiar method which gave what Delambre (1), pp. 400 ff., called 'false longitude' and 'false latitude'. They measured from the ecliptic to the equatorial pole and not to the ecliptic pole (cf. Kaye (4), p. 8). This, however, was not an invention of their own. H. Vogt (1) has shown that of the 471 spherical coordinate measurements preserved from the catalogue of Hipparchus, 64 are declinations, 67 are right ascensions, while 340 are ecliptic arcs determined by the intersection of the star's hour-circle with the ecliptic. The system therefore goes back to the period when the new Babylonian zodiacal astronomy was being transmitted to the Greeks, but when the old Babylonian equatorial astronomy still retained importance (−4th to −2nd centuries). Greek influence on Indian astronomy has of course been contested (cf. Das, 2) but seems well established, and this is a striking case of it. Neugebauer (9), p. 178, gives others as well. One need not suppose, however, that all Babylonian influence was transmitted through Greek intermediation; Persian culture must also be borne in mind.

[b] Cf. p. 372 below.

azimuth starts from the observer's south point). This system is never found in China.[a] It was particularly associated with the astrolabe,[b] an instrument of which no Chinese examples are known.

We now return to Ueta's analysis of the star-catalogues. In one of his most interesting contributions he was able to elucidate the level of accuracy aimed at in the measurements made. Besides giving the number of degrees in integral figures, the texts use also other technical terms which he recognised as meaning various parts of a degree. These are as follows:

tu[1]	the degree itself.
jo[2]	'weak'; one-eighth of a degree less than the degree given.
pan[3]	'half'; half a degree.
shao[4]	'small'; a quarter of a degree.
chhiang[5]	'strong'; an eighth of a degree.
shao chhiang[6]	'small, strong'; five-sixteenths of a degree.
pan jo[7]	'half, weak'; three-eighths of a degree.
pan chhiang[8]	'half, strong'; five-eighths of a degree.
thai[9]	'great'; three-quarters of a degree.
shao jo[10]	'small, weak'; three-sixteenths of a degree.

Ueta then proceeded to an attempt to date the epochs of the various observations by the use of the north polar distances given, finding graphically the most probable position of the pole for various groups of stars. Conclusions could thus be drawn as to the date of the observations.[c] The results may be summarised as in Table 27. This shows that many of the measurements still extant in texts such as the *Hsing Ching* and the *Khai-Yuan Chan Ching* do in fact go back to the time of Shih Shen and Kan Tê, but that others must be revised measurements made at the end of the Later Han period. The conclusion is important because it answers the question whether the measurements in degrees really date from the −4th century, or whether they were inserted by later copyists. But if the Warring States astronomers did in fact express their results in this comparatively accurate form, then it seems impossible that they could have obtained them without the use of armillary spheres or rings, equipped with alidades, and hence the conclusion of Maspero (4) that these instruments did not arise until the −1st century seems in need of revision.[d]

In any case we can see that the statement of Peters & Knobel that the star-catalogue in Ptolemy's *Almagest* was, until the time of Tycho Brahe, 'practically the only source

[a] For fuller explanations of the three systems, see R. Wolf (1), vol. 1, pp. 400, 437; Woolard (1).

[b] Cf. below, p. 375.

[c] Cf. the parallel work by Gundel (3), pp. 131, 148, on the dating of the observations in the star-catalogue of Hipparchus.

[d] See below, p. 343. But final judgment is premature, for new computations have been made by Hsi Tsê-Tsung (3), who reaches a date about +125, i.e. the time of Chang Hêng in the Later Han, for the measurements of the *hsiu*-determinatives. Of course this does not disprove the view that Shih Shen and his contemporaries made some of the other measurements.

[1] 度 [2] 弱 [3] 半 [4] 少 [5] 强 [6] 少强 [7] 半弱
[8] 半强 [9] 太 [10] 小弱

Table 27. *Probable dates of observations in the star-catalogues (after Ueta)*

	Probable epoch of observations
28 Hsiu	
6 (Chio, Hsin, Fang, Chi & Chang; perhaps also Tou)	−350
17 other *hsiu*	+200
2 data for N.P.D. lacking (Khang & Shen)	—
3 aberrant (Ti, Liu & Hsing)	—
62 Northern Hemisphere	
27 stars	−350
13 stars	+180
6 stars, N.P.D. data lacking	—
4 stars, remainder uncertain	+150
30 Southern Hemisphere	
10 stars	−350
16 stars	+200
4 aberrant	—

of information about the positions of the stars which the world possessed' (the exception being the +15th-century catalogue of Ulūgh Beg),[a] is wholly unjustified, though characteristic of the majority of historians of science hitherto in their neglect of East Asian contributions.[b] It is of much interest to compare the Chinese star-catalogues with those of Hipparchus and Ptolemy;[c] one finds that the latter were not only later (the catalogue of Hipparchus dates from −134), but also one-third smaller, since the total number of stars whose positions are given in the *Almagest* (in ecliptic coordinates) is 1028.[d] Now the celestial sphere contains about 41,000 'square degrees' and about 208,000 spaces corresponding to that occupied by the moon. The largest number of stars mentioned by any ancient astronomer is 14,020, given by Chang Hêng in the passage quoted on p. 265 above. This was almost one for every square degree, or one for every 5 'moon spaces'. It is interesting to note that the limit of magnitude visible to the average eye is somewhere between 6 and 7, so that no more than 14,300 and probably not many more than 10,000 are really actually to be seen by unaided sight.[e]

[a] See Knobel (3). Besides, there are many Arabic catalogues, based on original observations, dating from as far back as the +9th century; see Destombes (2, 3, 4).

[b] Misled by a remark of Gaubil's (2, p. 44; 3, p. 148) concerning the *Chin Shu*, Sédillot (2, pp. 587, 617) affirmed that no Chinese lists of star-coordinates existed before +1050.

[c] The exact relations between the catalogues of Hipparchus and Ptolemy have been the subject of dispute among historians of astronomy, some asserting that the latter simply reproduced Hipparchus' data with the addition of a correction for precession (e.g. Berry (1), p. 68), and others considering that Ptolemy did make personal observations (Dreyer, 1).

[d] According to Boll (2) the catalogue of Hipparchus contained only 850 stars. As late as +1602 Europeans were still worrying whether there could really be more than 1022 (Thorndike, 6).

[e] Spencer-Jones (1), p. 293.

There can be no doubt that the work of the Chou and Han Chinese in positional astronomy deserves a much more important place in the general history of science than it has so far been accorded.

It also seems worth while to lay emphasis on the fact that the system of celestial coordinates used throughout the modern world is essentially Chinese and not Greek—later, in dealing with astronomical instruments, we shall have occasion to examine what it was which led European astronomers in Tycho Brahe's time to abandon the Greek ecliptic coordinates, and to adopt equatorial ones. It may be, indeed, that the faithfulness of the Chinese to equatorial coordinates was not without its disadvantages; thus it is possible that the late discovery of the precession of the equinoxes in China was really due to the star coordinates used. In Greece, the discovery was made by Hipparchus in the −2nd century, i.e. at the same time as the construction of the first star-catalogue of Europe. In comparing his observations of certain stars with those which Timocharis and Aristyllus had made about 150 years earlier, Hipparchus found that their distances from the equinoctial points had changed. If he had been measuring their position solely on the equator this change might not have been so obvious.[a] Indeed, it was specifically masked by the fact that the Chinese equivalent of right ascension was the position within an individual *hsiu* and not the distance from the equinoctial node or first point of Aries. Only the slow movement of the solstitial and equinoctial points themselves brought Yü Hsi to his parallel discovery.

But when the Chinese astronomers did begin to pay more attention to positions on and near the ecliptic they found what looked like something new and interesting. In 1718 Edmund Halley discovered (1) that the 'fixed' stars have motions of their own.[b] Comparing the positions of Aldebaran, Arcturus and Sirius, as measured by Ptolemy and by himself, he found that they seemed to be advancing slowly southwards, an effect which could evidently have nothing to do with the apparent east–west shift due to precession. This 'proper motion', as it is called, has now been verified for thousands of stars, though even at its largest it is some ten to fifty times less than the annual 50·26″ of precessional movement. The fastest star known is Munich 15,040 (Barnard's Star) with a yearly motion of 10·25″, and there are about a hundred stars the displacement of which is more than 1″ per annum.[c] It has now been questioned whether Halley was not anticipated by the monk I-Hsing just about a thousand years earlier.[d] As explained elsewhere, I-Hsing had access to Hellenistic-Indian astronomy,[e] and with Liang Ling-Tsan constructed about +725 armillary spheres with ecliptically mounted sighting-tubes.[f] The Thang Histories, after giving the details of these, recount at some length[g] the observations of star positions then made, stating how they differed from the values in the ancient catalogues and maps (*chiu thu*[1]).[h] In more than

[a] I owe this point to Prof. F. J. M. Stratton. Cf. p. 200 above and p. 356 below.
[b] Cf. Grant (1), p. 554; Spencer-Jones (1), pp. 297ff.; O. Thomas (1), pp. 448ff.
[c] Becker (1), p. 225; Spencer-Jones (1), p. 299.
[d] The matter has been discussed most recently by Chhen Tsun-Kuei (5) and Hsi Tsê-Tsung (3).
[e] Cf. p. 202 above.
[f] Cf. p. 350 below.
[g] *Chiu Thang Shu*, ch. 35, pp. 4ff., esp. p. 5*b*; *Hsin Thang Shu*, ch. 31, pp. 3*b*ff., esp. p. 4*b*.
[h] Certainly the *Hsing Ching* and similar texts. The existence of star-charts is undoubtedly implied.

[1] 舊圖

ten of these cases I-Hsing found a north–south movement relative to the ecliptic just as Halley did afterwards. For example, an asterism in Taurus which had been 4° south of the ecliptic[a] in ancient times was now found to be directly on it.[b] I-Hsing himself wrote:[c] 'Thus the distances of (some of) the determinative and other calendar stars (from the ecliptic) differ from those measured by Lohsia Hung and others, while the 28 *hsiu* have remained the same.' The differences given in the text are, however, much too large, even in one instance 5°, but it is possible that I-Hsing's values were garbled by the scribes of the Bureau of Historiography who transmitted them to us, minutes being changed into degrees.[d] The Thang astronomers were at a disadvantage for only 850 years had elapsed since the time of Lohsia Hung, and 1100 since that of Shih Shen, while 1500 years separated Halley from Ptolemy. Still, it is just conceivable that I-Hsing observed a true effect, and (more important perhaps) it is clear that his mind was entirely open[e] to the possibility of such manifold stellar motions.[f]

(2) Star Nomenclature

The next question which arises is to what extent there was any similarity between Chinese and European recognition of asterisms and constellations. As will appear, the answer is that there was very little. The number of cases in which any parallelism of symbolic nomenclature can be made out is remarkably small, and the same groups of stars were not seen in the same patterns. Frequently a single European constellation appears on the Chinese planisphere as several different asterisms—for example, Hydra comprises the three *hsiu* Chang, Hsing and Liu, together with eight other star-groups having no similarity of symbolism with the European figure. From Houzeau[g] we may

[a] Chinese degrees.

[b] It is worth while to give the details of a few of these observations. Thus Chien hsing (S161), 7 stars in Sagittarius, had anciently been ½° N. of the ecliptic, but was now 4½° N. Thien kuan (S496), 2 stars in Taurus, had anciently been 4° S. of the ecliptic, but was now dead on it. Thien tsun (S601), 3 stars in Gemini, had anciently been a little to the N. of the ecliptic but was now on it. Hsü liang (S100), 4 stars in Aquarius, had anciently been just south of the ecliptic, but was now 4° N. of it. Finally, Chhang yuan (S376), 4 stars in Leo, had anciently been exactly on the ecliptic, but was now 5° N. of it. I-Hsing gave all his results in terms of asterisms, but the reference was probably to the determinative star of each.

[c] In his *Ta Yen Li Shu*, reproduced in *Hsin Thang Shu*, ch. 27A, p. 16*a*.

[d] Dr Arthur Beer has very kindly assembled the proper motions of the twenty stars mentioned above from the standard tables. Although most of them do not exceed 0·05″ per year, there are three (κ Aquarii, k Leonis and XI. 12 Leonis) which have proper motions well above the average: of the order of 0·137″ per year. In 1100 years this would give 2·5′. It is unlikely that I-Hsing could do as well as Tycho's (supposed) accuracy of 1′ of arc, but he may have been able to detect 2 or 3; medieval Chinese astronomers often gave fractions of $\frac{1}{16}$° or under 4′ of arc. Thus the possibility that he really saw some movement cannot quite be excluded. One notes the curious fact that of the nearest stars which seem to be moving very fast (see the list in Allen (1), pp. 205ff.) none appears in our examples, nor yet the brightest (Allen (1), pp. 209ff.), some of which have marked proper motions. Most of the stars in I-Hsing's sample above are faint ones, and about half are on the limit of visibility (mag. 5·5), though none would have been really impossible to see with naked eye and sighting-tube. We are much indebted to Dr Beer for his help in this.

[e] In contrast with the contemporary occidental belief in solid celestial spheres, which did not permit of random proper motions.

[f] That is to say, he accepted the reliability of the measurements of his far-distant predecessors and did not simply believe that he was correcting their values with up-to-date methods.

[g] (1), p. 820.

Table 28. *Recognition of constellations in the West* (*after Houzeau*)

Constellations (or rather, spherical quadrilateral zones or fields) recognised internationally in modern astronomy (Delporte, 1, 2)		Number of constellations
+2nd century, Ptolemy (138): zodiacal	12	
other[a]	39	
	51	51
+5th century, Proclus (Berenice's Hair)		1
+17th century, Keyser & Bayer (1603)	11	
Plancius (1605)	1	
Bartsch (1624)	2	
Habrecht (1628)	1	
Royer (1679)	1	
Hevelius (1690)	7	
	23	23
+18th century, La Caille (1752)		14
		89

make a table showing how the constellations grew up in the West (Table 28). We are now in a position to compare this list with the corresponding parts of the sky in the Chinese planisphere. From Table 29 it will be seen that only three zodiacal and seven extra-zodiacal constellations show any similarity of symbolism as between Chinese and European nomenclature. The former are Capricorn (Niu[1]); Leo (Hsüan yuan;[2] Dragon backbone,[b] connected with water-raising chain-pumps); and Scorpio (Fang,[3] Hsin[4] and Wei[5]). The latter are Auriga (Wu chhê,[6] the Five chariots[c]); Bootes, formerly of warlike significance (Hsüan ko,[7] the Sombre axe,[d] and nearby stars with military names); Canis Major (Thien lang,[8] Sirius[e]); Corona Australis (Pi,[9] a tortoise;[f] —here the similarity is only in the recognition of a circle of stars); Corona Borealis (Kuan so,[10] a coiled thong[g]—and the same remark applies); Orion (Shen,[11] with parallel human figure symbolism); Ursa Major (Pei tou,[12] the Northern dipper). The list is not at all impressive, and strongly suggests that the nomenclature of the Chinese constellations grew up in almost complete independence of the West. Minakata (1) made a good point when he said that one of the most striking characteristics missing from the

[a] Including the three into which Navis was split up, and the two Serpens instead of the original single one.

[b] S 93.

[c] S 716.

[d] S 96.

[e] S 509. Speaking of I-Hsing's determination of the coordinates of this star, Gaubil says (2, p. 86), 'Il faut avoüer que ce n'est pas un petit éloge pour *Y-hang* d'avoir l'an 725 de Jésus-Christ à la Chine mieux observé la latitude de *Sirius*, que les Astronomes des autres pays ses contemporains, et même que ceux qui lui furent postérieurs de plusieurs siécles.'

[f] S 284.

[g] S 188.

1 牛 2 軒轅 3 房 4 心 5 尾 6 五車 7 玄戈
8 天狼 9 鼈 10 貫索 11 参 12 北斗

Chinese skies is that of maritime names, for there is nothing corresponding to Cetus, Delphinus, Cancer, etc. And, on the other hand, the overwhelmingly agrarian-bureaucratic nature of ancient Chinese civilisation led to a multitude of star names in which the hierarchy of earthly officials found their counterparts.

Schlegel sought for as many correspondences as possible between the Chinese and the European nomenclatures because he was anxious to prove that Chinese astronomy was the origin of all astronomy. Most of his arguments, however, seem very far-fetched. Thus, for example, in considering the *hsiu* Chi[1] (Winnowing-basket)[a] he unearthed the fact that in some ancient texts this word signified a kind of wood used for making quivers—which suited him very well, since the stars concerned include γ, δ and ϵ Sagittarii. So also he found, near Virgo, the star with the Chinese name Thien ju[2] (Celestial milk, or nurse[b]), but it is in fact ω Serpentis[c] and separated from Virgo by Libra.[d] Sometimes the parallels are a little more striking, as in the case of the numerous stars with military significance (generals, army, beheaded prisoners, etc.) in the main field of Aquarius,[e] and the neighbouring Thien lei chhêng[3] (Celestial ramparts)[f] in Microscopium. Here there are distinct parallels with the Egyptian decan stars seen, for instance, on the famous planisphere of Denderah, as, indeed, de Paravey had pointed out before Schlegel. Another Egyptian–Chinese rapprochement is found in Aries, where in the *Hsing Ching* the 'Celestial prison' (Thien yü[4])[g] is given as a synonym for Lou[5] *hsiu*, and on Egyptian planispheres a man in chains is seen in this place.[h] Perhaps a re-examination of these matters in the light of the more accurate and extended knowledge now at our disposal, and with the collaboration of egyptologists, assyriologists and iranists, would justify some of Schlegel's comparisons; but the impression of one reader, at any rate, has been that his chapter on this subject is too full of special pleading to convince, especially when the role of probable coincidence is given due allowance. It seems safer to conclude, with Edkins (7), that on the whole the Chinese nomenclature of the constellations represents a system which grew up in comparative isolation and independence.

Such, too, was the mature conclusion of Bezold (1),[i] who pointed out that it does not exclude that transmission of a body of Babylonian astrological lore to China before the −6th century, which, as we saw above,[j] seems rather probable. Nor would it militate against the belief that certain basic ideas were transmitted about a thousand years earlier, e.g. the planispheric 'roads' which led to the system of the *hsiu*, the use

a (5), p. 661.
b S 453.
c Chhen Tsun-Kuei (*3*) gives 32 μ Serpentis.
d (5), p. 655.
e (5), p. 667.
f S 516.
g S 614.
h (5), p. 672. Another correspondence, suggested to us by Mr D. H. Kelley, might be seen in the names of Algol (β Persei). This star is called in Chinese Chi shih[6] (Heaps of corpses) and is surrounded by other stars in Perseus forming a constellation called Ta ling[7] (common grave); S 350, 650. Algol is of course al-gūl, a demon of death, but it was only what the Arabs made of the gorgon's head (see Ideler (2), p. 88).
i Yetts (5), p. 135, and Chao Yuan-Jen (*1*) agree.
j Sect. 14*a*, Vol. 2, p. 353.

1 箕 2 天乳 3 天壘城 4 天獄 5 婁 6 積尸
7 大陵

of the gnomon, the recognition of the position of the pole and the equinoctial points, and so on.

The last two lines in Table 29 record constellations which were not recognised and named by Europeans until the 17th and 18th centuries. These groups are all in the southern hemisphere and include the southern circumpolar stars. It has often been supposed that they were not known to the Chinese until the transmission of astronomical discoveries from Europe by the Jesuits.[a] Records exist, however, of the growth of Chinese knowledge about the stars and constellations of the southern hemisphere,[b] as witness, for example, the following passage in the *Chiu Thang Shu*:[c]

> In the 8th month of the twelfth year of the Khai-Yuan reign-period (+724) (an expedition was sent to the) south seas to observe Canopus (Lao jen[1]) at high altitudes and all the stars still further south, which, though large, brilliant and numerous, had never in former times been named and charted. These were all observed to about 20° from the south (celestial) pole. This is the region which the astronomers of old considered was always hidden and invisible below the horizon.

It is sad that the results of these investigations failed to survive, but the expedition remains a remarkable one for its time, indeed, unique in the early Middle Ages.[d] In order to observe stars as far south as α Trianguli Australis, which has an S.P.D. of about 21·5°, at altitudes higher than 15°, the party must have gone further south than Malaya, perhaps to somewhere near the southern tip of Sumatra (about 5° S. lat.). Sailing thence south another 10° the astronomers would have seen this star as high as 35° and Canopus at 52° altitude, but perhaps it is more likely that they made their observations from land as far south as they could get among the islands.[e]

It would be interesting to investigate the ancient and medieval descriptions of the stars in the light of modern astronomical knowledge with a view to determining their degree of exactness and penetration. For example, we know now that the Pleiades star-group is really an 'open cluster' of stars all moving with the same speed in the same direction,[f] and its brightness is largely due to the 1400 small stars of which it is composed, apart from the six or seven normally visible as pin-points of light to the naked eye. Hence the interest of Ssuma Chhien's description[g] of the Pleiades (Mao[2]) as Mao thou[3]—the hairy head,[h]—to which he adds the words 'the white-robed assembly' (*pai i hui*[4]). Chavannes[i] interpreted this astrologically as meaning that the

[a] So Schlegel (5), p. 553.

[b] Cf. further below, in Sect. 29*f* on navigation.

[c] Ch. 35, pp. 6*a*, *b*; tr. auct. Cf. *Thang Hui Yao*, ch. 42 (p. 755); *TCKM*, ch. 43, p. 51*b*; *TH*, p. 1407.

[d] It was organised by the two men, I-Hsing and the Astronomer-Royal Nankung Yüeh, who were also responsible for setting up a chain of nine stations along a meridian nearly 4000 km. long in order to measure solstice shadow lengths in different latitudes; on this see p. 293 below.

[e] The reading given suggests that they took with them armillary instruments, better used on land than at sea. The south coast of Java is quite a probable location for their work. In the +7th century the Chinese were not very familiar with Java and Sumatra, but by the +8th these countries were well-known way-stations on the routes between China and India (cf. Hirth & Rockhill (1), pp. 8ff.).

[f] Spencer-Jones (1), p. 376.

[g] *Shih Chi*, ch. 27, p. 11*a*.

[h] Cf. our Coma Berenices.

[i] (1), vol. 3, p. 351.

[1] 老人 [2] 昴 [3] 髦頭 [4] 白衣會

Table 29. *Relation between occidental constellations and Chinese star-groups*

	Occidental constellations, total no.	No. containing *hsiu*	No. containing only Chinese stars and star-groups other than *hsiu*	No. of *hsiu* contained	No. of other important Chinese stars and star-groups contained	Symbolic similarity of constellation names: none	very doubtful	doubtful	admissible if adjacent constellation fields allowed	positive
+2nd century, Ptolemaic:										
zodiacal	12	11	1	17½	146	5	1	3	0	3
extra-zodiacal	36	7	29	10½	290	21	1	3	4	7
				28						
+5th century	1	0	1	0	6	1	—	—	—	—
					442					
+17th century	23	0	23	0	50	23	—	—	—	—
+18th century	14	0	7	0	7	14	—	—	—	—

Pleiades preside over funerals, but Chu Wên-Hsin[a] conjectures that the phrase was a term, and a very appropriate one, for this star cluster.

Probably some of the old Chinese descriptions of the heavens contain mentions of nebulae also.[b] The Magellanic Clouds, which are the nearest of the extra-galactic nebulae, are quite visible to the naked eye as two detached portions of the southern Milky Way.

(3) Star Maps

Although there is no doubt that charts of the heavens were being constructed in China as early as the +3rd century,[c] and probably also in the Han,[d] none has come down to us from those times. We know, however, from Han carvings and reliefs, that the system of representing asterisms by patterns of dots or circles connected by lines goes back at least as far as that period. Fig. 102 (Pl. XXVI) shows a Han tomb-carving of the Weaving Girl (Chih nü;[1] Vega[e]) with a picture of a loom and weaver beside the star-group in question. In later times, this manner of depicting stars became associated, because of astrological connotations, with Taoism, and hence it is a frequent sight to see flags marked with various constellations hung out from Taoist temples; Fig. 103 shows such a flag photographed by the writer near Chungking, Szechuan, in 1946.[f] The *Shih Wu Chi Yuan*[g] describes an incident relating to +963 in which such flags figure prominently.

The MS. star-map from Tunhuang already described (p. 264) is a precious possession. Dating from about +940, it is almost certainly the oldest extant star-chart from any civilisation, if we exclude of course the highly stylised carvings and fresco-paintings of antiquity (e.g. Fig. 90).

[a] (5), p. 42.

[b] Perhaps we may so interpret the *yün chhi*[2] of *Chin Shu*, ch. 12, p. 7*b*. Strangely, the word 'spiral' (*chhün*[3]) is used of one of them, though the spiral nature of nebulae could not have been known by naked-eye observation.

[c] What is known about them has already been described (p. 264 above). Cf. p. 387 below, regarding celestial globes. The key date is +310 when Chhen Cho was doing his compiling work.

[d] The *Chhien Han Shu* bibliography lists a *Yüeh Hsing Po Thu*[4] (Silken Map of the Path of the Moon) which Kêng Shou-Chhang[5] presented to the emperor in −52. His *Yüeh Hsing Thu*[6] seems to have been an abridgement of it. The astronomical section of the bibliography also mentions a *Thu Shu Pi Chi*[7] of uncertain date. The *Hsü Han Shu* (quoted by Juan Yuan, *Chhou Jen Chuan*, ch. 3, p. 31) records a discussion in +92 by Yao Chhung,[8] Ching Pi[9] and others, in which it is said that star-maps always have methods of graduation, i.e. coordinates (*hsing thu yu kuei fa*[10]). About +100 Chang Hêng[11] produced a *Ling Hsien Thu*[12] (Map of the Spiritual Constitution of the Universe), which was probably a star-map (recorded in *Chiu Thang Shu*, ch. 47, p. 5*b*). Another example is the *Tzu Ko Thu*[13] (Chart of the Purple Palace) mentioned by Wang Mang in +19 (*Chhien Han Shu*, ch. 99c, p. 4*b*).

[e] S396.

[f] As a tentative identification of the asterism on the flag I suggest Thien chhün[14] (the Celestial granary), S491, which consists of numerous stars in Cetus. The Chinese ball-and-link convention has had some curious echoes; see for instance a map of the heavens due to the Kabbalist Jacques Gaffarel (+1601 to +1681; Richelieu's librarian) in *Curiosités Innouies*, 1637. It is reproduced in Seligmann (1), fig. 143, and was considered to be of 'oriental' origin.

[g] Ch. 3, p. 3*a*.

[1] 織女 [2] 雲氣 [3] 囷 [4] 月行帛圖 [5] 耿壽昌
[6] 月行圖 [7] 圖書祕記 [8] 姚崇 [9] 井畢 [10] 星圖有規法
[11] 張衡 [12] 靈憲圖 [13] 紫閣圖 [14] 天囷

Fig. 102. Han stone carving showing, on the left, the Weaving Girl constellation (Chih nü), with the stars α (Vega), η and γ Lyrae above her head; in the centre the sun, marked with its symbolical crow; and on the right, another constellation, probably one of the 'bird' *hsiu*, Fang, Hsin or Wei. All the asterisms are represented in the standard 'ball-and-link' convention. The whole composition would thus depict the sun at an hour-angle of about 260°, or $17\frac{1}{2}$ hr. From Bushell (1), vol. 1.

Fig. 103. Constellations on a Taoist flag, outside the Lao Chün Tung temple on the South Bank, Chungking (orig. photo. 1946); the main asterism depicted is probably Thien chhün in Cetus.

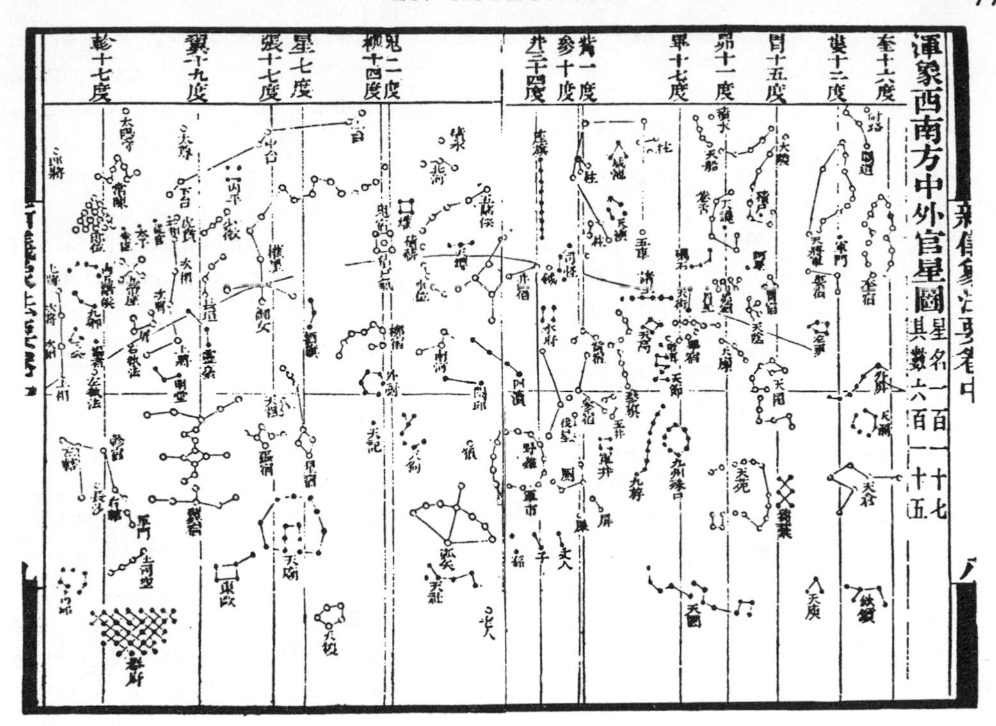

Fig. 104. The star-maps for the celestial globe in the *Hsin I Hsiang Fa Yao* of +1092; fourteen *hsiu* on 'Mercator's' projection. The equator will be recognised as the horizontal straight line; the ecliptic curves above it. Note the unequal breadth of the *hsiu* (cf. p. 239).

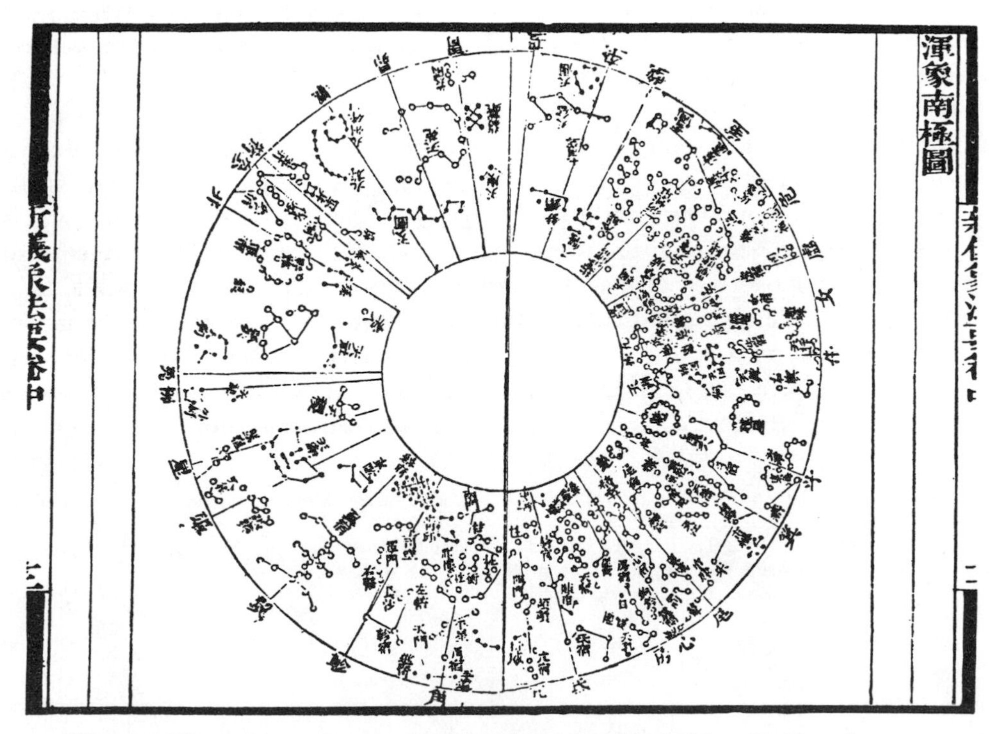

Fig. 105. The star-maps for the celestial globe in the *Hsin I Hsiang Fa Yao* of +1092; south polar projection.

If the maps incorporated in the *Hsin I Hsiang Fa Yao* (New Description of an Armillary Clock) by Su Sung[1] really date from the time when the book was written, they must be the oldest printed Chinese star-charts which we possess.[a] Begun in +1088, it was finished in +1094, and the maps which it contains are remarkable in several ways. They are five in number, one of the north polar region, two cylindrical orthomorphic 'Mercator' projections of the regions of declination about 50° N. and 60° S. (Fig. 104), and two polar projections, one of the north hemisphere and one of the south (Fig. 105). The space where the southern circumpolars should be is left blank. The drawing of the stars is much more carefully done than on the MS. map of rather more than a century earlier, but this also had the 'Mercator' projection.

Close examination reveals that while the first of these maps accepts the pole-star of −350 (Thien shu[2]) as the position of the true pole, the fourth places this position half-way between Thien shu and our Polaris (Thien huang ta ti[3]). This would suggest that Su Sung must have taken advantage of the observations of his contemporary Shen Kua, already referred to (p. 262 above). Moreover, the equinoctial points are located almost at Khuei[4] *hsiu*, and nearer Chio[5] than Chen,[6] positions which, according to precession, would be about right for +850, in the Thang. Su Sung was therefore distinctly up to date. But the later tradition of Chinese celestial cartography seems to have drawn from sources of an earlier time and to have perpetuated their observations, as we shall see in a moment.

The most famous of Chinese planispheres is that which was prepared in +1193 for the instruction of the young man who was to reign as Ning Tsung[7] from +1195 to +1224. The map was engraved on stone in +1247, and still exists as a stele in the Confucian temple at Suchow, Chiangsu. The inscription which accompanies it has been translated into French by Chavannes (8) and into English by Rufus & Tien (1). It can be recommended as one of the shortest (and most authentic) expositions of the Chinese astronomical system. After an introduction drawing upon Neo-Confucian philosophy,[b] the text describes the celestial sphere, with its 'red' and 'yellow' roads (equator and ecliptic). 'The Red Road', it says, 'encircles the heart of Heaven, and is used to record the degrees of the twenty-eight *hsiu*'; if this forthright statement had been known to modern scholars a great deal of 19th-century controversy would have been avoided. The text then mentions the 'white road', i.e. the path of the moon, crossing the ecliptic at an angle of 6°; and gives a correct account of lunar and solar eclipses. It is noted that there are 1565 named fixed stars.[c] The planetary portion is astrological, and the text ends with correlations between regions of the sky and the

[a] There is indeed no ground for doubting this; cf. the study of the transmission of the book in Needham, Wang & Price (1).

[b] Including the centrifugal cosmogony, already discussed above, Sect. 14*d*, Vol. 2, pp. 371 ff.; and an identification of the principle of organisation in Nature with that in man.

[c] Though only 1440 are shown on the chart.

1 蘇頌 2 天樞 3 天皇大帝 4 奎 5 角 6 軫
7 寧宗

Chinese cities and provinces supposed to be affected by celestial phenomena therein.[a] An interesting section refers to the role of the Great Bear as a seasonal indicator, and shows that the ancient system keying the circumpolars to the *hsiu* (cf. p. 233 above) had not been forgotten.[b]

The planisphere itself is reproduced in Fig. 106. Like all others of its kind, it shows the hour-circles of the determinative star of each *hsiu* as radiating straight lines. Differences in the representation of the asterisms from those adopted in other Chinese celestial maps are discussed by Rufus & Tien.[c]

Next requiring remark is the Korean star-chart prepared and engraved in +1395 at the order of Yi Tai Jo (Li Thai Tsu[1]), founder of the last Korean dynasty. According to the descriptions (Rufus (1, 2); Rufus & Chao, 1), it was based on an inscribed stele map dating from +672, and doubtless represents a somewhat older tradition than that of Su Sung. It is reproduced in Fig. 107. The galactic belt is given undue prominence, and there is a slight attempt to indicate magnitudes by variations in the size of the dots. The names of a number of astronomers associated with the making of this chart have come down to us.[d] Rufus (1) has given a translation of the accompanying inscription, which includes a review of the ancient Chinese cosmological theories (cf. p. 210 above) and tables giving the stars culminating at dawn and dusk for the twenty-four divisions of the year (cf. p. 405 below).

A third planisphere, of a rather different character, has been described by Knobel. It is a bronze bowl, 13½ in. in diameter, of provenance unknown save that it was taken from a Japanese junk probably at some time in the early 19th century. Depressions in the rim held two small compass needles, and the stars are represented by small raised dots on the floor of the bowl (Fig. 108).[e] It seems hard to assign a date to the chart, which any oriental navigator would have been glad to possess, but though not likely to be earlier than the +17th century, it is of fully traditional Chinese type.

[a] Rufus (3) believes that many of the asterisms in the Suchow planisphere were deliberately 'idealised' to suit the political theory contained in the attached text; I agree, however, with the comments of Chatley, who points out that the theory was as old as the Han, and would attribute inaccuracies rather to copyists' errors as well as to the difficulties of planispheric plotting without stereographic geometry.

[b] The passage is interesting since it shows that a line along the tail of the Bear was prolonged in both directions, and a line drawn perpendicular to it through γ Ursae Majoris, forming three divisions sweeping out six-hour intervals. It has been much discussed (Chalmers (1); de Saussure (11), (1), p. 391; Chavannes (8), p. 53; Rufus & Tien (1), p. 6) because the lines are said to point to each of the duodenary cyclical characters successively during the course of the year. These were in fact used not only as cardinal points of the horizon but analogously as stationary divisions of the equator, north and south being fixed by the meridian. Such a graduation can be seen round the rim of the planisphere. Cf. p. 240.

[c] Who also give a list of identifications of the stars.

[d] Kwon Keun (Chhüan Chin[2]), Ryu Pang-Taik (Liu Fang-Tsê[3]), Kwon Chung-Wha (Chhüan Chung-Ho[4]), Choi Yung (Tshui Jung[5]), No Eul-Chun (Lu I-Chün[6]), Yün In-Yong (Yin Jen-Lung[7]), Chi Sin-Won (Chhih Chhen-Yuan[8]), Kim Toi (Chin Tui[9]), Chün Yün-Kwon (Thien Jun-Chhüan[10]), Kim Cha-Yü (Chin Tzu-Sui[11]) and Kim Hu (Chin Hou[12]). We shall meet with Chhüan Chin again later on (p. 554).

[e] The specimen is in the possession of the Royal Scottish Museum.

1 李太祖 2 權近 3 柳方澤 4 權仲和 5 崔融
6 盧乙俊 7 尹仁龍 8 池臣源 9 金堆 10 田潤權
11 金自綏 12 金侯

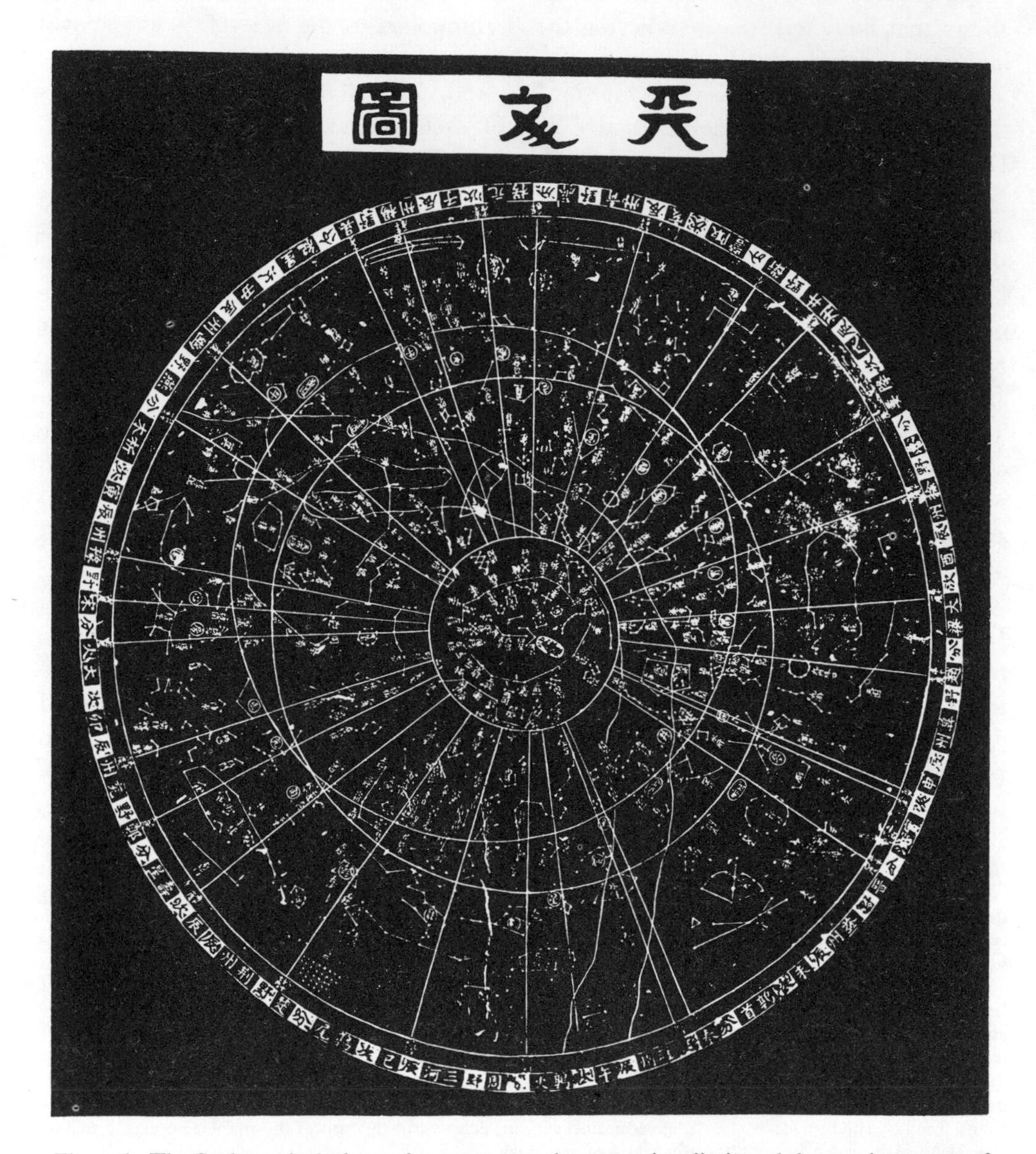

Fig. 106. The Suchow planisphere of +1193; note the excentric ecliptic and the curving course of the Milky Way (from Rufus & Tien). The planisphere with its explanatory text was prepared by the geographer and imperial tutor Huang Shang, and committed to stone by Wang Chih-Yuan in +1247.

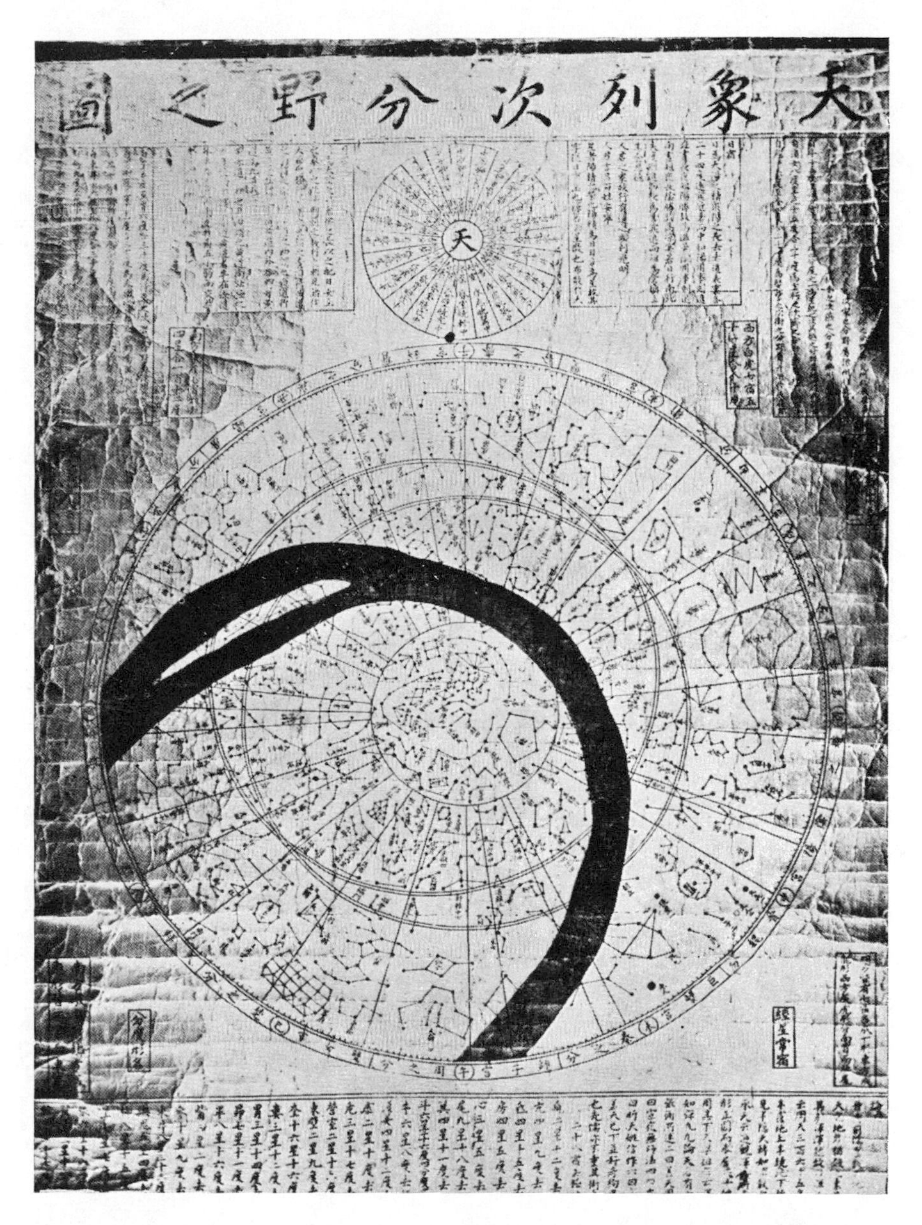

Fig. 107. Korean planisphere of +1395; *Thien Hsiang Lieh Tzhu Fên Yeh chih Thu* (Positions of the Heavenly Bodies and the Regions they govern) (from Rufus & Chao). This was based on an engraved stele of +672 and prepared by ten astronomers headed by Kwon Keun (Chhüan Chin); cf. Fig. 106. The Milky Way is represented by a black band.

Fig. 108. Planisphere in the form of a bronze bowl, used by navigators (photo. Knobel).

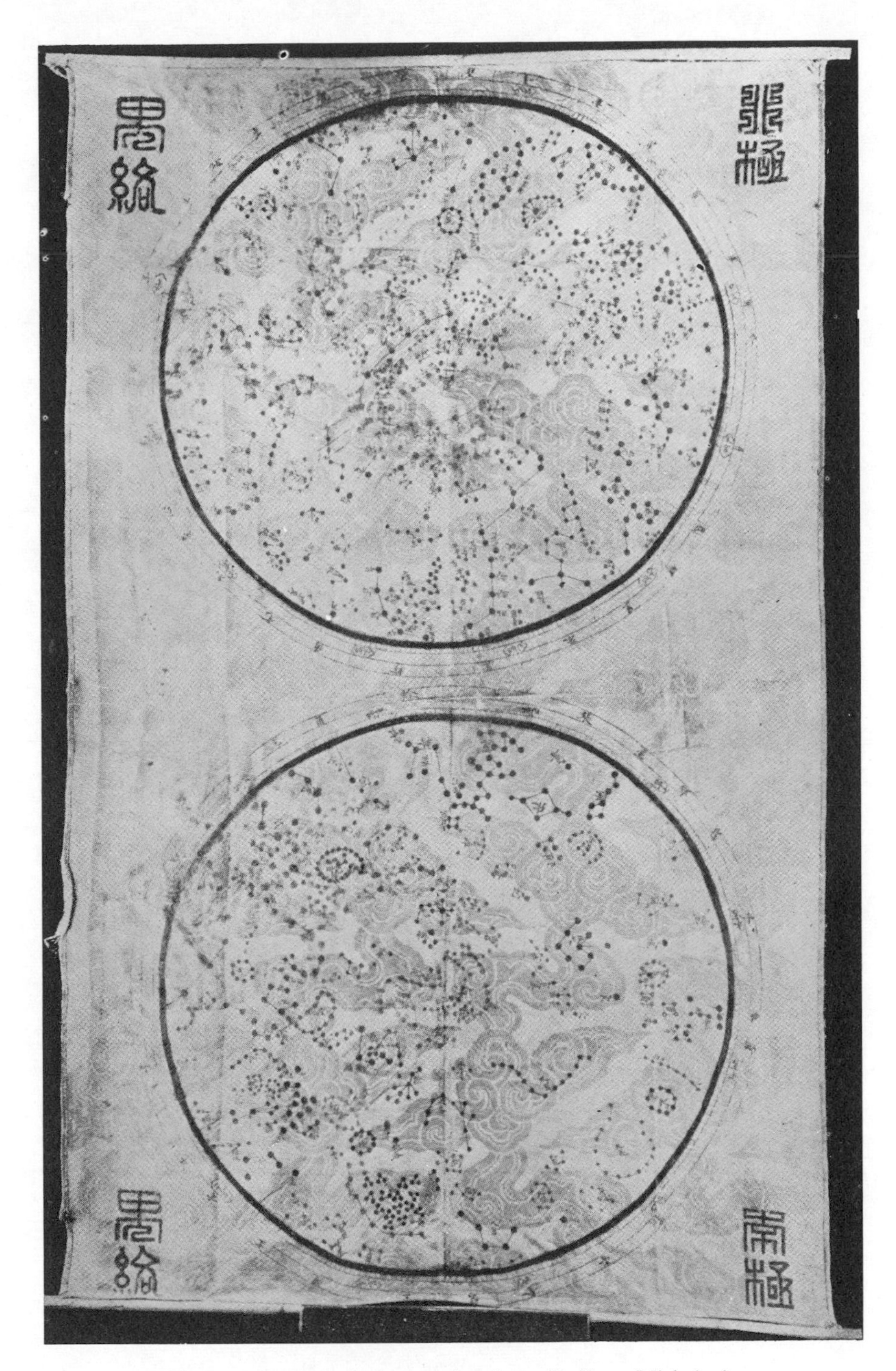

Fig. 109. Late planispheres painted on silk (from Michel, 2).

Later planispheres in MS. are not uncommon; one painted on silk has been figured by Michel (2) (Fig. 109). This must be of the Chhing, since it shows fully the south polar constellations.

The four planispheres just described all agree in maintaining Thien shu[1] as the pole-star, i.e. in perpetuating the system of −350, though the two former ones both mark Polaris (Thien huang ta ti[2]) by its imperial name. They further differ from Su Sung's best map in putting the equinoctial points between Chio[3] and Khang[4] on the one side, and between Khuei[5] and Lou[6] (and nearer Khuei than Lou) on the other. These would correspond (as Rufus noticed) with a date somewhere around +200, and may suggest, therefore, that the tradition from which Su Sung departed, and which these other maps retained, was in fact the situation fixed in Chhen Cho's map of approximately +300.[a] It seems tempting to connect this retrograde step with contemporary social and political trends. Su Sung and Shen Kua worked at a time when the reform movement associated with the name of the great minister Wang An-shih was in its prime.[b] The Suchow planisphere of a century later was made when conservative traditionalism had once again got the upper hand.

In the Sung period star-charts must have been fairly common, and it would be interesting to know what positions they showed. Ma Yung-Chhing,[7] in his *Lan Chen Tzu*[8] (Book of the Truth-through-Indolence Master), mentions[c] that about +1115 he often used to discuss astronomy with the monks of certain temples, who had, or made, celestial maps. In the +10th century there was a *Lieh Hsing Thu*[9] (Map of the Principal Stars) by some compiler now forgotten.[d] In the +12th Chêng Chhiao complained that the available printed star-charts were generally not to be relied upon and hard to correct. He advised chanting over portions of the *Pu Thien Ko* on clear nights to familiarise oneself with the celestial patterns.[e]

What is known of astronomical map-making in other parts of the world, however, suggests that we must not undervalue this whole tradition of Chinese star-charts from the Han time down to the Yuan and Ming, of which the MS. of about +940 is our oldest extant specimen. Thiele (1), B. Brown, and Sarton, in his *Introduction*, can offer no star-charts other than Chinese throughout the Middle Ages down to the end of the +14th century. Before that there were only the crude Egyptian representations and

[a] In the *Thu Shu Chi Chhêng* encyclopaedia of +1726, the astronomical parts of which were prepared under strong Jesuit influence, the position seems rather confused. Thien shu is shown as one of the stars of the Ssu fu group (cf. above, p. 261), and it is said to be also a name for the true pole. Polaris is not stated to be the pole-star (*Chhien hsiang tien*, ch. 44). It is central in the cubical Pardies polar projections made by Philippe Grimaldi in China in +1711. Those of the mid +18th century which illustrate the star-catalogue in which Ignatius Kögler participated: are quite modern in character, and Polaris takes its right place in them (see below, p. 454); Tsuchihashi & Chevalier (1); Rigge (1).

[b] It happens that we know both of them to have been associated personally rather with the Conservatives than the Reformers (see Needham, Wang & Price). But that did not prevent them from being progressive in scientific matters, and the spirit of their time favoured this.

[c] Ch. 2, p. 8.

[d] *TPYL* quotes from it twice, ch. 7, p. 4*a*; ch. 12, p. 11*a*.

[e] *Thung Chih Lüeh*, ch. 14, p. 1*b* (*Thung Chih*, ch. 38).

[1] 天樞 [2] 天皇大帝 [3] 角 [4] 亢 [5] 奎 [6] 婁
[7] 馬永卿 [8] 嬾眞子 [9] 列星圖

the mainly artistic Greek planispheres which showed allegorical constellation figures but not stars. We hear a little, indeed, of a silver table on which the stars were shown in their proper positions, said to have belonged about +850 to Charlemagne (Zinner, 1). But the scientific value of this is not sure, and it seems necessary to conclude that Europe had little or nothing to show before the Renaissance comparable with the Chinese tradition of celestial cartography.

The most important contemporary Chinese celestial atlas is that of Chhen Tsun-Kuei (*3*), which is an official publication of Academia Sinica. Good star-maps, with both Western and Chinese identifications marked on them, were also made, about ten years ago, by Shen Wên-Hou for the Fukien Meteorological Survey. These are to be found in the library of the Royal Astronomical Society, but the Chinese planispheres which J. Williams (1) presented in 1855 are no longer there, and had indeed already been lost by 1909, as Knobel (1) tells us.

(4) Star Legend and Folklore

At this point a few words may be said about the place of the heavenly bodies in Chinese legend and folklore, though the subject is an interlude in the present discussion. In view of the general importance of the stars as season signs in a predominantly agricultural civilisation, one would expect it to be considerable. Though there is no special monograph on the matter, much relevant information will be found in the usual sources, such as de Groot (2), Werner (1), Doré (1) and Hodous (1). A long paper by Solger (1) treats of a number of star-myths, e.g. Orion and Antares the quarrelling brothers,[a] Altair and Vega the herd-boy and the spinning-girl,[b] etc. The great treatise of Schlegel (5) on Chinese astrography deals systematically with these as they arise. Chinese literature itself would yield a mass of hitherto untranslated material from the innumerable books which deal with the strange and the mysterious, as also with folk-customs, e.g. the *Ching Chhu Sui Shih Chi*[1] (Annual Customs of Ching and Chhu) of Tsung Lin[2]—a Thang book. Sometimes people were supposed to be incarnations of the spirits of the planets: thus Huang Shih Kung[3] (Old Master Yellowstone), a legendary character connected with the origins of the Han dynasty,[c] was supposed to be a manifestation of Saturn;[d] while Tungfang Shuo,[4] one of Han Wu Ti's advisers, was thought to have been an incarnation of the planet Venus.[e]

In order to enliven what might otherwise be the tedium of so long a Section as this, I may perhaps be permitted to give one of the legends which gathered round the person of I-Hsing, the Buddhist astronomer of the Thang.[f] It comes from the *Ming Huang Tsa Lu* of +855.

[a] Cf. R. Wilhelm (7), p. 46.

[b] Cf. R. Wilhelm (7), p. 31.

[c] A particularly beautiful temple in his honour still exists on the road between Szechuan and Shensi, at Miao-thai-tzu.

[d] Schlegel (5), p. 629.

[e] *Fêng Su Thung I*, ch. 2; R. Wilhelm (7), p. 86. Similar stories were told of some of the deified persons depicted on the cover of the present book.

[f] Cf. pp. 119, 139, 202, 270, 274.

[1] 荊楚歲時記 [2] 宗懍 [3] 黃石公 [4] 東方朔

I-Hsing in his youth had belonged to a very poor family, and had had as a neighbour one Wang Lao, who often helped him. I-Hsing had tried to make some return, especially during the Khai-Yuan reign-period when he was in high favour with the emperor. Eventually Wang Lao, having killed someone, was imprisoned, and called upon I-Hsing for help. I-Hsing went to see him and said, 'If you want gold and silver I can give you all you want, but as for the law, I cannot change it.' Wang Lao reproached him, saying, 'What good was it to me that I ever knew you', and so they parted.

Later I-Hsing was in the Hun-Thien (Armillary Sphere) Temple, where there were several hundred workers. He ordered some of them to move a huge pot into an empty room. Then he said to two servants 'In a certain place there is a ruined garden. Do you hide there secretly tomorrow, from noon to midnight. Something will come—if it is seven in number, put them in the pot and cover them up, and if you lose one I shall give you a great beating.' About six o'clock in the evening, sure enough, a herd of seven pigs appeared, and they caught them all and put them in the pot; then they ran off and told I-Hsing, who was very pleased. Covering the pot with a wooden cover and matting, he wrote certain Sanskrit words upon it in red, the meaning of which his students did not understand.

Before long I-Hsing received a message to go urgently to a certain palace, where the emperor met him and said, 'The Head of the Astronomical Bureau has just informed me that the Great Bear has disappeared. What can it mean?' I-Hsing replied, 'This sort of thing has happened before. In the Later Wei dynasty they even lost the planet Mars. But there are no previous records of the disappearance of the Great Bear. Heaven must be giving you an important warning, perhaps of frost or drought. But your Majesty, with your great virtue, can influence the stars. What would most affect them would be a decision on your part in favour of life rather than death. So do we Buddhists preach forgiveness to all.' The emperor agreed, and issued a general amnesty.

Later the seven stars of the Great Bear reappeared in their places in the heavens. And when the pot into which the pigs had been put was opened, it was found to be empty.[a]

This mastery of I-Hsing over the stars of the Great Bear becomes perhaps more comprehensible when we remember that he did indeed write certain tracts or books about their astrological significances and astronomical relations.[b]

[a] Supplement, p. 3*a*, tr. auct. Greek and later European parallels in Boll (5), p. 84.

[b] E.g. the *Pei Tou Chhi Hsing Nien Sung I Kuei*[1] (Mnemonic Rhyme of the Seven Stars of the Great Bear, and their Tracks), TW 1305. Perhaps also the *Fo Shuo Pei Tou Chhi Hsing Yen Ming Ching*[2] (Sūtra spoken by a Bodhisattva on the Delaying of Destiny according to the Seven Stars of the Great Bear), TW 1307. On both these see Eberhard (12).

[1] 北斗七星念誦儀軌 [2] 佛說北斗七星延命經

(g) THE DEVELOPMENT OF ASTRONOMICAL INSTRUMENTS

(1) The Gnomon and the Gnomon Shadow Template

The most ancient of all astronomical instruments, at least in China, was the simple vertical pole.[a] With this one could measure the length of the sun's shadow by day to determine the solstices (called *chih*[1] from Shang times until now), and the transits of stars by night to observe the revolution of the sidereal year. It was called *pei*[2] or *piao*,[3] the meaning of the former being essentially a post or pillar, and the latter an indicator. *Pei* can be written with the bone radical[4] (as in the title of the *Chou Pei Suan Ching*, see p. 19 above), or with the wood radical,[5] in which case it means a shaft or handle. Ancient oracle-bone forms of the phonetic component show a hand holding what seems to be a pole with the sun behind it at the top (K 874 *b, j*), so that although this component alone came to mean 'low' in general, it may perhaps have referred originally to the gnomon itself. This is after all an object low on the ground in comparison with the sun, and shows the long shadow of a low sun at the winter solstice, the moment which the Chinese always took as the beginning of the tropic year.

K874

Other evidence that the Shang people were conscious of the casting of shadows is derived by Tung Tso-Pin (*1*) from the now purely literary word *tsê*,[6] which means the late afternoon when the sun is setting. Ancient bone forms of this show the sun and a man's shadow at different angles (see drawing).[b]

Probably the earliest literary reference to solstice observations which has come down to us is the passage in the *Tso Chuan* bearing the date −654, and it is therefore worth quoting.[c]

K924

In the fifth year of Duke Hsi, in spring, in the first month (December), on a *hsin-hai* day, the first of the month, the sun (reached its) furthest south point. The duke (of Lu), having caused the new moon to be announced in the ancestral temple, ascended to the observation tower (*kuan thai*[7]) in order to view (the shadow), and (the astronomers) noted down (its length) according to custom. At every equinox and every solstice, whether at the beginning (*chhi*[8]) of spring or summer, or at the beginning (*pi*[9]) of autumn or winter, it is necessary to note the appearance of the clouds and vapours, for (prognostication and) preparations against coming events.[d]

[a] See Fig. 110. The most extensive discussion of this, as of all other Chinese astronomical instruments, is contained in the remarkable monograph of Maspero (4).

[b] Though it may seem at first sight surprising, gnomon shadow measurements with a pole about 8 ft. high to determine solstices and so fix agricultural planning are still regularly carried out by a number of primitive peoples. A photograph (Fig. 111) showing two Borneo tribesmen measuring shadow length is given by Hose & McDougall (1); Anon. (26).

[c] Cf. Gaubil (3), p. 22.

[d] Tr. auct., adjuv. Couvreur (1), vol. 1, p. 247.

1 至 2 碑 3 表 4 髀 5 椑 6 仄 7 觀臺 8 啓 9 閉

Fig. 110. A late Chhing representation of the measurement of the sun's shadow at the summer solstice with a gnomon and a gnomon shadow template by Hsi Shu (the youngest of the Hsi brothers) in legendary antiquity. From *SCTS*, ch. 1, Yao Tien (Karlgren (12), p. 3).

From the words inserted in the translation it will be seen that the annalist was vague as to what exactly the prince did when he joined the party of astronomers, but this was a winter solstice, and there can be no doubt that they were measuring the length of the shadow. So important was this reading that the attendance of the ruler of the State in person is not surprising. Recently Dittrich (1) has made a new study of this particular solstice.

Though the *Huai Nan Tzu*[a] preserves a tradition that gnomons of 10 ft. length were anciently used (which would fit in with the strong evidence for a decimal metrology in the Chou period already mentioned),[b] this was early abandoned, presumably because it did not readily lend itself to simple calculations about the sides of right-angled triangles; and with certain exceptions, such as the 9 ft. gnomon of Yü Kuang[1] in +544, the length of 8 ft. is that generally mentioned in ancient and medieval texts. Even when in the Yuan period the interests of accuracy led to a much larger structure, a multiple of 8 ft., 40 ft., was chosen, as we shall see. The requirements of a perfectly horizontal base and a perfectly vertical pole were well understood before the Han, for the *Chou Li* speaks[c] of a water-level and of suspended cords. Han commentators took this to mean that cords of equal length were fixed one to each corner of the base, but Chia Kung-Yen[2] in the Thang supposed that four suspended plumb-lines were used; if this was so, the instrument was very like the *groma* of the Roman surveyors.[d]

The earliest measurements of the shadow's length were no doubt made with the foot-rules of the time, but as it was realised that these varied according to bureaucratic prescription and local custom, a standard jade tablet (*thu kuei*[3]), which may be called the Gnomon Shadow Template, was made for this purpose only. It is mentioned in the *Chou Li*,[e] and actual specimens, made of terra-cotta, one dated +164, are extant.[f]

> The Ta Ssu Thu[4] (a high official) [says the *Chou Li*],[g] using the gnomon shadow template, determines the distance of the earth below the sun, fixes the exact (length of the) sun's shadow, and thus finds the centre of the earth....The centre of the earth[h] is (that place where) the sun's shadow at the summer solstice is 1 ft. 5 in.

The moment of the solstice could thus be determined by placing the calibrated template at the base of the post due north for several days around the expected time, and taking noon of the day when the shadow most nearly coincided with it. One of the reasons why the winter solstice was determined indirectly by observing the summer

[a] In the passage referred to above, p. 225, with regard to cosmological theories.
[b] In the mathematical Section, pp. 89ff. above.
[c] *Khao Kung Chi* section, ch. 12, p. 15*a* (ch. 43, p. 19, tr. Biot (1), vol. 2, p. 554).
[d] Singer (2), p. 113; B & M, p. 616; D. E. Smith (1), vol. 2, p. 361.
[e] Ch. 12, p. 2*a* (ch. 42, p. 19, tr. Biot (1), vol. 2, p. 522).
[f] Maspero (4), p. 222. This was just after the time of Chang Hêng, who mentions the instrument in his *Tung Ching Fu* (see *Wên Hsüan*, ch. 3, p. 4*a*).
[g] Ch. 3, p. 13*b* (ch. 9, p. 16) tr. Biot (1), vol. 1, p. 200; eng. auct.
[h] Where exactly the Han people thought this was will shortly appear.

1 虞鄺 2 賈公彥 3 土圭 4 大司徒

Fig. 111. Two Borneo tribesmen measuring the sun's shadow length at summer solstice with a gnomon and a gnomon shadow template in recent times (photo. Hose & McDougall).

solstice was that a much shorter template was required for the latter. But a sidereal mark-point was also more easily found in this way, since for several centuries around −450, the determinative star of Niu[1] *hsiu* (β Capricorni) was exactly in opposition to the sun at the summer solstice. All the oldest template measures of the summer solstitial sun-shadow length agree around 1½ ft.; the *I Wei Thung Kua Yen*[a] gives 1·48 ft. and Liu Hsiang about −25 adopted 1·58, while Yuan Chhung[2] in +597 got 1·45 ft.—and most accounts give the place of measurement as Yang-chhêng,[3] a city[b] about fifty miles south-east of Loyang, for which latitude indeed the measurement would have been exact to within a small margin of error.

The template system was an attempt to overcome the chaos of primitive metrology, and did not persist. It does seem to anticipate, in some sense, modern devices such as the standard platinum metre. In +500 Tsu Kêng-Chih[4] made bronze instruments in which the gnomon and a horizontal measuring-scale were combined. About fifty years earlier Ho Chhêng-Thien[5] had proceeded to more careful observations of the winter solstice shadow. It was not until this was done that the inequality of the seasons could be discovered.[c] The lack of this knowledge was shown very clearly in two Han calendars discovered in Central Asia by Aurel Stein, dating from −63 and −39 (de Saussure, 33).

Study of the solstitial sun-shadow lengths furnishes exact knowledge of the obliquity (ε) of the ecliptic (Fig. 112). This is a fundamental astronomical datum which was probably known in the −4th century as soon as Shih Shen and Kan Tê began to measure right ascensions and declinations of celestial bodies. No precise figure, however, for the solar declination at the solstices is available before the *Hou Han Shu*. There, in his exposition of +89, Chia Khuei[6] said: 'At the winter solstice the sun is 115° from the pole; at the summer solstice it is distant by 67°'.[d] The difference divided by two gives a round figure of 24°.[e] The many further determinations of solstitial sun-shadow lengths made in subsequent centuries by Chinese astronomers[f]

[a] 'Verifications of the Powers of the Kua in the *Book of Changes*'; one of the Wei ('weft') Classics or apocryphal treatises, of Han date; see above, Vol. 2, p. 380.

[b] The same as the modern, and much dwindled, city of Kao-chhêng.[7]

[c] Failure to realise this had thrown the calendar of −105 two days out; there was no deliberate falsification such as de Saussure (25) suspected. Here again, the discovery took place much earlier in Greece, the inequality being known to Hipparchus in the −2nd century (Berry (1), p. 46). On Ho Chhêng-Thien's measurements, cf. Maspero (4), p. 258.

[d] Ch. 12, p. 5*a*; see Maspero (3), p. 276.

[e] Corresponding to 23° 39′ 18″ in our reckoning; see Gaubil (2), pp. 5, 8, 113, 114. Gaubil believed that this figure had been accepted in the days of Lohsia Hung[8] (*c.* −105), but if so it was then even less accurate. In any case it was never more than a round number. But an obliquity of $23\frac{15}{16}$ Chinese degrees is implied by a more accurate summer solstice declination given in the calendrical treatise o +178, written by Liu Hung[9] and Tshai Yung,[10] which now forms ch. 13 of the *Hou Han Shu* (pp. 21*b*, 23*b*). So also Liu Hung in his treatise of +200 on lunar eclipses, *Lun Yüeh Shih*,[11] now in ch. 12, p. 18*b*. Yet measurements of the two solstitial shadows in +173 by these same astronomers gave, when reduced by Gaubil (6), a value about 3′ 45″ higher (see Fig. 113). Unfortunately, Gaubil never gave exact references to the Chinese texts which he quoted. See Hartner (8).

[f] Cf. the summary given in the commentary by Li Shun-Fêng[12] on the *Chou Pei Suan Ching*, ch. 1, p. 10*b*.

[1] 牛 [2] 袁充 [3] 陽城 [4] 祖暅之 [5] 何承天 [6] 賈逵
[7] 告成 [8] 落下閎 [9] 劉洪 [10] 蔡邕 [11] 論月蝕
[12] 李淳風

were collected and computed by Gaubil (6) in +1734, but not printed until 1809,[a] when Laplace (1) accompanied them with a celebrated memoir on the saecular decline in the value of the obliquity. Together with a few Greek and Arabic determinations, all with Laplace's corrections, they will be found, beside the theoretical curve of Newcomb[b] and de Sitter, in Fig. 113. This curve, derived from celestial mechanics, gives us to some extent an opportunity of seeing how good the observations of the ancient and medieval astronomers were.

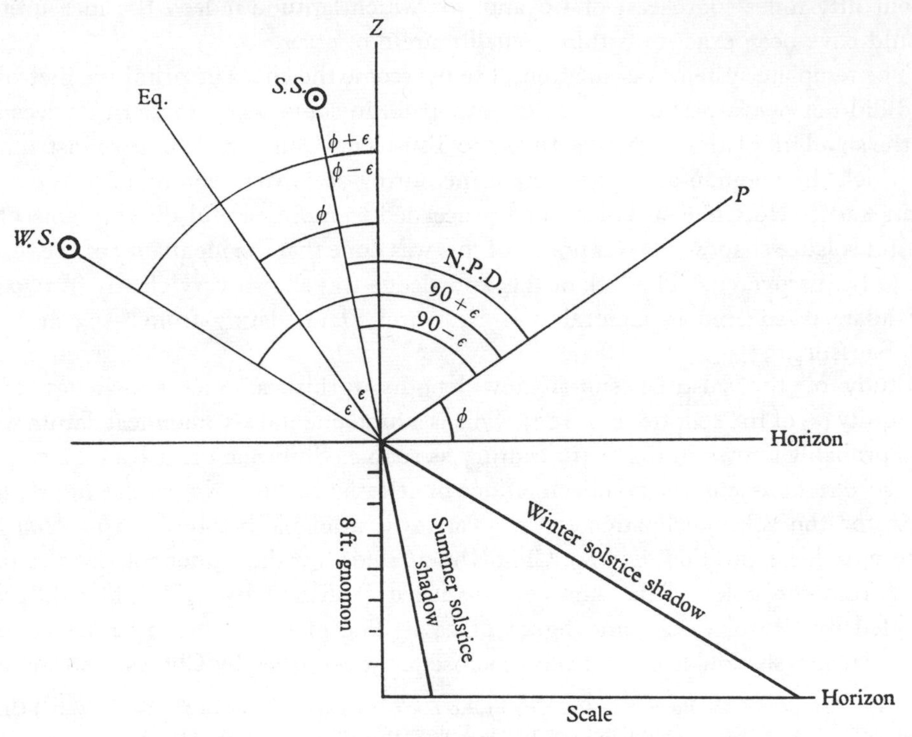

Fig. 112. Diagram to illustrate the obliquity of the ecliptic in relation to gnomon shadows (after Hartner).

Such at least would be the case if the theoretical curve could be relied upon as final. But it may require considerable revision. As will be seen from the graph, values obtained about the beginning of the Christian era and before tend to deviate widely on the higher side. The high figure obtained by Ptolemy (*c.* +143)[c] has long been puzzling. Now the observations of Liu Hung and Tshai Yung (+173), tabulated in the *Hou Han Shu*, seem to have been made with great care; and in their summer solstice N.P.D. for the ecliptic, the figure of 67° (Ch.) is followed by the character *chhiang*,[1] indicating that one-eighth of a degree must be added, making 67° 12′ 50″ (Ch.) (67·125°). Hartner (8) has shown that this reading must have referred to the sun's

[a] Gaubil (7). [b] From *Nautical Almanac*, 1931.
[c] *Almagest*, I, 12. Ptolemy's result was expressed as zenith distance ($\phi-\epsilon$), not N.P.D.

[1] 强

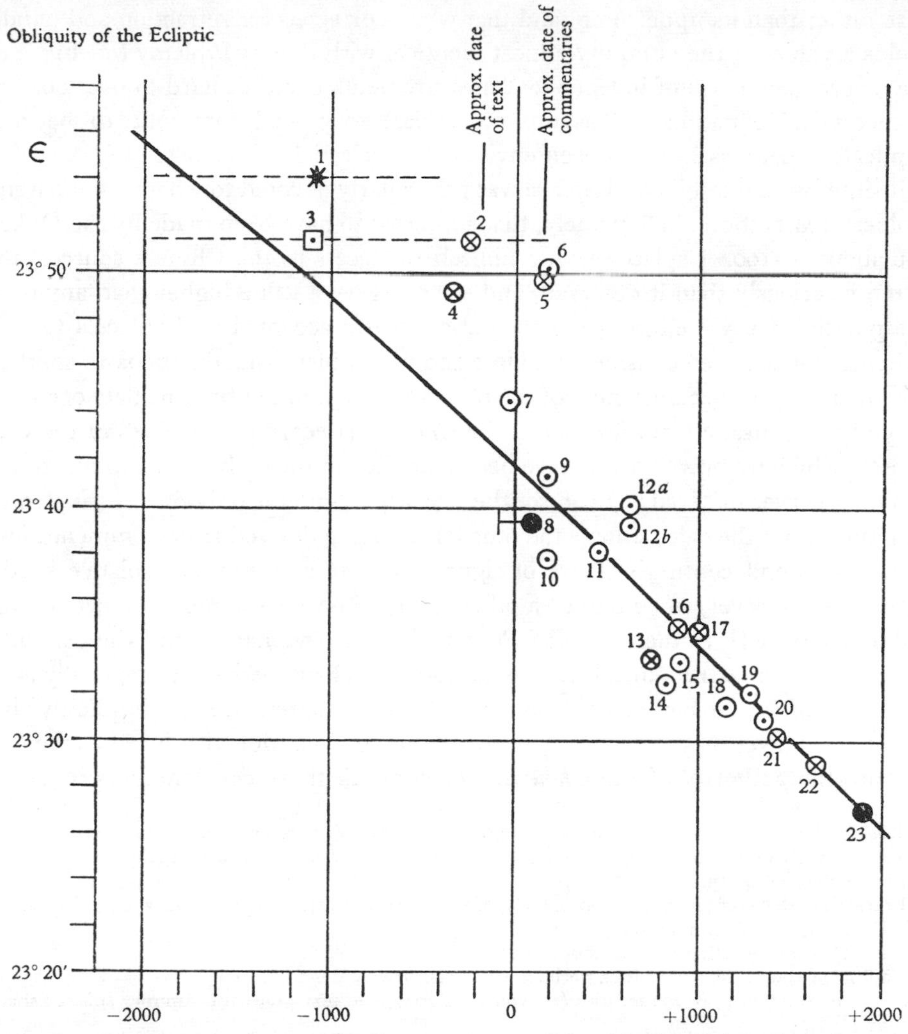

Fig. 113. Plot of ancient and medieval measurements of the obliquity of the ecliptic (see text).

1 *Chou Li* figure, subject to uncertainty as to place, length of units, and height of gnomon.
2–3 (dotted line) *Chou Li* figure recalculated by Hartner (8) assuming Yang-chhêng as the place.
2 Eratosthenes (corr. Nallino; Hartner).
3 Laplace's theoretical value.
4 Pytheas.
5 Ptolemy (Hartner).
6 Liu Hung & Tshai Yung (taking $67\frac{1}{8}$° Ch. value only; Hartner).
7 Liu Hsiang (calc. Gaubil; corr. Hartner).
8 Chia Khuei, with backward extension to Lohsia Hung, the range representing also values given in the *Chou Pei Suan Ching*.
9 Liu Hung & Tshai Yung (both solstices, calc. Laplace from Gaubil).
10 Liu Hung & Tshai Yung (both solstices, *Hou Han Shu* figures).
11 Tsu Chhung-Chih.
12*a*, *b* Li Shun-Fêng.
13 Value from al-Ma'mūn's observatory (Sind ibn 'Ali and Yaḥya ibn abi Manṣūr; Hitti (1), p. 375; Hartner).
14 Hsü Ang.
15 Pien Kang.
16 al-Battānī (corr. Nallino; Hartner).
17 Ibn Yūnus (corr. Nallino; Hartner).
18 Liu Hsiao-Jung.
19 Kuo Shou-Ching.
20 'Ali ibn al-Shāṭir.
21 Ulūgh Beg (corr. Nallino; Hartner).
22 Cassini.
23 Contemporary value.

centre rather than its upper limb, and that when corrected for refraction and parallax it yields a value for the obliquity almost identical with that of Ptolemy (cf. Fig. 113). Such a correspondence of independent measurements is rather hard to overlook, and the theoretical equation (an asymptotic series) may well turn out to be more complicated than has hitherto been envisaged in celestial mechanics.

The interest of Gaubil and Laplace was particularly devoted to what was apparently the oldest observation of all, namely, that supposed to have been made by the Duke of Chou about −1100. Laplace (having himself no access to the Chinese sources) took this more seriously than it deserved, and since it gave a value higher than any other, corresponding fairly well with the theoretical figure accepted in Laplace's time for that epoch,[a] it achieved considerable fame and was copied from one book to another.[b] Unfortunately, the circumstances of the observation are in doubt. The data occur, not in any of the dynastic histories, but in the *Chou Li* (Record of the Rites of the Chou Dynasty) which we now know to have been compiled in the early Han period,[c] though containing earlier material.[d] Even so, the text itself contains only one of the items of data required for the calculation,[e] the other three being derived from commentaries as late as the +2nd century.[f] Pairs of figures (winter and summer solstice shadow lengths) are, however, to be found in other ancient books,[g] the most important being the data contained[h] in the *Chou Pei Suan Ching*.[1] As we have seen,[i] the content of this mathematical-astronomical work is decidedly archaic, perhaps intrinsically as old as the time of Confucius, and the shadow-length measurements are implicitly those of Chou Kung, yet they differ from all the others just mentioned.[j] Nor has the *Chou Pei* quite the authority of the *Chou Li*. In sum, all these computations involve so

[a] Taking the modern formulae of Newcomb and de Sitter, it would now correspond rather to between −1900 and −1700. It is quite impossible to believe that any measurements in the *Chou Li* could be of that antiquity.

[b] Fréret (1); R. Wolf (1), vol. 1, pp. 421, 439, vol. 2, p. 96; (2), p. 7; (3), vol. 2, p. 79; Grant (1), p. 99; Berry (1), p. 11; Zinner (1), p. 288.

[c] Within a century either way of about −175.

[d] Gaubil, of course, accepted it as perhaps eight centuries older than this.

[e] Ch. 3, p. 14*b* (ch. 9, p. 17; tr. Biot (1), vol. 1, p. 201); the text gives the summer solstice shadow length as 1·5 ft.

[f] At ch. 3, p. 14*b*, the commentary of Chêng Chung[2] (*fl.* +50 to +83) gives the gnomon length as 8 ft., and the place as Yang-chhêng. At ch. 8, p. 29*b* (ch. 33, p. 60; tr. Biot (1), vol. 2, p. 279), the commentary of Chêng Hsüan[3] (+127 to +200) gives the winter solstice length of the shadow as 13 ft.

[g] In the *Hung Fan Wu Hsing Chuan*[4] of Liu Hsiang[5] the two shadow lengths are given as 1·58 and 13·14 ft. respectively. If this is genuine it would be of about −25. This book has been lost for many centuries, but the relevant quotation is preserved in various places, notably in the commentary by the Thang astronomer Li Shun-Fêng on the *Chou Pei Suan Ching*, ch. 1, p. 10*b*. Another pair of figures, again different, occur in the apocryphal treatise *I Wei Thung Kua Yen*[6] (1·48 and 13 ft.). This might also be dated as of the last quarter of the −1st century. Both these appear to refer rather to contemporary measurements than to those of Chou Kung.

[h] Ch. 2, pp. 6*b*, 7*a*.

[i] Pp. 210 and 257 above.

[j] Summer solstice 1·6 ft. and winter solstice 13·5 ft. The value which these figures give for the obliquity (comput. W. Hartner) is not very high, being almost the same as that mentioned by Chia Khuei, and suitable (according to the theoretical curve) for the +3rd century rather than the −1st, to say nothing of the −11th. But the numerical figures may have been interpolated in an ancient textual framework.

[1] 周髀算經 [2] 鄭衆 [3] 鄭玄 [4] 洪範五行傳 [5] 劉向
[6] 易緯通卦驗

many uncertainties—the date,[a] the place,[b] the height of the gnomon,[c] and the nature of the units of length[d]—that the high figure obtained by Gaubil and Laplace gave a fallacious appearance of precision.

Nevertheless, it is hard to reject entirely the figures given in the *Chou Li*.[e] Probably the best solution is to take them as traditional measurements which had been made somewhere during the Chou period and which had been afterwards carefully handed down through uninstructed intermediary hands. The margin of error seems great enough to justify us in choosing an epoch for the observations probable on quite other historical grounds, and this might be between the −9th and −3rd centuries. All that remains of the observation which so much interested Laplace is the residue that it may well have been the earliest in any nation of which the records have come down to us, and that it seems to have yielded a higher value than any subsequent estimation.[f]

In any case, the whole series of Chinese observations, culminating in the very accurate ones of Kuo Shou-Ching, proved of value to 18th-century astronomers in their discussions of what they called the mutability of the obliquity. In +1747 Euler had been able to derive a very small saecular diminution in obliquity from the perturbation effects to be expected from other planets, and it was in search of confirmatory observations that Laplace studied Gaubil's manuscripts.

A process of refinement continued gradually throughout Chinese history in this as in other astronomical matters. The *Chou Li* gave the shadow lengths only for the two solstices, not for each of the twenty-four fortnights of the tropic year. Lists of such figures dating from the Han exist, however.[g] Some of them, such as those contained in the *Chou Pei Suan Ching*[h] and the apocryphal treatise *I Wei Thung Kua Yen*,[i] seem to have been simply computed, but others, such as those in the calendrical chapters of

[a] Nothing but tradition links the figures in the *Chou Li* and the *Chou Pei* with Chou Kung.

[b] There is plenty of evidence that Yang-chhêng (=modern Kao-chhêng, lat. 34° 26′ N.), about fifty miles south-east of Loyang in Honan, was regarded by Chinese cosmographers in Han and even Chou times as the centre of the world, and that gnomon measurements were made there (Tung Tso-Pin *et al.*). Giant instruments of stone still exist at Yang-chhêng, as we shall shortly see (p. 296). But we have no certainty as to the locality of any of these ancient measurements. However, Dr W. Hartner (private communication to the author) has found by computation that if the latitude of Yang-chhêng is assumed, then most of the obliquity values move within reasonable limits if we suppose that the sun's centre, and not the upper or lower limb, was taken.

[c] Gnomons of 10 and 11 ft. are mentioned in the commentary of Li Shun-Fêng just mentioned. Dr W. Hartner further finds, however, that if the latitude of Yang-chhêng is assumed, then in all cases the height of the gnomon can be computed to fall between 7·9 and 8·5 ft. approximately, and more than 8·0 in nearly all cases.

[d] The units of length enter into the calculations only as relative quantities. But as the lengths of the standard foot were always changing in ancient times, serious confusion could have arisen if measurements from different dates were combined in pairs by unscientific editors or scribes, or if attempts at correcting individual values were made.

[e] Especially since the archaic *Chou Pei* also gives figures.

[f] If Thom (1, 2) is right, obliquity data for about −2000 may be derived from the siting of megalithic monuments in Europe, and these also yield high values (av. 23° 54·3′). The stones were aligned so as to indicate azimuth points of solstitial sunrise and sunset; no shadow length measurements were made.

[g] Cf. the similar Babylonian lists of *c.* −700 (the Mul-Apin tablets), given by Weidner (3).

[h] Ch. 2, pp. 6*b*, 7*a*.

[i] *Ku Wei Shu*, ch. 15, pp. 1*a* ff.

the *Hou Han Shu* and the *Chin Shu*, are probably actual records of observations. It will be remembered that the problem of the prediction of the daily motion of the sun was what led to the development of the method of finite differences by Tsu Chhung-Chih in the +5th century, and Li Shun-Fêng in the +7th.[a]

Another error arising from primitive computations was the long-standing idea[b] that the shadow length increased 1 in. for every thousand *li* north of the 'earth's centre' at Yang-chhêng and decreased by the same amount every thousand *li* south.[c] It is clear from the *Chou Li*[d] that observation of shadow lengths was used to determine latitude in fixing provincial and other territorial boundaries.[e] Numerous records of experiments in which this assumed relationship was disproved remain. In +445 Ho Chhêng-Thien had measurements made simultaneously at Chiaochow[1] (modern Hanoi; 5000 *li* south of Yang-chhêng) and at Lin-i in Indo-China.[f] The result was a value of 3·56 in. of shadow length for each thousand *li*. Similar observations were made in +508 and by Liu Chhuo[g] about +600.

It is interesting that a still earlier record exists of the observation of a military commander, Kuan Sui,[2] who in +349 led a successful expedition against the Champa people of Lin-I under Fan Wên.[3] Having pursued their army far into what is now Viet-Nam, 'he set up a gnomon in the 5th month, and found that the sun was to the north of it, casting a shadow to the south 9·1 in. long....Hence the inhabitants of that country', continues the account, 'have the doors of their houses facing north to turn to the sunshine'.[h] Thus the Chinese found what the Alexandrian Greeks had also known, namely, that south of the Tropic of Cancer (which passes just north of Canton) the sun will cast shadows at midday towards the south during part of the year. Kuan Sui's position has sometimes been located about 13° north, i.e. nearly as far down as Nha-trang; but Destombes (1) has recently computed from all the available texts that it must have been between 17°05′ and 19°35′. It was therefore about level with the most southerly point of Hainan Island.

But the most complete set of figures was that obtained by expeditions under the direction of Nankung Yüeh[4] and I-Hsing[5] in +721 to +725. We are told[i] that there were nine stations (including Yang-chhêng) with polar altitudes ranging from 17·4° at

[a] Cf. the mathematical Section, pp. 123 ff. above.

[b] E.g. in *Hun Thien Hsiang Shuo* quoted in *Chin Shu*, ch. 11, p. 6*b*; also in, for example, *Hsü Po Wu Chih*, ch. 1, p. 5*b*.

[c] The fallacy in this was the failure to realise the curvature of the earth's surface.

[d] See especially Biot (1), vol. 2, p. 279.

[e] At any rate for Han practice; how far back into the Chou this held good it would be difficult to say.

[f] The occasion was an expedition of the Liu Sung against this southern kingdom; cf. Cordier (1), vol. 1, p. 333. Cf. Gaubil (2), p. 50.

[g] Sources conflated in *Chhou Jen Chuan*, ch. 12. It is interesting that for this work Liu Chhuo is said to have enlisted the help of hydraulic engineers, surveyors and mathematicians.

[h] *Wên Hsien Thung Khao*, ch. 331, p. 16*a*, tr. Hervey de St Denys (1), vol. 2, p. 427. The translator's note erred in suggesting that Kuan Sui must have crossed the equator.

[i] *Chiu Thang Shu*, ch. 35, pp. 6*a* ff., abridged in *TCKM*, ch. 43, pp. 51*a* ff.; cf. *TH*, p. 1407; *Thang Hui Yao*, ch. 42 (p. 755).

1 交州 2 灌邃 3 范文 4 南宮説 5 一行

Lin-i[a] to 40° at Weichow.[b] Along this meridian line of 7973 *li* (just over 3500 km.)[c] simultaneous measurements of summer and winter solstice shadow lengths were made with standard 8 ft. gnomons. The text notes with interest that the +5th-century value for Chiaochow[d] was confirmed. The terrestrial distance corresponding to 1° was thus estimated as 351 *li*, 80 *pu*, and the difference in shadow length was found to be very close to 4 in. for each thousand *li*, or four times the amount accepted by the 'scholars of former times'.[e] This work must surely be regarded as the most remarkable piece of organised field research carried out anywhere in the early Middle Ages.[f]

It has been supposed that I-Hsing's meridian was as long as 13,000 *li* (5700 km.), its most northerly point being in the country of the Thieh-lê[1] (Tölös) horde of Turkic nomads beside Lake Baikal. But the text seems to indicate that the figures for this point,[g] though embodied by Gaubil in his table, were extrapolated, not observational. Five centuries later, in +1221, the work of I-Hsing's survey was completed by the Taoist Chhiu Chhang-Chhun and his party, who made gnomon observations at the summer solstice on the banks of the Kerulen river in northern Mongolia (about 48° N.), when travelling on a visit to Chingiz Khan at Samarqand.[h]

While the observed lengths of the shadows could serve for cosmological calculations, some of which were doomed to inaccuracy owing to the absence of other information (e.g. the curvature of the earth's surface), the most important observation of all was a null-point one, namely, the moment of minimum and maximum shadow length. That was essential for the construction of calendars. The great work of Tung Tso-Pin (*1*) on the astronomy and calendrical science of the oracle-bones has shown that the inscriptions contain references to the number 548; one such text is datable accurately as of −1210. This has much significance since $1\frac{1}{2}$ times 365·25 is 547·875; the meaning therefore is that 'the summer solstice recurs at 548 days after the winter solstice'. It can hardly be doubted, therefore, that the Shang people in the −13th and −14th centuries were using gnomons to determine the solstices, in the traditional calendar handed down as the 'Old Quarter-Remainder Calendar' (Ku ssu fên li[2]). Obviously regular and long-continuing observations were made, covering at least four tropic years, the shortest space of time in which the day-count could come to a whole

[a] Indrapura in Champa, capital of the Lin-I State, not far from Hué in modern Annam.

[b] An old city near modern Ling-chhiu, near the Great Wall in northern Shansi, almost on the same latitude as Peking.

[c] Stations were most numerous on the great plains north and south of the Yellow River. Only one was on the northern border of China proper, and two were in the far south (Indo-China). The points were not strictly on a north-south line.

[d] Hanoi in Tongking; the shadow was *southwards* 3·3 in.

[e] See the tabulation in Gaubil (2), p. 76. 'Quand le Bonze Y-hang', he wrote, 'n'auroit fait autre chôse que de procurer tant d'observations de la hauteur du Pôle, et de détèrminer la grandeur du Ly en le rapportant aux degrés de latitude, on lui auroit toujours une obligation infinie.'

[f] The chief observers, Ta-Hsiang[3] and Yuan-Thai[4] (monks apparently), were also responsible for the conduct of the expedition which went to map the constellations of the southern hemisphere (cf. p. 274 above).

[g] Polar altitude 51·8°. The place was taken as being as far to the north of Yang-chhêng as Lin-i was to the south of it.

[h] The incident is recorded by Li Chih-Chhang in his *Chhang-Chhun Chen Jen Hsi Yu Chi*, tr. Waley (10), p. 66.

[1] 鐵勒 [2] 古四分曆 [3] 大相 [4] 元太

Table 30. *Evaluation of the fraction* (sui yü) *of the tropic and the sidereal year*

	sui yü [1]	
	tropic year fraction (days)	sidereal year fraction (days)
True value	—	0·25637
True value (computed):		
Han +200	0·242305	—
Thang +750	0·242270	—
Yuan +1250	0·242240	—
Deduced from the oracle-bones (−13th century) [a]	—	0·25
Liu Hsin [2] (San Thung cal. −7) [b]	0·250162	
Tsu Chhung-Chih [3] (Ta Ming cal. +463) [c]	0·242815	—
Kuo Shou-Ching [4] (Shou Shih cal. +1281) [d]	0·242500	—
Han I [5] (Huang Chhu cal. +220) [e]	—	0·255989
Liu Chhuo [6] (Huang Chi cal. +604) [f]	—	0·257610
I-Hsing [7] (Ta Yen cal. +724) [g]	—	0·256250
Kuo Shou-Ching [4] (Shou Shih cal. +1281) [h]		0·257500

number. Attempts at exact evaluation of the odd fraction (*sui yü* [1]) of the tropic year (*sui shih* [8]) continued throughout Chinese history; Maspero [i] has tabulated twenty-three of them.[j] The nearest approaches, some quite early, may be seen from Table 30. In the original texts the numbers are given as fractions, of course, not as decimals, with the exception of those of Kuo Shou-Ching.

(2) Giant Instruments in Masonry

One of the most interesting chapters in the history of astronomy is concerned with a phase of development in which the search for accuracy led to the construction of instruments of extremely large size. This came about chiefly because of growing dissatisfaction with the degree of precision attainable by the metal-workers of the time, but the great advances in practical technology which accompanied the European

[a] Tung Tso-Pin (*1*).
[b] *Chin Shu*, ch. 18, p. 16*a*.
[c] *Sung Shu*, ch. 13, p. 26*b*.
[d] *Yuan Shih*, ch. 52, p. 15*a*; ch. 54, pp. 1*a*, *b*.
[e] *Chin Shu*, ch. 17, p. 2*a*.
[f] *Sui Shu*, ch. 18, p. 18*a*.
[g] *Hsin Thang Shu*, ch. 28A, pp. 1*a*, 2*a*, 4*a*.
[h] *Yuan Shih*, ch. 52, p. 15*a*; ch. 54, pp. 1*a*, *b*.
[i] (4), p. 233.
[j] The mass of information which he collected disposes completely of the categorical statement of Ginzel (3), p. 77, that the 'year-fraction' was not known anciently in China.

1 歲餘 2 劉歆 3 祖沖之 4 郭守敬 5 韓翊
6 劉焯 7 一行 8 歲實

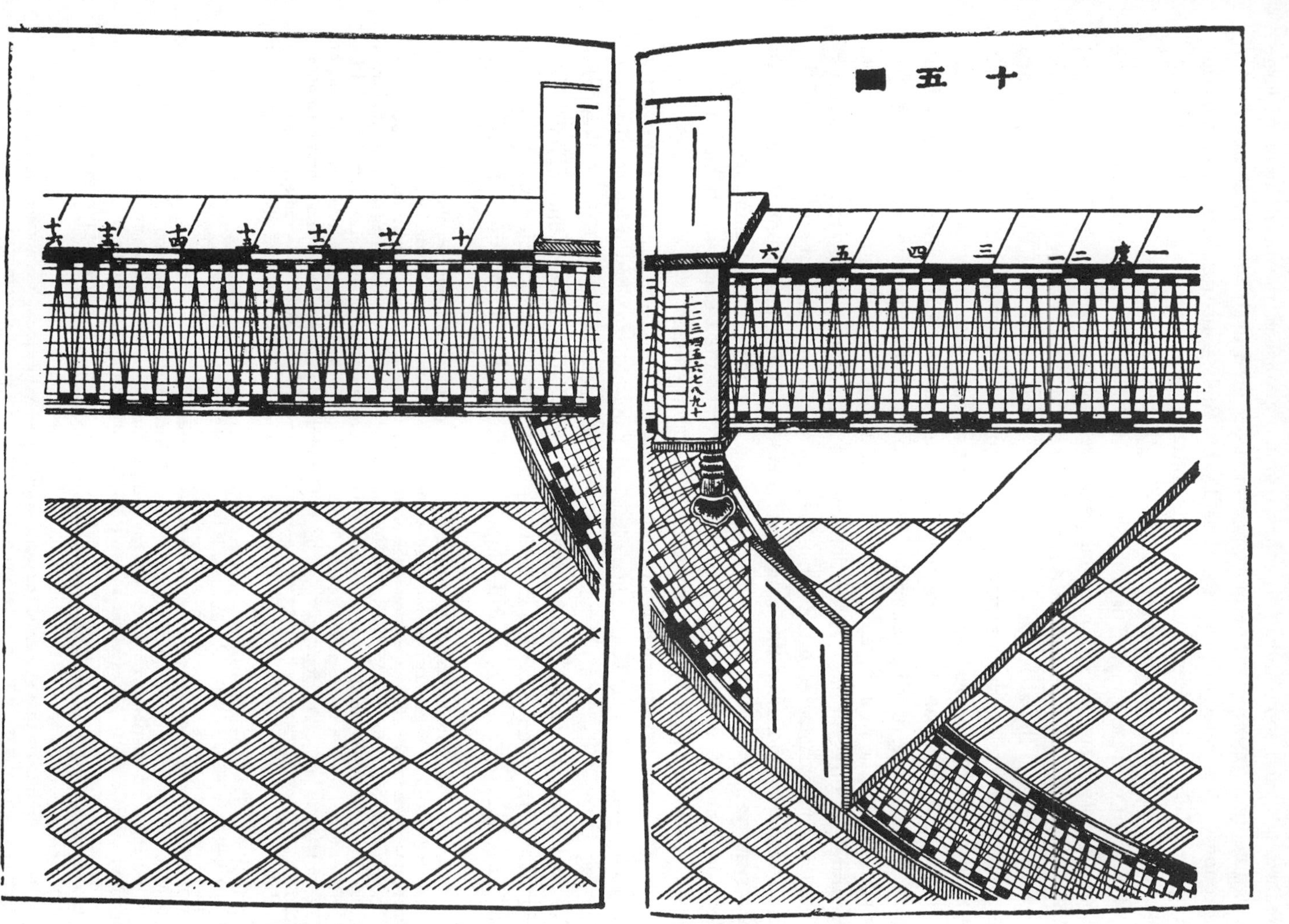

Fig. 114. Hommel 'vernier' graduations shown in the *Thu Shu Chi Chhêng*.

Renaissance, and soon spread to other regions, abolished the need for very large instruments of masonry.[a] The movement first declared itself in Chinese and Arabic astronomy, was perceptible in some of Tycho Brahe's instruments (especially his Augsburg 'Great Quadrant' and his later mural quadrant at Uraniborg),[b] and came to a very late climax in the observatories of 18th-century India.[c]

The giant Chinese gnomon of Kuo Shou-Ching[1] now to be described, set up in the Yuan dynasty about +1276, must be considered against this background. In +995 Abū'l-Wafā' al-Būzjānī[d] had used a quadrant with a radius of nearly 22 ft., and the sextant of Ḥāmid Ibn al-Khiḍr al-Khujandī (*d.* +1000)[e] had had a radius of as much as 57 ft. That of Ulūgh Beg, whose observatory at Samarqand began work about +1420, is said to have been as high as the dome of St Sophia in Byzantium, i.e. 180 ft.[f] Sédillot (3)[g] quotes an +11th-century Arabic astronomer[h] as having said that if he could make a circle of which one side was supported on the pyramids and the other on Mt Mocattam, he would gladly have done so, as the larger the instrument the more accurate the observations. Since the work of Kuo Shou-Ching, though markedly independent in character, was done, as we shall see,[i] in presence of astronomers of the Arab tradition and after the sending of models or diagrams from the Marāghah observatory in Persia, it seems almost certain that his giant gnomon, so natural a development of Chinese astronomy, was stimulated by the trend towards large instruments which had already started among the Arabic scholars.

The standard account of it is the publication of Tung Tso-Pin, Liu Tun-Chen & Kao Phing-Tzu (*1*).[j] At Kao-chhêng (the old Yang-chhêng), fifty miles south-east of Loyang, there still stands a remarkable structure, known as Chou Kung's Tower for the measurement of the Sun's Shadow (Chou Kung Tshê Ching Thai[2]) (see Figs. 115–17).[k] The tower is a truncated pyramid the sides of which measure some 50 ft. at the bottom and 25 ft. at the top. Two stairways lead from ground-level to the

[a] Apart from the introduction of better metal-working machines, it should be noted that the use of transversal dot lines in graduations was first introduced by Richard Chancellor (+1552 or +1553), chief pilot of the first English expedition to the north-east passage, when he was working with John Dee. This was a variant of the zigzag line system introduced by Johann Hommel (+1518 to +1562) from whom Tycho Brahe obtained it. Another plan, that of concentric circles, had also been described by P. Nuñez about +1542, but it was not so useful (cf. R. Wolf (3), vol. 1, p. 392). The 'vernier' scale of P. Vernier came in +1631 (Houzeau (1), p. 953). The Jesuit altitude quadrants and azimuth circles at Peking have Hommel graduations (personal observations, 1952; Fig. 114). The micrometer was invented by William Gascoigne about +1640 and telescopic sights were demonstrated by Robert Hooke about +1667 (Andrade, 1).

[b] Which had radii of about 20 and 8 ft. respectively (Raeder, Strömgren & Strömgren (1), pp. 29, 89).

[c] Descriptions by Kaye (4, 5), and first by W. Hunter in 1799.

[d] Mieli (1), p. 108; Hitti (1), p. 315.

[e] Sarton (1), vol. 1, p. 667.

[f] Judging from the photograph of the excavated portion of it given by Christie (opp. p. 184) this tradition seems correct. An excellent and fully illustrated account can now be found in the work of Kari-Niyazov (1).

[g] Pp. lvii, cxxix.

[h] Al-Qaraqa, quoted by al-Maqrīzī (Mieli (1), p. 270); Nolte (1), p. 14.

[i] P. 372 below.

[j] It was unfortunate that this excellent work, which appeared in 1939, was not available to Maspero for his great monograph on Chinese astronomical instruments (4), of the same year.

[k] The general similarity with the construction of a Babylonian *ziggurat* should not be overlooked.

1 郭守敬 2 周公測景臺

Fig. 115. The Tower of Chou Kung for the measurement of the sun's solstitial shadow lengths at Kao-chhêng (formerly Yang-chhêng), some fifty miles south-east of Loyang, considered by Chinese astronomers in ancient times to be the centre of the world (photo. Tung Tso-Pin *et al.*). In its present form the structure is a Ming renovation of the building erected by Kuo Shou-Ching about +1276 for use with the 40 ft. gnomon. This stood upright in the niche, and the shadow was measured along the horizontal graduated stone scale (for details see text). One of the rooms on the platform housed a water-clock (perhaps a water-driven mechanical clock), the other probably an armillary sphere.

Fig. 116. The Chou Kung Tower at Yang-chhêng; looking along the shadow scale with its water-level channels (photo. Tung Tso-Pin *et al.*).

Fig. 117. The Chou Kung Tower at Yang-chhêng; looking down at the shadow scale (photo. Tung Tso-Pin *et al.*).

platform,[a] upon which there stands, on the north side, a one-storey building of three rooms, the centre one having a wide opening to the north giving a good view of the top of the 40 ft. gnomon (now no longer there) and the shadow which it once cast. The top level was known as the Star Observation Platform (*kuan hsing thai*[1]); it probably once possessed a thin vertical rod for observations of meridian transits, and the records show that one of the rooms was equipped with a large clepsydra. Below the tower on the north side, lying on the ground and extending for over 120 ft., is the Sky Measuring Scale (*liang thien chhih*[2]). This carries, besides graduations, two parallel troughs communicating at the ends and forming a water-level, and it continues into the body of the pyramid, which is cut away to receive it, so that its base is perpendicularly below the wall of the central chamber of the observatory-house at the top. The gnomon was almost certainly an independent pole sunk into a socket at the near end of the horizontal graduated scale. The height of the platform itself above that of the graduated scale was only about 28 ft. The whole construction is of brick. According to the *Yuan Shih*, three other places (Tatu[3] (now Peking), Shangtu[4] (the Xanadu of the English poet, and the Mongol imperial summer capital in Chahar province), and Nanhai[5] in Kuangtung) were destined to receive gnomons of 40 ft., but only at Yang-chhêng and Peking was one actually set up, and Yang-chhêng alone (the 'centre of the earth') ranked so high as to have the tower as well.

Tung Tso-Pin and his collaborators bring much textual evidence to prove that Yang-chhêng was the place where the official astronomers, at least as early as the Han, made the 'standard' solstice measurements.[b] Although it is not certain whether there was any tower approaching the dimensions of the present one in times earlier than the Yuan, the Confucian temple surrounding it contains a stele set up by Nankung Yüeh in +723. This is so arranged that the sun at summer solstice casts no shadow beyond its heavy truncated pyramidal base, and the stele itself measures exactly 8 ft. according to the metrology in use during the Thang. Perhaps we can gain an idea of what the Thang tower would have looked like from a very interesting observatory tower still extant in Korea (Fig. 118). This is the Chan Hsing Thai[6] at Kyungju (Chhing-chow) near the south-east coast, which was built in the reign of Queen Sŏndŏk[7] of Silla (+632 to +647).[c] Standing about 30 ft. high, the stone bottle-shaped structure has

[a] The Jesuits found it quite impressive. In the *Novus Atlas Sinensis* of Martin Martini (1655), p. 62, we find: 'Ex hisce Têngfung non ita celeri pede ac calamo percurrenda, quippe quam item in ipso orbis centro ac meditullio constituunt Sinae, in ea spectatur etiamnum ingens regula supra aeneum planum ad perpendiculum erecta in certas divisa partes, uti et in ipsa plani superficie linea extensa in suas etiam partes distributa, quo instrumento Cheucungus magnus ille apud Sinas Astrologus et Mathematicus, summusque totius olim Imperii praefectus, umbram meridianam observabat, atque inde altitudinem poli, caeteraque quae ex ea colligi possunt, venabatur. Vixit is ante Chr. natum annis mille centum et viginti; ibidem visitur turris, in qua solitus syderum notare cursus ac conversiones, dicta Quonsingtai, hoc est adspiciendorum syderum turris....'

[b] For instance, the *Sui Shu* says, quoting Yü Hsi's *An Thien Lun* (*YHSF*, ch. 77, p. 4*b*), 'Lohsia Hung turned the armillary sphere (or rings) for the emperor Hsiao Wu Ti at the centre of the earth, so as to determine the seasons and the fortnightly periods...' (ch. 19, p. 14*b*).

[c] *Samguk Sagi*, ch. 5; see Anon. (5); Rufus (2), p. 13.

[1] 觀星臺 [2] 量天尺 [3] 大都 [4] 上都 [5] 南海 [6] 瞻星台
[7] 善德

one large window facing the pole-star and carried at the top a wooden platform for an armillary sphere and for the nightly observers.

In the *Yuan Shih* there is an account of the gnomon worth reproducing:

The graduated scale (*kuei piao*[1]) is made of stone 128 ft. long, 4′ 5″ wide and 1′ 4″ thick; its base is 2′ 6″ high. Circular basins are excavated at the north and south ends, each 1′ 5″ in diameter and 2″ deep. From 1′ north of the gnomon for 120′ a central strip of 4″ wide is marked off, 1″ on each side of which is divided into feet, inches and tenths, extending to the north end. One inch from the edge are water channels, 1″ deep, connected with the reservoirs at the ends, for the purpose of levelling.

Fig. 118. The astronomical observatory at Kyungju (Chhing-chow) in Korea, built between +632 and +647 (drawing from Anon. 5).

The gnomon is made 50 ft. long, 2′ 4″ wide and 1′ 2″ thick (of bronze?)[a] and is fixed in the stone base at the south end of the graduated scale. Inserted to a depth of 14′ in the earth, it rises 36′ above the scale. At the top the gnomon divides into two dragons which sustain a cross-bar (*hêng liang*[2]). From the centre of the cross-bar the measurement to the top of the gnomon is 4′ and hence to the top of the scale it is 40′. The cross-bar is 6′ long and 3″ in diameter, and carries a water-channel on the top for the purpose of levelling.[b] At its two ends and in the centre are transverse holes, $\frac{1}{5}$″ in diameter, through which are inserted rods 5″ long, having plumb-lines attached to them so that the correct position can be ascertained and lateral deflection prevented.

When a gnomon is short, the divisions on the scale have to be close together and minute, and most of the smaller divisions below feet and inches are difficult to determine. When a

[a] Tung Tso-Pin thinks brass.

[b] This was an application of the technique still used in the 'striding level' on modern transit instruments (Spencer-Jones (1), p. 79).

[1] 圭表 [2] 横梁

gnomon is long, the graduations are easier to read, but the inconvenience then is that the shadow is light and ill-defined, making it difficult to get an exact result. In former times, observers sought to ascertain the real point by using sighting-tubes, or a pin-point gnomon and a wooden ring, all devices for easier reading of the shadow mark on the scale. But now with a 40′ gnomon, 5″ of the graduation scale corresponds to what was only 1″ previously and the smaller subdivisions are easier to distinguish.[a]

So far there is nothing original except the increase of size, but the account continues with the description of a device, the Shadow Definer (*ying fu*[1]), which seems to have been an ingenious new invention.

The shadow definer is made of a leaf of copper 2″ wide and 4″ long, in the middle of which is pierced a pin-hole. It has a square supporting framework, and is mounted on a pivot (*chi chu*[2]) so that it can be turned at any angle, such as high to the north and low to the south (i.e. at right angles to the incident shadow-edge). The instrument is moved back and forth until it reaches the middle of the (shadow of the) cross-bar, which is not too well-defined, and when the pin-hole is first seen to meet the light, one receives an image no bigger than a rice-grain in which the cross-beam can be noted indistinctly in the middle. On the old methods, using the simple summit of the gnomon, what was projected was the upper edge of the solar disc. But with this method one can obtain, by means of the cross-bar, the rays from the centre of the disc without any error.

(In +1279) on the 30th May, I observed the summer solstice shadow as 12·3695 ft. in length; and on the 11th Dec. of the same year found that the winter solstice shadow was 76·7400 ft. long.[b]

This passage was long misunderstood. Gaubil (7), Wylie (7), and even Tung Tso-Pin (*1*),[c] thought that the instrument was placed at the top of the gnomon, but Maspero (4) showed rather convincingly that on the contrary it was moved along the horizontal graduated scale and had the effect of focusing, like a lens, the image of the cross-bar. That Kuo Shou-Ching should have utilised the principle of the pin-hole is not at all surprising, since, as will be seen in the Section on physics, it had been familiar to Chinese scientists at least three centuries earlier, and indeed the camera obscura may have passed from them to the Arabs. There is, moreover, contemporary supporting evidence for Maspero's interpretation, in a remark by the astronomer Yang Huan[3] (*d.* +1299).[d]

The observations of Kuo and his assistants were collected in a book with the title *Erh Chih Kuei Ying Khao*[4] (Studies on the Gnomon Shadows at the Two Solstices), but it has long been lost, and the calendrical chapters of the *Yuan Shih* are now our only resource. Laplace himself considered the work with the 40 ft. gnomon in the +13th century as perhaps the most accurate which had ever been done on solstice shadows.[e]

[a] Ch. 48, p. 8*b*, tr. Wylie (7), mod.

[b] Tr. Maspero (4), eng. auct.

[c] Who, however, appreciated the optical similarity to the diaphragm of the photographic camera.

[d] *TSCC, Li fa tien*, ch. 108, p. 3*b*.

[e] 'The observations made from +1277 to +1280 are valuable on account of their great precision, and prove incontestably the diminution of the obliquity of the ecliptic and the eccentricity of the earth's orbit between then and now' (Laplace (2), p. 398).

[1] 影符 [2] 機軸 [3] 楊桓 [4] 二至晷影考

The use of a hole in the top of the gnomon did not appear in China until the Ming, perhaps not till after the coming of the Jesuits. In a lengthy analysis, Maspero[a] has shown that the attribution of this to the Han people arose from a mistranslation by E. Biot[b] in his version of the *Chou Pei Suan Ching*; the passage really concerns an erroneous calculation of the sun's diameter based on the use of the sighting-tube. However, we have several descriptions of the gnomon dating from about +1440, but restored in +1744, which in 1900 was still in the enclosure of the old observatory (Kuan Hsiang Thai[1]) in Peking (see Fig. 119).[c] It was installed in a dark hall (the Kuei Ying Thang[2]), and the top of the 8 ft. vertical[d] had a hole which received sunlight from an opening in the roof.[e] This was the sort of instrument which paralleled the Chou Kung Tower among the massive Indian types, save that there the entering sun-rays were caught not upon a horizontal scale, but upon a curved quadrant.[f]

It is of interest to compare the Indian observatories with the Chou Kung Tower. Their builder was the Maharajah Jai Singh of Jaipur (+1686 to +1743), who set up more than forty instruments at Delhi, Jaipur, Ujjain and Benares.[g] Though a Hindu working with Hindu assistants, he was entirely in the Muslim-Arabic tradition of astronomy, and considered himself the continuator of the work of Ulūgh Beg.[h] Nevertheless, a great deal of European work was available to him, and his observers used Flamsteed and la Hire no less than Ptolemy.

The large instruments of masonry, plaster and metal which Jai Singh caused to be constructed (cf. Fig. 120) may be listed as of the following types:[i]

(1) *Samrāṭ yantra.* 'Supreme Ruler Instrument', huge equinoctial dial with gnomon sloping in the polar axis, flanked by quadrants in the plane of the equator (Fig. 121).

(2) *Miśra yantra.* 'Mixed Instrument', dial with a set of gnomons arranged in the form of arcs for two meridians west of Delhi and two meridians east.

(3) *Rāśivalaya yantra.* A battery of twelve zodiacal dials, oriented not on the equatorial but on the ecliptic pole, for the times of rising of each of the zodiacal signs; thus giving directly the sun's longitude.

[a] (4), pp. 273 ff.

[b] (4), p. 605.

[c] In the spring of 1946 I did not see it, but several buildings at the base of the tower on the city wall were locked at that time. Fabre's guide makes no mention of it. In 1952 I saw what might have been the marble base of it standing in the open air just south of the buildings. The references are Chhang Fu-Yuan (*1*) and Chhen Tsun-Kuei (*6*), p. 55.

[d] Lengthened to 10 ft. in the Chhing restoration.

[e] Good description in the *Ta Chhing Hui Tien*[3] (Imperial Statutes of the Chhing Dynasty), ch. 81, p. 1*b*.

[f] Other tall gnomons still exist. In 1956 I had opportunity to study that which is constituted by a hole in the roof of the Archiepiscopal Basilica of San Petronio at Bologna. Arranged by Domenico Cassini in 1695, the height of the 'gnomon' is 81 ft. and the distance between the solstice points on the meridian line marked in the floor about 168 ft.

[g] The summary which follows has been prepared from the elaborate works of Kaye (4, 5). I myself had the great pleasure of visiting the Delhi observatory (the Jantar Mantar) in 1942. Cf. Das (1). We are greatly indebted to Dr H. von Kluber for permission to use his unpublished photographs.

[h] Whose star catalogue has been made available by Knobel (3).

[i] Not all of these are found at any one of the four Indian observatories. I take the opportunity of listing them here as there will be later references to them. Cf. Soonawala (1).

[1] 觀象臺 [2] 圭影堂 [3] 大清會典

Fig. 119. Gnomon and bronze shadow scale of about +1440, repaired and provided with a new marble base in +1744; now at Peking but not intended for that location (photo. Whipple Museum Collection). Behind it, on the left, is a late small-scale reproduction of Stumpf's quadrant altazimuth, now at Nanking.

Fig. 120. The observatory of Jai Singh at Jaipur (*c.* +1725), seen from the top of the large *samrāṭ yantra*. In the centre foreground are two *scaphe* or bowl dials, the *jai prakāśa*, and to their right the *narivalaya yantra* or equatorial dial similar to the Chinese type. One of the white zodiacal dials (*rāśivalaya yantra*) is seen in the left foreground, and a small *samrāṭ yantra* stands behind the *scaphes*. (Unpub. photo. von Kluber, 1930).

Fig. 121. The observatory (Jantar Mantar) at Delhi (*c.* +1725), showing the great equinoctial sundial (*samrāṭ yantra*) from the north (unpub. photo. von Kluber, 1930).

Fig. 122. The observatory of Jai Singh at Jaipur; the two *chakra yantra* or equatorially mobile hour-angle circles set in the polar axis, clearly related to Kuo Shou-Ching's equatorial mounting (see Fig. 168) of *c.* +1275 (unpub. photo. von Kluber, 1930).

Fig. 123 (*a*). Korean *scaphe* sundial of the late +17th century (photo. du Bois Reymond).

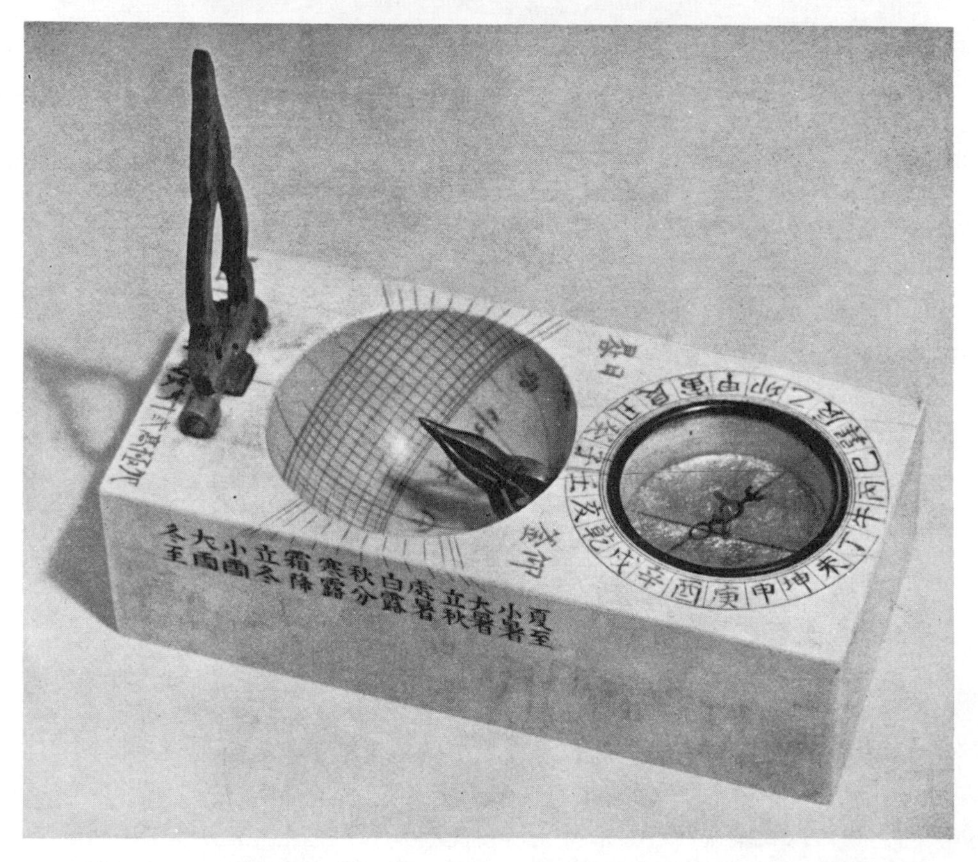

Fig. 123 (*b*). Japanese *scaphe* sundial (*hidokei*) with compass, made by Chiang Jun (probably a Korean). and dated 1810 (photo. Old Ashmolean Museum, Oxford).

(3*a*) *Krāntivṛitti yantra.* Identical with the 'torquetum' of late medieval European astronomers (see below, pp. 370 ff.), but now incomplete.

(4) *Dakshiṇovṛitti yantra.* Meridian quadrants on walls, like Tycho Brahe's, but larger, for star declinations at transit.

(5) *Shashtāṃśa yantra.* 'Sixty Degree Instrument', huge fixed sextants or graduated arcs in dark rooms built into the supports of the quadrants of the *samrāṭ yantra*, the sun's rays entering through an orifice in the roof—for solstice altitudes. Comparable with the Chinese gnomons.

(6) *Jai prakāś.* Inverted hemispheres marked with coordinates and surmounted by stretched cross-wires, which may indicate the sun's position by their shadows or may be used as alidades for star positions.

(7) *Kapāla.* 'Bowl', a variant of the last, in which the graduations correspond, not to the horizon but to the solstitial colure; shows rising zodiacal signs.

(8) *Rām yantra*, (9) *digaṃśa yantra.* Both of these really circular protractors for azimuth observations. The former a large cylindrical walled enclosure with a pillar in the centre, all graduated. The second similar but with two concentric walls, the outer one higher than the inner. 'Direction-Degree Instruments' performing the function of quadrant altazimuths.

(10) *Narivalaya yantra.* A horizontal cylinder set in the meridian line with dials on the faces in the equatorial plane, marked with equal radial graduations. This forms a double sundial exactly equivalent to the Chinese equinoctial types described below. The Jaipur example, illustrated by Kaye (5), pl. IX*b*, carries at the centre of one dial an inscription.[a] Chinese inspiration can hardly be absent.

Besides these, the observatories had equatorial armillary spheres or rings (*chakra yantra*, Fig. 122),[b] astrolabes (*yantra rāja*), and armillary rings in various mountings.[c] Some of the instruments, especially nos. 6, 8 and 9, exist in duplicate in order to allow of easy access by the observers to any desired point. As Kaye has pointed out, many[d] derive ultimately from the inverted hemisphere of Berossos,[e] which is so often figured,[f] and if correctly attributed, would date from about −270. The classical description of the *scaphe* or *hemicyclium excavatum* is in Vitruvius.[g]

It is thus of some interest that Dubois-Reymond has described an inverted hemisphere sundial of pale bronze kept in Korea in recent years but certainly of Chinese origin (Fig. 123). The hollow is the size of an ordinary soup-plate and the tip of the gnomon occupies the centre of the opening. An inscription states 'Polar height 37°, 39′, 15″' (the latitude of Seoul), and the instrument calls itself an 'Upward-looking Bowl Sundial' (*yang fu jih kuei*[1]). Numerous meridians and parallels are marked in the bowl, and there are the usual denary and duodenary characters on the rim. The

[a] The graduations are in hours, minutes, *ghaṭi* and *pala*.

[b] Up to 6 ft. diameter. Their derivation from Kuo Shou-Ching's 'Simplified Instrument' is obvious. See below, p. 371.

[c] Such as the *unnatāmsa yantra*, a 17½ ft. diameter altazimuth circle.

[d] Especially nos. 1, 2, 3, 6 and 7.

[e] Cf. Schnabel (1).

[f] As by B & M, p. 627; Zinner (1), p. 38. Cf. Drecker (1, 2).

[g] IX, 6–8. Cf. Diels (1), pp. 163 ff.

[1] 仰釜日晷

instrument incorporates a water-level. Though the writing is in seal character, the instrument, from its form, cannot be early, and may well be late +17th century—perhaps it was one of those which the Koreans An Kuo-Lin[1] and Pien Chung-Ho[2] took back from Peking about +1741.[a] But here for the first time we meet with the inclined gnomon in China,[b] an encounter which raises the question of the sundial as such, making it necessary that we should retrace our steps.

(3) The Sundial (Solar Time Indicator)

The determination of time by most sundials depends on noting the direction of the shadow, not the length.[c] Perhaps owing to its very ubiquity and familiarity, clear references to the sundial or sun-clock (horologium, as the medieval Europeans called it)[d] are rare in Chinese literature. The term *kuei piao*[3] is usually understood to refer to it. Here the character *kuei* combines the sun radical with an old word *chiu*[4] (K 1068), now meaning fault or blame, but which contains the divination symbol *pu*[5] and must have been originally concerned with the detection of auspicious and inauspicious times. An early reference is found in connection with the assembly (*po shih kung i*[6]) of calendar experts which Ssuma Chhien advised should be called together in −104. After meeting at the capital, the experts, says the *Chhien Han Shu*,

> determined the true east and west points, set up sundials and gnomons, and contrived water clocks (*li kuei i hsia lou kho*[7]). With such means they marked out the twenty-eight *hsiu* according to their position at various points in the four quarters, fixing the first and last days of each month, the equinoxes and solstices, the movements and relative positions of the heavenly bodies and the phases of the moon.[e]

The *Chhien Han Shu* bibliography mentions an *Jih Kuei Shu*[8] (Sundial Book) in thirty-four chapters which was included in a calendar compilation by Yin Hsien[9] (*fl.* −32 to −7).

[a] These bowl sundials afterwards became popular in Japan, where (like other types) they are called *hidokei*,[10] and occur in considerable numbers (Michel, 10).

[b] Actually, since the *scaphe* depends on sun shadow lengths and not directions, the fact that the gnomon of this instrument seems to be inclined in the polar axis, is immaterial to its function. All that matters is that something must cast a shadow from the centre of the opening of the bowl.

[c] Because they are oriented in the meridian and register the direction or azimuthal position of the sun. There are a certain number of types of sundial, however, which do measure the length of the shadow, and register the altitude of the sun. Besides the *scaphe* just mentioned, Michel (10) and Ward (1) describe, for example, cylindrical types (the 'shepherd's dial'), ring types and disc or 'astrolabe' types. Late Chinese versions of the two latter are known to exist, but there seems no reason for thinking them indigenous. The former (the 'chilindre') goes back to the +12th century at least, for an Arabic example of +1159 is known (Casanova, 1). A flat silver tablet and pin, used in the same way, was found at Canterbury in 1938; it is of the +9th century.

[d] Cf. Thorndike (1), vol. 5, p. 397; Sarton (1), vol. 3, p. 717.

[e] Ch. 21A, p. 16*b*, tr. Yetts (5). The calendar which resulted was the Thai Chhu calendar, and Têng Phing and Lohsia Hung must have been the two most important men among the experts.

[1] 安國麟 [2] 卞重和 [3] 晷表 [4] 咎 [5] 卜 [6] 博士共議
[7] 立晷儀下漏刻 [8] 日晷書 [9] 尹咸 [10] 日時計

Later, in the *Sui Shu*, there are numerous references to the same kind of instrument for the +6th century. But we here come up against a difficulty which we shall constantly meet with in the technological Sections, namely, that of determining the exact nature of an ancient instrument from the names loosely used for it. Here everything depends upon whether the gnomon was placed vertically, pointing towards the zenith (as it was for the measurements discussed in the previous sub-section), or whether it was inclined at an angle depending on the latitude, and pointing at the celestial pole[a]—only in the latter case are measurements of equal intervals of time possible. If one looks, for example, at the *kuei piao*[1] section in the *Thai-Phing Yü Lan* encyclopaedia[b] of +960, one finds that all the references are clearly to shadow-lengths of vertical gnomons and not shadow-directions of inclined ones.

The further elucidation of this subject brings up what was long one of the standard enigmas of Chinese archaeology, namely, the significance of the so-called 'TLV-mirrors' of the Han dynasty. A design which has been given this name is frequently found on the backs of Han metal mirrors,[c] as reference to any of the described collections will show, and one of them has had its cosmic symbolism minutely analysed by Yetts.[d] Fig. 124*a* shows the typical pattern, and the mirror from the Cull collection is reproduced in Fig. 126; it dates from the Hsin (+9 to +23). The earliest of such mirrors extant would be datable about −250.[e]

Similar markings were then noticed in some of the reliefs of Han times. A scene from the Wu Liang tomb-shrines (+147), shown in Fig. 125, depicts what was at first thought to be merely a banquet (Chavannes, 9), but which Laufer (7) and Nakayama (*1*) recognised to be a scene of magical operations. In the background a board which bears the TLV markings is apparently hanging on the wall, while on the ground, carefully drawn so as to indicate that it is horizontal, is a small table which may plausibly be identified as the diviner's plate or *shih*[2] (cf. Vol. 2, p. 361 above). This consists of a square representing earth surmounted by a rotating disc representing heaven. A diagram of the TLV-board is given in Fig. 124*b*.[f] Another stone relief from the Hsiao Thang shrine (+129) shows two men playing a game at a table the side of which again has a board with these symbols (Figs. 127 and 124*c*); probably this was meant to indicate what the surface of the table carried.

We thus come upon an association of divination with games which later, in the Section on physics, we shall find to have been of great significance in relation to the discovery of magnetic polarity. Yang Lien-Shêng (1, 2) has taken the matter further by describing a Han mirror (Fig. 130) which actually shows two immortals playing a game at a board with these markings (Fig. 124*d*). We know what they called it, for

[a] Strictly speaking, the word gnomon ought perhaps to be reserved for the vertical rod, while the pole-pointing one should be termed style or stylus; cf. Michel (8); K. Higgins (1).

[b] Ch. 4, pp. 4*b* ff.

[c] See, e.g., Koop (1), plates 71, 72.

[d] (5), pp. 116 ff.

[e] White & Millman (1).

[f] The Fêng brothers, in the *Chin Shih So*, were not very careful in getting the markings correct and Yetts' version should be preferred.

[1] 晷表 [2] 栻

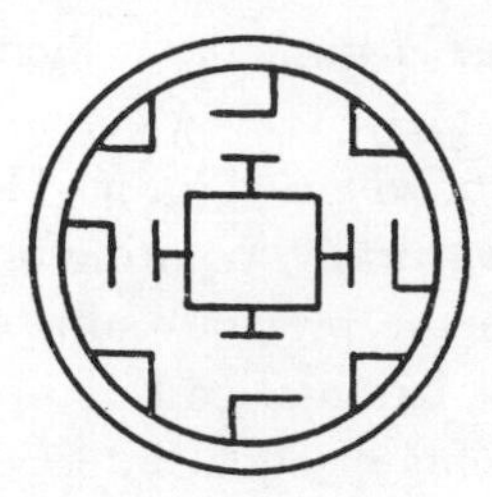

(*a*) Typical TLV design on Han mirrors (Yetts, 5).

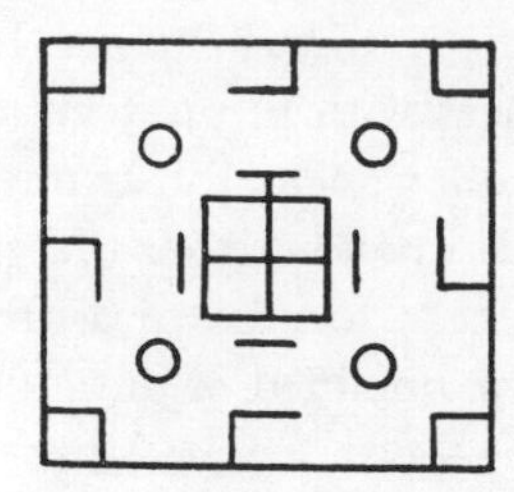

(*b*) Board for divination or the *liu-po* game in the Wu Liang tomb-shrine relief; Chavannes (9); Yetts (5). The carver forgot to put in three out of the four uprights of the T's, or else they became indistinct by weathering in course of time.

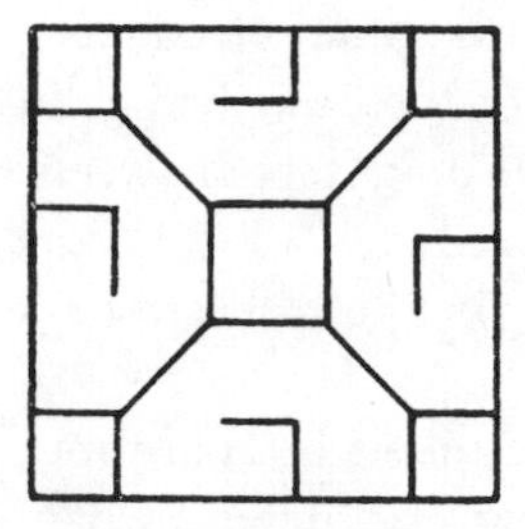

(*c*) Board for divination or the *liu-po* game in the Hsiao-Thang Shan tomb-shrine relief, *c.* +129 (White & Millman, 1). The carver made a mistake as to the direction of the right-hand and bottom L's.

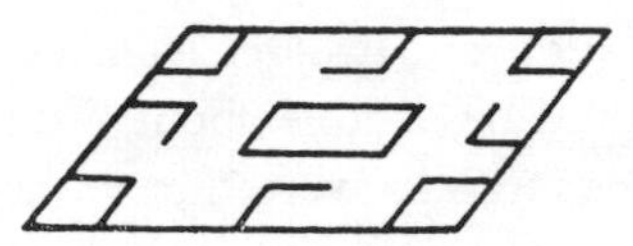

(*d*) Board for divination or the *liu-po* game in a Han mirror (Yang Lien-Shêng, 1).

Fig. 124. TLV designs on mirrors and divination boards.

Fig. 125. A scene from the Wu Liang tomb-shrines (*c.* +147) depicting magician-technicians at work. In the background, a TLV board hanging on the wall, in the centre a *shih* (diviner's board) on the floor. *Chin Shih So, Shih* sect. ch. 3, redrawn by Yetts (5).

Fig. 126. Bronze TLV mirror from the Hsin dynasty (+9 to +23). For the cosmic symbolism of these designs and their relation to the sundial, see text. This example (from the Cull Collection of Chinese Bronzes, cat. Yetts, 5) is inscribed with a poem beginning at the quincunx on the right (at 3 p.m.):

'The Hsin have excellent copper mined at Tan-yang,
Refined and alloyed with silver and tin, it is pure and bright.
This imperial mirror from the Shang-fang (State workshops) is wholly flawless:
Dragon on the east and Tiger on the west ward off ill-luck;
Scarlet Bird and Sombre Warrior accord with Yin and Yang.
May descendants in ample line occupy the centre,
May your parents long be preserved, may you enjoy wealth and distinction,
May your longevity endure like metal and stone
May your lot match that of nobles and kings.' (tr. Yetts, 16)

PLATE XXXVIII

Fig. 127. A scene from the Hsiao-Thang Shan tomb-shrines (*c.* +129) showing two men playing a game (perhaps *liu-po*) at a table, the side of which has a TLV board (photo. White & Millman).

the scene is inscribed *hsien jen liu-po.*[1] This game of the 'Six Learned Scholars' occurs in the −3rd-century poem 'Chao Hun'[2] (Calling Back the Soul) of Sung Yü,[3] and commentators say that it was played with chess-men (*chhi*[4]). Here one of the two immortals holds a cup for dice, and the other some sticks, while two more look on. Other representations have been found in carvings and tomb pottery models. Kaplan was on the right track in seeking to associate the TLV-boards with the *shih* (diviner's plate) itself, though his suggested identification may not be correct, for the plate was a divination instrument rather than a game.[a] The TLV-mirror, unlike the *shih*, has the circle outside, instead of inside, the square.

In view of all this then, it is extremely striking to find that the only existing objects which must have been Han sundials both bear TLV markings.[b] One of these was in the collection of the Manchu prince Tuan Fang;[c] it has been discussed by Thang Chin-Chu,[5] Chou Ching,[6][d] Maspero (4), Liu Fu (*1*), and Yetts (5). The other was acquired by Bishop White for the Toronto Museum, and has been discussed by White & Millman (1).[e] Both are identical in layout, though the former is of jade and the latter of limestone; I reproduce the latter here (Fig. 129, diagram in Fig. 128), since it is the more accurately graduated instrument.[f] The Chinese name for these dials is *tshê ching jih kuei.*[7]

The graduations appear on one face only. Around the hole or seating at the centre two circles are described with much accuracy, two-thirds of the annular space being divided into equal segments forming hundredths of the circumference. Where the lines meet the outer circle there are a series of small sockets numbered 1–69 in a clockwise direction by characters in 'small seal' style. This, and especially the old form of *chhi*,[8,9] the number 7, dates the instrument as Former Han, perhaps −2nd century. In addition, there are diagonal lines ending in the V's, four T's extending from the central square and circle, and four L's based on the cardinal point lines at the periphery.

Everything depends, in the interpretation of these dials, on whether they were used in the plane of the horizon or the equator. Chou Ching believed that they were placed horizontally for determinations of azimuth at sunrise and sunset, and this was the first interpretation adopted by Maspero (4), who supposed that the segmental graduations, corresponding to the sixty-nine quarters of an hour of the longest day, were placed to the north of the central gnomon. But noting that the shadow of whatever it was that was mounted in the central socket would have been too large for any accurate reading of the divisions, he rallied to the opinion of Thang Chin-Chu and concluded that a

[a] It is true that for primitive societies this distinction can hardly be made. Games were a form of divination which indicated the will of the gods, as we know from many Amerindian examples, and shall find abundantly in Sect. 26*i*.

[b] The connection between the dials and the mirrors was noted almost simultaneously by Yetts (5) and Karlbeck (1).

[c] And was described in his *Thao Chai Tshang Shih Chi.*

[d] In the notes to the same collection.

[e] A third exists in the form of a small fragment.

[f] The TLV markings were scratched in roughly on the jade dial, presumably by a later hand.

[1] 仙人六博 [2] 招魂 [3] 宋玉 [4] 棊 [5] 湯金鑄 [6] 周暻
[7] 測景日晷 [8] 七 [9] 十

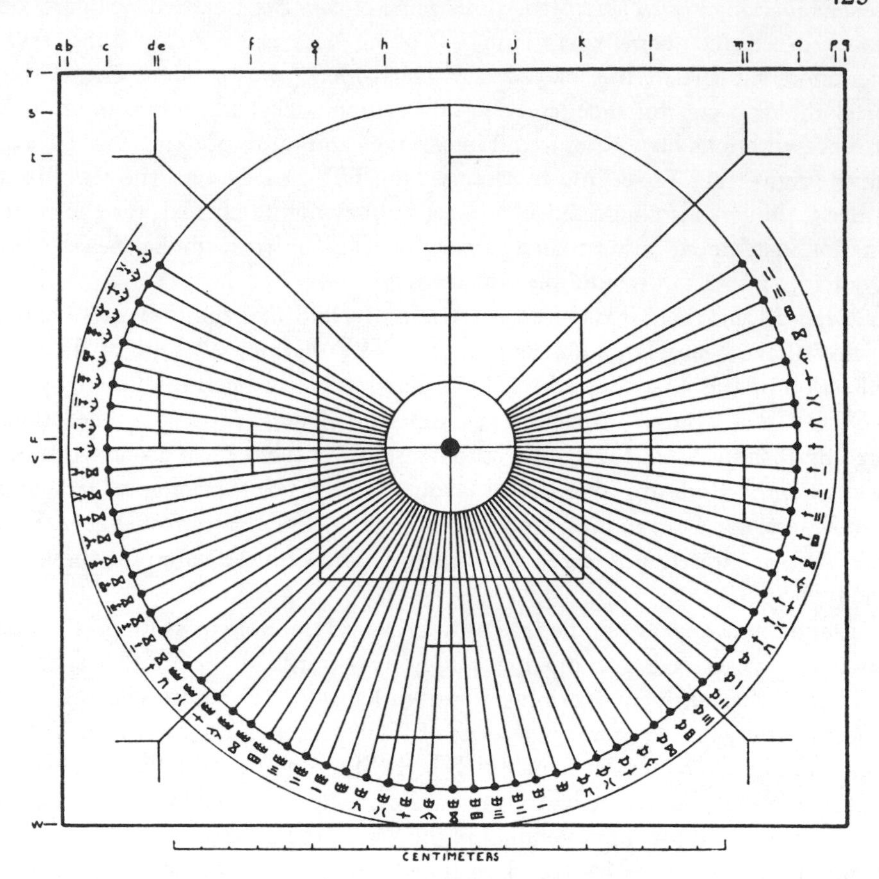

Fig. 128 Inscriptions on the Former Han sundial shown in Fig. 129 (from White & Millman, 1).

movable gnomon was placed in one or other of the peripheral sockets; the graduations thus being placed to the south of the central gnomon, and the time being given by the position at which the direction of the movable gnomon's shadow coincided with that of the fixed central gnomon. Since, however, the result of such a procedure would have been extremely inaccurate, he decided that these dials were not time-keepers at all, but regulators for the clepsydra. Its indicator-rods[a] would be selected by the official in charge according to the length of the day then dawning, and this would be known by the azimuthal position of the sun at sunrise. There is no doubt that what are sets of measurements of the sun's azimuthal position in relation to time have been preserved, e.g. those of Ho Jung[1] in +102[b] and Yuan Chhung[2] in +594,[c] but they were made with the 8 ft. gnomon, and tell us nothing about the mode of employment of these Han dials.

In view of the basically polar and equatorial character of Chinese astronomy, it

[a] See below, p. 321. [b] See *Hou Han Shu*, ch. 12, p. 8*b*. [c] *Sui Shu*, ch. 19, p. 26*b*.

[1] 霍融 [2] 袁充

PLATE XXXIX

Fig. 129. Plane sundial of the Former Han period, grey limestone, about 11 inches square (Collections of the Royal Ontario Museum, photo. White & Millman 1).

Fig. 130. Part of a Han mirror showing two feathered and winged Taoist immortals playing the *liu-po* game on a board with the TLV markings (photo. Yang Lien-Shêng, 1).

Fig. 131. A Taoist consulting the characteristic type of Chinese sundial; plate set in the equatorial plane, and pole-pointing gnomon (or stylus).

seems more likely that at an early time the discovery was made that if the base-plate was inclined in the equatorial plane, and the gnomon was made to point at the imperial pole of the heavens, a solar time-keeper would result.[a] This would occur very naturally to people who were accustomed to using the jade template of the circumpolar constellations (described below, p. 336), since this equinoctial star-dial must also be set in the plane of the equator. Accepting the suggestion of Thang Chin-Chu that the peripheral or secondary gnomon was movable, Liu Fu (*1*) proposed that the graduated surface was in fact inclined in the equatorial plane. Such a sundial receives the sun's rays on its upper face only for half the year, i.e. between March and September, when the sun is north of the equator. In the numerous examples of late (Ming and Chhing) sundials still found in China, the dial is graduated on both sides, and the pin-gnomon or style extends right through the dial so as to stick out below as well as above.[b] The dials in the imperial palaces at Peking are of this kind.[c] Possibly in the Han two dials were used, one facing upward and one down.

Liu's solution was that the movable peripheral gnomon was higher than the central one, so as to catch its shadow, the graduated part of the base-plate being located north of centre—he offered some evidence that this method was known in Yuan times, and referred to an obscure passage about gnomons in the *Chou Pei Suan Ching*.[d] But what is perhaps a better solution has been proposed by White & Millman (see Fig. 132). They suggest that the central object was not a gnomon but a rectangular bronze plate, connected with the peripheral T-shaped gnomon by a bronze bridge. Time would then be shown by that position in which the shadow of the peripheral gnomon fell on the upright plate in line with the bridge, and the season would be shown by the height of the cross-bar above the bridge, as in the annexed diagram.[e]

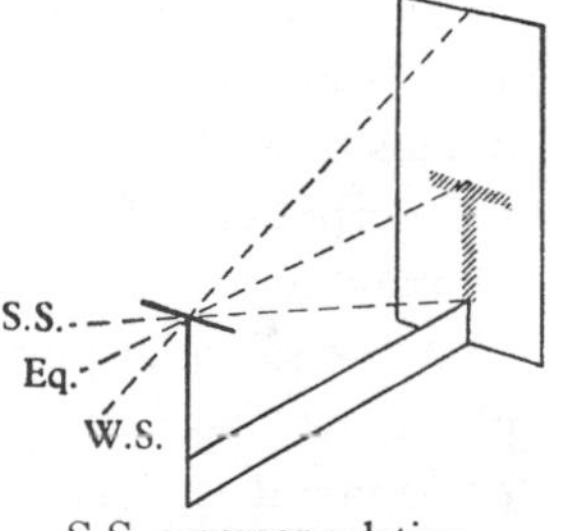

S.S. summer solstice
Eq. equinoxes
W.S. winter solstice

It remains only to explain the marks shaped like the letters T, L and V on these dials. Liu Fu suggested that they were records of measurements of the heights of the gnomons required, or of the shadows which they would be expected to throw at winter and summer solstices respectively. White & Millman (1) accept this view, but consider that the height of the peripheral gnomon was fixed by the distance from the circumference of the outer circle to the cross-bar of the T, while that of the central plate was fixed by the

[a] Neugebauer (3) says that this arrangement was, in a way, the converse of one which he thinks may have underlain the Greek development of the theory of conic sections (Menaechmus, *c.* −350; Apollonius, *c.* −220). His hypothetical type would have had the gnomon pointing to the sun at culmination, and the dial at right angles to the plane of the ecliptic. No such sundial has ever been found, but the British Museum possesses one Greek specimen of the characteristic Chinese type (see Drecker (3), ch. 5). Cf. p. 102 above.

[b] In the 17th century, Lecomte (p. 300) saw and described some 'antient' sundials centesimally graduated.

[c] Cf. Yetts (5), pl. XXXIV, and here Fig. 131. Cranmer (1) describes the construction of a replica.

[d] See E. Biot (4), p. 626.

[e] Later on (p. 365) we shall have occasion to refer to some remarkable applications of the principle of the peripheral gnomon in modern instruments used, for example, in navigation.

distance from the centre hole to the cross-bar of the L. At the least, this arrangement will give a system which actually works in practice. It should be noted that in this solution, the graduated part of the annulus is located south, not north, of the centre.

As for the V marks, Liu Fu adduced a number of cryptic passages in ancient texts which he felt justified the proposal that the most primitive sundial had been a square of cloth or oxhide stretched flat and placed in the equatorial plane. The V marks would be a reminiscence of four of the eight hooks originally used to make this arrangement stay in position, eventually restricted to four directions intermediate between the four cardinal points.[a]

Provisionally we may perhaps assume that the original purpose of the TLV markings was a practical and astronomical one. It was quite natural that they should have been reproduced on mirrors, especially those which embodied elaborate cosmical symbolism. The *liu-po* game board may have been an intermediate stage, or more probably an independent derivation. Doubtless connected with divination, nothing would have been more natural than to use for the game-board the sundial face, in which the form and march of the heavens were exhibited. The ornamental nature of the marks on the mirrors is shown by the fact that no example has yet been discovered with sockets in which gnomons could have been placed, and by the frequent dropping-out of the T or the L in the designs.

Uneasy about the sundial derivation, some modern scholars have sought further symbolism in the TLV mirrors. Cammann (2) wishes to link it with the Ming Thang complex[b] and with later Buddhist mandalas. Loewenstein has pointed out the similarity between the markings, especially the L's, and the swastika.[c] No doubt much remains to be discovered in these directions. One would expect that the design would have perpetuated itself in motifs of window lattices, and indeed some indications of this may be found in the compendium of Dye on the subject.[d]

If the equatorial-plane plate with both surfaces graduated and a gnomon extending both above and below was not in use in the Han, one would like to know when this development was made. The following passage, not before noted, may throw some light on the matter, for Tsêng Nan-Chung[1] himself, at any rate, evidently thought that he was introducing something new. The *Tu Hsing Tsa Chih*[2] (Miscellaneous Records of the Lone Watcher), written by his son or grandson, Tsêng Min-Hsing[3] (+1176), says:[e]

[a] What is perhaps more important is that all the Han sundials have a central square. This is distinctly suggestive of the central square on the circumpolar constellation template or equatorial star-dial, shortly to be described, from which they may have originated.

[b] Vol. 2, p. 287. See Granet (5), pp. 180ff.; Soothill (5); Maspero (25).

[c] The swastika appears clearly on Chinese Neolithic pottery and Shang bronzes. Loewenstein (1) brings evidence, not only from China, that it was a symbol both of fecundity and of death; he thinks therefore that it may have been connected in China with the origins of Yin-Yang dualism; if so its association with sundials would be understandable.

[d] (1), vol. 1, pp. 198, 201; vol. 2, p. 352.

[e] Ch. 2, p. 12*a*, tr. auct. The account of Tsêng Nan-Chung given in this book is interesting; he mastered astronomy from his youth, graduating in +1119. In order to watch for meridian transits of stars he took away the tiles from part of the roof of his house, and finally died owing to illness caused by falling asleep on an observation-platform in bitterly cold winter weather.

[1] 曾南仲 [2] 獨醒雜志 [3] 曾敏行

Fig. 132. Reconstruction of the mode of operation of the Han sundial shown in Fig. 129 (photo. White & Millman, 1).

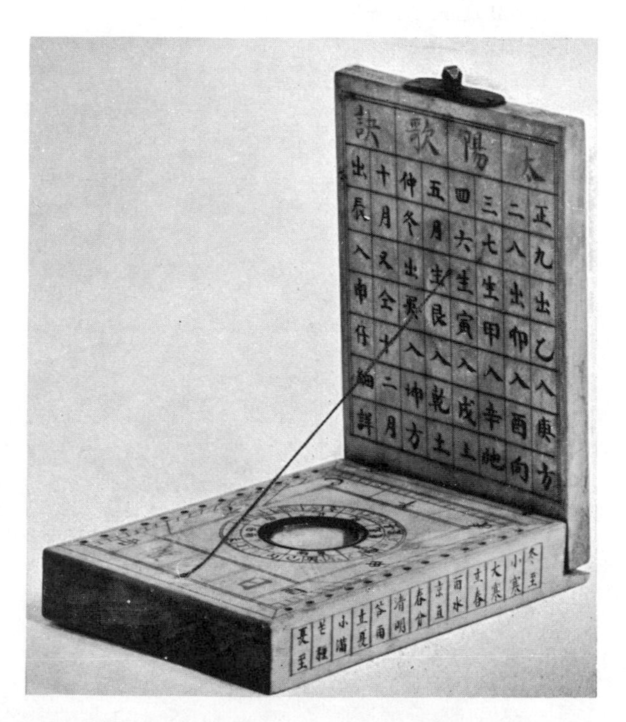

Fig. 133. Late Chinese portable sundial of Type *A*; the gnomon is formed by the stretched string and the horizontal plate has a simple stereographic projection. A compass-needle is provided for orientation. This type is probably not older than the end of the +16th century. From the collection of Mr W. G. Carey.

PLATE XLII

Fig. 134. Late Chinese portable sundial of Type *B*; the plate, which is in the equatorial plane and can be adjusted for latitude, carries a pole-pointing gnomon or style. A compass-needle, with detailed azimuth graduations, is provided for orientation. This type may go back at least as far as the Sung period. From the collection of Mr W. G. Carey.

Fig. 135. Equatorial moon-dial on the back of a Type *A* sundial; a pin in one of a set of holes gives adjustment for latitude. From the collection of Mr W. G. Carey.

Tsêng Nan-Chung used to say that while there was much in the ancient classics about recording the sun's shadow, it all concerned the length, and thus was not comparable with telling time by the water-clock. So at Yüchang he made a *Kuei Ying Thu*[1] (Diagram of the Sun Shadow), and constructed a sundial (*kuei*[2]). A round (wooden plate) was divided into four (equal divisions), and one of these segments was taken away (i.e. not graduated), so that (the graduated part) was shaped like a crescent moon. Hours and quarters were marked round the edge. The dial was supported on posts so that it was high towards the south and low towards the north (i.e. in the plane of the equator). The gnomon pierced the dial at the centre, one end pointing to the north pole and the other to the south pole. After the spring equinox one had to look for the shadow on the side facing the north pole, and after the autumn equinox, one found it on the other (the under) side. This instrument agreed more or less with the water-clock. Tsêng Nan-Chung was very proud of this device, and thought that he had got something which had not been achieved in ancient times. I, his descendant, once made one myself following his system. It was a strange thing to see how, at the equinoxes, the gnomon in the polar axis cast no shadow at all, the light falling directly on the edge of the dial. This is because its plane corresponds exactly with the equator. After the spring equinox the sun goes inside (i.e. north of) the equator, and after the autumn equinox it goes outside (i.e. south of) it. How exact and nice was this invention!

It would be impossible to have a clearer description of the typical equatorial sundial in which both faces are used according to the time of year. But further evidence would be needed to substantiate the attribution of the system solely to Tsêng Nan-Chung about +1130.

If, then, we are right in tracing the history of sun clocks in China to the discovery of the pole-pointing gnomon or style some time about the −4th century,[a] their development in Asia might be said to parallel that in the West.[b] The earliest ancient Egyptian unequal-hour shadow-clocks go back to the middle of the −2nd millennium (Ward), and the cup-dials already referred to, obvious inverted replicas of the celestial sphere with the unnecessary parts cut away, were probably not a new thing in Babylonia in the time of Berossos (−3rd century). The preparation of vertical or horizontal plane dials with gnomons in the polar axis required considerable knowledge of conic sections and they were therefore a natural product of Greek geometry.[c] Among the most famous are those on the walls of the Tower of the Winds at Athens, often figured,[d] but though the tower itself is of the −1st century, the dials may be later, even Byzantine. The art of gnomonics, and the stereographic projections which it involved, were particularly cultivated by the Arabs.[e]

[a] Mr Henri Michel (private communication) still prefers the view of Chou Ching that the Han dials so far discovered were goniometers for measuring the azimuth position of sunrise and sunset. If so, the nature of the Chhin and Han sundials remains to be determined.

[b] The book of Gatty, Eden & Lloyd (1) is mostly antiquarian and not of much help, but the encyclopaedia article of Godfray (1), which gives references to the literature, is excellent.

[c] Cf. Diels (1), pp. 179 ff. especially for the earliest 'double-axe' types.

[d] As by Napier Shaw (1).

[e] Cf. J. J. E. Sédillot (1) and Schoy (1).

1 晷影圖 2 規

(i) *Portable equinoctial dial-compasses*

In late times portable sundials which incorporated small magnetic compasses for orientation were manufactured in great numbers in China and are now commonly found in collections of scientific instruments. It does not seem to have been noted, however, that they fall into two quite distinct types, which we may call *A* and *B* (Figs. 133 and 134).[a] The first of these was described long ago by van Beek, Carus (2) and others. Upon opening the lid of the 'diptych', a fine cord is rendered taut; this is the gnomon, and it gives its shadow reading on the base-plate which carries the magnetic compass for orientation. This is therefore the characteristic occidental form in which the gnomon makes an acute angle with the plate and the graduations are unequal. One may feel confident that it was not known in China before the arrival of the Jesuits, and that it could not have been produced before the end of the Ming dynasty.[b] Indeed, its Chinese name is not only *phing mien jih kuei*[1] but also *yang kuei*[2] (the foreign sundial).[c]

Type *B*, however, is entirely different. Here also there is a plate carrying a magnetic compass, but the dial is graduated upon a separate plate which may be raised or lowered to any desired angle so that the gnomon, at right angles to the dial, may point at the pole whatever the latitude of the observer. The dial is then fixed in position by means of a prop at its back which is held in a toothed ratchet scale below. The curious thing is, however, that (in all specimens which I have seen) this ratchet scale is marked not with latitudes or names of cities,[d] but with the twenty-four fortnightly periods or *chhi*. The specimen of Type *B* figured here may be set at angles corresponding to a total range of from 15° to 50° N., which is just about the coverage required, Hainan being 20° N., Nanking 32° N. and Shangtu 42° N. It is therefore clear that each locality had its 'standard *chhi*'. On Hainan island, for instance, the shadow length at the equinoxes would be equal to that at Yang-chhêng (the centre) at summer solstice, while at Shangtu the equinoctial shadow would equal the winter solstice shadow at the centre of the whole region. Local calendars no doubt gave the 'standard *chhi*' setting. We see, therefore, that Type *B* is wholly in the Chinese tradition, and there is no reason whatever for regarding it as a Jesuit introduction—Shen Kua or even Chen Luan may well have been familiar with it.[e] And, indeed, we find a mention of

[a] The two specimens figured are from the collection of Mr W. G. Carey of Caius College, Cambridge.

[b] It will be seen further on (p. 374) that Jamāl al-Dīn brought two sundials to Peking in +1267, and they might have been of this type, though the earliest European example is of +1451 (Ward (1), p. 21). But in any case, it may be regarded as most unlikely that the necessary geometrical knowledge required to graduate the horizontal dial spread to any extent in China at that time.

[c] Matteo Ricci was a good gnomonist and made many sundials for his Chinese friends during the early years of his mission. Abundant references on this will be found in d'Elia (2).

[d] It will be noted that the *Yuan Shih*, ch. 48, pp. 12*b* ff., gives a table of polar altitudes at a number of different cities.

[e] The *Yuan Shih* says that Tsu Chhung-Chih in the +5th century fixed the methods of determining time by sun-shadow differences (*jih chha*[3]). But this probably concerns his work on the study of the sun's angular motion (see p. 123 above and p. 394 below). The only reference we have found in Chinese literature to a combination of sundial and magnetic compass occurs in relation to a book known as the *Pai Kung Phu*[4] (Record of the Hundred Works) about which we can find no information at all.

[1] 平面日晷 [2] 洋晷 [3] 日差 [4] 百工譜

a portable sundial in the *Shan Chü Hsin Hua* of Yang Yü written in +1360. He says[a] it was very convenient to use when travelling or on horseback, and he presented it to the emperor, who ordered it to be checked. At Shangtu it was about seven minutes slow and in Chekiang about seven minutes fast (difference 10° lat.). Apparently, therefore, this instrument had no adjustable plate, or perhaps insufficient adjustability.

Moreover, portable sundials even of Type *A* did not always completely abandon the plan of having the dial in the equatorial plane and the gnomon pointing at the pole. In our specimen of Type *A*, there is an arrangement on the outside of the lid destined to act as a moon-dial.[b] A pin is placed in an appropriate central hole, the lid is set at the required angle according to the latitude by two smaller pins fitted into one or other of a series of sockets in the base-plate (each one of these being again marked for a particular *chhi*),[c] the dial on the lid is turned so as to correspond with the day of the lunar month on which the instrument is being used, and the hour of the night then read off on a scale of equal graduations inscribed on the rotating disc (Fig. 135).

During the 17th century, dials of Type *B* became common in Europe; they are known as 'equinoctial', or better, 'equatorial', dials. The Whipple Museum at Cambridge has at least twenty-five specimens of these, all of which resemble exactly the antique Chinese type, save that the compass is mounted directly underneath the dial face and that there is usually a quadrant graduated in angles of co-latitude (Fig. 136). Now this type of sundial seems not to appear much before +1600,[d] and it may therefore be legitimate to suggest that while the Jesuits introduced to China the horizontal dial, they (or the Portuguese travellers who preceded them) brought back to Europe the simpler plan of the equatorial dial with its perpendicular pole-pointing gnomon or style, which they found there. But perhaps it is more likely that a transmission occurred earlier through Arabic or Jewish intermediation. In any case the occidental type (*A*) never displaced the old Chinese type (*B*) for sundials permanently set up, as in palaces, gardens or temples in China.[e]

Besides the two types so far mentioned, there is also found, more rarely, a third (*C*). As the example illustrated (Fig. 137) shows,[f] it consists of a series of ivory plates, each one ruled for a different latitude (the upper one in the photograph is for Thaiyuan in Shansi and the lower for Shenyang (Mukden) in Manchuria), and bearing a complex

[a] P. 16*b* (tr. H. Franke (2), no. 43).

[b] Cf. *Kao Hou Mêng Chhiu*, ch. 3, pp. 14 ff., and Schlegel (7*r*), p. 31.

[c] For van Beek, this graduation was an insoluble puzzle.

[d] Personal communication from Mr R. S. Whipple. Dr A. R. Hall tells me, however, that 'equinoctial' dials are described by Sebastian Münster in his *Compositio Horologiorum*, etc., of +1531 (see Thorndike (1), vol. 5, p. 331). Michel (10) also places the earliest European examples at the beginning of the +16th century. Moreover, Reich & Wiet (1) have described a remarkable portable equatorial dial made by 'Ali ibn al-Shāṭir at Damascus in +1366. Time by the sun's altitude was first determined with an alidade, then the instrument was set so that the equatorial dial with its equal divisions corresponded to the time, whereupon another semicircular dial gave the direction of the Qiblah. This was an elaborate way of doing without the magnetic compass (here the real limiting factor). Several kinds of equatorial sundials, like the Chinese but more complicated, are described in Bettini's *Apiaria* (1645), IX, progym. ii, pp. 38ff.; and *Rec. Math.* (1660), pp. 165ff.

[e] The equinoctial or equatorial dial, it is interesting to note, can be the most accurate of all sundials.

[f] From the collection of Dr R. Clay, to whom best thanks are due.

reticulum of lines stereographically projected. The gnomon, about an inch long, and dismountable for portability, fits into the central hole, and indicates by the position of the shadow of its tip both the time of the day and the period of the year. The orientation of the plates would be effected no doubt by the use of an auxiliary compass. While the specimens illustrated are certainly late, there is no decisive reason for assuming that this system goes back only to the time of the Jesuits, since, as we shall shortly see (p. 374), projects for dials of some such kind were brought to China from the Arabic world in +1267.

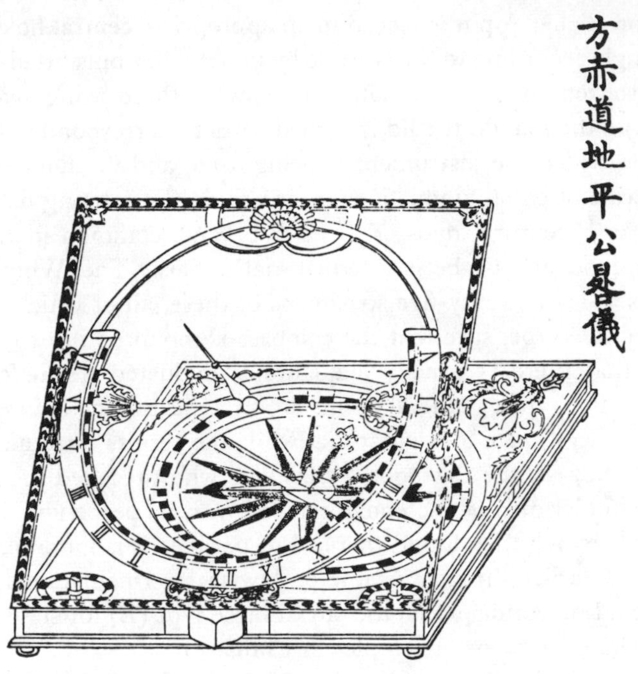

Fig. 136. Equatorial sundial of European design corresponding to the Chinese sundials of Type *B*, figured among the Jesuit astronomical instruments in the *Huang Chhao Li Chhi Thu Shih*.

The compasses in the Chinese portable sundials have been considered by Wang Chen-To;[a] one of Type *A* is extant bearing an inscription which shows that it was made by the Jesuit John Adam Schall von Bell himself in +1640. The notable thing about them is that they all have dry pivot suspensions. Although this way of mounting magnets had been devised in the Sung, floating magnets had continued more commonly in use, and the dry suspension of the 17th century was probably in large part a reintroduction from the West. In the early 18th century there grew up some famous schools of sundial and compass makers, for example, that of Yao Chhiao-Lin[1] in Shansi; later the art centred more on Canton.

[a] (5), pp. 153 ff.

[1] 姚喬林

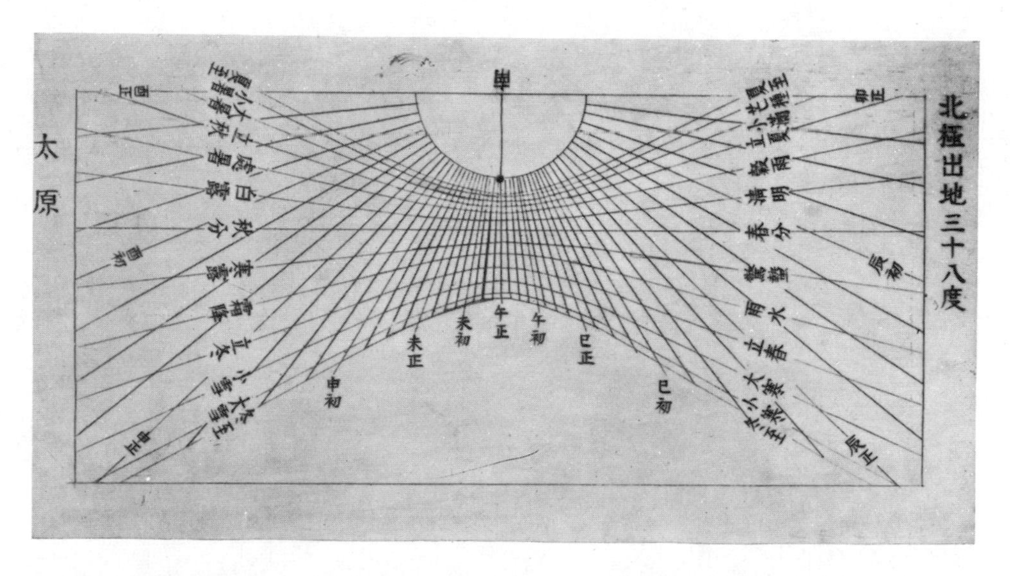

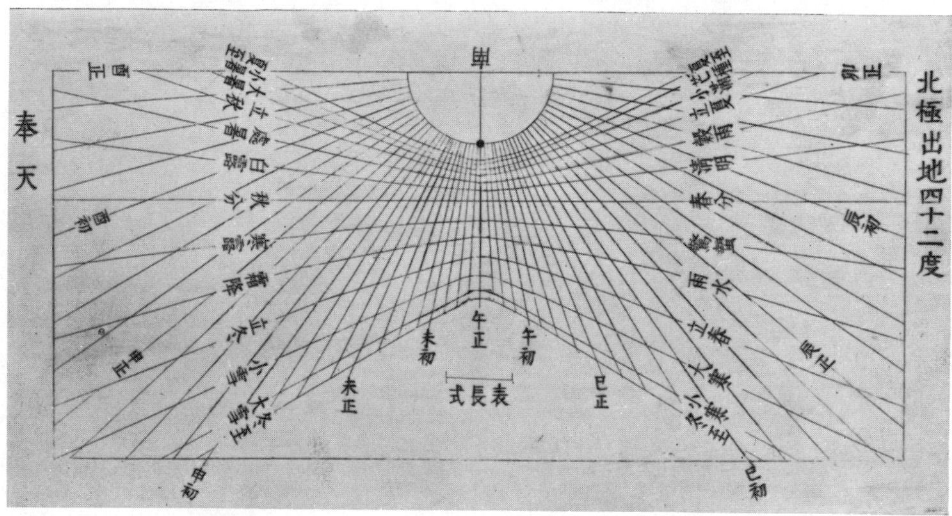

Fig. 137. Late Chinese portable sundial of Type *C*; two only (38° and 42°) of the series of plates for different latitudes are shown. From the collection of Dr R. Clay.

People in late Roman times had also thought of portable sundials,[a] and a few specimens have come down to us[b] arranged for different latitudes or climata from Britain through Narbonne to Aethiopia or Mauretania. Although the gnomon is obviously movable on the horizontal dial, and must have been intended to follow the polar axis, the mechanism is very obscure, and Diels[c] gave up the attempt to explain it.

Before leaving this subject it may be well to remember that the Chou and Han system of equal or equinoctial double-hours was extraordinarily modern. In Europe it was not until the coming of mechanical clocks in the +14th century that the old system of varying the length of the hours according to the length of the day[d] was abandoned. But in China by the −4th century at latest a permanent system of 12 equal double-hours, starting at 11 p.m., divided each day and night.[e] By a curious contrast, variable or temporal hours continued in Japan as late as +1873, and the pointer-readings of mechanical clocks were cleverly adapted to them.[f]

(4) The Clepsydra (Water-Clock)

What the sundial measures is true or apparent solar time, but owing to the eccentricity of the earth's orbit, which gives the sun its apparent unequal rate of motion, and owing to the tilt of the earth, which accounts for the obliquity of the apparent solar path (the ecliptic), true solar time and mean solar time do not coincide.[g] Hence the importance of the fact that from very early periods, methods of measuring time other than by the sun were employed. While in Europe reliance was placed mainly on the sundial until the development of mechanical clocks early in the +14th century, the Chinese paid great attention to the water-clock or clepsydra (κλεψύδρα, the 'water-stealer'), and in their culture it was brought to its highest perfection.

It is nevertheless quite certain that the clepsydra was not invented in China. Both in Babylonia[h] and in Egypt,[i] as we know from cuneiform texts and from actual objects and models recovered from Egyptian tombs, it had already been used for centuries by the time of the early Shang period (*c.* −1500). There were two obvious ways of arranging to measure time by the drip of water, either by noting the time taken for a vessel of specific shape to empty itself ('outflow type') or by seeing how long a vessel with no lower opening would take to fill ('inflow type'). The Babylonian clepsydras seem to have been mainly of the former sort, and cuneiform tablets show calculations about the adjustment of the amount of water put in to the varying length of the days. The Egyptians used both forms, but the inflow type is later and scarcer.

[a] Vitruvius (IX, viii, 1) says that Theodosius and Andrias invented dials adjustable for all latitudes.
[b] Descriptions by Baldini (1), Woepcke (1), Durand & de la Noë (1), and Drecker (1, 2).
[c] (1), pp. 187 ff.
[d] Sarton (1), vol. 3, pp. 716, 1125.
[e] See further, p. 398 below, and full details in Needham, Wang & Price (1).
[f] See the discussion of clockwork in Sect. 27*h* below.
[g] See p. 182 above, and p. 329 below. The difference is more than a quarter of an hour in November. Of the two components of this difference between dial time and clock time, the eccentricity factor vanishes only at perigee and apogee, but the obliquity factor vanishes at both equinoxes and both solstices. For further explanations, consult Barlow & Bryan (1), pp. 38 ff.
[h] See Neugebauer (2); M. C. P. Schmidt (1).
[i] See Borchardt (1); Sloley (1); Pogo (1), etc.

A particularly well-known one from Karnak, dating from *c.* −1400, and made of translucent alabaster about 14 in. high, bears on its outer surface an illustrated list of decan-stars, which has been described by Chatley (14). Both types had markings carved on the interior walls, from which time could be read off. An Assyrian clepsydra of the time of King Assurbanipal (*c.* −640) is described by Bilfinger (1).[a]

Not unnaturally the clepsydra attracted the interest of the Alexandrian physicists and engineers, who sought to combine the drip of water through narrow orifices with mechanical devices such as gear-wheels. Ctesibius (about the middle of the −3rd century—a contemporary of Chhin Shih Huang Ti)[b] seems to have been the first to introduce a float (*phellos sive tympanum*) into an inflow vessel; this carried an indicator-rod which showed the time by its height above the vessel's lid, or could be made to point to lines on a revolving drum. According to the reconstruction of Diels (1),[c] the inflow of water was regulated by a conical float valve.[d] It is to be doubted whether the Greeks knew of the clepsydra much before the −3rd century,[e] and Vitruvius strongly

[a] Books on clocks in general (e.g. Milham, 1; Britten, 1) generally open with brief and superficial remarks on the clepsydra; information of value must be sought in the assyriological and egyptological literature. Cf. Archibald (1), p. 60; Kubitschek (1), pp. 203ff.

[b] B & M, p. 483; Usher (1), p. 143.

[c] Reproduced by Neuburger (1), p. 228.

[d] The *locus classicus* is Vitruvius, IX, viii, 6; reproduced by B & M, p. 628. We thought at first that float valves did not occur in Chinese technique, but later noticed a passage in Chou Chhü-Fei's *Ling Wai Tai Ta* of +1178, in which he describes an interesting instrument used among the tribes-people of the south and south-west (ch. 10, p. 14*a*, *b*). The custom of drinking wine communally through long pipes of straw or bamboo has always been characteristic of them, and forms a central feature of some of their ceremonial dances, as may still be seen today. Chou Chhü-Fei now says that 'they drink wine through a bamboo tube two feet or more in length, and inside it has a "movable stopper" (*kuan li*[1]). This is like a small fish made of silver. Guest and host share the same tube. If the fish-float closes the hole the wine will not come up. So if one sucks either too slowly or too rapidly, the holes will be (automatically) closed, and one cannot drink.' Evidently the nodes of the bamboo had been fashioned into constricted orifices which were blocked by the float valve unless the sucking was done with moderation. Whether this principle was made use of in more advanced Chinese techniques, such as the mechanical automata described below (Sect. 27*b*), we do not know. A closely related term *kuan li*,[2] is met with in the Chhing (Sect. 27*c* below), when it means the pivots of the Cardan suspension. And yet another orthography (*kuan li*[3]) occurs in the *Chin Shu*, ch. 11, p. 7*a*, where the writer describes the proto-clockwork used by Chang Hêng in the +2nd century to drive his celestial globe by water-power. Reasons are given elsewhere (Needham, Wang & Price) for believing that here it should be translated 'trip-lug'. See further in Sect. 27*h* below. Finally, the Ctesibian float valve, or something very like it, appears in a clepsydra described in the +12th-century book of Wang Phu (see immediately below).

[e] A well-known passage in one of the fragments of Empedocles (mid −5th century) cannot be taken (as in B & M, p. 146) to refer to the water-clock (Diels-Freeman (1), p. 62). The word 'clepsydra' in the time of Empedocles meant not the water-clock but a kind of pipette for picking up small quantities of liquid out of a larger vessel; Powell (1); Last (1). Neuburger (1, p. 226) figures an ancient Egyptian painting of unknown date showing the use of the pipette, and points out how easily it could have been used as a time-measurer when provided with a bulb and small holes through which the liquid could escape. Very little has been written about pipettes in China, but as among the most widely used of all pieces of simple chemical apparatus they deserve attention. In the Thang they were called *chu tzu*.[4] 'Before the Yuan-Ho reign-period (+806 to +820) people ladling out wine used (only) cups and spoons....The name of the inventor is not known. In shape the *chu tzu* is like a (little) jar, with its mouth covered and a handle on its back. In the Thai-Ho reign-period (+827 to +835) the eunuch Chhiu Shih-Liang[5] took a dislike to its name because it was the same as that of Chêng Chu[6] (imperial physician and a great enemy of the eunuchs), so he had handles put on something like those of teapots, after which it was called *phien thi*[7] (the sideways-lifter)' (*Hsü Shih Shih*, in *Shuo Fu*, ch. 10, p. 52*a*, and *Shih Wu Chi Yuan*, ch. 8, p. 14*a*).

[1] 關捩 [2] 關棙 [3] 關戾 [4] 注子 [5] 仇士良 [6] 鄭注
[7] 偏提

implies that it was Ctesibius who invented the float. In the +1st century Heron of Alexandria also treated of water-clocks,[a] and one or other of the Alexandrians may have proposed a design whereby the water of the inflow vessel, on reaching a certain point, would siphon out into a rotating balance-wheel which its weight would turn, thus setting in motion a train of gears which would bring the recording drum to a new position so as to allow for the changes in the length of the day.[b] The first clepsydra in Rome is said to have been introduced in −159.[c] In the −1st century they must have been common (Vitruvius was writing about −28), and the Tower of the Winds at Athens (*c.* −50) had a large one.

In China the clepsydra was known as the 'drip-vessel' (*lou hu*[1]) or the 'graduated leaker' (*kho lou*[2]).[d] The problem of its first appearance there is not solved. The commentators of the Sui and Thang, such as Khung Ying-Ta (+574 to +648) in his *Shih Su*[3] (Studies on the Book of Odes), believed that there were references to it in these ancient folk-songs, but this interpretation may already have been doubted by the Sung (+960), for the *Thai-Phing Yü Lan* encyclopaedia includes none of these passages in its section on water-clocks.[e] Nevertheless, in view of the antiquity of the clepsydra in Babylonian civilisation, there is no reason why the Book of Odes (perhaps −7th century) should not contain references to its simpler forms.

(i) *Clepsydra types; from water-clock to mechanical clock*

The general development of clepsydra technique in China may be summarised as follows.[f] The most archaic type was no doubt the outflow clock, received in ancient times from the culture-centres of the Fertile Crescent. But it seems to have continued in occasional parallel use with inflow types down to quite late times.[g] The Chinese also knew another archaic device, the inverse variant of the outflow clepsydra, a floating bowl with a hole in its bottom so adjusted that it took a specific time to sink.[h] But from the beginning of the Han onwards the inflow clepsydra with an indicator-rod

[a] B & M, p. 498. He introduced the very ingenious idea of attaching a siphon to the float, so that the outflow of water occurred at an exactly constant pressure-head (Drachmann, 1), but it seems that this plan was generally ignored. So much so, indeed, that it was actually patented in 1912 by someone who thought it was a new idea (Horwitz, 2).

[b] This is often figured, as by Berthoud (1), pp. 35ff.; Baillie (1), pp. 89ff.; Britten (1). Doubt may be felt as to whether it was ever constructed. The real source may be Perrault's Vitruvius editions of 1673 and 1684.

[c] The authority is Cicero, *De Nat. Deorum*, II.

[d] The paper of McGowan (1) on the clepsydra in China is unsatisfactory, and most of the work on it is in Maspero (4). The *TSCC* devotes two whole chapters to it (*Li fa tien*, chs. 98, 99) the material in which has not yet been fully explored. Cf. Yabuuchi (*4*).

[e] Ch. 2, pp. 11*b* ff. Cf. *Yü Hai*, ch. 11, pp. 1*a*ff.

[f] We are much indebted to Dr Derek J. Price for assistance in the elucidation of this subject.

[g] As we shall shortly see, it frequently appears in Sung lists of clepsydra types.

[h] This has persisted until our own time in North Africa, where it is used as a kind of hour-glass for the control of irrigation-water sluices (Diels (1), p. 196, pl. XVI). It may have been used in medieval England (R. A. Smith). A Chinese example is afforded by the work of the Thang monk Hui-Yuan,[4] who arranged a series of lotus-shaped bowls to sink one after another during the twelve double-hours (*Thang Yü Lin*, ch. 5, p. 31*b*).

[1] 漏壺 [2] 刻漏 [3] 詩疏 [4] 惠遠

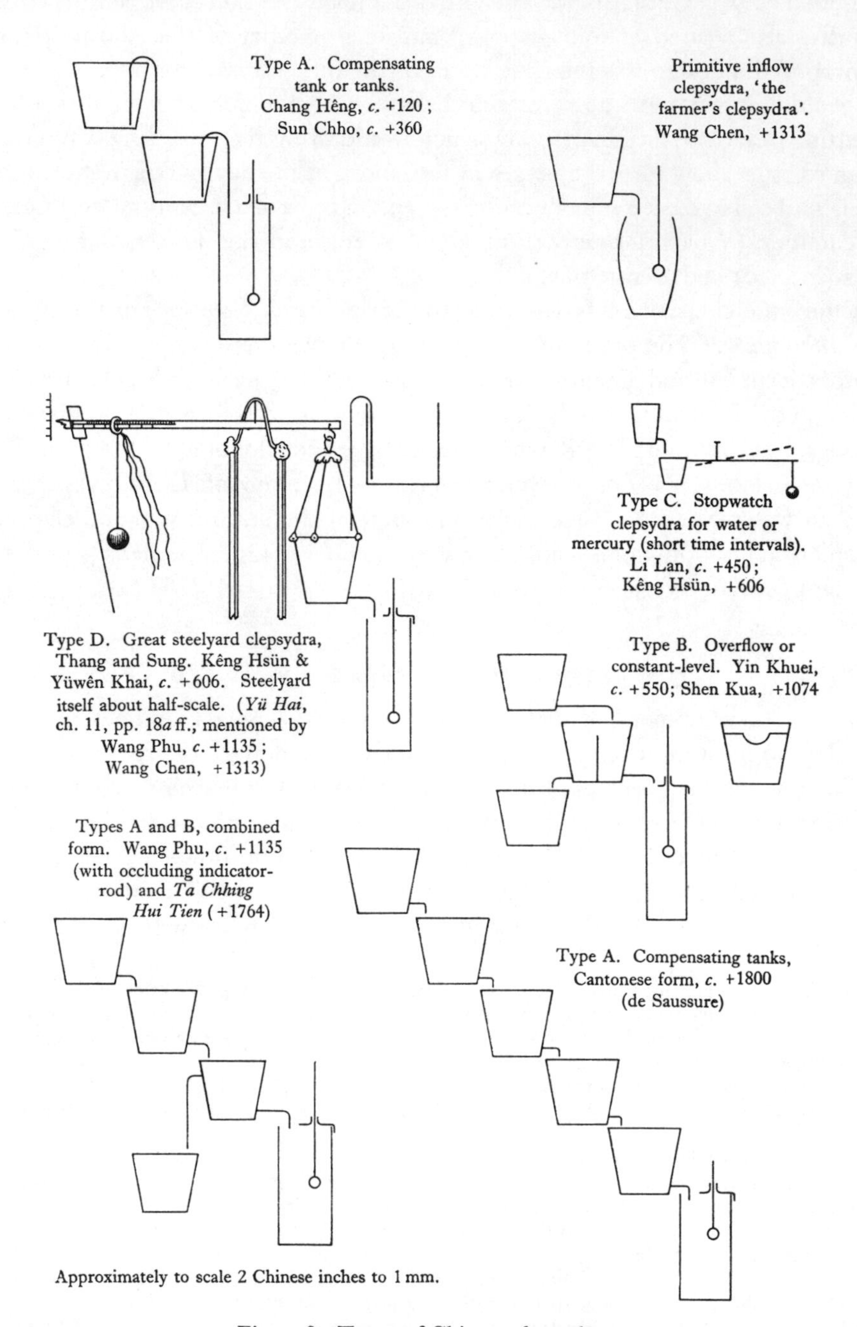

Fig. 138. Types of Chinese clepsydras.

borne on a float came into general, even universal, use. At first there was only a single reservoir, but it was soon understood that the falling pressure-head in this vessel greatly slowed the time-keeping as it emptied.

Throughout the centuries two principal methods were used to avoid this difficulty (see Fig. 138), on the one hand the extremely simple and ingenious plan (Type A) of interpolating one or more compensating tanks between the reservoir and the inflow receiver,[a] on the other hand (Type B) the insertion of an overflow or constant-level tank in the series.[b] The popularity of these methods and their modifications varied from time to time.[c] The first type, which we may call polyvascular, was an admirable means of cumulative regulation, for at each successive stage the retardation of flow due to diminishing pressure-head is more and more fully compensated. The level of an *n*th tank in a series (counting the initial reservoir), for instance, falls according to an expression something like

$$1 - \frac{1}{n!} t^n .$$

As many as six tanks above the inflow receiver are known to have been used.[d] The obscurity of some of the fragmentary descriptions which have come down to us led Maspero[e] to believe that the overflow tanks of Type B were in certain periods (e.g. Liang and Sung) provided with some kind of ball-valves, even incorporating the use of mercury, but we are now convinced that this was a mistake.[f] Where mercury entered in was in connection with clepsydras which involved weighing with balances.

These have hitherto been somewhat overlooked. They included at least two kinds, one in which the typical Chinese steelyard (the balance of unequal arms)[g] was applied to the inflow receiver (Type C), and another in which it weighed the amount of water in the lowest compensating tank (Type D). The former design naturally dispensed with the float and indicator-rod, and was usually made small and portable (*hsing kho lou*[1]).[h] Here mercury was sometimes used, with reservoir, delivery pipe and receiver all made of a chemically resistant material, jade. Such instruments, which depended on the approach of the steelyard arm to the horizontal,[i] were well adapted for the

[a] As we shall see immediately, there was at least one compensating tank already by the +2nd century.

[b] This practice seems to begin in the middle of the +6th century; see below, p. 324. First there was a double, i.e. partitioned, vessel, then an overflow pipe.

[c] E.g. in the extent to which siphons were used as against orifices and pipes at the base of the vessel.

[d] See de Saussure (29).

[e] (4), pp. 190, 200ff., 203.

[f] The confusion arose partly because it was apparently the custom among Chinese medieval clepsydra technologists to refer to the delivering ducts or pipes, made of jade or other very hard material, with carefully calibrated lumen, as 'weights' (*chhüan*[2]). So Shen Kua in *Sung Shih*, ch. 48, p. 15*b*: 'This is called the "weight", because it delivers (lit. weighs) a greater or lesser flow of water.' A good translation would perhaps be 'regulator'. The terminological confusion goes back, as we shall see, to the +6th century. The fact that there really were also clepsydras of various kinds embodying steelyards has naturally made confusion worse confounded.

[g] See Sect. 26*c* in Vol. 4.

[h] We hear first in detail about these in the middle of the +5th century, but they were probably used a good deal earlier than that.

[i] The significance of this in a general history of pointer-readings should not be forgotten. Cf. the discussion of the weathercock in Sect. 21*f* (p. 478 below), and of the magnetic compass in Sect. 26*i* (Vol. 4).

[1] 行刻漏 [2] 權

measurement of small time intervals, as by astronomers in studying eclipses, or by others for the timing of races, and we may reasonably call them 'stopwatch' clepsydras. They were termed *ma shang pên chhih*,[1] 'rapid water-clocks'.[a] The larger apparatus of Type D was used for public and palace clocks throughout the Thang and Sung. It made possible the seasonal adjustment of the pressure-head in the compensating tank by having standard positions for the counterweight graduated on the beam; and hence controlled the rate of flow for different lengths of day and night.[b] It also of course avoided the necessity of an overflow tank and warned the attendants when the clepsydra needed refilling.

Besides the above main types there were other less common ones. Most interesting were the so-called 'wheel clocks' (*lun lou*;[2] Type E) as to which more will be said in Section 27*h* on mechanical engineering under the head of clockwork. Our knowledge of these is rather meagre, but provisionally we may say that a clepsydra was arranged to drip into a vertical scoop-wheel (probably something like a noria working in reverse) bearing on its shaft a simple trip-lug. This pushed on a toothed wheel and ratchet by one tooth at each revolution, rotating thereby, with long pauses, a dial or some other form of indicator. As we shall see in Section 27, this was in all probability the kind of apparatus which rotated demonstrational armillary spheres and celestial globes from the time of Chang Hêng in the +2nd century to that of Kêng Hsün in the +6th. With this background in mind we can understand the lists of clepsydra types so often found in Sung literature:[c]

(1) Inflow float and indicator-rod clepsydra (*fou chien*[3] or *fou lou*[4]) (Types A and B).

(2) Sinking indicator-rod, i.e. outflow, clepsydra (*chhen chien*[5] or *hsia lou*[6]).

(3) Steelyard clepsydra, with balance and weights (*chhêng lou*[7, 8] or *chhüan hêng lou*[9]) (Types C and D).

[a] Clepsydras combined in various ways with steelyard balances were also well known in the Arabic culture-area. We meet with them notably in the *al-Kitāb Mīzān al-Ḥikma* (Book of the Balance of Wisdom), written by Abū'l-Fatḥ al-Manṣūr al-Khāzinī in +1122, parts of which have been translated by Khanikov (1). Cf. Winter (5). Al-Khāzinī seems to have used outflow vessels on the end of his steelyards (cf. Khanikov, pp. 17, 24, 105). Particularly interesting in this connection are the accounts in the *Chiu Thang Shu* (ch. 198, p. 16*a*) and *Hsin Thang Shu* (ch. 221B, p. 10*b*) of the great steelyard clepsydra on one of the gates of the city of Antioch. Each hour was sounded by the clanging of golden balls which dropped successively into a metal receiver. The first of these accounts (both translated by Hirth (1), pp. 53, 57, 213) was written about +945 and refers to the affairs of the +7th and +9th centuries. All this, therefore, was after the flourishing of steelyard clepsydras in China (+5th to +7th centuries). The Chinese recognised the Antioch device quite clearly as it was classified under the head of clepsydras in the *Yuan Chien Lei Han* encyclopaedia, ch. 369, p. 9*a*.

[b] This was an invention of the Sui, about +606, as we shall shortly see.

[c] For example, in the *Kuo Shih Chih*, quoted in *Yü Hai* (encyclopaedia of +1267), ch. 11, p. 18*a*, and in *Sung Shih*, ch. 76, p. 3*b*. According to Pelliot (41), pp. 44ff., this was the bibliographical part of the lost *Liang Chhao Kuo Shih*,[10] an unofficial history of the period +1023 to +1063, finished by Wang Kuei[11] in +1082. Less probably it was a part of an earlier *Liang Chhao Shih* which dealt with the period +960 to +998 and was presented by Wang Tan[12] in +1016. The fourfold classification appears again in Su Sung's account of the great astronomical clock of +1090 (*Hsin I Hsiang Fa Yao*, ch. 1, p. 5*a*; cf. p. 352 below). We have it also in the *Hsiao Hsüeh Kan Chu*[13] of Wang Ying-Lin[14] (about +1275), ch. 1, p. 32*b*.

1 馬上奔馳 2 輪漏 3 浮箭 4 浮漏 5 沈箭 6 下漏
7 稱漏 8 秤漏 9 權衡漏 10 兩朝國史 11 王珪
12 王旦 13 小學紺珠 14 王應麟

(4) 'Unresting' or 'continuous' (water-) wheel clepsydra (*pu hsi lou*[1] or *lun lou*[2]) (Type E).

Here we must not anticipate what our fourth volume will disclose concerning the origin and invention of the mechanical clock. Long before the association of the verge-and-foliot escapement with the falling weight drive in early +14th-century Europe, several forms of clepsydra technique had come together to produce a particular kind of mechanical clock in China from the +8th century onwards.[a] In this apparatus the constant-level tank of the clepsydra provided the major part of the chronometry, delivering water or mercury into the scoops of a water-wheel.[b] A minor part was provided by the adjustability of a steelyard or weighbridge which held up each scoop until it was full or nearly so. The essence of the new invention added by I-Hsing and Liang Ling-Tsan in +725 was the parallel linkage device which constituted the ancestor of all escapements. With this we must return to the history of the clepsydra proper. Diagrams of its four main types are given in Fig. 138.

(ii) *Clepsydras in history*

The most famous ancient allusion to the clepsydra occurs in the *Chou Li* (Record of the Rites of Chou),[c] where it is said:

> The Official in charge of Raising the Vessel (Chhieh Hu Shih[3]) hoists up a vessel to indicate where the well (of the encamping army) is, flies reins (like a flag) to mark the centre of the camp, and raises a basket as a signal to show where rations are being given out. On any occasion of army service, he elevates the (clepsydra) vessel so as to let the sentries know how many strokes they are to sound (during the hours of the night).[d] On any occasion of funeral rites, he elevates the (clepsydra) vessel so as to organise the relays of weepers. Invariably he keeps watch with fire and water, dividing the day and the night. In winter he heats the water in a cauldron, and so fills the vessel and lets it drip.

It is probable that we have here to do with a simple outflow type of clepsydra, in connection with which some attention was paid to temperature control.[e] Unfortunately, the date to which the passage refers is uncertain, since much of the *Chou Li* is of early Han origin; nevertheless, it could easily belong to the Warring States period, perhaps

[a] See the full account in Needham, Wang & Price.

[b] For further information on norias and water-wheels see Section 27*e*, *f* in Vol. 4.

[c] Ch. 7, p. 27*a* (ch. 28, p. 13, ch. 30, p. 28), tr. Biot (1), vol. 2, pp. 146, 201; Maspero (4), p. 205; eng. auct. mod. The oldest historical reference is in *Shih Chi*, ch. 64, p. 1*b*; an incident in the life of the military theoretician Ssuma Jang-Chü *c.* −500.

[d] One of the sounds most familiar to anyone who has lived in China is the strokes beaten by the night-watchmen on their bamboo drums (illustration in de Saussure, 29). The western analogy for the military use of the clepsydra is Aeneas Tacticus (*c.* −360), XXII, 10. For details on the unequal night-watches see Needham, Wang & Price.

[e] A Sung idea of what it was like is shown in Fig. 144. This is taken from the monograph of Wang Phu[4] on clepsydra technique about +1135 (see below, p. 326). He evidently took the *Chou Li* clepsydra to be an inflow one with an indicator-rod.

1 不息漏 2 輪漏 3 挈壺氏 4 王普

−4th century.[a] A surer point of reference is the assembly of calendar experts of −104 (see the excerpt already given, p. 302 above), but there is no indication as to the type of water-clocks they used.

The most primitive form of inflow clepsydra, consisting only of two vessels, the reservoir and the receiver, persisted as a rough time-measure in the countryside down to the +14th century or later. In +1313 Wang Chen refers to it[b] as the 'farmer's clock' (*thien lou*[1]), and shows the two pots being carried to the fields by a girl with a carrying-pole over her shoulder. He concludes his remarks by quoting a poem from the Horatian Mei Yao-Chhen[2] (*d.* +1060):

> The astronomers know of the rising and setting of stars,
> No error they make in predicting the heat and the cold;
> But the farmer too keeps time with his rustic pots
> And grudges the loss of a single inch on the dial.
> The drops of his sweat match those of the dripping clepsydra,
> The swathes of his cutting advance like the shadow itself.
> Who could disdain or regret this life-giving labour?
> Who can snatch back an hour from the realm of the past?

The first sure evidence for floats and indicator-rods in an inflow clepsydra is reached with Chang Hêng,[3] Pien Hsin[4] and Li Fan[5] about +85. In the fragment remaining of Chang Hêng's *Lou Shui Chuan Hun Thien I Chih*[6] (Apparatus for Rotating an Armillary Sphere by Clepsydra Water) he says:

> Bronze vessels are made and placed one above the other at different levels (*tsai tieh chha chih*[7]); they are filled with pure water. Each has at the bottom a small opening in the form of a 'jade dragon's neck' (*yü chhiu*[8]). The water dripping (from above) enters two inflow receivers (*hu*[9]) (alternately), the left one being for the night and the right one for the day.
>
> On the covers of each (inflow receiver) there are small cast statuettes in gilt bronze; the left (night) one is an immortal (*hsien*[10]) and the right (day) one is a policeman (*hsü thu*[11]). These figures guide the indicator-rod (lit. arrow, *chien*[12]) with their left hands, and indicate the graduations on it with their right hands, thus giving the time.[c]

Not only are the floats and indicator-rods here, but also apparently the siphon, for the term *yü chhiu* certainly meant this in later times.[d] The phrasing suggests three vessels in all; if this is the correct interpretation the clepsydra had at least one compensating tank. From this point therefore we can follow the development of Type A. The

[a] This is perhaps a modest estimate, as the compilers of the *Chou Li* were consciously trying to be as archaic as possible.

[b] *Nung Shu*, ch. 19, pp. 20*a*ff.

[c] Both fragments in *YHSF*, ch. 76, pp. 68*a*, *b*, appended to Chang Hêng's *Hun I*, tr. auct., adjuv. Maspero (4), p. 202. Both were preserved in Hsü Chien's *Chhu Hsüeh Chi* of +700 (ch. 25, pp. 2*a*, 3*a*); the second one also in the commentary by Li Shan on Lu Tso-Kung's *Hsin Kho Lou Ming* (Inscription for a New Clepsydra) in *Wên Hsüan*, ch. 56, p. 13*b*, i.e. *c.* +660. Cf. *Yü Hai*, ch. 4, p. 9*b*; ch. 11, p. 7*a*.

[d] Contributory support for this description of the Later Han clepsydra comes from the description in Hsü Shen's *Shuo Wên* dictionary, and from Chêng Hsüan's commentary on the *Chou Li*. The latter mentions the 100 'quarters of an hour' into which the indicator-rod was graduated.

1 田漏 2 梅堯臣 3 張衡 4 編訢 5 李梵
6 漏水轉渾天儀制 7 再疊差置 8 玉虬 9 壺 10 仙
11 胥徒 12 箭

reference to this kind of clepsydra by Wang Chhung in the *Lun Hêng* (cf. above, p. 214)[a] will take it back two or three decades before Chang Hêng, to +60 or +70.

Archaeological evidence shows us, however, that it can by no means have been new at the end of the +1st century. Two Former Han clepsydras survived, at least down to the Sung, at which time a drawing was made and a description given of one of them.[b] It was a small cylindrical bronze vessel with a pipe inserted at the bottom, and a hole in the lid for the indicator-rod to slide through (Fig. 139). The date in the inscription is given only in cyclical characters, so that an exact dating is not possible, but the latest would be −75 and the earliest −201. The name of the maker is on it—Than Chêng.[1]

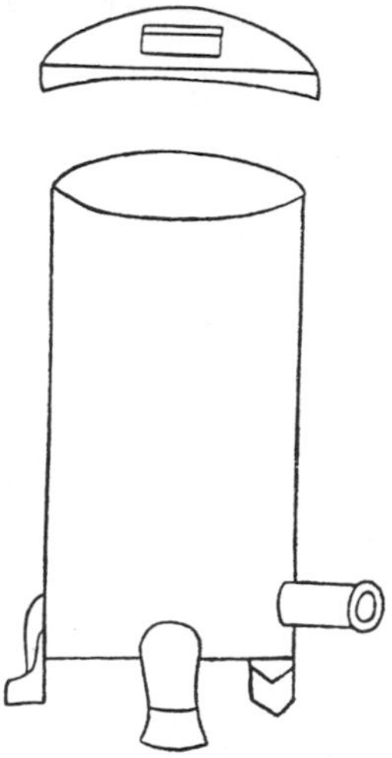

The relation of the Han float-and-pointer inflow-clock with that of the Alexandrians is therefore rather a difficult matter to determine. As we saw in an earlier Section,[c] there was contact between China and Roman Syria towards the end of the +1st century, though the series of 'embassies' from the West does not really begin until +120. There might, of course, have been earlier ones not recorded, but the possibility that a clepsydra of this type was in use much earlier, in fact at the beginning of the Han (−200), makes one hesitate to conclude that it was essentially an Alexandrian invention which found its way to China. For the present the problem must be left unsolved.[d] If it is hard to visualise two quite independent inventions of this technique, the possibility remains open that the transmission may have been in the opposite direction, for this type of clock may, as we see, go back in China before the opening of the overland route by Chang Chhien.[e] Most likely of all is a spread in both directions from the Fertile Crescent and ancient Egypt.

Fig. 139. Inflow clepsydra from the Former Han dynasty (between −201 and −75) made by Than Chêng. Described by Hsüeh Shang-Kung in his *Li Tai Chung Ting I Chhi Khuan Chih Fa Thieh* (+11th century), ch. 19, reproduced by Maspero (4).

The quotation from the *Chou Li* just given shows that the Han people were well aware of the change of viscosity of water with temperature,[f] and made rough efforts to keep its temperature constant in winter, or at least to prevent it from freezing. Huan Than[2] (−40 to +30) tells us:

> Formerly, when I was a Secretary at the Court I was in charge of the clepsydras (and found that) the differences in degrees varied according to the dryness and the humidity, the cold

[a] Forke (4), vol. 2, p. 84.

[b] In the *Li Tai Chung Ting I Chhi Khuan Chih Fa Thieh*[3] (Description of the Various Sorts of Bronze Bells, Cauldrons, Temple Vessels, etc., of all Ages) by Hsüeh Shang-Kung[4] (+11th century) (ch. 19, p. 2*b*); and in the *Khao Ku Thu*[5] (Illustrations of Ancient Objects) by Lü Ta-Lin[6] (+1092).

[c] Sect. 7 in Vol. 1. It was there pointed out that significance may attach to the mention of mechanical toys and acrobats in the presentations of these Western 'embassies'.

[d] We shall meet again with the inconvenience of finding important technological inventions appearing almost simultaneously in the Eastern Roman Empire and in China (cf. below, Sect. 27*f*).

[e] Cf. above, Sects. 6*b* and 7*e*; Vol. 1, pp. 107 and 173.

[f] It varies by a factor of 9 between 0° and 100° C.

[1] 譚正 [2] 桓譚 [3] 歷代鍾鼎彝器款識法帖 [4] 薛尙功
[5] 考古圖 [6] 呂大臨

and the warmth (of the surroundings). (One had to adjust them) at dusk and dawn, by day and by night, comparing them with the shadow of the sun (*kuei ching*[1]), and the divisions of the starry *hsiu*. In the end one can get them to run correctly.[a]

He had thus noticed variations in the evaporation-rate as well as the viscosity. The text is interesting, too, in that he speaks of the comparison of sidereal time with clock time.

From at least the later Chou period onwards the day and night was divided into twelve equal double-hours (*shih*[2])[b] and, by a parallel centesimal system, into one hundred 'quarters' (*kho*[3]).[c] The use of two receivers, one for the day and the other for the night, was an unnecessary complication. The varying lengths of light and darkness led to difficulties about the graduations of the indicator-rod which the Babylonians and Egyptians had already met with in marking their outflow clocks.[d] What the Chinese did was to have a series of indicator-rods differing by one quarter of an hour, and according to the season these were progressively substituted for one another. The earliest regulation concerning this[e] refers to the time of Han Wu Ti, about −130. Indicator-rods were changed every nine days.[f] In −58 a greater approach to accuracy was sought by making the change depend on the sun's observed declination, in successive stages of 2·4°; this was embodied in the Han Code,[g] and lasted through the +1st century. In +102 Ho Jung[4] linked the change with the calendrical fortnights (the twenty-four *chhi*[5]),[h] and this system endured for more than a millennium.

In the Chin dynasty we hear[i] of a clepsydra set up by Wei Phei[6] in the imperial palace in +332, and of another important one in +387. In the middle of the century Sun Chho[7] wrote a metrical inscription[j] for a clepsydra, in which the term *chhiu* (*ling*

[a] *CSHK*, *Hou Han* sect., ch. 15, p. 2*a*, tr. auct.

[b] See further below, p. 398, in connection with the calendar.

[c] The word persists in colloquial Chinese, as meaning a quarter of a modern hour, to the present day. Whether or not these two concurrent systems betray a difference of origin, the inconvenience of the discrepancy between them was always inviting attempts at rationalisation. Thus at the end of the Former Han the astrologer Kan Chung-Kho[8] and his disciple Hsia Ho-Liang[9] (*d.* −23 and −5 respectively) recommended a change from 100 to 120 quarters. But as this was coupled with subversive views about the dynastic succession, the reform lasted only a very short time (see *Chhien Han Shu*, ch. 11, pp. 5*b*ff.). Later on, in +507, the emperor Liang Wu Ti for some years reduced the number of quarters to 96, but this in turn was soon abandoned (*Chhou Jen Chuan*, ch. 9). Cf. Dubs (2), vol. 3, pp. 6, 7, 93; Maspero (4), pp. 208ff.

[d] For Vitruvius' ingenious method (IX, 8, 11), making use of different controlled hydrostatic pressures, see Diels (1), p. 209.

[e] It is recorded in Liu Hsiang's[10] *Hung Fan Wu Hsing Chuan Chi*;[11] quoted in *Sui Shu*, ch. 19, p. 25*a*, not, as Maspero (4), p. 208, says, *Hou Han Shu*, ch. 12, p. 9*a*.

[f] This was of course very inaccurate owing to the unequal rate of the sun's apparent motion (Maspero (4), p. 215).

[g] Ling chia, ch. 6, see *Hou Han Shu*, ch. 12, p. 8*b*.

[h] *Hou Han Shu*, ch. 12, p. 9*a*.

[i] *Yü Hai*, ch. 11, p. 7*b*; *Chhu Hsüeh Chi*, ch. 25, p. 3*a*.

[j] *CSHK* (Chin sect.), ch. 62, p. 5*b*; *TPYL*, ch. 2, p. 13*a*; *Yü Hai*, ch. 11, p. 8*b*.

1 晷景 2 時 3 刻 4 霍融 5 氣 6 魏丕 7 孫綽
8 甘忠可 9 夏賀良 10 劉向 11 洪範五行傳記

chhiu[1]) again appears, and also the term *yin chhung*[2] (Yin creeping animal).[a] These probably mean siphons. The epigram runs:

> Mysterious dragons spew forth water,
> Siphons receive it and let it escape,
> The vessels are set on their triple throne,
> The waters are massed like the vasty deep,
> In fullness and power the time abounds
> And riding on emptiness the flood descends.

This implies at least one intermediate compensating tank, and more probably two, since the receiving vessel would not require a base, and indeed in nearly all illustrations is shown without one. The polyvascular trend was fully established by the Thang, for later iconographic tradition ascribes to Lü Tshai[3] (*d.* +665)[b] the system of three compensating tanks between the reservoir and the receiver. Fig. 140, taken from the Sung encyclopaedia *Shih Lin Kuang Chi*,[4] shows the four vessels.[c] A figure of a Buddhist monk has taken the place of the immortal and the policeman. In the Sung period itself the constant-level overflow tank (Type B) prevailed, but the following +13th-century account, interesting because of its references to other sciences, seems to refer to the polyvascular apparatus. In his *Chhü I Shuo Tsuan* (Discussions on the Dispersal of Doubts), Chhu Yung[5] says,[d] about +1230:

> People talking about clepsydras have always asked how large in size the vessels should be, and how much water they should contain. There is also the question of the weight of the water (in steelyard clepsydras). The mouths of the siphons (*kho wu*[6]) deliver water in a stream as thin as a hair, and the usual concern was that it was not thin enough. Clepsydras have generally been used for determining the 'fire-times' (*huo hou*[7]) (appropriate times and durations of heating in alchemical and chemical operations).[e] But (the conditions at the points of) reception and delivery of the water are not the same. If a speck of dust is allowed to get into a siphon it will at once be blocked, and there are few (of such) clepsydras which will work properly for as many as three days running. But thinking it over by night I solved the problem. The lumen of the siphon must be enlarged to the size of a magnetic needle (*chung chen*[8])[f] so that small particles may follow the stream and pass through without causing any blockage—but at the same time the inflow receiver must be doubled in size.... I record this here for the benefit of those who like to busy themselves with miscellaneous matters.

[a] The term is found also in Lu Chi's[9] 'Essay on the Clepsydra' (Lou Kho Fu), written just before +300 (*CSHK* (Chin sect.), ch. 97, pp. 4*a* ff.). Cf. *Piao I Lu*, ch. 1, p. 14*a*.

[b] Cartographer and sceptical naturalist, expert on music, acoustics and Five-Element theory, conducted a controversy on logic with Hsüan-Chuang.

[c] This illustration comes from a Ming copy of +1478 in the Cambridge University Library (Pt. 1, ch. 2 (Li Hou), p. 3*a*). The work was first put together by Chhen Yuan-Ching[10] some time between +1100 and +1250, though not printed till +1325. The picture is often found elsewhere, as in the *Liu Ching Thu* (see immediately below), +1740 edition, Shih Ching sect., p. 43*a*.

[d] Ch. 1, pp. 17*a* ff., tr. auct.

[e] Cf. Vol. 2, pp. 330 ff. and Sect. 33 below.

[f] On this technical term see Sect. 26*i* below.

1 靈虬 2 陰蟲 3 呂才 4 事林廣記 5 儲泳 6 渴烏
7 火候 8 中針 9 陸機 10 陳元靚

Within living memory clepsydras for official time-keeping were in use in China. The most famous one surviving, now probably the oldest, is that at Canton (Fig. 141).[a] It is of the polyvascular type, with two compensating tanks. Made in the Yuan period (+1316)[b] by the bronze-founders Tu Tzu-Shêng[1] and Hsi Yün-Hsing,[2] it remained in continuous use until at least 1900, and still runs,[c] though now in the City Museum. But it was not the *ne plus ultra* of polyvascularity, for a Chinese almanac of 1831, reproduced by de Saussure (29), depicts no less than six vessels above the inflow receiver itself.[d] The last of these would have approached quite closely to a constant-level tank.

The overflow-tank clepsydra (Type B) naturally developed somewhat later than the simpler kind just described. Our oldest description comes from the *Lou Kho Fa*[3] (Clepsydra Technique) of Yin Khuei,[4] about +540, a few sentences of which were preserved in encyclopaedias.[e] One tank only intervened between the reservoir and the receiver; it contained a partition where the water 'hesitated' (*chhih-chhu*[5]), as if uncertain which way to go. Water issuing from a dragon-mouth on the reservoir was purified by a filter (*ching-wei*[6]), and left the overflow or constant-level tank by a delivery-pipe (*hêng-chhü*[7]) to reach the receiver, where an 'hour-jack' (*ssu chhen*[8]) guided the indicator-rod (*thien fou*[9]). These arrangements probably date from somewhat earlier, however, perhaps from the time of Tsu Chhung-Chih (late +5th century), whose son, Tsu Kêng-Chih,[10] was commissioned to repair the imperial standard clepsydra in +506. The new one which he made was honoured in the following year by a celebrated inscription.[f]

From the Sung many accounts of these clepsydras remain. Early in the dynasty the overflow principle was adopted in the 'lotus clepsydra' (*lien hua lou*[11]) of Yen Su,[12] so called because the top of the receiver was shaped in the form of this Buddhist emblem.[g] That ingenious man[h] first presented it in +1030, and after many tests it was officially

[a] Descriptions by Li Ping-Shui (1), McGowan (1) and Middleton Smith (1). Liu Hsien-Chou (*1*) quotes from a special monograph on it, the *Kuangchou Yen-Yu Thung Hu Chi*,[13] but we have not been able to identify the author or date of this. See, however, *Yang-chhêng Ku Chhao*, ch. 8, pp. 39*b* ff.

[b] Local tradition carries it back to the Nan Han kingdom (+917 to +971), but wrongly. For many years it was in the North Reverence Tower.

[c] According to a personal communication from Sardar K. M. Panikkar, who visited it in 1951. An interesting eye-witness account of its official use in 1899 will be found in A. M. Earle (1), p. 54.

[d] It also shows the stove for warming the water in winter, a rack of tablets for the five night-watches and a figure in western dress holding a compass.

[e] *Chhu Hsüeh Chi*, ch. 25, pp. 2*a*, *b*; *Yü Hai*, ch. 11, p. 6*b*. A full translation of the passage is given by Maspero (4), p. 193, but it needs modifying. He found Yin Khuei's words in *Sui Shu*, ch. 19, and in *Wên Hsüan*, ch. 56, but it must have been in commentaries not in the editions used by us. Yao Chen-Tsung (*1*) placed Yin Khuei in the Later Han, but this must be a mistake.

[f] This was the *Hsin Kho Lou Ming*[14] by Lu Chhui[15] (Lu Tso-Kung), which found a place in the *Wên Hsüan*, ch. 56, pp. 11*a* ff.; see also *CSHK* (Liang sect.), ch. 53, p. 7*a*.

[g] Description in *Chhing Hsiang Tsa Chi*,[16] ch. 9, pp. 8*a* ff.; also *Yü Hai*, ch. 11, pp. 15*a* ff.

[h] Scholar, painter, technologist and engineer; we shall meet him again in connection with the study of the tides (p. 491 below), and the construction of hodometers and south-pointing carriages (Sect. 27*c*).

1 杜子盛 2 洗運行 3 漏刻法 4 殷夔 5 踟躕
6 經緯 7 衡渠 8 司辰 9 天浮 10 祖暅之 11 蓮花漏
12 燕肅 13 廣州延祐銅壺記 14 新刻漏銘 15 陸倕
16 青箱雜記

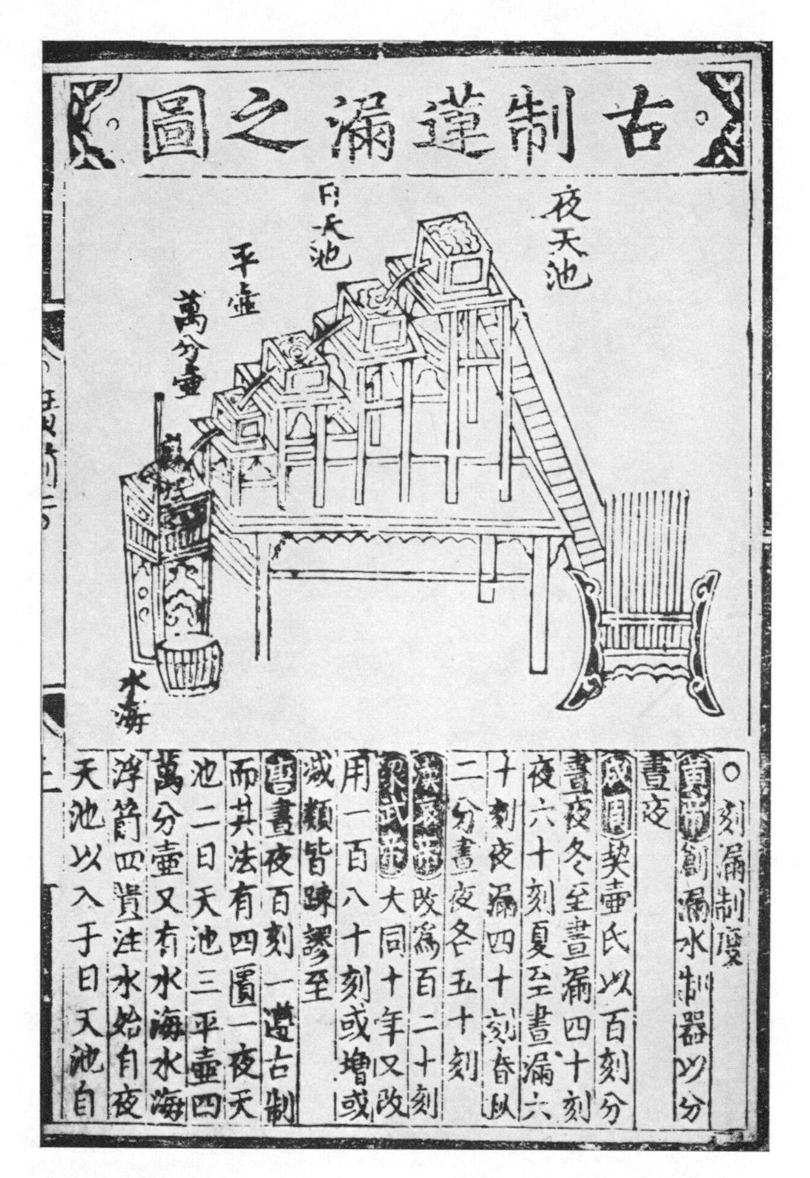

Fig. 140. The polyvascular type of inflow clepsydra associated with the name of Lü Tshai (*d.* +665); from a copy of the *Shih Lin Kuang Chi* encyclopaedia printed in +1478 (Pt. 1, ch. 2, p. 3*a*), in the Cambridge University Library. The text below refers to the legendary invention of the clepsydra by Huang Ti, the account of the clepsydra officials in the *Chou Li*, the temporary change to 120 quarters in the day and night by Han Ai Ti (−6 to −1), the temporary change to 180 quarters by Liang Wu Ti in +544, and the stabilisation of the 100-quarter system in the Thang.

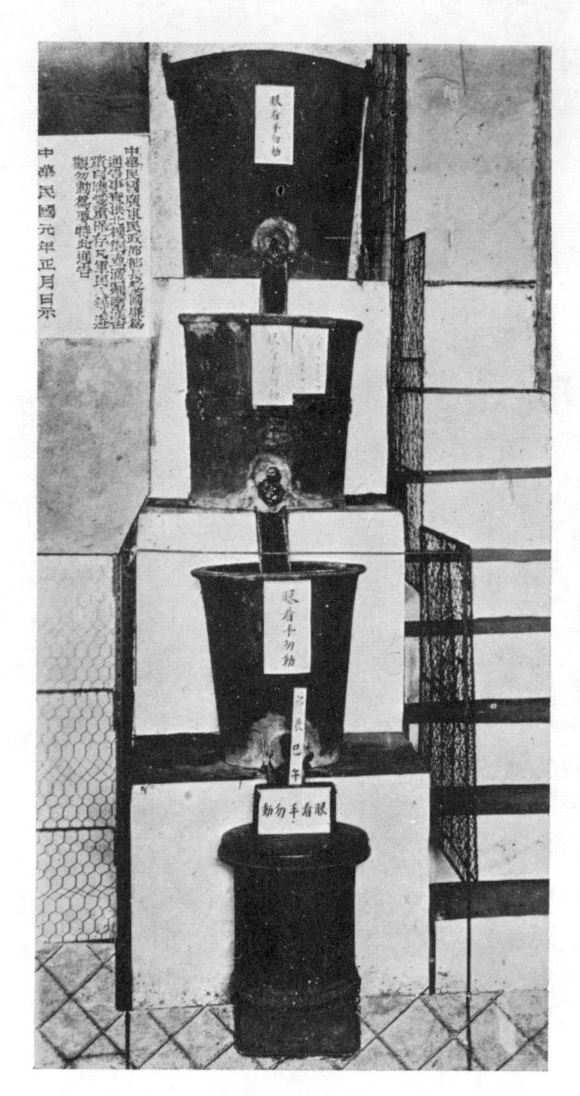

Fig. 141. The famous polyvascular inflow clepsydra at Canton made in the Yuan dynasty (+1316) by Tu Tzu-Shêng & Hsi Yün-Hsing, and in continuous use until after 1900. Photographed in the 1st year of the Republic (1912).

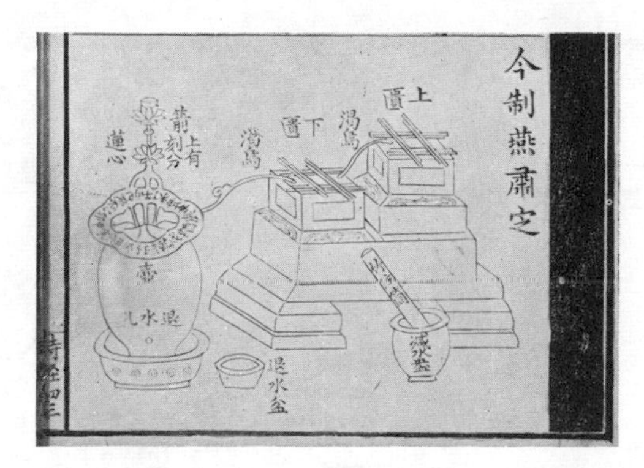

Fig. 142. The overflow type of inflow clepsydra associated with the name of Yen Su (+1030); from a Chhing edition (+1740) of the *Liu Ching Thu*.

Fig. 143. The combined type of inflow clepsydra; an example in the Imperial Palace at Peking (photo. Tung Tso-Pin *et al.*).

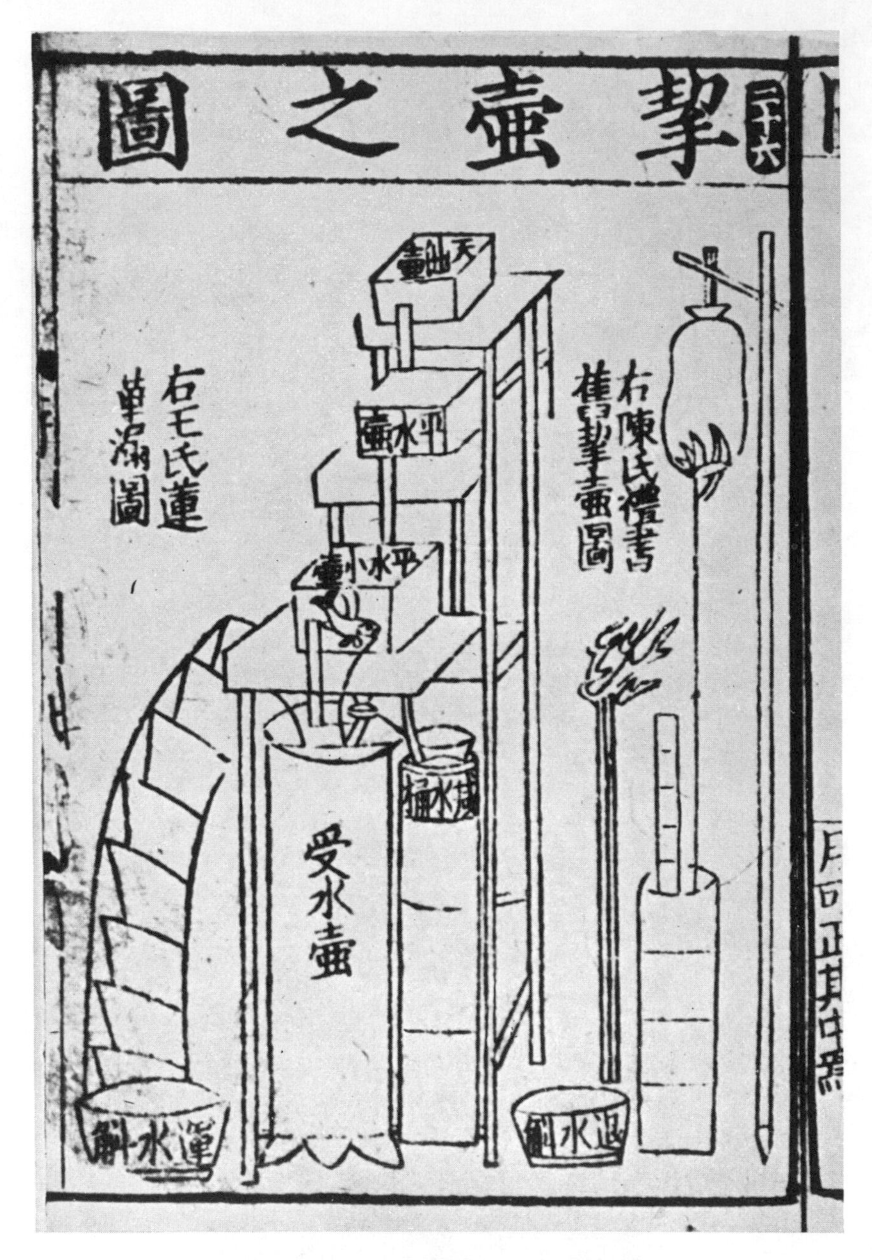

Fig. 144. The oldest printed picture of a clepsydra; an illustration from a Sung edition of Yang Chia's *Liu Ching Thu* (*c.* +1155). It shows the combined polyvascular and overflow type described by Wang Phu in his *Kuan Shu Kho Lou Thu* of *c.* +1135, which incorporated an automatic cut-off device acting when the indicator-rod rose to its fullest extent. On the right of the picture Yang Chia showed the oldest and simplest type of inflow clepsydra to illustrate the description in the *Chou Li*.

adopted six years later. At the same time other clepsydras were adapted to the overflow system.[a] Yen Su's model was frequently illustrated, as in Fig. 142 from the *Liu Ching Thu*[1] (Pictures of Objects mentioned in the Six Classics).[b] Indeed, it was taken as a prototype for centuries afterwards,[c] and the term 'lotus-clock' became proverbial.[d] Great efforts were made to improve it further in the +11th century by Shen Kua[e] whose memorial to the emperor of +1704 has survived,[f] indicating the same components with somewhat different names. Thus the overflow compartment in the constant-level tank[g] is termed the 'branch canal' (*chih-chhü*[2]), and the partition itself the 'water-leveller' (*shui kai*[3]). This had a semicircular piece cut out of it so that it looked like a worn whetstone,[h] and here the interesting word *shih*[4] is used, a rare technical term of ancient hydraulic engineering which means a division head between two separating canals. The pipes (*ching*[5]) were of jade,[i] and that which delivered water to the receiver was called the 'regulator' (lit. jade weigher; *yü chhüan*[6]). Shen Kua's[7] account in his own book[j] shows how closely the clepsydra was associated with other astronomical instruments, and kept in the Astronomical Bureau (Thien Wên Yuan[8]).

In the course of time, the two types (A and B) became (perhaps somewhat illogically) combined into one, so that the system consisted of a reservoir, a compensating tank, an overflow tank, an inflow receiver with rod-float, and an overflow-receiver. This is the apparatus described[k] in the *Ta Chhing Hui Tien* (Administrative Statutes of the Chhing Dynasty) of 1764. The vessels[l] can all be made out in Fig. 143 which shows

[a] Some of these were steelyards (Types C, D). Details in *Sung Shih*, ch. 76, p. 3*b*, and *Yü Hai*, ch. 11, pp. 16*a*, 19*a*. The examiners were the Astronomer-Royal, Wang Li,[9] two younger experts, Chang Tê-Hsiang[10] and Fêng Yuan,[11] with Yang Wei-Tê,[12] the man who twenty years later was to be the observer of the supernova of +1054 (see p. 427 below).

[b] Edition of +1740 edited by Wang Hao,[13] *Shih Ching* sect., p. 43*a*. There were many editions of this work in the Ming and Chhing periods, but the earliest was that of Yang Chia[14] in the Sung. Yen Su's clepsydra is also depicted in the *Shih Lin Kuang Chi* already mentioned (edition of +1478, Pt. 1, ch. 2, p. 3*b*).

[c] For example, Chang Hsing-Chien[15] made lotus clepsydras for the Chin Tartars between +1190 and +1208 (*Chin Shih*, ch. 22, pp. 32*b* ff.).

[d] 'In the wilderness there are no bells and drums, nor lotus clepsydra rods continuing day and night...'; so Tai Han, an +18th-century academician, in praise of solitude (*Yung Chhêng Shih Hua*, ch. 1, p. 10*b*).

[e] Cf. Vol. 1, p. 135. Biographical study by Chhien Chün-Hua (*1*).

[f] *Sung Shih*, ch. 48, pp. 14*b* ff. A translation of the passage is given by Maspero (4), pp. 188 ff., but it is very muddled and unsatisfactory. Our own version, omitted here, is available for consultation.

[g] Slight confusion is caused by Yen Su and Shen Kua who both speak of four vessels, but these were not in series, for one was the receiver and another a lower vessel to catch the overflow.

[h] Cf. the discussion of the shapes of lenses in Sect. 26*g* below.

[i] On the use of gem-stones (especially onyx) in Greek clepsydras cf. Drachmann (2), p. 18.

[j] *MCPT*, ch. 8, para. 8 (cf. Hu Tao-Ching (*1*), vol. 1, pp. 335 ff.).

[k] Ch. 81, p. 2*b*; tr. Maspero (4), pp. 185 ff.

[l] It is interesting that Chinese texts and illustrations of all periods invariably give the vessels a 'flower-pot' shape, while the receiver is cylindrical. The adoption of the flower-pot shape by the ancient Egyptians had undoubtedly been a rough attempt to compensate for the slower delivery as the pressure-head fell. A constant rate, at least for the first two-thirds of the outflow, can be attained by a paraboloid or a frustum of 77°; Borchardt (1). May not this Chinese tradition have derived originally from a knowledge of primitive Mesopotamian outflow clepsydras?

1 六經圖　2 枝渠　3 水槩　4 釃　5 頸　6 玉權
7 沈括　8 天文院　9 王立　10 章得象　11 馮元
12 楊惟德　13 王皜　14 楊甲　15 張行簡

the clepsydra still preserved in the Imperial Palace Museum in Peking. But we have a much older illustration of the same kind of clepsydra, dating from the +13th century, and this must be the oldest printed picture of a water-clock from any civilisation. For Fig. 144 is taken from a Sung edition of the *Liu Ching Thu* already mentioned.[a] The accompanying text identifies the 'Mr Wang' of the drawing as Wang Phu,[1] whose *Kuan Shu Kho Lou Thu*[2] (Illustrated Treatise on Standard Clepsydra Technique), long lost, must date from the neighbourhood of +1135. After speaking of steelyard clepsydras, he goes on to describe the type shown, and adds the very unusual information that when the indicator-rod rises to its full height it blocks up the orifice through which the water is delivered. It seems strange to find this seemingly Hellenistic device again in +12th-century China, though used for cutting off, rather than regulating, the flow of the water, and with just what object is not altogether clear.

Something must now be said of those types of clepsydra (C and D) which measured time not by the rise of an indicator-rod on a float, but by the actual weighing of the vessel. The simplest plan was to attach the receiving vessel to a balance, usually the characteristic Chinese steelyard.[b] Whether this practice started in the Han or San Kuo periods we do not know, but the oldest text we have concerning it speaks as if the device was well known—this is another *Lou Kho Fa*, written by the Taoist Li Lan,[3] in the Northern Wei dynasty about +450. The precious fragments have come down to us preserved in the *Chhu Hsüeh Chi*[4] encyclopaedia.[c] Li Lan says:

> Water is placed in a vessel (whence it issues by) a siphon of bronze shaped like a curved hook. Thus the water within is conducted to a silver dragon's mouth which delivers it to the balance vessel (*chhüan chhi*[5]). One *shêng*[d] of water dripping out weighs (*chhêng chung*[6]) one *chin*,[e] and the time which has elapsed is one quarter (-hour).

This is the small steelyard clepsydra (*chhêng lou*[7]). Li Lan then goes on to say:

> (It is by means of) jade vessels, jade pipes and liquid pearls (*liu chu*[8]) (that the) rapid stopwatch (*ma shang pên chhih*[9]) portable clepsydra (*hsing lou*[10]) (is constructed). 'Liquid pearls' is only another name for mercury.[f]

Thus he indicates that the short-interval stopwatch clepsydra which could easily be carried about for use in the field needed chemically resistant parts to allow of the use

[a] Yang Chia's first edition dates apparently from *c.* +1155, and was enlarged *c.* +1170 by Mao Pang-Han.[11] Yang Chia was thus incorporating up-to-date devices, but as we shall see later on the Southern Sung was far from equalling the Northern Sung in horological ingenuity (Sect. 27*h*).

[b] For detailed discussion of this, see Sect. 26*c* below.

[c] Ch. 25, p. 2*b*; this was compiled by Hsü Chien[12] in +700. And in Shen Yo's[13] *Hsiu Chung Chi*[14] (Sleeve Records).

[d] Usually translated 'pint'. The Han *shêng* is estimated as 0·36 U.S. dry pint.

[e] The catty or pound. At this period it corresponded to 0·22 kilo, or about half a pound.

[f] Tr. auct. The translation of Maspero (4), p. 192, involves preconceptions which we have abandoned.

1 王普 2 官術刻漏圖 3 李蘭 4 初學記 5 權器
6 枰重 7 稱漏 8 流珠 9 馬上奔馳 10 行漏
11 毛邦翰 12 徐堅 13 沈約 14 袖中記

of mercury, but worked on the same principle as all other ordinary steelyard clepsydras. It is hard to say how widely these forms were used, but by the Sui period (+605 to +616) a carriage bearing these portable time-keepers was considered an indispensable part of any imperial procession.[a]

The clepsydras of Type D were rather more complicated. They seem to have originated as the invention of two eminent technicians of the Sui, Kêng Hsün [1] and Yüwên Khai.[2] The *Sui Shu* says:[b]

At the beginning of the Ta-Yeh reign-period (+606) Kêng Hsün[c] made 'advisory inclining vessels' in the ancient style,[d] for filling with clepsydra water, and presented them to the emperor (Yang Ti) who was very pleased. The emperor asked him and Yüwên Khai[e] to follow the methods developed by the Northern Wei Taoist Li Lan. The old Taoist technique used a small steelyard clepsydra, with a vessel which weighed the water—thus it was portable. Accordingly (Kêng Hsün and Yüwên Khai) also set up (sun-) shadow measuring (gnomons by which to) graduate the scales of indicator-rods (or steelyard beams), and square vessels to hold the water above. These time-indicators were established below the drum-tower in front of the Chhien Yang Hall (of the Palace) at the eastern capital (Loyang). They also made portable stopwatch clepsydras for telling the time. These two instruments, the sun-computing dial and the clepsydra, can measure the (phenomena of) Heaven and Earth, and therefore they are the basic means of calibrating the armillary sphere and the celestial globe....

The new departure made at this time, which produced a large type of palace and public clock constantly in use until the end of the Sung dynasty (+14th century), can best be appreciated from the diagram in Fig. 138. A detailed description of it given in the *Kuo Shih Chih*[f] was fortunately incorporated in the *Yü Hai* encyclopaedia.[g] First the use of four indicator-rods[h] shows clearly that the compensating tank, not the inflow receiver, was attached to the steelyard. Since there were only four, not forty-eight,[i] as in the clepsydras of Yen Su and other AB-type designs, the variation for seasonal differences of day and night lengths must have been achieved by varying the rate of water inflow.[j] This depended on the pressure-head maintained in the compensating

[a] 'The *Sung Chhao Hui Yao* says that in the Sui dynasty there were bell and drum carriages accompanying the imperial carriage. The *Ta-Yeh Tsa Chi* says that the Praetorian Guard had a carriage with a portable clepsydra as well as the other two. Nowadays these are classed as imperial prerogative vehicles. The clepsydra carriage was introduced in the Sui period.' So *Shih Wu Chi Yuan*, ch. 2, p. 47*a*.

[b] Ch. 19, pp. 27*b* ff., tr. auct. Cf. *Hsü Shih Shuo*, ch. 6, p. 11*a*.

[c] An engineer and instrument-maker who had a very adventurous life in troubled times. His biography (*Sui Shu*, ch. 78, pp. 7*b* ff.) has been translated *in extenso* by Needham, Wang & Price.

[d] These vessels were hydrostatic puzzles which tipped over when full, to symbolise the ruin of princes and officials who greedily overreached themselves; see the discussion in Sect. 26*c*.

[e] Engineer, architect and Minister of Works under the Sui for thirty years. We shall often meet with him again, as in Sect. 27*c* on sailing carriages.

[f] As we have already seen, p. 318, this was probably part of the *Liang Chhao Kuo Shih*, finished by Wang Kuei in +1082.

[g] Ch. 11, pp. 18*a* ff.; a full translation of the passage will be found in Needham, Wang & Price.

[h] Almost certainly two for the day (a.m. and p.m.) and two for the night. Each would have been marked on one side with the 3 double-hours and 25 quarters correct for the equinoxes; and on the other sides with lesser and greater numbers of hours and quarters.

[i] The practice was to change one in the middle of each of the 24 fortnightly periods and one at its end.

[j] Thus it must run fast during the short summer nights and winter days.

[1] 耿詢　　[2] 宇文愷

tank, and that could be very exactly adjusted by moving a bronze ring carrying a suitable weight back and forth along the other arm of the steelyard. Indeed, this bore a series of graduations corresponding to the seasons. Attendants had to ensure that the arms did not vary from the horizontal, to adjust the pressure-head in the reservoir by moving the siphon, and fill tanks when necessary. A square iron guide with a handle was provided to steady the steelyard during adjustments.

Such were the main types of non-mechanical clepsydra which developed in China at different periods. A substantial literature grew up about them, but nearly all of it has perished. The *Thung Chih Lüeh* bibliography of +1150 lists[a] no less than fifteen such manuals, none of which is now available to us. The oldest was the *Lou Kho Ching*[1] of Ho Jung[2] who has already been mentioned, written about +100, but 'Clepsydra Manuals' were also written by Chu Shih[3] in the Chhen Dynasty about +563, and by an Astronomer-Royal of the same, Sung Ching,[4] a little later.[b] All the extant writings were digested into a treatise on the subject by Huangfu Hung-Tsê[5] in the Sui or early Thang. Yen Su himself presented an illustrated account of his methods (*c.* +1030). And finally, as we shall see in due course, the Sung period saw at least three books on mechanical water-clocks, only one of which has survived.[c]

In 17th-century Europe there was a certain revival of interest in water-clocks. John Bate described one of classical Chinese style, an inflow clepsydra with a figure pointing to the rising indicator-rod, in +1635, as well as other ancient types.[d] But most of them no longer worked on the simple and ancient drip principle, they consisted of a drum containing radial compartments separated by diaphragms each perforated with a small hole. The flow of water from one compartment to another partially counteracted the fall of the drum on chains or cords from a geared clock dial, or retarded its descent along rails on an inclined plane, thus acting as a liquid escapement. There can be no doubt that these designs all derived from medieval Arabic compartment-drum devices, some of which were elaborated by seekers after perpetual motion,[e] but which led to the liquid escapement mercury drum clock described in one of the parts of the *Libros del Saber de Astronomia*,[f] the corpus produced under the aegis of Alfonso the Wise, king of Castile, about +1275. The drum water-clocks did not spread very fast in the +17th century,[g] but became widely used in the +18th.[h] Some authors then

[a] Ch. 44, pp. 19*b*, 20*a*.

[b] Several famous astronomers also issued works with this title, e.g. Ho Chhêng-Thien[6] about +420 and Tsu Kêng-Chih about +507.

[c] See Sect. 27*h* below.

[d] A variant of the sinking bowl, a sinking glass bell with a small hole at the top, and an anaphoric clock (on which see Sect. 27*h*).

[e] Full details in Schmeller (1); there are close connections with India, which suggests the possibility of contacts still further East. See further in Sect. 27*h*.

[f] See Rico y Sinobas (1), vol. 4; Feldhaus (22), and Drover (1).

[g] Kircher (2) in +1643 makes no mention of them, and they were still a rarity for Graverol in +1691. They were described in his *Mathematical Recreations* by Ozanam, who attributed the invention to Domenico Martinelli in +1663.

[h] An elaborate description was given by Salmon in the *Descriptions of Arts and Trades* of the French Academy of Sciences, 1788, vol. 27 (tinsmith's work). At this time Sens was a great centre for their manufacture, according to Planchon (1).

1 漏刻經 2 霍融 3 朱史 4 宋景 5 皇甫洪澤
6 何承天

thought that they derived from what was considered a Chinese invention, namely, puppets containing pipes and reservoirs filled with mercury, which performed gymnastic evolutions in descending a set of steps (van Musschenbroek[a] in +1762; Beckmann[b]). One would like to know more about these; they could very easily have derived from the tradition of the trick hydrostatic vessels which goes back so far in China.[c] A link between these vessels and time-keeping we have already noticed—the fact that about +606 Kêng Hsün arranged for them to be filled slowly by clepsydra drip.[d] In any case vessels with compartments were very much in the Alexandrian tradition too, and the clock of Alfonso is chiefly important because it embodied a weight drive, and thus connects the Alexandrian tradition with the mechanical clocks of +14th-century Europe. But the substitution of weight for water drive belongs to another story, and we shall tell it in its place.

(iii) *Combustion clocks and the equation of time*

There has been much to say of sundials and clocks, but, as most people know, the time which they keep is not always the same. The fact that the apparent motion of the sun through the sky is irregular, sometimes faster and sometimes slower than its average motion, is one of the most important of ancient observations, going back to Greece and Babylonia (Neugebauer, 5). We have already seen (p. 123 above) Li Shun-Fêng, in +7th-century China, working out algebraic methods of dealing with it. Its most obvious consequence in ancient times was the inequality of the four seasons (the times between the passages of solstitial and equinoctial points), but ancient astronomers must have been aware that if they could make use of a clock keeping time regularly for long periods, it would not always agree with the sun. The effect is due (as already explained, p. 313) to two circumstances, the eccentricity of the earth's orbit, and the difference between the planes of the equator and the ecliptic; the greatest positive difference between clock-time and sundial time (14½ min.) occurring in February, and the greatest negative difference (16½ min.) in November. This is known as the Equation of Time.[e] A discrepancy of this order, just about one Chinese quarter (*kho*), might have been more easily detectable by a clepsydra than by the mechanical clocks of +11th-century China or +14th-century Europe, which could lose or gain 20 minutes a day.[f] After Kepler, its exact extent was easily calculated. By the middle of the 17th century, pendulum clocks had reduced their error to 10 seconds a day,[g] so that the difference could be measured accurately.

Though the future lay with mechanical clocks, there may have been other methods of time measurement which might, under certain circumstances, surpass water-clocks

[a] 'Ut dulce et jucundum addamus misceamus utili', he wrote (vol. 1, p. 143), 'describendam judicavi imagunculam a Chinensibus non longo ab hinc tempore inventam, se vivi hominis instar innumeris modis ex altiori gradu in gradum deorsum supinantem....'

[b] (1), p. 84.

[c] Cf. above, p. 327, and especially Sect. 26*c* in Vol. 4.

[d] Above, p. 327.

[e] Cf. Spencer-Jones (1), p. 44; Smart (1), pp. 42, 146; Barlow & Bryan (1), pp. 38 ff.

[f] Cf. below, Sect. 27*h*.

[g] Cf. Pledge (1), p. 70.

in accuracy. In a passage which we shall study later on for other reasons,[a] Hsüeh Chi-Hsüan,[1] a Sung scholar of the mid +12th century, said that apart from the water-clock and the sundial there was also the 'incense-stick' (*hsiang chuan*[2]) method of measuring time. It would seem very improbable that these were, or ever could have been, made so as to burn with any high degree of regularity.[b] But they have always been widely used in China,[c] and they made a great impression on one of the +17th-century Jesuits, Gabriel Magalhaens, who wrote:[d]

The Chinese have also found out, for the regulating and dividing the Quarters of the Night, an Invention becoming the wonderful Industry of that Nation. They beat to Powder a certain Wood, after they have peel'd and rasp'd it, of which they make into a kind of Past, which they rowl into Ropes and Pastils of several Shapes. Some they make of more costly Materials, as Saunders, Eagle, and other odoriferous Woods, about a fingers length, which the wealthy sort, and the Men of Learning, burn in their Chambers. There are others of less value, one, two and three Cubits long, and about the bigness of a Goose quill, which they burn before their Pagods or Idols. These they make the same use of as Candles to light them from one place to another. They make these Ropes of powder'd Wood of an equal Circumference, by the means of a Mould made on purpose. Then they wind them round at the bottom, lessening the circle till they come to be of a Conick figure, which enlarges itself at every Turn, to one, two and three handsbreadths in Diameter, and sometimes more;[e] and this lasts one, two and three days together, according to the bigness which they allow it. For we find some in their Temples that last ten, twenty, or thirty days. These Weeks (wicks) resemble a Fisher's Net, or a String wound about a Cone; which they hang up by the Middle, and light at the lower end, from which the Fire winds slowly and insensibly, according to the windings of the string of powder'd Wood, upon which there are generally five marks to distinguish the five parts of the Watch or Night. Which manner of measuring Time is so just and certain, that you shall never observe any considerable Mistake. The Learned Men, Travellers, and all Persons that would rise at a precise hour about Business, hang a little weight at the Mark, which shews the Hour when they design to rise; which when the Fire is come to that point, certainly falls into a Copper Bason, that is plac'd underneath, and wakes them with the noise of the fall. This Invention supplies the want of our Larum Watches, only with this difference, that this is so plain a thing and so cheap, that one of these Inventions, which will last four and twenty hours, does not cost above three pence; whereas Watches that consist of so many wheels and other devices, are so dear, that they are not to be purchas'd but by those that have store of Money.

Fig. 145 shows a metal incense-clock of the 18th century in which the burning point is made to wind its way through the maze of the strokes of a stylised seal character.

It may be asked whether the sand hour-glass was known in medieval China. Wang Chen-To (5) believes that this was a late introduction from Dutch and Portuguese

[a] Cf. below, Sect. 27*h*.

[b] Michel (11) illustrates one form, a straight stick. Another, seen in Fig. 145, conducts the glow through the maze-like channels of a stylised written character.

[c] A Sung reference is the *Hsiang Phu* of Hung Chhu, ch. 2, p. 5*a*. Cf. the King Alfred's candles, and 'auction by the candle' in England (Hough (2), p. 40).

[d] (1), p. 124.

[e] A striking photograph by Norman Lewis (1), p. 177, shows some of these incense spirals. For a fine colour photograph, see G. W. Long (1).

[1] 薛季宣 [2] 香篆

Fig. 145. An undated metal incense clock; the burning point of the tindery powder winds its way through the strokes of stylised characters. The maze of the tray at the back on the right is in the form of the word *shou* (longevity). *Shuang hsi* (double happiness) can be made out on the cover in the right foreground. Photo. Science Museum, London.

ships,[a] but Lin Yü-Thang[b] takes certain passages[c] in the writings of Su Tung-Pho to imply its use in the Sung. They seem to us to refer to clepsydras. In any case we know that from the beginning of the Ming, sand replaced water in some types of mechanical clocks with driving-wheels.[d]

None of these methods could have demonstrated the equation of time. Yet conceivably a lamp in a secluded place, if fed with oil of regular composition and quality, and well designed, might act as a time-keeper sufficiently good to detect the dial and clock discrepancy. For this reason there is much interest in something which Yang Yü tells us in his *Shan Chü Hsin Hua* of +1360, about the 'perpetual lights' which were kept burning in a certain temple.

Fan Shun-Chhen[1] was a Khaifêng man, and a famous physician in his time, of great ability and wide knowledge. He was particularly skilled in astronomy. In the Chih-Chun reign-period (+1330 to +1333) he was head of the Yung-Fu Buildings Upkeep Office. He once told me that the perpetual lights in the prayer-halls had each a reservoir which used twenty-seven *ko* of oil[e] annually. During the Chih-Yuan reign-period (+1264 to +1294) the issue had originally been fixed at thirteen *chin* to the *ko*, so the total amount was 351 *chin*.[f] In the course of a year, it was found, when a test was made, that there was an excess of fifty-two *chin*. So there was clearly a discrepancy between this and the time as measured by the sundial. This observation concerned the prayer-hall of the Blue Pagoda temple which was then under the supervision of the Yung-Fu Office.[g]

If this story is to the point it must be garbled, for the discrepancy cannot be seen on a yearly basis; but the fact that Yang Yü says distinctly that a difference between lamp-clock time and sundial time was noticed, suggests at least that Fan Shun-Chhen had been keeping careful comparative records for certain periods within the year. It is interesting that time-indicating lamps were used fairly widely in 18th-century Germany, according to Hough,[h] but unfortunately he supplies no details as to their accuracy.

[a] Cf. *Liu-Chhiu Kuo Chih Lüeh* (+1757), *Thu hui* sect., p. 34*b*. Wang Chen-To's evidence is rather convincing.
[b] (5), pp. 20, 229.
[c] E.g. one in the Mei Shan Yuan Ching Lou Chi[2] (*Tung-Pho Chhüan Chi, Chhien chi*, ch. 32) which speaks of the farmers breaking, at the harvest festival, the *lou*[3] with which they had timed their labours.
[d] See Sect. 27*h*. If there was any foreign stimulus here, it was Arabic or Persian, not Portuguese.
[e] A measure of fluid capacity.
[f] A measure of weight, the catty, 0·6 kilo. at this time.
[g] P. 22*a*, tr. H. Franke (2), no. 58, eng. auct. Cf. Franke (8).
[h] (2), p. 67.

[1] 范舜臣 [2] 眉山遠景樓記 [3] 漏

(5) THE SIGHTING-TUBE AND THE CIRCUMPOLAR CONSTELLATION TEMPLATE

Leaving gnomons and time-keepers, we now approach the instruments which the Chinese astronomers used for direct observations of the heavens. In an earlier sub-section (p. 262 above) we noted the use of the sighting-tube by Shen Kua in his +11th-century observations of the position of the true pole and the circular motion of the stars nearest to it.[a] We also noted (p. 300) the attempt recorded in the *Chou Pei Suan Ching* to determine the sun's diameter by means of a sighting-tube. There can be no doubt that such tubes (not, of course, equipped with lenses) were commonly employed in ancient Chinese astronomy; the later name for them was *wang thung*,[1] but they had other names, consideration of which will take us into unexpected regions. In the late Chou and Han periods their use by the official astronomers was so well known as to lead to a proverbial expression *i kuan khuei thien*[2]—looking at heaven through a tube.[b] The sighting-tube is probably referred to in *Huai Nan Tzu* (*c.* −120) where it is said:

> If people want to find out heights of objects and cannot do so they will be pleased if you teach them how to use the sighting-tube and the water-level (*kuan chun*[3]). If they want to know the weights of things and do not know how to do so, they will be pleased if you give them the balance and scales. If they want to determine distances far and near, they will be pleased if you show them how to 'shoot with the metal eye' (*chin mu*[4]) (i.e. to point the sighting-tube). From this one can see how much more should be provided to cope with that which has no directions and no bounds.[c]

This shows that the sighting-tube was used by surveyors as well as by astronomers, and indeed the Sung compendium of architecture, *Ying Tsao Fa Shih*, includes[d] a small drawing of an architect's sighting-tube (reproduced here in Fig. 146).

In a learned and brilliant paper, Eisler (2) traced the origin of the sighting-tube or *dioptra*, frequently mentioned by occidental classical writers and figured in medieval manuscripts, to the ancient Babylonians. He suggested that the Greek words *sphairophoros aulos* (σφαιροφόρος αὐλός), *speirophoros* (σπειροφόρος), etc., were perhaps corruptions of the term *sporophoros* (σποροφόρος), the seed-drill tube or funnel clearly represented on Babylonian cylinder-seals attached to the plough.[e] It may be, there-

[a] The sighting-tube was a commonplace in the Sung—philosophers like the Chhêng brothers often refer to it; cf. *Honan Chhêng Shih I Shu*, ch. 13, p. 1*a*; ch. 22A, p. 11*b*.

[b] Cf. *Chhien Han Shu*, ch. 65, p. 8*b*, but the *locus classicus* is *Chuang Tzu*, ch. 17 (Legge (5), vol. 1, p. 389), a reference which I owe to Mr W. Willetts. Chuang Chou is making fun of the Mohist and Logician Schools.

[c] Ch. 20, p. 15*a*, tr. auct. We owe the notice of this passage to the kindness of Dr Arthur Waley. In the +18th century Yao Fan[5] in *Yuan Chhun Thang Pi Chi*[6] suggested that *Huai Nan Tzu* was referring to some kind of spectacles, but the idea was not taken seriously and Naba (*1*) exploded it in a special paper. Yet it is still occasionally met with (H. T. Pi, 1).

[d] Ch. 29, p. 2*b*.

[e] Cf. Gustavs & Dalman (1); Meissner (1), vol. 1, p. 193, and others. For the history of the seed-drill plough, see below, Sect. 41.

1 望筒 2 以管窺天 3 管準 4 金目 5 姚範
6 援鶉堂筆記

fore, that the Old Babylonian astronomers applied an agricultural implement for their own purposes. Eisler also discussed the ancient and medieval belief that stars are visible by day when the sky is observed from the bottom of a deep well, excess light being thus excluded.[a] This interesting subject has since been reviewed by Aydín Sayili (1); there is much evidence that such wells were actually used, but not that they were useful. Much better established is the greater visibility of faint stars at night when extraneous light is cut off by a sighting-tube. Curtis has shown that stars as faint as the eighth magnitude can be seen at night through a $\frac{1}{4}$ in. hole in a black screen

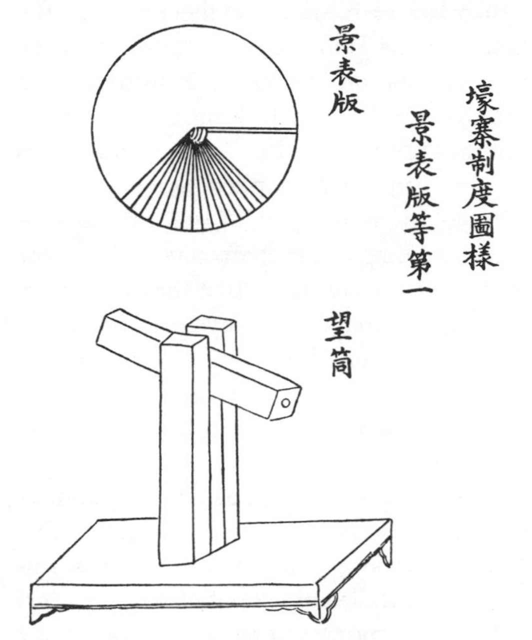

Fig. 146. Sighting-tube and quadrant from the *Ying Tsao Fa Shih* of +1103.

Fig. 147. A sighting-tube in use by a medieval Western astronomer; from a +10th-century codex of St Gall (reproduced by Eisler, 2).

15 ft. from the eye. The normal lower limit is the sixth magnitude. Whether ancient sighting-tubes were narrow enough to have much effect may be doubtful. Fig. 147, from Eisler, shows a +10th-century miniature from a codex of St Gall, in which Ptolemy is depicted as observing the pole-star through a sighting-tube. Later on it became a common attribute of pictured astrologers.[b] But though the sighting-tube was thus known in European as well as Asian culture, the incorporation of it in armillary spheres became a standard practice only in China.[c]

Much the most important ancient text on the sighting-tube is that contained in the Shun Tien chapter of the *Shu Ching* (Historical Classic). Its date is very uncertain,

[a] Hence perhaps the Platonic anecdote (*Theaet.* 174A) about Thales of Miletus falling into a well when observing the stars.

[b] See Michel (14) and Zinner (6).

[c] Perhaps because of the ready availability of the natural tubing of bamboo. We shall note later (p. 368) that Ricci was surprised by the sighting-tubes which he found instead of vane sights in Chinese astronomical instruments.

but might be within a couple of centuries more or less of the −6th century. '(Shun) examined the *Hsüan-Chi*[1] apparatus and the jade traverse (*Yü-Hêng*[2]), in order to bring into accord (the different cyclical periods of) the Seven Regulators (*i chhi chhi chêng*[3]).'[a] Legge[b] and Medhurst[c] translated *hsüan-chi*[d] as 'the pearl-adorned turning sphere', and the conception of it as some kind of armillary sphere dominated the minds of the Jesuit translators, who frequently reproduced pictures of Sung armillary instruments to illustrate it, though it is now clear that there was no justification for this. Legge, moreover, in his suggestion that the Seven Regulators were the seven stars of the Great Bear, was further from the truth than Karlgren, who took them to mean the sun, moon and five planets.[e] For the scholars of the Han the exact meaning of the term *hsüan-chi* was already lost, and there was a division of opinion as to whether it referred to an astronomical instrument or to a constellation: Ma Jung,[4] Tshai Yung[5] and Chêng Hsüan[6] maintained the former meaning, while Fu Shêng[7] and others maintained the latter.[f] Unquestionably the term came to be applied, as also did that of *yü-hêng*, to certain stars in Ursa Major (as we saw above, p. 238), and this attribution is again significant, as we shall see. There is nothing in the characters of the words *hsüan-chi* to justify their translation as 'sphere'; all one can say is that they must have signified some instrument made of jade. The solution of the mystery, extremely plausible if not definitive, is due to the recent work of Michel (1, 2). In order to explain it, we have to take another starting-point.

Among the most renowned objects of ancient Chinese jade work (Shang and Chou periods) are the objects which Wu Ta-Chhêng (*2*) and Laufer (8) call 'jade images and symbols of the deities of Heaven and Earth'. The former, known as *pi*,[8] are flat discs with large perforations in the centre, or (derivatively) rings of various dimensions. The latter, known as *tshung*,[9] are tubes or hollow cylinders to which four salient prisms are attached in such a way as to form a rectangular parallelepiped out of which a short length of tube projects at each end.[g] The whole is of course in one piece. Fig. 148*a*, *b* and *c* illustrate these two ancient ritual objects, which were doubtless carried in the imperial temple ceremonies. For centuries speculation has centred round their significance. Since the *tshung* was earthy, chthonic, feminine and Yin, it was proper to

[a] Ch. 2, tr. Karlgren (11), p. 77; (12), p. 4.

[b] (1), p. 38.

[c] (1), p. 14.

[d] These characters, and *hêng*, have been investigated etymologically by van Esbroeck (1). Without its jade radical, *chi* consists of pictographs for fine thread, dagger-point and knife (K 547). It may have meant originally the carving of straight lines like stretched threads on a piece of stone or jade, and it certainly acquired the meaning of careful quantitative observation using something finely adjusted. It forms, as we know, the phonetic of 'machine', and we have seen it used in the sense of the 'minute germs of things' (Sect. 10 in Vol. 2). *Hsüan* (K 269, 344) shows an eye examining markings on bones, hence scapulimancy and divination, hence any natural inquiry. But here it is mainly a homophone for *hsüan*,[10] 'turning'. *Hêng* (K 748*i*) shows a man with something straight coming out of his eye, surrounded by the pictograph for motion along a way, i.e. a man looking in a given direction.

[e] Cf. van Esbroeck (2).

[f] Cf. the discussion in Dubs (2), vol. 3, p. 328. Maspero (4), for once, was on the wrong track in deciding that constellations, and not an instrument, were meant (p. 293).

[g] These forms shade off into objects like napkin-rings.

1 璿璣 2 玉衡 3 以齊七政 4 馬融 5 蔡邕 6 鄭玄
7 伏勝 8 璧 9 琮 10 旋

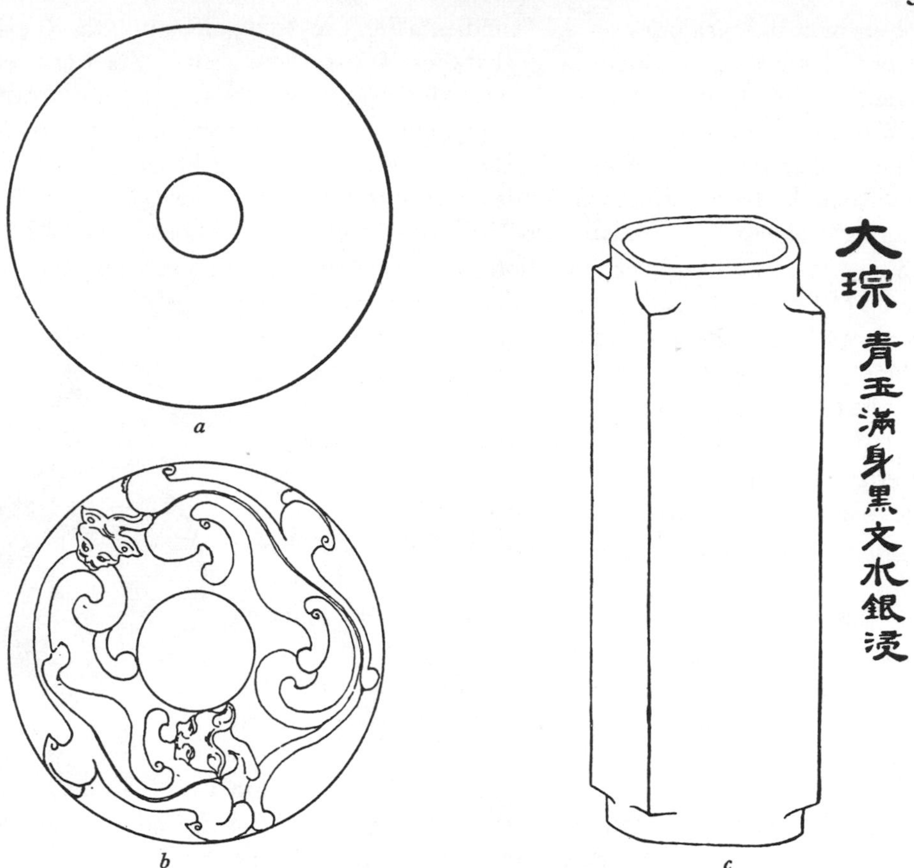

Fig. 148. The ancient ritual objects of jade, *pi* and *tshung*. *a*, a plain *pi*; *b*, *pi* ornamented with dragons; *c*, a *tsung*. From Laufer (8) and Michel (1).

the attendants of the empress, and its possible status as a womb-symbol has not been lost upon modern scholars,[a] who have contrasted it with the *kuei*[1] or elongated and pointed tablet, the jade symbol of sovereign power, heavenly, ouranic, male and Yang. Naturally the *tshung* was square outside and round inside (cf. the TLV mirrors conversely, p. 305 above; and the diviner's board similarly, Sect. 26*i* below). Since the *pi* was of valuable jade, it may have affected the traditional shape of Chinese coins.[b] Since it certainly had a male status one might well expect to find it combined with the *kuei*, and this is actually the case, for specimens exist in which the perforated disc of the *pi* is adorned with one, two or four projections shaped either in the pointed phallic

[a] Karlgren (9); Waley (7); Erkes (11); and Laufer (8), pp. 84, 99, 131. Following the theory that the shapes of ritual jade objects were generally derived from actual tool or weapon prototypes, Mr W. Willetts has suggested (private communication) that the *tshung* was patterned on the bearings of chariot wheels. There is a hint of this in the *Shuo Wên*.

[b] Laufer (8), p. 155.

[1] 圭

symbolism of the *kuei*, or as square 'handles' (Fig. 149).[a] Significantly, the *Chou Li* (Record of the Rites of Chou) says[b] that these *kuei-pi* were used in the sacrifices to heaven, the sun, the moon and the stars. Moreover, examples are not rare in which the *pi* discs carry rough incised representations of the seven stars of the Great Bear.

Meanwhile, another kind of *pi* had been under study by archaeologists. A good many examples are known in which the outer edge of the disc is very curiously carved (one might almost say graduated), being divided into three sections of equal length each beginning with a salient projection and a sharp indentation, and continuing with a series of teeth of variable shape until a plane circumference intervenes before the next set of graduations. Furthermore, on one side these strange discs bear two incised cross-lines almost at right angles which owing to their constancy and regularity cannot be dismissed as chance marks of the jade-cutter's saw. It was Wu Ta-Chhêng (*2*) who was the first to urge that these discs were certainly astronomical instruments, and to identify them with the *hsüan-chi* of the *Shu Ching*; Laufer (8) accepted this but was unable to explain their use. This has now been done by Michel (1, 2), who, pointing out that the protruding tube of the *tshung* seems as if made to fit into the perforation of the *pi*,[c] suggests that the *tshung* was originally nothing but a sighting-tube (and hence identical with the *yü-hêng*), while the *pi* was a degenerate ritualised form of what we may call the 'circumpolar constellation template'. For if the indentations on the circumference of the astronomical *pi* would fit the chief circumpolar constellations, then the true pole[d] would occupy the centre of the sighting-tube and an orientation among the *hsiu* and the outer constellations could readily be obtained. This the indentations do.

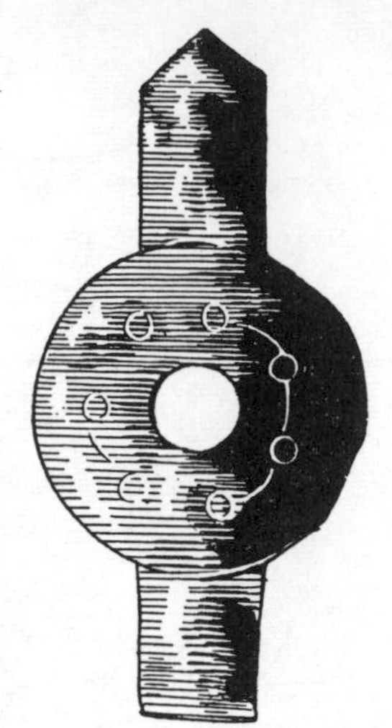

Fig. 149. The *pi* combined with the *kuei* (from Laufer, 8).

Fig. 150 shows the use of the *hsüan-chi*. When α and δ Ursae Majoris are taken by one of the three chief indentations, α Ursae Minoris will occupy the second, and the stars of the Tung fan[1] constellation in Draco and Cepheus will fit tolerably well along the third set of teeth. This would, about -600, centre the whole instrument on the pole, while β Ursae Minoris would be seen revolving in the field of view. If now the *tshung* is fitted into the *pi* so that one of its flat edges follows the double line engraved on the back of the *pi*, there will be four times in the year, for a given hour, at which the flat surface of the sighting-tube (*yü-hêng*) is either parallel to the horizon or at right angles to it. What then was the meaning of the single line at approximately right angles to the double one? Michel plausibly suggests that it represented the

[a] Laufer (8), pp. 167, 168 and plates XV and XXIII.

[b] Khao Kung Chi (ch. 42, pp. 17, 22); see Biot (1), vol. 2, pp. 522, 524.

[c] This of course destroys the phallic significance of the symbols, and indicates why the *kuei* forms had to be added on to the *pi* to indicate its masculinity.

[d] With its nearest star, then β Ursae Minoris (Kochab).

[1] 東藩

solstitial colure. It must be remembered that about −1250, in the Shang, α Ursae Majoris had a right ascension of approximately 90°, instead of the 163° of the present time. It should therefore, as he says, be possible to date extant specimens of the *hsüan-chi* by the angles which their single-line reference-marks make with the hour-circle of α Ursae Majoris (given on the instrument by the position of the principal

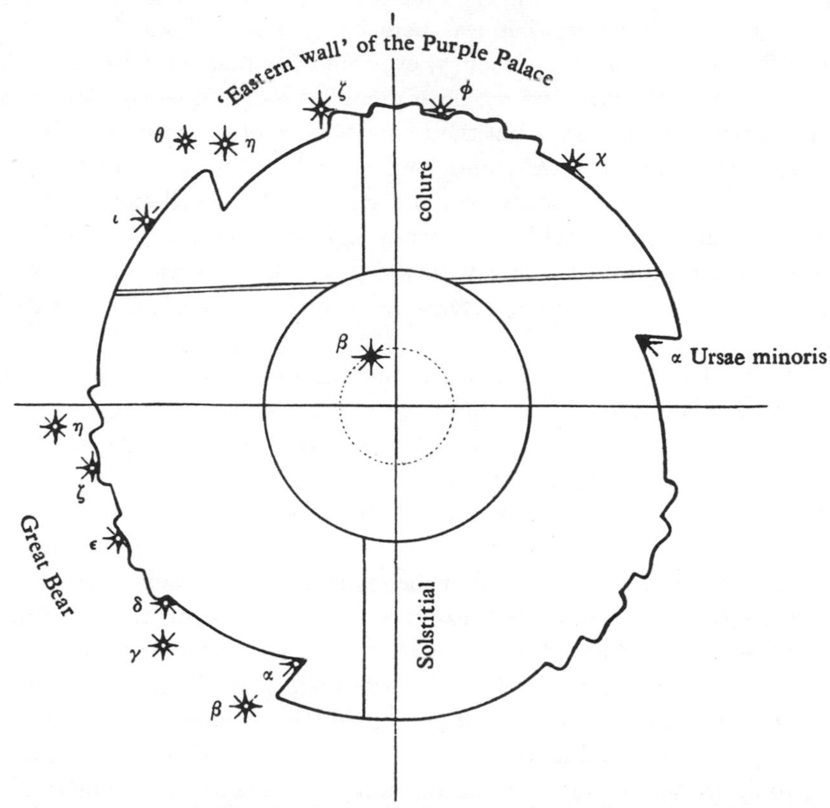

Fig. 150. Diagram to illustrate the use of the circumpolar constellation template (after Henri Michel).

adjacent notch). That shown in Fig. 151, taken from the Huet Collection, yields an angle of 20° corresponding to about −600; that in Laufer[a] gives one of $7\frac{1}{2}$° corresponding to about −1000.

The question arises as to the value of the instrument. What could it do which simple inspection of the stars would not? One answer is that it was primarily used for determining the position of the true celestial pole and hence the solstitial colure. This is shown by a famous passage in the *Chou Pei Suan Ching*, already alluded to:[b]

The true celestial pole (*chêng chi*[1]) is in the centre of the *hsüan-chi*....Wait and observe the large star at the middle of the north pole (area)....At the summer solstice, at midnight,

[a] (8), p. 107. So also an example at the Hague figured in Michel (2).
[b] P. 261 above. Biot himself (p. 622) failed to recognise that the *hsüan-chi* was an instrument.

[1] 正極

the southern excursion of this star is at its furthest point....At the winter solstice the northern excursion of this star is at its furthest point....On the day of the winter solstice, at the hour Yu (5–7 p.m.) the western excursion of this star is at its furthest point. On the same day, at the hour Mao (5–7 a.m.) the eastern excursion of this star is at its furthest point. These are the four excursions (*ssu yu*[1])[a] of the north polar star (as seen through the) *hsüan-chi*.[b]

There can be no doubt that what was here being observed was the circumpolar rotation of β Ursae Minoris.[c] Michel (7) even suggests that the four stars which on many Chinese star-maps surround the polar point in a rough square (cf. Fig. 106) were not meant to represent real stars but the four positions of β Ursae minoris.

Apart from this, the *hsüan-chi* could have served as a very rudimentary kind of orienting instrument for the study of more equatorial constellations. The various teeth on its circumference would give a series of right ascensions, as we should say, and this would be quite in keeping with the principle of 'keying' of circumpolar to equatorial constellations, so characteristic, as we have seen,[d] of ancient Chinese astronomy.[e]

The instrument of later ages which is at once called to mind by the circumpolar constellation template is, of course, the 'nocturnal'. This was a device for telling the time at night by means of the fixed stars, and consisted of a graduated disc or dial with a hole in the centre, and a long projecting arm capable of rotation (Fig. 152 and Fig. 153). The centre being placed in a direct line between the observer's eye and the pole-star, the arm was rotated until it came in line with α and β Ursae Majoris, whereupon the time was read off on the dial of the instrument, this having been previously adjusted to the right position for the season of the year. Nocturnals were first introduced about +1520 but enjoyed only a short vogue before the widespread production of portable watches sent them out of use.[f] Nevertheless, as Michel (4) points out, the principle of the polar sighting-tube still persists in the meridian prism of the modern theodolite. And it may well have given rise to the characteristic equatorial form of Chinese sundial (p. 307 above). But lacking accurate graduations, the *hsüan-chi* could hardly have fulfilled the function of a nocturnal time-keeper any better than direct observations of the position of the Great Bear itself.

If Wu Ta-Chhêng and Michel are right, the astronomical use of the sighting-tube and template may have preceded their adoption as ritual symbols of Earth and Heaven respectively, and therefore they must be very ancient indeed. Hence the interest of

[a] Note that the expression is the same as that used for the 'four displacements' of the earth; cf. above, p. 224.

[b] Ch. 2, p. 2*a*, tr. Biot (4), pp. 621, 622, eng. auct.; mod.

[c] One of the names of this star was Pei chi chung yang ta hsing[2] (the Great Star at the Centre of the North Pole), Maspero (3), p. 329.

[d] P. 232 above.

[e] A difficulty felt by many about the whole theory of Michel is that the construction of the templates would have necessitated star-maps more accurate than any of which we have record from those ages.

[f] One would expect nocturnals in China. What may be a Mongol specimen has been mentioned by Michel (10). Matteo Ricci presented one to the viceroy Liu Chieh-Chai at Nanchhang in 1596 (d'Elia (2), vol. 1, p. 377; Trigault, Gallagher ed. p. 285).

1 四游 2 北極中央大星

Fig. 151. Circumpolar constellation template from the Huet Collection (photo. Michel, 1, 7).

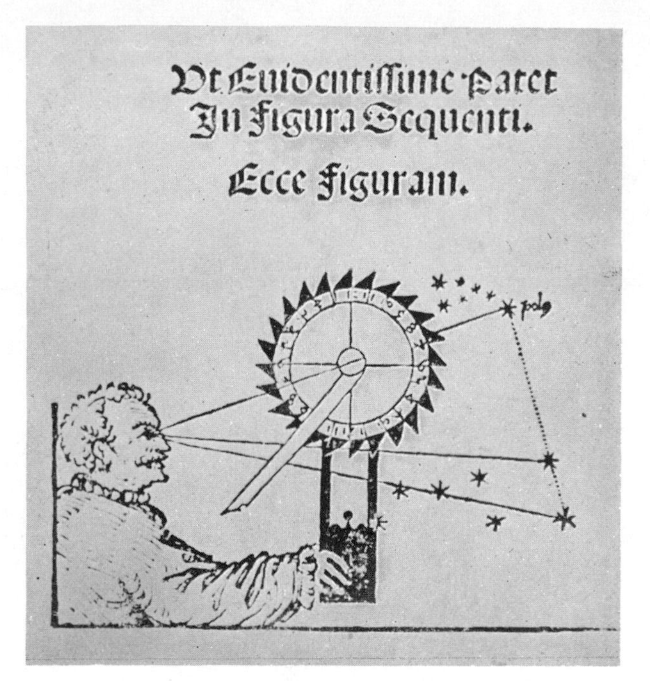

Fig. 152. Petrus Apianus' diagram (+1540) of the nocturnal; the descendant of the circumpolar constellation template.

Fig. 153. A late 17th-century European nocturnal; the descendant of the circumpolar constellation template (from Ward, 1).

the suggestion[a] of Hopkins (19) that the character *yang*[1] (K 720; the essential component of the Yang[2] of Yin and Yang) shows not the sun, but a *pi*, above a ceremonial table, while the related *yang*,[3] to raise, shows a human figure with extended arms taking up the jade template or the ritual symbol.

(6) The Armillary Sphere and other Major Instruments

The armillary sphere, or sphere composed of a number of rings (*armillae*; *huan*[4]) corresponding to the great circles of the celestial sphere, was the indispensable instrument of all astronomers for the determination of celestial positions before the invention of the telescope in the 17th century. In a sense, of course, the armillary rings still continue to exist, but modified and greatly reduced, in the form of those mountings which permit the modern telescope to be directed to any point of the sky; while the telescope itself is conversely a greatly magnified version of those sighting-tubes, alidades, fiducial markings or wires, by means of which the older astronomers fixed their star before reading off its position on the rings of their armillary spheres. It may be added that modern instruments for the determination of the components of terrestrial magnetism make use of the armillary principle, and indeed until recently looked at first sight rather like ancient spheres.

There are numerous expositions of the construction and use of the armillary,[b] but the clearest known to us are contained in the systematic book of Wolf.[c] An excellent monograph on its history is due to Nolte. From the beginning its Chinese name was *hun i*,[5] the celestial-sphere instrument. From early times it was used in China not only as an observational apparatus, but also as a kind of orrery[d] or demonstrational sphere by means of which calendrical computations could be assisted. For this purpose motion was imparted to it by the fall of water, and the movements of the heavenly bodies were reproduced in miniature with varying degrees of complexity and accuracy.[e] These power-driven armillary spheres are closely connected with the history of clockwork, discussion of which we must postpone till Vol. 4. In at least one case the automatic motion seems to have been used for more than purely demonstrational or computational purposes.

[a] For the generally accepted derivation, from sun and sunbeams, see Sect. 13*b* in Vol. 2, p. 227.

[b] E.g. Huggins (1); Gunther (1), vol. 2.

[c] (1), vol. 2, pp. 117, 118.

[d] It will be understood that we use the word 'orrery' in this Section with a sense wider than that which it ought really to bear. The mechanical model made for the Earl of Orrery by George Graham in +1706 was of course intended to illustrate the heliocentric system. A copy of this was sent to China before the century was out; on this see further in Sect. 27*h* below.

[e] Before the Renaissance this was much less common in the West. The chief European parallel is the planetarium or orrery ascribed by many Greek and Latin writers to Archimedes (late −3rd century). This was supposed to have been hydraulically operated (Schlachter & Gisinger (1), p. 51). In 1903 a remarkable chain of gear-wheels was fished out of the sea at Anti-Kythera and described by Svoronos and Rediadis (see Gunther (2) and Price (1) with illustrations). This seems much more likely to have been part of an orrery than, as Gunther thought, of an astrolabe, though rare examples of astrolabes with geared parts are known.

1 昜 2 陽 3 揚 4 環 5 渾儀

It may be said at the outset, in order to clarify the position, that there was a fundamental difference between the armillary spheres of China and those of the Europeans and Arabs, owing to the fact that the Chinese remained persistently faithful to what have become the 'modern' celestial coordinates (i.e. equatorial right ascension and declination) while the Westerners used ecliptic ones from the time of Hipparchus onwards.[a] This difference has already been explained (p. 266 above). It means that although most Chinese instruments had an ecliptic ring, and although much trouble was taken to determine the equivalence of degrees on the equator and ecliptic, the ecliptic ring was just as secondary to their main function as the equatorial ring was on European instruments.[b] Later we shall see how this led the Chinese in the +13th century to the invention of a polar axis mounting essentially identical with that used for all modern equatorial telescopes.

Nothing is known of the apparatus used by Aristyllus and Timocharis, the two Alexandrian astronomers, who, some sixty years later than Shih Shen and Kan Tê,[c] opened the European story of systematic observations of stellar positions. There seems good ground for thinking that Hipparchus, when he began his work in −134, made use at first of an equatorial armillary sphere, but then exchanged it for an ecliptic one.[d] The essentials of such instruments are, of course, similar. The main framework of the sphere will be constituted by the meridian circle, and to this the equatorial or ecliptic circle will be attached at right angles. Then, in the one case an hour-circle, in the other case a circle of celestial longitude, will be made to pivot upon the pole, or the ecliptic pole, respectively, and if this circle carries an alidade or a sighting-tube the position of any star will at once be found on the graduations of the two great circles. The form of armillary stabilised by Ptolemy in the *Almagest* (*c.* +150) is shown in Fig. 154. The solstitial colure was taken as the meridian, and inside it there was a second meridian ring pivoted in the axis of the celestial pole. To this inner ring there was fixed at right angles a circle in the plane of the ecliptic, while in the axis of the ecliptic pole it bore three movable latitude circles one without and two within. The innermost of these carried sights for determining star positions.[e] No further changes in the essential structure of the instrument then took place until the Renaissance, though gradually the Arab astronomers added all other possible great circles, such as the horizon, the equator, etc.[f] Their work was brought to the knowledge of Europeans largely through the remarkable body of translations made under the aegis of Alfonso X, King of Castile and Leon (+1252 to +1284),[g] the *Libros del Saber de Astronomia* (+1277). No less famous in medieval astronomy, however, was the treatise on the sphere, *Sphaera Mundi*, of Johannes Sacrobosco (John Holywood of Halifax), who was writing in Paris in +1229 and died in +1256.[h] All the astronomers of these

[a] They were abandoned only in the +16th century, with Tycho Brahe, as we shall see.
[b] This basic distinction was well appreciated by Maspero (4), p. 337.
[c] I.e. about −290.
[d] Wolf (1), vol. 2, p. 118; Grant (1), p. 438; B & M, p. 534.
[e] The description is given at the beginning of Bk. v of the *Almagest* (*Megale Syntaxis*). Cf. Dicks (1).
[f] Nolte (1); Sédillot (2, 3).
[g] Contemporary of Naṣīr al-Dīn al-Ṭūsī and of Kuo Shou-Ching; Sarton (1), vol. 2, p. 834.
[h] Tr. Thorndike (2).

centuries described armillary spheres—George Peurbach (*Theoricae Novae Planetarum*, +1472), Johannes Müller of Königsberg (Regiomontanus) (+1436 to +1476), and others.[a] Then with the work of Tycho Brahe as described in his *Astronomiae Instauratae Mechanica* (+1598)[b] the armillary sphere reached its highest degree of precision in construction (Fig. 155), and the Greek ecliptic mountings were abandoned in favour

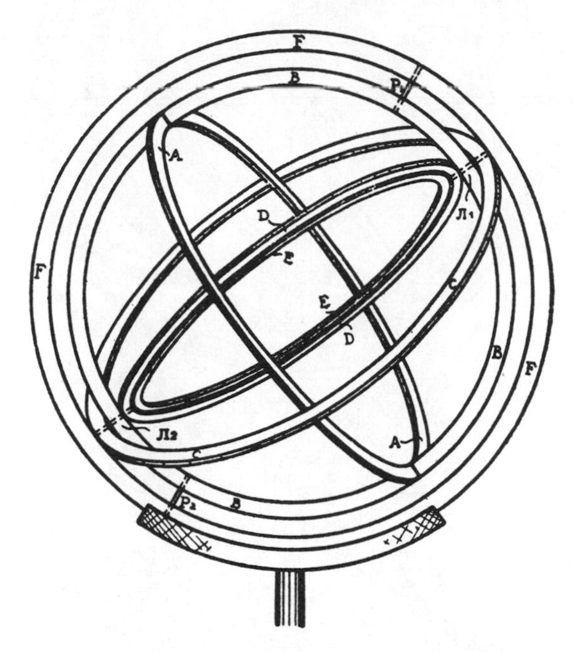

Fig. 154. Ptolemy's armillary sphere (*c.* +150), reconstructed by Nolte (1) after Manitius. *F*, outer fixed meridian (solstitial colure) ring, probably graduated; *B*, inner mobile meridian ring pivoted in the polar axis; *A*, ecliptic ring, graduated for observations; *C, D, E*, three mobile celestial latitude rings, the two outer ones graduated, the innermost one bearing sights; P_1, P_2, polar pivots; π_1, π_2, ecliptic polar pivots.

of the Chinese equatorial system. This is commonly regarded (e.g. by Zinner)[c] as one of the greatest technical advances of Renaissance astronomy. After the introduction of the telescope the sphere was relegated to purposes of demonstration rather than observation.[d] Let us now see how closely this development was paralleled in another civilisation, that of China.

[a] Their works were printed by Erhard Ratdolt in +1485 at Venice in books remarkable as being the first examples of colour printing in Europe (Redgrave, 1).

[b] Tr. Raeder, Strömgren & Strömgren (1).

[c] (2), p. 298.

[d] Apianus (1); Apianus & Gemma Frisius (1); Th. Blundeville (1).

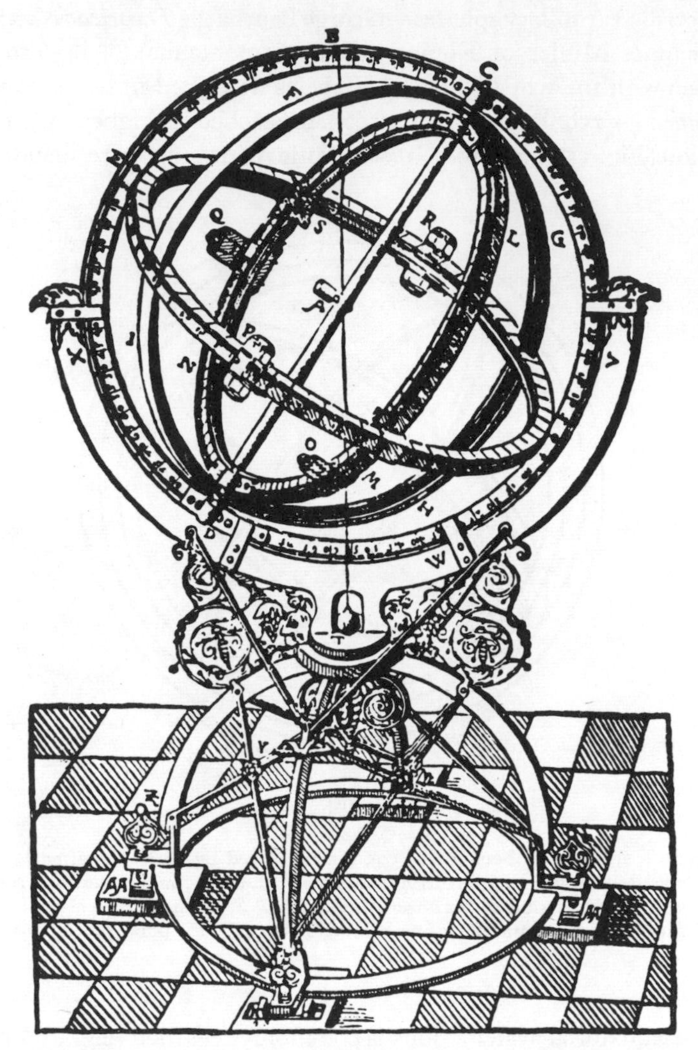

Fig. 155. Tycho Brahe's smaller equatorial armillary sphere (+1598); for detailed description see the translation and study of his *Astronomiae Instauratae Mechanica* by Raeder, Strömgren & Strömgren.

(i) *The general development of armillary spheres*

Maspero (4) devotes some sixty pages to the development and use of armillary spheres in China, but so much detail is impossible in a summary exposition,[a] and the course of events may be followed in Table 31, which presents what is known of the

[a] Maspero's monograph gives numerous synonyms of the various technical terms, and several important translations of passages about armillary spheres in the dynastic histories. In spite of his great, and deserved, authority, one cannot assume that all his conclusions are reliable; for instance, I believe that he was wrong in thinking (p. 325) that Khung Thing's armillary sphere of +323 had no equatorial circle (see *Sui Shu*, ch. 19, p. 16*a*). Moreover, there are errors in some of the textual references.

chief characteristics of the instruments of different periods. It makes no claim to be exhaustive; indeed, closer investigation might perhaps reveal as many instances again—but the most famous instruments are listed.[a]

The principal features of this development may be summarised as follows. In all probability the most primitive form of armillary was the simple single ring, with some kind of fiducial line or sights, which could be set up in the meridian or equatorial plane as desired. Measurement one way gave the north polar distance (the Chinese form of the declination), measurement the other way gave the position in a *hsiu* (the Chinese form of right ascension). This was very likely all that Shih Shen and Kan Tê (−4th century) had at their disposal.[b] It seems to have lasted down to the time of Lohsia Hung and Hsienyü Wang-Jen (late −2nd and early −1st centuries). But from the middle of the latter century onwards developments occurred rapidly. Kêng Shou-Chhang introduced the first permanently fixed equatorial ring in −52, Fu An and Chia Khuei added the ecliptic ring in +84, and with Chang Hêng about +125 the sphere was complete with horizon and meridian rings.[c] Chang Hêng was also the first to make a demonstrational type with a model earth at the centre and to accomplish the rotation of its rings by water-power, as well as the observational type, and he combined their use in a way we shall presently explain.

After this, spheres of both kinds continued to be made without much change for several centuries. But it was found inconvenient to have the ecliptic and equatorial rings immovably connected together, for while this sufficed for studying annual motion it did not help the study of diurnal motion. In +323 therefore Khung Thing[1] so arranged his sphere that the ecliptic ring could be pegged on to the equator at any point desired. This apparatus was the forerunner of all the most important later spheres. For a little more than three hundred years afterwards, Li Shun-Fêng made the radical innovation of building not two nests (*chhung*[2]) of concentric rings but three.[d] There was still the polar-mounted declination ring with its sighting-tube (the 'component of the four displacements', *ssu yu i*[3]) internally, but between it and the fixed outer nest (the 'component of the six cardinal points', *liu ho i*[4])[e] there was inserted an intermediate nest (the 'component of the three arrangers of time', *san chhen i*[5]). This new component had in the following centuries both equator and ecliptic,

[a] A large collection of quotations concerning the armillary spheres of all centuries is given in *TSCC*, *Li fa tien*, chs. 83, 84. We are particularly indebted to Dr Derek Price for studying with us the development of Chinese armillary technique; without his collaboration a clear picture would not have been attained.

[b] Such simple rings were also described by Ptolemy (*Almagest*, I, 12); cf. Dicks (1); Rome (1). The characters for the names of all these astronomers will be found incorporated in Table 31.

[c] It is noteworthy that the armillary sphere reached its definitive forms characteristic respectively of Chinese and Hellenistic civilisation just about the same time. For Chang Hêng and Claudius Ptolemy were contemporaries, the latter slightly the younger of the two, and the *Almagest* was finished about +144. Biographical memoirs of Chang Hêng have been written by Chang Yu-Chê (1, 2); Sun Wên-Chhing (*2, 3, 4*) and Li Kuang-Pi & Lai Chia-Tu (*1*).

[d] Cf. *Hsiao Hsüeh Kan Chu*, ch. 1, p. 9*b*.

[e] Cf. *Hsin I Hsiang Fa Yao*, ch. 1, p. 6*b*; *Hsiao Hsüeh Kan Chu*, ch. 2, p. 2*a*. The essential texts concerning Li Shun-Fêng's instrument were translated by Maspero (4), p. 321.

[1] 孔挺 [2] 重 [3] 四遊儀 [4] 六合儀 [5] 三辰儀

Explanation and Key to Table 31

Certain presence of a particular component is indicated by a +, certain absence by a ○; in cases of doubt a ? is inserted. Lack of information is indicated by blank spaces, except in cases where 'Absent' or some other explanation is given. Components may also be represented by numbers more or less than one. Technical terms for parts of the apparatus are based mainly on the *Hsin I Hsiang Fa Yao* of +1092, but many synonyms were used in other ages, some of which will be found in Maspero (4).

ABBREVIATIONS

CJC — *Chhou Jen Chuan*
CS — *Chin Shu*
CSHK(SK) — Yen Kho-Chün (*1*), San Kuo section
CSHK(C) — Yen Kho-Chün (1), Chin section
CTS — *Chin Thang Shu*
FY — *Fa Yen*
HHS — *Hou Han Shu*
HIHFY — *Hsin I Hsiang Fa Yao*
HSS — *Hsü Shih Shuo*
HTS — *Hsin Thang Shu*
KYCC — *Khai-Yuan Chan Ching*
MCPT — *Mêng Chhi Pi Than*
MSH — *Ming Shih*
NSH — *Nan Shih*
PSH — *Pei Shih*
SS — (Liu) *Sung Shu*
SSH — *Sung Shih*
SUS — *Sui Shu*
TH — Wieger (1)
THY — *Thang Hui Yao*
TPYL — *Thai-Phing Yü Lan*
TSCC — *Thu Shu Chi Chhêng* (*Li fa tien*)
WS — (Pei) *Wei Shu*
YH — *Yü Hai*
YHSF — Ma Kuo-Han (*1*)
YSH — *Yuan Shih*

Name of astronomer	Date	Dynasty	Scale (Chinese inches to degree)	Approx. circumference (Chinese feet)	Sighting-tube ring; Component of the Four Displacements (*ssu yu i*[55])		Demonstrational device (earth model) instead of observational apparatus	Outer nest; Component of the Six Cardinal Points (*liu ho i*[56])						Intermediate nest pivoted on polar axis; Component of the Three Arrangers of Time (*san chhen i*[57])						Application of motive power	References
					Sighting-tube (*wang thung*[35]) and Polar-mounted declination ring (*yu kuei*[36] or *hsüan i*[37])	Ecliptically mounted celestial latitude ring (*huang tao yu i*[38])		Equator (*thien chhang*[39])	Ecliptic (*huang tao*[40] or *jih tao*[41])	Prime meridian (*thien ching*[42] or *yang ching*[43])	Prime vertical (*thien ting huan*[44])	Horizon (*yin wei*[45] or *ti hun*[46])	Declination circles of perpetual apparition (*shang kuei*[47] and *hsia kuei*[48])	Equator (*chhih tao*[49])	Ecliptic (*huang tao*[50])	Solstitial colure circle (*san chhen i shuang huan*[51])	Equinoctial colure (*ssu hsiang huan*[52])	Lunar path (*pai tao yüeh huan*[53])	Diurnal motion gear ring (*thien yün huan*[54])		
Shih Shen[1] / Kan Tê[2]	*c.* −350	Chou			+[a]	Absent	○	Absent						Absent						○	*FY*, ch. 7, p. 2*b*; *SUS*, ch. 19, p. 14*b*; *SSH*, ch. 76, p. 1*a*; *TSCC*, ch. 83, p. 3*b*
Lohsia Hung[3]	−104	C/Han			+[a]		○													○	
Hsienyü Wang-Jen[4]	−78	C/Han			+[a]		○													○	
Kêng Shou-Chhang[5]	−52	C/Han	0·2	7	+		○	+	○	○	○	○	○							○	*HHS*, ch. 12, p. 6*a*
Fu An[6] / Chia Khuei[7]	+84 / +103	H/Han			+		○	+	+	○	○	○	○							○	*HHS*, ch. 12, pp. 4*a*, 5*a*; *CS*, ch. 11, p. 5*a*; *SUS*, ch. 19, p. 14*b*; *SSH*, ch. 76, p. 1*a*
Chang Hêng[8]	+125	H/Han	0·4	15	+ and ○		○ and +	+[b]	+[b]	+	○	+	?[c]							○ and +	*HHS*, ch. 13, p. 20*b*; *CS*, ch. 11, pp. 3*b*, 5*a*; *SUS*, ch. 19, pp. 2*a*, 7*b*, 14*b*, 15*a*; *KYCC*, ch. 1, p. 2*a*; *SSH*, ch. 48, p. 3*b*, ch. 76, pp. 1*a* ff.; *YHSF*, ch. 76, pp. 66*a* ff.; *TSCC*, ch. 83, pp. 4*a* ff.
Wang Fan[9]	+250	San Kuo (Wu)	0·3	11	○		+	+	+	+	○	+	+							○	*KYCC*, ch. 1, pp. 10*a* ff.; *CSHK* (*SK*), ch. 72, pp. 1*a* ff., 5*b*; *SS*, ch. 23, pp. 1*b* ff.; *TSCC*, ch. 83, p. 5*a*
Ko Hêng[10]	+260	San Kuo (Wu)			○		+													+	*SUS*, ch. 19, p. 18*a*; *YH*, ch. 4, p. 18*b*; *TPYL*, ch. 2, p. 10*a*; *CJC*, ch. 5

Table 31 (*continued*)

Name of astronomer	Date	Dynasty	Scale (Chinese inches to degree)	Approx. circumference (Chinese feet)	Sighting-tube ring; Component of the Four Displacements (*ssu yu i*[55]): Sighting-tube (*wang thung*[35]) and Polar-mounted declination ring (*yu kuei*[36] or *hsüan i*[37])	Sighting-tube ring: Ecliptically mounted celestial latitude ring (*huang tao yu i*[38])	Demonstrational device (earth model) instead of observational apparatus	Outer nest; Component of the Six Cardinal Points (*liu ho i*[56]): Equator (*thien chhang*[39])	Outer nest: Ecliptic (*huang tao*[40] or *jih tao*[41])	Outer nest: Prime meridian (*thien ching*[42] or *yang ching*[43])	Outer nest: Prime vertical (*thien ting huan*[44])	Outer nest: Horizon (*yin wei*[45] or *ti hun*[46])	Outer nest: Declination circles of perpetual apparition (*shang kuei*[47] and *hsia kuei*[48])	Intermediate nest pivoted on polar axis; Component of the Three Arrangers of Time (*san chhen i*[57]): Equator (*chhih tao*[49])	Intermediate nest: Ecliptic (*huang tao*[50])	Intermediate nest: Solstitial colure circle (*san chhen i shuang huan*[51])	Intermediate nest: Equinoctial colure (*ssu hsiang huan*[52])	Intermediate nest: Lunar path (*pai tao yüeh huan*[53])	Intermediate nest: Diurnal motion gear ring (*thien yün huan*[54])	Application of motive power	References
Liu Chih[11]	+274	Chin			○		+	+	+	○	○	+	○							○	*CSHK(C)*, ch. 39, pp. 5*a*ff.
Khung Thing[12]	+323	C/Chao	0·7	24	+		○	+	+[d]	+	○	+	○							○	*SUS*, ch. 19, pp. 16*a*, *b*
Chhao Chhung[13]	+402	N/Wei	0·7	24[e]	+		○	+	+[f]	+	+	+	○							○	*SUS*, ch. 19, p. 3*a*; *WS*, ch. 91, p. 1*b*; *TSCC*, ch. 83, pp. 8*a*, *b*
Hsieh Lan[14] (or Hu Lan[15])	+415	N/Wei	0·4	15[e]	+		○	+	+[f]	+	+	+	+							○	*CTS*, ch. 35, p. 2*a*; *HTS*, ch. 31, p. 2*a*; *THY*, ch. 42, p. 7*a*
Chhien Lo-Chih[16]	+436	L/Sung	0·5	18	○	Absent	+	+	+	+	○	+[g]	○	Absent						+	*SS*, ch. 23, p. 8*b*; *SUS*, ch. 19, pp. 17*b*ff.; *KYCC*, ch. 1, p. 23*b*; *YH*, ch. 4, p. 18*b*; *TSCC*, ch. 83, p. 7*b*; *TH*, p. 1111
Thao Hung-Ching[17]	+520	Liang		9	○		+					+[g]								+	*NSH*, ch. 76, p. 11*a*; *TPYL*, ch. 2, p. 10*a*
Kêng Hsün[18]	+590	Sui			○		+													+	*SUS*, ch. 78, pp. 7*b*ff.; *PSH*, ch. 89, p. 31*a*; *YH*, ch. 4, p. 26*a*; *HSS*, ch. 6, p. 11*a*

Li Shun-Fêng[19]	+633	Thang	0·4	15	○	+[h]	○	+		+	○	+	○	+	+	+	○	+	○	○	*CTS*, ch. 35, pp. 1*a*ff.; *HTS*, ch. 31, pp. 1*a*ff.; *THY*, ch. 42, p. 5*b*; *TSCC*, ch. 83, p. 9*a*
I-Hsing[20] & Liang Ling-Tsan[21]	+721	Thang	0·8	30	○	+[i] and ○	○ and +	○	Transferred	+	+	+	○	+	+	+	○	+	○	○ and +	*CTS*, ch. 35, pp. 1*a*ff.; *HTS*, ch. 31, pp. 1*b*ff. *THY*, ch. 42, p. 6*a*; *YH*, ch. 4, p. 24*a*; *TSCC*, ch. 83, pp. 9*b*ff.; *TH*, p. 1409
Chang Ssu-Hsün[22]	+979	Sung			○	○	+							+	+	?	○			+[j]	*SSH*, ch. 48, pp. 3*b*ff.; *YH*, ch. 4, pp. 29*a*ff.; *TSCC*, ch. 84, p. 1*b*
Han Hsien-Fu[23]	+995, +1010	Sung	0·5	18	+	○	○	+	+	+	+	+[k]	+	Absent						○	'Too simple', *MCPT*, ch. 8, p. 5*a*; *SSH*, ch. 48, pp. 4*b*ff., ch. 76, p. 1*b*, ch. 461, p. 6*b*; *YH*, ch. 4, pp. 30*a*ff., 32*b*ff.; *TSCC*, ch. 84, p. 2*a*
Chou Tshung,[24] Shu I-Chien[25] & Yü Yuan[26]	+1050	Sung			○	+	○	+	Transferred to intermediate nest	+	○	+	○	+	+	+	○	+	○	○	'Too complex', *MCPT*, ch. 8, p. 5*a*; *SSH*, ch. 76, pp. 1*b*ff.; *TSCC*, ch. 84, p. 3*a*
Shen Kua[27]	+1074	Sung			+	○	○	+						+	+	?	○	○		○	*YH*, ch. 4, pp. 35*b*ff.; *MCPT*, ch. 8, p. 5*a*; *TSCC*, ch. 84, p. 4*b*
Su Sung[28]	+1090	Sung	0·7	24[l]	+	○	○	+		+	○	+	○	+	+	+	+	○	+[m]	+	*HIHFY*, ch. 1; *TSCC*, ch. 84, pp. 8*b*ff.
Wang Fu[29]	+1124	Sung			○	○	+							+	+	?				+	*SSH*, ch. 80, pp. 25*b*ff.; *TSCC*, ch. 84, pp. 11*a*ff.
Li Chi-Tsung[30]	+1132	Sung			+	○	○														*SSH*, ch. 48, pp. 18*a*ff., ch. 81, pp. 13*b*ff.; *YH*, ch. 4, pp. 47*b*ff.; *TSCC*, ch. 84, p. 12*a*
Shao O[31]	+1150	Sung			+	○	○														*SSH*, ch. 81, pp. 15*a*ff.; *YH*, ch. 4, p. 48*b*; *TSCC*, ch. 84, pp. 13*a*ff.
Kuo Shou-Ching[32]	+1276	Yuan			+	○	○	+		+	○	+	○	+	+	+	+	○	○	○	*YSH*, ch. 164, p. 7*b* (not in ch. 48; not descr. in Wylie, 7); *TSCC*, ch. 84, pp. 15*b*ff. Cf. Johnson (1, 2)[n]
Huangfu Chung-Ho[33]	+1437	Ming		c. 18	+	○	○	+		+	○	+	○	+	+	+	+	○	○	○	*MSH*, ch. 299, p. 14*b*; *TSCC*, ch. 84, p. 18*b*[o]

Table 31 (*continued*)

Name of astronomer	Date	Dynasty	Scale (Chinese inches to degree)	Approx. circumference (Chinese feet)	Sighting-tube ring; Component of the Four Displacements (*ssu yu i*[55]): Sighting-tube (*wang thung*)[35] and Polar-mounted declination ring (*yu kuei*[36] or *hsüan i*[37])	Sighting-tube ring: Ecliptically mounted celestial latitude ring (*huang tao yu i*[38])	Demonstrational device (earth model) instead of observational apparatus	Outer nest; Component of the six Cardinal Points (*liu ho i*[56]): Equator (*thien chhang*[39])	Outer nest: Ecliptic (*huang tao*[40] or *jih tao*[41])	Outer nest: Prime meridian (*thien ching*[42] or *yang ching*[43])	Outer nest: Prime vertical (*thien ting huan*[44])	Outer nest: Horizon (*yin wei*[45] or *ti hun*[46])	Outer nest: Declination circles of perpetual apparition (*shang kuei*[47] and *hsia kuei*[48])	Intermediate nest pivoted on polar axis; Component of the Three Arrangers of Time (*san chhen i*[57]): Equator (*chhih tao*[49])	Intermediate nest: Ecliptic (*huang tao*[50])	Intermediate nest: Solstitial colure circle (*san chhen i shuang huan*[51])	Intermediate nest: Equinoctial colure (*ssu hsiang huan*[52])	Intermediate nest: Lunar path (*pai tao yüeh huan*[53])	Intermediate nest: Diurnal motion gear ring (*thien yün huan*[54])	Application of motive power	References
Extant Korean demonstrational armillary sphere (see p. 389)	+18th			4	○	○	+	+		+	○	+	○	+	+	+	?	+	○	+	Rufus (2)
Tycho Brahe (lesser)	+1598				+	○	○	○	○	+	○	○	○	+	○	○	+	○	○	○	I.e. simplified
Verbiest, equatorial }	+1673	Chhing		18	+	○	○	+[r]	○	+	○	○	○	Absent						○	I.e. simplified. *TSCC*, ch. 84, p. 20*a*
Verbiest, ecliptic }				18	○	+	○	○	○	+	○	○	○	○	+	+	○	○	○	○	I.e. simplified. *TSCC*, ch. 84, p. 20*a*
Kögler, equatorial	+1744	Chhing			+	○	○	+	○	+	○	○	○	+	○	½[p]	+	○	○	○	
Ptolemy	+2nd				○	3	○	○	○	+	○	○	○	○	+	+	○[q]	○	○	○	Cf. Nolte (1)
Jamāl al-Dīn	+1267				+	○	○	○	○	+	○	+	+[s]	Absent						○	*YSH*, ch. 48, pp. 10*b*, 11*a*

[a] It is conjectured that these single rings, which must have carried movable sights, were set up in the plane of meridian or equator as desired.

[b] Maspero (4), p. 336, believed that these two circles were movable, manually and mechanically, within the outer nest—in other words, that a mobile intermediate nest was already present at this date. If so (though he did not note it), some kind of colure ring would have been needed to unite them. But none is mentioned in the relevant texts.

[c] We are not sure whether the 'inner and outer rings' (*nei wai kuei*[58]) imply this.

[d] This ring is known to have been attachable to the equator circle at any desired point by means of pegs. Khung Thing's instrument was therefore the predecessor of all those with mobile intermediate nests.

[e] Both these armillary spheres were made of iron—striking testimony to the abundance of this metal in +5th-century China. Since cast iron had long been in common use, the rings may have been castings pegged together.

[f] In both instruments these rings were fixed to the inner polar-mounted declination ring—another attempt to achieve mobility for the ecliptic circle.

[g] These were not actual rings; the horizon line was given by a marking on the earth model at the centre of the sphere.

[h] Probably. It is not quite certain whether Li Shun-Fêng departed in practice from the traditional equatorially mounted sighting-tube.

[i] With steel bearings.

[j] Here the motive power was mercury. Freezing temperatures did not therefore interfere with the time-keeping of the mechanised armillary, as when water was used.

[k] A flat annulus with several sets of graduations.

[l] About twenty tons of bronze were necessary for this sphere.

[m] Here is the only place where this entry appears, because (so far as we know at present) the armillary of Su Sung was the only one in which an observational sphere was mechanised as well as a globe or demonstrational sphere. The transmission shaft for this 'clock drive' was carried upwards to connect with the gear ring through a hollow central columnar support. Such a support had been reintroduced into the design of armillary spheres by one of Su Sung's aides, Chou Jih-Yen,[34] who had himself constructed one about ten years before (*c.* +1080).

[n] This instrument lasted until the coming of the Jesuits (1600).

[o] This was an exact copy of Kuo Shou-Ching's armillary sphere, made in wood at Nanking, and afterwards cast in bronze at Peking after adjustment for the latitude. It still exists there.

[p] A semicircular solstitial colure ring supporting the equator circle in the southern hemisphere only.

[q] These were not fixed colure rings, but celestial latitude circles swinging in the same plane as the ecliptically mounted sighting-tube ring. Though the sphere of Ptolemy might therefore be said to have had five nests, it lacked any equatorial component. The inner meridian ring, however, was pivoted on the polar axis.

[r] Plus a semicircular equinoctial colure ring supporting the equator circle in the southern hemisphere only.

[s] Attached to the movable declination-ring.

1 石申 2 甘德 3 落下閎 4 鮮于妄人 5 耿壽昌 6 傅安 7 賈逵 8 張衡 9 王蕃 10 葛衡
11 劉智 12 孔挻 13 晁崇 14 解蘭 15 斛蘭 16 錢樂之 17 陶弘景 18 耿詢 19 李淳風
20 一行 21 梁令瓚 22 張思訓 23 韓顯符 24 周琮 25 舒易簡 26 于淵 27 沈括 28 蘇頌
29 王黻 or 黼 30 李繼宗 31 邵諤 32 郭守敬 33 皇甫仲和 34 周日嚴 35 望筒 36 遊規 37 旋儀
38 黃道遊儀 39 天常 40 黃道 41 日道 42 天經 43 陽經 44 天頂環 45 陰緯 46 地渾
47 上規 48 下規 49 赤道 50 黃道 51 三辰儀雙環 52 四象環 53 白道月環 54 天運環 55 四游儀
56 六合儀 57 三辰儀 58 內外規

and sometimes also a third circle for the moon's path, though this last was dropped after + 1050. These circles were held together by one, or sometimes two, colure rings. Meanwhile the outer nest, from which the mobile ecliptic had disappeared, retained the equator, together with meridian, prime vertical and horizon, just as in the older pre-Thang spheres.[a]

Li Shun-Fêng was working in the early part of the +7th century when Buddhist influence was strong and Hellenistic astronomical ideas were finding entry to China through Indian channels. He proposed, but perhaps the monk I-Hsing was the first to execute, the plan of having the sighting-tube mounted ecliptically in order that observation of celestial latitudes (so important in Hellenistic planetary theory) could more easily be made.[b] But the originality and particularity of Chinese astronomy was never easily modified, and with the exception of the instrument of Chou Tshung and Shu I-Chien in + 1050, ecliptic mountings were employed no more until the time of the Jesuits.[c]

A few further observations may be made. The spheres of Chhao Chhung in +402 and Hsieh Lan in +415 had the unusual feature of being in iron, not bronze, while I-Hsing and Liang Ling-Tsan used steel for their bearings about +720. Circles of perpetual visibility and invisibility were naturally more suited for demonstrational armillary spheres, such as those of Wang Fan (*c.* +250), but they seem to have been used occasionally on observational ones also, as in that of Hsieh Lan. As may be seen from the last column before the references in Table 31, most of the the demonstrational spheres from the time of Chang Hêng onwards, i.e. those in which the sighting-tube was replaced by a model earth at the centre, or later sunk in a flat-topped box to symbolise the terrestrial horizon, were rotated by water-power. The early forms of this drive must have been crude enough, but a turning-point came in +725 when I-Hsing and Liang Ling-Tsan invented a form of escapement and combining the rotation of the sphere with various kinds of jackwork constructed what was essentially the first of all mechanical clocks (see Sect. 27*h* in Vol. 4). A later clock, that of Chang Ssu-Hsün in +979, used mercury instead of water for the motive power.

It is interesting to compare the final form of the Chinese equatorial armillary sphere with those which issued from the Western tradition. Tycho Brahe's lesser equatorial armilla of + 1598 (Fig. 155) differed from those of Kuo Shou-Ching (*c.* + 1276, Fig. 156) and Su Sung[1] (+1090, Fig. 159) chiefly in being simpler, for it dispensed with the equator and horizon rings in the outer nest, and with the ecliptic in the inner. This was undoubtedly because Tycho's equipment included ecliptic armillary spheres as

[a] There was a reaction against the three-nest system about + 1000 when Han Hsien-Fu[2] went back to pre-Thang designs, but his instruments were pronounced much too simple by Shen Kua,[3] and had no successors.

[b] One cannot be too careful about terminology in this rather complex subject. When Hartner (8) says that Fu An in the + 1st century introduced 'the ecliptical armilla' into Chinese astronomy, he means 'the ecliptic ring in the equatorial armillary sphere', and not of course the Ptolemaic armillary sphere with the ecliptically mounted sights. That did not come in China until the +7th or the +8th century.

[c] With the possible exception of a special piece of apparatus used by Kuo Shou-Ching (p. 369 below). Shen Kua considered them too complicated (*MCPT*, ch. 8, para. 8; cf. Hu Tao-Ching (*1*), vol. 1, pp. 335 ff.).

[1] 蘇頌 [2] 韓顯符 [3] 沈括

Fig. 156. Equatorial armillary sphere of Kuo Shou-Ching (*c.* +1276), made for the latitude of Phing-yang in Shansi, but later kept at Peking, where this photograph was taken, and at Nanking. The instrument here shown may be one of the exact replicas made by Huangfu Chung-Ho in +1437. In this picture it is seen from the north-west (photo. Saunders).

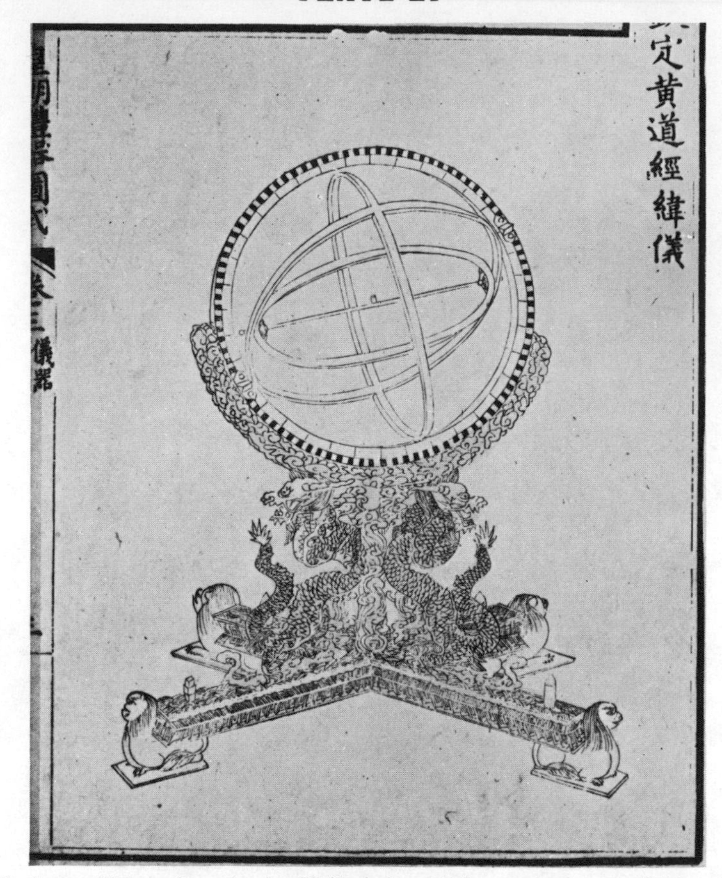

Fig. 157. Ecliptic armillary sphere in the Hellenistic tradition made for the Peking Observatory by Ferdinand Verbiest in +1673. This drawing, from the *Huang Chhao Li Chhi Thu Shih*, is similar to that in the French edition of Lecomte (1) but clearer.

Fig. 158. Equatorial armillary sphere in the Chinese tradition set up in the Peking Observatory by Ignatius Kögler and his collaborators in +1744 (see text). Through it can be seen (end-on) Ferdinand Verbiest's quadrant of +1673, and to the left his celestial globe (photo. Thomson, reprod. Bushell).

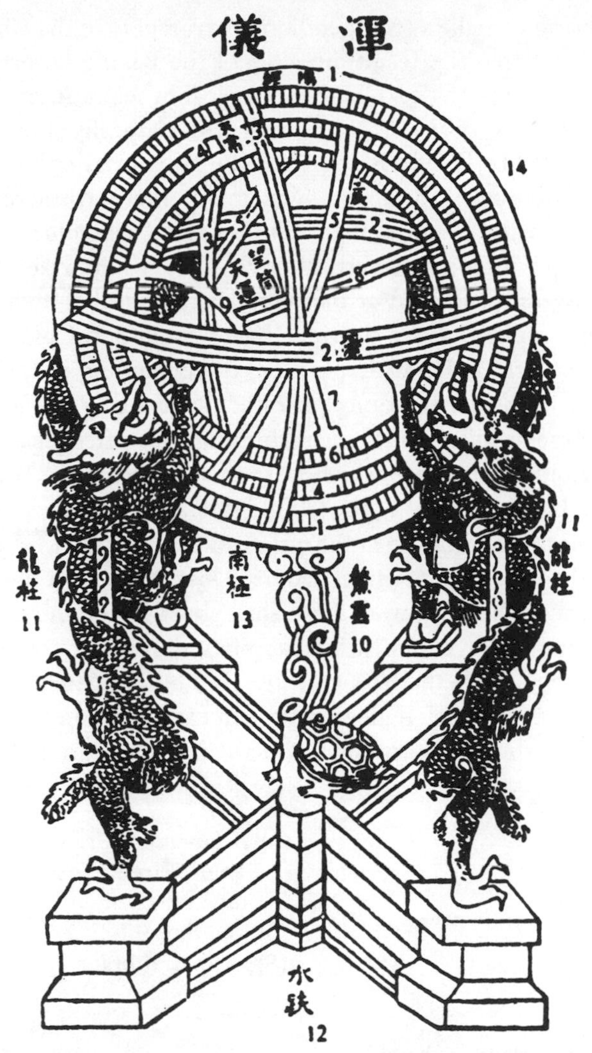

Fig. 159. Su Sung's armillary sphere (*hun i*) of +1090, described in the *Hsin I Hsiang Fa Yao* (redrawn and labelled by Maspero, 4); the first to be provided with a clock drive. For characters see also Table 31.

Outer Nest (*liu ho i*):
1 meridian circle (*yang ching huan*).
2 horizon circle (*ti hun huan*).
3 outer equator circle (*thien chhang huan*).

Middle Nest (*san chhen i*), see also Fig. 160*a*:
4 solstitial colure circle (*san chhen huan*).
5 ecliptic circle (*huang tao huan*).
9 diurnal motion gear-ring (*thien yün huan*), connecting with the power-drive.

Inner Nest (*ssu yu i*), see also Fig. 160*b*:
6 polar-mounted declination ring or hour-angle circle (*ssu yu huan*), with
7 sighting-tube (*wang thung*) attached to it and strengthened by a
8 diametral brace (*chih chü*).

Other Parts:
10 vertical column (*ao yün chu*), concealing the transmission shaft.
11 supporting columns in the form of dragons (*lung chu*).
12 cross-piece of the base, incorporating water-levels (*shui thieh*).
13 south polar pivot (*nan chi*).
14 north polar pivot (*pei chi*).

well; he was not trying to make a 'compendium instrument' in the Chinese style. It is also interesting to note that the Jesuit directors of the Peking observatory eventually made considerable concessions to Chinese usage. The ecliptic armillary sphere of Ferdinand Verbiest in +1674 (cf. Fig. 157) had naturally followed closely the Hellenistic pattern used by Tycho Brahe,[a] but the equatorial one set up by Ignatius Kögler and his collaborators in +1744 was much more Chinese than European (Fig. 158).[b] They added an equator to the outer nest and a half colure ring to the middle one, while dispensing with a prime vertical circle. More important they adopted the 'gun-barrel' sighting-tube[c] instead of the pinnules universal elsewhere, and the very practical split-ring armillary circles previously known only in China, where they had originated at least as early as Hsieh Lan's time, the beginning of the +5th century.

Perhaps we should consider the Sung (or rather specifically the Northern Sung) as the apogee of Chinese armillary instrument-making. We often hear in later times of the four famous armillary spheres[d] which were then constructed, between +995 and +1092. But the fall of the capital to the Chin Tartars in +1126 proved a heavy blow to astronomical science, and in spite of the efforts of devoted men,[e] the Southern Sung could never equal the instruments which had been lost. Although the armillary sphere of Su Sung,[1] the last of the four, was (as we shall shortly see) unique in an important respect, in its general construction it well typifies the 'great tradition' of Chinese armillary art. Su Sung's book, the *Hsin I Hsiang Fa Yao*, is particularly noteworthy for the numerous illustrations which it gives to explain each component part of the instrument. Fig. 159 shows the whole armillary sphere. The parts are indicated as follows:

Outer Nest
(1) meridian circle,
(2) horizon circle,
(3) outer equator circle.

Middle Nest
(4) solstitial colure circle,
(5) ecliptic circle,
(5*a*) inner equator circle (not shown),
(9) diurnal motion gear-ring connecting with the power-drive (see pp. 363 below).

Inner Nest
(6) polar-mounted declination ring or hour-angle circle, with diametral sighting-tube attached,
(7) sighting-tube,
(8) diametral brace.

[a] Cf. Chhen Tsun-Kuei (*6*), p. 23.
[b] Cf. Chhen Tsun-Kuei (*6*), p. 46.
[c] This, with its cross-strut, goes back to very ancient times in China (cf. pp. 336, 343 ff.). One cannot but surmise that the easy availability of bamboo tubes was a factor here when it came to the need of straight and sturdy piping from 4 to 8 ft. in length. This would be a parallel to the eminent role of bamboo in the development of gunpowder arms, on which see Sect. 30 in Vol. 6.
[d] They were those of Han Hsien-Fu, Chou Tshung & Shu I-Chien, Shen Kua and Su Sung. See *Hsiao Hsüeh Kan Chu*, ch. 1, p. 10*a*; *Yü Hai*, ch. 4, p. 47*b*; *Hsin I Hsiang Fa Yao*, ch. 1, p. 1*b*; *Sung Shih*, ch. 81, p. 13*b*; *Chhi Tung Yeh Yü*, ch. 15, p. 5*b*.
[e] Yuan Wei-Chi,[2] for instance, who had been a pupil and assistant of Su Sung, went south with the court, and about +1130 tried to construct new armillary apparatus, but in the absence of plans and documents, and at a time of dispersion of artisans, was not very successful. Three more instruments of indifferent quality followed before +1170.

[1] 蘇頌 [2] 袁惟幾

Other Parts

(10) vertical column (concealing the transmission shaft),
(11) supporting columns in the form of dragons,
(12) cross-piece of the base, incorporating water-levels,
(13) south celestial pole,
(14) north celestial pole.

Perhaps it is not surprising that the proper disposition of the components was beyond the power of the artist to limn; we must be thankful for having an +11th-century diagram of an armillary sphere at all. Parts of the instrument as depicted by Su Sung are reproduced in Fig. 160*a* and *b*, the former showing the middle nest bearing both equatorial and ecliptic rings, the latter showing the polar-mounted declination ring or hour-angle circle with its attached sighting-tube.[a]

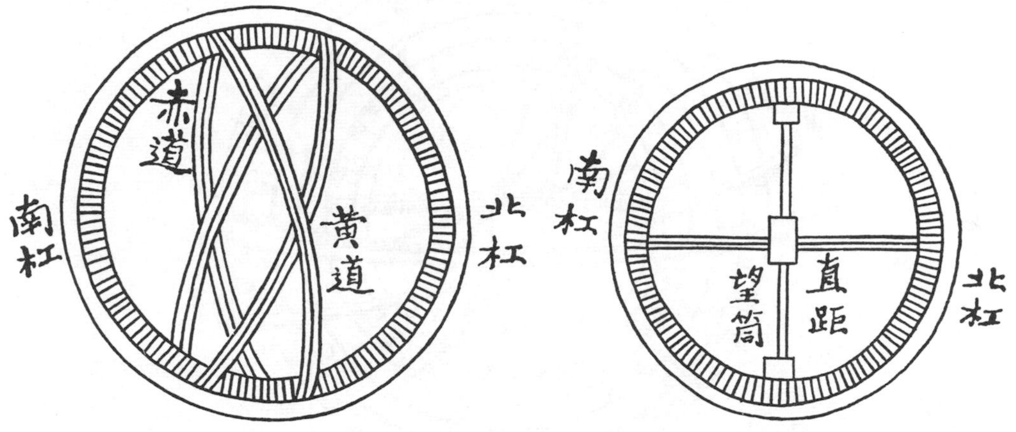

(*a*) The middle nest with its equator and ecliptic rings attached to the solstitial colure ring.

(*b*) The polar-mounted declination ring or hour-angle circle with its attached sighting-tube and diametral brace.

Fig. 160. Drawings of some of the component parts of the clock-driven equatorial armillary sphere of Su Sung, from the *Hsin I Hsiang Fa Yao* (+1090).

It is of interest to compare this instrument with that which was made for Alfonso X of Castile about two centuries later (Fig. 161).[b] Su Sung's armillary sphere was no less admirable, and in some ways superior to Alfonso's. The chief difference is that the Moorish one has special arrangements for making measurements of azimuth and altitude, according to the Arabic tradition. Though these were not important for Chinese methods Su Sung's horizon circle was duly graduated.[c] His treatise gives us

[a] Almost the whole of Su Sung's first chapter, which describes the armillary sphere, has been translated by Maspero (4), pp. 306ff. The whole of his memorial to the emperor, and of his third chapter, which describes his power-drive, has been translated by Needham, Wang & Price. His second chapter, which describes his celestial globe, has not yet been translated. The list of component parts of his sphere given above is not quite complete, but sufficiently so to elucidate the figure, in which we make use of the same numbering system as Maspero.

[b] See Nolte (1), p. 26, and of course the edition of Rico y Sinobas.

[c] The water-levels incorporated in the base of his apparatus will be noted. They were known to the Alexandrians, but how far they were used in the medieval Arabic and European culture-areas we are not sure. The bubble water-level was introduced by Thévenot in +1661 (Houzeau (1), p. 955).

an unrivalled picture of the achievements of Chinese mechanical engineering in the service of science about the time familiar to us as the period of William the Conqueror.

(ii) *Armillary instruments in and before the Han*

The greatest uncertainty of course concerns the earliest beginnings. The chief text is one to which reference has already been made (p. 216 above), namely, a few words in Yang Hsiung's *Fa Yen* (Admonitory Sayings)[a] of +5. To someone who asked him about the 'hun-thien'[1] he replied that Lohsia Hung had constructed it (*ying chih*[2]),

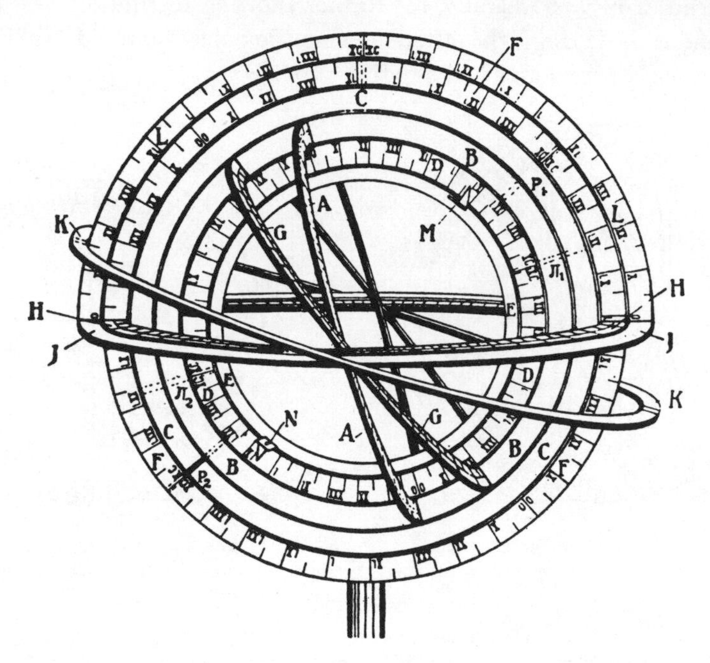

Fig. 161. Armillary sphere of Alfonso X, King of Castile (*c.* +1277), reconstructed by Nolte (1). *F*, outer fixed meridian (solstitial colure) ring, graduated; *B*, inner mobile meridian ring pivoted in the polar axis; *A*, graduated ecliptic ring attached to *B*; *C*, *D*, *E*, three mobile celestial latitude rings, the middle one graduated and the innermost one bearing *M*, *N* sights; P_1, P_2 polar pivots; π_1, π_2 ecliptic polar pivots; *G*, graduated equator ring attached to *B*; *H*, graduated horizon ring attached to *F*; *J*, horizon ring attached to *L*, half meridian graduated circle pivoted at the zenith, forming thus a pair of altazimuth quadrants, readings on which may be taken by *K*, a sighting ring for altitudes, pivoted on *J* and pierced by observation holes.

Hsienyü Wang-Jen had made calculations for it (*tho chih*[3]), and Kêng Shou-Chhang had checked it with actual observations (*hsiang chih*[4]). The commentators interpreted the answer as referring to the armillary sphere (*hun thien i*[5]), but Maspero (4), perhaps because the next question related to the Kai Thien[6] theory (cf. p. 210 above), believed

[a] Ch. 7, p. 2*b*.

[1] 渾天 [2] 營之 [3] 度之 [4] 象之 [5] 渾天儀 [6] 蓋天

that it referred to the celestial sphere theory (*hun thien shuo*[1]), and therefore that it was not valid as evidence for armillary rings as early as −104. But if Yang Hsiung had been speaking of the theory he would not have used the word *ying* which can only mean the construction of something material. Moreover, there is the further serious objection to the view of Maspero, which makes the first armillary rings start with Kêng Shou-Chhang in −52, that there are strong grounds[a] for believing that Shih Shen and Kan Tê in the −4th century did actually find star positions in degrees, and this would evidently have been quite impossible without armillary rings of some kind thus graduated. We seem therefore compelled to go back to the old opinion which Gaubil transmitted,[b] that Lohsia Hung did in fact set up armillary apparatus.

In the memorial of about +180 already mentioned[c] Tshai Yung spoke about armillary spheres as follows:

The bronze instruments used by the imperial astronomers are based upon (the Hun Thien theory).[d] A sphere eight feet (in circumference) represents the shape of the heavens and the earth. By means of it the ecliptic (graduations) are checked. The rising and setting of the heavenly bodies are observed, the movements of the sun and moon are followed, and the paths of the five planets traced. (Such instruments have been found to yield) wonderful and accurate results. This is a method which will remain unchanged for a hundred generations.[e] Although the official (astronomers) are in possession of the instruments, the original writings on the subject are no longer in existence. Moreover, no description (of the theory) is to be found in the previous official histories.[f]

The oldest text which we have on the armillary sphere is the remaining fragment[g] (about +125) of Chang Hêng's *Hun I*.[2] It opens as follows:

The equatorial ring goes round the belly of the armillary sphere $91\frac{5}{19}°$ away from the pole.[h] The circle of the ecliptic also goes round the belly of the instrument at an angle of 24° with the equator. Thus at the summer solstice the ecliptic is 67° and a fraction[i] away from the pole, while at the winter solstice it is 115° and a fraction[i] away.

Hence (the points) where the ecliptic and the equator intersect should give the north polar distances (*chhü chi*[3]) of the spring and autumn equinoxes.[j] But now (it has been recorded that) the spring equinox is $90\frac{1}{4}°$ away[k] from the pole, and the autumn equinox is $92\frac{1}{4}°$[k] away.[l] The former figure is adopted only because it agrees with the (results obtained

[a] P. 268 above.

[b] (2), p. 5. Forke (6), pp. 9ff., Wieger (3), p. 85, and Sarton (1), vol. 1, p. 195, all went far astray on these matters. [c] P. 210 above. [d] See p. 216 above.

[e] Since the equatorial mounting of telescopes and even radio telescopes is necessarily based on the same principles, he might comfortably have said a thousand generations.

[f] *Chin Shu*, ch. 11, p. 1*b*, tr. auct., adjuv. Ho Ping-Yü (1).

[g] In Ma Kuo-Han's collection of fragments (*YHSF*, ch. 76, pp. 66*a*ff.), which Maspero, strangely, never used. Also in *Hou Han Shu* (commentary), ch. 13, p. 20*b*. Tr. auct.

[h] I.e. 90° on our graduation, and theoretically $91\frac{5}{16}$ on theirs. Precisely this fraction (*shao chhiang*[4]) is given in Wang Fan's memorandum mentioned just below (*Chin Shu*, ch. 11, p. 5*b*).

[i] In both these cases, the word *chhiang* is used, which can mean (cf. p. 268 above) just $\frac{1}{8}°$, but the phrase runs *erh chhiang*[5] and *i chhiang*[6] respectively, and from the context $\frac{5}{19}°$ must be meant.

[j] I.e., they should both be the same as that of the equator itself.

[k] In both these cases, the word *shao*[7] is used (cf. p. 268 above).

[l] Both the texts read $91\frac{1}{4}°$ here, but the emendation is required for the passage to make sense, and the copyist's error would have originated very easily.

[1] 渾天說 [2] 渾儀 [3] 去極 [4] 少强 [5] 而强 [6] 亦强
[7] 少

by the) method of measuring solstitial sun shadows as embodied in the Hsia (dynasty) calendar.[a]

The upper one of the two circles is that which represents the ecliptic.[b]

This passage gives us a glimpse of a bronze armillary sphere of the Han period, but it is also interesting for another reason. Reference has already been made[c] to the discovery of the precession of the equinoxes by Yü Hsi[1] about +320, but here we see that Chinese astronomers were becoming uneasy about the relation of the sidereal and the tropic years at a much earlier date. An apparent inequality between the north polar distances of the equinoctial points could only have come about by the use of traditional values for the solstitial sun's shadow lengths,[d] and continued reliance upon a traditional pole-star,[e] at a time when new determinations ought to have been made of the former, and change ought to have been recognised in the latter. The sliding round of the equator on the ecliptic would naturally make one equinoctial point seem nearer to the pole than the other if no account were taken of the precessional movement of the pole itself. In Chang Hêng's time the old values seem to have been still orthodox, but clearly he doubted their validity. Half a century after his death, Liu Hung[2] knew that the 'year-differences' (*sui chha*[3])[f] were piling up, but it was Yü Hsi who first clearly distinguished[g] the 'celestial revolution' (*thien chou*[4]) from the 'year revolution' (*sui chou*[5]), obtaining for the precessional movement a value of about 1° in fifty tropic years.[h] Thus the saying was that he had 'dealt with the Heavens as the Heavens and with the Year as the Year' (*Shih thien wei thien, sui wei sui*[6]).[i]

But we must return to the text of Chang Hêng. It goes on to treat of the empirical graduation of the ecliptic by direct measurement from the equator of an armillary

[a] The reference is to the 'Ku Ssu Fên Li', cf. p. 293.

[b] Looking at the instrument from the south.

[c] Pp. 200, 220 and 270 above.

[d] See the discussion above (pp. 290ff.) on the 'Chou Kung' figures for the solstitial shadow lengths.

[e] See the discussion above (pp. 259ff.) on the pole and the pole-stars.

[f] The amount by which the sidereal year exceeds the tropic year is of the order of twenty minutes.

[g] This subject has been briefly discussed by Gaubil (2), p. 46; Chu Wên-Hsin (*1*), p. 105, and others, but the whole story has not so far received sufficient attention. Yü Hsi reached his conviction partly by a study of the apparent position of the solstices given in the Yao Tien chapter of the *Shu Ching*; cf. pp. 177 and 245 above.

[h] Later, Ho Chhêng-Thien[7] (+5th century) considered (as did Hipparchus long before) that 1° in 100 years was the best estimate (cf. *Sung Shu*, ch. 13, pp. 24*a*, 25*a*), while Liu Chhuo[8] (+6th century) gave 1° in 75 years. The true figure is about 71·6. It was not until the +10th century that the Hipparchan figure was definitively superseded in the west, by Ibn al A'lam (Destombes, 2) whose value was 70 years, and al-Battānī (Chatley (9); Mieli (1), p. 88; Hitti (1), p. 376), who gave one of just over 66. In the *Mêng Chhi Pi Than* (ch. 7, para. 11), Shen Kua recalls a legend that Lohsia Hung had foretold that his calendar would be inaccurate after 800 years, and that I-Hsing had duly arisen to correct it. On the contrary, says Shen Kua, it was already wrong when it was new. And besides, he adds, as Chang Tzu-Hsin showed (about +576), the *sui chha* was changing by 1° in 80 years, only a tenth of the time-span of the story. This was also the figure of Tung Ho in the Thang (*CJC*, 2nd. add., ch. 4).

[i] *Chhou Jen Chuan*, ch. 6, p. 76. The difference between the formulations of Hipparchus and Yü Hsi are interesting. The Greek noticed that as compared with earlier positional data, the longitudes of the stars had increased while their latitudes remained unchanged. They seemed therefore to be drifting eastwards, because the equator was moving westwards at constant obliquity with the ecliptic. The essential reference point was thus equinoctial. But the Chinese noticed that the winter solstice point seemed to be moving relatively to the stars westwards.

1 虞喜 2 劉洪 3 歲差 4 天周 5 歲周
6 使天爲天歲爲歲 7 何承天 8 劉焯

sphere. Being unacquainted with the work of his contemporary Menelaus of Alexandria[a] on spherical triangles, he had no resource but the following empirical graphic method.

> The difference in numbers, of degrees less or more, on the ecliptic should really be measured by observing the movements of the sun and moon with the bronze armillary throughout a year, but since there are often times of bad weather preventing observations, it is difficult to succeed completely in this. One therefore takes a small armillary sphere and graduates both equator and ecliptic in $365\frac{1}{4}°$ starting from the winter solstice point.... Next one fixes a thin ruler of flexible bamboo to each of the two poles, thus spanning exactly half the circumference of the sphere.... One then moves the ruler round, starting from the winter solstice point, and noting how many degrees less or more on the ecliptic correspond to those on the equator. These are the numbers of 'advance' (*chin*[1]) or 'retardation' (*thui*[2]). Moreover, by counting the number of degrees marked on the (semicircular) bamboo ruler, one obtains the north polar distance, at any point. The equator and the ecliptic are both divided into the twenty-four *chhi* (fortnightly periods), each being $15\frac{7}{16}°$ in length, and for each *chhi* there is an advance or a retardation of 1° on the ecliptic with regard to the equator. This is because the ecliptic, in accordance with the seasons, approaches at one time the northern, and at another the southern, pole, hence its position with respect to the equator is oblique (and its degrees differ).[b]

This procedure remained in use for one and a half millennia, until the coming of Arabic and Western influences in the Yuan and the Ming.[c] It was essentially a means of expressing movements along the ecliptic in equatorial coordinates. Traditional Chinese astronomy never extended the ecliptic degrees to form 'segments' of celestial longitude radiating from the pole of the ecliptic in the same way as they thought of the *hsiu* as segments of what we should call right ascension. The fragment of Chang Hêng concludes by comparing the degrees on the great circles with the days of the fortnightly periods, and by noting a saecular change in the positions of the solstices.

[a] D. E. Smith (1), vol. 2, p. 606.

[b] Tr. Maspero (4), p. 339, eng. auct. The use of a flexible ruler by Chang Hêng instead of a piece of string indicates that his instrument was indeed a computational armillary sphere and not a solid celestial globe.

[c] The earliest known list of *hsiu* extensions measured in degrees on the *ecliptic* is that of Tshai Yung[3] (+178) which was collected by Ssuma Piao[4] and came to be embodied in the *Hou Han Shu* (ch. 13, pp. 19*b*, 20*a*). The measurements had actually been made by Liu Hung.[5] What the Chinese wished to determine was not, of course, as Maspero well says, the celestial longitude of the observed star, but the longitude of the point of intersection of its hour-circle with the ecliptic (see also Eberhard & Müller, 2). Movements of sun and moon on the ecliptic were thus also reduced to terms of equatorial *hsiu*. Eventually, owing to the technical difficulties involved in accurate graduation and mensuration, the Chinese passed over to computation in the Sui calendar of +604 associated with the name of Liu Chhuo,[6] who used an arithmetical progression method. In +724, another method was employed by I-Hsing[7] for the same purpose. Finally Kuo Shou-Ching,[8] in the +13th century, gave a list of measurements to four places of decimals (*Yuan Shih*, ch. 54, pp. 9*b*, 14*b*; Gaubil (3), p. 145); he probably calculated them with the help of the trigonometry of Shen Kua and what the Arabs were then transmitting.

[1] 進 [2] 退 [3] 蔡邕 [4] 司馬彪 [5] 劉洪 [6] 劉焯
[7] 一行 [8] 郭守敬

A substantial account of the armillary sphere is contained in the *Hun Thien Hsiang Shuo* of Wang Fan[1] (*c.* +260).[a] This is particularly important in connection with the parallel development of the solid celestial globe (p. 386 below). Then comes (as already mentioned) that of Khung Thing (+323).[b] In the +6th century the mathematician Hsintu Fang[2] (cf. below, p. 632) wrote a book, the *Chhi Chun Thu*[3] (Specifications of Scientific Instruments), unfortunately long lost. Hsintu Fang was a man of the Northern Chhi dynasty, and had his tradition from the great Tsu family. We hear that when Tsu Kêng-Chih[4] (son of Tsu Chhung-Chih[5]) was captured by the Northern Chhi people, he was lodged in the palace of the Prince of An-Fêng, Kao Yen-Ming,[6] but not treated very well. Hsintu Fang, who also took great interest in mathematics and instrument-making, succeeded in obtaining better hospitality for him from the prince. Afterwards, when Tsu Kêng-Chih returned to Chhi, he taught Hsintu Fang everything he could of his father's knowledge.[c]

We get very few such glimpses of the human side of the story. A particularly interesting one is to be obtained from the fragments of the writings of Huan Than[7] (−40 to +30):

> Yang Hsiung[8] was devoted to astronomy and used to discuss it with the officials (*huang mên*[9]). He made an armillary sphere himself. An old artisan[d] once said to him: 'When I was young I was able to make such things following the method of divisions (graduations) to scale (lit. feet and inches) without really understanding their meaning. But afterwards I understood more and more. Now I am seventy years old, and feel that I am only just beginning to understand it all, and yet soon I must die. I have a son also, who likes to learn how to make (these instruments); he will repeat the years of my experience, and some day I suppose he in his turn will understand, but by that time he too will be ready to die.' How sad, and at the same time how comical, were his words![e]

This text, which escaped the eagle eye of Maspero, teaches us several things. Not only was Yang Hsiung himself (−53 to +18) concerned with the construction of an armillary sphere, but there was in his time a tradition of instrument-makers—this strengthens the case for the use of armillary rings by Lohsia Hung. Moreover, one can

[a] *Khai-Yuan Chan Ching*, ch. 1, pp. 10*a*ff.; *CSHK* (San Kuo sect.), ch. 72, pp. 1*a*ff., reconstructed from *Chin Shu*, ch. 11, pp. 5*b*ff. and similar sources. Tr. Maspero (4), p. 332 and Eberhard & Müller (2). They believed in a +4th-century remodelling; Maspero did not, and we agree with him.

[b] *Sui Shu*, ch. 19, p. 16*a*, *b*; tr. Maspero (4), p. 323.

[c] See *Pei Shih*, ch. 89, pp. 13*a*ff., *Pei Chhi Shu*, ch. 49, pp. 3*b*ff. This prince had in his household a great collection of scientific books and instruments, and it was these which Hsintu Fang described. By the +13th century his book was already lost, as noted by Chou Mi,[10] in his *Chhi Tung Yeh Yü*[11] (Rustic Talks in Eastern Chhi), ch. 15, p. 6*b*, who cites the *Pei Shih*.

[d] In the *Pei Thang Shu Chhao* and other texts quoted by Juan Yuan (*Chhou Jen Chuan*, ch. 2, p. 14) the name of Lohsia Hung or Lohsia Huang-Hung[12] appears instead of 'the officials' and the 'old artisan'. If this reading is correct it supports the view that Lohsia Hung made armillary apparatus, but from the dates of their lives as approximately known it is hard to believe that Yang Hsiung could have known him personally.

[e] *Yü Hai*, ch. 4, p. 29*a*; *CSHK* (Hou Han sect.), ch. 15, p. 2*a*, tr. auct.

1 王蕃 2 信都芳 3 器準圖 4 祖暅之 5 祖沖之
6 高延明 7 桓譚 8 揚雄 9 黃門 10 周密
11 齊東野語 12 落下黃閎

sense the lack of technical education, and the fewness of those who could explain the celestial sphere; besides it was regarded as a State secret, or at least 'restricted information'.

(iii) *The invention of the clock-drive*

The first sphere to be slowly rotated by means of a water-wheel powered by a constant pressure-head of water in a clepsydra was one of those built by Chang Hêng about +132. The *Chin Shu*, quoting from some writing of Ko Hung's, now lost, says:[a]

Though many have discoursed upon the theory of the heavens, few have been as well acquainted with the principles of the Yin and Yang as Chang Hêng[1] and Lu Chi.[2] They considered that in order to trace the paths and degrees of motion of the Seven Luminaries, to observe the calendrical phenomena and the times of dawn and dusk, and to collate these with the forty-eight *chhi*;[b] to investigate the divisions of the clepsydra and to predict the lengthening and shortening of the shadow of the gnomon, (finally) verifying all these changes by phenological observations[c]—there was no instrument more precise than the (computational) armillary (*hun hsiang*[3]).[d] (Thus) Chang Hêng made his bronze armillary sphere (*hun thien i*[4]) and set it up in a close chamber, where it rotated by the (force of) flowing water. Then, the order having been given for the doors to be shut, the observer in charge of it would call out to the watcher on the observatory platform, saying the sphere showed that such and such a star was just rising, or another star just culminating, or yet another star just setting. Everything was found to correspond (with the phenomena) like (the two halves of) a tally. (No wonder that) Tshui Tzu-Yü[5] wrote the following inscription on the (burial) stele of Chang Hêng: 'His mathematical computations exhausted (the riddles of) the heavens and the earth. His inventions were comparable even to those of the Author of Change. The excellence of his talent and the splendour of his art were one with those of the gods.' And indeed this was demonstrated by the armillary sphere and the seismographic apparatus[e] which he constructed.

Elsewhere in the same chapter we have another account in the words of the +7th-century author of the *Chin Shu*, Fang Hsüan-Ling, or one of his collaborators, possibly Li Shun-Fêng.

In the time of the emperor Shun Ti (+126 to +144) Chang Hêng constructed a (computational) armillary (*hun hsiang*[3]), which included the inner and outer circles,[f] the south and north celestial poles, the ecliptic and the equator, the twenty-four fortnightly periods, the stars within (i.e. north of) and beyond (i.e. south of) the twenty-eight *hsiu*, and the paths of the sun, moon and five planets. The instrument was rotated by the water of a clepsydra (lit. dripping water) and was placed inside a (closed) chamber above a hall. The transits, risings and settings of the heavenly bodies (shown on the instrument in the chamber) corresponded with (lit. resonated with)[g] those in the (actual) heavens, following the (motion of the) trip-lug, and the turning of the auspicious wheel.[h]

[a] Ch. 11, p. 3*b*, tr. auct., adjuv. Ho Ping-Yü (1). The passage may be dated about +330.
[b] I.e. the halves of the twenty-four fortnightly periods. Cf. p. 327 above. [c] See p. 463 below.
[d] See p. 383 below. [e] See below, p. 626 in Sect. 24.
[f] Probably the declination-circles of perpetual apparition and invisibility; cf. Table 31.
[g] Note the appearance of this philosophically significant technical term (cf. Vol. 2, p. 304).
[h] Ch. 11, p. 5*a*, tr. auct., adjuv. Ho Ping-Yü (1).

1 張衡 2 陸績 3 渾象 4 渾天儀 5 崔子玉

The general picture is thus quite clear. The procedure was for two observers to compare the indications of the mechanised sphere in a closed laboratory with the celestial phenomena actually occurring overhead. The concluding phrases give us certain clues about the nature of the power drive which we shall follow up in Section 27*h* on horological engineering; here we need only notice the existence of several pieces of evidence about Chang Hêng's arrangements contemporary with him. One of the most striking of these is a title or sub-title which Thang scholars carefully preserved when they quoted the only fragments of Chang Hêng's own writing on the clepsydra which have come down to us.[a] This title, which reads *Lou Shui Chuan Hun Thien I Chih*[1] (Apparatus for Rotating an Armillary Sphere by Clepsydra Water), was most probably that of a chapter in his *Hun I*[2] or *Hun I Thu Chu*[3] (Illustrated Commentary on the Armillary Sphere) from which we have already quoted.[b] The mechanisation technique was thus not something merely attributed to Chang Hêng by later generations. One of the apocryphal treatises of his time adds further confirmation, as we shall shortly see.

Indeed, the later tradition about Chang Hêng's achievement was so strong and widespread that there is no adequate reason for doubting it. Moreover, as we shall find in Section 27*h* on clockwork, nearly every succeeding century produced some astronomer or technician who accomplished the same thing. Table 31 already shows this.[c] The implication usually is that these spheres also were set up indoors and compared with the data of observational instruments. All this was before the invention of the first escapement by I-Hsing and Liang Ling-Tsan about +723, and after that time demonstrational armillary spheres and celestial globes continued to revolve by means of a power-drive down to the coming of the Jesuits.[d] From the beginning of the +8th century these instruments were thus nothing more nor less than great astronomical clocks, antedating the first mechanical clocks of the European +14th century.[e] Before the +8th century one can only call them mechanised orreries, or power-driven spheres for demonstration and computation, designed to show a rough approximation to time-keeping. In fact, throughout this long period of Chinese developments the form was often that of an observational astronomical instrument, but the essence was that of the clock. For the aim of the astronomer-engineers, from the time of Chang Hêng onwards, was to persuade a driving wheel to rotate sufficiently slowly to keep pace with the

[a] Hsü Chien in *Chhu Hsüeh Chi*, ch. 25, pp. 2*a*, 3*a*, in +700; Li Shan in *Wên Hsüan*, ch. 56, p. 13*b*, about +660. See *YHSF*, ch. 76, pp. 68*a*, *b*; *CSHK* (Hou Han sect.), ch. 55, p. 9*a*.

[b] P. 355 above.

[c] Note the work of Ko Hêng about +260 in the State of Wu, Chhien Lo-Chih in +436 in the Liu Sung dynasty, Thao Hung-Ching, *c.* +520, in the Liang, and Kêng Hsün, *c.* +590, in the Sui. Sometimes, as in this last case, the closed room system is extremely explicit (cf. Sect. 27*h* in Vol. 4).

[d] The Sung and Yuan produced many such devices from the mercury-driven machine of Chang Ssu-Hsün[4] in +979 to Su Sung's great apparatus of +1090, and the globe of Kuo Shou-Ching about +1276.

[e] We take this opportunity of withdrawing our statement (Vol. 1, p. 243) that clockwork was a distinctively European invention of the early +14th century, and referring to the monograph of Needham, Wang & Price for further details. The weight-driven mechanical clock with the verge-and-foliot escapement originated in Europe at that time, but for many centuries previously China had had water-driven mechanical clocks with another kind of escapement.

[1] 漏水轉渾天儀制 [2] 渾儀 [3] 渾儀圖注 [4] 張思訓

apparent diurnal revolution of the heavens, and after the beginning of the +8th century their problem was substantially solved. The question naturally presents itself—why did Chang Hêng and his successors want to do this?

The *Shang Shu Wei Khao Ling Yao*[1] (Apocryphal Treatise on the *Historical Classic*; the Investigation of the Mysterious Brightnesses), which dates from about the time of Chang Hêng, has a relevant passage of great interest. It runs as follows:

The motions of the 'jade instrument' are observed; dusks and dawns determine times and seasons—this means checking the culminations of stars. If the (computational armillary) sphere (*hsüan chi*[2]) indicates a meridian transit when the star (in question) has not yet made it, (the sun's apparent position being correctly indicated), this is called 'hurrying' (*chi*[3]). When 'hurrying' occurs, the sun oversteps his degrees, and the moon does not attain the *hsiu* in which it should be. If a star makes its meridian transit when the (computational armillary) sphere has not yet reached that point (the sun's apparent position being correctly indicated), this is called 'dawdling'[a] (*shu*[4]). When 'dawdling' occurs, the sun does not reach the degree which it ought to have reached, and the moon goes beyond its proper place into the next *hsiu*. But if the stars make their meridian transits at the same moment as the sphere, this is called 'harmony' (*thiao*[5]). Then the wind and rain will come at their proper time, plants and herbs luxuriate, the five cereals give good harvest and all things flourish.[b]

This gives us the explanation of the two spheres, the computational one in a closed room, the observational one on the terrace outside. Han astronomers, like their predecessors and their successors, were deeply concerned with all divergences or discrepancies between the indicated positions of the stars on the one hand and those of the sun and moon on the other. If, as is rendered highly probable by many texts,[c] the system was to have small objects representing sun, moon and planets attached somehow to the sphere or globe, yet freely movable thereon,[d] then the computer within would adjust their positions in accordance with the predictions of the calendar. If the star transits and other movements which he then announced[e] were not in accord with what the observer outside was seeing, then due corrections could be made in the calendrical computations.

In his memorial to the emperor in +1092, Su Sung quoted the above passage from the Han apocrypha, adding:[f]

From this we may conclude that those who make astronomical observations with instruments are not only organising a correct calendar so that good government can be carried

[a] Terms like 'hurrying' and 'dawdling' were, of course, used for the acceleratory and retardatory phases of planetary motions, but as can be seen from p. 399 below, they were quite different from these.

[b] *Sui Shu*, ch. 19, p. 13*b*, tr. auct. One late version of the text (*Ku Wei Shu*, ch. 2, p. 2*b*) made it read in the opposite sense so that the 'hurrying' and 'dawdling' referred to the stars and not to the instrument. This misled Maspero (4), p. 338, into thinking that the whole arrangement was astrological. On the contrary, its calendrical purpose is clear. For the elucidation of this we owe much thanks to Dr Derek Price.

[c] E.g. *Sung Shih*, ch. 48, pp. 3*b*ff., concerning Chang Ssu-Hsün; *Chiu Thang Shu*, ch. 35, pp. 1*a*ff.; *Sung Shu*, ch. 23, pp. 8*b*ff., and *Sui Shu*, ch. 19, pp. 17*b*ff., concerning Chhien Lo-Chih.

[d] For example, beads strung on threads. We know that this was the system used by Su Sung with his celestial globe of +1090 (*Hsin I Hsiang Fa Yao*, ch. 1, p. 4*b*).

[e] The *Chin Shu* says in two places that it was the computer within who was the first to speak.

[f] *Hsin I Hsiang Fa Yao*, ch. 1, p. 5*a*, tr. auct.

[1] 尙書緯考靈耀 [2] 璇璣 [3] 急 [4] 舒 [5] 調

on,[a] but also (in a sense) predicting the good and bad fortune (of the country) and studying the (reasons for) the resulting gains and losses.

This was an interesting rationalisation of the prognosticatory significance naturally attributed by the common people to the work of all medieval astronomers. We do in a way foretell the future, says Su Sung, because we know that if the calendar is always well adjusted the work of the farmers will keep perfect time with the seasons, and so (apart from special calamities) bring the best harvests. Nevertheless, it would probably be unwise to deny all astrological significance to the spheres and globes of medieval China, rotating more or less jerkily, with or without escapements, through nights of storm and days of rain and cloud. In our discussion of the history of clockwork[b] we shall bring forward evidence that some importance was attached to the positions of stars at conception. If the child conceived was of a rank which might qualify him later on for nomination as crown prince, such knowledge might take its place among affairs of State. There is therefore nothing surprising in the interest which the imperial family took, especially after the early +8th-century invention of the escapement, in the erection within the palace of astronomical clocks embodying steadily turning celestial globes. And here one cannot fail to recall that it was in China also (as we shall tell in its place)[c] that there occurred the invention of the magnetic compass—another device for orienting human beings in a world where the essential mark-points normally present had temporarily become obscured. I should like to believe that these discoveries were all manifestations of the uneasiness of the Chinese mind when separated from cosmic organic pattern. 'Wherever there is *Li*', we have already heard, 'east is east, and west is west.'[d] If the visible structure of the universe in space and time were hidden from sight, were there not perhaps invisible fields of force persisting which human ingenuity might reveal and use?

So much for clocks and pre-clocks. But there is another significance in these texts, which all the sinologists have missed. An armillary sphere used as a computer and kept in a closed room would need no sighting-tube, yet we find one description of an armillary so furnished which was made to rotate with due deliberation by the use of water-power. A practical astronomer sees at once the only obvious meaning of this.[e] Such a power-drive would be equivalent to the clockwork attached to all modern telescopes, which enables stars to be kept in view as they move along their circles (parallels) of declination. The movement ought to be a slow and steady one of 15° an hour. But if this was the object of the power-driven observational armillary sphere set up at Khaifêng in +1090, it was an invention which long antedated the similar arrangements of Europe. There the first proposal for an automatically adjusted clockwork-driven telescope was that of Robert Hooke in +1670,[f] and the system did not come into use until the decreasing focal length of telescopes permitted general adoption of strictly equatorial mountings. The name of Joseph Fraunhofer therefore (1824) is associated with the first effective clockwork drives.[g]

[a] That is to say, the administration of an agricultural society.
[b] Sect. 27*h* in Vol. 4.
[c] Sect. 26*i* in Vol. 4.
[d] P. 163 above.
[e] For insight into this question I am indebted to my friend Dr W. H. Steavenson.
[f] See Andrade (1).
[g] Cf. Houzeau (1), p. 949; von Rohr (1), p. 124; F. Meyer (1).

The astronomical clock-tower at Khaifêng was built by Su Sung, with a number of collaborators, between +1088 and +1090, and his description of it, the *Hsin I Hsiang Fa Yao*[1] (New Design for a (Mechanised) Armillary (Sphere) and (Celestial) Globe),[a] was presented to the throne in +1092. It was a building of two storeys surmounted by a platform, the whole about 35 ft. in height. The ground floor housed the power-drive, a water-wheel receiving the outflow of a constant-level tank and checked by a

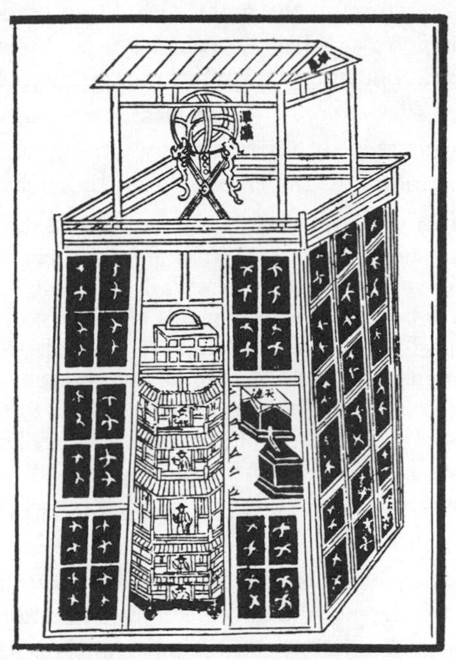

Fig. 162. General view of the astronomical clock-tower built at Khaifêng by Su Sung and his collaborators in +1090. On the top platform, some 35 ft. above the ground, there was a mechanised armillary sphere, in a chamber on the first floor there was a mechanised celestial globe, and below, in front of the machinery of a water-driven clock, numerous jacks appeared at the openings of a pagoda façade to announce the time. From the *Hsin I Hsiang Fa Yao*.

link-work escapement; it also exhibited to onlookers a façade in the form of a five-tiered pagoda at the doors of which an elaborate company of puppets gave visual and audible notice of the passage of time. Above this jackwork, on the first floor, a celestial globe, with its attached adjustable planetary models, automatically rotated. Finally, a suitable transmission shaft rose through the chamber and was carried up the central support of the armillary sphere on the open observation platform so as to rotate at will its inner nests of rings. As the sphere was fitted with a sighting-tube it was certainly intended for taking star positions. Now the system of placing the globe in the closed chamber below, and the armillary sphere on the open platform above, was a perfectly conscious allusion to the precedent established by Chang Hêng so long before. There is no cause for surprise in the appearance of the purely time-keeping function in the engine-room, but Su Sung's object in mechanising the armillary sphere seems at first

[a] See Fig. 162.

[1] 新儀象法要

sight a little puzzling. For this was some eight hundred years before the invention of the photographic plate, which must necessarily be carried round in accurate compensation for the movement of the earth in order to record the meagre lights of outer space.

The accuracy of Su Sung's clock drive was certainly not great. It must have moved forward jerkily at intervals of not less than five minutes, and one can hardly believe that it could have worked through the night without an error of at least 2°, yet this would not have been as good as what the naked eye could estimate. Even when the Chinese astronomers were following stars of low magnitude,[a] they were not at all likely to get lost among the constellations. Nevertheless, one must not underestimate the difficulties of precise adjustment of instruments weighing 20 tons or more. Su Sung gives us a hint in his memorial to the emperor that these caused much trouble:

> Your servant's opinion is that the system and principles of the armillary sphere and the celestial globe have been handed down to us (from former times) with detailed degrees and measurements. Yet the astronomical and calendrical officials continue to differ and argue among themselves. This is because the instruments of today no longer agree with those of old, and also the technical terms are not properly defined. Furthermore (the present instruments) for observations require the human hand to move them, and (as the motion of the) hand is sometimes too much and sometimes too little, so the degrees of motion of the celestial bodies (seem to) vary. And each (official) gets readings larger or smaller than those of his colleagues, and there is no definite conclusion.[b]

This suggests that the mechanisation was intended, at least in part, as a kind of coarse adjustment for the sphere during night observations. The fine adjustments could then be more accurately made by levering with wooden bars. Of course, if the motion had been more continuous, the drive could have been used for short periods only to permit of great accuracy in the measurement of interstellar distances by one observer. Since rotation would have continued during the time required to find and observe the second star, this allowance for the earth's motion would have greatly assisted star mapping. But such a use seems precluded by the evident discontinuity of the drive.

There is, however, reason for envisaging a further function of the apparatus. Su Sung himself has a statement that a celestial body could be held automatically in the field of the sighting-tube as it rotated. He speaks of maintaining the sun in this position by day.[c]

> (Our) armillary sphere can be used to observe the degrees of motion of the three luminaries (sun, moon, and stars). The ecliptic is added in the form of a single ring.[d] The sun is seen half on each side of this. The sighting-tube is made to point constantly at it so that its rays are always shining down the tube.[e] During one complete westward revolution of the heavens the sun moves eastward one degree. This device is a new invention.[f]

[a] We know that they were very interested in some of these; cf. p. 239 above.

[b] *Hsin I Hsiang Fa Yao*, ch. 1, p. 1*b*; tr. Needham, Wang & Price.

[c] *Hsin I Hsiang Fa Yao*, ch. 1, p. 4*b*; tr. Needham, Wang & Price.

[d] This contradicts Su Sung's description of an improved double, or split, ecliptic ring (ch. 2, pp. 15*a*, *b*; cf. Maspero (4), p. 314), but we have no explanation of the discrepancy. Perhaps these rings were dismountable and interchangeable.

[e] Lit. 'so that the body of the sun is always (sighted through) the lumen of the tube'.

[f] He means, the mechanising of an observational sphere. But this use of it is remarkably reminiscent

Presumably this procedure was intended as a check upon the time-keeping of the whole clock, though this is not expressly said, either in the memorial or in the body of the book. The jerky movement would simply have brought the sun's rays again directly down the sighting-tube each time the machinery advanced by a distance corresponding to a single scoop on the water-wheel.

Thus Su Sung, retaining the ancient mechanisation of the globe, extended it to the observational sphere above, and added machinery for giving time-signals below. But not all his contemporaries were convinced that the clock drive of the sphere was desirable, and even before it was finished one of the Han-Lin Academicians, Hsü Chiang,[1] was petitioning that another, non-mechanised, armillary purely for observations should be constructed.[a] Apparently the decision was taken to do this, but we hear no more of it. In any case Hsü's conservative objections seem to have prevented Su Sung's plan from ever being carried out again. For after ticking over for 36 years, his great astronomical clock was captured by the Chin Tartars when they stormed the capital and extinguished the Northern Sung. Though rebuilt in Peking, its parts gradually wore out and could not be replaced.[b] Meanwhile in the south the technicians were scattered and disorganised, and engineering did not revive for nearly a century, by which time attention shifted to other devices. Perhaps it was the growing skill of the bronze-founders which rendered the heavy construction of the Sung spheres unnecessary, and so lessened the need for the power-driven coarse adjustment.

We may conclude, therefore, that the mechanisation of the armillary spheres of ancient and medieval China, though generally applied to demonstrational or computational instruments, came near to the invention of the clock drive of the modern telescope. Su Sung, in coupling his clockwork to an observational sphere, fully accomplished this in principle, thereby anticipating Robert Hooke by six centuries and Fraunhofer by seven and a half. The question then arises, why do we hear of no such mechanised armillary spheres and celestial globes in ancient and medieval Europe?

of the Bumstead sun-compass, devised for aircraft in conditions where use of the magnetic compass is difficult. The Bumstead sun-compass was used around 1927 by Admiral Byrd in his flights from Spitzbergen to the North Pole and back, and also to and from the South Pole. It consists of a mechanical clock mounted in the equatorial plane, over the dial of which a perpendicular pin or gnomon, peripherally placed, is carried round on a pointer so that its shadow is held continually on a translucent plate moving round diametrically opposite. Since the timepiece, with its single pointer revolving once every 24 hours, is mounted upon a rotatable azimuth plate with graduations, the correct points of the compass can easily be ascertained—see the brief description in A. J. Hughes (1), pp. 117, 118. Of course, Bumstead knew no more of Su Sung than Su Sung could have known about his long-distant successor—and of course their objectives were different. A similar system is used in the McManigal sundial; the clock is mounted in the equatorial plane and the gnomon's shadow is made to fall upon a plate marked with the well-known figure-of-eight analemma correcting for the equation of time (Barton (1); Danjon (1), p. 76). This goes back to the beginning of the 18th century with Bedos de Celles and others. Instruments similar to that of McManigal, but without motive power, were made later in the century by P. M. Hahn, U. Adam and J. Engelbrecht (see Zinner, 8). The close relation between these modern devices and the equatorial sundials of the Han described above (p. 307) is evident.

[a] This was in +1089. See *Sung Shih*, ch. 80, pp. 25*a*ff.; the passage is fully translated in Needham, Wang & Price.

[b] See *Chin Shih*, ch. 22, pp. 32*b*ff., also fully englished in the same monograph.

[1] 許將

Surely the answer has already been given in the faithfulness of the Chinese to equatorial coordinates, in contrast to the ecliptic coordinates of the West. Was it not much more natural and easy to make an equatorial instrument demonstrate automatically the apparently uniform motion of the many stars on their declination parallels, than to make an ecliptic instrument demonstrate the obviously very complex motions of the few great luminaries on and near the ecliptic? A declination parallel (or circle) is a real line of daily apparent celestial motion, but on those circles of celestial latitude among which the Greeks constrained the stars no body ever moves.

The apparent diurnal revolution of the stars was always man's most fundamental natural clock, and, as we have seen, Chang Hêng in the +2nd century entitled his description 'Apparatus for Rotating an Armillary Sphere by Water from a Water-Clock' (*Lou Shui Chuan Hun Thien I Chih*). Now the most impressive part of Western mathematical astronomy was its theory of planetary motion in the ecliptic, so that similar technical developments were distinctly inhibited. There were of course the Anti-Kythera machine[a] and the anaphoric clocks,[b] but the features of ecliptic motion presented engineering problems so difficult that no real advances towards clockwork were made.[c] On the other hand, it is of particular interest to note how in China the mechanical clock and the mechanised astronomical instrument grew up together, combined in fact in the same machines. Europe received the results of this development in two stages. First came westwards, in all probability at the time of the Crusades, the water-driven mechanical clock, speedily to be transformed at the beginning of the +14th century into the more compact and accurate metal weight-driven clock with verge-and-foliot escapement. But this was a pure time-keeper and, in Price's celebrated phrase, 'a fallen angel from the world of astronomy'. Then at the end of the following two centuries (+1595) came the abandonment of the classical ecliptic coordinates by Tycho Brahe (perhaps as the result of some oriental stimulus), the proposal of Robert Hooke (+1670) implemented by J. D. Cassini (+1678),[d] and the beginning of equatorial telescope mountings in the work of Christopher Scheiner (+1625)[e] brought to perfection by James Short (+1732 to +1768).[f] Fraunhofer was thus able to follow unknowingly the precedent of Su Sung and thereby to restore the fallen angel of Western clockwork to that astronomical world in which it had had its Chinese origins, and which in China it had never altogether left.

[a] The remains of a planetarium of the +2nd century; cf. p. 339. See Rediadis (1); with other references and description in Price (1).

[b] Float-driven rotating dials typical of the Hellenistic world. See Diels (1); Drachmann (6); Price (4, 5).

[c] This does not exclude technical advances towards the component parts of clocks, e.g. the delicate gearing on some +11th-century astrolabes. See Wiedemann (13); Price (1).

[d] Cf. Daumas (1), p. 20.

[e] Scheiner (+1575 to +1650) was the Jesuit who quarrelled with Galileo about the priority of the discovery of sun-spots. In his *Rosa Ursina* (1630; Bk. III, pp. 347ff.), he attributes the construction to one of his brethren, Christopher Grünberger. It is interesting that he used it as a heliostat, just as Su Sung had done before and Fraunhofer did later. Cf. Bigourdan (1), p. 139; Danjon & Couder (1), p. 662; Daumas (1), p. 74; A. Wolf (2), pp. 136ff.; H. C. King (1), p. 41. Uccelli (3), p. 179, reproduces Scheiner's illustration of the apparatus.

[f] Short (+1710 to +1768) made a splendid series of telescopes between the dates mentioned; cf. Daumas (1), pp. 226, 232; H. C. King (1), p. 87. We are grateful to Mr P. A. Jehl for reminding us of the preceding work of Scheiner and Cassini. Cf. Chauvenet (1), p. 367.

(iv) *The invention of the equatorial mounting*

Of the armillary spheres of the Southern Sung little is known, but some of the +13th-century instruments of Kuo Shou-Ching[1] are still extant,[a] and have for long been kept at the Purple Mountain Observatory of Academia Sinica north-east of Nanking.[b] Made under the Yuan dynasty, they were still in use at the time of the arrival of the Jesuits in +1600. Here is what Matteo Ricci wrote about them:[c]

Not only in Peking, but in this capital also (Nanking) there is a College of Chinese Mathematicians, and this one certainly is more distinguished by the vastness of its buildings than by the skill of its astronomers. They have little talent and less learning, and do nothing beyond the preparation of almanacs on the rules of calculation made by the ancients—and when it chances that events do not agree with their computations they assert that what they have computed is the regular course of things, and that the aberrant conduct of the stars is a prognostic from heaven about something which is going to happen on earth. This something they make out according to their fancy, and so spread a veil over their blunders. These gentlemen did not much trust Fr. Matteo, fearing no doubt lest he should put them to shame; but when at last they were freed from this apprehension they came and amicably visited him in the hope of learning something from him. And when he went to return their visit he saw something that really was new and beyond his expectation.

There is a high hill at one side of the city, but still within the walls.[d] On the top there is an ample terrace, capitally adapted for astronomical observation, and surrounded by magnificent buildings erected of old. Here some of the astronomers take their stand every night to observe whatever may appear in the heavens, whether meteoric fires or comets, and to report them in detail to the emperor. The instruments proved to be all cast of bronze, very carefully worked and gallantly ornamented, so large and elegant that the Father had seen none better in Europe. There they had stood firm against the weather for nearly two hundred and fifty years,[e] and neither rain nor snow had spoiled them.

There were four chief instruments. The first was a globe, having all the parallels and meridians marked out degree by degree, and rather large in size, for three men with outstretched arms could hardly have encircled it. It was set into a great cube of bronze which

[a] The present writer had the great pleasure of examining two of them in 1946. Several were at Peking when the Jesuits first went there, but one was transported to Germany as loot after the Boxer Rebellion, whence in due course the German Government returned it to China from Potsdam. Good photographs of the Yuan armillary sphere and some of the Jesuit instruments which accompanied it, taken during their period at Potsdam, have been published by R. Müller (1).

[b] It is not certain that the instruments we have today were actually cast under the supervision of Kuo Shou-Ching, for it is known that exact replicas were made in +1437 by Huangfu Chung Ho.[2] See *Ming Shih*, ch. 25, pp. 15*a* ff. They were still on the Purple Mountain in April 1956 but it is expected that they will be moved to the National Planetarium and Museum of Astronomy in Peking.

[c] *Opere*, ed. Venturi, vol. 1, pp. 24, 315; *Fonti*, ed. d'Elia (2), vol. 2, pp. 56ff. Enlarged by Trigault (+1615), tr. Yule (1), vol. 1, p. 451; Gallagher (1), pp. 329ff., cit. Bernard-Maître (1), p. 59; Wylie (7), p. 14. These texts are here conflated and the translations modified where necessary.

[d] This was the hill called Pei Chi Ko[3] (Pole Star Pavilion), on the top of which now stands the Meteorological Research Institute of Academia Sinica. On the same spur of the Purple Mountain, also overlooking the Hou Hu lake, is the Cockcrow Temple, where the Liang emperor Wu Ti, a Buddhist, piously starved himself to death.

[e] Three hundred and fifty years would have been a nearer estimate.

[1] 郭守敬 [2] 皇甫仲和 [3] 北極閣

served as its pedestal, and in this box there was a little door through which one could enter to manipulate the works. There was, however, nothing engraved on the globe, neither stars nor terrestrial features. It therefore seemed to be an unfinished work, unless perhaps it had been left that way so that it might serve either as a celestial or a terrestrial globe.[a]

The second instrument was a great (armillary) sphere, not less in diameter than that measure of the outstretched arms which is commonly called a geometric pace. It had a horizon (-circle) and poles; instead of (solid) circles it was provided with certain double hoops (armillae), the void space between the pair serving the same purpose as the circles of our spheres.[b] All these were divided into 365 degrees and some odd minutes. There was no globe representing the earth in the centre, but there was a certain tube, bored like a gun-barrel, which could readily be turned about and placed at any degree or altitude so as to observe any particular star, just as we do with our vane sights[c]—not at all a despicable device.

The third piece of apparatus was a gnomon, the height of which was twice the diameter of the former instrument, erected on a rather long slab of marble which pointed to the north. This had a channel cut round the margin, to be filled with water in order to determine whether the slab was level or not, and the style was set vertical as in hour-dials.[d] We may suppose this gnomon to have been erected that by its aid the shadow at the solstices and equinoxes might be precisely noted, for both the slab and the style were graduated.

The fourth and last instrument, and the largest of all, was one consisting of, as it were, 3 or 4 huge astrolabes in juxtaposition; each of them having a diameter of such a geometrical pace as I have specified. The fiducial line, or Alhidada, as it is called, was not lacking, nor yet the Dioptra. Of these astrolabes, one having a tilted position in the direction of the south represented the equator; a second, which stood crosswise on the first, in a north and south plane, the Father took for a meridian, but it could be turned round on its axis; a third stood in the meridian plane with its axis perpendicular, and seemed to represent a vertical circle, but this also could be turned round to show any vertical whatever. Moreover, all these were graduated and the degrees marked by prominent (metal) studs, so that in the night the graduation could be read by touch without any light. This whole compound astrolabe[e] instrument was erected on a marble platform with channels round it for levelling.

On each of the instruments explanations of everything were given in Chinese characters, and there were also engraved the names of 28 zodiacal constellations which answer to our 12 signs. There was, however, one error common to all the instruments, namely that the elevation of the pole was taken to be 36°. Now there can be no question about the fact that the city of Nanking lies in lat. $32\frac{1}{4}°$, whence it would seem probable that these instruments were made for another locality and had been erected at Nanking, without reference to its position, by someone ill-versed in mathematical science.[f]

[a] The absence or faintness of the markings probably means that the instrument had suffered more from the weather than Ricci thought. For reasons which will shortly appear there can be no doubt that Kuo Shou-Ching's globe was a celestial one. Its 'parallels' were certainly not parallels in the Western sense of celestial latitude, but declination circles parallel with the equator. In Trigault's rewriting the phraseology gives the impression that the globe was rotated by hand, but Ricci's Italian clearly indicates that the entry was for the adjustment of machinery—presumably water-powered.

[b] The sighting-tube pointed between these split rings. It is notable that such a device was strange to Ricci.

[c] 'pinnulis' (Trigault).

[d] 'et stilus eo modo quo in horologiis ad perpendiculum collocatus' (Trigault).

[e] Ricci's use of the term 'astrolabii' in this paragraph was, of course, quite wrong and misleading; he was just writing a rough description. The instrument he saw had nothing to do with astrolabes.

[f] This point can be cleared up at once. The instruments which Ricci saw belonged to a set which had originally been made for the astronomical college founded by Yehlü Chhu-Tshai about +1200 at Phing-yang (modern Linfên) in Shansi (lat. just over 36°). In the scientific decadence of the Ming they

Some years afterwards Father Matteo saw similar instruments at Peking, or rather the same instruments, so exactly alike were they, insomuch that they had unquestionably been made by the same artists. And indeed it is known that they were cast at the period when the Tartars were dominant in China; so that we may without rashness conjecture that they were the work of some foreigner acquainted with our sciences.[a]

Thus Ricci was greatly impressed by the astronomical instruments of the Yuan dynasty, though holding a poor opinion of those Chinese contemporaries whom it was his strategy to supplant, and venturing a particularly erroneous guess about the origin of the instruments.

The authentic Chinese texts from which we gain information about the re-equipment of the imperial observatory in +1276 to +1279 include of course, the *Yuan Shih* (History of the Yuan Dynasty).[b] The instruments are there listed as follows:[c]

(1) *Ling lung i*[1] — Ingenious Armillary Sphere (Fig. 163). Ricci's 'second instrument'.

(2) *Chien i*[2] — Simplified Instrument (Figs. 164 and 165). Ricci's 'fourth instrument'.[d]

(3) *Hun thien hsiang*[3] — Celestial Globe.[e] Ricci's 'first instrument'.

(4) *Yang i*[4] — Upward-Looking Instrument.[f] A hemispherical sundial intermediate in size between that already described (p. 301 above) and the much larger *jai prakāś* instruments of the Indians (p. 301 above).

(5) *Kao piao*[5] — Lofty Gnomon.[g] Undoubtedly the 40 ft. gnomons, especially that at Yang-chhêng, already described (p. 297 above).[h]

(6) *Li yün i*[6] — Vertical Revolving Circle. The vertical circle in (2) described by Ricci, and seen in Figs. 163 and 165. This, with (13) below, would be equivalent to modern altazimuths and theodolites.[i]

(7) *Chêng li i*[7] — Verification Instrument. It is not clear what this was, but perhaps it was a component of (1) which permitted exact determinations of the positions of the sun and moon near the ecliptic; such at any rate was its stated purpose. Perhaps a sighting-tube ecliptically mounted.

(8) *Ching fu*[8] — Shadow Definer.[j] Already explained above (p. 299).

had been removed to Nanking. That the astronomers of Kuo Shou-Ching's time were very conscious of the importance of the latitude is shown by the fact that the *Yuan Shih* (ch. 48, pp. 12*b* ff.) reproduces a table of the latitudes of some twenty-five important centres, some of which undoubtedly had astronomical instruments; it is entitled 'Ssu Hai Tshê Yen'.[9] Cf. Gaubil (2), p. 110.

[a] Comment is hardly needed on this piece of European presumption.

[b] Chs. 48, pp. 1*a* ff. and 164, pp. 5*b* ff., paraphrased and partly translated by Wylie (7), who drew also on other sources such as the *Hsü Hung Chien Lu* and the *Chhou Jen Chuan*.

[c] The order varies in the two chapters; we adopt our own.

[d] *Yuan Shih*, ch. 48, p. 2*b*.

[e] P. 5*b*. Cf. pp. 382 ff.

[f] P. 6*a*. This description precludes the suggestion of d'Elia (2) that it was a gnomon. It was a bowl 12 ft. in diameter.

[g] P. 8*b*.

[h] Ricci's 'third instrument' must have been one of the usual 8 ft. gnomons.

[i] Spencer-Jones (1), p. 83.

[j] P. 9*b*.

1 玲瓏儀 2 簡儀 3 渾天象 4 仰儀 5 高表
6 立運儀 7 證理儀 8 景符 9 四海測驗

(9) *Khuei chi*[1] Observing Table.[a] Apparently an adaptation of the gnomon and shadow-definer to lunar shadows.

(10) *Jih yüeh shih i*[2] Instrument for Observation of Solar and Lunar Eclipses. Not explained.[b]

(11) *Hsing kuei*[3] 'Star-Dial.' Could this have been a forerunner of the +16th-century nocturnal (see p. 338 above)?

(12) *Ting shih i*[4] Time-Determining Instrument. Either the 'diurnal circle' of (2) above, or another name for (11).

(13) *Chêng fang an*[5] Direction-Determining Table.[c] This must have been an azimuthal circle, probably the 'earthly coordinate circle' of (2) above.

(14) *Hou chi i*[6] Pole-Observing Instrument. Presumably the polar sighting-tube embodied in (2) above, with the 'pole-determining circle' at the upper end.

(15) *Chiu piao hsüan*[7] Nine Suspended Indicators. Though details are not given, this probably refers to the *groma* or hanging plumb-lines whereby the trueness of the instruments, especially the gnomons, was checked (cf. p. 286 above).

(16) *Chêng i*[8] Rectifying Instrument. Purpose uncertain.

(17) *Tso chêng i*[9] Rectifying Instrument on a stand. Purpose uncertain.

Apart from the minor devices and those about which we have not sufficient information, the chief interest lies in the great armillary sphere (1) and the 'simplified instrument' (2). There is little need to describe the former more fully, since it did not differ in any fundamental way from that which Su Sung had already used in +1090, though no doubt of finer and exacter workmanship. But the second of these two, which Ricci could only describe as a collection of 'astrolabes' set in different axes, is something new. We may regard it as a simplification of the medieval instrument known as the 'torquetum',[d] which consisted of a series of discs and circles not, like those of the armillary sphere, placed concentrically. The Arabic and European versions had a disc mounted in the plane of the equator, and another one revolving at an angle to it in the plane of the ecliptic; this then carried a celestial latitude circle at right angles to itself.[e] The equipment was completed by a half-disc or protractor and plumb-line for reading off altitude. Probably the apparatus was used mainly for computations, as it permitted the direct conversion of ecliptic to equatorial coordinates and vice versa as well as other comparisons. The invention of this unwieldy instrument has often been ascribed to Kuo Shou-Ching's elder contemporary, Naṣīr al-Dīn al-Ṭūsī of Marāghah,[f]

[a] P. 10*a*.

[b] Perhaps this was the adjustable dioptra for determining disc diameters in partial eclipses described by Arabic writers (see L. A. Sédillot (1), p. 198).

[c] P. 7*b*.

[d] To be distinguished carefully from the 'triquetrum'; see below, p. 373.

[e] See R. Wolf (1), vol. 2, p. 117; Houzeau (1), p. 952; Gunther (1), vol. 2, pp. 35, 36; Anon. (25), p. 18, no. 348 and opp. p. 30; Rohde (1), pp. 79 ff.; Michel (12), p. 68 and pl. XIII. One of the best expositions is that of Michel (13), who showed that the Rectangulus of Wallingford, which Gunther (1), vol. 2, p. 32, had not understood, was a skeleton torquetum.

[f] Sarton (1), vol. 2, p. 1005. Cf. Sect. 7*j* above, in Vol. 1, pp. 217ff.

1 闚几 2 日月食儀 3 星晷 4 定時儀 5 正方案
6 候極儀 7 九表懸 8 正儀 9 座正儀

Fig. 163. Equatorial armillary sphere of Kuo Shou-Ching (+1276), copied by Huangfu Chung-Ho (+1437), seen from the north in the gardens of the Peking Observatory (photo. Thomson).

Fig. 164. The 'Simplified Instrument' (*chien i*) or equatorial torquetum devised by Kuo Shou-Ching *c.* +1270. It constitutes the precursor of all equatorial mountings of telescopes. Here the instrument, which may be one of the replicas of Huangfu Chung-Ho, is seen from the south-east. Its size may be gauged from the fact that the base-plate measures 18 × 12 ft. and that the revolving declination ring carrying the sighting-tube is 6 ft. in diameter (photo. Academia Sinica).

Fig. 165. Kuo Shou-Ching's equatorial torquetum seen from the south, so as to show the 'pole-determining circle' at the top, with its crossbars and central hole, attached to the 'normal circle' (photo. Academia Sinica).

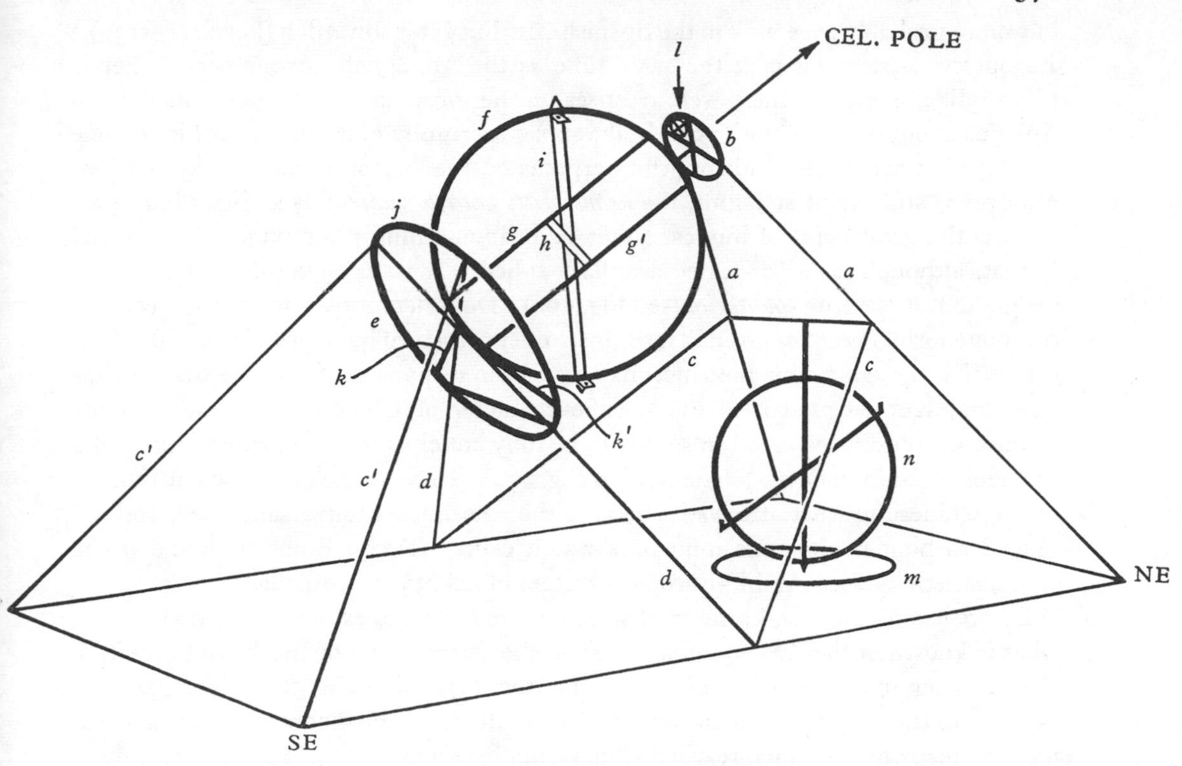

Fig. 166. Diagram of Kuo Shou-Ching's equatorial torquetum (*chien i*[1]), seen from the south-east, for comparison with Fig. 164. *a*, *a*, North pole cloud frame standards (*pei chi yün chia*[2]); *b*, normal circle (*kuei huan*[3]), diam. 2 ft. 4 in. (fixed); *c*, *c*, *c'*, *c'*, dragon pillars (*lung chu*[4]); *d*, *d*, south pole cloud frame standards (*nan chi yün chia*[5]); *e*, fixed diurnal circle (*pai kho huan*[6]), graduated in the 12 double-hours and the 100 quarters, each of the latter having 36 sub-divisions. This circle has four roller-bearings in its northern face to allow of the easy rotation of the equatorial circle (*j*) upon it. Diameter 6 ft. 4 in. *f*, Mobile declination ring or meridian double circle (*ssu yu shuang huan*[7]), 6 ft. in diameter, graduated on both sides in degrees and minutes, and carrying a sighting-tube or diametral alidade (*i*) for the determination of declination (actually N.P.D.); *g*, *g'*, stretchers (*chih chü*[8]); *h*, double brace (*hêng*[9]), also to prevent deformation; *i*, diametral alidade (*khuei hêng*[10]) with pointed ends (like ceremonial tablets) for accuracy (*tuan wei kuei shou*[11]), and sighting-vanes (*hêng erh*[12]). *j*, Mobile equatorial circle (*chhih tao huan*[13]), 6 ft. in diameter, graduated in degrees and minutes, and marked with the boundaries of the 28 *hsiu*. It is strengthened by cross-stretchers like the declination ring. *k*, *k'*, Independently movable radial pointers with pointed ends (*chieh hêng*[14]). It is not clear whether these carried sighting-vanes. Their name, 'boundary bars', indicates that they were used to mark off the boundaries of *hsiu*. *l*, Pole-determining circle (*ting chi huan*[15]), of diameter equivalent to 6°, attached to the upper part of the normal circle. It can be seen only in Fig. 165. This circle has a cross-piece inside it, with a central hole, and seems to have been used for determining the moment of culmination of the pole-star itself. Observation was effected through a small hole in a bronze plate attached to the south pole cloud frame standards. The main polar axis through the centre of the diurnal and equatorial circles and that of the normal circle was also provided with holes which constituted a polar sighting-tube. *m*, Fixed terrestrial coordinate azimuth circle (*yin wei huan*[16]); *n*, revolving vertical circle (*li yün huan*[17]) with alidades, for measurement of altitudes. For a translation of the description of the instrument in *Yuan Shih*, ch. 48, pp. 2*b*ff. see Wylie (7).

1 簡儀	2 北極雲架	3 規環	4 龍柱	5 南極雲架
6 百刻環	7 四游雙環	8 直距	9 橫	10 窺衡
11 端爲圭首	12 橫耳	13 赤道環	14 界衡	15 定極環
16 陰緯環	17 立運環			

but more probably goes back to the Spanish Muslim Jābir ibn Aflaḥ (born *c.* +1130).[a] Europeans were using it at the same time as the Marāghah astronomers in Persia (Thorndike, 3, 4), and there were treatises on the torquetum by Regiomontanus and Apianus about +1540,[b] but Tycho Brahe spoke scornfully of it,[c] and after him no one employed it except the Indians who perpetuated the Arabic tradition. An outdoor example is still extant at Jaipur, the *krāntivṛitti valaya yantra*[d] (see Fig. 168*a*).

Now the great point of interest about Kuo Shou-Ching's 'simplified instrument' is that, although (as a 'dissected armillary sphere') it is recognisably related to the torquetum, it is a true *equatorial* (see Fig. 166). Doubtless it was because the ecliptic components had been removed that the instrument received its name 'simplified'. This means that though Arabic influence may have been responsible for suggesting its construction, Kuo adapted it to the specific character of Chinese astronomy, namely equatorial coordinates. And in so doing, he fully anticipated the equatorial mounting so widely used for modern telescopes (cf. Fig. 170). Here we see again the faithfulness of the Chinese to what afterwards became the coordinates universally used, and are moved to ponder upon the influences which caused Tycho Brahe to abandon the characteristic Graeco-Arabic-European system of ecliptic coordinates.

Before examining this, however, it will be more logical to explore somewhat further what is known of the Arabic influence upon the astronomers of the Yuan.[e] The task is easy owing to the special studies which Hartner (3), Yabuuchi (5) and Tasaka[f] have devoted to the subject. They have been able to identify[g] the diagrams of seven astronomical instruments which reached China from Persia in +1267. In the *Yuan Shih*[h] a couple of pages are devoted to the 'Plans (or Models) of Astronomical Instruments from the Western Countries' (*Hsi yü i hsiang*[1])—these were sent by Hūlāgu Khan or his successor to Khubilai Khan through the hands of one of the Marāghah astronomers Cha-Ma-Lu-Ting[2] (Jamāl al-Dīn)[i] in person. The identity of this man is somewhat obscure, but he may have been the Jamāl al-Dīn ibn Muḥammad al-

[a] Sarton (1), vol. 2, p. 206. Cf. Repsold (1); Mieli (2), vol. 2, p. 144.

[b] The oldest European specimen is the one which Nicholas of Cusa bought in +1444 (see Hartmann, 1). It is preserved in the library of the hospital which he founded at his birthplace Kues in the valley of the Mosel below Trier (Fig. 167). Others may be seen in the Deutsches Museum at Munich, and one appears in the background of the famous painting by Holbein, 'The Ambassadors'.

[c] Raeder, Strömgren & Strömgren (1), p. 53.

[d] Kaye (5), pp. 32, 33 and fig. 58; Soonawala (1), p. 38. It now lacks several components.

[e] Wagner has described two interesting MSS. preserved in his time in the Pulkovo Observatory in Russia, one in Arabic or Persian, the other in Chinese. They are tables of the motion of sun, moon, and planets, calculated from an epoch starting at +1204 and written about +1261. As probable relics of the collaboration of Jamāl al-Dīn and Kuo Shou-Ching, they would be precious indeed, and it is to be hoped that they were not destroyed when the Observatory was burnt during the second world war. We have already mentioned (Vol. 1, p. 218) another MS. of about a century later (+1363). Preserved in Paris, it is an astronomical treatise with lunar tables written in Arabic with marginal notes in Mongol and a Chinese title-page. The author, 'Aṭā ibn Aḥmad al-Samarqandi, prepared it for a Yuan prince, Radna, son of Dukbal, a direct descendant of Khubilai and Chingiz. His Chinese name was Chen-Hsi-Wu-Ching Wang. On this interesting relic see Blochet (3), p. 169, Schefer (2), p. 24 and Destombes (4).

[f] Cf. Fuchs (1), p. 4. In the following identifications of Arabic terms and their interpretations we accept those of Hartner rather than those of Tasaka, some of which are certainly wrong.

[g] As Gaubil (2), p. 130, could not.

[h] Ch. 48, pp. 10*b* ff.

[i] Cf. above, p. 49. Sarton (1), vol. 2, p. 1021.

[1] 西域儀象 [2] 札馬魯丁

Fig. 167. Torquetum; this instrument once belonged to Nicholas of Cusa (+1444) and is still preserved in the library of his Hospice at Kues near Trier.

Fig. 168*a*. An Indian torquetum, the *krāntivṛitti valaya yantra*, at the observatory of Jai Singh at Jaipur (photo. Kaye). The instrument is incomplete but the two planes of equator and ecliptic can be seen. Cf. p. 301.

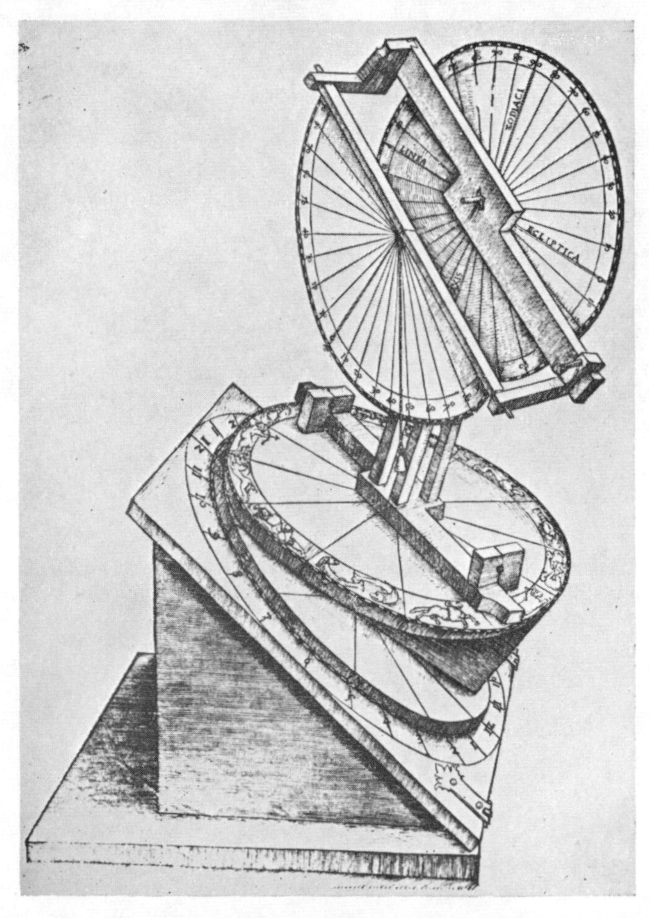

Fig. 168*b*. The torquetum of Petrus Apianus (+1540).

Najjārī[a] who had declined to take on the full responsibility of building the Marāghah observatory in +1258. The Chinese names of the instruments about which he was deputed to inform the Chinese astronomers, together with brief explanations of them, are given in the *Yuan Shih* text.

These instruments were the following:

Chinese transcription[b]	*Persian-Arabic original*	*Chinese translation and explanation*
(1) *Tsa-thu ha-la-chi*[1]	*Dhātu al-ḥalaq-i* ('the owner of the rings')	*Hun thien i.*[2] Armillary sphere, not Ptolemaic, but an equatorial with tropic circles or circles of perpetual visibility and invisibility attached to the movable declination ring carrying the sights[c]
(2) *Tsa-thu shuo-pa-thai*[3]	*Dhātu'sh-shu'batai (ni)* ('the instrument with two legs')	*Tshê yen chou thien hsing yao chih chhi*[4] ('instrument for observing and measuring the rays of the stars of the celestial vault'). The conjectures of Zinner (1), p. 236, that this was a divided circle, and of Fuchs (1), p. 4, that it was the Jacob's Staff of the surveyors, are not to be retained. Hartner (3) was essentially right in identifying it with Ptolemy's *organon parallacticon* (ὄργανον παραλλακτικόν), i.e. the 'long rulers' or triquetrum, for determining zenith distances of stars at culmination.[d] But it must have been the modification described about +840 by Ya'qūb ibn Isḥāq al-Kindī in a MS. translated by Wiedemann (9) from a copy of +1212, at which date the instrument was thus probably still in use. For the Chinese text of the *Yuan Shih* says that it had two sighting-tubes (*hsiao*[5]) (for determining interstellar distances in degrees), and this was not necessary when one limb was permanently fixed to point perpendicularly at the zenith

[a] Or more probably al-Bukhārī (note from Dr W. Hartner). A MS. catalogue of 72 stars dated +1285 by one Jamāl al-Dīn ibn Mahfuz exists in Paris (Destombes, 2).

[b] These are not the only extant transcriptions, but they are the most correct. Chhien-Lung editions of the *Yuan Shih* were subjected to revision by a commission of learned linguists who mongolicised all foreign words even when these had been transliterations from other languages such as Arabic (note from Dr W. Hartner).

[c] Its polar elevation was 36°, so it was probably planned for the observatory at Phing-yang; for this latitude would fit Teheran or Meshed but not Marāghah.

[d] *Almagest*, v, 12. The best description is Tycho Brahe's (Raeder, Strömgren & Strömgren (1), p. 44). See also Gunther (1), vol. 2, p. 15; Dicks (1); Drachmann (3).

[1] 咱秃哈剌吉 [2] 渾天儀 [3] 咱秃朔八台 [4] 測驗周天星曜之器
[5] 簫

	Chinese transcription	*Persian-Arabic original*	*Chinese translation and explanation*
(3)	*Lu-ha-ma-i miao-wa-chih*[1]	*Rukhāmah-i-mu'wajja*	*Tung hsia chih kuei.*[2] 'Solstitial Dial', i.e. plane sundial for unequal hours
(4)	*Lu-ha-ma-i mu-ssu-tha-yü*[3]	*Rukhāmah-i-mustawīya*	*Chhun chhiu fên kuei.*[4] 'Equinoctial Dial', i.e. plane sundial for equal hours
(5)	*Khu-lai-i sa-ma*[5]	*Kura-i-samā'*	*Hsieh wan hun thien thu*[6] ('obliquely set globe with map of the stars'). Celestial Globe
(6)	*Khu-lai-i a-erh-tzu*[7]	*Kura-i-arḍ*	*Ti li chih.*[8] Terrestrial Globe
(7)	*Wu-su-tu-erh-la*[9]	*al-Usṭurlāb*	Astrolabe. The text says: 'The Chinese name (for this) has not been worked out. The instrument is to be made of bronze, on which the times (hours) of the day and night are engraved.' Certainly not a clepsydra, as Zinner (1), p. 236, supposed.

The list is an interesting one. The first suggestion was certainly no novelty for the Chinese,[a] but Jamāl al-Dīn's instruments were surely adapted for ecliptic measurements, and as we saw above, Kuo Shou-Ching paid no attention to this. They would also have used a graduation of 360° but Kuo Shou-Ching retained the $365\frac{1}{4}$° system.[b] Nor was the fifth instrument, the celestial globe, anything new. On the other hand, the terrestrial globe was perhaps a novelty; there is no previous record of one before the time of Martin Behaim (+1492)[c] except the ancient globe of Crates of Mallos[d] in the −2nd century, which had been entirely forgotten. The Chinese text describes the new instrument as 'a globe to be made of wood, upon which seven parts of water are represented in green, three parts of land in white, with rivers, lakes, etc. Small squares are marked out so as to make it possible to reckon the sizes of regions and the distances along roads.' There seems no evidence, however, that the Chinese took this up.[e] As for the sundial, they were probably puzzled at the conception of unequal hours, and it is

[a] Hartner (3) went astray in his suggestion that armillary spheres were unknown in China until the +13th century—an underestimate of perhaps seventeen centuries. This is said not in criticism of him, since his facilities were lamentably inadequate, and the work of Maspero was not at that time available to him; but in order to correct a view which might otherwise have the authority of *Isis* in 1950. The error was fully rectified in Hartner (8).

[b] This seems the only justification, and not at all a strong one, for Johnson's emphasis (2) on what he calls the 'tragic conservatism' of the Chinese. Bosmans, himself a Jesuit, has on the contrary drawn attention to the imposition of the sexagesimal graduation of degrees and minutes on the Chinese by Verbiest at a later date. The Chinese had always graduated them decimally in tenths and hundredths. Bosmans freely admits that this change was retrograde.

[c] Ravenstein (1). Behaim's globe is now in the National Museum at Nuremberg.

[d] Sarton (1), vol. 1, p. 185; cf. Stevenson (1); Schlachter & Gisinger (1).

[e] There were of course Chinese terrestrial globes in later times, but we shall postpone the consideration of them till Sect. 29*f* on navigation.

[1] 魯哈麻亦渺凹只 [2] 冬夏至晷 [3] 魯哈麻亦木思塔餘
[4] 春秋分晷 [5] 苦來亦撒麻 [6] 斜丸渾天圖 [7] 苦來亦阿兒子
[8] 地理志 [9] 兀速都兒剌

fairly clear that their traditional type of sundial[a] persisted unchanged. The instrument most strikingly absent from the list is the torquetum, though it would have been expected more than any other, if Kuo Shou-Ching's 'simplified instrument' arose from the stimulus of contact with Arabic science, as on all the circumstantial evidence it did. Moreover, no torquetum is listed in al-'Urḍī's account of the equipment at the Marāghah Observatory.[b] But it would seem that Jamāl al-Dīn must have brought the idea with him.

As for the second and seventh of the instruments, if they were not adopted it was surely because they did not fit into the characteristic system of Chinese astronomy, polar and equatorial. The parallactic ruler for determination of zenith distance[c] could hardly interest astronomers in whose work the zenith played no part. And the astrolabe, so universal in Arabic and medieval European astronomy, was primarily intended to measure altitude and to compute ecliptic coordinate positions which the Chinese did not particularly want. Hartner considers that what the Marāghah astronomers offered was well chosen;[d] they did not send the apparatus for determining sines, azimuths and versed sines,[e] because it was probably known that the Chinese astronomers were unfamiliar with spherical trigonometry. Jamāl al-Dīn had an overwhelming task before him if he intended to explain to the Chinese the whole system of Arabic gnomonics and the mathematics required for the stereographic projection on which the astrolabe markings were based, and if he tried he certainly did not succeed. But what has been unperceived, even by Hartner, is that the measurements and computations which the rulers and the astrolabe could yield were simply not wanted in the Chinese polar-equatorial system.

The astrolabe is a very complex instrument upon which medieval Arabic and European astronomers lavished all their mathematical art. It might be called a 'flattened' armillary sphere,[f] combining the armillary rings of Hipparchus and the theodolite of Theon with a projection of the zodiac and the starry hemisphere. The luminous treatise of Michel (3), published a few years ago, makes recourse to older explanations of the theory and construction of astrolabes unnecessary, while a massive compendium by Gunther (2) gives elaborate details of the principal surviving instru-

a See above, pp. 308 ff. On the graduation of sundials for equal and unequal hours see Zinner (7).

b Seemann (1); Jourdain (1). Cf. Howorth (1), vol. 3, pp. 137ff.

c Illustration in Gunther (1), vol. 2, p. 15.

d It will be remembered that the Marāghah Observatory had on its staff at least one Chinese astronomer (Sect. 7*j* above, Vol. 1, p. 217). His name was apparently Fu Mêng-Chi, or as some think, Fu Mu-Chai[1] (cf. Sarton (1), vol. 2, p. 1005; Li Nien (*2*), p. 151). Dr Zaki Validi Togan transliterates from al-Banākitī (cf. above, Vol. 1, p. 221) rather Fu Mi-Chi.

e Cf. Jourdain (1) and Seemann (1).

f Hence the name 'astrolabium planisphaerium', R. Wolf (1), vol. 2, p. 45. A rare intermediate form also existed, in which flat rings were made so as to slide over a solid celestial sphere. One such instrument, made for Alfonso X, is figured in Singer & Singer (1), p. 227. The term 'astrolabe' was also that used by Ptolemy for his armillary sphere (ἀστρόλαβον ὄργανον), a fact which has caused some terminological confusion in modern times. It is impermissible now to refer to an armillary sphere as a 'spherical astrolabe' (with, e.g., Dubs (2), vol. 3, p. 328) for since the +13th century at least, this has meant something quite different.

1 傅穆齋

ments. The place of origin of the instrument is uncertain,[a] but its first known user was the Byzantine Ammonius (*c.* +500), though the earliest dated astrolabe is Persian, that of 'Aḥmad and Maḥmūd the sons of Ibrāhīm the astrolabist of Iṣfahān', +984. Venerable also is the Byzantine astrolabe of +1062, described by Dalton. The earliest extant treatise on its use derives from Joannes Philoponus (*c.* +525), the Byzantine physicist, a pupil of Ammonius, and in the next century the Syrian bishop Severus Sebokht[b] also wrote on it.[c] Not a single example is known from China, either by textual reference or actual preservation.

Fig. 169. The greater equatorial armillary instrument of Tycho Brahe (+1585); for detailed description see the translation and study of his *Astronomiae Instauratae Mechanica* by Raeder, Strömgren & Strömgren.

Although we are thus absolved from further discussion of the astrolabe itself, mention must be made here of the apparatus from which, in all probability, it originated. This was the anaphoric clock, a form of clepsydra not uncommon in Hellenistic times.[d] It consisted of a bronze dial,[e] marked with a star-map in planispheric projection, and made to rotate by the rise or fall of a clepsydra float attached to its drum by a counter-

[a] See Neugebauer (7) and Drachmann (6), who suspect that it was invented in or before the time of Ptolemy. It is mentioned in the *Tetrabiblos*, not in the *Almagest*.

[b] We have met with him before, see Sect. 7*j*, in Vol. 1, p. 220.

[c] Another old treatise, by the Jewish astronomer, Māshā'allāh of Baghdad (d. +815; Messahalla), has been translated in Gunther (1), vol. 5. See also J. Frank (1).

[d] See Diels (1), pp. 213 ff.; Usher (1), p. 97, 2nd edn. p. 145; Price (1, 4).

[e] Extant specimens discovered by Maass (1); Maxe-Werly (1). First explained by Benndorf, Weiss & Rehm.

weighted cord. The front of the disc was separated from the observer by a stationary network of wires representing meridian, equator, etc., and the position of the sun was indicated by a small stud plugged in by hand to one or other of a sufficient number of small holes in the disc along the ecliptic line. This arrangement is just the reverse of that of the astrolabe, where the star points are marked on the *aranea* or spider's-web, the coordinates being engraved on the various base-plates; and there is every reason to think that in fact the astrolabe arose by such a reversal.[a] The dial of the anaphoric clock, though mobile, was the first of all clock faces, and in Sect. 27*h* we shall discuss certain pieces of evidence which suggest that it was known and used in some Chinese clepsydras. If indeed the principle of the anaphoric clock travelled to China, the considerations already mentioned may suffice to explain why it did not generate the astrolabe there also. In any case the anaphoric dial is still with us, disguised as the 'rude star finder and identifier' of aeroplane navigation.[b]

It was Dreyer (2) who was perhaps the first to appreciate the historical importance of Kuo Shou-Ching's retention of the equatorial system in his 'simplified instrument'. 'We have here', he said, 'two remarkable instances of how the Chinese people often came into possession of great inventions many centuries before the western nations enjoyed them. We find here in the +13th century the equatorial armillae of Tycho Brahe, and better still, an equatorial instrument like those *armillae aequatoriae maximae* with which Tycho observed the comet of +1585 as also the fixed stars and planets.'[c] Johnson considers[d] that 'the Yuan instruments exhibit a simplicity which is not primitive, but implies a practised skill in economy of effort, and in this sense compares favourably with the Graeco-Muslim tendency to rely on separate instruments for each single coordinate to be measured—neither Alexandria nor Marāghah[e] exhibit any device so complete and effective and yet so simple as the "simplified instrument" of Kuo Shou-Ching. Actually our present-day equatorial mounting has made no further essential advance.'[f] This may be seen at once from Fig. 170 showing a 19th-century equatorial mounting, and Fig. 171 which is the Mount Wilson 100-inch reflector. The polar axis takes the form of a pivoted shaft in the former and a cradle in the latter. Johnson adds that the gun-barrel sighting-tubes rotating within the split rings were much preferable to the open alidades of the Arabs.[g]

[a] See Neugebauer (7); Drachmann (2) and especially (6). Drachmann thinks that Hipparchus used a plane astrolabe with the star-map on the base-plate (like the anaphoric clock), while by Ptolemy's time the reversal had taken place.

[b] This instrument consists of a ground-plate carrying a star-map in polar projection, together with a number of transparent plastic plates (one for every 10° of latitude) marked with coordinates, which fit over it. When adjusted for hour-angle from tables the altazimuths of all the main stars visible are at once shown.

[c] See Fig. 169. The difference was that Tycho retained a half-circle of the equator embracing the hour-circle centrally.

[d] (2), p. 104.

[e] For lists of the Marāghah equipment see Jourdain (1) and Seemann (1), summarised in Sarton (1), vol. 2, p. 1013.

[f] On the development of this in Europe details have already been given (p. 366 above). Cf. Olmsted (1).

[g] Another matter in which the Renaissance astronomers of Europe adopted Chinese practices was the greater use of meridian transit observations. There was Tycho's great mural quadrant (A. Wolf (1), p. 127), followed by the first mounting of a telescope permanently in the meridian by Roemer in +1681 (Dreyer (2); Grant (1), pp. 461 ff.; Spencer-Jones (2); A. Wolf (2), p. 137; H. C. King (1), pp. 105 ff.).

Thus Kuo Shou-Ching, who had no telescope, was the inventor of its equatorial mounting.[a] But his simplified torquetum did not die at this birth. It still lives on, unrecognised, among the most modern instruments of air and sea navigation. In the 'astro-compass' (shown in Fig. 172)[b] every component of the *chien i* can be seen, the equatorial circle (here adjustable for latitude), the declination ring (here a plate), the sighting device and the azimuth circle. Only now the latter is no longer separate but directly connected with the other parts to serve a new purpose, the determination of terrestrial bearings from celestial sightings according to prepared tables.[c]

Perhaps the most remarkable aspect of the whole story is the speed with which ideas could travel at that time. When Jamāl al-Dīn made his journey in +1267 to meet Kuo Shou-Ching in Peking, bringing with him his designs for the two *rukhāmah* dials, the greatest Arabic treatise on gnomonics had been finished only a dozen years before.[d] And it had been written by an astronomer from the other end of the Old World, Abū 'Alī al-Ḥasan al-Marrākushī, one of the ablest scientific men whom Morocco produced.[e] Hartner points out[f] that in sending on this new knowledge, the Marāghah astronomers were fulfilling the Ilkhan's desire that the Peking observatory should be equipped with the most modern instruments available anywhere in the known world. The complicated dials, as we have seen, awoke little interest, but the suggestion which was outstandingly successful, that of the torquetum, came from just as far away. If Jābir ibn Aflaḥ was really the inventor of this instrument, then within the span of two generations (*c.* +1130 to *c.* +1270), what started as an original, though not very practical, idea in Andalusia far in the West, had evoked a really important practical invention in Peking.[g]

An urgent question thus arises. What was it that led Tycho Brahe in the 16th century

[a] It is interesting to reflect that Marco Polo was in China just at the time when all this was going on. But he noticed only the astrological aspects of Chinese state-supported astronomy. Some of his texts contain a chapter (ch. 33; see Yule (1), vol. 1, p. 446) 'Concerning the Astrologers in the City of Cambaluc'. He says that no less than five thousand of them were entertained by the Great Khan with annual maintenance and clothing, and that 'they have a kind of astrolabe on which are inscribed the planetary signs, the hours and critical points of the whole year'. All three sects of star-clerks, Cathayan, Saracen, and Christian (presumably Uighur Nestorian), used these instruments for prognostications, on which they were consulted by many people. Moreover, they prepared 'certain little pamphlets called Tacuin', Ar. *taqwīm*, i.e. calendars or ephemerides, which were published by the government in surprising numbers. For example, in +1328 more than three million copies were printed and issued. Some of these (e.g. one for +1408) later came into the possession of such men as Robert Boyle, Robert Hooke, and Samuel Pepys, arousing their curiosity concerning Chinese astronomy (cf. p. 391 below). On *taqwīm* and *taquinum* see Thorndike & Sarton (1). The word meant anything drawn up in tables; cf. the term *li chhêng* (pp. 9, 36, 107 above).

[b] Mentioned rather than described in A. J. Hughes (1), pp. 116, 117.

[c] For acquaintance with this instrument and its uses we are much beholden to Dr Martin Hinton, F.R.S.

[d] It was the *Jāmi' al-Mabādī wa'l-Ghāyāt* (Book of the Beginnings and Endings). This work was translated by J. J. E. Sédillot, and formed the basis of the treatise of Schoy (1) on Arabic gnomonics.

[e] Mieli (1), p. 210; Suter (1), no. 363.

[f] (3), p. 188.

[g] Practical not only in its long-term uses but also because Kuo Shou-Ching made good use of it. Among his lost works there are two significant titles: the *Hsin Tshê Erh-shih-pa Hsiu Tsa Tso Chu Hsing Ju Hsiu Chhü Chi*[1] (New Catalogue of Star Positions of Hsiu and other Constellations in Right Ascension and Declination), and the *Hsin Tshê Wu Ming Chu Hsing*[2] (New List of Positions of hitherto unnamed Stars); *Yuan Shih*, ch. 164, p. 12*b*. Apparently he himself described his *chien i* and other instruments but his *I Hsiang Fa Shih*[3] (Designs of Astronomical Apparatus) also failed to survive.

[1] 新測二十八宿雜坐諸星入宿去極 [2] 新測無名諸星 [3] 儀象法式

Fig. 170. A nineteenth-century equatorial telescope mounting (from Ambronn, 1).

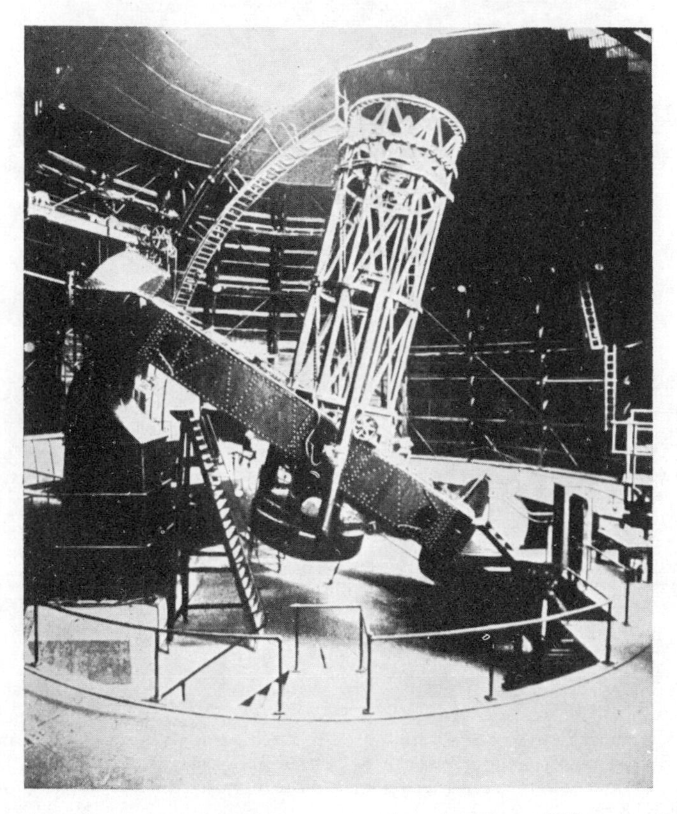

Fig. 171. A modern equatorial telescope mounting, the Mount Wilson 100-inch reflector.

PLATE LVII

Fig. 172. A modern form of the torquetum; the astro-compass used in aerial navigation (see text). Kuo Shou-Ching's components can all be seen in it, the equatorial circle (here adjustable for latitude), the declination ring (here a plate), the sighting device, and the azimuth circle. Height approx. 8 in.

Fig. 173. Ecliptic armillary sphere erected by Ferdinand Verbiest at the Peking Observatory in +1674 (photo. Nawrath).

Fig. 174. Celestial globe made at the Marāghah Observatory about +1300 (the time of Kuo Shou-Ching) by Muḥammad ibn Mu'ayyad al-'Urḍī (Dresden Museum, photo. Stevenson).

to abandon the age-old Graeco-Muslim ecliptic coordinates and ecliptic armillary spheres in favour of the equatorial coordinates which the Chinese had had all along? Gunther[a] expressed the greatest surprise that the Chinese had used equatorial armillae, which were considered one of the chief advances of European Renaissance astronomy, and concluded that Kuo Shou-Ching had anticipated Tycho by three centuries. Now Tycho tells us himself in his book on instrument construction[b] that he found zodiacal or ecliptic armillary spheres very unsatisfactory because since they are not always in a position of equilibrium (i.e. since their centre of gravity shifts according to the position of the equinoctial junctions) they become deformed owing to the weight of the metal, introducing errors of as much as a couple of minutes of arc. For this reason, he preferred to construct equatorial armillary spheres (Fig. 155). But a purely technical reason of this kind seems insufficient for a change in the basic method of expressing celestial positions. Dreyer (3) therefore raised the question of whether there had been some Arabic influence sapping the assured convention of ecliptic methods. L. A. Sédillot suggested[c] that equatorial armillary spheres had been known to the Arabs, and provided more evidence later, quoting, among other sources, Bettini's seventeenth-century view that Ibn al-Haitham had used them in the late +10th or early +11th century.[d] In such a book as this it is impossible to pursue the matter further, but its importance would seem to justify a special investigation. There exists the possibility that, in spite of the general trend of Muslim astronomy, there were isolated instances of the use of equatorials,[e] and that these might have been derived from Arab-Chinese contacts (at any time indeed back to the Han).[f] The idea might then have stimulated a few European astronomers, such as Gemma Frisius, who in +1534 first described a small portable equatorial armillary sphere,[g] and then in turn Tycho Brahe himself, whose reasons for making the change can hardly have been confined to the purely technical one mentioned above.[h]

How paradoxical it is, in the light of all this, that when the Jesuits proceeded to enlighten the Chinese in scientific matters, they erected in +1674 an *ecliptic* armillary sphere at the Peking observatory (Fig. 173).[i] And Verbiest, in his book of 1687 about

[a] (1), vol. 2, pp. 145 ff.

[b] Tr. Raeder, Strömgren & Strömgren (1), pp. 55 ff.

[c] (1), p. 198.

[d] (3a), p. cxxxiv. Bettini (*Apiaria*, VIII, progym. iii, prop. vii, p. 41) had written: 'Adhibuit Tycho armillare quoddam instrumentum quod tamen comperi ego positum et adhibitum olim fuisse ante Tycho ab Alhazeno.' Alfraganus and Albategnius were in Tycho's library (Kleinschnitzová, 1).

[e] We have just seen (p. 373) that the instrument (or plan) which Jamāl al-Dīn took to Peking in +1267 was an equatorial, but this may have been purposely designed for a known Chinese tradition. It also had a graduated horizon ring, but the Chinese text (*Yuan Shih*, ch. 48, pp. 10b, 11a) does not speak of any component for taking altitudes as well as azimuths. In any case, the instrument was strikingly un-Greek. Cf. Seemann (1), pp. 57 ff.

[f] If we may include pre-Muslim times and the Syrian and Persian predecessors of Arabic science.

[g] Dreyer (3), p. 316.

[h] The knowledge of the third anomaly (variation) of the moon's motion may be a parallel case. L. A. Sédillot (2), vol. 1, p. 70, suggested that Tycho might have come across some hint of Abū'l-Wafā's +10th-century work on this. But there is doubt about the Arabic discovery.

[i] Lecomte (1), p. 67; Chhen Tsun-Kuei (6), p. 23. Some of the instruments still possess (1952) their sighting-tubes, square-sectioned externally and tubular internally, with the fiducial cross-wires in place. Some writers even inform us that the Jesuit instruments were modelled on Tycho's (e.g. Thiel (1), pp. 99 ff.).

astronomy in China, has nothing to say of the Yuan instruments, save that they had been the products of a 'ruder Muse'.[a]

The serene assurance of the Jesuits in their astronomical superiority, misplaced though it may now seem in true historical perspective, was part of that general occidental self-confidence which impelled later historians of science to assume that the Muslims had exerted a powerful influence on Chinese astronomy. The facts, as we have just seen them, are quite otherwise.[b] But the custom was to say, with Reinaud & Favé in 1845, 'l'on réforma a Pékin l'astronomie nationale d'après les travaux exécutés en Occident'.[c] This statement was based on a summary of Gaubil's in his history of the Mongol dynasty, but he, who knew his material well, spoke much more cautiously.[d] It is interesting to see what he said.

A la onzième Lune [+1280] on fit publier l'astronomie à laquelle Hiuheng,[e] Vangsun,[f] Yangkongy[g] et Cocheouking[h] travailloient depuis long-temps; Cocheouking eut la meilleure part à ce grand ouvrage.[i]

[J'ai parlé au long de l'astronomie publiée l'an 1280 par ordre de Houpilay[j] dans le traité que j'ai envoyé sur l'astronomie Chinoise, et qui a été imprimée en 1717.[k] Yelutchoutsay[l] fut chargé par Gentchiscan[m] de ce qui regardoit l'astronomie. Cet astronome rectifia beaucoup ses idées en Occident à la la suite de Gentchiscan, et au retour publia une astronomie.[n]

[a] After Verbiest's reconstruction of the Peking Observatory in 1672 they were put away in storerooms, where Mei Ku-Chhêng, the historian of mathematics, often examined them in 1713 and 1714. But in 1715 the Jesuit father B. K. Stumpf (Chi Li-An[1]) melted down several of them to make a bronze quadrant. By 1744 the Simplified Instrument, the Armillary Sphere and a Terrestial Globe were the only ones left, and for some unknown reason not open to inspection in Gaubil's time (+1723 onwards); see his (2), p. 108. Mei Ku-Chhêng rightly deplored the action of Stumpf, who had authority as Director of the Bureau of Astronomy and Calendar.

[b] Yabuuchi (1) and Chhien Pao-Tsung (7) strongly concur.

[c] (1), p. 200. Cf. L. A. Sédillot (2), vol. 2, p. xii. Even the most estimable of modern writers are capable of describing Kuo Shou-Ching's *chien i* as 'constructed by Persian astronomers' (Cable & French (2), p. 40; Cronin (1), p. 141).

[d] (12), p. 192.

[e] Hsü Hêng[2] (+1209 to +1281), better known as Hsü Lu-Chai,[3] the Neo-Confucian philosopher.

[f] Wang Hsün[4] (*c.* +1255 to +1282). He was Astronomer-Royal at the time of the re-equipment of the observatory in +1279.

[g] Yang Kung-I[5] (*fl. c.* +1255 to +1318).

[h] Kuo Shou-Ching, Deputy Astronomer-Royal in +1279.

[i] This was the *Shou Shih Li*[6] (Shou Shih Calendar). The *I*[7] (Explanations) now occupy chs. 52 and 53 of the *Yuan Shih* (History of the Yuan Dynasty), starting from p. 2*b* of the former. The *Ching*[8] (Manual of Computations) now occupies chs. 54 and 55 of the *Yuan Shih*. The former begins with a description of gnomon solstice observations (see p. 299 above), and proceeds to the determinations of tropic and sidereal year fractions, winter solstice positions, *hsiu* extensions; then adds solar and lunar ephemerides, gives an account of the moon's nodes, states the principles of eclipse computations used, and ends with a register of all eclipses noted since ancient times and a history of all the preceding calendars. The latter gives the formulae for all the calculations involved, including ecliptic as well as equatorial *hsiu* extensions, and planetary ephemerides. The whole work was completed by Kuo Shou-Ching alone after the death of Wang Hsün in +1282 (*Yuan Shih*, ch. 164, p. 12*a*).

[j] Khubilai Khan.

[k] This was Gaubil (3). Much of it is a translation (pp. 69ff.) of the *Shou Shih Li I Ching*, including the tables.

[l] Yehlü Chhu-Tshai[9] (+1190 to +1244).

[m] Chingiz Khan.

[n] This was the (*Hsi Chêng*) *Kêng-Wu Yuan Li*[10] (Calendar of the Kêng-Wu Year (+1210) made during the Western Expedition). It now occupies chs. 56 and 57 of the *Yuan Shih*. It was used from about +1220 onwards, but never promulgated.

1 紀理安　2 許衡　3 許魯齋　4 王恂　5 楊恭懿
6 授時曆　7 議　8 經　9 律楚材　10 西征庚午元曆

Au commencement du règne d'Houpilay, les Astronomes du pays d'Occident publièrent deux astronomies, l'une selon la méthode d'Occident,[a] l'autre selon la méthode Chinoise, mais corrigée.

Cocheouking prit un milieu, et suivant dans le fonds la méthode d'Occident, il conserva tant qu'il pût les termes de l'astronomie Chinoise, mais il la réforma entièrement sur les époques astronomiques, et sur la méthode de réduire les tables à un méridien, et d'appliquer ensuite les calculs et les observations d'autres méridiens.[b] J'ai déjà dit que les Princes Mongous avoient à leur Cour des Médecins[c] et des Mathématiciens d'Occident; ils faisoient des corps separés des médecins et des mathématiciens Chinois. D'ailleurs ils vivoient très-bien ensemble, et les livres où est l'histoire de ces temps-là louent fort en général l'habileté de ces Etrangers, et avouent en particulier que c'est d'eux que Cocheouking prit ce qu'il avoit de meilleur.[d]][e]

Les Mathématiciens d'Occident en grand nombre et en grand crédit à la Cour avoient déjà beaucoup travaillé sur l'astronomie et ils avoient fait de très-beaux instrumens.[f] Cocheouking, homme d'un génie et d'un travail extraordinaire, secouru des trois Seigneurs Chinois que je viens de nommer, et parfaitement au fait sur les méthodes que les Occidentaux avoient fait connoître à la Cour, mit la dernière main à l'Astronomie Chinoise. Outre cela il fit de grands instrumens de léton, sphères, astrolabes,[g] armilles, boussoles,[h] niveaux, gnomons, dont un étoit de 40 pieds. J'ai parlé ailleurs de l'ouvrage de Cocheouking, et j'ai expliqué la méthode qui se voit dans ce qui nous reste de son astronomie.

Thus Gaubil gave liberal credit to the greatness of Kuo Shou-Ching, though his term 'Occidentaux' for the Arabs and Persians was rather misleading for later western readers. But the facts do not support his belief in deep Muslim influence. The Chinese did not adopt the advanced geometry[i] needed for the stereometric projections on astrolabes and sundials, nor were they interested in the unequal-hour system, nor the division of the circle into 360°, nor did terrestrial globes appeal to them. They remained untouched by the Arabic altazimuth star-coordinate system and the European ecliptic coordinates alike, remaining faithful to those equatorial coordinates which

[a] This refers to the Wan Nien Li[1] (Ten Thousand Year Calendar) prepared by Jamāl al-Dīn, after his arrival in Peking, and given official recognition in +1267. But it only lasted nine years, for in +1276 Kuo Shou-Ching, with many colleagues and assistants, was commissioned to produce the Shou Shih calendar, promulgated in +1281. This lasted more than a century, till +1382, when there was again an Arabic calendar for two years, followed by a Chinese one which continued till as late as +1554. On Kuo Shou-Ching's calendar see Chhien Pao-Tsung (7).

[b] This cannot be true, for the meridian of Yang-chhêng (see p. 297 above) had for centuries been accepted as China's Greenwich.

[c] This is a reference to Ai-Hsüeh[2] (+1227 to +1308), an eminent physician indeed, but not a Byzantine (as d'Ohsson implied, (1), vol. 2, p. 377), rather a Nestorian Arab from Syria. He reached high office in the Mongol service, and had been Director of the Astronomical Bureau a few years before Jamāl al-Dīn arrived. His biography is in *Yuan Shih*, ch. 134, p. 7*a*.

[d] No text in any way indicating this has ever been met with by us.

[e] The square brackets indicate what was a footnote in Gaubil (12).

[f] We know of no evidence whatever that at this time there were many of them, or that they made any important instruments at any time.

[g] This is of course quite erroneous. See above, p. 268 and p. 375.

[h] If this is true, one would very much like to have details of them. There is no reason why it should not be true.

[i] As we have already seen, however, there may have been an acceptance of some trigonometry, especially spherical.

[1] 萬年曆 [2] 愛薛

were to become universal in modern astronomy; thus came the invention of the equatorial mounting. They were unaffected by Ptolemaic geometric epicyclic planetary theory, continuing to prefer the non-committal 'Babylonian' algebraic formulations. How far Chinese calendrical computations were affected by Arabic methods, only further research will reveal.

(7) Celestial Globes

The idea of representing the constellations and stars on the surface of a material globe goes back rather far in Greek antiquity, and may just possibly have been derived from Babylonian practice.[a] The first scientifically designed celestial globe for which there is fair evidence seems to have been that of Eudoxus of Cnidus,[b] and the astronomical poem of Aratus[c] was almost certainly written (about −270) with a celestial globe in mind. A +3rd-century mosaic preserved at Trier shows him before such an instrument receiving inspiration from the muse Urania. Only one globe from approximately this period still exists; the famous Farnese marble now at Naples.[d] The tradition was continued by Ptolemy in the +2nd century, and it lasted on into the Middle Ages; for example, in the +7th century the Byzantine Leontius Mechanicus wrote a small treatise on the construction and use of celestial globes.[e] The globe as a symbol of the universe, or of the power of earthly rulers, was of course frequent in works of art and stamped on coins. From the time of Leontius for some centuries onwards the making of celestial globes passed into the hands of the Arabs,[f] and the return to the West is typified by the globe which an Arabic astronomer made for Frederick II of Sicily in the +13th century. This was some fifty years before the time when the observatories of Naṣīr al-Dīn al-Ṭūsī at Marāghah and Kuo Shou-Ching[1] at Peking were flourishing. A Marāghah globe of about +1300 is still preserved at Dresden (Fig. 174); its inscription says that it was made by Muḥammad ibn Mu'ayyad al-'Urḍī.[g] It has pivots both at the pole and the ecliptic pole.[h]

Although it has long been generally accepted that the classical Chinese technical term for a celestial globe was *hun hsiang*,[2] the history of the thing itself has never been properly clarified. Since the star-position measurements of Shih Shen and Kan Tê antedated those of Hipparchus by two centuries (they were working about the time of the death of Eudoxus), there is no reason why 'images of the spherical heavens' (*hun thien hsiang*[3]) in the form of solid globes should not have been made in the Chhin or

[a] See the monographs of Schlachter & Gisinger (1) and Stevenson (1).

[b] Younger than Mo Ti and older than Mencius: −409 to −356. But Sarton (1) accepts no celestial globe before the time of Hipparchus (−2nd century).

[c] The *Phaenomena*; cf. p. 201 above. See Böker (1). Translation by G. R. Mair.

[d] Stevenson (1), pp. 14ff. It shows only constellation symbols, not individual stars.

[e] There is not much evidence that any of these were mounted equatorially.

[f] A beautiful one of +1080 is preserved at Florence (Bonelli (1), no. 2712); it is the earliest known Arabic globe. On Arabic globes in general see Destombes (2, 3); Winter (6).

[g] The son, in fact, of the al'Urḍī whose precise description of the instrumental equipment at Marāghah has come down to us; cf. Jourdain (1); Seemann (1).

[h] Which explains the statement made in the *Yuan Shih* about the globe sent from Marāghah to Peking in +1267, that it would 'revolve obliquely'; see p. 374 above. This globe of al-'Urḍī's son is illustrated in Pope (1), vol. 6, pl. 1403; we reproduce the photograph of Stevenson.

[1] 郭守敬 [2] 渾象 [3] 渾天象

early Han. As we have seen,[a] this is the time from which date our oldest literary references to celestial sphere theory and to armillary rings. The problem really depends on the interpretation of certain technical terms.

Traditionally the celestial globe (*hun hsiang*[1]) has been contrasted with the observational armillary sphere (*hun i*[2]) equipped with its sighting-tube. But Needham, Wang & Price, in their study of the clockwork movements of Chinese astronomical instruments, became convinced that account must be taken of three, not two, types of apparatus. Besides the two just mentioned there was also the demonstrational armillary sphere,[b] with or without a model earth supported on a pin at the centre,[c] and with or without arrangements for rotating continuously the nest of rings. These workers reached the conclusion that in early texts, especially before the +4th century, the term *hun hsiang*[1] often, if not indeed always, means the demonstrational armillary sphere. The existence of such instruments with earth models is proved by the statement[d] of Wang Fan[3] about +250 explaining his plan of removing the model from the centre of the sphere and replacing it by a horizontal box-top outside to symbolise the earth. It would seem that the demonstrational armillary sphere half sunk in a box-like casing persisted at least until the time of I-Hsing and Liang Ling-Tsan in the early +8th century. On the other hand, the true solid celestial globe seems not to have originated before the time of Chhien Lo-Chih[4] (+435), and it is quite significant that this occurred only after Chhen Cho[5] had constructed about +310 a standard series of star-maps based on the *Hsing Ching* catalogue which had been begun in the −4th century. Thus the boundary between the terms *hun hsiang* and *hun i* is, at least in earlier ages, rather fluid, for either may be employed to refer to the demonstrational armillary sphere. Sometimes[e] a *hun i* is contrasted with a *hou i*[6] ('observational instrument'), showing that the former could also mean the demonstrational sphere. It is probably best to take as true celestial globes only those instruments of which it is said that they depicted the stars of all the constellations according to the colours of the three schools of astronomers.

These remarks may be illustrated by a couple of key quotations. The first is from the astronomical chapter of the *Sui Shu*,[f] written about +654 probably by Li Shun-Fêng:

> *Uranographic Models* (*Hun Thien Hsiang*[7]):
>
> The characteristic of uranographic models is that they have the round parts, but not the straight part (*yu chi erh wu hêng*[8]).[g]

[a] Pp. 216 and 343 above.

[b] In some cases the adjective 'computational' might be more appropriate; cf. p. 361 above.

[c] One would very much like to know whether this was made spherical or flat, square and platelike, but no clue to this has been given by any text which we have seen. Ancient cosmological statements (pp. 213, 217 above) could have authorised either form.

[d] *CSHK* (San Kuo section), ch. 72, p. 5*b*; quoted in full below.

[e] As in *Hsin I Hsiang Fa Yao*, ch. 1, p. 4*a*, tr. Needham, Wang & Price (1).

[f] Ch. 19, pp. 17*a*ff., tr. auct.

[g] The allusion here is of course to the *hsüan chi yü hêng* instrument referred to in the *Shu Ching*. We have identified it (following Michel, cf. pp. 336 ff. above) with the jade circumpolar constellation template and the jade sighting-tube. In the time of Li Shun-Fêng and Su Sung, who knew much less about it than modern archaeologists think they do now, it was interpreted as some kind of armillary sphere, and the early Jesuits followed this tradition.

1 渾象 2 渾儀 3 王蕃 4 錢樂之 5 陳卓 6 候儀
7 渾天象 8 有機而無衡

At the end of the Liang dynasty (*c.* +550) there was one in the Secret Treasury (Pi Fu[1]).[a] It was made of wood, as round as a ball, several arm-spans in circumference, and pivoted on the south and north poles, while round the body of it were shown the twenty-eight *hsiu*, as also the stars of (each of) the Three Masters,[b] the ecliptic, the equator, the milky way, etc. There was also an external horizontal circle surrounding it, at a height which could be adjusted,[c] to represent the earth. The southern extremity of the polar axis penetrated below this to represent the south (celestial) pole, while the northern extremity rose above it to represent the north (celestial) pole. When the globe rotated from east to west, the stars which made their meridian transits morning and evening corresponded exactly (*ying*[2])[d] with their degrees, and the equinoctial and solstitial points, as well as the fortnightly periods, all also checked—there was absolutely no difference from the heavens.

This was not at all like the (observational) armillary sphere (*hun i*[3]) which must have a sighting-tube for measuring and computing the motions of the sun and moon and the positions of the stars in degrees.

The Astronomer-Royal of the Wu Kingdom (+222 to +277), Chhen Miao,[4] said:[e] 'The worthies of old made an instrument of wood which was called *hun thien*[5] ("celestial sphere").' Was he not referring to this (i.e. a solid celestial globe such as that used at the end of the Liang dynasty)? According to these words, the (armillary) sphere (*i*[6]) and the (celestial) globe (*hsiang*[7]) were two different things, only distantly related to each other. Thus what Chang Hêng made (about +125) was a celestial globe (*hun hsiang*[8]) with the (models of the) seven luminaries (attached to it).[f] But Ho Chhêng-Thien[9] could not distinguish the difference between the (armillary) sphere (*i*[6]) and the (celestial) globe (*hsiang*[7]); therefore he went astray.[g]

In the (Liu) Sung dynasty, in the 13th year of the Yuan-Chia reign-period (+436), the emperor Wên Ti ordered the Bureau of Astronomy to construct new armillary instruments (*hun i*[3]). The Astronomer-Royal Chhien Lo-Chih, following the old theories, explained them by means of a uranographic model apparatus (*i hsiang*[10]), (i.e. a demonstrational armillary sphere)[h] made of bronze, taking 0·5 inch to the degree, with a diameter of 6·0825 ft. and a

[a] Perhaps simply another name for the Imperial Observatory.

[b] Shih Shen, Kan Tê, and Wu Hsien; cf. pp. 197, 263 above.

[c] This must surely mean that arrangements were included for simulating different degrees of geographical latitude by varying polar altitude.

[d] Note the philosophical significance of this term, 'response by resonance'; cf. Vol. 2, p. 304. Li Shun-Fêng himself wrote a *Kan Ying Ching*.[11] What is left of it shows that it was a treatise on unexplained natural phenomena and action at a distance.

[e] This is almost his only appearance on the stage of Chinese astronomical history.

[f] The writer, perhaps Li Shun-Fêng, was thus inclined to attribute to Chang Hêng a solid celestial globe. But the evidence indicates rather that it was a demonstrational armillary sphere. For we are never told that all the constellations were marked on it, and if it had been a globe he could have used a string instead of a flexible ruler to mark out the *hsiu*-extensions on the ecliptic.

[g] This statement is particularly interesting. Ho Chhêng-Thien (+370 to +447) was of the Liu Sung dynasty, and an old man when Chhien Lo-Chih's instruments were being made. If this was indeed the time when the solid celestial globe began to supplant the demonstrational armillary sphere, this confusion would explain itself, for evidently the terminology did not change when the thing changed. The fragments of Ho Chhêng-Thien's *Hun Thien Hsiang Shuo*[12] in *CSHK* (Sung section), ch. 24, p. 2*b*, throw no further light on the matter.

[h] This phrase could equally well be translated (as the sequel will show) 'a (demonstrational armillary) sphere and a (solid celestial) globe'.

1 祕府 2 應 3 渾儀 4 陳苗 5 渾天 6 儀
7 象 8 渾象 9 何承天 10 儀象 11 感應經
12 渾天象說

circumference of 18·2625 ft.[a] The earth was (shown) fixed immovably within the sphere of the heavens. Besides a meridian circle through the south and north poles, there were two rings, one for the ecliptic and one for the equator, on which were marked the divisions of the 28 *hsiu*; and the positions of the Great Bear and the Pole Star (were also shown). The sun, moon, and five planets were indicated on the ecliptic.[b] There was an axle (in the polar axis) permitting rotation like the real heavens, and so exhibiting 'the (determination of times and seasons by) dusks and dawns as also the culminations of stars'.[c] Everything corresponded with the heavens like (the two halves of) a tally.

Towards the end of the Liang dynasty (+557) this instrument was placed in front of the Wên Tê Hall (of the Imperial Palace). As for its system of construction, you might regard it as an armillary sphere (*hun i*[1]) but it had no sighting-tube inside it; or you might regard it as a celestial globe (*hun hsiang*[2]) but it had no earth horizon outside it. Thus it combined the two systems to make a third type of instrument.[d] Accordingly if you studied its function, it was more or less like a celestial globe, but it did not distort the pattern of the universe for it represented the earth occupying its proper position within the concavity of the heavens.

In the San Kuo (Three Kingdoms) period (+222 to +280), in the State of Wu, there was also Ko Hêng[3] (*fl.* +250),[e] who was a perfect master of astronomical learning and capable of making ingenious apparatus. He altered the astronomical instrument (*hun thien*[4]) in such a way as to show the earth fixed at the centre of the heavens, and these were made to move round by a mechanism (*i chi tung chih*[5]) while the earth remained stationary.[f] (This demonstrated) the correspondence (*ying*[6]) of the (shadows on the) graduated sundial with the motions of the heavens. This it was that Chhien Lo-Chih imitated.

In the seventeenth year of the Yuan-Chia reign-period (+440), he (Chhien) also made a small astronomical instrument (*hun thien*[4]) taking 0·2 inch to the degree, with a diameter of 2·2 ft. and a circumference of 6·6 ft. The 28 *hsiu*, and all the constellations both north and south of the equator, were indicated by pearls of three colours, white, green and yellow, according to the three schools of astronomers. The sun, moon, and five planets were attached to the ecliptic, and the rotation of the heavens demonstrated with the earth (-horizon) across the middle.

This sphere and globe, made in the Yuan-Chia reign-period, were both transported to Chhang-an (modern Sian) in the ninth year of the Khai-Huang reign-period (+589) (of the Sui dynasty) after the conquest of the Chhen.[g] Then, at the beginning of the Ta-Yeh reign-period (+605) they were moved to the astronomical observatory (*kuan hsiang tien*[7]) at the eastern capital (i.e. Loyang).

[a] The term *shao*[8] used here, one of the standard astronomers' terms for fractions, allows the easy expression of four decimal places (cf. p. 268 above). In view of the advanced knowledge of the value of π at this time (cf. p. 101 above), it is curious that here it should be taken as 3.

[b] Presumably in the form of models the positions of which were adjustable thereon.

[c] The quotation comes from the Han apocryphal treatise *Shang Shu Wei Khao Ling Yao*, in *Ku Wei Shu*, ch. 2, p. 2*a*, cf. p. 361 above. From the parallel passage in *Sung Shu*, ch. 23, pp. 8*b* ff., we know that Chhien Lo-Chih's sphere was driven by water-power.

[d] Li Shun-Fêng speaks as if the demonstrational armillary sphere was the youngest or derivative type of instrument, but the historical evidence is clear that the solid celestial globe occupies this position.

[e] We know almost nothing about this astronomer, but are prompted to wonder whether he could have been a relative of Ko Hsüan,[9] a key figure in the development of the Taoist Church (cf. Vol. 2, p. 157; *fl.* +238 to +255), and his great-nephew, Ko Hung[10] the famous alchemist (+280 to +360).

[f] A parallel passage from another book, quoted in *TPYL*, ch. 2, p. 10*a*, attests this.

[g] The Chhen dynasty had inherited them from the Liang thirty years earlier at Nanking.

1 渾儀 2 渾象 3 葛衡 4 渾天 5 以機動之 6 應
7 觀象殿 8 少 9 葛玄 10 葛洪

Before drawing our conclusions from this very informative passage we had better put beside it the words of Wang Fan (*c.* +250),[a] and the best setting for them is the introduction written by Su Sung in +1090 for his second chapter,[b] that on the celestial globe. Su Sung says:

The celestial globe (*hun hsiang*[1]) has not traditionally been kept in the Bureau of Astronomy and Calendar. Ours is based on the description in the (astronomical chapter of the) *Sui Shu*, with some modifications.... According to this memoir the celestial globe has the round part (of the instrument mentioned in the *Historical Classic*) but not the straight part. At the end of the Liang dynasty there was one in the Secret Treasury, made of wood, as round as a ball, and several arm-spans in circumference....

What we have now made is the same apparatus, simplifying the methods of Liang Ling-Tsan and Chang Ssu-Hsün regarding the (automatic rotation of the) sun, moon, and five planets,[c] which are tied around the 265 degrees (of the ecliptic) and rotate following the celestial movements.

Wang Fan (in his *Discourse on Uranographic Models*) said: 'In the (demonstrational armillary) sphere method (*hun hsiang*[1]) the earth should be placed at the centre of the heavens. But as this is not very convenient, we may look at the matter in the opposite sense, and simply represent the earth by a (flat-topped) box outside. The perspicacious will see that there is no real difference—it depends on the point of view. Although the appearance may look unusual and strange, it is in perfect agreement with the principles (*li*[2]) involved. This is indeed an ingenious thing.'[d] Thus the plan of having the earth sphere (*ti hun*[3]) represented (as a flat horizon) outside the celestial globe (or demonstrational armillary sphere) originated with the technique of Wang Fan.

This brings us to the following picture. When the first fully developed armillary spheres were made by Chang Hêng[4] about +125 he made them in two forms (for reasons we now know), one for observation, the other for demonstration and computation. It does not seem likely that he placed an earth model at the centre. Nor did Lu Chi[5] in +225.[e] This appears to have been the contribution of Ko Hêng in the middle of this century, and almost immediately afterwards Wang Fan adopted the alternative of sinking the sphere in a box the top of which represented the earthly horizon. Liu Chih about +274 followed Ko Hêng.[f] The demonstrational armillary sphere then continued in use through the centuries.[g] We must glance in a moment at its last representative.

[a] His work was also called *Hun Thien Hsiang Shuo*[6] (Discourse on Uranographic Models). It was preserved in part in three of the dynastic histories (the *Chin Shu*, *Sung Shu* and *Sui Shu*) as also in *TPYL* and *KYCC*, whence it was reconstituted as fully as possible in *CSHK* (San Kuo section), ch. 72, pp. 1*a* ff.

[b] *Hsin I Hsiang Fa Yao*, ch. 2, pp. 2*a*, *b*; tr. auct.

[c] On this subject, see Sect. 27*h* below.

[d] See *CSHK* (San Kuo section), ch. 72, p. 5*b*. Cf. p. 218 above.

[e] See *Chin Shu*, ch. 11, p. 7*a*; *Sung Shu*, ch. 23, p. 5*b*; *Sui Shu*, ch. 19, p. 15*b*; and the previous reference, p. 5*a*.

[f] For references and characters see Table 31. Liu Chih's earth model is particularly interesting because it was square, and therefore must have been flat.

[g] As by Thao Hung-Ching (with earth model) in +520, Kêng Hsün in +590, I-Hsing (sunk in a box) in +725, and Chang Ssu-Hsün in +979. References and characters will be found in Table 31. It

[1] 渾象 [2] 理 [3] 地渾 [4] 張衡 [5] 陸績 [6] 渾天象說

Meanwhile, at the beginning of the +5th century the true solid celestial globe had started upon its career. We have seen that Chhien Lo-Chih made at least two[a] instruments, one of the traditional armillary type with earth model, and one true celestial globe. As already suggested, the limiting factor was probably the preparation of standard star-maps based on a conflated *Hsing Ching*. Chang Hêng[b] had made something of this kind, but the manuscript was soon lost, and Chhen Cho, who had been Astronomer-Royal of the Wu State like Chhen Miao, compiled the first standard maps in +310. We are expressly told[c] that Chhien Lo-Chih's globe or globes were based on these.[d] After this time there is a long interval, and the next great celestial globe of which we can be sure is that placed in his astronomical clock-tower by Su Sung[1] in +1090 (Fig. 175). There then followed the globe of Kuo Shou-Ching (+1276) already mentioned,[e] and finally the tradition was inherited by the Jesuits, whose clockwork-driven bronze instrument beautifully inscribed (Fig. 176) can still be seen on the observatory terrace of the city wall at Peking.[f]

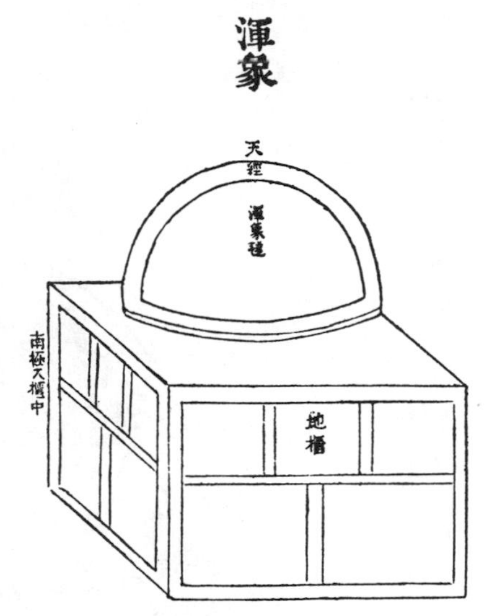

Fig. 175. Su Sung's celestial globe, sunk in a casing and rotated by a water-driven mechanical clock; from the *Hsin I Hsiang Fa Yao* of +1092. It was placed in a chamber on the first floor of the clock-tower at Khaifêng.

The celestial globe about which we have the fullest information is certainly that commissioned in +1086 and completed by Su Sung and his collaborators in +1090, for the whole of the second chapter of his *Hsin I Hsiang Fa Yao* is devoted to it. Attached to his star-maps[g] is an interesting series of diagrams showing the precessional changes in the positions of the equinoctial and solstitial points since the times

has been suggested to us by Dr Lu Gwei-Djen that the fretted balls within balls which were so favourite a product of the Chinese ivory-carvers (cf. Holtzapfel & Holtzapfel (1), vol. 4, p. 426) were perhaps a popularised and fanciful reminiscence of demonstrational armillary spheres. Equally related to armillary spheres, and much more important, may be the gimbals or 'Cardan suspension', the invention of which in China (as we shall see in Section 27*c* below) took place remarkably early.

[a] We say at least two because it will be seen from the foregoing texts that the dimensions given in the *Sui Shu* for the globe of the Liang and the small one made by Chhien Lo-Chih do not correspond. He probably made more than one globe.

[b] Judging from the extant titles of some of his lost books.

[c] *Sui Shu*, ch. 19, pp. 2*a*, *b*.

[d] On star-maps see above, pp. 276 ff. The latent period between Chhen Cho and Chhien Lo-Chih may seem rather long, but historical circumstances go some way to explaining it. When the Chin dynasty was defeated in +317 by the northern barbarians, the instruments of Chang Hêng, Lu Chi and Wang Fan were all left behind at Chhang-an. It was not until +418 that the old capital was recaptured by the first emperor of the Liu Sung dynasty. The astronomical equipment then recovered must have acted as a strong stimulus on Chhien Lo-Chih and his colleagues. See *Sung Chu*, ch. 23, pp. 8*b* ff., tr. Needham, Wang & Price (1).

[e] P. 367 above.

[f] Cf. p. 452 below.

[g] On these see p. 278 above.

[1] 蘇頌

Fig. 176. Celestial globe made for the Peking Observatory in 1673 by Ferdinand Verbiest (from the *Huang Chhao Li Chhi Thu Shih*).

of the earliest records accessible to him, in connection with the varying numbers of quarter-hours of day and night as sounded by his clock.[a] It is very regrettable that none of the earlier medieval Chinese celestial globes has survived.[b] To sum the matter up, one may say that as compared with European developments the Chinese were in some respects tardy, in others advanced. For although the early +5th century was rather late for the first appearance of the solid celestial globe, the practice of setting a model earth at the centre of a demonstrational armillary sphere from at least the middle of the +3rd century was extremely enlightened. Europeans did not come to this until the end of the +15th century.[c] Fig. 178 shows such a sphere made at Rome by Carolus Platus and dated +1588. Fig. 177 shows a Jesuit orrery on the same plan from 18th-century China.

To conclude this chapter let us examine one of the last representatives of the demonstrational armillary sphere tradition. This exceedingly interesting piece of mechanism, though of workmanship probably as late as the 18th century, incorporates many features of the Asian clockwork traditions of two millennia. It is a demonstrational armillary constructed in Korea.[d] Its mechanical drive includes a time-telling and striking device to complete its clock function. The sphere itself (Fig. 179)[e] is 1 ft. 4 in. in diameter, and the diameter of the earth model within it is about 3½ in. The position of the sun is given on the band representing the ecliptic, which is marked with the 24 divisions of the year. The motion of the moon is indicated on the ring representing its orbit (i.e. its apparent path), which is divided by pegs to mark the 28 *hsiu* (lunar mansions). And now at every step we find echoes of the ancient traditions, for the model earth itself stems from Ko Hêng and Chhien Lo-Chih, while the presence of a horizon circle recalls Wang Fan and his external box. Moreover, the drive is transmitted through a rotating polar axis, just as in all the powered instruments from Chang Hêng to Kuo Shou-Ching, and the middle nest of rings is made to rotate exactly as in the great sphere of Su Sung. The special ring for the path of the moon recalls the designs of Li Shun-Fêng (early +7th century). The earth model is marked with the chief continents, resembling that terrestrial globe brought to Peking in +1267 by Jamāl al-Dīn.[f] Lastly, within the casing of the mechanism, which has two weights and is regulated by a pendulum with a simple escapement, there are still further devices of classical Chinese and Arabic type (balls periodically released,[g] norias to raise them up again, and time-telling discs appearing like jacks at a window) which we shall better appreciate in the engineering Section (27*h*). Although we do not know the

[a] What seems to have been a dissectable globe or hemisphere was invented in +1137 (*Yü Hai*, ch. 1, p. 35*a*, *b*); cf. p. 582 below and Sect. 27*b*.

[b] We have no concave representations either. The most remarkable monument surviving of star-maps painted in the concave surface of a cupola is constituted by the frescoes of the Umayyad desert palace of Quṣair 'Amrah (cf. Hitti (1), p. 271), built about +714. Their astronomy has been studied by Beer (1). The cupola was, of course, rather rare as an architectural form in China, but it is not impossible that ceiling star-maps may be found in one or other of the cave-temples which Chinese archaeological discoveries are now adding to the well-known glories of Tunhuang.

[c] See Price (3).

[d] It was preserved in private hands until at least 1936. Whether it has survived the holocaust of the 'fifties we do not know.

[e] From Rufus (2), p. 38; reproduced and discussed also in Rufus & Lee (1).

[f] See p. 374.

[g] Cf. Sect. 7*h*, in Vol. 1, pp. 203 ff. and Fig. 33.

exact date of this astronomical clock, it seems to derive from a long antecedent Korean tradition, for a demonstrational armillary sphere moved by water-power was successfully constructed for King Hyojong (Hsiao Tsung) in +1657 by Choi Yoo-Chi (Tshui Yu-Chih[1]). The history of science in the Chinese culture-area will always remain incomplete so long as thorough studies of the scientific and technological achievements of the neighbouring peoples have not been made.[a]

(*h*) CALENDRICAL AND PLANETARY ASTRONOMY

Although there is a very large literature, still growing almost daily, on the Chinese calendar, its interest is, we suggest, much more archaeological and historical than scientific. A calendar is only a method of combining days into periods suitable for civil life and religious or cultural observances. Some of its elements are based on those astronomical cycles which have obvious importance for man, such as the day, the month and the year; others are artificial, such as the week and the subdivisions of the day. The complexity of calendars is due simply to the incommensurability of the fundamental periods on which they are based. 'The supply of light by the two great luminaries', writes Fotheringham (1), in one of the best general accounts of the subject,[b] 'is governed by the periods known to astronomers as the solar year and the synodic month, while the return of the seasons depends on the tropical year. The length of the synodic month at the present time is 29·5305879 days, while that of the tropical year is 365·24219 days.' Calendars based on the former figure, depending only on lunations, make the seasons unpredictable, while calendars based on the latter cannot predict full moons, the importance of which in ages before the introduction of artificial illuminants was considerable.[c] The whole history of calendar-making, therefore, is that of successive attempts to reconcile the irreconcilable, and the numberless systems of intercalated months (*jun yüeh*[2]),[d] and the like, are thus of minor scientific interest. The treatment here will therefore be deliberately brief. As the *Chin Shu* says:[e]

The movements of the three luminaries are not bound to follow unvarying rules (of relation and proportion) such as are sought for by the technicians (of calendar-making) with their

[a] See Addendum below, p. 683, on the role of Korean culture in the history of science.

[b] Other general discussions of the calendar worth consulting are, apart from the erudite and exhaustive work of Ginzel (1), Ionides & Ionides (1) and Philip (1).

[c] The ancient peoples were also much impressed by the obvious influences which the moon seemed to have on living things from women to sea-urchins (cf. Vol. 1, p. 150), and which before long it was found to have on the tides. A discussion on the relation of lunar calendars to matriarchal customs in ancient China will be found in the papers of Hirschberg (1); Pancritius (1); Koppers (1); and Shirokogorov (2). There is also a curious connection between sexology and calendrical science; thus Jung Chhêng, one of the famous experts on sexual matters in ancient times (cf. Vol. 2, p. 148), was also handed down in some sources as the originator of the calendar (see Kaltenmark (2), p. 56).

[d] The presence of the radical for 'king' in this character well indicates the arbitrary social element in calendar-making. It is indeed 'the king standing in a gateway' and Soothill (5), p. 62, makes the suggestion that it refers to the emperor's station on the threshold between one room of the Ming Thang and another when an intercalary month intervened in the normal cycle of his perambulation.

[e] Ch. 11, p. 6*a*, tr. auct., adjuv. Ho Ping-Yü. The passage is a quotation from Wang Fan's *Hun Thien Hsiang Shuo* (see *CSHK*, San Kuo sect., ch. 72, p. 3*a*), and therefore dates from about +260.

[1] 崔攸之 [2] 閏月

PLATE LX

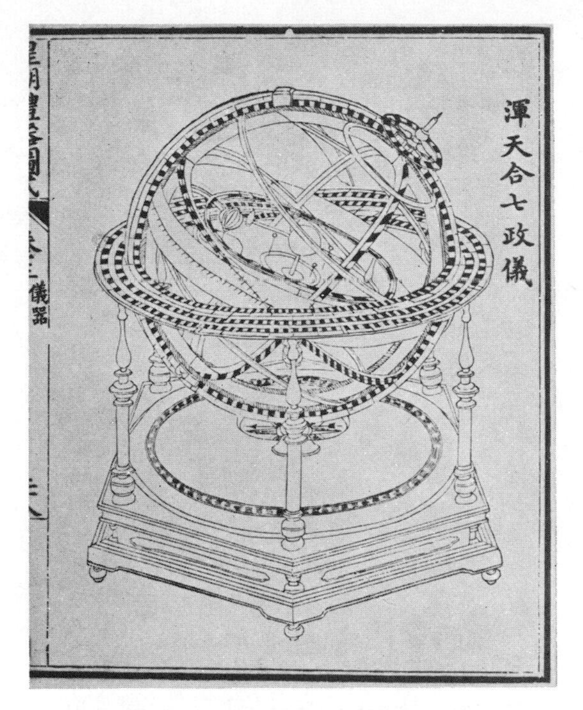

Fig. 177. Jesuit instruments of the 18th century; an orrery (demonstrational armillary sphere) from the *Huang Chhao Li Chhi Thu Shih* (+1759).

Fig. 178. Demonstrational armillary sphere made by Carolus Platus, +1588 (Price, 3).

Fig. 179. Clockwork-driven demonstrational armillary sphere made in Korea in the 18th century, but incorporating numerous special features of the Chinese and Arabic traditions of instrument-making (see text). Apparatus of this kind was made for the court of Seoul in +1657 by Choi Yoo-Chi (Tshui Yu-Chih).

calculations. The calendars of different schools (of astronomers) thus show differences and similarities. When their methods are compared, divergences are (naturally) found among them.

But real scientific interest attaches to the accuracy of the estimates of the periods of revolution of the planets which were formed in ancient and medieval times.

Perhaps the most important source for all Chinese calendrical questions is the *Ku Chin Lü Li Khao* of Hsing Yün-Lu (*c.* +1600). The older European disquisitions on the Chinese calendar,[a] though not useless or wholly erroneous, have been superseded as regards their historical conclusions by the new investigations of the Shang oracle-bones reported by Tung Tso-Pin (*1*) and others. Useful descriptions of more recent date are those of de Saussure (16*d*) and Hartner (4), but the most exhaustive discussions are in Japanese.[b] A comprehensive work in Chinese on the successive approximations represented by the hundred-odd official calendars[c] used in the various Chinese dynasties is that of Chu Wên-Hsin (*1*).[d] Lastly we may mention the treatises of Li Jui (*1*) on seven ancient calendars; work which still retains its value today.

Some estimates were given above[e] of the accuracy to which the Chinese ascertained the 'year-fraction' (*sui yü*[1]) of the tropical and sidereal year (*sui shih*[2]) respectively.

[a] E.g. Ideler (1); Fritsche (1); Kingsmill (2); Kühnert (1, 2). The early Fellows of the Royal Society were interested in the Chinese calendar. The British Museum possesses a copy of a Chinese ephemeris (Birch MSS. 4394, fol. 26) on which are written the words 'Lent me by Mr Robert Boyle, Oct. 29th, 1671'. Thus it was probably studied by Robert Hooke, whose paper on China appearing some years later in the *Phil. Trans.* includes an attempt to copy some of the characters, and a short discussion on the measurement of time by the Chinese. We headed our first volume with an excerpt from this interesting work, which was reproduced in abridged form in the *Miscellanea Curiosa* for 1707 (see Anon. 33). Another calendar, for the year +1408, is in the Pepysian Library at Magdalene College, Cambridge; it was catalogued wrongly by Pepys himself as a manuscript.

[b] For example there are monographs and papers by Hashimoto (*2*); Yabuuchi (*8*); Shinjō (*2, 4, 5*) and Iijima (*7*).

[c] The names and details of these will be given in the glossaries forming part of the concluding volume of the present work.

[d] Among special studies of different ages may be mentioned the work of Chavannes (13), Liu Chao-Yang (*5, 6, 7,* 2) and Iijima (*6*) on the Shang and the Chou. Eberhard (14) believes that it is possible to trace several local calendars in very ancient times and that it was this which gave rise to the idea that there had been official calendars of the Hsia, Shang and Chou 'dynasties'. Jao Tsung-I (*1*) has described an interesting calendrical diagram, with text, inscribed on silk and of Chhu State provenance, dating from about the −4th century. Han calendars have been studied by Shinjō (*6*) and by Eberhard & Henseling, while the reforms of −104 and +85 have been examined by Yabuuchi (*7*) and Yü Phing-Po (*1*). For the Northern Chhi we have Yen Tun-Chieh (*12*); for the Sui and Thang Yabuuchi (*1, 9*) and Tung Tso-Pin (*2*). On the entry at that time of Buddhist elements see Eberhard (12, 15); and for Manichaean influence Chavannes & Pelliot; Wylie (15) and Yeh Tê-Lu (*1*). For the Sung calendars see Yabuuchi (*14*), and for those of the Yuan and Ming Yabuuchi (*15*). That of the Jurchen Chin has been investigated by Yen Tun-Chieh (*11*). Chhien Pao-Tsung (*1*) has written an interesting chapter (ch. 8) on the relation of mathematics to calendrical science in medieval China.

Japanese calendrical science, on which see, for instance, Kanda (*3*) and Yabuuchi (*16*), derived, like that of Korea (p. 683), from China. Before +604 the traditional Hi-oki[3] calendar had been purely lunar, but in this year the Yuan Chia Li[4] (Genka-reki) calendar due to Ho Chhêng-Thien (+443) was introduced by a Korean monk Chhüan-Lê[5] (Kwanroku), also called Sêng-Tu[6] (Sōzu), and officially adopted. There followed a succession of calendars until +861 when the Hsüan Ming Li[7] (*Semmei-reki*) was inaugurated, to last right down to +1684. This had been prepared by Hsü Mao in +822, but was in force only seventy-one years in its own country. Copies of the *Semmei-reki* are still not infrequently seen.

[e] Table 30, p. 294.

[1] 歲餘 [2] 歲實 [3] 日置 [4] 元嘉曆 [5] 觀勒 [6] 僧都 [7] 宣明曆

From the +5th century onwards both were rather precisely known. Maspero has collected[a] similar data (Table 32) for the length of the lunation (*shuo shih*[1]). Even in the Yuan period (+13th century) the approximation was not better than this, and one of the reasons for the adoption of the calendar of the Jesuits was the greater accuracy of their evaluations of the length of years and lunations. However, as no calendar can make the two periods fit, the scientific value of the change was not great. The extra fraction of a day was termed the *hsiao yü*[2] ('little excess') and was placed as numerator over the number of divisions in the day (*jih fa*[3]), giving a fraction equivalent to the decimals shown.

Table 32. *Values for the length of the lunation*

	Lunation (days)
True value	29·530588
Deduced from the oracle-bones (−13th century)[b]	29·53
Yang Wei[4] (Ching Chhu cal. +237)[c]	29·530598
Ho Chhêng-Thien[5] (Yuan Chia cal. +443)[d]	29·530585
Tsu Chhung-Chih[6] (Ta Ming cal. +463)[e]	29·530591

(1) Motions of the Sun, Moon and Planets

Indigenous Chinese astronomy never attempted an advanced geometrical analysis of the moon's motions. One of the earliest literary mentions of them is in the Chou Yüeh[7] chapter (The Revolutions of the Moon) in the *I Chou Shu*[8] (Lost Books of Chou), dating perhaps from some time before −300. *Huai Nan Tzu* gives[f] an account of them, stating that the daily eastward motion is 13°, a figure long afterwards accepted. Shih Shen knew that the rate of motion varied, and that the moon diverged from the ecliptic north and south.[g] Periods of faster speed were called *thiao*,[9] and those of slower motion *tshê ni*.[10] The first mention of the 'Nine Roads of the Moon' (*chiu hsing*[11]) is in the *Hung Fan Wu Hsing Chuan* of Liu Hsiang about −10, where a detailed description is given.[h] The roads were traditionally assigned colours (like

[a] (4), pp. 238 ff.
[b] Tung Tso-Pin (*1*).
[c] *Sung Shu*, ch. 12, pp. 10*b*, 14*a*; *Chin Shu*, ch. 18, pp. 2*a*, 3*b*.
[d] *Sung Shu*, ch. 13, pp. 1*b*, 3*a*.
[e] *Sung Shu*, ch. 13, pp. 26*b*, 27*b*.
[f] Ch. 3, pp. 7*b*, 8*a*.
[g] *Khai-Yuan Chan Ching*, ch. 11, p. 3*a*.
[h] Cf. *Chhou Jen Chuan*, ch. 2, p. 16. We noticed above (p. 276) Kêng Shou-Chhang's description of the moon's path in −52.

1 朔實 2 小餘 3 日法 4 楊偉 5 何承天 6 祖沖之
7 周月 8 逸周書 9 朓 10 側匿 11 九行

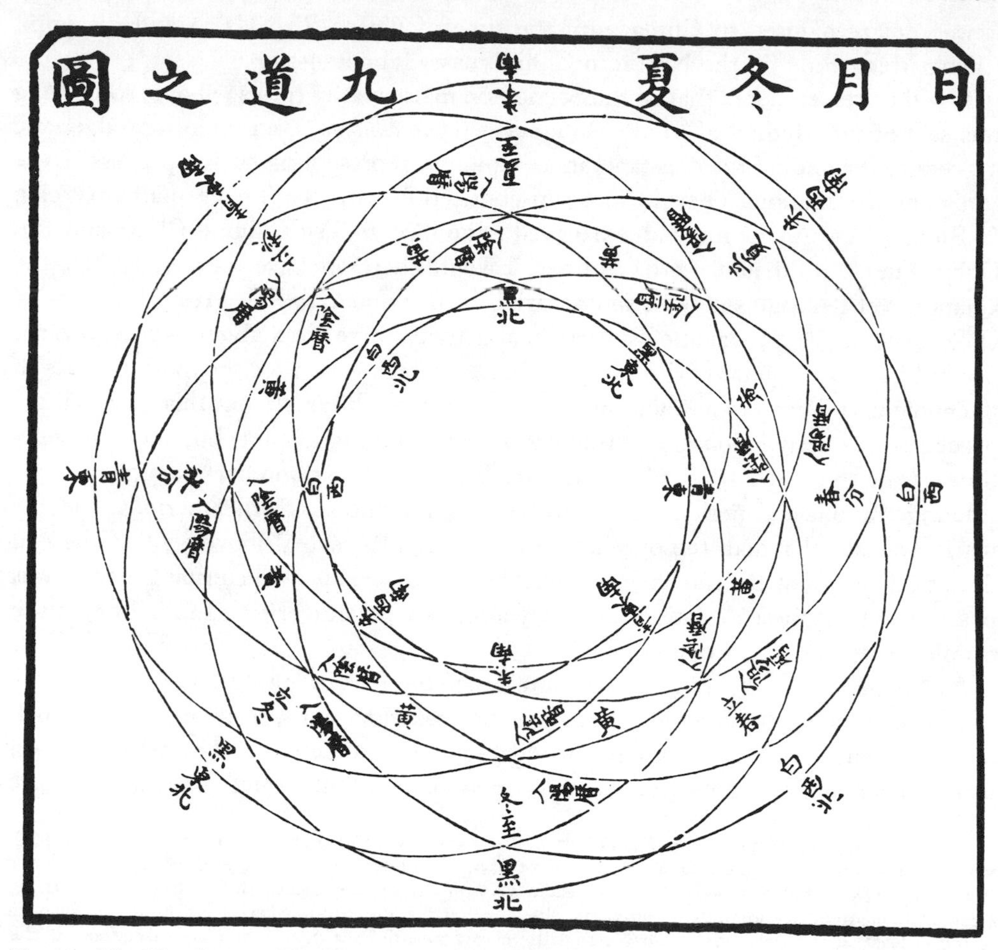

Fig. 180. The Nine Roads of the Moon, a late Chhing representation. The diagram shows the progressive forward motion of the apsidal line (the major axis of the moon's orbit), eight different positions of apogee (represented by the outermost bulges on the diagram) being passed through in 8–9 years (actually 3232·575 days). The 'road' should of course be drawn as one single interweaving line, not as nine separate lines, but that was the old Han tradition.

the equatorial red and the ecliptic yellow), green, white, red and black.[a] An examination of this Han theory will be found in Chhien Pao-Tsung (*4*); Eberhard & Müller (2) have discussed its simplification in the +3rd century. The roads are seen in Fig. 180; a late Chhing diagram given by Medhurst.[b]

The inequality of the seasons was a remarkably late discovery in China. Though the fact that the equinoxes are not placed at exactly equal intervals between the solstices may have been implicit in data available to the Han astronomers, and had already led Hipparchus in the −2nd century to his eccentric circle theory of the sun's motion,[c]

[a] This is still reproduced in late books such as the Ming *Li Hai Chi*, p. 5*a*.
[b] Cf. Barlow & Bryan (1), p. 163; Thomas (1), p. 314.
[c] Berry (1), p. 43.

it was not recognised in China until the time of Chang Tzu-Hsin[1] and his pupil Chang Mêng-Pin[2] (in the Northern Chhi dynasty, about +570).[a]

It will be remembered that in the Section on mathematics (p. 123 above) something was said of the Method of Finite Differences (*chao chha fa*[3]), an important algebraic procedure for finding the constants in an equation representing such a process as the sun's angular motion. There it arose in connection with the Thang mathematician Li Shun-Fêng. As the method is believed to go back to Tsu Chhung-Chih[4] and Liu Chhuo[5] in the +5th and +6th centuries, it would therefore have been available to the Changs. Yet its origins may be much earlier in Babylonia, to judge from the work of O. Neugebauer.[b] Systematic observational activity there from about −700 onwards provided the astronomers with fairly accurate average values for the main periods of phenomena (motions of sun and moon; planetary revolutions). On this basis simple processes of interpolation and extrapolation could be carried out, but then, at some time during the −4th or −3rd century, some ingenious astronomer[c] thought of considering complicated periodic phenomena as summations of other periods, each of which was simpler than the one in question. Thus the actual movement of the sun deviates from what the movement would be if the velocity was constant, and it was possible to treat these deviations as if they also were periodic, rising and falling, in the simplest case in a linear manner.[d]

As Neugebauer says, we have here in essence the idea of 'perturbations', which is so fundamental in all phases of celestial mechanics, whence it spread into every branch of exact science. But the treatment was algebraic, not geometrical. 'The determination of characteristic constants (for example, period, amplitude, and phase in periodic

[a] Gaubil (2), p. 59; Chhen Tsun-Kuei (5), p. 117. See also *MCPT*, ch. 7, para. 11; Hu Tao-Ching (*1*), vol. 1, p. 294; and *Chhou Jen Chuan*, ch. 11. Biot (1), p. 71, thought that the work of these two men was the origin of what the *Sūrya Siddhānta* in India says on the same subject. But in view of the strong Hellenistic element in Indian astronomy, and the fact that the core of the *Sūrya Siddhānta* goes back to +400 at least, it is much more probable that there was an influence in the opposite direction—if the discovery was not an independent one. Gaubil noted that the brief account of Chang Tzu-Hsin in *Pei Chhi Shu*, ch. 49, p. 9*b* (cf. *Pei Shih*, ch. 89, p. 20*b*) says nothing of foreign influence, but it tells us very little about him anyway save that he lived much in mountain retirement. By the +6th century good armillary spheres were available in China, and Chang and his pupils are said to have made many observations (*Sui Shu*, ch. 17, p. 4*a*; cf. ch. 19, pp. 27*a*, *b*, *Hsin Thang Shu*, ch. 27B, p. 1*a*) with 'round instruments'. Perhaps therefore what should be inferred from all this is that they determined the times of equinoxes much more accurately than was possible with classical gnomon shadow measurements only. Though the stimulus may well have come from abroad, the new knowledge was not as fundamental for the Chinese astronomical system as it had been for the geometrical theory of the Greeks. Yet the time of the statement of the inequality of the seasons can be important in the study of calendrical questions in China. De Saussure (33) was thus able to clear up a difficulty of Pelliot's arising from a calendar on a Han tablet of −63.

[b] (1), pp. 28ff.; (8); (9), pp. 98ff. Full publication in (10). The most important work which revealed this was that of Kugler (1, 2).

[c] It may have been Naburiannu (Schnabel). On him see Neugebauer (9), p. 130.

[d] Tables of these 'linear zigzag functions' are found on cuneiform ephemerides running from about −250 to about −50 (the time of Lohsia Hung and Ssuma Chhien). See Neugebauer (9), pp. 110ff. The Seleucid Babylonians could by these means predict lunar eclipses (and their magnitudes) quite satisfactorily, but not of course solar eclipses, for that would have needed an insight into the actual dimensions of the solar system. At the same time they were in a position to know when these eclipses were possible, and other times when they could be quite excluded.

1 張子信 2 張孟賓 3 招差法 4 祖沖之 5 劉焯

motions) not only requires highly developed methods of computation but inevitably leads to the problem of solving systems of equations corresponding to the external conditions imposed by the observational data.[a] Moreover, the Seleucid 'Chaldeans' constructed planetary ephemerides upon the same principles as those for the moon and the sun. Their results were just about as good as those obtained in the Ptolemaic planetary system, and yet they involved no interpretations in terms of circular motions or any other mechanical models. These Babylonian algebraic concepts undoubtedly influenced Diophantus (late +3rd century),[b] but not so much the Greeks of the geometrical schools, to which Hipparchus had belonged.[c] It seems probable that they made their way, through channels still exceedingly obscure,[d] to East Asia, where they would have helped to fix the predominantly algebraic character of Chinese mathematics, and even stimulated specifically the algebraic methods in astronomy which come to light there in the +5th century. Analogous radiation carried the Babylonian methods to India.[e]

Parallels between Chinese and Babylonian ephemerides for lunar and solar motion have indeed long been suspected. Thus Sédillot,[f] who knew of the 'Chaldean' methods through Greek sources such as Geminus of Rhodes (*fl.* −70),[g] recognised 'linear zigzag functions' in the ephemerides contained in some of the dynastic histories[h] and accessible to him in the transcriptions of Gaubil.[i] Kuo Shou-Ching in his short historical introduction to the subject[j] tells us that the first to prepare such tables was Liu Hung in +206. He also gave the technical terms used for the accelerating and retarding phases of the anomalous motion, *chi*[1] and *chhih*[2] respectively. The equations were additive during the former and subtractive during the latter. Each phase was divided into two halves, a beginning (*chhu*[3]) and an ending (*mo*[4])[k].

[a] Neugebauer (1), p. 33.

[b] Neugebauer (1), p. 114.

[c] This is not to say that Hipparchus did not utilise Babylonian empirical data; cf. Neugebauer (1), p. 118; (9), p. 151. The teaching activity of the 'Chaldean' Berossos (Berussu), who founded a school on the island of Cos about −270, is generally regarded as an important factor in the transmission of Babylonian astronomical knowledge to the Greeks. One would like to know the names of the Babylonian star-clerks who travelled east rather than west.

[d] Hence the importance of ascertaining something about science in Bactria if possible. We have already discussed Sogdia as a way-station (pp. 204 ff. above).

[e] It is rapidly becoming clear that much of Indian astronomy will only become comprehensible in the light of Mesopotamian researches. 'We can now understand', writes Neugebauer (9), pp. 165, 169, 'whole sections in Varāha-Mihira's *Pañca Siddhāntikā* by means of Babylonian planetary texts.' A stimulating survey of the present position will be found in Neugebauer (12), who gives justified credit to Schnabel (2) for having been the first to realise the Babylonian-Indian parallels thirty years ago. He emphasises that Indian astronomy is twofold, for Babylonian arithmetical-algebraic methods are found largely in Tamil (southern) sources (cf. Neugebauer, 11), while the Greek geometrical influence is evident in the (northern) *Sūrya Siddhānta*. It is tempting to associate the former with the Roman trading stations such as Arikamedu near Pondicherry, datable by coins and pottery between the −1st and +4th centuries (Wheeler, 4), just as one naturally associates the latter with the Bactrian Greeks and the Sassanian Persians. Cf. Vol. 1, pp. 178, 200, 232 ff.

[f] (1), p. 11; (2), pp. 483, 492, 617, 624 ff.

[g] Sarton (1) vol. 1, p. 212.

[h] E.g. *Yuan Shih*, ch. 54, pp. 19*a* ff.

[i] (3), pp. 131, 133, 141.

[j] *Yuan Shih*, ch. 52, p. 29*b*.

[k] It will be remembered that Sédillot (7) maintained that the third inequality in lunar motion (variation) was discovered by Abū'l-Wafā' in the +10th century.

[1] 疾 [2] 遲 [3] 初 [4] 末

Thorough study of Chinese ephemerides and their relations with Babylonian astronomy is much needed;[a] Yabuuchi (*9*) and Eberhard (12) have begun it. Here we would only mention the existence of elaborate planetary ephemerides in the *Chhi Yao Jang Tsai Chüeh*[1] (Formulae for avoiding Calamities according to the Seven Luminaries), beginning for the year +794. This is a book with Sogdian connections, put together by Chin Chü-Chha[2] in the +9th century. But Thang astronomers had other tables also, such as those of the *Pai Chung Ching*,[3] which seem[b] to have started from +657, the *Wu Hsing Hsing Tsang Li*[4] and the *Chin Kuei Ching*[5] (Golden Box Manual). The Iranian connection is of great interest in view of transmission of computation techniques from the Mesopotamian culture area some time during the first centuries of the +1st millennium.

(2) Sexagenary Cycles

The most ancient day-count in Chinese culture did not depend on the sun and moon at all. It was that sexagesimal cyclical system to which reference has already several times been made,[c] a series of twelve characters (the so-called 'branches'; *chih*[6])[d] being combined alternately with a series of ten (the so-called 'trunks' or 'stems'; *kan*[7])[e] so as to make sixty combinations at the end of which the cycle started all over again. These characters are among the commonest on the oracle-bones of the mid −2nd millennium, and in the Shang period they were used strictly as a day-count. The practice of using them for the years as well did not come in until the end of the Former Han in the −1st century,[f] but thenceforward both uses continued till modern times.[g] The etymology of these characters, in their capacity as cyclical signs, is lost in the mists of antiquity; modern scholars such as Karlgren (1), for example, are far less confident than de Saussure was (8, 9) on this subject.[h]

a For example, Sédillot claimed, (6, 8), (2) p. 80, that the Hakemite Tables of Ibn Yūnus (+1007) were the basis of those adopted by Kuo Shou-Ching and incorporated in the *Yuan Shih*, but this requires substantiation.

b *Khai-Yuan Chan Ching*, ch. 104, p. 2*a*; and see Eberhard (12), p. 228.

c In the Introduction (Sect. 5*a*, in Vol. 1, p. 79), the Section on the pseudo-sciences (Sect. 14*a*, in Vol. 2), and that on mathematics (p. 82 above).

d These are: *tzu*,[8] *chhou*,[9] *yin*,[10] *mao*,[11] *chhen*,[12] *ssu*,[13] *wu*,[14] *wei*,[15] *shen*,[16] *yu*,[17] *hsü*,[18] *hai*.[19]

e These are: *chia*,[20] *i*,[21] *ping*,[22] *ting*,[23] *wu*,[24] *chi*,[25] *kêng*,[26] *hsin*,[27] *jen*,[28] *kuei*.[29]

f Chatley (19) thinks it was in +4. Dubs (2), vol. 3, p. 330, notes an instance in +12. The two systems ran concurrently so that the Chinese new year's day could occur on any one of the days of the day-count cycle (cf. Fritsche, 1).

g Both at Marāghah and at Samarqand (Ulūgh Beg) they were taken very seriously (Sédillot (3*a*), pp. 32 ff.). On the relations of the Chinese and Muslim calendars see recent papers by Yen Tun-Chieh (*10*) and Liu Fêng-Wu. In the early 18th century the Chinese sexagenary cycles were introduced to western scholars by Noel (1) and des Vignoles (1).

h A few meanings, however, may be considered established, as, for example, that *ssu*[13] was originally a drawing of an embryo (cf. Hopkins, 6, 27). *Hsü*,[18] a drawing of a man with an axe, may represent Orion (Shinjō, 1; Hashimoto, *2*).

1 七曜攘災决 2 金俱吒 3 百中經 4 五星行藏曆
5 金匱經 6 支 7 干 8 子 9 丑 10 寅
11 卯 12 辰 13 巳 14 午 15 未 16 申 17 酉
18 戌 19 亥 20 甲 21 乙 22 丙 23 丁 24 戊
25 己 26 庚 27 辛 28 壬 29 癸

Whether Babylonian or not, the sixty-day count was convenient enough in that six cycles made approximately a tropical year. Both words for 'year' (*sui*,[1] *nien*;[2] K 346, 364) seem to derive from pictograms connected with harvest grain and annual sacrifices. Similarly, the sixty-day cycles broke down into six periods of ten days each (*hsün*[3]; K 392, apparently a graph for a 'little number' of days),[a] and therefore approximately two lunations. The graph for 'moon' (*yüeh*[4]) always stood for 'month' as well. When necessary, one to three ten-day units were intercalated, or occasionally a whole sixty-day cycle. The ten-day period (*hsün*[3]) lasted down to our own time and still exists in rural places.[b] The seven-day week[c] is an introduction to China not earlier than the Sung,[d] as may be seen from a discussion about +1205.[e]

The Chinese sexagesimal cycle can be thought of in the image of two enmeshed cogwheels, one having twelve and the other ten teeth, so that not until sixty combinations have been made will the cycle repeat. It is one of the most interesting parallels between ancient Chinese and ancient Amerindian civilisation that similar, though more complicated, cycle systems are found in the latter.[f] Morley, in his book on the Mayas,[g] explains that the 260 days of their religious year (*tzolkin*) were made up by combining cyclically the numbers 1 to 13 with the 20 day-glyphs, and that this year was then in turn combined with the calendar year (*haab*) which was composed of the 20 day-glyphs and the 19 months,[h] or 365 days. In this case, the two cogwheels were of 260 and 365 teeth respectively, giving a total period of 18,980 days, or 52 years. The cyclical period so formed was adopted by the Aztecs[i] and other Central American cultures.[j] But as its multiples were awkward to handle arithmetically, the Mayas developed a parallel system, based on vigesimal progression (*tuns*, *katuns*, *baktuns*, etc.) which gave them an extremely precise chronology.[k]

The usual view about the ten Kan was that they had been developed by combining the Five Elements with the Yin-Yang dualism (de Saussure, 8), but this cannot be true seeing that they long ante-date these late Chou theories. It is much more probable, as Shinjō (1) thinks, that they were the names of the days of the ten-day 'week' (*hsün*[3]). By the early Han they had come to be associated with a series of ten astrological names

[a] See Shinjō (*1*).

[b] An abortive attempt was made to introduce it to Europe during the French Revolution (Anon., 26).

[c] The excellent monograph of Colson shows that while in the West the seven-day week was originally Hebrew, the planetary day-names were not adopted until about the beginning of our era. On their introduction to China see Wylie (15) and Chavannes & Pelliot (1). Cf. Schiaparelli (1), p. 260.

[d] There is, however, some evidence for an early Chou system of seven-day weeks which did not persist (Wang Kuo-Wei (*3*), see Shinjō (*1*), p. 8). This has been related by Tung Tso-Pin (*7*) to the terms *sêng-pa*[5] and *ssu-pa*[6] which occur in Chou bronze inscriptions with reference to the phases of the moon, each both alone and preceded by the word *chi*,[7] thus making four periods in each lunation. The system must have been used between the −10th and −8th centuries. Perhaps it was dropped because of the early Chinese predilection for decimal computations.

[e] *Lü-Chai Shih Erh Pien*, ch. 2, p. 3*a*.

[f] This has also been noted by Hartner (10).

[g] (1), pp. 265 ff. And J. E. S. Thompson (1).

[h] Eighteen of the months had twenty days each, and one had five only.

[i] The *tonalamatl*, see Spinden (1).

[j] Burland (1). The oldest Amerindian calendrical inscriptions are in fact Zapotec, perhaps three centuries before the Mayas, but less complex.

[k] There were about 2½ *katuns* in one 52-year period.

1 歲 2 年 3 旬 4 月 5 生霸 6 死霸 7 既

of obscure origin.[a] The twelve Chih were already in very ancient times applied to the lunations (months) of the tropical year, but also served in other ways, particularly as azimuth direction points (compass points)[b] and as names for the double-hours of each sidereal day.[c] Some think that the twelve cyclical signs derive from rites proper for each lunation (Bulling, 1). The Chinese also had a parallel list of astrological names which seem to have been borrowed from the Jupiter cycle of years.[d] Planetary cycles thus demand consideration next.

(3) PLANETARY REVOLUTIONS

The most convenient account of ancient Chinese planetary astronomy is in Maspero (3).[e] The *Khai-Yuan Chan Ching*[f] gives the essential technical terms which came down from the −4th-century astronomical schools. As has already been noted[g] each planet was associated with one of the Five Elements and one of the cardinal points, thus:

(1) Jupiter	Sui hsing[1] (the Year-star)	W	East
(2) Mars	Ying huo[2] ('Fitful Glitterer')	F	South
(3) Saturn	Chen hsing[3] (the Exorcist)	E	Centre
(4) Venus	Thai pai[4] (the 'Great White One')	M	West
(5) Mercury	Chhen hsing[5] (the Hour-star)	w	North

The apparent forward motion of a planet was termed *shun*;[6] its retrograde motion *ni*.[7][h] It thus rose (heliacally or otherwise) (*chhu*[8]), advanced (*chin*[9]), changed its

[a] Chavannes (1), vol. 3, p. 652; de Saussure (1), p. 220.

[b] See Table 34, and more extensively in Sect. 26*i* below. Cf. also pp. 240, 279 above.

[c] In the early Shang, before about −1270, there were no equal subdivisions of the day and night, and a set of some half-dozen terms sufficed for dawn, midday, dusk, etc. But from that time onward the day was divided into 6 equal double-hours by the use of 7 special terms, and there was one also for midnight. These usages have been established by the researches of Tung Tso-Pin (*1*) and others (*Yin Li Phu*, Pt. 1, ch. 1, pp. 4*b* ff.). The full cycle of 12 equal double-hours is found stabilised in the Han (−2nd century onwards) but it probably goes well back into the Chou. A passage in the *Tso Chuan* under date −534 (Couvreur (1), vol. 3, p. 129) seems to show that at that time only 10 double-hour names were used, two of the periods lasting four hours (cf. *Hsiao Hsüeh Kan Chu*, ch. 1, p. 23*b*). A passage in the *Kuo Yü* (*Chou Yü*), ch. 3, pp. 36*a*, *b*, giving what purports to be a discussion of −519, indicates that by the −4th or −3rd century the 12 cyclical characters (*chih*) had become attached to the 12 equal double-hours. This duodecimal system, from which modern practice is derived, goes back no doubt to the Old Babylonian division of the day and night into 12 equal double-hours (*kas.bu*): see Ginzel (1), vol. 1, p. 122. China may well have received it from this source. Like the Babylonians also the Chinese had, and long indeed retained, a parallel system of 5 unequal night watches. On the whole subject see further de Saussure (9); Chavannes (7); and Needham, Wang & Price (1).

[d] See Table 34.

[e] A wealth of information on the planets, most of it reliable, will of course also be found in Schlegel (5); and de Saussure (1), pp. 430 ff., is worth reading.

[f] Ch. 64, pp. 11*b* ff.

[g] In the discussion of the symbolic correlations, Sect. 13*d*, in Vol. 2, p. 262 above. Cf. *Huai Nan Tzu*, ch. 3, p. 3*a*.

[h] This was a term of some philosophical importance, see Sect. 46 below. Ssuma Chhien observes that in ancient times only the retrogradation of Mars, which is much more marked than that of the other planets, had been noticed (*Shih Chi*, ch. 27, p. 42*b*; Chavannes (1), vol. 3, p. 409).

1 歲星 2 熒惑 3 鎭星 4 太白 5 辰星 6 順
7 逆 8 出 9 進

direction of movement (*fan*[1]), or retired (*thui*[2]), and ultimately set (*ju*[3]). A short retrogradation was called a 'retreat' (*so*[4]), and an unexpectedly rapid advance a 'gaining' (*ying*[5]) or an 'urgency' (*chi*[6] or *su hsing*[7]). When it appeared to remain stationary in one place, it was said to 'dwell' there (*chü*[8] or *liu*[9]); if this phase took a particularly long time, more than twenty days, the terms *su*[10] or *shou*[11] were employed. It was thus said to 'guard' the constellation to which it was near; if it actually entered it was said to be 'trespassing' (*fan*[12]). Unexpectedly slow motion was a 'tarrying' (*chhih hsing*[13]). Collective terms for the planets were the Five Wefts (Wu Wei[14]) or the Five Walkers (Wu Pu[15]). All these technical terms, which first appear in the writings of the −4th-century astronomers, were conserved through subsequent times.

According to the means at their disposal the observers of the Warring States period must have mapped the planetary motions.[a] In the *Shih Chi*, Ssuma Chhien has much to say about retrogradations.[b] Here reproduced is a diagram of the retrogradations of Mercury (Fig. 181), from a very late source,[c] but illustrating facts which must have been familiar in China.[d] By comparing the figures for the estimates of the periods of revolution dating from various times between −400 and +100, a clear indication of the astronomical refinement during that time can be obtained (Table 33). It will thus be seen that before the end of the +1st century the estimates of the synodic periods had become quite reliable.[e] This was the case a little later also in the West, for about +175 Cleomedes the Stoic[f] gave figures of about the same degree of precision. Naturally Mercury caused the most difficulty as it is the hardest to observe.[g]

In spite of so much accurate observation, Chinese study of planetary motion remained purely non-representational in character. Unlike that of the Greeks, in which the geometry of circles and curves was so prominent, it perpetuated the algebraic treatment of the Babylonian astronomers such as Naburiannu and Kidinnu,[h] and never sought a geometrical theory of planetary motions.[i] If this gave it, at the time of the arrival of the Jesuits, a somewhat archaic character, one must also remember that the attachment of the Greeks to circular movements[j] had been almost too great,

[a] For Hellenistic knowledge of planetary movements, cf. Bouché-Leclercq (1), pp. 111 ff.

[b] E.g. ch. 27, pp. 19*a*, 22*a*, 43*a*; Chavannes (1), vol. 3, pp. 365, 371, 409; Chu Wên-Hsin (5), p. 28. We had particular occasion to notice this in Vol. 2, p. 553. Yet Narrien (1), p. 349, could write: 'it does not appear that previous to the introduction of European astronomy the Chinese had even observed that the planetary movements were alternately direct and retrograde'.

[c] *TSCC*, *Chhien hsiang tien*, ch. 27, p. 16*a*.

[d] Just such annular ecliptic graphs are found in +9th-century astronomical MSS. (Zinner (2), p. 56).

[e] What the Han histories have to say about planetary cycles has been analysed by Eberhard & Henseling (1); and the Buddhist texts of the Thang which deal with the subject have been investigated by Eberhard (12).

[f] Sarton (1), vol. 1, p. 211.

[g] Among Ptolemy's errors in planetary longitudes the greatest were those for Mercury.

[h] Cf. Neugebauer (9), p. 130.

[i] See the expositions of Dreyer (4); Pannekoek (1); and Zinner (2).

[j] The multiple spheres of Eudoxus, the excentric circles of Hipparchus and the epicycles of Ptolemy.

1 返 2 退 3 入 4 縮 5 贏 6 疾 7 速行
8 居 9 留 10 宿 11 守 12 犯 13 遲行 14 五緯
15 五步

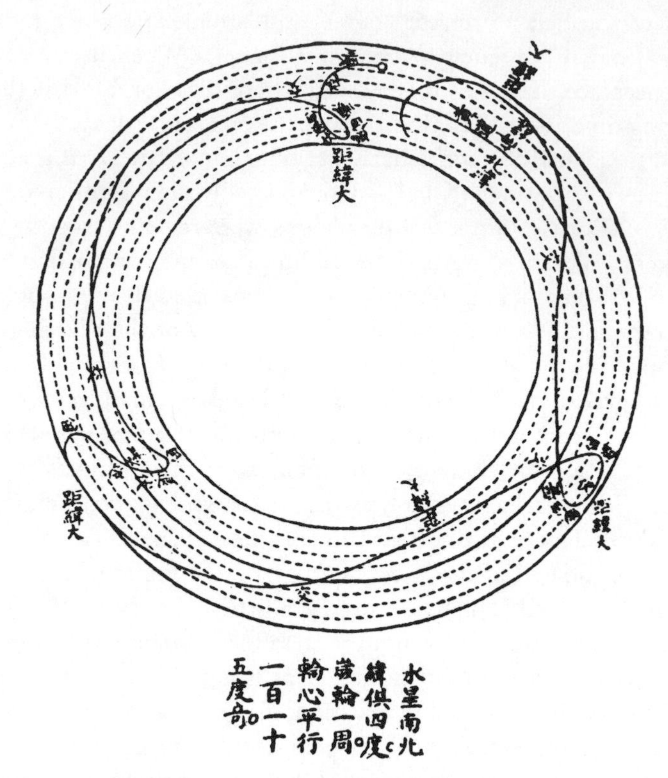

Fig. 181. Annular ecliptic plot to show the retrogradations of Mercury (from the *Thu Shu Chi Chhêng*, +1726). Many of the technical terms for the motions inserted in the diagram are ancient, according with their definitions in the *Chin Shu* (+635).

as we know from the titanic efforts which Kepler and his successors, and indeed all the scientific men of the Renaissance, had to make to get away from them.

Nevertheless, it would be rash to assume that Chinese astronomers never visualised planetary orbits. In the interesting astronomical chapter of the *Chu Tzu Chhüan Shu* (Collected Works of Chu Hsi),[a] which records conversations taking place about +1190, the philosopher speaks about 'large and small circular tracks' (*ta lun*,[1] *hsiao lun*[2]), i.e. the small 'orbits' of the sun and moon, and the large 'orbits' of the planets and the fixed stars. Particularly interesting is his recognition of the fact that retrogradation (*ni*[3]) was only an apparent phenomenon, depending on the relative and different speeds of the various bodies. He suggested that the calendar experts ought to realise that all *ni*[3] and *thui*[4] movements were in fact progressive movements (*shun*[5] and *chin*[6]).[b]

[a] Ch. 50. It fully deserves a complete translation.

[b] All on p. 12*b*. Chang Tsai[7] said something similar (*Chêng Mêng* in *Sung Ssu Tzu Chhao Shih*, ch. 4, p. 6*b*).

[1] 大輪 [2] 小輪 [3] 逆 [4] 退 [5] 順 [6] 進 [7] 張載

Table 33. *Estimates of periods of planetary revolutions*

	Mercury	Venus	Mars	Jupiter		Saturn	
	days	days	days	days	years	days	years
True values:							
Sidereal Period[a]	88·97	224·7	686·98	4333	11·86	10759	29·46
Synodic Period[b]	115·877	583·921	779·936	398·884		378·092	
Shih Shen (−4th)[c]	—	736	780	—		—	
Kan Tê (−4th)[c]	68+	585	—	400		—	
Eudoxos of Cnidos[d] (−408 to −355)	110	570	260	390		390	28
Huai Nan Tzu[e] (−2nd)	—	635	—	—		—	
The author of the *Lo Shu Wei Chen Yao Tu*[1] (perhaps −2nd)[f]	74+	—	—	—		—	
Ssuma Chhien (−1st) (*Shih Chi* figures)[g]	—	626	—	395·7	12	360	28
Liu Hsin[2] (end −1st)[g] (*Chhien Han Shu* figures)[g]	115·91	584·13	780·52	398·71	11·92	377·93	29·79
Li Fan[3] (+85) (*Hou Han Shu* figures)	115·881	584·024	779·532	398·846	11·87	378·059	29·51

[a] The sidereal period is the time of the planet's revolution in its orbit about the sun relative to the stars.

[b] The synodic period is the interval between two conjunctions with the earth relative to the sun, i.e. the time during which the planet makes one whole revolution as compared with the line joining the earth to the sun.

[c] Figures preserved in the *Khai-Yuan Chan Ching* and drawn therefrom by Maspero (3). The motions of the planet Mars are also referred to in the probably −4th-century *Yen Tzu Chhun Chhiu*[4] (Master Yen's Spring and Autumn Annals), ch. 1, p. 22*a*.

[d] Sarton (1), vol. 1, p. 117. The figures come from Simplicius' commentary on Aristotle (see Balss (1), pp. 107, 109, 271 and Zinner (2), p. 5).

[e] Ch. 3, p. 4*a*.

[f] In *Ku Wei Shu*, ch. 36, pp. 1*a* ff. and *Khai-Yuan Chan Ching*, ch. 53, p. 2*b*.

[g] The figures from the *Shih Chi* and the two Han histories have been extracted and converted to decimals by Chhien Pao-Tsung (*1*), p. 15; Chu Wên-Hsin (*1*), pp. 47 ff.; Chatley (16, 17); and Eberhard (13). There are minor variations in the decimal places. Chu gives the figures accepted in all the calendars throughout the dynasties.

1 洛書緯甄曜度　2 劉歆　3 李梵　4 晏子春秋

(4) Duodenary Series

The Jupiter cycle must have attracted attention at a very early date on account of the fact that the sidereal period of this planet is almost exactly 12 years (actually 11·86), and therefore invited correlation with the 12 cyclical characters and the 12·37 lunations in the tropical year. The *Chi Ni Tzu* book,[a] transmitting a southern −4th-century naturalistic tradition, says:

When the Great Yin (Thai Yin[1]) is at the position of the element Metal during the first three years, there will be abundant harvests. When it is at Water, there will be damage to crops for three years. When it stands in Wood, for another three years, there will be prosperity; when it stands in Fire, again for three years, there will be drought. So some times are suitable for storing agricultural products while at other times rice may be given away. Accumulations need not be for more than three years. Deciding wisely and with judgment, and helped by the Tao (of Nature) one may draw on surpluses to mend shortages. In the first year there will be a double harvest, in the second a normal one, in the third a poor one. During floods chariots should be built, during droughts boats should be prepared. Bumper harvests come every six years, and famines every twelve. Thus the sage, predicting the recurrences of Nature, prepares for future adversity.[b]

Here Thai Yin is not, as might be supposed, the moon, but an invisible 'counter-Jupiter' which moved round in the opposite direction to the planet itself.[c] Jupiter (Sui hsing[2]), with the other planets, appears to move eastwards or anti-clockwise through the stars, so a 'shadow-planet' (Thai sui[3] or Sui yin[4]) was invented to move with them, accompanying the sun. Wang Chhung, in the *Lun Hêng*, devotes a whole chapter[d] to this peculiar theory.[e] The twelve Jupiter-stations were named *tzhu*[5] and the whole cycle of years a *chi*.[6] The names for the years have survived in two forms, one set astronomical[f] another astrological;[g] and it was natural therefore that they should sometimes also have been used for the months of a single year and[h] for the double-

[a] In Ma Kuo-Han's fragment collection (*YHSF*, ch. 69); *Chi Ni Tzu*, ch. 1, p. 3*a*; ch. 2, p. 1*b*. Cf. above, Sect. 18, in Vol. 2, p. 554.

[b] Tr. auct. Censorinus (+3rd century) has an astonishingly similar passage: 'Proxima est hanc magnitudinem quae vocatur dodekaeteris, ex annis vertentibus duodecim. Huic anno Chaldaico nomen est, quem genethliaci non ad solis lunaeque cursus sed ad observationes alias habent accomodatum quod in eo dicunt tempestates frugumque proventus sterilitates item morbosque circumire' (*De Die Natali*, ch. 18). I am indebted to the late Dr H. Chatley for drawing my attention to this.

[c] See *Shih Chi*, ch. 27, p. 15*b*; Chavannes (1), vol. 3, pp. 357–62, 653 ff.; Forke (13), p. 502; Liu Than (*1*).

[d] Ch. 73 (tr. Forke (4), vol. 2, pp. 302 ff.).

[e] Much, also, in *Huai Nan Tzu*, ch. 3, pp. 13*a*, 13*b*, 15*a*, 18*a*. Cf. *Yen Thieh Lun*, ch. 36. Parallels in Greek thought will be recalled, notably the 'counter-earth' (*antichthon*) of Philolaus of Tarentum, a Pythagorean of the late −5th century (see Freeman (1), p. 225; Schiaparelli (1), pp. 334, 357). Could the imaginary planet in Arabic astronomy, al-Kaid, have been in some way an echo of the Chinese counter-Jupiter? It has been studied recently by Hartner (9).

[f] See Maspero (3), p. 284; de Saussure (12).

[g] See de Saussure (9), in (1), p. 289.

[h] After +507, when for a time the number of 'quarters of an hour' in the day and night was reduced from 100 to 96; cf. Maspero (4), p. 210.

1 太陰 2 歲星 3 太歲 4 歲陰 5 次 6 紀

Table 34. *The twelvefold series*

	Cyclical *chih*[1]	Astronomical term (*tzhu*[2])	Astrological or calendrical term	Associated *hsiu*	Associated compass-point
1	*tzu*	*hsüan-hsiao*[3]	*shê-thi-ko*[15]	Nü, Hsü, Wei (10, 11, 12)	N.
2	*chhou*	*hsing-chi*[4][a]	*tan-o*[16]	Tou, Niu (8, 9)	N.
3	*yin*	*hsi-mu*[5]	*chih-hsü*[17]	Wei, Chi (6, 7)	E.
4	*mao*	*ta-huo*[6]	*ta-huang-lo*[18]	Ti, Fang, Hsin (3, 4, 5)	E.
5	*chhen*	*shou-hsing*[7]	*tun-tsang*[19]	Chio, Khang (1, 2)	E.
6	*ssu*	*shun-wei*[8]	*hsieh-hsia*[20]	I, Chen (27, 28)	S.
7	*wu*	*shun-huo*[9]	*thun-than*[21]	Liu, Hsing, Chang (24, 25, 26)	S.
8	*wei*	*shun-shou*[10]	*tso-o*[22]	Ching, Kuei (22, 23)	S.
9	*shen*	*shih-chhen*[11]	*yen-mao*[23]	Tshui, Shen (20, 21)	W.
10	*yu*	*ta-liang*[12]	*ta-yuan-hsien*[24]	Wei, Mao, Pi (17, 18, 19)	W.
11	*hsü*	*chiang-lou*[13]	*khun-tun*[25]	Khuei, Lou (15, 16)	W.
12	*hai*	*chhü-tzu*[14]	*chhih-fên-jo*[26]	Shih, Pi (13, 14)	N.

[a] This series was considered to start with *hsing-chi*, a term which refers to the group of stars Chien hsing[27] (π Sagittarii, and neighbouring fainter stars; S 161). This point was taken as the starting-point of all the planetary cycles (de Saussure (1), pp. 450 ff.).

1 支 2 次 3 玄枵 4 星記 5 析木 6 大火 7 壽星 8 鶉尾 9 鶉火 10 鶉首 11 實沉 12 大梁 13 降婁 14 娵訾 15 攝提格 16 單閼 17 執徐 18 大荒落 19 敦牂 20 協洽 21 涒灘 22 作噩 23 淹茂 24 大淵獻 25 困敦 26 赤奮若 27 建星

hours of a single day, as well as for the years of the Jupiter cycle.[a] The astronomical terms were used to designate the positions of Jupiter, the astrological or calendrical terms applied to the positions of counter-Jupiter.[b] In Table 34 the names and numbers of the associated *hsiu*, *chih*, and compass-points, are also given.

It will be noticed that the astrological-calendrical series begins with the curious term *shê-thi-ko*[1] which we have already encountered.[c] Some Chinese historians of astronomy (e.g. Ting Wên-Chiang; Chu Kho-Chen, 1) have believed that this and some of the other terms are transliterations of the Indian names for the years of the Jupiter cycle,[d] but de Saussure (11) and Chavannes (1) always contested this, and it seems unconvincing. While the meanings of the astronomical terms are obvious enough, those of the astrological series remain very obscure. There has also been much discussion of chronology based on the Jupiter cycle, since the year-terms are found not infrequently in ancient texts. For all such questions the reader is referred to de Saussure (1) and to Eberhard, Müller & Henseling (1), who believe that the *Tso Chuan* dates were systematically 'corrected' by Liu Hsin on the basis of the Jupiter cycle. After the coming of the Jesuits, this twelvefold equatorial division was of course thoroughly confused with the twelve ecliptic signs of the Graeco-Egyptian zodiac, with which it had nothing in common.[e]

Month-names, however, were not the most important in the old Chinese calendar. The twelve months of the tropical year (*chhi*,[2] a very meteorological conception) were divided into twenty-four fortnights, of which twelve were '*chhi*-centres' (*chung chhi*[3]) and twelve were '*chhi*-nodes' (*chieh chhi*[4]—the analogy being with the nodes of a bamboo). Though the names for these are listed in many dictionaries, and are still in common use in China today, they are given in Table 35.[f] They are undoubtedly very ancient; one of the earliest literary references to them occurs in the *Mu Thien Tzu Chuan* (Account of the Travels of the Emperor Mu) which is of late Chou time (cf. de Saussure, 26, 26*a*). Before then, in the Shang, there were months of twenty-nine days (*hsiao yüeh*[5]) and thirty days (*ta yüeh*[6]); sometimes two of the latter came successively (*phin ta yüeh*[7]).[g]

[a] The subject has been investigated by Shinjō (*1*) and de Saussure (12). Cf. Chavannes (1), vol. 3, p. 663.

[b] Evidences of their use from ancient texts and inscriptions are given by Chavannes (1), vol. 3, pp. 656ff.

[c] P. 251 above.

[d] It will be remembered that Chalmers derived *shê-thi-ko* from *Bṛhaspati-cakra*. Chu Kho-Chen sees in it rather the *nakshatra Kṛttikā* where the full moon was in the month *Kārttika* (cf. E. Burgess (1), pp. 315, 317) corresponding to the year *mahākārttika* of the Jupiter cycle. He would also derive in a parallel way *ta-yuan-hsien* from the *nakshatra Dhaniṣṭhā* corresponding to *mahāśrāvaṇa*. But there are no other equations even plausible.

[e] Cf. de Saussure (16*d*), p. 337.

[f] Further details about them, and the way in which intercalation was performed, will be found in de Saussure (16*d*), and Chatley (18), based of course on Gaubil and Biot. Cf. Kühnert (4). J. F. Davis (2) discussed them in the *Phil. Trans.* for 1823.

[g] Tung Tso-Pin (*1*), *Yin Li Phu*, Pt. 1, ch. 1, pp. 9*a* ff., ch. 2, pp. 2*a* ff. But he thinks that the Shang people had something of the fortnightly period system (Pt. 11, ch. 3, pp. 27*a* ff., ch. 5, pp. 14*a* ff.).

1 攝提格 2 氣 3 中氣 4 節氣 5 小月 6 大月
7 頻大月

Table 35. *The twenty-four fortnightly periods* (chhi)

	Chinese name		Translation	Beginning
1	Li Chhun	立春	Beginning of Spring	5 Feb.
2	Yü Shui	雨水	The Rains	20 Feb.
3	Ching Chê (or Chih)	驚蟄	Awakening of Creatures (from hibernation)	7 March
4	Chhun Fên	春分	Spring Equinox	22 March
5	Chhing Ming	清明	Clear and Bright	6 April
6	Ku Yü	穀雨	Grain Rain	21 April
7	Li Hsia	立夏	Beginning of Summer	6 May
8	Hsiao Man	小滿	Lesser Fullness (of Grain)	22 May
9	Mang Chung	芒種	Grain in Ear	7 June
10	Hsia Chih	夏至	Summer Solstice	22 June
11	Hsiao Shu	小暑	Lesser Heat	8 July
12	Ta Shu	大暑	Greater Heat	24 July
13	Li Chhiu	立秋	Beginning of Autumn	8 Aug.
14	Chhu Shu	處暑	End of Heat	24 Aug.
15	Pai Lu	白露	White Dews	8 Sept.
16	Chhiu Fên	秋分	Autumn Equinox	24 Sept.
17	Han Lu	寒露	Cold Dews	9 Oct.
18	Shuang Chiang	霜降	Descent of Hoar Frost	24 Oct.
19	Li Tung	立冬	Beginning of Winter	8 Nov.
20	Hsiao Hsüeh	小雪	Lesser Snow	23 Nov.
21	Ta Hsüeh	大雪	Greater Snow	7 Dec.
22	Tung Chih	冬至	Winter Solstice	22 Dec.
23	Hsiao Han	小寒	Lesser Cold	6 Jan.
24	Ta Han	大寒	Greater Cold	21 Jan.

NOTE. Each of these twenty-four periods corresponds to 15° motion of the sun in longitude on the ecliptic. The average fortnightly period is 15·218 days; the half (lunar) month, 14·765. The names of the periods suggest that the list was first established in, or north of, the Yellow River valley.

At some quite early time[a] the twelvefold division became associated with a cycle of animals, the ox, sheep, dragon, pig and so on. We have already come across a reflection of this when the symbolic correlations were under discussion.[b] There has been great debate among scholars, both Eastern and Western, about the origin of these associations, some maintaining, with Chavannes (7) and Boll (4), that the Chinese took them over from neighbouring Turkic peoples or from the ancient Middle East, and others

[a] Boodberg (1) thinks as early as the −6th century.
[b] Sect. 13*d*, in Vol. 2, p. 266.

labouring to prove, with de Saussure (10, 16*c*), that they were essentially Chinese in origin.[a] Most of the arguments need re-examination in the light of modern knowledge about the dates and probable reliabilities of texts, but in any case, so far as the historian of science is concerned, whoever invented the animal cycle is welcome to it. Its interest seems purely archaeological and ethnological.[b] The naming of years for animals still continues in numerous Asian cultures, e.g. those of the Mongols and Tibetans as well as the Chinese.

(5) Resonance Periods

In view of the incommensurability of their two fundamental periods (cf. above, p. 390). calendar-makers in all cultures have attached importance to certain 'resonances', i.e. to those rather longer lapses of time which end in approximate concordance of phase. Thus nineteen tropical years are almost equivalent to 235 lunations. In the West this period is associated with the name of Meton of Athens (*fl.* −432);[c] in China it was known[d] as the *chang*.[1] Similarly, four such periods are equivalent to exactly 76 years or 27,759 days, and this unit is associated with the name of Callippus of Cyzicos (*fl.* −370 to 330, and therefore a contemporary of Shih Shen),[e] who obtained a rather satisfactory calendar if one day was dropped out of each such period. The 76-year cycle was known in China[f] as the *pu*.[2] It was then found that 27 *chang* were equivalent to 47 lunar eclipse periods (about 135 months each), i.e. 513 years, and this was termed a *hui*.[3] Three *hui* (or 81 *chang*) constituted a *thung*[4] (1539 years), periods less than this not giving a round number of days. The smallest concording period of sexagenary cycles, lunations, tropical years and eclipse periods, was found to be 3 *thung*, or 4617 years. Another unit was the *chi*[5] (already met with as a term for Jupiter's synodic revolution) or the *sui*[6] or *ta chung*,[7] which was 20 times the *pu*, i.e. 1520 years, or 19 × 487 sexagesimal day cycles.[g] Three *chi* constituted a *ta pei*,[8] *yuan*[9] or *shou*.[10][h] The *Chou Pei Suan Ching* goes on to say[i] that seven *shou* equalled one *chi*[11] or 31,920 years, after which time 'all things come to an end and return to their original state'. This 'grand period' is interesting, for it equals exactly four Julian cycles.[j] As is

[a] Opinion now inclines to the latter view, especially since the finding of references in Han texts by Pelliot (42). Professor Louis Bazin informs us that the calendrical knowledge of the Turks of Central Asia is more and more clearly seen to have derived from Chinese culture.

[b] Of course an adequate solution of the problem of the animal cycle might throw much light on the travel of proto-scientific ideas among ancient and medieval peoples.

[c] Sarton (1), vol. 1, p. 94.

[d] E.g. *Huai Nan Tzu*, ch. 3, p. 8*a*; *Chou Pei Suan Ching*, ch. 2, p. 13*a*.

[e] Sarton (1), vol. 1, p. 141.

[f] E.g. *Huai Nan Tzu*, ch. 3, p. 5*a*; *Chou Pei Suan Ching*, ch. 2, p. 13*a*. Cf. Kühnert (3).

[g] Cf. *Shih Chi*, ch. 27, p. 38*b* (Chavannes (1), vol. 3, p. 403); *Huai Nan Tzu*, ch. 3, p. 5*a*; *Chhien Han Shu*, ch. 26, p. 23*a*; *Chou Pei Suan Ching*, ch. 2, p. 13*b*.

[h] *Chou Pei Suan Ching*, ch. 2, p. 14*a*. More elaborate details on these terms, taken largely from the *Chhien Han Shu* and the *Hou Han Shu*, will be found in Chatley (15, 16).

[i] Ch. 2, p. 14*a*.

[j] As pointed out by Chatley (15, addendum). In the beginnings of modern geology, Lyell (1), vol. 1, p. 23, cited a 17th-century Latin translation from an Arabic source, by Abraham Ecchellensis, according to which the 'Gerbanites', a 'sect of astronomers who flourished some centuries before the

1 章 2 蔀 3 會 4 統 5 紀 6 遂 7 大終
8 大備 9 元 10 首 11 極

well known, the Julian cycle (7980 years) was proposed by Scaliger many centuries after the text which we are considering. It is the L.C.M. of the Metonic *chang*, the Callippic *pu*, the 28-year Victorinus sabbatical cycle, the 15-year Roman indiction, the 80-year period in which the 60-day cycle recurs, and the 60-year cycle itself. Eberhard & Müller (1) and Eberhard, Müller & Henseling (1) have given abundant translations of the formulae used with much exactness by the Han astronomers for calculating cyclical dates.[a]

Planetary cycle constants were termed *chi mu*[1] and day-month-year cycle constants *thung mu*.[2] With the Han value for the length of the lunation, the smallest number of lunations which would give a round number of days was 81 (i.e. 2392 days), and when this was combined with the lunar eclipse cycle of 135 months, the former being multiplied by 5 and the latter by 3, both giving 405 lunations or 11,960 days, the shortest period of whole days in which the eclipse cycle could be completed was found. The *dénouement* here, pointed out by Eberhard & Henseling (1), is that this period is identical with the great *tzolkin* of the Mayas.[b]

As already mentioned,[c] the parallel between the shorter of these cycles and those known to the Greeks has seemed to some scholars rather strong evidence of transmission from the West to China. But Tung Tso-Pin (*1*) now brings forward reasons[d] for thinking that the *chang* and the *pu* were known to the people of the Shang dynasty, not indeed by those names, which have not so far been found on the oracle-bones, but because from certain dates which were evidently considered of great importance by the inscribers of the bones, precisely those intervals can be derived. Two of these are eclipse dates (−1311 and −1304), which have been confirmed by modern computation methods as correct, and the other two are new moon dates (*shuo*;[3] −1313 and −1162).[e] The time elapsing is exactly 2 *pu* or 8 *chang*. If this should be confirmed we can hardly suppose that these periods were not known to the Babylonians[f] as well as to the Chinese of the Shang.[g]

Christian era', taught that the world was re-made and all species created anew, every 36,420 years. It would take a whole book, which we do not have time to write, to follow out all the ramifications of such beliefs. For this reference we are indebted to Sir Ronald Fisher.

[a] From ch. 21B of the *Chhien Han Shu*. This chapter forms a systematic, though empirical, treatise which, as Chatley (9) rightly emphasises, antedated the *Almagest* by over a century. Its independent status has been much obscured by the fact that it became incorporated so soon in a dynastic history. Largely due to Liu Hsin,[4] the chapter embodies his San Thung Li[5] (Three Sequences Calendar) of −7. Some appreciation of the further implications of this title may be gained from Fêng Yu-Lan (1), vol. 2, p. 58.

[b] Forty-six ordinary *tzolkins*, cf. Spinden (1), p. 141.

[c] Above, p. 186.

[d] *Yin Li Phu*, Pt. II, ch. 1, pp. 11*a*, 12*a*; ch. 4, p. 5*a*.

[e] The fixing of the Shang calendar involves working backwards from the epoch of the Chuan Hsü[6] calendar of −366 or −370, when the beginning of the sexagenary cycle of days happened to coincide with the winter solstice and a new moon (*Yin Li Phu*, Pt. I, ch. 2, p. 2*a*; see also Shinjō, 1).

[f] Cf. Fotheringham (1); and Heath (4) on Naburiannu (*fl.* −500) and Kidinnu (another contemporary of Shih Shen; *fl. c.* −380).

[g] Tung Tso-Pin's discovery would invalidate the argument of de Saussure (22, 22*a*) that the calendar table in ch. 26 of the *Shih Chi* is a later interpolation by Chhu Shao-Sun,[7] because the *pu* period is mentioned in it. Chhu did, of course, carry the table down to −29, the last of the 76 years starting from −104, and long after the death of Ssuma Chhien.

[1] 紀母 [2] 統母 [3] 朔 [4] 劉歆 [5] 三統曆 [6] 顓頊
[7] 褚少孫

It was natural that the Han astronomers should wish to incorporate into these systems the synodic and sidereal periods of planetary revolutions. Their thought was that at the 'beginning' (of the world, or of a world-cycle) there had been a general conjunction (*chü*[1]) of planets, and at the end of the world this would happen again.[a] There are conjunctions of Jupiter and Saturn at almost the same place in the sky every 59·5779 years—by coincidence very close to the sexagesimal year cycle (Chatley, 15, 20). Conjunctions of Jupiter, Saturn and Mars recur every 516·33 years, which may account for the 500+ year period given in *Mêng Tzu*[b] for the recurrence of great sages. Planetary conjunctions (*ho*[2])[c] and occultations (*shih*,[3] *ju*[4])[d] were closely watched. Occasionally there may have been a distortion of date for political reasons, as in the conjunction[e] of −205 investigated by Dubs (21) and Nōda (*4*)[f] but more often the records, when recalculated today, are found to be quite reliable, e.g. the occultation of Mars by the moon in −69 and of Venus in +361 (Hind). We hear from time to time of predictions proved to be accurate, such as one by Tou Yen[5] in the Wu Tai period, *c.* +955. The notion of the 'general conjunction', found also in Greek and Hellenistic literature, seems to derive, according to Schnabel, from Berossos and the Babylonians. Now 3 *thung* periods would be equivalent to 9 times the lunar eclipse cycle of 135 lunations, to 4617 years, and to 28,106 sexagesimal day cycles; and this was the unit which the Han scholars, notably Liu Hsiang and Liu Hsin, combined with the planetary cycles. It was thought that in 138,240 years, all the planets would exactly repeat their motions, so that by combining this 'cogwheel' with the 4617-year period, a whole 'world-cycle' was taken to be 23,639,040 years. The beginning of it was known as the 'Supreme Ultimate Grand Origin' (*Thai Chi Shang Yuan*[6]). This idea continues, in various forms, throughout the later history of Chinese astronomy.[g]

[a] Such a thing is of course impossible. Planets may agree in longitude or in latitude, but never in both at the same time. Cf. what was said above in Section 16*d* on the Neo-Confucian school of philosophers, about 'world-periods', Vol. 2, pp. 485ff.

[b] VII (2), xxxviii.

[c] One of the most famous is of course that referred to in the *Chu Shu Chi Nien* (Bamboo Books—semi-apocryphal, cf. Vol. 1, pp. 74, 79, 165), which may be that of −1059, but was perhaps computed backwards in later times to suit theory (Chatley, 15, 21). There was a technical term for each type of conjunction. When Jupiter met Venus it was a 'combat' (*tou*[7]), but when Mars did so it was a 'fusion' (*shuo*[8]). Another metallurgical term, 'quenching' (*tshui*[9]), applied to conjunctions of Mars with Mercury, while irrigation engineering gave 'blocked channels' (*yung chü*[10]) for meetings of Mercury with Saturn. These terms arose quite naturally from Five-Element theory.

[d] Another term *hsi*,[11] 'raiding', was used for occultation preceded by rapid approach.

[e] *Chhien Han Shu*, ch. 26, p. 23*b*; cf. *Shih Chi*, ch. 27, pp. 15*b*, 42*a* (Chavannes (1), vol. 3, pp. 357, 407).

[f] Another case occurred in March 1725, when a conjunction of Mars, Venus, Mercury and Jupiter was represented to the throne by some of the Chinese astronomical officials as an auspicious general conjunction. But it had been carefully observed by Gaubil, Jacques and Kögler (see Souciet, vol. 1, pp. 101ff. and opp. p. 208), who protested (Pfister (1), p. 646). We thank Mr P. A. Jehl, who studied this incident with the late Fr. Hersart de la Villemarqué, S.J., for bringing it to our notice.

[g] It was discussed, for instance, by Kirch (2) in 1727; des Vignoles (2) in 1737; and by Costard in his rather hostile account of Chinese astronomy in the *Phil. Trans.* for 1747. The reader will remember from p. 120 above the efforts made by I-Hsing in the +8th century to calculate the date of the 'Grand Origin'. It is significant of the amplitude of his conception of the universe that it worked out, not to −4004, but to some 97 million years before our era.

[1] 聚 [2] 合 [3] 蝕 [4] 入 [5] 竇儼 [6] 太極上元
[7] 鬬 [8] 爍 [9] 焠 [10] 壅沮 [11] 襲

(i) RECORDS OF CELESTIAL PHENOMENA

(1) Eclipses

The conception of an eclipse of the sun or moon as the gradual eating of the luminary by some celestial dragon manifests itself in the earliest term for eclipse, *shih*,[1] which appears on the Anyang oracle-bones.[a] This word[b] simply means 'to eat', and the original pictograph showed a food vessel with a lid. The formation of a more precise technical term (*shih*[2]) by the addition of the 'insect' radical (Rad. 142), did not take place till comparatively late, after the Han.

How far back in Chinese history eclipses were observed is a question which has been involved in dispute for centuries, ever since growing confidence in eclipse cycles induced I-Hsing in the +8th century to make computations about the eclipses of the Hsia and Shang dynasties. Traditionally, the solar eclipse recorded in the *Shu Ching* (Historical Classic)[c] was regarded as of the −3rd millennium. 'On the first day of the month, in the last month of autumn, the sun and moon did not meet (harmoniously) in Fang (*hsiu*)—(*Nai chi chhiu, yüeh shuo, chhen fu chi yü Fang*[3])', so runs the text. Identifications of this have varied from −2165 to −1948 according to the choice of successive students,[d] but since the period in question is clearly legendary in character, and the text almost certainly late,[e] attempts to fix the year were abandoned.[f] The opinion then grew up that the earliest recorded solar eclipse was that in the *Shih Ching* (Book of Odes),[g] which, on the basis of intensive studies by Hartner (5), Hirayama & Ogura (1), and Nōda,[h] has been identified as having occurred in −734.[i] This used to be accepted as the most ancient verifiable eclipse in the history of any people (Fotheringham, 2) since the Babylonian records were all lost after the time of Ptolemy. Copies of some of these, dating from −747 (Nabonassar's time) or even perhaps −763, had apparently been sent to Greece by Aristotle's nephew Callisthenes after the capture of Babylon by Alexander the Great in −331.[j] Recent researches on the oracle-bones, however, have greatly increased the priority of the Chinese as to extant records, since

[a] For parallels in other cultures cf. Forke (6), p. 98. We have already come across a recognition by Hsün Chhing in the −3rd century of the conventional folklore character of the idea in his time (Sect. 14*b* in Vol. 2, p. 365).

[b] K921.

[c] Ch. 9 (Yün Chêng), tr. Legge (1), p. 82; Medhurst (1), p. 127.

[d] Gaubil; Williams (2); v. Oppolzer (1); Schlegel & Kühnert (1); S. M. Russell (1); Rothman (1); Liu Chao-Yang (*2*).

[e] As was noted above (p. 189) it is generally regarded as a forgery of the +4th century. Yet the sentence occurs in a quotation in the *Tso Chuan* under date −524, Duke Chao, 17th year (Couvreur (1), vol. 3, p. 275). But this may not mean that it is earlier than the −3rd century, since the *Tso Chuan* text was interpolated by Han Confucians. The same applies *a fortiori* to the quotation in the *Chu Shu Chi Nien*, ch. 3, p. 16*b*.

[f] See Hirayama & Ogura (1).

[g] Minor Odes, Chhi Fu decade, no. 9, in Legge (8); Karlgren (14), p. 138; Mao no. 193.

[h] (4), p. 365; (*8*).

[i] Replacing the date −776 accepted by Fotheringham (2). Hartner (5) and Waley (21) found that if the later year is accepted the allusions to contemporary events in the group of poems concerned make sense.

[j] Heath (4), p. xiv.

1 食 2 蝕 3 乃季秋月朔辰弗集于房

Tung Tso-Pin (*1*) was able to identify six lunar eclipses and one solar eclipse, recorded by the Shang people on bones which have come down to us.[a] These dates are, for the moon, −1361, −1342, −1328, −1311, −1304 and −1217 and, for the sun, −1217.[b] To these he joined a seventh lunar eclipse, that of −1137, mentioned in the *I Chou Shu*,[c] apparently correctly. In a later paper (*3*) he added the deciphered records of three more eclipses, one lunar, one solar and one which might be either, all undatable.

An interesting point is that the Shang people were watching lunar eclipses for calendrical rather than astrological reasons, since they can only occur on or very near the date of the full moon (*wang*[1]), and the bone inscriptions indicate the desire of their writers to check as closely as possible the dates of the lunations. We also have a hint of the degree of State organisation at the time, since the first and fifth of the eclipse-records just mentioned are followed by the word *wên*[2]—i.e. 'this was reported from the provinces, and not actually seen at the capital'. This technical term continued in use in the Han, a millennium and a half later. It is interesting that the solar eclipse in the *Shih Ching* is termed *chhou*[3] (ugly, abnormal), while lunar eclipses are termed *chhang*[4] (usual), which suggests that the latter were expected but the former were not.

(i) *Eclipse theory*

By what date did the Chinese arrive at an approximately correct understanding of the nature of eclipses? The question[d] is bound up with the methods used for eclipse prediction. In the West the first correct prediction of a solar eclipse is associated with the name of Thales of Miletus, who foretold the eclipse of −585,[e] presumably using the Babylonian *saros* period of 223 lunations.[f] The difficulty which the ancients encountered in dealing with eclipses simply arises from the fact that the moon's path does not follow the ecliptic exactly; if it did there would be a solar eclipse at every new moon, and a lunar one at every full moon. Hence the gradual development of various arbitrary periods permitting approximate predictions of when eclipses would be likely to recur.

Of the recognition of the reflection of sunlight from the moon, something has already been said.[g] Though the statement that the moon is illuminated by the sun is attributed to Parmenides (−6th century), it seems perhaps more likely that the true discoverer of this fact, necessary for the understanding of the nature of lunar eclipses, was Anaxagoras (*b.* −500).[h] In China, however, though the observation of eclipses had

[a] See also the papers of Liu Chao-Yang (*1, 2, 8, 9*).

[b] These dates are in historical, not astronomical, reckoning; the latter will of course be one year less. All agree with the Canon calculated by Tung Tso-Pin (*Yin Li Phu*, Pt. II, ch. 3, pp. 7*b* ff.), and all but one (−1328) agree with the Canon calculated by Dubs (22), who, however (26), prefers a series of dates about 150 years later.

[c] Cf. Vol. 1, p. 165.

[d] The early work of Wylie (14) on this is not now very helpful.

[e] 18th May; Herodotus, I, 74.

[f] Clement of Alexandria, *Stromata*, I, 65; Pliny, *Hist. Nat.* ch. 12, sect. 53.

[g] P. 227 above.

[h] Heath (4), p. 27.

1 望 2 聞 3 醜 4 常

evidently been systematically pursued in an age contemporary with pre-Hellenic and Old Babylonian culture, the understanding of their nature seems to have come later than in Greece.[a]

Shih Shen in the −4th century was certainly well aware that the moon had something to do with solar eclipses, because he gave instructions for predicting them based on the relative position of the moon to the sun.[b] When the moon was in conjunction (*chiao*[1]) with the sun, during the dark nights (*hui*[2])[c] at the beginnings and ends of lunations, such eclipses could, he believed, occur at any time. According to Maspero,[d] Shih Shen probably thought, not that the moon's substance was interposed between the sun and the earth, but that a Yin influence radiating from it overcame the Yang influence of the sun.[e] This explains why Kan Tê (perhaps considering sun-spots as incipient eclipses), could speak of eclipses beginning at the centre of the sun and spreading outwards.[f] It probably also explains why the late Chou and Han astronomers spoke of 'veilings' (*po*[3]) of the sun, which were not full eclipses—e.g. Ching Fang[4] in the −1st century.[g] The Babylonians, too, had believed that eclipses could occur on days of the month other than the 1st and the 15th (Sayce, 2). Possibly these were effects produced by high fog. As for lunar eclipses, the −4th-century astronomers seem to have been less well informed than those of the Shang, since they supposed that they could occur at any time during the lunation.

In Ssuma Chhien's time (*c.* −100), it was thought that the eclipses of the moon could be due to influences emanating from almost any heavenly body, including not only the five planets but also Arcturus and Antares.[h] Ssuma Chhien knew that there were definite periods for lunar eclipses, but he did not say this for solar ones, and ventured upon no theory for the causation of the latter.

The 'radiating influence' theory was still alive in the +1st century, because Wang Chhung mentions it. Yet from what he says it is clear that by then the correct view had long been known. He himself argues against this, preferring to adopt a theory of periodical 'shrinkage' or fading of the *chhi* of the two great luminaries. Why he thus opposed the correct view is an interesting problem in itself, and the whole passage is so valuable that it is worth reproducing in full. He says:

According to the scholars, solar eclipses are brought about by the moon. It has been observed that they occur at times of new moon (lit. on the last days (*hui*[2]) and first days (*shuo*[5]) of months), when the moon is in conjunction (*ho*[6]) with the sun, and therefore the moon can eclipse it. In the 'Spring and Autumn' period there were many eclipses, and the *Chhun Chhiu* says that at such and such a month at new moon there was an eclipse of the

[a] Mythological theories such as the eating of the sun by the three-legged crow supposed to inhabit it, and of the moon by its toad or hare, can be found in such texts as *Kuan Tzu*, ch. 55 (see Haloun, 5), and Chhu Shao-Sun's[7] ch. 128 of the *Shih Chi*.

[b] *Khai-Yuan Chan Ching*, ch. 9, p. 2*a*.

[c] For the discussion between Kuo Mo-Jo and Hopkins as to whether this word occurs in the oracle-bones and as to its relation with *shih* (eclipse), see Hopkins (30).

[d] (3), p. 292.

[e] Cf. *Huai Nan Tzu*, ch. 3, p. 2*a*.

[f] *Khai-Yuan Chan Ching*, ch. 9, p. 9*a*.

[g] *Khai-Yuan Chan Ching*, ch. 17, p. 1*a*.

[h] *Shih Chi*, ch. 27, p. 30*a*, tr. Chavannes (1), vol. 3, pp. 386 ff.

1 交 2 晦 3 薄 4 京房 5 朔 6 合 7 褚少孫

sun, but these statements do not imply that the moon did it. Why should (the chroniclers) have made no mention of the moon if they knew that it was really responsible?[a]

Now in such an abnormal event the Yang would have to be weak and the Yin strong, but (this is not in accord with) what happens on earth, where the stronger subdue the weaker. The situation is that at the ends of months the light of the moon is very weak, and at the beginnings almost extinct; how then could it conquer the sun? If you say that eclipses of the sun are due to the moon consuming it, then what is it that consumes (in a lunar eclipse) the moon? Nothing, the moon fades of itself. Applying the same principle to the sun, the sun also fades of itself.[b]

Roughly speaking, every 41 or 42 months there is a solar eclipse, and every 180 days there is a lunar eclipse.[c] The reason why the eclipses have definite times is not (as the scholars say) that there are (recurring) abnormal events due to the periods (of the moon's cycle), but because it is the nature of the *chhi* (of the sun) to change (at those times). Why should it be said that the moon has anything to do with the times of (changing of) the sun('s *chhi*) at the first and last days of months?[d] The sun should normally be full; if there is a shrinkage (*khuei*[1]) it is an abnormal event (and the scholars say that) there must be something consuming (the sun). But in such cases as landslides and earthquakes, what does the consuming then?[e]

Other scholars say that when the sun is eclipsed, the moon covers it (*yüeh yen chih*[2]). The sun is further away (lit. above), but the moon being nearer (lit. below) obstructs (*chang*[3]) the view of its shape. The sun and moon being in conjunction, they intrude (*hsi*[4]) upon each other. If the moon were further away (lit. above) and the sun nearer (lit. below), then the moon could not cover the sun. But since the opposite is true, the sun is obstructed, its light is covered by that of the moon, and therefore a solar eclipse is caused. Just as, in gloomy weather, neither sun nor moon can be seen. When the edges contact, the two consume each other; when the two are concentric they face each other exactly covered and the sun is nearly extinguished. That the sun and moon are in conjunction at times of new moon is simply one of the regularities of the heavens.

But that the moon covers the light of the sun in solar eclipses—no, that is not true. How can this be verified? When the sun and moon are in conjunction and the light of the former is 'covered' by the latter, the edges (*yai*[5]) of the two must meet at the beginning, and when the

[a] Wang Chhung thus begins with a very bad argument, especially for him, the appeal to written authority.

[b] In this paragraph Wang Chhung is led away by too much reliance on the Yin and Yang theory. Moreover, later on he argues in contradiction of the view that what happens on earth is any guide to understanding what appears to happen in the heavens.

[c] Cf. *Lun Hêng*, ch. 53 (Forke (4), vol. 2, p. 14), where other short periods are given. Wang Chhung certainly did not realise that solar eclipses are more frequent than lunar ones, but less likely to be seen from any particular station. Presumably his figure of 180 days is connected with the time taken by the sun to pass from one node to another, 173 days; or with the 177 days in 6 lunations (cf. Spencer-Jones (1), p. 180). Ssuma Chhien had spoken of periods of 113 and 121 months (*Shih Chi*, ch. 27, p. 30*b*; Chavannes (1), vol. 3, p. 388). This was probably a groping after the Chinese 135-month period (see below, p. 421), established later in the −1st century.

[d] Forke thought that this sentence referred to the fact that new moons constantly recur without necessarily involving eclipses. This would have been a very suitable argument for Wang Chhung to have brought forward, but it does not seem quite clear that this was what he meant.

[e] Wang Chhung has now stated his own view, namely, that nothing 'consumes' the sun or comes in front of it, but that it shrinks or fades according to a rhythm of its own. And the moon has a similar rhythm. His arguments, weak though they are, support rather strikingly just the kind of conception which would naturally arise in an organic world-outlook (cf. p. 287 above, in Vol. 2, Sect. 13*f*). The moon's proximity at times of solar eclipse was for him just a coincidence.

[1] 虧 [2] 月掩之 [3] 障 [4] 襲 [5] 崖

light reappears they must have changed places. Suppose the sun is in the east and the moon is in the west. The moon falls back[a] (lit. moves) quickly eastwards and meets the sun, 'covering' its edge. Soon the moon going on eastwards passes the sun. When the western edge (of the sun), which has been 'covered' first, shines again with its light, the eastern edge which was not 'covered' before, should (now) be 'covered'. But in fact we see that during an eclipse of the sun the light of the western edge is extinguished, yet when (the light) comes back the western edge is bright (but the eastern edge is bright also). The moon goes on and covers the eastern (inner) part as well as the western (inner) part. This is called 'exact intrusion' (*ho hsi*[1]) and 'mutual covering and obscuring' (*hsiang yen chang*[2]). How can these facts be explained (by astronomers who believe that the moon covers the light of the sun in solar eclipses)?[b]

Again, the scholars assert that the bodies of the sun and moon are quite spherical. When one looks up at them, their shape seems like that of a ladle or a round basket, perfectly circular. They are not the *chhi* of a light seen from far off, for *chhi* could not be round. But (my opinion is) that in fact the sun and moon are not spherical; they only appear to be so on account of the distance. How can this be verified? The sun is the essence of fire, the moon the essence of water. On earth, fire and water never assume spherical forms, so why should they become spherical only in the heavens? The sun and moon are like the five planets, and these in turn like the other stars. Now the other stars are not really round, but only appear to be so in their shining, because they are so far away. How do we know this? In the 'Spring and Autumn' period, stars fell down (upon the earth) at the capital of the State of Sung.[c] When people went near to examine them, it turned out that they were stones, but not round. Since these (shooting) stars were not round, we may be sure that the sun, the moon, and the planets are not spherical either.[d]

To summarise, it is clear that the correct theory was already widely held in Wang Chhung's time (*c.* +80). Yet he preferred the view that the sun and moon had intrinsic rhythms of brightness of their own,[e] an idea almost identical with some of the speculations of Lucretius on the subject.[f] He supported this with objections drawn, for instance, from observations of annular eclipses, and from rather sophisticated doubts about the shapes of the heavenly bodies.[g] But perhaps the real interest of his position lies in the fact that a Confucian sceptic[h] was criticising the theories of

[a] Cf. the statement in the *Chin Shu*, quoted on p. 219 above, that the moon falls back 13° eastwards each day, while the sun falls back 1°.

[b] We feel convinced that the objection of this paragraph is based upon observations of annular solar eclipses. With no knowledge of the dimensions or distances of the two bodies, the phenomenon of annularity might have been puzzling to Wang Chhung and his contemporaries, whose ideas of perspective were not very developed.

[c] *Tso Chuan*, Duke Hsi, 16th year (Couvreur (1), vol. 1, p. 310). This incident is dated −643.

[d] *Lun Hêng*, ch. 32, pp. 5*b* ff., tr. auct., adjuv. Forke (4), vol. 1, pp. 269 ff.

[e] Somewhat like the pulsating variable stars of modern astronomy. Variable stars were noted in medieval China and Japan (Iba (1), pp. 141, 143).

[f] *De Rer. Nat.* v, 719 ff. and esp. 751 ff.

[g] Ko Hung's criticisms of the views of Wang Chhung on this are in the *Chin Shu*, ch. 11, p. 5*a*. Ko Hung rightly enough adduced the shapes seen in eclipses to show that the heavenly bodies were really spherical. Cf. Aristotle, *De Caelo*, 297 b 25.

[h] At the beginning of the chapter from which we have quoted, Wang Chhung made one of his sceptical attacks on the folklore stories of animals visible in the sun and moon. But his conservatism is shown by the fact that on the whole he supported Kai Thien views (cf. p. 218 above).

[1] 合襲 [2] 相掩障

practical astronomers not unconnected with Taoism. Wang Chhung was probably not inclined to draw any great distinctions between magicians like Luan Ta and star-clerks like Lohsia Hung, or between Ching Fang the prognosticator and Liu Hsin the calendar-calculator. The fact that the imperial court employed them all merely discredited the court. He, Wang Chhung, was against them all, and all their theories—but sometimes the theories happened to be right.

The correct view is to be found in Liu Hsiang's *Wu Ching Thung I*[1] (The Fundamental Ideas of the Five Classics), *c.* −20: 'When the sun is eclipsed it is because the moon hides him as she moves on her way.'[a] It would seem likely, then, that the right view was reached at some time between the early Warring States period and the middle of the Former Han; perhaps the naturalists of Tsou Yen's school had something to do with it. After Wang Chhung, it was well established; Chang Hêng says in his *Ling Hsien* about +120:

> The sun is like fire and the moon like water. The fire gives out light and the water reflects it. Thus the moon's brightness is produced from the radiance of the sun, and the moon's darkness (*pho*[2]) is due to (the light of) the sun being obstructed (*pi*[3]). The side which faces the sun is fully lit, and the side which is away from it is dark. The planets (as well as the moon) have the nature of water and reflect light. The light pouring forth from the sun (*tang jih chih chhung kuang*[4]) does not always reach the moon owing to the obstruction (*pi*[3]) of the earth itself—this is called '*an hsü*',[5] a lunar eclipse.[b] When (a similar effect) happens with a planet (we call it) an occultation (*hsing wei*[6]); when the moon passes across (*kuo*[7]) (the sun's path) then there is a solar eclipse (*shih*[8]).[c]

Wang Chhung's pulsation theory is an admirable example of the inhibitory effect of a world-view otherwise good in itself—organic naturalism. That this was the *philosophia perennis* of China we sought to show in the second volume of the present work. Even the simple conceptions of converging paths and mutual shadowing were too mechanistic for Wang Chhung; he preferred to believe in a rhythm emanating from the intrinsic natures of the celestial organisms in question. A similar inhibition of thought can be seen in the *Lun Thien*[9] (Discourse on the Heavens)[d] of a somewhat later thinker Liu Chih,[10] written about +274. Though quite aware of the intersection of the paths of the moon and sun, he cannot believe that in an organic hierarchic universe the inferior Yin could ever normally cover and obstruct the superior Yang, so as to cause a solar eclipse. Moreover, a lunar one cannot be due to the earth's shadow only since Yang and Yin being inseparable the moon has some Yang and would still shine by a lesser light of its own.

[a] Cit. in *Khai-Yuan Chan Ching*, ch. 9, p. 3*a*.

[b] The technical term *an-hsü* continued in use at least until the Ming (cf. the *Li Hai Chi* of Wang Khuei, *c.* +1400; p. 5*a*).

[c] *YHSF*, ch. 76, p. 63*b*, tr. auct.; cit. also in comm. *Hou Han Shu*, ch. 20, p. 4*a*. Cf. *Khai-Yuan Chan Ching*, ch. 1. See below, p. 421, concerning the technical term *kuo*.

[d] *CSHK* (Chin sect.), ch. 39, pp. 5*a* ff.

[1] 五經通義 [2] 魄 [3] 蔽 [4] 當日之衝光 [5] 闇虛 [6] 星微
[7] 過 [8] 食 [9] 論天 [10] 劉智

The Yin and the Yang respond (*ying*[1]) to one another; what is pure receives light, what is cold receives warmth—such communication needs no intermediary. They can mutually respond in spite of the vast space which separates them. When a stone is thrown into water, (the ripples) spread forth one after another; this is the propagation of the *chhi* of the water. Mutual echoing means mutual receptivity; there is no bound which (the mutual influence of things) cannot attain, there is no barrier which can stand in its way. (Thus it is that) the purest substance (i.e. the moon) receives the light of the Yang (the sun).... Yin and Yang receiving from each other, when one flourishes the other must decay.... If there were only a reflection of light (between sun and moon) and no question of the mutual radiation and reception of *chhi*, then the brightness of the Yin ought to flourish when the Yang is flourishing (yet it is eclipsed at times when there is no solar eclipse), and it ought to decay when the Yang decays (yet new moons accompany solar eclipses). There would then be no explanation of the differences between the sun and moon.[a]

Thus here again we see a mind prevented from accepting the simplest explanation of the facts by a preconceived theory of an organic universe of inevitably interacting parts. It is true that elsewhere, e.g. in the understanding of the cause of the tides, this deep belief in action at a distance could prove beneficial.[b]

It is of interest to listen to a Chinese astronomer of the Sung dynasty discussing eclipses in the +11th century. Shen Kua wrote as follows (+1086):

When I was compiling and editing books in the Chao Wên (Kuan[2])[c] I participated in detailed discussions concerning the equipment of the Astronomical Observatory with new armillary spheres....

The Director[d] asked me about the shapes of the sun and moon; whether they were like balls or (flat) fans. If they were like balls they would surely obstruct (*ai*[3])[e] each other when they met. I replied that these celestial bodies were certainly like balls. How do we know this? By the waxing and waning (*ying khuei*[4]) of the moon. The moon itself gives forth no light, but is like a ball of silver; the light is the light of the sun (reflected). When the brightness is first seen, the sun (-light passes almost) alongside, so the side only is illuminated and looks like a crescent. When the sun gradually gets further away, the light shines slanting, and the moon is full, round like a bullet. If half of a sphere is covered with (white) powder and looked at from the side, the covered part will look like a crescent; if looked at from the front, it will appear round. Thus we know that the celestial bodies are spherical.

(On the other hand) the sun and moon are made of *chhi*, having form but no solid substance; hence they could meet without obstructing one another.

He also said: 'Since the sun and moon are in conjunction (*ho*[5]) and in opposition (*tui*[6]) once a day,[f] why then do they have eclipses only occasionally?' I answered that the ecliptic and the moon's path are like two rings, lying one over the other (*hsiang tieh*[7]), but distant by a small amount.[g] (If this obliquity did not exist), the sun would be eclipsed whenever the two bodies were in conjunction, and the moon would be eclipsed whenever they were

[a] Tr. auct.
[b] Cf. p. 483.
[c] The 'College of the Glorification of Literature', attached directly to the Imperial Chancery.
[d] Doubtless a lay official charged with administration.
[e] We shall meet with this term again in the Section on optics (26*g*) below.
[f] Of twenty-four hours.
[g] Actually $5\frac{1}{4}°$.

1 應 2 昭文舘 3 礙 4 盈虧 5 合 6 對 7 相疊

exactly in opposition. But (in fact) though they may occupy the same degree,[a] the two paths are not (always) near (each other), and so naturally the bodies do not (always) intrude upon one another. When at conjunction (in right ascension) and also near (in declination) (*thung tu erh yu chin*[1]), that is to say, when the ecliptic and the moon's path cross each other (at one or other of the nodes), then the sun and moon invade and cover each other. The node (*chiao chhu*[2]) is (the point of) complete eclipse. If the conjunction is not central and symmetrical (*pu chhüan tang chiao tao*[3]), the eclipse will be more or less partial according to the extent of the mutual interference.

In a solar eclipse (I continued), if the moon's path enters from outside and crosses over to the inside (of the ecliptic), first contact will be in the south-west and last contact in the north-east.[b] If it starts from the inside and crosses outwards, the contrary will be the case. If the sun is east of the node, it will be eclipsed from the inner side; if it is to the west, the eclipse will begin on the outer side. A total eclipse which has started due west will end due east....[c]

The node retreats more than one degree a month,[d] so 349[e] days[f] make up the whole period (*chi*[4]) (of the synodic revolution of the node).[g]

The Western (i.e. Indian) conception of 'tracing Rahu and Ketu[h] backwards' is only a name for what we call the path (of retreat) of the nodes. Rahu is the ascending one (*chiao chhu*[5]) and Ketu the descending one (*chiao chung*[6]).[i]

Later in the Sung period (about +1180) the philosopher Chu Hsi gave quite a clear account of eclipses in commenting on one of the *Shih Ching* songs. He said:

At the end of a lunar month, a solar eclipse occurs when there is a conjunction (*ho*[7]) of the sun and moon at the same degree (*thung tu*[8]) of east and west (right ascension), and on the same line (*thung tao*[9]) of north and south (declination). The moon then covers (*yen*[10]) the sun and so causes the eclipse. Similarly, at the full moon, when it is in opposition (*tui*[11])with the sun at the same degree and along the same line, it is 'protected' (*khang*[12]) from the sun, and there is a lunar eclipse.[j]

When the Taoist Chhiu Chhang-Chhun was on his way from Peking to the Mongol court and to visit Chingiz Khan at Samarqand, he and his party observed a total solar

[a] In right ascension or celestial longitude.

[b] Because of the 5° inclination of the moon's path to the ecliptic. Since the moon is falling back through the stars eastwards faster than the sun (though the apparent motion of both is westwards), eclipses must begin at the west side.

[c] There follows a repetition of similar information, *mutatis mutandis*, for lunar eclipses.

[d] The correct figure is 19° 21′ per annum.

[e] Some texts write 249 here, but this must be a copyist's error.

[f] Lit. conjunctions and oppositions.

[g] I.e. the time required for the sun to come back to the nodes. The correct figure is 346·62 days.

[h] See p. 228 above.

[i] *MCPT*, ch. 7, paras. 13–16, tr. auct., incl. Liao Hung-Ying. Cf. the comments of Hu Tao-Ching (*1*), vol. 1, pp. 308ff.

[j] *Shih Chi Chuan*[13] (Collected Commentaries on the Book of Odes), cit. in *Ko Chih Ku Wei*, ch. 1, p. 8*a*, *b*, tr. auct. See also *Chu Tzu Chhüan Shu*, ch. 50, pp. 1*b*, 12*a*; *Chu Tzu Yü Lei*, ch. 1, p. 9*b*.

1 同度而又近 2 交處 3 不全當交道 4 朞 5 交初
6 交中 7 合 8 同度 9 同道 10 揜 11 對 12 亢
13 詩集傳

eclipse. This was on 23 May +1221, at the Kerulen River in Northern Mongolia.[a] After they had arrived at Samarqand in 1222, Chhiu Chhang-Chhun had a discussion with an astronomer there concerning the information which they had systematically gathered along their route about the differing times and degrees of totality noted in different places. Li Chih-Chhang, the secretary of the expedition, recorded this.[b] 'It is just as though one covered a candle with a fan', said Master Chhiu. 'In the direct shadow of the fan there is no light, but the further one moves to the side, the greater the light becomes'. This must be one of the earliest investigations in history of the path of an eclipse shadow on the earth's surface.

(ii) *Extent, reliability and precision of the records*

Thirty-seven eclipses, from −720 onwards, are recorded in the *Tso Chuan*; their identifications, naturally of much importance for chronology, have often been discussed.[c] By a remarkable coincidence, the list of lunar eclipses given by Ptolemy in the *Almagest* started from −721. From the beginning of the Han onwards there are systematic records in all the dynastic histories. Wylie (8) collected data from them for 925 solar[d] and 574 lunar eclipses, down to +1785, but the fullest list is that of Huang (3).[e] Recent statistical summaries which may be mentioned are those of Wên Hsien-Tzu (2), Kao Lu (*1*) and Chu Wên-Hsin (*4*). The latter author has also devoted a monograph especially to eclipses of the sun in Chinese records (*2*).[f]

The eclipses of the Han dynasty have been studied with particular care.[g] Attempts have been made to assess the reliability of the official astronomers of ancient times; thus Eberhard (6), Airy, and Dubs (2) give figures shown in Table 36. As to the last class, bad weather for observation might account for some gaps, but this is not likely, since the missing ones are evenly distributed over the months of the year.[h] Eberhard suggests that the 'impossible' ones were calculated afterwards according to some cycle and interpolated in the records; if this were so, some of the unidentified class might also be accounted for in this way. Alternatively, eclipses may have been put in for

[a] *Chhang-Chhun Chen Jen Hsi Yu Chi*, ch. 1, p. 10*a*.

[b] In his *Chhang-Chhun Chen Jen Hsi Yu Chi*, ch. 1, p. 22*a*, tr. Waley (10), pp. 66, 94. Goodrich (9), to whom we owe a reminder of this interesting incident, makes the astronomer at Samarqand Chinese, but this does not seem to follow absolutely from the wording of the text, though a few pages later (p. 23*b*, tr. p. 97) a Mr Li is mentioned, who was 'in charge of the Observatory'. It would indeed be interesting to know more about him and his colleagues. He can hardly be identical with the Li Ta-Hsi (or Ta-Chi or Na-Hsi), the Chinese scholar who, according to al-Banākitī, assisted Rashīd al-Dīn al-Hamdānī in the preparation of the *Jāmi' al-Tawārikh* of +1305, for that would be at least a generation later (information kindly supplied by Dr Zaki Validi Togan).

[c] Chu Wên-Hsin (*2*); Williams (3); S. J. Johnson (1); Eberhard, Müller & Henseling (1); Schjellerup (1); Fotheringham (2).

[d] Chhen Tsun-Kuei (2) counts 985, and Chu Wên-Hsin (*2*) 921.

[e] These are checked against the Canon of v. Oppolzer (2) and may be compared with the Canon of Ginzel (2) for the Mediterranean region, and the tables of P. V. Neugebauer.

[f] Japanese eclipse records will be found in the monograph of Kanda (*1*); all are taken from the *Nihongi* chronicle which begins about +620 (see Snellen, 1). For Korean eclipse records see Rufus (2), pp. 16, 44.

[g] Dubs (23, 24) and relevant chapters in (2). That of +31 received a special study by Kirch (1) in 1723.

[h] Remembering China's monsoon climate.

political reasons (praise or blame; *pao pien*[1]), in order to criticise the government, or omitted (twenty-three for both periods) when the rulers were doing better. Thus during the reign of the unpopular and criminal empress of Kao Tsu, there occurred an eclipse announcement (−186), though none could have taken place. Then there are long blank periods when nothing was recorded (e.g. from −177 to −160, from −68 to −56, and from −54 to −42). The opinion of Dubs,[a] however, is very favourable to the Han astronomers; he believes that the 'impossible' class can be explained by textual errors, and finds little evidence that eclipses were fabricated for political purposes.

Table 36. *Observations of solar eclipses by ancient Chinese astronomers*

	Chhun Chhiu (Tso Chuan)		*Chhien Han Shu*	
Identifiable and verified by modern computation				
Very striking	21		12	
Visible	5		9	
Not striking	2		6	
Hardly visible	3		6	
Partial	1		5	
		32		38
Identifiable if slight textual errors are assumed	0		14	
Unidentified	3		0	
Impossible	2		3	
Total number recorded	37		55	
Computed by modern methods as striking, but of which no observation was recorded	14		28	

Recently, Bielenstein has re-examined the problem, representing graphically the statistics of eclipses recorded by reigns as a function of those which should have been recorded. It appears that the records, while never falsified (except in the case of the empress just mentioned), were often left incomplete, and the extent of their incompleteness tallies with the popularity of the reign.[b] If 'warnings from Heaven' seemed not to be needed, the astronomers may have noted eclipses but refrained from memorialising about them, hence they were not recorded by the historiographers. Bielenstein shows that a similar relationship holds good for all other kinds of portents, and brings evidence in favour of the view (obviously plausible) that 'popularity' meant popularity with the high officials of the court rather than with the mass of the people. The Han records are therefore most likely to be complete in the reigns most unsatisfactory to the Confucian bureaucracy.

[a] (2), vol. 1, pp. 212, 288, vol. 3, pp. 551, 559.

[b] Perhaps the weakness of Bielenstein's argument is the lack of evidence on which he (or anyone else, as yet) can base an estimate of this.

[1] 褒貶

In his examination of the Yuan period, H. Franke (1) confirms the view that the association of eclipses with other portents must always be taken into consideration. During the reign of the last Yuan emperor, Shun Ti, a great number of portents were recorded, but it may be doubted whether they were really more frequent then. There was a stylised pattern of what was supposed to take place at the catastrophic close of a dynasty.

Many passages must exist, if one could only find them, throwing light on the customs and mental processes of the members of the Astronomical Bureau in the various dynasties, and Franke discovered an excellent one in the *Shan Chü Hsin Hua* (New Discussions from the Mountain Cabin), written by Yang Yü in +1360. It throws a flood of light on what went on. Yang Yü says:

> When I was a Co-signatory Observer in the Bureau of Astronomy, there came a special imperial edict that we were to pay particular attention to celestial presages. On the first day of the seventh month in the sixth year of the Chih-Yuan reign-period (+1340), there came (to my house) one of the Senior Observers, a Mr Chang, who asked me to go to the Observatory as quickly as possible. When we arrived there together, we were met by Commissioner Li, dressed up in state apparel, who said: 'Last night there appeared the Ching Hsing phenomenon.[a] That is a very auspicious omen. I consider that it ought to be memorialised immediately. I suppose we shall be richly rewarded.' So I looked up the files which contained the records of earlier memorials, and came to a very different conclusion. I said, 'Although the phenomenon has occurred on the last day of the month (i.e. at the new moon), its shape was slightly different from what it ought to be. Besides, if the Ching Hsing appears, there ought to be reports coming in of wine-sweet springs, phoenixes, purple herbs, and felicitous clouds, in order to corroborate (lit. assist, *fu*[1]) (the celestial omen). But (on the contrary) there are epidemics and catastrophes in Shensi, brigands and robbers in the central provinces, and in Fukien rebels are active. I am afraid it won't do. Why should the Tao of heaven be proclaiming the opposite (to the Tao of earth)?' But Mr Li was most obstinate, and stuck to his opinion. So I said 'Up to now, only the six Observers here have seen the phenomenon. In the unlikely possibility of its having been generally seen by people throughout the country,[b] will they not have taken it as an omen of evil?' Finally he agreed to wait and see if it appeared again (that night), before we memorialised about it. And indeed only nine days later the planet Venus 'crossed the meridian'.[c] All this shows how careful one has to be not to take lightly responsibilities like these.[d]

It is thus fairly clear that before the Chinese records can be made full use of by modern astronomers or meteorologists interested in periodicities, a good deal more historical analysis and research will be needed. Nevertheless, if they were not more accurate than would appear from some of their severest critics, it would have been

[a] What this was will be explained a few pages below (p. 422).

[b] This reminds us that in the absence of artificial light available on any considerable scale, nearly everyone went to bed at dusk and rose at dawn, so the official Observers on duty were not likely to be checked by any educated person, and others whose trade might keep them abroad at night were unlikely to know very much about the heavens.

[c] This was a very bad sign; cf. Schlegel (5), p. 635; Chavannes (1), vol. 3, p. 374.

[d] P. 14*a*, tr. H. Franke (2), no. 35, eng. auct. Cf. Franke (8).

[1] 副

impossible to find known periodicities in them, as has in fact been done, e.g. in the case of the sun-spot cycle (p. 435 below). A striking example of their correctness has been given by Dubs[a] with regard to the solar eclipse of −96. This had been among the unidentified group until Chavannes studied some of the calendar tablets discovered by Stein in a Han watch-tower in the desert, and found that Huang (3) had assumed wrongly the position of a certain intercalary month—then when the necessary correction was made, the statement of the *Chhien Han Shu* was shown to be accurate. Around −16 the astronomers were, for some reason or other, watching for solar eclipses with particular care, and Dubs thinks that they used some special means, such as mirrors, for detecting partial eclipses.[b] For one of these it is said: '(Heaven) caused the capital alone to know of it; the kingdoms in the four (directions) did not perceive it'. Conversely, the provinces had their observers. In −145 there was an eclipse visible only at sunrise at the tip of the Shantung peninsula; yet it was duly reported and recorded.

It is interesting to see the gradual advance in precision in the records. Already in the *Chhun Chhiu* there are three cases in which the word *chi*[1] occurs, showing that the eclipse was total. With regard to the solar eclipses of −442, −382 and −300, the *Shih Chi* says that the daylight was so darkened that the stars could be seen (*chou hui hsing chien*[2]). Han records have technical terms such as *chi chin*,[3] nearly total, and *pu chin jo kou*,[4] crescent-shaped; in addition to *chi*,[1] total. A partial eclipse of three-tenths (*san fên*[5]) is also once mentioned. The degree of partiality is recorded in all subsequent dynasties, and the Thang records have a phrase *ta hsing chieh chien*[6]—all the great stars could be seen. The Han records sometimes mention the duration of eclipses and the times of their onset and ending, correct to a quarter of an hour. Records of the Thang and Sung frequently have very exact details, though not always. The celestial positions of eclipses (number of degrees in what *hsiu*, etc.) are generally recorded in the Han and always in the Thang. Chu Wên-Hsin's (*2*) laborious check of all these eclipses with the Oppolzer Canon shows that the great majority of them were faithfully recorded.

(iii) *Eclipse prediction*

Chinese astronomers naturally devoted much attention throughout the centuries to the prediction of eclipses, though like all such efforts before the Renaissance, this could only be empirical. As is well known, the Babylonians had identified a period generally known as the *saros*[c] (18 years 11 days, i.e. 223 synodic), at the end of which eclipses recur in the same relative positions of sun and moon. We see now that this depends simply on the periods of revolution of the moon and its nodes relative to the

[a] (2), vol. 2, p. 141, vol. 3, p. 557.

[b] (2), vol. 2, p. 420. I think more probably the equivalent of smoked glass, i.e. a piece of almost transparent jade, mica or rock-crystal (cf. below, p. 436).

[c] The idea that the Babylonians used this term in this sense has been shown by Neugebauer (9), p. 135, to be a pure historical myth. For Berossos, the *saros* meant 3600 years. The association of the term with the eclipse period seems to have arisen from an injudicious emendation of Suidas by Halley.

1 旣　2 晝晦星見　3 幾盡　4 不盡如鉤　5 三分
6 大星皆見

sun. One of the difficulties in the empirical prediction of eclipses was that those of the sun are hard to observe from any one centre, each solar eclipse being visible over only a small path on the earth's surface. Yet there are more solar than lunar eclipses; in each *saros* about forty-one of the former and twenty-nine of the latter.[a] For solar eclipses to recur in approximately the same place three *saroi* (one *exeligmos* of Ptolemy) are necessary. It does not seem that the Han people recognised either of these periods, but they developed one of their own, the *shuo wang chih hui*[1] (later called *chiao shih chou*[2]) of 135 months,[b] during which twenty-three eclipses took place. This was much used by Liu Hsin[3] in the San Thung[4] (Three Sequences) calendar[c] of −7 (cf. Eberhard & Henseling, 1); and must have been developed during the −1st century. Some scholars have believed that Liu Hsin adjusted the dates of eclipses in the *Chhun Chhiu* to correspond with this cycle (Eberhard, Müller & Henseling).

By the early years of the +3rd century the path of the moon was analysed more clearly. Liu Hung's[5] methods of eclipse prediction recognised the nodes (the points where the moon's path crosses the ecliptic), terming them *kuo chou fên*,[6] and assessed the angle of the path with the ecliptic (*chien shu*[7]) as 6° approximately.[d] This was in the Chhien Hsiang calendar[e] of +206. In the same century Yang Wei[8] was able to predict the directions of first and last contact (*khuei chhi chio*[9] and *chhü chiao hsien*[10]) for solar eclipses.[f] This was much refined by Chiang Chi[11] (*c.* +390), who could apparently predict the extent of partiality.[g] Early in the +7th century Liu Chhuo[12] and Chang Chou-Yuan[13] gave the times of first and last contact (*chhi chhi*[14]), the position in the heavens (*so tsai*[15]), and the probable extent of partiality (*shih fên*[16]). Terms used in the Thang by I-Hsing[17] and other astronomers in their predictions were *chhu khuei*[18] for time of onset, *shih shen*[19] for time of greatest immersion, and *fu yuan*[20] for time of last contact. Annular eclipses were called *huan shih*.[21] At this time attempts were also made to predict the geographical path along which solar eclipses would be visible.

In the Sung the prediction of eclipses was sometimes assigned to one bureau,[h] the Thai Shih Chü,[22] and their observation to another, the Ssu Thien Chien.[23] Shen Kua's

[a] Cf. Spencer-Jones (1), p. 177; Berry (1), pp. 19, 56; Dubs (2), vol. 1, p. 163.

[b] The most accurate period is the 19th century one of Simon Newcomb—358 months.

[c] *Chhien Han Shu*, ch. 21B, p. 1*b*; *Hou Han Shu*, ch. 12, p. 18*a*.

[d] Liu Hung's figure in Chinese degrees is equivalent to 5° 54′. Hipparchus (−2nd century) had fixed it at 5°. The actual value is 5° 8′.

[e] *Chin Shu*, ch. 17, pp. 1*b* ff., esp. 7*a* ff. Cf. his treatise *Lun Yüeh Shih* (On Lunar Eclipses) preserved in *Hou Han Shu*, ch. 12, pp. 17*a* ff.

[f] *Chin Shu*, ch. 18, pp. 6*a*, *b*.

[g] In the middle of the +6th century there was a famous dispute about the prediction of an eclipse. Five leading astronomers of the Northern Chhi named four different hours on the same day, but none of them was exactly right. 'And so their disputing continued, without reaching any conclusion, till before long the dynasty was extinguished' (*Sui Shu*, ch. 17, p. 4*b*; *Chiu Thang Shu*, ch. 32, p. 1*a*).

[h] See p. 191 above, concerning the system of two parallel observatories at the capital.

1 朔望之會　2 交食週　3 劉歆　4 三統　5 劉洪
6 過周分　7 兼數　8 楊偉　9 虧起角　10 去交限
11 姜岌　12 劉焯　13 張胄元　14 起訖　15 所在
16 食分　17 一行　18 初虧　19 食甚　20 復圓
21 環食　22 太史局　23 司天監

Mêng Chhi Pi Than of the +11th century contains[a] an interesting passage about eclipse computation in his time, including a tribute to the work of his friend Wei Pho,[1] whose canon of eclipses accounted for nearly all those in the *Tso Chuan*, thus excelling even that of I-Hsing. The following passage (from the *Fêng Chhuang Hsiao Tu*[2] (Maple-Tree Window Memories) written in the early +13th century) shows that at that time there must have been widespread interest in eclipse calculations, and that the official astronomers apparently had no monopoly of accuracy in prediction.

> In the 4th year of the Chhing-Yuan reign period (+1198) the Astronomer-Royal predicted a solar eclipse for the first night of the ninth month, but persons who were not in office said that it would occur during the day time, and they proved to be right. In the 2nd year of the Chia-Thai reign period (+1202) the Astronomer-Royal predicted an eclipse for the second half of the noon double-hour on the first day of the fifth month, but Chao Ta-Hsien,[3] who had no official position, said that it would occur in the first half after three quarters, and would be only partial to the extent of three *fên*. The emperor ordered the Staff Writer Chang Ssu-Ku[4], Chu Chhin-Tsê[5] and others to supervise the checking of the matter by the Assistant Armillary Observer and Chao proved to be right. The astronomical officials were found guilty of negligence and severely punished. Since the capital moved to the south of the river, the calendar has been full of mistakes.[b]

Although the methods of eclipse prediction were maintained at a high, though still empirical, level in the time of Kuo Shou-Ching[c] (the last two decades of the +13th century), there was a steady decline in the Ming dynasty, and the earlier methods were forgotten. It will be remembered that the prediction of eclipses was one of the most important reasons for the credit which the Jesuits were able to obtain at the imperial court.

(iv) *Earth-shine and corona*

Among celestial phenomena allied to this subject, mention may be made of the effect known as 'earth-shine', seen when the sunlit earth illuminates by reflection the unlighted part of the moon. As the 'Ballad of Sir Patrick Spens' puts it:

> I saw the new moon yestereen
> Wi' the old moon in her arms,
> And if we gang to sea, Master,
> I fear we'll come to harm.

Kühnert (5), in an elaborate paper, has shown that this phenomenon was recognised by the Chinese under the terms Tê-hsing[6] or Ching-hsing.[7] Ssuma Chhien says:

> When the sky is serene, then the Ching-hsing (Resplendent orb) appears. It is also called

[a] Ch. 18, para. 11; cf. Hu Tao-Ching (*1*), vol. 2, pp. 604ff.

[b] Ch. 2, p. 18*b*, tr. auct. Later on, as in the clockwork part of Section 27, we shall have further evidence of the damage done to Chinese science and technology by the fall of the Northern Sung and the retirement of the capital to Hangchow.

[c] His *Ku Chin Chiao Shih Khao*[8] (Studies of Eclipses Old and New) unfortunately failed to survive.

[1] 衛朴 [2] 楓牕小牘 [3] 趙大猷 [4] 張嗣古 [5] 朱欽則
[6] 德星 [7] 景星 [8] 古今交食考

Tê-hsing (Orb of virtue). It has no constant form, but it appears (to the people of) countries which follow the Tao.[a]

This was the tradition which caused premature rejoicings in the Peking Astronomical Observatory in +1340, as reported by Yang Yü in the passage quoted on p. 419 above. It is strange that the European interpretation was quite contrary in character.

Liu Chao-Yang (*1*) has suggested that the oracle-bones of the −2nd millennium may contain the first recorded observation of the solar corona during an eclipse. This is quite visible to the naked eye, and was discussed by Plutarch (contemporary of Wang Chhung and Chang Hêng) in the +1st century,[b] as also by Kepler later. The date of the bone fragment studied by Liu must be either −1353, −1307, −1302 or −1281. The eclipse concerned was therefore not one of those established by Tung Tso-Pin (1).[c] The bone bears characters which have been deciphered as *san yen shih jih, ta hsing*[1]—three flames ate up the sun, and a great star was visible.[d] It seems not unreasonable, therefore, to suppose that this was a record of especially striking solar prominences or coronal streamers.

Another which may be relevant is that in the *Tso Chuan* for −490, where it is said[e] that 'a cloud like a flock of red crows was seen flying round the sun' (*yu yün jo chung chhih wu chia jih i fei*[2]). The term *jih erh*,[3] which means a kind of solar halo,[f] may also have been used to refer to the corona. Loewenstein suggests that corona observations may have been at the origin of the 'winged sun' symbol, so characteristic of Assyria and Persia, but not unknown in ancient China.

(2) Novae, Supernovae, and Variable Stars

Eclipses are not the only celestial phenomena for which a wealth of records is available to us in Chinese texts. The total number of stars in the heavens visible to the naked eye is not constant; we know now that from time to time stars rise into visibility while others disappear and that the magnitude or brightness of stars often changes. Stars which before were but faintly visible may undergo suddenly an increase in brightness of a million-fold. Such stellar explosions give rise to what are called 'novae', or, if the cataclysm is exceptionally great, 'supernovae'. Other stars may show regular periodical variations of brightness, hence the term 'variables'. All these phenomena are of the highest importance for current cosmological theories, as may be seen from many excellent expositions.[g]

[a] *Shih Chi*, ch. 27, p. 33*a*, tr. auct. Chavannes (1), vol. 3, p. 392, noted that the word Ching was that used many centuries later for designating Nestorian Christianity; he wondered whether this had any connection with the story of the star of the Magi. Later histories report occurrences of the phenomenon, e.g. *Chin Shu*, ch. 12, p. 4*a*; *Sung Shih*, ch. 56, p. 21*a*.

[b] Berry (1), p. 390.

[c] See above, p. 410.

[d] Presumably a planet, perhaps one of the brighter fixed stars (cf. Chhen Tsun-Kuei (*5*), p. 59).

[e] Duke Ai, 6th year; Couvreur (1), vol. 3, p. 631.

[f] See p. 475 below.

[g] Such as that of Spencer-Jones (1), pp. 323ff.

[1] 三焰食日大星 [2] 有雲如衆赤烏夾日以飛 [3] 日珥

Fig. 182. The oldest record of a nova. The inscription on this oracle-bone, dating from about −1300, reads (in the two central columns of characters): 'On the 7th day of the month, a *chi-ssu* day, a great new star appeared in company with Antares.'

What must certainly be the most ancient extant record of a nova is contained in one of the oracle-bones, dating from about −1300, studied by Tung Tso-Pin (*1*) (Fig. 182). The inscription says, 'on the 7th day of the month, a *chi-ssu* day, a great new star appeared in company with Antares (*hsin ta hsing ping Huo*[1])'.[a] Another bone-inscription of the same period says: 'On the *hsin-wei* day the new star dwindled (or disappeared) (*hsin-wei yu hui hsin hsing*[2]).' That this refers to the same phenomenon seems probable, for the second date is only two days after the first, and some such rise and

[a] *Yin Li Phu*, Pt. II, ch. 3, p. 2*a*. The exact year is not given, but it must have been between −1339 and −1281. 'Fire-star' here cannot refer to the planet Jupiter, for the planets did not receive their element names until the elaboration of Five-Element theory in the −4th century. This supernova position agrees closely with the radio-star 2C. 1406 (private communication from Dr A. Beer).

[1] 新大星並火 [2] 辛未有毀新星

fall of brightness would be expected. This term *hsin hsing* was used for novae until the middle of the Han period, when it was replaced by the better known technical term *kho hsing*,[1] guest-star.[a]

At the end of the +13th century Ma Tuan-Lin devoted the 294th chapter of his *Wên Hsien Thung Khao* to a list of the extraordinary stars which had appeared since the beginning of the Han,[b] and this was translated and annotated in 1846 by E. Biot (8). Some of these got into the comet register of Williams (6), and all were brought together in the important paper of Lundmark (1), though the new catalogue of Hsi Tsê-Tsung (*1*) now supersedes previous listings. Another old Chinese collection is that in the *Thu Shu Chi Chhêng* encyclopaedia,[c] under the heading *Hsing pien pu*[2] or 'Records of Unusual Occurrences in the Heavens'. The venerable records extracted by Ma Tuan-Lin from the dynastic histories do not always make a clear distinction between novae and comets, but generally the description given is sufficient to identify the phenomenon. The texts usually state the time and duration of appearance, the position in the heavens,[d] and the brightness and colour of the star.[e] For example, in +185:

> In the 2nd year of the Chung-Phing reign-period, in the tenth month, on a *kuei-hai* day, a guest-star (*kho hsing*[1]) appeared in the midst of the constellation Nan Mên[3] (α, β Centauri); it was as big as the half of a bamboo mat and showed the five colours in turn, now beaming now lowering. It diminished in brightness little by little and finally disappeared about July of the following year.[f]

It is interesting to find that Ma Tuan-Lin's list begins with that same star of −134 which stimulated Hipparchus to embark upon his general stellar catalogue;[g] he recorded it in Scorpio, and the Chinese duly noted it in their equivalent, the *hsiu* Fang.[4] But this star was more probably a comet, not a nova,[h] for Pliny (our only source for the story) distinctly says that it moved.[i] The Chinese data will also accept this interpretation. Ma was copying the *Chhien Han Shu*,[j] which uses the term 'guest-star', but the longer account in the *Shih Chi*[k] refers to an appearance at this

[a] Cf. Chou Kuang-Ti (*1*). An early form, *pin hsing*,[5] occurs in *Lü Shih Chhun Chhiu*, ch. 30 (vol. 1, p. 61) (−3rd century); cf. R. Wilhelm (3), p. 78, who explains it wrongly.

[b] For the novae, −134 to +1203.

[c] *Shu chêng tien*, chs. 27–59. Here the records are all intermingled (novae, comets, meteors, conjunctions, etc.).

[d] We believe that the astronomers, probably from the Han onwards, measured positions in degrees. But in the official histories they are generally given only roughly in relation with this or that constellation; presumably the historiographers simplified the information they received.

[e] Cf. Chu Wên-Hsin (*4*).

[f] *Hou Han Shu*, ch. 22, p. 6*a*; *Wên Hsien Thung Khao*, ch. 294, p. 2326·3, tr. auct., adjuv. Biot (8).

[g] Berry (1), p. 51.

[h] The question has been especially studied by Fotheringham (4), cf. Merton (2). There is also an association with the birth of Mithridates Eupator, king of Pontus. His biographer, Reinach (1), not knowing of the Chinese observations, dismissed all the statements about a comet at his birth as legendary. This was the same Mithridates whose water-mills, the first recorded in the occident, fell into the hands of the Romans at his death in −63 (see Sect. 27*f* in Vol. 4).

[i] *Hist. Nat.* II, 26 (24), 95.

[j] Ch. 26, p. 27*b*; tr. Fotheringham (4).

[k] Ch. 27, p. 42*a*; tr. Chavannes (1), vol. 3, p. 408.

1 客星　2 星變部　3 南門　4 房　5 賓星

time of the 'Standard of Chhih-Yu' (Chhih-Yu chhi [1]) which is always defined as a particular kind of comet.[a] In any case it was carefully watched at both ends of the Old World.

Lundmark (1) made a very interesting observation. When the suspected novae are plotted according to galactic coordinates, they show a spatial distribution quite similar to known novae observed in modern times. The 'guest-stars' were therefore true novae, appearing, as they should have done, within our own galaxy, and not only in the brightest regions but sometimes where the star density is quite low. This result has a bearing on the arguments about the reliability of Chinese records, for if 'guest-stars' had been invented to criticise the government, it is highly unlikely that they would all have been placed in the right part of the sky.

It was natural and inevitable that the appearances of novae should be interpreted astrologically. An example of this may be seen in the Thang book *Thai Pai Yin Ching* [2] (Manual of the White and Gloomy Planet of War, i.e. Venus), an important treatise on military affairs by Li Chhüan [3] written in +759. Here the appearance of a 'guest-star' is regarded as a portent of great military significance.[b] And of course the *Khai-Yuan Chan Ching* has a learned disquisition[c] on such prognostications.

The giant stellar explosions which give rise to supernovae are now thought to occur about once in every one or two centuries in our own galaxy, and the frequency of the phenomenon in other galaxies is about the same.[d] Gamow has recently told the story of the only three supernovae of which there is historical record.[e] One was the 'New Star' of Tycho Brahe observed in +1572,[f] a second was that seen by his pupil Kepler in +1604,[g] and the third, that of +1054, was recorded only by the Chinese.[h] This was the origin of the so-called Crab Nebula, which appears today as a somewhat shapeless and diffuse bright cloud (Fig. 183) and which is still expanding. Measurements of the rate of expansion show that it must have started from a central point some eight centuries ago. Since the Chinese records say that at its maximum apparent brightness the guest-star was as bright as Venus, it can easily be calculated that at the time of the explosion the star was several hundred million times as bright as our sun.

The value of the Chinese records for modern astronomy has been underlined by Baade.[i] Sinologists and astronomers have collaborated in the careful examination of

[a] *Shih Chi*, ch. 27, p. 32*b*, tr. Chavannes (1), vol. 3, p. 392, and more elaborately in *Chin Shu*, ch. 12, p. 4*b*, tr. Ho Ping-Yü (1). Nevertheless Hsi Tsê-Tsung (*1*) still admits the apparition of −134 as a nova in his catalogue.

[b] Ch. 8, p. 13*b*. Cf. *Chin Shu*, ch. 12, p. 6*b*.

[c] Ch. 77, pp. 1*a* ff.

[d] Cf. Stratton (1), p. 259.

[e] Cf. also Wattenberg (1).

[f] In Cassiopeia. Hsi Tsê-Tsung (*1*), no. 82, gives the Chinese record.

[g] In Ophiuchus. This nova was also recorded in China (E. Biot (9); Williams (6), p. 93; Hsi Tsê-Tsung (1), no. 85) and Korea (Iba).

[h] And the Japanese.

[i] They have all kinds of applications. Thus attempts have been made from time to time to identify the 'star in the East' of the Magi of Christian tradition. Lundmark (2) suggests that it was a nova recorded in China in −5. He also suggests that the rebellion of Simon bar Kochba ('son of the Star'; a pun on his real name, ben Koseba) may have had some connection with the nova which was recorded in China in +123.

[1] 蚩尤旗 [2] 太白陰經 [3] 李筌

texts and computations so that little or no doubt now remains[a] that the Crab supernova was in fact the 'guest-star' of +1054. Five texts have been assembled which describe the phenomenon,[b] but only one need be quoted:

In the fifth month of the 1st year of the Chih-Ho reign-period,[c] Yang Wei-Tê[1] (Chief Calendrical Computer) said, 'Prostrating myself, I have observed the appearance of a guest-star; on the star there was a slightly iridescent yellow colour. Respectfully, according to the disposition for emperors,[d] I have prognosticated, and the result said, "The guest-star does not infringe upon Aldebaran; this shows that a Plentiful One is Lord, and that the country has a Great Worthy." I request that this prognostication be given to the Bureau of Historiography to be preserved.'[e]

This was done, and the emperor was congratulated. In the month of April +1056, it was reported that the guest-star had become invisible, which was an omen of the departure of guests.

Originally this star had become visible in June (+1054) in the eastern heavens in Thien-kuan[2] (ζ Tauri). It was visible by day, like Venus; pointed rays shone out from it on all sides. The colour was reddish-white. Altogether it was visible for twenty-three days.[f]

These observations were made at Khaifêng, the Sung capital, but the Liao astronomers at Peking did not lag behind Yang Wei-Tê; they also reported the phenomenon.[g] Nor did it escape the perspicacity of the Japanese, two of whose chronicles contain a virtually identical passage on the subject (Iba).[h] Indeed, the date of first observation given in these texts is some ten days earlier than that of the Chinese records.

Attention has been directed by several authors to the great historical importance of Tycho Brahe's new star.[i] It was one of the events which shook to its foundations the Aristotelian theory of the 'perfection' of the heavens and prepared the way for the acceptance of the cosmology of Copernicus. In his acute analysis of trends of 17th-century thought, Willey (1) has quoted Galileo's words:

Comets have been observed which have been generated and dissolved in parts higher than the Lunar Orb, besides the two New Stars, Anno 1572 and Anno 1604—without contradiction much higher than all the Planets. And in the face of the Sun itself, by help of the Telescope, certain dense and obscure substances, in substance very like to the foggs about the Earth, are seen to be produced and dissolved.[j]

As Willey says, these demonstrations of change and imperfection in the heavens

[a] Hubble (1); Oordt (1); Duyvendak (17); Duyvendak, Mayall & Oordt (1).
[b] Including *Sung Shih*, ch. 56, p. 25*a*, and ch. 12, p. 10*b*.
[c] The exact date corresponded to 27 August +1054.
[d] Because of the imperial yellow colour, suggests Duyvendak.
[e] *Sung Hui Yao*, ch. 52, p. 2*b*, tr. Duyvendak (17). One of the most interesting features of the Chinese descriptions of this supernova is that their accounts of its diminution in brightness closely agree with the rate of decay which, as we now know, it must have followed. The significance of this was pointed out to us by Professor Fred Hoyle.
[f] Tr. Duyvendak (17).
[g] *Chhi-Tan Kuo Chih*, ch. 8, p. 6*b*.
[h] *Meigetsuki*, entry under the eighth day of the eleventh month, +1230; and *Ichidai Yōki*.
[i] His *De Nova Stella* of +1573 was reprinted in facsimile by the Royal Danish Science Society in 1901.
[j] *Mathematical Collections and Translations*, ed. T. Salusbury (1661), p. 25.

1 楊惟德 2 天關

powerfully contributed to the general overthrow of medieval scholasticism.[a] Sarton has said, in another connection, that the failure of medieval Europeans and Arabs to recognise such phenomena was due, not to any difficulty in seeing them, but to prejudice and spiritual inertia connected with the groundless belief in celestial perfection. By this the Chinese were not handicapped.

The extent to which the Chinese records of 'guest-stars' remain of living interest to current astronomical research may be seen in the field of radio-astronomy, where during the past few years great additions to knowledge have been made.[b] In 1932 studies of the 'noise' in radio-wave receivers showed that it followed regular variations, the period of which was the sidereal day, hence that it must be originating beyond the solar system. Since that time radio aerials of interferometer type have been developed and employed in systematic surveys of the sky, with the result that a great number of radio-wave sources of great intensity ('radio-stars') are now known. The nature and origin of these, and how far they may be identified either with faintly luminous bodies now visible, or possibly with novae which have ceased to be visible, are among the questions which are the subject of intensive research at the present time. In this context the relevance of the Chinese novae observations (the only ones we have for more than a millennium before the Middle Ages) is obvious.[c] By 1950, when the positions of some fifty radio-stars had been determined, only seven of them were found to coincide with visual objects, but one of these was the Crab nebula in Taurus (the Chinese supernova of +1054), and the others were nebulosities which might once have been novae, some in other galaxies. More recently, a search with the 200-inch telescope in California has permitted the visual identification of some fifteen radio-stars, but these are extremely few in relation to all the radio-stars now known. It is interesting that other supernovae, that of Kepler (+1604) and perhaps that of Tycho Brahe (+1572), find their place among them as radio-wave emitters, though now practically invisible.

The rapid upsurge of this new and powerful[d] method of study of the birth and death of stars in the remoter parts of the universe, with all that it implies for our understanding of cosmology, makes urgently necessary the reduction of the information contained in the ancient and medieval Chinese texts to a form utilisable by modern astronomers in all lands. For this purpose, however, collaboration between competent sinologists and practical astronomers and radio-astronomers is indispensable. The most important radio source in the sky, that in Cassiopeia, is in our own galaxy, probably not farther away than 1000 light-years, and seems to be a nebulous bundle

[a] The words of Tycho ring like its death knell: 'Concludo igitur hanc stellam, non esse ullam Cometarum speciem, vel aliquod igneum Metheoron sive infra Lunam sive supra generentur: sed lucentem in ipso firmamento esse stellam, nulla aetate a mundi exordio ante nostra tempora prius conspectam.' Cf. M. H. Nicolson (1), p. 23.

[b] See, e.g., Shakeshaft (1); Ryle & Ratcliffe (1).

[c] There is however the technical difficulty that the Chinese novae observations have come down to us, through the hands of the official historians, without precise positions in degrees. On the other hand, the radio-stars are now extremely numerous. Many possibilities of misinterpretation are thus open.

[d] It may well be that some of the radio-stars at present being mapped are farther away from the earth than man will ever be able to see by visible light.

of hot filaments of a type previously unknown deriving from a supernova.[a] A tentative identification of it with a nova recorded in China in +369 has recently been made, but unfortunately this seems to rest upon a misunderstanding.[b] However, it seems likely that a powerful radio source near Antares may be identifiable with the supernova which the Chinese recorded on their oracle-bones about −1300.

Search should be made in Chinese and Japanese records for mentions of variable stars.[c] Bobrovnikov has pointed out that the variations of at least two of them (Algol in Perseus and Mira in Cetus) are readily visible to the naked eye,[d] and one was indeed discovered[e] before the use of the telescope.

[a] Thought by some to have exploded about 280 years ago.

[b] The circumstances of this case, and the lessons they teach regarding what is meant by adequate sinological-astronomical collaboration, are so interesting that details must be given. The nova of +369 was duly included in the list of Biot (8), p. 21, who of course obtained his notice of it from Ma Tuan-Lin's *Wên Hsien Thung Khao* (+1319), ch. 294. But the basic source is easy to find. The *Chin Shu* (ch. 13, p. 20*b*) says: 'In the fourth year of the Thai-Ho reign-period (+369), in the second month, a guest-star was seen in the Western Wall of the Purple Palace (*tzu kung, hsi yuan*[1]). In the seventh month it disappeared.' The text goes on to say how the diviners interpreted its significance. Biot translated correctly and added only the remark that the Western Wall was about equivalent to the circle of perpetual apparition, which is true enough for terrestrial latitude 34° N., i.e. that of Yang-chhêng, the traditional central observatory of China (cf. p. 297). The nova was next incorporated in the list of Williams (6), p. 29 (no. 132), who seems to have thought that Tzu Kung meant the circle of perpetual apparition, and whose sinology we may estimate (without undervaluing his great services) by the fact that he gave the name 'She Ke' (i.e. *Shih Chi*) to the entire set of dynastic histories. Now the Tzu Kung (Purple Palace) is a synonym (*Shih Chi*, ch. 27, p. 1*b*) for the northern circumpolar region (more correctly the Tzu Wei Yuan[2]); and its Western Wall extends from 70° to 210° R.A., following approximately 70° decl. N. It runs through the constellation fields of Camelopardalis, Ursa Major, and Draco. The position of the nova was thus described very vaguely in the official history, though the Chin astronomers themselves may have been more precise. But when we look in the most modern novae lists of Lundmark (1) and Hsi Tsê-Tsung (*2*), we find the position for that of +369 tabulated as 0° R.A. and 60° decl. N. How could so great a difference have come about? As his chief authority Lundmark relied upon Zinner (3), who must have made some effort, perhaps without sinological help, to get back to the sources. Most probably he was misled by the fact that in the index of Chinese stars and constellations drawn up by Schlegel (5) no 'Tzu Kung' appears, except in the combination Tzu Kung Chhi[3] (the Flag(staff) of the Purple Palace; no. 693) which is, in fact, a small constellation in Cassiopeia also known as Ko Tao[4] (the Approach to the Hall). It runs along the meridian of R.A. 10° between 45° and 57° decl. N. Tzu Kung Chhi is not a modern name for Ko Tao, as Schlegel (5), p. 327, thought, for one finds it in the *Khai-Yuan Chan Ching* (+718), ch. 66, p. 1*a*. Zinner presumably did not realise that Tzu Kung is an alternative name for the whole circumpolar region, and simply had to disregard the words 'Western Wall'. Tzu Kung Chhi also satisfied him, because, as he said, it is approximately on the circle of perpetual apparition (for a mid-European latitude). Then, in 1952, this confusion, preserved by Lundmark, led Shklovsky & Parenago to suggest that the Cassiopeia radio source, the position of which is about 352° R.A. and 58° decl. N., should be identified with the nova of +369. And the irony of the situation reached its height when the doyen of Chinese meteorologists, Chu Kho-Chen, in an article (*6*) on the value of researches in the history of science in China (1954), cited *en passant* this identification. But he went on to record that in the previous year the Moscow Academy of Sciences had requested the Chinese astronomers to carry out the task, in the interests of all, which we have referred to above. The firstfruits of this most desirable work are now available in a paper by Hsi Tsê-Tsung (*2*). He accepts four out of six of Shklovsky's identifications of 'Chinese' novae with radio sources and suggests alternatives for the others, adding eleven further novae which were recorded in positions quite close to those of radio-stars under study today.

[c] See the preliminary remarks of Iba (1) who indicates relevant texts.

[d] The only evidence of an ancient observation of this kind is a Babylonian cuneiform text of very doubtful significance (Kugler (2), Schaumberger's *Ergänzungsheft*, p. 350). It deals with a constellation the stars of which are said to be sometimes close together and sometimes wide apart. Several statements of this kind occur in the astrological treatise in the *Chin Shu* (ch. 12). A special investigation might decide whether observations of variable stars lie behind them.

[e] By Fabritius in +1596.

[1] 紫宮西垣 [2] 紫微垣 [3] 紫宮旗 [4] 閣道

(3) COMETS, METEORS, AND METEORITES

While there exist a few Babylonian cuneiform records of comets as far back as −1140,[a] and observations of them were quite frequent in ancient and medieval Europe,[b] the Chinese records are by far the most complete—as Olivier points out at the beginning of his excellent monograph on cometary phenomena. The computation of approximate orbits for some forty comets which made appearances earlier than about +1500 has been based almost entirely on the Chinese observations. As in the case of novae, the first compilation of these events, as noted in the dynastic histories, was made by the Chinese themselves. Ma Tuan-Lin (*fl.* +1240 to +1280) incorporated them in his *Wên Hsien Thung Khao*. The 286th chapter of this work, which dealt with comets down to +1222, was translated by Gaubil (10) in a manuscript still kept at the Paris Observatory, and he added to it the 212th chapter of the supplement, which carried the record down to the end of the Ming (+1644). Another translation of Ma Tuan-Lin's record was made in 1782 by C. L. J. de Guignes (2), and E. Biot (9) completed it by a list drawn from the official histories of the Sung, Yuan and Ming (+1222 to +1644). All these catalogues of comets, however, were found to be incomplete by Williams, who in 1871 published (6) what remains the fullest list, giving abundant details of no less than 372 comets, from −613 to +1621.

To give an idea of the care with which the Chinese astronomers described their comets, we may select the record of the comet of +1472, studied also in Europe by Johannes Müller of Königsberg (Regiomontanus).[c]

In the 7th year of the Chhêng-Hua reign period (+1472), in the twelfth month, on a *chia-hsü* day (of the sexagenary cycle), a comet was seen in the star-group Thien thien[1] (σ, τ Virginis). It pointed towards the west. Suddenly it went to the north, touched the star 'Right conductor' (Yu shê-thi[2]; η, ι, τ Bootis), and swept through the Thai Wei Yuan[3] (the 'Enclosure' of stars in Virgo, Coma Berenices and Leo), touching Shang chiang[4] (ν Comae Berenices), Hsin chhen[5] (2629 Comae Berenices), Thai tzu[6] (*E* Leonis), and Tshung kuan[7] (2567 Leonis). Its tail now pointed directly towards the west. It swept transversely across the Lang wei[8] (*a–k* Comae Berenices) of the Thai Wei Yuan. On a *chi-mao* day its tail had greatly lengthened. It extended from east to west across the heavens. The comet then proceeded northwards, covering about 28°, touched Thien chhiang[9] (ι, θ, χ Bootis), swept through the Great Bear (Pei tou[10]), and passed near the San Kung[11] (three small stars at the north of Canes Venatici) and Thai Yang[12] (χ Ursae Majoris), finally entering the Tzu Wei Yuan[13] (Circumpolar Enclosure).[d] It was now perfectly visible in full daylight. At various times it was seen in the Khuei[14] (the 'box' or 'body' of the Great Bear), and near Thien ti hsing[15] (β Ursae Minoris), Shu tzu[16] (5 Ursae Minoris), Hou fei[17] (b3162

[a] Sayce (2), p. 52, referring to Brit. Mus. Western Asiatic Inscriptions, vol. 3, p. 52, no. 1.
[b] For the astrological significance attributed to them, see Bouché-Leclercq (1), p. 357.
[c] Thorndike (1), vol. 4, pp. 359, 422, 442. The following passage is in *Ming Shih*, ch. 27, p. 10*b*.
[d] See p. 259 above.

[1] 天田 [2] 右攝提 [3] 太微垣 [4] 上將 [5] 幸臣 [6] 太子
[7] 從官 [8] 郞位 [9] 天槍 [10] 北斗 [11] 三公 [12] 太陽
[13] 紫微垣 [14] 魁 [15] 天帝星 [16] 庶子 [17] 后妃

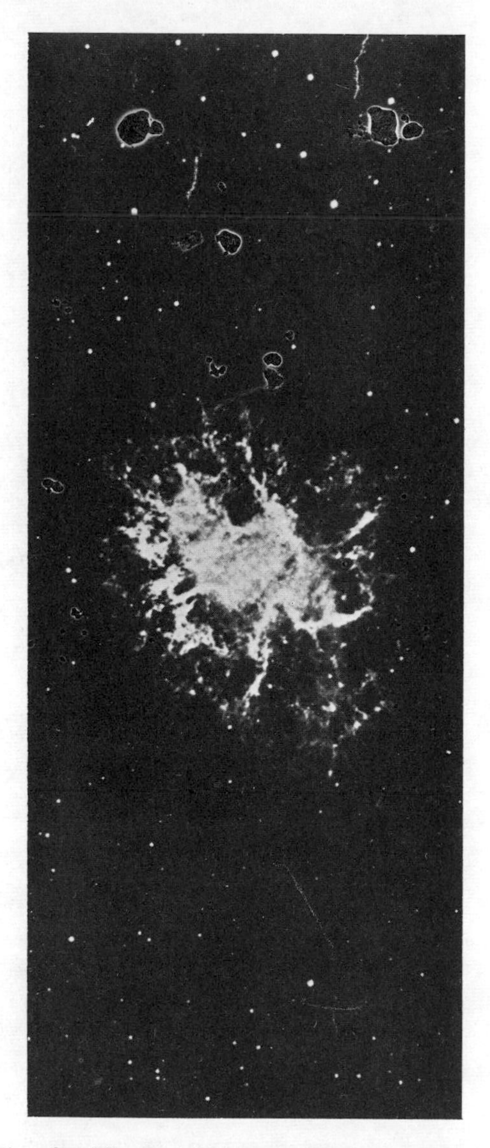

Fig. 183. The Crab Nebula in Taurus, remains of the supernova of +1054 observed only by Chinese and Japanese astronomers. Photograph on red-sensitive plate showing the central small but dense hot star illuminating the nebulosity still expanding around it (photo. Gamow).

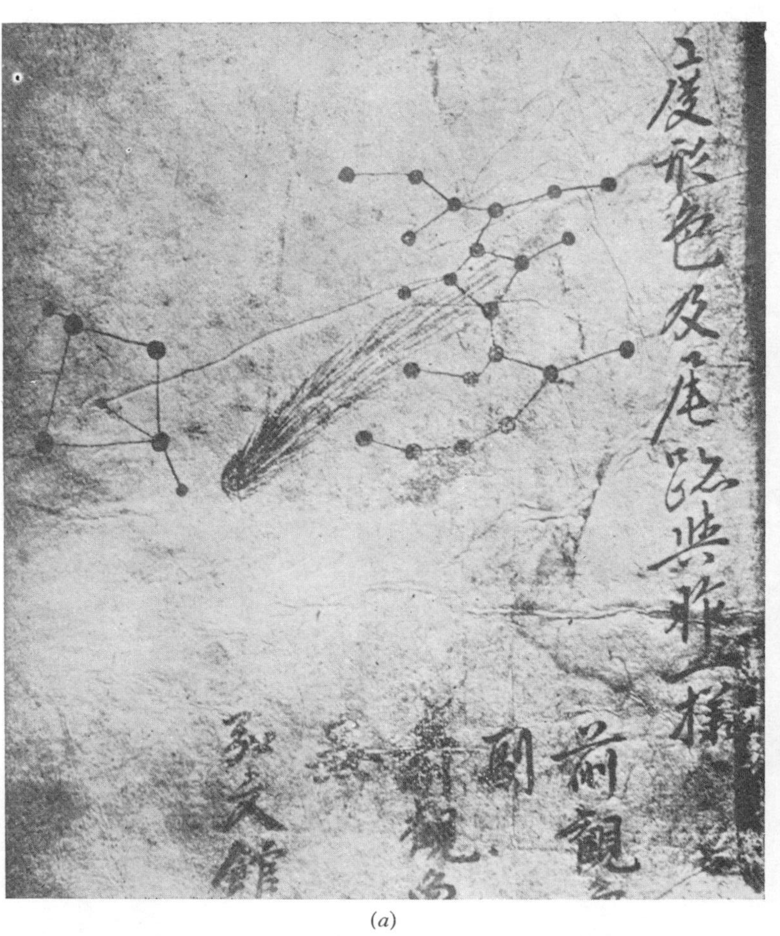

(*a*)

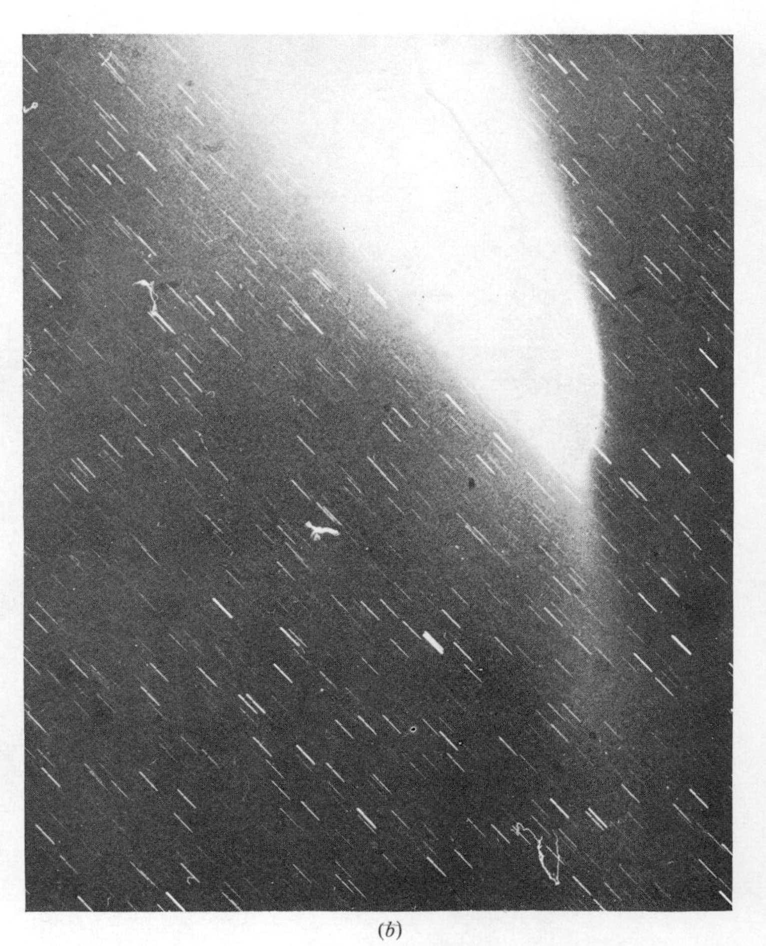

(*b*)

Fig. 184. (*a*) MS. drawing of a comet passing between the *hsiu* I and Chen, on the night of 28 October 1664, from the records of the Korean Astronomical Bureau (photo. Rufus, 2). The annotation at the side says that the shape and colour of the comet and its tail were the same as the previous night (*tu hsing sê chi wei chi yü tso i yang*). The notes below refer to previous observation and the last line on the left signs for the Bureau (*Hung Wên Kuan*).

(*b*) Comet Arend-Roland 1956 *h* photographed at Cambridge in yellow-red light about midnight, 28 April 1957. The notable 'spike' pointing away from the tail is clearly shown (photo. Argue & Wolf). Though 'spiked' comets are rare, references to them exist in ancient and medieval Chinese sources (see Needham, Beer & Ho Ping-Yü).

Ursae Minoris), Kou chhen[1] (ζ, ϵ, δ and other stars in Ursa Minor),[a] San shih[2] (ρ 2006, σ 2027 and 2031 Ursae Majoris), Thien lao[3] (Xh 80, 101, 133, 163, 177 and 170 Piazzi, Ursae Majoris), Thien huang ta ti[4] (α Ursae Minoris, our pole-star), Shang wei[5] (χ Cephei), Ko tao[6] (ξ, o, π, θ, ϕ, ν Cassiopeiae), Wên chhang[7] (θ, υ, ϕ and other stars in Ursa Major), Shang thai[8] (ν, ι Ursae Majoris), etc. On an *i-yu* day it moved to the south, touched the *hsiu* Lou[9] (α, β, γ Arietis), and passed through Thien a[10] (e 602 Arietis), Thien yin[11] (δ, ζ, τ etc. Arietis), Wai phing[12] (α, δ, ϵ, ζ, μ, ν and ξ Piscis), and Thien yuan[13] (γ, δ, ϵ, ζ, η, π, τ etc. Eridani). In the first month of the 8th year, on a *ping-wu* day, it was going towards Wai phing[12] in the *hsiu* Khuei.[14] Gradually it faded, and it was a long time before it finally disappeared.[b]

From such a description the path of the comet is easily traceable. First seen in Virgo, it goes northwards, becomes circumpolar and almost polar, then descends to the south through Cassiopeia and Cepheus, and ends beyond Aries. The use of the expression 'swept through' was particularly appropriate, since the technical term for comets from very early times in China was *hui hsing*[15] or *sao hsing*[16], 'brush-stars'.[c] Chhen Tsun-Kuei (*4*), discussing Han comets, gives a number of synonyms, such as *thien chhan*,[17] 'edging-in stars'; *phêng hsing*,[18] 'sailing stars';[d] *chhang hsing*,[19] 'long stars'; *chu hsing*,[20] 'candle-flame stars', etc. Confusion with novae is of course always to be checked, for comets do not necessarily have tails. When a comet comes into line with the earth and the sun its tail is no longer visible and its light may appear nebulous. The Chinese had a special term for a comet in opposition, *po hsing*[21], clearly distinguishing it, at least theoretically, from a nova.[e] Whether any manuscript drawings of comets still exist in the records of the Astronomical Bureau at Peking we do not know, but Fig. 184 shows a late one from Korean records. The comet is seen passing between the *hsiu* I[22] and Chen.[23] The *Thien Wên Ta Chhêng Kuan Khuei Chi Yao*, a compilation first made at the end of the +14th century, illustrates many different kinds of comets and novae, still according with the definitions in the +7th-century *Chin Shu*.[f]

'Of all comets', says Olivier, 'there is no doubt that Halley's Comet has had the most important influence on astronomy. This comes not only from the fact that its periodicity was established before that of any other, but also because its history can be traced accurately for over two thousand years.' That this is so is due to the careful way in which the Chinese observations were recorded. Halley's own observations were made in 1682. He recognised his comet as the same which Apianus had seen in 1531 and Kepler in 1607, and predicted that it would return, as it did, in 1758. The

[a] See p. 261 above.
[b] Tr. Biot (9), eng. auct. Identifications ours.
[c] Cf. the expression 'stellae comatae', hairy stars.
[d] Or perhaps 'withy-stars' if *phêng*[24] is correct.
[e] The definition is in *Chin Shu*, ch. 12, p. 4*b* (tr. Ho Ping-Yü). Ch. 13, p. 17*a* describes a case of a *po* changing into a *hui* in November and December, +236.
[f] Chhen Tsun-Kuei (*5*), pp. 63, 69, 71, reproduces these illustrations, though poorly.

1 勾陣	2 三師	3 天牢	4 天皇大帝	5 上衛	6 閣道
7 文昌	8 上台	9 婁	10 天阿	11 天陰	12 外屏
13 天苑	14 奎	15 彗星	16 掃星	17 天攙	18 蓬星
19 長星	20 燭星	21 孛星	22 翼	23 軫	24 蓬

importance of this return can, as Olivier says, hardly be over-estimated; it proved that some comets at least are definitely members of the solar system, and that Newton's laws fitted their motions as well as those of the planets.[a] After the next reappearance, in 1835, astronomers and sinologists set to work together to produce a complete recomputation of all the recurrences of the comet.[b] The earliest Chinese observation which may have been Halley's comet[c] is that of −467, but the data are insufficient for certainty; there can, however, be no doubt that the comet of −240 (the 7th year of Chhin Shih Huang Ti) was Halley's. The reappearance in −163 is doubtfully identifiable, but those of −87 and −11 are quite definite, and from the details for the latter, which the Chinese observed minutely for 9 weeks, Hind (2) was able to calculate the approximate orbit, the elements of which approach those of Halley's comet so closely as to leave no doubt. He was able to do the same for the data of Wang Chhung's time (+66). After this, every reappearance in the 76-year cycle is found in the Chinese texts,[d] including that of +1066, so familiar in Anglo-Norman history.

The Chinese were also the first to observe that the tails of comets always point away from the sun. E. Biot (12) reported an interesting statement,[e] concerning a comet observed on 22 March and following days in +837: 'In general, when a comet appears in the morning, its tail points towards the west, and when it appears in the evening, its tail points towards the east. This is a constant rule'.[f] But it was left for Apianus in 1532 to state, not only that the tail points away from the sun, but that its direction coincides with the radius vector.[g]

Few elaborate theories about comets seem to have been produced in China; naturally some early writers ascribed them to derangements of the Yin and Yang (e.g. *Huai Nan Tzu*[h]). But as if in premonition of a modern theory,[i] others associated different kinds

[a] For a good general account, see Plummer (1).

[b] The results may be seen in the work of Cowell & Crommelin (1); E. Biot (10); Hind (2, 3); Hirayama (1) and others, summarised conveniently by Wên Hsien-Tzu (1) and Chu Wên-Hsin (*4*). Cf. Schove (8, 9, 10); Kamienski (1).

[c] *Shih Chi*, ch. 15, pp. 4*b*, 5*a*.

[d] There is some confusion in the +7th and +8th centuries.

[e] As so often, he gave no reference, but from Biot (10) it is almost certain that he read it in *Wên Hsien Thung Khao*, ch. 286, p. 2270.3. When was the generalisation first made? Ma Tuan-Lin was verbally copying *Hsin Thang Shu*, ch. 32, p. 7*a*, but it is interesting to find that the statement is absent from *Chiu Thang Shu*, ch. 36, p. 13*b*, which gives a much longer description of the same comet, though differing considerably in detail. This might suggest that the rule was known about +1050 when Ouyang Hsiu & Sung Chhi were writing the *Hsin Thang Shu*, but not about +950, the time of composition of the *Chiu Thang Shu* by Liu Hsü. Such an inference, however, is quite precluded by the fact that the same statement occurs in the *Chin Shu*, ch. 12, p. 4*b*, finished by Fang Hsüan-Ling in +635. Here indeed we find it in its fullest form, for the text adds that 'if the comet is south or north of the sun its tail always points following the same direction as the light (radiating) from the sun'. In this source, the statement is attributed to the official astronomers in general, who, we are told, explain it by saying that comets shine by reflected light like the moon. Besides, the circumstances of the +837 comet (almost certainly Halley's; cf. Schove (8, 10) and Kamienski) show that the rule must by then have been well known. The director of the Bureau of Astronomy at the time was Chu Tzu-Jung,[1] a man whose name only survived because the emperor appealed to him for advice on this occasion; the comet was particularly disturbing because it seemed to violate this very rule. The first statement of the generalisation must go back at least to the +7th century and probably to the +6th.

[f] *Fan hui hsing chhen chhu tsê hsi chih, hsi chhu tsê tung chih; nai chhang yeh.*[2]

[g] Sarton (1), vol. 3, p. 1122.

[h] Ch. 3, p. 2*a*.

[i] Olivier (1), p. 211.

[1] 朱子容 [2] 凡彗星晨出則西指夕出則東指乃常也

of comets with the different planets. This system was worked out in the *Fêng Chio Shu*[1] of Ching Fang[a] about −50 and, though his text is lost, a parallel in one of the apocryphal treatises of the same period remains.[b] Each comet was thought to originate from a particular planet.[c]

Between comets and meteor streams there is a close connection. It is known that the August Perseid meteors move in the same orbit as Tuttle's Comet, while the November Leonids (which used to give such great showers every thirty-three years) follow that of Tempel's Comet (1866, I), and the May Aquarids that of Halley's. They are almost certainly remains of disintegrated comets. In the Chinese literature there is a vast mass of information about meteors, known as *liu hsing*[2], 'gliding' or 'shooting' stars, or *pên hsing*[3], 'energetic' stars, and meteor showers (*liu hsing yü*[4]), as well as the actual fall of meteorites upon the earth's surface (*hsing yün*[5]).[d] Ma Tuan-Lin summarised it in the +13th century, and chapters 291 and 292 of his *Wên Hsien Thung Khao* were translated, first by Rémusat (4, 5) and then by E. Biot (11). Many of the observations are highly detailed, and the whole list fills over 200 quarto pages. The material would bulk even larger if the records in the local histories were drawn upon, as has been done by Chu Wên-Hsin (*4*) for one province in a preliminary way, as also by Hsieh Chia-Jung (*1*). The earliest date from −687 and −644 (subject to the usual reservations about the adjustment of Chou texts by Han scholars).[e] A very great Leonid shower was reported in +931. So complete were the records that Biot was able to make a statistical account of frequencies for the Sung period +960 to +1275, including the directions towards the cardinal points, and the colours of the displays.[f] At this time the maximum frequencies were in July and October.[g] Before the Sung there are 149 records of meteor showers, during it 272, and in the Yuan and Ming 74. The period of recurrence mentioned above (33·11 years, actually 33·25) is clearly shown by the Chinese records.[h]

Here is an example of careful description of the fall of a meteorite, from Shen Kua's *Mêng Chhi Pi Than*:

In the 1st year of the Chih-Phing reign period (+1064), there was a tremendous noise like thunder at Chhang-chou[6] about noon. A fiery star as big as the moon appeared in the

[a] Summarised in *Chin Shu*, ch. 12, p. 6*a*, *b*.

[b] This is the *Ho Thu Wei Chi Yao Kou*[7] (Apocryphal Treatise on the River Chart; Investigation of the Full Circle of the (Celestial) Brightnesses), in *Ku Wei Shu*, ch. 33, pp. 1*a* ff.; cf. *Chin Shu*, ch. 12, pp. 5*b*, 6*a*.

[c] Cf. Boll (5), pp. 46ff. on Babylonian and Greek ideas about the associations of the fixed stars with different planets.

[d] Cf. *Chin Shu*, ch. 12, p. 7*a*, *b*; *Khai-Yuan Chan Ching*, ch. 2, p. 9*b*.

[e] There was a rationalisation of the old meteor records in the Confucian editing of the *Chhun Chhiu*; see Wu Khang (1), p. 174.

[f] Note the remarks made elsewhere about the Sung as the flowering-period of Chinese science (Vol. 2, p. 493). Note also that the very complete description of the comet of +1472, given above, was made during the Ming, when the physical sciences were at a low ebb. On meteor frequency throughout the ages see Paneth (1).

[g] The difference in the months is purely calendrical; these must have been the Perseids and Leonids.

[h] Cf. T. Fu.

1 風角書 2 流星 3 賁星 4 流星雨 5 星隕 6 常州
7 河圖緯稽耀鈎

south-east. In a moment there was a further thunderclap while the star moved to the south-west, and then with more thunder it fell in the garden of the Hsü family in the I-hsing district. Fire was seen reflected in the sky far and near, and fences in the garden round about were all burnt. When they had been extinguished, a bowl-shaped hole was seen in the ground, with the meteorite glowing within it for a long time. Even when the glow ceased it was too hot to be approached. Finally the earth was dug up, and a round stone as big as a fist, still hot, was found, with one side elongated (i.e. pear-shaped). Its colour and weight were like iron. The governor, Chêng Shen,[1] sent it to the Chin Shan temple[2] at Jun-chou,[3] where it is still kept in a box and shown to visitors.[a]

Meteorites had many other names in Chinese books besides the *yün*[4] already mentioned, or *yün shih*.[5] Further information is contained in a valuable chapter by Chang Hung-Chao,[b] who points out that one of the oldest names must be that contained in the *Shan Hai Ching* (ch. 16), namely, *thien chhuan*,[6] 'hounds of heaven'. He also notes that meteorites were often confused (as in other civilisations) with stone axes of the Neolithic period.[c] There is a reference to this in the *Chiu Thang Shu* (Old History of the Thang Dynasty), where, about +660, a meteorite presented to the emperor was called 'the stone axe of the thunder-god' (*Lei Kung shih fu*[7]).[d] Other names were 'the thunder-god's ink-block' (*lei mo*[8]) or 'thunder-lumps' (*phi li chen*[9]), and it is these which formed the headings under which Li Shih-Chen in the +16th century treated of meteorites in his *Pên Tshao Kang Mu*.[e]

(4) Solar Phenomena; Sun-Spots

The outstanding example of a celestial phenomenon which in general escaped the attention of Europeans because of their preconceived idea of the perfection of the heavens, is that of the spots on the sun.[f] In Europe their discovery was one of the advances made by Galileo in his use of the telescope. According to his own statement,[g] he first saw them towards the end of +1610, but he did not publish his results till +1613,[h] by which time the same discovery had been independently made by Harriot in England, Fabricius in Germany, and Scheiner in Holland.[i] Previously, dark spots on the sun seen with the naked eye had been ascribed to the passage of planets across it. Scheiner thought that the spots might be small satellites of the sun. But Galileo was able to refute all such views; he demonstrated that the spots must be on or close

[a] *MCPT*, ch. 20, para. 3, tr. auct. Attention was drawn to this by Vacca (2). Cf. Hu Tao-Ching (*1*), vol. 2, p. 649. Other good descriptions will be found in *Kuei Hsin Tsa Chih, Hsü Chi*, ch. 1, p. 35*b*; *Pieh Chi*, ch. 1, p. 7*b*. Perhaps the oldest record is for −662; *Kuo Yü* (Chou Yü), ch. 1, p. 22*a*; cf. Fêng Yu-Lan (1), vol. 1, p. 24.

[b] (*1*), pp. 372 ff.

[c] See further de Mély (5); Belpaire (2).

[d] Cf. Ennin's +9th-century diary; Reischauer (2), p. 128.

[e] Ch. 10, pp. 45*b* ff.; Read & Pak (1), nos. 113, 114.

[f] Cf. Pelseneer (2).

[g] Berry (1), p. 154.

[h] *Istoria e Dimonstrazioni intorno alle Macchie Solari e loro Accidenti* (Mascardi, Rome, 1613).

[i] See details in W. M. Mitchell; Wohlwill.

1 鄭伸 2 金山寺 3 潤州 4 隕 5 隕石 6 天犬
7 雷公石斧 8 雷墨 9 霹靂碪

to the surface of the sun, as if they were clouds, and that their motion was such as to indicate that the sun revolved on its axis with a period of about a month.[a]

Sarton (3) has shown that observations of sun-spots in the occident go back rather further than has usually been thought, but they were very fragmentary and rare. The earliest mention seems to be in Einhard's *Life of Charlemagne*, referring to about +807 and this was interpreted as a transit of Mercury. Another observation, by Abū al-Faḍl Ja'far ibn al-Muqtafī in +840, was explained as a transit of Venus. Other references date from about +1196 (Ibn Rushd) and +1457 (the Carraras).

In the 19th century a cyclical recurrence of sun-spot activity was established by Schwabe, with a period of about 11 years, but fluctuating between 7·3 and 17 years.[b] There are also superposed irregular variabilities of much greater period. It was then realised by Sabine and others that the 11-year sun-spot period corresponded with a similar period found in terrestrial magnetism (magnetic 'storms', etc.).[c] Today the connection between sun-spot activity and the conditions of the ionosphere, which profoundly affect the transmission of radio messages, is fully accepted, and there may be an effect on the meteorological situation too (Napier Shaw). The International Astronomical Union has a permanent commission at work on solar-terrestrial phenomena.

But the Chinese records are by a long way the most complete which we have. They start nearly a thousand years before the first reference in the west, that is, in the time of Liu Hsiang, −28.[d] Between that time and +1638 there are 112 descriptions of outstanding sun-spots in the official histories, but there are also numerous mentions in local topographical records, volumes of memoirs, and other kinds of publications, which have not so far been fully collected.[e] The black spots are referred to as *hei chhi*,[1] *hei tzu*[2] or *wu*,[3] and their size is often described 'as big as a coin', 'as big as a hen's egg', a peach, a plum, etc. Extensive lists of observations are in the +13th-century *Wên Hsien Thung Khao* and the *Thu Shu Chi Chhêng* encyclopaedia,[f] but the most convenient one is in the book of Chu Wên-Hsin.[g] In the western-language literature, after an early mention by Kirkwood, lists were published by Williams (5), Lovisato (1), Hosie (1) and de Moidrey. The last-named author, who enjoyed the collaboration of P. Huang, was able to find a fair approximation to the 11-year cycle from these ancient observations.[h] This has been confirmed by S. Kanda (1), who found a period of 10·38 to 11·28 years, and could also detect a superimposed fluctuation of much longer period (975 years). Probably the most complete list so far is in Japanese, that of S. Kanda (*2*).[i]

[a] The true nature of the spots was of course not known until quite modern times, but Chu Hsi in the +12th century had refuted the idea that they were earth-shadows (*Chu Tzu Yü Lei*, ch. 1, p. 10*a*).

[b] Spencer-Jones (2); Oppenheim (1).

[c] Berry (1), p. 385; Spencer-Jones (1), p. 153.

[d] 10th May; Dubs (2), vol. 2, p. 384, tr. *Chhien Han Shu*, ch. 27, p. 17*b*.

[e] For example there is abundant material in the *Chin Shu*, ch. 12, pp. 1*a*, 15*b* ff. (finished in +635), and the *Thung Chih Lüeh* of +1150 also discusses sun-spots (ch. 50, p. 7*b*).

[f] *Shu chêng tien*, ch. 24.

[g] (*4*), p. 80.

[h] See also Schove (2, 3, 6, 7, 11), who finds the mean period of the rhythm to be 11·1 years, with a longer cycle of 78 years.

[i] For Korean records, starting early in the +11th century, see Rufus (2), p. 19.

1 黑氣　2 黑子　3 烏

The use of the term *wu*,[1] which means 'crow' as well as 'black', raises the question whether some references long before −28 might be based on sun-spot observations. As Chhen Wên-Thao has pointed out, the existence of a crow in the sun (the colleague of the rabbit in the moon) was part of traditional Chinese mythology in Chou and early Han times.[a] This we find in the *Lun Hêng*, where Wang Chhung says, 'The scholars hold that there is a three-legged crow in the sun...' (*Ju yüeh jih chung yu san tsou wu*[2]),[b] after which he goes on to argue, in his sceptical way, that the thing is impossible. But it may mean that the black spots had been observed, perhaps as early as the time of Tsou Yen. Again, Dubs notes medieval traditions that in −165 the character *wang*[3] appeared in the sun;[c] Schove (6) inclines to take this as the earliest dated sun-spot.

As in the case of eclipses, it seems likely that from an early date the sun must have been observed through smoky rock-crystal or semi-transparent jade.[d] Indeed, this technique is specifically referred to by Li Shih-Chen,[e] who says that the 'books on jade' mention certain kinds which were used for looking at the sun (*kuan jih yü*).[4] A large piece of this was brought, according to one account, as tribute by an embassy from Fu-Sang about +520; we shall examine the passage, which may refer to observations of sun-spots, in Sect. 26*g* below. But haze due to dust-storms from the Gobi would also have permitted the observation of sun-spots.

As Ionides & Ionides have said,[f] Chinese astrological conclusions were for once perhaps right in maintaining a connection between celestial and terrestial phenomena. Though not at present generally admitted, it is not impossible that the sun-spot period may be connected, through meteorological effects, with events of social importance, such as good and bad harvests. Such a correlation is indeed maintained for Japanese rice crop famines since +1750 by Arakawa (1). Conceivably the basis of the association made in the *Chi Ni Tzu* book (p. 402 above) between the 12-year Jupiter cycle and an agricultural production cycle, may really have been, as Chatley (1) suggests, the 11-year sun-spot period. Correlations between the periodicity of sun-spot maxima and other phenomena such as aurorae, using in part Chinese data, have been attempted by several workers.[g]

[a] Cf. Granet (1).
[b] Ch. 32, tr. Forke (4), vol. 1, p. 268. Cf. *Fu Hou Ku Chin Chu*, cit. *TPYL*, ch. 764, p. 4*b*.
[c] (2), vol. 1, p. 258; *Yü Hai*, ch. 195, p. 2*a*.
[d] There is no reason why sun-spot observations could not have been made very early. Many sun-spots are easily visible with the naked eye at sunrise or sunset, or reflected on the surface of still waters. They can be seen well through thin haze. A spot about 30,000 miles, or four earth-diameters, across, is large enough to be seen without telescopic aid, and many spots are much larger, even as large as twenty earth-diameters (Sarton (1), vol. 3, p. 1856).
[e] *Pên Tshao Kang Mu*, ch. 8, p. 48*b*.
[f] P. 79.
[g] See Fritz (1), Schove (1), etc.

[1] 烏 [2] 儒曰日中有三足烏 [3] 王 [4] 觀日玉

(j) THE TIME OF THE JESUITS

In the history of intercourse between civilisations there seems no parallel to the arrival in China in the 17th century of a group of Europeans so inspired by religious fervour as were the Jesuits, and at the same time so expert in most of those sciences which had developed with the Renaissance and the rise of capitalism. Although for our present plan the year 1600 is the turning-point, after which time there ceases to be any essential distinction between world science and specifically Chinese science, yet the part played by the Jesuits in Chinese astronomy has so many links with the Asian astronomy of former centuries, and so much to teach us about the mutual impact of Chinese and Western thought, that it is impossible to avoid a brief account of what happened during the 17th century.[a]

From hints which have already been dropped on the way, it will have been suspected that the coming of the Jesuits was by no means (as it has often been made to appear) an unmixed blessing for Chinese science. Let us make a provisional balance-sheet of the merits and demerits of their contribution, before illustrating some of its items from contemporary documents. In the first place, the European methods for the prediction of eclipses were greatly superior to the traditional empirical Chinese methods.[b] This was first demonstrated for the solar eclipse of 15 December 1610, when Sabbathin de Ursis was acting as the principal Jesuit astronomer after the death of Matteo Ricci.[c] Secondly, the Jesuits brought a clear exposition of the geometrical analysis of planetary motions, and of course the Euclidean geometry necessary for applying it.[d] This had many other uses, as (thirdly) in gnomonics and the stereographic projections[e] of the astrolabe, and in surveying.[f] A fourth contribution was the doctrine of the

[a] The literature on the Jesuit period is voluminous but diffuse. We have already cited the works of Bernard-Maître (*1*, *5*) on Ricci's own contribution, and the encyclopaedia of Pfister (*1*). Among general summaries may be mentioned those of Yabuuchi (*17*) and Bernard-Maître (*16*).

[b] But Trigault's pretension (Gallagher ed. p. 325) that the true nature of eclipses was unknown in China before the time of the Jesuits, was, as we have seen, quite baseless.

[c] Comedy sometimes attended Jesuit predictions. In 1636 they stated that Jupiter would pass between two stars in Cancer and then begin a phase of retrogradation. Some of the Chinese officials thought this presaged a calamity by fire so they falsified the observations, but the explosion of a large powdermill near Peking confirmed strikingly the exactness of the Jesuits (Bernard-Maître (*7*), p. 457).

[d] Trigault informed his readers (Gallagher ed. p. 326) that Chinese astronomers knew nothing of the two celestial poles. How far this deviated from the truth will be apparent from the whole history of Chinese astronomical instruments, which we have outlined in the foregoing pages. Moreover, the Jesuits brought the full Ptolemaic theory of epicycles, but the victory of Copernicus was so near that Chinese astronomers could surely well have done without them.

[e] Cf. the *Hun Kai Thung Hsien Thu Shuo*[1] (On Plotting the Coordinates of the Celestial Sphere and Vault) by Li Chih-Tsao[2] of +1607. He was a collaborator of the Jesuits.

[f] In some cases, the Jesuits were really only reminding the Chinese of things which they themselves had developed long before, but which the degenerate science of the Ming had forgotten. Thus the Euclidean definition of a geometrical point caused much admiration (Bernard-Maître (*1*), p. 51), neither the Jesuits nor their Chinese friends being aware that the Mohists had discussed such matters before the Han (cf. Vol. 2, p. 194 above). Similarly, Ricci's demonstrations of the use of geometrical quadrants and other means of measuring heights, distances and depths, were attended with enthusiasm (Bernard-Maître (*1*), p. 57), no one apparently being aware of the survey techniques already explained in the +3rd-century *Hai Tao Suan Ching* (p. 31 above). Trigault least of all (cf. Gallagher ed. p. 326).

[1] 渾蓋通憲圖説 [2] 李之藻

spherical earth[a] and its division into spaces separated by meridians and parallels. Fifthly, the new 16th-century algebra of the time of Vieta was made available to the Chinese, with many new computing methods, and ultimately mechanical devices such as the slide-rule. Sixthly, by no means the least valuable transmission was the most up-to-date European technique of instrument-making, graduating of scales, micrometer screws, and the like. The spread of the telescope was the climax of this.[b]

(1) China and the Dissolution of the Crystalline Spheres

On the other hand, the world-picture which the Jesuits brought was that of the closed Ptolemaic-Aristotelian geocentric universe of solid concentric crystalline spheres.[c] Hence they opposed the indigenous Hsüan Yeh doctrine of the floating of the heavenly bodies in infinite space, and the irony was that they did so just at a time when the best minds in Europe were breaking away from the closed Aristotelian system. Hence also (second) they obstructed the spread of the Copernican heliocentric doctrine in China, for after all they could not but be sensitive to the condemnation of Galileo by the Church. It followed, thirdly, that they substituted an erroneous theory of the precession of the equinoxes for the cautious Chinese refusal to form any theory at all about it.[d] Fourthly, they completely failed to appreciate the equatorial and polar character of traditional Chinese astronomy, and therefore, confusing the *hsiu* divisions with the zodiac, equalised the duodenary equatorial divisions when there was no need to do so.[e] Fifthly, in spite of the advance to equatorial coordinates which was just being made by Tycho Brahe, the Jesuits imposed the less satisfactory Greek ecliptic coordinates upon Chinese astronomy, which had always been primarily equatorial, and actually constructed in Peking, as we have seen, an ecliptic armillary sphere (cf. pp. 379, 451).

A fascinating glimpse of this paradoxical situation is seen in letters which Ricci wrote on 28th October and 4th November 1595, enumerating the 'absurdities', as he called them, of the Chinese.[f] They say, he wrote, that

(1) The earth is flat and square, and that the sky is a round canopy; they did not succeed in conceiving the possibility of the antipodes.[g]

[a] This again was not new to China, as the Jesuits thought (cf. Trigault, Gallagher ed. p. 325) for it had been part of the ancient Hun Thien theory (cf. p. 217 above). It is strange that Chinese cosmology had so little affected Chinese cartography. The Chinese may perhaps be pardoned for a certain reluctance in accepting Ricci's presentation of terrestrial sphericity, since he reserved the centre of the earth for the hell of the damned.

[b] Between 1634 and 1638 the Jesuits made many gifts of astronomical apparatus to the last Ming emperor (details in Bernard-Maître (7), p. 450). The monarch himself observed the eclipse of 20 Dec. 1638.

[c] See especially Aristotle, *De Caelo*, Bk. 2. There is a charming story of one of the friends of the Jesuits, Chhü Thai-Su,[1] accepting a prism of rock-crystal and decorating its case with an inscription to the effect that this was a piece of the very material of which the sky is composed (Trigault, Gallagher ed. p. 318). On the dissolution of the closed cosmos in Europe see Koyré (5); Dingle (2).

[d] Bernard-Maître (1), p. 88. The true explanation was reserved for Newton (Berry (1), p. 235).

[e] De Saussure (16*d*); Bernard-Maître (1), pp. 49, 76, 86, etc.

[f] Venturi (1), vol. 2, pp. 175, 184, 185, 207; arranged by Bernard-Maître (1), p. 48.

[g] Ricci had certainly met with a survival of the Kai Thien theory (p. 210 above).

[1] 瞿太素

(2) There is only one sky (and not ten skies). It is empty (and not solid). The stars move in the void (instead of being attached to the firmament).[a]

(3) As they do not know what the air is, where we say that there is air (between the spheres) they affirm that there is a void.

(4) By adding metal and wood, and omitting air, they count five elements (instead of four)—metal, wood, fire, water and earth. Still worse, they make out that these elements are engendered the one by the other; and it may be imagined with how little foundation they teach it, but as it is a doctrine handed down from their ancient sages, no one dares to attack it.[b]

(5) For eclipses of the sun, they give a very good reason, namely that the moon, when it is near the sun, diminishes its light.[c]

(6) During the night, the sun hides under a mountain which is situated near the earth.[d]

Here we see the elements of superiority in European science at the turn of the 16th and 17th centuries imposing a fundamentally wrong world-picture, that of the solid spheres, on the fundamentally right one which had come down from the Hsüan Yeh school, of stars floating in infinite empty space.[e]

The point is worth looking at a little more closely. Five years after these words had been written by Ricci in his letter home, William Gilbert was saying in his *De Magnete*:

Who has ever made out that the stars which we call fixed are in one and the same sphere, or has established by reasoning that there are any real, and, as it were, adamantine spheres? No one has ever proved this, nor is there a doubt but that just as the planets are at unequal distances from the earth, so are these vast and multitudinous lights separated from the Earth by varying and very remote altitudes—they are not set in any sphaerick frame or firmament. The intervals of some are from their unfathomable distance matter of opinion rather than of verification; others less than they are yet very remote, and at varying distances, either in that most subtile quintessence, the thinnest aether, or in the void....It is evident then that all the heavenly bodies set as if in destined places are there formed into spheres, that they tend to their own centres, and that round them there is a confluence of all their parts. And if they have motion, that motion will rather be that of each round its own centre, as that of the Earth is, or a forward movement of the centre in an orbit, as that of the Moon is....But there can be no movement of infinity and of an infinite body, and therefore no diurnal revolution of that vastest Primum Mobile.[f]

[a] As has already been mentioned (p. 198 above), an echo of the Greek spheres may have penetrated to China about −300, but it played no part in Chinese astronomical thought. See also p. 603 below. 'The Chinese had never heard', wrote Trigault later, 'that the skies are composed of solid substance, that the stars are fixed and not wandering round aimlessly, and that there are ten celestial orbs, enveloping one another, and moved by contrary forces' (Gallagher ed. p. 326). Indeed, Kho Chung-Chiung wrote against these ideas in his *Hsüan-Yeh Ching* of +1628.

[b] The arguments in favour of the Aristotelian elements were not a whit better. Trigault, however, related with pride the triumph of the Four elements over the Five in the lectures and pamphlets of the Jesuits (Gallagher ed. p. 327). Yet this was almost contemporary with the epoch-making work of Jean Rey and John Mayow in Europe, and within half a century the whole edifice of element-theory was finally exploded by Robert Boyle.

[c] Ironical. It will be remembered that this is the ancient Yin-Yang influence theory (p. 412 above), which one or other of Ricci's interlocutors must have fished out of the *Lun Hêng*.

[d] A relic of the legendary cosmology which Ricci must have obtained from some uneducated acquaintance rather than from anyone skilled in astronomy.

[e] It is unfortunate that Cronin (1), in a widely-read and moving book published in 1955, still seeks to perpetuate the Jesuit legend of the backwardness of Chinese cosmology and astronomy (cf. e.g. p. 138).

[f] Thompson tr., p. 215.

Twenty years earlier, Giordano Bruno, in his *De Infinito Universo*, had been pointing the same moral, in his usual more violent way:

> The difficulty proceedeth from a false method and a wrong hypothesis—namely of the weight and immovability of the Earth, and the position of the Primum Mobile, with the other seven, eight, nine, or more, spheres, on which stars are implanted, impressed, plastered, nailed, knotted, glued, sculptured or painted—and that these stars do not reside in the same space as our own star, named by us Earth.[a]

Thus was the 'false method' and 'wrong hypothesis' in cosmology introduced to China. But did any stimulus come back in exchange?

Most of the scholastics, following Aristotle, had held that a plurality of worlds was an impossibility.[b] But in the 17th century the doctrine rapidly gained ground,[c] and it was accompanied by a great proliferation of 'scientific' romances of interplanetary travel, which have been reviewed in an excellent book by M. H. Nicolson (1).[d] Among the literary works in which these themes were set forth there are certain coincidences which hint that the scepticism of the Chinese as to the solid spheres had not been without influence. Thus, for example, Francis Godwin,[e] in *The Man in the Moone; or, A Discourse of a Voyage Thither, by Domingo Gonsales, the Speedy Messenger* (1638), one of the earliest scientific romances, makes his narrator fly to the moon in a machine propelled by wild geese. From there the earth looks like any other planet, and after some time the narrator, 'free of that tyrannous lodestone, the Earth', and acquiring another sort of lodestone as antidote to the earth's attraction,[f] floats down safely, arriving precisely in China, where he meets both mandarins and missionaries.[g] Both the Chinese and the lunar people speak a tonal language. In less romantic form, the same idea was urged by John Wilkins,[h] in his *Discovery of a World in the Moon, tending to prove that 'tis probable that there may be another habitable World in that Planet* (also 1638), and Christian Huygens (1698). But the Chinese theme occurs again in the

[a] Gentile, *Op. Ital.*, vol. 1, p. 402; Lagarde, *Op. Ital.*, vol. 1, p. 388; D. Singer (1), p. 71.

[b] *De Caelo*, especially Bk. 1. See Duhem (1), vol. 2, pp. 55 ff. Pietro d'Abano, however, had suggested in +1310 that the heavenly bodies might not be borne on spheres, but rather move freely in space (Duhem (3), vol. 4, pp. 241 ff.; Crombie (1), p. 202). On him see especially Thorndike (1), vol. 2, pp. 874 ff.; Sarton (1), vol. 3, pp. 439 ff.

[c] See McColley (1) and D. Singer (1).

[d] This trend began with the first English translation (by Francis Hicks) of the works of Lucian of Samosata (*b.* +120) which had dealt with voyages to the moon. There were other ancient works, such as Cicero's (−106 to −43) *Somnium Scipionis* (Keyes, 2), and Plutarch's (+48 to +123) *De Facie in Orbe Lunae*, which had contained similar ideas, but which had lain dormant through the long period of dominance of 'solid sphere' conceptions. It will hardly be believed that Lucian's contemporary, Chang Hêng the great astronomer, mathematician and engineer, also wrote an imaginary journey beyond the sun—his *Ssu Hsüan Fu*,[1] an essay contained in the *Wên Hsüan* collection (ch. 15): cf. Hughes (7), pp. 79, 87, 117, 118; (8).

[e] +1562 to +1633. He had known Bruno at Oxford, and later became Bishop of Llandaff and Chester. Cf. M. H. Nicolson (1), p. 71, for a fuller account.

[f] It will be recalled that just such a device was imagined by H. G. Wells in his *First Men in the Moon*.

[g] McColley (2) conjectures that 'Domingo Gonsales' partly conceals Matteo Ricci—both were born in 1552 and reach Peking in 1601; both have a subordinate called Diego (in Ricci's case, Didace de Pantoja) and use Macao as their depot for dispatches to Europe. It is certain that Godwin borrowed much from Trigault's *De Christiana Expeditione apud Sinas*.

[h] +1614 to +1672, Bishop of Chester. For a fuller account see M. H. Nicolson (1), p. 93.

[1] 思玄賦

amusing political satire of Daniel Defoe, *The Consolidator; or Memoirs of Sundry Transactions from the World in the Moon* (1705), who mentions both Godwin and Wilkins. So far as China is concerned the satire is double-edged, for while many exaggerated stories are told of Chinese inventions and discoveries,[a] the Chinese devotion to natural law and good custom is used to contrast the political absolutism of European governments.[b] The Consolidator, or 'Apezolanthukanistes', is a real anticipation of the aeroplane; it has wings and carries enough fuel to last out the voyage to the moon. As Chhien Chung-Shu (2) has pointed out, the ambivalent attitude of European to Chinese science is well seen, for while Defoe speaks of the 'gross and absurd ignorance of the Chinese of the motions of the heavenly bodies', the interplanetary aeroplane could only be supposed to have been made and used by people well versed in astronomy and mechanics. Yet a third recurrence of the Chinese theme is found in Miles Wilson's *History of Israel Jobson, the Wandering Jew*, who visited all the planets and allegedly wrote the account of his travels in Chinese (1757).[c] Indeed, it is hardly possible to take up any seventeenth-century book dealing with the idea that there are other peopled solar systems besides our own, without finding some reference to China. For example, in the *Entretiens sur la Pluralité des Mondes* written by B. de Fontenelle in 1686, the subject is discussed in the guise of evening conversations with a great lady at a country house. After many passing mentions of China,[d] the narrator says:[e]

Je viens de vous dire, repondis-je, toutes les nouvelles que je sçay du Ciel, & je ne croy pas qu'il y en ait de plus fraîches. Je suis bien fâché qu'elles ne soient pas aussi surprenantes et aussi merveilleuses que quelques Observations que je lisois l'autre jour dans un Abregé des Annales de la Chine, écrit en Latin, & imprimé dépuis peu. On y voit des mille Etoiles à la fois qui tombent du Ciel dans la Mer avec un grand fracas, ou qui se dissolvent, et s'en vont en pluye, & cela n'a pas esté veu pour une fois à la Chine. J'ay trouvé cette Observation en deux temps assez éloignez, sans compter une Etoile qui s'en va crever vers l'Orient, comme une fusée, toujours avec grand bruit. Il est fâcheux que ces spectacles-là soient reservez pour la Chine, & que ces Pays-cy n'en ayent jamais eu leur part. Il n'y a pas longtems que tous nos Philosophes se croyoient fondez en expérience pour soutenir que les Cieux et tous les corps Celestes estoient incorruptibles & incapables de changement, & pendant ce temps-là d'autres hommes à l'autre bout de la Terre voyoient des Etoiles se dissoudre par milliers; cela est assez différent.

Mais, dit-elle (Mme la Marquise), n'ay-je pas toûjours oüy dire que les Chinois estoient de si grands Astronomes?

Il est vrai, repris-je, mais les Chinois y ont gagné à estre separez de nous par un long espace de Terre, comme les Grecs et les Romains à en estre separez par une longue suite de siecles; tout éloignement est en droit de nous imposer....

[a] Printing and gunpowder are of course referred to, and the construction of very large ships, but there is also mention of storm and tide prediction, and imaginary inventions such as typewriters and dictaphones are described.

[b] The Chinese polity is said to be based on the belief that 'Natural Right is superior to Temporal Power', a heretical doctrine for the West, 'long since exploded by our learned doctors, who have proved that kings and emperors came down from heaven with crowns on their heads, and that all their subjects were born with saddles on their backs.' On Defoe's book in general see M. H. Nicolson (1), p. 183.

[c] M. H. Nicolson (1), p. 177.

[d] Pp. 46, 95, 165, 182, 236 ff.

[e] P. 294.

So here again, with bantering tone, a European expositor of post-Renaissance cosmology, while gently making fun of the Chinese reputation for wisdom in such matters, hints that indeed they had long known of the birth and death of stars (novae, meteors, etc.) which European tradition had made immutable; and of the almost infinite distance and diversity of those lights which European tradition had supposed truly fixed to a revolving crystal sphere. In a word, specific investigations might be worth making to ascertain whether the complete disbelief of the Chinese in the solid celestial spheres of the Ptolemaic-Aristotelian world-picture, which became evident to the Jesuits as soon as they began to discuss cosmology in China, was not one of the elements which combined in breaking up medieval views in Europe, and contributed to the birth of modern astronomy. For such a suggestion there is contemporary evidence in the words of Christopher Scheiner, seeking about 1625 to show that the realm of the stars had a fluid nature.

> The peoples of China [he said] have never taught in any of their innumerable and flourishing academies that the heavens are solid; or so we may conclude from their printed books, dating from all times during the past two millennia. Hence one can see that the theory of a liquid heavens is really very ancient, and could be easily demonstrated; moreover one must not despise the fact that it seems to have been given as a natural enlightenment to all peoples. The Chinese are so attached to it that they consider the contrary opinion (a multiplicity of solid celestial spheres) perfectly absurd, as those inform us who have returned from among them.[a]

And we may take as another example the words of Nieuwhoff (1665) who found the idea of a plurality of worlds very Chinese.[b]

In any case, the height of irony is reached when we find Wells Williams, in 1848, reproaching late Chhing popular writers for their belief in solid celestial spheres, under the impression that this was a primitive Chinese doctrine still persisting.[c]

(2) The Imperfect Transmission

The sincere (and well-justified) admiration of Ricci himself for the instruments of Kuo Shou-Ching has already been noted.[d] But such was the decadence of the late Ming period, and so convinced were the Europeans of their scientific superiority, that the accounts of Chinese astronomy which got through to 17th-century Europe were mainly unfavourable. One or two relevant passages are worth looking at. Thus Trigault wrote:

> They have some Knowledge also of Astrologie and the Mathematikes: In Arithmetike and Geometry antiently more excellent, but in learning and teaching confused. They reckon

[a] *Rosa Ursina* (1630), p. 765 (Bk. 4, Pt. 2, ch. 29, Pro Caelo Liquido Auctoritates Astronomorum), tr. Bernard-Maître (7), p. 57, eng. auct. Scheiner's informant must have been Trigault.

[b] Nieuwhoff had accompanied the Dutch ambassadors to Peking in +1656 (Cordier (1), vol. 3, p. 262). In the second part of the book which he wrote about this embassy and about China he said: 'Et les opinions qu'ont quelques-uns (des Chinois) dans la physique, conformes à celles de Democrite et de Pythagore touchant la pluralité des Mondes, monstrent assés combien ceux de cette Nation se plaisent à l'étude des choses naturelles' (p. 14).

[c] (1), vol. 2, p. 73.

[d] P. 367 above.

four hundred Starres more than our Astrologers have mentioned, numbring certaine smaller which do not always appeare. Of the heavenly Appearances they have no rules: they are much busied about foretelling Eclipses, and the courses of Planets, but therein very erroneous; and all their Skill of Starres is in a manner that which we call Judiciall Astrology, imagining these things below to depend on the Starres. Somewhat they have received of the Westerne Saracens, but they confirme nothing by Demonstration, only have left to them Tables, by which they reckon the Eclipses and Motions.

The first of this Royal Family (the Ming) forbad any to learne this Judiciall Astrology but those which by Hereditary right are thereto designed, to prevent Innovations. But he which now reigneth mayntayneth divers Mathematicians, both Eunuches within the Palace and Magistrates without, of which there are in Pequin two Tribunals, one of Chinois, which follow their owne Authors, another of Saracens which reforme the same by their rules, and by conference together. Both have in a small Hill a Plaine for contemplation where are the huge Mathematicall Instruments of Brasse before mentioned: One of the Colledge nightly watcheth thereon as is before observed. That of Nanquin exceeds this of Pequin, as being then the Seat Royall. When the Pequin astrologers foretell Eclipses, the Magistrates and Idoll Ministers are commanded to assemble in their Officiary Habits to helpe the labouring Planets, which they think they do with beating brazen Bels, and often kneelings, all the time that they thinke the Eclipse lasteth, lest they should then be devoured (as I have heard) by I know not what Serpent.[a]

Another passage, from Lecomte, is of much interest as giving an eye-witness account of that nightly observation which had been going on for perhaps three millennia.

They still continue their Observations. Five Mathematicians spend every Night on the Tower in watching what passes over head; one is gazing towards the Zenith, another to the East, a third to the West, the fourth turns his eyes Southwards, and a fifth Northwards, that nothing of what happens in the four Corners of the World may scape their diligent Observation. They take notice of the Winds, the Rain, the Air, of unusual Phenomena's, such as are Eclipses, the Conjunction or Opposition of Planets, Fires, Meteors, and all that may be useful. This they keep a strict Accompt of, which they bring in every Morning to the Surveyor of the Mathematicks, to be registered in his Office. If this had always been practised by able and careful Mathematicians, we should have a great number of curious Remarks;[b] but besides that these Astronomers are very unskilful, they take little care to improve that Science; and provided their Salary be paid as usual, and their Income constant, they are in no great trouble about the Alterations and Changes which happen in the Sky. But if these Phenomena's are very apparent, as when there happens an Eclipse, or a Comet appears, they dare not be altogether so negligent.[c]

Now the two most important features in European astronomy at the time the Jesuits began their work in China were (*a*) the invention and use of the telescope, and (*b*) the acceptance of the heliocentric theory of Copernicus. The former they transmitted, but the latter, after some hesitations, they held back. The reform of the Chinese calendar, which usually looms so large, and of which some criticisms have already been made (pp. 258, 404 above), has been exhaustively described by Bernard-Maître (7); in

[a] The passage is taken from the recension of his *De Christiana Expeditione* (+1615) englished in *Purchas his Pilgrimes*, vol. 3, p. 384. Also Gallagher ed. p. 31.

[b] Lecomte evidently had no idea as to the relative value of ancient and medieval records of celestial phenomena in China and Europe.

[c] *Memoirs and Observations*, etc., p. 71.

reality, it was much less significant than the two developments just mentioned. Owing to the researches of d'Elia (1, 3) we are now fairly well informed about what really happened. Szczesniak (2) has said that the Copernican conflict had an even more tragic history in China than in Europe, since it lasted down to the end of the 18th century. Duyvendak (6) has underlined the importance of the failure of the Jesuits to transmit the heliocentric system of the universe.

Ricci died in Peking in 1610, the same year in which Galileo published his *Sidereus Nuntius*. In the following winter, Christopher Clavius and other Jesuits of the Roman College repeated his telescopic observations[a] and confirmed them. But this made the Jesuits anti-Aristotelian rather than anti-Ptolemaic. Clavius, the teacher and friend of Ricci, died in 1612. The two condemnations of Galileo's Copernican views[b] were in 1616 and 1632, and must have had considerable effect on the China Mission. The first reference to the telescope in Chinese is in the *Thien Wên Lüeh*[1] (Explicatio Sphaerae Coelestis)[c] by Emanuel Diaz (Yang Ma-No[2]) of 1615, where Galileo is said to have devised it because he 'lamented the weakness of the unaided eye' (*ai chhi mu li*[3]). Venus was seen as big as the moon, and Saturn looked as if it had a hen's egg on each side (Fig. 185), while Jupiter's satellites were clearly visible (Fig. 187). In 1618 Johannes Terrentius (Johann Schreck; Têng Yü-Han[4]) arrived in China; he had been the seventh member of the Cesi Academy, having been elected next after Galileo, and was an astronomer and physicist of great gifts. He brought with him a telescope, which was eventually given to the emperor in 1634, and remained in touch with Galileo, who was not very helpful, and with Kepler, who took more interest.[d] Kepler sent out with the Polish Jesuit Michael Boym (Pu Mi-Ko[5])[e] a set of the (Copernican) Rudolphine Tables in 1627, and Boym, who stayed at Macao, passed them on with enthusiastic praise to Peking.[f] In the previous year, Adam Schall von Bell (Thang Jo-Wang[6])[g] had pub-

[a] Especially on the lunar mountains, the sun-spots, the phases of Venus, the shape of Saturn, the moons of Jupiter, the nebula in Orion, the *Praesepe* cluster in Cancer, and many stars not before visible. John Adam Schall von Bell, later to be the first European Director of the Chinese Bureau of Astronomy, was present as a young man in the hall of the Roman College in May 1611 when Galileo received a triumphant welcome from Clavius and his 'mathematicians' after their confirmation of his discoveries.

[b] See Banfi (1) and de Santillana (1), two recent valuable contributions to this long discussed subject.

[c] Cordier (8), p. 18. On the coming of the telescope to Japan see Mikami (*17*).

[d] 'Father Terrentius', says Gaubil (5), p. 285, 'wrote to the celebrated Kepler, telling him what the Yao Tien (chapter of the *Shu Ching*) reports about the stars. Undoubtedly it was on this occasion that he sent him the Chinese method for computing solar eclipses. Kepler was also informed of the eclipses in the *Shu Ching* and the *Shih Ching*, as well as of others taken from the *Chhun Chhiu* and the (dynastic) histories. But now we have not been able to find here the copies of the letters, nor of the replies which Kepler doubtless made to them.' Actually the letters and the answers were printed in 1630 in Silesia (Pfister (1), p. 157), and reprinted in the Frisch edition of Kepler's works, vol. 7, pp. 667 ff. Cf. Bernard-Maître (7). On the general question of the relations of Terrentius (Schreck) and other Jesuit scientists with Galileo and Kepler, see Gabrieli (1, 2) and Bernard-Maître (12). Schreck knew them both personally.

[e] See Szczesniak (6).

[f] See Szczesniak (3).

[g] See his biography by Väth (1). Though Terrentius and then Rho had been in charge of calendar reform from 1629 onwards, Schall von Bell was the first to receive official rank as Director (Chien Chêng[7]) of the Bureau of Astronomy (Chhin Thien Chien[8]). No Jesuit was ever President of it. The succession

1 天問畧 2 陽瑪諾 3 哀其目力 4 鄧玉函 5 卜彌格
6 湯若望 7 監正 8 欽天監

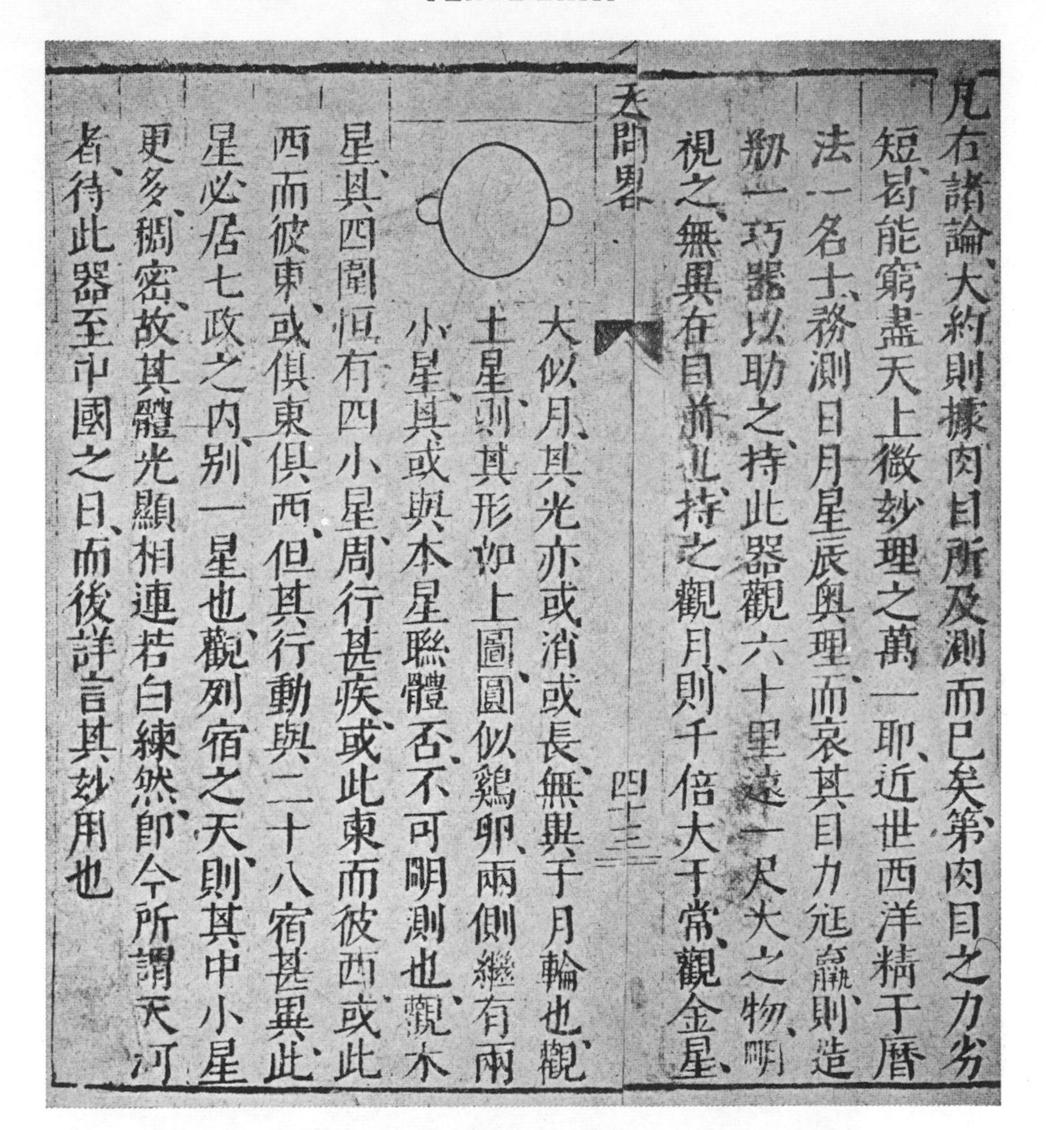

凡右諸論大約則據肉目所及測而已矣第肉目之力劣
短曷能窮盡天上微妙理之萬一耶近世西洋精于曆
法一名士務測日月星辰奥理而哀其目力尩羸則造
剏一巧器以助之持此器觀六十里遠一尺大之物明
視之無異在目前也持之觀月則千倍大于常觀金星

天問畧　四十三

大似月其光亦或消或長無異于月輪也觀
土星則其形如上圖圓似雞卵兩側繼有兩
小星其或與本星聯體否不可明測也觀木
星其四圍恒有四小星周行甚疾或此東而彼西或此
西而彼東或俱東俱西但其行動與二十八宿甚異此
星必居七政之內別一星也觀列宿之天則其中小星
更多稠密故其體光顯相連若白練然即今所謂天河
者待此器至中國之日而後詳言其妙用也

Fig. 185. Two pages from the *Thien Wên Lüeh* (Explicatio Sphaerae Coelestis) of Emanuel Diaz (+1615) relating for the first time in Chinese the discoveries made with Galileo's telescopes.

The right-hand page reads: 'Most of the above observations (on lengths of days and nights, eclipses, etc.) were made with the naked eye. But such vision is short; how could it attain to even one ten-thousandth part of the mysterious principles of the vast Heavens? Quite recently, however, a famous scholar of the West, particularly learned in calendrical science and himself an observer of the sun, moon and planets, deploring the weakness of his eyes, has made an ingenious instrument which helps them. With this device an object one foot in size can be seen at a distance of 60 *li* as clearly as if it was in front of one's eyes. The moon, seen through (this telescope) seems a thousand times larger than usual.'

The left-hand page reads: 'Venus appears as large as the moon; its light increases and decreases just like the moon's. Saturn, as shown in the above diagram, seems to have a rounded shape like that of an egg, with two small stars on each side, but whether or not they are attached to it we do not know. Jupiter can be seen always surrounded by four small stars, continually revolving around it at great speed, one at the west and one at the east, or vice versa, and sometimes all on the west side or all on the east side—(in any case) their movement differs greatly from that of the 28 *hsiu*. For these stars must remain in the regions of (each of) the seven planets, and form (indeed) a special sort of star. Then, when one looks (with the telescope) at the great constellations in the firmament one sees an immense multitude of small stars closely crowded together, hence the light from their bodies seems to form a white stream, which we call the Milky Way. As soon as one of these instruments arrives in China we shall give more details on its marvellous uses' (tr. d'Elia (3), mod. auct.).

The 'triple' appearance of Saturn was first seen by Galileo in +1610 but he could never interpret it; the disposition of the rings and satellites was not clarified until Huygens' *Systema Saturnium* of 1659.

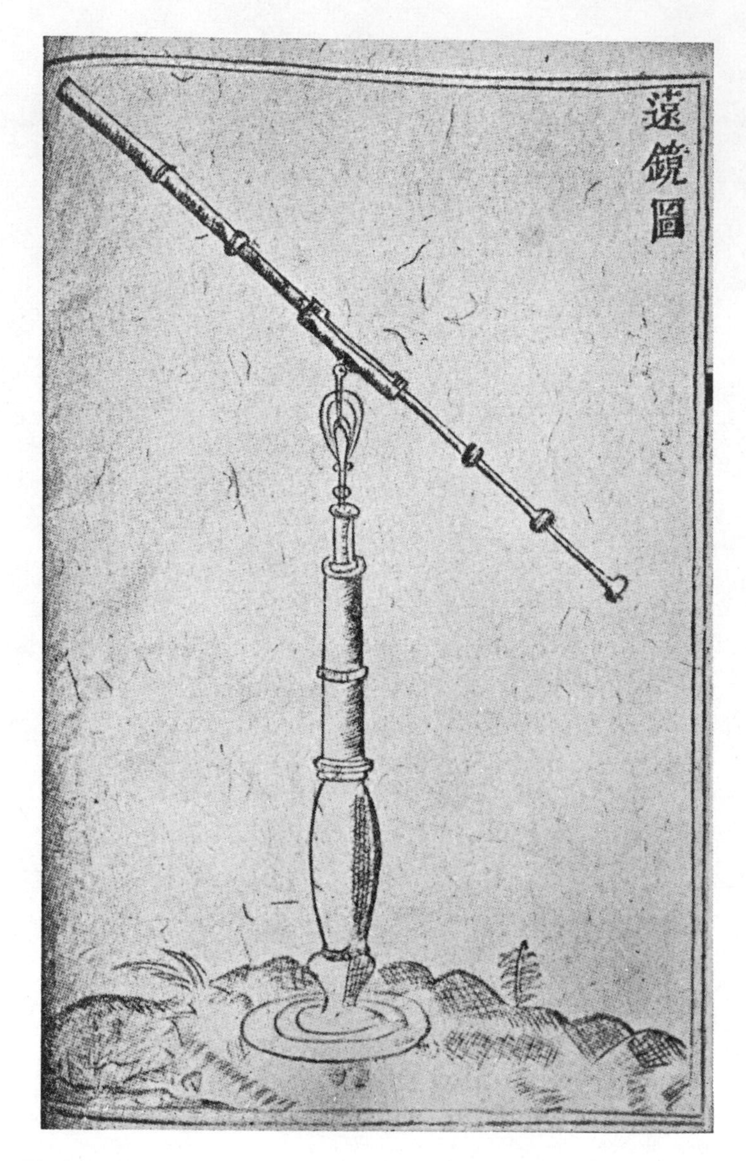

Fig. 186. The first Chinese picture of the telescope, the 'far-seeing optick glass', from Adam Schall von Bell's *Yuan Ching Shuo* of +1626.

lished a Chinese treatise on the telescope, *Yuan Ching Shuo*[1] (The Far-Seeing Optick Glass) (cf. Fig. 186).[a] But not until +1640, in Schall's history of Western astronomy in Chinese, were the names of Galileo (Chia-Li-Lüeh[2]), Tycho Brahe (Ti-Ku[3]), Copernicus (Ko-Pai-Ni[4]) and Kepler (Kho-Pai-Erh[5]) actually mentioned.[b]

It is clear that in this early period, especially before the condemnation of Galileo, the missionaries were not at one on Copernicanism. It was favoured by Boym, and taught by Nicholas Smogułęcki (Mu Ni-Ko[6]), another Pole, at Nanking, while Wenceslaus Kirwitzer, who went out with Terrentius, was definitely a Copernican, but

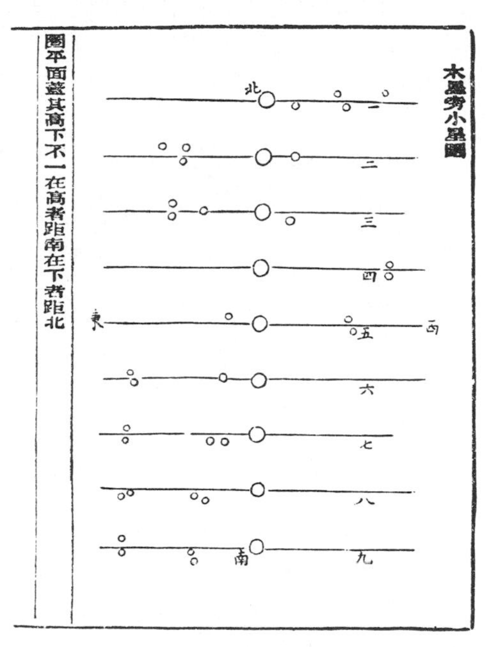

Fig. 187. Diagrams of the moons of Jupiter (from the *Thu Shu Chi Chhêng*).

died young in 1626. In general it may be said that Chinese books between 1615 and 1635 described the telescopic discoveries, but did not mention Copernicanism, then for a short time the heliocentric theory was described, but after news of the condemnation had reached China a curtain descended and a return to the Ptolemaic view took

was as follows: Schall (1645–66), Verbiest (1669–88), P. M. Grimaldi (1688–1706 and 1710–12), A. Thomas (Acting, 1686–94), C. Kastner (1707–9), B. Kilian Stumpf (1712–20), I. Kögler (1720–46), A. von Hallerstein (1746–74), Felix da Rocha (1774–81), J. d'Espinha (1781–3), and J. B. d'Almeida (1783–1805). Pfister (1), p. 886, says that d'Almeida closed the line, but from Huc (1), himself a Lazarist, we learn that the last missionary to be employed in the Bureau of Astronomy, perhaps not as Director, was the Lazarist Gaetan Pires-Pereira (d. 1838).

[a] Cordier (8), p. 37. This book included a rough picture of the Crab Nebula, though neither Schall nor his readers knew what its later importance for cosmology would be, nor that it had come from the +1054 supernova observed only in China and Japan.

[b] *Li Fa Hsi Chuan*;[7] see Pfister (1), p. 180. Such transliterations were long in use, cf. *Chhou Jen Chuan*, chs. 43–6.

[1] 遠鏡說 [2] 伽離畧 [3] 弟谷 [4] 歌白泥 [5] 刻白爾
[6] 穆尼閣 [7] 曆法西傳

place.[a] This had long before been clearly expounded in the *Chien Phing I Shuo*[1] (Elementary Explanations of Astronomical Instruments)[b] of 1611 by Sabbathin de Ursis (Hsiung San-Pa[2]).

Humanistic colleagues have sometimes expressed surprise that the Jesuits could have been so successful at preparing for the Chinese court a calendar of 'Renaissance' type while at the same time adhering to the Ptolemaic world-system and rejecting the Copernican.[c] The first answer is that on the purely calendrical level there is nothing to choose between them. The geocentric and the heliocentric hypotheses were in strict mathematical equivalence; whether the earth or the sun was at rest, the lengths and angles were identical, and similar triangles had to be solved.[d] Decision lay far beyond the frame of reference of the calendar-computer; it needed the relatively accurate

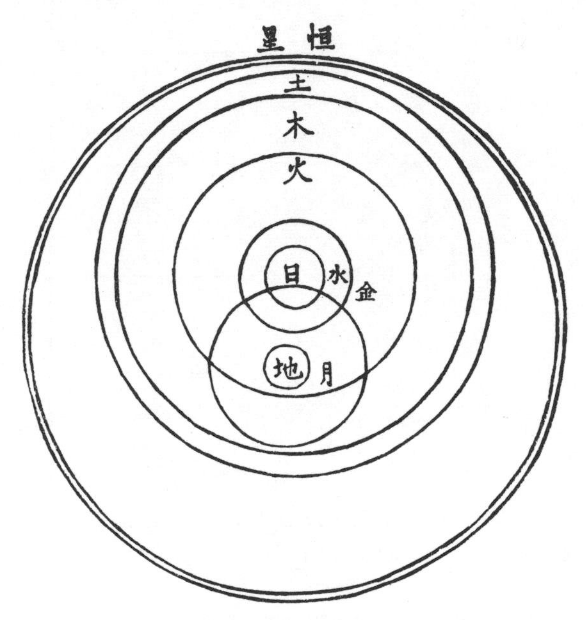

Fig. 188. Diagram of the Tychonic theory of the solar system (from the *Thu Shu Chi Chhêng*).

observational data of the age of Cassini and Flamsteed. Secondly, the Chinese themselves had produced very good calendars for centuries before the time of the Jesuits without employing any geometrical model of the solar system at all. A calendar is only a method of reconciling terrestrial-celestial periodicities observed as carefully as possible, predicting their recurrences, and adjusting conventional human time units

[a] In the *Thu Shu Chi Chhêng* encyclopaedia there is a diagram of the Tychonic theory of the solar system (Fig. 188 from *Li fa tien*, ch. 65, p. 3*b*); cf. Berry (1), p. 137. We have not found a diagram of the Copernican.

[b] Pfister (1), p. 105.

[c] I am much indebted to Dr Victor Purcell for raising this point in challenging form.

[d] See the discussion of Price (2), p. 94.

[1] 簡平儀說 [2] 熊三拔

(months, days, etc.) to the best fit.[a] Where the Jesuits scored was in the more advanced character of their instruments and the superiority of their mathematics (old geometry indeed but quite new algebra). But it took them nearly a century to learn to profit by the great wealth of recorded Chinese celestial observations.

A curious result of the Jesuit failure to make use of the Chinese records until the time of Gaubil was that when Terrentius in 1628 expounded the telescopic discovery of sun-spots in his *Tshê Thien Yo Shuo*[1] (Brief Description of the Measurement of the Heavens),[b] there was no mention of the fact that they had been known for a dozen centuries before the Europeans discovered them.

Szczesniak (1) has contrasted the situation of China with that of Japan. The effect of the closure of Japan between 1616 and 1720 was to emphasise the contribution of Dutch traders rather than that of Roman Catholic missionaries. When the first modern observatory in Japan was founded about 1725 under the direction of Nakane Genkei,[2] Copernican ideas were fully admitted there.[c] But in China it was not until the early 19th century, with the contributions of the Protestant missionaries, such as Joseph Edkins, Alex. Wylie and John Fryer, that Copernican views really spread. Some detail of their work will be found in Szczesniak (2). But it is impossible to accept his contention that the main reason of the Jesuits for not propagating Copernicanism was the resistance of the Chinese to any abandonment of the geocentric world-view; this can have been only a part of the story. On the whole, one concludes that the Jesuit contribution was not an unmixed blessing.

(3) 'Western' Science or 'New' Science?

Nevertheless, the later work of the Society was indeed impressive. Between 1629 and 1635 the second generation of missionaries, including, besides Terrentius and Schall von Bell, James Rho (Lo Ya-Ku[3]), and, to a minor extent, Nicholas Longobardi (Lung Hua-Min[4]),[d] collaborating with Hsü Kuang-Chhi, Li Chih-Tsao and Li Thien-Ching,[5] produced a monumental compendium of the scientific knowledge of the time. This was entitled, upon its presentation in the latter year, the *Chhung-Chên Li Shu*[6] (Chhung-Chên Reign-Period Treatise on (Astronomy and) Calendrical

[a] A great deal of nonsense has been written about the calendar reform of the Jesuits in China. Cronin (1), pp. 142, 230, 231, writes as if the Gregorian Calendar (partly the work of Matteo Ricci's teacher Christopher Clavius) adopted in Europe in 1582 was a corollary of the Ptolemaic theory. This, he suggests, had been introduced to China by the Arabs, but rejected there, so that its absence caused calendrical disorder. On the contrary, the Gregorian Calendar was simply another system of intercalation, more ingenious than its forerunners, but still arbitrary, as any such system of reconciling incommensurables must be. Cf. Fotheringham (1), p. 743; Philip (1), p. 20.

[b] Pfister (1), p. 157.

[c] The growth of modern science in Japan lies outside our field. But attention may be drawn to an interesting series of papers by Mikami (8, 9, 10, 11, 14). A Japanese physician (Petrus Hartsingius Japonensis, perhaps identical with Hatono Sōha) studied in 17th-century Leiden, and managed to return to his own country. An Italian missionary (Giuseppe Chiara) abandoned his mission, took a Japanese name (Sawano Chūan[7]) and settled down to spend his life translating Dutch scientific books, especially on astronomy. On Nakane Genkei see Hayashi (2), pp. 354ff.

[d] We have met with him before in connection with Leibniz (Sect. 16*f*, in Vol. 2, p. 501 above).

[1] 測天約說 [2] 中根元圭 [3] 羅雅谷 [4] 龍華民 [5] 李天經
[6] 崇禎曆書 [7] 澤野忠庵

Science).[a] After the Manchus came in, ten years later, in 1645, Schall von Bell attained greater favour, and the encyclopaedia was reissued as the *Hsi-Yang Hsin Fa Li Shu*[1] (Treatise on (Astronomy and) Calendrical Science according to the New Western Methods).[b] Eventually it formed the basis for the *Yü-Ting Li Hsiang Khao Chhêng*[2] (Complete Studies on Astronomy and Calendar),[c] edited by Ho Kuo-Tsung[3] and Mei Ku-Chhêng[4] and printed in 1723. Later on, in 1738, after much of it had been incorporated in the *Thu Shu Chi Chhêng* encyclopaedia,[d] it was improved by the addition of astronomical tables embodying the new observations of Cassini and Flamsteed. These were the work of Ignatius Kögler (Tai Chin-Hsien[5]) and Andrew Pereira (Hsü Mou-Tê[6]).[e]

Here we must halt a moment. The reader will probably have noticed nothing especially significant in the preceding paragraph, seemingly concerned only with the recitation of fact. But actually it raises certain points of extreme importance in these culture-contacts of the two great civilisations, and we must look at the facts more closely. It is vital today that the world should recognise that 17th-century Europe did not give rise to essentially 'European' or 'Western' science, but to universally valid world science, that is to say, 'modern' science as opposed to the ancient and medieval sciences. Now these last bore indelibly an ethnic image and superscription. Their theories, more or less primitive in type, were culture-rooted, and could find no common medium of expression. But when once the basic technique of discovery had itself been discovered, once the full method of scientific investigation of Nature had been understood, the sciences assumed the absolute universality of mathematics, and in their modern form are at home under any meridian, the common light and inheritance of every race and people. Of argument about elements and humours, Yin and Yang, or 'philosophical sulphur', there could be no end, the disputants could reach no common ground. But the mathematisation of hypotheses led to a universal language, an oecumenical medium of exchange, a reincarnation of the merchants' single-value standard on a plane transcending merchandise. And what this language communicates is a body of incontestable scientific truth acceptable to all men everywhere. Without it

[a] It has 100, 110 or 137 chapters (not of course 'volumes' as so often said), according to different editions and sources. Cf. Pfister (1), p. 156; Bernard-Maître (7), p. 452; Li Nien (*4*), vol. 1, p. 167, (*21*), vol. 3, p. 37.

[b] It now contained an appendix entitled *Hsin Fa Piao I*[7] (Differences between the Old and the New Astronomical Systems), and later on, in 1656, Schall von Bell added his *Li Fa Hsi Chuan* (History of Western Astronomy), already mentioned, as another.

[c] This formed, with the treatise on acoustics and music, *Lü Lü Chêng I*, and the companion one on mathematics, *Shu Li Ching Yün*, the three parts of the *Lü Li Yuan Yuan* (Ocean of Calendrical and Acoustic Calculations), an official and imperial publication. There is room for some doubt about the true authorship of the *Li Hsiang Khao Chhêng*, though it was certainly Chinese and not Jesuit; cf. Hummel (2), pp. 93, 285, 922. Elaborate description by Bernard-Maître (7). Cf. p. 53 above.

[d] *Li fa tien*, chs. 51–78.

[e] Andrew Pereira is of particular interest to us as he was the only Englishman among all the Jesuits of the China Mission. He came of a family of the name of Jackson, settled in Oporto, doubtless connected with the wine trade and naturalised Portuguese. He seems to have been a very sympathetic character, and was a particular friend of the Yung-Chêng emperor, not otherwise an amateur of missionaries. Cf. Pfister (1), p. 652.

1 西洋新法曆書 2 御定曆象考成 3 何國宗 4 梅瑴成
5 戴進賢 6 徐懋德 7 新法表異

plagues are not checked, and aircraft will not fly. The physically unified world of our own time has indeed been brought into being by something that happened historically in Europe, but no man can be restrained from following the path of Galileo and Vesalius, and the period of political dominance which modern technology granted to Europeans is now demonstrably ending.

In their gentle way, the Jesuits were among the first to exercise this dominance, spiritual though in their case it was meant to be. To seek to accomplish their religious mission by bringing to China the best of Renaissance science was a highly enlightened proceeding, yet this science was for them only a means to an end. Their aim was naturally to support and commend the 'Western' religion by the prestige of the science from the West which accompanied it. This new science might be true, but for the missionaries what mattered just as much was that it had originated in Christendom. The implicit logic was that only Christendom could have produced it. Every correct eclipse prediction was thus an indirect demonstration of the truth of Christian theology. The *non sequitur* was that a unique historical circumstance (the rise of modern science in a civilisation with a particular religion) cannot prove a necessary concomitance. Religion was not the only feature in which Europe differed from Asia. But the Chinese were acute enough to see through all this from the very beginning. The Jesuits might insist that Renaissance natural science was primarily 'Western', but the Chinese understood clearly that it was primarily 'new'.

Thus the 'Chhung-Chên treatise' of 1635 reappeared ten years later as the 'Western treatise according to New Methods'. Schall von Bell had been wanting to use the word 'Western' a long time previously. In a letter to Francis Furtado (Fu Fan-Chi[1]) of November 1640, he said he was aiming at a Hsi Kho[2] (Western Bureau) within the Li Kho[3] (Department of the Calendar), but that the disadvantage of this was that it put it only on a level with the Muslim Bureau (Hui-Hui Kho[4]) already existing. He wrote: 'The word Hsi (Western) is very unpopular (with the Chinese), and the emperor in his edicts never uses any word other than Hsin (New); in fact the former word is employed only by those who wish to depreciate us.'[a] But after the change of dynasty Schall evidently felt that he could freely use the term 'Western'; after all, the Manchus were foreign too. So for many years printed calendars bore the title '...i Hsi-Yang Hsin Fa' (according to the New Western Methods). For this he was taken to task in 1661 by Yang Kuang-Hsien,[5] and three years later formally condemned by the President of the Ministry of Rites for having used a formula 'injurious for the dignity of the empire'.[b] However, before long, Schall having died in 1666, his Belgian

[a] Bernard-Maître (7), p. 463.

[b] Yang Kuang-Hsien was a scholar, amateur astronomer, and pertinacious anti-Jesuit controversialist. Associated with him was the Muslim astronomer Wu Ming-Hsüan.[6] Schall could not remember who had originally authorised the phrase 'Western', but it seems to have been agreed upon in the first year of the Manchu régime, when Rho was still in charge of the calendar reform, having inherited this position from Terrentius. These early days of the scientific activity of the mission are referred to in the *Ming Shih*, ch. 326, pp. 17*b* ff. (tr. Bretschneider, 7). On the prosecution and imprisonment of Schall, see Bernard-Maître (7), p. 477.

[1] 傅汎濟 [2] 西科 [3] 曆科 [4] 回回科 [5] 楊光先
[6] 吳明烜

successor Ferdinand Verbiest (Nan Huai-Jen[1])[a] was called to the Khang-Hsi emperor (who had succeeded in 1662) and spent no less than five months daily with him teaching and explaining the new mathematics and astronomy.[b] It was then about 1669 that the encyclopaedia was reissued with again a new title, the *Hsin Fa Suan Shu*[2] (Treatise on Mathematics (and Astronomy) according to the New Methods).[c] The emperor's insistence united him unknowingly with that group of men at the other end of the world who exactly at the same time were meeting in the Royal Society to work out the implications of the 'new, or experimental, philosophy'—just as new for Europe as for China.[d]

Down to the very end of the mission the Jesuits were the prisoners of their limited motive and the Chinese sought persistently to emphasise the continuity of the new science with the old. For example, in 1710 Jean-François Foucquet (Fu Shêng-Tsê[3]) and others of the Society wished to make use of the new planetary tables of P. de la Hire,[e] but the Father-Visitor would not permit it, for fear of 'giving the impression of a censure on what our predecessors had so much trouble to establish, and occasioning new accusations against our religion'.[f] Any acceptance of Copernicanism would equally have raised doubts about all Ricci's teachings. In fact the penalty of enlisting live science in the service of fixed doctrine was to inhibit its development—Urania's feet were bound. Only in certain cases could the Jesuits move forward; for instance, the armillary sphere of 1744 was a Chinese (and therefore 'modern') equatorial, not a Greek ecliptic, instrument; the old European coordinates were quietly given up. Meanwhile the Chinese were much concerned to show that the study of Nature had a continuous history. This is transparent from passages such as the following:[g]

In the Wan-Li reign-period (+1573 to +1619) the western foreigner Li Ma-Tou (Matteo Ricci) made designs for an armillary sphere, a celestial and a terrestrial globe, etc. Li Chih-Tsao of Jen-ho wrote a discussion on the discovery, construction and use of the armillary sphere, which, though of some length, did not include diagrams. For (the new design) was not essentially different from the (long-known) apparatus constituted by the 'component of the six cardinal points', the 'component of the three arrangers of time' and the 'component of the four displacements'.[h] The main improvement was that whereas formerly the polar altitude was fixed in the casting, the new model was so arranged as to be adjustable for different latitudes—a very convenient thing....

The design and construction of astronomical instruments, and the making of observations have always been the first duty of astronomers. Those who are technically skilful can devise

[a] See Bernard-Maître (11), and de Burbure (1) from whom we reproduce Fig. 189.

[b] Large scrolls and wall-diagrams of planispheres, planetary motions, and astronomical instruments, well printed in Chinese, both by Schall von Bell in the thirties and by Verbiest in the sixties, are still extant, though very rare. I had the opportunity of examining a number of them in the possession of Messrs P. R. Robinson, Ltd., in London in 1956. All published bibliographies are inadequate regarding these charts.

[c] Cf. Bernard-Maître (7), p. 481.

[d] Mr Jean Chesneaux was the first to see the importance of the successive changes in the titles of the Jesuit books, and I owe to him an appreciation of it. We have emphasised it together elsewhere (Chesneaux & Needham, 1).

[e] *Tabulae Astronomicae*, Paris, 1702 (R. Wolf (3), vol. 2, p. 288).

[f] Pfister (1), p. 551.

[g] *Ming Shih*, ch. 25, pp. 17*a*, 19*b*; tr. auct.

[h] Cf. pp. 343ff. above.

[1] 南懷仁 [2] 新法算書 [3] 傅聖澤

Fig. 189. Ferdinand Verbiest, habited as a Chinese official, with his sextant and his celestial globe. The picture is a Japanese print by Utagawa Kuniyoshi (1797 to 1861), one of a series of 108 entitled *Tsugoku Suikoden Koketsu hyaju hachinin ikkō* (The Hundred and Eight Heroes of the Popular Novel *Shui Hu Chuan*). The legend reads: 'Chioku Seigōyō [Chih-To-Hsing, Wu Yung], a man from the village of Tōkei, having the cognomen Gāgaku-kyudō, and the literary name Karyu-sensei. In military science he was not inferior to Kōmei [i.e. Chuko Liang] and Taikohō [i.e. Chiang Tzu-Ya], and for wiles and stratagems he equalled Hanrei [i.e. Fan Li]. He was a general at Ryosanhaku [i.e. Liang Shan Po, the headquarters of the rebellion described in *All Men are Brothers*].' None of these names resembles in the least Verbiest's real Chinese name, Nan Huai-Jen, but the allusion must be to his casting of cannon for the Chhing government in +1675 against Wu San-Kuei. His conflation with one of the heroes of popular literature in this way is remarkable (photo. Michel, reproduced by de Burbure).

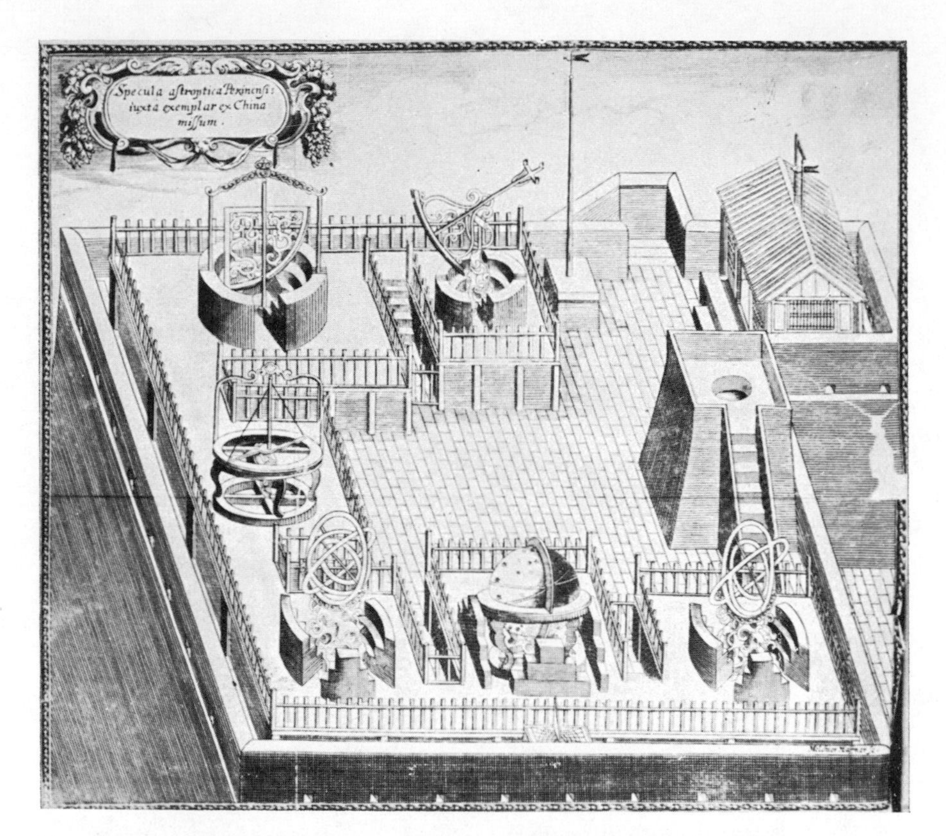

Fig. 190. The Peking Observatory refitted by Ferdinand Verbiest (+1674), according to the engraving made by Melchior Haffner for Verbiest's *Astronomia Europaea* of +1687 and often copied elsewhere both in Chinese and Western books (e.g. *Thu Shu Chi Chhêng*). The view is taken from the south. The instruments shown are as follows, proceeding from the north-east corner of the platform (background, right): sextant, quadrant, horizon circle, ecliptic armillary sphere, celestial globe, equatorial armillary sphere. The eastern wall of the city, not shown, runs along the right-hand side of the picture.

Fig. 191. The Peking Observatory, photographed about 1925 from the north-east corner of the platform. On the right the armillary sphere of Kögler & von Hallerstein (+1744) and Verbiest's quadrant. In the centre, Verbiest's celestial globe. At the back, in the southern row, from right to left, Verbiest's ecliptic armillary sphere and horizon circle, then Stumpf's quadrant altazimuth (+1714) (photo. Whipple Museum Collection). The background is formed by the roofs and trees of the city of Peking.

Fig. 192. The sextant of +1673 at the Peking Observatory (photo. Whipple Museum Collection).

ingenious improvements. Thus the Westerners have many instruments with various names of which we cannot speak here, but among them the two instruments *hun kai*[1] and *chien phing*[2] are the most refined.[a] Reference must be made to the complete accounts for all cannot be recorded herein.

And as a pendant to this consider the air-pump and its name. When Michel Benoist (Chiang Yu-Jen[3]) demonstrated this to the Chhien-Lung emperor in 1773 he termed it the *yen chhi thung*[4] ('air-testing pipe'), but the monarch decided next day that its name should be changed to *hou chhi thung*[5] ('air-observing pipe'), 'this word being more elegant, having served in classical writings for natural observations on the celestial bodies as well as on agricultural matters'.[b] He was of course thinking of the antique names of the weather-vane and the seismograph,[c] and of one of the terms for the armillary sphere.[d]

(4) The Integration of Chinese Astronomy into Modern Science

In 1669 there began the great refitting of the Peking observatory (Figs. 190, 191)[e] under the care of Ferdinand Verbiest (Fig. 189). The instruments of Yuan or Ming time were taken down from the astronomical platform on the eastern wall of the city,[f] and a new set installed in their place, where they have remained until the present time.[g] The Jesuit and later instruments are as follows:[h]

(1) Simple ecliptic armillary sphere, *huang tao ching wei i*,[6] supported on four dragon heads. Verbiest, 1673.

[a] These are not standard names but probably refer to instruments of quadrant altazimuth type.

[b] Pfister (1), p. 823, from *Lettres Ed. & Curieuses*, vol. 4, p. 224.

[c] Cf. pp. 469, 478, 627 below. [d] Cf. p. 383 above.

[e] The literature on the instruments and their arrangement on the observatory platform is scattered but copious. The best drawings of them in a Western work are those in Lecomte (French edition only). Starting from the general view given by Verbiest himself (Fig. 190), which appears also in *TSCC*, *Li fa tien*, ch. 93, pp. 2*b*, 3*a*, and was often copied (as by du Halde), the observatory can be studied in the descriptions and illustrations of many successive authors from the middle of the last century—J. Thomson (1), Mouchez (1), Bosmans (2), Damry (1), Planchet (1), Kao Lu (1), F. B. Robinson (1). Very recently Chhen Tsun-Kuei (*6*) has produced a standard account of the instruments.

Intensive study has been devoted to the observatory's history by Mr P. A. Jehl of Paris, to whom I am indebted for much interesting information.

[f] They long remained in and among the buildings belonging to the Bureau of Astronomy below the platform, but in recent times two at least were kept at the Purple Mountain Observatory at Nanking (cf. p. 367 above).

I have had the good fortune to be able to visit the Peking Observatory twice (1946 and 1952). Since on the latter occasion the site was in a military area, I take this opportunity of thanking Captain Chou Li-Kung for the very warm welcome which he accorded to our party. The instruments were in good condition.

[g] Unless they have already been removed to indoor positions at the National Planetarium and Museum of Astronomy first opened in Peking in May 1956. Dr Chu Kho-Chen has informed us that some such better protection for them is planned. Apart from this, four of the Jesuit instruments (nos. 3, 4, 6 and 8), together with the armillary sphere of Kuo Shou-Ching (see p. 369 above), were carried away in 1901 by the Germans as part of their Boxer Rebellion indemnity, but returned to their original places after the First World War, in 1920. Their interim location had been the Sans Souci Palace at Potsdam. Photographs of them taken there were published by R. Müller (1).

[h] Most of the best known enumerations are confused and erroneous, e.g. Couling (1), p. 402; Fabre (1), pp. 76ff.; Arlington & Lewisohn (1), pp. 155ff.

[1] 渾蓋 [2] 簡平 [3] 蔣友仁 [4] 驗氣筒 [5] 候氣筒
[6] 黃道經緯儀

(2) Simple equatorial armillary sphere, *chhih tao ching wei i*,[1] supported on the arched back of a dragon. Verbiest, 1673.
(3) Large celestial globe, *thien thi i*,[2] encased in a horizon framework with four pedestals. Verbiest, 1673.
(4) Horizon circle for azimuth measurements, *ti phing ching i*,[3] table supported on four pedestals, pointers slung from an overhead bearing. Verbiest, 1673.
(5) Quadrant, *ti phing wei i*,[4] or *hsiang hsien i*,[5] supported on a vertical shaft with upper and lower bearings. Verbiest, 1673.
(6) Sextant, *chi hsien i*,[6] on a single pedestal (Fig. 192). Verbiest, 1673.
(7) Quadrant altazimuth,[a] *ti phing ching wei i*.[7] Stumpf, 1713–15.
(8) Elaborate equatorial armillary sphere,[b] *chi hêng fu chhen i*.[8] Kögler, 1744, assisted by von Hallerstein and Gogeisl, perhaps also by Gaubil and de la Charme (Sun Chang[9]).
(9) Smaller celestial globe,[c] *hun hsiang*[10].

Illustrations of the first six of these were published by Verbiest under the title of *I Hsiang Thu*[11] (Designs of Astronomical Instruments) with their description *I Hsiang Chih*[12] (1673),[d] and finally incorporated into the *Yü-Ting I Hsiang Khao Chhêng*[13] (Complete Studies of Astronomical Instruments) edited by Kögler & von Hallerstein in 1744. A great deal of this, richly illustrated, had appeared in the *Thu Shu Chi Chhêng*,[e] but the most beautiful drawings are those published by Tung Kao[14] in the *Huang Chhao Li Chhi Thu Shih*[15] (ch. 3) of 1759 and 1766. We have reproduced from it the drawing of Verbiest's globe and other pictures (Figs. 136, 176, 178). We now add two showing the preparation of other instruments (Figs. 193 and 194). The western

[a] This instrument is of a rather clumsy design, differing from all the others in having graduations inlaid in brass, figures in Arabic numerals, but no founder's inscription and no dragon decorations (cf. Chhen Tsun-Kuei (*6*), p. 45).

It has commonly been supposed that King Louis XIV of France presented it in these years to the Khang-Hsi emperor, but no record of this remains in Chinese sources, as far as we know, and Mr P. A. Jehl informs me that no evidence of manufacture in Paris, or shipment, can be found. On the other hand, Bernard Kilian Stumpf (Chi Li-An[16]), who was Director of the Astronomical Bureau from 1712 to 1720, was afterwards said by his Chinese colleagues to have melted down some old bronze instruments to make new ones (Pfister (1), p. 645). If this was so, the quadrant altazimuth must be a work of Stumpf's, and its bronze must stem from instruments of the Yuan or Ming—which we would much rather have had today than his.

[b] For the constructional details of this see p. 352 and Table 31 above.

[c] There is considerable mystery about this instrument. Its casing and pedestal, though apparently of 18th-century style, differ from those of the Verbiest globe. It appears in photographs of the observatory taken during the Potsdam interregnum and sometimes earlier, but disappears after 1920. One wonders whether it could be the globe of Kuo Shou-Ching described in *Yuan Shih*, ch. 48, p. 5*b*, presumably seen both by Ricci and Lecomte (cf. Needham, Wang & Price), and eventually given an 18th-century housing?

[d] Pfister (1), p. 354.

[e] *Li fa tien*, chs. 85ff. Most of the illustrations of apparatus are in chs. 93–5, with explanations in chs. 89–92.

[1] 赤道經緯儀 [2] 天體儀 [3] 地平經儀 [4] 地平緯儀
[5] 象限儀 [6] 紀限儀 [7] 地平經緯儀 [8] 璣衡撫辰儀
[9] 孫璋 [10] 渾象 [11] 儀象圖 [12] 儀象志 [13] 御定儀象考成
[14] 董誥 [15] 皇朝禮器圖式 [16] 紀理安

Fig. 193. Making the Jesuit astronomical instruments for the Peking Observatory in +1673; a page from the *Thu Shu Chi Chhêng* showing the grinding of a bronze armillary ring.

Fig. 194. Testing the trueness of a bronze armillary ring (from the *Thu Shu Chi Chhêng*).

counterpart to these publications was the *Astronomia Europaea sub Imperatore Tartaro-Synico Cam Hy* (Khang-Hsi) *appellato, ex Umbra in Lucem Revocata*, of 1687, from which we have already seen the illustration of the whole observatory.[a] In the same period comes the book by another Jesuit, Francis Noel (Wei Fang-Chi[1]), published at Prague in 1710, in which he gave for European readers much information on the stems, branches, *hsiu*, etc., with a rough correlation of Chinese and European star-catalogues, and a discussion of Chinese metrology.[b] During the 18th century many observations on eclipses were made, largely by Jacques-Philippe Simonelli (Hsü Ta-Shêng[2]) and published jointly by him with Kögler and Melchior della Briga between 1744 and 1747. A great deal of positional work was done with the new instruments, and a catalogue of 3083 stars was included in the *I Hsiang Khao Chhêng* edition of 1757, under a preface written by the emperor himself. The astronomers responsible were Kögler and Felix da Rocha (Fu Tso-Lin[3]) with Augustin von Hallerstein (Liu Sung-Ling[4]) and Anton Gogeisl (Pao Yu-Kuan[5]). These observations have all been reduced to modern expression and published in translated form by Tsuchihashi & Chevalier. Another event of importance in this period was the *Huang Tao Tsung Hsing Thu*[6] (Star-Maps on Ecliptic Co-ordinates)[c] by Kögler, published a few years after his death in 1746.

The transmissions of the Jesuits seem to have affected a number of Chinese scholars who were more or less outside their circle. The works of Wang Hsi-Shan,[7] for instance, deserve a special investigation.[d] His *Wu Hsing Hsing Tu Chieh*[8] (Analysis of the Motions of the Five Planets), published in 1640, proposed essentially what had been the system of Tycho Brahe, namely, that the sun moves round the earth but all the other planets move round the sun (+1583).[e] One of his diagrams is reproduced in Fig. 195. There is no evidence that this was not independently thought out, perhaps from a bare hint that someone in the West had conceived this idea. He followed it up three years later with a larger work, the *Hsiao-An Hsin Fa*[9] (Wang Hsi-Shan's New (Astronomical) Methods), which was an attempt to synthesise Western and Chinese ideas. So far as I can see, this astronomer was a capable man; at least he understood the Chinese system, and knew that the *hsiu* were equatorial divisions, which was more than the Jesuits did.

His contemporary Hsüeh Fêng-Tsu[10] was more closely connected with the Jesuits, since he was a collaborator of Smogułęcki at Nanking,[f] and therefore probably a

[a] A work with a very similar title, the *Liber Organicus Astronomiae* etc., had been printed by Verbiest in Peking as early as 1668. It has 125 plates of astronomical instruments (see Cordier (8), p. 46; Pfister (1), p. 358).

[b] Cf. Slouka (1). Noel was also an accomplished linguist, sinologist and philosopher whose other books (2, 3, 4) played a part in the Rites Controversy. He was strongly on the sinophile side.

[c] Pfister (1), p. 647.

[d] So also does the rare book *Han Yü Thung*[11] (General Survey of the Universe) of +1648 by Hsiung Ming-Yü[12] and his son Hsiung Jen-Lin[13] on general astronomy (see Hummel, 6).

[e] Berry (1), p. 137. Cf. Fig. 188.

[f] Cf. p. 52 above.

1 衛方濟 2 徐大盛 3 傅作霖 4 劉松齡 5 鮑友管
6 黃道總星圖 7 王錫闡 8 五星行度解 9 曉菴新法
10 薛鳳祚 11 函宇通 12 熊明遇 13 熊人霖

Copernican. His *Thien Hsüeh Hui Thung*[1] of +1650 was again a conciliation of Chinese and Western astronomy, and his treatise on eclipses *Thien Pu Chen Yuan*[2] (True Origins of the Celestial Movements) was the first book in Chinese to make use of logarithms. Other scholars followed tradition in being more interested in chronology, e.g. Hsü Fa[3], who in +1682 supported the unorthodox 'Bamboo Books' datings

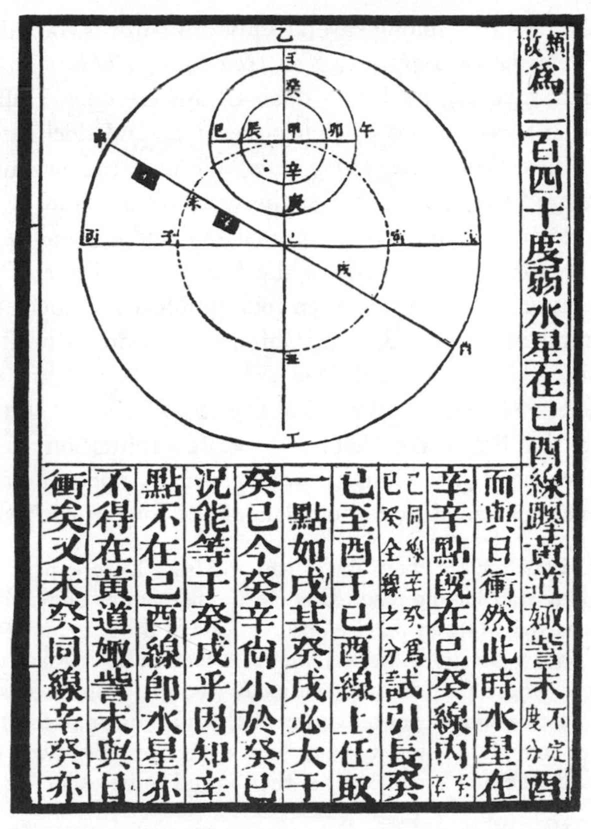

Fig. 195. A geometrical construction from Wang Hsi-Shan's *Wu Hsing Hsing Tu Chieh* (+1640) to explain the Tychonic theory of the solar system.

in his *Thien Yuan Li Li Chhüan Shu*[4] (Complete Treatise on the Thien Yuan Calendar).[a] About this time also was Shao Ang-Hsiao's[5] *Wan Chhing Lou Thu Pien*[6] (Study of Star-Maps from the Myriad Bamboo Tablet Studio).

As the 18th century went on, Chinese astronomers and mathematicians emancipated themselves more and more from the spell which the Jesuit apparition had woven during the decadent Ming and early Chhing times. The *Li Suan Chhüan Shu* (Complete Works on Calendar and Mathematics) of Mei Wên-Ting[7] (1723), of which we have

[a] Cf. Pinot (2), p. 63; Puini (1). This was the book to which Schlegel was principally indebted.

[1] 天學會通 [2] 天步眞原 [3] 徐發 [4] 天元曆理全書
[5] 邵昂霄 [6] 萬青樓圖編 [7] 梅文鼎

before spoken, included much astronomy.[a] His work stimulated a younger man,[b] Chiang Yung,[1] whose *Shu Hsüeh*[2] (Mathematical Astronomy) and *Thui Pu Fa Chieh*[3] (Analysis of Celestial Motions) both appeared about the middle of the century. This was contemporary with Shêng Pai-Erh's[4] defence of the Tychonic against the Ptolemaic system in his *Shang Shu Shih Thien*[5] (Discussion of the Astronomy in the Historical Classic).[c]

At the end of the century, among several important works, mention may be made of the treatise on celestial cartography, *Kao Hou Mêng Chhiu*[6] (Investigation of the Dimensions of the Universe), by Hsü Chhao-Chün[7] (1800).[d] Half a century later, when Fêng Kuei-Fên[8] gave tables of right ascensions and declinations of 100 stars in his *Hsien-Fêng Yuan Nien Chung Hsing Piao*,[9] Chinese astronomical science might be said to have merged at last with that of the world as a whole.

It was not to be expected that the over-emphasised, and in many respects erroneous, claims of the Jesuits for the superiority of the European science of their time would escape a strong reaction. Though this often took political and social forms, as may be read in many accounts of the period, some Chinese astronomers of the old school were actively in opposition. Thus in 1631 Wei Wên-Khuei[10] and his son Wei Hsiang-Chhien[11] published two books on calendrical science (*Li Yuan*[12] and *Li Tshê*[13]) which were so important[e] that Schall von Bell had to write a refutation, the *Hsüeh Li Hsiao Pien*.[14] On the other hand, the Chinese were generally open to conviction, for the following statement was signed by ten officials of the Bureau of Astronomy:

> At first we also had our doubts about the astronomy from Europe when it was used in the *chi-ssu* year (1629), but after having read many clear explanations our doubts diminished by half, and finally by participating in precise observations of the stars, and of the positions of the sun and moon, our hesitations were altogether overcome. Recently we received the imperial order to study these sciences, and every day we have been discussing them with the Europeans. Truth must be sought not only in books, but in making actual experiments with instruments; it is not enough to listen with one's ears, one must also carry out manipulations with one's hands. All (the new astronomy) is then found to be exact.[f]

Unexpectedly, the Jesuit intervention led in due course to a rediscovery on the part of the Chinese of the achievements of their own civilisation before the Ming decadence.

[a] Wylie (1), p. 90, gives a list of the component parts of this collection. Cf. p. 48 above.
[b] Who was destined to be the teacher of Tai Chen, see above, Sect. 17*d*, in Vol. 2, p. 513.
[c] It is to be regretted that the otherwise excellent biographical encyclopaedia of Hummel (2) is distinctly weak on the scientific men of the Chhing, and recourse must be had to special studies, such as that of Chhang Fu-Yuan (2) on Hsü Po-Chêng[15] and Than Yün,[16] which, however, are as yet insufficient.
[d] This contains the *Ching Thien Kai*[17] (Comprehensive Rhymed Catalogue of Stars) by Ricci himself, with Li Wo-Tshun.[18]
[e] It would be interesting to re-investigate these controversies in the light of modern knowledge.
[f] Tr. Bernard-Maître (7), p. 445; eng. auct.

1 江永 2 數學 3 推步法解 4 盛百二 5 尙書釋天
6 高厚蒙求 7 徐朝俊 8 馮桂芬 9 咸豐元年中星表
10 魏文魁 11 魏象乾 12 曆元 13 曆測 14 學曆小辯
15 許伯政 16 譚澐 17 經天該 18 李我存

For mathematics this story has already been briefly described (p. 53 above); Mei Wên-Ting and his grandson Mei Ku-Chhêng were prominently associated with it. In astronomy there was the book of Shêng Pai-Erh just mentioned; and in +1819 the treatise of the Taoist Li Ming-Chhê[1] *Yuan Thien Thu Shuo*[2] (Illustrated Discussion of the Fields of Heaven), while also still Tychonic, referred to the achievements of the ancients in China. In Japan a similar movement was connected with Buddhism, as in the *Bukkoku Rekishō-hen* (The Astronomy of Buddha's Country) by Entsū (1810) described by Mikami (12).[a] Mention was made at the beginning of our book[b] of the works composed by late Chhing scholars in order to prove that all important inventions and discoveries had originally been made in China, for example, the *Ko Chih Ku Wei* of Wang Jen-Chün, and sometimes this was taken to considerable lengths. For instance, in his *I Shu Pien*[3] (The Antheap of Knowledge; miscellaneous essays), Wang Ming-Shêng[4] (+1722 to +1798) maintained[c] that much occidental calendrical science and astronomy had originated from the work of Tsu Chhung-Chih in the +5th century, which had been preserved under the Liao dynasty, and, upon its break-up in 1125, taken to Arabia (Thien-Fang[5]) by the academicians (*lin ya*[6]) of the West Liao State (Ta-Shih[7])[d] whence it passed to Europe.

All in all, the contribution of the Jesuits, chequered though it was, had qualities of noble adventure. If the bringing of the science and mathematics of Europe was for them a means to an end, it stands for all time nevertheless as an example of cultural relations at the highest level between two civilisations theretofore sundered. Truly the Jesuits, with all their brilliance, were a strange mixture, for side by side with their science went a vivid faith in devils and exorcisms.[e] Though some superstitions wilted in their presence,[f] philosophers might opine that they brought as many with them. As for their judgment of the sciences of China, we know now that it was vitiated for two reasons; the Ming dynasty was a period of decline which had few exponents of indigenous tradition, and such men the Jesuits rarely met. Nor must the linguistic difficulties and the scarcity of old books be forgotten. Secondly, since the Jesuits desired to convince the Chinese of the superiority of Western religion by demonstrating the superiority of Western science in their day and age, they were hardly tempted to think like historians when they came upon Chinese achievements in science or technology. Nevertheless many of the Jesuits conceived a warm enthusiasm for Chinese culture,

[a] Entsū was remarkably learned in the history of Chinese astronomy from Lohsia Hung through I-Hsing and Chhüthan Hsi-Ta to Kuo Shou-Ching. But the discoveries of modern science temporarily overshadowed all earlier contributions.

[b] Vol. 1, p. 48.

[c] Ch. 72. Cit. in *Ko Chih Ku Wei*, ch. 2, p. 10*b*.

[d] The name of the first ruler of the West Liao (Qarā-Khitāi) State, a short-lived colony of Chhi-tan exiles in Turkestan (+1125 to +1211) was Yehlü Ta-Shih. The second part of his name was later used to designate the State itself (Eberhard (9), p. 230; Wittfogel, Fêng Chia-Shêng *et al.* pp. 619ff.). Cf. Vol. 1, p. 133.

[e] Cf. Trigault (Gallagher ed. p. 552).

[f] Such as the reliance upon lucky and unlucky days (Trigault, Gallagher ed. p. 548).

[1] 李明徹 [2] 圜天圖說 [3] 蛾術編 [4] 王鳴盛 [5] 天方
[6] 林牙 [7] 大石

and with the Renaissance behind them they successfully achieved a task which had proved beyond the powers of their Indian forerunners in the Thang, namely, to open communications with that world-wide universal science of Nature into which the Chinese achievements would also be built.

(*k*) SUMMARY

An epilogue to a long Section should have the grace of being short. It will by now be abundantly evident that the Chinese contribution to the development of astronomical science was a very remarkable one (see Table 37).[a] Without running over again all the specific points to which attention has been drawn, we may mention: (*a*) the elaboration of a polar and equatorial system strikingly different from that of the Hellenistic peoples, though equally logical; (*b*) the early conception of an infinite universe, with the stars as bodies floating in empty space; (*c*) the development of quantitative positional astronomy and star-catalogues two centuries before any other civilisation of which comparable works have come down to us; (*d*) the use in these catalogues of equatorial (i.e. essentially modern) coordinates, and a faithfulness to them extending over two millennia; (*e*) the elaboration, in steadily increasing complexity, of astronomical instruments, culminating in the +13th-century invention of the equatorial mounting, as an 'adapted torquetum' or 'dissected' armillary sphere; (*f*) the invention of the clock drive for that forerunner of the telescope, the sighting-tube; and of a number of ingenious mechanical devices ancillary to astronomical instruments;[b] and (*g*) the maintenance, for longer continuous periods than any other civilisation, of accurate records of celestial phenomena, such as eclipses, novae, comets, sun-spots, etc.

The most obvious absences from such a list are just those elements in which occidental astronomy was strongest, the Greek geometrical formulations of the motions of the celestial bodies, the Arabic use of geometry in stereographic projections, and the physical astronomy of the Renaissance. We often hear of 'the Greek genius for inquiry—the desire to know not only the facts, but the reasons for the facts...', but this is surely a false antithesis. It was not necessary that the reasons should be conceived either geometrically or mechanically. The Chinese did not feel the need for these forms of explanation—the component organisms in the universal organism followed their Tao each according to its own nature, and their motions could be dealt with[c] in the essentially 'non-representational' form of algebra. The Chinese were thus

[a] Cf. the judgment of Lecomte in 1685: 'As for Astronomy, it must be confest that never did People in the World addict themselves so constantly to it. This Science is beholding to them for abundance of Observations; tho' the History that reports them in general, hath not been careful to descend to particulars, which would be necessary for the reaping all the benefit such Elucubrations seem to promise. However, it hath not been unprofitable to Posterity. We have above Four hundred Observations, as well of the Eclipses and Comets, as Conjunctions, that make good their Chronology, and may conduce to the perfecting of ours' (p. 222).

[b] Outstandingly, the application of water-power to such apparatus for the rotation of celestial globes and demonstrational armillary spheres from the +2nd century onwards, and then the nvention of the first mechanical clock escapement at the beginning of the +8th (cf. Needham, Wang & Price). These achievements will be fully described in Vol. 4 (Sect. 27).

[c] As the Babylonians, too, had dealt with them.

Table 37. *Chart to show the comparative development of astronomy in East and West*

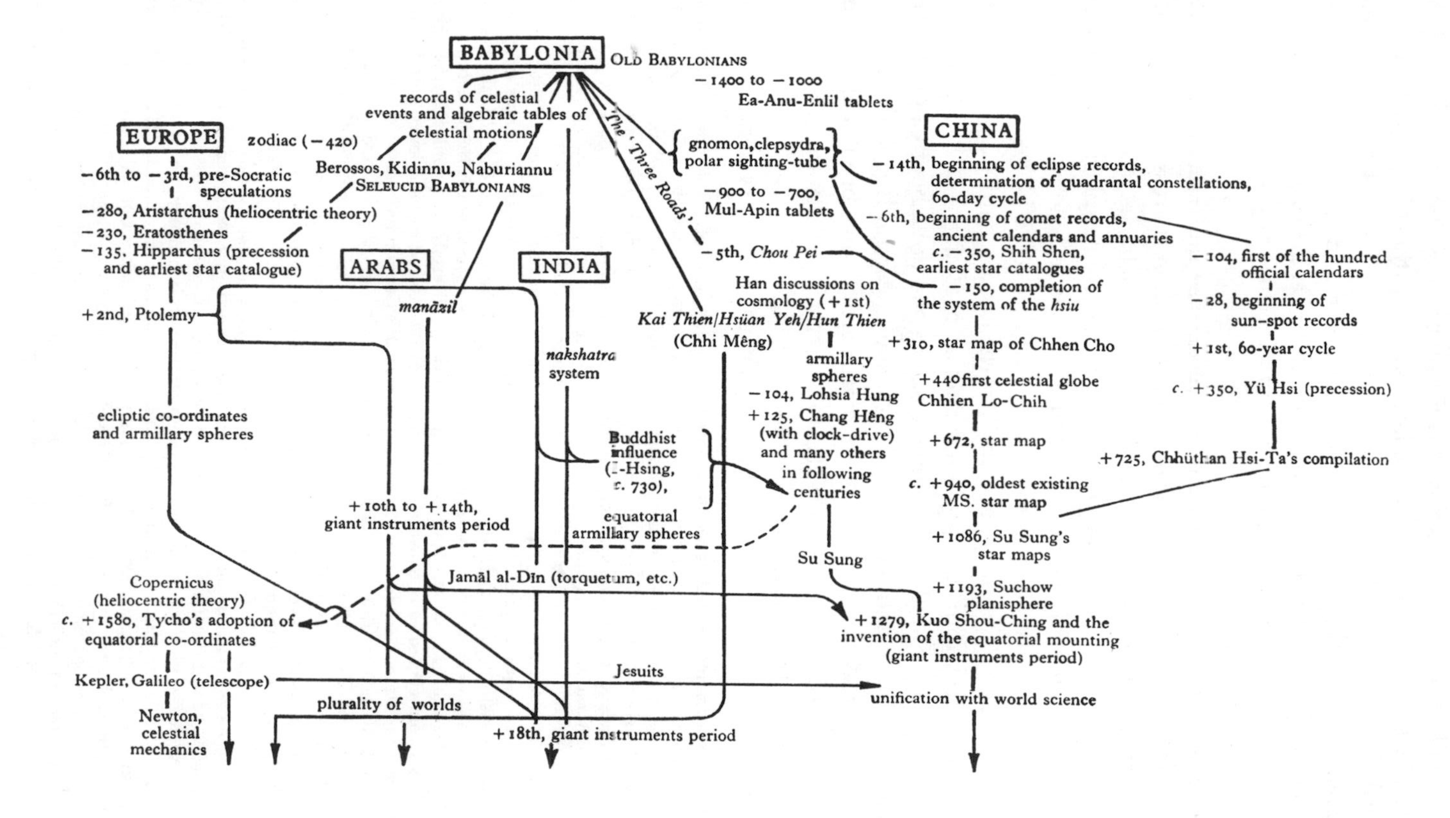

free from that obsession of European astronomers for the circle as the most perfect figure, an obsession from which it took a Kepler to escape. Nor did they experience the medieval prison of the crystalline spheres, those unexpectedly adamantine materialisations of the spirit of Greek geometry. If, like all Chinese science, Chinese astronomy was fundamentally empirical and observational, it was spared the excesses and aberrations, as well as the triumphs, of occidental theorising. But clearly it requires a much more important place in the history of science as a whole than historians have been wont to give it.[a]

Hardly any other subject known to me has been so much distracted by the difficulty of disentangling facts from a maze of arguments often at cross-purposes and based on insecure foundations.[b] The Jesuits, followed by the scholars of the 19th century, began by accepting all the astronomical content of the Chinese legendary material. Their successors, sinologically better informed, scrapped it wholesale, and Maspero (15)[c] summarised the situation by denying that anything could be known of Chinese astronomy before the −6th century at best. But the ink was hardly dry on his paper before it became evident from the profound studies of Tung Tso-Pin and others on the oracle-bones that it would be possible to say a good deal about Chinese astronomy in the middle of the −2nd millennium, not from uncertain legend, but from very concrete inscriptions.[d] The study of this material is only now beginning,[e] and more finds must be expected. Possibly they will throw light on the eastward radiation of primitive astronomy from the lands of the Fertile Crescent.

All in all, it seems that de Saussure has been justified, by evidence of which he never

[a] At an earlier point we had occasion to note the presumption of a Master of Trinity who dismissed all Chinese contributions to astronomy without being able to read a word of any. We owe a comparable and entertaining statement to a Parisian scholar contemporary with him, L. P. Sédillot. 'C'est assez nous occuper de ces aberrations d'un peuple qui n'a jamais su s'élever de lui-même à la moindre spéculation scientifique; esclave de pratiques superstitieuses ou astrologiques, dont il ne s'est jamais entièrement dégagé, il n'a tenu aucun compte des anciennes observations éparses dans ses annales, si tant est qu'elles soient réelles; et, au lieu d'examiner les apparences de la voute étoilée avec cette vive curiosité qui s'attache aux phénomènes, jusqu'à ce que les lois et la cause en soient parfaitement connues, les Chinois n'ont appliqué cette persévérance caractéristique dont on leur fait honneur, qu'à des rêveries sans portée en astronomie, triste fruit d'une routine barbare', (2), p. 603. Sédillot knew no more Chinese than Whewell, but conceived that the substantial services which he rendered as an Arabist to the history of Arabic astronomy authorised him to deny all credit to peoples further East. The only value of these old comedies is to warn us of the necessity of modesty and open-mindedness.

[b] Hence some strange echoes. In the light of all that this Section has reported it is astonishing to read the following statement in an authoritative history of astronomy published in 1954. 'It seems that around −1100 the Chinese had accurately established the obliquity of the ecliptic and the position of the winter solstice in the sky, but after the −5th century the study of astronomy was abandoned in China and many of its findings even destroyed. All the later conceptions were imported afterwards by the Arabs and Europeans' (Abetti (1), p. 24).

[c] I opened this Section by mentioning his summary as perhaps the best available today. But so many advances have been made since 1930 that it is nevertheless very unsatisfactory. If we compare the conclusions of this Section with what Maspero wrote, we could say that (*a*) Chinese astronomy may be studied in its earliest forms in the −13th, not the −6th century; (*b*) it was not independent of Babylonian; (*c*) Chinese mathematics was not so inadequate as he supposed; (*d*) the solar dates were not first known in the −7th century; (*e*) the gnomon was probably not the oldest instrument; (*f*) the sundial was not unknown; (*g*) Chang Hêng's sphere was not the earliest Chinese armillary; etc.

[d] Even the names of many scapulimantic diviners of the −14th century are known (see the Biographical Glossary in the concluding volume). It is very probable that they were star-clerks also.

[e] Of the 10,000 bone fragments in the possession of Academia Sinica, not more than 600 have yet been studied, as by Tung Tso-Pin (*1*), from the point of view of the history of astronomy.

knew, de Saussure, that practical navigator, who, though in the service of alien arms, warmed to the genius of the Chinese people, perhaps precisely because his trade gave him the necessary understanding. Allowing for a certain exaggeration of time, we can still read with appreciation the words with which he closed one of his early papers (4) forty years ago:

> Amidst the darkness which veils the mysterious antiquity of China, the text of the *Yao Tien* opens a scene before us. One of the terraces of the imperial palace distinctly appears; it is the Tower of the Mathematics. A flickering rushlight reveals what is going forward, and by the light it casts on the graduations of the clepsydras, we can discern the astronomers choosing four stars, at that time equally located at the four quarters of the celestial equator, but destined by their movements to reveal to later ages the extent of a history saecular by more than forty times.

21. METEOROLOGY

(*a*) INTRODUCTION

Meteorology is a word which has undergone much change of meaning since the time when it was born in ancient Greece. For Aristotle (as in his *Meteorologica*[a]) it included the study of many phenomena which are now known to be celestial, such as meteors and meteorites[b] (from which it took its name), comets, and the Milky Way, though at that time they were classed as belonging to the 'sublunary' world. Ancient meteorology also included much which would now be called physical geography, such as the origin and nature of rivers, the distribution of land and sea, etc., and also mineralogical matters such as the formation of metals and rocks. In the modern sense, meteorology has become essentially the study of climate, weather and all the events which go on within the earth's atmosphere, together with tidal phenomena.

There is no work in Chinese literature similar in scope to Aristotle's *Meteorologica*, but that does not mean that the Chinese were not deeply interested in weather matters. In this Section we must glance at some of their contributions; for example, they were long in advance of the West in certain methods of meteorological measurements, and kept records of a more complete nature over a much longer time.[c] As regards the understanding of the tides they were also sometimes considerably in advance of Europeans. Nevertheless, the principal works on the history of meteorology[d] make hardly any use of Chinese material.

(*b*) CLIMATE IN GENERAL

This is a subject to which we shall have to return in the concluding part of the book. In considering the differences between Chinese and Occidental civilisation, it will be necessary to compare the climatic conditions in which they developed. Here it need only be said that an excellent short account of the Chinese climate will be found in the books of Sion (1) and Cressey (1), whose words are the basis of the following brief description.[e]

China's succession of passing weather is the product of the seasonal monsoon

[a] Tr. E. W. Webster. On Greek meteorological theory in general see O. Gilbert (1).

[b] Already dealt with in the astronomical Section (p. 433 above).

[c] These are now serving in current statistical investigations, as we shall see.

[d] Napier Shaw & Austin (1); Hellmann (1). I know of nothing in any Western language on the history of meteorology in China except the paper of Chu Kho-Chen (2), which is very brief. Nor is there any monograph in Chinese specifically devoted to the subject. Kimble (1), pp. 151–60, gives a good description of medieval European meteorology.

[e] Cf. also Roxby (2, 3, 4). Unfortunately, Roxby's best work, done for the British Government, remains unpublished, though printed, and cannot be cited; nor has it been available for the present study, though a request for access to it was directed to the proper official quarter. I am now informed that the book may be consulted, under restricted conditions, in certain British libraries.

circulation,[a] the occasional tropical cyclones, and the procession of continental cyclonic storms, all modified by the relief of the subcontinent. Though China's weather is conditioned more by influences from the land-mass to the west than from the Pacific Ocean to the east, yet the aridity of central Asia contends with the moisture of the south-east Asian seas over a perennial Chinese battlefield. During the summer the air masses over inner Asia become heated, expand, rise and overflow towards the encircling oceans. The consequent reduced pressure causes warm moist oceanic air to be drawn in along the surface, thus setting up a very large-scale convectional circulation. Winter reverses the process. The resulting winds are the monsoon winds, less regular in China than India, but equally the basic background of the climate. Hence the fact, familiar to everyone who has lived in China, that the rainfall occurs mainly in a distinct rainy season, occupying usually the three summer months.

The prevailing tendency for monsoonal indraught in summer and outflow in winter is complicated by the eastward movement of high- and low-pressure systems migrating from time to time. The low-pressure areas, or depressions, are most common in the transition season (spring and early summer), and give rise to the characteristic unsettled spells of cloudy and showery weather in central and north China associated with changes in the air mass prevailing over this area.[b] The third element in the Chinese climate is the tropical typhoon, a small but very intense disturbance with extremely low pressure at the centre, steep barometric gradients, and wind velocities up to 165 miles an hour. Though the typhoon as a whole moves fast (often several hundred miles a day), the total area under the sway of its damaging winds is frequently not more than 100 miles in diameter. The typhoon originates in the Pacific, and after travelling westward tends northward as it strikes the China coast, eventually dying out in the interior provinces.

A climatic zonation of the provinces has been devised by Chu Kho-Chen (5).

To what extent has there been a saecular change in the Chinese climate? The question has received a good deal of discussion, and the consensus of opinion has been that China (or at least north China) was formerly both warmer and moister than at present. This conclusion has been drawn mainly from phenological data contained in ancient texts. Information about climate may be deduced from statements concerning the recurrences of the annual phenomena of plant and animal life. One of the first modern efforts in this direction[c] was that of E. Biot (13), who noted, for instance, the references in the *Shih Ching* (Book of Odes) to rice culture and the growing of mulberry, jujube-tree and chestnut in the Yellow River area and north of it. His conclusion was that if there had been any change it had not been a great one. Chu Kho-Chen (4), however, compared the phenological observations contained in books

[a] Mahdihassan (1) derives our word monsoon from the characters *mao-chan*,[1] regular soaking rains, but the phrase is so uncommon, if it ever existed at all, that the argument is unconvincing. The usual terms in Chinese are *shih ling fêng*[2] (old), *hai fêng*,[3] *chi chieh fêng*,[4] and *chi hou fêng*[5] (more modern).

[b] Chu Kho-Chen (6).

[c] A gradual cooling of the Chinese climate had long been suspected by Chinese historians, e.g. Chin Lü-Hsiang[6] (+1232 to +1303) in his discussion of the *Yüeh Ling*, on which see p. 195 above.

[1] 卯霑 [2] 時令風 [3] 海風 [4] 季節風 [5] 季候風
[6] 金履祥

of the Warring States period[a] with the contemporary observations summarised by Gauthier, and found that such annual events as the blooming of the peach-tree (*Prunus persica*), the commencement of song by the cuckoo (*Cuculus micropterus*), and the first appearance of the house-swallow (*Hirundo rustica*), were all placed by these ancient books from a week to a month earlier than the records of the present day. The weather must have been warmer. Many other writers have supported this view. Mêng Wên-Thung (*1*) collected references to bamboos and mulberry-trees during the Han to demonstrate it. Hsü Chung-Shu (*2*) and Yao Pao-Hsien (*1*) pointed out that elephants existed at the Yellow River latitude in the Shang and Chou periods, but are not mentioned thereafter, and that down to the Sung, but not since, crocodiles were indigenous in south China.[b] The richness of the Shang landscape in bamboos is indicated by the bones of the bamboo rat (*Rhizomys sinensis*), as de Chardin & Phei Wên-Chung noted.

Very laborious efforts were made, independently, by Wittfogel (3) and by Hu Hou-Hsüan (*1*), to deduce something conclusive about the climate of the Shang dynasty (−14th century) by a study of oracle-bone inscriptions. Their results supported the above views. But apart from the obvious difficulties of the assumptions which it was necessary to make as to why questions about rain or snow should have occurred at certain seasons more than others, Tung Tso-Pin (*4*) strongly criticised their statistical methods, and the subject must be regarded as still unsettled. The general impression among archaeologists, however, seems to be that China in Shang times was moister and warmer than now (Chêng Tê-Khun, 4).

As for the growth of weather-lore in China, one can only agree with Chu Kho-Chen (2), that prediction never advanced beyond the stage of peasant proverbs.[c] Of course it did not do so in Europe either before the Renaissance.[d] Such proverbs form a part of the 'omen-text' which was one of the main components of the *I Ching* (Book of Changes),[e] and are often met with in ancient books, for example *Tao Tê Ching*, ch. 23: 'A hurricane never lasts a whole morning, and a sudden rainstorm will not go on for the whole day.' In the *I Ching*, haloes foretell storms, and eastward-travelling clouds (the opposite direction to the monsoon rains) bode well for travellers. Weather observation and prediction must have played a large part in the study of Nature by the Taoists; one quotation from a +12th-century source showing this has already been given.[f] The supposed influence of climate on human beings in health and disease was also discussed, as in *Kuan Tzu*,[g] in passages quite reminiscent of the Hippocratic book, *Airs, Waters and Places*.[h]

[a] Yüeh Ling in *Li Chi*; *Huai Nan Tzu*; *Lü Shih Chhun Chhiu*, and *I Chou Shu*.

[b] Cf. Read (4), nos. 104, 105. *Crocodilus porosus* (*chiao lung*[1]) is perhaps not yet quite extinct in Cantonese rivers. *Alligator sinensis* (*tho lung*[2]) is still found in the Lower Yangtze and its lakes, so much so, indeed, as to constitute a nuisance by damaging dams and dykes (*NCNA*, 4th April 1954). All evidence indicates that formerly these species, and probably others, were commoner, and had a much more northerly and westerly range.

[c] A mass of meteorological material is contained in chs. 65–94 of the *Chhien hsiang tien* in the *Thu Shu Chi Chhêng* encyclopaedia. This has not yet been digested on the basis of modern knowledge.

[d] Cf. Inwards' book (1).

[e] See above, Vol. 2, p. 308.

[f] See above, Vol. 2, p. 86.

[g] Ch. 39; see above, Vol. 2, p. 44, and also p. 650 below.

[h] Cf. B & M, pp. 160, 173, 177.

[1] 蛟龍 [2] 鼉龍

(c) TEMPERATURE

It has already been noted that excessive seasonal cold and heat (*han shu i*[1]) were among the happenings which the Han 'phenomenalists' interpreted as 'heavenly reprimands' for deficiencies on the part of the emperor or his administration.[a] It was thus for essentially 'astrological' purposes that records were kept of summers especially hot or winters especially severe, and these found their way into the official histories. We also noted[b] that temperature was important for the ancient Chinese on account of difficulties in making the clepsydra keep accurate time. Records of excessive cold and heat occupy four chapters in the *Thu Shu Chi Chhêng* encyclopaedia,[c] and from these

Table 38. *Correlation of severe winters with sun-spot frequency*

Century	No. of severe winters per century		Sun-spot frequency (Chinese records)
	Europe (Brückner, 2)	China (Chu Kho-Chen)	
+ 6th	—	19	7
+ 7th	—	11	0
+ 8th	—	9	0
+ 9th	11	19	8
+ 10th	11	11	1
+ 11th	16	16	3
+ 12th	25	24	16
+ 13th	26	25	6
+ 14th	24	35	9
+ 15th	20	10	0
+ 16th	24	14	2

dynastic history data Chu Kho-Chen (3, 4) has inferred the existence of long-period climatic pulsations (Table 38). A rough correlation with the sun-spot frequency can also be seen. Chu Kho-Chen was able to illustrate these figures by the witness of contemporary diarists, such as Kuo Thien-Hsi,[2] who kept a weather journal from +1308 to +1310, which was one of the coldest periods.[d] In China from the +11th century at least there was a custom of keeping weather records for nine nine-day periods following the winter solstice; this was called *shu chiu han thien*.[3] In Ming

[a] See above, Vol. 2, pp. 379, 381.
[b] Above, p. 321.
[c] *Shu chêng tien*, chs. 103–6.
[d] By another of those numerous coincidences of simultaneity, it was just about the same time that the first continuous meteorological observations were made in Europe—by the English parish priest William Merle, between 1337 and 1344 (Sarton (1), vol. 3, pp. 117, 674). Cf. Thorndike (5); Hellmann (3, 4); Kimble (1), p. 160.

1 寒暑異 2 郭天錫 3 數九寒天

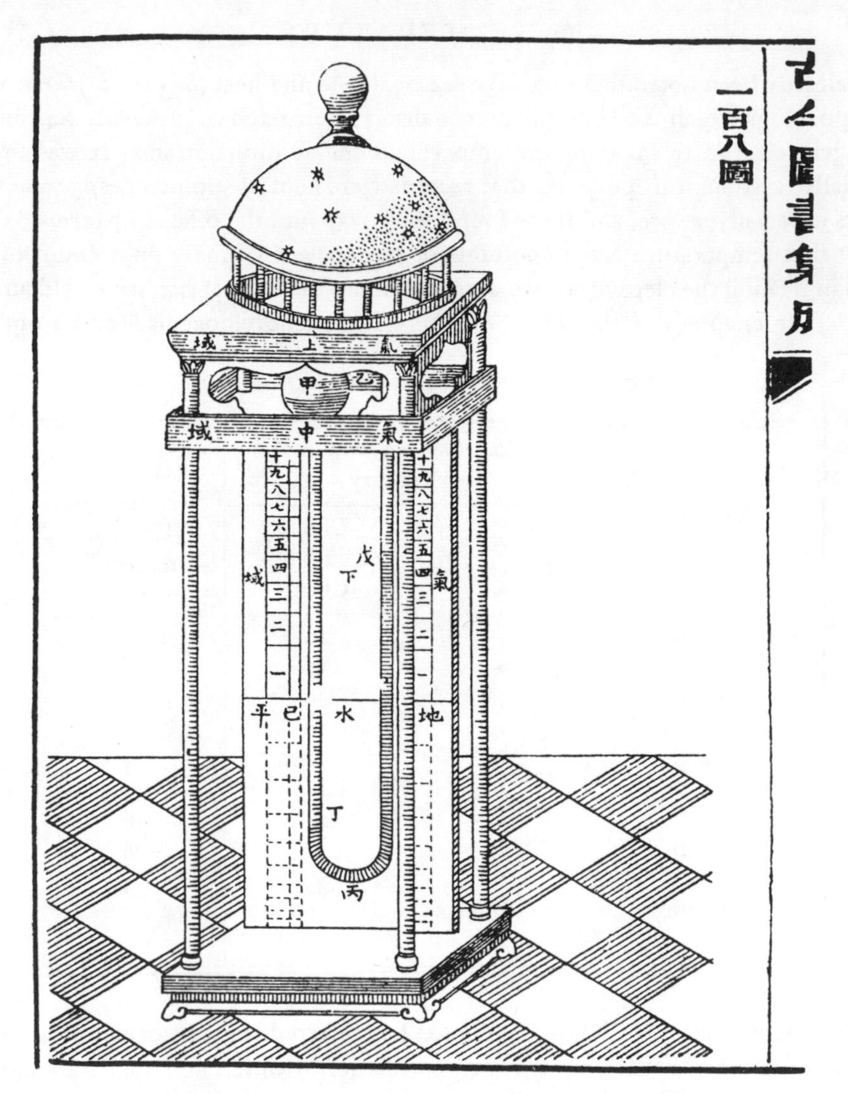

Fig. 196. Ferdinand Verbiest's air thermometer, *c.* +1670 (from the *Thu Shu Chi Chhêng*).

and Chhing times people often used to record the daily weather during this period on special charts which were made for the purpose and filled in according to well-known conventions.[a]

Renaissance quantitative thermometry was introduced to China by Verbiest, whose instrument, similar to the air thermometer of Galileo and affected by barometric pressure, is figured in the *Thu Shu Chi Chhêng*[b] (Fig. 196).

[a] Details and references in Yang Lien-Shêng (6).
[b] *Li fa tien*, ch. 95, p. 40*b*, explained in ch. 92, pp. 1*a* ff.

(*d*) PRECIPITATION

The study of the Anyang oracle-bones has shown that as far back as the −13th century rather systematic meteorological records were being kept. Tung Tso-Pin has analysed[a] a series of bone inscriptions from −1216 in which rain, sleet, snow, wind and direction of rain and wind are all mentioned for many ten-day periods. Many successful predictions are recorded because of the habit of the scribes of writing an additional note on the bone that, in fact, for example, it did snow, after the divination process had said that it would. As in other ancient civilisations, early meteorology was closely connected with divination. In the *Tso Chuan* under date −654, it is stated[b] that particularly careful observations of cloud forms and other atmospheric phenomena were made at the times of the solstices and equinoxes. We have already seen[c] that in the *Chou Li* there is mention of a special official whose duty it was to observe phenomena such as clouds and wind and to make prognostication from them. The *Chhien Han Shu* bibliography lists several books on clouds, rain and the rainbow, one by a certain Thai I.[1] This continued throughout Chinese history; in the +12th century, for example, the *Wei Lüeh*[2] of Kao Ssu-Sun[3] has a section on divination by clouds,[d] and the *Thung Chih Lüeh* bibliography of about +1150, which has often before been quoted, lists no less than twenty-three books on meteorological forecasting. There was also of course the desire to control as well as to foresee, hence much 'rain magic' in ancient and medieval China. Ma Kuo-Han reconstructed[e] a fragmentary Han *Chhing Yü Chih Yü Shu*[4] (Prayers for and against Rain).[f]

Of greater scientific interest is the recognition of the meteorological water-cycle in China. Perhaps the oldest indication that this was understood occurs in the *Chi Ni Tzu* book, a naturalist work probably of the late −4th century. There it is said that

> the wind is the *chhi* of heaven, and the rain is the *chhi* of earth. Wind blows according to the seasons and rain falls in response to wind. We can say that the *chhi* of the heavens comes down and the *chhi* of the earth goes upwards.[g]

In the following century the *Lü Shih Chhun Chhiu* speaks similarly:

> The waters flow eastwards from their sources, resting neither by day nor by night. Down they come inexhaustibly, yet the deeps are never full. The small (streams) become large and the heavy (waters in the sea) become light (and mount to the clouds). This is (part of) the Rotation of the Tao.[h]

[a] (*1*), Pt. II, ch. 9, pp. 44*a* ff.; (5).
[b] Duke Hsi, 5th year; Couvreur (1), vol. 1, p. 248.
[c] Above, p. 190. Cf. pp. 284, 476.
[d] Ch. 8, p. 14*b*.
[e] *YHSF*, ch. 78, p. 46*a*.
[f] Cf. Sect. 10*h* above, and Schafer (1).
[g] See above, Sect. 18*f*, Vol. 2, p. 554. The words quoted are found in ch. 2, p. 3*b* in the text as given in the Ma Kuo-Han collection, *YHSF*, ch. 69, p. 19*a*; cited also *Thai-Phing Yü Lan*, ch. 10, p. 9*a*, tr. auct. Parallel statements, perhaps equally ancient, in *Yo Chi*, para. 3 (in *Shih Chi*, ch. 24, p. 14*a*, tr. Chavannes (1), vol. 3, p. 253), and in *Nei Ching*, ch. 5, p. 2*a* (tr. Veith (1), p. 115).
[h] Ch. 15, tr. R. Wilhelm (3), p. 38, eng. auct.; vol. 1, p. 31.

[1] 泰壹 [2] 緯略 [3] 高似孫 [4] 請雨止雨書

So also the Han book *Ho Thu Wei Kua Ti Hsiang*,[1] one of the 'weft classics',[a] says that the clouds are water from the Khun-lun mountains evaporating and rising[b] (*Khun-lun shan yu shui, shui chhi shang chêng*[c] *wei hsia*[2]). This statement would be about −50. In the +1st century, however, the clear distinction between the circulation in the atmosphere and the vast distances of the starry firmament had still not become fully accepted, since Wang Chhung, in his *Lun Hêng*, has an interesting passage on the subject. As already mentioned, the Greeks were subject to similar confusions.[d] Wang Chhung says:

The Confucians also maintain that the expression that the rain comes down from heaven means that it actually does fall from the heavens (where the stars are). However, consideration of the subject shows us that rain comes from above the earth, but not down from heaven.

Seeing the rain gathering from above, people say that it comes from the heavens—admittedly it comes from above the earth. How can we demonstrate that the rain originates in the earth and rises from the mountains? Kungyang Kao's commentary on the *Spring and Autumn Annals* says;[e] 'It evaporates upwards through stones one or two inches thick, and gathers. In one day's time it can spread over the whole empire, but this is only so if it comes from Thai Shan.' What he means is that from Mount Thai rain-clouds can spread all over the empire, but from small mountains only over a single province—the distance depends on the height. As to this coming of rain from the mountains, some hold that the clouds carry the rain with them, dispersing as it is precipitated (and they are right). Clouds and rain are really the same thing. Water evaporating upwards becomes clouds, which condense into rain, or still further into dew. When the garments (of those travelling on high mountain passes) are moistened, it is not the effect of the clouds and mists (through which they are passing), but of the suspended rain water.

Some persons cite the *Shu Ching* which says, 'When the moon follows the stars, there will be wind and rain'[f]—or the *Shih Ching*, which says, 'The approach of the moon to Pi *hsiu* will bring heavy rain showers'.[g] They believe that according to these two passages of the classics it is heaven itself which causes the rain. What are we to say to this?

When the rain comes from the mountains, the moon passes the (other) stars and approaches Pi *hsiu*. When it approaches Pi *hsiu* there must be rain. As long as it does not rain, the moon has not approached, and the mountains have no clouds. Heaven and Earth, above and below, act in mutual resonance. When the moon approaches above, the mountains steam below, and the embodied *chhi* meet and unite. This is (part of the) spontaneous Tao of Nature. Clouds and fog show that rain is coming. In summer it turns to dew, in winter to frost. Warm, it is rain, cold, it is snow. Rain, dew, and frost, all proceed from the earth, and do not descend from the heavens.[h]

[a] 'Apocryphal Treatise on the River Chart; Examination of the Signs of the Earth.'

[b] *Ku Wei Shu*, ch. 32, p. 4*a*.

[c] Note that this is the word which was afterwards applied to distillation.

[d] For their views of the relations of clouds and mountains, cf. the monograph of Capelle.

[e] *Kungyang Chuan*, ch. 12, p. 15*a*, Duke Hsi, 31st year.

[f] Ch. 24 (Hung Fan); Karlgren (12), p. 35.

[g] Pt. II, bk. 8, no. 8; Karlgren (14), p. 184, no. 232; Waley (1), p. 120. Pi *hsiu* is of course the Hyades.

[h] *Lun Hêng*, ch. 32, tr. auct., adjuv. Forke (4), vol. 1, p. 276.

[1] 河圖緯括地象 [2] 崑崙山有水水氣上蒸爲霞

The passage is interesting not only on account of its clear understanding of the water-cycle, but also because of the appreciation of mountain ranges in the precipitation process. As to the seasonal lunar and stellar connections, the thought of Wang Chhung (about +83) is that in some way or other the cyclical behaviour of the *chhi* on earth, where water is distilled into mountain clouds, is correlated with the behaviour of the *chhi* in the heavens, which brings the moon near to the Hyades at certain times.[a]

In later times the water circulation was well understood. The *Shuo Wên*, completed soon after Wang Chhung's time, defines clouds as the 'moisture evaporated from marshes and lakes' (*jun chhi*[1]). In the +3rd century the cycle was discussed by Yang Chhüan in his *Wu Li Lun* already mentioned.[b] Many statements of it can be found in the Sung[c] and Ming,[d] where the views of a Yeh Mêng-Tê[2] and a Wang Khuei correspond closely with those of the Arabic encyclopaedists such as Abū Yaḥyā al-Qazwīnī, *c.* +1270.[e]

In Europe the recognition of the water-cycle goes back to the −6th century, with Anaximander of Miletus.[f] Aristotle built his *Meteorologica* round the idea of two terrestrial emanations (*anathumiasis*, ἀναθυμίασις), one aqueous (*atmidodestera* ἀτμιδωδεστέρα) and the other gaseous (*pneumatodestera*, πνευματωδεστέρα).[g] The former corresponds rather closely to the Chinese conception of the ascending aqueous *chhi* of the earth; the latter may have originated from observations of such things as the sulphur-depositing gases of fumaroles, and was called upon to explain the formation of minerals and metals in the rocks. We shall meet with its close analogue in what the *Huai Nan Tzu* book has to say on the same subject.[h] One can only leave open the possibility of some transmission in either direction, but the very early date at which these ideas were developing seems to make it most unlikely. The general similarity between *chhi*,[3] *pneuma* (πνεῦμα) and *prāna*, need not imply that there was anything but a community of Mesopotamian origin and later independent developments. The word *chhi* was from the beginning applied to the weather,[i] for which the oldest combined phrase was *chhi hou*.[4] *Hou* has the general meaning of waiting, hence of some particular time or duration,[j] and was also used for prediction of rain or wind, *hou yü*,[5] *hou fêng*.[6] *Chhi* still persists in the modern term for meteorology, *chhi hsiang hsüeh*.[7]

[a] As usual with Wang Chhung, these ideas exemplify well the organic naturalist view of the universe; cf. Sect. 13*f* above, in Vol. 2.

[b] Cf. p. 218 above. Quoted in the mid-12th century *Hsü Po Wu Chih*[8] (Supplement to the Record of the Investigation of Things), ch. 1, p. 4*b*.

[c] E.g. Yeh Mêng-Tê's *Pi Shu Lu Hua*[9] (Summer Holiday Discussions), ch. 1, p. 23*b*, of about +1150—or Chhen Chhang-Fang's[10] *Pu Li Kho Than*[11] (Discussions with Guests at Pu-li), ch. 2, p. 4*b*, of about +1110.

[d] *Li Hai Chi* (probably late +14th), p. 50*b*; also in *Chhi Hsiu Lei Kao*.

[e] Mieli (1), p. 150.

[f] K. Freeman (1), p. 62.

[g] B & M, p. 242.

[h] Below, p. 640.

[i] And hence to the fortnightly calendrical periods, already mentioned, p. 405 above.

[j] Cf. the expression referred to above, Sect. 13*g* in Vol. 2, p. 330, concerning the correct times in alchemy for the application of heat to reactions.

1 潤氣 2 葉夢得 3 氣 4 氣候 5 候雨 6 候風
7 氣象學 8 續博物志 9 避暑錄話 10 陳長方 11 步里客談

In different ages individuals naturally acquired particular fame in rain forecasting. One such was the magician Ching Fang[1] of the −1st century; his lore was recorded to some extent in his *I Chang Chü*[2] (Commentary on the Book of Changes).[a] Another was Lou Yuan-Shan[3] of the Sung. The +10th-century *Thai-Phing Yü Lan* encyclopaedia,[b] the *Wei Lüeh*,[c] etc., quote an otherwise mysterious lost book by one Huang Tzu-Fa[4] of uncertain date, probably Han, on rain prediction, the *Hsiang Yü Shu*.[5] Lacking our cloud classification (cirrus, cumulus, etc.[d]) the medieval Chinese devised many technical terms which have not yet been properly analysed in the light of modern knowledge.[e] Scudding black clouds, yellow clouds like covered chariots (*fu chhê*[6]),[f] shuttle-shaped (*shu chu*,[7] lenticular) clouds, fish-scale (*yü lin*,[8] *cho yü*[9]) clouds,[g] grass-shaped (*tshao mang*[10])[h] clouds, etc. are mentioned. Thunderstorm alto-cumulus may be meant by 'cap' (*kuan*[11]) clouds, and by clouds like flocks of sheep, or pigs or water buffaloes. The 'anvil' shape of typical thunderclouds appears in Ching Fang's description of clouds like drums and raised drumsticks. Alternatively, it seems to have been designated by the term 'catapult-carriage cloud'[i] (*phao chhê yün*[12]), the shape being that of a truncated triangle seen upside down. Another weather sign to which much attention was paid was the lunar halo (*yüeh yün*[13]), caused by cirro-nebula at great heights, and regarded by the Chinese as a sure sign of wind.[j] Lunar haloes showing rainbow colours were termed *chü mu*,[14] 'typhoon-mothers'. Solar haloes were closely studied.[k] In the Sung period a great weather expert was Liu Shih-Yen,[15] whose successes are described[l] in the *Hou Ching Lu*[16] of Chao Tê-Lin.[17] Rhyming weather-lore of sailors is preserved in the *Tung Hsi Yang Khao*[m] and the *Kuang Yü Thu*[n] of +1579. Sung material of this kind is to be found in the *Mêng Liang Lu*.[o] Records of the phenomenon of red rain are preserved in the *Thu Shu Chi Chhêng* encyclopaedia.[p]

Even Chu Kho-Chen (2) thought that the first hygrometer in China was the deer-gut instrument of Verbiest (Fig. 197).[q] But in fact the Chinese had taken advantage of the hygroscopic properties of such things as feathers and charcoal from quite early times. In the *Huai Nan Tzu* book there are at least two mentions of a practice of

[a] And often quoted, for example, by the *Thai-Phing Yü Lan*, ch. 8, p. 2*b*; *Hsü Po Wu Chih*, ch. 1, pp. 6*a*, 8*b*.
[b] Ch. 8, p. 7*b*.
[c] Ch. 8, p. 15*b*.
[d] Due to Luke Howard, cf. Napier Shaw (1).
[e] Cf. *Chin Shu*, ch. 12, pp. 10*a* ff.; tr. Ho Ping-Yü (1).
[f] Perhaps the conical tops of cumulo-nimbus.
[g] Presumably dappled cirro-cumulus.
[h] Presumably cirrus striae.
[i] *Kung Chhi Shih Hua*, ch. 5, p. 3*a*; *Piao I Lu*, ch. 1, p. 6*b*; *Thang Yü Lin*, ch. 8, p. 24*a*.
[j] *TPYL*, ch. 4, p. 12*b*.
[k] Cf. p. 474 below.
[l] Ch. 4, p. 5*a*.
[m] Ch. 9, p. 11*b*.
[n] Ch. 2, p. 74*b*.
[o] Ch. 12, p. 15*b*.
[p] *Shu chêng tien*, ch. 143.
[q] *TSCC*, *Li fa tien*, ch. 95, p. 41*a*, explained in ch. 92, pp. 3*a* ff.

1 京房 2 易章句 3 婁元善 4 黃子發 5 相雨書
6 覆車 7 杼軸 8 魚鱗 9 濯魚 10 草莽 11 冠
12 砲車雲 13 月暈 14 颶母 15 劉師顏 16 候鯖錄
17 趙德麟

Fig. 198. A Chinese rain gauge of +1770 preserved in Korea (photo. Wada).

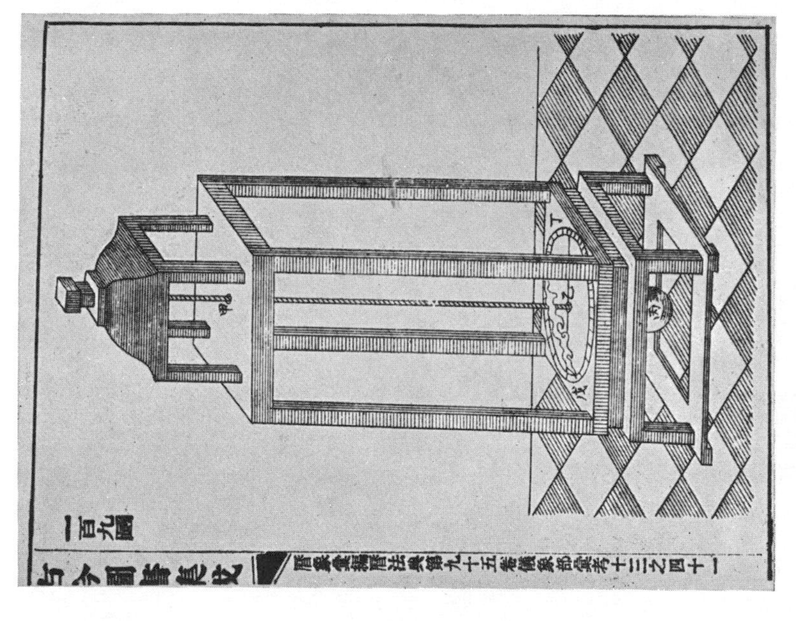

Fig. 197. Ferdinand Verbiest's deer-gut hygrometer, *c.* +1670 (from the *Thu Shu Chi Chhêng*).

weighing elm charcoal[a] against a tared pan of earth with a view to rain prediction by testing the moisture in the atmosphere. The first[b] simply says that when there is dryness the charcoal will be light, and when there is damp it will be heavy (*sao ku than chhing, shih ku than chung*[1]). The second[c] says that by suspending (*hsüan*[2]) a feather or some charcoal one can detect the *chhi* of dryness or damp, adding 'Thus by small things one can observe great ones, and know what is far away from what is near' (*i hsiao chien ta, i chin yü yuan*[3]). We meet with the same test in the *Shih Chi*[d] in but three words *hsüan thu than*,[4] to which the +5th-century commentator (Phei Yin) adds that it was performed especially at the solstices, perhaps with the object of foretelling the coming weather for the whole season. This is elaborated in the *Kan Ying Lei Tshung Chih*[5] (Record of the Mutual Resonances of Things) by Chang Hua[6] (+232 to +300),[e] where the fullest details are given.[f] It is interesting that in the 15th century in Europe, Nicholas of Cusa used exactly the same method, weighing wool against stones (Cajori, 5). The technique is often mentioned in later Chinese texts.[g]

So much for the forecasting of rain. Since it had its inevitable sequel in the rising of rivers and canals, with the danger of floods, always so serious in China, it would not be surprising to find that the Chinese made use of rain gauges from an early period. In Europe the very simple idea of catching rain in some kind of container so as to permit measurement dates only from +1639, when it was introduced by Benedetto Castelli (a friend of Galileo) at Perugia. In recent years, meteorologists have become familiar with Korean rain gauges of the +15th century, through the work of Wada (1, *1*) and Lyons. Among the earliest of these, known as *tshê yü chhi*[7] (rain-measuring instruments), were several in bronze set up in +1442, and concerning which the relevant imperial decrees are preserved in Korean historical texts. They stood on Cloud-Watching Platforms (*yün kuan thai*[8]). In +1770 meteorological observatories with these rain gauges were set up in all the chief cities of Korea; one of them is here reproduced (Fig. 198).

What has hitherto been less generally realised, however, is that the rain gauge was not a Korean invention, but goes back a good deal earlier in China.[h] The chief

[a] If it really was charcoal. Peek has described a particularly hygroscopic kind of graphite (*chhing hui*[9] or *hei hui*[10]) which might have been what was used. Chang Hung-Chao (*1*), p. 202, agrees.
[b] Ch. 3, p. 6*a*.
[c] Ch. 17, p. 5*a*.
[d] Ch. 27 (Thien Kuan), p. 37*a*; Chavannes (1), vol. 3, p. 400.
[e] In *Shuo Fu*, ch. 24, p. 19*a*. Cf. Li Shun-Fêng's *Kan Ying Ching*, in *Shuo Fu*, ch. 9, p. 2*a*.
[f] What may have been an ancient form of hygrometer is seen in the story of the skin of some kind of sea-animal the hairs of which rose and fell in accordance with the tides. The reference is in the +12th-century record of an embassy to Korea referred to below (pp. 492, 511); see Moule (3), p. 152. Cf. *Shan Chü Hsin Hua*, p. 53*a* (H. Franke (2), no. 146).
[g] E.g. *Piao I Lu*,[11] ch. 1, p. 14*a*. Iron counterweights in *TPYL*, ch. 871, pp. 4*a* ff.
[h] It may even be of Babylonian origin. Passages which seem to show the use of rain gauges have been noted by Sammadar in the *Arthaśāstra* (Shamasastry ed. p. 127), and in +2nd-century Hebrew texts by Vogelstein (cf. Hellmann, 2).

1 燥故炭輕，溼故炭重　2 縣 for 懸　3 以小見大以近喩遠
4 縣土炭　5 感應類從志　6 張華　7 測雨器　8 雲觀臺
9 青灰　10 黑灰　11 表異錄

evidence for this is that the mathematical book *Shu Shu Chiu Chang* by Chhin Chiu-Shao, of +1247, contains[a] problems on the shape of rain gauges,[b] here called *thien chhih tshê yü*.[1] At that time there would seem to have been one in each provincial and district capital. Chhin discusses the determination of the rain falling on a given area of ground from the depth of rain-water which is collected in vessels of conical or barrel shape.

Still more remarkable, the same book shows us that snow gauges were also in use, *chu chhi yen hsüeh*.[2] These were large cages made of bamboo, and Chhin gives sample problems concerning them.[c] They were doubtless placed beside mountain passes and on uplands. If local magistrates in the Sung really transmitted to the capital readings on rain and snow fall, the high officials must have been greatly assisted in making their calculations concerning the maintenance and repairs likely to be required for dykes and other public works.

Needless to say, Chinese records have a very long and abundant list of floods and droughts. The data are in all the chief historical compendia as well as the official histories and encyclopaedias.[d] A great digest has been made of this information by Chhen Kao-Yung *et al.* (*1*). In Western languages the pioneer work of Hosie (2, 3) has been superseded by elaborate studies.[e] While keeping in mind certain insecurities in the recording, such as failure to keep records in periods of chaos and struggle, some periodicities can be made out. Table 39 is from Chu Kho-Chen (3). Chhen Ta (1) has made the correlation that the migration to Malaysia took place in the very dry +15th century, and that the migration to the Pescadores and Formosa accompanied the very dry +7th. The formation of Chinese colonies in Hawaii, North America and South Africa was a 19th-century phenomenon. One can hardly overlook the fact that the extremely dry centuries from +300 to +600 correspond with the long period of political fragmentation terminated only by stable unification under the Sui and Thang. Conversely the centuries of Thang and Sung stability were characterised by relatively abundant rainfall.[f]

Naturally the Chinese records of bad weather contain references to damaging storms of hail (*pao*[3]). Horwitz (8) was moved to investigate the history of the practice of firing mortar shells at hailstorm clouds in order to start the precipitation or deflect it from the crops under protection. This is widely current today among vineyard cultivators in the south of France, who use rockets. Horwitz found the first European mention in the autobiography of Benvenuto Cellini, written at the end of the

[a] Ch. 4, pp. 107ff.

[b] This fact was first brought to my attention by my friend Dr Yeh Chhi-Sun.

[c] Ch. 4, p. 110. Remarkable work on the forms of snowflakes was published by a Japanese daimyo, Doi Toshitsuru (+1789 to 1848), early in the nineteenth century (*Sekka Zusetsu*).

[d] E.g. *TSCC*, *Shu chêng tien*, chs. 76–80, 86–94, 124–32, etc. But no attempt has been made to conflate them for world use.

[e] Those of Chu Kho-Chen (3), Yao Shan-Yu (1, 2, 3), Ting Wên-Chiang (2), K. Y. Chêng (1), Schove (4) and others. For a general background to their work see Brooks (1).

[f] Cf. above, Vol. 1, pp. 184ff. Li Chi (2) has also made correlations between climatic pulsations and political disturbances.

[1] 天池測雨 [2] 竹器驗雪 [3] 雹

16th century,[a] but also noted, from the travel account of Bastian,[b] that during the Khang-Hsi period (+1662 to +1722) it was the practice of Lamas to have guns fired off at rain clouds in Kansu. The civil officials were asked to take steps to obtain the pardon of the gods of mountains and rivers on these occasions. Horwitz questioned, therefore, whether the idea might not have originated on purely animistic grounds in China and have been transmitted to the West—but we do not know the answer, and the subject should be further investigated.

Table 39. *Ratio of droughts to floods by centuries*

Century	Raininess ratio droughts per century / floods per century
+2nd	1·98
+3rd	1·60
+4th	8·20
+5th	2·06
+6th	4·10
+7th	3·30
+8th	1·32
+9th	1·80
+10th	1·80
+11th	1·70
+12th	1·04
+13th	1·80
+14th	1·05
+15th	2·25
+16th	1·95

(*e*) RAINBOWS, PARHELIA AND SPECTRES

There are references to the rainbow[c] in the Anyang oracle-bone inscriptions.[d] The modern term for it, *hung*,[1] perpetuates, because of the 'insect' radical which it contains, the idea of the Shang people that it was a visible rain-dragon. In a special paper Hopkins (26) has discussed the ancient form of the pictogram[e] (see drawing). Other terms, *i* or *ni*[2,3] use either the 'insect' or the 'weather' radical.

In the Sung period, Shen Kua described a double rainbow which he saw when on a diplomatic mission about +1070 to the Chhi-tan Tartars in Kansu.[f] Both he and

[a] Ed. Bettoni, vol. 2, p. 56. [b] (1), p. 410.
[c] Cf. *TSCC*, *Chhien hsiang tien*, ch. 76; *Shu chêng tien*, ch. 73.
[d] Tung Tso-Pin (*1*), Pt. 1, ch. 1, p. 6*a*.
[e] One frequently sees such a double-headed animal (amphisbaena) sculptured over niches in the Buddhist cave-temples of Northern Wei date at Tunhuang (cf. Fig. 19 in Vol. 1). Cf. Bosch (1).
[f] *MCPT*, ch. 21, para. 1; cf. Hu Tao-Ching (*1*), vol. 2, p. 670.

1 虹 2 蜺 3 霓

Sun Yen-Hsien[1] considered that the rainbow was due to the reflection of the sunlight from suspended water-drops. Two centuries had yet to elapse before Quṭb al-Dīn al-Shīrāzī (+1236 to +1311) gave in Persia the first satisfactory explanation of the rainbow (essentially that of Descartes) in which the ray is refracted twice and reflected once, through a transparent sphere.[a]

But the rainbow in all its beauty is far outdone by the strange complex of phenomena which includes concentric haloes and 'mock suns' or parhelia. Under certain atmospheric conditions, when clouds of hexagonal or pyramidal ice-crystals, columnar or

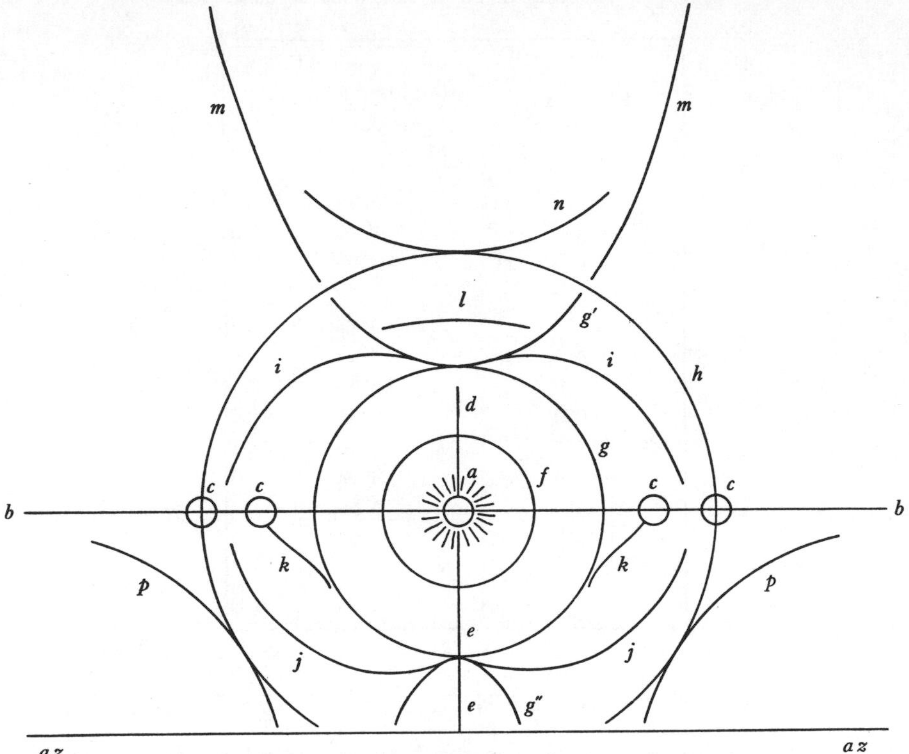

Fig. 199. Halo-components; cylindrical azimuth projection.

Key: *a*, sun; *az*, horizon; *b*, parhelic circle; *c*, parhelia (mock suns); *d*, supra-solar column; *e*, infra-solar column; *f*, Hall's halo; *g*, 22° halo; *g'*, upper tangent arc of 22° halo; *g"*, lower tangent arc of 22° halo; *h*, 46° halo; *i*, upper arcs of circumscribing oval halo; *j*, lower arcs of circumscribing oval halo; *k*, oblique arcs of Lowitz; *l*, Parry's arc; *m*, extensions of *g'* enclosing zenith; *n*, upper tangent arc of 46° halo (circumzenithal arc); *p*, infra-lateral tangent arcs of 46° halo (the supra-lateral counterparts are predicted by theory but have never been seen).

plate-like, are falling slowly through the air at considerable heights, the sun is seen surrounded by haloes, and flanked by as many as four centres of bright light ('mock suns', Fig. 199*c*, *c*) at the same altitude. These are located at or near the junctions of

[a] Sarton (1), vol. 2, pp. 23, 1018; vol. 3, p. 141; Pledge (1), p. 67; Mieli (1), p. 151; Wiedemann (2); Crombie (1). A few years later the same explanation was given by Theodoric of Freiburg in far-away Europe, and by al Shīrāzī's pupil, Kamāl al-Dīn al-Fārisī. Cf. p. 162 above. See also Boyer (3).

[1] 孫彥先

a bright horizontal line (the parhelic circle, Fig. 199*b*, *b*) with an inner and an outer halo (Fig. 199*g*, *h*).[a] In addition, there may be a fifth sun image (the anthelion) at 180° on the horizontal line, i.e. due opposite to the true sun, two more (the paranthelia) at about 120°, and (very rarely) two at 90°. The haloes may be brightly coloured (with red to the inside), but the parhelic circle is white. A vertical column or pillar may cross it at the sun.[b] Haloes and mock suns are observable in all latitudes, but though not commoner in northern and polar regions they are there more splendid and various in form. The first European description was that of Christopher Scheiner in +1630, who saw a display at Rome, the second that of Hevelius at Danzig in +1661, but the most complex effect ever seen was probably that described by Tobias Lowitz at St Petersburg in +1794.

In China, however, astrologically minded star-clerks had been devoting meticulous observation to halo phenomena centuries earlier, and so impressive they found them that an emperor himself did not disdain to write an illustrated book which dealt, *inter alia*, with such phenomena.[c] This is the *Thien Yuan Yü Li Hsiang I Fu*,[1] by Chu Kao-Chih,[2] who reigned (for one year only) as Jen Tsung,[3] in +1425. We reproduce here two of its pictures (Fig. 200*a* and *b*).

Even more surprising perhaps is the recent discovery of Ho Ping-Yü (1) that the pages which the *Chin Shu*[4] (History of the Chin Dynasty) devotes[d] to the 'Ten Haloes' (*Shih Yün*[5]) yield technical terms for almost every component of the solar halo system. Complete haloes were known as *yün*,[6] and the multiple suns (*shu jih*[7]) were strung along the *mi*[8] ('complete' or parhelic circle). Partial lateral arcs of the 46° halo were termed *erh*[9] ('ear-rings'), those of the 22° halo *pao*[10] ('embracements'), and of Hall's halo *chüeh*[11] ('thumb-rings', Fig. 199, *f*). Partial upper arcs of the 46° halo were *hsü*[12] ('arrays'), of the 22° halo *kuan*[13] ('bonnets'), while the *jih tai*[14] (sun crown) can clearly be identified as Parry's arc (Fig. 199, *l*). Even the curious triangular effects known as oblique Lowitz arcs (Fig. 199, *k*) are described as *thi*[15] ('supporting brackets'). The lower tangent arc of the 22° halo (Fig. 199, *g″*) was called *ying*[16] (the 'tassels'), and the infra-lateral tangent arcs of the 46° halo (Fig. 199, *p*) had the very appropriate name of *chi*[17] ('leaning lances'). Of course all kinds of presages were

[a] The inner halo is at 22° from the sun, the outer at 46°. Both have inverted tangent arcs (Fig. 199, *g′*, *g″*, *n*) touching them at their highest and lowest points, though naturally the lower tangent arc of the outer halo can rarely be seen.

[b] The most elaborate description of the phenomena will be found in the treatise of Pernter & Exner (1), pp. 242 ff. The descriptions in standard works such as those of R. W. Wood (1), 2nd ed. p. 437, 3rd ed. p. 394, or R. S. Heath (1), pp. 339 ff., are often too brief to be fully comprehensible, and indeed the accounts in some of the best-known reference books are wrong. The classical geometrical explanation was given by Bravais in 1847. There is an abridged account in Mascart (1), vol. 3, pp. 472 ff. The most recent work on the subject is that of Liljequist in 1956, who illustrates many displays and relates particular halo-system components to particular types of ice-crystal.

[c] It was never printed, but a manuscript copy, with coloured illustrations, is preserved in the Cambridge University Library.

[d] Ch. 12, pp. 8*a* to 9*b*; tr. Ho Ping-Yü (1).

1 天元玉曆祥異賦　2 朱高熾　3 仁宗　4 晉書　5 十煇
6 暈　7 數日　8 彌　9 珥　10 抱　11 璚　12 序
13 冠　14 日戴　15 提　16 纓　17 戟

inferred from these appearances,[a] but the precision of the observations is astonishing. We thus have in hand some twenty-six technical terms current in the +7th century (the text dates from about +635), and cannot but conclude that the Europeans of the 17th century were long anticipated in the close study of solar halo phenomena.

Indeed, some of these terms go much further back. In Section 20*c*, when discussing the 'official' character of Chinese astronomy, we had occasion to mention the Shih Chin, one of the prognosticatory officials mentioned in the *Chou Li*. According to the description of his duties there given:[b]

> The Shih Chin[1] is concerned with the method of studying the Ten Haloes (*shih yün*[2]) in order to observe extraordinary phenomena (in the heavens) and thence to prognosticate good fortune or bad. The first is called *chin*[3] ('invasive' haloes),[c] the second is called *hsiang*[4] (the 'image'),[d] the third *hsi*[5] (or *hui*,[5] 'metal ornaments'),[e] the fourth *chien*[6] ('overshadowings'),[f] and the fifth *an*[7] ('darkenings', solar or lunar eclipses). The sixth is called *mêng*[8] (gloom of fog, etc.), the seventh is called *mi*[9] (the 'complete' parhelic circle), the eighth *hsü*[10] ('ordered arrays'),[g] the ninth *chi*[11] (lit. rainbow, but explained as a synonym for *yün*,[12] full haloes), and lastly the tenth *hsiang*[13] ('suggestive' cloud-forms).[h]
>
> The Shih Chin is charged with tranquillising the people, and explaining to them what heaven sends down upon earth.[i] At the beginning of the year he starts his observations, and at the end of the year he analyses them.

This clearly shows that the star-clerks of the Chhin and Han were interested in halo phenomena and carefully recorded them. We would in any case know this from the short description of Ssuma Chhien,[j] which contains half a dozen of the technical terms used in the *Chin Shu* and later writings.[k]

One cannot but wonder whether haloes and parhelia were not the basis of the Chinese mythological story, to which there are many references in ancient works,[l] of the appearance of ten suns in the sky simultaneously in the time of the emperor Yao.[m]

[a] A list of some 28 such displays, between the years +249 and +420, is given later on in the same chapter, pp. 15*a*–17*b*, together with the political events which followed, and which they were thought to have portended. For further details see Ho & Needham (1).

[b] Ch. 6, p. 29*b* (ch. 24, p. 30), tr. auct., adjuv. Biot (1), vol. 2, p. 84. For his staff see ch. 5, p. 7*b* (ch. 17, p. 24), tr. Biot (1), vol. 1, p. 411.

[c] The *Chin Shu* defines these as the *erh*, the *pao* and the *chüeh* already mentioned, together with the *pei*,[14] the upper tangent arc of the 22° halo and the circumscribing oval halo (Fig. 199, *g′*, *i*) forming a band convex to the sun.

[d] The same as the *pei* just described, but with the Parry arc (Fig. 199, *l*), looking like a great flying bird with the sun in its claws.

[e] Vertical columns (Fig. 199, *d*, *e*).

[f] Probably partial upper arcs of Hall's halo (Fig. 199, *f*).

[g] Partial upper arcs of all the haloes.

[h] For example, says the *Chin Shu*, a red cloud which takes the shape of a hunter. Cf. pp. 190, 284.

[i] By this time the seasonal occurrence of meteor showers, for example, was certainly known.

[j] *Shih Chi*, ch. 27, p. 29*a*, *b*, tr. Chavannes (1), vol. 3, p. 385.

[k] So also some of the apocryphal treatises, e.g. the *Ho Thu Wei Chi Yao Kou*, in *Ku Wei Shu*, ch. 33, p. 1*b*.

[l] E.g. *Tso Chuan*, *Chu Shu Chi Nien*, *Li Sao*, *Chuang Tzu*, etc. Cf. Granet (1).

[m] As Schlegel (7*a*) suggested long ago.

1 眡祲 2 十煇 3 祲 4 象 5 鑴 6 監 7 闇
8 瞢 9 彌 10 敍 11 隮 12 暈 13 想 14 背

This was discussed more than once by Wang Chhung in the *Lun Hêng*[a] about +80. The suggestion seems at any rate more probable than that which would refer the legend to the days of the ten-day week, all the suns of which would thus have been brought upon the stage by the story-tellers at one and the same time.[b]

Fully circular coloured haloes are sometimes seen today from aircraft. Similar phenomena have long been observed on mountains, notably in China the famous 'Buddha Light' (*Fo kuang*[1]) nimbus on Omei Shan in Szechuan.[c] In +1177 Fan Chhêng-Ta recognised this as being a perfectly natural kind of rainbow; the same, he thought, was also true of iridescent colours seen in certain waterfalls.[d]

(*f*) WIND AND THE ATMOSPHERE

Observations on wind, more or less fragmentary, were being made throughout Chinese history.[e] The *Huai Nan Tzu* book (*c.* −120) contains[f] a list of eight wind 'seasons' during the year. Later, the classification of winds was much elaborated; for example, the *Thang Yü Lin*[g] (collected in +1107) and the Ming *Shu Ssu Shuo Ling*[2] (Discussions of a Bookseller) by Yeh Ping-Ching[3] list twenty-four seasonal winds. The monsoon winds from the south-east were of course early named[h]—*hsin fêng*[4] or *hua hsin fêng*.[5] A special wind preceding the hurricane or typhoon (*chü*[6]) was called the 'preparatory' wind (*lien fêng*[7]),[i] and so on.

In the Ming book *Li Hai Chi*, Wang Khuei mentions[j] experiments with kites to test the behaviour of winds. This raises the question of the antiquity of the weather cock or vane in China—not as unimportant a matter as it might seem, since this simple

[a] Chs. 84 and 32, tr. Forke (4), vol. 1, pp. 89, 271. It is interesting that Wang Chhung came to the conclusion that the supernumerary suns were not true suns, but his arguments were not very meritorious.

[b] It is curious that the total number of suns in a great display will amount to ten (four near the true sun, two at 90°, the two paranthelia, and the anthelion). If the true sun be excluded, a parhelion at the lowest point of Hall's halo will make up the number.

These paragraphs were first drafted in March 1954, at a time when minor but interesting displays of parhelia had just been visible in Cambridge. We happened to have been consulting Wang Chhung's chapter on the sun.

[c] Cf. for example, Franck (1), p. 579, for a description. The phenomenon is analogous to the 'Brocken Spectre' of Europe (Pernter & Exner (1), pp. 446ff.). See a recent case described by D. M. Black (1).

[d] *Wu Chhuan Lu*, ch. 1, pp. 13*b*, 17*b*, 18*a*.

[e] Cf. *TSCC*, *Chhien hsiang tien*, chs. 65–9; *Shu chêng tien*, chs. 60–4.

[f] Ch. 3, p. 4*a*.

[g] Ch. 8, p. 24*a*.

[h] E.g. in *Lü Shih Chhun Chhiu*, ch. 62 (vol. 1, p. 121), tr. R. Wilhelm (3), p. 159; and discussed in the *Yen Fan Lu*[8] of Chhêng Ta-Chhang[9] of +1175.

[i] See the early +9th-century *Ling Nan I Wu Chih*[10] (Record of Strange Things South of the Passes) by Mêng Kuan.[11] 'Wind' had many proto-scientific significances, as in the 'hard wind' of the Taoists, already mentioned (Vol. 2, pp. 455, 483 above), which had a part to play in the prehistory of aeronautics (Sect. 27 below). In parallel with *prāna*, and like *chhi*, other 'winds' were extremely important in medical theory (Sect. 44) and in Buddhist embryology (Sect. 43).

[j] P. 4*b*.

1 佛光 2 書肆說鈴 3 葉秉敬 4 信風 5 花信風
6 颶 7 鍊風 8 演繁露 9 程大昌 10 嶺南異物志
11 孟琯

instrument is perhaps the oldest of all pointer-reading devices, the importance of which in the philosophy of the natural sciences requires no emphasis. In Europe it does not seem possible to trace it back beyond the Tower of the Winds at Athens (*c.* −150) which Andronicus of Cyrrha fitted with a weather vane.[a] In China it must be of about the same date, since the *Huai Nan Tzu* book speaks[b] of a thread or streamer (*wan*[1]), which a Han commentator explains as 'a wind-observing fan' (*hou fêng shan*[2]).[c] Military treatises from the San Kuo onwards call it *wu liang*[3] ('five ounces'), alluding to the weight of feathers to be used in it. Later books bring out the bird-like form which it was given, for example the mid +4th-century *Ku Chin Chu* (Commentary on Things Old and New), which calls it by its usual name (*hsiang fêng wu*[4]). Another term is *hsien*.[5] Its invention is variously ascribed to legendary personages such as Huang Ti.[d]

Chu Kho-Chen (2) conjectures that some attempt was made in the Han to construct an anemometer. This view is based[e] on passages in the *San Fu Huang Thu*[6] (Description of the Palaces at Chhang-an), a late +3rd-century book attributed to Miao Chhang-Yen.[7] Discussing towers and pavilions, he writes:[f]

> The Han 'Ling Thai'[8] (Observatory Platform) was eight *li* north-west of Chhang-an. It was called 'Ling Thai' because it was originally intended for observations of the Yin and the Yang and the changes occurring in the celestial bodies, but in the Han it began to be called Chhing Thai.[9] Kuo Yuan-Sêng,[10] in his *Shu Chêng Chi*[11] (Records of Military Expeditions),[g] says that south of the palaces there was a Ling Thai, fifteen *jen* (=120 feet) high, upon the top of which was the armillary sphere made by Chang Hêng. Also there was a wind-indicating bronze bird (*hsiang fêng thung wu*[12]), which was moved by the wind; and it was said that the bird moved (only, or faster?) when a 1000-*li* (very strong?) wind was blowing. There was also a bronze gnomon 8 feet high, with a (horizontal scale) 13 feet long and 1 foot 2 inches broad. According to an inscription, this was set up in the 4th year of the Thai-Chhu[13] reign-period (i.e. −101).[h]

The evidence of this passage does not seem very clear, though unless the movements of the bird had something to do with the strength of the wind there would hardly be any point in referring to them at all. Apart from gusts and turbulence, the stronger the wind the more fixed an ordinary weathercock would be. Moreover, elsewhere in

[a] See the discussions of Beckmann (1), vol. 2, p. 281, and Hellmann (2). The subject is closely related to the development of the compass-card, on which see S. P. Thompson (1). All these writers discuss the Hellenistic and medieval names of the winds. Cf. below, Sect. 26*h*, *i*.

[b] Ch. 11, p. 17*b*.

[c] Cf. *Lun Hêng*, ch. 43 (Forke (4), vol. 1, p. 111).

[d] As in *Shih Wu Chi Yuan*, ch. 2, p. 46*b*.

[e] Personal communication from Dr Chu Kho-Chen.

[f] Ch. 4, tr. auct.

[g] This seems to be another name for Kuo Hsiang,[14] the Taoist commentator, elsewhere referred to (*d.* +312). The book is apparently a lost one.

[h] This was just after the date of the astronomical work and calendar reform with which the names of Lohsia Hung and Têng Phing are associated (cf. p. 302 above).

1 綄 2 候風扇 3 五兩 4 相風烏 5 俔 6 三輔黃圖
7 苗昌言 8 靈臺 9 清臺 10 郭緣生 11 述征記
12 相風銅烏 13 太初 14 郭象

the same book (Chapter 2, on the Han palaces), there is mention of a bronze phoenix set on a tower roof 'which faced the wind on a turning axle above and below, as if flying'; this suggests the continuation of the axis to a lower floor where it could have been fitted with a device to indicate, if not to record, the speed of rotation by the wind. It is not out of place to reflect that the cup anemometer is a version of the paddle-wheel theme, and that it was in the Han (as we shall later see [a]) that the first water-wheels appeared. Besides, the bronze phoenix is stated to have been 5 ft. high, which seems rather excessive for a weather-cock, but not for a device where wind resistance was sought. If this interpretation is justified, the Han anemometer may have been an anticipation of the modern four-cup type, for the early Renaissance instruments of Egnatio Danti (1570) and Robert Hooke (1667) were of the pendulum pattern not used today.[b]

A good description of a whirlwind with waterspouts is given by Yang Yü in his *Shan Chü Hsin Hua* (New Discourses from the Mountain Cabin) of +1360.

On the 15th day of the twelfth month, in the 8th year of the Chih-Chêng reign-period (+1348), at about 3 o'clock in the afternoon, there appeared in the south four black 'dragons' coming down from the clouds, and taking up water. Shortly afterwards another one appeared in the south-east, and lasted a considerable time before it disappeared. This was seen at Chia-hsing city.[c]

As to effects of the earth's atmosphere, we have already seen an example of an appreciation of their importance (p. 226 above). About +400 the astronomer Chiang Chi [1] undertook to explain why the sun rises and sets as a large red globe but appears small and white at noon:[d]

The terrestrial vapours (*ti chhi* [2]) do not go up very high into the sky. This is the reason why the sun appears red in the morning and evening, while it looks white at midday. If the terrestrial vapours rose high into the sky, it would still look red then.[e]

He thus understood that when at a low altitude, the sun was seen through a thicker layer of the earth's atmosphere than when it was high in the heavens. A century earlier Kuo Pho had spoken [f] of the dawn and sunset mists as *fên chin*;[3] and in the +11th century Shen Kua and others [g] used expressions such as *cho*,[4] *cho fên* [5] and *yen chhi chhen fên*,[6] referring to misty atmospheres caused by small suspended particles.

As for mirages, an explanation substantially correct was given by Chhen Thing in the Ming period.[h]

a Sect. 27*f*, in Vol. 4 below.
b Cajori (5), p. 47.
c P. 27*b*, tr. H. Franke (2), no. 71.
d According to the *Sui Shu*, ch. 19, p. 11*b*; cf. Chu Wên-Hsin (*1*), p. 108.
e Tr. auct. Cf. Vol. 2, p. 81.
f *TPYL*, ch. 771, p. 6*b*.
g *Sung Shih*, ch. 56, p. 21*a*, for example.
h *Liang Shan Mo Than*, ch. 11, p. 4*b*.

1 姜岌 2 地氣 3 氛祲 4 濁 5 濁氛 6 煙氣塵氛

(g) THUNDER AND LIGHTNING

It was natural that the ancient Chinese should have conceived of thunder (*lei*[1]) and lightning (*tien*[2])[a] as the result of a clash between the two deepest physical forces which they could imagine, the Yin and the Yang. In the broadest sense, when one bears in mind what this theory contributed to the conception of the positive and negative in Nature, including ultimately the phenomena of electricity and chemical combination, there was a great element of truth in the old Chinese ideas.[b] The *Huai Nan Tzu* book (*c.* −120) says:

> The Yin and Yang hurl themselves upon one another and this is the cause of thunder. Their forcing their way through each other produces lightning.[c]

But such naturalistic theories did not still the superstitious fears of the Han people, who, in accordance with the theory of phenomenalism, viewed the electrical discharges of the heavens as 'heavenly reprimands' for improper governmental or private proceedings. Hence the firmly naturalistic statement of Wang Chhung in his *Lun Hêng* (+83). We see him at his best in this passage, part of a special chapter on the subject.

> At the height of summer, thunder and lightning come with tremendous force, splitting trees, demolishing houses, and from time to time killing people. The common idea is that this splitting of trees and demolishing of houses is due to Heaven setting a dragon to work. And when the lightning strikes a person and kills him, this is attributed to some secret faults which he must have committed, such as eating unclean things. The roar and roll of thunder, they say, is the voice of Heaven's anger, like men gasping with rage. Ignorant and learned alike talk thus, making inferences from the ways of men (to those of Heaven) in order to make sense of what happens.
>
> But this is all nonsense. The genesis of thunder is one particular kind of energy (*chhi*), and one particular kind of sound. Its splitting of trees and demolishing of houses is one with the rushing on men and the killing of them. All these effects happen at the same time. Are we then to ascribe to a dragon the effects on trees and houses, calling in the supposed hidden faults as an additional explanation where a person is concerned? A dragon at work would be auspicious, and could not give rise to an inauspicious event. This would not be in accord with its Tao. For the two things (the non-ethical and the ethical) to happen instantaneously and with the same sound, is unreasonable....
>
> The *Li Chi* speaks of goblets with thunder patterns carved on them. One thunder rushes forth, the other reverts, one is coiled up and the other stretched out. They make a noise when they rub against one another. Their crossing and colliding causes the deep rumbling sound. (This representation is quite correct.) The crashing noise is the shooting forth of the *chhi*. If this *chhi* hits a man, he dies.

[a] Cf. *TSCC*, *Chhien hsiang tien*, chs. 77–9; *Shu chêng tien*, chs. 74–5.

[b] In this connection, the role played by thunder and other meteorological phenomena as symbols in the *I Ching* (Book of Changes) may be recalled (above, Sect. 13*g*, in Vol. 2, pp. 312ff.).

[c] Ch. 4, p. 12*b*, tr. auct. Parallel words in the −1st-century text, *Chhun Chhiu Wei Yuan Ming Pao* (in *Ku Wei Shu*, ch. 6, p. 3*b*). We find the same idea in *Chuang Tzu*, ch. 26 (Wai Wu), and in the *Kuliang Chuan* (cit. *TPYL*, ch. 13, p. 2*a*), so it was certainly current in the late Chou period.

1 雷 2 電

To speak truly, thunder is the explosion (*chi*[1]) of the *chhi* of the solar Yang principle. This may be understood by the fact that in January, when the Yang begins to grow, we hear the first thunders, while in May, when it is predominant, the thunder is continuous and severe. In autumn and winter, when the Yang is dying away, thunderstorms decline. In the summer, when the Yang is reigning, the Yin disputes its supremacy, so that there is collision, friction, explosions and shootings. What happens to be struck will be harmed. By pure chance a person sheltering under a tree or in a house may be hit and killed.

How can we test this? Throw a ladle of water into a smelting-furnace. The *chhi* will be stirred and will explode like thunder. Those who are near may well be burned. You may consider heaven and earth as like a furnace, the Yang *chhi* as the fire, and clouds and rain as abundant water. Hence violent disturbances must arise. How could people who are hit not be harmed and die?

When founders melt iron, they make moulds (*hsing*[2]) of earth and when these are dry the metal is allowed to run down into them. If they are not, it will skip up, overflow and spurt about. If it hits a man's body it burns his skin. Now the fiery Yang *chhi* is hotter than molten iron, and the exploding Yin is damper than earth or clay. When the Yang *chhi* hits a man it does more than cause a pain of burning.[a]

Lightning is essentially fire. Such *chhi*, burning a man, leaves a mark. If the mark looks like some writing, people, seeing it, are tempted to regard it as a statement concerning his guilt written by Heaven. This again is empty nonsense. If Heaven struck men dead with lightning (of set purpose) it would take care to write its characters clearly in order, instead of indistinctly and unclearly. Marks made by lightning are certainly not characters written by Heaven....[b]

The parallel with Lucretius springs to the mind:

> Hoc est igniferi naturam fulminis ipsam
> perspicere et qua vi faciat rem quamque videre,
> non Tyrrhena retro volventem carmina frustra
> indicia occultae divum perquirere mentis,
> unde volans ignis pervenerit aut in utram se
> verterit hinc partem, quo pacto per loca saepta
> insinuarit, et hinc dominatus ut extulerit se,
> quidve nocere queat de caelo fulminis ictus.[c]

Other Han scholars, such as Huan Than (before +30), had urged the same point of view.[d]

[a] In passing we may note the metallurgical significance of this paragraph as an indication of the knowledge of cast iron technique in the +1st century, antedating European practice by twelve centuries; see below, Sects. 30 and 36.

[b] Ch. 23, tr. auct., adjuv. Forke (4), vol. 1, p. 285; Hughes (1), p. 324.

[c] *De Rer. Nat.* VI, 379 ff. Leonard's translation runs:

> This, this it is, O Memmius, to see through
> The nature of the fire-fraught thunderbolt;
> O this it is to mark by what blind force
> It maketh each effect, and not, O not,
> To unwind Etrurian scrolls oracular,
> Inquiring as to the occult will of gods...
> Or what the thunderstroke portends of ill... (p. 265).

On the fulgural divination of the Etruscans see Bouché-Leclercq (2) and R. Berthelot (1).

[d] *TPYL*, ch. 13, p. 8*a*.

1 激 2 形

In later centuries the denial of prognosticatory significance in lightning and thunder became a commonplace of Confucian sceptical rationalism. In the early +11th century Su Hsün,[1] the father of the famous poet Su Tung-Pho, alluding to a popular belief that death by lightning was a punishment for lack of filial piety, remarked that it would have its work cut out to punish all those who deserved it, and evidently it did not do so.[a] In the same century, about +1078, Shen Kua recorded a description, so carefully written that it might have been destined for the columns of *Nature*, of certain lightning effects.

A house belonging to Li Shun-Chü was struck by lighting. Brilliant sparkling light was seen under the eaves. Everyone thought that the hall would be burnt, and those who were inside rushed out. After the thunder had abated, the house was found to be all right, though its walls and the paper on the windows were blackened. On certain wooden shelves, certain lacquered vessels with silver mouths had been struck by the lightning, so that the silver had melted and dropped to the ground, but the lacquer was not even scorched. Also a valuable sword made of strong steel had been melted to liquid, without the parts of the house nearby being affected. One would have thought that the thatch and wood would have been burnt up first, yet here were metals melted and no injury to thatch and wood. This is beyond the understanding of ordinary people. There are Buddhist books which speak of 'dragon fire' which burns more fiercely when it meets with water[b] instead of being extinguished by water like 'human' fire. Most people can only judge of things by the experiences of ordinary life, but phenomena outside the scope of this are really quite numerous. How insecure it is to investigate natural principles using only the light of common knowledge, and subjective ideas.[c]

Similar precise and objective descriptions were given in later centuries, as by Yang Yü about +1360.[d] But the explanation of the true nature of thunder[e] and lightning had to await the full flood of post-Renaissance science.

(*h*) THE AURORA BOREALIS

In the Chinese historical records there are a number of descriptions of 'strange lights' (*kuang i*[2])', 'coloured emanations', etc.,[f] which can only refer to aurora observations. E. Biot (14) collected forty of these from the *Wên Hsien Thung Khao* and its supplement, and about sixty are contained in the *Thu Shu Chi Chhêng* encyclopaedia.[g] The earliest dates from −208 and the latest from +1639. The *Khai-Yuan Chan Ching* (+718) quotes[h] lost books, such as the *Yao Chan*[3] (Divination by Weird Wonders) of the Han

[a] See Forke (9), p. 139.

[b] Could this have been an echo of one of the conventional descriptions of Byzantine 'Greek fire'? In Sect. 30 we shall show that 'Greek fire' or something very like it was widely used in China in the +10th and +11th centuries.

[c] *MCPT*, ch. 20, para. 10, tr. auct. Cf. Hu Tao-Ching (*1*), vol. 2, p. 656.

[d] *Shan Chü Hsin Hua*, p. 50*b* (tr. H. Franke (2), no. 137).

[e] On the confusion of Neolithic axes ('thunder-axes') with meteorites, a few words have been said above (p. 434).

[f] Cf. above, pp. 190, 284, 476, on the duties of observers given in the *Tso Chuan* and the *Chou Li*.

[g] *Shu chêng tien*, ch. 102.

[h] Ch. 2, p. 3*a*.

[1] 蘇洵 [2] 光異 [3] 妖占

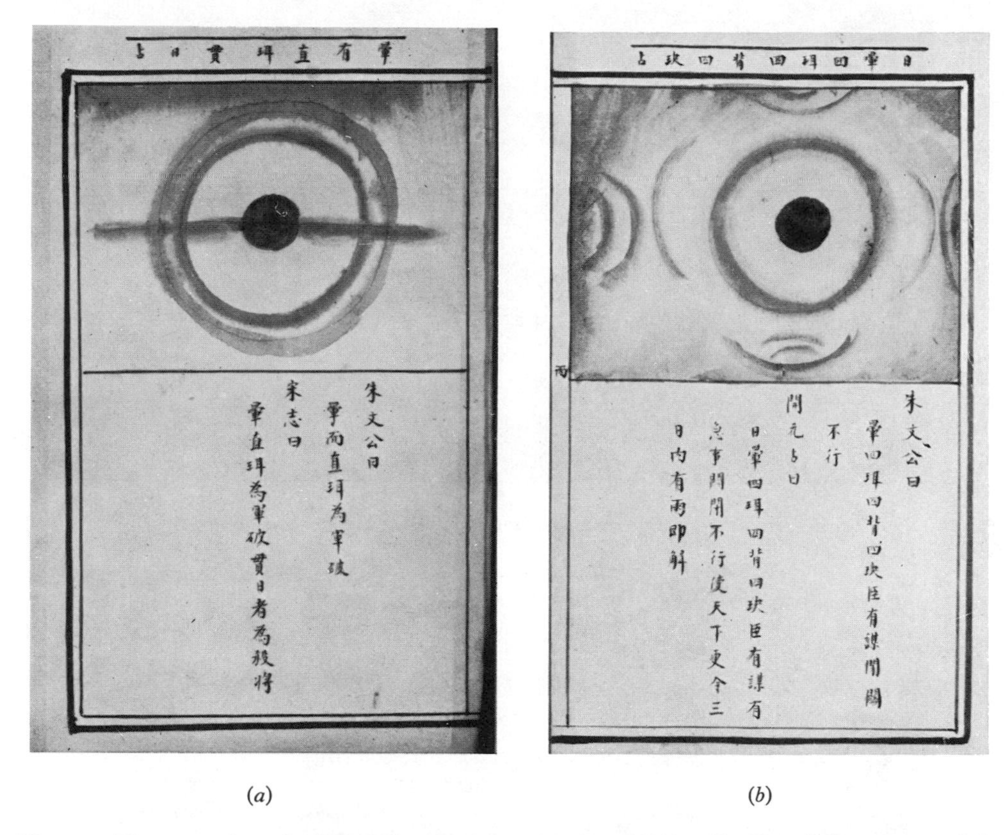

(a) (b)

Fig. 200. Two pages from the MS. *Thien Yuan Yü Li Hsiang I Fu* by Chu Kao-Chih, emperor of the Ming, *c.* +1425 (Cambridge University Library), showing parhelic phenomena.

Translation:

(*a*) '*Haloes* (*Yün*) *having straight Erh and threading the sun.* Chu Wên Kung says that when there is a *yün* with vertical *erh* an army will be defeated. The *Sung History* Memoirs also say that it means this, adding that if the sun is threaded through, a commander will be killed.'

(*b*) '*A solar halo* (*Yün*) *with four Erh, four Pei and four Chüeh.* Chu Wên Kung says that this signifies conspiracies on the part of ministers; let the gates be shut and (the emperor) not stir forth. The *Khai-Yuan Chan Ching* says exactly the same, further advising the issue of (emergency) orders throughout the empire. If within three days there is rain, the orders are to be cancelled.'

Thus although the motive was often astrological, the observations of the haloes were precise. In the above drawings one can recognise the parhelic circle, the 22° and 46° haloes, and a variety of tangent arcs.

PLATE LXXI

Fig. 201. The bore on the Chhien-Thang River near Hangchow (photo. Beer).

Fig. 202. Lin Chhing's picture of the bore on the Chhien-Thang River (from *Hung Hsüeh Yin Yuan Thu Chi*, 1849).

prognosticator Ching Fang,[1] mentioning great displays in −193 and −154, which were attributed, like earthquakes, to excessive Yin. No complete listing of the Chinese observations has yet been made,[a] since the phenomenon was not recognised very clearly as an entity, and went by many names, e.g. *chhih chhi*[2] (red vapour), *pei chi kuang*[3] (north polar light), etc. A vast literature would thus have to be searched by a sinologist and a meteorologist in collaboration. We noticed, for instance, a description of a purple aurora in one of the Five Element chapters of the *Hsin Thang Shu* (New History of the Thang) for +882, but this particular one does not seem to occur in the encyclopaedia's list.[b] An earlier aurora, covering the whole northern sky with shimmering red light in +763, was also recorded in Irish and Anglo-Saxon chronicles.[c] One of the oldest Chinese descriptions[d] is that for −30, when 'at night there were seen in the sky luminous vapours, yellow and white, with (streamers) more than 100 feet long, which brightly lit up the ground. Some said that these were cracks in the heavens (*thien lieh*[4]); others said that they were the "swords of heaven" (*thien chien*[5]).' This terminology is strangely reminiscent of the *χάσματα* (chasms) of Aristotle, who thus refers to aurorae in the *Meteorologica*,[e] but any connection between the terms is hard to believe.[f] Of course no explanation of the phenomena was possible until modern times. Their close association with sun-spots has been realised since 1859.

(*i*) SEA TIDES

Until modern times there was, on the whole, more knowledge of, and interest in, the phenomena of tides in China than in Europe. As has often been pointed out, this was probably due to the fact that the tides in the Mediterranean are slight and did not attract much notice by the ancient naturalists. The coasts of China, however, have tides of considerable range, for example, about twelve feet in spring off the mouth of the Yangtze. Moreover, China possessed one of the only two great tidal bores or eagres in the world, that on the Chhien-Thang[6] River near Hangchow (Fig. 201), the other being in the northern mouth of the Amazon, far away from any ancient civilisation. The bore on the Severn is much smaller. From very early times, therefore, a natural phenomenon of great impressiveness, obviously closely allied to coastal tides in general,

[a] But see Kanda (1, 2), whose reference to, and translation of, the Fu Hou passage are, however, both wrong.

[b] 'In the seventh month of the 2nd year of the Chung-Ho reign-period, on a *ping-wu* day, at night the whole of the north-western sky was red; the vapour was like purple or deep crimson borders or boundaries in the heavens', ch. 34, p. 11*b*. Cf. the translations of these chapters by Pfizmaier (67).

[c] See *Hsin Thang Shu*, ch. 34, p. 11*a*, which however makes it late in +762. A beginning has been made by Schove (1, 6) at correlating the Chinese and European data. Cf. Størmer (1).

[d] *Fu Hou Ku Chin Chu*, in *YHSF*, ch. 73, p. 50*b*; cit. *TPYL*, ch. 2, p. 6*b*. Compiled about +140. There is nothing exactly corresponding in the annals of the *Chhien Han Shu*, but in the previous year (−29) 'divine lights' had been seen in the first month 'in all three districts' (ch. 10, p. 3*b*). Perhaps Fu Hou was one year out in his reckoning. The term *thien lieh* goes back to the −1st century at least, for the *Chin Shu* (ch. 11, p. 7*a*) quotes a passage from Liu Hsiang which uses it.

[e] 1, 5, 342 b. Cf. Seneca (Clarke & Geikie tr. p. 38).

[f] We are indebted to Mr Francis Celoria for drawing our attention to this similarity.

1 京房 2 赤氣 3 北極光 4 天裂 5 天劍 6 錢塘

invited explanation by Chinese thinkers.[a] We owe to A. C. Moule (3) an exhaustive monograph on the Hangchow bore and the history of tidal theory in China; this includes a bibliography of descriptions of the bore,[b] the most recent being that of Chatley (22). No medieval Chinese drawing of the bore is known to exist, but I reproduce here a picture from Lin Chhing's[1] *Hung Hsüeh Yin Yuan Thu Chi*[2] of 1849 (Fig. 202).[c]

The normal behaviour of tides is of course a rise, slow at first, fastest at mid-sea-level, and diminishing to high water; then a fall returning to the original low level. The total cycle occupies about 12 hr. 25 min., this being the interval between successive meridian transits of the moon. The rise is greatest at the beginning of each lunation, and least about two weeks after it. The range of the tides, the intervals between transits and high waters, and the difference between syzygy and spring tide, however, greatly depend upon local conditions, such as the coastal topography, the configuration of the littoral sea-bottom, etc. Places in north China may have but one tide in each twenty-five hours, while Southampton may have four in a day. As one proceeds up a tidal estuary, such as that of the Yangtze or the Thames, the period of rise shortens and the period of fall increases. A bore is simply an extreme example of this in which the first half of the rise is almost instantaneous.[d] In the Chhien-Thang River the bore usually has a front of about twelve feet followed by a further rise of six feet in the first hour, and a total range of some twenty feet. Out at sea, where the bore commences, the speed of the waves is as much as seven knots; in the river, after the passage of the bore, a flood current of ten knots occurs. A standing wave of perhaps thirty feet high is generated at the point where two wave fronts intersect, and if this happens near the shore, it may flood over the very massive sea walls.[e] A thunderous noise is heard long before the bore arrives, and after it has passed junks proceed upstream in the strong current hardly under control.

Elaborate discussions have been entered upon by Moule (3) and Gillis (himself a sailor) on the usual words for tides, *chhao*[3] and *hsi*,[4] the former meaning morning, flood, or spring tide, and the latter evening, ebb, or neap tide. The terms are, it will be seen, somewhat vague. Moreover, their etymology is very obscure. Giles (8) drew attention to a phrase common in ancient writings, *chhao tsung yü hai*,[5] meaning that rivers do homage to the sea, as if at an imperial court. Moule thinks, therefore, that

[a] The far more terrifying and destructive tidal waves of the Sea of Japan were naturally not so familiar to the Chinese, but Schlegel (7 *r*), p. 50, noted what is quite probably a reference to them in the +4th- or +5th-century *Hai Nei Shih Chou Chi* (entry under Phêng Lai). Waves rising to a hundred feet in height, with no wind, need by no means be purely imaginary (cf. the description of Bernstein). Tidal waves, of course, have nothing to do with tides, but are caused by underwater earthquakes or volcanic explosions. At a distance, tidal regularities may be affected, as in the occurrence of +1347 reported in *Shan Chü Hsin Hua*, p. 26*a* (H. Franke (2), no. 64).

[b] E.g. McGowan (3). Its first modern scientific analysis was that of Commander W. U. Moore in 1888. Cf. Moule (15), pp. 19 ff.

[c] Lin Chhing was a great expert in water conservancy and hydraulic works; we shall meet with him again in connection with that subject (Sect. 28 *f*).

[d] For the theory of bores, see Doodson & Warburg (1); Allen (1).

[e] Chatley (22). Perhaps 'shock wave' would be a better term.

[1] 麟慶 [2] 鴻雪因緣圖記 [3] 潮 [4] 汐 [5] 朝宗于海

the word *chhao*[1] was derived not directly from 'sunrise' but indirectly from this analogy, since the court assembled at dawn. The word *hsi*[2] is connected more directly with the evening, since the phonetic is an ancient pictogram of the crescent moon.

Already in the early −2nd century a good bore was expected at full moon, as appears from the poetical composition Chhi Fa[3] (The Seven Beguiling Tales)[a] by Mei Shêng[4] (*d.* −140):[b]

> The guest said: 'At full moon in the eighth month I hope to go to Kuang-Ling[5] with the feudal lords and with companions from far away, and with my brothers, to see the bore on the River Chhü.[6] When we reach the spot, before we see the bore itself, but only the places where the strength of the water has come, it will be alarming enough to make us tremble at what it has over-ridden, uprooted, thrown into confusion, and swept away....'[c]

But a causal connection with the moon is not stated, and when the sick prince asks the guest what the force is which drives the wave, he replies that, though not supernatural, it is not recorded. In the +1st century, however, the dependence of the tides on the moon was clearly indicated, and by none other than Wang Chhung in his *Lun Hêng*. The whole passage affords so remarkable an example of how that great sceptic would take a popular belief and tear it limb from limb, that I shall hope to be excused for citing it in full.[d] Wu Tzu-Hsü,[7] a virtuous minister[e] of the State of Wu in the feudal period, had been unjustly killed or driven to death by his prince[f] Fu Chhai,[8] and thrown into the river about −484. The popular belief was that the vengeful spirit of the minister regularly thereafter roused the waves to their periodical wrath and havoc:

The story.

> (1) It has been recorded that the King of Wu, Fu Chhai, put Wu Tzu-Hsü to death, and had (his body) cooked in a cauldron, sewn into a leather sack, and thrown into the River. Wu Tzu-Hsü, incensed, lashed up the waters, so that they rose in great waves and drowned people. In later times temples were erected to him on the Yangtze at Tan-thu[9] in Kuei-Chi[10] as well as on the Chê[11] river of Chhien-Thang,[12] for the purpose of appeasing his anger and stopping the wildness of the waves. Now the statement that the King of Wu put Wu Tzu-Hsü to death and had him thrown into the river, is reliable enough, but the idea that out of spite Wu Tzu-Hsü lashes up the waters, is absurd.

Contradictory negative examples.

> (2) Chhü Yuan,[13] full of disgust, threw himself into the Hsiang[14] River, but (boisterous) waves did not arise on it. Shen Thu-Ti[15] jumped into the Yellow River and died, but waves did not appear in that case either.

[a] *CSHK* (Chhien Han sect.), ch. 20, p. 6*b*; *Wên Hsüan*, ch. 34. [b] See Nagasawa (1), p. 148.
[c] Tr. Moule (3). The Chhü seems to have been a river in Northern Chiangsu north of the Yangtze.
[d] The division of the argument into sections is ours.
[e] G2358. Biography in *Shih Chi*, ch. 66.
[f] G576. This king was one of the pioneer canal-builders, cf. Sect. 28*f* below.

1 潮 2 汐 3 七發 4 枚乘 5 廣陵 6 曲
7 伍子胥 8 夫差 9 丹徒 10 會稽 11 浙 12 錢塘
13 屈原 14 湘 15 申徒狄

Objection

(3) People will certainly say that as to violence and wrath Chhü Yuan and Shen Thu-Ti did not equal Wu Tzu-Hsü.

refuted,
leading to a new argument.

(4) But in the State of Wei, (Chung) Tzu-Lu's[1] body was salted in pickle. And in the Han time the body of Phêng Yüeh[2] was boiled. The valour of Wu Tzu-Hsü certainly did not exceed that of Tzu-Lu and Phêng Yüeh. Yet these two men could not vent their anger when their bodies were being cooked in the cauldrons, and raise a disturbance to bespatter the bystanders.

Why wait to show annoyance till thrown in the river?

(5) Moreover, Wu Tzu-Hsü was first put into the cauldron and only afterwards thrown into the river. What was his spirit doing when his body was in the cauldron? Why was it so meek in the cauldron and so bold in the river? Why was its indignation not the same in both places?

And what river was it, anyway?

(6) Furthermore, when he was thrown into the river, what river was it? There is the Yangtze at Tan-thu, the Chê river of Chhien-Thang, and the Ling[3] river at Wu-thung.[4] If you say he was thrown into the Yangtze at Tan-thu (you have to explain why) it never rises in boisterous waves. If you say he was thrown into the Chê river at Chhien-Thang (then you have to admit that) not only the Chê river but also the Shan-Yin[5] and the Shang-Yü[6] rivers have waves. Perhaps you will maintain that the body was taken out of the sack and cut up into three pieces, one piece being cast into each of these three rivers?

A pointless story, in Wang Chhung's opinion.

(7) For human hatred there is still some justification as long as the deadly enemy is alive, or some of his descendants still left. But now the Wu State has long ceased to exist, and Fu Chhai had no sons. What was once Wu is now the Prefecture of Kuei-Chi. Why should the spirit of Wu Tzu-Hsü still resent the wrong once done him, and continue to excite the waves? What is he still after?

Geographical objection.

(8) When the States of Wu and Yüeh existed, the present Kuei-Chi circuit was divided, so that Yüeh governed Shan-Yin and Wu had the present Wu city as its capital. South of Yü-Chi all the land belonged to Yüeh, and north of it to Wu, the Chhien-Thang river thus forming the frontier between the two kingdoms. The Shan-Yin and Shang-Yü rivers both come out of Yüeh territory. So if Wu Tzu-Hsü's anger causes the violence of the waves they ought to do their damage on the Wu side only—why should they affect the Yüeh side? To harbour a grudge against the King of Wu and yet to wreak malice also on Yüeh is contrary to reason (*wei shih tao li*[7]) and does not show the distinction which a spirit is (surely) capable of making.

[1] 仲子路 [2] 彭越 [3] 陵 [4] 吳通 [5] 山陰 [6] 上虞 [7] 違失道理

Objection comparing mental and physical action.

(9) Besides it is difficult to excite waves, but easy to move men. The living rely on the strength of their nerves, the dead must use their essence and soul (*ching hun*[1]). Alive, Wu Tzu-Hsü could not move the living, or preserve his body, and himself caused its death (by suicide when the King presented him with a sword). When the strength of his nerves was lost, and his essence and soul evaporated and dispersed (*fei san*[2]), how could he excite waves?

Return to argument of (2).

(10) Hundreds and thousands of people have found themselves in the predicament of Wu Tzu-Hsü—crossing rivers in boats, they have not reached the other shore. But Wu Tzu-Hsü was the only one who had his body boiled in a cauldron first. After such treatment what harm could he do? (He was not even drowned in the water directly.)

Parallel legends.[a]

(11) King Hsüan[3] of Chou killed his minister, the Lord of Tu.[4] Lord Chien[5] of Chao killed his officer Chuang Tzu-I.[6] Years later (the ghost of) Tu shot and killed King Hsüan, and (the ghost of) Chuang felled the Lord Chien. Although these stories seem (at first sight) true, they are really all empty nonsense. But even supposing them true, Wu Tzu-Hsü, whose body was no longer intact, having been broken down by the boiling, was not in a position to have acted in the same way against King Fu Chhai. So why consider the beating of the waves a revenge? Or even as proof of Wu Tzu-Hsü's retaining his consciousness?

(12) (The trouble is that) popular stories, though not true, are represented in paintings, so that even scholars and wise men are deceived by them.

A scientific view.

(13) Now the rivers in the earth are like the pulsating blood-vessels of a man. As the blood flows through them they throb or are still in accordance with their own times and measures. So it is with the rivers. Their rise and fall, their going and coming are like human respiration, like breath coming in and out.[b]

The natural processes of Heaven and Earth have remained the same from the most ancient times (*shang ku*[7]). When the *Shu Ching* says, 'The Yangtze and the Han pursue their common course to the sea', that was in the time before Yao and Shun (the legendary emperors). When the rivers fall into the sea, there is nothing but rapid motion (as a rule), but when the sea enters the three rivers (i.e. the Chhien-Thang, the Shan-Yin and the Shang-Yü already mentioned), the waters begin to roar and foam, doubtless because their channels are small, shallow, and narrow.

[a] Cf. *Mo Tzu*, ch. 31, tr. Mei (1), p. 161.

[b] This is a microcosm-macrocosm theory (cf. above, Sect. 13*f*, in Vol. 2, p. 294). Respiratory theories of tides were also held in Europe; see below, p. 494.

[1] 精魂 [2] 飛散 [3] 宣 [4] 杜 [5] 簡 [6] 莊子義 [7] 上古

Another example of a river with a tidal wave; cf. (6).

(14) The River Chhü[1] in Kuang-Ling[2] also has such great waves. As the poet says: 'Vast is the expanse of the Yangtze, and great the waves of Chhü.' This is certainly due to the passage being narrow and impeded. If after having been done to death in Wu, Wu Tzu-Hsü's spirit were to go and produce boisterous waves at Kuang-Ling, that would certainly not be a sign of its intelligence. (No harm had been done to it there.)

And what about rapids in general?

(15) When streams are deep they are still and broad, but when they are shallow, with much sand and many rocks, they boil up into rapids. A tidal bore and a rapid are much the same thing. If Wu Tzu-Hsü is to be held responsible for the tidal waves, who is it that lives in the rivers and makes falls and rapids?[a]

And ordinary storms?

(16) Besides, sometimes a storm excites the waters of the three rivers, so that they drown people. Wu Tzu-Hsü's spirit would have to be made responsible for the wind too.

Localisation of the effects.

(17) Moreover, I remark that when the wave enters the three rivers (of the Chhien-Thang estuary) it foams and rages along the banks, but in the middle of the stream there is no noise. So if Wu Tzu-Hsü is responsible, his body must (in some way) be concentrated along each of the banks.

The moon is the answer to the problem; cf. (13).

(18) Finally the rise of the wave follows the waxing and waning moon, smaller and larger, fuller or lesser, never the same. If Wu Tzu-Hsü makes the waves, his wrath must be governed by the phases of the moon! (*ju Tzu-Hsü wei thao, Tzu-Hsü chih nu i yüeh wei chieh yeh*[3]).[b]

Thus does Wang Chhung prowl round the traditional animistic belief, attacking it from a variety of angles before giving it the knock-out blow based on the correlation of the tides with the moon.

The earliest reference to the sea walls built along the Chhien-Thang river to withstand the tides and the bore comes from Wang Chhung's time. His contemporary, Hua Hsin,[4] who was governor of the district from +84 to +87, seems to have been the first to organise the construction of dykes of sufficient strength.[c]

In Wang Chhung's mind, the influence of the moon on the tides must have been combined with a microcosm-macrocosm respiration theory, as we have seen. But these naturalistic views had to contend in his time with others more primitive. The *Shan Hai Ching* (Classic of the Mountains and Rivers)[d] stated that the tides were the

[a] Note again Wang Chhung's appreciation of the role played by the topography of the sea and estuary bottom.

[b] *Lun Hêng*, ch. 16, tr. auct., adjuv. Forke (4), vol. 2, pp. 247 ff.; Moule (3), p. 149.

[c] *Hou Han Shu*, ch. 101, p. 10*b*; *Thai-Phing Huan Yü Chi*, ch. 93; *Thung Tien*, ch. 182, p. 966·2.

[d] This is generally regarded as a Former Han book built round material from Warring States times, and perhaps going back to the school of Tsou Yen.

[1] 曲 [2] 廣陵 [3] 如子胥爲濤子胥之怒以月爲節也 [4] 華信

result of a sea-serpent or whale (*hai chhiu*[1] or *chhing ni*[2]) coming out from and going into its cave.[a] The Buddhist books, with their stories of *nāgas*, encouraged this idea. However, in the +3rd century, Yang Chhüan, in his *Wu Li Lun* (Discourse on the Principles of Things),[b] reaffirmed that the moon, being itself of the purest aqueous principle, influenced the tides, which were great or small according to her waxing and waning.[c] His older contemporary Yen Chün[3] had written the first monograph specially devoted to tidal theory. Yang Chhüan's view was supported early in the +4th century by Ko Hung, whose *Pao Phu Tzu* says:

> The waves of the sea heave up and down with the waning and waxing of the moon. The *chhao*[4] is that which comes in the morning, the *hsi*[5] is that which arrives in the evening. The influence of the moon produces water, so when the moon is full the *chhao* is large.[d]

And Ko Hung goes on to correlate, not very successfully, the positions of the sun (and hence the moon) in the four seasons with the behaviour of the tides at those times.[e] He also offers a strange alternative theory that the tides are due to the 'overflowing' of the Milky Way, which passes under the sea in the diurnal revolution of the heavens. It seems that what he had in mind was the noria or water-wheel which raises water in buckets attached to its rim.[f] In a third passage he emphasises the importance of the topography and configuration of estuaries and rivers with regard to the Chhien-Thang bore, and scoffs at the story of Wu Tzu-Hsü, saying that the phenomenon had existed since the foundation of heaven and earth, long before the quarrel between the King of Wu and his minister.

The next advances were made in the Thang. Tou Shu-Mêng,[6] about +770, who wrote the *Hai Thao Chih*[7] (or *Hai Chiao Chih*[8]) (On the Tides), seems to have been one of the first to deal with the lunar theory of the tides with any scientific detail. When the moon is passing through *hsi-mu*[9] and *ta-liang*,[10] he said, the water rises higher. These two terms, it will be remembered, are two of the twelvefold divisions of the equator and its *hsiu*, corresponding to the autumn and spring equinoxes respectively.[g] Apparently Tou Shu-Mêng thought that the moon actually caused the

[a] It seems impossible to find this passage in the text of the book as we now have it (cf. the relevant Sino-French index), but it was quoted early, as in the *Fêng Thu Chi*[11] (Record of Airs and Places) by Chou Chhu[12] in the Chin (+3rd century), see Pelliot's note in Moule (3), p. 148; and *TPYL*, ch. 68, p. 4*b*.

[b] Already referred to, p. 218 above.

[c] Klaproth (1) seems to be the only sinologist who has noticed this passage, which he took from a quotation given in the Thang encyclopaedia *Thang Lei Han*[13] by Yü An-Chhi.[14]

[d] We have not found the exact location of these remarks; as yet there is no index to *Pao Phu Tzu*, but they are quoted in *TPYL*, ch. 68, p. 5*b*, tr. Moule (3).

[e] He says that the tides are higher in summer than in winter, which is true only for the day tides. Natural, therefore, is his statement that they slowly increase during spring and diminish during autumn. In his seasonal correlation he introduces the old erroneous calculations about the distance of the sun from the earth (cf. pp. 21, 225, 257 above).

[f] See below, Sect. 27*e*. The passage might well be taken as evidence of the existence of this machine in Ko Hung's time.

[g] Cf. Table 34 above. See also de Saussure (16*a*).

1 海鰌 2 鯨鯢 3 嚴畯 4 潮 5 汐 6 竇叔蒙
7 海濤志 8 海嶠志 9 析木 10 大梁 11 風土記
12 周處 13 唐類函 14 俞安期

water to swell and contract. He was clear, however, on the monthly springs and neaps, saying, 'In one night and day there are two *chhao* and two *hsi*, in one new and full moon two springs and two neaps (*i hui i ming tsai chhao tsai hsi*[1])'.[a]

In the following decades Fêng Yen,[2] in his *Chien Wên Chi*[3] (Records of Things Seen and Heard) (+800), accurately described daily changes in the time of high water. Li Chi-Fu,[4] in his *Yuan-Ho Chün Hsien Thu Chih*[5] (General Geography of the Yuan-Ho reign-period) of +814, was equally precise, saying that the tide is smallest on the tenth and twenty-fifth days of the lunar month, and largest on the third and eighteenth days. This was in particular reference to the Chhien-Thang bore, which he graphically described. In +850 came a short tractate which was to be famous, Lu Chao's[6] *Hai Chhao Fu*[7] (Essay on the Tides).[b] From this we know that regular tide tables (*thao chih*[8]) were already in use, and that the association of neaps with the moon's quarters was fully accepted. At the same time, he retained some attachment to the respiration theory, if this term may be applied to the somewhat vague conception of the original *chhi* puffing out and drawing in (*yuan chhi i yüeh*[9]). But he introduced a new theme,[c] namely, that the sun had a part to play in the matter[d] (*hai chhao chih sêng hsi tzu jih, erh thai yin tshai chhi hsiao ta yeh*[10]); the sun produced tides, but the moon determined their dimensions. His essay was often afterwards quoted[e] and criticised, as we shall see.

Of course, he did not think in terms of the gravitational attraction of the sun on the mass of water, but supposed a kind of periodical explosion and consequent tidal wave to occur each time that the white-hot sun entered the water. With such basically erroneous theories in mind, it was not surprising that Lu Chao considered that the farther away the moon (the Yin force) was from the sun (the Yang force) the larger the tide would be—a conclusion exactly opposite to what we know now to be the truth.[f]

Another point of view was taken by Chhiu Kuang-Thing[11] about +900, in his *Hai Chhao Lun*[12] (Discourse on the Tides), which took the form of a dialogue between two characters, the Old Fisherman of the Eastern Sea (Tung Hai Yü Ong[13]) and the Recluse of the Western Hill (Hsi Shan Yin Chê[14]). He thought it was absurd to

[a] Tou Shu-Mêng was frequently afterwards quoted, in Sung books, such as the *Yün Lu Man Chhao*[15] (Random Jottings at Yün-lu) by Chao Yen-Wei,[16] and other books shortly to be mentioned.

[b] To be found in *TSCC*, *Shan chhuan tien*, ch. 315, p. 6*b*.

[c] Or perhaps built further on views which Ko Hung had held.

[d] Today of course we know that this is true, the attractive force of the sun being comparable with that of the moon by the ratio 3 to 7 roughly.

[e] As in the *Chiu Jih Lu*[17] (Daily Journal), a Sung book by a Mr Chao[18] who adopted the pseudonym Kuan Yuan Nai Tê Ong[19]—The Old Gentleman of the Water-Garden who attained Success through Forbearance. There is much on the tides in this book, which is abridged in *Shuo Fu*, ch. 14, p. 3*a*.

[f] The largest tides occur when the gravitational forces of the moon and the sun pull in the same direction.

1 一晦一明再潮再汐 2 封演 3 見聞記 4 李吉甫
5 元和郡縣圖志 6 盧肇 7 海潮賦 8 濤志 9 元氣噫噦
10 海潮之生兮自日而太陰裁其小大也 11 邱光庭 12 海潮論
13 東海漁翁 14 西山隱者 15 雲麓漫抄 16 趙彥衛
17 就日錄 18 趙 19 灌園耐得翁

suppose that the sea could expand and contract, as the land would do so too, and there would be no difference of water-level; what happened was that the land periodically rose and sank in accordance with cosmic respiration.

Notable advance was made with Yü Ching's[1] *Hai Chhao Thu Hsü*[2] (Preface to Diagrams of the Tides) of +1025. This work[a] was perhaps intended to introduce the *Hai Chhao Thu Lun*[3] (Illustrated Discourse on the Tides) by a celebrated engineer[b] Yen Su,[4] who had been studying the tides for the previous ten years. Yü Ching established that the day tide was larger in spring and summer, while the night tide was larger in autumn and winter. He also knew the association of the tidal bore with new and full moon. The most important passages of Yen Su's book were afterwards repeatedly quoted.[c] Here is one of them:

The original forces (*yuan chhi*[5]) breathe in and out, and the sky following the forces expands and contracts,[d] while the tide going and coming in the seas follows the sky and flows and ebbs. Since the sun is the mother of the double Yang, and Yin is born of Yang, the tide is subject to the sun; and since the moon is the essence of the Yin, and water is Yin, the tide follows the motions of the moon. For this reason, following the sun and responding to the moon, copying the motions of the Yin and yet subject to the Yang, the tide is at its highest at new and full moon, contracts as the moon waxes or wanes, is at its lowest at the first and last quarter, and grows just before full and new moon. That is the reason why there are large and small tides.

Now beginning at new moon at the *tzu* double-hour (11 p.m. to 1 a.m.) in the middle of the night, the tide is high at the earth's *tzu* position, 4·165 *kho*.[e] When the moon is removed from the sun at the earth's *chhen* point by the daily movement of 3·72 *kho*, which tallies with the positions which the moon has reached in relation to the position of the sun, the tide is bound to respond. After the moon has passed the full, it still travels eastward, and the tide is governed by the sun and again responds to it westward, until, on reaching 4·165 *kho* of *tzu* at the next new moon, sun, moon, and tide all meet again at the *tzu* position. Thus we know that the tide is bound to be governed by the sun and turn to the west, so that when the moon approaches (the meridian at) *tzu* (about midnight) or *wu* (noon), the spring tide is certain to be at its highest, and when the moon reaches (the meridian at) *mao* (about 6 a.m.) or *yu* (about 6 p.m.) the neap tide is certain to be at its lowest. There may be small differences of lateness or earliness, but on the whole the flow and ebb, the springs (*ying*[6]) and neaps (*hsü*[7]), do not miss their proper times.[f]

[a] Another work of about the same time was the *Chhao Shuo*[8] (Discussion of the Tides), interesting because its author, an official of Hangchow, was the distinguished Taoist who edited the *Yün Chi Chhi Chhien* collection—namely Chang Chün-Fang[9] (*Mo Chi*[10] (Things Silently Recorded), p. 59*a*).

[b] Cf. above, p. 324, and below, Sect. 27*c*.

[c] Especially in the *Hsi Chhi Tshung Hua*[11] (Western Pool Collected Remarks) of Yao Khuan[12] later in the same century. This in its turn was drawn on for the *Hsien-Shun Lin-An Chih*[13] (Hsien-Shun reign-period Records of the Hangchow District) by Chhien Yüeh-Yu[14] of +1274. This was the source translated by Moule (3).

[d] One cannot escape being reminded here of the ancient theories of expansion and contraction, dissipation and congelation (above, Sects. 10 and 16).

[e] A *kho* (quarter) was rather less than 15 minutes because there were generally 100 in the day-and-night of 12 double-hours, not 96.

[f] Tr. Moule (3), mod.

1 余靖 2 海潮圖序 3 海潮圖論 4 燕肅 5 元氣
6 盈 7 虛 8 潮說 9 張君房 10 默記 11 西溪叢話
12 姚寬 13 咸淳臨安志 14 潛說友

From this passage, though somewhat obscurely expressed, it is clear that attempts were being made at quite accurate observation early in the +11th century. It is also clear that a distinct notion of 'influence' on the part of the celestial bodies was in mind. Exactly how near this came to a formulation in terms of 'gravitational' attraction or force depends on the interpretation of the terms used. For example, Chang Tsai (the Neo-Confucian), about the same time, spoke of the *ching*[1] of the moon, a radiating 'seminal virtue', and of the *hsiang kan*[2] of the water, i.e. its response.[a] It is not clear why the time of two minutes past midnight in the above passage is given with such refinement, but the moon's transit at new and full would be approximately midnight. The average daily change of fifty-six minutes in the time of the transit is also not far wrong.[b] Part of Yen Su's work consisted of a detailed tide-table for Ningpo.

Other tide-tables of this date were those of Lü Chhang-Ming[3] in +1056, which are contained in the *Shun-Yu Lin-An Chih*[4] (Shun-Yu reign-period Records of the Hangchow District)[c] by Shih Ê.[5] The *Cho Kêng Lu* of +1368 says[d] that these tables were inscribed by Hsüan Chao[6] on the walls of the Chê-chiang Thing[7] pavilion which stood on the banks of the Chhien-Thang river.[e] Compare with these facts the +13th-century tide-table giving the time of 'fflod at london brigge' which exists in a British Museum manuscript.[f] This is the earliest in Europe.

In the latter half of the century Shen Kua took Lu Chao to task for his theory that the sun was involved in the causation of the tides. In the *Mêng Chhi Pi Than* we read:

> Lu Chao says that the tide of the sea is formed because it is stirred up by the rising and setting of the sun. This has not the slightest basis. If the tide were due to this cause it would have a diurnal regularity. How could it happen that it sometimes comes in the morning and sometimes in the evening?
>
> I have myself given much study to its periodical motion, and found that the tide comes to high water whenever the moon makes its meridian transit. If you wait for this moment you will never miss the tides. Here I am referring to the tide on the coast itself; the farther you are away from the sea (i.e. up an estuary or the like) there will be a delay varying according to the place....[g]

Thus in +1086 Shen Kua clearly defined what we now call the 'establishment of the port', i.e. the constant interval between the theoretical time of high tide and the time when it occurs at the place in question, or in other words the degree of continental retardation.[h]

In September +1124, Hsü Ching[8] wrote the preface to his *Hsüan-Ho Fêng Shih Kao-Li Thu Ching*[9] (Illustrated Record of an Embassy to Korea in the Hsüan-Ho

[a] *Chêng Mêng*, ch. 4, p. 6*b*. On stimulus and response see Vol. 2, p. 304, and of course Sect. 39 below.
[b] It should be 51.
[c] Ch. 10, p. 5*a*.
[d] Ch. 12, p. 12*a*.
[e] His exact date is obscure.
[f] Cotton MSS., Julius D, 7.
[g] Pu appendix, ch. 2, para. 3, tr. auct. Cf. Hu Tao-Ching (*1*), Vol. 2, pp. 931 ff.
[h] Cf., for example, Barlow & Bryan (1), p. 347.

1 精 2 相感 3 呂昌明 4 淳祐臨安志 5 施諤
6 宣昭 7 浙江亭 8 徐兢 9 宣和奉使高麗圖經

reign-period), though the book was not printed till 1167. Hsü had accompanied a Chinese ambassador to that country upon the accession of a new Korean king, and we shall meet with his book again in connection with the history of the magnetic compass.[a] The relevant parts on the tides have been translated by Moule (3), and will not here be given as they only set forth at greater length what had already been said by Yen Su and Shen Kua.

Such the position remained until the coming of modern science in the 18th century. In the Yuan, Liu Chi[1][b] and Chêng Ssu-Hsiao[2][c] supported the respiratory theory. During the Ming decadence there had been little save the discussion in Wang Khuei's *Li Hai Chi*,[d] which made no advance on the Sung. In 1781 Yü Ssu-Chhien[3] made a collection of the medieval writers on tidal theory, the *Hai Chhao Chi Shuo*.[4] It now only remains to glance at the comparable development in Europe.

Although, as has been said, the weak or imperceptible tides of the Mediterranean did not invite study by the Greek and Hellenistic scientists, it cannot be said that they were unknown. Indeed, the first formulations of lunar influence in Europe anticipated those of the Chinese.[e] One of the earliest mentions of tides refers to that at Suez on the Red Sea, where there is a variation of some six feet; it was made by Herodotus[f] (almost the contemporary of Wu Tzu-Hsü). When Tsou Yen was telling of the encircling seas, Pytheas of Marseilles at the other extreme bound of the Old World was experiencing the tides of the English Channel (*c.* −320). Just at this time, too, the sailors of Alexander the Great, upon reaching the mouth of the Indus near Karachi, were much surprised not only by the tides but by some kind of tidal bore.[g] And Aristotle's pupil, Dicaearchos of Messina, conjectured (like Ko Hung later) that the sun was responsible in some way for the tides.

It seems that Antigonus of Carystos (*c.* −200) was the first of the Greeks to suggest that the moon was the primary influence. His position was thus similar to that of the poet Mei Shêng, but advance was more rapid since Seleucos the Chaldean of Seleuceia on the Persian Gulf[h] correlated the tides with the moon's motions about −140, and half a century later Poseidonius of Apameia[i] (contemporary of Lohsia Hung) stated the meridian rule, the quarter rule of springs and neaps, and the combined actions of moon and sun.[j] This level was not reached in China until the Thang and Sung, as we have seen, but while the Chinese were advancing, Europe (apart from a few repetitions in Latin authors, often incorrect) forgot the progress which had been made. With the exception of Bede the Venerable (*fl.* +700, just before Tou Shu-

[a] Sect. 26*i*.
[b] See Forke (9), p. 307.
[c] See Huang Chieh (*1*).
[d] Pp. 5*b*, 51*b*.
[e] Brunet & Mieli (1) devote a special chapter to tidal theory in antiquity. Cf. also the monograph of Almagià.
[f] II, 11.
[g] Bunbury (1), vol. 1, p. 447.
[h] Sarton (1), vol. 1, p. 183; Tarn (1), p. 43.
[i] Bunbury (1), vol. 2, p. 97. The relation of his thought to Stoic organism is well pointed out by Sambursky (1), pp. 142 ff.
[j] The speculation that some breath of these discoveries might have reached Wang Chhung has already been touched upon (Vol. 1, p. 233).

[1] 劉基 [2] 鄭思肖 [3] 俞思謙 [4] 海潮輯說

Mêng), there was nothing done between −100 and +1500.[a] Bede, stimulated, like the Chinese, by the close presence of striking tidal phenomena, noted local differences and gave a good account of the tides in general in his *De Temporum Ratione*. Curiously, he made the same mistake as Tou about the tides at the equinoxes.

Meanwhile other theories had been competing with the moon in Europe as in China. The microcosmic-macrocosmic respiration idea, first found perhaps in Strabo,[b] had great success throughout the Middle Ages, and captivated as remarkable a mind as Leonardo da Vinci, who actually tried to calculate the size of the world lung. There is also a parallel to the sea-cavern story of the *Shan Hai Ching*; it seems to originate in Wang Chhung's older contemporary, the geographer Pomponius Mela (*fl.* +43), and found supporters such as another geographer in the same century as Tou Shu-Mêng—Paul Warnefrid (*d.* +797).[c] There is no clue here as to any transmission of ideas.

The conception of the 'establishment of the port', adumbrated by Bede, came to full expression in Giraldus Cambrensis' *Topographia Hibernica*[d] of +1188. We met with it in China in +1086. But the Chinese had a clear priority as to the systematic preparation of tide-tables, which, as we have seen, go back to the +9th century at least. In the +11th they were much more enlightened on the theory of the subject than the Europeans until the Renaissance. The crowning irony was that Galileo rejected Kepler's lunar theory of tides on the ground that it was astrological.[e] It was not until the time of Newton that the true gravitational explanations of tidal phenomena were worked out and accepted. Yet it would never have occurred to any Chinese observer that the moon could not have an effect on terrestrial events—such a separation would have been contrary to the whole world view of organic naturalism.[f]

[a] For the ideas on the tides current in medieval Europe see Kimble (1), pp. 161–8.

[b] *Geogr.* III, 8.

[c] Sarton (1), vol. 1, p. 539.

[d] Sarton (1), vol. 2, p. 418.

[e] Berry (1), p. 167; Pledge (1), p. 62.

[f] See above, Sect. 13*f*, in Vol. 2, pp. 287, 293.

THE SCIENCES OF THE EARTH

22 GEOGRAPHY AND CARTOGRAPHY

(*a*) INTRODUCTION

After the expanses of the stars and the deeps of sky and sea, it is quite natural to turn to the homelier features of the mundane world, to the realm of the explorers and geographers. But geography is a subject which lies on the borderline between the natural sciences and the humanities. Any systematic treatment of the cumulative growth of geographical knowledge in China would far exceed the limits of our plan. On this subject there is a vast literature, both in Chinese and Western languages, but it belongs rather to history itself than to the history of science. So does the identification of ancient place-names which has generated untold erudition both in China and the West. Sinologists have always been especially intrigued by the knowledge of foreign countries possessed by the Chinese in various periods; this field, therefore, has been a favourite one. All such detail will have to be omitted.[a] What does seem really relevant to the present book is the development of scientific cartography in China, and here there is a tale to unfold, the neglect of which has greatly vitiated all the standard works on the history of geography in general which we have seen.

A summary of it has nevertheless been available for Western scholars for nearly half a century in the fundamental paper of Chavannes (10). What approximates to a monograph on the history of Chinese cartography has also been contributed by Herrmann (8), though it was buried in the reports of an expedition. In the Chinese language we owe a valuable introduction, mainly bibliographical in character, to Wang Yung (*1*), and besides this one may occasionally find shorter articles on the history of geography in China, such as that by Huang Ping-Wei.

Whoever sets out to write on the history of geography in China faces a quandary, however, for while it is indispensable to give the reader some appreciation of the immense mass of literature which Chinese scholars have produced on the subject, it is necessary to avoid the tedium of listing names of authors and books, some of which indeed have long been lost. Only a few examples can be given, but it should be understood, even when it is not expressly said, that they must often stand simply as representative of a whole class of works.

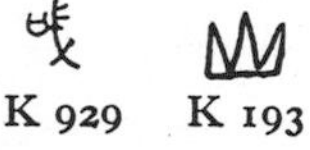

Many geographical symbols of great antiquity are embedded in the Chinese language. The character for river (*chhuan*[1]) is an ancient graph of flowing water, the character for mountain (*shan*[2]) was once an actual drawing of a mountain with three peaks (K193), and that for fields (*thien*[3]) shows enclosed and divided spaces. Political

[a] There is no Western-language treatise doing for geography what the book of Schlegel (5) did for Chinese astronomy in cataloguing the names and synonyms of individual stars and constellations.

1 川 2 山 3 田

boundaries are seen in the character for country (*kuo*[1]) where the frontier encloses the symbols for 'mouths' and 'dagger-axes', the eaters and the defenders (K929*c*).[a] Bone and bronze forms of the character which came to mean 'map' (*thu*[2]) actually show a map. Unfortunately, this word acquired a general signification covering any kind of diagram or drawing, so that in cases where a book disappeared at an early time it is not possible to be sure whether the *thu* which it was said to have had were really maps.[b] In any case it would not be far off the mark to guess that the pictographic character of Chinese encouraged the idea of mapping.[c]

As for the ideas about the shape of the earth current in ancient Chinese thought, mention has already often been made[d] of a prevailing belief that the heavens were round and the earth square (cf. Forke, 6). But there was always much scepticism about this. Thus in the *Ta Tai Li Chi* Tsêng Shen, replying to the questions of Shanchü Li, admits[e] that it was very hard to see how, on the orthodox view, the four corners of the earth could be properly covered (*ssu chio chih pu yen yeh*[3]). We saw also, in the astronomy Section,[f] that it was repeatedly stated (as by Yü Sung and Chang Hêng) in the +1st and +2nd centuries that the universe was like a hen's egg, and the earth was like the yolk in the midst of it.[g] Chinese thinkers of all ages joined Yü Hsi (*c.* +330)[h] in expressing scepticism about the square and flat earth: if it was square, said Li Yeh,[i] the movements of the heavens would be hindered (*chih ai*[4]). In his view, it was spherical, like the heavens, but smaller; and all supporters of the Hun Thien theory[j] must have tended to believe this.[k] The influence of these views on Chinese cartography, however, remained slight, for, as we shall see, it revolved around the basic plan of a quantitative rectangular grid, taking no account of the curvature of the earth's surface.[l] At the same time Chinese geography was always thoroughly naturalistic, as witness the passage about rivers and mountains already quoted from the *Lü Shih Chhun Chhiu*.[m]

[a] Cf. 'wapentake'.

[b] Cf. pp. 206 above, and 518, 536 below.

[c] Mention should be made somewhere here of the forgotten glossary of Chinese geographical technical terms by Stanley. A glossary of topographical terms such as are met with on Chinese maps was prepared by L. Giles (12).

[d] Pp. 211, 213, 220, 383 above.

[e] Ch. 58, tr. R. Wilhelm (6), p. 126, and given in full above, Vol. 2, p. 269.

[f] Pp. 211 and 217 above.

[g] Cf. Ko Hung: 'According to the commentary on (Chang Hêng's) *Hun Thien I* (Manual of the Armillary Sphere), the heavens resemble an egg, while the earth is like the yolk within the egg, and is situated alone in the heavens. The heavens are vast and the earth small' (*Chin Shu*, ch. 11, p. 3*a*). He would have been speaking about +300.

[h] See p. 220 above.

[i] *Ching-Chai Ku Chin Chu* (+13th century) cited in *Ko Chih Ku Wei*, ch. 4, p. 12*a*.

[j] Above, pp. 216 ff.

[k] In the +11th century one of the Chhêng brothers (I-Chhuan or Ming-Tao) wrote 'Below the earth there must be heaven (also). What we call earth is nothing but an object inside heaven. It has assembled like a mist, and just because over a long period it has not dispersed, it is considered the counterpart of heaven' (*Honan Chhêng shih I Shu*, ch. 2B, p. 5*a*). Cf. Yao Ming-Hui (*1*).

[l] We noted above how slow the Chinese were to take account of this in connection with the differing lengths of the gnomon's shadow (p. 225).

[m] Sect. 10*c* above in Vol. 2, p. 55 (R. Wilhelm (3), p. 112).

1 國 2 圖 3 四角之不揜也 4 窒礙

圓地總無方隅

地球在天中圓如彈丸海水附土爲氣所褁皆是圓形圓則無隅無方東極成西南觀成北如泛海者二舶俱從大洋一處開[illegible]一舶往東一舶往西俱可至中國元往東者從西面到元往西者從東面到理勢不

格致草 地圓 函宇通

Fig. 203. A page from the *Ko Chih Tshao* of Hsiung Ming-Yü (+1648), with a diagram to explain the sphericity of the earth. He speaks of it in the classical phrase of Han astronomers—'as round as a crossbow-bullet'. But Western influence is seen in the graduation of the sphere into 360° rather than $365\frac{1}{4}$°. Moreover, while one continent bears pagodas, and the circumnavigating ships are of Chinese rig, the antipodes are equipped with a large European building like a cathedral (photo. Hummel).

In a rare 17th-century treatise on astronomy and geography, the *Ko Chih Tshao*[1] of Hsiung Ming-Yü[2] (+1648), which has been discussed by Hummel (6), there is a picture (Fig. 203) showing that, as the accompanying text says, 'The Round Earth certainly has no Square Corners', and explaining that a ship could return to its port of origin after circumnavigating the earth. This could well be taken as evidence of how strange the idea of a spherical earth was to the decadent scholars of the Ming, if significantly Hsiung Ming-Yü had not taken care to use, in his explanation, the very same words which had been employed in ancient times by the Hun Thien theorists—'as round as a crossbow bullet' (*yuan ju tan wan*[3]).[a]

[a] See p. 217 above. The antiquity of this conception in China was emphasised by the Khang-Hsi emperor in his studies of +1711.

[1] 格致草 [2] 熊明遇 [3] 圓如彈丸

In this connection it would be interesting to make a study of the steps by which the expression *ti li*[1] (lit. 'earth pattern') came to have its present meaning of geography. This was certainly being used in the +1st and +2nd centuries as we use it now. Earlier there was no doubt a close connection with geomancy,[a] but the significance of the term *li*, about which so much has already been said,[b] and which fundamentally implied pattern and organisation as well as rational principle, will not escape us.

In what follows we shall have to compare rather carefully the parallel march of scientific geography in the West and in China. It may be said at the outset that both in East and West there seem to have been two separate traditions, one which we may call 'scientific, or quantitative, cartography', and one which we may call 'religious, or symbolic, cosmography'. The European tradition of scientific map-making was completely interrupted for centuries by a dominance of the latter, though originally it was older than the Chinese, but the parallel Chinese tradition, once it had begun, was not so interrupted. Before taking up these interesting comparisons, however, it is necessary to say something about the geographical classics and treatises of China through the centuries.

(*b*) GEOGRAPHICAL CLASSICS AND TREATISES

(1) Ancient Writings and Official Histories

Presumably the oldest Chinese geographical document which has come down to us is the Yü Kung[2] (Tribute of Yü) chapter of the *Shu Ching* (Historical Classic), which after having been granted any date back to the end of the −3rd millennium, is now considered to be probably −5th century, approximately contemporary with the pre-Socratic philosophers in Greece.[c] It will be remembered that Yü the Great was the legendary hero-emperor who 'mastered the waters' and became the patron of hydraulic engineers, irrigation experts and water-conservancy workers in after ages. The chapter is of great interest for many reasons; it lists the traditional nine provinces, their kinds of soils, their characteristic products, and the waterways running through them. It is thus important for the early history of soil science and hydraulic engineering, and constitutes a primitive economic geography.[d] The accepted view is that the part of China covered by the Yü Kung chapter included the lower valleys of the Yangtze and the Yellow Rivers, with the plain between them and the Shantung peninsula;[e] to the west the upper reaches of the Wei and Han rivers were known, together with the southern parts of the provinces of Shansi and Shensi. This was hardly the half of the region which Chinese civilisation was ultimately to occupy.

To give an idea of the text, we may read the account of the first two provinces.

[a] See Sect. 14*a* in Vol. 2, pp. 359ff. [b] See Sect. 18*f*. in Vol. 2, pp. 557ff.

[c] That is to say, ch. 6 is not one of the sixteen chapters which are regarded as certainly pre-Confucian nor is it among the seven more which are probably so.

[d] It has often been translated, as by Legge (1), Medhurst (1), Karlgren (12).

[e] Cf. Herrmann (8, 10).

1 地理 2 禹貢

Yü disposed the lands (in order). Going along the mountains, he put the forests to use, felling the trees. He determined the high mountains and the great rivers.

In Chi[1] (the first province)[a] he started work at Hu-khou, regulated the Liang and Chih (defiles), adjusted Thai-yuan and came to Yo-yang. At Than-huai he effected achievements and came to the Hêng and Chang rivers. The soil of this province is white and loose.[b] Its revenues are of the upper class, first grade; with an admixture (of lower class grades). Its fields are of the middle class, second grade. The Hêng and Wei rivers were made to follow their courses. Ta-lu (the Great Plain) was cultivated. The Niao-I barbarians had garments of skin (presumably as tribute). Yü closely followed to the right the Chieh-shih rocks, and arrived at the Ho (the Yellow River).

Between the Chi river and the Yellow River is the province of Yen[2] (the second province).[c] The nine branches of the Yellow River were led into their proper channels. The Lei-hsia district was made into a marsh, and the Yung and Chü rivers joined it. The mulberry grounds were stocked with silkworms, and the people descended from the hills and dwelt in the plains. The soil of this province is black and fat,[b] its grass is luxuriant and its trees are tall. Its fields are of the middle class, third grade. Its revenues were proportionate, after thirteen years' work on improvements had been done. Its tribute is lacquer and silk, and in the baskets presented there are textiles of various colours and patterns. Sailing on the Chi and the Tha rivers, Yü returned to the Ho (the Yellow River)....[d]

As has often been pointed out, this ancient inventory of the Chou 'empire' is essentially in terms of physical geography; the boundaries of the 'provinces' (*chiu chou*[3]) are natural, not political. It is also completely devoid of magic, and even of fantasy or legend, apart from the appearance of Yü himself. The majority of details concern the three 'inner' provinces (Chi,[1] Yü[4] and Yung[5]), while the six outer ones are less fully described. The words *tao shan*,[6] 'tracing out the mountains', seem to show that the idea of mountain-ranges, called *shan mo*[7] ('veins of mountains') by later scholars, was present in the mind of the writer.[e]

It is usual to hold[f] that the Yü Kung implies a naïve map of concentric squares (Fig. 204). This is based on the concluding sentences of the chapter, where it is said that throughout a zone 500 *li* (presumably in all directions) from the capital there are the 'royal domains' (*tien fu*[8]), within the next concentric zone of 500 *li* are the 'princes' domains' (*hou fu*[9]), then come the 'pacification zone' (*sui fu*[10]), the zone of allied barbarians (*yao fu*[11]), and lastly the zone of cultureless savagery (*huang fu*[12]). There is nothing in the text, however, to justify the traditional view that these zones were concentric squares; this was probably just assumed on the basis of the cosmological doctrine of the square earth. The point is more important than it may seem, for if the zones were thought of as concentric circles, this ancient gradient system might have

[a] Approximately the North China plain north of the Yellow River (Herrmann, 1).

[b] We must defer till Section 41 on agriculture the exact interpretation of the pedological terms used here.

[c] North-western Shantung (Herrmann, 1).

[d] Tr. Karlgren (12), p. 12; mod.

[e] Cf. Ong Wên-Hao (*1*).

[f] De Saussure (16*e*); Rey (1), vol. 1, p. 402; copied by J. O. Thomson (1), p. 43. Medhurst (1), p. 118, gives such a diagram.

1 冀 2 兗 3 九州 4 豫 5 雍 6 導山 7 山脈
8 甸服 9 侯服 10 綏服 11 要服 12 荒服

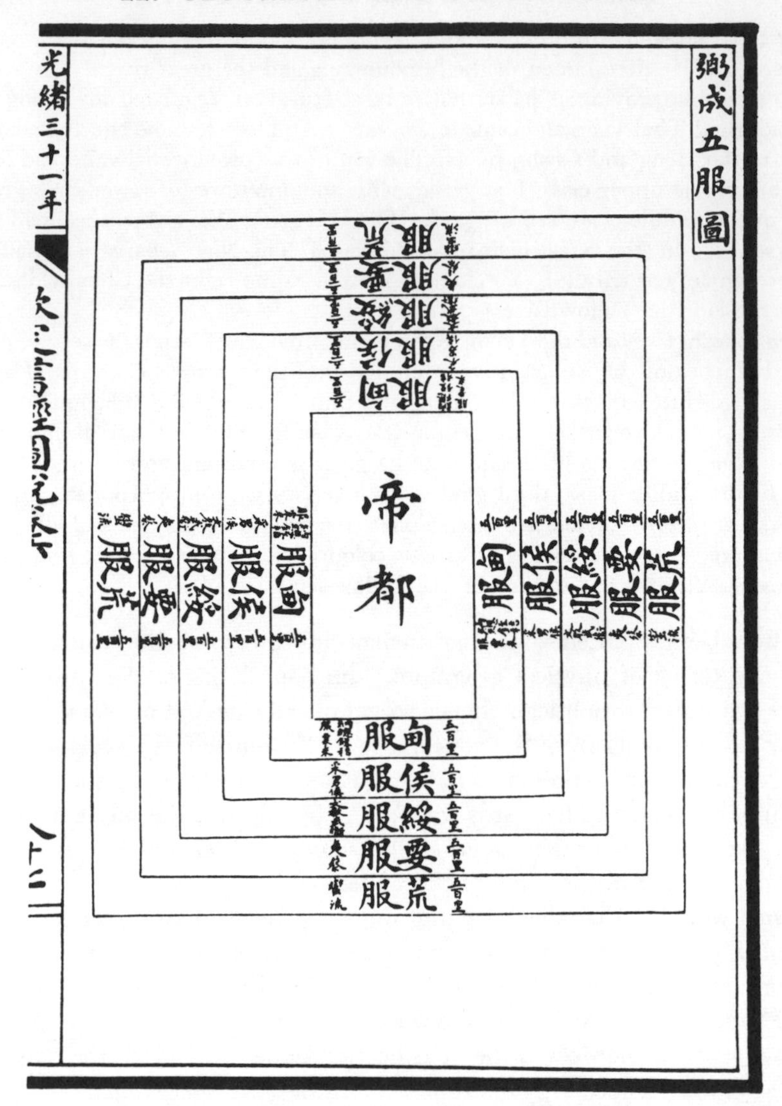

Fig. 204. The traditional conception of the radiation of ancient Chinese culture from its imperial centre. From *SCTS*, ch. 6, Yü Kung (Karlgren (12), p. 18). Proceeding outwards from the metropolitan area, we have, in concentric rectangles, (*a*) the royal domains, (*b*) the lands of the tributary feudal princes and lords, (*c*) the 'zone of pacification', i.e. the marches, where Chinese civilisation was in course of adoption, (*d*) the zone of allied barbarians, (*e*) the zone of cultureless savagery. The systematisation can never have been more than schematic but Egypt and Rome might have used a similar image, all unconscious of the equally civilised empire at the eastern end of the Old World.

been one of the sources of the East Asian discoidal tradition of 'religious cosmography' which we shall study in its place.[a] On the other hand concentric squares would foreshadow a rectangular grid.

In general, it may be said that the Yü Kung, the first naturalistic geographical survey in Chinese history, is approximately contemporary with the first map-making

[a] See p. 565 below.

in Europe. This is associated with Anaximander (*fl.* –6th century).[a] But the Chinese document is much more detailed and elaborate than anything which has come down to us from Anaximander's time. Throughout Chinese history the influence of the Yü Kung was enormous; all Chinese geographers worked under its aegis,[b] drew the titles of their books from it, and tried unceasingly to reconstruct the topography which it contained.[c]

Brief consideration must now be given to the curious subject of the Nine Cauldrons of the Hsia (*Hsia ting*[1]) and the book known as the Classic of the Mountains and Rivers (*Shan Hai Ching*[2]). The cauldrons were supposed to have been cast with pictures or maps of some kind illustrating the various places or regions of the country and the strange things found in them.[d] The book has generally been regarded as the text relating to these pictures. The *locus classicus* concerning them is in the *Tso Chuan*, and purports to relate to the year –605.

The prince of Chhu (State) attacked the barbarians of Lu-hun. In the course of this, he arrived at Lo and held a review of his troops on the (royal) territory of Chou. The High King (of Chou), Ting, sent Wangsun Man[3] to present his compliments to the prince. The latter (who had already conceived the design of usurping the overlordship) questioned the envoy as to the size and weight of the Nine Cauldrons (symbols of the sovereign authority). Wangsun Man replied 'The sovereignty depends upon the virtue of the king, not upon the cauldrons. Formerly, when the dynasty of the Hsia had attained the height of its greatness, the distant regions made pictures of the (material and spiritual) things natural to them (*yuan fang thu wu*[4]), and the governors of the nine provinces made tribute offerings of metal. With this (Yü the Great) caused cauldrons to be cast on which these pictures were represented (*chu ting hsiang wu*[5]). In this way the people were instructed so that they could recognise all things and spirits both good and evil (*shen chien*[6]). And thus when they travelled over the rivers and marshes, and through the mountains and forests, they did not meet with any adversities. (They did not go in fear of the weird spirits and genii of mountains and waters). Moreover, spirits such as the Chhih,[7] the Mei[8] and the Wang-liang[9] did not come to meet them (were not offended at their intrusions). Thus concord reigned between men and spirits, and the people received the favour of Heaven. But Chieh (the last High King of the Hsia) having fallen into viciousness, the cauldrons passed to the dynasty of the Shang, which kept them six hundred years. Chou[10] (the last High King of the Shang) having reached the bounds of cruel despotism, they passed to the dynasty of the Chou.[11] Whatever size the cauldrons may be, they are heavy if the High King has great virtue, and light if he is unjust and perverse. True virtue is protected by Heaven with constancy. When the High King Chhêng installed the cauldrons at Chia-ju, the oracle predicted that his House would reign for thirty generations and seven centuries. Such is the will of Heaven. It is true that the

[a] Bunbury (1), vol. 1, p. 122; K. Freeman (1), p. 63.
[b] Cf. below, pp. 514, 538, 540, 547 and 577.
[c] Inscribed steles dealing with the geographical and hydraulic work of Yü the Great, and dating from many different times, were set up (and still exist) on many famous mountains in China (cf. C. T. Gardner, 2).
[d] The European parallel here would appear to be the 'maps' which Aristagoras of Miletus caused to be cast on bronze tablets about –500 (Bunbury (1), vol. 1, p. 122).

1 夏鼎 2 山海經 3 王孫滿 4 遠方圖物 5 鑄鼎象物
6 神姦 7 螭 8 魅 9 罔兩 10 紂 11 周

virtue of the Chou is not what it was, but the mandate of Heaven has not changed, and the time has therefore not yet come to make enquiries about the weight of the cauldrons.'[a]

We are certainly here in the presence of an ancient tradition, perhaps magical and ritualistic rather than geographical. In an interesting and learned monograph Chiang Shao-Yuan has developed the view that the pictures on the urns or cauldrons were primarily those of spirits, analogous to the fauns, nymphs and nereids of Greek mythology, but more terrifying, which the ancient Chinese believed to haunt the wilds. These numina of localities would be what the text refers to as the *pai wu*,[1] the 'hundred beings', though the less familiar wild animals would also be included. Travelling officials or envoys would know what god should be honoured with sacrifice in what place.[b]

A similar atmosphere pervades the *Shan Hai Ching*, which, however, bears also a resemblance to the Yü Kung in that it often mentions the existence of quite reasonable minerals, plants and animals.[c] An elaborate table of these, together with the fabulous animals, plants and semi-human races and peoples, has been drawn up by Ho Kuan-Chou & Chêng Tê-Khun (*1*). These authors, with Maenchen-Helfen, also discuss the very difficult problem of the date of the book. It was certainly current in some form in the Former Han period (Ssuma Chhien refers to it),[d] and a good deal of the material, on internal evidence, goes back to the time (and probably the school) of Tsou Yen[e] (late −4th century). Some of the contents are likely to be much older even than that, for Wang Kuo-Wei (*2*) pointed out that one of the personages mentioned in the *Shan Hai Ching*, Wang Hai,[2] was already a god of some kind in the Shang period (−13th century) and appears as such on the oracle-bones.[f] On the other hand, the later chapters (6–18) may be of Later Han or even Chin date. As Wang Yung says, many of the topographic features mentioned in the book can be approximately identified, and it forms a veritable mine of information concerning ancient beliefs about natural things such as minerals and drugs.[g]

The chief discussion has centred round the fabulous beings and peoples described. Taking the view that the *Shan Hai Ching* is the oldest 'traveller's guide' in the world,[h] Schlegel attempted a number of naturalistic identifications—thus (7*b*) the *wên shen*

[a] Duke Hsüan, 3rd year, tr. Couvreur (1), vol. 1, p. 575; Chiang Shao-Yuan (1), p. 130; eng. auct., adjuv. Chavannes (10).

[b] Wang Chhung, in his usual manner, scoffs at the whole story (Forke (4), vol. 1, p. 505). But it had great veneration throughout Chinese history, and in +697, for example, urns were cast with representations of the provinces on them in imitation of those of Yü the Great (see Chavannes, 10). Again, in +1104 Wei Han-Chin,[3] a thaumaturgist from Szechuan, was entrusted with the casting of nine new urns by the tottering government of the Northern Sung (*Sung Shih*, ch. 462, p. 10*b*; for the context see Needham, Wang & Price).

[c] There is a translation, now old, of the first (mountain) section by de Rosny (1). A brief account of the contents of the whole book is given by Maspero (2), p. 611. Finsterbusch (1) has recently studied it with relation to Han art motifs.

[d] *Shih Chi*, ch. 123, p. 21*b*.

[e] Cf. above, Sect. 13*c* in Vol. 2, pp. 232ff.

[f] Cf. Chiang Shao-Yuan, p. 42 n.

[g] Tokarev (1) has identified the minerals mentioned in it.

[h] With which Chiang Shao-Yuan concurs, p. 273.

[1] 百物 [2] 王亥 [3] 魏漢津

kuo[1] were probably barbarian tribes of the Kuriles which practised tattooing, the *pai min kuo*[2] and *mao jen*[3] (hairy white people) were probably the Ainu (7*h*), the *yü i kuo*[4] must have been the 'malodorous barbarians' of the Siberian coast from whom the Chinese imported fish glue for bows in very early times (7*o*), and so on. He fortified his identifications, which are still of interest, by passages from many other ancient and medieval Chinese books. But a large proportion of the peoples mentioned are clearly fabulous—heads that fly about alone, winged men, dog-faced men, bodies with no heads, and the like. Since a great many of these appear also in Greek mythology, the problem of transmission at once presents itself. De Mély (2) collected from late encyclopaedias some seventy kinds of these fabulous beings (nearly all of which appear in the *Shan Hai Ching*),[a] and in all but very few cases could point to their analogues in Greek and Latin authors. Herodotus (−5th century) is one of the earliest sources, but there is much similar material in Strabo[b] and Pliny. It was assembled and concentrated by Gaius Julius Solinus in the +3rd century in his *Collectanea Rerum Memorabilium*, which was essentially a compilation of the 'nonsense' in Pliny, and which, with its title changed to *Polyhistor* in a +6th century revision, supplied abundant 'marvels' for geographers throughout the European Middle Ages.[c] It is interesting to compare a couple of illustrations from the *Shan Hai Ching* with parallels from Solinus (Fig. 205*a*, *b*, *c*, *d*).

The transmission question has been discussed by Wang I-Chung (*1*); Maspero (2); de Mély (2) and Laufer (9) among others. Occidental scholars have been strongly inclined to regard this Chinese body of mythical teratology as of Greek origin. In certain cases they may be right. The story of the battles of the pygmies with the cranes, which occurs in many ancient Greek authors, is first found in the *Wei Lüeh*[5] of Yü Huan,[6] a +3rd-century book (the time of Solinus).[d] But it is going too far to derive all the fabulous beings of the *Shan Hai Ching* from Greek sources; some of them may well go back in China beyond the time of Herodotus. Attempts, such as that of Wei Chü-Hsien (*3*), to trace them to Indian mythology, are not convincing either, yet it may well be that some primary Indian or Iranian (or even Mesopotamian) source may have radiated them in both directions (cf. Kennedy, 1). That Babylonian

[a] Among them may be mentioned: (*a*) flying heads; (*b*) acephali; (*c*) cynocephali; (*d*) winged men (cf. the winged genii often seen in Han tomb-reliefs); (*e*) men with holes through their chests; (*f*) sirens or unipeds (also frequent in Han tomb-reliefs; cf. Fig. 28 in Vol. 1, p. 164); (*g*) men with interlaced legs; (*h*) long legs; (*i*) long arms; (*j*) giants; (*k*) men with no bellies; (*l*) long ears; (*m*) pygmies at war with cranes; (*n*) steganopods or sciopods which use their feet to shade themselves from the sun; (*o*) tailed men; (*p*) centaurs; (*q*) one-armed men; (*r*) cyclops or monoculi; (*s*) beings with eyes or whole faces in their chests; (*t*) three-headed men; (*u*) three-bodied men; (*v*) worms with human heads; (*w*) shapeless golem with wings; (*x*) tiger-headed centaurs; (*y*) lions with human heads; (*z*) men with bifurcated tongues or limbs. Somewhere in the network of legends a place has to be found for the three-legged crow supposed to reside in the sun (p. 436 above); for the one-legged Khuei, the thunder-drum musician (Granet (1), p. 507); and other specialists.

[b] Who himself was very sceptical (II, i, 9).

[c] Lloyd Brown (1), p. 86; Bunbury (1), vol. 2, p. 675. Solinus mentions the Seres, but only repeats Pliny respecting them. If there was transmission it would have been long before his time.

[d] Cf. *TPYL*, ch. 796, p. 7*a*.

1 文身國 2 白民國 3 毛人 4 鬱夷國 5 魏略
6 魚豢

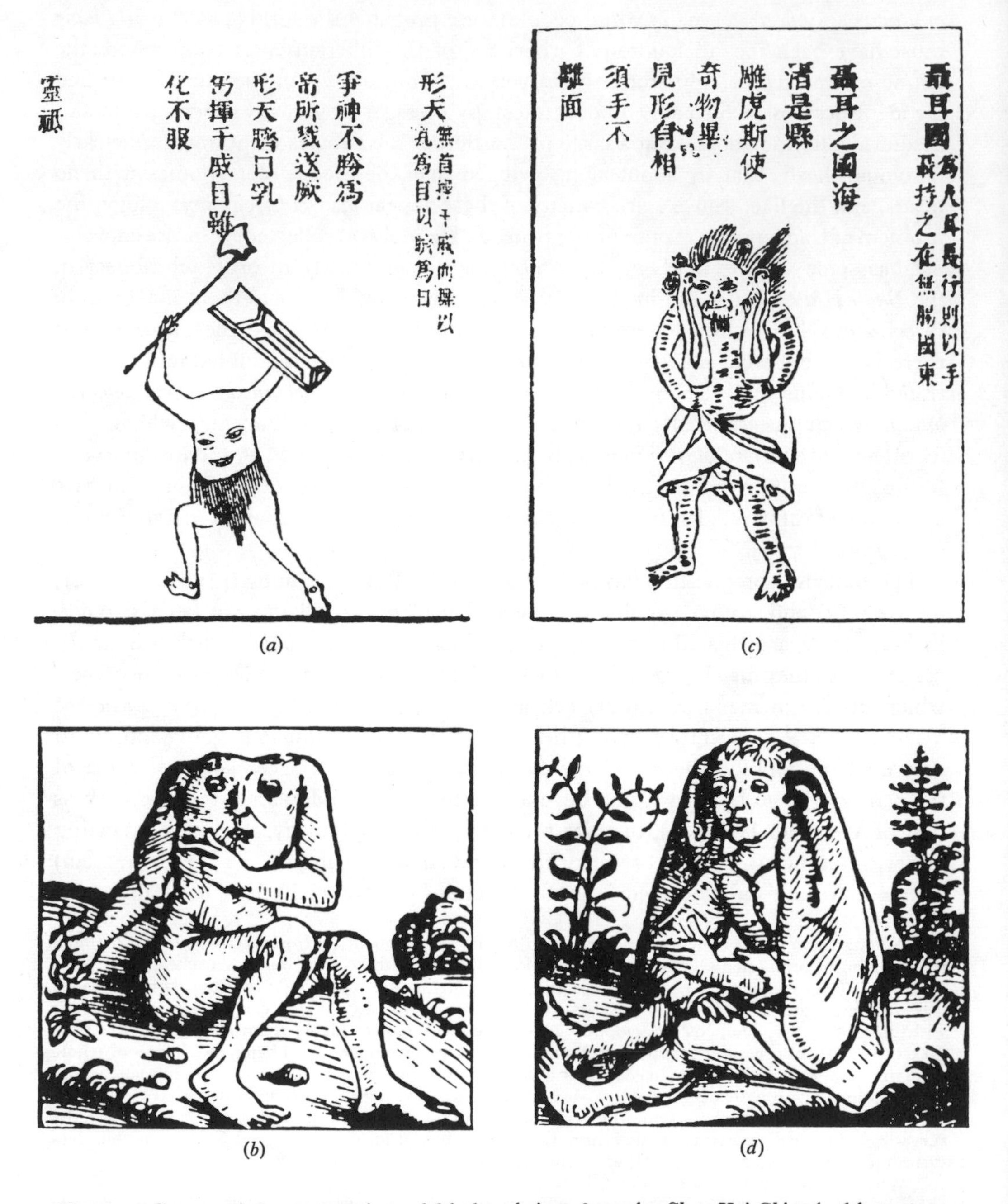

(a) (c)

(b) (d)

Fig. 205. Comparative representations of fabulous beings from the *Shan Hai Ching* (−6th to +1st centuries) and from the *Collectanea Rerum Memorabilium* of Solinus (+3rd century). *a*, *b*, acephali; *c*, *d*, long-eared men.

diviners were extremely interested in terata we know from the special works of Oppert (1), Lenormant (1, 2) and R. C. Thompson (1).[a] Maenchen-Helfen shows fairly convincingly that the passages of the book which describe paradise-like places[b] derive from earlier Indian legends. Wittkower identifies Ctesias (−4th century) and Megasthenes (−3rd) as the chief sources of transmission westwards.

What is curious is that the material has not yet been examined from the point of view of the history of biology. To an embryologist it seems obvious that most of the different abnormalities which form the basis of the corpus of legend could have been derived from human and animal monstrosities naturally occurring. Reference may be made to the book of C. J. S. Thompson on the history of teratology, and the classical experimental work of Dareste. So Cordier (4) traced the Eastern and Western legends of cynocephali to hirsute-faced human beings, citing living Burmese examples, as well as to cynocephalic monkeys. If this point of view should prevail, there would be no reason to assume any transmission at all, at any rate to account for origins. Such an interpretation would, however, not be inconsistent with the more social appraisal suggested by Sinor (1), in which the existence of these myths would be related to a kind of xenophobia present in all ancient peoples.

Another book of semi-legendary type which has some geographical interest is the *Mu Thien Tzu Chuan*[1] (Account of the Travels of the Emperor Mu). This was translated long ago by Eitel (1) and more recently by Chêng Tê-Khun (2). It was one of the documents discovered in +281 in the tomb of a ruler of the Wei[2] State, Hsiang Wang[3] or An Li Wang,[4] who had died in −245. There has thus always been some doubt about the authenticity of the book, but the modern view is that although it has nothing to do with the period in which this High King of the Chou is supposed to have lived (−10th century) it is doubtless a Warring States composition.[c] The book gives an account of three peregrinations of the king, including all kinds of festivities, hunts and side-trips. Rough identifications of the mountains visited and deserts and rivers crossed are possible (de Saussure, 26, 26*a*). But there is much legendary material in the book, especially the visit of the king to the fabulous imperial goddess of the west, Hsi Wang Mu.[5]

While a considerable legendary element still persists in the books of the Chhin and early Han time, such as the *Lü Shih Chhun Chhiu* and the *Huai Nan Tzu*, concrete geographical information is more and more taking its place.[d] Especially chapter 4 of the latter book (tr. Erkes, 1), which concerns the earth's surface, is of geographical interest. In this connection it is interesting that we are told that one of the accusations against Liu An, Prince of Huai Nan, was that he and his group of Taoist magician-

[a] Our very word 'monsters' comes from *monstrare*, to show 'because', as St Augustine said, 'they betoken somewhat' (*City of God*, II, 303).

[b] Cf. the Taoist versions (Sect. 10*i* in Vol. 2, p. 142).

[c] Cf. Chavannes (1), vol. 2, pp. 6 ff.; vol. 5, pp. 446 ff.

[d] The views of Tsou Yen in the late −4th century on the nine continents of the world, of which China formed only one, have already been the subject of discussion in Vol. 2 (Sect. 13*c*, pp. 233, 236). They remained well known in the Han, for Wang Chhung criticises them favourably towards the end of ch. 31 of the *Lun Hêng* (tr. Forke (4), vol. 1, p. 256).

1 穆天子傳 2 魏 3 襄王 4 安釐王 5 西王母

scientists had made use of maps (*ti thu*[1]).[a] In this his chief associate seems to have been the otherwise unknown Tso Wu.[2] But the story of Chinese cartography must be postponed for a few pages.

From the +1st century onwards, each of the official histories contains a geographical section (Ti Li Chih[3]), the whole forming an immense compilation concerning the changes in place-names and local administrative divisions controlled by the dynasty, descriptions of mountain ranges, river systems, taxable products etc. The first of these, that in the *Chhien Han Shu*,[b] is supposed to have been written by Liu Hsiang about −20 and incorporated later,[c] with additions by Chu Kan.[4] There is an index to it by Sargent (2). Very little of the geographical chapters in the official histories has ever been translated.[d]

We may now consider the other main classes of Chinese geographical literature as developed during the centuries. They may perhaps be summarised as follows:

(1) Anthropological geographies;
(2) Descriptions of the southern regions;
(3) Descriptions of foreign countries;
(4) Accounts of travels;
(5) Hydrographic books;
(6) Descriptions of the coast;
(7) Local topographies: (*a*) districts controlled from a walled city; (*b*) famous mountains; (*c*) cities and palaces;
(8) Geographical encyclopaedias.

(2) ANTHROPOLOGICAL GEOGRAPHIES

The human geographies developed rather naturally from the descriptions of fabulous peoples of the type of the *Shan Hai Ching*. The generic name for this literature, Chih Kung Thu[5] (Illustrations of the Tribute-Bearing Peoples), does not, however, appear until the Liang (about +550), when it occurs as the title of a book[e] prepared by Chiang Sêng-Pao[6] or Hsiao I.[7] But Chuko Liang is said to have written a *Thu Phu*[8] on the southern tribes already in the San Kuo period (+3rd century).[f] In the Thang and other dynasties it was customary for foreign tribute-bearers to attend at an office known as the Hung Lu,[9] where the officials took notes of the geography and customs of their countries, and committed these to writing as part of an ever-growing body of government intelligence[g] (Fig. 206). In due course this gave rise to numerous books.

[a] *Chhien Han Shu*, ch. 44, p. 11*a*; *Chih Lin Hsin Shu*[10] (by Yü Hsi the astronomer), in *YHSF*, ch. 68, p. 38*b*.
[b] Ch. 28B.
[c] Maspero (16).
[d] But there are some partial translations of the separate chapters on foreign countries.
[e] *Hsüeh Chai Chan Pi*, ch. 2, p. 9*b*. The locations and customs of more than thirty peoples were said to have been described and illustrated. Hsiao I was the personal name of the Emperor Liang Yuan-Ti.
[f] *Hua Yang Kuo Chih*, ch. 4.
[g] *Thang Hui Yao*, ch. 63 (pp. 1089 ff.).

1 地圖 2 左吳 3 地理志 4 朱贛 5 職貢圖
6 江僧寶 7 蕭繹 8 圖譜 9 鴻臚 10 志林新書

Fig. 206. A late Chhing representation of the attendance of barbarian envoys presenting tribute at the Hung Lu Department where details of their countries and products were recorded. From *SCTS*, ch. 25, Li Ao (Medhurst (1), p. 209).

Thus in +628 there was the *Wang Hui Thu*[1] (Illustrations of the Princely Assembly)[a] by Yen Li-Pên,[2] who also produced a *Hsi Yü Chu Kuo Fêng Wu Thu*[3] (Illustrated Account of the Strange Customs and Products of the Western Countries). Another famous work of the kind was the *Ka-Hsia-Ssu Chhao Kung Thu Chuan*[4] (Illustrated Account of the Kirghiz) by Lü Shu[5] about +844. In the Sung the genre was continued in several notable books, for example, the *Hua I Lieh Kuo Ju Kung Thu*[6] (Illustrated Enumeration of the Tribute-Bearing Regions, both Chinese and Barbarian) by Tshui Hsia.[7] Few of them now remain, but there is the *Huang Chhing Chih Kung Thu*[8] (Chhing Dynasty Illustrated Records of Tributary Peoples) by Fu Hêng,[9] one of Chhien-Lung's generals who campaigned in Sinkiang and Burma.

(3) Descriptions of Southern Regions and Foreign Countries

There were no very sharp lines of distinction between this ethnological geography and what might be described as ethnology and folklore proper, since there were in all centuries, and still are, large enclaves of territory occupied by tribal peoples only very partially affected by Chinese culture—the Miao, the Yao, the Yi, the Chia-jung, etc. Moreover, as Chinese civilisation expanded to the south, more and more descriptions of the strange things, as well as the topography, of the southern regions, were written. Works on folk customs and their geographical distribution were termed Fêng Thu Chi,[10] and descriptions of unfamiliar regions I Wu Chih.[11] Of the former, the oldest are the *Chi-chou Fêng Thu Chi*[12] (Customs of the Province of Chi)[b] by Lu Chih[13] about +150, and the *Fêng Su Thung I* of Ying Shao about +175, already mentioned.[c] Chu Kan (whom we have just met) also wrote one, but it has not survived. The customs of the tribal peoples aroused a continuing interest on the part of Chinese scholars, and in the +17th-century Chhen Ting[14] gave an account of them, with a critical analysis of previous works on the same subject, in his *Thung Chhi Hsien Chih*[15] (Brief Notes from Thung-chhi).[d] Reference has already been made[e] to the important early descriptive books about the southern regions—Yang Fu's +2nd-century *Nan I I Wu Chih* (Strange Things from the Southern Borders), and Wan Chhen's +4th-century *Nan Chou I Wu Chih* (Strange Things of the South). Sometimes these works were devoted entirely to the plants and animals; we shall therefore meet with them again in the Sections on the biological sciences. In the Sung the distinguished scholar Fan Chhêng-Ta[16] wrote an important book on the topography and products of the southern provinces, *Kuei Hai Yü Hêng Chih*.[17]

[a] See *Hsüeh Chai Chan Pi*, ch. 2, p. 9*b*. Yen Li-Pên is better known as a painter.

[b] The first province mentioned in the Yü Kung inventory, see above, p. 501.

[c] Vol. 2, p. 152.

[d] Manuscripts with coloured illustrations of the tribal peoples dating from the 18th century and later are not uncommon; there is one in the Wellcome Medical Library. This literature has been reviewed by Jäger (3).

[e] Vol. 1, p. 118.

1 王會圖 2 閻立本 3 西域諸國風物圖 4 戞黠斯朝貢圖傳
5 呂述 6 華夷列國入貢圖 7 崔峽 8 皇淸職貢圖
9 傅恆 10 風土記 11 異物志 12 冀州風土記 13 盧植
14 陳鼎 15 峝谿纖志 16 范成大 17 桂海虞衡志

An enormous literature still exists, after all losses, on the geography of foreign countries. Among its earliest representatives was the *thu shu*[1] (illustrated account) of the Huns mentioned as having been presented in −35.[a] But it is unlikely that Chang Chhien in the −2nd century had failed to provide some kind of charts and illustrations of the peoples whom he had visited in the West. In the +2nd century Tsang Min[2] produced an elaborate memorandum dealing with fifty-five countries.[b] Between the Han and the Thang some twenty books of importance are listed by Hsiang Ta (*1*). There was, for example, the *Wu Shih Wai Kuo Chuan*[3] (Account of Foreign Countries in the time of the Wu Kingdom) by Khang Thai[4] about +260. All this was greatly intensified in the Thang period when intercourse with other countries, especially India, was so widespread. In an earlier Section[c] the works of the Buddhist pilgrims, largely geographical, have already been described, for example, the *Fo Kuo Chi* of Fa-Hsien in the +5th century, and the *Ta Thang Hsi Yü Chi* of Hsüan-Chuang in the +7th. But religious pilgrims were not the only men who wrote of foreign countries; as time went on ambassadors and envoys made a great contribution to this literature. The reader may remember Wang Hsüan-Tshê, the bold Chinese representative at the court of Magadha in the +7th century, who took back to China the Indian alchemist apparently able to make mineral acids.[d] Now Wang Hsüan-Tshê himself wrote a book, *Chung Thien-Chu Kuo Thu*[5] (Illustrated Account of Central India), which only remains in the form of fragments. In the Sung this source of travel opportunities led to a number of books of great importance. In +1124 Hsü Ching, an official in the train of a Chinese ambassador to Korea, wrote an account of the voyage and the country, the *Hsüan-Ho Fêng Shih Kao-Li Thu Ching* (Illustrated Record of an Embassy to Korea in the Hsüan-Ho reign-period).[e] Similarly, in +1297 a description of Cambodia, the *Chen-La Fêng Thu Chi*,[6] was given by Chou Ta-Kuan,[7] a counsellor of the Chinese ambassador there.[f] This was in the Yuan, but interest was also directed towards the North at the time, very naturally, and we find, for instance, a Mongol geographer, Na-Hsin,[8] producing an archaeological topography of all regions known to the Chinese north of the Yellow River, the *Ho Shuo Fang Ku Chi*.[9]

The climax of this literature was reached in the Ming, with the +15th-century voyages of Chêng Ho, the famous admiral,[g] which gave rise to a number of books; but

[a] *Chhien Han Shu*, ch. 9, p. 12*b* (Dubs (2), vol. 2, p. 332). There is more in this than meets the eye, and we shall shortly discuss it further (p. 536).

[b] *Hou Han Shu*, ch. 88, p. 16*b*. It dealt mostly with central and south Asia.

[c] That on Contacts, Sect. 7*i* in Vol. 1, pp. 207ff.

[d] Vol. 1, pp. 211ff. Cf. Lévi (1).

[e] We have met with this before (pp. 471, 492) and shall do so again (Sects. 26*i*, 29) in connection with the history of the magnetic compass, navigation, etc. The book originally included maps, afterwards lost.

[f] Here may be mentioned the work of Pelliot (16) on the Chinese knowledge of Cambodia (Fu-Nan[10]) as an example of the close investigations which sinologists have made of certain aspects of Chinese geographical learning. The translations by Pelliot (9, 33) of the *Chen-La Fêng Thu Chi* replace the older one of Rémusat (3). In Chinese there is a monograph by Fêng Chhêng-Chün on the relations of China with the South Seas.

[g] Cf. above, Vol. 1, p. 143, and below, pp. 557ff.

1 圖書　2 臧旻　3 吳時外國傳　4 康泰　5 中天竺國圖
6 眞臘風土記　7 周達觀　8 納新　9 河朔訪古記　10 扶南

we shall defer mention of these until a little later on account of their close connection with the history of cartography. Two earlier books, however, cannot be omitted. One is the *Ling Wai Tai Ta*[1] (Information on what is Beyond the Passes), by Chou Chhü-Fei,[2] written in +1178, and dealing with the geography of many Asian kingdoms, even far to the west. The other is the *Chu Fan Chih*[3] (Records of Foreign Peoples), by Chao Ju-Kua[4] in +1225, of which there is a famous translation by Hirth & Rockhill.[a] Mention of it reminds us of yet another source of information about foreign countries, namely, trade, for Chao's work was mainly inspired by contacts with Arab merchants who came and went with their goods in the Chinese ports. Four centuries later, the same stimulus was still effective, as we see from Chang Hsieh's[5] *Tung Hsi Yang Khao*[6] (Studies on the Oceans East and West), in which the geography of thirty-eight kingdoms, mostly islands in south-east Asia, is described (+1618). Military expeditions also gave rise to geographical literature, as in the case of the *Chiao Li Chiao Phing Shih Lüeh*[7] (Materials on the Pacificatory Expedition to Annam during the Li Dynasty) of +1551. Finally, mention may be made of the *I Yü Thu Chih*[8] of about +1430, earlier in the same dynasty (Illustrated Record of Strange Countries), to which sinologists have given much attention, partly on account of its rarity.[b] The book is anonymous, but there are indications of authorship which point to the Ming prince Ning Hsien Wang (Chu Chhüan[9]) whom we often meet in other connections.[c] Hummel (8) has sketched the state of Chinese geographical knowledge of foreign countries at the time when the Jesuits arrived, referring to the *Hsien Pin Lu*[10] (Record of All the Guests) compiled by Lo Yüeh-Chhiung[11] in +1590.

How does this general picture compare with the development of descriptive geography in the West? The Chinese had nothing of the quality of Herodotus or even of Strabo at times contemporary with them, but during the gap between the +3rd and the +13th centuries, when European learning sank so low, the Chinese were far more advanced, and steadily progressing. The floor was held in Europe by Solinus and his myths, almost as if the *Shan Hai Ching* had continued to dominate in China without competition from diplomats like Khang Thai, pilgrims like Fa-Hsien, ethnographers like Ying Shao, and trade superintendents like Chao Ju-Kua. In the Thang period, almost the only reasonable representative that the West could produce was the Syrian bishop Jacob of Edessa (+633 to +708).[d] As Sarton points out, however,[e] the Arabs match up better. By the Sung, about +950, they were laying the foundations of later Western geography, with al-Ya'qūbī, Ibn Khurdādhbih, al-Ma'sūdī, Ibn al-Faqīh,

[a] Others also tried their hand, e.g. de Rosny (4).

[b] Moule (4); Hummel (7); see also Sarton (1), vol. 3, p. 1627. The only copy known is in the Cambridge University Library (see Fig. 207*a,b*). It was, however, reprinted in the *Wan Yung Chêng Tsung Pu-Chhiu-Jen Chhüan Pien*[12] (The 'Ask No Questions' Complete Handbook for General Use), an encyclopaedia issued by Yü Wên-Thai[13] in +1609.

[c] Vol. 1, p. 147.

[d] A pupil of our old friend Severus Sebokht (Vol. 1, p. 220). Cf. Sarton (1), vol. 1, p. 500.

[e] (1), vol. 2, p. 41.

1 嶺外代答 2 周去非 3 諸蕃志 4 趙汝适 5 張燮
6 東西洋考 7 交黎剿平事略 8 異域圖志 9 朱權
10 咸賓錄 11 羅曰褧 12 萬用正宗不求人全編 13 余文台

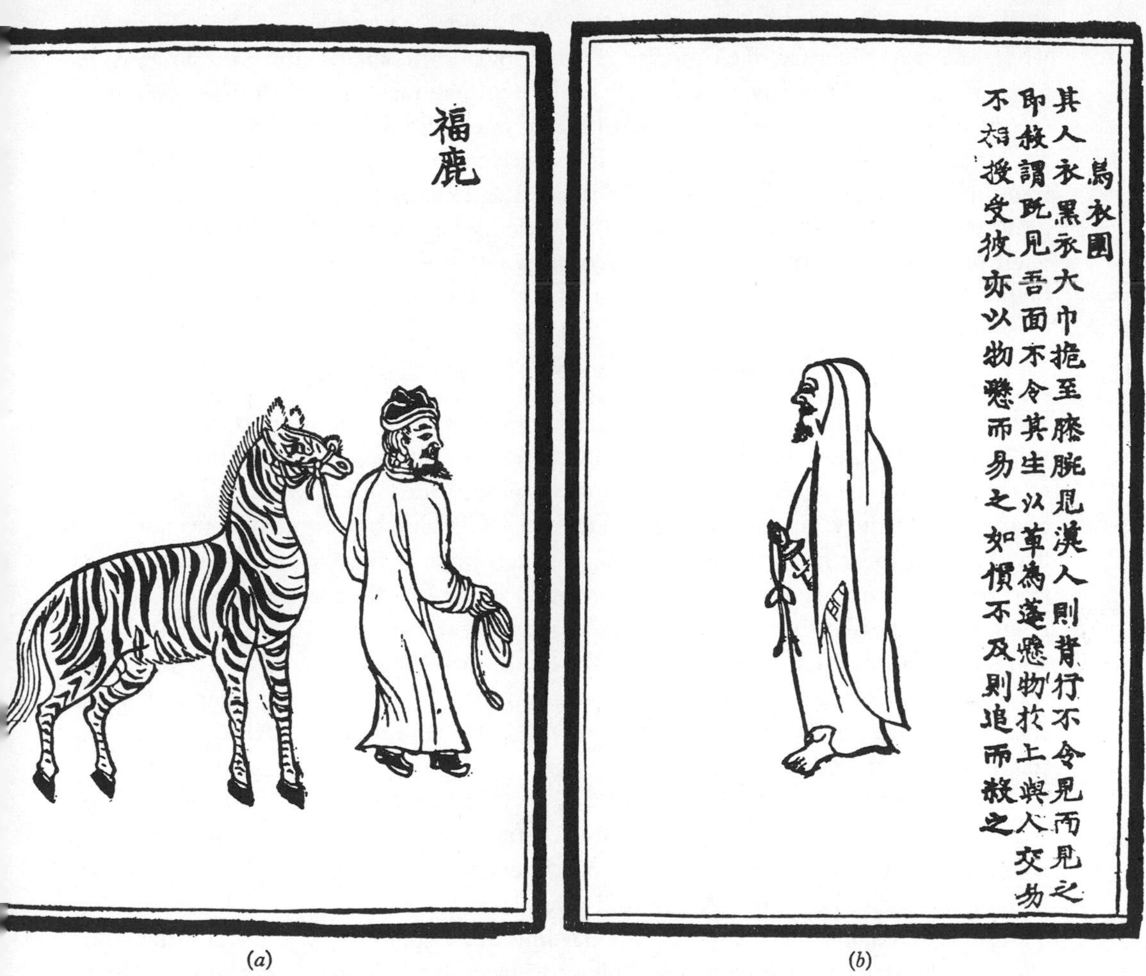

(*a*) (*b*)

Fig. 207. Two pages from the *I Yü Thu Chih* (Illustrated Record of Strange Countries) of *c.* +1430, perhaps written by the alchemist, mineralogist and botanist Chu Chhüan, a prince of the Ming (Ning Hsien Wang), and almost certainly profiting from the zoological and anthropological knowledge gained in the expeditions of Chêng Ho.

(*a*) A zebra (*fu lu*).

(*b*) An inhabitant of the country of Black-clothed People, no doubt some outpost of Arab culture. The description says that they hide their faces from Chinese visitors, and anyone who sees them is killed. Bargainers in trade are separated by a curtain, but one has to be careful for if the native merchants are dissatisfied with the deal one is likely to be pursued and slain.

Photographs from the unique copy in the Cambridge University Library.

al-Iṣṭakhrī and Ibn Ḥauqal. Arabic geography reached its climax with al-Idrīsī in the +12th century but still yields many good names in the +13th.[a] Of course, the West had had its pilgrim literature, analogous to that of the Chinese Buddhists, beginning with 'the first of the Christian guide-books',[b] the *Itinerary from Bordeaux to Jerusalem* of +333; and its records of trading voyages, such as the *Christian Topography* of Cosmas

[a] See Mieli (1), pp. 79, 114, 158, 198, 210, 301, and the collection of de Goeje (1).

[b] Beazley (1), vol. 1, pp. 26, 57. This genre long continued, as in the itinerary to the Holy Land written by the Irish monk Symon Simeonis just a thousand years later (Sarton (1), vol. 3, p. 787).

Indicopleustes written about +540, when the Liang were in power at Nanking.[a] But when one reads the careful chronicles of the Renaissance, such as the *True Story of the Conquest of New Spain* by Bernal Diaz del Castillo about 1520, or the *Relación de las Cosas de Yucatán* of Diego de Landa (1566), one feels that the West was only now beginning to follow a path of objective description which the Chinese had been treading for the previous millennium and a half. Already once we have met with this pattern, when examining the growth of knowledge of the tides,[b] and it will manifest itself again throughout the present Section, with especial clarity in the cartographic field.

(4) Hydrographic Books and Descriptions of the Coast

We may pass to the next class of writings. In view of the great importance of waterways for the Chinese social and economic system at all times, it was natural that close attention should be paid to them. The first treatise of the kind was that of Sang Chhin[1] of the −1st century, the *Shui Ching*[2] (Waterways Classic), but the text as we now have it is thought to be from the hand of some geographer of the San Kuo period, at any rate before +265. It gives a brief description of no less than 137 rivers. About the beginning of the +6th century it was enlarged to nearly forty times its original size by a great geographer Li Tao-Yuan[3] and given the title *Shui Ching Chu*[4] (The Waterways Classic Commented).[c] This constitutes a work of the first importance.[d] From the titles of several other books (*Chiang Thu*[5]), it would seem that rivers were being mapped from the Chin onwards.

Among treatises of this kind in the Sung may be mentioned the *Wu Chung Shui Li Shu*[6] (The Water-Conservancy of the Wu District) by Shan O[7] (+1059). Shan spent more than thirty years exploring the lakes, rivers and canals in the region of Suchow, Chhangchow and Huchow. A hundred years later Fu Yin[8] wrote the *Yü Kung Shuo Tuan*[9] (Discussions and Conclusions regarding the Geography of the Tribute of Yü), in which he dealt mainly with the Yellow River valley. Some diagrammatic charts, presumably of the +12th century, are still included in his book (Fig. 208).

In the Chhing dynasty much larger works were produced. The *Hsing Shui Chin Chien*[10] (Golden Mirror of the Flowing Waters)[e] is owing to Fu Tsê-Hung[11] and appeared in +1725. It includes many panoramic maps showing the locations of rivers and lakes (Fig. 209). Then there was the *Shui Tao Thi Kang*[12] (Complete

[a] Beazley (1), vol. 1, pp. 41, 273.

[b] P. 493 above. Cf. p. 101.

[c] Hu Shih (5) has written on the involved later history of the text.

[d] It deals incidentally with certain foreign countries, e.g. northern India, but this is not its best side (Petech, 1). One has to wait until +1535 for anything like it in Europe, if indeed the book of Corrozet & Champier may sustain a comparison.

[e] Its size was more than doubled by its continuation, the *Hsü Hsing Shui Chin Chien*, prepared by Lei Shih-Hsü[13] and Yü Chêng-Hsieh[14] before 1832. All these men were high officials in charge of river conservancy.

[1] 桑欽 [2] 水經 [3] 酈道元 [4] 水經注 [5] 江圖
[6] 吳中水利書 [7] 單鍔 [8] 傅寅 [9] 禹貢說斷 [10] 行水金鑑
[11] 傅澤洪 [12] 水道提綱 [13] 黎世序 [14] 俞正燮

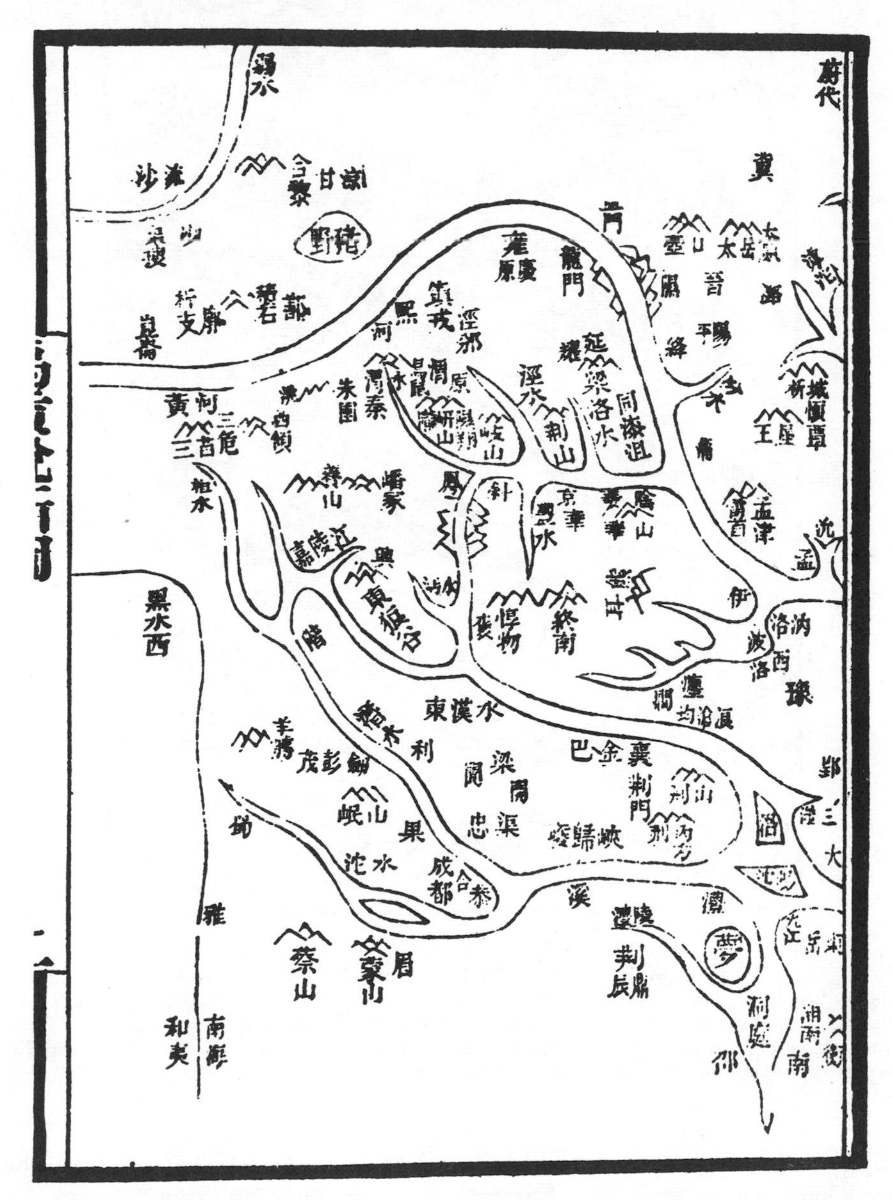

Fig. 208. A diagrammatic chart of the river systems of West China, from Fu Yin's *Yü Kung Shuo Tuan* (Discussions and Conclusions regarding the Tribute of Yü) of *c.* + 1160. The great bend of the Yellow River round the Ordos Desert and its passage through the Lung Mên gorges will be seen at the top, while lower down the old road and canal connecting the Wei River with the Han River (i.e. the Yellow and Yangtze River systems) through the Chhinling mountains can be made out. In Szechuan, Chhêng-tu and Mt Omei are marked.

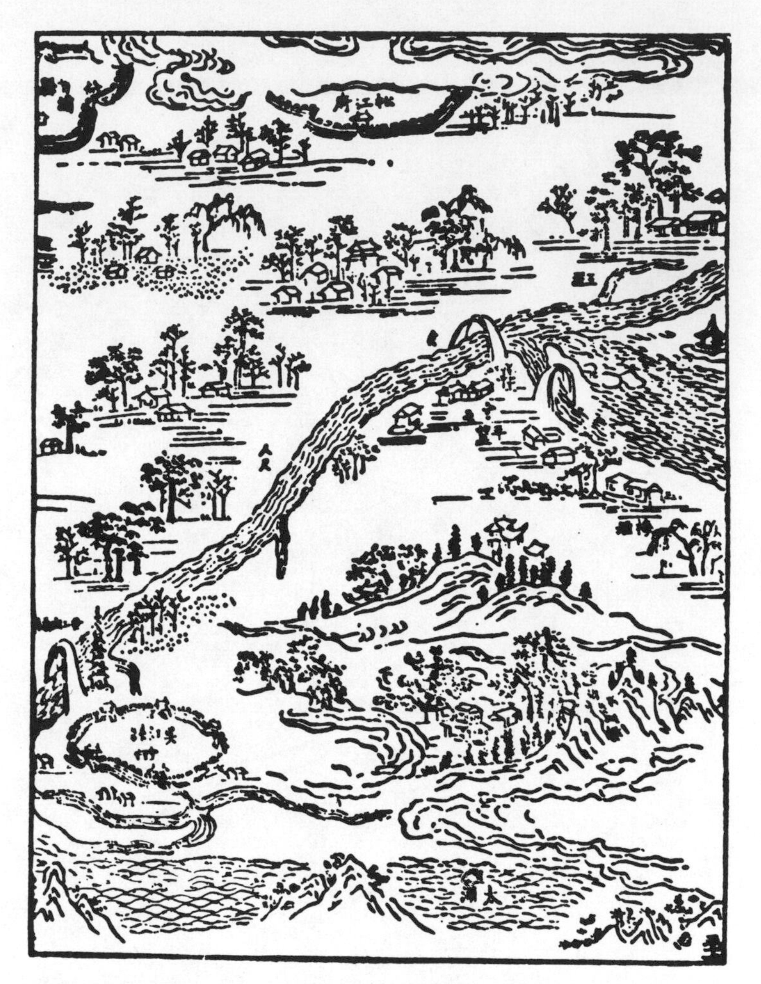

Fig. 209. A panoramic map from Fu Tsê-Hung's *Hsing Shui Chin Chien* (Golden Mirror of the Flowing Waters) of +1725. The lake in the foreground is Lake Thaihu in Chiangsu and the round walled city to the left is Wuchiang; the vista is thus looking east. The Grand Canal is shown crossing the panorama towards Hangchow past the village of Wang-ching. Beside the city of Wuchiang the famous 'Drooping Rainbow Bridge' (Chhui Hung Chhiao) is marked, and two other bridges are shown. A reproduction such as this cannot do justice to the delicacy of the original.

Description of Waterways)[a] by Chhi Shao-Nan[1] in +1776. There seems to be no class of geographical literature in Europe quite corresponding to this.

Allied to the group of books on the greater and lesser waterways is another on the geography of the Chinese coast. But most books on this subject are late. In +1562 Chêng Jo-Tsêng[2] published a large work on it, the *Chhou Hai Thu Pien*[3] (Illustrated

[a] There is no sharp line of distinction between these works on river geography and works on river conservancy as a technique, which we shall study in Sect. 28*f* below.

[1] 齊召南 [2] 鄭若曾 [3] 籌海圖編

Seaboard Strategy),[a] with maps 'in the rudest style of art'.[b] The stimulus had been the marauding raids of Japanese pirates which were a great affliction to all the maritime provinces during this century. But coastal protection against encroachment of the sea also gave rise to a special literature, as an example of which may be mentioned the *Chhih Hsiu Liang Chê Hai-Thang Thung Chih*[1] (Historical Account of the Coastal Protection Works of Chekiang Province, prepared by Imperial Command) written by Fang Kuan-Chhêng[2] in +1751.

One of the finest monuments of this period is a silk scroll which has been carefully studied by Mills (8). It derives from the same origins as the series of maps of the Chinese coast found in Chhen Lun-Chhiung's[3] *Hai Kuo Wên Chien Lu*[4] (Record of Things Seen and Heard about the Coastal Regions) of +1744. Wylie points out[c] that Chhen's father was one of those concerned with the subjugation of Formosa at this time, and that he collected his information among the sailors into whose company he was consequently thrown. Scroll maps of this character continued to be produced for well over a hundred years afterwards, and Mills (7) has recently described two more which have been acquired by the British Museum.

(5) Local Topographies

The series of local topographical writings, wrote Wylie (1), are probably unrivalled by any nation for extent and systematic comprehensiveness. Anyone at all acquainted with Chinese literature is familiar with the host of 'gazetteers', as they came to be called, which are really local geographies and histories (termed in general *fang chih*,[5] if of provinces *thung chih*,[6] and of smaller districts *fu chih*,[7] *chou chih*[8] or *hsien chih*[9]). In other literatures there is little comparable to this forest of monuments which the industry of provincial scholars erected over the centuries.[d] The earliest of the genus, it might be considered, is the *Yüeh Chüeh Shu*[10] (Book of the Former State of Yüeh), by an unknown author about +52. But the work which is generally regarded as the beginning of the story is the *Hua Yang Kuo Chih*[11] (Historical Geography of Szechuan),[e] written by Chhang Chhü[12] in +347. In this there is much about rivers, trade-routes, and the various tribes.[f] Szechuan had also been subjected to mapping, for in the book of Chhang we read of a *Pa Chün Thu Ching*[13] (Map of Szechuan) apparently made in the Later Han about +150.

[a] This geographer left many valuable works in MSS., ten of which were published in 1932 with the title *Chêng Khai-Yang Tsa Chu*.[14] In Fig. 210 we reproduce a portion of a Ming map of the Chekiang coast, the photograph of which we owe to the kindness of Mr R. Alley.
[b] Wylie (1); Hummel (12).
[c] (1), p. 48.
[d] Cf. Hummel (9).
[e] Literally 'Record of the Country south of Mt Hua'.
[f] There is a description by S. H. Fêng (1), but the book merits translation in full.

[1] 勅修兩浙海塘通志 [2] 方觀承 [3] 陳倫炯 [4] 海國聞見錄
[5] 方志 [6] 通志 [7] 府志 [8] 州志 [9] 縣志 [10] 越絕書
[11] 華陽國志 [12] 常璩 [13] 巴郡圖經 [14] 鄭開陽雜著

Such local map-making, however, was slow to spread, for this kind of title does not appear much until the bibliography of the *Sui Shu* (late +6th century) when suddenly many are mentioned.[a] With the growth of a stable bureaucracy, in which men were generally sent to serve in places far from their homes, the local topographies acquired social importance. About +610 the emperor ordered officials all over the country to compose records of customs and products, illustrated by maps or diagrams, and to present these to the imperial secretariat. Hence arose the *Chu Chün Thu Su Wu Chhan Chi*[1] (Record of the Local Customs and Products of all the Provinces), which apparently included maps, though not certainly, since all was later lost. The Sung dynasty continued energetically this compilation and collection of local records. Not long after it had come to power, in +971, the emperor ordered Lu To-Sun[2] to 're-write all the Thu Ching in the world', and this official, pursuing his colossal task, travelled through the provinces collecting all the relevant available texts. They were then worked together by Sung Chun,[3] and by +1010 no less than 1566 chapters were finished. Some kind of cartographic survey was clearly involved in these activities, for the *Sung Shih* says:

Yuan Hsieh[4] (*d.* +1220) was Director-General of government grain stores. In pursuance of his schemes for the relief of famines he issued orders that each *pao*[5] (village) should prepare a map which would show the fields and mountains, the rivers and the roads in fullest detail. The maps of all the *pao* were joined together to make a map of the *tu*[6] (larger district), and these in turn were joined with others to make a map of the *hsiang*[7] and the *hsien*[8] (still larger districts). If there was any trouble about the collection of taxes or the distribution of grain, or if the question of chasing robbers and bandits arose, the provincial officials could readily carry out their duties by the aid of the maps.[b]

Chu Shih-Chia (*1*) and other scholars have shown that less than 10 per cent of the Sung local topographies date from the earlier or northern period of the dynasty. But, on the other hand, though map-making was practised all through the Sung, after the move to the south, geography apparently became less literary-descriptive and more practical and cartographic. As to the kind of material which was collected by the officials, one might mention the interesting *Sha-Chou Thu Ching*[9] and *Tun-huang Lu*[10] from the extreme north-west of the empire, dating from +886, studied and translated by L. Giles (8, 9), while from the south-east there was Fan Chhêng-Ta's *Wu Chün Chih*[11] of about +1185.[c] In another connection[d] we have already referred to two famous Sung topographies of the Hangchow district in the +12th and +13th centuries. A typical topography such as the *Chiang-yin Hsien Chih*[12] goes back to +1194.

[a] Geographers of importance in this movement were Lang Wei-Chih,[13] Yü Shih-Chi[14] and Hsü Shan-Hsin.[15]

[b] Ch. 400, p. 8*a*, tr. auct.

[c] Not printed till +1229.

[d] The discussion of the tides above, pp. 491, 492. Cf. Chu Shih-Chia (*2*).

1 諸郡土俗物產記 2 盧多遜 3 宋準 4 袁燮 5 保
6 都 7 鄉 8 縣 9 沙州圖經 10 燉煌錄 11 吳郡志
12 江陰縣志 13 郎蔚之 14 虞世基 15 許善心

ig. 210. Part of a late Ming panoramic map of the Fukien coast, looking eastwards, about lat. 27° N.; one of a set of photographs
 Mr R. Alley. The first large walled city on the left is Fên Shui Kuan on the Chekiang border, then comes the Phu-mên inlet and
e fort of Thung Shan Ying. The large walled city on the right is Fu-ning (mod. Hsia-phu); as this ranked as a *chou* only in the
te Ming period the scroll is approximately dated. Out to sea on the right is the island of Ta Yü Shan with its profile correctly
awn (unlike the two small islands on the left which have been turned at right angles). This is probably the kind of coastal map
hich was made and used by the geographers of the late +16th century, the time of Chêng Jo-Tsêng (see Mills, 8).

Wu Chhi-Chhang (*2*) estimates that even before the end of the Sung the number of titles amounted to 220. The quantity of these books then increased from the Ming onwards, till today there is hardly a town, however small, which does not have its own historical geography. Than Chhi-Hsiang's catalogue of the local topographies in the Peking National Library lists 5514, and registers of other private and public collections by Chu Shih-Chia (*3*, *4*) give 8771.[a] In earlier times they were not considered a very meritorious type of literature, but rather as 'perfunctory documents the compilation of which supplemented the income of impecunious provincial literati' (Hummel, 9). But in the +17th century their value was officially recognised; eminent scholars such as Tai Chen and Chang Hsüeh-Chhêng wrote some themselves, and in due course these books became invaluable source-material.[b]

The *fang chih* were not only confined to cities and their administrative districts. Similar books were devoted to famous mountains.[c] A special stimulus to this was the location of Taoist and Buddhist monasteries in mountain regions.[d] Wylie (1) refers to more than a dozen examples of these books, among which the +11th-century *Lu Shan Chi*[1] (about a famous mountain near the Poyang Lake), by Chhen Shun-Yü,[2] and the +18th-century *Lo Fou Shan Chih*[3] (the mountains north of Canton, where Ko Hung had his alchemical laboratory), by Thao Ching-I,[4] may be mentioned. But there must be many dozens of them.

At the other extreme of urban interest come the books which dealt with cities alone.[e] Here the type-specimen is perhaps the *Lo-Yang Chhieh Lan Chi*,[5] a description of Buddhist establishments at the Northern Wei capital, by Yang Hsüan-Chih,[6] about +500. Of greater interest, however, are the two books *Tu Chhêng Chi Shêng*[7] (The Wonder of the Capital) and *Mêng Liang Lu*[8] (The Past seems a Dream), both about Hangchow under the Sung.[f] There are others besides.

What is there in the West to compare with this vast mass of literature? Greek and Hellenistic antiquity supplies no real parallel,[g] and there seems little to adduce

[a] Here there may be some duplication—the probable overall figure is some 6500.

[b] Aurel Stein (7) once had occasion to verify a point which had seemed most improbable to him before he knew the locality. He wrote 'This...illustrates afresh the risk run in doubting the accuracy of Chinese records without adequate local knowledge'.

[c] We have spoken already of the sacred mountains of China in Vol. 1, p. 55. Their numinous quality goes back, no doubt, to the mountain worship of primitive times; see K. Kanda (*1*).

[d] Thus when I visited Chin-Yün Shan in Szechuan the abbot kindly presented me with a copy of the *Chin-Yün Shan Chih*[9] (Cloud-Girdled Mountain Records), by Thai-Hsü and Chhen-Khung.

[e] They are among the earliest to develop. Chao Chhi's[10] *San Fu Chüeh Lu*[11] (Verified Records of Three Cities) was written about +153 but it did not survive. The three cities were Chhang-an (Sian), Fêng-i (Ta-li) in Shensi, and Fu-fêng (Hsiang-yang). We still possess, however, the *San Fu Huang Thu*[12] (Description of the Three Districts in the Capital, Chhang-an), attributed to Miao Chhang-Yen,[13] probably of the San Kuo period, late +3rd century.

[f] These books preserve many fascinating details of the life of the city a short while before Marco Polo visited it. Cf. below, p. 551. The former was by a Mr Chao (The Old Gentleman of the Water-Garden who attained Success through Forbearance) in +1235; the latter was by Wu Tzu-Mu[14] in +1275. Parts have been translated by Moule (5, 15). On the plans of the city see Moule (15), pp. 12ff.

[g] There was, of course, Pausanias' *Description of Greece*, about +168.

[1] 廬山記 [2] 陳舜俞 [3] 羅孚山志 [4] 陶敬益 [5] 洛陽伽藍記
[6] 楊衒之 [7] 都城紀勝 [8] 夢粱錄 [9] 縉雲山志 [10] 趙岐
[11] 三輔決錄 [12] 三輔皇圖 [13] 苗昌言 [14] 吳自牧

from the early Middle Ages. By the +13th century we have the anonymous French *L'Estat de la Citez de Jherusalem* (1222),[a] which might offer a comparison with the description of Loyang six hundred years before, and the works of Giraldus Cambrensis (*Topographia Hibernica*, 1188; *Itinerarium Cambriae*, 1191), which it would be interesting to contrast in detail with the *Kuei Hai Yü Hêng Chih* of his exact contemporary Fan Chhêng-Ta. But this kind of geography did not develop in Europe as it did in China during the Sung, and one has to wait for the Ming, that is to say, for Renaissance Europe, to find good parallels. The *Historia de Gentium Septentrionalium* of Olaus Magnus the Swede (1567) is then comparable to the ethnological type of Chinese geographies which we have noticed. And England produces at the opening of the modern period the *Itinerary* of John Leland (1506 to 1552, curiously Chinese in his appointment as Antiquary-Royal) and the *Britannia* of Wm. Camden (1551 to 1623), though both these were as much archaeological as geographical. On the whole there is, as regards detailed local topography, the same gap of a millennium in Europe as we noted before and shall shortly see again, with the difference that the ancient Western world had not made much progress in this field beforehand.

(6) Geographical Encyclopaedias

It only remains to mention some of the great compendia of geographical information which the Chinese produced from the Chin dynasty (+3rd and +4th century) onwards. They were mainly descriptive, in the style of Strabo rather than Eratosthenes, though during the Thang and Sung they certainly contained some maps, now long lost. Perhaps one might name as the earliest of the kind the *Shih San Chou Chi*[1] (Record of the Thirteen Provinces) by Khan Yin,[2] written between +300 and +350. In the Chhi period (+479 to +502) there was Lu Chhêng's[3] *Ti Li Shu*[4] (Book of Geography), which was apparently based on no less than 160 geographical works of earlier date, and enlarged later in the Liang (before +557). At the same time a large *Yü Ti Chih*[5] (Memoir on Geography) was produced by Ku Yeh-Wang.[6] We have just mentioned the compilations of local topographies which were made fifty years later by imperial order in the Sui. Throughout the Thang, new geographies were appearing, such as the *Kua Ti Chih*[7] of Wei Wang-Thai[8] and others in +638, and the *Ti Li Chih*[9] of Khung Shu-Jui[10] a little later; besides these there were the important books of the great cartographer Chia Tan[11] about +770 which will be mentioned below.

The oldest extant general work, however, is the *Yuan-Ho Chün Hsien Thu Chih*[12] (Yuan-Ho reign-period Illustrated Geography), by Li Chi-Fu,[b] written about +814. This was towards the end of the Thang. The period between Thang and Sung pro-

[a] Sarton (1), vol. 2, p. 672; Beazley (1), vol. 2, p. 208. The brief and anonymous Latin tract *On the Houses of God in Jerusalem* (Beazley (1), vol. 1, p. 36) of +808, also invites comparison with the *Lo-Yang Chhieh Lan Chi*.

[b] It will be remembered that Li Chi-Fu has a place in the history of tidal theory (p. 490 above).

1 十三州記 2 闞駰 3 陸澄 4 地理書 5 輿地志
6 顧野王 7 括地志 8 魏王泰 9 地理志 10 孔述睿
11 賈耽 12 元和郡縣圖誌

duced one important geographer, Hsü Chiai,[1] but shortly after the beginning of the Sung there appeared (between +976 and +983) a great compendium which is still consulted, the *Thai-Phing Huan Yü Chi*[2] (General Description of the World in the Thai-Phing reign-period), by Yüeh Shih,[3] in 200 chapters. In the +11th century there were further books by Wang Chu[4] (+1051) and Li Tê-Chhu[5] (+1080), and in the +12th by Chhen Khun-Chhen[6] (+1111) and Ouyang Wên[7] (+1117). But during this time the treatment was becoming more and more literary and historical-biographical and less geographical in the scientific sense. The +13th century, however, saw two valuable geographies, the *Yü Ti Chi Shêng*[8] (The Wonders of the World), by Wang Hsiang-Chih[9] (cf. Haenisch, 1), and the *Fang Yü Shêng Lan*[10] (Triumphant Vision of the Great World), by Chu Mu,[11] both about +1240.

The succeeding centuries produced less of this description, and we need only notice the three great comprehensive geographies of the Empire.[a] The first of these was started about +1310 but never printed, and now only some thirty-five chapters remain; the second was published in +1450;[b] while the third appeared in the +18th century, having been initiated under the direction of Hsü Chhien-Hsüeh[12] in +1687. This last includes every conceivable kind of geographical detail for all the provinces, and also a mass of information about dependencies and tributary states beyond the frontiers of China proper.[c] Also early in the +18th century appeared the *Thu Shu Chi Chhêng* encyclopaedia, in which a great quantity of geographical material is included. This is found in the second division (Fang Yü Hui Pien[13]) which includes four sections. The first (*khun yü*[14]), though dealing also with mineralogy and geology, has some ninety chapters on historical topography and cartography. The second (*chih fang*[15]) is by far the longest of all the sections in the encyclopaedia, having no less than 1544 chapters, and consists of a detailed description of all provinces, cities, towns and the areas controlled by them. The third (*shan chhuan*[16]) deals similarly with mountains and rivers. The fourth (*pien i*[17]) is concerned with foreign peoples and countries.

By this time, of course, the Chinese work has to face what is in many ways a damaging comparison with post-Renaissance Europe. Again we see that it was during the earlier medieval time that China was more advanced. Between the Han and the Thang, the West had nothing to set beside the Chinese geographers, and during the Sung there was no competition except on the part of the Arabs. Only with the Ming decadence and the rise of modern science in Europe did the West draw considerably ahead.

[a] *Ta Yuan I Thung Chih*,[18] *Ta Ming I Thung Chih*,[19] *Ta Chhing I Thung Chih*.[20] They did not prevent the appearance of geographical encyclopaedias by private persons: e.g. the *Huang Yü Khao*[21] by Chang Thien-Fu[22] in +1588. Cf. Goodrich (11).

[b] This Ming geography perhaps stimulated a parallel official Korean work of much merit, the *Tongguk Yeji Seungnam*[23] by Sŏ Kŏ-Jŏng,[24] begun about +1470 and completed in +1530.

[c] Small parts of it have been translated or abstracted by Biot (15, 16).

1 徐鍇 2 太平寰宇記 3 樂史 4 王洙 5 李德芻
6 陳坤臣 7 歐陽忞 8 輿地紀勝 9 王象之 10 方輿勝覽
11 祝穆 12 徐乾學 13 方輿彙編 14 坤輿 15 職方
16 山川 17 邊裔 18 大元一統志 19 大明一統志
20 大清一統志 21 皇輿考 22 張天復 23 東國輿地勝覽
24 徐居正

(c) A NOTE ON CHINESE EXPLORERS

The immense mass of geographical knowledge represented by the literary works which have just been described, as well as the achievements of scientific cartography to which the following subsections will be devoted, was certainly not gained except by the accumulated observations of countless travellers and explorers. Some were engaged upon official or diplomatic missions, others were travelling in the cause of religion, but they all added their store of experience and their observations, more or less accurate and complete, to the growth of knowledge about the terrestrial world. On account of their great importance for contacts between China and Europe, we have already described the exploits of Han officials such as Chang Chhien and Kan Ying in another place,[a] as also the work of later pilgrims such as Hsüan-Chuang and Fa-Hsien.[b] On account of their close connection with advanced cartography we shall examine later[c] the voyages of the eunuch admiral Chêng Ho.[d] But we may take advantage of the present opportunity to pause for a moment and survey a few other outstanding aspects of Chinese travel and exploration.

So great has been the fame of Marco Polo and the other European travellers of the +13th century[e] that their Chinese counterparts who also made important journeys have generally been overlooked.[f] The most accessible account of them in a Western language we owe to Bretschneider (2). Bretschneider translated parts of the *Hsi Yu Lu*[1] (Record of a Journey to the West) by the statesman and patron of astronomers, Yehlü Chhu-Tshai,[2] who accompanied Chinghiz Khan on an expedition to Persia (+1219 to +1224). Here we find also the *Pei Shih Chi*[3] (Notes on an Embassy to the North) by Wukusun Chung-Tuan,[4] a Chin Tartar who returned from a mission to Chinghiz Khan in +1222.

Another famous journey was that of the Taoist adept and alchemist Chhiu Chhang-Chhun[5] (+1148 to +1227), who was summoned to the court of the Mongol emperor, Chingiz Khan, then in Afghanistan, and made the journey there and back from Shantung between +1219 and +1224. The record of it, the *Chhang-Chhun Chen-Jen Hsi Yu Chi*[6] (Western Journey of the Taoist Chhang-Chhun),[g] by his secretary Li Chih-

[a] Sect. 7*e*, *g*, in Vol. 1, pp. 173, 196. Mem. Hirth (2).

[b] Sect. 7*i* in Vol. 1, pp. 207ff. The collection of Hennig (4) includes most of the Chinese voyages of exploration. Cf. also the diaries of the Japanese monks Ennin (+838 to +847) and Jōjin (+1072 to +1073).

[c] P. 556 below.

[d] The Chinese sailors of the +15th century who enlarged the geographical knowledge of their countrymen had some interesting successors in the +18th. Chhen Kuan-Shêng (3) has made a study of Hsieh Chhing-Kao,[7] who between 1783 and 1797 visited all the principal trading ports of Europe, Asia, Africa and America. Himself illiterate, his *Hai Lu*[8] was written down for him by a Rusticianus, Yang Ping-Nan;[9] it is purely factual and objective.

[e] For a convenient résumé see Sykes (1) or Komroff (1).

[f] Indeed Europeans are accustomed to think that they themselves throughout the centuries discovered the inhabitable globe. It would be just as true to say that Europe was first discovered by China, in the days of Chang Chhien's visit to Bactria.

[g] *TT* 1410.

[1] 西遊錄 [2] 耶律楚材 [3] 北使記 [4] 烏古孫仲端 [5] 邱長春
[6] 長春眞人西遊記 [7] 謝淸高 [8] 海錄 [9] 楊炳南

Chhang,[1] has several times been translated,[a] the most recent version being that of Waley (10). As the preface says, Li Chih-Chhang 'kept a record of their experiences throughout the journey, noting with the greatest care the nature and degree of the difficulties—such as mountain-passes, river-crossings, bad roads and the like—with which they had to contend; also such differences and peculiarities of climate, clothing, diet, vegetation, bird-life and insect-life as they were able to observe'. Indeed, during the journey through Mongolia and Central Asia, Chhiu Chhang-Chhun's party made observations of a solar eclipse and took gnomon solstice-shadow measurements at a more northerly point than any Chinese astronomers had reached before them.[b] Then in +1259 Chhang Tê[2] went on an embassy from Mangu Khan to his brother Hūlāgu and left a record of it in the *Hsi Shih Chi*[3] (Notes on an Embassy to the West).[c] The last translation of Bretschneider[d] was the biography of the grandson of Yehlü Chhu-Tshai, Yehlü Hsi-Liang,[4] who travelled widely in Central Asia between +1260 and +1263. Moule (6) has analysed the long journey from south to north of Yen Kuang-Ta[5] of +1276. But though all these records are of great interest, their authors were mostly official travellers whose notes were incidental. Of greater scientific interest were journeys the main motives of which were geographical.

For example, there was the question of the source of the Yellow River. In the time of Chang Chhien (−2nd century) it was supposed that the Khotan River, issuing from the Khun-Lun Mountains (which form the northern escarpment of the Tibetan plateau), and making its way round the north of the Tarim basin to fall into Lop Nor, was the source of the Yellow River. Between Lop Nor and a pass near Lanchow the river was supposed to flow underground. But by the time of the Thang the true situation was clearly understood. In +635 a general named Hou Chün-Chi,[6] in charge of a punitive expedition against a Tibetan tribe, pushed west as far as Djaring Nor and 'contemplated the sources of the Yellow River'.[e] The river does in fact take its origin near this lake.[f] The identification was confirmed not long afterwards by Liu Yuan-Ting,[7] who was sent as ambassador to the Tibetans in +822. Starting from Sining above Lanchow, he took a south-westerly route to Lhassa which must have led him to cross both the Yellow River below Ngaring Nor and the Yangtze somewhere near Jyekundo.[g] Hou and Liu probably did not realise that the Yellow River makes an enormous detour around the Amne Machin range before doubling

[a] As by Palladius into Russian in 1866, and by Bretschneider (2), pp. 35ff. A recent appreciation is due to Fedchina (1).

[b] Cf. pp. 416 and 293 above.

[c] Tr. Bretschneider (2), pp. 109 ff.

[d] (2), pp. 157 ff. It is ch. 180 of the *Yuan Shih*.

[e] *Hsin Thang Shu*, ch. 221 A, p. 7*a*.

[f] So remote are these parts of the world, however, that discoveries still remained to be made in 1953. Chou Hung-Shih (1) has given an account of current explorations which show that the Yellow River rises neither from Djaring Nor nor from Ngaring (Oring) Nor, nor yet from Khotun Nor (Pools of the Stars) still farther west, but from the Yoko-chung-lieh stream in the Yaho-Latahotzu mountains.

[g] *Chiu Thang Shu*, ch. 196 B, p. 15*a*; *Hsiu Thang Shu*, ch. 216 B, p. 6*b*. The itinerary is in the *Hsin Thang Shu*, ch. 40; tr. Bushell (3).

1 李志常 2 常德 3 西使記 4 耶律希亮 5 嚴光大
6 侯君集 7 劉元鼎

back again, but otherwise they were perfectly correct. In +1280 Khubilai Khan sent out a scientific expedition under Tu Shih[1] to clear up this question, and the results were embodied in a book *Ho Yuan Chi*[2] by the geographer Phan Ang-Hsiao.[3] Chu Ssu-Pên[a] made use of Tibetan sources as well as this work, and there is a special chapter on the upper course of the Yellow River in the *Yuan Shih*.[b]

Elsewhere[c] we touch upon the road guides which are extant in Chinese literature, and were prepared primarily for practical purposes. Bretschneider (4) has translated such an itinerary of the Ming period, from Chiayükuan (Fig. 14 in Vol. 1, facing p. 103), the then 'gate of China' on the Old Silk Road north-west of Lanchow, to Istanbul. It is, indeed, in the Ming that we come to the prince of all Chinese itinerary-makers, the traveller Hsü Hsia-Kho[4] (+1586 to +1641), a man whose interests were neither official nor religious, but scientific and artistic. There is a good account of him by Ting Wên-Chiang (3), and a collective work on him by several authors recently appeared under the editorship of Chang Chhi-Yün. For more than thirty years he perambulated the most obscure and wildest parts of the empire, exposed to all kinds of difficulties and sufferings, often dependent on the patronage of local scholars who helped him after he had been robbed of all his belongings, or local abbots who were willing to pay him for composing a history of their monastery. On sacred mountain and snowy pass, beside the rice-terraces of Szechuan and in the semi-tropical jungles of Kuangsi, there inevitably was to be seen Hsü Hsia-Kho with his notebook. His biographers agree in saying that he thoroughly disbelieved in the theories of the geomancers, and wished to go and see for himself the dispositions of the great mountain regions radiating from the Tibetan massif. His notes, typical excerpts of which have been translated by Ting Wên-Chiang, read more like those of a 20th-century field surveyor than of a 17th-century scholar. He had a wonderful power of analysing topographical detail, and made systematic use of special terms which enlarged the ordinary nomenclature, such as staircase (*thi*[5]), basin (*phing*[6]), etc. Everything was noted carefully in feet or *li*, without vague stock phrases.[d]

Ting (himself an outstanding geologist, and one of the best scientific minds which China has produced in the present century) gave an example of Hsü Hsia-Kho's perspicacity. In Yunnan true crystalline schists are rare, and Ting while on a journey in 1914 thought that he had discovered in the Yuan-Mo Valley a curious outcrop of typical mica schists among the red sandstone of that province. But he found that Hsü had noted it three centuries earlier. On 6th December 1639 he observed that at this place 'the rocks shine like gold sand, brilliantly yellow in the sunlight like flakes of mica pressed together'.

Hsü Hsia-Kho's chief scientific achievements were, first the discovery of the true source in Kweichow of the West River (Hsi Chiang) of Kuangtung. Secondly, he

[a] See below, p. 551.
[b] Ch. 63, p. 18*a*.
[c] P. 544.
[d] It has been suggested that Hsü Hsia-Kho was influenced by the Jesuits, but the evidence is against this. Some of his findings were however incorporated in Martini's Atlas.

[1] 都實 [2] 河源記 [3] 潘昂霄 [4] 徐霞客 [5] 梯 [6] 坪

established that the Mekong and the Salween were separate rivers. Thirdly, he showed that the Chin Sha Chiang[1] river was none other than the upper waters of the Yangtze; this had long been very confusing on account of the enormous bend which it makes to detour the Lu Nan Shan[2] mountains south of Ningyuan.[a]

A curious factor which led to the increase of geographical knowledge in the Chhing dynasty was the custom (of course not new) of banishing scholars to remote parts of the empire for real or alleged misdemeanours. Thus in 1810, Hsü Sung,[3] a good scholar, then Commissioner of Education for Hunan, was accused of using his office to promote the sales of his own books, and of failing to assign topics for essays on the classics by degree aspirants. Two years later he was banished to Sinkiang, where he remained for seven years. But the result was that Chinese geographical literature was enriched by four excellent books on the remote parts of that dominion,[b] notably the *Hsi Yü Shui Tao Chi*[4] (Account of the River Systems of the Western Regions),[c] and commentaries on the chapters in the Han dynastic histories on Central Asia.

(d) QUANTITATIVE CARTOGRAPHY IN EAST AND WEST

(1) Introduction

The history of scientific geography and cartography is usually presented as containing an unaccountable gap between the time of Ptolemy (+2nd century) and about +1400. If we take up any standard work on the subject[d] we find what seem to be certain conventions as to the participation of China—there are discussions of medieval European knowledge of China, what the Arabs said about it, and the stimulus of the visits made by the merchants and the religious-diplomatic envoys in the +13th century—but never by any chance the story of Chinese cartography itself.[e] Kimble, indeed, is so good as to say that 'fifty years of Cathay had wrought a bigger change in the outlook of the Western world than a whole cycle of medieval Europe',[f] but he refers only to the widening of Western horizons. Yet during the whole of the millennium when scientific cartography was unknown to Europeans, the Chinese were steadily developing a tradition of their own, not strictly astronomical, but as quantitative and exact as they could make it.

Nor can it be claimed that the material has been unavailable to occidental historians

[a] According to Than Chhi-Hsiang's paper in the collective work already mentioned, Hsü Hsia-Kho was not the first to suggest this latter identification; it had been mooted by Pan Ku in the Han and by Li Tao-Yuan in the +6th century. Chu Ssu-Pên seems also to have known the separateness of the Mekong and the Salween in the Yuan dynasty.

[b] See Fuchs (3).

[c] Analysed by Himly (8).

[d] E.g. Beazley (1), Brown (1), Bunbury (1), Kimble (1), Nordenskiöld (1), Wright (1).

[e] Beazley, indeed, has a section on Chinese geography (1), vol. 1, pp. 468 ff., and gets wind, through Gaubil, of Chia Tan's great map in the Thang, but adds: 'It is not worth our while to enter any further into a subject which has so little relation with Western and Christian thought.' The recent book of Tooley (1) deserves commendation for a short section on Chinese map-making (pp. 105 ff.), based on Chavannes, but the romanisations are badly mauled.

[f] (1), p. 147.

1 金沙江 2 魯南山 3 徐松 4 西域水道記

of science. It is now nearly half a century since Chavannes (10) published his fundamental paper in which he sketched the growth of accurate mapping in China.[a] Before that time the limitations of a Santarem[b] or a Huttmann, who thought that Chinese cartography began in the Yuan period, or even that Chinese maps had never been graduated at all, were perhaps pardonable, but now it is time that the picture was appreciated as a whole.[c] We shall follow Chavannes, adding to his material as we go, for there were certain aspects of which he did not treat. But first we must glance at the vicissitudes of map-making in the West.

(2) Scientific Cartography; the Interrupted European Tradition

The development of Greek cartography has so often been expounded that it is only necessary here to remind ourselves of its essential features in very few words.[d] It began with Eratosthenes (−276 to −196),[e] the contemporary of Lü Pu-Wei, whose application of a coordinate system to the earth's surface originated from his determination of the earth's curvature. The famous observations of the gnomon shadows at summer solstice at Syene and Alexandria led to the approximately correct figure of 25,000 geographical miles for the earth's circumference.[f] It is to be noted that the spherical earth was as much at the basis of Greek cartography as the flat earth was at the basis of Chinese. But in practice it made less difference than would seem at first sight, for the Greeks never developed satisfactory projections for describing the spherical surface on a flat sheet of paper.

The *oikoumene*, or inhabited world, of Eratosthenes was oblong, 78,000 stadia (about 7800 geographical miles) in length, and 38,000 stadia from north to south. This was crossed by a series of parallels (of latitude), chosen according to solstitial gnomon shadow-lengths, and another series of meridians, chosen arbitrarily. The fundamental parallel was that of Rhodes, which began at the Sacred Promontory in west Spain, touched the Sicilian strait and the tip of Greece, and after passing through Rhodes ran along the south edge of the Taurus mountains. The fundamental meridian was that of Syene, Alexandria, Rhodes and Byzantium; Syene being supposed to lie exactly on the tropic. This meridian was even more inaccurate than the parallel, both involving considerable distortion of the true positions of the places named. Another meridian passed through Carthage, Sicily and Rome. There was no way of determining the distances between these meridians except dead reckoning, mostly by sea voyages, and the length of the Mediterranean consequently came out about one-fifth too great.[g]

[a] Some years after Chavannes, the same field was covered independently (in Japanese) by Ogawa (*1*, *2*) but the influence of this in the West has been negligible. The papers of Soothill (4) and Cressey (2) add little to Chavannes.

[b] (1), p. 359.

[c] I fear that this has still not been achieved even in the latest general history of cartography (Bagrow (1) in 1951).

[d] The account of Bunbury, though old, still remains one of the fullest and best.

[e] B & M, p. 474.

[f] Bunbury (1), vol. 1, pp. 615 ff.

[g] Among special studies on Greek parallels and meridians, those of Heidel (1) and Diller (1) may be consulted.

Hipparchus (*fl.* −162 to −125),[a] the contemporary of Liu An and his school, criticised the work of Eratosthenes and introduced various rectifications, including the term *climata* for the areas between parallels. The parallels of Eratosthenes had been arbitrary, but Hipparchus made them equal and astronomically fixed. In the *oikoumene* there were eleven, the southernmost one being half-way between the equator and the tropic, the next corresponding to a solstitial day of 13 hours, the next to one of 13½ hours, etc. The northernmost one, passing through north Britain, corresponded to a solstitial day of 19 hours. For longitude he made no new advance.

With Ptolemy (*fl.* +120 to +170),[b] who was working at the same time as Tshai Yung, the accurate or scientific cartography of the ancient world reached its greatest height. No less than six out of the eight books of his *Geography* are occupied with tables of latitude and longitude of specific places, given to a precision of one-twelfth of a degree.[c] But the longitudes were really only guesswork. Hipparchus, indeed, had suggested a way of measuring them by observations at different stations of the onset of lunar eclipses, but only one or two experiments of this kind were available to Ptolemy. The ancient world was not able to organise scientific observations on the scale required. However, Ptolemy greatly reduced the estimate of the length of Asia which had been given by Marinus of Tyre (the distance from the 'Stone Tower'[d] to Sera Metropolis), and in this he was fully justified.[e] On his largest map, which covered 180° of longitude and 80° of latitude, he made an attempt to show the meridians and parallels as curved lines.[f]

Here, however, we have to take notice of a point which will prove of particular interest to us, namely, that in his maps of smaller areas or individual countries, Ptolemy used a simple rectangular grid. In this he followed the example of Marinus of Tyre, whom we have just mentioned.[g] Marinus (*fl. c.* +100) has perhaps had less credit than is his due in the history of cartography, for like that of Eratosthenes, his work is known to us only at second hand.[h] It will be worth remembering that he was especially interested in the extension of geographical knowledge towards the East, and made use of the data supplied by Maës Titianus, a Syrian engaged in the silk trade with the Chinese (Seres).[i] It will also be worth while bearing in mind that Marinus of Tyre was an exact contemporary of the astronomer Chang Hêng. Marinus was content, then, to draw his latitude parallels and longitude meridians at right angles to each other.

None of the maps made in Ptolemy's time has come down to us. What the people of

[a] Bunbury (1), vol. 2, pp. 2 ff.; B & M, p. 544.
[b] Bunbury (1), vol. 2, pp. 546 ff.; B & M, pp. 769, 787.
[c] B & M, pp. 802 ff.
[d] Cf. the map in Vol. 1, p. 171 (Fig. 32).
[e] On Ptolemy's knowledge of East Asia there are monographs by Gerini and by A. Berthelot.
[f] Historians of geography (e.g. Bunbury (1), vol. 2, p. 544) usually consider that, although Ptolemy made great efforts to represent more accurately the curvature of the earth's surface, the accuracy of the information available from the remoter regions was hardly worth it.
[g] Bunbury (1), vol. 2, p. 543.
[h] Bunbury (1), vol. 2, pp. 519 ff.; B & M, p. 634.
[i] Cf. above, Sect. 7*e* in Vol. 1, p. 172. The gross over-estimation of the length of Asia seems to have been due to him. A date of −20 to −1 is now thought likely for his travels; M. Cary (1).

the Renaissance thought they must have looked like is seen in Fig. 211 from Ruscelli's +1561 edition of Ptolemy.[a] Many manuscript codices state that the maps were drawn by Agathodaemon of Alexandria, who remains for us a somewhat mysterious character, since he may have lived at any time between the +2nd and the +13th centuries; the latter period being that of the oldest known manuscripts.[b] The only other ancient map which may be mentioned here is the very distorted road-map of the Roman world (giving mileages) discovered by Conrad Peutinger in 1507 and now bearing his name.[c] Though it had been copied from some unknown source by a monk of Colmar in +1265, it is quite pre-christian in spirit, and must have been first made sometime between about +20 and +370. Though in a sense quantitative, it is diagrammatic and makes no attempt at coordinates.

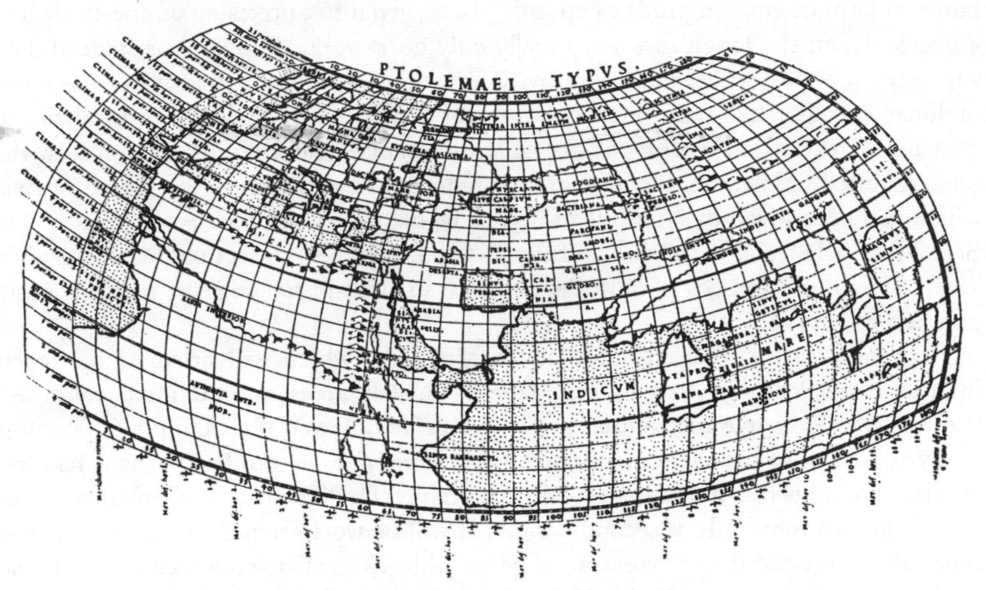

Fig. 211. Ptolemy's world map as reconstructed by the Venetian Ruscelli (+1561). Longitude is expressed in fractions of hours east of the Fortunate Islands, latitude designated by the number of hours in the longest day of the year (from Lloyd Brown).

(3) Religious Cosmography in Europe

After the time of Ptolemy, the Great Interruption sets in. European map-making suffers a degeneration so extreme that it would hardly be believable if we were not so familiar with it. Scientific cartography is drastically replaced by a tradition of religious cosmography.[d] All attempt at coordinates is abandoned, and the world is represented as a disc[e] divided by a few partitions into continents, across which rivers and mountain ranges stray in wild disorder. There is no lack of these 'maps' or 'Mappaemundi', to

[a] Cf. B & M, pp. 792ff.; Lloyd Brown (1), opp. p. 54.
[b] Lloyd Brown (1), p. 73.
[c] Beazley (1), vol. 1, p. 381; B & M, p. 1038; K. Miller (2).
[d] On the comparative background of this, Eliade (2), pp. 315 ff., is worth reading.
[e] Or sphere, if Taylor (2) is right.

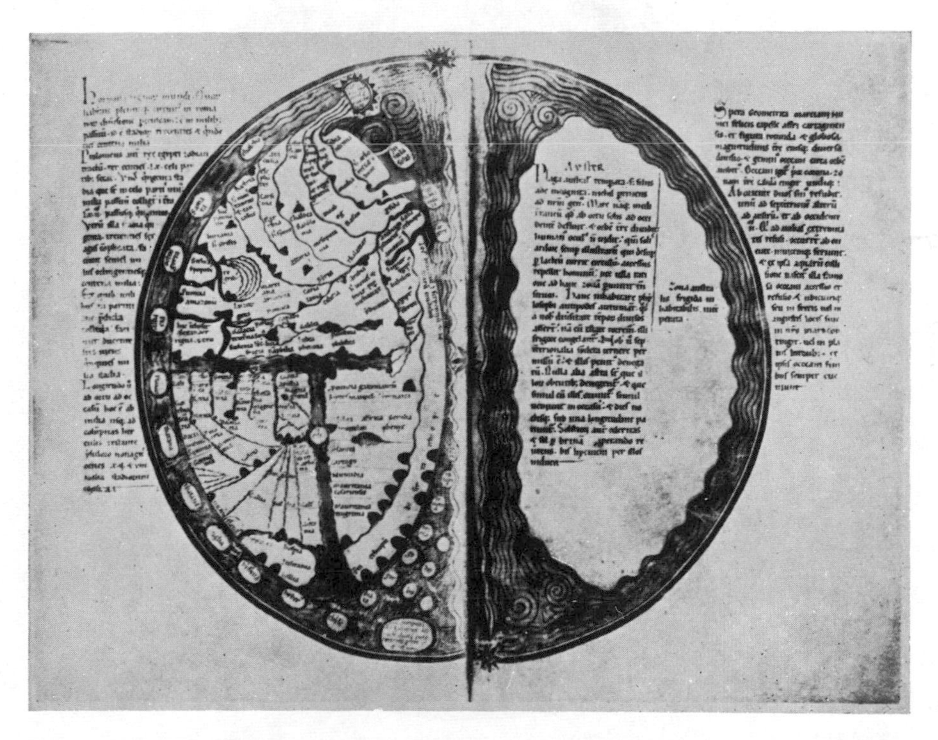

Fig. 212. World map of Martianus Capella (*fl.* +470), from a MS. of the *Liber Floridus*, *c.* +1150 (from Kimble). The northern hemisphere to the left is drawn in the T-pattern, the southern has simply an unknown continent.

PLATE LXXIV

Fig. 213. The 'Psalter Map' of the mid +13th century, a T-O design centred on Jerusalem. Note the 'Wall of Gog and Magog' to the left at the top, built (according to legend) by Alexander the Great to keep out the 'barbarians' of East Asia, but probably a European echo or rumour of the real Great Wall. The Indian Ocean appears as the dark wedge to the right at the top (from Beazley).

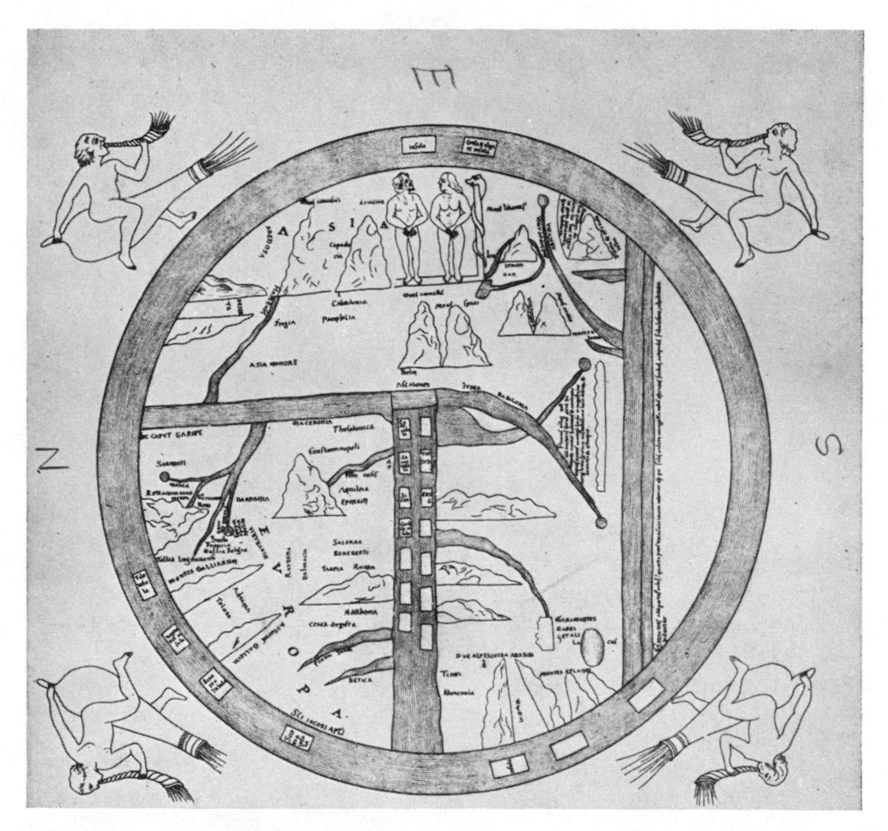

Fig. 214. A world map of the type due to Beatus Libaniensis (*d.* +798), from a Turin MS. of +1150 (from Lloyd Brown). Here again the T-O shape is seen. Paradise and Eden are placed in the Far East at the top; below the Mediterranean with its many islands separates Europe from Africa.

use the medieval term,[a] which may be found in numerous collections.[b] Their classification has been the subject of an interesting paper by M. C. Andrews.[c] The usual name for them is 'wheel-maps' or 'T-O maps'—the significance of the latter we shall see in a moment.

These strange representations are divisible into various families among surviving manuscripts. They are found as illustrations to the texts of numerous medieval authors, among whom the following may be mentioned. First comes the *In Somnium Scipionis* of Ambrosius Theodosius Macrobius (+395 to +423),[d] who is accompanied by Orosius (*fl.* +410) and his *Historia adversus Paganos*,[e] and followed by Martianus Capella (*fl.* +470) and his *Satyricon*.[f] These men still show the influence of Ptolemy (Fig. 212), but after the +5th century Ptolemy is hardly ever mentioned. With Isidore of Seville (*fl.* +600 to +636)[g] further degeneration has taken place, as we see from the diagrams attached to his *Etymologiae*. At the end of the +8th century, the work of the Spanish priest Beatus Libaniensis (*d.* +798)[h] in his *Commentary on the Apocalypse* set the style for a large number of medieval wheel-maps, the earliest extant example dating from +970 and the last from +1250, approximately.[i] In the map which Henry of Mainz[j] added in +1110 to the *Imago Mundi* of Honorius of Autun, there is a certain improvement, but not much, and that of William of Conches (*d.* +1154) is also notable.[k] The so-called Psalter Map of the mid +13th century (Fig. 213) is important on account of the emphasis which it gives to the orbocentric position of Jerusalem, but this is seen also extensively in other medieval wheel-maps.[l] The significance of this we shall appreciate later. To illustrate another example, I give a map of the Beatus type from a Turin MS. of +1150 (Fig. 214).[m] The degree of schematisation which could be reached may be appreciated from Fig. 215, taken from a Venetian edition of Isidore of Seville of +1500.[n]

Let us look at the accompanying sketches (Fig. 216), which summarise the main forms of European religious cosmography. The earliest of them, which has been called Macrobian, still retains the Ptolemaic recognition of a southern sub-equatorial half of the world, an unknown antipodean continent. But for the *oikoumene* the geographer is content with a T, the Mediterranean forming its vertical stroke, and the two parts of the horizontal stroke being represented by the Rivers Tanais (Don) and Nile. The position of Jerusalem, if marked, is always orbocentric. Later traditions, such as that associated with the name of Beatus, dispense altogether with the antipodes or southern hemisphere, and draw the *oikoumene* as occupying the whole of the world

[a] More than 600 are known, but few or none date from before the +10th century.

[b] Such as those of Santarem (1, 2), K. Miller (1), Yusuf Kamal (1), Beazley (1), Lloyd Brown (1), Kimble (1), Taylor (2), etc.

[c] Cf. also Taylor (1).

[d] Kimble (1), p. 8; Beazley (1), vol. 1, p. 343.

[e] Kimble (1), p. 20; Beazley (1), vol. 1, p. 353.

[f] Kimble (1), p. 9; Beazley (1), vol. 1, p. 340.

[g] Kimble (1), p. 23; Beazley (1), vol. 1, p. 366.

[h] Lloyd Brown (1), pp. 94, 119, 126; Kimble (1), p. 183; Beazley (1), vol. 2, pp. 549, 550, 554.

[i] Santarem (2), pl. XII; Yusuf Kamal (1), vol. 3, pp. 871, 947.

[j] Beazley (1), vol. 2, p. 563.

[k] Yusuf Kamal (1), vol. 3, pp. 868, 921.

[l] E.g. the Hereford Mappamundi (Moir & Letts). Cf. Beazley (1), vol. 3, p. 528.

[m] Beazley (1), vol. 2, p. 552.

[n] Lloyd Brown (1), p. 103.

disc. A third tradition preserved a vague memory of the Hellenistic climata, showing them simply as parallel lines without meridians and superimposed on no geographical features. This is found, for example, in the +1110 map (Fig. 217) of Petrus Alphonsus of Huesca, a Spanish Jew.[a] But it goes back to Macrobius.

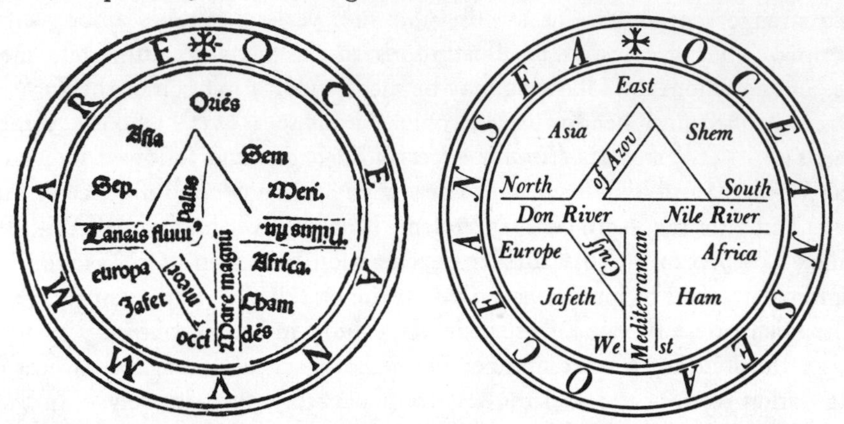

Fig. 215. The extremely schematised world map of Isidore of Seville (+570 to +636) in a Venetian edition of his *Etymologiae* of +1500. Interpretation on the right (from Lloyd Brown).

The T-O maps received their classical explanation in a +15th-century poem by Leonardo Dati (+1365 to 1424), *La Sfera* (*c.* +1420).[b]

> Un T denttro adun O mostra ildisegno
> chome inttre partti fu diviso ilmondo
> elasuperiore emagor rengno
> chequasi pigla lameta delmondo
> asia chiamatta elgrenbo ritto segno
> chepartte iltterzo nome dalsechondo
> africho dicho daleuropia elmare
> mediteraneo traese imezzo apare.

(A T in an O gives us the division of the world into three parts. The upper part and the greatest empire take nearly the half of the world. It is Asia; the vertical bar is the limit dividing the third from the second, Africa, I say, from Europe; between them appears the Mediterranean Sea.)

The T-O map tradition did not die out till as late as the +17th century, as may be seen from a book such as the *Variae Orbis Universi*, by Petrus Bertius (+1628).[c] Before leaving it, however, a word may be said about certain aberrant representations in the occident. There was, for instance, the *Christian Topography*[d] written by Cosmas Indicopleustes between +535 and +547. This was essentially a controversial work, intended to expose 'the wicked folly of the Greeks in Geography'.[e] It therefore has

[a] Beazley (1), vol. 2, p. 575.

[b] III, 11; Yusuf Kamal (1), vol. 4, p. 1436. In the year of Dati's birth, Andrea di Bonaiuto painted a T-map in the frescoes of the Chapter House (the 'Spanish Chapel') of Sta Maria Novella at Florence. There it can still be seen. Cf. Bargellini (1), pp. 30, 32.

[c] Lloyd Brown (1), pl. XV.

[d] Tr. McCrindle (7). Cf. p. 513 above.

[e] Beazley (1), vol. 1, p. 297. Beazley gives a very full account of Cosmas. Cf. B & M, p. 1045.

importance in connection with the exact reasons why Christian Europe, under the influence of the Fathers, threw away all the Hellenistic achievements in quantitative cartography.[a] The principal picture which Cosmas gives of the universe has often been reproduced, but as we are here comparing occidental with oriental levels of attainment, we may not omit it (Fig. 218). There is Heaven with its walls and barrel-vault, while below is the flat surface of Earth, with some schematic sea and land hard to interpret.[b] We are particularly to note, however, the high mountain in the north (or in the centre?) round which the sun is pictured both rising and setting, for this was an idea well known and widespread in Asian thought, and Cosmas himself, as his cognomen indicated, had made the voyage to India, and perhaps to Ceylon.[c] We shall come back later to this mountain of Cosmas, as to a focal point in the intercourse between peoples.[d] Among other European maps which deviated from the general sort may be mentioned the +8th-century map preserved at Albi, which is perhaps the most degenerate of all, showing merely a horseshoe-shaped continent surrounding the Mediter-

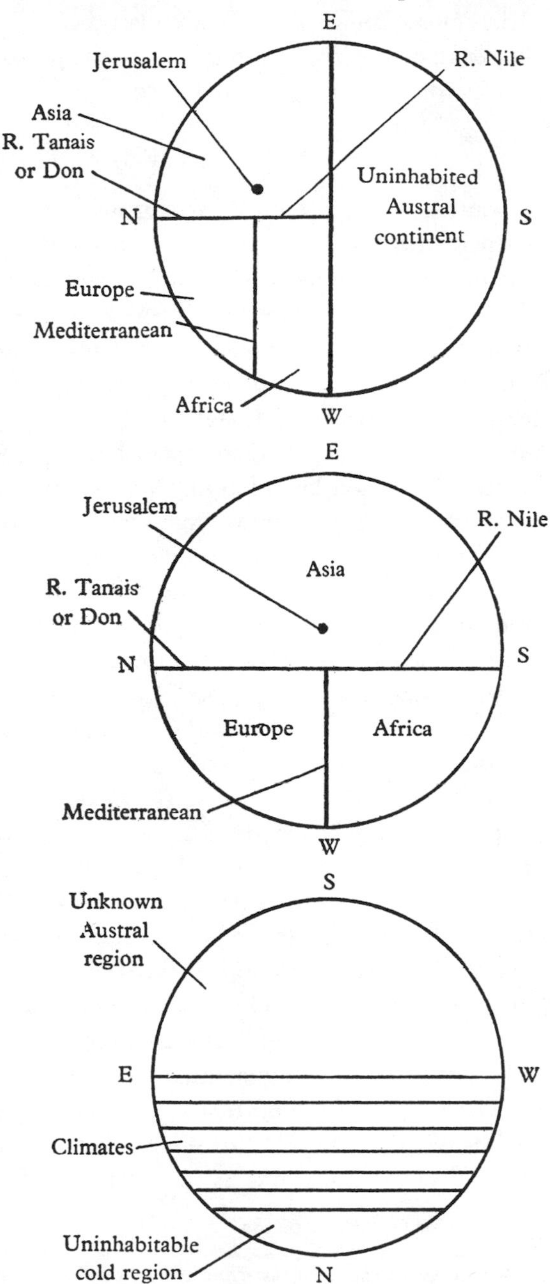

Fig. 216. Diagrams to illustrate the chief forms of the European tradition of religious cosmography.

[a] This point is discussed at some length by Kimble (1), pp. 14 ff., and by Beazley. St Augustine's vigorous attack on the idea of the antipodes (*Civitas Dei*, XVI, 9), for instance, is well known.

[b] From other diagrams it is fairly clear that he meant the large inlet for the Mediterranean and the two smaller ones for the Red Sea and the Persian Gulf.

[c] Beazley (1), vol. 1, pp. 190 ff. Cf. the discussion in *Chin Shu*, ch. 11, p. 4*b*, where Ko Hung is arguing (*c.* +300) against Wang Chhung's support of the Kai Thien theory, and says that it would involve a vertical not a horizontal half-disc of the sun at rising and setting. Cf. p. 218 above. The passage has been translated by Ho Ping-Yü (1).

[d] Cf. Thomson (1), p. 387. It must be, presumably, Mount Meru (see below, p. 568).

ranean gulf.[a] Conversely the late +10th-century 'Cottoniana' map, which seems to have been prepared in the household of an Archbishop of Canterbury by an Irish monk, shows on a square plan the main contours of the European land-masses better than any Western map before the portolans.[b] It is to the 'portolans', or handy sea-charts, that we must now turn.

(4) The Role of the Navigators

About +1300 there began to appear in the Mediterranean region, first slowly and then in increasing numbers, sea-charts for essentially practical use. These showed mainly the contours of the land-masses, along the edges of which the names of all havens and coast towns were marked, often with the addition of flags to indicate their political allegiance (a point which it might be very useful for a pilot to know). The first of these which is dated (Fig. 219) is a Vesconte portolan of +1311,[c] but some extant specimens are thought to go back perhaps thirty years previously. All the examples known are Italian, Spanish, Portuguese or Catalan—no Byzantine or Arabic versions have yet been found. Many publications showing these portolan charts are available.[d] Fig. 220 shows Spain on the portolan of Angelino Dulcerto (+1339).

The characteristic graduation of the portolans was not meridians and parallels, nor even a rectangular grid, but an interlocking network of rhumb-lines or loxodromes radiating from a series of compass wind-roses placed in positions arbitrarily chosen.[e] The connection of the re-introduction of quantitative cartography with the use of the mariner's compass is thus obvious; magnetic polarity had become known in Europe just before +1200, so that about a century later charts based on the compass were becoming general. The manner of using the portolans is described in a work of +1542, written by Jan Rotz of Dieppe, the *Boke of Ydrography*; the pilot selected a loxodrome line in the neighbourhood of the ports or points of departure and arrival of the day's run, such that it was as parallel as possible to the line directly joining them; hence the bearings required could be read off.[f] But the fact that the coastal outlines were now so well drawn shows that a great deal of dead-reckoning information had accumulated, and it was therefore not long before the rhumb-lines or loxodromes were superimposed on a rectangular grid. This is seen in the example taken from Dulcerto (Fig. 220), and occurs even earlier.[g] Nevertheless, the earliest portolans are not without scales at the edges of the maps graduated in 'portolan miles'.

The origins of the portolans have been much discussed, as by Uhden. Bunbury[h] points out that the pilots of the Mediterranean had always had their written *periploi*, and refers to such works as the treatise of Timosthenes of Rhodes[i] on ports, which

a Beazley (1), vol. 1, p. 385.

b Beazley (1), vol. 2, p. 559.

c Beazley (1), vol. 3, p. 513.

d One may mention, apart from the vast collection of Yusuf Kamal (1), the works of Stevenson (2), Cortesão (1), de Reparaz-Ruiz (1), Nordenskiöld (1), Hinks (1), etc.

e Lloyd Brown (1), p. 139; Kimble (1), p. 191; Beazley (1), vol. 3, pp. 513 ff.

f Kimble (1), p. 192; Taylor (3), p. 70.

g Nordenskiöld (1), pls. VII, VIII and IX.

h (1), vol. 1, p. 587.

i Admiral of the Egyptian fleet under Ptolemy Philadelphos, and apparently an authority much followed by Eratosthenes.

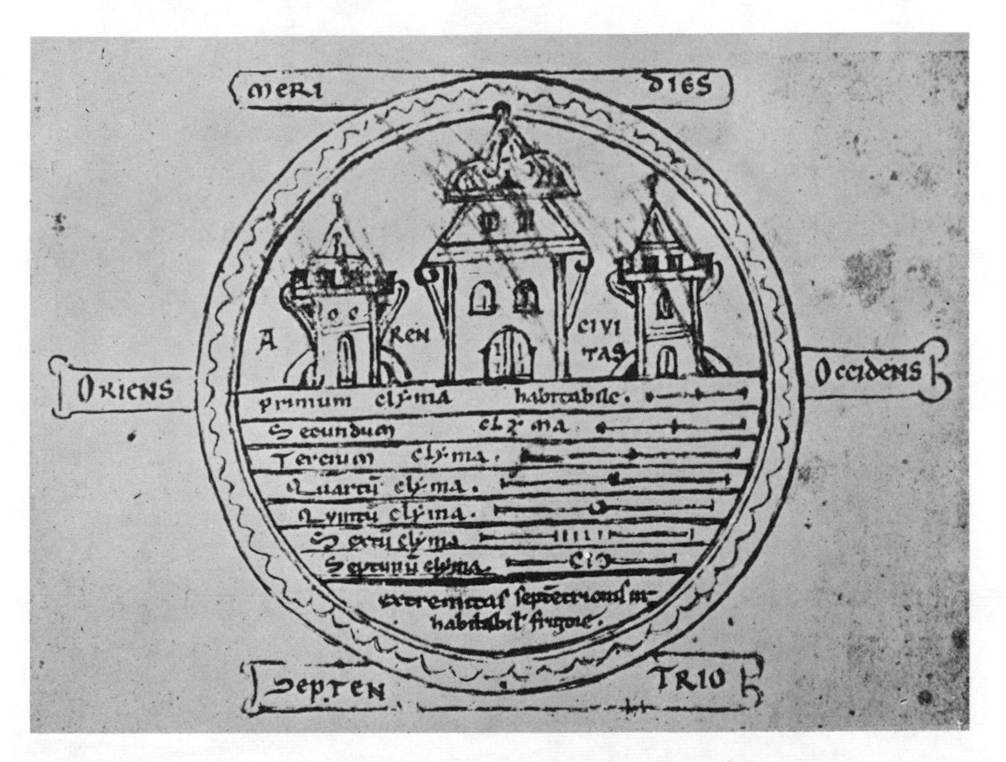

Fig. 217. The climate map of Petrus Alphonsus, *c.* +1110 (from Beazley). Seven *climata* are given until at the bottom the cold and uninhabitable northern region is reached. Instead of a southern continent in the upper hemisphere we are shown the towers of 'Aren Civitas', i.e. Arīn, the mythical prime meridian city of the Arabic geographers, which has been identified as Ujjain in India, capital of the +5th-century Gupta kings (see p. 563 below).

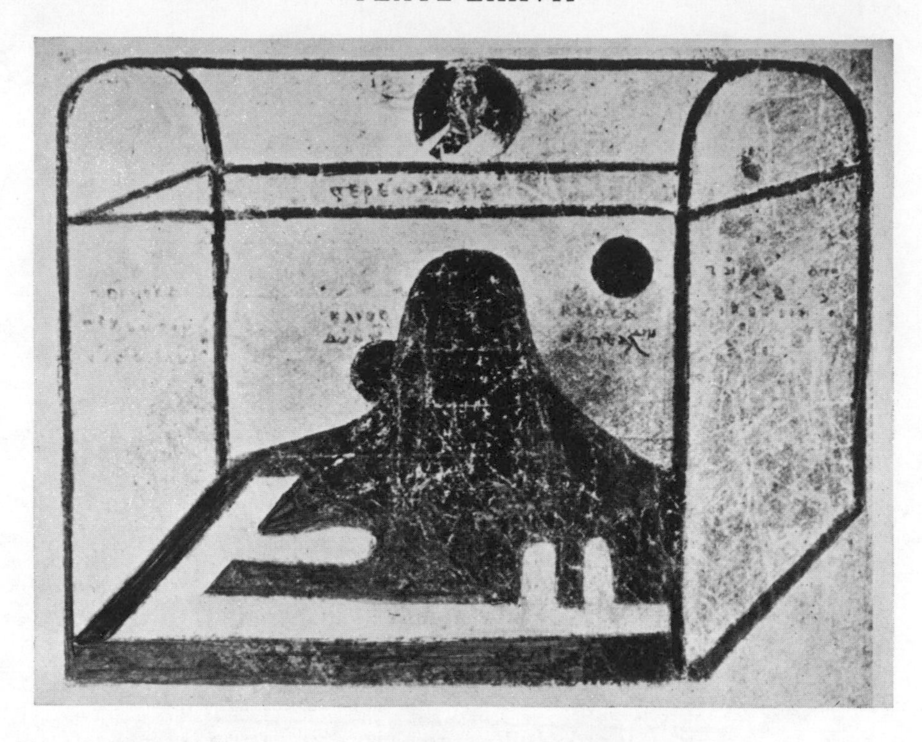

Fig. 218. The universe according to Cosmas Indicopleustes' *Christian Topography*, *c.* +540. The rising and setting suns move round the great mountain in the north; the inlets of the Mediterranean, Red Sea and Persian Gulf are shown below, the heavens have the shape of a barrel vault, and within them the Creator surveys his works (from Beazley).

Fig. 219. One of the oldest extant portolan charts, the P. Vesconte map of +1311 (from Beazley). Rhumb-lines, rectangular grid, and graduated edge can all be seen, and at the top is the cartographer's signature and date. The part of the map shown portrays some coasts in the Levant.

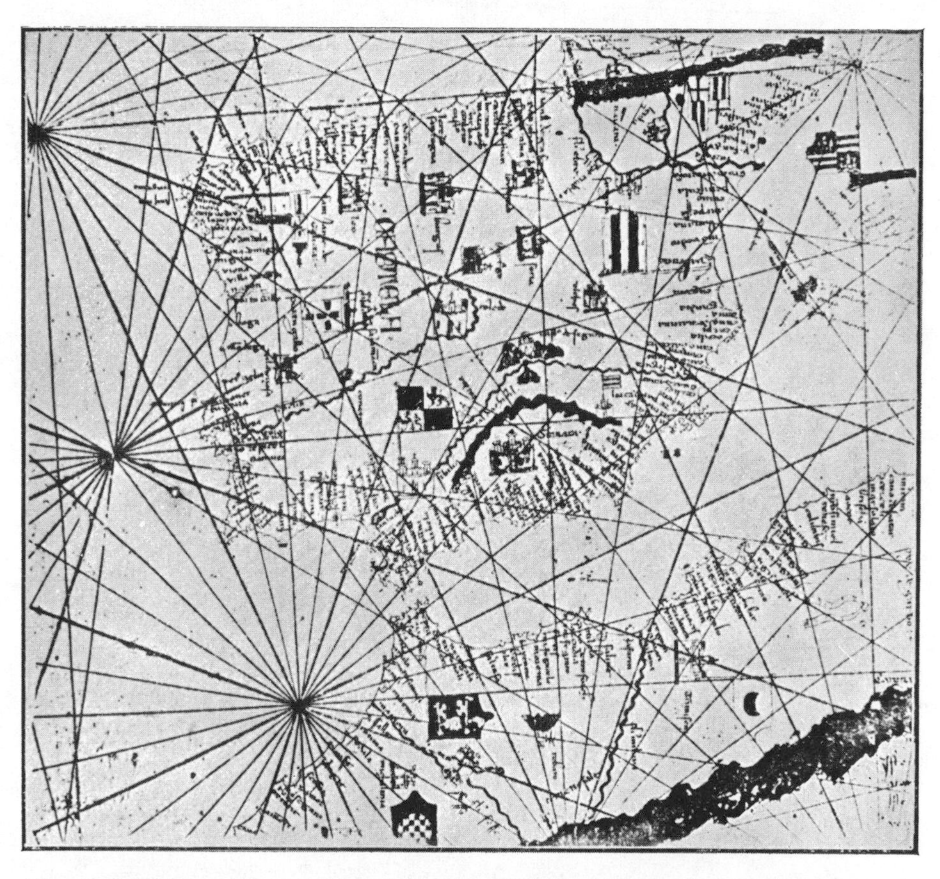

Fig. 220. Spain on the portolan of Angelino Dulcerto, +1339 (from de Reparaz-Ruiz). Behind the rhumb-lines and grid the Straits of Gibraltar and the whole outline of the peninsula clearly appear, with an abundance of names of ports and coastal features, marked by flags to indicate territorial authorities.

Fig. 221. Ceylon on a Byzantine grid-map in a Mt Athos MS. of *c.* +1250 (from Langlois). The name Taprobane can be seen in the second row of compartments from the top. This grid system was presumably a survival of Ptolemaic coordinates and existed only in limited parts of the Greek world during the Middle Ages.

gave the distances separating them (between −285 and −247). But for the most part this knowledge grew up in the technical patrimony of seafaring men through the centuries and was not committed to writing by scholars. It was the coming of the magnetic compass which liberated it from this obscurity. As for the immediate stimulus of the portolans, Arabic influence has long been suspected,[a] and there is now a tendency to look still further east.[b]

The developments following the portolan period, and the effect of the Renaissance on cartography, are well known.[c] Prince Henry the Navigator, of Portugal (+1394 to +1460), was one of the greatest figures not only in the organisation of celebrated expeditions, but also in the revival of the knowledge of Ptolemy's geographical ideas.[d] During the previous centuries the Ptolemaic coordinates had been known and reproduced only in the Greek-speaking world. Greek manuscripts showing them are known, for example, the Urbino MS. of about +1200 (reproduced by J. Fischer) and the Mount Athos MS. of about +1250 (reproduced by Langlois). Here Fig. 221 shows the representation of Taprobane (Ceylon)[e] in the latter. The earliest Latin manuscripts date from about +1415. By +1475 Ptolemy was firmly back again. Then came the time of Gerard Mercator (+1512 to +1594) and the great world map on his cylindrical orthomorphic projection of +1538.[f] With the *Mariner's Mirror* of Waghenaer (+1584) and the contemporary maps of Ortelius we are fully in the modern period.[g]

(5) Scientific Cartography; the Continuous Chinese Grid Tradition

The essential point which is now to be made is that just as the scientific cartography of the Greeks was disappearing from the European scene, the same science in different form began to be cultivated among the Chinese. A tradition which continued without interruption down to the coming of the Jesuits was born in the work of Chang Hêng (+78 to +139), the contemporary, as has been said, of that Marinus of Tyre who was one of Ptolemy's principal sources. But before we can mention him it will be more logical to glance at the evidence for Chinese map-making before his time.

[a] Vernet (1, 2).

[b] Bagrow, Mazaheri & Yajima (1).

[c] See Santarem (1), Kimble (1), Taylor (3, 4), Yule (2), Lloyd Brown (1), and others. After the time of Marco Polo there was a flourishing of world maps made in Europe which gave much more information about East Asia than had been possible in earlier times. We shall come back to them again in Section 29 on nautical technology because of the light they throw on medieval ship-building. For identifications of the place-names given on them, see the lexicon of Hallberg (1).

[d] Lloyd Brown (1), pp. 108 ff.

[e] The name Ceylon itself comes to us from the Chinese transliteration of Siṁhala-dvīpa through Arabic, according to Mahdihassan (4).

[f] Lloyd Brown (1), pp. 158 ff.

[g] Lloyd Brown (1), pp. 144 ff.

(i) *Origins in Chhin and Han*

The *Chou Li* (Record of the Rites of Chou) can be taken as telling us something about the ideas of the Former Han period, if not about the Chou. Its references to maps are remarkably numerous. The Director-General of the Masses (Ta Ssu Thu[1]) has the function of preparing the maps of the feudal principalities (*chang chien pang chih thu ti chih thu*[2]) and of registering their populations.[a] The Directors of Regions (Chih Fang Shih[3]) are placed in charge of the maps of the empire (*chang thien hsia chih thu*[4]) in accordance with which they supervise the lands in its different districts.[b] The Geographer-Royal (Thu Hsün[5]) takes care of the maps of the provincial circuits (*chang tao ti thu*[6]), and when the emperor is making a tour of inspection he rides close to the imperial vehicle in order to explain the characteristics of the country and its products.[c] The Antiquary-Royal (Sung Hsün[7]) is likewise in close attendance on such occasions, armed with notes on historical geography (*fang chih*[8])[d] in order to explain points of archaeological interest.[e] Then there are Frontier Agents (Hsing Fang Shih[9]) who register the boundaries of principalities and domains.[f] Maps for special purposes are also mentioned. Thus the Superintendents of Mining (Kung Jen[10]) observe the occurrence of ores of metals, and make maps of the locations (*tsê wu chhi ti thu*[11]) which they give to the miners.[g] Military maps are also spoken of.

That the Han people should have said so much about maps in their idealised picture of the perfect imperial bureaucracy is not surprising, since the first historical reference to a map in China goes back to the −3rd century. In −227 the crown prince of the State of Yen (Yen Tan Tzu[12]) induced a certain Ching Kho[13] to make an attempt to assassinate that king of Chhin who was afterwards to become the first emperor of the Chhin dynasty.[h] This man came into the king's presence with a map of the district of Tu-Khang (*Tu-Khang chih ti thu*[14]) on the pretext that this domain would be offered to him. The map was painted on silk, and contained in a case (*hsia*[15]), but when it was taken out a dagger appeared as well; this Ching tried to use, but failed.[i] The scene was a favourite one with the decorators of Han tombs.[j] Fig. 222 shows a version of it

[a] Ch. 3, p. 10*b* (ch. 9, p. 1; Biot (1), vol. 1, p. 192).
[b] Ch. 8, p. 24*b* (ch. 33, p. 1; Biot (1), vol. 2, p. 263).
[c] Ch. 4, p. 34*b* (ch. 16, p. 21; Biot (1), vol. 1, p. 368).
[d] This must be one of the earliest uses of this term, which, as noted above, came to be used for local topographies.
[e] Ch. 4, p. 34*b* (ch. 16, p. 22; Biot (1), vol. 1, p. 369).
[f] Ch. 8, p. 30*b* (ch. 33, p. 64; Biot (1), vol. 2, p. 282).
[g] Ch. 4, p. 37*a* (ch. 16, p. 33; Biot (1), vol. 1, p. 377).
[h] The story is told in Ching Kho's biography in *Shih Chi*, ch. 86, which has been translated by Bodde (15). A somewhat embroidered version (which, however, contains certain features confirmed by epigraphic evidence) has come down to us as the *Yen Tan Tzu*,[12] reconstituted from quotations (tr. Chêng Lin, 1); this dates probably from the +2nd century. Passages in the *Chan Kuo Tshê* relating to the *attentat* of Ching Kho have been translated by Margouliès (3), p. 99.
[i] *Shih Chi*, ch. 86, pp. 10*a* ff., 17*b*.
[j] Rudolph (1); Edwards (1).

1 大司徒 2 掌建邦之土地之圖 3 職方氏 4 掌天下之圖
5 土訓 6 掌道地圖 7 誦訓 8 方志 9 形方氏
10 卝人 11 則物其地圖 12 燕丹子 13 荊軻 14 督亢之地圖
15 匣

from the Wu Liang tomb-shrine reliefs of the +2nd century. The term *ti thu*[1] also occurs several times in the *Chan Kuo Tshê* (Records of the Warring States), i.e. writings which are most probably of Chhin or very early Han time. In the *Kuan Tzu* book, too, there is a chapter on military maps with these words as title (ch. 27), but this may not be pre-Han.

When Chhin Shih Huang Ti became emperor, he assembled all available maps of the empire. This is clear from the description which Ssuma Chhien gives of the taking of Hsien-yang city by the leader who was later to become first emperor of the Han.

Fig. 222. Ching Kho's attempt to assassinate the King of Chhin, afterwards Chhin Shih Huang Ti, in −227, with a dagger hidden in a rolled silk map taken from a wooden case. The king is seen leaping away on the left and in the centre the dagger with its tassel is deflected by a pillar at the foot of which is the map-case. Beside it is another case containing the head of Fan Yü-Chhi which Ching Kho had also presented to the King of Chhin. Above is Ching Kho's 13-year old assistant, Chhin Wu-Yang, too frightened to do anything, and on the right the King's physician, Hsia Wu-Chü, has secured Ching Kho with his medicine-bag. The relief is one of those in the Wu Liang tomb-shrines (+147). For the full story see Bodde (15).

This was in −207. While the other generals were engaging in pillage, Hsiao Ho[2] sought out the secretarial offices and took charge of all ordinances, reports and maps. These were of inestimable advantage to the Han dynasty.[a] They must have lasted till the end of the +1st century, for Pan Ku, who died in +92, refers to them at least twice in the *Chhien Han Shu*.[b] But by Phei Hsiu's time (+3rd century) they had disappeared—they were probably incised on wooden boards. All through the Han there are references to maps. When Chang Chhien had returned from the West (−126), the emperor consulted, we are told,[c] ancient maps and books (*ku thu shu*[3])

[a] *Shih Chi*, ch. 53, p. 1*b*.
[b] Ch. 28A, p. 15*a*, and ch. 28B, p. 4*a*.
[c] *Chhien Han Shu*, ch. 61, p. 3*b*.

[1] 地圖 [2] 蕭何 [3] 古圖書

and decided that the mountain from which the Yellow River took its source should be called Khun-Lun. The first use of the expression *yü ti thu*,[1] derived from the conception of earth as a chariot and heaven as a chariot-roof,[a] comes in −117, when maps of the whole empire were submitted to Han Wu Ti in connection with the investiture of three of his sons as feudal princes.[b] There was a famous military map-making in −99, when the general Li Ling[2] was campaigning against the Huns. He made a complete chart of the mountains and steppes as far as thirty days' journey north of the frontier, and sent a copy of it back for presentation to the emperor.[c] As might be expected, Wang Mang was aware of the importance of maps, and delegated a special official, Khung Ping,[3] to assemble and study them in connection with problems of fiefs.[d]

Much discussion has centred upon the *thu shu*[4] (charts and documents)[e] which were presented to the emperor in −35. These have already been mentioned above (p. 511). The incident is one which has numerous implications.[f] In the previous year the Protector-General of Central Asia, Kan Yen-Shou,[5] and his assistant, Chhen Thang,[6] had made an unauthorised expedition as far west as the Talas River, where their forces stormed the city of the Hun leader (Shanyu) Chih-Chih[7] and killed him. This was in retaliation for his murder of a Chinese diplomatic envoy in −43. When they returned to China it was at first uncertain whether they would be punished for having forged an imperial order, or honoured for their signal military success. What concerns us is that after presentation 'his (or its) *thu-shu* were (or was) shown to the honoured ladies of the imperial palace (*i chhi thu shu shih hou kung kuei jen*'[8]). The problem of what these documents were has considerable importance because this battle was the one in which it is believed that the Chinese soldiers fought against some 150 Roman legionaries who had taken mercenary service, after the famous defeat of Crassus in Parthia (−54), with the Sogdians or the Huns.[g] The difficulty lies in the ambiguity of the word *thu* which could mean either maps or pictures.

Duyvendak (16) formed the view that the account of the battle given in the biography of Kan Yen-Shou in the *Chhien Han Shu*[h] constitutes in fact a series of captions and explanations for a set of pictures afterwards lost. They could have been analogous to the battle pictures found in some of the Later Han tomb-shrines. Dubs (30) raised upon this a further hypothesis, that these pictures were similar to the representations carried in Roman triumphal processions, and that their existence was directly due to information about the practice provided by the Roman legionaries captured by Kan Yen-Shou and Chhen Thang.

[a] See above, p. 214, in the sub-section on cosmological theories.
[b] *Shih Chi*, ch. 60, p. 5*b*. The presentation of a map was part of the ceremonial.
[c] *Chhien Han Shu*, ch. 54, p. 5*a*.
[d] *Chhien Han Shu*, ch. 99B, p. 23*a*.
[e] The reference is *Chhien Han Shu*, ch. 9, p. 12*b*.
[f] Cf. above, Vol. 1, pp. 236 ff., and in Vol. 6 below, Sect. 30 on military technology.
[g] The military aspects of this will be deferred till the appropriate Section.
[h] Ch. 70, pp. 9*b* ff.

[1] 輿地圖 [2] 李陵 [3] 孔秉 [4] 圖書 [5] 甘延壽 [6] 陳湯
[7] 郅支 [8] 以其圖書示後宮貴人

The foundations of this story seem rather weak. That Roman legionaries actually were in mercenary service with the Shanyu (Dubs, 6, 29) may indeed be accepted; that they were captured alive by the Chinese and settled in Kansu is possible, even probable, though not proven. But really the only reason for thinking that 'battle-pictures' were meant at all, is that 'maps and charts' would not have been shown to 'the ladies of the imperial harem', as Duyvendak calls it.[a] I cannot help feeling that both these scholars underrated the intelligence of the empress, the concubines and the court beauties. No one would hesitate to discuss important and interesting matters with some of the women depicted on the Han frescoes which the reproductions of Fischer (2) have made familiar.[b] Besides, if the decorations of medieval Western maps (cf. Bagrow, 1) could be so exciting, why should those of the Han have been less so for their time? Modern mathematical cartography is not in question. And the only authority who can be quoted for the view that the documents were definitely pictures and not charts, lived more than a thousand years after the event.[c] Maps, doubtless much embellished with drawings, still seem the preferable interpretation.

Cartography continued to interest the people of the Later Han. In +26, when Kuang Wu Ti was fighting to establish the new dynasty, he opened a large map, probably painted on silk, on one of the gate-towers of a city which his forces had just taken, and said to Têng Yü,[1] one of his generals: 'Here are all the commanderies and feudal domains of the empire; what we have just taken is only a very small part. How could you have thought it easy to conquer the whole of it?'[d] After Kuang Wu Ti had become securely established on the throne, a special ceremony was held annually from +39 at which the Minister of Works (Ta Ssu Khung[2]) presented a map of the empire.[e] Again, in +69, when Wang Ching[3] was charged with repairing the breaches in the Yellow River dykes at Khaifêng, he was given[f] a set of maps illustrating the treatise of Ssuma Chhien's on the Rivers and Canals (*Ho Chhü Shu*[4]).[g]

(ii) *Establishment in Han and Chin*

This brings us to the time of Chang Hêng.[5] Much has already been said of his work in astronomy (pp. 216 ff., 343, 359 ff.), and we shall soon meet with him again in the field of seismology. None of the fragments of his writings which survive deals with cartography, but that it was he who originated the rectangular grid system seems very probable from the pregnant phrase used about him by Tshai Yung.[h] He is said to have

[a] He even says 'Documents of that sort were much too precious for the administrators of the Han empire to be toyed with by women'.

[b] And as if to strengthen the interpretation here preferred, there was the girl map-maker of the first emperor of the Wu State in the San Kuo period; whom we shall meet a page or two hence. The Lady Fêng (Fêng Lao[6]) may also be remembered (*Chhien Han Shu*, ch. 96B, p. 6*a*). Having 'gained a reputation as a historian, calligrapher, and manager of business' (*nêng shih, shu, hsi shih*[7]) (Wylie (11), p. 90), she was entrusted, just before −51, with important diplomatic functions in Sinkiang.

[c] Liu Tzu-Hui[8] (*fl.* +1127). Commentators of much earlier date said that they were charts.

[d] *Hou Han Shu*, ch. 46, p. 1*b*.

[e] *Hou Han Shu*, ch. 1B, p. 11*b*.

[f] *Hou Han Shu*, ch. 106, p. 7*b*.

[g] Now ch. 29 of the *Shih Chi*.

[h] *Hou Han Shu*, ch. 89, p. 2*a*.

1 鄧禹 2 大司空 3 王景 4 河渠書 5 張衡
6 馮嫽 7 能史, 書, 習事 8 劉子翬

'cast a network (of coordinates) about heaven and earth, and reckoned on the basis of it' (*wang lo thien ti erh suan chih*[1]). The celestial coordinates must have been the *hsiu*;[a] unfortunately, we cannot tell exactly what the terrestrial ones were. The title of one of his books was: 'Discourse on Net Calculations' (*Suan Wang Lun*[2]), and there was also a 'Flying Bird Calendar' (*Fei Niao Li*[3]), but if the word *li* were a mistake for *thu*,[4] as seems possible (Sun Wên-Chhing),[b] the title may have referred to a 'Bird's-Eye Map'.[c] That Chang Hêng occupied himself with map-making is sure, for a *Ti Hsing Thu*[5] (physical geography map) was presented by him in +116. At a later point the question of possible connections with the Greek cartographers will be raised.

But the San Kuo and early Chin periods were even more important than the Han for the attainment of the definitive style of Chinese cartography. The following story is reported in the *Shih I Chi*[6] (Memoirs on Neglected Matters) by Wang Chia[7] (late +3rd century).[d]

> Sun Chhüan[8] (reigned +222 to +248 as first emperor of the Wu State) searched for an expert painter to draw a map with mountains, rivers, and all physical features, for military purposes. The younger sister of the prime minister was presented to him (as a suitable person), and he asked her to limn the mountains, rivers and lakes of the nine continents. She suggested that as the colours of a drawing would fade it would be better to make the map in embroidery, and this was accordingly done....

Perhaps it may be doubted whether the result was very scientific. But there was living at the time a young man who was destined to be, as Chavannes (10) calls him, the father of scientific cartography in China. This was Phei Hsiu[9] (+224 to +271).

In +267 Phei Hsiu was appointed Minister of Works (Ssu Khung[10]) by the first emperor of the unifying dynasty of the Chin. The 35th chapter of the *Chin Shu* preserves[e] particulars of the map-making in which he then engaged, together with the text of his preface to the maps. It is of great interest.

> Considering that his office concerned the land and the earth; and finding that the names of mountains, rivers and places, as given in the Yü Kung, had suffered numerous changes since ancient times, so that those who discussed their identifications had often proposed rather forced ideas, with the result that obscurity had gradually prevailed; Phei Hsiu made a critical study of ancient texts, rejected what was dubious, and classified, whenever he could, the ancient names which had disappeared; finally composing a geographical map of the Tribute of Yü in 18 sheets, with the title *Yü Kung Ti Yü Thu*.[11] He presented it to the emperor, who kept it in the secret archives.

[a] See p. 266 above. [b] (*4*), pp. 124, 126, 130, 133.

[c] At any rate, Shen Kua seems so to have interpreted it (*MCPT*, Pu appendix, ch. 3, para. 6), in the +11th century. Cf. Hu Tao-ching (*1*), vol. 2, pp. 991 ff.

[d] Ch. 8, p. 1*b*, tr. auct. This was certainly the lady Chao, sister of Chao Ta;[12] she was a most talented person and we shall meet with her again (Sect. 31).

[e] Pp. 3*a* ff.

1 網絡天地而算之 2 算罔論 3 飛鳥曆 4 圖 5 地形圖
6 拾遺記 7 王嘉 8 孫權 9 裴秀 10 司空
11 禹貢地域圖 12 趙達

The preface said:

'The origin of maps and geographical treatises goes far back into former ages. Under the three dynasties (Hsia, Shang and Chou) there were special officials for this (Kuo Shih[1]). Then, when the Han people sacked Hsien-yang, Hsiao Ho collected all the maps and documents of the Chhin. Now it is no longer possible to find the old maps in the secret archives, and even those which Hsiao Ho found are missing; we only have maps, both general and local, from the (Later) Han time. None of these employs a graduated scale (*fên lü*[2])[a] and none of them is arranged on a rectangular grid (*chun wang*[3]). Moreover, none of them gives anything like a complete representation of the celebrated mountains and the great rivers; their arrangement is very rough and imperfect, and one cannot rely on them. Indeed some of them contain absurdities, irrelevancies, and exaggerations, which are not in accord with reality, and which should be banished by good sense.

'The assumption of power by the great Chin dynasty has unified space in all the six directions. To purify its territory, it began with Yung and Shu (Hupei and Szechuan), and penetrated deeply into their regions, though full of obstacles. The emperor Wên then ordered the appropriate officials to draw up maps of Wu[4] and Shu.[5] After Shu had been conquered and the maps were examined, with regard to the distances from one another of mountains, rivers and places, the positions of plains and declivities, and the lines of the roads, whether straight or curved, which the six armies had followed; it was found that there was not the slightest error.[b] Now, referring back to antiquity, I have examined according to the Yü Kung the mountains and lakes, the courses of the rivers, the plateaus and plains, the slopes and marshes, the limits of the nine ancient provinces and the sixteen modern ones, taking account of commanderies and fiefs, prefectures and cities, and not forgetting the names of places where the ancient kingdoms concluded treaties or held meetings; and lastly, inserting the roads, paths, and navigable waters, I have made this map in eighteen sheets.

'In making a map there are six principles observable:

(1) 'The graduated divisions (*fên lü*[2]), which are the means of determining the scale to which the map is to be drawn (*so i pien kuang lun chih tu yeh*[6]).[c]

(2) 'The rectangular grid (*chun wang*[3]) (of parallel lines in two dimensions), which is the way of depicting the correct relations between the various parts of the map (*so i chêng pi tzhu chih thi yeh*[7]).[d]

(3) 'Pacing out the sides of right-angled triangles (*tao li*[8]), which is the way of fixing the

[a] It is interesting that Phei Hsiu should have employed this term (*lü*), for its most ancient form (K498) was a pictogram of a *net* for catching birds. It has already been encountered as a technical term for ratio in the mathematics Section, p. 99 above.

[b] Chavannes (10) suggests that Phei Hsiu himself had begun by being cartographer to the Chin general staff, as one might say.

[c] Note the use of an expression here which compares the distance between the parallel lines with the ruts made according to the gauge of chariot wheels.

[d] Our interpretation here differs from that of Chavannes, who translated *chun wang*[3] by 'exact orientation'. *Chun* is surely the horizontal lines. Its basic meaning is 'water-level' (cf. *Mêng Tzu*, IV (1), i, 5). Wang surely carries an undertone of verticality, and means the vertical lines. As an astronomical technical term, it means the opposition of moon and sun at full moon; when the full moon is making its transit, the sun is directly below it. The term *wang thung*, as we saw above (p. 332), means the polar sighting-tube, for use in the (vertical) meridian. And in the *Tso Chuan* there is a character with a deformed neck which made him always look upwards (*wang shih*[9]) (Duke Ai, 14th year; Couvreur (1), p. 694).

[1] 國史 [2] 分率 [3] 準望 [4] 吳 [5] 蜀
[6] 所以辨廣輪之度也 [7] 所以正彼此之體也 [8] 道里 [9] 望視

lengths of derived distances (i.e. the third side of the triangle which cannot be walked over) (*so i ting so yu chih shu yeh*[1]).[a]

(4) '(Measuring) the high and the low (*kao hsia*[2]).

(5) '(Measuring) right angles and acute angles (*fang hsieh*[3]).

(6) '(Measuring) curves and straight lines (*yü chih*[4]). These three principles are used according to the nature of the terrain, and are the means by which one reduces what are really plains and hills (lit. cliffs) to distances on a plane surface (*so i chiao i hsien chih i yeh*[5]).

'If one draws a map without having graduated divisions, there is no means of distinguishing between what is near and what is far. If one has graduated divisions, but no rectangular grid or network of lines, then while one may attain accuracy in one corner of the map (*i yü*[6]),[b] one will certainly lose it elsewhere (i.e. in the middle, far from guiding marks). If one has a rectangular grid, but has not worked upon the *tao li* principle, then when it is a case of places in difficult country, among mountains, lakes or seas (which cannot be traversed directly by the surveyor), one cannot ascertain how they are related to one another. If one has adopted the *tao li* principle, but has not taken account of the high and the low, the right angles and acute angles, and the curves and straight lines, then the figures for distances indicated on the paths and roads will be far from the truth, and one will lose the accuracy of the rectangular grid (*shih chun wang chih chêng i*[7]).[c]

'But if we examine a map which has been prepared by the combination of all these principles, we find that a true scale representation of the distances is fixed by the graduated divisions. So also the reality of the relative positions is attained by the use of paced sides of right-angled triangles; and the true scale of degrees and figures (*tu shu*[8]) is reproduced by the determinations of high and low, angular dimensions, and curved or straight lines. Thus even if there are great obstacles in the shape of high mountains or vast lakes, huge distances or strange places, necessitating climbs and descents, retracing of steps or detours—everything can be taken into account and determined. When the principle of the rectangular grid is properly applied, then the straight and the curved, the near and the far, can conceal nothing of their form from us.'[d]

Although Phei Hsiu left us so clear an account of his methods, his actual maps did not survive in any form, though modern scholars have made attempts at reconstructing them—Herrmann (8, 9), for instance, who considers Phei Hsiu quite worthy to be compared with Ptolemy. In +1697 Hu Wei[9] had already made such a reconstruction in his *Yü Kung Chui Chih*[10] (A few Points in the Vast Subject of the Yü Kung).[e] There was a tradition among later scholars[f] that the map of Phei Hsiu had been constructed

[a] Again the interpretation differs from Chavannes, who took *tao li*[11] to mean simply distances measured out on roads. He was therefore unable to make sense of the sentence about the third principle in the succeeding paragraph. In connection with this and the three following principles, we may recall that Phei Hsiu was a contemporary of Liu Hui, whose work on surveying by the use of similar right-angled triangles was discussed in the mathematics Section (p. 31 above).

[b] Note the use of a technical term in geometry, related to the extraction of roots, discussed above (p. 66) in the mathematics Section.

[c] Chavannes confessed his inability to translate this phrase. It was because he had misinterpreted the second principle. Note the sorites form of the whole paragraph.

[d] Tr. auct., adjuv. Chavannes (10), Vacca (6).

[e] *Huang Chhing Ching Chieh*, ch. 27, p. 53*b*.

[f] So Chhüan Tsu-Wang in *Huang Chhao Ching Shih Wên Pien*, ch. 79, p. 12*b* (*c*. +1750).

[1] 所以定所由之數也 [2] 高下 [3] 方邪 [4] 迂直
[5] 所以校夷險之異也 [6] 一隅 [7] 失準望之正矣 [8] 度數
[9] 胡渭 [10] 禹貢錐指 [11] 道里

on a scale of two inches to the 1000 *li*, but this is uncertain. They fully recognised[a] that the rectangular grid was at least as old as Phei Hsiu.[b]

There is room for some speculation as to the source which may have suggested the coordinate system to Phei Hsiu and Chang Hêng. The characters combined in the ancient phrase *ching thien*[1] (well-field system), which had been the subject of social and economic debate since feudal times, both plainly betray an ancient manner of thinking of land-allotment in coordinate terms. Then there was the system of concentric squares (Fig. 204) described in the *Shu Ching* (Yü Kung chapter) and already mentioned (p. 502 above). Nor should the origins of the abacus be forgotten; it may be significant that the first name in Chinese history to be associated with this device was that of the Taoist mathematician Hsü Yo, whose *floruit* was just between the time of Chang Hêng and that of Phei Hsiu.[c] At what time the terms *ching*[2] and *wei*[3] were first used for the *chun-wang* coordinates (as Phei Hsiu called them) is hard to say, but they certainly meant textile warp and weft before they were employed by map-makers.[d] The very fact of drawing maps on silk, from the Chhin onwards, would invite the suggestive idea that the position of a place could be fixed by following a warp and a weft thread to their meeting-place. This was perhaps the significance of Sun Chhüan's girl cartographer, and we know that both Phei Hsiu and Chia Tan (the great Thang map-maker shortly to be mentioned) used silk for their maps; the former used as much as eighty rolls of fine silk (*chien*[4]), and the latter had his maps drawn on very smooth white silk (*hsien kao*[5]).[e] Shen Kua in the Sung used thin lustrous silk (*chüan*[6]) for the same purpose.

Another set of related ideas and techniques was that of the diviner's board (*shih*[7]), the magnetic compass and chess. Evidence that these things belong together will be offered below in Section 26*i* on physics. Here I would only like to refer to the compass indications given on the diviner's board (cf. Fig. 223), and the similar markings on Han 'cosmic mirrors' (cf. Yetts, 5). Describing these mirrors, the Sung archaeologist, Wang Fu,[8] says[f] that they had, at the various compass points, names of

[a] For instance, Liu Hsien-Thing[9] in his *Kuang Yang Tsa Chi*[10] (*c.* +1695).

[b] Chhen Phan (*2*) has recently drawn attention to another version of Phei Hsiu's complaint against the maps of earlier times and their inaccuracy. It is a fragment quoted in the *Chiu Chia Chin Shu Chi Pên*[11] (Collected Texts of Nine Versions of the History of the Chin Dynasty) by Thang Chhiu[12] (ch. 5). Phei Hsiu specifically names the maps (*thu*) attached to the *Ho Thu Wei Kua Ti Hsiang*,[13] one of the 'weft' classics (*Ku Wei Shu*, ch. 32).

[c] Cf. what was said above (pp. 77, 107) about the origins of coordinate geometry.

[d] Actually the use of the word *ching* for the geographical north-south direction and of *wei* for the east-west direction, goes back to the Han in the *Ta Tai Li Chi* (ch. 81; cf. R. Wilhelm (6), pp. 250 ff.). This is found also in the commentaries of Kao Yu[14] of the Later Han (on *Lü Shih Chhun Chhiu*, chs. 62 ff., and on *Huai Nan Tzu*, ch. 4). It may be noted that this is the opposite of the modern usage, where *ching* means latitude and *wei* longitude. The change seems to have come about in the Thang.

[e] Wang Yung (*1*), pp. 60, 68.

[f] In his *Hsüan-Ho Po Ku Thu Lu*[15] (Hsüan-Ho reign-period Illustrated Record of Ancient Objects), cit. *KCCY*, ch. 56, p. 1*a*.

[1] 井田 [2] 經 [3] 緯 [4] 縑 [5] 纖縞 [6] 絹 [7] 栻
[8] 王黼 [9] 劉獻廷 [10] 廣陽雜記 [11] 九家晉書輯本 [12] 湯球
[13] 河圖緯括地象 [14] 高誘 [15] 宣和博古圖錄

divinities (*ling*[1]), trigrams (*kua*[2]), the duodenary cyclical characters (*chih*[3]), the fortnightly periods (*chhi*[4]), and so on. Each side of the square being divided into three by the eight radiating divisions which separate the twenty-four azimuth points into groups of three, then if coordinate lines were drawn[a] across the space so enclosed, nine small squares (*yeh*[5] or *chiu kung*[6]) would be formed.[b] The word *yeh*[5] will be met with again in connection with the astronomical-geographical work of Li Shun-Fêng in the Thang (see below, p. 544). The central division was called *chen*.[7][c] Wang Fu adds that there are intersecting points of the coordinates (*tsho tsung ching wei*[8]), and that on the background of the mirror the five sacred mountains (*wu yo*[9]) are sometimes represented.[d] Although we are doubtless dealing here with very symbolic or diagrammatic representations, they did go back, on boards and mirrors, to the cosmological speculations of the Han people, and could not have been unknown to Chang Hêng and Phei Hsiu. Moreover, the ancient methods of divination in which 'pieces' like chess-men (*chhi*[10]) were thrown on to a marked board, perhaps the *shih* itself, seem to have been connected with another old system of coordinates, namely, the multiplication table, for which the characteristic expression is *li-chhêng*.[11] It will be remembered (pp. 9, 36, 107 above) that one of these was found in the Tunhuang manuscripts (Li Nien, 7). Another Tunhuang manuscript[e] has the title *Chiu Kung Hsing Chhi Li-Chhêng*[12]—'Tabulation of Data in the Calendar of the Nine Palaces in which the Chess Pieces Move'. This calendar had been introduced by the astronomer Li Yeh-Hsing in +548, and the tables were drawn up by Wang Chhen.[f] But the exact connections between the coordinate system of chess (or rather 'pre-chess', if the term be allowable) and that of cartographic practice require much further investigation.

The mention of astronomical aspects of the map grids raises at once the question to what extent the Chinese cartography of Phei Hsiu and Chang Hêng was keyed to celestial phenomena. In this respect there would seem to have been little difference between the Chinese and the Greeks, for while the latter used the gnomon shadow and the length of the solstitial day to determine latitude, the former were also perfectly aware that the shadow length varied continuously in the north-south line.[g] The *Chou Li* says[h] that the Surveyors (Thu Fang Shih[13]) concern themselves with the method of the gnomon shadow template[i] (*chang thu kuei chih fa*[14]) for determining the sun's shadow length, and by its aid measure the earth, constituting fiefs and principalities, i.e. presumably fixing their boundaries. As for longitude, the Chinese were no worse

[a] The expression used is *thou*,[15] lit. 'thrown across', which may be significant; see on, Sect. 26*i*.

[b] Cf. the nine divisions of heaven (*yeh*[5]) in *Huai Nan Tzu*, ch. 3. The other term persisted for many centuries as *chiu kung ko*[16] to mean the squared paper on which children practised writing characters.

[c] Again a significant word, which will occur in connection with the Buddhist-Taoist wheel-maps, see below, p. 567.

[d] Cf. below, p. 566, in connection with these wheel-maps.

[e] British Museum, Stein no. 6164.

[f] See p. 107 above.

[g] Cf. pp. 292 ff. above.

[h] Ch. 8, p. 29*b* (ch. 33, p. 60; Biot (1), vol. 2, p. 279).

[i] Cf. above, pp. 286 ff.

[1] 靈 [2] 卦 [3] 支 [4] 氣 [5] 野 [6] 九宮 [7] 鎮
[8] 錯綜經緯 [9] 五岳 [10] 棊 [11] 立成 [12] 九宮行棊立成
[13] 土方氏 [14] 掌土圭之法 [15] 投 [16] 九宮格

PLATE LXXX

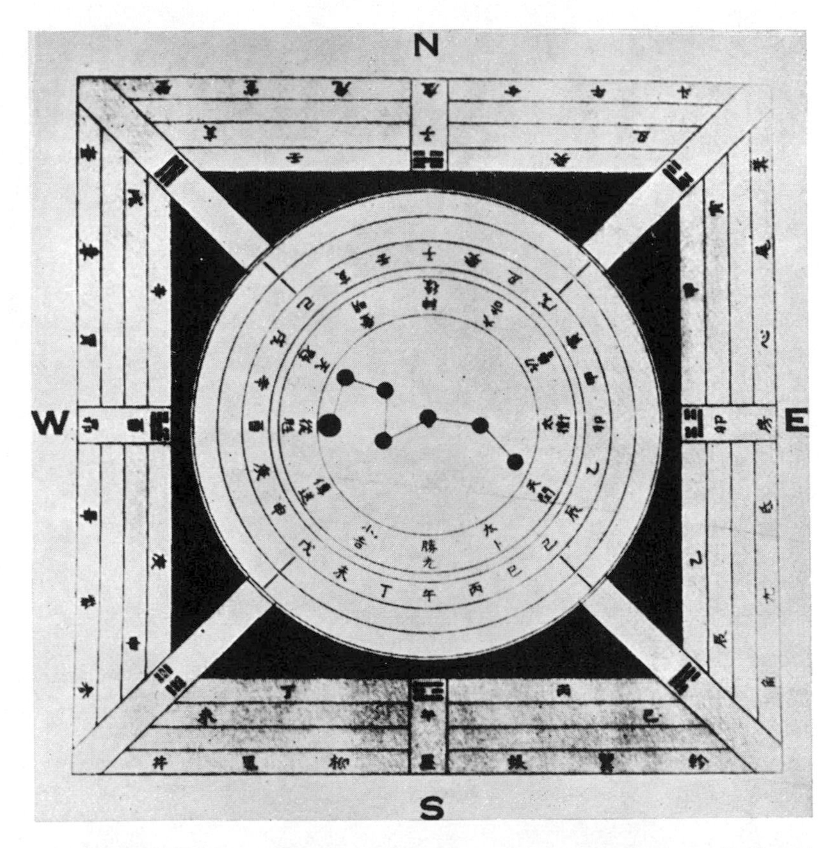

Fig. 223. A reconstruction of the *shih* or diviner's board of Han times (Rufus (2) after Harada & Tazawa), a symbolic representation of heaven and earth which is connected with the origins of the magnetic compass. The square plate of the earth was surmounted by a rotating round plate signifying the heavens. The former was marked with cyclical characters and *hsiu*, the latter carried the figure of the Great Bear, with cyclical characters and prognosticatory signs. Fragments of these boards were first found in two Han tombs in Korea, the tomb of Wang Hsü (*d.* +69), and the Painted Basket tomb.

off than the Greeks. Its measurement with any degree of accuracy did not become possible until the 18th century, with the invention of the marine chronometer.[a] Throughout antiquity and the middle ages, dead reckoning was the only way.[b]

(iii) *Development in Thang and Sung*

Between the Chin and the Thang, maps continued to be produced, but we know little about their scientific value. The 97th chapter of the *Pei Shih* (History of the Northern Dynasties) gives a description of the Western (i.e. Central Asian) peoples and kingdoms, which, from internal evidence, also took the form of a map (+437), and which has been thus reconstructed by Herrmann (11). In the Liang (early +6th century) we hear reports of maps inscribed on stone steles,[c] which is interesting in connection with the magnificent (and still existing) Sung examples, shortly to be examined. In +605 a famous grid map was made by Phei Chü,[1] who was Commercial Commissioner for the Sui dynasty with his headquarters at Chang-yeh (Ganchow) on the Old Silk Road in Kansu. This also has been reconstructed by Herrmann (11), while Jäger (1) has translated both Phei Chü's biography[d] and the few fragments which remain of his *Sui Hsi Yü Thu*[2] (Map of the Western Kingdoms in the Sui Dynasty).[e]

The Thang saw a great development of map-making. The wide extensions of the empire which were added early in the dynasty stimulated the mapping of Central Asia, and no doubt account for the books and maps of Hsü Ching-Tsung[3] in +658,[f] Wang Ming-Yuan[4] in +661,[g] and Wang Chung-Ssu[5] in +747.[h] Unfortunately, all these were afterwards lost. Then came the greatest Thang cartographer, Chia Tan[6] (+730 to +805), and although it is true that his work did not survive either, we nevertheless have a good deal of information about it. It was in +785 that he was entrusted by the emperor (Tê Tsung) with the preparation of a map of the whole empire, but the work was not completed till +801, smaller sections, as of Kansu and Szechuan, being presented in the meantime.[i] Its title was *Hai Nei Hua I Thu*[7] (Map of both Chinese and Barbarian Peoples within the (Four) Seas). The map was 30 ft. long and 33 ft. high, constructed on a grid scale of 1 in. to 100 *li*. It must therefore have covered an *oikoumene* of 30,000 *li* from east to west and 33,000 *li* from north to south; which is much more than the 7000 *li* square shown in Fig. 226 (the map of +1137), and must

[a] Lloyd Brown (1), pp. 208 ff.
[b] For the possible use of hodometers in the time of Phei Hsiu and Chang Hêng, see below, p. 577.
[c] *Shu I Chi*,[8] ch. 2, p. 6*b*.
[d] *Sui Shu*, ch. 67, p. 9*a*.
[e] The work of Phei Chü was known unusually early to occidental scholars, e.g. Ritter (1), vol. 7, p. 560, and K. F. Neumann (1), p. 187. There was a map of the Yantgze River at this time which is thought to have used the grid.
[f] *Hsin Thang Shu*, ch. 221 A, p. 10*a*, and biogr. ch. 223, pp. 1*a* ff.
[g] Herrmann (8), p. 249.
[h] *Hsin Thang Shu*, ch. 43 B, p. 14*a*, and biogr. ch. 133, pp. 4*b* ff.
[i] Some authorities give these dates as +784 and +796.

1 裴矩 2 隋西域圖 3 許敬宗 4 王名遠 5 王忠嗣
6 賈耽 7 海內華夷圖 8 述異記

therefore have been, to all intents and purposes, a map of Asia.[a] Herrmann (8) notes that this scale is about 1 : 1,000,000. One is tempted to compare the great map of Chia Tan with that of Cassini (*c.* +1680) at the Paris Observatory,[b] which had a diameter of 24 ft. and used a polar azimuthal projection.

Chia Tan's contemporary Li Chi-Fu was also a geographer of importance.[c] One of his maps was connected with his *Yuan-Ho Chün Hsien Thu Chih* already mentioned, and another was concerned with all the fortified points and strategic locations north of the Yellow River. This was set up in the imperial baths and the emperor (Hsien Tsung) consulted it daily.[d] Li Tê-Yü[1] (the son of Li Chi-Fu) continued geographical work when governor of Szechuan, making military maps to facilitate the control of the barbarian tribes.[e] All these maps were connected with itineraries. At about the same time Yuan Chen[2] had a road-map prepared on the occasion of the marriage of a princess to the Khagan of the Uighurs in +821.[f] Chia Tan prepared itineraries from China to Korea, Tongking, Central Asia, India, and even Baghdad;[g] others are extant in the excellent *Man Shu*[3] (Book of the Barbarians) by Fan Chho[4] (*c.* +862), and most of these have been analysed by Pelliot (17) and Chavannes (14).[h]

It is possible that in the Thang a further effort was made to link geographical with celestial coordinates. The hint arises from an obscure group of writings which Chavannes overlooked but to which Wang Yung has drawn attention. In this period we hear of certain maps (*fang chih thu*[5] or *fang yü thu*[6]) which yet do not seem to be, as their names would at first sight imply, associated with local topographies; and which are associated with Taoist and Buddhist scholars, especially Li Shun-Fêng, I-Hsing and Lü Tshai.[i] The astronomical chapters of both the *Thang Shu* say that Li Po,[7] the father of Li Shun-Fêng, resigned his official position to live the life of a Taoist, and produced one of these maps.[j] The text continues:

The celestial bodies are suspended in the heavens without changing as time passes, but the names of provinces and districts have always changed, and thus have made difficulties for the scholars of later generations. So in the Chên-Kuan reign-period (+627 to +649) Li Shun-Fêng wrote his *Fa Hsiang Chih*[8] (The Miniature Cosmos),[k] in which all the

[a] *Chiu Thang Shu*, ch. 138, p. 7*b*. Ancient place-names were inserted in black, and those of Chia Tan's own time in red.

[b] Lloyd Brown (1), p. 219.

[c] Cf. above, pp. 490, 520.

[d] *Chiu Thang Shu*, ch. 148, p. 6*a*; *Hsin Thang Shu*, ch. 146, p. 5*b*.

[e] *Hsin Thang Shu*, ch. 180, p. 3*a*.

[f] *TSCC*, *Ching chi tien*, ch. 429, *i wên*, p. 3*a*.

[g] *Hsin Thang Shu*, ch. 43B, p. 14*a*.

[h] This was something in common with European geographical literature, which was full of itineraries, from the *Bordeaux Pilgrim* of +333 onwards (Beazley (1), vol. 1, pp. 26, 57). Cf. pp. 513, 524 above.

[i] All have been encountered before, pp. 350, 202, and 323.

[j] We met him already as an astronomer, p. 201 above.

[k] From other passages we know that this lost book discussed the merits and defects of various types of armillary spheres (*Chiu Thang Shu*, ch. 35, p. 1*a*; *Hsin Thang Shu*, ch. 31, p. 1*b*; *Yü Hai*, ch. 4, p. 21*a*), but it must have done much more than this.

1 李德裕 2 元稹 3 蠻書 4 樊綽 5 方志圖
6 方域圖 7 李播 8 法象志

provinces and districts of the Thang were included. At the beginning of the Khai-Yuan reign-period (+713 onwards), I-Hsing remodelled this work and reduced its size.[a]

The biographies of Lü Tshai and of Shang Hsien-Fu[1] in the *Chiu Thang Shu* speak[b] of similar maps made about +630 and +695 respectively. Significantly, the latter was an astronomer in charge of the imperial observatory. Then there was an official, Lü Wên,[2] who wrote a preface[c] about +800 to a geographical work *Ti Chih Thu*[3] by Li Kai[4] in which he said:

> The boundaries of every square inch correspond to the celestial divisions on high. The (relation between the) phenomena in the heavens and the aspect (i.e. physical features) of the earth can thus be brilliantly seen (*Fang tshun chih chieh erh shang tang hu fên yeh; chhien hsiang khun shih ping jan kho kuan*[5]).[d]

The problem now is, what exactly was this cartographic movement in the Thang? There are three obvious possibilities. First, *fên yeh* certainly included straight astrology. The association of particular earthly regions with sections of the sky was a very old idea in China, found in early Han and pre-Han texts.[e] But it may also have laid a new emphasis on the importance of showing physical features as opposed to political territorial divisions. To the fixed stars would correspond the fixed mountains and rivers, not the changeable city names. As we shall see in a moment, this might be the background of the greatest extant achievement of +12th century map-making. Thirdly, the efforts of these Taoists and Buddhists may have been directed towards a really astronomical coordinate system. It would have been just as easy to draw meridian lines parallel to the hour-circles separating the different *hsiu*, as it had been for the Greeks to draw them on the model of celestial longitudes. And it would have been just as difficult for the Thang people to have fixed their terrestrial longitudes in any way, as it was for everyone else before the construction of Mr Harrison's marine chronometer.[f] I am not sure that this would have implied (as would seem at first sight) a spherical earth, though by the time of the Thang there had been so many foreign contacts that men such as Li Shun-Fêng could hardly have been unaware that there was such a hypothesis.[g] It will be remembered that already for the +10th century documentary evidence exists[h] that a projection analogous to that of Mercator was used, the *hsiu* being represented as long rectangles centred on the equator and of course very distorted towards the poles. It would, then, have been possible to fit the system of the *hsiu* into the traditional terrestrial rectangular grid.[i]

[a] *Chiu Thang Shu*, ch. 36, p. 1*a*, tr. auct.; cf. *Hsin Thang Shu*, ch. 31, p. 8*b*.
[b] Chs. 79 and 191.
[c] He had been ambassador to Turfan, which may have encouraged his geographical interests.
[d] *Lü Ho-Shu Wên Chi*, ch. 3, tr. auct.
[e] Cf. p. 200 above. *Lü Shih Chhun Chhiu*, ch. 62. *Huai Nan Tzu*, ch. 3, pp. 15*a*, *b*, systematised in *Chin Shu*, ch. 11, pp. 19*a* ff.
[f] Cf. Gould (1).
[g] Besides, they would have known the ideas of the founders of the Hun Thien cosmology long before.
[h] Cf. pp. 264, 276 above regarding the Tunhuang manuscript star-map.
[i] The chief study of this *fên yeh* system is due to Hsü Wên-Ching,[6] who produced it in +1723 (*Thien Hsia Shan Ho Liang Chieh Khao*[7]).

[1] 尙獻甫 [2] 呂溫 [3] 地志圖 [4] 李該
[5] 方寸之界而上當乎分野乾象坤勢炳然可觀 [6] 徐文靖
[7] 天下山河兩戒考

The second alternative above may serve to alert us to the possibility of a primitive kind of contour map originating in the Thang. Ogawa (*1*) has drawn attention to a very interesting chart of Thai Shan[1] (the eastern sacred mountain) which appears in +17th-century editions of an anonymous work known as the *Wu Yo Chen Hsing Thu*[2] (Map of the True Topography of the Five Sacred Mountains). As Fig. 224 shows, the system of delineating the mountain ranges compares not unfavourably with the modern style placed beside it.[a] Quotations from this book in the *Thai-Phing Yü Lan*

Fig. 224. Early attempt at contour mapping; on the right a representation of the Thai Shan mountain range from a +17th-century edition of the *Wu Yo Chen Hsing Thu*, the text of which is of much earlier but uncertain date. On the left, for comparison, a contour map of modern type. From Ogawa (*1*).

encyclopaedia[b] show that at some time or other quite precise measurements were made of the breadth and depth of gorges. The date of the text is uncertain, but it may be significant that the title is mentioned in the *Han Wu Ti Nei Chuan*[3] (The Inside Story of Emperor Wu of the Han), an anonymous Taoist romance probably of the +4th century attributed to Ko Hung. This contains a cartographical passage which will be examined later in connection with the Chinese wheel-map tradition.

As has been indicated above (p. 521) a great deal of geographical work was done in the Sung, and it is from this period that the oldest examples of Chinese cartography still extant derive. Early in the dynasty (before +1000) there is an account[c] of a map

[a] Cf. the geomantic diagram already reproduced in Fig. 45 (Vol. 2, opp. p. 360). There can be no doubt that for many centuries in China specialists in particular sciences or pseudo-sciences had their own quite effective conventions in physical geographic mapping.

[b] E.g. *TPYL*, ch. 44, p. 5*a*; ch. 663, p. 1*b*.

[c] *Sung Shih*, ch. 292, p. 6*b*.

[1] 泰山 [2] 五嶽眞形圖 [3] 漢武帝內傳

of the Western countries, and strategic maps of Kansu, prepared by a geographer in the imperial service, Shêng Tu.[1] A Commissioner of the Tibetan Borderlands, Liu Huan,[2] made a map of the mountains and presented it to the emperor in +1040.[a] Remarkably accurate plans of cities had already been made in the Thang. In +1080 Liu Ching-Yang[3] and Lü Ta-Fang were instructed to make a historical map of the city of Chhang-an (Sian), which they did to scale (*chê fa*[4]) at 2 in. to the mile, identifying and marking sites of ancient palaces and the like.[b] This is the basis of the plans preserved in the *Chhang-an Chih Thu*[5] of Li Hao-Wên.[6]

But all other records of the +11th century are overshadowed by the two magnificent maps which still exist, carved in stone in +1137,[c] and now in the 'Forest of Steles' (Pei Lin[7]) at Sian. To these the papers of Chavannes (10) and Aoyama (*4*) were primarily devoted. The first is entitled *Hua I Thu*[8] (Map of China and the Barbarian Countries), the second *Yü Chi Thu*[9] (Map of the Tracks of Yü (the Great)). Both are about 3 ft. square (see Figs. 225 and 226).

Though the dates of the present examples are the same within a few months,[d] the first seems a good deal more archaic than the second. It has no grid, the coast-line is sketchy, the Shantung peninsula is not shown, and the river-systems are imperfect. The barbarian countries are represented rather by texts than by geographical markings, but within China both places and mountains are shown. As the latest date in the textual material is +1043, this date may be taken as very probable for the first composition of the map. On internal evidence, the anonymous geographer had the Thang map of Chia Tan at his disposal.[e] The inscriptions mostly concern the barbarian peoples, and have been fully translated by Chavannes. Neither map shows Formosa though both have Hainan.

The second map (Fig. 226) is one which does the greatest credit to the Sung cartographers. Its inscription says that the grid scale is 100 *li* to each square. Its coastal outline is much firmer, and comparison of the network of river systems with a modern chart shows at once the extraordinary correctness of the pattern. Anyone who compares this map with the contemporary productions of European religious cosmography (e.g. Figs. 212, 214, 217) cannot but be amazed at the extent to which Chinese geography was at that time ahead of the West.[f] But although the Map of the Tracks of Yü has a more modern look than the first of these two maps, it seems rather to have belonged to a different and not much younger tradition. On the copy preserved at

[a] *Shêng Shui Yen Than Lu*[10] (Fleeting Gossip by the River Shêng) by Wang Phi-Chih,[11] ch. 2, p. 9*b*.
[b] *Yün Lu Man Chhao*, ch. 2, p. 11*a*.
[c] The reign-period given is that of a transient Chinese buffer-state in a puppet relation to the Liao.
[d] One was engraved in the 4th, the other in the 10th month.
[e] Soothill (4) actually urges that this *was* Chia Tan's map, from which the sheets for the barbarian lands had been lost, and that the texts were a substitute for them. Sung bibliographies show that versions of Chia Tan's map were still extant.
[f] Heawood (in Soothill, 4) pronounced it superior to even the best Hellenistic efforts. There was nothing like it in Europe till the Escorial MS. map of about +1550 described by de Reparaz-Ruiz (2).

1 盛度 2 劉渙 3 劉景陽 4 折法 5 長安志圖
6 李好文 7 碑林 8 華夷圖 9 禹跡圖 10 澠水燕談錄
11 王闢之

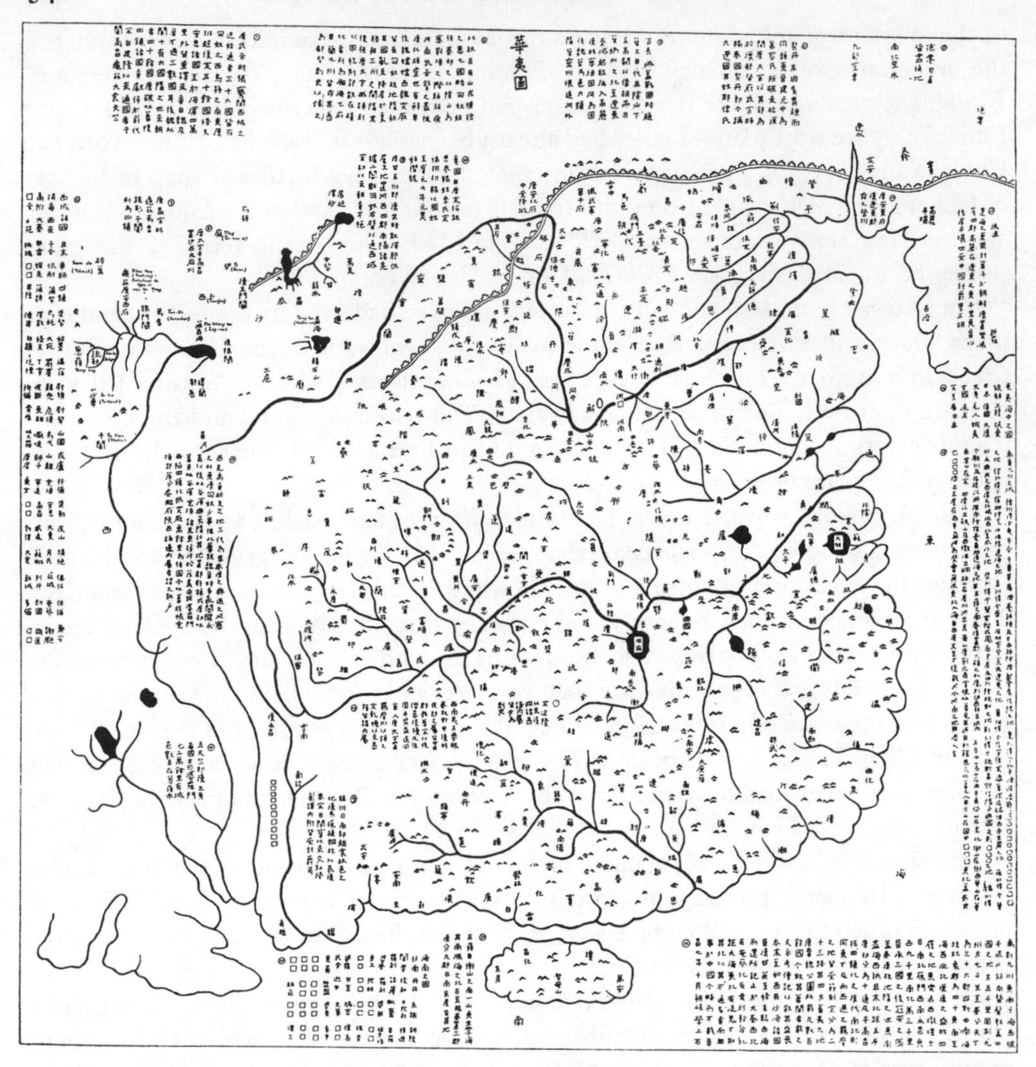

Fig. 225. The *Hua I Thu* (Map of China and the Barbarian Countries), one of the two most important monuments of medieval Chinese cartography, carved in stone in +1137 but probably dating from about +1040 (from Chavannes). The size of the original, which is now in the Pei Lin Museum at Sian, is about 3 ft. square. The name of the geographer is not known.

Chen-chiang in Chiang-su, which Yü Chhih[1] (Prefectural Director of Studies) caused to be engraved in +1142, there is mention of an earlier copy of +1100 which itself was based on the 'old Chhang-an version'.[a] In any case there is no doubt that the purpose of the second map was to instruct students in the geography of the Yü Kung.[b]

[a] There is nothing impossible in Wang Yung's suggestion that it may go back to Yüeh Shih, the geographer of the *Thai-Phing Huan Yü Chi* in the late +10th century.

[b] This is why the second map is less sound than the first on the south-western rivers (Salween, Mekong, etc.); these were not known to the Yü Kung. Similar mistakes in it were pointed out to Chu Hsi by a scholar from Szechuan (*CTCS*, ch. 50, p. 28*b*).

[1] 俞篪

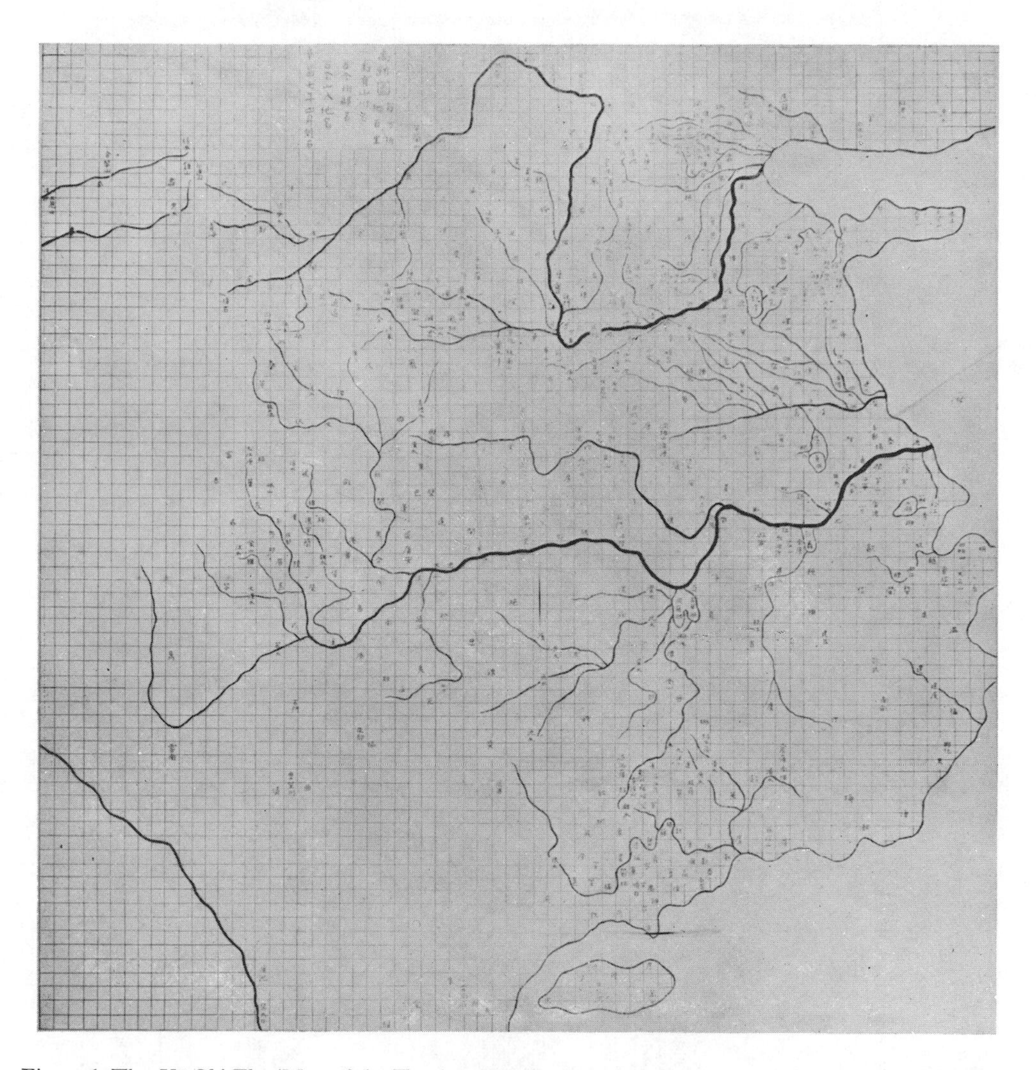

Fig. 226. The *Yü Chi Thu* (Map of the Tracks of Yü the Great), the most remarkable cartographic work of its age in any culture, carved in stone in +1137 but probably dating from before +1100 (from Chavannes). The scale of the grid is 100 *li* to the division. The coastal outline is relatively firm and the precision of the network of river systems extraordinary. The size of the original, which is now in the Pei Lin Museum at Sian, is about 3 ft. square. The name of the geographer is not known.

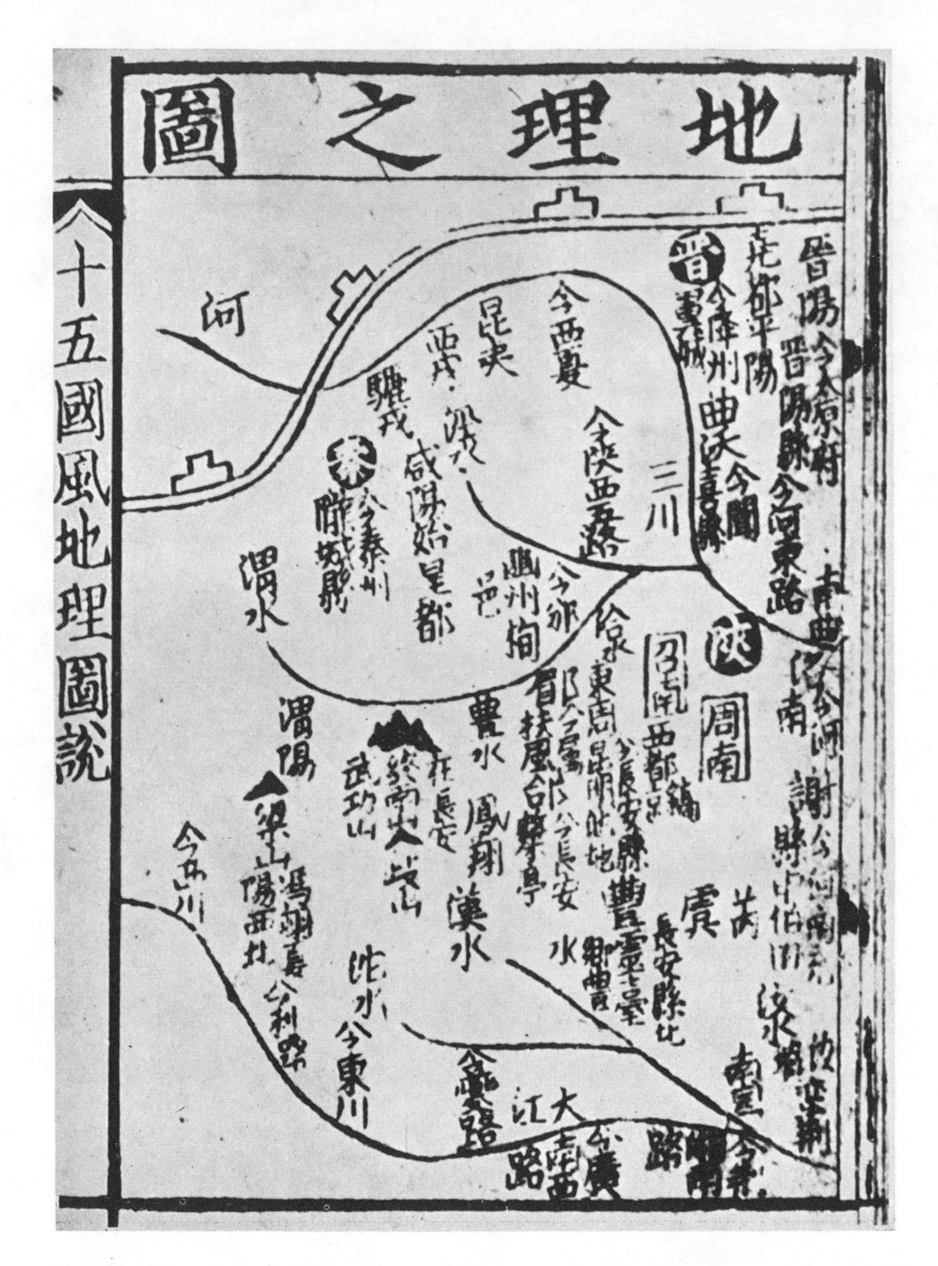

Fig. 227. The oldest printed map in any culture, a map of West China (*Ti Li chih Thu*) in the *Liu Ching Thu* (Illustrations of Objects mentioned in the Six Classics), an encyclopaedia edited by Yang Chia in the close neighbourhood of +1155 (copy in the Peking National Library). This map should be compared with that in Fig. 208; though it does not reach as far south. Provincial names are indicated here in white on black, and the line of the Great Wall is prominent.

It is of interest that in both these maps the north is at the top, and this is true for all the Sung maps which have survived. The practice of placing the south at the top seems to have originated among the Arabs rather than the Chinese[a] and to have become known only later in China. There is a Sung map included in the late Chhing book of Chang Chien[1] on the Hsi-Hsia State, *Hsi-Hsia Chi Shih Pên Mo*[2] (Rise and Fall of the Hsi-Hsia), which must date from before +1125, and in which the north is at the top (Pelliot, 18). We illustrate another from a Sung edition of the *Liu Ching Thu* (Illustrations of Objects mentioned in the Six Classics) in the Peking National Library (Fig. 227). This was edited by Yang Chia[3] and printed in the close neighbourhood of +1155; here the interest is that it gives us our first Chinese *printed* map. The first European printed map does not come until about two centuries later, in the *Rudimentum Novitiorum* (the Lübeck Chronicle, an account of the history of the world),[b] printed in that city by Lucas Brandis in +1475. If the two are compared (Fig. 228), the advantage seems to lie much with the Chinese cartographer, whose work, though crude, shows no tendency to compress the topography into an artificial disc-like form, and confines itself to the practical setting-down of the characters of place-names instead of scattering artistic and mythological drawings over the board.[c]

Of the many other Sung maps which have not survived, and of their makers, details will be found in Wang Yung.[d] Some of the most distinguished men of the time cultivated cartography, for instance Chu Hsi the Neo-Confucian philosopher, Shen Kua, Huang Shang[4] the astronomer,[e] and Lu Chiu-Shao[5] the learned brother of the idealist philosopher Lu Chiu-Yuan. An historical map, *Chih Chang Thu*,[6] was made[f] by Shui An-Li.[7] A passage from Shen Kua indicates the high strategic value which was placed on maps in the Sung period, and the disinclination to allow copies of them to leave the country. He says:

In the Hsi-Ning reign-period (+1068 to +1077) ambassadors came from Korea bringing tribute. In every *hsien* city or provincial capital which they passed through they asked for local maps, and these were made and given to them. Mountains and rivers, roads, escarpments and defiles, nothing was omitted. When they arrived at Thiehchow they asked for maps, as usual, but Chhen Hsiu,[8] who was then prefect of Yangchow, played a trick on them. He said that he would like to see all the maps of the two Chekiang provinces with which

[a] Cf. the remarks below, p. 563, on al-Idrīsī's world map.

[b] Cf. Hind (1), vol. 2, p. 363; Pollard (1), p. 50; Klebs (1), no. 867 (i); Santarem (1), vol. 3, p. 230, (2), pl. 18; Anon. (27); Bagrow (2), pp. 61, 62.

[c] Of course, Brandis' map was soon followed by the first printed edition of Ptolemy (Bologna, 1477).

[d] Mori (*1*) has described a map of about +1270 preserved in Japan. Masuda (*1*) has written on the political significance of cartography in this period.

[e] Huang Shang's maps have been studied by Aoyama (*5, 8*). He was working between +1190 and +1194, when his *Yü Ti Thu*[9] was carved on wood. A copy of it is still preserved in a Japanese cloister. Another of his maps was the basis of the great Suchow stele map shortly to be described.

[f] Cf. *Liang Chhi Man Chih*, ch. 6, p. 8*a*.

1 張鑑 2 西夏紀事本末 3 楊甲 4 黄裳 5 陸九韶
6 指掌圖 7 税安禮 8 陳秀 9 輿地圖

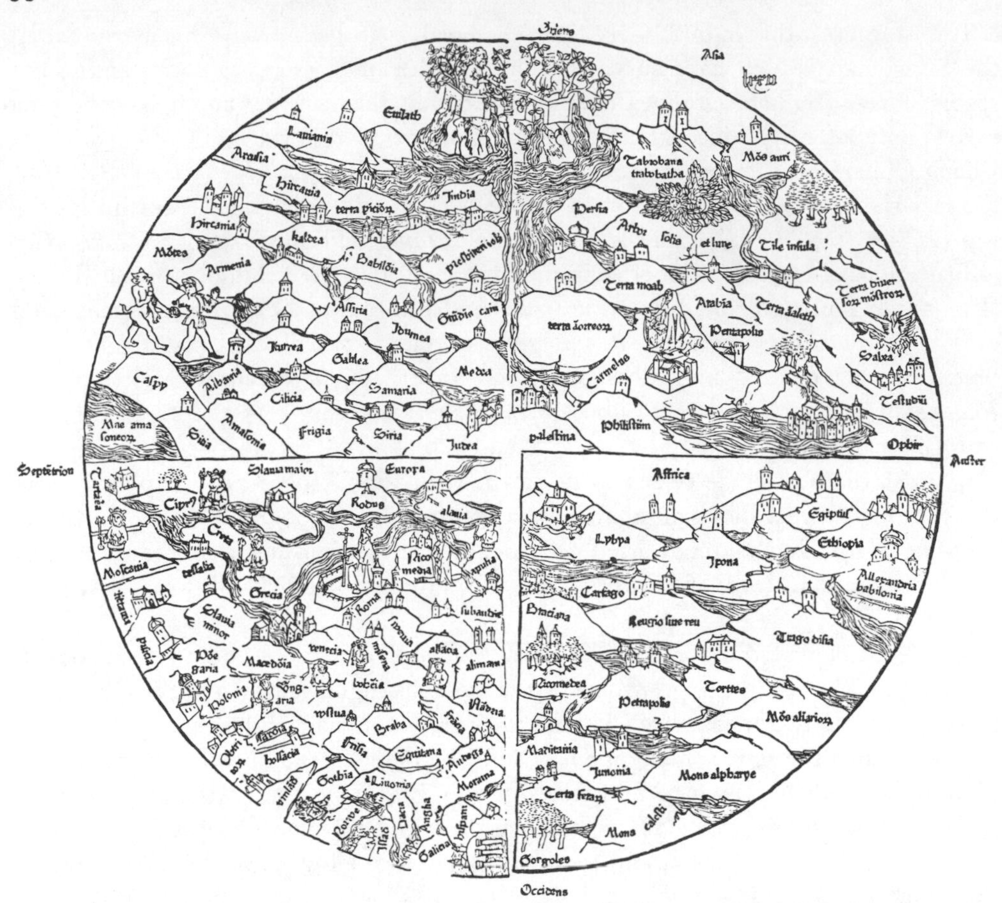

Fig. 228. The first printed European map, for comparison with Fig. 227; Lucas Brandis' woodcut for the *Rudimentum Novitiorum* of +1475. Here the T-O influence is still at work, but hardly succeeds in organising the confusion; Babylonia appears twice, Dacia adjoins Norway, and the Earthly Paradise is still in the Far East.

they had been furnished, so that he could copy them for what was now wanted, but when he got hold of them, he burnt them all, and made a complete report on the affair to the emperor.[a]

The other most important epigraphic monument of the Sung time is the companion stele to that which shows the stellar planisphere already examined (p. 278 above). This is still at Suchow, where it has been seen and studied by many modern scholars.[b]

[a] *Mêng Chhi Pi Than*, ch. 13, para. 11, tr. auct., adjuv. Chavannes (10). The *Liang Chhi Man Chih* (ch. 5, p. 4*b*) amplifies this by saying that a previous ambassador to Korea, Lu To-Sun,[1] had brought back complete maps of that country about +970; so the Koreans were only trying to 'get their own back'. Cf. Addendum below.

[b] Chavannes (8), Vacca (5), Aoyama (5), Wang Yung (*1*). It was one of four which Huang Shang had presented to the throne in +1194, but Aoyama considers that in all its essentials it goes back to Shen Kua's time a century or more earlier. Indeed it purported to represent China as it was before the loss of the capital (Khaifêng) to the Chin Tartars in +1126.

[1] 盧多遜

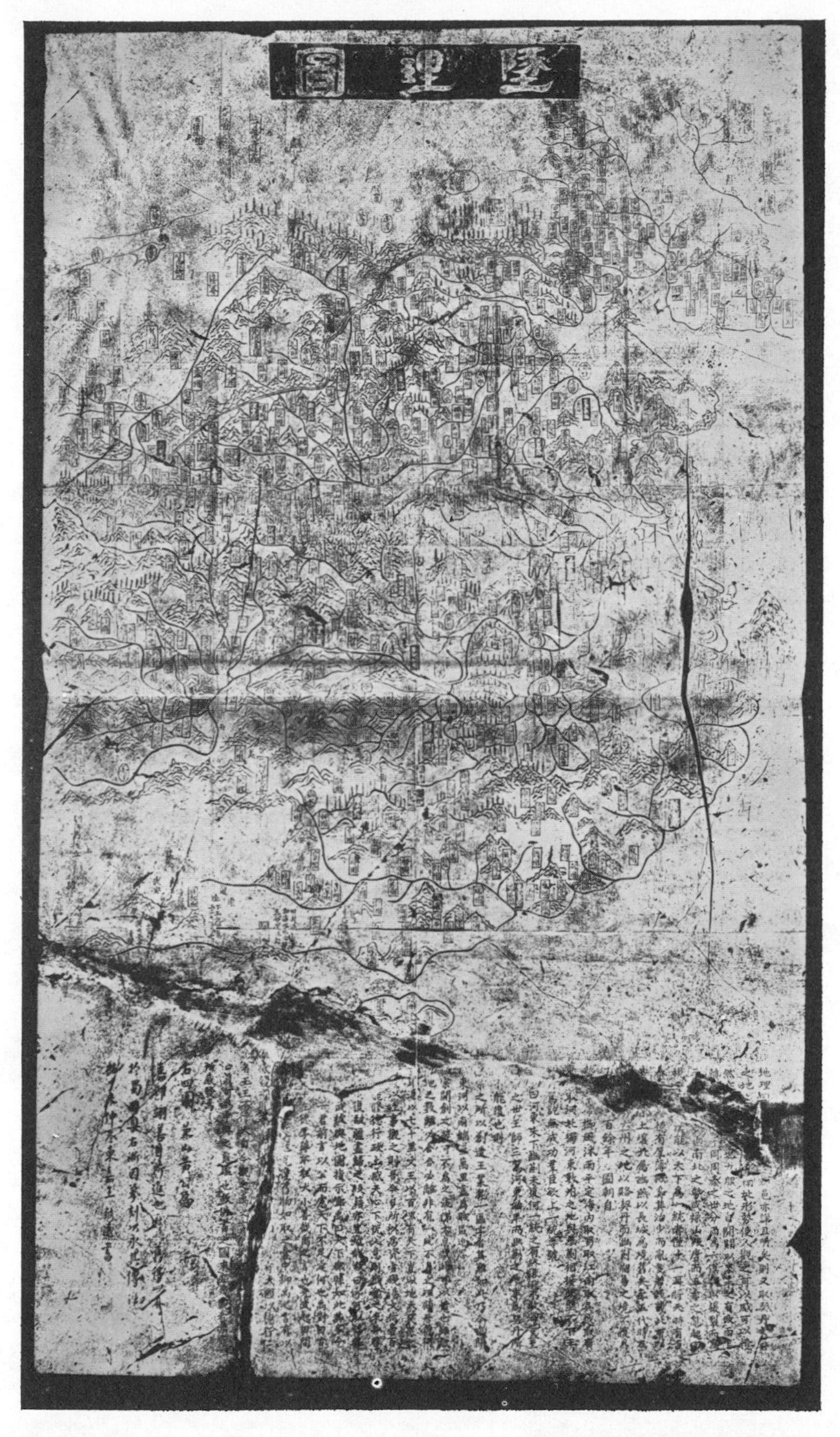

Fig. 229. The *Ti Li Thu* (General Map of China), a map in the *Hua I Thu* tradition (cf. Fig. 225), made by Huang Shang about +1193 and engraved on a stone stele at Suchow by Wang Chih-Yuan in +1247 (from Chavannes). The size of the original map is about $3\frac{3}{4} \times 3\frac{1}{4}$ ft. The character *chui* in the title is an old alternative form for *ti*.

Fig. 230. One of the narrative frescoes in the Chhien-Fo-Tung cave-temples near Tunhuang in Kansu. This painting is of early Thang date

Made in +1193, it was engraved on stone in +1247 by Wang Chih-Yuan.[1] It is in the 'China and Barbarians' rather than the 'Tracks of Yü' tradition (see Fig. 229), mountains and forests being more naturalistically drawn in, with no grid, and with place-names inserted in cartouches, just as on the fanciful topographic story-pictures of the Thang so abundant on the walls of the cave-temples at Tunhuang (see Fig. 230). The coast-line and the Shantung peninsula are, however, better than in the +1137 map. The inscriptions have been fully translated by Chavannes (8). The same series of steles has also an important city-map of Suchow, engraved in +1229 (Moule, 15), interesting in connection with those made for Sian four centuries earlier. Thus we have a plan of the city some fifty years before it was visited by Marco Polo.

(iv) *Climax in Yuan and Ming*

With the Yuan period we reach a man who might be regarded as the focal figure in the history of Chinese cartography, Chu Ssu-Pên[2] (+1273 to +1337).[a] Inheriting the tradition of Chang Hêng and Phei Hsiu, he was able to summarise by its aid the large mass of new geographical information which the Mongol unification of Asia had added to earlier knowledge possessed by the Thang and Sung. Chu Ssu-Pên was a younger contemporary of the astronomer Kuo Shou-Ching, and he in his turn was succeeded by two geographers hardly less brilliant than himself, Li Tsê-Min[3] and the monk Chhing-Chün,[4] about whom more will shortly be said. Doubtless all these men benefited by the contacts with Western Muslims, Persians and Arabs, such as Jamāl al-Dīn[b] and the others whose lives and work have been studied by Chhen Yuan (*3*).[c] The views of Chu Ssu-Pên on the geography of China's border countries were perpetuated in edition after edition of his great atlas, and lasted almost undisputed until the beginning of the 19th century.[d]

Chu Ssu-Pên prepared his map of China between +1311 and +1320, using older maps, literary sources, and the results of personal travel.[e] His map was a large one, not dissected, and bore the simple title *Yü Thu*[5] (Earth-Vehicle (i.e. Terrestrial) Map).[f] He himself was rather chary of mapping the more distant regions, saying, in sceptical words which deserve to be remembered:

> Regarding the foreign countries of the barbarians south-east of the South Sea, and north-west of Mongolia, there is no means of investigating them because of their great distance, although they are continually sending tribute to the court. Those who speak of them are

[a] Biography by Naito (*1*).

[b] See above, p. 372.

[c] There were geographers among them, as witness Shan-Ssu[6] (Shams al-Dīn) and his *Hsi Yü Thu Ching*[7] (Geography of the Western Countries), now lost.

[d] We owe to Fuchs (1) a valuable monograph on the geographical work of Chu.

[e] It is of interest that Chu Ssu-Pên was himself a Taoist, and studied under famous Taoists such as Chang Jen-Ching[8] and Wu Chhüan-Chieh.[9] He says in his preface that among other sources he consulted the *Yü Chi Thu* of +1137.

[f] Rare copies in manuscript and inscribed on stone in a Taoist temple lasted till the +19th century but have all been lost.

1 王致遠　2 朱思本　3 李澤民　4 清濬　5 輿圖
6 贍思　7 西域圖經　8 張仁靖　9 吳全節

unable to say anything definite, while those who say something definite cannot be trusted; hence I am compelled to omit them here.[a]

As we shall see, however, Chu's successors had access to better information, though their cartography did not equal his.

For about two centuries this great map existed only in manuscript or epigraphic[b] form, but in +1541 it was revised and enlarged by Lo Hung-Hsien[1] (+1504 to +1564), and printed about +1555 under the title *Kuang Yü Thu*[2] (Enlarged Terrestrial Atlas). The preface reads:

> Chu Ssu-Pên's map was prepared by the method of indicating the distances by a network of squares (*yu chi li hua fang chih fa*[3]), and thus the actual geographic picture was faithful. Hence, even if one divided (the map) and put it together again, (the individual parts) in the east and west fitted faultlessly together....His map was 7 ft. long and therefore inconvenient to unroll; I have therefore now arranged it in book form on the basis of its network of squares.[c]

Apart from the general map (see Fig. 231), there were sixteen sheets of the various provinces, sixteen of the border regions, three of the Yellow River, three of the Grand Canal, two of sea routes, and four devoted to Korea, Annam, Mongolia and Central Asia. The scale was usually 100 *li* to the division, but sometimes also 40, 200, 400 or 500.[d] Lo Hung-Hsien naturally drew on many other Yuan and Ming sources[e] in his revision, including the schematic grid-map of the north-western countries (*Hsi-Pei Pi Ti-Li Thu*[4]) in the *Yuan Ching Shih Ta Tien*[5] (History of Institutions of the Yuan Dynasty), +1329 (Fig. 232).[f]

In spite of Chu Ssu-Pên's caution about far-distant regions, it is a remarkable fact, as Fuchs (1) has pointed out, that he and his contemporaries already recognised the triangular shape of Africa. In European and Arabic maps of the +14th century the tip of Africa is always represented as pointing eastwards, and this is not corrected until the middle of the +15th century; the Chinese atlas of +1555, however, has it pointing south, and other evidence shows that Chu Ssu-Pên must have drawn it in this way as early as +1315.[g] From the middle of the +16th century, further editions of the *Kuang Yü Thu* continued to appear, the last being that of +1799.[h]

[a] Tr. Fuchs (1).

[b] It is known that an epigraphic copy was still extant in +1715. We now have no access to the original form of the map.

[c] Tr. Fuchs (1).

[d] The +16th-century printed editions employ quite modern symbolism to indicate physical features and sizes of settlements. Thus cities of the first order are indicated by a white square, those of the second rank by a white lozenge, and those of the third by a white circle; post-stages by a white triangle and forts by a black square, etc. (see Fig. 233). Colours to indicate areas and frontiers are known to have been used in a military map as early as +1084.

[e] Fuchs (1), p. 13.

[f] There is a special study on this by Ting Chhien (cf. Pelliot, 19); but the most comprehensive identification of the place-names on it is that of Bretschneider (2), vol. 2, pp. 1–136, who includes a map with transliterations. See below, p. 564.

[g] I am indebted to Dr W. Fuchs personally for the establishment of this fact. Wang Yung (*1*), p. 91, is wrong in supposing that the map containing Africa was introduced only in the edition of +1579 edited by Chhien Tai.[6]

[h] Cf. Klaproth (2); Himly (1). The study of these editions, some of which are very rare, is a complicated matter, for which the reader is referred to Fuchs (1) and Hummel (10). The later ones are not

[1] 羅洪先 [2] 廣輿圖 [3] 有計里畫方之法 [4] 西北鄙地理圖
[5] 元經世大典 [6] 錢岱

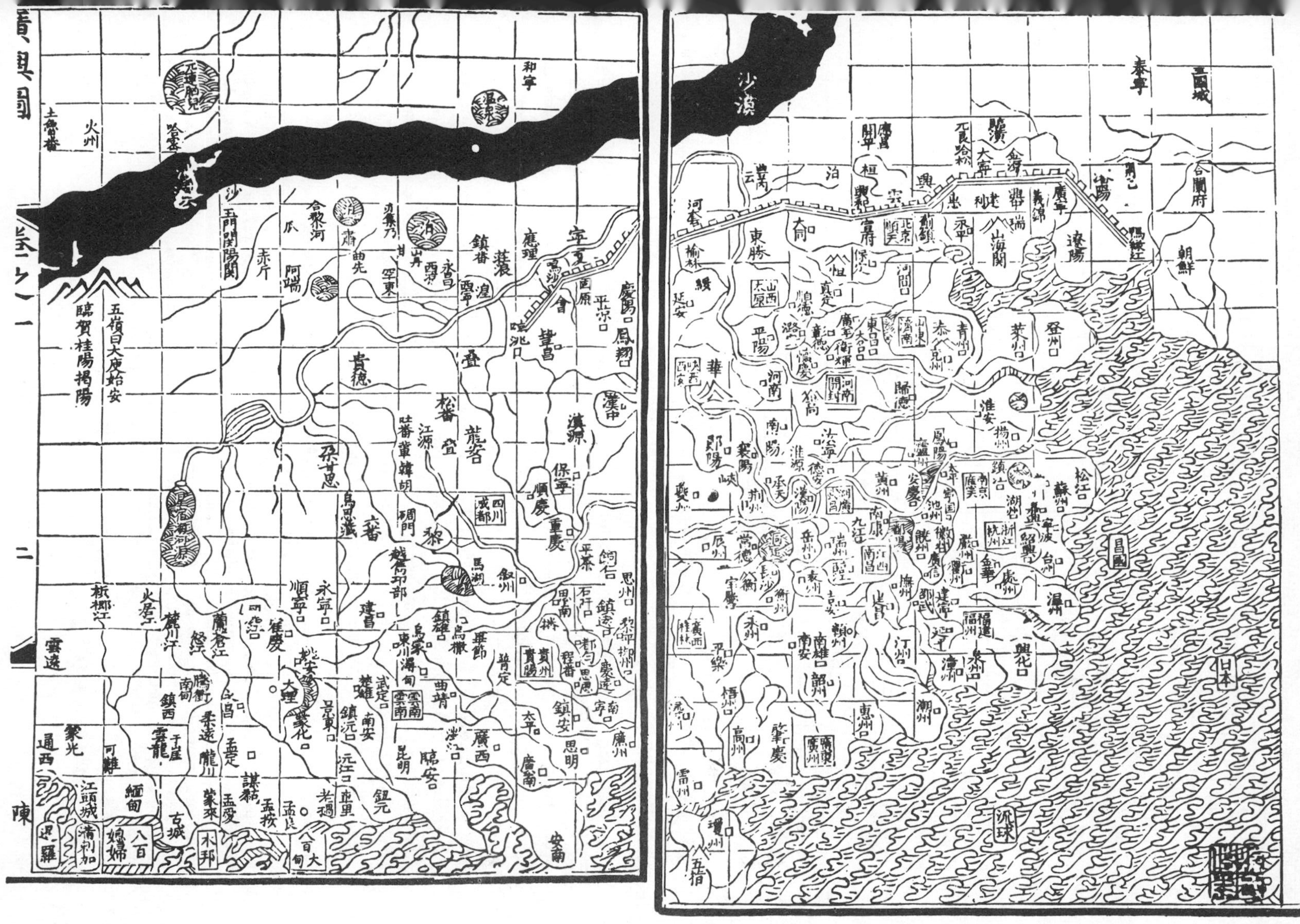

Fig. 231. Two pages from the *Kuang Yü Thu* (Enlarged Terrestrial Atlas), begun by Chu Ssu-Pên about +1315 and enlarged by Lo Hung-Hsien about +1555; the general map of China on a grid scale of 400 *li* to the division. The black band in the north-west represents the Gobi Desert. A map deriving from this was printed in *Purchas his Pilgrimes*.

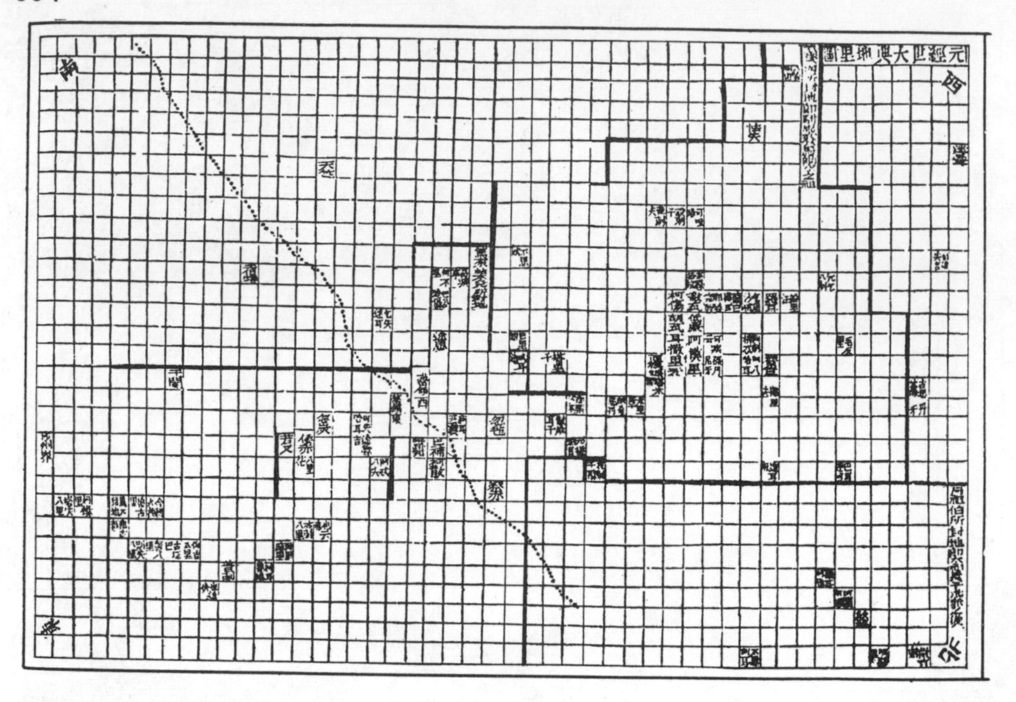

Fig. 232. The schematic grid-map of the north-western countries in the *Yuan Ching Shih Ta Tien* (History of Institutions of the Yuan Dynasty), +1329. North is at the right-hand bottom corner. This system of marking nothing but place or tribe names on a scale grid has been called the 'Mongolian style' in cartography.

The other two geographers of whom we have already spoken specialised rather on the construction of maps of the maximal known world, such as that of al-Idrīsī two centuries before them. Li Tsê-Min, of whom we know nothing save that he flourished around +1330, produced a *Shêng Chiao Kuang Pei Thu*[1] (Map for the Diffusion of Instruction). That of the monk Chhing-Chün (+1328 to +1392) must have been made some forty or fifty years later; it was called *Hun-I Chiang-Li Thu*[2] (Map of the Territories of the One World). Both of these maps got to Korea in +1399 through the agency of the Korean ambassador, Chin Shih-Hêng,[3] and were there combined in +1402 by Li Hui[4] and Chhüan Chin[5],[a] with the title *Hun-I Chiang-Li Li-Tai Kuo Tu chih Thu*[6] (Map of the Territories of the One World and the Capitals of the Countries

so good as the earlier, and some even distorted the grid system to correspond with the size of the page. On the other hand, other maps, for example one of Japan, were gradually added. Statistical data on population, etc. were also expanded. A copy of one edition reached Florence in +1606; for its history see Moule (14) and Frescura & Mori (1). A manuscript translation of the text was made in +1701 by Francesco Carletti with a Chinese collaborator, and the identity of this was recognised after still a further hundred years by Klaproth (1). Rightly he called it 'un des monumens géographiques les plus curieux que je connaisse, aprés le livre de Ptolémée'. Meanwhile the *Kuang Yü Thu* in China had been the model from which the Jesuit Michael Boym made some fine manuscript maps still extant, about +1654 (see Fuchs, 5), and a little later it was the basis of the *Atlas Sinensis* of Martin Martini (below, p. 586).

[a] Cf. p. 279 above.

[1] 聲教廣被圖 [2] 混一疆理圖 [3] 金士衡 [4] 李薈 [5] 權近
[6] 混一疆理歷代國都之圖

in Successive Ages). A copy of this, dating from about +1500, has been preserved in Japan,[a] and the Chinese *Ta Ming Hun-I Thu*[1] of about +1585 is in the same tradition.

Ogawa[b] was the first to draw attention to the Korean map, which is in size about 5 × 4 ft., and has been carefully described by Aoyama (*1*, *2*, *3*), though only so far as the Asian parts are concerned.[c] The fact that the names of the Chinese cities are all the same as in +1320 suggests that the map as a whole must go back to Chu Ssu-Pên's own time. The treatment of the Western parts is very interesting; it includes about 100 place-names for Europe and about 35 for Africa,[d] which has its correct triangular shape and points in the right direction. In the northern part of Africa the Sahara is shown black, like the Gobi in so many Chinese maps (including the *Kuang Yü Thu*), and the position of Alexandria is indicated by a prominent pagoda-like object representing the famous Pharos. The outline of the Mediterranean is well drawn, but the cartographers failed to black it in, perhaps not being quite sure that it was an ordinary sea. Germany and France are marked phonetically (A-lei-man-i-a and Fa-li-hsi-na), and the Azores are shown. One of the two largest capitals in the world, judging from the symbols adopted, is Pyongyang in Korea, but Europe has the other city of equal importance, the position of which would indicate approximately Budapest. Such extensive knowledge of the occident, which was much clearer than that of Europeans about Chinese geography at the time, must no doubt have been gained from Arab, Persian and Turkish contacts,[e] notably perhaps

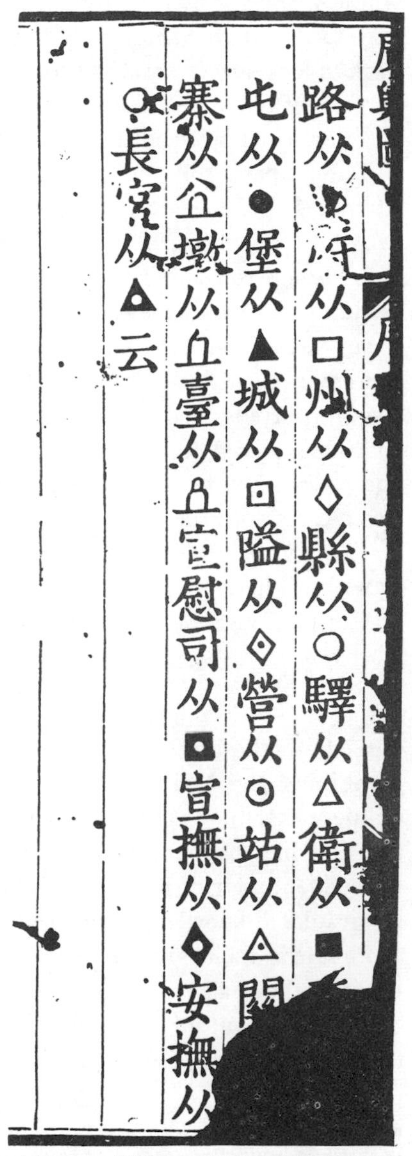

Fig. 233. The key to the symbols in the *Kuang Yü Thu*. Cities of the first order were indicated by a white square, those of the second order by a white lozenge, and those of the third by a white circle; post-stages by a white triangle and forts by a black square, etc. This must be one of the earliest occasions on which systematic symbolism of this kind was employed.

[a] Neither of the +14th-century components has survived, unfortunately. [b] (*1*), p. 606.

[c] Curiously, Japan was not added until the version of +1500.

[d] So far it has not been possible to identify many of them. For the information contained in the remainder of this paragraph I am under much obligation to Dr W. Fuchs, who is engaged in the study of the occidental parts of this map, and with whom I have seen a photographic copy of it.

[e] A check on al-Idrīsī's forms for Germany and France, kindly made by Mr D. M. Dunlop, gives al-Lamānīyah and

[1] 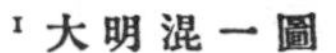

that terrestrial globe which came with Jamāl al-Dīn to Peking in +1267, and was marked with a grid (presumably of latitudes and longitudes), though the Korean map has none. Essentially, however, it was a concrete result of the conquest of almost the whole *oikoumene* by the Mongols. A final point of interest concerning this remarkable map is that it could not include the information which Chêng Ho certainly brought back five years later about the peninsular nature of India. Only in a subsequent version of about +1580 (in the Imperial Palace Museum at Peking) is India shown as a peninsula between south-east Asia and Africa.

The maps of this type are rightly regarded by Fuchs (1) as the most magnificent examples of Yuan cartography, completely overshadowing all contemporary European or Arabic world maps. Parts of the Korean map are here reproduced (Figs. 234, 235). One should remember that the maps of Chu Ssu-Pên, on which it was based, were contemporary with the early portolans in the West. The extent of the lead which the Yuan cartographers had, however, may perhaps best be appreciated by comparing the Korean map with the fantastic Catalan map of +1375, reproduced by Yule,[a] which also purported to show Asia as well as Europe. Of course that too was based on +13th-century material, for example, the information collected by Marco Polo, but when one compares what the two groups of cartographers did with their material, the advantage lies clearly with the Chinese.

The *Kuang Yü Thu* tradition gave rise to other atlases towards the end of the Ming dynasty. There was, for instance, after +1600, the *Huang Ming Chih-Fang Ti Thu*[1] of Chhen Tsu-Shou,[2] a worthy successor.

(6) Chinese Sailing Charts

Just about the time when Chhüan Chin and his group of Korean geographers were combining the Yuan world-maps into one, there began that remarkable series of Chinese maritime explorations which have already been mentioned above.[b] During the first half of the +15th century a series of expeditions under a eunuch admiral ranged far and wide in the South Seas and the Indian Ocean, greatly adding to Chinese geographical knowledge and bringing back all kinds of rarities to the imperial court. Here it is in their cartographical aspect that we must consider them, but we can hardly do so without a slightly fuller explanation than they before received.[c]

Two texts will suffice to give an idea of the scale of the expeditions. The first is from the *Li-Tai Thung Chien Chi Lan*[3] (Essentials of History):[d]

al-Afransīyah. Apparently neither al-Idrīsī nor Ibn Khaldūn mentions the Azores. That the Korean world-map of +1402 does mention them is quite extraordinary, for they were not rediscovered by the Portuguese until after +1394, and not generally known till about +1430.

[a] (2), vol. 1, p. 299, and vol. 2. Cf. the valuable name-index of Hallberg (1).

[b] E.g. in the historical Section (6*j*), Vol. 1, p. 143.

[c] General accounts are available in the papers of Mayers (3), who was the first to study them; Pelliot (2), Duyvendak (8, 9, 10, 11), Mills (3), etc. Recent Chinese monographs are those of Fan Wên-Thao (*1*); Fêng Chhêng-Chün (*1*), and Chêng Hao-Shêng (*1*); there is also the valuable paper of Liu Ming-Shu (*2*).

[d] By Lu Hsi-Hsiung[4] (editor and compiler), +1767.

[1] 皇明職方地圖 [2] 陳組綬 [3] 歷代通鑑輯覽 [4] 陸錫熊

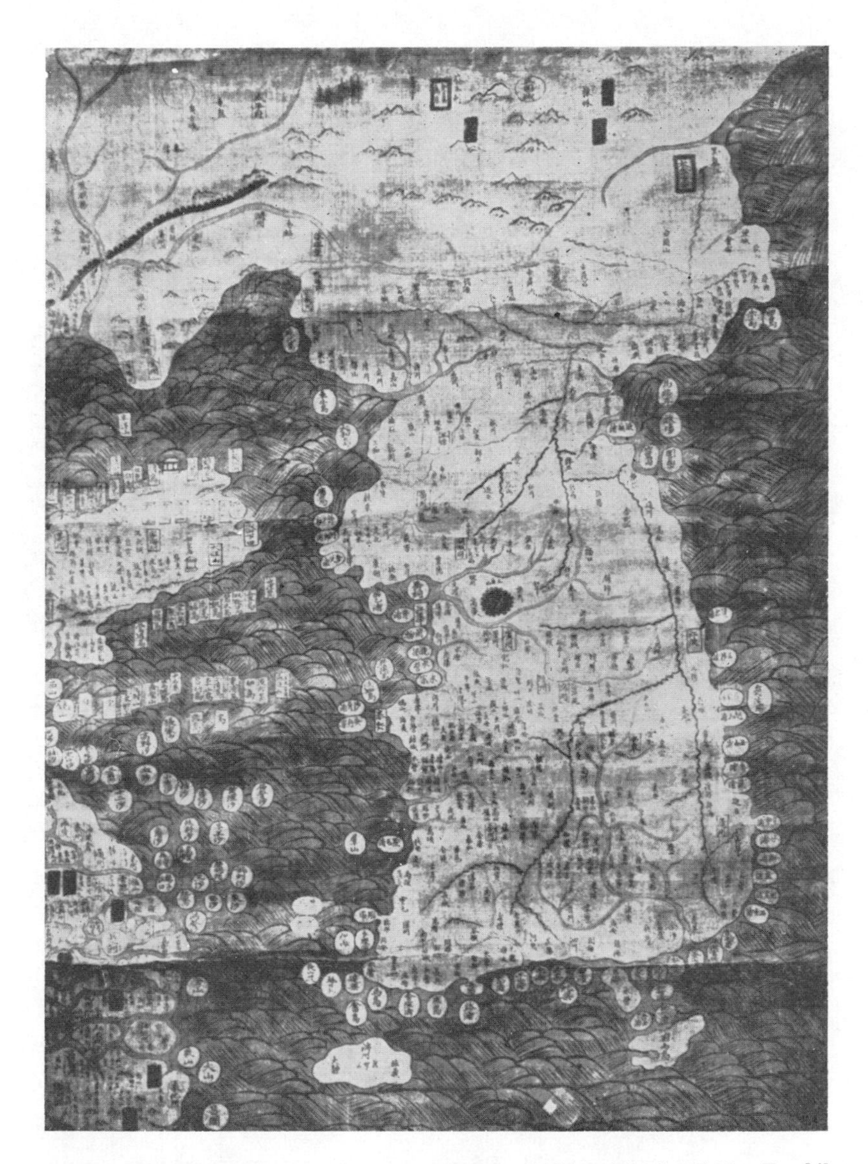

Fig. 234. The Korean world map of +1402, *Hun-I Chiang-Li Li-Tai Kuo Tu chih Thu* (Map of the Territories of the One World and the Capitals of the Countries in Successive Ages), by Yi Hwei (Li Hui) and Kwon Keun (Chhüan Chin), cf. Fig. 235. This portion shows the Korean peninsula and, to the left, the coasts of China (from Aoyama).

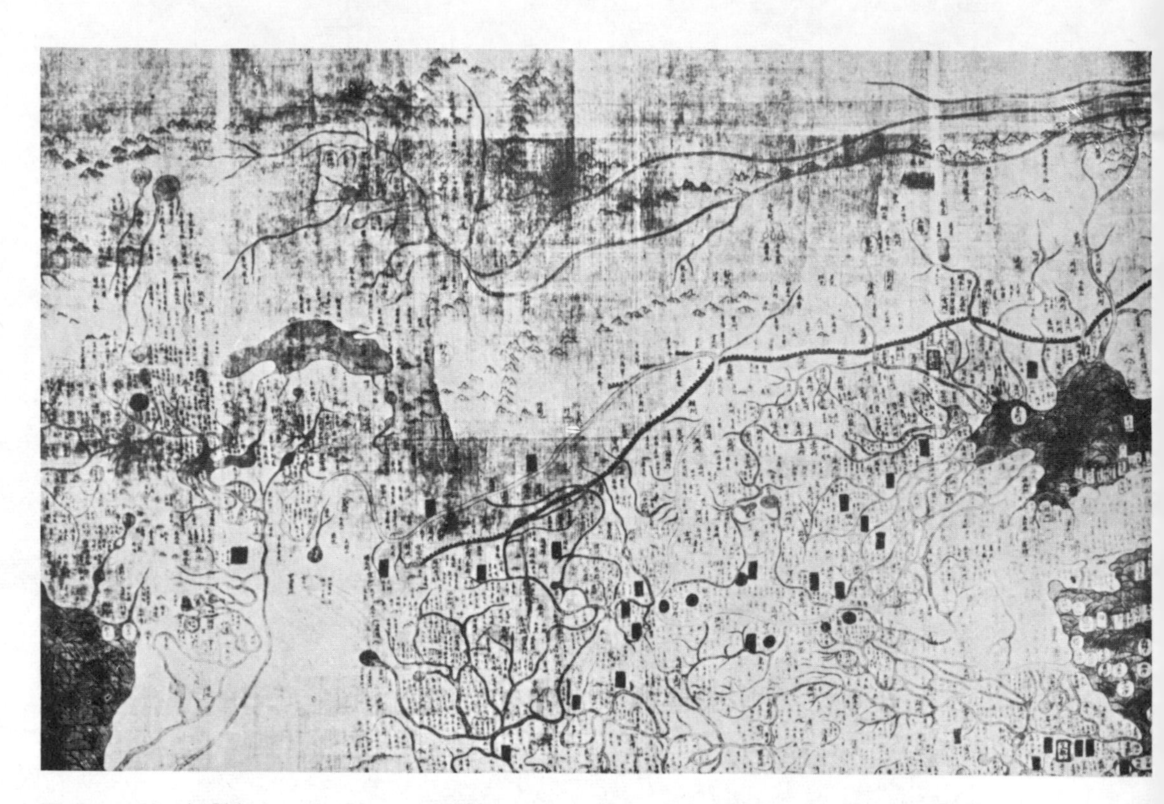

Fig. 235. North China on the Korean world map of +1402. The great bends of the Yellow River, marked in white, can be seen, and the serrated black line of the Great Wall. The large lake in Sinkiang may have indicated Lop Nor (from Aoyama).

In the third year of the Yung-Lo reign-period (+1405), the eunuch Chêng Ho,[1] commonly known as the 'Three-Jewel Eunuch', a native of the province of Yunnan, was sent on a mission to the Western Ocean.

The emperor, under the suspicion that the last emperor of the Yuan dynasty might have fled beyond the seas, commissioned Chêng Ho, Wang Ching-Hung[2],[a] and others, to pursue his traces. Bearing vast amounts of gold and other treasures, and with a force of more than 37,000 soldiers under their command, they built great ships, sixty-two in number, and set sail from Liu-chia-chiang[b] in the prefecture of Suchow, whence they proceeded by way of Fukien to Chan-Chhêng, and thence on voyages throughout the western seas.

Here they made known the proclamations of the Son of Heaven, and spread abroad the knowledge of his majesty and goodness. They bestowed gifts upon the kings and rulers, and those who refused submission they overawed by force. Every country became obedient to the imperial commands, and when Chêng Ho turned homewards, sent envoys in his train to offer tribute. The emperor was highly gladdened, and after no long time commanded Chêng Ho to go overseas once more and scatter largesse among the different States. On this, the number of those who presented themselves before the throne grew ever greater. Chêng Ho was commissioned on no less than seven embassies, and thrice he made prisoners of foreign chiefs. His exploits were such as no eunuch before him, from the days of old, had equalled. At the same time, the different peoples, attracted by the profit of Chinese merchandise, enlarged their mutual intercourse for purposes of trade, and there was uninterrupted going to and fro. Thus it came to pass that in those days 'the Three-Jewel Eunuch who went down into the West' (*San Pao Thai-Chien hsia hsi-yang*[3]) became a proverbial expression; and all who, in after times, were sent as bearers of commissions to the countries by sea, were wont to impress the outer nations with the name of Chêng Ho. Yet, as regards China, the treasure that was lavished on these undertakings brought no profit in return; while, of the soldiers in the expeditions, many perished by shipwreck or were cast away in distant lands, so that the number who returned, after nearly a score of years had elapsed, was not more than one or two in ten.[c]

By a fortunate discovery of recent years, we are enabled to gain a glimpse of how these things seemed to Chêng Ho and his companions. In 1937 Wang Po-Chhiu discovered a stele in Chhang-lo, Fukien, which bears an inscription of sailors' gratitude to a Taoist goddess, dedicated at the New Year in +1432, just before the seventh and last expedition weighed anchor.

The imperial Ming dynasty, in unifying seas and continents, surpasses the Three Dynasties, and goes even beyond the Han and the Thang. The countries beyond the horizon and at the ends of the earth have all become subjects, and to the most westerly western, or to the most northerly northern countries, however far they may be, the distances and routes may be calculated. Thus the barbarians from beyond the seas, though their lands are truly distant, have come to audiences bearing precious objects and presents.

The emperor, approving of their loyalty and sincerity, has ordered us (Chêng Ho) and others, at the head of several tens of thousands of officers and troops, to embark upon more than a hundred large ships, in order to go and confer presents on them, thus to make manifest

[a] Chêng Ho's vice-admiral. [b] Near present Shanghai.

[c] Ch. 102, tr. Mayers (3), mod. The last sentence is, of course, orthodox Confucian anti-eunuch propaganda.

[1] 鄭和 [2] 王景弘 [3] 三寶太監下西洋

the transforming power of the (imperial) virtue, and to show kind treatment to distant peoples. From the third year of the Yung-Lo reign-period (+1405) till now, we have seven times received the commission of ambassadors to the countries of the Western Ocean.

The barbarian countries which we have visited are: by way of Chan-Chhêng (Champa), Chao-Wa (Java), San-Fo-Chhi (Palembang) and Hsien-Lo (Siam), crossing straight over to Hsi-Lan-Shan (Ceylon) in South India, Ku-Li (Calicut) and Kho-Chih (Cochin),[a] we have gone to the western regions Hu-Lo-Mo-Ssu (Hormuz),[b] A-Tan (Aden), and Mu-Ku-Tu-Shu (Mogadishiu, in Africa).[c] All together more than thirty countries large and small. We have traversed more than one hundred thousand *li* of immense water spaces, and have beheld in the ocean huge waves like mountains rising sky-high. We have set eyes on barbarian regions far away hidden in a blue transparency of light vapours, while our sails, loftily unfurled like clouds, day and night continued their course with starry speed, breasting the savage waves as if we were treading a public thoroughfare. Truly this was due to the majesty and the good fortune of the Court, and moreover we owe it to the protecting virtue of the Celestial Spouse (Thien Fei[1]).

The power of the goddess, having indeed been manifested in previous times, has been abundantly revealed in the present generation. In the midst of the rushing waters it happened that, when there was a hurricane, suddenly a divine lantern was seen shining at the mast-head, and as soon as that miraculous light appeared the danger was appeased, so that even in the peril of capsizing one felt reassured and that there was no cause for fear....[d]

The discovery of this and other similar inscriptions is all the more valuable as for some reason or other the official records in the imperial archives subsequently disappeared.

Nevertheless, the expeditions gave rise to four books of great importance, which have been analysed by Pelliot (2). The earliest was the *Hsi-Yang Fan Kuo Chih*[2] (Record of the Barbarian Countries in the Western Ocean) by Kung Chen[3] in +1434; quickly followed by the *Hsing Chha Shêng Lan*[4] (Triumphant Visions of the Starry Raft)[e] by Fei Hsin[5] in +1436—both of these men had been among Chêng Ho's officers. Then came the turn of one of the Chinese Muslim interpreters, Ma Huan,[6] who in +1451 produced the *Ying Yai Shêng Lan*[7] (Triumphant Visions of the Boundless Ocean).[f] The presence of Ma Huan was particularly natural since it is now established that Chêng Ho's real family name was Ma, and that he also was of Yunnan Muslim extraction.[g] The last of the four books is the *Hsi-Yang Chhao Kung Tien Lu*[8]

[a] All these places were visited on the first three expeditions (+1405 to +1407, +1407 to +1409, +1409 to +1411). On the third, the ruler of Ceylon (Vīra Alakēsvara, sometimes wrongly styled Vijaya Bāhu VI) made difficulties, and was taken in temporary captivity to China.

[b] The fourth voyage (+1413 to +1415) added Hormuz and the Persian Gulf. On this occasion Chinese Muslim interpreters were attached to the staff; hence the presence of Ma Huan.

[c] The last three expeditions (+1417 to +1419, +1421 to +1422, +1431 to +1433) explored the east coast of Africa, including Melinda, and brought back, among other things, the giraffe. The elucidation of all these dates is due to Duyvendak (9).

[d] St Elmo's Fire. Tr. Duyvendak (8, 11), mod. auct.

[e] The usual term for ships carrying ambassadors.

[f] Translations of this famous work are available by Groeneveldt (1); Rockhill (1); Phillips (2); and Duyvendak (10). It is reproduced in part in *TSCC*, *Pien i tien*, chs. 58, 73, 78, 85, 86, 96, 97, 98, 99, 101, 103, 106.

[g] Liu Ming-Shu (2); Duyvendak (8).

1 天妃 2 西洋番國志 3 鞏珍 4 星槎勝覽 5 費信
6 馬歡 7 瀛涯勝覽 8 西洋朝貢典錄

(Record of the Tribute-Paying Western Countries) of +1520[a] by Huang Shêng-Tshêng.[1]

Now while all these works are full of information about the people and products of the places visited by the expeditions they do not contain (and seem never to have contained) any maps. Huang, however, speaks of 'sailing directions' (lit. compass bearings, *chen wei*[2]), which he used as one of his sources.[b] And in fact we do possess certain maps which give the routes followed by Chêng Ho and his ships; these have been preserved at the end[c] of a book on military technology, the *Wu Pei Chih*[3] (Treatise on Armament Technology) written by Mao Yuan-I[4] before +1621 and presented to the emperor seven years later. In an introductory note it is stated that these maps were derived from the records left by Chêng Ho.[d] Duyvendak (10) and Wang Yung have noted that the grandfather of Mao Yuan-I was Mao Khun,[5] lifelong collaborator of a man whose chief work was the defence of the coast against Japanese pirates (Hu Tsung-Hsien,[6] governor of Fukien) and the friend of an outstanding authority on coastal geography (Chêng Jo-Tsêng, already mentioned, p. 516). Very probably Mao Khun found these maps in the governor's *yamên* and handed them down to his grandson.[e]

Part of one of these maps is here reproduced (Fig. 236). It can be seen at once that it does not derive from the tradition of Phei Hsiu's grids, but is a true mariner's chart giving elaborate compass-bearings in the legends along the lines of travel, adding the distances in 'watches',[f] and noting all points along the coast which could be of importance to sailors.[g] It is distorted by an extreme schematism, which Mills (1) has compared to diagrammatic 'underground tube' maps or steamship companies' voyage-schedule maps of today. In the portion reproduced, we see the opening of the Persian Gulf to the left, with Hormuz (86) and Muscat (81), while to the right is the entrance of the Red Sea by Socotra (60) and Aden (62). The land towards us is therefore Arabia, and the farther land India, where Bombay is marked (67).[h] Neither south nor north

[a] A significant date, since it saw the beginning of the Spanish conquests of the American continent. These, together with the previous voyages of Columbus, would have been impossible but for two essential Chinese inventions, the magnetic compass and the sternpost rudder, available also to Chêng Ho. Huang's book has been translated by Mayers (3).

[b] His expression, *Chen Wei Pien*,[7] may be taken as the title of a specific book. This was noticed by Pelliot (9), p. 139. Though Rockhill (1), p. 77, was inclined to think that the expression was used only in a general sense, Pelliot always believed that evidence of a specific book would some day come to light, cf. (2*a*), p. 345, (2*b*), p. 308, (33), p. 79.

[c] Ch. 240.

[d] We also possess a MS. of +15th-century sailing directions, in the Bodleian Library, which has been analysed by Duyvendak (1). On this see Sects. 26*i* and 29*f* below.

[e] Before the end of the 17th century they were reproduced again, less well than in the *Wu Pei Chih*, by Shih Yung-Thu[8] in his *Wu Pei Pi Shu*[9] (Confidential Treatise on Armament Technology). On this edition see Duyvendak (10), p. 18, and Pelliot (33), p. 78.

[f] Ten *kêng*[10] or *ching*[10] was equivalent to the distance covered in one day and one night with a favourable wind. One *kêng* equalled 2 h. 24 min. sail, or about 60 *li*. On these matters see S. Wada (1) who deals with Chinese navigation to the Philippines before the Ming.

[g] E.g. half-tide rocks and shoals as well as ports and havens.

[h] Actually Mahaim.

[1] 黄省曾 [2] 鍼位 [3] 武備志 [4] 茅元儀 [5] 茅坤
[6] 胡宗憲 [7] 鍼位編 [8] 施永圖 [9] 武備秘書 [10] 更

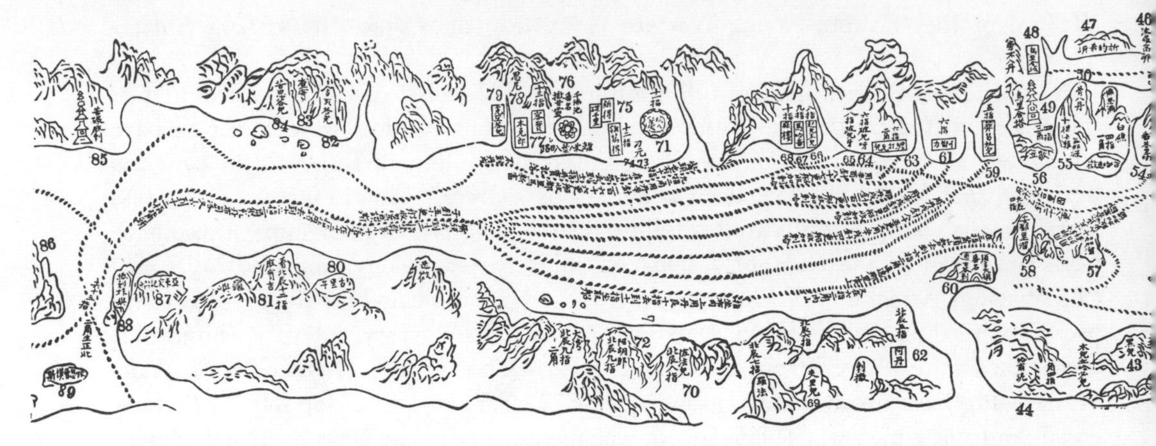

Fig. 236. One of the sea-charts from the *Wu Pei Chih*. Though not printed until +1621, these charts date from the expeditions of Chêng Ho (+1405 to +1433). In the above map the coast at the top is that of western India, that at the bottom representing Arabia. To the left is the opening of the Persian Gulf, to the right is the entrance to the Red Sea. The Indian Ocean is thus compressed to a schematic corridor in which sailing tracks are marked with precise compass-bearings and other instructions. Places can be identified by the numbers Hormuz (86), Muscat (81), Socotra (60), Aden (62), Bombay (Mahaim) (67). Alongside each one is given the altitude of the pole-star in *chih* (finger-breadths); further details on this and other points of navigation will be found in Sect. 29*f*. Such charts as these may be considered Chinese portolans but they are quite different in type from the portolans of the West (from Phillips).

is at the top, but east, and part of the distortion is in order to squeeze the Indian Ocean sufficiently to allow of representation within the limits of the size of the book. These maps have been exhaustively analysed and their place-names identified by G. Phillips (1); and it is not possible to accept the low estimate of them entertained by Liu Ming-Shu (*2*).[a] Indeed, they correspond, not only in nature, but also in date (early +15th century) with the portolan charts of Europe, the only difference being that they give their compass bearings in words instead of drawing rhumb-lines from arbitrarily chosen centres.[b] As to their accuracy, we are indebted for a close examination to Mills (1) and Blagden, who were both familiar with the coastline of the whole Malayan peninsula, and formed a high admiration for the precision of the Chinese sailing directions. Moreover, Mulder (1) has recently considered the material from the pilot's point of view. Routes are generally given for inner and outer passages where islands are concerned, sometimes with preferences if outward- or homeward-bound. Bearings are given by cyclical characters, sometimes doubled, and sometimes with the addition of the word 'red' (*tan*[1]) which probably meant 'due' south, east or north as

[a] Rockhill (1) preferred the maps in the +1564 *Yü Ti Tsung Thu*[2] (General World Atlas) by Shih Ho-Chi,[3] but they are not comparable. Shih's maps belong to the tradition of Chu Ssu-Pên, in which the grid covers land but not sea, and though he shows South Africa just as Chu did, correctly enough, his disposition of islands is very arbitrary.

[b] 'The Chinos', wrote Mendoza (+1585), 'doo governe their ships by a compass divided into 12 partes, and doo use no sea cardes, but a brief description of Ruter (Routier), wherewith they doo navigate and saile....' Of course some of the later Western portolans are of very superior execution.

[1] 丹 [2] 輿地總圖 [3] 史霍冀

the case might be.[a] An accuracy of 5° was general, which is to be considered excellent for a pilot of +1425.[b]

There has been some discussion as to possible foreign influence on the *Wu Pei Chih* charts, which it now seems safe to take as associated with Chêng Ho's expeditions. Duyvendak (10) and Pelliot (2) suspect Arabic influence.[c] Fan Wên-Thao (*1*) notes that Shih Pi[1] brought home from Java in the early Yuan period (before +1297) a map of the country, and in +1372 the king of a place which Fêng Chhêng-Chün (*1*) has identified as Coromandel sent one. The difficulty is, however, that no Arabic portolan has so far been discovered, and we are therefore very ill-informed about the nautical charts which the Arabic merchant captains must surely have had. Until this is elucidated it will be hard to decide to what extent the makers of the Chinese charts received stimuli from Arab sources. We shall return to this subject in the sub-section on navigation (29 *f* below). In the meantime we must be content, with Mills, to take leave of Chêng Ho as we watch him 'crashing north at a steady six knots' on his last homeward voyage.

(7) The Role of the Arabs

The extensive Arab–Chinese contacts of which we have just been speaking make this a convenient place to consider the next portion of the whole picture, namely, the role of the Arabs in the history of cartography.[d] Their close relation with the Byzantine world gave them an early contact with the remains of the Hellenistic geographers, and Ptolemy was available to them from the middle of the +9th century. In the time of al-Ma'mūn's Caliphate (+813 to +833) new lists of latitudes and longitudes were prepared.[e] It was therefore natural that the tradition of quantitative cartography was never quite lost among the Arabs.

Nevertheless, for the earlier centuries (+8th to +11th) the other tradition, that of religious cosmography, was very strong if not entirely dominant. Many examples of Arabic T-O maps and wheel-maps with climates[f] are known. There was also a tendency to greater geometrical schematisation, so that the appearance of the wheel-maps lost all resemblance to the actual contours of sea and land. These things may be seen in the albums of K. Miller (4) and of Yusuf Kamal. I reproduce one of the highly geometrised maps in Fig. 237; it is that of Abū Isḥāq al-Fārisī al-Iṣṭakhrī about +950.[g]

[a] Probably standing for *tan*,[2] 'single', i.e. 'dead on'. Some markings on the compass were actually painted red; cf. Sect. 26*i* below.

[b] Cf. the studies of Chang Li-Chhien (*1*) and S. Wada (1). Dumoutier (1) has described an Annamese 'portolan', but we have not had access to his reproduction of it.

[c] But as Mr J. V. Mills points out to us, the tradition could have been purely Chinese. Chu Ssu-Pên already showed the sea-routes between Fukien and Manchuria as paths running horizontally across his charts (see Fuchs (1), pls. 37 and 38).

[d] This may be studied in Kimble (1), who devotes a special chapter to the 'rise and fall' of Muslim geography; in Beazley (1), vol. 3, and in the special monograph of Nafis Ahmad (1).

[e] K. Miller (3) has compared the Arab and Hellenistic values for the degree and the earth's circumference.

[f] See pp. 529ff. above.

[g] Sarton (1), vol. 1, p. 674; Hitti (1), p. 385; Mieli (1), p. 115.

[1] 史弼 [2] 單

The great Sicilian geographer al-Idrīsī made a T-map and a climate-map about +1154.[a] The most ancient Turkish world-map, by Maḥmūd ibn al-Ḥusain ibn Muḥammad al-Kāshgharī in +1074, which shows Khanbaliq (Peking) and Kashgar, is also of the wheel-type.[b] Wheel-maps with climate 'latitude' lines were still current in the +13th century, for example, that of Zakarīyā' ibn Muḥammad al-Qazwīnī[c] (+1203 to +1283) figured by Yusuf Kamal.[d] They compare very poorly with the contem-

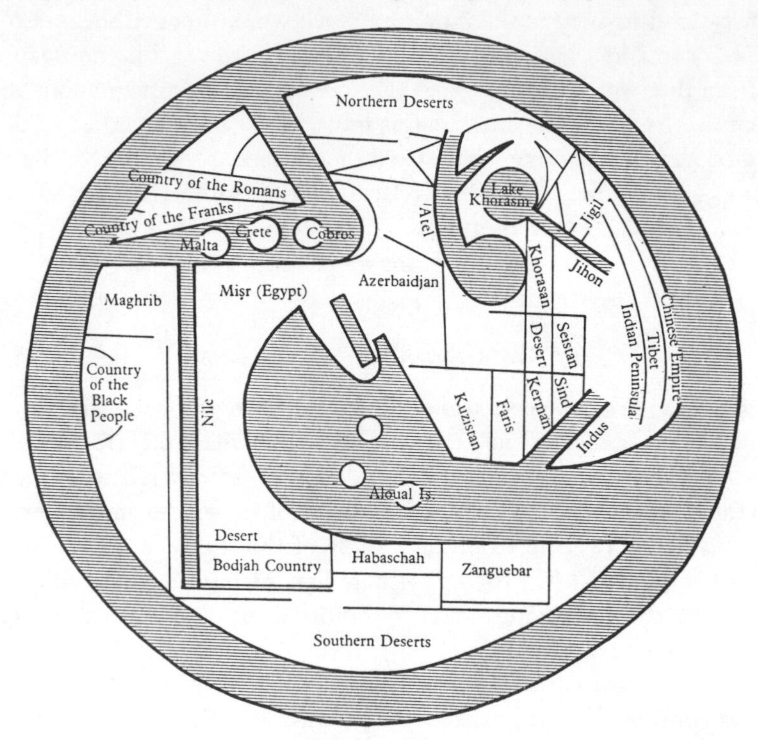

Fig. 237. An Arabic wheel-map, that of Abū Isḥāq al-Fārisī al-Iṣṭakhrī and Abū al-Qāsim Muḥammad ibn Ḥauqal (+950 to +970). It clearly shows the strong tendency to geometrical stylisation characteristic of the second period of Arab cartography (Mieli (1), pp. 115, 201). In the original, east was at the top, just as in the T-O maps of contemporary Latin Europe, but instead of the Earthly Paradise the Arab scholars knew enough to place in the Furthest East both China and Tibet. Note also how the tip of Africa points eastwards, a mistake which the Chinese geographers were the first to correct (from Beazley after Reinaud).

porary Chinese maps which we have been examining. By the +14th century, however, the general configuration of the known world was becoming fairly clear. When in the Lebanon in 1948, I had the privilege of examining a manuscript book written in his own hand by Naṣīr al-Dīn al-Ṭūsī in +1331, and belonging to the collection of Professor Sami Haddad at Beirut. Entitled *Memoranda on Astronomy*,[e] it gives a world-map on a disc or sphere, with climates for the northern half only (Fig. 238). Like the

[a] Miller (4), vol. 3, p. 160. For other similar ones see his pp. 131, 135, 116.
[b] Miller (4), vol. 3, p. 142.
[c] Sarton (1), vol. 2, p. 868.
[d] Yusuf Kamal (1), vol. 3, p. 1050.
[e] It must be identical with the work known to Brockelmann (2), vol. 1, p. 511, as *al-Tadhkirah al-Nāṣirīyah*. We are grateful to Mr D. M. Dunlop for checking this.

world-map of Marino Sanuto (+1306 to +1321)[a] which it closely resembles, it shows the tip of Africa pointing east and not south. Alone of the three contemporaries, Chu Ssu-Pên (cf. above, p. 552) got this point right. But all three of these early +14th-century cartographers failed to appreciate the peninsular character of India. The chief point to be made, however, is that the Arabs did participate in the tradition of religious cosmography, making their wheel-maps centre on Mecca rather than on the Jerusalem, Mt Meru or Mt Khun-lun of other civilisations.[b]

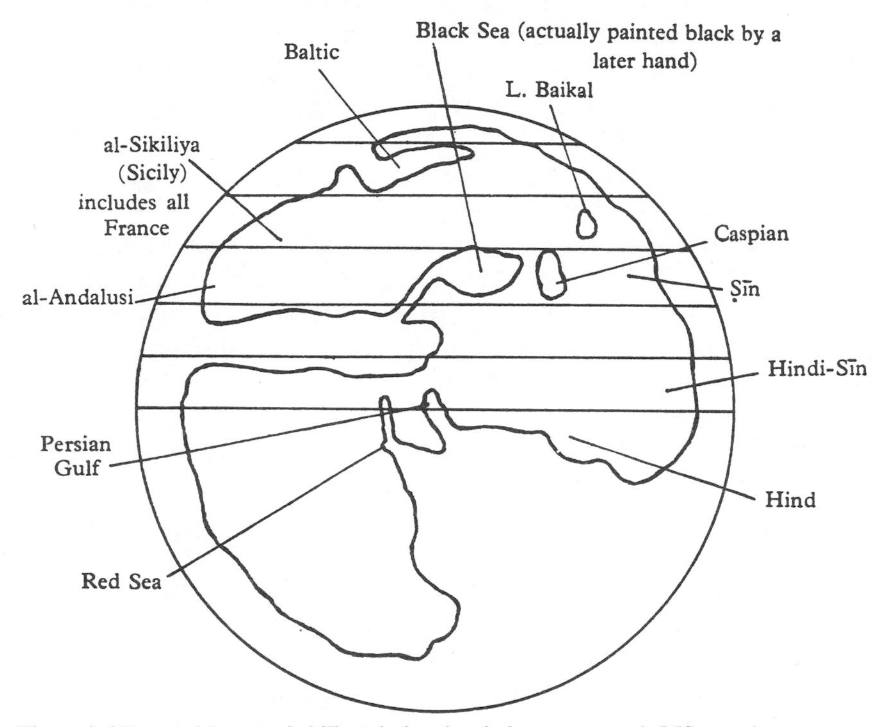

Fig. 238. The world-map of al-Ṭusī (orig. sketch from autograph MS. copy), +1331.

In comparing the scientific cartography of the Muslims with the Chinese, there are three chief maps to keep in mind. The first and most famous of these was the world-map of Abū 'Abdallāh al-Sharīf al-Idrīsī (+1099 to +1166),[c] made about +1150 for Roger II, the Norman king of Sicily, and often reproduced in modern works.[d] This was fully in the Ptolemaic tradition, using nine parallels of latitude (climates) and eleven meridians of longitude, but arranged on a projection like Mercator's and

[a] Beazley (1), vol. 3, opp. p. 521.

[b] Actually the Arab cartographers invoked a mythical city, Arim or Arym, which was supposed to be situated on the central meridian of the *oikoumene* some 10° east of Baghdad. In spite of this, Mecca was often placed orbocentrically (Wright (1), p. 86). The name Arīn seems to derive from the Indian city of Ujjain in Mālwā which had been one of the capitals of the +5th-century Gupta kings and the site of a famous observatory. Ptolemy had known it as Ozene. Arabic geographers spoke of Arīn as 'the cupola of the earth', a phrase which suggests connections with the central mountain of ancient Indian cosmology.

[c] Sarton (1), vol. 2, p. 410; Mieli (1), p. 198.

[d] K. Miller (4); Yusuf Kamal (1), vol. 3, p. 867; de Reparaz-Ruiz (1).

making no attempt to allow for the earth's curvature.[a] In this respect it resembled the Chinese grid-maps. Al-Idrīsī made use of a great variety of sources for his information about further Asia and Africa, for example, Ibn Khurdādhbih's[b] itineraries of the Muslim world (+9th century) and al-Mas'ūdī's[c] descriptions of the East (+10th). Kimble expresses surprise at the failure of Latin scholars to make any use of al-Idrīsī's work, composed though it was at the chronological and geographical focal point of Islamic and Christian civilisation. From the reproduction here given (Fig. 239), which may be compared with Fig. 226, the advantage seems to rest with the contemporary Chinese grid-map of +1137. Yet early in the next century Abū 'Alī al-Ḥasan ibn 'Alī al-Marrākushī[d] gave a list of 134 coordinate reference-points which connected geography with astronomy more closely than was ever the case in China. For although the altitude of the sun or the pole-star fixed the latitude, the Arabs made more thorough use of the old Hellenistic suggestion of determining the longitude by comparing the onset times of lunar eclipses.

The two other sets of maps of importance for our argument are essentially grid-maps constructed not by Chinese but by Muslims. The first of the kind were those made by Hamdallāh ibn abū Bakr al-Mustaufī al-Qazwīnī (+1281 to +1349)[e] to illustrate his *Ta'rīkh-i-Guzīda* (Select Chronicle).[f] Al-Mustaufī al-Qazwīnī was certainly in contact with East Asia, since in other works he gives the Mongol equivalents of plant and animal names. His three maps of Iran are on rectangular grids, and his two world-maps, though discoidal, have grids (Fig. 240). The manner of inserting nothing but place-names, without symbols for any physical feature, is identical with that seen in the *Yuan Ching Shih Ta Tien* of +1329 (see above, p. 554). This 'Mongol style', if such we might call it, appears again in the maps of Ḥāfiẓ-i Abrū[g](*d.* +1430).[h]

Most interesting of all, the non-astronomical grid system has even a contemporary Latin representative. The world-map of Marino Sanuto has already been mentioned,[i] but this Italian was responsible also for a map of Palestine, which illustrated his *Liber Secretorum Fidelium Crucis* (a kind of Crusader's geography). This map[j] dates from +1306 and shows twenty-eight 'spatia' ruled from north to south, and eighty-three 'spatia' ruled from east to west (Fig. 241). But this is the solitary example of the use of a grid by any European before the later portolans and the revival of Ptolemy. It may not be without significance that it was a map of an Arab country.

The outlines of a possible general scheme of transmission are now becoming clear.[k] How far did the Arab attempts at quantitative cartography at the beginning of the +14th century influence the later development of the European portolans? Was Arab quantitative cartography entirely due to the knowledge of Ptolemy, or was there some

[a] Kimble (1), p. 57.
[b] Hitti (1), p. 384; Sarton (1), vol. 1, p. 606.
[c] Sarton (1), vol. 1, p. 637.
[d] Mieli (1), p. 210; Sarton (1), vol. 2, p. 622.
[e] Sarton (1), vol. 3, p. 630.
[f] K. Miller (4), pls. 83, 84, 85, 86; Yusuf Kamal (1), vol. 4, pp. 1255 ff.
[g] Sarton (1), vol. 3, p. 1855.
[h] K. Miller (4), vol. 3, p. 178 and pls. 72 and 82.
[i] On him see T. Fischer (1) and Kretschmer (1); Sarton (1), vol. 3, p. 769.
[j] Reproduced by Yusuf Kamal (1), vol. 4, pp. 1162 ff., esp. p. 1173; Nordenskiöld, pl. VII. Described by Beazley (1), vol. 3, pp. 309, 391, 521, 529.
[k] Vernet (1) has also perceived this.

PLATE LXXXVII

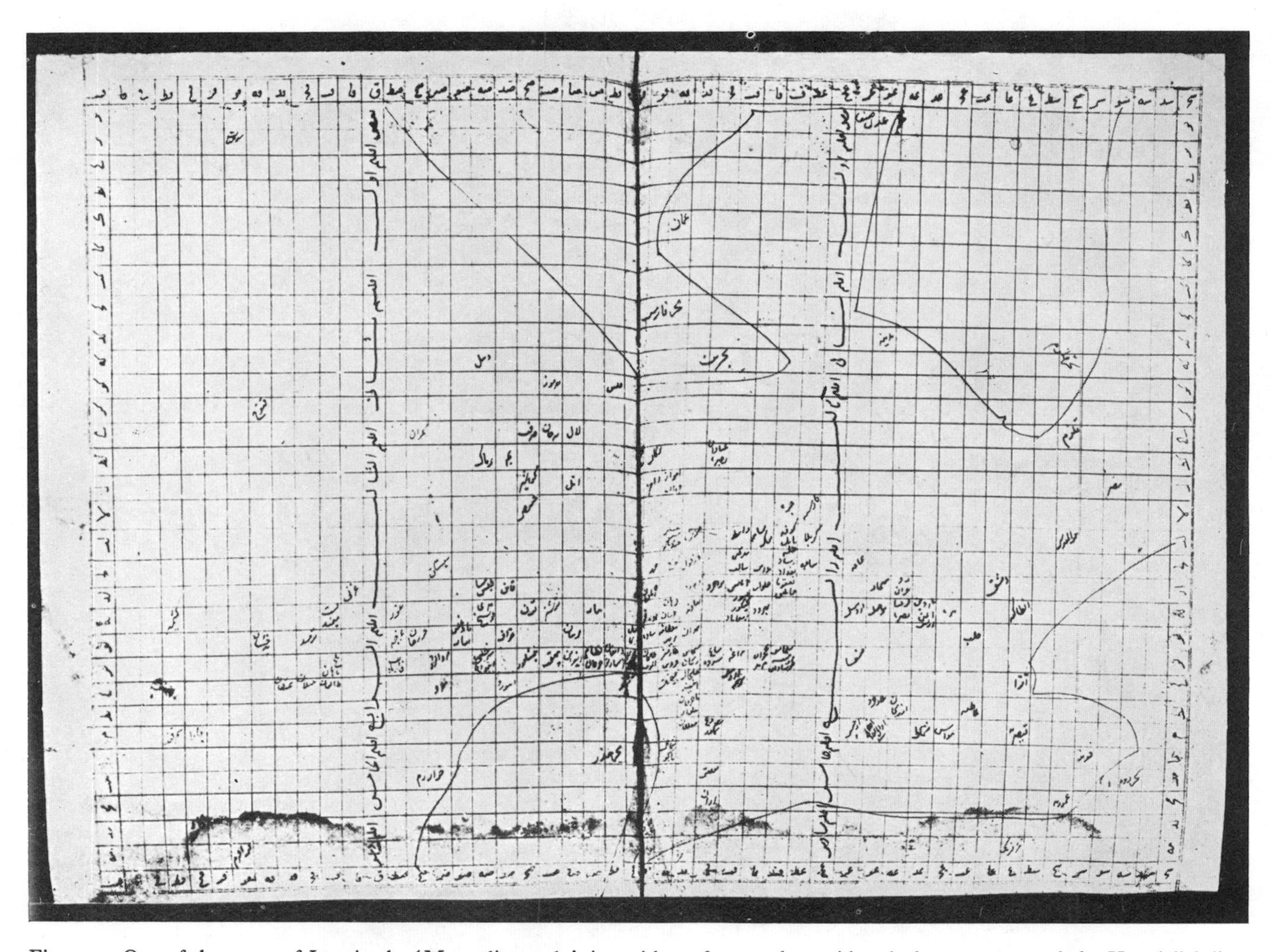

Fig. 240. One of the maps of Iran in the 'Mongolian style', i.e. with no features but grid and place-names, made by Hamdallāh ibn abū Bakr al-Mustaufī al-Qazwīnī, *c.* +1330, to illustrate his *Ta'rīkh-i-Guzīda* (Select Chronicle) (from K. Miller and Yusuf Kamal).

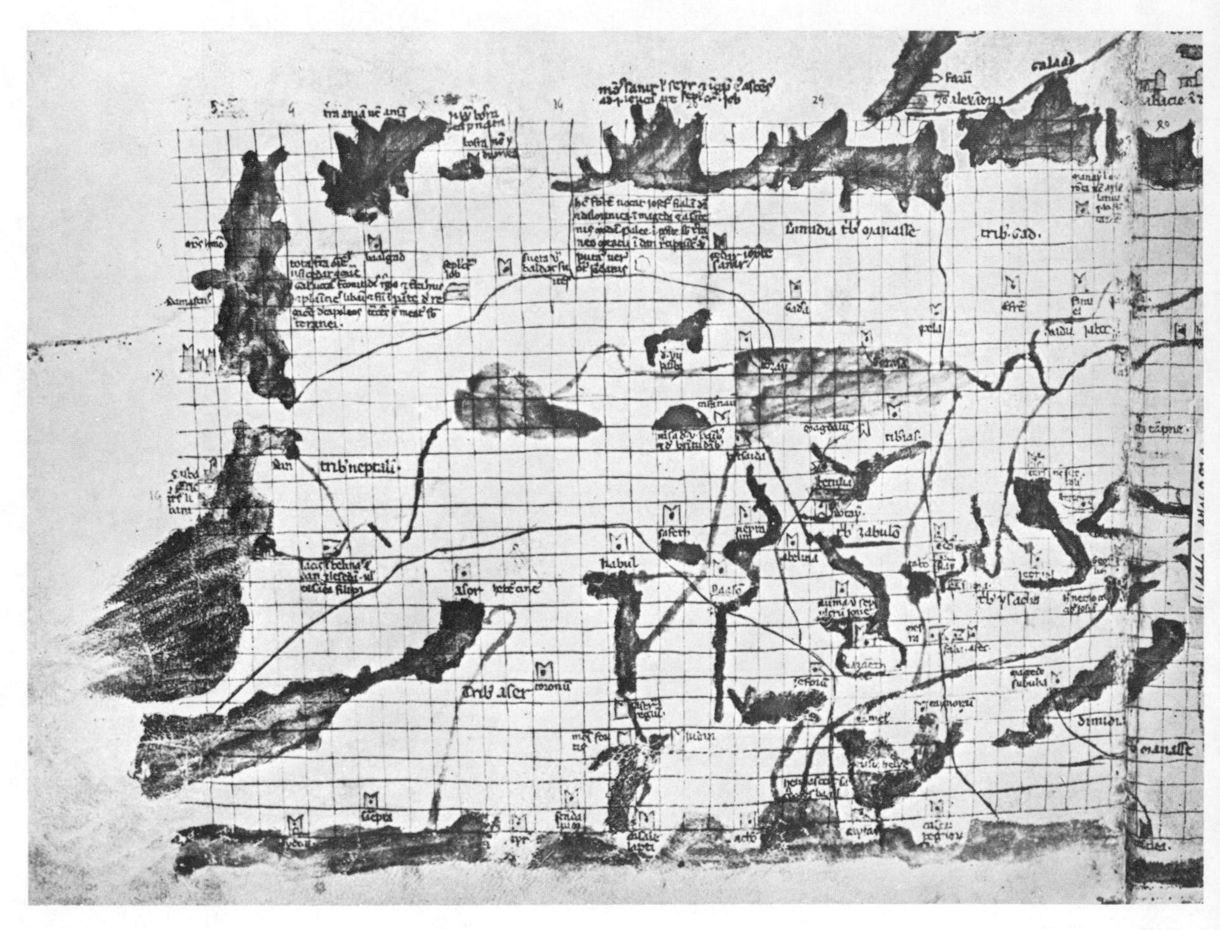

Fig. 241. The map of Palestine on the Chinese grid system made by Marino Sanuto to illustrate his *Liber Secretorum Fidelium Crucis* (+1306) (from Yusuf Kamal).

stimulus from the grid-maps of China, where the tradition had long been current? The Arab colony at Canton was certainly well established by the middle of the +8th century, and the two following centuries saw the extensive travels in China of Sulaimān the Merchant and of Ibn Wahb al-Baṣrī.[a] Then the Mongol conquests of the +13th century brought the Arab and Chinese worlds in closest contact. Perhaps Marino Sanuto was employing a characteristic Chinese practice without knowing it.

(8) Religious Cosmography in East Asia

Only one major component of the overall picture remains to be described, namely, the existence of a tradition of religious cosmography in East Asia. This was not recognised by former sinologists but has been established in recent researches, especially by Nakamura (1). Essentially, this type of map is centred on the legendary mountain in Central Asia or the north of Tibet, Mt Khun-Lun;[1] west of which are unknown regions, while to the east Korea, China, Indo-China and India form a series of promontory-continents extending into the eastern oceans. Around these outer seas there is a ring-continent itself surrounded by further ocean. Quite a number of these maps have been published, as by Courant,[b] Cordier[c] (Fig. 242), Yi Ik-Seup (1), Hulbert (1) and Rosetti (1), as well as Nakamura himself. They are particularly frequent in Korea, where they occur as woodcuts, manuscripts and paintings on screens. Though they undoubtedly represent archaic tradition, paradoxically most of them are of late date (17th and 18th centuries). Nakamura showed, however, that many legendary names of countries taken from the *Shan Hai Ching* appear on them (110 out of 145), while others come from the Yü Kung chapter of the *Shu Ching*, the *Mu Thien Tzu Chuan*, *Lieh Tzu*, etc. Often there are annotations in the corners copied straight out of *Huai Nan Tzu*. No place-name later than the +11th century has been seen on any of them.

There can be no reason for doubting that the Koreans received this tradition from China, though it seems never to have been so popular there. Herrmann reproduces[d] a Chinese map of exactly this type dating from +1607 by Jen Chhao.[e] There is another in the *Thu Shu Pien*[2] (On Maps and Books)[f] of +1562 by Chang Huang,[3] entitled *Ssu Hai Hua I Tsung Thu*[4] (Complete Map of China and the Barbarian Lands within the Four Seas). This again has Mt Khun-Lun in the centre.[g] The compiler says:

> This map is copied from a Buddhist work. It represents Jambūdvīpa (the *oikoumene*) within the four oceans of the universe.... Although this map, taken from the *Fo Tsu Thung*

[a] Renaudot (1); Reinaud (1); Ferrand (2).

[b] (1), vol. 2, pl. X.

[c] (3), pl. V.

[d] (8), pl. XII. Cf. his p. 244.

[e] We cannot identify this man or his work as Herrmann omitted the characters of his name and gave no further details.

[f] Ch. 27, pp. 39*b*, 40*a*.

[g] It also includes Ming place-names, however, as well as earlier ones which come mostly from the *Ta Thang Hsi Yü Chi* of Hsüan-Chuang. There was thus some sort of connection with Buddhist maps showing pilgrim itineraries (cf. Julien, 1). Nakamura discusses the relationship with a very queer map of +1194 preserved in Japan.

[1] 崑崙 [2] 圖書編 [3] 張潢 [4] 四海華夷總圖

Chi, does not clearly represent the shape of the world, I give it here. Such Buddhist attempts are not as a rule convincing.[a]

This note is of much interest, since it indicates the Buddhist-Taoist nature of the tradition, and the low value in which it was held by scholars who knew the scientific tradition of Phei Hsiu. The *Fo Tsu Thung Chi*[1] (Records of the Lineage of Buddha and the Patriarchs) still exists in the Tripiṭaka;[b] it is due to a monk Chih-Phan[2] and dates from +1270. Its cosmographical content seems to go back, through +7th-century intermediates of which only the names are known, to the *Ssu Hai Pai Chhüan Shui Yuan Chi*[3] (Record of the Sources of the Four Seas, and the Hundred Rivers) with maps, prepared by another monk, Tao-An,[4] in +347.

It is just about this time that we have a text which is perhaps the principal literary reference to this wheel-map tradition. It occurs in the *Han Wu Ti Nei Chuan* (Inside Story of the Emperor Han Wu Ti), a Taoist work of the Chin period attributed to Ko Hung. Ogawa (*1*) drew attention to this, but without seeing its proper relevance. The emperor is on a visit to the legendary goddess of the West, Hsi Wang Mu.

On this occasion the emperor also saw certain baskets in which there were small books bound in purple silk. He asked whether these contained the techniques of the holy immortals, and if he could be allowed to glance at them. Hsi Wang Mu took them out, and showed one of them to him, saying, 'This is the *Wu Yo Chen Hsing Thu*[5] (Map of the True Topography of the Five Sacred Mountains). Only yesterday all the immortals from Chhing-Chhêng[6] Mountain[c] came to me asking for it, so I shall have to give it to them. It was made by San-Thien Thai-Shang Tao-Chün[7] (The Highly Exalted Taoist Adept of the Three Heavens)[d] and it is extremely secret and important. How could so gross a person as you be fit to carry it? I will bestow upon you instead the Mysteriously Shining Scripture of Life (*Ling Kuang Sêng Ching*[8]) so that you may communicate with the spirits and strengthen your will (to immortality).'

But the emperor, prostrating himself, insisted on having the map. Hsi Wang Mu then said: 'Formerly, in the 1st year of the Shang-Huang Chhing-Hsü[9] reign-period,[e] San-Thien Thai-Shang Tao-Chün went down to see the wide world, and investigated the differences of lengths and breadths among the rivers and seas. He also observed the differences of height among the hills and mountains. Then he established the (position of the) Pillar of Heaven (Thien Chu[10])[f] and arranged the geographical features in their positions all around it. Then he placed (on the map) the five mountains (in a manner) imitated from the method

[a] Tr. Nakamura (1).

[b] N 1661; TW 2035.

[c] Near Kuanhsien in Szechuan.

[d] The second person of the Taoist trinity (cf. Vol. 2, p. 160).

[e] An imaginary reign-period.

[f] Another +4th-century work, the *Shen I Ching* (Book of the Spiritual and the Strange) says (para. 9, Chung Huang Ching): 'On Mt Khun-Lun there is a bronze pillar which goes high up to the heavens. This is called Thien Chu. It has a circumference of 3000 *li*, as round and smooth as if cut with a knife. At the base of it are the dwellings of the immortals.' Again, the *Shui Ching* opens with the statement that Mt Khun-Lun is orbocentric. All this was a commonplace of legend, presumably signifying the Kai Thien axis (p. 211 above).

[1] 佛祖統紀 [2] 志盤 [3] 四海百川水源記 [4] 道安
[5] 五獄眞形圖 [6] 青城 [7] 三天太上道君 [8] 靈光生經
[9] 上皇清虛 [10] 天柱

PLATE LXXXIX

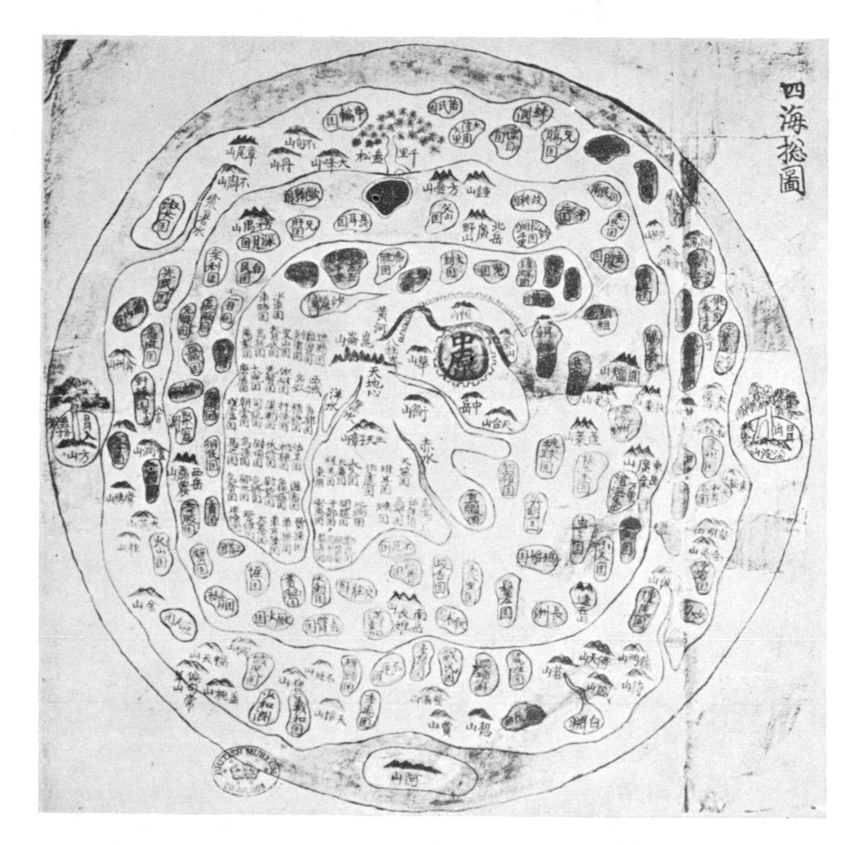

Fig. 242. Religious cosmography in East Asia; an 18th-century Korean MS. wheel-map in the Buddhist tradition centred on Khun-Lun Shan (equivalent of Mt Meru) and entitled *Ssu Hai Tsung Thu* (Comprehensive Chart of the Four Seas). Numerous islands are shown in the oceans, and many countries in the ring-continent. At east and west there are two islands each with a Tree of the Sun and Moon; cf. the 'Arbores Solis et Lunae' appearing in the east at the top of Figs. 213 and 228 above. Cf. the 'Insula Solis' in the same position in Fig. 212. North is here at the top, and north of the central plain of China (Chung Yuan) the line of the Great Wall can be seen crossing the Yellow River (from Cordier, 3).

of concentric zones (i.e. taking one as central and disposing the others symmetrically) (*chih wu yo erh ni chu chen fu*[1]). The position of honour (the central position) was given to Mt Khun-Lun, where the dwellings of the immortals were depicted. An important position was also given to the island mountain of Phêng-Lai[a] where the halls of the spirits were shown. The Water of An[2] (the moon?) was written in as very holy and sublime, since it is the origin of the extreme Yin influence. The Thai Ti[3] (Great Emperor, i.e. the sun?) was similarly placed in the wilderness where the Fu-Sang tree is.[b] A little hill no bigger than ten feet square can be the place where human fates are decided. A small island in the stormy sea has enough room for the Nine Sages. Each continent[c] has its proper name....All of them are located in an orderly manner in the huge environing ocean with its tremendous waves. The rivers are also seen flowing (on the map), some green and some black. The assembly of spirits is depicted cradled on the waves; all the immortals and jade girls are gathered there. Some of their names are hardly known, but their real features are distinctly displayed. Thus by the use of the compass and the square the rivers and their upper reaches were measured, and the mountains drawn with circularly curving lines (*phan chhü*[4]). The mountain-ranges bend back upon themselves and the smaller hills wander back and forth (*ling hui fou chuan*[5]). The height of the mountains and the extent of their slopes (*shan kao lung chhang*[6]) are shown by lines turning and curving (*chou hsüan wei shê*[7]). Indeed, they look like written characters (*hsing ssu hua tzu*[8]). Thus the written names of the mountains were determined by their respective natural shapes, and the reality of the mountains is enshrined in symbols.[d] The diagrams of the shapes (of the mountains) were kept secret in Yuan-Thai,[9] but when they were taken out they served as charms and talismans among the immortals. By their aid, Taoists can pass safely over all mountains and rivers. The hundred gods and the assembly of immortals showed the greatest respect for this map and acclaimed it. Although you are not their equal, you have often visited mountains and marshes, you have the heart of an eager seeker after truth, and you have not lost sight of the Tao. I rejoice that this is so, and now therefore I will give you the map. Keep it, I enjoin you, with as much care as men show to their lords and their parents. For if the secrets of it should leak out to ordinary people, evil fortune will overtake you.'[e]

There can thus be no doubt that we have to do with a mixed Buddhist–Taoist tradition of religious cosmography, analogous to that of Europe, but with Mt Khun-Lun as orbocentric instead of Jerusalem. Its connection with the protective charms of the Taoists is interesting—*Pao Phu Tzu*, chapter 17, is entirely concerned with this subject, and I reproduce one of Ko Hung's numerous diagrams (*ju shan fu*[10]) which seems to depict at least four of the five symmetrical mountains (Fig. 243). Among the Tunhuang manuscripts at the British Museum, there is a *Shou Shou Wu Yo Yuan Fa*[11] (The Received Method of Drawing the Circles of the Five Mountains).[f] But it would take us too far from the present argument to follow this further.

[a] Cf. Sect. 13*c* in Vol. 2, p. 240.
[b] Cf. Granet (1). Suns hung like fruits on a mythical tree.
[c] Cf. the ideas of Tsou Yen (Sect. 13*c* in Vol. 2, p. 236).
[d] Now we can understand the idea behind the Taoist 'tallies' and charms which would protect adepts who had to cross wild mountains. Bearing in their hands the very essences of the topographic pattern, they would be immune from any evils which the spirit of the mountain, if not understood, might do them (cf. Sect. 10 in Vol. 2, esp. p. 140).
[e] Tr. auct.
[f] British Museum, Stein no. 3750.

[1] 植五嶽而擬諸鎮輔 [2] 安水 [3] 太帝 [4] 盤曲 [5] 陵回阜轉
[6] 山高隴長 [7] 周旋委蛇 [8] 形似畫字 [9] 元臺 [10] 入山符
[11] 授受五嶽圓法

The cosmography of Tsou Yen in the −4th century had, as we have seen,[a] an environing ocean, though not a central mountain. So also did that of the Kai Thien astronomical theorists.[b] When Yung Hêng,[1] therefore, in the Sung, says in his *Sou Tshai I Wên Lu*[2] (Collection of Strange Things Heard)[c] that 'in the east, north and south there are seas with different names, but actually it is all one sea', he was not necessarily influenced by the wheel-map tradition, though this may be a reference to it. However, the *Ho Thu Wei Kua Ti Hsiang* (one of the apocryphal or weft classics) says[d] that the Khun-Lun Mountain is in the centre of the earth, corresponding to heaven, and that the eighty regions are scattered all around it (*pu jao chih*[3]). China is in the south-east and occupies only one of these regions. If this was written towards the end of the later Han, one can perhaps see in it the combination of the indigenous world-picture deriving from Tsou Yen with that which came from India.[e]

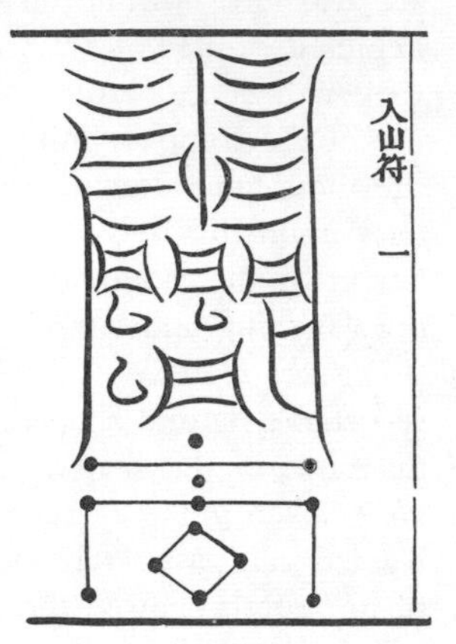

Fig. 243. A Taoist talisman from Ko Hung's *Pao Phu Tzu*, c. +300, intended to protect those who wandered through wild mountains, and perhaps a crude diagrammatic representation of their disposition.

Obviously everything points to an Indian origin of these wheel-maps; and perhaps of all wheel-maps a Babylonian one.[f] A glance at the work of Kirfel on the cosmography of the Indians shows immediately that in Buddhist and Jain ideas, long before our era, there were four continents centring upon Mt Meru (Uttarakuru in the north, Pūrvavideha in the east, Jambūdvīpa in the south, and Aparagoyāna in the west). Brahmanic tradition also has the ring-continent, modified by what were perhaps echoes of Greek 'climates'. The classical literary exposition of this cosmography is that of the *Abhidharma-kośa* (bk. 3),[g] probably composed about +370. Crystallised in stone by the Khmer people, builders of the (+9th century) Phnom Bakhen between Angkor Vat and Angkor Thom, it is still one of the wonders of the world.[h] One can hardly doubt that this cosmographic tradition came into China with Buddhism, perhaps joining earlier indigenous conceptions of Mt Khun-Lun as central. But never did it triumph over the tradition of scientific cartography there, as its analogue did in Europe.

[a] See Sect. 13*c* in Vol. 2, p. 236.
[b] See above, p. 212.
[c] Ch. 1, p. 1*a*.
[d] *Ku Wei Shu*, ch. 32, p. 4*a*.
[e] F. W. K. Müller (1) has published a late, but interesting, pictorial representation of Mount Meru with its environing continents, from a Japanese source. For an earlier Chinese Buddhist version see Feer (1).
[f] It would be interesting to know whether they were pre-Buddhist in China. The *Hou Han Shu*, ch. 112B, p. 18*b*, gives an account of the magician Fêng Chün-Ta (cf. Vol. 2, p. 148) who begged unsuccessfully from another Taoist Lu Nü-Sêng[4] for a 'Five-Mountain Map'. But the story is in the Thang commentary, not the text, so it does not prove much. On sacred mountains in all old Asian religion, cf. Quaritch Wales (2).
[g] Tr. de la Vallée Poussin (7). Cf. Renou & Filliozat (1), vol. 2, pp. 377, 380.
[h] See Filliozat (9).

[1] 永亨 [2] 搜采異聞錄 [3] 布繞之 [4] 魯女生

(e) CHINESE SURVEY METHODS

It is to this tradition of scientific cartography that we must now briefly return. Before we can draw the threads of this Section together, there are three more subjects which must be touched upon, first the probable survey methods which were used by the Chinese cartographers who made the grid-maps; secondly, the origin of relief and other special maps; and thirdly, the coming of Renaissance cartographic science to China. We shall then be able to round off the whole by a comparative retrospect and a final surprise.

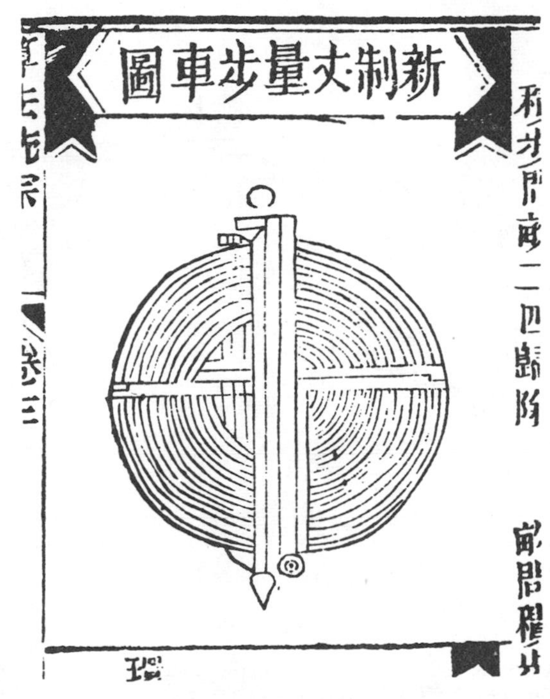

Fig. 244. Chinese survey methods; a reel of measuring tape (from the *Suan Fa Thung Tsung* of +1593).

It seems quite safe to assume that by the beginning of the Han, the Chinese were in possession of the simple and ancient survey instruments which had been known to the Babylonians and Egyptians.[a] As we have seen,[b] the astronomical use of the gnomon goes back to the earliest Chou or Shang times, and it could not have been set up without the use of the water-level and the plumb-line. Ropes, cords (Fig. 244) or chains must also have been used, together with graduated poles (Fig. 245). One of the earliest literary references to calculations of surveyors in the field may be that in the

[a] From the considerable literature on ancient surveying methods, I will refer only to the paper of Lyons (2) on ancient Egyptian, and of Walters (1) on Greek and Roman, instruments. For the Middle Ages and the Renaissance there is now a valuable book by Kiely (1) and an excellent study of topographical maps by Price (7).

[b] Pp. 284ff. above.

Fig. 245. Chinese survey methods; apparatus figured in the *Wu Ching Tsung Yao* (Collection of the most important Military Techniques) by Tsêng Kung-Liang (+1044). To the left a graduated vertical pole (*tu kan*) and a surveyor with his sighting-board (*chao pan*); to the right a water-level (*shui phing*) with three floating sights and two plumb-lines (*Chhien chi*, ch. 11). The text of the description of these instruments in Tsêng's book follows verbally that of a much earlier work, the *Thai Pai Yin Ching* (Manual of the White and Gloomy Planet of War) written by Li Chhüan in +759.

Shan Hai Ching,[a] where Yü the Great orders two of his legendary assistants (Ta Chang[1] and Shu Hai[2]) to pace out the size of the world. Shu Hai, says the text, held the counting-rods in his right hand (*yu shou pa suan*[3]) and pointed to the north with his left. This, at any rate, gives a glimpse of late Chou or early Han surveyors. Similarly, the *Chou Li* speaks[b] of the compass and square, the plumb-line and water-level (*kuei*,[4] *chü*,[5] *hsien*,[6] *shui phing*[7]) in more than one place.[c]

[a] Ch. 9, p. 3*a*.

[b] Ch. 11, p. 15*b* (ch. 40, p. 52; tr. Biot (1), vol. 2, p. 481).

[c] Cf. *Mêng Tzu*, IV (1), i, 5.

1 大章 2 竪亥 3 右手把筭 4 規 5 矩 6 縣 7 水平

One of the developed forms of the plumb-line was the *groma*,[a] associated with Rome and Hellenistic Egypt, but possibly of older origin. It consisted of two sets of plumb-lines fixed at right angles and arranged to turn round a vertical axis. One pair could be used for sighting and the other to determine the direction at right angles. It seems that the Chinese of the Han also used this instrument—the Khao Kung Chi section of the *Chou Li* says[b] that the builders (Chiang Jen[1]) 'level the ground (by the use of) water (-levels), and suspend (plumb-lines) (*shui ti i hsien*[2]).' Then 'they test the verticality of posts, gnomons or poles, by the plumb-line (*chih yeh i hsien*[3]).' To the first of these phrases the +2nd-century commentary of Chêng Hsüan says that 'at the four corners (of an instrument) four straight (lines) hang over the water, and (the surveyors) observe with this the high and the low; when this has been decided the ground can be levelled'.[c]

Other names for the water-level were *chun*[4] (which may be an ancient pictogram of the instrument), *shui hêng*[5] (the 'water balance'),[d] and *shui nieh*.[6] Fig. 245, taken from the *Wu Ching Tsung Yao* (Collection of the most important Military Techniques), by Tsêng Kung-Liang in +1044, shows what seems at first to be a kind of plane table or altitude theodolite. But the description shows that it is a trough with three floats, each bearing a fiducial sight.[e] The other important invention of Hellenistic antiquity, besides the *groma*, was the *dioptra* of Heron of Alexandria (*c.* +65). This was the ancestor of the altazimuth theodolite of today, but how nearly its actual use approximated to that of the latter is a point of some doubt (Lyons, 2). That the Chinese used sighting-tubes (*wang thung*[7]) from the Han period or earlier has already been indicated above.[f] I have not, however, come across any early descriptions which would suggest that it was mounted on a graduated quadrant, except, of course, in the special case of armillary rings. If these were already coming into use in the −4th century, it is hard to believe that the advantages of an azimuthal quadrant mounting for surveying would not have been appreciated. The surveyor's sighting-tube shown in Fig. 146 is mounted for altitudes.

A good deal has already been said[g] of the practical geometry set forth by Liu Hui (cf. Fig. 246) in his *Hai Tao Suan Ching* (+263),[h] and we have noted[i] the significant fact that he and Phei Hsiu were contemporaries. Têng Ai,[8] a general of the Wei State, was also living at that time, and it is recorded of him that whenever he saw a high mountain or a broad marsh, he always 'estimated the heights and distances, measuring

[a] Singer (2), p. 113.
[b] Ch. 12, p. 15*a* (ch. 43, p. 19; Biot (1), vol. 2, p. 553). Cf. p. 231 above.
[c] Chiang Yung, in his commentary on the Khao Kung Chi section (*Chou Li I I Chü Yao*[9]), deals with the matter in some detail (ch. 7, p. 6*b*).
[d] Cf. Schlegel (5), p. 408.
[e] Another good illustration of this is in the late Chhing *Hsiu Fang So Chih*[10] (Brief Memoir on Dyke Repairs), by Li Shih-Lu.[11]
[f] P. 332. The reference to it in *Huai Nan Tzu* will be remembered.
[g] P. 31 above.
[h] Tr. van Hée (7, 8).
[i] P. 540 above.

1 匠人 2 水地以縣 3 置槷以縣 4 準 5 水衡
6 水臬 7 望筒 8 鄧艾 9 周禮疑儀要舉 10 修防瑣志
11 李世祿

Fig. 246. An illustration of survey geometry from the *Thu Shu Chi Chhêng*; measuring the height of an island crag (*Li fa tien*, ch. 122).

by finger-breadths, before drawing a plan of the place and fixing the position of his camp'.[a] The account adds that people used to laugh at him for being so particular. But the remark is interesting as showing how widespread in the +3rd century were survey techniques involving, for the most part, the properties of similar right-angled triangles.

Perhaps the most important survey instrument of the European Middle Ages was

[a] *Wei Chih*[1] (in *San Kuo Chih*), ch. 28, p. 17*b*, quoted in *TPYL*, ch. 335, p. 2*a*.

[1] 魏志

the *baculum*, cross-staff, or Jacob's staff.[a] In its simplest form it was a graduated rod about 4 ft. long, of rectangular cross-section, and having a cross-piece which could slide along this while remaining at right angles to it. Measurements of a distant line which could not be reached and paced were made by standing approximately half-way along it at different distances from it, and sighting the two ends of the line past each end of the cross-piece. The dimensions of the staff being known, and also the distance between the observer's positions, the length of the line could be calculated[b]

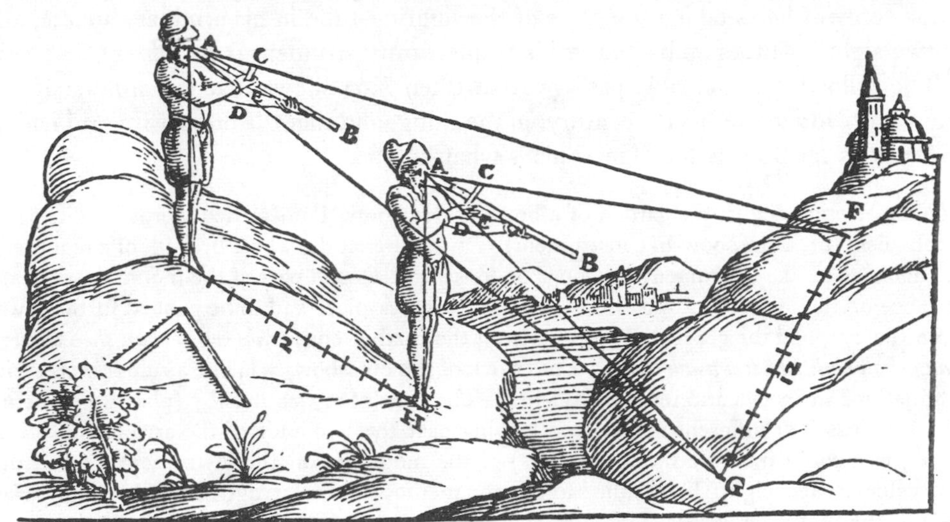

Fig. 247. The use of Jacob's Staff; an illustration from the *De Re et Praxi Geometrica* by Oronce Finé (Paris, +1556).

(see Fig. 247). The accepted view is that this simple device was first described by the Provençal Jewish scholar, Levi ben Gerson (+1288 to +1344), in 1321.[c] It certainly

[a] D. E. Smith (1), vol. 2, p. 346; Lloyd Brown (1), p. 182; Feldhaus (1), col. 543; Taylor (3, 4, 5). Some, like the writer, may have come across Thomas Bancroft's epitaph on William Holorenshaw:

> Lo in small closure of this earthly bed
> Rests he who heaven's vast motions measured,
> Who, having known both of the land and sky,
> More than famed Archimed, or Ptolemy,
> Would further press, and like a Palmer went,
> With Jacob's Staff, beyond the firmament.

[b] The instrument had other uses, for instance, it could measure altitudes of celestial bodies.

[c] See Sarton (1), vol. 3, pp. 129, 600. One form of it had of course been described by Ptolemy (*Almagest*, v, 14) and Pappus (*Commentary*, v), under the name of the *dioptra*, but it was only capable of measuring very small angles. The diameter of the sun was obtained by its aid. We are indebted to Dr Derek Price for a reminder of this. Ptolemy's 'triquetrum' (cf. above, p. 373) was also an allied instrument. When one of its limbs was freed from the vertical zenith-pointing position by al-Kindī in +9th-century Baghdad, it became essentially a quadrant without a graduated arc, and could be used to measure the angular distances of any two stars, not only the zenith distances of stars at culmination. Al-Kindī's description has been translated by Wiedemann (9), and the fact that the manuscript in question was copied as late as +1212 suggests that the instrument was then still being found useful. By making two observations from different positions in a straight line, the heights or lengths of distant terrestrial objects could also be ascertained, and this use was also described by al-Kindī. But all these instruments had actual material limbs along the lines of sighting; the turning-point came when someone appreciated the simple fact that the cross-piece of a cross-bow, if equipped with sights at each end and made to slide on the butt, would do just as well. On Levi ben Gerson and the cross-staff see also Steinschneider (4), pp. 248, 270.

lasted a long time, till about +1594, when it was replaced by the backstaff of John Davis.[a] This in its turn gave place in the +18th century to reflecting instruments, which were the direct forerunners of the modern sextant. It would be natural to assume, then, that knowledge of the cross-staff reached the Chinese through Arab contacts in the +15th century, if not indeed still later through the Jesuits.[b] Similarly, it has been supposed[c] that the use of cross-wires in sighting-tubes or alidades was introduced to China by the visit of Jamāl al-Dīn in +1267. The *Yuan Shih*, however,[d] speaks only of holes taking the place of the sighting-tube in his armillary sphere, and of two sighting-tubes embodied in his triquetrum instrument (*tshê chhi*[1]).[e]

The following remarkable passage from Shen Kua shows that the cross-staff was known already in the +11th century, in the Sung, and that the cross-wire grid had an ancestry as far back as the Han. This is what he says:

When I once dug in the garden of a house at Haichow, I unearthed a crossbow trigger-mechanism (or, a crossbow-like instrument). On looking at the whole breadth of a mountain, the distance on the instrument was long; on looking at a small part of the mountain-side, the distance on the instrument was short (because the cross-piece had to be pushed further away from the eye, and the graduation started from the further end) (*chhi wang shan shen chhang, wang shan chih tshê wei hsiao tuan*[2]). The (stock of the crossbow) was like a rule (*chhih*[3]) with graduations in inches and tenths of an inch. The idea of it was that by (placing) an arrow (*tsu*[4]) (across it at different points) and looking past the two ends of the arrow (*mu chu tsu tuan*[5]) one could measure the degree (*tu*[6]) of the mountain on the instrument, and in this way calculate its height. This is the same as the method of similar right-angled triangles (*kou ku*[7]) of the mathematicians.

Now the Thai Chia[8] chapter[f] (of the *Shu Ching*) speaks of a man with his bow drawn and his finger on the trigger 'aiming at the target (*wang hsing*[9]) embraced in the graduations (*tu*[6]) (of his sights), and so letting fly'. I suppose that this degree system (*tu*[6]) was something like the graduations on the crossbow-instrument just described.

Then in the Han time (Liu) Chhung, Prince of Chhen (Chhen Wang Chhung[10]) was very skilful with the crossbow; he could hit the bulls-eye with a hundred per cent score. His method was:[g] '*Thien fu ti tsai, tshan lien wei chi, san wei san hsiao, san wei wei ching, san hsiao wei wei, yao tsai chi ya.*'[11] These words are (at first sight) rather obscure and difficult to understand. '*Thien fu ti tsai*' (Heaven covers and the Earth sustains) probably refers to the motions of the hands in holding the crossbow, one in front and one behind. The next four characters ('*tshan lien wei chi*') refer to the position of the arrow (*tsu*[4]) (used as cross-piece)

[a] Cf. Cooper (1).

[b] Mikami has a paper (15) on the study of the Dutch art of surveying by the Japanese. Cf. Bernard-Maître (1), p. 57.

[c] *Ko Chih Ku Wei*, ch. 2, p. 27*a*; apparently following an opinion of Mei Wên-Ting.

[d] Ch. 48, p. 11*a*.

[e] This was the modified form described by al-Kindī (see previous page).

[f] Ch. 14; cf. Medhurst (1), p. 147; Legge (1), p. 97. Thai Chia was the fourth Shang emperor in the traditional lists, and famous as an archer.

[g] As recorded by Hua Chhiao,[12] quoted in *Hou Han Shu*, ch. 80, p. 2*b*; later cit. e.g. *TPYL*, ch. 348, p. 3*a*.

1 測器 2 其望山甚長望山之側爲小短 3 尺 4 鏃
5 目注鏃端 6 度 7 句股 8 太甲 9 往省 10 陳王寵
11 天覆地載參連爲奇三微三小三微爲經三小爲緯要在機牙 12 華嶠

in relation to the degrees marked (on the stock), and this position in turn depends upon the distance of the target. This is the 'Triple Connection' (*tshan lien*[1]); it is like a steelyard or balance (and with it the proper elevation of the crossbow can be assessed). It is exactly the same principle as the use of similar right-angled triangles for the measurement of heights and depths. The '*san ching san wei*' (three lengthwise and three crosswise; i.e. a grid or cross-wire sights) are set up on the frame (*phêng*[2]) (or, as if a framework), and by means of them (the archer) can mark his target, whether high or low, to the right or to the left.

I once arranged such a 'three lengthwise and three crosswise' (grid) on a crossbow, and also sighted the target with the cross-piece arrow, and the result was that my shots were successful seven or eight times out of ten. If graduations are added on the trigger-mechanism (itself), the accuracy will be still further improved.[a]

The significance of this passage, written about +1085, is that in Europe the cross-staff was also known as the crossbow or arbalest, and may plausibly be supposed to have developed from that particular propulsive mechanism. Now, as we shall see later,[b] the crossbow was much more ancient and widespread in China than in Europe. It seems to have been introduced from the East twice, first for artillery use only, in Hellenistic and early Byzantine times, and then in the +11th century for what became a period of crossbow dominance. Hence an application to surveying at the time of Levi ben Gerson or somewhat before, would be natural. In China, on the other hand, it formed the standard weapon of the Han armies, and was in continuous use from the −4th century onwards.[c] Could this fact have had something to do with a superiority in survey methods which assisted the continuity of grid cartography after Phei Hsiu? In any case, Shen Kua certainly had in his hands a crossbow-like instrument with graduated divisions for the measurement of heights, breadths and distances. This is what we saw that Phei Hsiu considered so necessary as one of his 'Essentials of Map-Making'.[d] Shen Kua further believed that the sighting techniques of the Prince of Chhen nine hundred years before had also involved an instrument with a stock graduated for use as a cross-staff. That Liu Chhung and other Han marksmen used cross-wire grid sights is well assured from other evidence[e] but the epigram does not attest the use of the cross-staff at that time, as Shen Kua thought, and if it was not known then it was certainly not known in the *Shu Ching* period. The time of Phei Hsiu himself would be a reasonable guess for its introduction.

If the cross-staff was really first developed in China, and afterwards found its way to Europe, the association with Levi ben Gerson suggests that its transmission may have been effected through Jewish circles.[f] The probable role of Hebrew travellers and merchants in carrying ideas and techniques between East Asia and Western Europe

[a] *Mêng Chhi Pi Than*, ch. 19, para. 13, tr. auct. Cf. Hu Tao-Ching (*1*), vol. 2, p. 635.
[b] In Sect. 30*e* on military technology.
[c] It was even developed in the Sung, as we shall see, into a repeating weapon of machine-gun type (Sect. 30*e*).
[d] P. 540 above.
[e] On this see Sect. 30*e* below.
[f] See Addendum on p. 681 below.

1 參連 2 棚

has already been noted,[a] and we shall find another possible example of their influence in a topic in Section 27*c* on engineering.[b]

A passage on the application of survey methods to cartography is also contained in the *Mêng Chhi Pi Than*:

As for the geographical books (writes Shen Kua) there was formerly a *Fei Niao Thu*[1] (Bird's-Eye Map)[c] but we do not know who its author was. The Fei Niao system may be explained as follows. As roads and paths are sometimes winding and sometimes straight, without any definite rule, if a walker starts out in any one of the four directions from a given point along a path, his pacing will not help us to get the direct distance. Therefore what we call 'straight lines in the four directions' (*ching chih ssu chih*[2]) have to be measured by other methods, just as a bird can fly in a straight line unaffected by the convolutions of mountains and rivers.

I recently made a map of the counties and prefectures (*Shou Ling Thu*[3])[d] on a scale (*fên lü*[4]) of 2 inches for 100 *li*. I used the methods of *chun wang*[5] (rectangular grid), *hu yung*[6] (mutual inclusions),[e] *phang yen*[7] (checking from the side),[f] *kao hsia*[8] (heights and depths), *fang hsieh*[9] (right angles and acute angles), and *yü chih*[10] (curved and straight lines). With these seven methods one can work out the distances as a bird would fly them. The finished map had *fang yü*[11] (four-cornered, square, divisions) strictly to scale (*yuan chin chih shih*[12]). Then the four (azimuth) directions and the eight positions may be increased to twenty-four, these being designated by the twelve cyclical signs, eight of the denary series of cyclical signs, and four of the *kua* (trigrams). Thus later generations with the help of my recorded data, and using the twenty-four directions, will be able to reconstruct the map showing the positions of districts and towns without the slightest mistake, even if the original copies should be lost.[g]

This statement is of particular interest, for it suggests rather strongly that towards the end of the +11th century Chinese cartographers were recording compass-bearings, as in modern ordnance surveys. It will be remembered that elsewhere in the same book we have the earliest definite reference to the magnetic needle in any literature.[h] Maps of the kind here described by Shen Kua may therefore have resembled the Mediterranean portolans of three centuries later, where a series of compass roses are accompanied by a network of approximate meridians and parallels.[i] But in +1085

[a] In Sect. 13*e*, *f* (Vol. 2, pp. 278, 297 above), concerning the passage of philosophical ideas, tables of categories, and the like.

[b] The invention of the Cardan suspension.

[c] Or 'As-the-Crow-Flies' Map, suggested the late Prof. A. C. Moule. Shen Kua is perhaps alluding to the maps of Chang Hêng (see above, p. 538).

[d] Other texts assembled by Hu Tao-Ching (*1*), vol. 2, pp. 991 ff. show that this map was imperially commissioned in +1076 and presented in +1087.

[e] This technical term is not one of Phei Hsiu's, but it may well refer to similar right-angled triangles, and is perhaps an extension of his *tao li* (cf. p. 539 above). Cf. *Hai Tao Suan Ching*, problems 1–6. Hu Tao-Ching's text reads *ya yung*.[13]

[f] Again, this technical term is not one of Phei Hsiu's. Cf. *Hai Tao Suan Ching*, problems 7 and 9.

[g] Pu Appendix, ch. 3, para. 6, tr. auct.

[h] Cf. Sect. 26*i* below.

[i] Chu Kho-Chen (*4*) and Wang Chen-To (*5*), p. 114, concur.

[1] 飛鳥圖 [2] 徑直四至 [3] 守令圖 [4] 分率 [5] 準望
[6] 互融 [7] 傍驗 [8] 高下 [9] 方斜 [10] 迂直 [11] 方隅
[12] 遠近之實 [13] 牙融

Europe was still far indeed from having reached this point. Unfortunately, none of the Chinese 'land-portolans', such as this, has survived. The practice, however, is relevant to the question of the indigenous origin of the *Wu Pei Chih* maps of Chêng Ho's time.[a] If Shen Kua's methods were at all widespread, one might expect to hear about the same time of the establishment of 'trigonometrical stations' from which the bearings would be taken. Something of this sort may lie behind the military map which Liu Chhang-Tsu[1] presented to the emperor in +1083.[b]

For determining the lay of sloping ground, graduated poles were used in Shen Kua's time. This we know from one of his autobiographical sections, where he describes work which he once did as a government hydraulic engineer (between +1068 and +1077).

The Pien Canal (Pien Chhü[2]) had brought down so much silt that the bottom of it had become higher than the ground outside by about 12 ft., all the way between the eastern water-gate of the capital to Yung-chhiu and Hsiang-yi. Looking down from the embankments one could see the inhabitants as if they were living in a deep valley. In the Hsi-Ning reign-period the government were considering a plan for clearing out the Lo River and leading it into the Pien Canal. My duty as a government conservancy official was to measure the canal from Shang-shan-mên near the capital to the Huai-khou at Ssu-chou (the place where the canal debouched on the Huai River). The distance was 840 *li*, 130 *pu* (about 175 miles). The ground at the capital was found to be higher than that at Ssu-chou by 194 ft. 8 in. A test boring was made in the Po river-bed several *li* east of the capital, and the old bed was not seen until it had gone down 30 ft. (such was the accumulation of silt).

The instruments used (to determine the slope) were the *shui phing*[3] (water-level), the *wang chhih*[4] (sighting-tube with graduated scale, presumably an altitude theodolite), and *kan chhih*[5] (graduated pole). But the measurement of the fall could not be made with perfect accuracy, so the following steps were taken. As the silt was piled outside the bed of the Pien Canal or on its embankments, dams were built to collect the water in the bed itself. After water had been let in, another dam was built, the whole series being in a succession like steps. The differences between the water levels were then measured, and their sum gave the total fall.[c]

This serves to remind us of another point of social importance, namely, the stimulus which the characteristic water conservancy works of China must have given to the art of cartographic survey.[d] Many problems of this kind are contained in Chhin Chiu-Shao's mathematical work of +1247, the *Shu Shu Chiu Chang* (Figs. 248 and 249).

The question may be raised whether the hodometer or taximeter carriage was used to any extent by the ancient and medieval Chinese cartographers. The simple application of gear-wheels involved has already been alluded to in Section 7*l* on culture contacts,[e] where it was pointed out that the first references to hodometers in China are at least as early as those in Europe. They go back to Prince Tan of Yen (−240 to −226)

[a] Cf. p. 560 above.

[b] *Yü Hai*, ch. 14, p. 37*b*.

[c] *Mêng Chhi Pi Than*, ch. 25, para. 8, tr. auct. Cf. Hu Tao-Ching (*1*), vol. 2, pp. 795ff.

[d] Cf. the abiding interest shown by Phei Hsiu and others in everything connected with the Yü Kung.

[e] Vol. 1, pp. 152, 195, 229, 232 above.

[1] 劉昌祚 [2] 汴渠 [3] 水平 [4] 望尺 [5] 幹尺

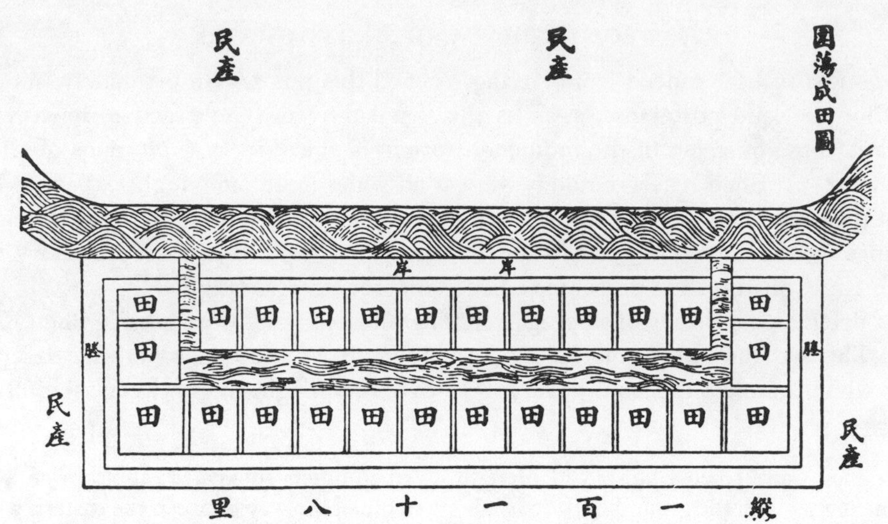

Fig. 248. Plan of an irrigation survey problem by Chhin Chiu-Shao, +1247 (*Shu Shu Chiu Chang*, ch. 6). The total length of the parallel canal is given as 118 *li* or 59 km. The irrigated fields are marked with the character *thien*, the major river dyke is called *an*, and the minor dykes *chhêng*. Cartographical accuracy comes second in these representations to geometrical formalisation.

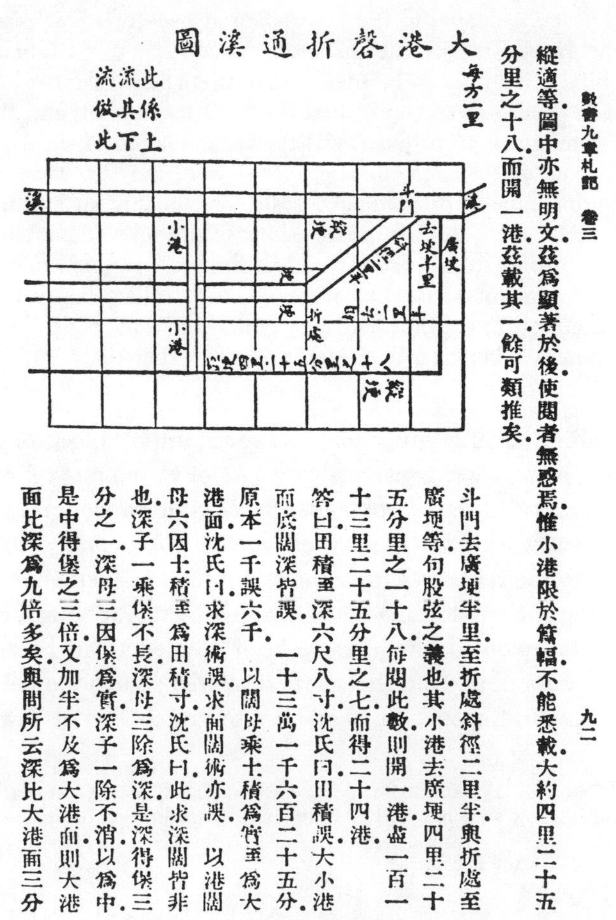

數書九章札記　卷三

縱適等圖中亦無明文。茲爲顯著於後。使閱者無惑焉。惟小港限於篇幅。不能悉載。大約四里二十五分里之十八。而開一港。茲載其一。餘可類推矣。

斗門去廣埂半里。至折處斜徑二里半。與折處至廣埂等。句股弦之義也。其小港去廣埂四里二十五分里之一十八。每閱此數。則開一港。盡一百一十三里二十五分里之七。而得二十四港。

答曰。田積亟深六尺八寸。沈氏曰。田積誤大小港面底闊深皆誤。　一十三萬一千六百二十五分。原本一千誤六千。以闊母乘土積爲實亟爲大港面。沈氏曰。求深術誤。求面闊術亦誤。以港闊母六因土積亟爲田積寸。沈氏曰。此求深闊皆非也。深子一乘保不長。深母三除爲深。是深得保三分之一。深母三因保爲實。深子一除不消。以爲中是中得保之三倍。又加半不及。爲大港面則大港面比深爲九倍多矣。與問所云深比大港面三分

九二

Fig. 249. Sung Ching-Chhang's reconstruction of the preceding problem in his *Shu Shu Chiu Chang Cha Chi* (1842), ch. 3. He draws the map with closer approach to reality and inserts further details such as the sluice-gate (*tou mên*) at the entrance to the parallel canal, and the smaller lateral derivates (*hsiao chiang*). The main canal runs along ½ *li* away from the river and takes 2½ *li* to reach that position.

and to Han Yen-Shou (between −140 and −70). The proper place for the main discussion of the hodometer (called *chi li ku chhê*,[1] 'mile-counting drum carriage') will be in Section 27*c* on engineering, but in view of the fact that Chinese map-making must have involved so much dead-reckoning, one cannot help wondering whether such men as Chang Hêng, Phei Hsiu and Chia Tan did not make some use of it. The Renaissance cartographers seem to have done so. Lloyd Brown reproduces[a] an illustration of a surveyor's carriage with a dial-recording hodometer from Paul Pfintzing's *Methodus Geometrica* of +1598. The French physician Jean Fernel had in the same century measured a meridian line between Amiens and Paris by the use of a hodometer.[b] However, no Chinese text has yet come to light which demonstrates the employment of this device by the cartographers. In steppe or plain country, if not in mountainous regions, it could have been helpful.

We shall also examine in the appropriate place the famous 'south-pointing carriage' (*ting nan chhê*[2]), which was a more elaborate application of gearing principles. This was probably invented in the Han, and could conceivably have been of cartographic value in open regions where the ground was not too broken, though one has the impression that it belonged to the mechanical arcana of the emperor and was not capable of much practical use. Again, we have not found any text which would relate it to map-making.

(*f*) RELIEF AND OTHER SPECIAL MAPS

The idea of representing the surface of the earth in modelled form on an exaggerated scale, so that the contours of mountain and river systems may clearly be seen, seems to us obvious enough. And, indeed, it has a much longer history than is usually suggested by historians of geography. Let us, in this case, start from recent times and work backwards.

For John Evelyn the idea was not so obvious, as may be seen by a paper of his which appeared in the *Philosophical Transactions of the Royal Society* for +1665, 'Of a Method of making more lively Representations of Nature in Wax than are extant in Paintings, and of a New Kind of Maps in Bas-Relief; both practised in France'. Lloyd Brown believes[c] that the earliest known relief map was one of the Swiss canton of Zürich in +1667. Feldhaus[d] and other writers attribute the priority to Paul Dox, who in +1510 thus represented the neighbourhood of Kufstein. Sarton[e] takes the matter further back by pointing out that Ibn Baṭṭūtah (+1304 to +1377) describes a relief or raised map which he saw at Gibraltar.[f] I shall now show that relief maps of a scientific character were well known in China in the +11th century.

[a] (1), p. 243.
[b] Lloyd Brown (1), p. 290.
[c] (1), pp. 273, 337, 362, 364.
[d] (1), col. 552.
[e] (1), vol. 3, pp. 1157, 1620.
[f] Defrémery & Sanguinetti (1), vol. 4, p. 359.

[1] 記里鼓車 [2] 定南車

Our main evidence for this is contained in the *Mêng Chhi Pi Than* (+1086), in which Shen Kua wrote:

When I went as a government official to inspect the frontier, I made for the first time a wooden map upon which I represented the mountains, rivers and roads. After having explored personally the mountains and rivers (of the region), I mixed sawdust with wheat-flour paste (modelling it) to represent the configuration of the terrain upon a kind of wooden base. But afterwards when the weather grew cold, the sawdust and paste froze and was no longer usable, so I employed melted wax instead. The choice of these materials was dictated by the necessity of making something light which would not be difficult to transport. When I got back to my office (in the capital) I caused (the relief map) to be carved in wood, and then presented it to the emperor. The emperor invited all the high officials to come and see it, and later gave orders that similar wooden maps should be prepared by all prefects of frontier regions. These were sent up to the capital and conserved in the imperial archives.[a]

Shen Kua must therefore take an important place in the history of relief maps, as of other aspects of geography.[b] But there are further records of Sung scholars interested in the relief technique. About +1130, Huang Shang[1] made a wooden relief map,[c] and this attracted the attention of the great Neo-Confucian, Chu Hsi, who requested his friend Li Chi-Chang[2] in a letter to try and find it.[d] Chu Hsi himself constructed relief maps in sticky clay (*chiao ni*[3]) as well as in wood. Lo Ta-Ching[4] records in his *Ho Lin Yü Lu*[5] (Jade Dew from the Forest of Cranes) a conversation with Chao Shih-Shu[6] (i.e. Chao Chi-Jen[7]) who told him of the philosopher's love for mountain scenery, which he would always go out of his way to visit. Then he added:

Chu Hsi also made a wooden Map of the Countries of the Chinese and Barbarians (*Hua I Thu*[8]), upon which the convexities and concavities of mountains and rivers (*wa tieh chih shih*[9]) were carved. Eight pieces of wood were used, with hinges to connect them together. The map could be folded up and one person could carry it. Whenever he travelled, he took this along with him. But it was never really completed.[e]

How far back the idea of relief maps goes before the time of Shen Kua is not clear, but there are several features in ancient Chinese art which may well have given rise to it. One of these is the custom of representing sacred mountains in sculptured relief upon incense-burners and jars. Laufer has described in detail[f] the well-known bronze 'hill-censers' or Vast Mountain Stoves (*po shan hsiang lu*[10]) of the Han dynasty, some examples of which have survived. These are mentioned in all Chinese archaeological works, such as the Chhing *Hsi Chhing Ku Chien*[11] (Hsi Chhing Catalogue of Ancient

[a] Ch. 25, para. 22, tr. auct., adjuv. Chavannes (10). Cf. Hu Tao-Ching (*1*), vol. 2, p. 813.
[b] See the special article on him by Chu Kho-Chen (*4*).
[c] *Yü Hai*, ch. 14, p. 38*a*.
[d] *Chu Tzu Wên Chi*, replies to Li Chi-Chang.
[e] Ch. 3, p. 5, tr. auct. The passage is quoted in *Shuo Fu* (1652 edition of Thao Thing), ch. 21 (*Ho Lin Yü Lu*), p. 1*b*.
[f] (3), pp. 174 ff.

1 黃裳 2 李季章 3 膠泥 4 羅大經 5 鶴林玉露
6 趙師恕 7 趙季仁 8 華夷圖 9 凹凸之勢 10 博山香爐
11 西清古鑑

Mirrors and Bronzes) of +1751.[a] One is shown in Fig. 250*a*. The cover of these incense-burners, which were also made in pottery, is always shaped in the form of a realistic hill or mountain, with holes through which the perfume escaped, and some arrangement by which the mountain is surrounded by water.[b] This art motif must be at least as early as the Former Han, for the censers are referred to by Chang Chhang,[1] who died in −48, in his *Tung Kung Ku Shih*[2] (Stories of the Eastern Palace), as well as in other Han books.[c] The *Hsi Ching Tsa Chi*, which, though written later, is well informed on Han matters, attributes[d] the construction of such censers to the famous mechanic[e] Ting Huan,[3] who incorporated in them 'many queer birds and strange animals, all of which could turn round by themselves'—activated, perhaps, by the

Fig. 250. Two kinds of vessel from which the earliest relief maps may have originated. On the left (*a*), a pottery 'hill-censer' (*po shan hsiang lu*) of the Han period, its cover moulded so as to represent one of the magic mountain islands of the Eastern Sea. On the right (*b*), a pottery mortuary jar the lid of which also represents modelled mountains. After Laufer (3), pls. LV and LVII.

ascending hot-air current. In later times, these hill-censers, fitted with long handles, were adopted in Chinese Buddhism for liturgical purposes, and many of the donors depicted on the frescoes at the Tunhuang caves may be seen carrying them. But as Chinese archaeological books often suggest a Taoist identification, the usual view is that the hill-censers were originally supposed to represent the famous sacred island-mountains in the Eastern Sea, of which Phêng-Lai[4] was the most important. The representation of these mountains long persisted in other materials, such as the large blocks of jade thus carved, many of which were in the imperial collections in recent

[a] Especially ch. 38, pp. 40 ff. The 'Western Retreat' was a library and museum in the imperial palace.
[b] Illustrations and comment in Laufer (3), p. 192; Siren (1), vol. 2, pls. 35, 36, 37; Hentze (1), p. 203; R. L. Hobson, vol. 1, p. 7; Koop (1), pl. 57.
[c] On stylistic evidence Wenley (1) places at least one existing specimen in the late Warring States period.
[d] Ch. 1, p. 8*a*.
[e] Cf. below, Sect. 27.

[1] 張敞 [2] 東宮故事 [3] 丁緩 [4] 蓬萊

times.[a] The identification with Phêng-Lai seems probable enough, as the tradition seems to have begun about the time of Taoist dominance under Han Wu Ti. Moreover, Han mortuary jars with lids representing a mountain in relief are also common,[b] and here the connection with ideas of immortality would be evident (Fig. 250*b*). It may well be that the idea of relief maps originated in this way.

This is, however, not the only possibility. Wang Yung has pointed out that the earliest Chinese maps (as indeed we have already seen)[c] were carved on wood, citing the 'Fu Pan Chê'[1]—Bearers of the Tables of Population—to whom Confucius always used to bow.[d] There seems here a very old connection between geographical maps and population statistics.[e] Among the many other meanings of the word *pan* (also written *pan*[2]) is that of the mould with branching channels used for casting coins,[f] as also, of course, the carved board from which each page of a book was printed. No doubt in China the carving and incising of characters was particularly important from the oracle-bones and the bamboo slips onwards, and so might have given rise to the idea of the material representation of the earth's surface in relief. Of course carved inscriptions were also a feature of the Babylonian and Egyptian civilisations. In any case, there is one strange account of what may have been a relief map in the −3rd century, which cannot be omitted. In describing the tomb of Chhin Shih Huang Ti, the *Shih Chi* says:

In the tomb-chamber the hundred water-courses, the Chiang (the Yangtze River) and the Ho (the Yellow River), together with the great sea, were all imitated by means of flowing mercury, and there were machines which made it flow and circulate. Above (on the roof) the celestial bodies were all represented; below (presumably on the floor or on some kind of table) the geography of the earth was depicted.[g]

Here, at the least, there must have been channels for the mercury to flow in, so that a relief map is implied. This was in −210.

Then, in the Han, we find in +32 a mention of strategic maps made by a general, Ma Yuan,[3] in which the disposition of mountains and valleys was represented by modelling in rice.[h] He demonstrated this before the first emperor of the Later Han at a critical moment in the campaigns which established this dynasty. The same technique persisted in the Thang, for in the +9th century Chiang Fang[4] wrote a special essay on the subject, *Chü Mi Wei Shan Fu*.[5]

The use of wood for maps led to one rather remarkable development in the +5th century, namely, the making of what seems to have been a 'jigsaw map'. The story comes from the official history of the Liu Sung dynasty, which says:

Hsieh Chuang[6] (+421 to +466) made a wooden map ten feet square, on which mountains, water-courses and the configuration of the earth were all well shown. When one separated

[a] Hansford (1), pl. XXIX.
[b] Laufer (3), pp. 198 ff.
[c] P. 535.
[d] *Lun Yü*, X, 16, iii.
[e] This we see in its late form in the *Kuang Yü Thu*.
[f] Cf. below, Sect. 36.
[g] Ch. 6, p. 31*a* (tr. Chavannes (1), vol. 2, p. 194; Wieger, *TH*, p. 225).
[h] *Hou Han Shu*, ch. 54, p. 6*b*. The passage has been translated by Bielenstein (2), p. 50.

[1] 負版者 [2] 板 [3] 馬援 [4] 蔣防 [5] 聚米爲山賦 [6] 謝莊

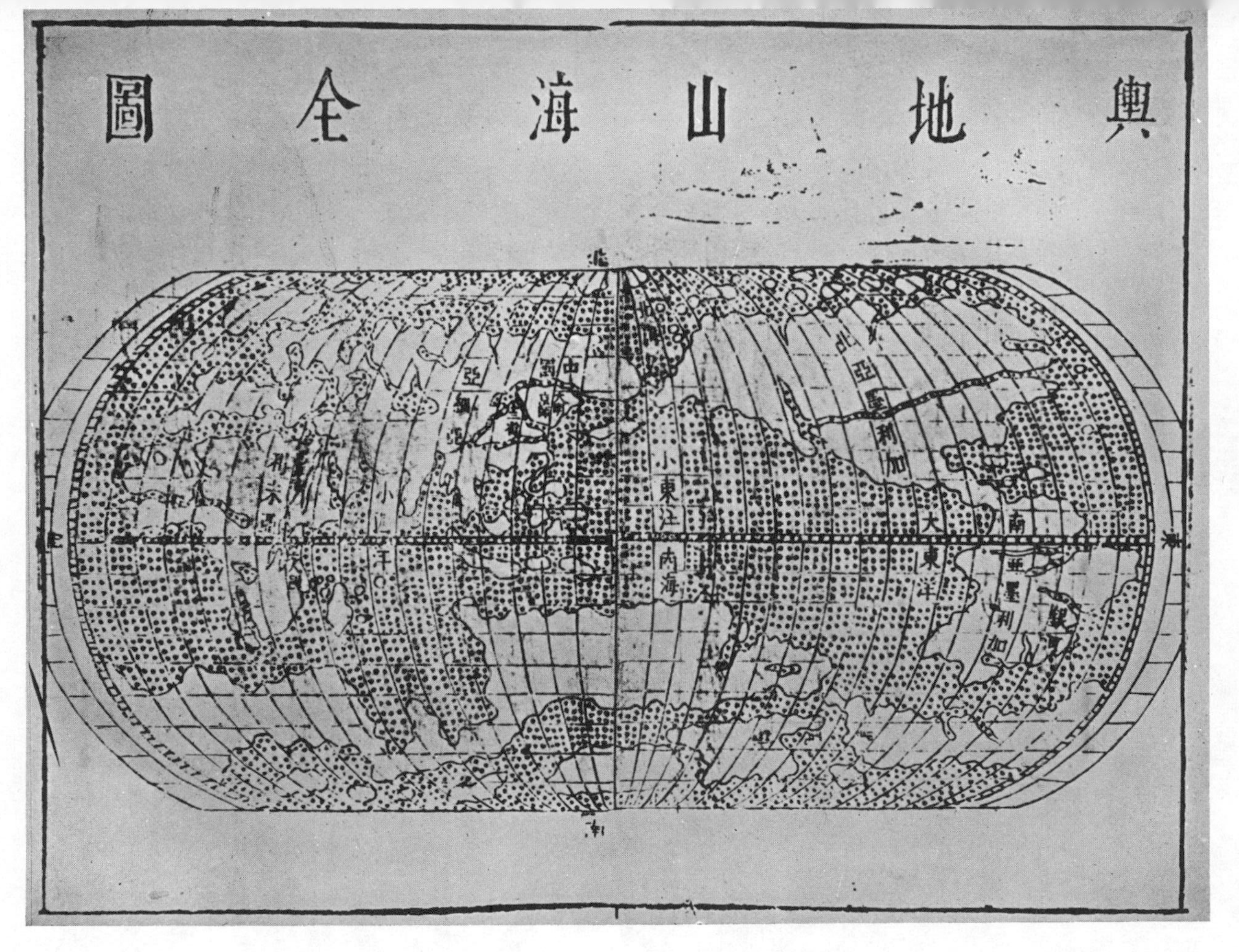

Fig. 251. The first edition of Matteo Ricci's world-map (+1584), reproduced by Chang Tou-Chin in his *Thu Shu Pien* of +1623 (from d'Elia (2), vol. 2, Pl. VIII). Entitled *Yü Ti Shan Hai Chhüan Thu*, it has straight parallels and curving meridians; China is represented almost at the centre, with 'the capital of the Great Ming Dynasty' in prominent characters. A large and straggling antarctic continent is shown, perhaps to balance the land masses of the northern hemisphere. Was the part of it south of the East Indies meant for more than New Guinea? This raises the question of an Asian tradition (which the Jesuits may have encountered) of a great unknown southern continent; see Sect. 29 hereafter.

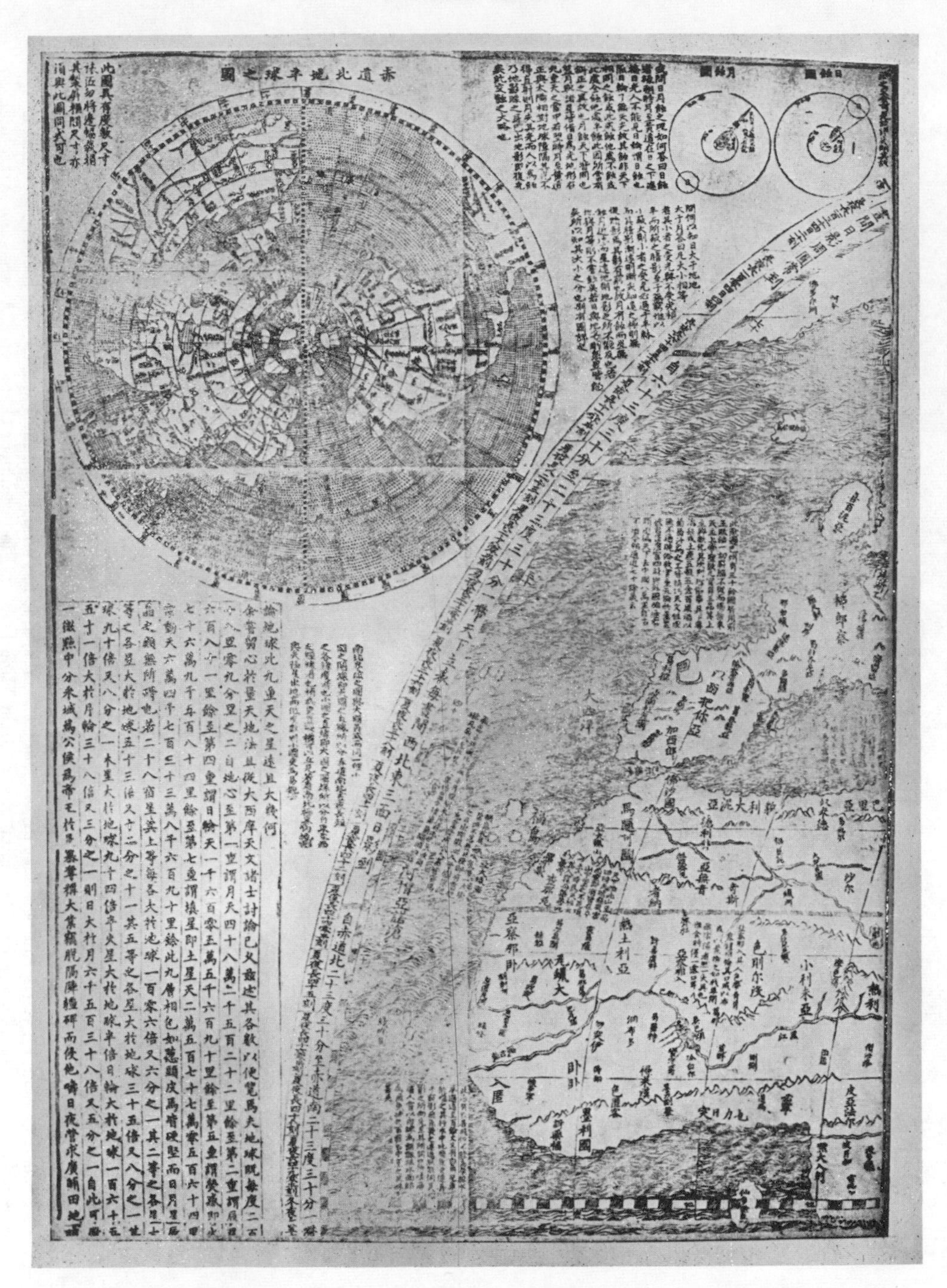

Fig. 252. A corner (one-twelfth) of the definitive world map of Matteo Ricci (+1602) entitled *Khun Yü Wan Kuo Chhüan Thu*. The projection was similar, but the larger format permitted the inclusion of far more information. Here can be seen the western part of Africa, Spain, France, Ireland and the western coasts of England and Scotland. A north polar projection occupies the left-hand top corner, and at the top on the right are two small eclipse diagrams (from d'Elia (3), vol. 2, Pl. II).

(the parts of the map) then all the districts were divided and the provinces isolated; when one put them together again, the whole empire then once more formed a unity.[a]

But there seems to be no further mention of this in Chinese history. Though the date of the first map of this kind in the West is not easy to determine, it would be hard to believe that it was as early as this.

As for terrestrial globes, none is mentioned in Chinese texts until the Yuan time, when there is a description of the model brought by Jamāl al-Dīn in +1267.[b] Apparently it had little effect, for such globes were not again made until the Jesuit period.

(g) THE COMING OF RENAISSANCE CARTOGRAPHY TO CHINA

In 1583 Matteo Ricci established himself at Chao-chhing[1] (then the capital of the two provinces Kuangtung and Kuangsi), and it was there that he was asked by Chinese scholarly friends, especially Wang Phan,[2] to prepare for them a map of the world.[c] This was the beginning of his famous world-map (*Khun Yü Wan Kuo Chhüan Thu*[3])[d] of 1602. It was on a flattened sphere projection, with parallel latitudes and curving longitudes, and it was certainly based on the world-map of Ortelius (+1570).[e] For example, like the map of Ortelius[f] before it, and that of Kepler (1630)[g] after it, it marks the entirely imaginary isle of 'Friesland' in the Atlantic south of Iceland.[h] Ricci's map has been carefully analysed by Baddeley and Heawood, while the text on it has been fully translated by L. Giles (10) and Goodrich (8). There has been much discussion as to Ricci's Chinese sources,[i] and his influence on subsequent Chinese

[a] *Sung Shu*, ch. 85, p. 1*b*, tr. auct., adjuv. Chavannes (10).

[b] See above, p. 374. We shall return to the matter in Sect. 29*f*.

[c] Bernard-Maître (1), pp. 40, 50, 60–2, 70; Trigault (Gallagher tr.), pp. 166 ff., 301 ff., 331, 397, 536. See Fig. 251.

[d] Photographically reproduced by the Chinese Society of Historical Geography at Peiping in 1936. And two years later, in sumptuous form, by d'Elia (5). Cf. Bernard-Maître (13); Szczesniak (7); Vacca (10).

[e] It is usually held that the contribution of Ricci to Chinese geography was enormous. 'He gave to the Chinese', it is said, 'their first complete map of a round world showing the western hemisphere and the five continents in their relative positions.' This is true in so far as America was not previously known to them. But to say that he 'demonstrated how to describe and locate places by longitude and latitude' is not quite fair, for, as we have seen (p. 292), the different lengths of gnomon shadows had been known in China for many centuries, and longitude was very difficult to determine accurately before 'Mr Harrison's Number One' of +1735. The measurement of the degree of latitude had been going on in China at least since the time of I-Hsing who determined it about +725 as 351 *li* 80 *pu* (Gaubil (2), pp. 77, 78). Ricci certainly assigned to many places previously unknown to the Chinese transliterations which they bear to this day; but his services have sometimes been proclaimed rather uncritically. For example, d'Elia (5) gives a particularly bad Chinese map of China and Korea of +1555 as a foil to Ricci's world map, but makes hardly any reference to Chu Ssu-Pên and none at all to Li Hui and Chhüan Chin, the great summarisers of the cosmographical knowledge of the Mongol age. Cf. Szczesniak (7). See Fig. 252.

[f] Cf. Lloyd Brown (1); Tooley (1).

[g] Nordenskiöld (1), pp. 153, 198.

[h] We get this again in Giulio Aleni's *Chih Fang Wai Chi*[4] of +1623 (Fig. 253). Dunlop (2) suggests an origin from al-Idrīsī's 'Raslānda' and a confusion with the Faröes.

[i] Puini (3), Chhen Kuan-Shêng (2).

[1] 肇慶 [2] 王泮 [3] 坤輿萬國全圖 [4] 職方外記

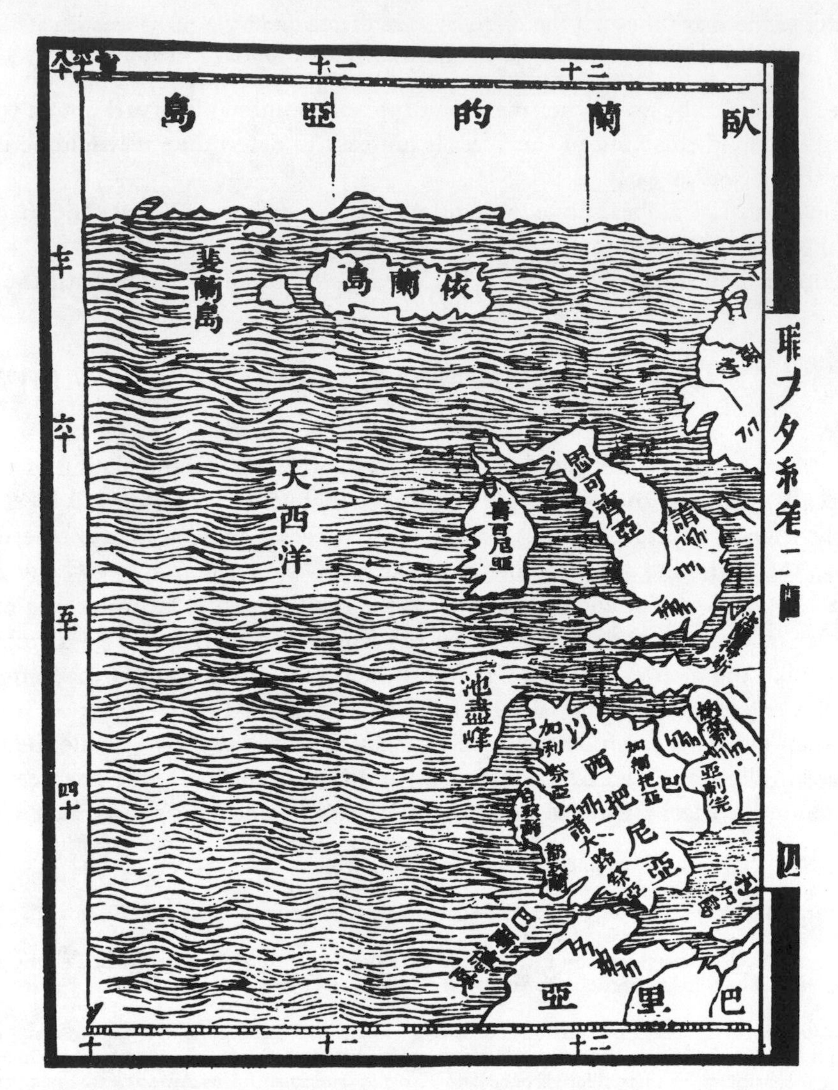

Fig. 253. One of the maps illustrating the *Chih Fang Wai Chi* (On World Geography) of Giulio Aleni, +1623, one of Matteo Ricci's Jesuit cartographic successors in China. This portion shows Spain, France and the British Isles. East Anglia is labelled *An*, Scotland *Ssu-kho-chhi-ya*, and Ireland *Hsi-pai-ni-ya*, Spain being *I-hsi-pa-ni-ya*. Iceland appears as *I-lan Tao*, with further west the mythical island of *Fei-lan Tao* (Friesland). The adjacent sheet has a spouting whale and a large ship of European rig—exotic reading for a Chinese scholar.

cartography has been described by Chhen Kuan-Shêng (1) and Hummel (11). To Bernard-Maître (8) we owe an exhaustive discussion on all aspects of Chinese geography in the +17th and +18th centuries, which, however, belongs to the history of science in general rather than to that of the characteristic Chinese contributions. Subsequently the art and science of map-making was cultivated by several of the Jesuits—Sanbiasi's 'Ricci-type' map of +1648 has been described by Mills (2), and Verbiest's stereographic projection map of +1680 by Ahlenius. For these later works the reader is referred to Fuchs (3, 5), Bernard-Maître (8), and Herrmann (8).[a]

Great geographical and cartographic activity took place in the Khang-Hsi reign-period (+1662 to +1722) when the emperor was personally interested in extending the scientific knowledge of his vast dominions. Much use was made of Manchu and Mongol travellers and explorers. In +1677 the Manchu officials Umuna[b] and Sabsu[c] explored the Long White Mountain region in Manchuria, and between +1712 and +1715 Tulišen[d] made a long journey to visit the Torguts on the Lower Volga. In continuation of the Thang and Yuan expeditions to the sources of the Yellow River, and the wanderings of Hsü Hsia-Kho (cf. above, p. 524), Lasi[e] and Sulun[f] conducted a five-month expedition in Tibet in +1704, which led, after the further investigation of Amida[g] in +1782, to the official publication *Chhin-Ting Ho Yuan Chi Lüeh*.[1]

Meanwhile an elaborate programme of work had led to the Khang-Hsi Jesuit Atlas, the *Huang Yü Chhüan Lan Thu*,[2] which has been reproduced and exhaustively studied by Fuchs (2, 3). The original idea seems to have been that of Jean François Gerbillon (Chang Chhêng[3]), who persuaded the emperor to undertake a complete survey of the empire. The project lasted from +1707 to +1717, and when complete was not only the best map which had ever been made in Asia, but better and more accurate than any European map of its time. The individual sheets,[h] dissected, and without the coordinates, were incorporated in the *Thu Shu Chi Chhêng* encyclopaedia (+1726). The Jesuit who bore the chief responsibility was Jean-Baptiste Régis (Lei Hsiao-Ssu[4]); among his assistants were Joachim Bouvet (Pai Chin[5]) and Pierre Jartoux (Tu Tê-Mei[6]), together with Chinese scholars such as Ho Kuo-Tung.[7] The atlas was engraved on copper plates in +1718 by Matteo Ripa (Ma Kuo-Hsien[8])[i] and afterwards reproduced many times, in Europe as well as in China.

The history of this period would have to deal, not only with the bringing of Renaissance cartography to China, but also with the advances made by Western geographers in knowledge of Asia which the new access to Chinese sources permitted.[j] João de Barros first published at Lisbon (+1563) his *Terceira Decada da Asia*, which consisted of translations from Chinese geographical texts.[k] The first map of China to appear in a European atlas was that of Fr. Ludovico Georgio, a Portuguese Jesuit, in the

[a] Pp. 287 ff. The parallel history of cartography in Japan may be approached through the papers of Ramming (1); Dahlgren (1).

[b] Wu-Mo-Na.[9]

[c] Sa-Pu-Su.[10]

[d] Thu-Li-Shen.[11]

[e] La-Hsi.[12]

[f] Shu-Lan.[13]

[g] A-Mi-Ta.[14]

[h] Of China proper and Manchuria, not of Tibet and Mongolia.

[i] A secular priest, not a Jesuit. His memoirs are not without interest. A copy of his map, preserved in the Chinese College at Naples, has been described by Petech (3). On the whole project see also Bernard-Maître (14).

[j] Tenri University has made available a collection of Western maps of Japan between 1552 and 1840 Anon., 6). See also the review of Szczesniak (8).

[k] Cf. Boxer (1), p. lxxxvi.

1 欽定河源記畧 2 皇輿全覽圖 3 張誠 4 雷孝思 5 白晉
6 杜德美 7 何國棟 8 馬國賢 9 武默訥 10 薩布素
11 圖理琛 12 拉錫 13 舒蘭 14 阿彌達

third addendum of Ortelius (+1584).[a] The first to attain correct views on the Chinese coastline was J. H. van Linschoten (+1596), and it is well known that Fr. Martin Martini's *Atlas Sinensis* of +1655 was largely based on the *Kuang Yü Thu*, which European geographers such as d'Anville greatly admired.[b] Vacca (5) goes so far as to say that up to the end of the 17th century European geographers simply copied from Chinese atlases, and not always very accurately, never improving on them. Purchas' *Pilgrims* (+1625) contains[c] an interesting 'Map of China, taken out of a China Map printed with China characters, etc. gotten at Bantam by Capt. John Saris'.[d] This is entitled *Huang Ming I Thung Fang Yü Pei Lan*[1] and seems to be derivative from the *Kuang Yü Thu*.[e]

Meanwhile the stream of Chinese geographical scholarship continued. In the +17th century fine contributions were made such as the *Thien Hsia Chün Kuo Li Ping Shu*[2] (Merits and Drawbacks of all the Countries in the World)[f] by Ku Yen-Wu.[3] The following century produced a very great work, the Chhien-Lung Atlas of China. This was based on Jesuit surveys carried out between 1756 and 1759, and the maps were executed on a scale of 1:1,500,000. By 1769 the wood-cut edition was ready, and by 1775 it was engraved on copper plates. Once again China was ahead of all other countries in the world in map-making, and the enlightened imperial patronage of the Jesuits and their Chinese co-workers was well rewarded.[g] Much of the surveying was done by Felix da Rocha (Fu Tso-Lin[4]) and Joseph d'Espinha (Kao Shen-Ssu[5]), while Michel Benoist (Chiang Yu-Jen[6]) worked on the cartography at Peking.[h] A good 'Geography of the Empire' (*Chhien-Lung Fu Thing Chou Hsien Chih*[7]) appeared in +1787, and a 'Historical Geography of the Sixteen Kingdoms' (i.e. +5th century) (*Shih Liu Kuo Chiang Yü Chih*[8])[i] in +1798, both by Hung Liang-Chi.[9] But all this lies outside the scope of the present book.

It is interesting that there was much reluctance on the part of the Chinese geographers to give up the rectangular grid system. During the first half of the 19th century many atlases showed both grid-lines and latitude-longitude lines on the same map (for example, the *Huang Chhao I Thung Ti Yü Chhüan Thu*[10] produced by Li Chao-Lo[11] in 1832). In such maps the crossing of a given latitude and longitude was taken as the central point, and the system of rectangular spaces laid down in relation to it, but some other latitudes were also shown.[j]

[a] It is not nearly so good as the grid-map of +1137. On Georgio see Szczesniak (7).
[b] Huttmann (1); Bernard-Maître (15); Tooley (1).
[c] Pt. III, Bk. ii, ch. 7, Sect. 7, p. 401.
[d] Whose journeys to the East are described in Purchas, Pt. I, Bk. iv, ch. 2, p. 384.
[e] The general disposition of river and lake systems is the same, thus Erh Hai in Yunnan is crescent-shaped in both, and the relations of Yellow River and Great Wall in Shensi are identical. The map in Purchas reproduces the symbols for city sizes, but their meaning seems not to have been understood. It also inverts the names of Szechuan and Kweichow provinces.
[f] Cf. Forke (9), p. 479.
[g] See Bernard-Maître (8) and Mills (7).
[h] Cf. Pfister (1), pp. 774, 776, 818, 821, 865.
[i] Translation by des Michels (1). Cf. Forke (9), p. 563.
[j] I am grateful to Mr J. V. Mills and Dr W. Fuchs for drawing attention to this.

1 皇明一統方輿備覽 2 天下郡國利病書 3 顧炎武 4 傅作霖
5 高愼思 6 蔣有仁 7 乾隆府廳州縣志 8 十六國疆域志
9 洪亮吉 10 皇朝一統地輿全圖 11 李兆洛

(*h*) COMPARATIVE RETROSPECT

It is now possible to look back over the mass of material which has been analysed, and to gain a provisionally comprehensive view such as that summarised in Table 40. All that has now to be done is to recapitulate the main points in it, and to add a few further considerations in the nature of cement for which there was no opportunity before. As suggested at the opening of this Section, we may distinguish between scientific cartography and religious cosmography. The former tradition begins early in the West, with the great Greek and Hellenistic geographers, but then dies out completely, leaving the field to the orbocentric wheel-maps of the latter tradition. But just as this occurs in Europe, we find the Chinese initiating a long line of scientific map-makers which is never interrupted until it blends with the cartography of the Renaissance.

The first question which arises is whether there could have been any transmission of the idea of meridians and parallels eastwards from Marinus and Ptolemy to Chang Hêng and Phei Hsiu. At present there is no way of answering this. Facts which have been given already in the Section on culture contacts,[a] however, are sufficient to indicate that such a transmission would have been possible through the early travels of traders and 'ambassadors' from Roman Syria. The parallel case of the passage of Ptolemy's geography to India has been much discussed.[b] Cosmas Indicopleustes, for instance, relates that the 'Brahmans' held views very like those of Ptolemy.[c] But even if we could prove that an idea was transmitted, it would not help us much with regard to the detailed local sources of Chang Hêng and Phei Hsiu, and we have no need to assume any lack of originality on their part.

In the West, Hellenistic quantitative cartography was conserved in dormant state only in Byzantine civilisation, but after it became available to the Arabs, in the +9th century, led to much greater geographical achievements in their culture, such as the work of al-Idrīsī in the +12th. Meanwhile, the Latin world was dominated by the religious wheel-maps in the style of Macrobius, Orosius and Beatus, and awakening came, not from scholarly geographers, but from practising sailors and navigators at the beginning of the portolan period about +1300. That this was at least partly dependent upon the transmission of the magnetic compass from China cannot be questioned.[d] But it was not long before the loxodromes of the portolans began to be supplemented by a rectangular grid, and the second question which arises, therefore, is to what extent this too could have been a transmission in the westward direction. Arabic map-makers, and even a Latin, Marino Sanuto, were using rectangular grids almost from the beginning of the portolan period. Now the Arabs were well established at Canton from the middle of the +8th century onwards, and the two following centuries were marked by many travels in China, those of Sulaimān the Merchant and of Ibn Wahb al-Baṣrī (Ferrand, 2); those of Abū Dulaf ibn al-Muhalhil (Ferrand, 1)

[a] Section 7*h* in Vol. 1, pp. 191 ff.
[b] Cf. Kennedy (1), pp. 498 ff.
[c] McCrindle (7), vol. 2, p. 48.
[d] Cf. Sect. 26*i* below.

Table 40. *Chart to show the comparative development of cartography in East and West*

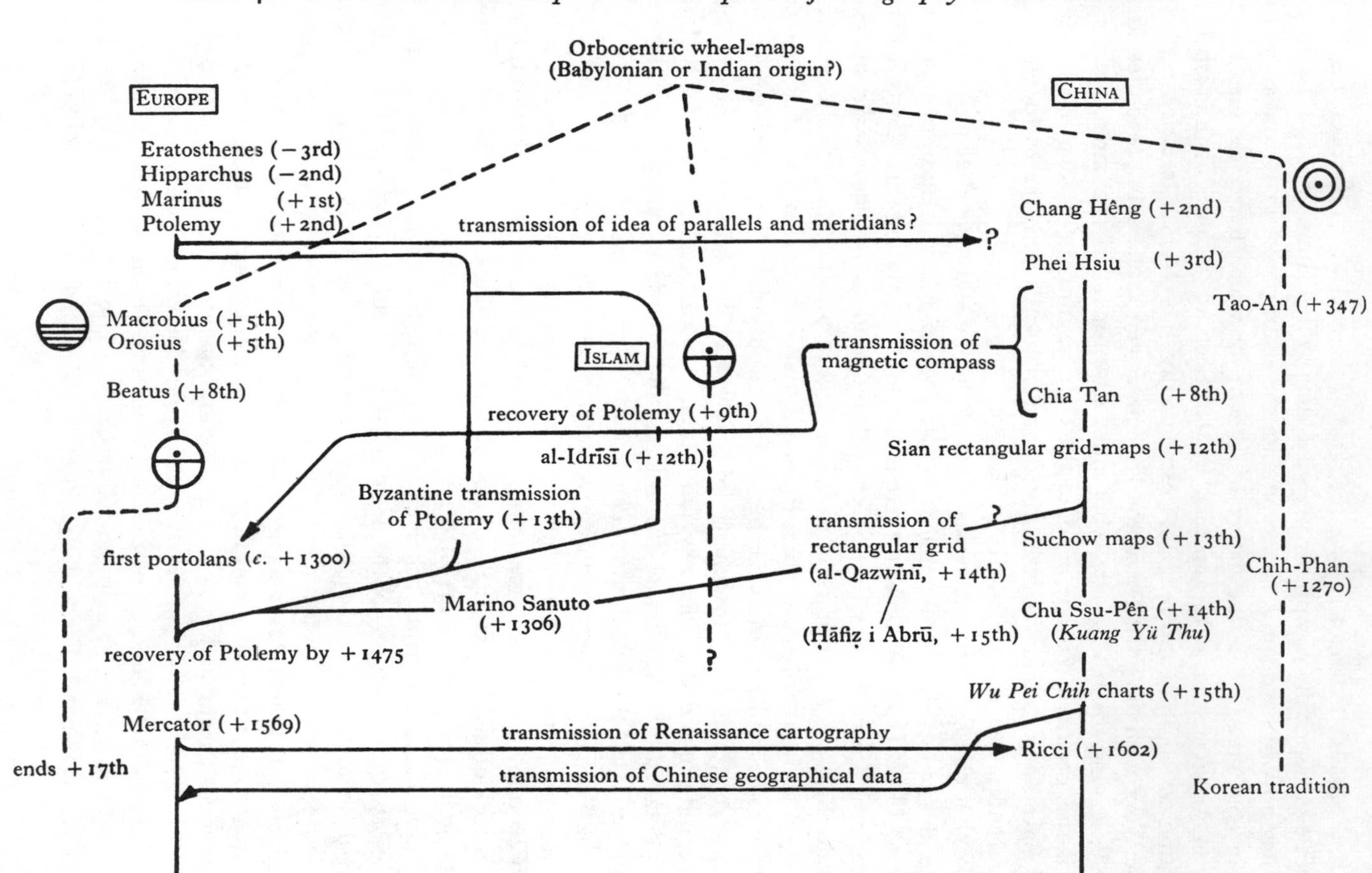

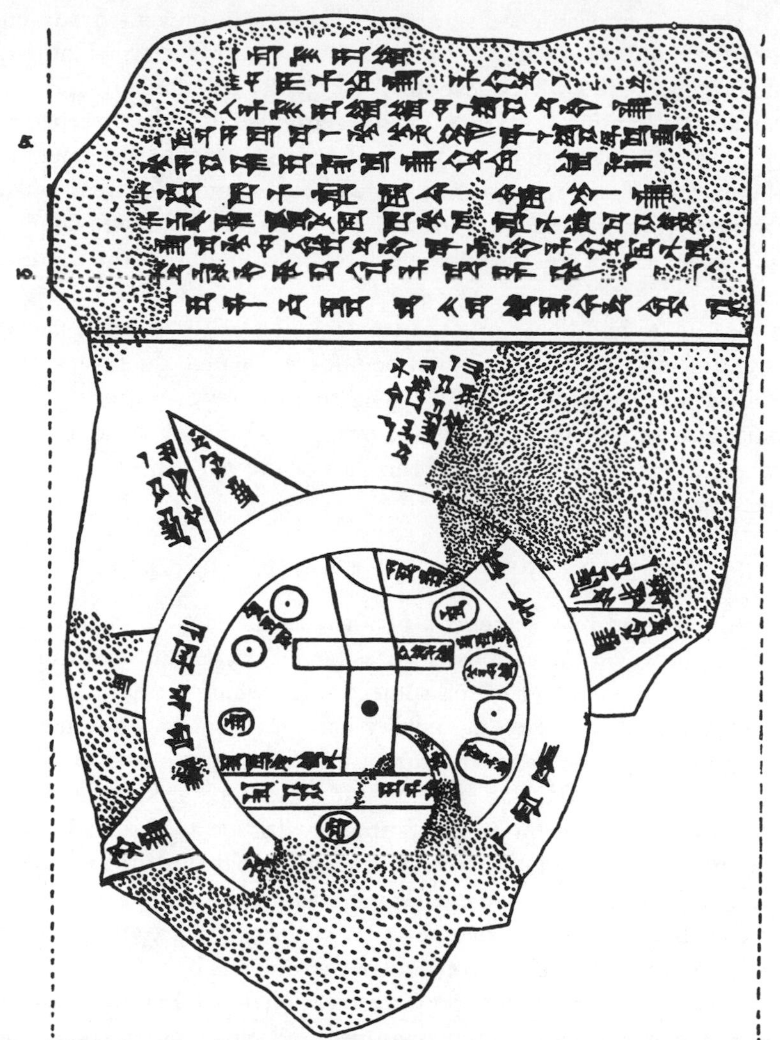

Fig. 254. Babylonian discoidal map on a clay tablet of the -7th century (*Cuneiform Texts in the British Museum*, vol. 22, pl. XLVIII).

and the Christian monk al-Najrānī (Ibn al-Nadīm). It would almost be surprising if men such as these had brought back with them no knowledge of Chinese quantitative cartography. Moreover, al-Mustaufī al-Qazwīnī was working in the period of Mongol dominance (early 14th) and his type of map (a grid with place-names only) is identical with contemporary Chinese maps. The suggestion is therefore permissible that the advances of cartography in $+14$th- and $+15$th-century Europe were due, not only to the study of Ptolemy by the Arabs, but also to some transmission of quantitative cartographic principles from China.

We may now return for a moment to the religious cosmographic tradition. This was quite clearly present among both the Arabs and the Chinese, though with the latter, at any rate, it never dominated in geography to the extent which it did in medieval

Europe. Perhaps Confucian good sense and Taoist skill prevented any enthusiastic reception of the Buddhist world-conception centring on Mt Meru (Khun-Lun). But its ultimate origins invite a third question. Was it really Babylonian? In India it seems to be of considerable antiquity. Winter (1) has reminded us of its presence in a part of the Jain Canon (the *Sūrya-prajñapti*)[a] which may go back to the later part of the −1st millennium. Hence it may not be without significance that Akkadian and Babylonian maps (Fig. 254) ranging from about −2500 to −500 are known, and that they are all discoidal.[b] In any case, there can be little doubt that the Chinese received it from India.

It only remains to add that while the transmission of Renaissance cartography to China in the time of Matteo Ricci cannot be underestimated, the reverse transmission of geographical information about East Asia to the +17th-century geographers of Europe must also be remembered. It was owing to the solid work of generations of Chinese map-makers that knowledge of this part of the world became incorporated in modern geography.

(*i*) THE RETURN OF THE RECTANGULAR GRID TO EUROPE

These concluding lines may serve as a pendant to the long story of cartography in East and West, in which the various cultures have been seen to be so interwoven, as much, perhaps, as in any other of the subjects with which we have to deal.

At the beginning of the present century cartographers had become so adept in methods of representing on paper the curvature of the earth's surface, and the astronomical determination of latitude and longitude had become so refined, that no one would have taken anything but a very superior attitude to the traditional Chinese rectangular grid. It would not have been imagined that this grid would again have been required. Yet in the First World War it was found, about 1915, that in the co-operation of artillery with survey engineers for accurate gunfire an orthomorphic projection was necessary. And thus rectangular coordinates, or 'carroyage', returned to the field of activity, a conical orthomorphic projection due to Lambert being eventually adopted.[c] For convenience of immediate recognition, without corrections, of the distance between two points, the rectangular grid once more became normal in military maps. More recently still, it has become essential in trans-polar flying, where any course taken (except in the north-south direction) rapidly crosses the converging meridians at a constantly changing angle, and navigation is made even more difficult by the proximity of the magnetic pole.[d] A particular meridian is therefore accepted as standard and a parallel grid network based on this is employed to keep the angle of course constant. But we do not suggest that the similarity with the age-long Chinese tradition was known to those who made these innovations in modern Europe.

With this we bring to an end our study of Chinese cartography and geography.

[a] Masson-Oursel, de Willman-Grabowska & Stern (1), p. 174; Renou & Filliozat (1), vol. 2, p. 613; Thibaut (2), p. 21.

[b] Unger (1, 2); Meek (1); *Cuneiform Texts from Babylonian Tablets in the British Museum*, vol. 22, pl. XLVIII.

[c] Cf. Steers (1); Hinks (2); Bramley-Moore (1).

[d] Cf. p. 365 above.

23. GEOLOGY

[AND RELATED SCIENCES]

(a) INTRODUCTION; GEOLOGY AND MINERALOGY

The study of the Chinese contributions to geology and mineralogy labours under the great difficulty that there exists as yet no monograph on the subject either in Chinese or a Western language—in other words the basic work has not yet been done. In the course of a general survey such as this it is naturally impossible to do it, and we must therefore be content with what can only be a rough first approximation. Moreover, geology and mineralogy are largely modern, post-Renaissance, sciences. The former really begins with Nicholas Stensen's[a] *Prodromus* of +1668, which introduced conceptions such as folded strata, faulting, volcanic intrusions, eroded forms, etc. Not until the latter half of the 18th century did geology as we know it take shape, with the Neptunian theories of A. G. Werner[b] and his followers opposing the Plutonian ideas of J. Hutton[c] and his school. Ancient and medieval times had contributed relatively little to geology, and the general backwardness in investigating the structure of the earth's crust[d] may broadly be said to have been shared by the Chinese, though they were responsible, as we shall see, for several empirical discoveries and inventions of great interest. Mineralogy goes back further, in the sense that catalogues of different kinds of stones, ores, gems and minerals were being made already in antiquity—here the ancient and medieval 'Lapidaries' (corresponding to the Herbals and Bestiaries) of Europe can be matched with (and sometimes are found to have been surpassed by) those of India and China. But again modern mineralogy as we know it developed from the concern of the 18th century about the classification of mineral substances. Some, such as Berzelius, sought a system based purely on chemical composition; others, such as Linnaeus, preferred a natural history classification; Werner adopted a mixed system, and it was not until the crystallographic work of R. J. Haüy[e] that modern systematic mineralogy could arise.

In order to assess Chinese contributions, one needs good general histories of geology and mineralogy in Europe. Even these are not too numerous. The most complete and useful which I have found is that of F. D. Adams, but certain smaller papers are

[a] +1638 to +1686; Pledge (1), pp. 91, 124; Adams (1), p. 358. Cf. the source-book of Mather & Mason (1).

[b] +1749 to +1815; Adams (1), p. 209.

[c] +1726 to +1797; Pledge (1), p. 97; Adams (1), p. 238.

[d] A factor here which is not usually taken into account is the necessity of really stable political conditions before geologists can make field trips on a scale sufficient to allow of the verification of hypotheses. Members of the Chinese Geological Survey in our own time, for example, have lost their lives owing to the incomprehension of backward tribesfolk. But such conditions are rapidly becoming a thing of the past.

[e] +1743 to +1821; Adams (1), p. 205. The best study of the development of mineralogy since Haüy is that of Groth (1).

valuable, such as that of Bromehead (1). The well-known book of Geikie deals essentially only with the post-Renaissance development of geology, and the same applies to that of Zittel, now rather old. The most important Chinese work is that of Chang Hung-Chao, *Lapidarium Sinicum*, which is a thorough discussion of some of the most important mineral substances mentioned in Chinese texts, including many disputed questions of nomenclature.[a] The mineralogical parts of the books of the *Pên Tshao*[1] type have given rise to several works of importance in Western languages, notably those of de Mély (1), Read & Pak (1), and Geerts (1). These, together with smaller contributions covering more specialised parts of the field, will be discussed below in their proper place. For the mineralogical knowledge of classical antiquity in Europe, there is the old and forgotten, but still useful, compendium of selected texts and comments by Lenz (1), besides recent work by Hiller (1), Wellmann (1) and others. Monographs by Berger and Sagui parallel these for geology. On medieval mineralogy Mieleitner may be consulted. In working on this subject one has to bear in mind the literature on mining in China;[b] it is enormous in size and scope, and we have not been able to check more than a small part of it for points of mineralogical interest.[c]

(b) GENERAL GEOLOGY

(1) Pictorial Representations

In the first place the capacity for accurate observation and representation of geological forms as shown in Chinese pictures and book illustrations should not be underestimated. If one takes up a modern book such as the *Hsia Chiang Thu Khao*[2] (Illustrations of the Yangtze Gorges), typical of many produced during the last century, but with drawings and text in purely Chinese traditional style showing no trace of Western influence, one finds that many geological formations are very clearly represented. In the copy which I saw in Chungking[d] in 1946 I noted scarps, flat-tops and peneplains,[e] granite boulder tors,[f] a picture showing very fine dipping strata,[g] a mesa hill cap of hard rock or a monadnock,[h] and a red sandstone hogback anticline.[i] It is probable that a whole text-book of geology could be illustrated largely from Chinese pictures

[a] Among subsidiary articles in Chinese, I may refer to Yin Tsan-Hsün (*1*). There is a forgotten glossary of Chinese geological and mineralogical terms by Muirhead. Though the book of Ko Li-Phu (A. W. Grabau) has not been accessible to us, I believe that it deals mainly with quite modern times.

[b] Most of it of course concerns mineral deposits in China as described by modern geologists, both Chinese and Western, or the modern exploitation of these deposits; but accounts of traditional mining practice are embedded in it. Among the books on the mineral wealth of China, those by W. E. Wang and by Torgashev may be mentioned.

[c] Comparisons with the mineralogy of the Arabs will of course be desired; here I can only indicate a few pointers to that literature—Wiedemann (1), Ruska (1), Mullet (1). For comparative mining traditions see Sébillot (1).

[d] It included the text of the *Hsia Chiang Chiu Sêng Chhuan Chih*[3] (Record of the Yangtze Gorges Lifeboat Service); cf. Worcester (1), p. 28. Dr Michael Bolton kindly brought it to my attention.

[e] Pp. 1, 13, 24.

[f] P. 30.

[g] P. 32.

[h] P. 54.

[i] P. 63.

[1] 本草 [2] 峽江圖考 [3] 峽江救生船志

and book illustrations, mountain scenery having always been one of the favourite subjects of artists. For a study of this observational side of geology in China abundant material would be found in the local histories and geographies, the *hsien chih*,[1] a class of literature which has already been described.[a] Numerous illustrations of this kind are scattered through the late encyclopaedias, such as the *Thu Shu Chi Chhêng* of +1726, in the Shan chhuan[2] division. So also the late geographies, such as the *Ta Chhing I Thung Chih* of +1730,[b] would have to be drawn upon; it was from this that Biot (15) selected a number of fragmentary geological notes, as, for example, on the basaltic columns (*shih sun sên pu*[3]) at Shih-Mên Shan[4] in Yunnan.[c]

If we look through the section of the *Thu Shu Chi Chhêng* just mentioned, we have no difficulty in finding many scenes of geological interest. Fig. 255 shows an example of the rejuvenation of a valley, the previously stabilised floor having been cut sharply through by new river erosion, forming sharp river terraces. Another effect of water is seen in the superficial deposit of water-rounded boulders (Fig. 256), which the artist, however, has strewn rather too freely in unlikely positions. In Fig. 257 we have a platform of marine denudation with a wave-cut arch; following this level to the right the eye detects further relics of the same platform. The manner of representing the crags on the right at the top is a conventional one, seen in many such pictures, and perhaps related to calligraphic traditions and techniques; its geological meaning is not obvious.[d] Fig. 258 is another example of extreme rejuvenation. The 'crag and platform' motif was a favourite one. Dipping strata are faithfully reproduced in Fig. 259, and the Permian basalt cliffs of O-Mei Shan are seen in Fig. 260. Fig. 261 is interesting as it shows in the distance a very typical U-shaped glacial valley, as well as dipping strata in the foreground. Finally, the karst limestone pinnacles, rising with extreme abruptness out of the flat plain, so typical of the landscape of Kuangsi, are shown in Fig. 262, which represents the neighbourhood of Kweilin.[e]

Perhaps the most striking of all such pictures is that which the Sung painter Li Kung-Lin[5] (*fl. c.* +1100) made of an exposed anticlinal arch at Lung-Mien Shan[6] (Fig. 263).[f]

The fact that so many identifications are possible is in itself a remarkable testimony of the faithfulness to Nature with which Chinese painters manipulated their brush-strokes. To illustrate geological structures was indeed far from their thoughts, and the accurate description of such forms may well have little to do with the aesthetic

[a] P. 517 above.

[b] Cf. p. 521 above. It would be of interest to follow back these iconographic traditions as far as possible; they could certainly be traced to the Sung.

[c] The remarkable descriptive work of Hsü Hsia-Kho at the end of the Ming has already been noticed above, p. 524.

[d] One can see it applied elsewhere in the illustration of the crystals of calcite given in the *Pên Tshao Kang Mu*. On all such conventions Petrucci (3) is to be consulted.

[e] I am much indebted to Mr Brian Harland for his advice in these geological identifications.

[f] These mountains are north-west of Thung-chhêng in Anhui, just north of the Yangtze between Hankow and Nanking.

1 縣志 2 山川 3 石笋森布 4 石門山 5 李公麟
6 龍眠山

Fig. 255. Geology in Chinese art: (*a*) the rejuvenation of a valley at Li-Shan, near Fei-hsien, in Shantung (*TSCC*, *Shan chhuan tien*, ch. 23).

Fig. 256. Geology in Chinese art: (*b*) deposit of water-rounded boulders at I-Shan in southern Shantung (*TSCC*, *Shan chhuan tien*, ch. 26).

Fig. 257. Geology in Chinese art: (*c*) a platform of marine denudation with a wave-cut arch at Lao-Shan near Tsingtao on the Shantung coast (*TSCC, Shan chhuan tien*, ch. 29).

Fig. 258. Geology in Chinese art: (*d*) extreme rejuvenation of a valley at Kuang-Wu Shan in northern Honan (*TSCC, Shan chhuan tien*, ch. 51).

appreciation of Chinese painting itself, yet surely the old empirical closeness to Nature of the Taoists was still there, and the world depicted was the real world. After we had made the study of landscapes on which the preceding paragraph was based, we found that others had already approached the matter in a different way. For many centuries Chinese literature has contained handbooks or guides for painters which include a great variety of type-forms or standard components of pictures, and among these,

Fig. 259. Geology in Chinese art: (*e*) dipping strata at Hsiang-Shan south of Khaifêng near the tomb of Pai Chü-I (*TSCC*, *Shan chhuan tien*, ch. 64).

Fig. 260. Geology in Chinese art: (*f*) the Permian basalt cliffs of O-Mei Shan in western Szechuan (*TSCC*, *Shan chhuan tien*, ch. 173). This region is the site of famous Buddhist abbeys, and from the top of the mountain, one of the westernmost peaks of the Tibetan massif, the Fo Kuang (Buddha Light), a kind of Brocken Spectre, can often be seen (cf. p. 477 above).

geological structures (hills, mountains, waters) naturally have a conspicuous place.[a] Among the handbooks perhaps the most famous is the *Chieh-Tzu Yuan Hua Chuan*[1] (Mustard-Seed Garden Guide to Painting), issued by Li Li-Ong[2] & Wang Kai[3] in +1679, a book which in some editions included examples of very early polychrome block printing, afterwards to be raised to so famous an art by Japanese masters. Tentative geological identifications of the type-forms shown in chapter 3 of this work[b] have been made by Taki (1) and March (3).[c]

[a] Late Japanese popular publishing abounded in manuals of this kind, which are thus among the commonest of all Japanese books in occidental collections and private libraries.

[b] In the *Chhu chi* (first series).

[c] With the collaboration of D. McLachlan. We thank Dr Chêng Tê-Khun for these references.

[1] 芥子園畫傳　[2] 李笠翁　[3] 王槪

Apart from certain general terms, such as *lun khuo*[1] (outline mountain ranges like a pile of cut-off wheel-rims) and *chang kai*[2] (the last enveloping outline of a peak far away), the forms are termed *tshun*[3] (mountain wrinkles), i.e. standard representations of mountain curvatures. Of these the teachers distinguished rather more than twenty variations of what was possible for a brush. Glaciated or maturely eroded slopes, sometimes steep, are shown by the technique called 'spread-out hemp fibres' (*phi ma tshun*[4]), and mountain slopes furrowed by water into gullies are drawn in the *ho yeh*

Fig. 261. Geology in Chinese art: (*g*) a typical U-shaped glacial valley in the distance with dipping strata to the right; Chhi-Chhü Shan near Pao-ning in northern Szechuan (*TSCC, Shan chhuan tien*, ch. 178). This scenery is typical of the Szechuan-Shensi border and the road through the Chhin-Ling mountains.

Fig. 262. Geology in Chinese art: (*h*) limestone karst masses and pinnacles near Kweilin in Kuangsi (*TSCC, Shan chhuan tien*, ch. 193). Cf. Vol. 1, Fig. 4.

tshun[5] manner ('veins of a lotus-leaf hung up to dry'). 'Unravelled rope' (*chieh so tshun*[6]) indicates igneous intrusions and granite peaks; 'rolling clouds' (*chüan yün tshun*[7]) suggest fantastically contorted eroded schists. The smooth roundness of exfoliated igneous rocks is seen in the 'bullock hair' method (*niu mao tshun*[8]), irregularly jointed and slightly weathered granite appears in 'broken nets' (*pho wang tshun*[9]), and extreme erosion gives *kuei mien*[10] or *khu-lou*[11] *tshun* ('devil-face' or 'skull' forms). The horizontal stratification of sedimentary rocks is easily to be seen in the *chê tai*[12] ('folded belts') *tshun*, and faulted angular rocks are unmistakable in the *hsiao fu phi*[13] and *ta fu phi*[14] *tshun* ('hacked or cleaved with a small, or large, axe'). Lastly,

1 輪廓 2 嶂蓋 3 皴 4 披麻皴 5 荷葉皴 6 解索皴
7 卷雲皴 8 牛毛皴 9 破網皴 10 鬼面 11 骷髏
12 折帶 13 小斧劈 14 大斧劈

cleavages across strata, with vertical jointed upright angular rocks, looking somewhat like crystals, are depicted in the 'horse-teeth' (*ma ya tshun*[1]) technique. These are only a few of the technical terms, but they show the degree of systematisation that was reached.

(2) The Origin of Mountains; Uplifting, Erosion, and Sedimentary Deposition

As regards statements concerning the origin of mountains, Chinese literature proves unexpectedly rich. Here the most famous text is that of the Neo-Confucian Chu Hsi. In the *Chu Tzu Chhüan Shu* we find the following passage (the philosopher has just been talking about the periodical cataclysmic destruction of the world and its re-creation at stated very long intervals of time):[a]

> In the beginning of the heavens and the earth, before the chaotic matter was divided, there was, I think, nothing but fire and water. Earth was formed as a deposit from the water. Even today, when we stand on a high elevation and look far and wide, the company of the hills looks like the waves of the sea. The waters must have moulded them into these shapes, but as to when their solidification (*ning*[2]) took place, we have no idea. That which must have been at first extremely soft, became hard and solidified. Some one remarked that these processes resembled the moulding of beaches by the tides, and the philosopher agreed. The heaviest part of the original waters, he said, formed the earth, and the finest parts of the original fire formed the wind, the thunder, the lightning, the sun, the stars, and so on....[b]
>
> The small is the sample (lit. the shadow) of the great. Only one day and one night is enough to see it.[c] The Five Peaks[d] are all but one Chhi, (blown by) a vast breathing. The waves roar and rock the world boundlessly, the frontiers of sea and land are always changing and moving, mountains suddenly arise (*pho*[3]) and rivers are sunk and drowned (*yin*[4]). Human things become utterly extinguished and ancient traces entirely disappear; this is called the 'Great Waste-Land of the Generations'. I have seen on high mountains conchs and oyster shells (*lo pang*[5]), often embedded in the rocks. These rocks in ancient times were earth or mud, and the conchs and oysters lived in water. Subsequently everything that was at the bottom came to be at the top, and what was originally soft became solid and hard. One should meditate deeply on such matters, for these facts can be verified.[e]

The great importance of this passage for the history of palaeontology will be referred to shortly. But, as Grabau has pointed out, the main geological interest of the conversation (which would have taken place about +1170) is that Chu Hsi recognised the fact that the mountains had been elevated since the day when the shells of the living animals had been buried in the soft mud of the sea-bottom.[f] Three centuries later, in the time

[a] Cf. above, in Sect. 16*d*, Vol. 2, pp. 485ff.

[b] On the 'centrifugal cosmogony' see above, in Sect. 14*d*, Vol. 2, pp. 371ff.

[c] For example, the changed shape of sand dunes or beaches after a severe storm.

[d] Probably a reference to the five sacred mountains.

[e] Ch. 49, pp. 19*b*, 20*b*, tr. auct., adjuv. le Gall (1), p. 121, Forke (6), p. 112, (9), pp. 182 ff., Bodde (6).

[f] The parallel between the wave forms of mountains or folded strata and the waves of the sea was an associated, but really quite distinct, idea. It could easily have originated from the observation of ripple marks on sand beaches, the great pressures in the earth's crust not yet being visualised. The

1 馬牙皴 2 凝 3 勃 4 湮 5 螺蚌

Fig. 263. Geology in Chinese art (*i*): an exposed anticlinal arch at Lung-Mien Shan near Thung-chhêng in Anhui, just north of the Yangtze between Hankow and Nanking; a painting by Li Kung-Lin (*fl. c.* +1100).

of Leonardo da Vinci, who takes an important place, as we shall see, in the history of palaeontology, it was still supposed that the fact that shells were found in the Apennine mountains indicated that the sea had once stood at that level. Leonardo himself, however, attributed the origin of mountains to elevation and distortion after the formation of their fossil-containing rocks.[a]

It might be thought that this was a flash of insight on the part of the great Sung philosopher, and stood alone. In fact, however, it was only part of a train of thought which had begun centuries before and was to continue long afterwards. The pursuit of this leads us to a peculiar technical term of the Taoists—*sang thien*,[1] the 'mulberry-grove'. Originally it was a place-name and perhaps a constellation-name, but by the Thang it had come to be used for land which had once been covered by the sea, or would be in the future. Li Pai so refers to it in poems.[b] At this time Yen Chen-Chhing[2] wrote an essay, *Ma-Ku Shan Hsien Than Chi*[3] (Notes on the Altars to the

view that the mountains were actually formed by the effects of the waters under the sea, afterwards rising in violent cataclysms high above sea-level, seems to have been particularly associated with the Neo-Confucian philosopher Tshai Yuan-Ting,[4] the exact contemporary of Chu Hsi. For Mr J. R. McEwan informs us that Japanese 18th-century books frequently give as the authority for this view a work entitled *Tsao-Hua Lun*[5] (Discourse on the Creation) attributed to Tshai. Unfortunately this is not mentioned among his other books as listed in *Sung Shih*, ch. 434, p. 6*b*, or *Sung Yuan Hsüeh An*, ch. 62, nor is it referred to by Forke (9), p. 203. On the other hand Tshai did write a book on geomancy (the *Fa Wei Lun*,[6] Effects of Minute Causes) and from his sayings recorded in the above sources one can see that the expression *tsao-hua*—the Author or Foundation of Change—was always on his lips. Indeed, this expression parallels the favourite phrase *Ko Chih*[7] (cf. Vol. 1, pp. 48–9) and signals titles of interest to us. Perhaps the *Tsao-Hua Lun* was an opuscule preserved only in Japan. Now it may be significant that the *Tao Tsang* contains (*TT*318) a *Tsao-Hua Ching* (Creation Canon; full title *Tung-Hsüan Ling-Pao Chu Thien Shih-Chieh Tsao-Hua Ching*[8]) of unknown date and authorship, which 'explains the structure of the universe, heavens, earth, hells, sun, moon, continents, upheavals by water, fire and wind, renewed creations' and so on. Buddhist influence is obvious, but there is a geological relevance too. We think that this was probably compiled in the Thang period, for the *Tsao-Hua Chhüan Yü*[9] (The Beginnings of the Creation) by Chao Tzu-Mien[10] (now lost), which appears in the *Thang Shu* bibliography, was very probably connected with it. A Ming commentary, the *Tsao-Hua Ching Lun Thu*[11] by Chao Chhien,[12] still exists. In another direction, geological thought links up with alchemy. Below, p. 638, we shall study some remaining fragments of a *Tsao-Hua Chih-Nan*[13] by Thu Hsiu Chen Chün[14] (Guide to the Creation, by the Earth's Mansions Immortal), especially in connection with the theory of the growth of metals in the womb of the uprisen mountains. To accelerate such growth artificially was one of the alchemists' principal aims. Hence, presumably, the *Tsao-Hua Chhien Chhui*[15] (The Hammer and Tongs of Creation) by the Ming prince Ning Hsien Wang[16] (cf. Vol. 1, p. 147, and Sect. 33 below). And even the title of a lost Ming book makes sense in this connection—the *Tsao-Hua Fou Hung Thu*[17] (Diagrams of the Natural Incubation of Mercury), i.e. the formation of its ores in the earth; by an alchemist who called himself Shêng Hsüan Tzu[18] (the Ascending-Mercury Master). In using the word 'Creation' for the translation of all these titles, we must remind the reader that the conception of creation in the full sense *ex nihilo* was entirely un-Chinese (see Vol. 2, p. 581). The Author of Change was no person, but a numinous allegory of the Tao.

[a] The importance of Chu Hsi's statement has been fully realised by modern Chinese scientists, as, for example, by the doyen of living Chinese geologists, Li Ssu-Kuang (*1*). Also by western writers, e.g. Silcock (1).

[b] Waley (13), p. 44.

1 桑田 2 顏眞卿 3 麻姑山仙壇記 4 蔡元定 5 造化論
6 發微論 7 格致 8 洞玄靈寶諸天世界造化經 9 造化權輿
10 趙自勔 11 造化經論圖 12 趙謙 13 造化指南
14 土宿眞君 15 造化鉗鎚 16 寧獻王 17 造化伏汞圖
18 昇玄子

Immortals on Ma-Ku Mountain),[a] in which he quoted some remarks attributed to the legendary Taoist woman immortal Ma Ku:

Ma Ku said, 'Since I was last invited here I have seen that the Eastern Sea has turned into groves and fields (*sang thien*[1]). This change has occurred three times. The last time I arrived at Mt Phêng-Lai[b] (for an assembly of Hsien immortals) I noticed that the sea was only half as deep as it had been at the previous meeting. It looks as if the sea will again be turned to mountains and dry land.'

Fang Phing[2] laughed and said, 'The sages all maintained that where the sea is now the dust will one day be flying.'

And Yen Chen-Chhing goes on to say:

Even in stones and rocks on lofty heights there are shells of oysters and clams to be seen. Some think that they were transformed from the groves and fields once under the water (*sang thien*[1]).[c]

This essay must have been written about +770.[d]

But Ma Ku's views on the *sang thien* were by no means new in the +8th century. Yen was quoting from the *Shen Hsien Chuan*[3] (Lives of the Divine Hsien) attributed to Ko Hung.[4] If this attribution were right, the book would date from about +320, but though it is probably not as old as this it is certainly pre-Thang. The same words of Ma Ku and Fang Phing appear in it in a section devoted to her in the 7th chapter, where other women immortals are also discussed. One of the oldest mentions must be that ascribed to Kuo Pho about +310 in the *Shih Shuo Hsin Yü*;[e] he defended the site of a grave which some had criticised as being too near the water's edge by saying that it would all turn into mulberry groves as the centuries went by. Moreover, other pre-Thang references may be adduced. It will be remembered that in the Section on mathematics we met with an interesting book entitled *Shu Shu Chi I* (Memoir on some Traditions of Mathematical Art) which purports to have been written by Hsü Yo[5] about +190, but which cannot in any case have been later than about +570 when Chen Luan[6] wrote the commentary on it. Here we find, in the text itself, the following curious passage:

Those who do not recognise the 'three' yet boastfully claim to know the 'ten', are like the River People who lost the direction of their destination, and blamed it on the unskilful hand of the direction-finder.[f] If one does not know the length or shortness of small moments

[a] *TSCC, Shan chhuan tien*, ch. 149, *i wên*, p. 2*a*. Also in Yen's collected works, ch. 13. He was a Taoist (*Thang Yü Lin*, ch. 6, pp. 2*b* ff.).

[b] The magic island in the Eastern Sea, cf. above, Sect. 13*c* in Vol. 2, p. 240.

[c] Tr. auct.

[d] Yet Sarton (1), vol. 3, p. 213, calls 'astounding' the early +10th-century statement of al-Mas'ūdī that sea could turn into dry land, and land again be covered by the sea.

[e] Ch. 3A, p. 31*a*.

[f] The significance of this passage, together with others in the same book, for the history of the magnetic compass, will be referred to below, Sect. 26*i*.

1 桑田 2 方平 3 神仙傳 4 葛洪 5 徐岳 6 甄鸞

of time (*chha-na*[1]),[a] how can one appreciate what Ma Ku meant by the '*sang thien*' (i.e. the long periods of centuries during which the sea is turned into dry land)? If one does not understand how to distinguish between very small quantities (*chi-wei*[2]),[b] how can one understand the hundreds of millions (of units) which form the whole universe?[c]

It is therefore safe to say that at some time between the +2nd and the +6th centuries, the expression *sang thien* had a familiar connotation equivalent to what we should now call 'geological time'. Another reference to these ideas occurs in the *Chin Shu* (History of the Chin Dynasty), which speaks of Tu Yü[3] (+222 to +284) as follows:

Tu Yü often used to say that the high hills will become valleys and the deep valleys will become hills. So when he made monumental steles recording his successes he made them in duplicate. One was buried at the bottom of a mountain, and the other was placed on top. He considered that in subsequent centuries they would be likely to exchange their positions.[d]

Here again there was a long lapse of time between Tu Yü's own period and the time when the *Chin Shu* was written, just before +635, yet the *Chin Shu* is considered well informed and its writers must have had access to reliable sources. Besides, an earlier book, the *Shih Shuo Hsin Yü*,[e] describes how Huan Wên[4] said about +360 that the whole Chinese continent would sink beneath the sea. Whether or not we should really refer these ideas to the +3rd century or before, there can be no doubt that they are pre-Thang, and thus long before Chu Hsi. In the Thang, moreover, they were widespread; there is, for instance, a famous poem of Chhu Kuang-Hsi,[5] one of Yen Chen-Chhing's contemporaries, which says that 'the sea will be changed into groves of mulberry-trees'.[f] After the +8th century the term *sang thien* came to be applied poetically to the sea in general.[g] In +1210 the poet Chang Tzu,[6] for example, brought the idea into several of his writings.[h] In the +14th century it must have been very common, for a writer such as Chhen Thing[7] has recourse to it in his *Liang Shan Mo Than*[8] (Jottings from Two Mountains)[i] to explain why an iron chain which was

[a] I.e. Skr. *kṣaṇa*, 'atomic' instants of time.

[b] I.e. piled-up particles, 'atoms', Skr. *paramāṇu*. We shall return to these terms in Sect. 26*b*.

[c] Pp. 3*b*, 4*a*, tr. auct.

[d] Ch. 34, p. 11*b*, tr. auct.; cit. in *TPYL*, ch. 589, p. 2*b*. The story is often quoted, as in the +9th-century *Lin Chio Chi*[9] (The Unicorn Horn Collection of Examination Essays), p. 12*a*. Tu Yü, a minister of State, was also a geographer, astronomer, and patron of engineers.

[e] Ch. 3B, p. 19*a*; also *Yü Chien*, ch. 3, p. 7*a*.

[f] Cf. the Thang inscriptions recorded in the *Ku Kho Tshung Chhao*, pp. 32*a* to 61*b*. And a poem by a Buddhist abbot of the +9th century in one of the Tunhuang scrolls (Bib. Nat. Pelliot no. 2104); for this reference we thank Mr Wu Chhi-Yü.

[g] The same chapter of the *Thu Shu Chi Chhêng* which quotes Yen's essay, gives a number of others by men such as Tsêng Ying-Hsiang[10] of the Ming; and Huang Ju-Hêng,[11] Hsiung Jen-Lin[12] and Hsieh Chao-Shen[13] of the Chhing—all expressing the same idea of geological transformations by changes in the relative level of sea and land. Cf. *Thang Chhüeh Shih*, ch. 1, p. 24*a*; *Ao Yü Tzu* (*c.* +1040), ch. 1, p. 4*b*; *Sung Chhuang Pai Shuo* (+1157), p. 11*a*; *Chhing Po Pieh Chih* (+1194), ch. 3, p. 1*b*; *Tung Hsiao Shih Chi* (+1302), ch. 4, p. 7*a*, ch. 11, p. 6*b*.

[h] *Nan Hu Chi*[14] (Southern Lake Collection of Poems), ch. 2, p. 14*a*, ch. 3, p. 1*b*.

[i] Ch. 9, p. 4*b*.

1 刹那 2 積微 3 杜預 4 桓溫 5 儲光羲 6 張鎡
7 陳霆 8 兩山墨談 9 麟角集 10 曾應祥 11 黃汝亨
12 熊人霖 13 謝兆申 14 南湖集

supposed to have been lost in a river centuries before was found in a well at the top of a hill.[a]

Though the ideas about mountain formation as we have so far described them had a distinctively Taoist and Neo-Confucian connection, there can be little doubt that they originally derived from the Indian notion, brought to China probably by the Buddhists, of periodical cataclysms in which the world was destroyed and formed anew.[b] Hence some interest attaches to a curious story preserved in the *Yeh Kho Tshung Shu*[1] (Collected Notes of the Rustic Guest), written by Wang Mou[2] in +1201. It relates to an incident alleged to have occurred in −120 in the reign of Han Wu Ti. During digging operations near the Khun-ming Lake 'black ashes' (*hei hui*;[3] perhaps asphalt, coal, peat, or lignite) were found, and to a question of the emperor regarding these Tungfang Shuo replied, 'You had better ask the Taoists from the Western Countries' (*kho wên hsi yü Tao jen*[4]), i.e. Buddhist monks. The *Phien Tzu Lei Pien*[5] (Collection of Phrases and Literary Allusions) has a quotation[c] from the Wu Ti chapter of the *Chhien Han Shu* (which is not there now) to the effect that the Buddhist monks replied: 'These are the remains of the last cataclysm of heaven and earth' (*Tzhu shih thien ti chieh hui chih yü yeh*[6]). Modern researches on the time when Buddhism entered China do not allow us to accept the story as it stands,[d] since monks were certainly not there in the Former Han dynasty; but the story must be quite old, for it is also found in the *San Fu Huang Thu*[7] (Description of the Palaces in the Capital), a Chin book;[e] and in the *Kao Sêng Chuan* (Biographies of Famous Buddhist Monks) of the +6th century. It therefore probably dates from the earlier years of Buddhism in China. If it had been pressed a little further it could have brought the scientific thinkers of China to an early appreciation of the role of volcanic phenomena in mountain formation. We shall return to this subject in a moment.

Adams points out that the references to mountain building in European classical times are extremely scarce.[f] As may be seen from Kretschmer's account, European medieval views were dominated by the Noachian Flood legend, essentially an echo of very ancient Babylonian local happenings. This was the background of the explanation of fossils on high mountains given by Leonardo and already mentioned. Long before him, however, there had been advanced geological thinking in Islamic civilisation. In the mid +10th century the Brethren of Sincerity,[g] in their *Rasā'il Ikhwān al-Ṣafā'*, had discussed changes in sea-level, denudation, peneplanation, river evolution, and so on.[h] As we know now from the work of Holmyard & Mandeville, the *De Congelatione et Conglutinatione Lapidum*, which used to be ascribed to Aristotle, was really part of

[a] Cf. *Yü Yin Wên Ta*, p. 7*a*; *Kuei Chhien Chih*, ch. 14, p. 10*b*.
[b] Cf. Sect. 16*d* in Vol. 2, p. 420 above.
[c] Taken from *TPYL*, ch. 871, p. 6*a*.
[d] Cf. Franke (5), Maspero (5).
[e] Ch. 18. It is quoted by the +3rd-century *Nan Fang Tshao Mu Chuang*.
[f] (1), p. 330.
[g] See above, Sects. 10*f* and 13*f*, Vol. 2, pp. 95ff., 296.
[h] Paraphrased translation by Dieterici (1). See Rushdi Said (1).

[1] 野客叢書 [2] 王楙 [3] 黑灰 [4] 可問西域道人 [5] 騈字類編
[6] 此是天地刧灰之餘也 [7] 三輔皇圖

Avicenna's *Kitāb al-Shifā'* (Book of the Remedy) written in +1022. After dealing with petrifactive processes, Avicenna ascribes the formation of heights to natural convulsions such as earthquakes which suddenly raise a part of the land, or to the erosive action of winds and floods which carry away part of a formation while leaving the rest. He speaks of agglutination and petrifaction from the original flood waters in terms which distinctly resemble the ideas of Chu Hsi rather over a century later. A true explanation of the nature of fossils would follow, and Avicenna gives it. The geology of the Brethren and of Avicenna leaves a much more modern impression than some European suggestions of later date, such as the theory of Ristoro d'Arezzo (*c.* +1250)[a] that the configuration of the earth's surface corresponded to the positions of the fixed stars in the eighth sphere (whether nearer to or farther from the earth), an idea which received the support of Dante Alighieri. There seems to be a certain community between medieval geology in Islam and in China; it is much less fanciful than that of Europe. Not until a book such as the *De Montium Origine* of Valerius Faventies (+1561)[b] do the more sober considerations of the effects of earthquakes, volcanic action, and sedimentary deposition, enter into the growing geology of the Renaissance. For the full distinction between primary (volcanic) and secondary (stratified eroded) mountains, we have to await the work of Lazzaro Moro (+1687 to +1740).[c]

The slow erosion of uplifted strata was quite clearly conceived in medieval Chinese thinking. There was an old expression *ling chhih*[1] or *ling i*[2] which probably meant originally the gradual slope of hills, as in *Huai Nan Tzu*[d] and *Hsün Tzu*,[e] but which commentators in the Chin and later interpreted in the sense of reduction in course of time. River-valley erosion is referred to by the Han poet Chia I,[3] who speaks of the way in which rivers change their courses, and sometimes return to their old beds;[f] and by Emperor Yuan of the Liang (*c.* +550).[g] In the Sung, Shen Kua gives, in the *Mêng Chhi Pi Than*, a number of interesting accounts of observations. About +1070 he wrote:

Yen-Tang Shan[4] (mountain) near Wênchow is very beautiful but there is no mention of it in the old books.... It seems to be referred to first by a Thang monk Kuan-Hsiu,[5] who wrote a poem about another monk No Chü-Lo[6] who lived there.... Now I myself have noticed that Yen-Tang Shan is different from other mountains. All its lofty peaks are precipitous, abrupt, sharp and strange; its huge cliffs, a thousand feet high, are different from what one finds in other places. Its peaks are hidden by foothills so that from outside one cannot see much, but when you get near the peaks themselves, they seem to pierce the sky. Considering the reasons for these shapes, I think that (for centuries) the mountain torrents have rushed down, carrying away all sand and earth, thus leaving the hard rocks standing alone.

In places like Ta Lung Chhiu,[7] Hsiao Lung Chhiu,[8] Shui Lien,[9] and Chhu Yüeh Ku,[10] one can see in the valleys whole caves scooped out by the force of the water. Standing at the

[a] Adams, (1), pp. 335, 341.
[b] Tr. Adams (1), pp. 348 ff.
[c] Adams (1), p. 368.
[d] Chs. 11 and 20.
[e] Ch. 28, p. 4*b*.
[f] Quoted in *Shih Chi*, ch. 84, p. 11*b*.
[g] *Chin Lou Tzu*, ch. 4, p. 17*a*.

[1] 陵遲 [2] 陵夷 [3] 賈誼 [4] 雁蕩山 [5] 貫休 [6] 諾矩羅
[7] 大龍湫 [8] 小龍湫 [9] 水簾 [10] 初月谷

bottom of the ravines and looking upwards, the cliff face seems perpendicular, but when you are on the top, the other tops seem on a level with where you are standing. Similar formations are found right up to the highest summits. So also we find even in small waterways that the banks become scooped out like a rounded shrine or oratory (*khan*[1]) wherever the force of the swirling water strikes against them. This is the same thing happening. Now in the large gorges of Chhêng Kao[2] and Shensi we can see the earth standing straight up as much as a hundred feet (loess canyons). They are, indeed, a small model of Yen-Tang Shan though this is simply earth while Yen-Tang Shan is hard rock.[a]

These eroded cliffs which Shen Kua saw are shown in Fig. 264 taken from the *Thu Shu Chi Chhêng*.[b] Shen Kua also describes sedimentary deposition.

When I went to Hopei on official duties I saw that in the northern cliffs of the Thai-Hang Shan[3] mountain-range, there were belts (strata) containing whelk-like animals, oyster-shells, and stones like the shells of birds' eggs (fossil echinoids). So this place, though now a thousand *li* west of the sea, must once have been a shore. Thus what we call the 'continent' (*ta lu*[4]) must have been made of mud and sediment which was once below the water. The Yü Mountain,[5] where Yao[c] killed Kun, was, according to ancient tradition, by the side of the Eastern Sea, but now it is far inland.

Now the Great (i.e. the Yellow) River, the Chang Shui,[6] the Hu Tho,[7] the Cho Shui[8] and the Sang Chhien[9] are all muddy silt-bearing rivers. In the west of Shensi and Shansi the waters run through gorges as deep as a hundred feet. Naturally mud and silt will be carried eastwards by these streams year after year, and in this way the substance of the whole continent must have been laid down. These principles must certainly be true.[d]

Thus Shen Kua fully understood in the +11th century those conceptions which, when stated by James Hutton in 1802, were to be the foundation of modern geology.[e]

Again, Tu Wan, in the *Yün Lin Shih Phu* of +1133, makes clear references to weathering and erosion processes.[f] We can therefore certainly not follow Hoover & Hoover in their statement that Agricola's writings in +1546 were 'the first adequate declaration of the part played by erosion in mountain sculpture'.[g] Exposed fossil-containing strata had been described by Tu Mu[10] thirty years earlier as something long known and familiar.[h]

There is plenty of material in Chinese literature on other interesting questions of general geology, but as it has not been systematically worked up, I shall have to be content with simply drawing attention to some of the subjects on which much more could be found.

[a] *MCPT*, ch. 24, para. 14, tr. auct. Cf. Hu Tao-Ching (*1*), vol. 2, p. 762.

[b] *Shan chhuan tien*, ch. 132.

[c] Legendary emperor; cf. Vol. 2, p. 117, for Kun, the father of Yü the Great.

[d] *MCPT*, ch. 24, para. 11, tr. auct. Cf. Hu Tao-Ching (*1*), vol. 2, p. 756.

[e] Shen Kua's contemporaries had no thought of denouncing his views as heretical and irreligious. But neither was any Geological Society founded to search for facts which should test them.

[f] E.g. ch. 1, pp. 6*b*, 9*a*. Cf. also *Hua Man Chi*,[11] ch. 1, p. 5*b*; *Tung Hsiao Shih Chi*,[12] ch. 11, p. 6*a*, *b*, ch. 14, p. 2*b*.

[g] (1), p. xiii. Nor was erosion and sea encroachment first expounded by John Ray in +1692, as is often said.

[h] *Nan Hao Shih Hua*, p. 13*a*.

1 龕 2 成皐 3 太行山 4 大陸 5 羽山 6 漳水
7 滹沲 8 涿水 9 桑乾 10 都穆 11 畫墁集 12 洞霄詩集

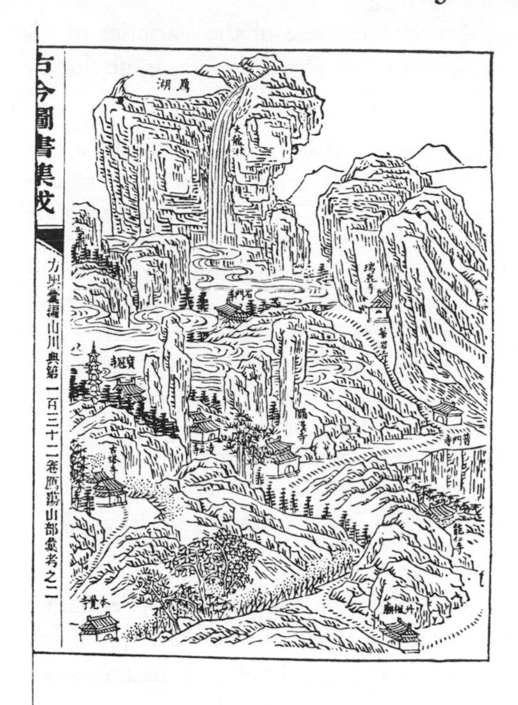

Fig. 264. Geology in Chinese art: (*j*) the eroded cliffs of Yen-Tang Shan near Wênchow on the coast of southern Chekiang (*TSCC*, *Shan chhuan tien*, ch. 132). These were some of the mountains which stimulated Shen Kua in the +11th century to consider erosion and sedimentation, and to state the basic geological principles concerning them.

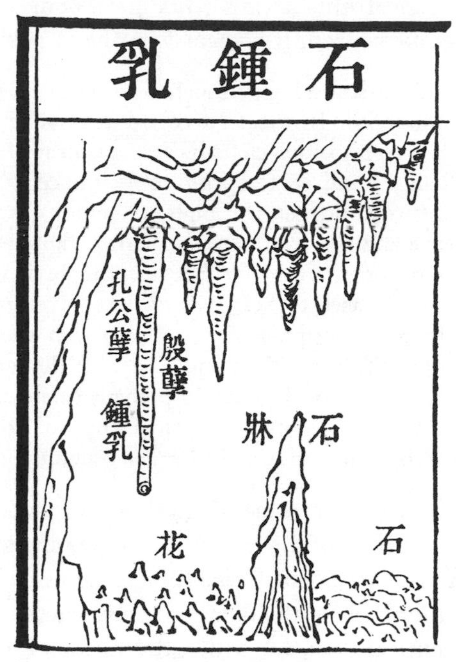

Fig. 265. Stalactites (*khung kung nieh*, *yin nieh* or *chung ju*), stalagmites (*shih chhuang*), and crystalline deposits (*shih hua*); from Li Shih-Chen's *Pên Tshao Kang Mu* of +1596.

(3) Caves, Underground Waters, and Shifting Sands

The study of caves (*tung*[1]) and the formations found in them has been pursued throughout Chinese history; they were of particular interest to the early Taoist hermits, and the word continued always to have a numinous significance in the Taoist religion. Stalactites (Fig. 265) are included in the earliest lists of inorganic and chemical substances and drugs which have come down to us, namely, the *Chi Ni Tzu* book (probably −4th century)[a] and the *Shen Nung Pên Tshao Ching*[2] (Pharmacopoeia of the Heavenly Husbandman), the first of the *Pên Tshao* series[b] and in date either Former Han or +1st century. In these lists they bear the name which they retained till the end, *shih chung ju*[3] (stone bell milk).[c] Another ancient name[d] was *khung kung nieh*.[4] The *Shen Nung Pên Tshao Ching* also distinguishes stalagmites as *yin nieh*.[5]

[a] Cf. above, p. 402 and in Vol. 2, Sect. 18*f*, p. 554.
[b] Cf. below, Sect. 38, where the series is tabulated.
[c] See below, Sect. 33, with its tabulations of the mineral and chemical substances recognised in the old Chinese literature.
[d] Cf. Chang Hung-Chao (*1*), pp. 213 ff.

[1] 洞 [2] 神農本草經 [3] 石鍾乳 [4] 孔公孽 [5] 殷孽

Ko Hung about +300 refers to them in connection with one of the varieties of his magical mushrooms (*chih*[1]).[a] Though the passage in the *Pao Phu Tzu* is obviously embroidered, it is clear what he is talking about.

The 'stone-honey mushroom' (*shih mi chih*[2]) grows in the stone caves of the Shao-Shih mountain. In those caves are deep valleys through which one cannot pass. If a stone is thrown into such clefts its sound is heard for half a day. There is a stone pillar standing more than a hundred feet from the entrance to the caves, having on top of it what looks like an upturned lid, ten feet high. One may observe the stone-honey *chih* fall from the roof of the cave and slip into this cover. At definite intervals drops invariably appear and fall like rain-water off the eaves of houses. Single drops fall continually. Though the stone-honey *chih* never ceases to fall, the lid never overflows. Above the cave characters are inscribed in 'tadpole' script, saying, 'Those who succeed in eating one *tou* of stone-honey *chih* will enjoy life for a thousand years.' Although all the Taoists long to visit this place, few are able to do so. One could perhaps catch the drops with a wooden bowl attached to the end of a strong bamboo rod, but I know of no one who has succeeded. Still, since we have the inscription, someone in former generations presumably did.[b]

In the later mineralogical books, such as the *Yün Lin Shih Phu*[3] (Cloud Forest Lapidary) of the Sung,[c] stalactites and stalagmites are always discussed.[d] Other names found are *chiang shih*[4] (ginger-stone), *shih phi*[5] (stone spleen) and *wei shih*[6] (stomach stone). Like all other products, whether inorganic, herbal or animal, they had pharmaceutical uses, for which the various *Pên Tshao* books may be consulted.

The existence of underground streams was well recognised. There is mention of one in a Sung book, the *Hsien Chhuang Kua I Chih*[7] (Strange Things seen through the Barred Window),[e] by Lu Ying-Lung.[8] A connection of some wells with the sea was suspected in certain cases, where the water-level seemed to rise and fall with the tides, hence the name *hai yen*[9]—eye of the sea. A Thang book, the *Yu-Yang Tsa Tsu* (+8th century), speaks of this, as also two Sung books, the *Mo Kho Hui Hsi* (+1080) and the *Yo-Yang Fêng Thu Chi*[10] (Customs and Notable Things of Yo-yang), by Fan Chih-Ming,[11] of later date.[f]

A great deal of information about springs is to be found in the encyclopaedias[g] and geographical collections. The role of petrifying springs in the history of geological and mineralogical thought is so important[h] that it is not surprising that these attracted

[a] These have never been identifiable, but were probably stone concretions of unusual shapes and colours, which were thought likely to confer immortality if powdered and swallowed.

[b] Ch. 11, p. 3*b*, tr. Feifel (3), p. 6; mod.

[c] Ch. 3, p. 8*a*. See below, p. 645.

[d] Cf. *Ling Wai Tai Ta*, ch. 7, p. 13*b*.

[e] P. 11*a*.

[f] Cf. the curious idea, widespread in +16th- and +17th-century Europe, that sea-water was conveyed subterraneously to the mountains and there distilled by internal heat as if in an alembic (Adams (1), p. 440). See also on mineral waters *Yung Chhêng Shih Hua*, ch. 1, p. 11*a*; *Yü Chien*, ch. 9, p. 3*b*, *Thang Yü Lin*, ch. 8, p. 23*a*.

[g] E.g. *TSCC*, *Khun yü tien*, chs. 31–42.

[h] See Adams (1).

1 芝 2 石蜜芝 3 雲林石譜 4 薑石 5 石脾
6 胃石 7 閑窗括異志 8 魯應龍 9 海眼 10 岳陽風土記
11 范致明

much attention in China also. A good description of one occurs in the early +6th-century *Shu I Chi* (Records of Strange Things).

The Yang Chhüan[1] (spring) is located north of Thien-Yü Shan[2] (mountain). A clear stream flows forth several dozen yards. Herbs or pieces of wood placed in the water are all turned to stone (*hua wei shih*[3]), clear and hard.[a]

Ancient Chinese writings refer constantly to two place-names which indicate phenomena of geological interest. Just as authors of the Warring States time or early Han are always speaking of Mt Khun-Lun,[b] so they speak also of the *liu sha*[4] (shifting sands) and of the *jo shui*[5] (weak water), both in the far west or north-west. The former occurs twenty-two times in the *Shan Hai Ching* and the latter five times; both are mentioned frequently in *Lü Shih Chhun Chhiu*, *Huai Nan Tzu*, etc.

For our purpose the identification of the places is of secondary importance; the interest of the names lies in the phenomenon concerned. The shifting sands or quick-sands were, I think, exactly what they seemed. Shen Kua discusses the matter in the *Mêng Chhi Pi Than*:

The Thang encyclopaedia, *Thang Liu Tien*, discussing the Five Elements, has, among various other sections, one on a 'Ping River' (Ping Ho[6]). People now do not know what this means. But when I used to talk with the generals in Fu and Yen[7] at the An-Nan[8] camp, I found in the army records that many soldiers had been lost at the Fan River. They said that the Yüeh people called it *nao sha*[9] and the northern people called it *huo sha*[10].[c] At the Wu Ting river, going over with a horse, the sand shifted a hundred feet away from where one was, like people treading on curtains, though where we trod was firm. In some places, however, they would sink in. Horses, camels, and waggons have all been engulfed, and hundreds of soldiers have disappeared there. This is indeed *liu sha*[4]—flowing sand. *Ping* may also be written *pan*[11] meaning 'deep mud'. This is why 'Ping Ho' came to mean, in the divination books, bad luck, like the present term, *khung wang*,[12] i.e. 'emptiness and death'.[d]

There are certainly localities in north-west China today which still present the same difficulties. A recent experience is to be found in the book of Band & Band,[e] who describe an awkward passage in 1943 near the north bend of the Yellow River by saying that it was like 'walking on chocolate blancmange'.

The phenomenon of 'singing sands' is also referred to in Chinese texts. This is particularly associated with the sand dunes surrounding the temple and lake of Yüeh-Ya Chhüan[13] some miles west of Tunhuang (Cable & French).[f] The nature of the phenomenon has been much discussed.

[a] Tr. auct.
[b] Cf. above, p. 565.
[c] Exactly the same as our 'quick'-sands.
[d] Ch. 3, para. 11, tr. auct. Cf. Hu Tao-Ching (*1*), vol. 1, p. 128.
[e] (1), p. 209.
[f] (1), p. 63. When I myself visited this place the sands refused to sing.

1 陽泉 2 天餘山 3 化爲石 4 流沙 5 弱水 6 湴河
7 鄜延 8 安南 9 淖沙 10 活沙 11 埿 12 空亡
13 月雅泉

(4) Petroleum, Naphtha, and Volcanoes

The second ancient term of interest, *jo shui* ('weak water'),[a] is often taken simply as the name of a river. But ancient authors repeatedly state that wood will not float in it. Kuo Pho (about +300) says, in his commentary on the *Shan Hai Ching*, that it will not bear the weight even of a wild-goose feather (*pu shêng hung mao*[1]). In a passage from his contemporary, Ko Hung, already quoted,[b] we have met with the same idea before. So also Chang Shou-Chieh,[2] the great Thang commentator of the *Shih Chi*, says that 'without the help of a boat made of feathers you will not be able to cross it' (*fei mao chou pu kho chi*[3]).[c] While some may prefer to believe that these were fancies derived from an ancient name, the suggestion may be put forward that in all these cases the foundation for the stories was the natural occurrence of petroleum seepages.[d] The reference is often to various parts of Central Asia now hard to identify, but such seepages certainly exist and have existed in the western provinces, especially Kansu and Szechuan. Fractions of fairly low boiling-point can occur (paraffins and naphthas) as well as the thick black natural oils, and it could easily be imagined that in some circumstances these would attract attention as 'waters' in which even wood or feathers would sink.[e] The natural petroleum which occurred at Lao Chün Miao[4] in Kansu (near Suchow) was locally known and used for greasing cart-axles many years before the present oilfield with its refineries was established there. It is mentioned in many old records such as the *Yuan-Ho Chün Hsien Thu Chih*, which records that in the Northern Chou period, between +561 and +577, it was used successfully in war against the Turks who were besieging Chiu-chhüan (Suchow).[f] Chang Hung-Chao discusses these natural seepages,[g] and mentions some of the other terms used for rock-oil or petroleum, such as *shih yu*,[5] *shih chhi*[6] ('stone lacquer'), *shih chih shui*[7] ('stone fat water'), and *mêng huo yu*[8] ('fierce fire oil').[h] Stone lacquer was probably one of the earliest terms; as such the *Po Wu Chi*,[9] a book[i] written by Thang Mêng[10] about +190, knows it.

[a] Besides the references mentioned above, we have it in many descriptions of Western Asia, e.g. *Hou Han Shu*, ch. 118, the Nestorian Stone, the *Wei Lüeh*, the *Wên Hsien Thung Khao*, ch. 339, the *Chu Fan Chih*, etc. all translated and discussed by Hirth (1); see esp. p. 291.

[b] Above, Vol. 2, p. 438.

[c] Ch. 123, p. 6*b* (the Ferghana chapter).

[d] It is interesting that the Indians told Megasthenes of a river in the north, everything thrown into which sank like stone. It was called Silas, presumably from Skr. *silā*, stone (Megasthenes, Frag. 19; Arrian, *Indica* 6, 2; Strabo, xv, c, 703; Bevan (1), p. 404). Cf. Marco Polo, ch. 22 (Moule & Pelliot ed.).

[e] For Western and Arabic references, cf. Forbes (4*a*). Cf. John Eldred's account of Mesopotamian seepages in +1583 (in Forbes (4*b*), p. 26) and Tardin's book of +1618.

[f] In Sect. 30 we shall give a detailed account of its military use about +1100.

[g] (*1*), pp. 205 ff.

[h] As to this last term, cf. Sect. 30 below on military technology.

[i] 'Notes on the Investigation of Things.' Not to be confused with the *Po Wu Chih* of a century later, as Wittfogel, Fêng *et al.* do (1), p. 565, probably through trusting to *TSCC*.

1 不勝鴻毛 2 張守節 3 非毛舟不可濟 4 老君廟 5 石油
6 石漆 7 石脂水 8 猛火油 9 博物記 10 唐蒙

This Later Han record of miscellaneous scientific matters says:

In the mountains south of Yen-shou[1a] there are certain rocks from which springs of 'water' arise. They form pools as big as bamboo baskets, and the stuff flows away in small streams. This liquid is fatty and sticky like the juice of meat. It is viscous like uncongealed grease. If one sets light to it, it burns with an extremely bright flame. It cannot be eaten. The local people call it 'stone lacquer'.[b]

Another version of the story, cited by Li Tao-Yuan in the +6th century,[c] adds that it was used for greasing cart-axles,[d] and the bearings of water-power trip-hammers.[e] Apart from the use of naphtha in warfare,[f] natural petroleum was later used, in the Sung, for obtaining carbon for ink. Shen Kua says:

Petroleum (*shih yu*) is produced in Fu[2] and Yen[3] (places in Shensi and Kansu). This 'oil-water' (*chih shui*[4]) is the same substance as that which old accounts[g] describe as coming from Kao-nu Hsien.[5] It comes out mixed with water, sand and stones. In the spring the local people collect it with pheasant-tail brushes,[h] and put it into pots where it looks like lacquer. It can easily be burnt, but its smoke, which is very thick, makes the curtains all black. I once thought that this smoke might be useful, and tried to collect its deposit for making ink. The black colour was as bright as lacquer and could not be matched by pine-wood resin ink. So I made a lot of it and called it Yen-Chhuan Shih I[6] (Yen River Stone Juice). I think that this invention of mine will be widely adopted. The petroleum is abundant, and more will be formed in the earth while supplies of pine-wood may be exhausted. Pine-forests in Chhi and Lu have already become sparse. This is now happening in the Thai-Hang mountains. All the woods south of the Yangtze and west of the capital are going to disappear in time if this goes on, yet the ink-makers do not yet know the benefit of the petroleum smoke. The smoke of 'stone charcoal' (*shih than yen*[7]—coal) also makes cloths black (and could be used for ink).[i]

The striking feature of this passage, written about +1070, is not that Shen Kua used petroleum to produce a black smoke,[j] just as is done in physiological laboratories today for making kymograph papers, but that he foresaw the de-forestation of his country, and thought that the 'inexhaustible' supplies of oil in the earth could be used as a substitute for wood. It is rather characteristic of the Chinese pre-industrial era that petroleum presented itself to an exceptionally intelligent man (as Shen Kua certainly was), not as a new and immensely significant source of power, but only a new way of making ink.

a In Shensi, mod. Yenan.
b In Ma Kuo-Han's collection, *YHSF*, ch. 73, p. 4*b*; tr. auct.
c *Shui Ching Chu*, ch. 3, p. 28*b*.
d So also *Yu-Yang Tsa Tsu*, ch. 10, p. 2*b*.
e Cf. below, Sect. 27*f*.
f Cf. below, Sect. 30.
g Cf. *Chhien Han Shu*, ch. 28B, p. 6*a*, which says that this place had 'water which would burn'. This is the oldest reference so far found.
h This technique is also mentioned by Agricola in the +16th century (ch. 7).
i *MCPT*, ch. 24, para. 2, tr. auct. with Chang Tzu-Kung. Cf. Hu Tao-Ching (*1*), vol. 2, p. 745.
j The same idea had occurred to Ibn Ḥauqal a century earlier (Forbes (4*a*), p. 18, (4*b*), p. 28; Mieli (1), p. 115). He was placed nearer to good sources of the material.

1 延壽 2 鄜 3 延 4 脂水 5 高奴縣 6 延川石液
7 石炭煙

As for natural gas, we must reserve a discussion of its very long practical use by the Chinese to Section 37 on the salt industry, where opportunity will arise for describing the fire-wells and salt-wells of Szechuan and other provinces. The use of the gas for evaporating the brine goes back certainly to the Han, and knowledge of it is probably older still. The following story, however, concerns a seepage of naphtha. The *Fa Yuan Chu Lin*[1] (Forest of Pearls in the Garden of the Law), a Thang Buddhist encyclopaedia by Tao-Shih[2] (+668), describes a pool of fire somewhere near Nepal, which was visited by the Chinese ambassador Wang Hsüan-Tshê about +650:[a]

> There was a little lake of water which would burn. If one set light to it, a brilliant flame would appear all over its surface as if coming forth from the water. If one poured water on it to extinguish it, the water changed to fire and burned. The Chinese ambassador and his suite cooked their meal on it. Later he asked the king of that country about it...and was told that it protected a gold casket containing the diadem of Maitreya.[b]

This brings us to volcanic phenomena. As there are in China proper no volcanoes, all information on this subject had to come from outside.[c] One of the first references occurs in the late +3rd-century *Shih I Chi*:

> In Tai-Yü Shan[3] there is an abyss a thousand miles deep, in which water is always boiling. Metal or stones thrown into it are attacked and reduced to mud. In winter the water dries up and yellow smoke billows forth from the ground many yards high. People who live among these mountains dig down to the depth of several tens of feet and get scorched stone like charcoal, which will burn with flames. It can be ignited by a candle, and the flames are blue. The deeper they dig the more fire they get.[d]

This does not sound like an eye-witness account, and some believe that it was a description of Mt Etna brought by Syrian traders, though it seems at least equally likely that it may derive from reports of the south seas brought back by official travellers like Khang Thai, who was envoy to Indo-China about +260.[e] There is, moreover, another text in one of the apocryphal classics, which Chhen Phan (*4*) thinks goes back to the time of Tsou Yen (−4th century), but is more probably late Han. Here belongs also the mention of a volcano in the spurious Yün Chêng chapter of the *Shu Ching*;[f] the fire consumed even jade and stone.

Of hot springs, however, there were always plenty in China.[g] In the Thang, Hsü

[a] We have met him before in several connections (Vol. 1, p. 211; Vol. 2, p. 428).

[b] Ch. 16, p. 15*b*, tr. Lévi (1), p. 314; eng. auct. The common-sense approach of the Chinese is most characteristic.

[c] The Europeans had volcanoes at much closer quarters, hence the larger interest shown by them anciently in this (cf. Adams (1), pp. 399 ff. and elsewhere).

[d] Ch. 10, p. 6*a*, tr. auct. For other early references see *TPYL*, ch. 869, p. 5*b*, ch. 871, p. 3*a*.

[e] His account of Cambodia is in *Liang Shu*, ch. 54, p. 6*b*; *Nan Shih*, ch. 78, p. 5*a*.

[f] Ch. 9, tr. Medhurst (1), p. 128.

[g] The author has very pleasant personal remembrances of the baths at Fuchow (Fukien), Peiwên-chhüan (Szechuan), and An-ning (Yunnan). He takes this opportunity of recording his deep indebtedness in many ways to the friends in whose company he visited and enjoyed them, Dr Huang Hsing-Tsung and Dr Wu Su-Hsüan.

1 法苑珠林 2 道世 3 岱輿山

Chien,[1] in his encyclopaedic *Chhu Hsüeh Chi*[2] (Entry into Learning), noted[a] that if the upper reaches of a stream smell sulphurous, its springs are likely to be warm or hot.[b] About +800 Li Ho[3] described an arseniferous spring. In the Sung, the author of the *Chhi Tung Yeh Yü* believed that the heat was caused by the combustion of sulphur and alum underground.[c] In the Ming, Wang Chih-Chien[4] listed many curious kinds of mineral-laden waters and their effects.[d] A whole monograph could of course be written on Chinese ideas concerning volcanic phenomena.[e]

(c) PALAEONTOLOGY

Though the term 'fossils' is now reserved for the remains of plants and animals embedded in the strata of the earth's surface, its original meaning was much wider and included everything of interest which could be obtained by digging. We shall glance later on at the medieval and Renaissance theories of 'lapidifying' emanations or juices, in connection with the early ideas on the generation of minerals. Here we are concerned only with the gradual recognition of the fact that some of the life-like forms discovered in the rocks were in fact the remains of ancient animals, perhaps of species now extinct. If it be held that palaeontology did not begin until after the appearance of classificatory systems of Linnean character, it has nevertheless what must then be called a pre-history going back for many centuries. This is what we have to trace in comparing Chinese with occidental statements on the subject.

Sometimes it is convenient to begin at the modern end and work backwards. Andrade (1) informs us that Robert Hooke, during the latter part of the 17th century, was 'the first to recognise the true nature of fossils and their importance as a record of the earth's history'. According to Raven,[f] John Ray, the great systematist (+1627 to +1705) was 'one of the first to assert that fossils were the remains of living organisms'. The quotations already given from Shen Kua (+11th century) and Chu Hsi (+12th) would alone suffice to show that these statements cannot be true. In any case, the progress of correct views in the 16th and 17th centuries may readily be followed in the discourse of Adams[g] on the 'lapides figurati', as they were then called. Von Zittel[h] also shows that while some of the most industrious collectors (such as Martin Lister and Edward Lhwyd)[i] held erroneous ideas, the right explanations were given by Palissy (+1580), Alessandro degli Alessandri (+1520) and Girolamo Fracastoro (+1517). All these followed[j] the achievement of Leonardo da Vinci, who in +1508 wrote his famous passage about the shells of oysters found in high mountains.[k] But Avicenna (Ibn Sīnā) in +1022 had said the same thing, in almost the same words,[l]

[a] Ch. 7, pp. 8*a* ff. He attributed the statement to Chang Hua's *Po Wu Chih*, *c.* +290.
[b] So also Thang Tzu-Hsi's[5] *I Chao Liao Tsa Chi*, ch. 1, p. 40*b*.
[c] Ch. 1, p. 6*a*.
[d] *Piao I Lu*, ch. 2, p. 7*a*.
[e] Cf. *Chhih Ya*, ch. 2, p. 5; *I Chao Liao Tso Chi*, ch. 1, p. 26*a*.
[f] (1), p. 170.
[g] (1), pp. 250 ff.
[h] (1), p. 17.
[i] Gunther (1), vol. 14.
[j] And, if Duhem is right (1, vol. 2, pp. 283 ff.), derived from him.
[k] Ed. McCurdy (1), vol. 1, p. 330.
[l] Holmyard & Mandeville (1), p. 28.

1 徐堅 2 初學記 3 李賀 4 王志堅 5 唐子西

while al-Bīrūnī in +1025 had deduced the former presence of a sea from the finding of fossil fishes.[a] This takes us back to the time of Shen Kua and Chu Hsi. There are no immediately previous contributions from Europe or Islam, and it will therefore be convenient at this point to see how far back the recognition and study of fossil remains goes in China, before returning to the contributions, mostly afterwards forgotten, of Greek and Hellenistic writers.

Recapitulating what was said a few pages above, it will be remembered that Yen Chen-Chhing in about +770 clearly associated the occurrence of fossils with what we should now call sedimentary strata afterwards lifted into mountains. The leading idea under which he did so was that of the 'mulberry-grove', a Taoist technical term which could have sprung, one feels, straight from the nature-mysticism of Chuang Tzu, with its conviction of the mutability of all things. It is in the light of this, which we were able to trace back at any rate to the end of the Han (+2nd century), that we may read the quotations to be given on the following pages concerning particular types of fossils.

We shall begin with fossil plants, and then deal briefly in succession with the remains of extinct invertebrates—brachiopods (a group in which the fossil forms far surpass the living ones in number and variety), cephalopods (ammonites, nautiloids) and other molluscs; and arthropods (trilobites, an extinct group; and fossil crustacea). Lastly something will be said about fossil reptiles and mammals.

(1) Fossil Plants

China certainly contributes to the beginnings of palaeobotany.[b] Knowledge of the petrification of pine-trees may go back to the +3rd century, since encyclopaedias quote Chang Hua as saying that all such trees turn into stone after three thousand years.[c] But in the Thang a great interest was taken in the matter. The *Hsin Thang Shu* says[d] that in one of the regions of Central Asia inhabited by the Uighurs, there is a river called Khang-Kan[1] which in three years will turn into blue-coloured stone any pieces of wood dropped into it. In +767 the painter Pi Hung[2] made a famous fresco depicting fossilised pine-trees.[e] At the end of the +9th century Tu Kuang-Thing said in his *Lu I Chi*[3] (Records of Strange Things):

> In a pavilion on a mountain in Yung-khang[4] Hsien near Wuchow[5] there are what look like rotten fir-trees. But if you break a piece off, you will find that it is not decayed in the water, but a substance changed into stone, which previously was not yet transformed in that manner. On examining (other) pieces in the water (at that place), they turn out to be transformations of the same character. These metamorphoses do not differ from fir-trees as to branches and bark; they are simply very hard.[f]

[a] Prostov (1).

[b] Of course we cannot now be sure, in any individual case, such as those here to be mentioned, that the author was really dealing with 'petrified wood' or plant remains. Many other things have often been mistaken for it, such as fibrous calcite, asbestos, various banded concretions, metamorphosed schists, inorganic cylindrical casts, etc.; even by modern geologists. The point is, what did the medieval Chinese writers think they were observing? Fossil plants.

[c] Laufer (13), p. 22. The passage cannot be found in the present text of the *Po Wu Chih*, however.

[d] Ch. 217B, p. 7*b*.

[e] *Phei Wên Yün Fu*, ch. 100A, p. 21*b* (3903·3).

[f] Cit. in *Ko Chih Ching Yuan*, ch. 7, p. 11*b*, tr. Laufer (13); mod.

1 康干 2 畢宏 3 錄異記 4 永康 5 婺州

This was written probably towards the end of the +9th century. Half a century later these petrified pines of Yung-khang attracted the attention of the Taoist poet Lu Kuei-Mêng,[1] who wrote two poems on them.[a]

The fossilisation of pine trees was appreciated also in the Sung. In the *Mo Kho Hui Hsi* (+1080) we read:

In Hu Shan[2] there are Pu[3] tree (-trunks)[b] several feet in length, of which half has been changed into stone and half is still the hard wood. Tshai Chün-Mo,[4] seeing these and thinking them particularly strange brought one to his house and I myself[c] saw it there.[d]

The *Chêng Lei Pên Tshao*[5] of about +1110 remarked, in words quoted textually in subsequent pharmacopoeias:

Now in Chhuchow[6] there is a kind of 'pine-stone' (*sung shih*[7]) which looks like the trunk of a pine, but which is actually stone. Some say that these really were once pine-trees, which in the course of ages have been transformed into stone. People polish them and keep them. They are of the same nature as asbestos.[e]

It was good observation that only part of a tree-trunk may petrify.[f] But there may have been some confusion with another kind of stone called *sung lin shih*[8] (pine-forest stone), which was found in Szechuan; this simply has elaborate dendritic markings due to crystallisation of manganese dioxide.[g] One of the first descriptions of this is in the *Tung Thien Chhing Lu Chi*,[9] by Chao Hsi-Ku[10] (+13th century).[h]

In other cases also it is not easy to be sure whether true fossil plants are being described, or whether the writers really mean something else. Thus the *Kuei Hai Yü Hêng Chih* (Topography and Products of the Southern Provinces), about +1175, speaks of 'stone plum-trees' (*shih mei*[11]) and 'stone cypresses' (*shih po*[12]), but as it also says that they grow in the southern seas, they were probably different kinds of coral.[i] Yet if this is so it is odd that Fan Chhêng-Ta did not use the usual word for coral, *shan hu*,[13] which is at least as old as the +7th century.[j] He also distinctly says that they are of the same nature as the fossil crabs (*shih hsieh*[14] and *shih hsia*[15]).

[a] According to the *Yün Lin Shih Phu* of +1133; ch. 2, p. 3*a*.
[b] Unidentifiable, cf. B II, 202.
[c] Phêng Chhêng.
[d] Ch. 5, p. 6*b*, tr. auct.
[e] Entry *Pu hui mu*,[16] asbestos, tr. auct. Cf. Chang Hung-Chao (*1*), p. 198. Cf. also de Mély (1), p. 86; his text, p. 83; *PTKM*, ch. 9, p. 44*a*.
[f] Other Sung and Yuan references in *Hsieh Chhuan Chi*, ch. 2, p. 19*b*; *Wu Li Pu Shih Hua*, p. 11*a*; *Tung Hsiao Shih Chi*, ch. 10, p. 7*a*.
[g] Figured by Chang Hung-Chao (*1*), pl. XIII.
[h] In *Shuo Fu*, ch. 12, p. 18*b*.
[i] In *Shuo Fu*, ch. 50, p. 10*b*; *Ling Wai Tai Ta*, ch. 7, p. 15*b*.
[j] RP 33.

1 陸龜蒙 2 壺山 3 布 4 蔡君謨 5 證類本草
6 處州 7 松石 8 松林石 9 洞天清錄集 10 趙希鵠
11 石梅 12 石柏 13 珊瑚 14 石蟹 15 石蝦 16 不灰木

Yao Yuan-Chih,[1] in his +17th-century book, the *Chu Yeh Thing Tsa Chi*[2] (Miscellaneous Records of the Bamboo Leaf Pavilion), says: 'Not only wood but also herbs can change into stone. When this happens to plants the "water-absorbing stone" (*shang shui shih*[3]) is produced'.[a] The usual name for this stone was *han shui shih*,[4] and Chang Hung-Chao has been able to identify it[b] as calcareous tufa formed from some species of *Chara*. It seems that recognition of this stone goes back to the Thang.[c]

Fossil bamboo-shoots are much spoken of in the Sung period. Lu Yu's[5] *Lao Hsüeh An Pi Chi*[6] (Notes from the Hall of Old Learning)[d] describes the finding of these *shih sun*[7] at Chhêngtu, but he talks of caves and wells in the same passage, and may perhaps have referred to stalagmites. On the other hand the following words of Shen Kua seem to indicate fairly clearly some kind of fossil plants:

> In recent years [*c.* +1080] there was a landslide on the bank of a large river in Yung-Ning Kuan[8] near Yenchow.[9] The bank collapsed, opening a space of several dozens of feet, and under the ground a forest of bamboo shoots was thus revealed. It contained several hundred bamboos with their roots and trunks all complete, and all turned to stone. A high official happened to pass by, and took away several, saying that he would present them to the emperor. Now bamboos do not grow in Yenchow. These were several dozens of feet below the present surface of the ground, and we do not know in what dynasty they could possibly have grown. Perhaps in very ancient times the climate was different so that the place was low, damp, gloomy, and suitable for bamboos. On Chin-Hua Shan[10] in Wuchow[11] there are stone pine-cones, and stones formed from peach-kernels, stone bulrush roots, stone fishes, crabs, and so on, but as these are all (modern) native products of that place, people are not very surprised at them. But these petrified bamboos appeared under the ground so deep, though they are not produced in that place today. This is a very strange thing.[e]

One cannot help being reminded of the providential landslide which turned Roderick Murchison into a geologist.

(2) Fossil Animals

With the fossils of animals we are on much surer ground. The brachiopods are a very ancient phylum of invertebrates bearing a superficial resemblance to bivalve molluscs. Owing to the fact that the many extinct species of *Spirifer* and related genera had shells which look like the outstretched wings of a bird, the Chinese name for them was *shih yen*[12] (stone-swallows). These 'stone-swallows' were identified as fossil brachiopods by Davidson in +1853.[f] The first important reference to them seems to be

[a] Ch. 8, tr. auct.
[b] (*1*), p. 309.
[c] RP 102.
[d] Ch. 5, p. 13*a*.
[e] *MCPT*, ch. 21, para. 17, tr. auct. They might however have been corals, crinoids, fulgurites or various other things. One cannot expect too much from contemporaries of William the Conqueror, and the main interest is that Shen Kua thought they were fossil plants. Cf. Hu Tao-Ching (*1*), vol. 2, p. 692.
[f] Cf. Hanbury (1), p. 274; RP 107.

1 姚元之 2 竹葉亭雜記 3 上水石 4 含水石 5 陸游
6 老學庵筆記 7 石筍 8 永甯關 9 延州 10 金華山
11 婺州 12 石燕

towards the end of the +5th century, when Li Tao-Yuan said in the *Shui Ching Chu* (Commentary on the Waterways Classic):

In Shih-Yen Shan[1] there are a sort of stone oysters (*shih kan*[2]) which look like swallows. Hence the name of the mountain. There are two varieties of these stone shapes, one large and one small, as if they were parents and offspring. During thunderstorms these 'stone-swallows' fly about as if they were real swallows. (Yet) Lo Han said: 'Now the (stone-) swallows do not fly about any more.'[a]

Li Tao-Yuan was quoting here from the *Hsiang Chung Chi*[3] of Lo Han,[4] who had been an official of the Chin (*fl. c.* +375),[b] so it seems likely that the fossils had been recognised at least as early as the +4th century.

From the *Thang Pên Tshao* (+660) onwards, these fossils were incorporated into the pharmaceutical compendia. Fig. 266, which shows a drawing of them, is taken from the *Pên Tshao Kang Mu*.[c] From the centuries following the time of Li Tao-Yuan more than thirty texts are known which emphasise the value and beauty of these stone-swallows and describe their collection as imperial tribute. These include the geographical chapters of the Thang dynastic histories.[d] It is noteworthy that Li Tao-Yuan, who spoke of 'stone oysters', came nearer the truth than the many later writers who accepted the technical term 'stone-swallows' without question. Yen Chen-Chhing, in the important passage quoted above (p. 600) on mountain formation, used the same words as Chu Hsi (*lo pang*,[5] molluscs), yet what he actually saw may well have been brachiopod fossils. They were recognised by Tsêng Min-Hsing in +1176 as having once been in the sea.[e]

It was a curious idea that the stone-swallows came out of the rocks and flew about in windy and stormy weather. But the legend at any rate stimulated a Sung scholar to disprove it experimentally in the +12th century. It had considerable authority; not only had Lo Han mentioned it, but his +4th-century contemporary, the famous painter Ku Khai-Chih,[6] had definitely affirmed it,[f] as also many later writers. But Tu Wan was not content to believe the story, and wrote (+1133) in his *Yün Lin Shih Phu* (Cloud Forest Lapidary):

The stone-swallows are produced at Ling-ling[7] in Yungchow.[8] It was said in ancient times that when it rained they flew about. In recent years, however, I have climbed up high

[a] Tr. auct.
[b] Biogr. *Chin Shu*, ch. 92, p. 20*a*.
[c] It is a curious fact that many Chinese Devonian brachiopods (especially *Spirifer* spp.) have been described on the basis of pharmaceutical material, no outcrop sources of these species having yet become known to geologists (personal communication from Dr K. P. Oakley and Dr Helen Muir-Wood). In other cases the sources are well known.
[d] Most of the texts are given in the thorough discussion of Chang Hung-Chao (*1*), pp. 251 ff.
[e] *Tu Hsing Tsa Chih*, ch. 4, p. 5*b*.
[f] If we are to believe what Chiang Yü[9] says in the +17th century in his *Hsiao-Hsiang Thing Yü Lu*[10] (Listening to the Rain at Hsiao-hsiang).

[1] 石燕山 [2] 石蚶 [3] 湘中記 [4] 羅含 [5] 螺蚌
[6] 顧愷之 [7] 零陵 [8] 永州 [9] 江昱 [10] 瀟湘聽雨錄

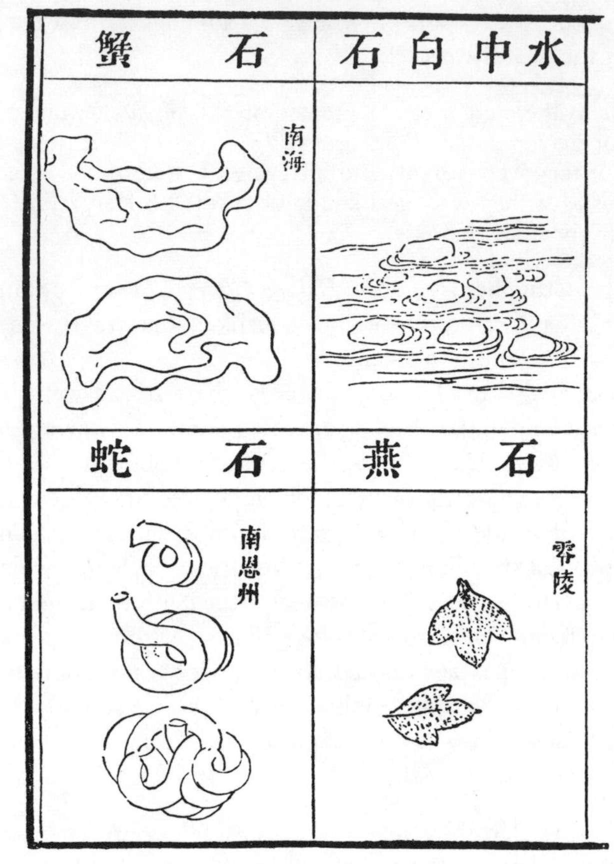

Fig. 266. Drawings of fossil animals from Li Shih-Chen's *Pên Tshao Kang Mu* of +1596. At the top to the left, *shih hsieh*, i.e. stone crabs. Below, to the left, *shih shê*, i.e. stone serpents; and to the right *shih yen*, i.e. brachiopods such as *Spirifer* and related genera.

cliffs, and found many of these stone shapes with the form of swallows. Some of them I marked with my pen. As the rocks were exposed to the blazing sun they cracked and weathered when thunder showers came, and the ones which I had marked fell to the ground one after another. It was because of the expansion in the heat and contraction in the cold that they fell flying through the air. They cannot really fly. The local people have in their houses many stone plates which show these stone-swallow shapes.[a]

The texts frequently say that the stone-swallows were dissolved or preserved in vinegar for use as medicine. The *Lung-Chhüan Hsien Chih*[1] (Lung-chhüan Local Topography) of +1762 says: 'People who like to do unnecessary things catch them in nets, cut each one into two, and put them into a vessel with vinegar in it. Those which can move after being put in are supposed to be the best ones.'[b] No doubt the pharmacists deceived themselves in thinking that spontaneous splittings of the stone were real movement, but perhaps what they did was not so unnecessary after all. It is

[a] Ch. 2, p. 3*b*, tr. auct.

[b] Tr. auct.

[1] 龍泉縣志

at any rate interesting that geologists still use acetic acid today for removing calcareous matrix from phosphatic fossils so as to bring out fine detail. Tu Wan seems to allude to this very practice when he says[a] that the workers use 'drugs' to etch certain stones.[b] But even more to the point is the fact that the traditional Chinese diet was always lacking in calcium (owing to the absence of milk products from the diet), and a good source of assimilable lime was greatly needed. Even if a large part of the calcium in fossil bones and other mineral materials was in the form of phosphate rather than carbonate, some calcium acetate would certainly be formed, and this would be more suitable for pharmaceutical use, corresponding indeed to the employment of calcium lactate and other similar organic salts today. The *Pên Tshao Kang Mu* prescribes fossils for dental and other troubles likely to be due to hypocalcaemia.[c]

By the time of Li Shih-Chen in the +16th century a certain confusion had come about between the stone-swallow fossils and a bird, the cave-dwelling martin, *Chelido dasypus*, which had the same name (*shih yen*[1]).[d] This had been introduced into the pharmacopoeia by the +10th-century *Jih Hua Pên Tshao*,[2] and of course used for different purposes. Li Shih-Chen clears up any doubts, however, and treats of the fossil brachiopod and the bird in two quite different chapters of the *Pên Tshao Kang Mu*.[e]

True fossil lamellibranchiate molluscs are also mentioned, as we have already seen from several quotations ('stone oysters'), and gastropods too ('conchs'), or what were thought to be so. The *Thang Pên Tshao* notes the resemblance of the former to the stone-swallows. Chou Chhü-Fei, about +1178, in his *Ling Wai Tai Ta*,[f] speaking of Hsiangchow[3] in Kuangsi, says that apart from the stone-swallows, there are many fossils of another kind resembling sea oysters (*hai kan*[4]) which, once living, have become inlaid (*khan*[5]), as it were, into the stone. He adds that these are not the real stone-swallows, so that a clear distinction must have been made in his time. But fossil lamellibranch molluscs did not acquire an important place in the lapidary parts of the pharmaceutical compendia, though the *San Tshai Thu Hui* encyclopaedia (Ming) gives a *shih ko*[6] (fossil clam) along with the true stone-swallows.[g]

Some very impressive coiled fossil shells of soft-bodied animals which possessed an external skeleton were recognised by the Chinese, at any rate from the Sung onwards, the first Pên Tshao in which they appear being Su Sung's[7] *Pên Tshao Thu Ching*[8] (Illustrated Pharmacopoeia) of +1070. I illustrate these 'stone-serpents' (*shih shê*[9]) in Fig. 266, but an even cruder drawing[h] is contained in the +1468 edition

a *Yün Lin Shih Phu*, ch. 2, p. 2*a*.
b *Kung jen yung yao tien hua chien chih.*[10]
c De Mély (1), pp. 131, 237; text, p. 126.
d Read (3), no. 287.
e Chs. 10 and 48 respectively.
f Ch. 7, p. 15*a*.
g De Mély (1), p. 130.
h The modern edition of the *Thu Ching Yen I Pên Tshao* in the *Tao Tsang* (*TT* 761) has illustrations, but I do not feel confident that these go back to the +11th and +12th centuries in their present form.

1 石燕 2 日華本草 3 象州 4 海蚶 5 嵌 6 石蛤
7 蘇頌 8 本草圖經 9 石蛇 10 工人用藥點化鐫治

of the *Chêng Lei Pen Tshao*.[1] An alternative name was *yang chio lo*[2] (ram's horn conchs).[a] Su Sung[b] wrote:

The stone-serpent appears in rocks which are found beside the rivers flowing into the southern seas. Its shape is like a coiled snake with no head or tail-tip. Inside it is empty. Its colour is reddish purple. The best ones are those which coil to the left. It also looks like the spiral shell of a conch. We do not know what animal it was which was thus changed into stone. But (the petrifaction) must have been similar to that which occurred in the case of the 'stone-crabs'.[c]

Here it is interesting that sinistral and dextral coiling was noticed, and fossils showing it would have been almost certainly gastropods rather than cephalopods. Then Khou Tsung-Shih,[3] in the *Pên Tshao Yen I*[4] of +1115, acutely observed that in his opinion the stone-serpent had never been a snake like modern snakes.

The stone-serpent is in colour like earthen plates on old walls or like hanging hawthorn berries.[d] Both ends are the same size. The stone-serpent is a very different kind of thing from the stone-crabs, for they were once real crabs subsequently changed (into stone), but the stone-serpent was never a snake (such as we find nowadays). People use it very little.[e]

Shen Kua alludes, about +1080, to the finding of such fossils:[f]

In the Chih-Phing reign-period (+1064 to +1067) a man of Tsêchow[5] was digging a well in his garden, and unearthed something shaped like a squirming serpent, or dragon. He was so frightened by it that he dared not touch it, but after some time, seeing that it did not move, he examined it and found it to be stone. The ignorant country people smashed it, but Chhêng Po-Shun,[6] who was magistrate of Chin-chhêng[7] at the time, got hold of a large piece of it on which scale-like markings were to be seen exactly like those on a living creature. Thus a serpent or some kind of marine snake (*chhen*[8]) had certainly been turned to stone, as happens with the 'stone-crabs'.

And about the same time the author of the *Hsi Chhi Tshung Yü* (Western Pool Collected Remarks) attributes[g] to Wang Chin-Chhen[9] the idea that there were particular places under the sea where serpents and crabs were petrified.

Another kind of cephalopod with external skeleton gave rise to conspicuous fossils, the orthoceratids, or straight-shelled nautiloids. Ordovician limestone containing them was known by the time of the Ming at any rate as 'pagoda-stone' (*pao tha shih*[10]),[h]

[a] RP 109; de Mély (1), p. 132.

[b] Shen Kua's contemporary, whose astronomical maps and instruments we have already admired (pp. 277, 351 above).

[c] Tr. auct. Quoted in subsequent compendia, such as *PTKM*, ch. 10, p. 43*b*.

[d] What both these authors said about the colour of the ammonites was presumably due to impregnation of the fossils with iron salts, giving a brownish pink colour.

[e] Ch. 5, p. 7*b*; tr. auct. Quoted also in subsequent compendia, such as *PTKM*, ch. 10, p. 43*b*.

[f] *MCPT*, ch. 21, para. 18, tr. auct. Cf. Hu Tao-Ching (*1*), vol. 2, p. 693.

[g] Ch. 2, p. 34*a*.

[h] Li Ssu-Kuang (*1*), p. 104.

[1] 證類本草 [2] 羊角螺 [3] 寇宗奭 [4] 本草衍義 [5] 澤州
[6] 程伯純 [7] 晉城 [8] 蜃 [9] 王蓋臣 [10] 寶塔石

the septa of the fossil encouraging such an analogy. Other names were '*Thai Chi shih*'[1] on account of a fanciful resemblance to the circles which the Neo-Confucians liked to embody in their philosophical diagrams,[a] and also 'straight-horn stone' (*chih chio shih*[2]). Chang Hung-Chao figures a fine specimen of *Orthoceras sinensis* from a locality in Hupei.[b] The nautiloids were not used in medicine, nor was it realised that they were animal remains.

Fossil arthropods had more of a success in medieval China than the fossil cephalopods just mentioned. Stone containing trilobites was called 'bat-stone' (*pien fu shih*[3]), from the resemblance which the cross-sections of these animals have to the wings of bats.[c] In +1637 Wang Shih-Chên[4] gave a detailed description of these fossils in his *Chhih Pei Ou Than*[5] (Chance Conversations North of Chhih-chou). But already fourteen centuries earlier trilobite-containing rocks had been prized, for Kuo Pho, in his commentary on the *Erh Ya* dictionary, says that another name for bats was *chih mo*[6] and that the men of Chhi use such 'bat-stone' for inkstones (*chih mo yen*[7]). The fossils were probably identical with what was called *shih tshan*[8] ('stone silkworms'), and as such they entered the pharmacopoeias.[d] The earliest in which they were mentioned was the *Khai-Pao Pên Tshao*[9] of +970.

Pleistocene crabs were much more widely known. The same early Sung lapidary introduced them into medicine.[e] The most common fossil species of China is *Macrophthalmus latreilli*, frequently found in Hainan and Kuangsi; and the name by which they were always known was *shih hsieh*,[10] 'stone crabs'. Fan Chhêng-Ta mentions them several times in his *Kuei Hai Yü Hêng Chih* (+1175),[f] and in one place suggests that they were formed from sea-foam, which is a little reminiscent of the 'succus lapidificus' of European medieval writers. He also says there was a different variety (*shih hsia*[11]). But plenty of other writers from Sung to Chhing[g] affirm that they were real crabs which had been caught in the mud and afterwards petrified. As in the case of the stone-swallows, there was in late times a confusion with a living species,[h] which Li Shih-Chen had to clear up.

Fossil and sub-fossil vertebrates in which the Chinese were interested included the remains of fishes, reptiles and mammals. The earliest reference to fossil fish has recently been the subject of a discussion started by Sarton (4), who reported the examination of such specimens by King Louis IX in +1253, as given in an account by

[a] Cf. Sect. 16*d* above (Vol. 2, p. 461).

[b] (*1*), pl. XI.

[c] Chang Hung-Chao (*1*), figures masses of trilobites (*Drepanura premesnili*) in Cambrian rock from Shantung, pl. VIII.

[d] For the *Pên Tshao Kang Mu* see RP 110; for *San Tshai Thu Hui*, see de Mély (1), p. 132, text, p. 127.

[e] RP 108; de Mély (1), p. 132, text, p. 127.

[f] P. 7*b* (in *Shuo Fu*, ch. 50, pp. 10*b*, 18*b*); also *Ling Wai Tai Ta*, ch. 7, p. 15*a*.

[g] E.g. Yao Khuan in his late +11th century *Hsi Chhi Tshung Yü*, ch. 2, p. 34*a*; Kao Ssu-Sun in *Hsieh Lüeh* of +1184 (in *Shuo Fu*, ch. 36, p. 18*a*); Chhu Jen-Huo[12] in *Hsü Hsieh Phu*[13] (Chhing).

[h] See Read (5), no. 214*c*; *Telphura* spp.

1 太極石 2 直角石 3 蝙蝠石 4 王士禎 5 池北偶談
6 蟙䘃 7 蟙䘃硯 8 石蠶 9 開寶本草 10 石蟹
11 石蝦 12 褚人穫 13 續蟹譜

the chronicler Joinville in 1309. Pease (1) then adduced the famous passage from Xenophanes at which we shall glance briefly later, and Eisler (3) conjectured that certain terms on Babylonian cuneiform inscriptions might refer to fossil fish. Of a more certain character was the passage from al-Bīrūnī (+1025) translated by Prostov. Goodrich (6) drew attention to the words of Chu Hsi already quoted, though they are not strictly relevant to the point at issue, since shell-fish are not fish. It was left to Rudolph (2) to draw on the excellent discussion of Chang Hung-Chao,[a] which shows that, so far as China is concerned, we can start definitely in the +6th century. The first reference to 'stone fishes' (*shih yü*[1]) comes in the *Shui Ching Chu* of Li Tao-Yuan (*d.* +527).

At Stone-Fish Mountain in Hsiang-hsiang Hsien[2] there is a lot of non-magnetic iron ore (*hsüan shih*[3]). It is of dark colour and veined like mica. If you split open the outer layer there are always shapes of fishes inside, with scales and fins, heads and tails, just as if they were carved or painted. They are several inches long, and their details are perfect. If burnt they will even give off a fishy smell.[b]

The *Yün Lin Shih Phu* of +1133 has a long passage on the subject.

At the top of the mountains in Hsiang-hsiang Hsien in Thanchow[4] there are horizontal stones buried in the earth. Digging down several feet one can unearth slabs of blue stone; this is called the 'cover of the fish-stone'. Below this the stone is bluish grey or whitish, and removing layer by layer, you come upon the shapes of the fishes. They look like 'false carp' (*chhiu chi*[5]).[c] Their scales and antennae are all as perfect as if drawn with ink. Digging down twenty or thirty feet blue stone is again seen; this is called the 'support of the fish-stone'. Underneath is sand and earth. In the stone the fishes seem to follow one another as if they were swimming. Sometimes their shapes are injured or indistinct, obscured by spots, as if the river-weeds had been petrified with them. Then among hundreds of specimens hardly one or two are clear. The fishes in the stone have no definite orientation, but lie in all directions; sometimes they are coiled like dragons. Occasionally one can get them out complete so that both sides can be seen in all perfection.

The local people make falsifications of these fish by painting stone with lacquer. But if you scrape it and burn it you can easily distinguish by the smell what is true and what is false.

Moreover, in Lung-Hsi[6] (Kansu) there is a place called Yü-lung (Fish-Dragon). Here also if you take out the stone and split it you can obtain many fish shapes, just the same as what is produced in Hsiang-hsiang. I wonder whether in very ancient times the mountains fell down upon the rivers in which these fishes lived, so that after many ages they were condensed into stone (*sui chiu thu ning wei shih*[7]), as we see them now.

[a] (*1*), pp. 258 ff.

[b] Ch. 38, p. 3*a*, tr. auct. Substantially the same passage is repeated in the *Yu-Yang Tsa Tsu* of +860 (ch. 10, p. 4*a*). Other Thang books, the *Chhao Yeh Chhien Tsai*[8] (Records of the Doings of Gentle and Simple) and the *Lin Chio Chi*[9] (Unicorn Horn Collection of Examination Essays), p. 15*a*, also mention these fossils. The latter distinctly recognises them as petrifactions. For a late reference see *Chhih Pei Ou Than*, ch. 21, p. 10*b*, ch. 22, p. 17*b*, ch. 23, p. 15*b*.

[c] Read (6), no. 147.

[1] 石魚 [2] 湘鄉縣 [3] 玄石 [4] 潭州 [5] 鰍鯽 [6] 隴西
[7] 歲久土凝爲石 [8] 朝野僉載 [9] 麟角集

As Tu Fu says, writing about Yü-lung: 'At night there is nothing but the sound of the water; the birds and mice of the empty mountain are all hiding away.'[a]

Here we find an unusually precise description of the geological location of the stone, and the account of a palaeontological test. This seems to have been necessitated by some considerable demand for the fossil fish, which encouraged forgeries, but it was aesthetic rather than medical, for they were not prominent in the *Pên Tshao* books. Or else it was because they were regarded as charms for good harvests, and placed in cupboards to keep away 'silver-fish' (*Lepisma*), bookworms, and other harmful insects.[b] The commonest fossil fishes in Chinese rocks are *Lycoptera* spp.[c]

The bones and teeth of fossil reptiles, birds and mammals were always known in China as 'dragon's bones' and 'dragon's teeth' (*lung ku*[1] and *lung chhih*[2]). What Li Shih-Chen says about them in the *Pên Tshao Kang Mu* has been translated by Read.[d] While most of them were undoubtedly of mammalian origin, and from a great variety of species (*Rhinoceros*, *Mastodon*, *Elephas*, *Equus*, *Hippotherium*, etc.) some of them were bones of deinosaurs and pterodactyls. The great esteem in which 'dragon's bones' were held medicinally[e] aided modern palaeontologists to their discoveries of fossil man in China (*Sinanthropus pekinensis*), as Davidson Black describes. Moreover, it was through the examination of drug-store material that the first discovery of the inscribed oracle-bones[f] was made just before the beginning of the present century.[g] It is an interesting fact that the incorporation of vertebrate fossils in the pharmacopoeia occurred earlier than in the case of others already described; for the bones are mentioned in works of Han or San Kuo period such as the *Shen Nung Pên Tshao Ching* and the *Ming I Pieh Lu*, while the teeth appear first in the Chin *Li shih Yao Lu*[3] (Mr Li's Record of Drugs). The *Chhien Han Shu* (*c.* +100) says in its chapter on irrigation and water-conservancy: 'A canal was dug at Chêng[4] to introduce the waters of the Lo[5] River to Shang-yen[6]....During the excavations dragon bones were found and therefore the canal was named the "Dragon-Head waterway".'[h] This was approximately in −133.[i] So also Wang Chhung says, about +83, that 'when the floods had been controlled and were flowing to the east, strange bones were found where the serpents and dragons had been; thus there was evidence of these queer creatures.'[j]

[a] Ch. 2, p. 1*a*, tr. auct.

[b] We know this from the *Lu-Ling I Wu Chih*[7] (Strange Things of Lu-ling), which refers to a stone-fish monument (probably a particularly fine section) seen in +971, which bore the date +763.

[c] Figured by Chang Hung-Chao (*1*), pls. IX, X.

[d] (4), nos. 102*a*, *b*. Cf. Hanbury (1), p. 273.

[e] Schlosser (1) describes the trade.

[f] Cf. above, pp. 83 ff. in Vol. 1, Sect. 5*b*.

[g] See Creel (1), p. 2, for the story.

[h] Ch. 29, p. 5*b*, tr. auct.

[i] Similar finds are being made by the hydraulic engineers of the great Huai River Conservancy Works nearly two thousand years later. The eminent palaeontologist Dr Phei Wên-Chung is leading a field survey of them (*NCNA*, 7 April 1954).

[j] *Lun Hêng*, ch. 9. Forke (4), vol. 1, p. 173, entirely misunderstood this passage. Our attention was drawn to it by Mr D. Leslie.

1 龍骨 2 龍齒 3 李氏藥錄 4 徵 5 洛 6 商顏
7 廬陵異物志

The common name shows that the fossils were always regarded as being the remains of animals long dead, even if somewhat mythological, but the existence of prehistoric monsters is clearly stated by Khou Tsung-Shih early in the +12th century.[a] Presumably the only pharmaceutical value which the bones ever had was as a source of lime, so it is curious that among the various methods of preparation which Li Shih-Chen gives, digestion with vinegar to produce an assimilable salt is not mentioned.

Some pages previously, we dropped the occidental story at the time of Ibn Sīnā and al-Bīrūnī. Since then we have seen that from the Han (about −1st century) onwards, the Chinese accumulated a mass of information about various fossil remains. In comparing their records with those of Western pre-Renaissance writers, one has the impression that they defined and recognised their material with a good deal more precision and comprehension, as well as appreciating something of the geological significance of fossils. What remains to be filled in is the contribution of the ancient Mediterranean civilisations. At this we can only glance, referring the reader to books such as those of Geikie and Adams, and to papers such as those especially devoted by v. Lasaulx and McCartney to 'fossil lore in Greek and Latin literature'.

The oldest Greek reference is certainly that ascribed to Xenophanes (*fl. c.* −530 but only known through a quotation in Hippolytus (+3rd century):

> Xenophanes says that the sea is salt because a great variety of mixed materials flow into it. He further says that the land and sea were once mixed up, and even thinks that the land is dissolved in course of time by moisture. For this he says he has the following proofs. Shells (of sea-animals) are found far inland and on mountains, and he tells us that in the stone-quarries at Syracuse imprints (remains) of fish and a certain kind of seaweed have been found, while at Paros in the depths of the rock there are impressions of sardines, and at Malta similar moulds of all kinds of marine creatures. He says that it follows from this that at one time all these lands were under water. After all things had been turned into mud, the impressions were dried out and consolidated. And men must all perish when the earth has been carried down into the sea and (again) become mud; and from that point generation will begin anew, and these changes take place in all the world.[b]

Here the echo of the Indian periodical cataclysms is interesting, but one can see that if the passage really belongs to Xenophanes the pre-Socratics were able to think along lines which seem very advanced when we meet with them in a Chu Hsi or a Leonardo. As for later references to fossils in Greek authors, some have proved to be mere misunderstandings,[c] but others are quite definite, as, for example, in Strabo,[d] Pausanias,[e]

[a] *Pên Tshao Yen I*, and copied in later compendia. The mythological aspect of the subject, about which there is, of course, a large literature (e.g. Hornblower, 2), is not really relevant here. The point is that the medieval Chinese scientists associated the animal in their minds with the genuine fossilised material.

[b] *Refutationes*, I, 14, 15, tr. Bromehead (1), Heath (4), K. Freeman (1), p. 103; mod. Although all western classical scholars seem to accept with little hesitation the genuineness of such quotations given after a lapse of eight centuries, one cannot help remarking that sinologists would show great reluctance to do so in a parallel Chinese case. The passage comes from a rather corrupt text, but the references given will indicate some of the emendations and the reasons for them.

[c] Bromehead (1), p. 104.

[d] *Geogr.* I, iii, 4.

[e] I, xliv, 6.

Pliny[a] and so on. Conclusions of a geological character, however, were not usually drawn; the facts were simply noticed as curiosities. Theophrastus (−370 to −287) started the idea of fish-spawn being scattered among the rocks and then petrifying, or of a special *vis plastica* which was able to bring about morphogenesis but not to confer life and movement on the product.[b] After the patristic period, the presence of fossil sea-shells in rocks was taken simply as evidence for the flood, as, for example, in Priscian's Latin translation of Dionysius Periegetes (*c.* +436)[c] or Isidore of Seville's *Etymologiae* (*c.* +630).[d] Finally, interest in fossils completely died out until the Renaissance.

The picture thus presented is very reminiscent of the Great Interruption which was so prominent in the history of quantitative cartography.[e] Brilliant insights or great achievements were attained by the Greeks, but from about the +2nd to the +15th century China was much more advanced than Europe, until modern science begins to appear. The pre-history of palaeontology also illustrates this, and it does not seem at all likely that any stimulus was received by the Chinese from the West at the beginning of their best period.

[a] *Hist. Nat.* XXXVII, x, 11.
[b] Geikie (1), p. 16.
[c] Bunbury (1), vol. 2, p. 685.
[d] XIII, 22, ii; cf. Kimble (1), p. 155.
[e] P. 587 above. And we can recognise a pattern which we have also seen in mathematics (e.g. p. 101), in astronomy (pp. 366, 379), in tidal theory (p. 493) and in human geography (pp. 514, 520).

24. SEISMOLOGY

(a) EARTHQUAKE RECORDS AND THEORIES

If China had no active volcanoes it nevertheless formed part of one of the world's greatest areas of seismic disturbance from the earliest times. It was natural, therefore, that the Chinese should have kept extensive records of earthquakes, and these indeed now constitute the longest and most complete series which we have for any part of the earth's surface.[a] Earlier lists, taken from the Chinese histories, have all been incorporated in those of Huang (2) and Anon. (*8*), but it is still good to read some of the original texts. The *Thu Shu Chi Chhêng* encyclopaedia, for example, devotes six chapters to excerpts from the dynastic annals concerning earthquakes.[b] From Omori's study we learn that up to +1644 there had been 908 recorded shocks for which we have precise data. Among the earliest was that of −780 mentioned in the *Shih Chi*,[c] when the courses of three rivers were interrupted. At Nanking between +345 and +414 there were 30 shocks, and between +1372 and +1644 no less than 110. But the main area always lay north of the Yangtze and in all the western provinces. From the records there emerge twelve peaks of frequency between the end of the Sung and the beginning of the Chhing, showing a 32-year periodicity. Earthquakes sometimes affected several provinces, but different regions did not usually have them simultaneously, and there seems to have been no time correlation between Chinese and Japanese earthquakes. One of the worst Chinese ones was that of 25 September +1303 in Shansi, while that of 2 February +1556 is said to have killed more than 800,000 people in Shansi, Shensi and Honan. A famous siege of Sian in +1128 was ended by an earthquake.

No great progress was made in ancient or medieval China regarding the theory of earthquakes, which indeed in Europe also had to await post-Renaissance conceptions of the nature of the earth's crust. However, in connection with the −8th-century earthquake already mentioned, we find an early statement of ideas concerning them. The *Shih Chi* says:

In the second year of the reign of King Yu[1] (of Chou), the three rivers of the western province were all shaken and their beds raised up. Poyang Fu[2] said: 'The dynasty of the Chou is going to perish. It is necessary that the *chhi* of heaven and earth should not lose their order (*pu shih chhi hsü*[3]); if they overstep their order (*kuo chhi hsü*[4])[d] it is because there

[a] Anyone with intimate acquaintance in China will have come across personal experience of seismic phenomena. One of the writer's best friends was made an orphan by the loss of all his family when their cave-dwelling home in Honan was destroyed in an earthquake about 1930.

[b] *Shu chêng tien*, chs. 115–21. Among other accounts of earthquakes may be mentioned those in chs. 34–6, 88 and 89 of the *Hsin Thang Shu*, translated by Pfizmaier (67).

[c] Ch. 4, p. 24*b*. Huang (2), vol. 2, p. 45.

[d] Cf. Heraclitus on the sun 'not transgressing his measures', in Vol. 2, pp. 283, 533 above.

1 幽　2 伯陽甫　3 不失其序　4 過其序

is disorder among the people. When the Yang is hidden and cannot come forth, or when the Yin bars its way and it cannot rise up, then there is what we call an earthquake (*ti chen*[1]). Now we see that the three rivers have dried up by this shaking; it is because the Yang has lost its place and the Yin has overburdened it. When the Yang has lost its rank and finds itself (subordinate to) the Yin, the springs become closed, and when this has happened the kingdom must be lost. When water and earth are propitious the people make use of them, when they are not, the people are deprived of what they need. Formerly when the rivers I and Lo dried up, the dynasty of the Hsia perished. When the Ho dried up, the dynasty of the Shang perished. Now the virtue of the Chou is in the same state as that of these dynasties was in their decline.... The Chou will be ruined before ten years are out; so it is written in the cycle of numbers[a] (*shu chih chi yeh*[2]).[b]

Apart from the usual aspect of State prognostication[c] in Poyang Fu's discourse, its interest lies in what it says about the Yin and Yang.[d] Similar ideas were prevalent throughout the Chhin and Han periods,[e] as we know from the *Lü Shih Chhun Chhiu*[f] and the *Lun Hêng*;[g] it was thought, too, that earthquakes could be predicted astrologically. The theory of the imprisoned Yang was embodied in the *I Ching* (Book of Changes) under the fifty-first hexagram *chen*,[3] which could refer either to thunder or earthquake shocks, and will still be found in Sung books, such as the *Yü Thang Chia Hua*[4] (+1288)[h] by Wang Yün.[5]

Let it not be thought that these theories were more primitive than those which the ancient Mediterranean world entertained on the subject of earthquakes. Lones has summarised what was thought in Aristotle's time and before it. The *Meteorologica*[i] recapitulates previous explanations—Anaxagoras believed that earthquakes were caused by excess of water from the upper regions bursting into the under parts and hollows of the earth; Democritus thought that this happened when the earth was already saturated with water, and Anaximenes suggested that the shocks were caused by masses of earth falling in cavernous places during the processes of drying. Aristotle himself in the −4th century attributed the instability to the vapour (*pneuma*, πνεῦμα) generated by the drying action of the sun on the moist earth, and to difficulties met with by the vapour in escaping.[j] This is closely parallel to Chinese ideas of the

[a] Cf. p. 402 above; *chi* as a technical term for the Jupiter Cycle, which is perhaps what Poyang means here.

[b] Tr. Chavannes (1), vol. 1, p. 279; eng. auct. Parallel texts, *Kuo Yü*, Chou Yü, ch. 1, pp. 20*a* ff.; *Chhien Han Shu*, ch. 27C, p. 5*b*.

[c] Cf. *Khai-Yuan Chan Ching*, ch. 4, p. 3*a*.

[d] In view of what has been said above (Sect. 13*c*, *e*, Vol. 2, pp. 241 ff.), it seems unlikely that anyone in the −8th century would have appealed to Yin and Yang in quite so definite a way, but the speech attributed to Poyang Fu by Ssuma Chhien would be quite characteristic of the time of Tsou Yen or somewhat before.

[e] The *Khai-Yuan Chan Ching* (+718) quotes (ch. 2, p. 3*a*) the *Hsing Ching* as saying that earthquakes were due to excess of Yin, and menaced princes.

[f] Ch. 29, tr. R. Wilhelm (3), p. 74.

[g] Chs. 17, 42, 43, 58; tr. Forke (4), vol. 1, pp. 112, 127; vol. 2, pp. 160, 211.

[h] 'Agreeable Talks in the Jade Hall', ch. 2, p. 2*a*.

[i] II, vii, viii (365*a* 14 ff.).

[j] The lively statement of this theory by Shakespeare will be remembered (*King Henry IV* (Pt. I), Act. III, sc. 1, ll. 25 ff.).

1 地震 2 數之紀也 3 震 4 玉堂嘉話 5 王惲

imprisonment of *chhi*. A few pages below, we shall have occasion to notice an even closer parallel between Aristotelian and Chinese ideas on the generation of rocks and ores, both being expressed in pneumatic terms. No further progress was made in seismic theory until modern times.

But the importance of pneumatic ideas equally in China and Europe is quite striking. Seneca, writing just over ten years before the birth of Chang Hêng, says:

> The chief cause of earthquake is air, an element naturally swift and shifting from place to place. As long as it is not stirred, but lurks in a vacant space, it reposes innocently, giving no trouble to objects around it. But any cause coming upon it from without rouses it, or compresses it, and drives it into a narrow space...and when opportunity of escape is cut off, then 'With deep murmur of the Mountain it roars around the barriers', which after long battering it dislodges and tosses on high, growing more fierce the stronger the obstacle with which it has contended.[a]

One might be reading Ching Fang or any other Han mutationist.

(*b*) THE ANCESTOR OF ALL SEISMOGRAPHS

But if theory did not advance, China has unquestionably the credit of having produced the ancestor of all seismographs. This was due to the brilliant mathematician, astronomer and geographer, Chang Hêng[1] (+78 to +139), whom we have already met with in these other connections.[b] Many modern Western seismologists, such as Milne,[c] Sieberg and Berlage, have freely recognised[d] the great merit which belongs to Chang Hêng in this field.[e] Modern Asian scholars, themselves geophysicists, such as Li Shan-Pang and Imamura, have studied his work, while others have given detailed biographies of him.[f]

The basic text which describes Chang Hêng's instrument is contained in the biographical chapter devoted to him in the *Hou Han Shu*.[g] The information which it gives is fortunately rather detailed, and the passage has been repeatedly translated or paraphrased into Western languages.[h] The following version attempts to combine the

a *Quaestiones Naturales*, tr. Clarke & Geikie (1), p. 247. The verse quoted is Virgil, *Aeneid*, 1, 55–6.

b Cf. pp. 100, 343, 363 and 537 above.

c But the description of Chang Hêng's instrument was suppressed from later editions; one must use that of 1886.

d Davison's work on the founders of seismology, however, does not mention Chang Hêng, and it is particularly strange that Knott's does not do so either, since Knott (like Milne) worked in Japan.

e Here a strange circumstance must be mentioned. Sarton (1), in his great *Introduction*, refers the invention of the seismograph to Chhao Tsho,[2] who lived in the −2nd century (vol. 1, p. 196). The only authority cited is Sieberg, who describes (p. 211) Chang Hêng's instrument, erroneously referring it to −136 instead of +132, and giving only his Japanese name (or rather a distortion of it) 'Chiocho'. So Milne and Shaw refer only to 'Choko', and a writer in *Nature* in 1939 knows only 'Tyoko', which is Imamura's version. Now the biography of Chhao Tsho in *Chhien Han Shu*, ch. 49, says nothing about a seismograph or anything remotely connected with it. We believe, therefore, that Homer must here have nodded, and that the whole entry should be omitted from the next edition of the *Introduction*.

f Sun Wên-Chhing (*2, 3, 4*); Chang Yin-Lin (*4*); Chang Yü-Chê (1, 2).

g Ch. 89, p. 9*b*, cit. *TPYL*, ch. 752, p. 1*b*.

h Pelliot & Moule (1); Imamura (1); Forke (6); H. A. Giles (5); Waley (11); Milne (1).

1 張衡 2 鼂錯

best elements in previous renderings, all of which have been carefully compared with the text:

In the first year of the Yang-Chia reign-period (+132) Chang Hêng also invented (*tsao*[1]) an 'earthquake weathercock'[a] (*hou fêng ti tung i*[2]—i.e. a seismograph).[b]

It consisted of a vessel of fine cast bronze, resembling a wine-jar, and having a diameter of eight *chhih*[3].[c]

It had a domed cover, and the outer surface was ornamented with antique seal-characters, and designs of mountains, tortoises, birds and animals.

Inside there was a central column capable of lateral displacement along tracks in the eight directions, and so arranged (that it would operate) a closing and opening mechanism (*Chung yu tu chu, pang hsing pa tao, shih kuan fa chi*[4]).[d]

Outside the vessel there were eight dragon heads, each one holding a bronze ball in its mouth, while round the base there sat eight (corresponding) toads with their mouths open, ready to receive any ball which the dragons might drop.

The toothed machinery (*ya chi*[5])[e] and ingenious constructions were all hidden inside the vessel, and the cover fitted down closely all round without any crevice.

When an earthquake occurred the dragon mechanism of the vessel was caused to vibrate so that a ball was vomited out of a dragon-mouth and caught by the toad underneath. At the same instant a sharp sound was made which called the attention of the observers.

Now although the mechanism of one dragon was released, the seven (other) heads did not move, and by following the (azimuthal) direction (of the dragon which had been set in motion), one knew (the direction) from which the earthquake (shock) had come (lit. where the earthquake was). When this was verified by the facts there was (found) an almost miraculous (*shen*[6]) agreement (i.e. between the observations made with the apparatus and the news of what had actually happened).

Nothing like this had ever been heard of before since the earliest records of the *Shu* (*Ching*).

On one occasion one of the dragons let fall a ball from its mouth though no perceptible shock could be felt. All the scholars at the capital were astonished at this strange effect occurring without any evidence (of an earthquake to cause it). But several days later a messenger arrived bringing news of an earthquake in Lung-Hsi (Kansu).[f] Upon this every-

[a] Pelliot was puzzled here by the reference to winds, and thought that they might have something to do with earthquakes, but the point is really quite clear; Chang Hêng's apparatus determined the azimuth direction in which the earthquake's epicentre lay, just as the weathercock determined the direction of the wind. Chu Kho-Chen (2) has urged that two separate instruments are referred to here, a weathercock (*hou fêng i*)—which he interprets as some kind of anemometer—and the seismograph (*ti tung i*). Dr Chu feels that this is justified by passages from the *San Fu Huang Thu* which we have already noticed (see p. 478 above) in connection with weathercocks (private communication), but we, at any rate, have not been able to find any reference to the seismograph in the text of this Chin book.

[b] Others have carefully written 'seismoscope' instead of 'seismograph', but the following description will, I think, justify the use of the latter term.

[c] As the Later Han *chhih* was just on nine of our inches, the diameter would be six feet.

[d] This is the key sentence of the whole. We have assumed that *kuan* and *fa* are antitheses, but *fa-chi* was the technical term for the contemporary crossbow trigger, a beautiful piece of mechanism in cast bronze (cf. below, Sect. 30). If this sense is preferred the interpretation of Wang Chen-To is strengthened.

[e] This need not mean toothed gearwheels. Both Wang Chen-To and Imamura need teeth or pins in one way or another, as will be seen.

[f] About 400 miles away to the north-west. Ong Wên-Hao (2) has made a study of the earthquakes in Kansu province, and gives a register of their recorded occurrences (−780 to +1909).

1 造 2 候風地動儀 3 尺 4 中有都柱傍行八道施關發機
5 牙機 6 神

one admitted the mysterious (power of the instrument). Thenceforward it became the duty of the officials of the Bureau of Astronomy and Calendar to record the directions from which earthquakes came.[a]

It will be agreed that this passage is one of the greatest interest. Let us see in what way it is possible to reconstruct the instrument of Chang Hêng. We owe to Wang Chen-To (*1*) an interesting attempt, but what is particularly valuable in these matters is the approach of men who are themselves practical experimentalists in the fields concerned—in this case Milne and Imamura.

First, Milne saw correctly that the 'central column' was essentially a pendulum. Both he and Wang Chen-To assumed it to be a suspended pendulum, while Imamura

Fig. 267. A reconstruction of the external appearance of the first of all seismographs, that of Chang Hêng (+132), by Li Shan-Pang.

believes that it must have been an inverted one. All the seismologists have appreciated the technical difficulty of constructing an apparatus in which only one of the balls should drop out (and thus 'write' its record of the phenomenon), since there are always, besides the main longitudinal shock-wave, other components, mostly lateral, some of which may be of great force. It was necessary therefore to include some arrangement whereby the apparatus would be immobilised immediately after giving its first response.

A reconstruction of the external appearance of the apparatus by Li Shan-Pang is given in Fig. 267. Fig. 268 then shows the reconstruction of the internal mechanism proposed by Wang Chen-To. The pendulum, heavy and fat, carries eight arms radiating in the eight azimuthal directions. Each of these bears at its end a vertical pin

[a] Tr. auct.; adjuv. Pelliot & Moule (1); Imamura (1); Forke (6), p. 19; Pfizmaier (92), p. 148; H. A. Giles (5), p. 277.

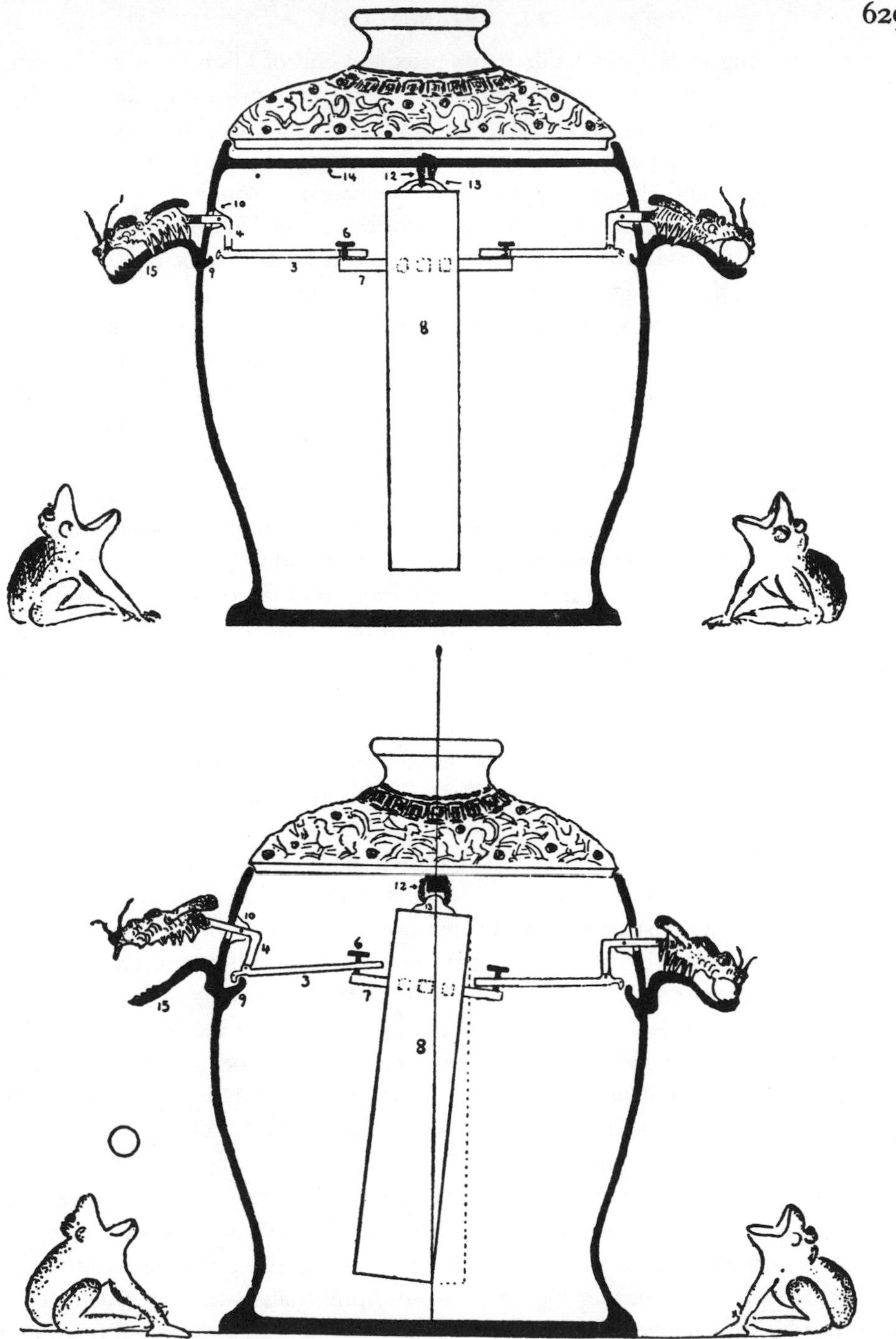

Fig. 268. An attempted reconstruction of the mechanism of Chang Heng's seismograph by Wang Chen-To (*1*). The pendulum carries eight mobile arms radiating in as many directions and each connected with cranks which are provided with catch mechanisms at the periphery. Any one of the cranks which raises a dragon head and so releases a ball is thus at the same time caught and held, thus immobilising the instrument. 3, crank; 4, right-angle lever for raising the dragon's head; 6, vertical pin passing through a slot in the crank; 7, arm of pendulum; 8, pendulum; 9, catch; 10, pivot on a projection; 12, sling suspending the pendulum; 13, attachment of sling; 14, horizontal bar supporting the pendulum; 15, lower jaw of dragon supporting ball.

loosely engaging with a small slot at the proximal end of a long crank. The crank is pivoted at the further end, so that when the pendulum swings in that direction the dragon's upper jaw is raised, releasing the ball from the mouth, while at the same time a hook on the crank catches on to a projection or annular rim attached to the inside of the vessel's wall, and thus immobilises the mechanism. Much of Wang Chen-To's paper is devoted to showing that the people of the Later Han were able to construct such a mechanism, as is proved by the levers and cranks used in contemporary devices such as the crossbow trigger. Here he is certainly right.

Imamura's contention would be, however, that the immobilisation mechanism proposed by Wang would not have sufficed to prevent further motion due to transverse shock components. He favours a pendulum of inverted type (see Fig. 269).[a] A pendulum of circular cross-section with a heavy bob stands on a base-plate; its diameter is 3 cm., its centre of mass 17 cm. high, and the minimum acceleration required to overturn it is 8·7 gals.[b] Above the bob the pendulum is prolonged into a pin (perhaps the 'tooth' of the text), which passes through the central hole of two diaphragms fixed horizontally in the upper part of the vessel. Both of these are provided with eight slots or guides (perhaps the 'tracks' implied in the text), the important point being that once the pin has entered one of these as a result of the initial shock, it is unable to respond to subsequent shock components in such a way as to knock out any further balls. The transmission of the pin's movements to the ball in question is effected simply by laying sliders on the lower slotted plate; these are each available to give a push to each of their respective balls, which would normally be held in place by very small indentations in the bronze or some similar device. The only linguistic difficulty about Imamura's view is that the text distinctly says 'the seven (other) heads did not move', implying that one of them did; but perhaps we should not insist upon so strict an interpretation of Fan Yeh's words. In any case, it is of much interest that Imamura and Hagiwara constructed a model instrument on this pattern and tested it out at the Seismological Observatory of Tokyo University. It was found that the ball was usually released not by the initial longitudinal wave, but by secondary transverse waves; though if the first shock was very strong it would do. It may well have been necessary, therefore, for Chang Hêng to calibrate his instrument empirically. The historiographers would hardly be likely to have taken note of the fact (if such it was) that in some circumstances one had to take the direction at right angles to the dropped ball as that of the earthquake's epicentre. Imamura points out that by estimating the duration of the preliminary tremors, and consequently the focal distance, one could give a rough estimate, not only of the direction, but also of the distance, of the earthquake. It will have been noticed that the text is a little confused as to these two determinations.

The principle of recording by means of dropping balls was one in which Chang Hêng had been anticipated by Heron of Alexandria (*fl.* +62), who used them in some of his hodometers[c] (cf. Sect. 27*c* below). It is interesting to note, however, that they

[a] Subsequently Wang Chen-To (*6*) also adopted the hypothesis of an inverted pendulum, as may be seen from his latest model (Fig. 270).

[b] 1 gal. = 1/1000 g.

[c] Cf. Diels (1); Beck (1), p. 53.

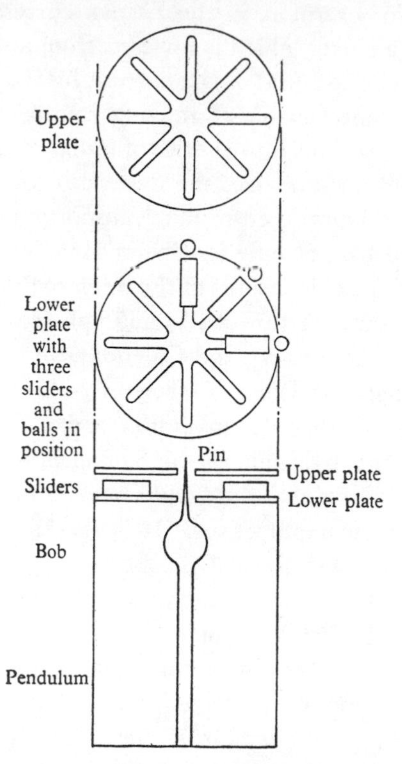

Fig. 269. Imamura Akitsune's reconstruction of the seismograph of Chang Hêng employing the principle of an inverted pendulum. When knocked over by an earth tremor the pin at the top must enter one or other of the slots provided and expel a ball by pushing one or other of the eight sliders. It can then no longer leave the slot and the instrument is immobilised.

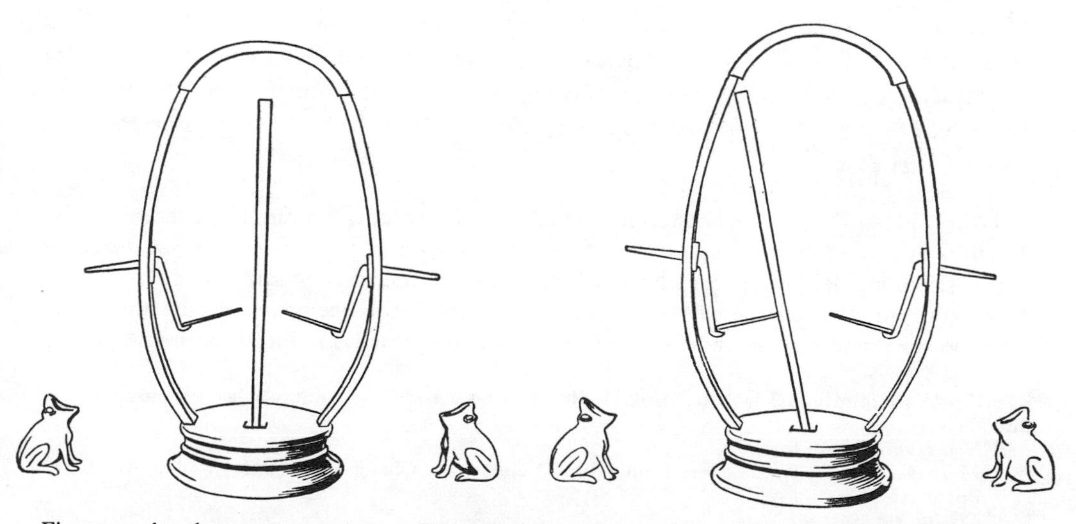

Fig. 270. Another reconstruction of Chang Hêng's seismograph by Wang Chen-To (*6*), accepting Imamura's principle of the inverted pendulum. Only the skeleton of the instrument is shown here, Wang Chen-To still favouring lever systems rather than slots and sliders.

are still used today, as, for example, in the Ekman current meter. This consists of a mechanism attached to a sinker which is lowered from a ship on the end of a cable. A propeller (idling) records the speed of the current by the number of its revolutions in a given time, while a vane capable of moving round a vertical axis controls the periodical fall of small bronze balls into a series of radial compartments. The statistical curve of their distribution in these gives the mean direction of the current.[a]

That his instrument was considered of much importance at the time is clear from the fact that it is referred to, not only in Chang Hêng's biography, but also in the 'Annals' chapter in which the chief events of the reign of the emperor Shun are given.[b] It is merely said that for the first time a seismograph (*hou fêng ti tung thung i*[1]) was installed in the Bureau of Astronomy and Calendar, and the commentary adds that Chang Hêng, who was Thai Shih Ling[2] (Astronomer-Royal) at the time, made it.[c] We ought not to miss the point that the invention had a certain connection with the centralisation of government; by its aid the high officials would have advance notice of an earthquake in a distant province, and would be able to take measures to deal with needs or disturbances which might ensue. It is therefore a parallel with the rain-gauges and snow-gauges noticed above.[d]

From the *Hou Han Shu*[e] we know something of the seismic shocks which occurred during the period of Chang Hêng's lifetime. Since +46 there had been twenty-five important earthquakes in more than fifty commanderies and provinces. In the decades before Chang Hêng's invention was perfected, the capital itself (Sian) had been rocked thrice. In +133 and +135, just afterwards, there were again two earthquakes which did damage at the capital. A bad one centred in Lung-Hsi (Kansu) came in +138, and this (or its associated tremors) may well have been the occasion of the striking test of the delicacy of the instrument.

Those who mention the seismograph of Chang Hêng generally imply that it was a chance achievement which had no subsequent history. This does not seem to be true, however, for we can find at least two references to similar instruments in later centuries. Hsintu Fang[3] we have already met as a mathematician;[f] he flourished in the latter half of the +6th century and served the Northern Chhi dynasty. The *Pei Chhi Shu* tells us that

Hsintu Fang when young showed mathematical ability and was praised by everyone in his locality. A man of great ingenuity, he was often absorbed in high contemplation, forgot to eat or sleep, and fell into holes when walking about. Sometimes he said to his friends that when concentrating on the mysterious refinements of the mathematical art he would not even notice thunder. His knowledge of techniques (eventually) induced the emperor Kao

[a] For this information I have to thank Dr Ronald Fraser and Rear-Admiral Day, Hydrographer of the Navy.
[b] *Hou Han Shu*, ch. 6, p. 8*a*.
[c] Moreover, it was mentioned in Tshui Yuan's[4] epitaph on Chang Hêng (*Chin Shu*, ch. 11, p. 3*a*).
[d] Pp. 471 ff.
[e] Ch. 26, pp. 1–5.
[f] P. 358 above.

[1] 侯風地動銅儀 [2] 太史令 [3] 信都芳 [4] 崔瑗

Tsu (of the Sui) to entertain him as a client....He wrote a book called *Chhi Chun*[1] (Specifications (or Calibration) of Instruments) which included descriptions of armillary spheres, the 'earthquake (weathercock)' (seismograph), hydrostatic vessels[a] (*wo chhi*[2]), water-clocks, and so on, all illustrated by diagrams....[b]

Thus if he was able to explain the principles of the seismograph and enjoyed imperial favour, the presumption that he made and used one is not unreasonable. Indeed the fuller account in the *Pei Shih* explains[c] that in earlier life Hsintu Fang was a client in the household of a certain prince An-Fêng Wang.[3] This man, whose personal name was Kao Yen-Ming,[4] had collected a large library of books on mathematics and science as well as the classics, and had in his palace all kinds of scientific instruments. He asked Hsintu Fang to collaborate in making calculations, but before long Kao Yen-Ming had to flee to the south and Hsintu was obliged to write the book alone.[d]

In the succeeding generation Lin Hsiao-Kung[5] took up the subject. The *Sui Shu* says:

Lin Hsiao-Kung was adept at astronomy and mathematics....The emperor Kao Tsu treated him intimately and with all hospitality. He proved very successful in predicting fortune and misfortune, and the emperor entrusted him with all Yin-Yang matters....He wrote the *Wo Chhi Thu*[6] (Diagrams of Hydrostatic Vessels) in three chapters, the *Ti Tung Thung I Ching*[7] (Manual of the Bronze Earthquake (-Indicating) Instrument) in one chapter, and the *Thai I Shih Ching*[8] (Great Unity Divining-Board Manual) in thirty chapters....[e]

Lin was working during the period +581 to +604. As regards the actual construction of the seismograph the same remarks apply as in the case of Hsintu half a century earlier. It should, moreover, be remembered that this was just the period when Kêng Hsün and others were making particularly complicated water-clocks with balance adjustments,[f] so that there is no need to doubt mechanical ingenuity and ability at this time, though before the unification by the Sui (+581) conditions were doubtless unsettled.[g]

The seismograph seems not to have outlasted the Thang, however. In a passage from the Yuan book *Chhi Tung Yeh Yü* by Chou Mi (*c.* +1290), quoted elsewhere,[h] it is interesting to find that, while the principles of all other instruments had been

[a] See below, Sect. 26*c*.
[b] Ch. 49, p. 4*a*, tr. auct.
[c] Ch. 89, p. 13*a*.
[d] It will be remembered (p. 358) that Hsintu Fang was a pupil of Tsu Kêng-Chih, whom he had befriended when an exile at the prince's court.
[e] Ch. 78, p. 16*a*, tr. auct. Cf. *Pei Shih*, ch. 88, p. 37*a*.
[f] Cf. p. 327 above.
[g] In the Sui bibliography we find a *Ti Tung Thu*.[9] This may have been either another description of the instrument, or possibly some kind of divinatory diagrams for earthquake prediction, or even a map on which seismic disturbances were plotted.
[h] Ch. 15, p. 6*b*, tr. Needham, Wang & Price. Cf. p. 671 below.

1 器準 2 欹器 3 安豐王 4 高延明 5 臨孝恭
6 欹器圖 7 地動銅儀經 8 太一式經 9 地動圖

preserved, that of the seismograph had been lost and could not be recovered. Chou Mi mentions the two books of Hsintu Fang and Lin Hsiao-Kung, but says that both had disappeared and that no example of the instrument had survived. He goes on:

> Judging from general principles, I see very well that celestial phenomena have definite (quantitative) rules to follow (*thien wên yu kuei tu kho hsün*[1]) so that (it is possible to make instruments to measure) hours and quarters succeeding one another without the slightest inaccuracy. For example, the armillary sphere can measure (these movements). But earthquakes come from (unpredictable and) unmeasurable (*pu tshê*[2]) collisions of the Yang and the Yin. Take the case of the body of a man; the blood and the *chhi* are sometimes in accord (*shun*[3]) and sometimes in opposition (*ni*[4]), hence the flesh responds, and he winks his eyes or moves his ears. If the *chhi* reaches (the vital point) he moves; if it does not, he is motionless. And yet this instrument (the seismograph) is said to have been placed in the capital far away from the place where the earthquake occurred. How could the collision of the two *chhi* make the bronze dragons vomit forth the balls? I cannot at all make out the principle, and would very much like to find someone who understands it.[a]

This passage is particularly interesting because the difficulty felt by Chou Mi was perhaps what we should call the distinction between phenomena which show great natural regularity, and others which require statistical methods for their treatment. He could not understand how an instrument could, as it were, 'lie in wait' for an irregularly recurring phenomenon, especially at a distance. In other words he could only visualise repetition as given in celestial recurrences. He did not realise that man himself (as we saw on p. 160 above) must organise repetitions if there was to be fruitful experimentation on earth, and only a recording machine planned and tested in this way could register fitful phenomena. Perhaps significantly Chou Mi was a scholar but Hsintu Fang was a practical man much nearer to the artisans. The passage also shows, by its significant biological and microcosmic analogy, that the ancient pneumatic theory was exerting a positively inhibitory influence on scientific thinking. Finally, the account of Chang Hêng's seismograph came to be frankly disbelieved, as we may see from the Chhing book *Chhiao Hsiang Hsiao Chi*[5] (Woodsmoke Perfume Jottings)[b] of Ho Hsiu,[6] who reproduces, in the +18th century, exactly the same argument as Chou Mi in the +13th.[c]

Possibly the seismographic pendulums of +6th-century China found their way somehow to the West, for we hear of seismographic instruments at the Marāghah observatory in +13th-century Persia.[d] Then again there is a great gap until 1703 when de la Hautefeuille set up the first modern seismograph.[e] The principle adopted was

a Tr. auct.

b Ch. 2, p. 9*a*.

c 'A fixed thing cannot measure that which is essentially non-fixed' (*yu ting pu nêng tshê wu ting*[7]).

d Köprülü (1); Godard (1).

e Technical terms for types of shock and tremor date only from Galesius, *De Terraemotu Liber*, +1571 (Adams (1), p. 405). First estimate of velocity of propagation of the wave was that of Michell, +1761 (Adams (1), p. 415). First intensity scale was that of Domenico Pignataro, +1788 (Adams (1), p. 420).

1 天文有晷度可尋 2 不測 3 順 4 逆 5 樵香小記
6 何琇 7 有定不能測無定

the spilling of an overfilled dish of mercury—this was used throughout the 18th century, and by Cacciatore as late as 1848.[a] Only since that time have the extremely delicate and complex mechanisms current today been developed.

It would be interesting to know whether the Thang seismographs persisted in Japan. Instruments of apparently traditional type there in which a pendulum carries pins projecting in many directions and able to pierce a surrounding paper cylinder, have been described.[b]

[a] Von Zittel (1), p. 282; Berlage (1). A good account of the development will be found in Boffito (1), pp. 143 ff., who illustrates (pl. 97) the mercury seismograph of Atanasio Cavalli (+1785). It is strange that Chang Hêng's principle was more like that of modern than of 18th-century seismographs.
[b] Horwitz (7); Kreitner (1), p. 293.

25. MINERALOGY

(a) INTRODUCTION

As was stated at the beginning of this Section, the systematic study of minerals and rocks is largely a modern and post-Renaissance development; it could not flourish until the basic principles of crystallography had been worked out, and the microscopical study of rock sections had become possible. Nevertheless, in pre-Renaissance times the Chinese contributed at least as much as Europeans to what there was of scientific and dispassionate study of stones and mineral ores.

It is at this point that we come, for the first time in this book, to a science which was, at least in its earlier phases, purely descriptive. This confronts us with special difficulties, for it would obviously be impossible, within the limits of so restricted a survey as ours, to deal exhaustively with all the individual members of the class of inorganic mineral substances which were studied. At the same time they cannot be dismissed in too summary a way, since it is necessary to give some idea of the extent to which mineralogy (as also later on, botany and zoology) was cultivated. I shall therefore select a certain number of substances for brief discussion, choosing those on which historians of science have thrown most light. Something will also be said, whenever a descriptive science has to be dealt with, of the principles and methods of the classificatory systems which were employed. Then it will be necessary to attempt some comparison of the total number of substances or species which were recognised and named, in the lapidaries, herbals, bestiaries, or pharmaceutical compendia of East and West. This implies short surveys of the surviving literature. Moreover, the practical use of such knowledge in applied arts such as mining and medicine must not be overlooked. But first of all, let us glance at the most important of the general theories about minerals which were entertained in ancient and medieval China.

(b) THE THEORY OF *CHHI* AND THE GROWTH OF METALS IN THE EARTH

It has already been pointed out (p. 469 above) in the Section on meteorology, that for Aristotle[a] there were two emanations or exhalations (*anathumiaseis*, ἀναθυμιάσεις) of the earth, given off under the influence of the sun. One kind, derived from the moisture within the earth and on its surface, is a moist vapour 'potentially like water' (indeed the evaporating water which forms the clouds and returns as rain). The other, which comes from the earth itself, is hot, dry, smoky, highly combustible 'like a fuel', 'the most inflammable of substances', and 'potentially like fire'. Aristotle's doctrine of the dry gaseous exhalation (*pneumatodestera*, πνευματωδεστέρα), and its relation to the formation of minerals, has lately been elucidated by Eichholz. Metals were sup-

[a] *Meteorologica*, I, iv (341 b 6 ff.).

posed to be congealed from the moist exhalation,[a] and all other minerals and rocks by the dry. The exact nature of the processes involved, as Aristotle describes them, remains obscure, but it seems that while the moist exhalation was thought to be the material cause of the metals, the dry one was rather the efficient cause of the minerals. However this may be, the important point for us is the 'pneumatic' character of the doctrine; the *chhi*[1] of Chinese writers seems a conception quite parallel to Aristotle's 'exhalations'.[b]

The preface to chapter 8 of the *Pên Tshao Kang Mu* opens thus:

Stone is the kernel of the *chhi* and the bone of the earth. In large masses it forms rocks and cliffs, in small particles it forms sand and dust. Its seminal essence (*ching*)[2] becomes gold and jade; its poisonous principle becomes arsenolite (*yü*[3]) and arsenious acid (*phi*[4]). When the *chhi* becomes congealed (*ning*[5]) it forms cinnabar and green vitriol. When the *chhi* undergoes transformation, it becomes liquid and gives rise to alums and to mercury. Its changes (are manifold), for that which is soft can become hard, as in the case of milky brine which sets to rock (-salt); and that which moves can become immobile, as in the case of the petrifaction of herbs, trees, or even of flying or creeping animals, which once had animation yet turn into that which has it not. Again, when thunder or thunderbolts turn to stones, there is a transformation of the formless into that which has form....[c]

This is surely the same doctrine as that of Greek 'exhalations'. Probably both derive from sources more ancient still, perhaps Babylonian, and certainly finding expression in the *prāṇa* of Sanskrit texts anterior to both Greece and China. The quotation just given was of course from the +16th century, but we can go back through the +11th[d] to the +5th, and then to the −2nd. In thus surveying this field of ideas we find at once that the theory of exhalations was intimately connected with a twin theory of the gradual growth of minerals and metals in the earth, and of their transformation into one another. 'The theory of exhalations', says Berthelot,[e] 'was the point of departure for later ideas on the generation of metals in the earth, which we meet with in Proclus,[f] and which reigned throughout the middle ages.' In so far as they recognised what we should now call processes of slow chemical change in the constituents of the earth's crust, they were not so far wrong; where they erred was in the assumption that metallic and other elements themselves changed into one another. This was what Ibn Sīnā was one of the first to doubt, in his famous passage against alchemy.[g] But nevertheless the conviction that minerals and metals did change into one another while slowly

[a] They all had the element of water, as their fusibility showed.

[b] The exhalations persisted until the +16th century and even later in the miners' lore of 'Witterung' or 'shimmering', an atmospheric phenomenon which was supposed to indicate the position of ore veins (Sisco & Smith (1), p. 33; Adams (1), p. 301). It would be interesting to compare the concepts revealed by the new dictionary of colloquial and dialect rock terms used by miners and quarrymen (Arkell & Tomkeiev, 1) with those of parallel Chinese practical men and writings. See below, pp. 673 ff.

[c] Tr. auct., adjuv. de Mély (1), pp. xxiii, 3. Often quoted elsewhere, as in *San Tshai Thu Hui*, ch. 59.

[d] Cf. the passage from Shen Kua, *MCPT*, ch. 25, para. 6, quoted above (Sect. 13*d*, Vol. 2, p. 267) in connection with the Five Elements.

[e] (2), p. 260.

[f] +410 to +485.

[g] Holmyard & Mandeville (1), p. 41.

[1] 氣 [2] 精 [3] 礜 [4] 砒 [5] 凝

growing in the earth was powerfully effective in encouraging the belief that by suitable methods alchemists could succeed in accelerating similar changes under laboratory conditions.

Li Shih-Chen wrote:

> It is said in the *Ho Ting Hsin Shu*[1] (New Book of Poisonous Substances)[a] that copper, gold and silver have a common root and origin. That which obtains the *chhi* of the purple Yang[b] produces green (matter),[c] and after two hundred years this becomes stone; in the midst of this the copper grows. It is because its *chhi* possesses a Yang (nature) that it is so hard and tough.[d]

This takes us back to the +12th century or somewhat before, a time when poets such as Chang Tzu[2] were making use of these ideas in their writings.[e] Similar transformations were also thought to occur with the sulphides of arsenic. Li Shih-Chen says:

> Orpiment[f] is formed on the Yin side of mountains, therefore it is called 'female yellow' (*tzhu huang*[3]). The *Thu Hsiu Pên Tshao*[4] (Earth's Mansions Pharmacopoeia)[g] tells us that if the petrifying *chhi* of the Yang is not sufficient, a female (*tzhu*[5]) mineral is formed; if it is sufficient, a male (*hsiung*[6]) mineral is formed. They take five hundred years to consolidate and form minerals. During this time they react (lit. carry out the role of husband and wife), which is why they are called male and female.[h]

That this goes back at least to the +5th century we know from a quotation in the *Thai-Phing Yü Lan* encyclopaedia from a lost book, the *Tien Shu*[7] (Management of all Techniques), by a prince of the Liu Sung Dynasty, Chien Phing Wang[8] (*fl.* +444).[i]

> The most precious things in the world are stored in the innermost regions of all. For example, there is orpiment (*tzhu huang*[8]). After a thousand years it changes into realgar[j] (*hsiung huang*[9]). After another thousand years the realgar becomes transformed into yellow gold.[k]

[a] This is a very obscure book. Li Shih-Chen quotes it among his authorities (ch. 1A, p. 37*b*), but its date is not known to us. Chêng Chhiao in the +12th century wrote a *Ho Ting Fang*[10] (Prescriptions involving Poisons). *Ho ting* was a proverbial term for a poisonous substance.

[b] I suspect this may refer to cinnabar. It was a term of alchemical importance sometimes taken as a personal pseudonym (see below, Sect. 33).

[c] Copper carbonate, no doubt.

[d] *PTKM*, ch. 8, p. 12*a*, tr. auct., adjuv. de Mély (1); cf. de Mély (1), pp. xxv 21; text, p. 18.

[e] *Nan Hu Chi*, ch. 3, p. 1*b*.

[f] Arsenious sulphide.

[g] This again is a very obscure book, not mentioned in any of the official bibliographies, and not in the *Tao Tsang*. Li Shih-Chen cites it in his own bibliography as the *Thu Hsiu Chen Chün Tsao-Hua Chih-Nan*[11] (Guide to the Creation, by the Earth's Mansions Immortal), ch. 1A, p. 17*a*. It was presumably a Taoist book of Sung or Ming.

[h] *PTKM*, ch. 9, p. 39*b*, tr. auct. Cf. de Mély (1), pp. xxiv, 80.

[i] Biography in *Sung Shu*, ch. 72, p. 5*a*. A contemporary of Proclus.

[j] Arsenic sulphide.

[k] Ch. 988, p. 3*a*, tr. auct.

[1] 鶴頂新書 [2] 張鎡 [3] 雌黃 [4] 土宿本草 [5] 雌 [6] 雄
[7] 典術 [8] 建平王 [9] 雄黃 [10] 鶴頂方
[11] 土宿眞君造化指南

There are other quotations in Li Shih-Chen's *Pên Tshao Kang Mu* from these Taoist authors. We read[a] that according to the *Ho Ting* books, when cinnabar (*tan sha*[1]) is acted upon by 'the *chhi* of the green Yang' it forms ores, which after 200 years become green ('the girl is then pregnant'), and gives birth after 300 years to lead which in a further 200 years changes to silver; lastly, after 200 years, the '*chhi* of the Great Unity' is obtained, and gold appears. Gold, they say, is the son of cinnabar, the Yang within the Yin; before its formation the Yang must die and the Yin be condensed. We also read[b] that, according to the Earth's Mansions Immortal, rock-salt (*lu shih*[2]) is first in 150 years changed into magnetite, then after 200 years into iron,[c] which, if not dug up and smelted, will turn into copper (by means of the sun's *chhi*),[d] that into 'white metal', and that finally into gold. 'Thus iron, silver and gold have a common root and origin.'

We may detect here more than one chemical theory. The root and origin may be an undifferentiated *chhi* or exhalation, its products depending on the balance of Yang and Yin, or else perhaps cinnabar (mercuric sulphide) or salt, or (in another text)[e] lead. But whatever the starting-point, it was agreed on all hands that there was a series of chemical changes in the earth whereby minerals and metals were produced in definite order. Berthelot (3) was fully justified, in reviewing de Mély's book, when he set beside these Chinese quotations the following words from Vincent of Beauvais (*c.* +1246):

> Gold is produced in the earth with the aid of strong solar heat, by a brilliant mercury united to a clear and red sulphur, concocted for more than 100 years....White mercury, fixed by the virtue of incombustible white sulphur, engenders in mines a matter which fusion changes to silver....Tin is generated by a clear mercury and a white and clear sulphur, concocted for a short time subterraneously. If the concoction is very prolonged, it becomes silver....[f]

Berthelot supposed that this theory must have entered Chinese proto-scientific literature from Western sources. Yet we can find a classical passage in Chinese from the −2nd century (and therefore anterior to most of the Greek chemical writings,[g]

[a] *PTKM*, ch. 8, p. 11*a*. Cf. de Mély (1), pp. xxv, 158.

[b] *PTKM*, ch. 8, p. 36*a*. Cf. de Mély (1), pp. xxvii, 35.

[c] This belief was frequently appealed to by those who tried to explain the polarity and declination of the magnetic needle (cf. Sect. 26*i* below). The change from magnetite to iron was supposed to occur under the influence of the Yang *chhi* of the south; therefore the iron would naturally 'follow a maternal principle' and point there. But metal was associated with the west, and declination was therefore brought about by what might be called a 'conflict of loyalties'. Cf. Wang Chen-To (*4*), p. 214.

[d] Here is a hint of an astrological correlation. As may be seen from Berthelot (1, 2) this was extremely important in occidental mineralogical thought. But it seems rather less emphasised in parallel Chinese writings.

[e] *PTKM*, ch. 8, p. 16*b*. Cf. de Mély (1), pp. xxvi, 27. This is Thu Hsiu Chen Chün again, whom Li Shih-Chen quotes as saying that lead is the ancestor of all metals (*chhien wu chin chih tsu*[3]). The same idea is present in Hellenistic chemical texts such as those due to Bolos of Mendes (Berthelot (1), p. 229; Berthelot & Ruelle (1), p. 167).

[f] *Speculum Maius* (see Sarton (1), vol. 2, p. 929).

[g] The oldest author is Bolos of Mendes, the pseudo-Democritus, whose *floruit* must be placed in the −2nd century, i.e. the same time as that of Liu An, prince of Huai Nan (Festugière (1), p. 222).

1 丹砂 2 鹵石 3 鉛五金之祖

which centre on the +2nd) stating it quite clearly. This is the conclusion of the fourth chapter[a] of *Huai Nan Tzu*, studied attentively by Erkes (1) and Dubs (5). It runs as follows:

When the *chhi* of the central regions (lit. the 'main-lands') ascends to the Dusty Heavens (*ai thien*[1]), they give birth after 500 years to *chüeh*[2] (an unknown mineral, perhaps realgar). This in its turn produces after 500 years yellow mercury, yellow mercury after 500 years produces yellow metal (gold), and yellow metal in 1000 years gives birth to a yellow dragon. The yellow dragon, penetrating to the treasuries (of the earth) gives rise to the Yellow Springs.[b] When the dust from the Yellow Springs ascends and becomes yellow clouds, the Yin and Yang beat upon one another, produce peals of thunder, and fly out as lightning. The (waters) which were above thereupon descend (as rain), and the running streams flow downwards uniting in the Yellow Sea.

When the *chhi* of the eastern regions (lit the 'edge-lands') ascend to the Caerulean Heavens, they give birth after 800 years to *chhing tshêng*[3] (azurite or malachite, i.e. copper carbonate). This in its turn produces after 800 years green mercury, green mercury after 800 years produces blue metal (presumably lead), and blue metal in 1000 years gives birth to a Caerulean dragon. The Caerulean dragon, penetrating to the treasuries (of the earth) gives rise to the Green Springs. When the dust from the Green Springs ascends and becomes blue clouds, the Yin and Yang beat upon one another, produce peals of thunder, and fly out as lightning. The (waters) which were above thereupon descend (as rain), and the running streams flow downwards uniting in the Caerulean Sea.[c]

It would be tedious to prolong this passage, which continues its parallelism to the end. So far we have had the yellow Centre and the blue-green East. The pattern of symbolic correlations unrolls itself as follows:

South	'bull-lands'	Red	700 yrs. *chhih tan*[4] (red cinnabar)	red mercury	copper
West	'weak lands'	White	900 yrs. *pai yü*[5] (arsenolite)	white mercury	silver
North	'cow-lands'	Black (or dark)	600 yrs. *hsüan chih*[6] (dark grindstone)	black mercury	iron

The passage is certainly archaic. After what was said earlier[d] concerning the association of colours, compass-points, etc., we cannot be surprised to meet with it again here. But what is quite surprising is to find that same juxtaposition of meteorological and mineralogical exhalations which, as we saw, was characteristic of Aristotelian doctrine. I am inclined to think that the expression 'heavens' may be partly metaphorical; perhaps the real sky and the real rain were meant in the case of the centre, but for the other four regions, the reference may be rather to those upper layers of the earth's crust in which the mineral transformations were to occur. In any case, the text

[a] Pp. 12*a* ff.
[b] A proverbial expression for the Plutonic regions.
[c] Tr. Erkes (1), Dubs (5); slightly mod.
[d] Sect. 13*d* above (Vol. 2, pp. 261 ff.).

[1] 埃天 [2] 砄 [3] 青曾 [4] 赤丹 [5] 白礜 [6] 玄砥

mentions many of the inorganic substances which later became so important in alchemy—arsenic sulphide, sulphur, arsenious acid,[a] mercuric sulphide, mercury, and the metals. Dubs is surely right in his view that the association of all these with the five elements (manifested by the colours and the directions) indicates strongly that the alchemical-mineralogical doctrine goes back to the −4th-century school of Tsou Yen.[b] He points out, moreover, that the order in which the elements are here listed is the same as that found in the surviving fragments of Tsou Yen. We shall see further evidence connecting the School of Naturalists with alchemy below.[c] Here it is only necessary to note the link between the theory and the practice. The aim of alchemy was the artificial production of gold, which it was thought would confer material immortality; this mineralogical idea made the possibility more plausible by describing the natural production of gold as the spontaneous metamorphosis of certain minerals. I agree fully with Eliade[d] and others when they suggest that the early alchemists believed that with the aid of divine beings they could hasten this natural metamorphosis and thereby bring about alchemical transformation. The Chinese theory of the metamorphosis of minerals is then fully developed by −122, and probably goes back to −350 or before. It is extremely difficult to believe that Tsou Yen and his School derived it from Aristotle or the pre-Socratics.[e] The future will probably show that it came from some intermediate and older source whence it radiated in both directions.

(*c*) PRINCIPLES OF CLASSIFICATION

A few words may here be in place concerning the graphic tools which the Chinese had at their disposal for classifying stones and minerals. The two radicals which were most important were of course *shih*[1] for all stones or rocks, and *chin*[2] for all metals and alloys. Oracle-bone forms of the first character show what may be a mouth (signifying a person) under an overhanging projection, and though the experts such as Karlgren (1) do not commit themselves, the graph certainly suggests a man sheltering in a rock cave (K795*c*). The second has often been taken as a drawing of a mine, with a cover over the shaft, and two nuggets of ore (K652*b*), but this seems perhaps rather sophisticated for the Shang. One of the factors which prevented or hindered a logical use of these radicals in classification was the fact that their derivatives, very numerous, included all kinds of stone and metal implements, as well as noises made by stones and metals, and relevant verbs and adjectives. A third radical, *yü*,[3] applied to jade and all kinds of precious stones.[f] A

K795*c*

K652*b*

[a] In +1637 Sung Ying-Hsing took the trouble to deny that tin ore was generated from arsenic trioxide (*TKKW*, ch. 14, p. 10*b*). In our own time it was still believed in Annam that gold grew in the earth from 'black bronze' (Przyłuski, 6).

[b] Cf. above, Sect. 13*c* (Vol. 2, pp. 232 ff.).

[c] Sect. 33.

[d] (4), and especially a more recent monograph (5), of luminous insight and deep learning.

[e] In view of what we know of possible inter-cultural contacts in those times (Sect. 7 above).

[f] According to Hopkins (12) and other scholars, its ancient forms were pictographs of a girdle pendant composed of several pieces of jade.

[1] 石 [2] 金 [3] 玉

fourth, *lu*,[1] was dedicated to salt, but unfortunately this was not given any general application until modern times, only eleven derivatives being listed in Couvreur (2). The lexicographers disclaim any interpretation of the Shang graph (K71*b*), but one may suggest that it was a bird's-eye picture of the salterns in which the brine was evaporated. Or it may have been, as Baas-Becking thought, an attempt to draw a large crystal of salt.

K71*b*

Colours played an important part in the classification of minerals, as was natural. *Tan*,[2] which came to mean red, was a great witchword in Chinese alchemy and mineralogy, having had the significance of cinnabar (mercuric sulphide) for as far back as it can be traced.[a] Ultimately, on account of the use of cinnabar as a drug of immortality by the Taoists, *tan* came to mean any medicine, pill or prescription. An alternative way of writing *tan*[3] suggests the mineral in a furnace or container, but the bone form (K150*b*) is more probably a picture of an oracle-bone, or part of a human skeleton, with a spot of magic red applied to it. Similarly, other minerals and salts were given names ending in yellow (*huang*[4]) or blue-green (*chhing*[5]), for example (as we have just seen) 'male yellow' and 'female yellow', in just the same way as in our own time dyes of complex organic structure are known as Nile Blue or Brilliant Green.

K150*b*

Another word which ought to have been capable of extended classificatory use was *fan*,[6] used for alum and, with qualifiers, all related substances. The upper part of this character means a fence or a hedge, and the philologists have of course renounced all hope of understanding its origin; but for anyone who has seen the methods employed in traditional Chinese chemical technology (as at the Tzu-liu-ching brine-field), it will seem quite plausible to suppose that the character refers to the method used for evaporating solutions at 'room temperature', namely, by pouring them continuously in the open air over a large structure like a hedge made of dry thorn branches, thus increasing enormously the air-water surface (Fig. 271).

Powders were known as *sha*[7] ('sands', lit. small particles of rock), *hui*[8] ('ashes'), *fên*[9] (from finely divided rice meal), *ni*[10] ('muds'), *shuang*[11] ('frosts', if white in colour), and *thang*[12] 'sugars' (cf. 'sugar of lead'). Inorganic substances were termed 'fats' if of greasy or viscous consistency, such as clays or soapstone, *chih*[13] or *kao*.[14] In the quotations which have been given on the preceding pages we have seen several indications that the Chinese thought of the earliest stages in the growth of ores as being of a soft, plastic or viscous consistency; and exactly the same idea (natural enough on any conception of slow petrifaction) occurs in late medieval and early Renaissance Western writers, who spoke of these materials as 'Bur' or 'Gur'.[b]

[a] There was, of course, confusion later with litharge and other red substances.

[b] Adams (2), pp. 283, 290. Though I have not seen it suggested, van Helmont may have been simply pushing these ideas a stage further in his invention of the terms 'Blas' and 'Gas', the latter of which was destined to make its fortune.

1 鹵 2 丹 3 日 4 黃 5 青 6 礬 7 砂
8 灰 9 粉 10 泥 11 霜 12 糖 13 脂 14 膏

Fig 271. Evaporator of traditional type at the Tzu-liu-ching brine-field (orig. photo.). Solutions of salts are continuously poured over these large thatch structures in the open air; the air-water surface is thus enormously increased and the brine concentrated. If this method is ancient it may explain the structure of the character *fan*, alum.

(d) MINERALOGICAL LITERATURE AND ITS SCOPE

Of what does the old Chinese mineralogical literature consist? First and foremost of that remarkable series of compendia, stretching from the Han to the Chhing, comprised in the term *pên tshao*[1]—the pharmacopoeias. Some two hundred collections, many of them very large, are included under this rubric. It would be logical to give our description of them here, since the study of minerals brings us in real contact with them for the first time,[a] but as the majority of the objects described in them are plants and plant drugs, it may be more logical to postpone their bibliography until Section 38 on botany. The reader is therefore requested to refer to the table in that chapter. The classical description of this mass of systematised knowledge is that of Bretschneider (1), but his interest was wholly on the botanical side, and no one has properly studied as yet their mineralogical contents. This overlaps into chemistry, since it is not easy to draw any sharp line of distinction between inorganic substances occurring naturally as ores or minerals and thus serving as the starting-points of chemical or alchemical preparations, and other inorganic substances which it was customary for the technicians to prepare artificially. In Section 33 on chemistry, therefore, will be found a Table of the principal substances used by ancient and medieval Chinese mineralogists and chemists, with the principal names and synonyms current in different periods. Before we can consider this, however, we must look at the literature more closely, and the size of the field which it covered.

The *Shen Nung Pên Tshao Ching*[2] (Pharmacopoeia of the Heavenly Husbandman)[b] is certainly a work of Han date; it no longer exists separately but is incorporated with commentaries in later *Pên Tshao* books.[c] Important is the fact that mineral remedies were included in the pharmacopoeia from its earliest beginnings, and this first *Pên Tshao* described forty-six inorganic substances, arranging them in three ranks according to their believed therapeutic value.[d] Examples follow:

First Rank: cinnabar, mica, mineral waters, stalactites, alum, saltpetre, steatite, copper carbonate, haematite, quartz, amethyst, a variety of clays.

Second Rank: orpiment, realgar, sulphur, mercury, magnetite, actinolite, marble, felspar, copper sulphate, azurite.

Third Rank: stalagmites, iron oxides, iron, lead tetroxide, lead carbonate, tin, salt, agate, arsenolite, lime, fuller's earth.

But this list, which will date from the −1st or +1st century, is not the oldest we have. The *Chi Ni Tzu*[3] book, which may be of the −4th century, and is almost surely pre-

[a] But cf. pp. 434 and 436 above. For bibliography see Lung Po-Chien (*1*).

[b] The name is taken, of course, from the legendary agronomic-medical culture-hero.

[c] Its text was reconstituted and commented on by Miu Hsi-Yung[4] in +1625. There is an unpublished translation of part of it by Hagerty (1), which I have been privileged to use by the kindness of its maker.

[d] *PTKM*, ch. 2, p. 34*a*. Cf. Tokarev (1) on the minerals in the *Shan Hai Ching*.

1 本草 2 神農本草經 3 計倪子 4 繆希雍

Han, gives a list of some twenty-four inorganic substances, most of which are included in the lists of the *Shen Nung Pên Tshao*. It seems, however, to be a record of tribute or economic geography rather than medical in aim.

All the subsequent *Pên Tshao* books open with chapters on mineral substances, though some treat them much more fully than others. Thus the *Chêng Lei Pên Tshao*[1] (Reorganised Pharmacopoeia), edited by Thang Shen-Wei[2] about +1115, deals with 215, while the *Shih Liao Pên Tshao*[3] (Nutritional Medicine Pharmacopoeia)[a] of Mêng Shen[4] about +670, concentrating attention on substances of nutritive value from animals and plants, discusses only two or three. The climax of the series, the *Pên Tshao Kang Mu*[5] (The Great Pharmacopoeia) of Li Shih-Chen[6] (+1596), gives a very elaborate treatment to 217 substances.[b] We are able to identify these with considerable precision, since J. F. Vandermonde in +1732 returned to France from Macao bearing labelled specimens of each of the minerals described by Li Shih-Chen, together with a translation which he had made of the mineral chapters of the *Pên Tshao Kang Mu* with the aid of Chinese pharmacists. The eighty specimens and the translation were deposited in the care of Jussieu at the Paris Museum of Natural History, where after being lost for many years they were studied first by E. Biot (17) and then by de Mély (1). Analyses were made, where necessary, by the French mineralogist A. Brongniart. The translation (or rather, abridged paraphrase) of Vandermonde will be found as an appendix to de Mély's book; it is still not entirely without value, but in view of the date at which it was made it contains, not surprisingly, misunderstandings. De Mély himself (1896) gives a translation of the corresponding chapters of a Japanese edition of the *San Tshai Thu Hui* encyclopaedia, originally compiled by Wang Chhi in +1609, reproducing the text itself. The *Pên Tshao Kang Mu* was also studied by another worker, Geerts, who used the Japanese commentary of Ono Ranzan,[7] and intended to describe the entire compendium, with identifications and notes mostly of Japanese relevance, but the task was not carried beyond the chapters on minerals (1878); it is a work of some merit. Lastly, there is the indispensable catalogue and commentary of Read & Pak already mentioned.

It will naturally be asked whether there were in the Chinese literature any books specifically on minerals apart from those in which they figure as materia medica. 'The few works which the Chinese possess, touching the subject of mineralogy', said Wylie,[c] 'are scarcely deserving a claim to the designation of science.' But this judgment was much too severe. We are fortunate in possessing, from the Thang dynasty, an important treatise on minerals by Mei Piao,[8] the *Shih Yao Erh Ya*[9] (Synonymic Dictionary of Minerals and Drugs),[d] prepared in the close neighbourhood of +818. This is a veritable key to the language of the Thang alchemists, for it lists 335 synonyms of

[a] Edited from a Tunhuang manuscript by Nakao.
[b] Cf. Fig. 29 in Vol. 1, facing p. 164, for some of its illustrations of plants.
[c] (1), p. 118.
[d] *TT* 894. Its second and third parts deal with vegetable drugs and standard prescriptions.

1 證類本草 2 唐愼微 3 食療本草 4 孟詵 5 本草綱目
6 李時珍 7 小野蘭山 8 梅彪 9 石藥爾雅

sixty-two chemical substances.[a] It has hitherto been too much overlooked, at least by Western scholars. It may be said to parallel the alchemical lists given in the Marcianus MS. 299 of the +11th century, the Paris MS. 2327 of +1478 (Berthelot),[b] or Ruhland's *Lexicon Alchemiae* of +1612. Then, with the Sung flowering of scientific monographs devoted to special subjects, came a whole group of books devoted to stones and minerals, the starting-point of which was aesthetic rather than medical. One of the first of these was the *Yüyang Kung Shih Phu*[1] (Treatise on Stones by the Venerable Mr Yüyang),[c] which must have been written towards the close of the +11th century, since it speaks of Mi Fu[2] (whom we shall meet again immediately) as one of the great officials of the age. Not much of it is left. Then between +1119 and +1125 came the *Hsüan-Ho Shih Phu*[3] (Hsüan-Ho reign-period Treatise on Stones)[d] by the Szechuanese monk Tsu-Khao;[4] this discussed sixty-three kinds of stones, but only the contents table remains. Complete still, however, is the *Yün Lin Shih Phu* (Cloud Forest Lapidary) by Tu Wan,[5] dating from +1133. Some quotations from this excellent author have already been given.[e] He is one of those who deserve the praise of de Mély's remark:[f] 'If, as in the occidental lapidaries, each mineral or stone is furnished with magico-medical formulae and improbable fables, the spirit of observation and analysis, which is totally lacking in the European treatises, brings forth many intelligent comments.' Other Sung books there are, of course, which devote considerable space to minerals from a non-medical point of view, for example the *Tung Thien Chhing Lu Chi* of the +13th century, already mentioned.[g]

The Ming and the Chhing did not live up to this standard, and one can only mention the unsystematic book of Yu Chün[6] in +1617, the *Shih Phin*[7] (Hierarchy of Stones), as well as smaller works such as *Kuai Shih Tsan*[8] (Strange Rocks), by Sung Lo[9] (+1665), and Kao Chao's[10] *Kuan Shih Lu*[11] (On Looking at Stones) of +1668.

Whoever undertakes to write the much-needed monograph on the history of mineralogy in China will have to take into account also another class of books, namely those on inkstones. From the time of the introduction of carbon-black ink and brushes it was the delight of Chinese scholars to select stones suitable for grinding their ink-blocks with water. This led to the description, at least, of a considerable variety of rocks. The most famous treatise on the subject is that of Mi Fu, a high official of the

[a] The careful listing of synonyms in pharmaceutical codices had begun much earlier, and is found in the *Hsin Hsiu Pên Tshao* of +659, as well as the *Pên Tshao Thu Ching* of +1070 and all later works of the same kind. A century after the *Shih Yao Erh Ya* a Japanese physician, Fukane no Sukehito,[12] produced a famous synonymic materia medica with Japanese equivalents added, the *Honzō-Wamyō*[13] (+918). This has come down to us, and Karow (1) has given a detailed description of it. Lost for centuries, it was not printed until 1796.

[b] (2), pp. 92 ff.

[c] In *Shuo Fu*, ch. 16, p. 25*a*.

[d] In *Shuo Fu*, ch. 16, p. 24*b*.

[e] Above, pp. 615, 620. There is an unpublished translation by H. Bendig (1).

[f] (1), p. xi; and de Mély knew his Occidental and Islamic lapidaries.

[g] P. 613.

1 漁陽公石譜 2 米芾 3 宣和石譜 4 祖考 5 杜綰
6 郁濬 7 石品 8 怪石贊 9 宋犖 10 高兆
11 觀石錄 12 深根輔仁 13 本草和名

Sung, the *Yen Shih*[1] of about +1085; this has been translated by van Gulik (2). It was not the first, however, for the *Yen Phu*[2] (Inkstone Record) of the late +10th century, perhaps by Shen Shih,[3] but also attributed to Su I-Chien,[4] describes thirty-two sorts of stones. The *Hsi-Chou Yen Phu*[5] was also famous; it was from the brush of Thang Chi[6] who finished it by +1066;[a] some fifty varieties (if we include those in the appendices) are mentioned. Particular places renowned for the production of inkstones had books of their own, such as the *Tuan-Chhi Yen Phu*[7] by an unknown author about +1135.[b] There seems to be no close parallel to this literature in the occident, and it has not so far been examined by a mineralogist, except for the interesting chapter of Chang Hung-Chao, who states[c] that limestone and clay shales were generally preferred on account of their smooth surface.

Our future author would also have to take into account the numerous dictionaries and encyclopaedias, the mineralogical material in which can obviously not be summarised here. Some of this, such as the portions of the *Thai-Phing Yü Lan* translated by Pfizmaier (95), is anecdotal and of no value except as an index of the dates at which various terms were in use. Another source would be that of the archaeological books, such as the *Ko Ku Yao Lun*[8] (Ming).[d]

We are now in a position to construct a comparative table showing the coverage of the successive lapidaries of East and West (Table 41).

A fuller treatment would enable a surer conclusion, but it is clear first that the number of stones and inorganic substances distinguished and named by the Chinese was certainly no less than that of the West. Again, we find that while the Greeks were perhaps somewhat more advanced than their Chinese contemporaries the advantage was lost in the Middle Ages, so that by the beginning of the +11th century Chinese systematism was a couple of hundred years ahead. It is striking that before the Renaissance no people recognised more than about 350 items. With the *Gemmarum et Lapidum Historia* of de Boodt (+1550 to +1632)[e] the pre-modern period may be said to come to an end.

[a] Printed, probably for the first time, together with other minor works on the same subject, *c.* +1160, by Hung Ching-Po.[9]

[b] The titles of other books of the kind will be found in the bibliographies of the official histories. Cf. also Wylie (1), p. 116.

[c] (*1*), p. 276.

[d] This book, by Tshao Chao[10] (+1387), devotes ch. 7 to ancient stone objects, and to the stones from which they were made.

[e] Tr. Hiller (2).

1 硯史 2 硯譜 3 沈仕 4 蘇易簡 5 歙州硯譜
6 唐積 7 端溪硯譜 8 格古要論 9 洪景伯 10 曹昭

Table 41. *The coverage of stones and substances in Western and Eastern lapidaries*

	Western lapidaries	No. of stones or substances described	Eastern lapidaries	No. of stones or substances described
c. −350			*Chi Ni Tzu* (Chi Jan)	24
c. −300	*De Lapidibus* (Theophrastus)[a]	*c.* 70		
−1st century			*Shen Nung Pên Tshao Ching*	46
+60	*De Mat. Med.* (Dioscorides)[b]	100		
c. +220	*Rivers and Mountains* (pseudo-Plutarch)[c]	24		
+300	*Cyranides* (Hermetic)[c]		*Pao Phu Tzu* (Ko Hung)	*c.* 70
+818			*Shih Yao Erh Ya* (Mei Piao) (total synonyms	62 335)
+1022	*Kitāb al-Shifā'* (Ibn Sīnā)[d]	72		
c. +1070	*Lapidarium* (Marbodus)[e]	60	*Pên Tshao Thu Ching* (Su Sung)	58
+1110			*Chêng Lei Pên Tshao* (Thang Shen-Wei)	215
+1120			*Hsüan-Ho Shih Phu* (Tsu-Khao)	63
+1133			*Yün Lin Shih Phu* (Tu Wan)	110
c. +1260	*De Mineralibus* (Albertus Magnus)[f]	70		
c. +1278	*Lapidarium* (Alfonso X)[g]	280		
+1502	*Speculum Lapidum* (Camillus Leonardus)[h]	279		
+1596			*Pên Tshao Kang Mu* (Li Shih-Chen)	217

(*e*) GENERAL MINERALOGICAL KNOWLEDGE

Some general remarks may now precede the few special entries on individual minerals for which we shall have space. The oldest classification of minerals, found already in Theophrastus, was simply whether or not they were changed by heat. Fusibility, therefore, distinguished stones from minerals, and this we duly find in *Pao Phu Tzu*, who says:

> Other substances decay when buried under the ground, or are melted when subjected to fire. But if the five kinds of mica are put right into a blazing fire, they will never be destroyed, and if buried they never decay.[i]

Many other texts reveal the same fundamental test, as we shall find in Section 36 on metallurgy.

The term for mica[j] throughout Chinese history was 'cloud-mother', *yün mu*.[1] The description of it given in the *Pên Tshao Thu Ching* of +1070 is not bad. Su Sung says:

[a] Eng. tr. J. Hill (1).
[b] Eng. tr. Goodyer (1).
[c] Fr. tr. de Mély (3).
[d] Eng. tr. Holmyard & Mandeville (1).
[e] Fr. tr. Pannier (1); cf. Evans (1), pp. 33 ff.
[f] Cf. Thorndike (1), vol. 2, ch. 59; Evans (1), pp. 84 ff.
[g] Cf. Evans (1), p. 42.
[h] See bibliography.
[i] Ch. 11, p. 8*b*, tr. Feifel (3), p. 16.
[j] The special study of Schafer (5) is now available.

[1] 雲母

Mica grows in between earth and rocks. It is like plates in layers which can be separated, bright and smooth. The best kind is white and shining....Its colour is like purple gold. The separate laminae look like the wings of a cicada. When they are piled up they look like folded gauze. It is said that this belongs in the category of glass. It can be used in the preparation of medicines. Thao Hung-Ching [+5th century] says [in his version of the *Ming I Pieh Lu*]: 'According to the *Hsien Ching*[1] (Manuals of the Immortals) there are eight kinds of mica, which can be differentiated by looking through them at the sun. Those with a bluish white colour but mainly dark are called *yün-mu*,[2] those with white and yellow markings but mainly blue are called *yün-ying*,[3] those with blue and yellow markings but mainly red are called *yün-chu*;[4] those like ice or dew but with yellow and white specks are called *yün-sha*,[5] those which are yellow and white and very crystalline are called *yün-i*,[6] and those which are bright with pure transparent spots are called *lin-shih*.[7] These six varieties can be eaten, but each has its suitable time of year. Dark and black mica with iron-like spots is called *yün-tan*;[8] that which is opaque and fatty is called *ti-chêng*[9] ('earth-steam'). These two should not be eaten. In any case (the powder) should be prepared with every care otherwise the greatest harm could be done when such things enter the stomach.[a]

Here the description of the thin transparent plates is clear enough, and there is, moreover, a systematic attempt to distinguish between the many different kinds of mica. Though it might be hard to be quite sure what Thao Hung-Ching and Su Sung meant, it would seem that the first samples in their series are muscovite, that the 'mainly red' variety would be lepidolite, the most transparent kind phlogopite, and the dark or black kind certainly biotite. The speck-like inclusions in biotite gave rise to the later name of *chin hsing shih*.[10] Elsewhere, in his entry on *yü hsieh*[11] (jade fragments, or traces), Su Sung quotes a *Pieh Pao Ching*[12] (Manual of the Distinctions between Precious Stones) as recommending observation of stones with both reflected lamplight and transmitted sunlight.

The descriptions in the *Pên Tshao* books are so therapeutic in orientation that they will be more suitably deferred until Section 45 on pharmacology. All the old Chinese mineralogical works, however, paid attention to crystal form, noticing which substances had crystals of hexagonal, needle-like, pyramidal, and other types. The *Pên Tshao Thu Ching* describes quartz crystals with six faces (*liu mien*[13]) 'as sharp as if cut by a knife', and calcite crystals 'square and angular' (*fang lêng*[14]). The word *mien*[15] for crystal face occurs regularly in Tu Wan's lapidary of +1133. Li Shih-Chen emphasises the needle-shaped crystals of sublimed arsenious acid,[b] quoting in this connection the *Pên Tshao Pieh Shuo*[16] (Informal Remarks on the Pharmacopoeia) by Chhen Chhêng[17] of +1090. Long before Li Shih-Chen, Su Sung, in the +11th-century *Pên Tshao Thu Ching*, had spoken of the subconchoidal fracture of native cinnabar. He said:

Cinnabar is found several dozens of feet deep in the ground. The local people find it by

[a] *Thu Ching Yen I Pên Tshao*, ch. 1, pp. 7*a*, 8*a*, *b*, tr. auct. Parallel passages in *Pao Phu Tzu*.

[b] Ch. 10, p. 27*a*. Almost all the world's arsenic is still obtained as a by-product in the flue dust of smelters treating arsenious ores for the recovery of gold, copper, etc.

1 仙經 2 雲母 3 雲英 4 雲珠 5 雲砂 6 雲液
7 磷石 8 雲膽 9 地涿 10 金星石 11 玉屑 12 別寶經
13 六面 14 方稜 15 面 16 本草別說 17 陳承

means of the 'sprouts' (*miao*[1]) (i.e. signs drawn from the presence of other kinds of stone or even herbs).[a]

Cinnabar (*tan sha*[2]) is thus found in association with a kind of white stone (*pai shih*[3]) which is known as the 'cinnabar bed' (*chu sha chhuang*[4]). The mineral grows on this stone....

Upon breaking the lumps of the mineral, it is seen to form precipitous slopes (with surfaces) like walls, as smooth inside as plates of mica....

The *Thang Pên Tshao* (by Su Kung) says that there are more than ten different kinds. The best kind, *kuang ming sha*[5] (brilliant sand) grows in (a rock-formation) called 'stone shrine' (*shih khan*[6]). The largest lumps are as big as eggs, the small ones like jujube fruits, chestnuts or hibiscus berries. The broken surfaces shine smooth like mica.

The *Lei Kung Yao Tui*[7] (Lei Kung's Answers concerning Drugs) says that one can find lumps (crystals)...with fourteen surfaces, each looking as bright as a mirror. On gloomy or rainy days, humidity like a red juice forms on the broken surfaces.[b]

This is interesting for several reasons. We catch a glimpse of the great attention paid to signs of ore beds,[a] and to the special characteristics of the country rock. The characteristic fracture of the mineral is noted, and there is an attempt to describe crystal form; cinnabar has hexagonal symmetry and occurs in rhombohedra which are often twinned. It was presumably some such crystals which caught Lei Kung's attention.[c] Brelich, indeed, has described penetration twin rhombohedra cinnabar crystals which he saw when he visited the mercury mines in Kweichow; these would have just fourteen faces.

Effects due to the contiguity of mineral deposits are not infrequently noted. For instance, Tu Wan, speaking of the green colour of a certain stone (*shao shih*[8]),[d] says that it arises from the distillation of the vapours of the 'sprouts' of copper, which are not far off, since this stone always occurs in the neighbourhood of copper ores.[e] This is not very different from saying that the stone contains inclusions of malachite or other copper-bearing mineral. It is, moreover, reminiscent of the views put forward by Agricola in the +16th century, regarded by Hoover & Hoover[f] as the foundation of modern theory, according to which ore channels were formed by the circulation of ground waters ('succi') in fissures subsequent to the deposition of the surrounding rocks.[g] Similar statements concerning native copper sulphate occur in the *Jih Hua Pên Tshao* (+970) and concerning copper carbonate in the *Pên Tshao Thu Ching* (+1070).[h] Sometimes expressions such as 'turbid yellow water' are used instead of

[a] See below the special sub-section on geobotanical prospecting (pp. 675 ff.).
[b] Tr. auct. The last two quotations also come from *Thu Ching Yen I Pên Tshao*, ch. 1, pp. 2*a*, 3*a*, 4*a*.
[c] The compiler of the *Lei Kung Yao Tui* was probably the physician Lei Hsiao.[9]
[d] Not identified, but probably malachite or some other copper-containing mineral.
[e] *Yün Lin Shih Phu*, ch. 1, p. 17*b*: *Yin thung miao chhi hsün chêng, chieh tzhu shih kung chhan chih yeh.*[10] Cf. *TPYL*, ch. 988, p. 4*b*, quoting a passage which is not in the present text of the *Shen Nung Pên Tshao*. Cf. the Buddhist doctrine of 'perfuming' (*hsün*), Vol. 2, p. 408.
[f] (1), pp. xiii, 52.
[g] Cf. a classical paper by Posepny, and the recent book of Bateman.
[h] *Thu Ching Yen I Pên Tshao*, ch. 2, pp. 6*a*, 7*b*.

1 苗 2 丹砂 3 白石 4 朱砂牀 5 光明砂 6 石龕
7 雷公藥對 8 韶石 9 雷斅 10 因銅苗氣薰蒸即此石共產之也

chhi, as in the *Thu Ching*'s entry under *shih chung huang tzu*[1] (brown iron ore; haematite).[a]

Perhaps the most vigorous description of the conception of the deposition of ore beds from the circulation of ground waters in rock fissures comes from the pen of Chêng Ssu-Hsiao,[2] who died in +1332, about two centuries before Agricola was in his prime.[b]

In the subterranean regions there are alternate layers of earth and rock and flowing spring waters. These strata rest upon thousands of vapours (*chhi*) which are (distributed in) tens of thousands of branches, veins and thread (-like openings). (There are substances there) both soft and firm, ever flowing back and forth, and undergoing transformations. (The veins are) slanting and delicate, like axles interlocking and communicating. (It is like a) machine (*chi*[3]) rotating in the depths, (and the circulation takes place as if the veins had) intimate mutual connections (and as if) there were piston bellows (*tho yo*[4]) (at work). The mysterious network (*hsüan kang*[5]) spreads out and joins together every part of the roots of the earth. The (innermost parts of the earth are) neither metal nor stone nor earth nor water (as we know them). Thousands and ten thousands of horizontal and vertical veins like warp and weft weave together in mutual embrace. Millions of miles of earth are as if hanging and floating on a sea boundlessly vast. Taking all (including land and sea) as earth, the secret and mystery is that the roots communicate with each other. The natures, veins, colours, tastes and sounds, both of the earth, the waters, and the stones, differ from place to place. So also the animals, birds, herbs, trees and all natural products, have different shapes and natures in different places.

Now if the *chhi* of the earth (*ti chhi*[6]) can get through (the veins), then the water and the earth (above) will be fragrant and flourishing...and all men and things will be pure and wise....But if the *chhi* of the earth is stopped up (*sai*[7]), then the water and earth and natural products (above) will be bitter, cold and withered...and all men and things will be evil and foolish....

The body of the earth is like that of a human being. In men there is much heat in and under the watery abdominal organs (*shui tsang*[8]); if this were not so, they could not digest their food nor do their work. So also the earth below the aqueous region is extremely hot; if this were not so, it could not 'shrink' all the waters (*so chu shui*[9]) (i.e. evaporate them and leave mineral deposits), and it could not drive off all the (aqueous) Yin *chhi* (*hsiao chu Yin chhi*[10]). Ordinary people, not being able to see the veins and vessels which are disposed in order within the body of man, think that it is no more than a lump of solid flesh. Likewise, not being able to see the veins and vessels which are disposed in order under the ground, they think that the earth is just a (homogeneous) mass. They do not realise that heaven, earth, human beings, and natural things, all have their dispositions and organisations (*wên li*[11]). Even a thread of smoke, a broken bit of ice, a tumbledown wall or an old tile, all have their dispositions and organisations. How can anyone say that the earth does not have its dispositions and organisations?[c]

[a] RP no. 81. *Thu Ching Yen I Pên Tshao*, ch. 2, p. 24*b*.

[b] The passage is quoted in abridged form by Huang Chieh in his essay on Chêng Ssu-Hsiao. The full text appears in a letter 'Reply to Mr Wu the hermit, who had asked him about Field Expeditions to observe Geomantic Matters (*Ta Wu Shan-Jen wên yuan yu kuan ti li shu*[12])', in *So-Nan Wên Chi*[13] (Collected Writings of Chêng Ssu-Hsiao), p. 12*a*. Another short biography of Chêng Ssu-Hsiao occurs in *Sung I Min Lu*, ch. 13, pp. 2*b* ff.

[c] Tr. auct.

[1] 石中黃子 [2] 鄭思肖 [3] 機 [4] 槖籥 [5] 玄綱 [6] 地氣
[7] 塞 [8] 水臟 [9] 縮諸水 [10] 消諸陰氣 [11] 文理
[12] 荅吳山人問遠遊觀地理書 [13] 所南文集

The passage is certainly a fine one. Though the idea of some kind of circulation going on in the earth was quite common in geomantic circles, Chêng applies it here with clarity to the deposition of minerals by evaporation of (or precipitation from) ground waters in ore channels.[a] It is interesting, too, that he adapts to his purpose the ancient medical theory of pathogenesis by the stopping up of pores, in which connection, naturally enough, he draws a conscious analogy of macrocosm-microcosm type. Of this mode of thinking we have already treated in Sect. 13*f* above. And he ends by a noble affirmation of those organic conceptions so characteristic of all Chinese thought.

The old books mention innumerable practical uses of the various minerals.[b] At periods, said de Mély,[c] when very few Europeans were clear as to the distinction between sal ammoniac, saltpetre, alum and the like, the industrial processes of the Chinese, such as tanning, dyeing, painting and firework-making, had led them to make the necessary identifications. The distinction between the two sulphides of arsenic (realgar and orpiment) appears in the second oldest list of chemicals which we have, i.e. from the Former Han onwards (−1st century); they were *hsiung huang*[1] and *tzhu huang*[2] respectively.[d] Ferrous sulphate (*lu fan*;[3] *tsao fan*[4]) served in dyeing; mercuric sulphide (vermilion, cinnabar; *shui yin chu*[5]) was used for red inks and paints as well as for alchemical preparations; powdered steatite (*hua shih*[6]) was added to paper as a filler. Litharge (*mi tho sêng*[7]) was a constituent of varnish. Skins were dried with saltpetre (potassium nitrate; *phu hsiao*[8]), treated with ammonium chloride (*nao sha*[9]) and dyed with ferrous sulphate (*huang fan*[10]). What saltpetre yielded when mixed with sulphur (*liu huang*[11]) and carbon (charcoal, *than*[12]) was known to the firework-makers and the military technicians. Oxides of cobalt (*pien chhing*[13]) and copper (*thung fên*[14]) were used by the porcelain-makers and enamellers.[e] Cobaltiferous ore (*yen shou*[15]) and salts of lead (oxides, carbonate or acetate) served for colours and glazes. Sometimes a substance had value in medicine as well as in technology, for instance kaolin (*kao ling thu*[16]) essential in ceramics but also used therapeutically as an antacid, like kieselguhr (*shih mien*[17]). Calcium sulphate (*shih kao*[18]) played an important part in the preparation of the bean curd (*tou fu*[19]) so universal in Chinese diet. One of the oldest chemical industries was the preparation of lead acetate (*chhien shuang*[20]) for cosmetic use as white paint. Arsenolite (As_4O_6) (*yü shih*[21]), while poisonous to rats, was found to have an accelerating effect upon the growth of silkworms.[f] Copper

[a] We know now (cf. Bateman, 1) that mineral deposition is more often due to cooling and release of pressure than to the evaporation of solutions by heat, but this does not detract from the insight of Chêng Ssu-Hsiao and Agricola as to chemical changes in circulating mineral-bearing waters.

[b] Although this touches upon fields to which other Sections are devoted, we glance at it here since so many of the substances used were obtained from 'native' minerals without much preparation.

[c] (1), p. xi.

[d] See now Schafer (6).

[e] We shall give closer attention to subjects such as this in Sect. 35 on ceramics.

[f] Arsenolite is mentioned in the *Shan Hai Ching* (cf. de Rosny (1), p. 55), and it is Kuo Pho's +4th-century commentary on it which starts the silkworm story (*PTKM*, ch. 10, p. 23*b*). The phenomenon

1 雄黃 2 雌黃 3 綠礬 4 皂礬 5 水銀硃 6 滑石
7 密陀僧 8 朴消 9 硇砂 10 黃礬 11 硫黃 12 炭
13 扁青 14 銅粉 15 岩手 16 高嶺土 17 石麪 18 石膏
19 豆腐 20 鉛霜 21 礜石

sulphate (*shih tan*,[1] lit. petrified bile) found employment as a fungicide as well as in medicine, and stalactitic limestone (*shih chung ju*[2]) was used as a chemical fertiliser. White arsenic (*phi shuang*[3]), obtained as a by-product from copper smelting, was applied to the roots of rice-plants during replanting, to protect them from insect pests; similarly wood soaked in copper acetate (*thung chhing*[4]) was protected from decaying, especially under water.[a]

(*f*) NOTES ON SOME SPECIAL MINERALS

The best way to pursue the subject will now be to offer a few notes on particular mineral substances such as alum, asbestos, and diamonds.[b] It will be more convenient, however, to postpone the discussion of certain important mineral substances to other Sections. Thus coal will be discussed under mining and metallurgy (36); fuller's earth under textiles (31); rock crystal under physics (optics) in Sect. 26*g*; magnetite and amber under physics (magnetism) in Sect. 26*i*; salt and natural gas under salt mining (37); saltpetre under martial technology (gunpowder) in Sect. 30; kaolin under ceramics (Sect. 35).

(1) Aetites

Aetites, or the 'Eagle-Stone', an object of interest to the ancients, but of minor significance in modern mineralogy, was simply a nodular mass (geode) of haematite or other mineral containing a loose centre produced by the leaching out of intermediate more soluble layers. Bromehead (2) has devoted an interesting paper to it. If this is what Theophrastus had in mind when he spoke of stones which are pregnant and beget young, then specimens had become known already in Aristotle's time, but the name itself does not appear until Dioscorides. The *shih nao*,[5] which seems to correspond to it, does not appear in the oldest Chinese lists, but is mentioned in the *Ming I Pieh Lu* and by Ko Hung. The occidental fables about its reputed value in childbirth could probably be paralleled in Chinese literature, but it does not seem that it was ever regarded as of much importance there.[c]

itself was probably due to the greater toxicity of arsenic compounds for viruses than for their lepidopteran hosts. Speyer, for example, found that administration of sublethal doses of arsenic compounds to silkworms infected with polyhedral virus gave a higher percentage of pupations than in the control series.

[a] This goes back to Ko Hung in the +4th century (*PTKM*, ch. 8, p. 15*b*).

[b] After this Section had been completed two notes of just this kind were published by Schafer, one on mica (5) and one on orpiment and realgar (6). Had their learning been available to us in time we should have included appropriate entries for these minerals here; as it is we shall draw upon them in Sections 33, 34 and elsewhere.

[c] Cf. Laufer (12), p. 9.

[1] 石膽 [2] 石鐘乳 [3] 砒霜 [4] 銅青 [5] 石腦

(2) ALUM

Alum, by contrast, was a substance of very great industrial importance. The history of its production and use in Europe has recently been the subject of a brilliant and monumental study by Singer (8). Its main application was as a dyer's mordant,[a] for which it had to be pure, and there is no doubt that it was purified rather thoroughly in several ancient civilisations (cf. Lucas (1) for Egypt). It was also used for making hides supple, for sizing paper and finishing parchment, for glass-making, clearing of natural waters, and fire-proofing of wood. Its styptic, emetic and astringent properties were greatly appreciated in medicine. The term *fan*,[1] for which an etymology was suggested above,[b] occurs in the oldest Chinese lists, including *Chi Ni Tzu* of the −4th century. Deposits of the native hydrated double sulphates of aluminium and iron, magnesium or manganese, were worked by the Egyptians, Greeks and Romans, but later alunite was used. This is a basic alum insoluble in water, but upon calcination it yields alumina and soluble potash alum; the Chinese distinguished native alum (*sêng fan*[2]) from that prepared by roasting alunite (*khu fan*[3]). Highly crystalline products were known as *fan ching*.[4] It would require a special investigation to determine from what date these two sources developed; we only know a little about the localities of origin. For example Wu Phu[5] says, in his *Wu shih Pên Tshao*[6] of +225, that alum came from Kansu and Chiangsi.[c] Singer (8) thinks that in Asia Minor the roasting of alunite began about the +10th century, and certainly by late medieval times granular alum was imported to China from the West. The third chief method appeared in Arabia in perhaps the +12th century; it consisted of boiling rocks containing aluminium sulphate with urine, so as to form ammonium alum.[d] When this was extended to boiling aluminous shales, the product rivalled the Papal monopoly of alunite, to the economic consequences of which so much of Singer's book is devoted. I cannot find that this third method was used in China. There exists a description of traditional alum-working there (Anon. 3), which it is interesting to compare with a parallel description for India by Ray.[e] One of the oldest names for alum in China was *shih nieh*;[7] as such it is mentioned in the *Shan Hai Ching*. Discussing this, Chang Hung-Chao (*1*) seems to think that the alunite-roasting process was very ancient, and that exploitation of native alum was not carried on by the Chinese until relatively modern times.

As regards the western locality from which alum was imported, Laufer (1) offers evidence that the term Po-Ssu[8] meant somewhere in the south seas, as well as Persia. The *Hai Yao Pên Tshao*[9] (Drugs of the Southern Countries beyond the Seas) by

[a] *Huai Nan Tzu*, ch. 2, p. 6*a*, refers to this use.

[b] P. 642.

[c] *TPYL*, ch. 988, p. 4*a*.

[d] This process was doubtless discovered by finding that alunite from near a latrine gave a much better yield. Agricola mentions all three methods (Hoover & Hoover (1), p. 564).

[e] (1), 2nd edn. p. 230.

1 礬　2 生礬　3 枯礬　4 礬精　5 吳普　6 吳氏本草
7 石涅　8 波斯　9 海藥本草

Li Hsien[1] (*c.* +775), says that there were two kinds of alums, one made in Malayan Po-Ssu, and the other coming through there from Ta-Chhin[2] (Arabia). A significant western connection is the use of alum with henna in China in the +13th century as reported by Chou Mi in his *Kuei-Hsin Tsa Chih*.[a]

(3) Sal Ammoniac

Ammonium chloride (sal ammoniac) was also of importance both medically (a stimulating expectorant and mild cholagogue) and chemically.[b] The 'hammoniac' salt of Pliny cannot have been ammonium chloride, since its volatility and deliquescence are not mentioned, nor does the salt appear among the lists of substances given by the Alexandrian chemists of the +3rd century.[c] Stapleton (1) thought that it was probably introduced into chemistry by the Arabic alchemists towards the +10th century, partly because they were able to prepare it by the dry distillation of hair.[d] Nevertheless, it occurs naturally in volcanic situations in Central Asia, and was probably collected there from an early date. Stapleton suggested that the Arabic word *nūshādur* was perhaps derived from Chinese *nao sha*,[3] a suggestion which Laufer[e] somewhat cavalierly dismissed. Laufer's view that the borrowing was in the inverse direction was so far plausible in that no one (including Chang Hung-Chao[f] subsequently) had been able to find any reference to *nao sha* in Chinese texts earlier than the +6th century, when the *Wei Shu*[4] was written (+572). It had not been noticed that it occurs in Wei Po-Yang's +2nd-century *Tshan Thung Chhi*,[g] where it bears a correct reference to the refrigerant effect of the salt on boils. Whether Ko Hung's *lu yen*[5] means sal ammoniac is not clear, and the next main reference in a technical book is apparently the *Thang Pên Tshao* of +660. Now it is indeed remarkable, as Laufer pointed out, that the orthography of the Chinese term is so fluctuating, a fact which would suggest a phonetic transcription. In works of the +7th century, such as the *Sui Shu*, variants such as *nao*,[6] *nung*[7] and *jao*[8] (all probably homophones of *nao*) are found. In the +9th century Mei Piao writes *niu*[9] (probably also then pronounced *nao*). In the +6th century Wei Shou (in the *Wei Shu*) had written *kang* or *wang*[10].[h] This word persisted through Thang and Sung books, and *nao*[6] seems to have become definitely stabilised only in the Ming.

All the Chinese books agree (for example, the *Hsi Yü Thu Chi*[11] (Illustrated Record of Western Countries) of Phei Chü,[12] *c.* +610; the *Pên Tshao Thu Ching* of Su Sung,

[a] *Hsü chi*, ch. 1, p. 17*a*.

[b] E.g. as a flux, and for making chlorides of silver and mercury.

[c] Partington (1), p. 147.

[d] In such a process, ammonium chloride remains behind, ammonium carbonate sublimes in crystalline state, and hydrogen sulphide is given off.

[e] (1), pp. 503 ff.

[f] (*1*), p. 221.

[g] Ch. 30 (tr. Wu & Davis (1), p. 257). We have checked this in the *Tao Tsang* text.

[h] Graphic corruptions. And there are even more variant forms.

1 李珣 2 大秦 3 硇砂 4 魏書 5 磠鹽 6 硇
7 磠 8 鐃 9 砠 10 碙 11 西域圖記 12 裴矩

c. +1070; and the *Yeh Huo Pien*[1] (Memoirs of a Mission achieved in the Wilds) of Shen Tê-Fu,[2] *c.*+1398)[a] that native sal ammoniac came from the west, i.e. Szechuan, Kansu, Sinkiang, and Tibet, where it was collected from the neighbourhood of volcanic fumaroles. Towards the end of the +18th century the Manchu geographer Chhi-shih-i Lao-jen[3] said, in his *Hsi Yü Wên Chien Lu*[4] (Things Seen and Heard in the Western Countries), concerning Kucha and Turfan:

Nao sha is produced in the mountains of that name which are north of the city of Kucha. In spring, summer and autumn, the caves there are full of fire. From a distance they look like thousands of lamps, and approach is difficult because of the heat. In winter, due to the excessive cold and heavy snow, the fires die down. Local people go there to collect the sal ammoniac, entering the caves naked because of the heat.[b]

In his *Chu Yeh Thing Tsa Chi* (Miscellaneous Records of the Bamboo Leaf Pavilion) a century earlier, Yao Yuan-Chin wrote:

The mountains where sal ammoniac is produced near Kucha were called in the Thang the 'Great Magpie Mountains'. No one dares go near them in spring or summer. Even in the cold weather, the people take off their ordinary clothes, and wear leather bags with holes through which they can see. They enter the caves to dig up (the sal ammoniac), but come out after one or two hours and could not possibly stay longer than three; even then the leather bag is scorching hot. The *nao sha* sparkles on the ground, but not much of it can be obtained. The product has to be kept in earthen jars with their mouths tightly closed, and kept cool, otherwise it will all disappear. It will also disappear if subjected to wind, wetness or damp; leaving only a white residue of granular appearance. Though this is the least valuable part, it is probably the only kind which finds its way to the central parts of China.[c]

These descriptions have close parallels in Arabic authors.[d] On account of this volatility, so well recognised, ammonium chloride acquired another Chinese name, *chhi sha*;[5] and doubtless because of its western origin, it was also known as *ti yen*,[6] barbarian salt. It was probably always impure, often mixed with sulphur or sulphates. If, as seems likely, its collection in volcanic Central Asian regions goes back very far, the earliest term may well have been Sogdian or Persian, giving rise later both to *nūshādur* and to *nao sha*.

(4) Asbestos

Asbestos is unusual among minerals in having a fibrous structure; it is composed of calcium magnesium silicate. Though it occurs in many forms (tremolite, actinolite, other kinds of amphibole, etc.) and with various admixture of other elements, the fibres are usually separable and look like flax. Chrysotile, or fibrous serpentine, is a hydrous magnesium iron silicate which resembles asbestos and has similar properties.

[a] Cf. Schott (2).
[b] Tr. auct.
[c] Tr. auct.
[d] Cf. de Meynard & de Courteille (1), vol. 1, p. 347; Ouseley (1), p. 233.

[1] 野獲編 [2] 沈德符 [3] 七十一老人 [4] 西域聞見錄 [5] 氣砂
[6] 狄鹽

From very early times it was discovered that these fibres could be woven so as to form a kind of cloth which was indestructible by fire. It was natural that such a thing should be considered a wonder and a work of great art, and in fact there are numerous references to asbestos cloth from the centuries just before the beginning of our era onwards. Those in European authors have long been known, but we owe an analysis of the Chinese literature to Wylie (9) and Laufer (11), whose learned essays, however, did not exhaust the subject. Although the intrinsic interest of asbestos minerals is not so great, the nature of the substance set a scientific puzzle to the people of ancient and medieval times, and the study of their ideas about it throws some unexpected light on the growth of scientific thinking. This justifies a treatment rather less brief than might at first sight be called for.

Theophrastus in the −4th century did not know of asbestos; neither did Chi Yen. But Strabo (−63 to +19) speaks of 'fire-resisting napkins',[a] and in this he is followed by many later authors, such as Dioscorides[b] in the +1st century, and Apollonius Dyscolus the grammarian,[c] in the +2nd. Pliny knew of it as coming from Arcadia and India.[d] The idea that it was of vegetable origin is Hellenistic, found for the first time in the 'Alexander Romance' of Pseudo-Callisthenes.[e] The Greek word asbestos, which we still use, meant simply 'inextinguishable' or 'unquenchable', presumably from the use of the substance in the wicks of lamps.

Laufer (11) thought that the Chinese first knew of asbestos through trade with Roman Syria. Certainly the *Wei Lüeh*[1] of Yü Huan[2] (+239 to +265) lists it as a product of Arabia (Ta-Chhin), as also does the *Hou Han Shu*,[f] where it is called by its commonest name, *huo wan pu*[3] ('cloth washable in fire' or fireproof cloth). But there is some evidence that the Chinese knew of it before the turn of the era, and before Syrian or Arabic trade with China had begun. As we do not know which parts of the *Lieh Tzu* book are of pre-Han date and which are not, we cannot be sure whether the following passage was written in the −3rd century or thereabouts, but the probability of this seems greater than Laufer was willing to admit. We read:

When King Mu of the Chou dynasty made a great expedition against the Western Jung people (Hsi-jung[4]) they presented (to propitiate him) a Khun-Wu[5] sword and some fireproof cloth (*huo wan pu*[3]). The sword was one foot eight inches long, a red blade of fire-transformed (or refined) steel (*lien kang*[6])[g] which would cut jade like clay.[h] The fireproof cloth was cleaned by being thrown into a fire, when the cloth became the colour of the fire, and the dirt assumed the colour of the cloth.[i] When taken out of the fire and shaken, it became

[a] *Geogr.* x, i, 6.
[b] v, 156.
[c] *Historiae Mirabiles*, xxxvi; cf. Sarton (1), vol. 1, p. 286. Contemporaries of Chang Hêng.
[d] *Hist. Nat.* xxxvii, 54, 156; cf. Thorndike (1), vol. 1, p. 214.
[e] +2nd or +3rd century; cf. Thorndike (1), vol. 1, p. 551; Cary (1).
[f] Ch. 118, p. 10*b*; cf. Hirth (1), p. 249.
[g] On the development of iron and steel technology in China see Sect. 30*d* below.
[h] The association of the 'jade-cutting knife' (which we shall discuss shortly in connection with that mineral) and the fireproof cloth is repeated in other ancient texts.
[i] This is an excellent description of the glowing red asbestos, with the dirt particles on it not yet brought to red heat.

1 魏略 2 魚豢 3 火浣布 4 西戎 5 錕鋙 6 煉鋼

as white as snow. A certain prince did not believe it, and thought that those who brought news of it must be mistaken, but Hsiao Shu[1] said, 'Must the prince insist on maintaining a preconceived idea, and deny the (demonstrable) truth?'[a]

In this passage the keynote of many subsequent discussions is struck, and Hsiao Shu appears as the prototype of all those Taoists who used the example of fireproof cloth to convince Confucian sceptics that there was more in heaven and earth than was dreamt of in their philosophy. Another somewhat uncertain evidence for knowledge of asbestos in pre-Han times is the quotation from the *I Chou Shu* (Lost Books of Chou)[b] given by Chang Hua about +290, in which a tribute of fireproof cloth is mentioned.[c] Such tributes are recorded frequently in books of the San Kuo and Chin periods, such as the *Hai Nei Shih Chou Chi*[2] (Record of the Ten Sea Islands), or the *I Wu Chih*[3] (Memoirs of Marvellous Things) by Hsüeh Yü,[4] or the *Shih I Chi* (Memoirs on Neglected Matters)[d] by Wang Chia (*c.* +300).

The last-named book, however, has another story which is of interest, referring to the year −598, or more probably −308.[e]

In the second year of King Chao[5] of Yen, the sea-people brought oil in ships, having used very large kettles for extracting it, and presented it to him. Sitting in the Cloud-Piercing Pavilion he enjoyed the brilliant light of the (lamps in which the) dragon blubber (was burnt). The light was so brilliant that it could be seen a hundred *li* away; and its smoke was coloured red and purple. The country people, seeing it, all said, 'What a prosperous light!', and worshipped it from afar. It was burnt with wicks (*chhan*[6]) of asbestos (*huo wan pu*[7]).[f]

Whatever the date to which this really refers, it must surely imply that some kind of primitive sealing or whaling was going on in Han or pre-Han times, and that the oil or blubber was consumed in the courts of the coastal princes with unburning wicks.[g] Rather good evidence for asbestos in the Later Han is the story found in Fu Hsüan's[8] *Fu Tzu*[9] (+3rd century) that Liang Chi,[10] a famous general (*d.* +159), had an incombustible gown, which he used to throw in the fire at parties. To this there are many later European parallels.[h] A reliable statement of a presentation of fireproof cloth by an unnamed western people occurs in the *San Kuo Chih*[i] where such a gift was received by the Wei emperor between +240 and +253.

[a] Ch. 5, p. 27*a*, tr. Wylie (9), mod. et add. Cf. Wieger (7), p. 149. The words about the sceptical prince may well be an addition of the +3rd century, for a reason which will appear immediately below.

[b] Cf. Vol. 1, p. 165 above.

[c] *Po Wu Chih*, ch. 2, p. 6*b*.

[d] Ch. 9, p. 3*b*.

[e] Because there were two rulers of Yen with this same name. At least six hundred years had thus elapsed between the event and our text; yet this is really no worse than the time-gap of the generally accepted Xenophanes text on fossils, p. 622 above.

[f] Ch. 10, p. 3*b*, tr. auct.

[g] Under the heading of *lung kao têng*[11] (dragon oil lamps) the quotation is reproduced in the *Chhi Kuo Khao*[12] (Investigations on the Seven States) by Tung Yüeh[13] (a Ming book), ch. 14, p. 11*a*.

[h] E.g. Marco Polo, ch. 60 (Moule & Pelliot ed.).

[i] Ch. 4, p. 1*b*.

1 蕭叔 2 海內十州記 3 異物志 4 薛珝 5 昭王
6 纏 7 火浣布 8 傅玄 9 傅子 10 梁冀
11 龍膏燈 12 七國考 13 董說

So far there had been no theories about the origin and nature of the fireproof cloth. But Ko Hung about +300 provided several, all wrong. What he said was as follows:[a]

> There are three kinds of fireproof cloth. It is said that in the ocean there is a (volcanic) mountain, Hsiao Chhiu,[1] with fire that burns of itself.[b] This fire rises in the spring and is extinguished in autumn. On the island grows a tree the wood of which is non-inflammable, but only scorches slightly, assuming a yellow colour. The inhabitants use it for fuel, but this fuel is not transformed to ashes. When their food has been cooked they put out the fire by water, and use it again and again—an inexhaustible supply. The barbarians also gather flowers from these trees and weave cloth from them. (This is the first kind of fireproof cloth). Further, they also peel the bark, boil it with lime, and so weave cloth, coarse and not so good as the former. (This is the second kind.) Moreover, there are white rodents (*pai shu*[2]), covered with hairs each three inches long, which live in hollow trees; these can enter fire without being burnt, and their hair may be collected and woven into cloth. This is the third kind.[c]

It is probable that one source of this description was the bark cloth or 'tapa', both beaten and woven, made by various peoples of the south Pacific. There is reason for thinking that Ko Hung took it from the reports of the south seas, particularly an island called 'Natural Fire Island' (Tzu-Jan Huo Chou[3]), which Khang Thai[4] had brought back from his diplomatic mission to Cambodia about the middle of the +3rd century.[d] Many other writers of the same period used this material, generally incorporating the fire-rat story into it, for example the *Wu Lu*[5] (Record of the Kingdom of Wu) by Chang Pho[6] (+3rd century);[e] Kuo Pho's commentary on the *Shan Hai Ching*,[f] and the *Shen I Ching* (Book of Strange Spiritual Manifestations),[g] probably of the +4th century.

We are here in the presence of the salamander legend. In the history of science it is simply an index of the difficulty which ancient minds had in believing it possible that textile fabrics could be anything other than animal or vegetable. Fibres of plastics and glass, modern congeners of asbestos, were still in the womb of time. Ko Hung deserves little credit for launching the fire-rat idea, but Laufer deserves still less for refusing to admit that Ko Hung could have thought of it himself.[h] In fact, the early history of the 'salamander' is obscure. The authenticity of the passage in Aristotle,[i] which speaks of flies engendered in copper furnaces, and of the salamander putting

[a] This passage seems not to be in the *Pao Phu Tzu* now, but is quoted in *TPYL*, ch. 820, p. 8*b*, ch. 869, p. 5*a*, and in Kao Ssu-Sun's *Wei Lüeh*[7] (+12th century), ch. 4, p. 3*a*. A parallel passage occurs in the +6th-century *Shu I Chi*; and *Hsüan Chung Chi* (*TPYL*, ch. 868, p. 8*a*).

[b] Cf. Vol. 2, p. 438 above.

[c] Tr. Laufer (11).

[d] See the discussion of Pelliot (16), p. 74, on the narrative as it eventually found its way into the *Liang Shu*, ch. 54, compiled in the +7th century.

[e] Quoted in *TPYL*, ch. 820, p. 8*a*, and in *Wei Lüeh*,[7] ch. 4, p. 3*a*.

[f] Ch. 16, p. 7*a*.

[g] Ch. 3 (*Nan huang ching*); *TPYL*, ch. 869, p. 7*a*.

[h] 'At first sight it is striking', he says, 'that Ko Hung's notice precedes in time any Western version of the legend, yet this can rationally be explained.' A remarkable example of the *idée fixe* of some Western scholars that no Chinese could ever originate anything.

[i] *Historia Animalium*, 552 b 11, 17.

1 蕭邱 2 白鼠 3 自然火洲 4 康泰 5 吳錄 6 張勃
7 緯略

out fires, is in grave doubt. Most of the references are late, such as that in Augustine[a] (end of the +4th century). There may also be a connection with the phoenix legend,[b] which occurs first in the *Physiologus* book[c] of Christian allegories (end of the +2nd century). In any case the salamander was fully established in Islamic literature by the +10th century[d] and in Europe by the +12th.[e] Since the Chinese references are all earlier than the Arabic ones, and since it would be unthinkable that an idea, even a silly one, could have originated in China, some common source in West Asia about the beginning of our era must be assumed. Thus Laufer.

About the time of the birth of St Augustine, Kan Pao wrote his *Sou Shen Chi* (Reports on Spiritual Manifestations). It contained[f] a passage of much interest on asbestos.

> Within the wastelands surrounding the Khun-Lun mountains there is a burning fiery hill. Upon it there are beasts, birds, plants and trees, which all thrive in the midst of the fire; hence fireproof cloth is either a textile made from the bark of the plants and trees on the hill, or else from the coverings of the birds and beasts.
>
> In the time of the Han dynasty, at a remote period, there were offerings of this cloth from the western regions, but during the long interval which elapsed between that and the beginning of the Wei dynasty, people came to doubt of its existence. The emperor Wên Ti[g] (+220 to +226), considering that the fierce nature of fire was incompatible with the preservation of life, wrote a book entitled *Tien Lun*[1] (Discourses on Literature) in which he showed the absurdity of the whole thing and warned intelligent people against giving any credence to it. When the emperor Ming Ti (+227 to +239) ascended the throne, he issued an edict to the three dukes, saying 'The maxims in the essay by my imperial predecessor are imperishable.' He caused it to be carved in stone outside the door of the ancestral temple, and also among the stone-carved classical texts in the Great College, to be a perpetual testimony to coming generations.
>
> Not long afterwards an envoy from the western regions arrived with an offering of fireproof cloth, whereupon the emperor ordered that the inscription should be obliterated. It thus became a subject of general ridicule.[h]

Although this passage was quoted in the commentary of a dynastic history, it has nevertheless a Taoist flavour of enjoyment at the discomfiture of the imperial Confucian

[a] *De Civ. Dei*, ch. 21.

[b] Cf. Hubaux & Leroy (1); Rundle Clark (1).

[c] Sarton (1), vol. 1, p. 300.

[d] There are some remarkable parallels. For instance, a story of non-inflammable birds from a volcanic region being kept in a palace, told by Su Ê[2] in his *Tu Yang Tsa Pien*[3] (ch. 2, p. 1*a*) at the end of the +9th century with reference to +805, recurs in almost identical words in the writings of Abū 'Ubayd al-Bakrī of Cordova (+1040 to +1094); Hitti (1), p. 568; Mieli (1), p. 185; see de Slane (1), p. 43.

[e] Cf. Robin (1), pp. 136ff.

[f] I say contained, because the passage is not in the text as we now have it. It is quoted by Phei Sung-Chih[4] in his +429 commentary on the *San Kuo Chih* text of +290 (ch. 4, p. 1*b*); and also by the *TPYL*, ch. 820, p. 9*a*. Phei Sung-Chih adds some archaeological observations of his own on these inscriptions made when he himself visited Loyang.

[g] Tshao Phei,[5] son of Tshao Tshao.

[h] Tr. Wylie (9).

1 典論　2 蘇鶚　3 杜陽雜編　4 裴松之　5 曹丕

sceptic; it is thus not surprising that it occurs in a parallel version in *Pao Phu Tzu*.[a] There is no reason to doubt the general lines of the story, however.

We come now to the point of greatest scientific interest, namely the date at which the mineral nature of asbestos was first clearly recognised. Strabo, Dioscorides, and Plutarch were sound on the subject,[b] but Pliny hesitated, being inclined to regard it as a variety of linen—then vegetables flourished and salamanders pranced throughout Islam and the European Middle Ages. It is sometimes said that Marco Polo was the first to report the truth once again,[c] but in fact he had long been anticipated by Chinese writers. Probably the earliest text which can be cited is the *Tung Ming Chi*[1] (Light on Mysterious Things), ascribed to Kuo Hsien[2] of the Han, but more probably of the +5th or +6th century; in this book[d] asbestos is called *shih ma*[3] (stone hemp) and *shih mo*[4] (stone veins). The writer says that 'stone veins' are woven to make string and cord. The stuff comes from Phu-Tung country and is as fine as silk; it will support a weight of 10,000 catties. It is derived from a stone which one must beat in order to separate its fibres. One can plait it into ropes like hemp and ramie, and one can also make cloth of it. It is called 'stone hemp'. In the early Sung there was clarity on the subject—Su Sung says in his *Pên Tshao Thu Ching* (+1070):

> The 'incombustible tree' grows at Shang-tang. It is now to be found in the Tsê-Lu mountains. It is of the nature of a mineral, bluish-white in colour resembling rotten wood. When exposed to the fire it does not burn—hence its name (*pu hui mu*[5]).[e] Some call it 'the root of soapstone (or talc)' (*hua shih chih kên*[6]),[f] for wherever soapstone (steatite) is found, there it is likely to be also. There is no particularly suitable time for collecting it.[g]

So also a younger contemporary, Tshai Thao,[7] discussing asbestos in his *Thieh-Wei Shan Tshung Than*[8] (Collected Conversations at Iron-Fence Mountain, *c.* +1115), remarked, 'It has nothing to do with the hair of rats! (*fei shu mao yeh*[9])'.[h]

Chang Yüeh[10] of the Thang, in his *Liang Ssu Kung Chi*[11] (Tales of the Four Lords of Liang), gives an account of an experimental differentiation of asbestos products in the +6th century, i.e. at a time about the same as that of the *Tung Ming Chi* mentioned above.

> Master Chieh,[12] passing through a market, noticed traders offering three rolls of fireproof cloth. Recognising (the true kind) from afar, he exclaimed, 'This is fireproof cloth indeed—those other two pieces are made from twisted bark, but this one is made from the hair of a

[a] Ch. 2 (tr. Feifel (1), p. 149). Cf. ch. 8 (tr. Davis & Chhen (1), p. 309). And it was often repeated afterwards, e.g. in the *Tu I Chih*[13] (Things Uniquely Strange) of Li Jung[14] in the Thang (ch. 1, p. 7*a*).
[b] Thorndike (1), vol. 1, p. 213; K. C. Bailey (1), vol. 2, p. 256.
[c] Ch. 39 (Yule & Cordier ed., vol. 1, p. 213).
[d] Ch. 3.
[e] This was the traditional pharmaceutical name.
[f] This was the remark which struck de Mély (1), xxiii, 85, 106, 220, as showing such remarkable mineralogical acuity.
[g] *Thu Ching Yen I Pên Tshao*, ch. 6, p. 13*a*; quoted also in *PTKM*, ch. 9, p. 43*b*, tr. Wylie (9).
[h] Ch. 5, p. 20*b*.

1 洞冥記 2 郭憲 3 石麻 4 石脈 5 不灰木
6 滑石之根 7 蔡絛 8 鐵圍山叢談 9 非鼠毛也 10 張說
11 梁四公記 12 杰 13 獨異志 14 李冗

rodent.' On making enquiry of the merchants, their statement agreed exactly with that of Master Chieh. When he was asked the difference between the cloth of vegetal and animal origin, he replied, 'That made from trees is stiff, that from the rodent hair is pliable; such is (one) way to distinguish them. Moreover, if you take a burning-mirror and ignite the bark of *chê*[1] trees[a] on the north side of a hill, you will find that it will soon be changed (i.e. bark-cloth is not fireproof).' The experiment was made, and turned out just as he said.[b]

Here then, although the salamander theory was accepted, it was understood that bark-cloth was not properly *huo wan pu*.

Many later rejections of the fable can be found; Wylie instanced two Ming books, the *Shu Wu I Ming Su*[2] (Disquisition on Strange Names for Common Things), by Chhen Mou-Jen;[3] and the *Tan Chhien Tsung Lu*[4] (Red Lead Record) by Yang Shen.[5] On the other hand numerous writers upheld either the vegetable[c] or the animal[d] theory, or both at once.[e] In +1430 Chang Ning[6] wrote:

The first fireproof cloth I ever saw was at the house of Chang Hsing-I at Suchow, and at the 'Pure Unity' Buddhist cloister at Jen-ho near Hangchow, in both instances about the size of a two-cash piece. Recently I saw some at the house of Chu Ming-Yu, long and narrow like a sash. When saturated with oil it could be used as a candle; placed over the fire it might be used to burn incense. When the oil was exhausted and the fire extinguished, the cloth was as perfect as before. The statements regarding the handkerchief of Liang Chi, the tribute in the time of Emperor Wu of the Wei, and the record in the Yuan history of the stone tissue at Chhieh-Chhih Mountain, which could be woven, are all credible therefore, and no fables.[f]

Here he refers to a memorial by a Muslim official in Mongol service, A-Ho-Ma,[7] in +1267, which described asbestos by yet another name, *shih jung*,[8] 'stone silk-floss'.[g]

In spite of anything that Dioscorides or Su Sung had ever said, the Oxford Philosophical Society in +1684 was still about on the level of Ko Hung. On 10 September in that year, Nicholas Waite wrote from London to Edward Tyson, in a letter still preserved in the Society's correspondence:[h]

The greate respect and honour I bore to those most learned and ingenious professors of your Society I did in few days after my arrivall into this Citty (Oxford) expose to your sight and tryall a piece of Cloth which by your experiment of the fire consumed nott, and you

a *Cudrania* spp. (R nos. 599, 600; B II, 501).
b Quoted also in *Wei Lüeh*, ch. 4, p. 3*b*, tr. Laufer (11).
c E.g. the *Hsüan Chung Chi*,[9] by a Mr Kuo,[10] of uncertain, but pre-Sung, date (*YHSF*, ch. 76, p. 30*a*, *b*).
d E.g. the *Fa Mêng Chi*,[11] by Shu Hsi.[12] If the authorship is genuine, this mention of the 'fire-rat cloth' may antedate Ko Hung's, being of about +280. The passage was preserved in *Chhu Hsüeh Chi*, ch. 29, p. 30*a*, and *TPYL*, ch. 911, p. 6*a*; whence *YHSF*, ch. 62, p. 21*a*.
e E.g. the *Yeh Kho Tshung Shu* of +1201, ch. 30, p. 12*b*.
f In his *Fang-Chou Tsa Yen*[13] (Reminiscences of (Chang) Fang-Chou), tr. Wylie (9).
g *Yuan Shih*, ch. 205, p. 2*a*.
h Gunther (1), vol. 12, p. 226; letter no. 126.

1 柘 2 庶物異名疏 3 陳懋仁 4 丹鉛總錄 5 楊慎
6 張寧 7 阿合馬 8 石絨 9 玄中記 10 郭
11 發蒙記 12 束晳 13 方洲雜言

being then desirous I should give you a narrative of its substance and in whatt partes of India made, will now communicate the same accompt I received from one Conco, a naturall Chynees resident in the Citty of Batavia in the North East partes of India (*sic*), who by means of Keay Arear Sukradana likewise a Chynees and formerly Cheife Coustomer to the Old Sultan of Bantam, did after several year's dilligence procure from a greate Mandarin in Lanquin, a province in China, nere $\frac{3}{4}$ of a yard of said Cloth, declaring thatt he was Credibly informed thatt the Princes of Tartaria and others adjoyning, use it for burning their dead, and thatt itt was believed by them to be made of the underparte of the roote of a Tree growing in the province of Sutan; and suposed in like manner of the Todda Trees in India and of the upper parte of said roote nere the surface of the ground, was made a finer sorte, which in three or fower times burning I have seen diminish almost halfe; and thatt out of said Tree distills a Liquour which nott consuming is used with a weeke of the aforesaid Cloth to burne in their Temples to posteritie. Now if the nature of this Subject as a vegittable correspond not with your greate experiments or judgments, your Commands obleidging me to render the same relation I received, hope there will nott be any ill construction made thereof, which is all the favour and kindness desired by, Gentlemen, Your most assured humble Servant, etc.

One hopes, indeed, that they put no ill construction on it; in any case, by +1701, when Ciampini's letter was published in the *Philosophical Transactions of the Royal Society*, the matter was set at rest once and for all, and asbestos was acknowledged mineral. Yet by a crowning irony, the Jesuit Ferdinand Verbiest contributed to the *Thu Shu Chi Chhêng* encyclopaedia in +1726 a description of the *sa-la-man-ta-la*[1] with an appropriate legend.[a]

Once again we meet with the same pattern of advance; first the Greeks record accurate information, but then between Hellenistic times and the Renaissance the Chinese are more advanced than Europeans.[b]

(5) Borax

Borax (native sodium tetraborate, or tincal) is apparently not mentioned in ancient Chinese writings, and first appears in the pharmacopoeia in the *Jih Hua Pên Tshao*[2] of *c.* +970.[c] This is about the same date as its first mention in the West—'Armenian borax', in a Coptic manuscript.[d] It probably came to China, at that time as later, from the natural deposits in what is now Chhinghai province (north-eastern Tibet), around Lake Kokonor. It may also have been produced from certain of the brines at the Tzu-liu-ching field (see below, Sect. 37), as it is today. The use of borax (*phêng sha*[3]) as a preparatory agent for soldering and brazing (in the molten state it cleans metal

[a] *Chhin chhung tien*, ch. 125, *hui khao*, 3, pp. 8*b*, 9*a*.

[b] It would be interesting to know more of Persian and Indian literature, i.e. that of those countries where asbestos was probably actually produced. Yet Finot's Indian lapidaries make no mention of asbestos.

[c] Laufer (1), p. 503, suggests that the *hu lo*[4] mentioned as a Persian product in the *Sui Shu* (ch. 83, p. 7*b*) may be a transliteration of the Persian word *būrak*, which is a forerunner of our borax. In this case, the Thang people may have received some of it from the west (Persia or Tibet).

[d] Partington (1), p. 193.

1 撒辣漫大辣 2 日華本草 3 硼砂 4 呼洛

surfaces by dissolving metallic oxides)[a] goes back in China to the +11th century, for it is mentioned by Su Sung (*kho han chin yin*[1]),[b] and the Chinese chemists of this time must have been quite interested in it, as it is referred to by Tuku Thao[2] and the 'Earth's Mansions Immortal'. Li Shih-Chen says that borax 'kills' the five metals, as saltpetre does; presumably this refers to the preparation of metallic salts.[c] The mild and non-irritant antiseptic quality of borax, which has given it such wide use in Western, and even modern, medicine, was appreciated by the Chinese pharmacists, who prescribed it for all kinds of external, including ophthalmic, affections.

(6) Jade and Abrasives

Jade calls for more than passing remark, since it is no ordinary mineral; the love of it was one of the most characteristic features of Chinese civilisation, and its texture, substance and colour gave inspiration to carvers, painters and poets for more than three thousand years.[d] Jade has given rise to an immense literature both in Chinese and Western languages, but this is mostly concerned with aesthetic appreciation and social uses. What is of interest to us here is rather the technological aspects of the mineral, its mining, and above all the manner in which it was worked, no easy achievement in view of its great hardness. To such a study the recent book of Hansford has been a valuable contribution.

The word which we use, jade, is a corruption of the Spanish *ijada*, meaning the flank or the loins, the full form being *piedra de ijada*. During the course of the Spanish conquests, green stones much prized by the Mexicans as amulets against diseases of the kidneys were brought back to Europe together with repute of their worth. The alternative name, *piedra de los riñones*, was Latinised as *lapis nephriticus*, hence the modern term nephrite. The ancient Chinese word, which had none of these connotations, has been discussed above.[e] It is a remarkable fact that the appreciation of jade and the art of working it was a feature also of ancient Central American civilisation,[f] and of the New Zealand Maoris.[g] The Amerindian material was jadeite and that of the Maoris nephrite. Current researches in Siberian archaeology are beginning to suggest a common prehistoric origin for the jade-loving cultures on both sides of the Pacific.

The largest book on jade (and probably the largest on any subject in any language, for it is a mammoth work) is that by Bishop, Bushell, Kunz, Li Shih-Chhüan, Lilley, Thang Jung-Tso *et al.* which describes the Bishop Collection, much of which was subsequently presented to the Metropolitan Museum, New York. The text and translation of a specially-written 'Discourse on Jade' (*Yü Shuo*[3]) by Thang Jung-Tso appears in it, with a series of pictures in Chinese style by Li Shih-Chhüan (*Yü Tso Thu*[4])

[a] This invention can apparently be traced back to Mycenae (Partington (1), p. 351). Cf. Maryon & Plenderleith (1), pp. 649ff.; Forbes (8), pp. 47ff.

[b] Cit. *PTKM*, ch. 11, p. 35*b*.

[c] Cf. Vol. 1, p. 212 above.

[d] In the Section on Taoism, Vol. 2, p. 43 above, a text was given which showed an early attempt to describe the mineralogical qualities of jade, though only ethical and aesthetic terms were available.

[e] P. 641.

[f] See Joyce (1); Kraft (1).

[g] See Chapman (1).

1 可銲金銀 2 獨孤滔 3 玉說 4 玉作圖

illustrating the processes of working jade. It also contains a mass of material, edited by Kunz, on jade mineralogy. Better known is the monograph on jade by Laufer (8),[a] almost purely archaeological in content, however. Laufer's work should not be used without recourse to Pelliot (20), whose introduction to the Lu Catalogue is considered the most acute, critical and scholarly discussion of ancient jades which we have. Recent books which may also be mentioned are those of Pope-Hennessy (1, 2) and Salmony (1).[b] Needless to say, there is a mass of material in the Chinese encyclopaedias, eight chapters being devoted to jade, for example, in the *Thu Shu Chi Chêng*.[c] Chang Hung-Chao (*1*) gives five useful sections to the subject.

True jade (*chen yü*[1]), or nephrite, is a crypto-crystalline silicate of calcium and magnesium belonging to the amphibole class of minerals and related to fibrous actinolite (amphibole asbestos). Jadeite, though similar in appearance, is a silicate of sodium and aluminium classed among the pyroxene minerals.[d] It is also crypto-crystalline, but generally composed of small grains rather than minute fibres,[e] hence, though slightly harder than nephrite, it does not offer quite such a tough resistance to the tools of carvers. The colours of jade are due mainly to the presence of compounds of iron, manganese, and chromium. The first are the most important, giving all shades from the rich greens through yellows and browns to blacks. Greyish tints may be due to manganese, and some blacks to chromite but the chief significance of chromium is in the apple and emerald greens of jadeite. Pale yellows may sometimes be due to titanium and blue and lavender tints to vanadium.

There are varieties of other minerals resembling jade fairly closely; the Chinese knew them as 'false jade' (*fu yü*[2]). Some specimens were certainly serpentine,[f] others pyrophyllite[g] or even greenish steatite.[h] All of these are considerably softer than true jade. Indeed, one of the most interesting features of jade is its great hardness—on the Mohs scale jadeite is 6·5 to 7, nephrite 6 to 6·5, quartz is 7 and felspar 6.[i] This region is all in the upper half of the whole range reaching from talc at the soft end to diamond at the hard. It should be realised that this level of hardness is much harder than any pure metal; so that the working of jade must have been a great problem for the ancients. True jades can also often be distinguished from false by measurements of density and refractive index.

[a] This drew much on the greatest Chinese study of jades, the work of Wu Ta-Chhêng (*2*). But it was somewhat impaired by reliance on the 18th-century forgery which purported to be a description of the imperial collection of jades in +1176 (see Vol. 2, p. 394 above).

[b] Cf. Chêng Tê-Khun (3); H. A. Giles (5), vol. 1, p. 312.

[c] *Shih huo tien*, chs. 325–32.

[d] Cf. Yoder.

[e] I say generally, because fibrous samples are known.

[f] Magnesium silicate. The 'jade' of north-west Kansu is known to be green serpentine (Pelliot, 21). When at Yümên[3] (Jade Gate) in 1943, I obtained four beautiful wine-cups of this stone. There is a curious historical confusion here; the location of the Han and Thang Yümên was much farther west, in Sinkiang, nearer to the principal source of true jade, and the present Yümên obtained its name only in the Ming. But its jade is only 'jade'.

[g] Aluminium silicate.

[h] Soapstone, magnesium silicate.

[i] The statement often seen, that jade is harder than quartz, is not correct.

[1] 眞玉 [2] 砆玉 [3] 玉門

After elaborate discussions on the sources of jade, it is now agreed that the rivers and mountains of Khotan (Yü-Tien[1]) and Yarkand in Sinkiang were the principal, perhaps the only, centres of production of the mineral for over two millennia. Already in the −4th century, as Haloun (4) has shown, the Yüeh-chih people were intermediaries in the trade. The *Chhien Han Shu* noted the mines.[a] Particulars about the region given in Chinese books have several times been translated.[b] Khotan is an oasis surrounded to the north, east and west by the desert of the Tarim basin, and to the south by the lofty Khun-Lun mountains; it has always been on one of the loops of the Old Silk Road. The jade was found in the valleys of two rivers, the Karakash and the Yurungkash, either mined (*shan liao*[2]) or collected as lumps in the river-bed (*tzu yü*[3]). Tu Wan, in his lapidary of +1133, speaks of both methods, but Sung Ying-Hsing's[4] *Thien Kung Khai Wu*[5] (The Exploitation of the Works of Nature) of +1637 describes only the collection of water-worn pieces (cf. Fig. 272).[c] We possess descriptions of the jade mines from two modern travellers, Cayley, who visited them in +1870, and Stoliczka, who was there four years later; at that time they were quite deserted owing to local rebellions.[d] From their descriptions, which did not give much praise to the mining methods used, it seems that fires were set against the veins to cause cracks, these being afterwards opened by wooden wedges.[e] The veins are remarkable, however; Stoliczka saw some of pale green jade amounting in thickness to ten feet.

Jadeite was not known in China before the 18th century, at which time it began to be imported through Yunnan from the Burmese deposits. An ancient term, *fei tshui*,[6] which had originally meant kingfisher plumage and subsequently been applied (as by Tu Wan) to certain fine green nephrites, was revived to describe it.[f] There are descriptions of the Burmese mines by Griffith[g] and Chhibber.[h] The question whether there were ever any sources of nephrite or jadeite within China proper is discussed by Hansford at length, with negative conclusions. It was for long supposed that jade weapons and implements found at neolithic sites in Europe must also have come from Central Asia, but European deposits are now known to exist.

As for the ancient history of jade in China, we know that it was carved already by the Shang people (−13th century) as objects have been found at Anyang (Shih Chang-Ju, *1*, *2*). Jade pieces of Chou date have been described by Kuo Pao-Chün, and Han ones by Sekino *et al.* (*1*).[i] Here we are not concerned with problems of stylistic

[a] Ch. 96A, p. 8*a* (tr. Wylie, 10).

[b] Notably the version of Rémusat (7) of the chapter on Khotan in *TSCC*, *Pien i tien*, ch. 55; and a paraphrase by Ritter (2), vol. 5, p. 401, of the relevant parts of Chhi-shih-i Lao-jen's 18th-century *Hsi Yü Wên Chien Lu*.

[c] In this picture we see women and girls at work in the river; they were supposed to go in naked, the idea being that their Yin-ness would attract the Yang jade.

[d] In the 17th century Benedict Goes (see Vol. 1, p. 169) also made some observations as he passed through Sinkiang on his way to China (Trigault, tr. Gallagher, pp. 506 ff.). Goes was given a rich present of jade by his friend the Princess of Kashgar to sell in China and to support himself with on the way (p. 502).

[e] Cf. Hoover & Hoover (1), p. 118.

[f] Cf. Hansford (1), p. 45.

[g] (1), p. 132.

[h] (1), p. 24.

[i] Perhaps one of the earliest literary references to jade is in the Yü Kung, chapter 6 of the *Shu Ching* (under Yung-Chou province); cf. Karlgren (12), p. 15; Demiéville (2).

[1] 于闐 [2] 山料 [3] 子玉 [4] 宋應星 [5] 天工開物 [6] 翡翠

Fig. 272. Collection of water-worn jade nuggets by women and girls in the Karakash and Yurung-Kash rivers at Khotan (Sinkiang); a picture from Sung Ying-Hsing's *Thien Kung Khai Wu* of +1637, ch. 18.

chronology, but with the ancient methods of working the jade. The neolithic jade tools described by Andersson (5)[a] give evidence of having been made without the help of metal, but abrasive sand must have been used, probably with laminae of sandstone or slate as cutters, and abrasion is the secret of all subsequent jade working.[b] The ancient name for these cutting or grinding stones was *chih-li*;[1] this is frequently found in the classics.[c] The *Shuo Wên* gives other words, such as *lan*[2] and *tsho*[3] with the same meaning. *Huai Nan Tzu* speaks of *chien-chu*,[4] saying that it was used for working jade. Similar achievements were made in the earliest periods of ancient Egypt, as Partington points out,[d] for example, vases of diorite and bottles of rock crystal with narrow necks,

[a] Cf. Laufer (8), pl. II.

[b] Early references to this are very rare, but Chou Mi at the end of the Sung describes the use of river-sand by the jade workers (*Chhi Tung Yeh Yü*, ch. 16, p. 11*a*).

[c] E.g. Yü Kung chapter of *Shu Ching*; *Shih Ching*; also *Shan Hai Ching*; see Chang Hung-Chao (*1*), p. 178.

[d] (1), p. 96.

[1] 砥礪 [2] 厱 [3] 厝 [4] 礛[石+諸]

when iron tools were certainly not available. He imagines abrasive powders, obsidian cutters and 'almost infinite patience'. Hence the interest of the 'Khun-Wu' sword or knife, which 'would cut jade like clay', and which comes into prominence about the −3rd century. We met with it in the quotation from *Lieh Tzu* a few pages back,[a] but many other references could be adduced.[b] One feels that just as asbestos was a real thing, so also there was something behind this curious term. Laufer (12) built up an elaborate argument to prove that it was the diamond point, but Hansford gives good reasons for rejecting this view. The phrase 'Khun-Wu knife' became proverbial in later times for any steel knife of fine quality.[c] The only connection with the diamond is that the term for the latter, *chin kang*,[1] was also in late times applied to other hard substances such as corundum, and too much emphasis cannot therefore be placed on Li Shih-Chen's comment on a passage which he quotes[d] from the +4th-century *Hai Nei Shih Chou Chi*.[e] This says that Khun-Wu stone is found in the western regions near the Shifting Sands, and is smelted like iron to make knives which shine like rock crystal and cut jade like mud. Li Shih-Chen simply says that this is 'the largest kind of *chin-kang*'. Moreover, while in traditional jade-cutting the diamond is a useful auxiliary, as Hansford puts it, all the important cutting tools are of iron or steel.

The central invention here was the rotary disc-knife (*cha tho*[2]); see Fig. 273. The oldest reference to it seems to be indirect, namely the mention in the *Chin Shih* (History of the Chin (Tartar) Dynasty)[f] of abrasive sand for use with jade-grinding wheels (*nien yü sha*[3]); this refers to the +12th century, though compiled about +1350. The *Yuan Shih* says that in +1279 an official Abrasives Depot was established at Ta-thung. The first specific reference to the rotary disc-knife is in the *Thai-Tshang Chou Chih*[4] (Topography of Thai-tshang, in Chiangsu), the earliest edition of which is +1500; here it is called *sha nien*,[5] with reference to a famous worker, Lu Tzu-Kang.[6] These are relatively late texts, but perhaps it is not surprising that the methods used in ancient times were kept secret by the technicians, in this case as in so many others. The possibility in fact arises that what lay behind the Khun-Wu story was the first application of a treadle-driven rotary steel knife and abrasive sand.[g] Hansford has found no evidence that rotary tools were used for this purpose in the bronze age, apart from the

[a] Ch. 5, p. 27*a*.

[b] E.g. *Po Wu Chih*, ch. 2, p. 6*b*; *Shih I Chi*; *Khung Tshung Tzu*; *Hsüan Chung Chi*; *Hou Han Shu*, ch. 112B, p. 18*b*, in the commentary on the biography of Kan Shih.[7]

[c] E.g. the +13th-century *Sui Yin Man Lu*,[8] ch. 4, p. 7*a*.

[d] *PTKM*, ch. 10, p. 37*a*.

[e] It is in the entry for Liu-chou.

[f] Ch. 24, pp. 12*b*, 13*a*. It was tribute from places in what is now part of Inner Mongolia.

[g] The usual view is that rotating knives were not used till the +5th or +6th century (Hansford (1), p. 64), but it lacks any convincing basis. From what we know in general of anthropology and the history of technology, there seems no reason for thinking that a treadle-mechanism could not have been used as early as the −3rd century; cf. Leroi-Gourhan (1). On the other hand Chêng Tê-Khun (4) maintains that rotary disc-knives as well as point and tube drills, possibly of bronze, were used in the Shang period. Usher (1), p. 107, and Middleton (1) believe that disc-knives rotated by a bow-drill mechanism (see on, Sect. 27*a*, *b*) were regularly used by Hellenistic gem-cutters and carvers.

1 金剛 2 鍘鉈 3 碾玉砂 4 太倉州志 5 砂碾
6 陸子剛 7 甘始 8 隨隱漫錄

bamboo drill. By the use of the microscope, however, he has discovered traces of rotary tools on late Chou perforated jade discs (*pi*[1]), and this would be in agreement with our tentative interpretation. It is also in fair conformity with the rather late date for the general use of iron in China.[a] Whether jades of earlier periods show any similar traces has yet to be determined with certainty. In any case, whoever introduced the rotary disc-knife deserves an important place in the history of technology when we remember the cutting wheels of today, made of rubber-bonded silicon carbide and rotating at 20,000 r.p.m. (170 m.p.h. at the cutting edge), wheels which will cut almost anything at any temperature.[b]

Grease was used as a medium for the abrasives. In +1092, Lü Ta-Lin, describing a San Kuo belt-hook (*tai kou*[2]),[c] wrote: 'It is said that with toad-grease (*chhan fang*[3]) and a Khun-Wu knife, jade can be worked like wax.'[d] About +1470 Tu Ang,[4] in his *San Yü Chui Pi*,[5] says the same thing, using the words *hsia ma fang*,[6] and citing Thao Hung-Ching of the +5th century.[e] Of course many literary writers in all centuries failed to realise that the abrasive contained in the grease was the important factor, and thought that the organic material must exert some softening effect on the stone.

Abrasive grinding (*cho*,[7] *cho mo*[8]) was carefully studied in Peking by Hansford just before the second world war. His illustration of the large grinding disc (Fig. 274) may be compared with the drawing of Li Shih-Chhüan (Fig. 273). He found six abrasives in use, the softest being quartz sand (*huang sha*[9]) and the hardest being diamond points on drills. There were also crushed almandine garnets (i.e. silicate of calcium and iron; *hung sha*,[10] *chieh yü sha*,[11] *tzu sha*[12]), crushed black corundum (i.e. emery; oxides of aluminium and iron; *hei sha*[13]), and modern silicon carbide (cf. Fig. 275). The polishing medium (*pao yao*[14]) is a mixture of carborundum and very fine loess calcite. It is fairly clear that the most ancient abrasive was the quartz sand, and that the corundum was a discovery of the +12th century about the time referred to in the Chin and Yuan Histories quoted above, while the garnet had come into use rather earlier, in the +10th. If Chang Hung-Chao[f] is right in identifying *wên shih*[15] with garnet powder this can be traced back to the Thang. The general succession is exactly the same as the order of hardness of the three abrasives.[g]

For an account of modern traditional jade-working Hansford will not be found

[a] Cf. Sect. 30*d* below.

[b] It is almost bizarre that the late Dr H. W. Dickinson should have regarded the Taylors of Southampton about +1760 as the first mechanics to have made the circular saw into a practical workshop tool

[c] See e.g. Lemaître (1); Alley (4).

[d] *Khao Ku Thu*, ch. 8, p. 10*a*.

[e] Hansford did not recognise here one of the technical terms for gear-wheels (*hsia-ma*) so that 'toad-grease' may not mean quite what it says. In fact, it suggests the use of toothed wheels for mechanical advantage in some jade-cutting techniques, perhaps now abandoned.

[f] (*1*), p. 129.

[g] Laufer (12), p. 51, gives evidence to show that Chinese corundum was exported to the Islamic countries during the Sung.

1 璧 2 帶鈎 3 蟾肪 4 都昂 5 三餘贅筆 6 蝦蟆肪
7 琢 8 琢磨 9 黃砂 10 紅砂 11 解玉砂 12 紫砂
13 黑砂 14 寶藥 15 文石

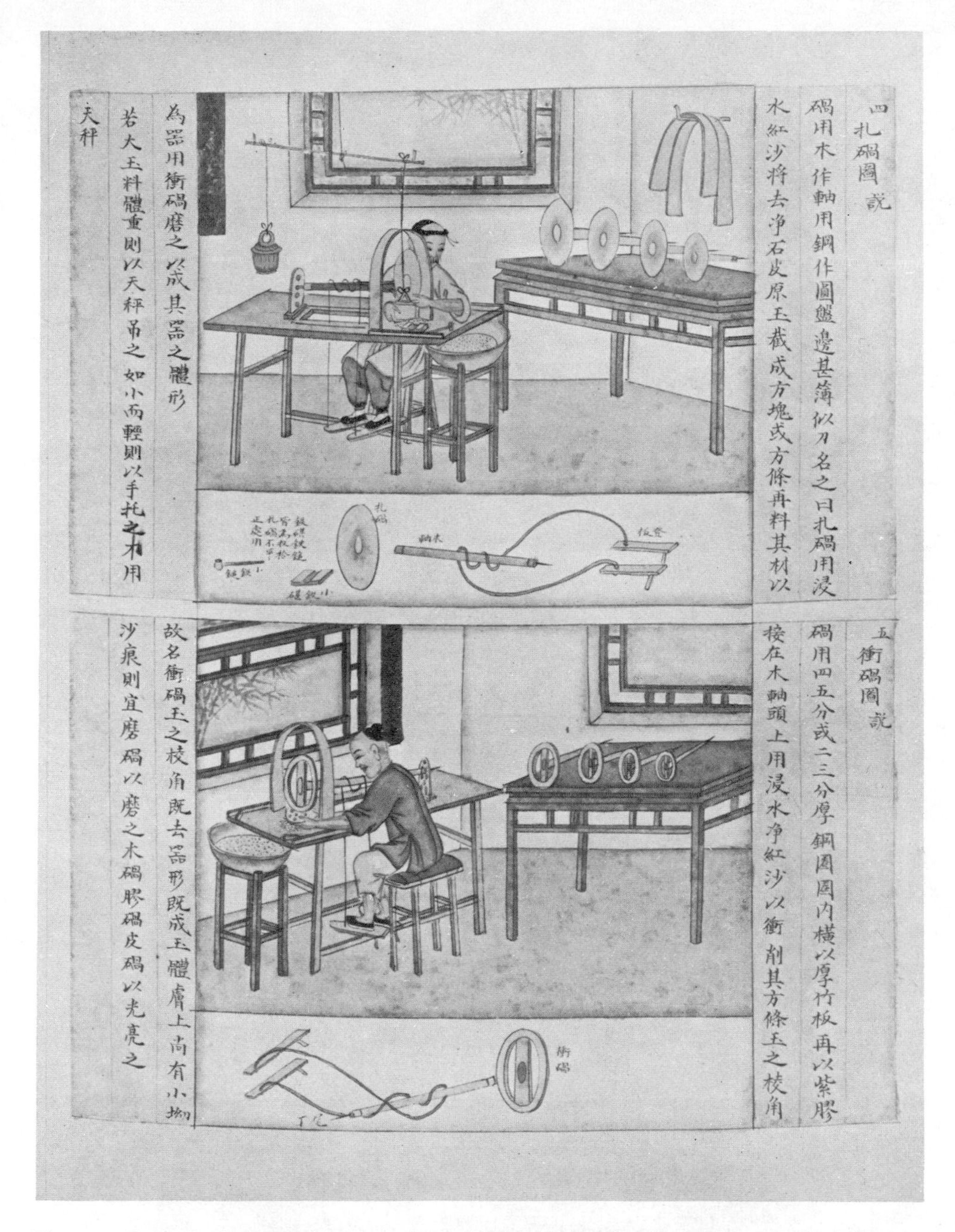

Fig. 273. Rotary tools for working jade (from Li Shih-Chhüan). Above, the steel disc knife (*cha tho*, here called *cha thuo*) with its treadle mounting and protective shield; the accompanying text mentions that *hung sha* (crushed almandine garnets) should be used with it. It is also said that when the piece of jade being worked upon is heavy, a balance suspension should be rigged up, and indeed one is shown in the picture. Below, the steel grinding wheel (*mo tho*, here called *chhung thuo*), also rotated by a treadle and operating with crushed garnets as abrasive. The text adds a reference to the polishing wheels such as the *chiao thuo*, made of a mixture of shellac (*tzu chiao*) and carborundum, and the leather buffing wheels (*phi thuo*).

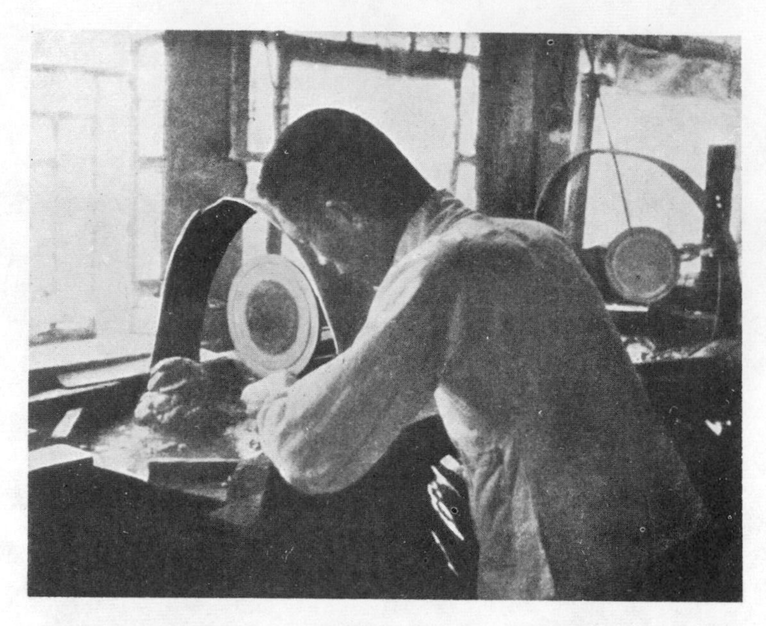

Fig. 274. A Peking jade worker using the large steel grinding wheel (*mo tho* or *chhung thuo*); from Hansford (1).

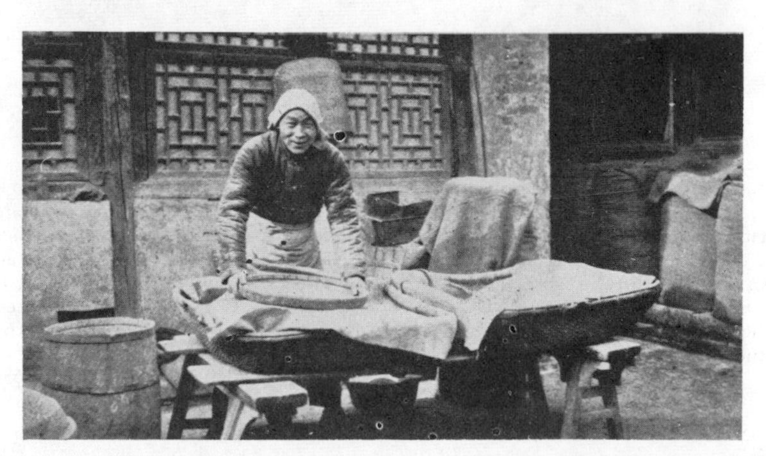

Fig. 275. Preparation of the abrasives in a Peking jade workshop; above, crushing garnets in a roller-mill; below, sifting the crushed material (from Hansford, 1).

disappointing.[a] Apart from the wire saw (*la ssu tzu*[1]) most of the tools (*tho tzu*[2]) are mounted on treadle lathes. These include not only the discoidal knives but also all kinds of borers (*la tsuan*,[3] the tubular drill, and *ta yen tsuan*,[4] the fine diamond drill).[b]

(7) Precious Stones, including the Diamond

Precious stones demand a brief reference. There comes a point, however, where it is hard to distinguish between the history of science and the history of substances and objects which have some scientific interest. Since gems were prized for purely aesthetic or superstitious (medical)[c] reasons, and since it was impossible to classify them rationally before modern times, their history is of primarily archaeological concern and we cannot expect it to throw much light on the development of scientific thought or techniques. The subject is moreover bedevilled by uncertainties in the identification of the various terms both in East and West, and if this is serious in the case of European history[d] it is far worse when the technical terms for precious stones in India,[e] Persia and China[f] have to be considered.[g] I do not know of any special work in Chinese medieval times devoted solely to precious stones,[h] but of course there is a mass of references to them in the lapidaries, antiquarian books, and encyclopaedias already mentioned. Fig. 276 shows a Chinese gem miner descending a shaft (+17th century). Bretschneider translated[i] a short section of the *Cho Kêng Lu* of +1366 about the gems of foreign origin which the author, Thao Tsung-I, knew at that time. Pfizmaier (95) translated a chapter of the *Thai-Phing Yü Lan* encyclopaedia (+980) on gems.

Two monographs were devoted by Laufer to the subject. In the more important one (12) he investigated Chinese knowledge of the diamond. We have already had to deal with the linear or chain-like macromolecules of asbestos and with a planar or layer-lattice structure—mica. The diamond is fully three-dimensional, with a cross-linked infinite lattice. There can be little doubt that the main ancient source of this remarkable crystalline carbon, the hardest of 'stones', was India, and among the earliest references is that in the −1st-century *Arthaśāstra*,[j] though Dioscorides also knew it in the +1st. The *Periplus* mentions its export from India.[k] The earliest

[a] Consult also Hildburgh (1).

[b] Referred to by Chou Mi at the end of the +13th century (*Chhi Tung Yeh Yü*, ch. 16, p. 11*a*).

[c] There is a good review of this aspect by Ponder (1).

[d] Cf. the standard works such as King (1, 2); Bauer (1); Spencer (1); Kunz (1, 2); Evans (1); McLintock & Sabine (1), etc.

[e] Cf. Finot (1).

[f] Cf. Laufer (1, 12, 13); Chang Hung-Chao (*1*); T. Wada (1).

[g] Gem-stones and their tentatively identified Chinese names are included in the Table in Sect. 33 below.

[h] The *Thu Ching Yen I Pên Tshao* (ch. 1, p. 13*b*) refers to a *Pieh Pao Ching*[5] (Manual of the Distinctions between Precious Stones), but this is not in the bibliographies of the official histories, and it has not been possible to trace any further information about it.

[i] (2), vol. 1, pp. 173, 194. Reprinted by de Mély (1), p. 251.

[j] Shamasastry ed., p. 89. According to the latest views (Kalyanov, 1) this work did not attain its present form until the +3rd century.

[k] Schoff ed., p. 45. On the medieval Levantine and Asian gem trade see Mugler (1).

[1] 拉絲子 [2] 鉈子 [3] 拉鑽 [4] 打眼鑽 [5] 別寶經

Fig. 276. A gem miner descending a shaft (from the *Thien Kung Khai Wu* of +1637, ch. 18). Sung Ying-Hsing says that the gem miners of Yunnan and other remote provinces often meet with dangerous, even lethal, gases in the workings, so that they rope themselves together in case one of them is overcome. The pits are generally very deep but free from water. The words *phou mien* on the right simply mean 'cut away so as to show the shaft'.

reference to the diamond in China was thought to be that in the *Chin Chhi Chü Chu*[1] (Daily Acts of the Chin Emperors)[a] which says that the Tunhuang region presented diamonds in +277, adding that they are derived from gold, come from India, can be used to carve jade, and though scoured many times, do not dissolve.[b] In this passage they bear the name ever afterwards retained, *chin-kang*.[2] Recently, Maenchen-Helfen (2) has been able to adduce a still earlier passage, dating from +114. The term, which means literally 'gold-hard' or 'metal-hard', was used by the Han and Chin translators to render the Buddhist term *vajra*, or thunderbolt of Indra, so that it meant in general anything which was firm, hard or indestructible. The close association of

[a] Quoted in *TPYL*, ch. 813, p. 8*b*.
[b] Laufer (12), p. 35; Hansford (1), p. 109.

[1] 晉起居注 [2] 金剛

the diamond with gold occurs also in Pliny, but the origin of the idea, apart from some general similarity of very precious and brilliant things, has not yet been explained.

For the folk-lore of the diamond the reader is referred to Laufer's monograph. Stories about its collection were similar in East and West. There was a legend that pieces of flesh were thrown into valleys where the diamonds lay, and were then picked up with the stones adhering to them by eagles or other birds so that the gems could be collected. This is related by Epiphanius, a Cypriot bishop, in the latter part of the +4th century,[a] and appears also in the *Liang Ssu Kung Chi* with reference to about +510.[b] The story may have originated in sacrifices of animals at the openings of mines, or perhaps also in a method later used in China[c] in which people walked up and down over diamond-containing sands with grass-sandals, these being later collected and burnt, and the stones collected from the ashes.

Presumably it was the exceptional hardness of the diamond which gave rise to the notion of its indestructibility.[d] Ko Hung mentions this[e] about +300, and reproduces another story common to both East and West, that the diamond was sensitive to the horn or blood of rams.[f] Lead was also thought to be capable of breaking it, as we find both in Dioscorides, in Arabic alchemical writings, and in the *Tan Fang Chien Yuan* of Tuku Thao (+11th century).[g] This certainly originated from the practice of wrapping up diamonds in lead foil so that their splinters should not be lost when diamond powder was being made. The use of diamonds for cutting hard stones such as jade has been read (probably wrongly) into early Chinese references such as the *Hsüan Chung Chi* (+6th century)[h] where a roller-cutter may possibly be referred to; but in later texts, such as Chou Mi's *Chhi Tung Yeh Yü* (*c.* +1290)[i] it is to be accepted. Apparently, however, the Chinese were not acquainted with cut and polished diamonds until the Portuguese brought some to Macao.[j]

As regards other stones, Farrington & Laufer (1) have given a short, but excellent, paper on the agate, which includes Chinese material. A more exhaustive study was that of Laufer (13) on the turquoise (hydrated aluminium phosphate). Our name for it derives simply from Turkey, whence we first knew it, but anciently it seems to have originated from Persia, spreading east as well as west and reaching India and China. Especially in Tibet was it prized, forming, with coral, the two commonest decorative stones on all Tibetan works of art, and taking a place almost equivalent to jade in China.

[a] Laufer (12), p. 9. Marco Polo was one of the many who told the story (ch. 175, Moule & Pelliot ed.).
[b] Laufer (12), pp. 7, 19, who considers, plausibly enough, that the story migrated from West to East because the *Liang Ssu Kung Chi* refers to Byzantium (Fu-Lin) in connection with it.
[c] Laufer (12), p. 6.
[d] The diamond's combustibility was not demonstrated till +1777 by Bergmann (v. Kobell (1), p. 388).
[e] Cit. by Li Shih-Chen, *PTKM*, ch. 10, p. 36*b*.
[f] An idea probably astrological in origin (Laufer (12), p. 24).
[g] Cit. by Li Shih-Chen in the same entry.
[h] In Ma Kuo-Han's collection (*YHSF*, ch. 76, p. 34*b*), also *TPYL*, ch. 813, p. 8*b*.
[i] Ch. 16, p. 11*a*, *b*.
[j] So Fang I-Chih[1] in the *Wu Li Hsiao Shih*[2] (Small Encyclopaedia of the Principles of Things), +1664, ch. 8, p. 28*a*.

[1] 方以智 [2] 物理小識

Since the Tibetan word for it, *gyu*, is indigenous, it must have been known there since very remote times. The Chinese term *sê-sê*,[1] obviously a transcription of some foreign word, and thought by previous scholars to mean turquoise, is identified by Laufer with the rubies of Badakshan. There are numerous references to them in the official histories of the Thang. What seems to be certainly turquoise is first mentioned in the +14th-century *Cho Kêng Lu* as *tien tzu*.[2] Its modern name, *lu sung shih*,[3] goes back only to the +18th century, but the stone has been mined in China since the Yuan time.

(8) The Touchstone

The touchstone deserves a notice on account of its connection with an interesting piece of technique. It was usually a coarse-grained or flinty velvet-black jasper (basanite), i.e. crypto-crystalline silica associated with quartz and coloured by iron and other oxides. The use of it was in the assaying of alloys of gold and silver. Medieval metallurgists had a set of 24 'needles', or little bars of gold, each of a known and marked standard from 1 carat up to 24 carats. Taking the gold to be assayed, they rubbed it on the stone, and then rubbed one or other of the standard bars alongside the streak which it had left, choosing first the standard apparently nearest in composition to the unknown. In ancient times the colour of the streak was the only guide, but after the introduction of mineral acids, a little aqua regia was applied and the effect of the solvent on the two streaks noted.[a] If any differences were seen, other standard bars were used until an exact coincidence was obtained. The method depends on differences in reflectivity and colour which cannot be detected from the metal surfaces themselves but which become visible when the metal is finely divided on the dark abrasive surface of the stone. According to Sisco & Smith (1), it can give tolerable accuracy in the hands of a skilled person, provided that the match is carefully made and that there are no unknown components in the alloy under test. The technique has considerable historical interest since it is the ancestor of all those colorimetric and nephelometric methods so widely used in modern chemistry and biochemistry.

The use of the touchstone in Greece is evidently very old, since there are references to it in the poets back to the −6th century (Pindar, and Bacchylides[b]). Plato mentioned it.[c] It was minutely described by Theophrastus[d] and (with misunderstandings) by Pliny.[e] Agricola in +1556 goes into its use at some length, giving tables for the standard compositions of gold-silver, gold-copper, gold-silver-copper and silver-copper bars.[f] So also the *Probierbüchlein* of +1524.[g] The earliest Chinese reference yet found seems to be in the archaeological book, *Ko Ku Yao Lun* of +1387,[h] where the 'gold-testing stone' (*shih chin shih*[4]) is said to come from Szechuan, to be pure black and delicately smooth, and to be used for testing alloys of silver as well as of gold. Another

[a] Cf. Lord (1).
[b] Frag. 10.
[c] *Gorgias*, 486D.
[d] *Hist. Plant.* paras. 78–80.
[e] *Hist. Nat.* XXXIII, 43.
[f] Hoover & Hoover (1), p. 252.
[g] Sisco & Smith (1), pp. 86, 181.
[h] Ch. 7, p. 16*a*.

[1] 瑟瑟 [2] 甸子 [3] 綠松石 [4] 試金石

mention is in the *Yün-nan Thung Chih*[1] of about +1730, the revised version of older historical geographies of that province which was made under the superintendence of the Manchu official O-Erh-Thai[2] during his governorship.[a] Here the stone is said to be the colour of a new ink-block, and to be greatly prized by the merchants who test alloys with it. Chang Hung-Chao, however, suggests[b] that a much older term, *tsung-chhü*,[3] refers to the touchstone. This is given in old dictionaries such as the *Kuang Ya*[4] by Chang I[5] of the Three Kingdoms period (about +230)[c] and *Kuang Yün*[6] commenced by Lu Fa-Yen[7] in the Sui (+6th).[d] A much older book, the *Thung Su Wên*[8] by Fu Chhien[9] of Later Han or Chin, says that 'just as the *chien-chu*[10] is used to master jade, so the *tsung-chhü*[3] is used to master gold'.[e] But one can hardly be sure that these were not simply stones for grinding. The exact date of the first use of the touchstone in Chinese metallurgy therefore remains somewhat obscure.[f]

(g) THE SEARCH FOR MINERAL DEPOSITS

(1) Geological Prospecting

The methods used by ancient miners for finding the locations of deposits of ores and minerals must have been primarily based on traditional geological lore, the observation of the 'lie of the land', the direction of strata, and the knowledge of what kinds of rock were likely to be associated with the mineral sought.[g] The mining works of the European 16th century give hints of such information. There must have been similar traditions in China, as we saw above.[h] But new methods of prospecting have now been introduced, some depending on geophysics (as in the measurement of gravity and of

[a] The scientific interests of O-Erh-Thai (Ortai) were shown later in his editorship of two imperially commissioned compendia, one on agriculture, the *Shou Shih Thung Khao*, and the other on medicine, the *I Tsung Chin Chien*. See his biography in Hummel (2), p. 601.

[b] (*1*), p. 179.

[c] *Kuang Ya Su Chêng*, ch. 8A, p. 14*b*.

[d] *Kuang Yün Chiao Pên*, *Shang phing* sect., p. 15*a*.

[e] Ma Kuo-Han's collection, *YHSF*, ch. 61, p. 23*a*.

[f] It might perhaps be worth suggesting that an old, almost proverbial, phrase, *fei-tshui hsieh chin*[11]—the *fei-tshui* stone can reduce gold to minute particles—may conceal the use of the touchstone kept as a professional secret of the metallurgists. This is mentioned in that curious compendium of Taoist lore, the *Lu Shih*[12] of Lo Pi[13] of the Sung (*fide Ko Chih Ku Wei*, ch. 2, p. 32*a*). I should not have suggested this were it not for the fact that Tu Wan (*Yün Lin Shih Phu*, ch. 2, p. 6*a*) wrote that he had tested the statement repeatedly and proved it true. In +1067 Ouyang Hsiu gave an eye-witness account of such a test (*Kuei Thien Lu*, quoted in *Yün Shih Chai Pi Than*, ch. 1, p. 11*b*). As we saw above, the *fei-tshui* was some kind of jade or stone related to jade; Tu Wan in +1133 certainly did not mean what the term means now, i.e. jadeite.

[g] Modern geology and geochemistry have of course given a theoretical foundation to these old observations. For example, all the world's tin is obtained from ore-bodies formed originally in or near acid igneous rocks of the granite type, while chromium always occurs in intimate association with basic igneous rocks such as gabbro. Without knowledge of the chemistry of rocks, the search for minerals could never have been put upon a rational basis.

[h] Pp. 649ff.; 660.

1 雲南通志 2 鄂爾泰 3 磫礭 4 廣雅 5 張揖
6 廣韻 7 陸法言 8 通俗文 9 服虔 10 礛礑
11 翡翠屑金 12 路史 13 羅泌

the results of experimental detonations), others depending on the observation and chemical analysis of plants growing in the area under investigation. Thus geobotanical prospecting involves the study of the oecology of plants which are known either to accumulate unusually large quantities of the element concerned, or to be capable of living in environments containing amounts of it toxic to most other plant species. Similarly, bio-geochemical prospecting involves the chemical or spectrographic analysis of plants and soils. The latter could evidently not have been possible before the rise of modern science, but the former, in so far as it consisted of an empirical association of certain plants with certain minerals, goes back in China a surprisingly long way, as we shall see.

Prospecting by the empirical association of one rock or mineral with another was presumably part of the most ancient traditional Chinese miner's lore, since echoes of it find their way into Han or pre-Han texts. Thus in the *Kuan Tzu* book:

> Huang Ti[a] said, 'I should like to know about these things.' Po Kao[1] answered,[b] 'Where there is cinnabar above, yellow gold will be found below. Where there is magnetite above, copper and gold will be found below. Where there is *ling shih*[2] above,[c] lead, tin, and red copper will be found below. Where there is haematite (*chê*[3]) above, iron will be found below. Thus it can be seen that the mountains are full of riches.'[d]

And on the next page, when Duke Huan of Chhi puts the same question to Kuan Chung, the reply confirms the significance of haematite, cinnabar and magnetite, adding that lead above is an indication of silver ore beneath it.

There is also a good deal of this in the *Shan Hai Ching*, though its descriptions of mountains are ambiguous in that you cannot tell whether one material was found at the top of a mountain and another at its base, or whether the deposits lay at different depths in the same location. An example[e] states that jade is above and copper ore below. This kind of association may be instanced from many following centuries; thus the *Pên Tshao Shih I*[4] of +725 states:

> Generally one sees those who search for gold dig down into the earth for several feet until they come to a stone called *fên tzu shih*[5] ('tangle-stone') (which accompanies the gold).[f] This is always in black lumps, as if charred, and underneath it is the gold-bearing ore, also in lumps, some as large as one's finger, others as small as beans, and coloured a mulberry yellow. When first dug out it is friable.[g]

So sure was this association that later books (such as the *Pên Tshao Yen I*[6] of +1115) call the black stone the 'gold companion stone' (*pan chin shih*[7]). One catches glimpses

[a] Legendary emperor.
[b] Legendary minister of Huang Ti.
[c] RP no. 135*i*; unidentifiable.
[d] Ch. 77, p. 2*a*, tr. auct. The translation of Than Po-Fu *et al.*, p. 146, appears to be confused. The passage is quoted also in *TPYL*, ch. 810, p. 5*b*; hence Pfizmaier (95), p. 234.
[e] Ch. 3, p. 16*a*; de Rosny (1), p. 131.
[f] This country-rock is unidentifiable.
[g] Quoted by Li Shih-Chen, *PTKM*, ch. 8, p. 4*b*, tr. auct. Cf. de Mély (1), pp. xxiv, 14.

1 伯高 2 陵石 3 赭 4 本草拾遺 5 紛子石
6 本草衍義 7 伴金石

also of disagreements; thus Su Sung (+1070) reports that realgar is an indicator for the presence of gold, but Khou Tsung-Shih (+1115) specifically denies it.[a] Thao Hung-Ching (*c.* +500) relates[b] that sulphur is considered by some to be the *i*[1] (essential juice?) of alum, so that they ought to be found together, but considers this theory not always borne out, because in the south there are places where alum is found but no sulphur. Actually there is often in fact some connection between them, both being related to volcanic phenomena (Singer, 8). Another association is that between iron pyrites and alum (both containing iron and sulphur), and the *Shan Hai Ching* in fact mentions this, as Chang Hung-Chao[c] points out. Lastly, Sung Ying-Hsing wrote in +1637 'Where iron ore lies on the north of a hill lodestone is likely to occur on the south side, but there are many exceptions to this'.[d]

(2) Geobotanical and Bio-geochemical Prospecting

But the Chinese not only recognised the associations of ores and rocks, they also noted certain connections between plants and ores. That the ore and the plants might be separated by many feet or yards, or greater distances, was not felt by them as a difficulty; the conception of 'action at a distance' appears in many ancient texts.[e] In Section 38 on botany we shall take cognisance of a correlation which was made in the −3rd century between the *fu-ling*,[2] a fungus parasitic on pine-tree roots, and the dodder (*thu-ssu*[3]), a plant parasitic on the branches. In Section 39 on zoology we shall find that at least as early the effect of the moon on the reproductive cycle of certain marine invertebrates was discovered.[f] A whole chapter of the *Lü Shih Chhun Chhiu* is devoted to examples of this action at a distance.[g] And as we saw in Section 21 on meteorology, the Chinese were perhaps less hindered than Europeans in the recognition of the true cause of the tides by their appreciation of the organic oneness of the universe.[h]

Among the *miao*,[4] or indications, of ore deposits, plants played a considerable role. The best known text which shows this is perhaps that in the *Yu-Yang Tsa Tsu*, written by Tuan Chhêng-Shih about +800. There we find the following:

When in the mountains there is the *tshung*[5] plant[i] (the ciboule onion), then below silver will be found. When in the mountains there is the *hsiai*[6] plant[j] (a kind of shallot), then below gold will be found. When in the mountains there is the *chiang*[7] plant[k] (ginger), then below

[a] *Thu Ching Yen I Pên Tshao*, ch. 3, pp. 2*b*, 7*a*.
[b] Quoted in *Thu Ching Yen I Pên Tshao*, in its entry under sulphur (ch. 3, p. 16*a*).
[c] (*1*), p. 171. Cf. his introduction.
[d] *TKKW*, ch. 14, p. 9*b*.
[e] Cf. Vol. 2, pp. 293, 355, 381, 408.
[f] Cf. Vol. 1, p. 150.
[g] Ch. 45, tr. R. Wilhelm (3), p. 114.
[h] Cf. in this connection Sect. 26 below on physics and acoustics, and all that was said on organic philosophy in Vol. 2.
[i] R666, *Allium fistulosum*.
[j] R663, *Allium bakeri*.
[k] R650, *Zingiber officinale*.

1 液 2 茯苓 3 菟絲 4 苗 5 葱 6 薤 7 薑

copper and tin will be found. If the mountain has precious jade, the branches of the trees all around will be drooping.[a]

The significance of this will be appreciated in the light of modern findings described below. It is by no means isolated in Chinese literature. The association of *tshung*[1] (*Allium*) with silver deposits occurs again[b] in the *Pên Tshao Kang Mu* as also does that of the *hsiai*[2] (another species of *Allium*) with gold.[c] But Li Shih-Chen quotes these statements from a source which must have been that of Tuan Chhêng-Shih also, namely the *Ti Ching Thu*[3] (Illustrated Mirror of the Earth). This work, which survives only in the form of fragments collected by Ma Kuo-Han,[d] is mentioned in the Sui bibliography and, though its writer's name is lost,[e] was evidently concerned with mining and geobotanical prospecting.[f] Ma Kuo-Han believed that it was written in the Liang dynasty (first half of the +6th century).

The *Ti Ching Thu* has some interesting statements which Tuan Chhêng-Shih did not copy. For example, it says that 'if the stalk of (a certain) plant is yellow and elegant (*tshao hêng huang hsiu*[4])' copper will be found below.[g] This indicates that not only the presence or absence of plants was looked for but also their physiological condition. The text goes on to say that the essence (*ching*[5]) of copper transforms itself into *ma*[6] and *thung*,[7] words which literally mean 'horses' and 'boys', but which may well have referred to names of plants no longer identifiable.[h] Similarly the *Ti Ching Thu* says that 'if the (leaves) of (a certain) plant are green, and the stalks red, much lead will be found below',[i] adding that the essence of lead and tin transforms itself into the 'old-woman', again perhaps a plant-name. Though we can hardly say what species were meant, modern knowledge confirms most strikingly that the presence or absence of metallic elements in the soil affects the appearance of plants. Exactly how far back this accumulated knowledge goes is hard to determine, but another piece of information occurs in the *Thu Ching Yen I Pên Tshao* (*c.* +1120) which quotes[j] Chang Hua (*c.* +290) as saying[k] that 'where the *liao*[8] (smartweed) grows abundantly,[l] there must be plenty of haematite (*yü yü liang*;[9] ferric oxide) below'. Moreover, the *Shan Hai Ching*, in

[a] Ch. 16, p. 3*b*, tr. auct.

[b] *PTKM*, ch. 8, p. 8*a*; cf. de Mély (1), pp. xxvii, 19.

[c] Ch. 8, p. 5*a*; cf. de Mély (1), pp. xxvii, 13.

[d] *YHSF*, ch. 78, pp. 31*a* ff.

[e] We are told by the Emperor Yuan of the Liang that in his time there were three manuals of this kind, one by Shih Khuang[10] (obviously a pseudonym taken from the Chou period), a second by Pai Tsê,[11] and a third by Liu Chia.[12] Of these men nothing is now known (*Chin Lou Tzu*, ch. 5, p. 21*a*). A fragment of a *Pai Tsê Thu*, mostly about spirits which lead miners to ore sites, is preserved in *YHSF*, ch. 77, pp. 58*a* ff. Chhen Phan (*1b*) has discussed the *Ti Ching Thu* and related writings.

[f] In this it differs from two other books, also of the Liang, the *Ti Ching*[13] (Mirror of the Earth) and the *Thien Ching*[14] (Mirror of the Heavens), which are almost entirely divinatory in character. What survives of both will also be found in *YHSF*, ch. 78.

[g] Quoted by Li Shih-Chen in *PTKM*, ch. 8, p. 12*a*; cf. de Mély (1), pp. xxvii, 22.

[h] About twenty Chinese plant names have the prefix *ma*;[6] cf. our 'horsetail'.

[i] Quoted by Li Shih-Chen in *PTKM*, ch. 8, p. 16*b*; cf. de Mély (1), pp. xxvii, 27.

[j] In its entry under haematite (ch. 2, p. 12*a*).

[k] We have not, however, been able to find the statement in the present version of the *Po Wu Chih*.

[l] R573, *Polygonum hydropiper* (ch. 2, p. 12*a*).

1 葱 2 薤 3 地鏡圖 4 草莖黃秀 5 精 6 馬
7 僮 8 蓼 9 禹餘糧 10 師曠 11 白澤 12 六甲
13 地鏡 14 天鏡

spite of its usual ambiguity, seems to speak[a] of an association between gold ore and *hui-thang*.[1 b] There is also a reference in *Wên Tzu*[2] to lush growth and drooping branches in the neighbourhood of jade.[c] In general, therefore, we should not be far wrong in thinking that this kind of empirical lore was growing steadily from the Han time onwards until the Sui. Naturally there was magic in it as well as true knowledge, for no plant could betray the presence of a mineral such as magnesium silicate, no matter how much men might value it as jade.

We have now, finally, to glance at the present position of knowledge in the light of which these achievements must be evaluated. When Jules Raulin, a great contemporary of Pasteur, began the first successful experiments in the culture of the lower plants on synthetic media, it became clear that they needed for their growth the presence of many metallic and other elements, though only in traces. Javillier and Schöpfer have recently reviewed the wide repercussions of this work. The importance of 'trace-elements' or 'oligo-elements' in all branches of biology may be seen from the books of Brewer (1) and Stiles (1).[d] Plant diseases caused by deficiencies in these may be diagnosed by inspection (Wallace, 1).[e] But on the other hand modern research has made clear that plants vary greatly in the extent to which they will accumulate within themselves metallic elements present in the soil. The role of plants as biological concentrators and indicators was first brought out in a classical lecture by the Norwegian geochemist V. M. Goldschmidt in 1935. Starting from a curious observation that certain coal ashes were extremely rich in germanium, it was found that there is a circulation from subsoil to surface layers, with absorption of elements by the plants and accumulation in their maximally transpiring organs, the leaves. Later, from leaves undergoing decay there is a leaching process in which the most soluble salts go down into the subsoil, while the less soluble ones remain above. Nearly twenty elements are now known to participate in this Goldschmidt Enrichment Principle, as it is called.[f] It will follow, therefore, that if there are rich deposits of a certain element in or below the subsoil, this will accumulate in the soil itself and in the plants which grow on it, and this process may reach a point which many plant species may find intolerable, thus leading to oecological changes. But besides this, there are active processes by which plants accumulate elements in their tissues, for example, boron.[g]

During the past fifteen years this new information has been actively applied to mineral prospecting, Swedish and Italian workers being in the forefront. The con-

[a] In the entry for Chung-Huang Shan in ch. 2.

[b] This combined plant name (if it is a combined one) is hard to identify. The *hui* could be a kind of orchid (B II, 406), as de Rosny (1), p. 67, took it, or *Ocimum* the basil (B III, 60; R 134a). *Thang* alone could be the wild pear (R432) or plum (R446) or hawthorn (R422).

[c] Cf. *Piao I Lu*, ch. 2, p. 7*a*, quoting the *Yün Chhi Yu I* of the Thang. See also *Hsün Tzu*, ch. 1 p. 5*b* (tr. Dubs (8), p. 36); *Po Wu Chih*, ch. 1, p. 7*b*.

[d] Cf. also a report (Anon., 4).

[e] Hence the interest of the old Chinese statements concerning the *appearance*, not only the presence or absence, of plants in relation to minerals.

[f] Notably Au, Be, Zn, Cd, Sc, Tl, Ge, Sn, Pb, As, Mn, Co, Ni.

[g] For an excellent review of this subject see Rankama (1).

1 蕙棠 2 文子

tributions of Palmqvist & Brundin,[a] reviewed by Landergren (1), are regarded as of particular merit.[b] Following the order of metals in which the medieval Chinese were interested, we may glance at a few details, astonishing indeed in themselves. Silver does not seem very prominent so far, but gold is known to accumulate in *Equisetum arvense* and *palustre*, the horsetails, which may contain as much as 4 oz. Au/ton.[c] As regards copper, tin and zinc, there have been chemical analyses by Warren & Delavault and Warren & Howatson of parts of various trees which may act as indicators for copper deposits. Copper is specifically accumulated in *Polycarpaea spirostylis*; and two other plants, *Viscaria alpina* and *Melandrium dioecium*, can endure copper concentrations in the soil which leave them without competitors.[c] The 'zinc pansy', *Viola calaminaria*, accumulates zinc up to 1% of the ash, and flourishes in the neighbourhood of its ores, but it is outdone by *Thlaspi* spp. which go as high as 16% of the ash.[d] *Holosteum umbellatum*, when growing on mercury-rich soils, actually shows intracellular droplets of metallic mercury (Rankama & Sahama, 1). *Astragalus* spp. seem quite insensitive to selenium, otherwise a highly poisonous element. Lead, mentioned by the old Chinese writers, accumulates in *Amorpha canescens* and *Panicum crusgalli* (Sinyakova), but many families have no detectable traces. Nickel indicator plants, such as *Alyssum bertolonii*, have been studied by Rankama (2) and Minguzzi & Vergnano (1), following the classical work on cobalt accumulators of Bertrand & Mokragnatz (1). Some of the *Lycopodium* species accumulate so much aluminium that they were used traditionally in Europe as sources of mordant instead of alum, as well as sources of yellow dye.[e] Of course in modern geobotanical technique it is necessary not only to study the oecological distribution and the physiological condition of the flora, but also to make chemical analyses of the plants, soils, waters, and even animals, of the region.

In view of these facts it is remarkable that Chinese literature contains a long-standing assertion that metals could actually be obtained from certain plants. The *Wu Li Hsiao Shih* of +1664 quotes[f] a passage from the *Kêng Hsin Yü Tshê* of +1421 saying that the *chhi* of gold was present in the rape-turnip (*man chhing*;[1] *Brassica rapa-depressa*, R477), that of silver was present in a kind of weeping willow (*shih yang liu*[2]), those of lead and tin in mugwort (*ai hao*;[3] *Artemisia vulgaris*, R9), chestnut, barley and wheat; and that of copper in the Indian sorrel (*san yeh suan*;[4] *Oxalis corniculata*, R367). In at least one case practical use was made of such knowledge in the extraction of mercury, as we shall see in a moment. It would be very interesting to put these old statements to the test by modern methods.

One would think that miners in all civilisations must have acquired some empirical knowledge of plant-mineral associations. But we have found no trace of medieval

[a] Though imperfectly available, owing, apparently, to their connection with the commercial firm Svenska Prospektering Aktiebolaget.

[b] Geobotanical prospecting is in actual practical use for Cu, Ni, Sn, Cr and W.

[c] Rankama (1); Vogt & Bergh (1).

[d] Rankama (1); Robinson, Lakin & Reichen (1).

[e] Analyses by Hutchinson & Wollack (1).

[f] Ch. 7, p. 12*b*.

[1] 蔓菁 [2] 石楊柳 [3] 艾蒿 [4] 三葉酸

European ideas on the subject. Agricola says very little about it; he notes only the lie of frost, and a certain 'sickliness' about trees or bushes.[a] Not till the *Flora Saturnisans* of the German mining chemist J. F. Henckel (+1760) do we come upon the traces of geobotanical prospecting. At first he seems not to go beyond Agricola, for with reference to arid patches near mines he says (p. 66):

Je sçais qu'on attribue ces effets aux exhalaisons qui s'élevent des fentes des mines qui détruisent tout ce que le terrein qui est au-dessus de ces fentes, et qu'elles traversent, peut contenir de propre à la végétation, empêchent que les rosées ne fertilisent ces endroits, comme les endroits voisins sur lesquels cette cause de destruction n'agit point; il ne faut cependant pas adopter entiérement les soupçons des Mineurs.

But later he quotes 'un passage de Cluvier qui mérite qu'on y réfléchisse' (p. 254):

Nous croyons avoir raison de croire et de dire que chaque plante participe d'un métal particulier; aussi les plantes peuvent-elles servir à faire connoître les différentes especes de minéraux qui sont dans la terre. Le *saffranum* ou le saffran bâtard indique où il y a des mines d'or; et le houx, où il y a de l'alun. On voit au-dessus des filons des mines des plantes qui ont quelque connexion avec ces mines qui y croissent très-bien, tandis que d'autres ont peine à y venir, et y dépérissent. Le vitriol se trouve en grande quantité dans le chêne et dans les bigarades d'où l'on peut aisément l'en retirer. Les Orientaux sçavent faire du salpètre avec une certaine espece d'osier....

The reference is to the *Passe-tems* of Cluvier, i.e. Philip Cluverius (+1580 to +1623), the eminent German historical geographer. This is one of those points at which we seem to be standing at the watershed between the old and the new in European science, for while one sentence evokes the doctrine of signatures and the ancient Kabbalistical and 'Chinese' symbolic correlations (every plant corresponds to a certain metal),[b] another indicates how Henckel and his contemporaries were in fact beginning to open up the subject of plant biochemistry, especially in those analyses of the inorganic constituents of plants, to which much of the *Flora Saturnisans* is devoted. It is noteworthy that there are certain references to China in Henckel, especially discussion of a plant found in that country which contained so much mercury that it would give rise to an amalgam as soft as butter when ground with copper filings.[c] But the successful pursuit of geobotanical prospecting in the European tradition would require a special investigation.

[a] Hoover & Hoover (1), p. 38. It is interesting that the Jesuits, urged on by Li Thien-Ching, started to make a translation of Agricola into Chinese in +1639. Within a year eight chapters were presented to the throne, but then for some reason the work stopped and the Chinese text was never published (Bernard-Maître (7), p. 462).

[b] Cf. Sect. 13*f* in Vol. 2, pp. 261 ff.

[c] Pp. 196, 245, 252. Though Henckel gave no Chinese reference, the texts are readily identifiable. Li Shih-Chen (*PTKM*, +1596, ch. 9, p. 10*b*) quotes Su Sung as giving in the *Pên Tshao Thu Ching* (about +1070) a method for getting mercury from a kind of purslane, the *ma chhih hsien*[1] (*Portulaca oleracea*, R554) by careful pounding, drying and autolysis. He could obtain 8 or 10 ounces from 10 catties of the dried plant. The account was quoted again in the *Wu Li Hsiao Shih* (+1664), ch. 7, p. 12*b*, but Sung Ying-Hsing (*Thien Kung Khai Wu*, ch. 14, p. 10*b*), supposing that the story referred to tin not mercury, disbelieved it.

[1] 馬齒莧

It would also be interesting to make a search of the Chinese literature in the light of present knowledge of plant species which are known to be ore-indicators. None of those, indeed, which have just been mentioned is the same as any of the plants referred to in the texts, but the identification of the latter is perhaps too uncertain for that to be significant. Conversely, if the plants which the Chinese seem to have used could ever be identified with sufficient exactness, it would be interesting to examine them in the field on the basis of modern knowledge. We can hardly follow the subject further here, but enough has been said to make it evident that the medieval Chinese observations were the forerunners of a vast and rapidly growing body of modern scientific theory and practice.[a] To what extent this holds true, not only in geophysics, but also in physics itself and throughout the realm of the technologies of the physical world, will appear from the discussions collected in the following volume.

[a] Meanwhile the Chinese themselves do not always appreciate the acuity of their forefathers; for example, in Phan Chung-Hsiang's otherwise excellent review of modern prospecting methods, geobotanical techniques are mentioned as if they were purely of modern Western origin. Even a special (and admirable) monograph on geobotanical prospecting and indicator-plants, published by Hou Hsüeh-Yü at Peking in 1952, is silent regarding the Chinese pioneers. However, after the publication of a preliminary draft of the present sub-section two years later (Needham, 30), we were happy to see a good account of the same material appear in Chinese (Yen Yü, *3*).

ADDENDA

ISRAEL AND KHAZARIA

(to p. 575 above, and to Vol. 1, Sect. 7)

Insufficient justice was done in the first volume of this work to the part played by Israel in the traffic of ideas and techniques between East and West. Opportunity may be taken to repair the omission at this point.

For the +9th century we have a considerable amount of information about the group of Jewish merchants known as the Radhanites (al-Rādhānīyah). According to Abū al-Qāsim ibn Khurdādhbih (+825 to +912; Mieli (1), p. 79; Hitti (1), p. 384), in his *Kitāb al-Masālik w'al-Mamālik* (Book of Roads and Provinces) of +846, the Radhanites travelled regularly between China and Provence. Besides Arabic and Greek, they spoke Frankish, Slavonic, Spanish and Persian. Journeying both by land and sea, they took eastwards eunuchs, girl slaves, boys, brocades, furs and swords; returning with musk, aloes, cinnamon, camphor and medicinal plants. Damascus, Oman, Hindustan and the (Jewish) Khazar kingdom in the Crimea were among their way-stations. It will be remembered that this was in the latter part of the Thang period, and that the Arab merchant Sulaimān al-Tājir, who wrote about China, was Ibn Khurdādhbih's exact contemporary. That porcelain was among the wares which the Radhanites brought back is very likely, since Thang fragments have been found in Egyptian rubbish heaps dating from that time (cf. Sect. 6*f* in Vol. 1, p. 129). But if we judge from correspondences of dates, the Radhanites may well have been responsible for other introductions. Independent lines of evidence point to the +9th and +10th centuries as the time when the efficient collar harness (Sect. 27*d* below) came from Chinese Central Asia to Europe, and the only other efficient harness, the breast-strap type, had come not very long before. Similarly, it is about this period that we have to look for the transmission of the art of deep drilling from Szechuan to Artois (Sect. 37 below), and for the second westward introduction of the crossbow (Sect. 30*d* below). The passage of Ibn Khurdādhbih is translated in E. N. Adler (1), p. 2; and further information about the Radhanites will be found in the works of Heyd (1), vol. 1, pp. 125 ff., Mez (1), Rabinowitz (1), and Muntner (1), pt. 2. Some have thought that the Radhanites took their name from their western headquarters in the Provençal towns along the Rhône, but others derive it (no more plausibly) from the name of the Persian town of al-Raiy, near modern Teheran, one of the key points on the great east-west trunk road known as the Khurāsān highway (cf. Fig. 32, in Vol. 1, p. 171, and Hitti (1), p. 323).

In connection with the Radhanites mention must also be made of the Khazar kingdom, for its Khan and many of its people accepted Judaism as their national religion about +740. The Khazars were a Turkic group (see Barthold, 1) related to the White (Ephthalite) Huns (Grousset, 1) and to the Uighurs (Klaproth, 3). The words *ketzer* (heretic) and 'hussar' probably derive from the Khazars, and the Karaim of Poland may be their descendants. They occupied for many centuries the lands north of the Caucasus, including the lower valleys of the Don and the Volga, as well as the westward outlying region of the Crimea. In this way they had a strategic position on the trade routes, and merited the name which has been given to them of 'the Venetians of the Euxine and the Caspian'. Although the Khazar kingdom can be traced from the +6th to the +11th century, its period of greatness was between

+640 and +960. The Chinese knew of the Khazars as the Kho-Sa[1] branch of the Turks (*Hsin Thang Shu*, ch. 221 B, tr. Hirth (1), p. 56; *Thung Tien*, ch. 193, p. 1041.3; *Wên Hsien Thung Khao*, ch. 339, p. 2659.2, both tr. Hirth (1), p. 83, also pp. 1044.3, 2663.2 respectively). One of the sources of their information was undoubtedly Tu Huan, the officer captured by the Arabs at the battle of the Talas River (+751), who brought back information about the Chinese technologists settled at Baghdad (cf. Vol. 1, p. 236). Khazaria was in close trade relations with Byzantium. Evidently we ought to know more about this interesting people; until recently the only monograph on the Khazars was in Hebrew (Poliak, 1), but we now have an excellent book in English (Dunlop, 1). According to Dr A. N. Poliak (in a private communication), Ibn al-Nadīm (*d.* +995) says that Chinese was understood in Khazar lands, and that the khanate used Chinese court etiquette. For the Khazar term *tudun*, governor, a Chinese origin has been suggested (see Dunlop (1), p. 174), perhaps from *tu thuan shih*,[2] imperial commissioner for provincial troops. The Khazars had an institution like the shogunate, with a Beg as well as a Khaqan (Dunlop (1), p. 208). Arab geographers, Poliak tells us, speak of mausolea suspended over rivers (see Sect. 28*e* below on suspension-bridges) though this may be a misunderstanding for royal tombs excavated underneath flowing rivers; of synagogues with pagoda-like towers, and of military flame-throwers (cf. Sect. 34 below), these last perhaps really Arabic (cf. Poliak, 2). The Volga delta certainly produced rice. As an example of the role which Khazaria may have played in scientific and technological transmissions, we remember the embassy there of the great algebraist al-Khwārizmī between +842 and +847 (cf. pp. 107, 147 above).

The historian of science must also take note of the community of Jews established in Khaifêng from the beginning of the +12th century onwards (their first synagogue dated from +1163). They may well have started as Radhanites who settled in China. Throughout the centuries they produced eminent physicians. About +1415, one of their number, Chao Yen-Chhêng,[3] was a close collaborator of the Ming prince Chou Ting Wang, who studied pharmaceutical botany and maintained a botanic garden. Yen-Chhêng seems to have been part author with the prince of the famous *Chiu Huang Pên Tshao* (Famine Herbal) of +1406, and more certainly of the *Phu Chi Fang*[4] (Simple Prescriptions for Everyman), issued by the same group. In +1423 he was honoured by the emperor, and in rebuilding the synagogue aided by the prince. One wonders whether there could have been any mutual influences of Hebrew and Chinese science or medicine in this partnership, and one would like to know something of Yen-Chhêng's Hebrew library. At the end of the following century the Jewish scholar Ai Thien[5] (+1573 to +1605) visited Matteo Ricci (the story may be read in Trigault, Gallagher tr. pp. 107ff.), and again in the early Chhing the community produced an eminent physician, Chao Ying-Chhêng.[6] Most of the information in this paragraph is due to the researches of White & Williams (see esp. (1), vol. 3, pp. 110ff.).

KOREA

The conviction has developed in the minds of the author and his collaborators during the writing of this book, that of all the peoples inhabiting the Chinese culture-area, the Koreans were for many centuries those most interested in all kinds of scientific matters. In Section 20 we have noted (p. 302) Korean interest in Jesuit sundials in the 18th century, as well as a remarkable astronomical clock from Korea (p. 389), and we have reproduced a drawing of a Korean observatory tower of the +7th century (Fig. 118) as well as a picture of a comet taken

[1] 可薩 [2] 都團使 [3] 趙俺誠 [4] 普濟方 [5] 艾田 [6] 趙映乘

from late Korean records (Fig. 184(*a*)). But there is abundant evidence from much earlier times, as may be seen from the perusal of Ma Tuan-Lin's account of Korea (Kao-Chü-Li[1]) in chapter 325 of the *Wên Hsien Thung Khao* (*c.* +1280; tr. Hervey de St Denys, 1). The people were known to be fond of books and learning; there were schools at almost every crossroads. In +951 they presented certain astrological-astronomical books, including one of the Chhan-Wei, *Hsiao Wei Tzhu Hsiung Thu*[2] (Apocryphal Treatise on the Filial Piety Classic; Diagrams of Male and Female (Influences)), the text of which had presumably been lost in China (existing fragments are in *YHSF*, ch. 58, p. 58*a*). In +1016, the Korean ambassador, Kuo Yuan,[3] departed for home laden with calendrical books and a medical work *Shêng Hui Fang*[4] (Glory-of-the-Sages Prescriptions). This was probably the work now known as *Shêng Hui Hsüan Fang*. Koreans repeatedly took doctorates at the imperial university, e.g. Chin Hsing-Chhêng[5] in +980 and Khang Chien[6] in +1021. In the latter year, the embassy requested, and obtained, geographical books; and in +1075 one of the Korean kings sent for physicians, pharmacists, painters and sculptors. But towards the end of the +11th century cultural relations between China and Korea were impaired because the suspicion grew up that the Koreans were coming to gather intelligence on behalf of the Chin (Jurchen) Tartars. Consequently in +1085 and +1092 Korean embassies were not given all the books for which they had asked, and already before +1080 (as we saw in Sect. 22, p. 549) there was great reluctance to provide them with maps which previously had been freely given. After the retreat to Hangchow early in the +12th century, scientific intercourse was reduced, but in later times it revived again. We have studied above (pp. 554 ff.) the outstanding achievement of Korean cartographers about +1400.

General histories of Korean literature (e.g. Trollope, 1) regard the *Yun Tai Rok*,[7] a calendrical work by Tshui Chih-Yuan[8] (Choi Chi-Won +858 to +910), who was trained at the Thang capital, as perhaps the oldest Korean astronomical book. No less famous, however, is the *Sip Chong Rok*,[9] written by Chin Chhêng-Tsê[10] (Kim Song-Taik) of the +11th century. The register of strange celestial phenomena, *Tyentung Sanguiko*[11] of +1708, is an important work. In the early nineteenth century, Korea produced two prolific astronomical writers, the brothers Nan Ping-Chê[12] and Nan Ping-Chi[13] (Nam Pyeng-Chul and Pyeng-Kil). There is a general history of science in Korea by Hung I-Pyǒn̄ (*1*).

1 高句麗 2 孝緯雌雄圖 3 郭元 4 聖惠方 5 金行成
6 康戩 7 年代曆 8 崔致遠 9 十精曆 10 金成澤
11 天東象緯考 12 南秉哲 13 南秉吉

BIBLIOGRAPHIES

A Chinese books before +1800

B Chinese and Japanese books and journal articles since +1800

C Books and journal articles in Western languages

In Bibliographies A and B there are two modifications of the Roman alphabetical sequence: transliterated *Chh-* comes after all other entries under *Ch-*, and transliterated *Hs-* comes after all other entries under *H-*. Thus *Chhen* comes after *Chung* and *Hsi* comes after *Huai*. This system applies only to the first words of the titles. Moreover, where *Chh-* and *Hs-* occur in words used in Bibliography C, i.e. in a Western language context, the normal sequence of the Roman alphabet is observed.

When obsolete or unusual romanisations of Chinese words occur in entries in Bibliography C, they are followed, wherever possible, by the romanisations adopted as standard in the present work. If inserted in the title, these are enclosed in square brackets; if they follow it, in round brackets. When Chinese words or phrases occur romanised according to the Wade-Giles system or related system, they are assimilated to the system here adopted without indication of any change. Additional notes are added in round brackets. The reference numbers do not necessarily begin with (1), nor are they necessarily consecutive, because only those references required for this volume of the series are given.

ABBREVIATIONS

A	*Archeion*
AAA	*Archaeologia*
AAEEG	*Annuaire de l'Assoc. pour l'Encouragement des Etudes Grecques*
A/AIHS	*Archives Internationales d'Histoire des Sciences* (contin. of *Archeion*)
AAN	*American Anthropologist*
AAR	*Art and Archaeology* (Washington)
ABAW/MN	*Abhandlungen d. bayerischen Akademie d. Wissenschaften, München* (Math.-nat. Klasse)
ABAW/PH	*Abhandlungen d. bayerischen Akademie d. Wissenschaften, München* (Phil.-hist. Klasse)
ACASA	*Archives of the Chinese Art Society of America*
ACLS	American Council of Learned Societies
ADVS	*Advancement of Science* (British Association, London)
AE	*Ancient Egypt*
AEO	*Archives d'Etudes orientales* (Upsala)
AGMNT	see *QSGNM*
AGMW	*Abhandlungen z. Geschichte d. Math. Wissenschaft*
AGNT	see *QSGNM*
AGSB	*Bulletin of the American Geographical Society*
AGWG/MP	*Abhandlungen d. Gesellschaft d. Wissenschaften z. Göttingen* (Math.-phys. Klasse)
AGWG/PH	*Abhandlungen d. Gesellschaft d. Wissenschaften z. Göttingen* (Phil.-hist. Klasse)
AHAW/PH	*Abhandlungen d. Heidelberger Akademie d. Wissenschaften* (Phil.-hist. Klasse)
AHOR	*Antiquarian Horology*
AHR	*American Historical Review*
AHSNM	*Acta Historia Scientiarum Naturalium et Medicinalium* (Copenhagen)
AI	*Ars Islamica*
AIEO/UA	*Annales de l'Institut des Etudes orientales* (*Université d'Alger*)
AJ	*Asiatic Journal and Monthly Register for British and Foreign India, China and Australia*
AJA	*American Journal of Archaeology*
AJP	*American Journal of Philology*
AJSC	*American Journal of Science*
AJSLL	*American Journal of Semitic Languages and Literature*
AKG	*Archiv f. Kulturgeschichte*
AKML	*Abhandlungen f. d. Kunde des Morgenlandes*
AM	*Asia Major*
AMG	*Annales du Musée Guimet*
AMM	*American Mathematical Monthly*
AMN	*Archiv för Math. og Naturvidenskab* (Christiania/Oslo)
AMP	*Archiv d. Math. u. Physik*
AN	*Anthropos*
ANHGN	*Abhandlungen d. Naturhistorischen Gesellschaft zu Nürnberg*
ANP	*Annalen d. Physik*
ANS	*Annals of Science*
ANSSR/AC	*Academy of Sciences of the U.S.S.R., Astronomical Circular*
AOF	*Archiv f. Orientforschung*
APAW	*Abhandlungen d. preuss. Akad. Wiss. Berlin*
APAW/MN	*Abhandlungen d. preuss. Akad. Wiss. Berlin* (Math.-nat. Klasse)
AQ	*Antiquity*
ARAB	*Arabica*
ARC	Agricultural Research Council (U.K.)
ARLC/DO	*Annual Reports of the Librarian of Congress* (Division of Orientalia)
ARSI	*Annual Reports of the Smithsonian Institution*
ARUSNM	*Annual Reports of the U.S. National Museum*
AS/BIHP	*Kuo-Li Chung-Yang* (now *Chung-Kuo Kho-Hsüeh*) *Yen-Chiu Yuan, Li-Shih Yü-Yen Yen-Chiu So Chi-Khan* (*Bulletin of the Institute of History and Philology, Academia Sinica*)
AS/CJA	*Chung-Kuo Khao Ku Hsüeh Pao* (*Chinese Journal of Archaeology, Academia Sinica*)
ASAW/PH	*Abhandlungen d. Sächischen Akad. Wiss. Leipzig* (Phil.-hist. Klasse)
ASI	*Actualités scientifiques et industrielles*
ASPN	*Archives des Sciences physiques et naturelles*
ASR	*Asiatic Review*
ASSB	*Annales de la Société scientifique de Bruxelles*
ASTNR	*Astronomische Nachrichten*
ASTRO	*Astrophysics Journal*
ASTSN	*Atti d. Soc. Toscana d. Sci. Nat.*
AT	*Atlantis*
AUL	*Annales de l'Université de Lyon*

AUON *Annali dell'Istituto Universitario Orientale di Napoli*

BA *Baessler Archiv (Beiträge z. Völkerkunde herausgeg. a. d. Mitteln d. Baessler Instituts, Berlin)*

BAN *Bulletin of the Astronomical Institutes of the Netherlands*

BAS *Bulletin astronomique*

BAU *Bulletin of Ankara University*

BBSHS *Bulletin of the British Society for the History of Science*

BBSSMF *Bollettino di Bibliografia e di Storia delle Scienze Matematiche e fisiche* (Boncompagni's)

BCGF *Bulletin de la Commission géologique de Finlande*

BCP *Bulletin catholique de Pékin*

BCS *Chung-Kuo Wên-Hua Yen-Chiu Hui Khan (Bulletin of Chinese Studies*, Chhêngtu)

BEFEO *Bulletin de l'Ecole française de l'Extrême Orient* (Hanoi)

BG *Bulletin de Géographie*

BGHD *Bulletin de Géographie historique et descriptive*

BGSA *Bulletin of the Geological Society of America*

BGTI *Beitr. z. Gesch. d. Technik u. Industrie* (changed to *Technik Geschichte BGTI/TG* in 1933)

BIFAO *Bulletin de l'Institut français d'Archéologie Orientale* (Cairo)

BLSOAS *Bulletin of the London School of Oriental and African Studies*

BM *Bibliotheca Mathematica*

BMFEA *Bulletin of the Museum of Far Eastern Antiquities* (Stockholm)

BMON *Bulletin Monumental*

BMQ *British Museum Quarterly*

BMRAH *Bulletin des Musées royaux d'Art et d'Histoire* (Brussels)

BNGBB *Berichte d. naturforsch. Gesellschaft Bamberg*

BNI *Bijdragen tot de taal- land- en volken-kunde v. Nederlandsch-Indië*

BNLP *Kuo-li Pei-phing Thu Shu Kuan Khan (Bulletin of the National Library of Peiping*, Peking)

BOR *Babylonian and Oriental Record*

BSEIC *Bulletin de la Société des Etudes indochinoises*

BSG *Bulletin de la Société de Géographie* (cont. as *La Géographie*)

BSMF *Bulletin de la Société mathématique (de France)*

BU *Biographie universelle*

BUA *Bulletin de l'Université de l'Aurore* (Shanghai)

BUSNM *Bulletin of the U.S. National Museum*

BV *Bharatiya Vidya*

C *Copernicus: International Journal of Astronomy* (Dublin and Göttingen)

CA *Chemical Abstracts*

CAM *Communications de l'Académie de Marine* (Brussels)

CEN *Centaurus*

CET *Ciel et Terre*

CH *Chih-Hsüeh (Learning)*

CHER *Chhing-Hua (University) Engineering Reports*

CHHP *Chhi-Hsiang Hsüeh Pao (Meteorological Magazine)*

CHI *Cambridge History of India*

CHJ *Chhing-Hua Hsüeh Pao (Chhing-Hua* (Ts'ing-Hua University) *Journal)*

CIMC/MR *Chinese Imperial Maritime Customs* (Medical Report Series)

CJ *China Journal of Science and Arts*

CLHP *Chin-Ling Hsüeh Pao (Nanking University Journal)*

CLJ *Classical Journal*

CLTC *Chen-li Tsa Chih (Truth Miscellany)*

CMJ *China Medical Journal*

CNCK *Chhing-Nien Chung-kuo Chi Khan (Young China Magazine)*

CP *Classical Philology*

CQ *Classical Quarterly*

CR *China Review* (Hong Kong and Shanghai)

CRAIBL *Comptes Rendus de l'Académie des Inscriptions et Belles Lettres* (Paris)

CRAS *Comptes Rendus de l'Académie des Sciences* (Paris)

CRAS/USSR *Comptes Rendus de l'Académie des Sciences* (U.S.S.R.)

CREC *China Reconstructs*

CRR *Chinese Recorder*

CRRR *Chinese Repository*

CS *Current Science*

CSPSR *Chinese Social and Political Science Review*

CST/HIJ *Chung-Shan Ta-Hsüeh Yü Yen Li-Shih Yen-Chiu So Tsou Khan (Sun Yat-Sen University Journal of Linguistics and History)*

CT *Connaissance du Temps*

CTE *China Trade and Engineering*

CZ *Chigaku Zasshi (Journal of the Tokyo Geographical Society)*

D *Discovery*

DHA *Dock and Harbour Authority*

DUZ *Deutsche Uhrmacher-Zeitung*

DWAW/PH *Denkschriften d. k. Akademie d. Wissenschaften, Wien* (Vienna) (Phil.-hist. Klasse)

EB *Encyclopaedia Britannica*
EG *Economic Geology*
ENB *Ethnologisches Notizblatt* (Kgl. Mus. f. Völkerkunde, Berlin)
END *Endeavour*
EOL *Jaarboek; Ex Oriente Lux*
ERE *Encyclopaedia of Religion and Ethics* (ed. Hastings)
EW *East and West* (Quart. Rev. pub. Istituto Ital. per il Medio e Estremo Oriente, Rome)
EXP *Experientia*

FEQ *Far Eastern Quarterly*
FF *Forschungen und Fortschritte*
FJHC *Fu-Jen Hsüeh-Chih* (*Journal of Fu-Jen University*, Peking)
FMKP *Fortschritte d. Mineralogie, Kristallographie u. Petrologie*
FMNHP/AS *Field Museum of Natural History* (Chicago) *Publications*, Anthropological Series
FMNHP/GLS *Field Museum of Natural History* (Chicago) *Publications*; Geological Leaflet Series

G *Geimon* (*Art Journal*)
GB *Geibun* (*Arts Journal*)
GHA *Göteborgs Högskolas Årsskrift*
GJ *Geographical Journal*
GM *Geological Magazine*
GP *Geophysics*
GR *Geographical Review*
GTIG *Geschichtsblätter f. Technik, Industrie u. Gewerbe*

HANL *Harvard Astronomical Newsletter*
HCLT *Hui-Chiao Lun-Than* (*Review of Muslim Affairs*)
HH *Han Hiue* (*Han Hsüeh*); *Bulletin du Centre d'Etudes sinologiques de Pékin*
HITC *Hsüeh I Tsa Chih* (*Wissen und Wissenschaft*)
HITH *Hsüeh I Thung Hsün* (*Science and Art Correspondent*)
HJAS *Harvard Journal of Asiatic Studies*
HMSO Her Majesty's Stationery Office (London)
HORJ *Horological Journal*

IC *Islamic Culture*
IHQ *Indian Historical Quarterly*
ILN *Illustrated London News*
IM *Imago Mundi: Yearbook of Early Cartography*
IPR Institute of Pacific Relations
ISIS *Isis*
ISP/WSFK *I Shih Pao; Wên Shih Fu Khan* (*Literary Supplement of the People's Betterment Daily*)

JA *Journal asiatique*
JAOS *Journal of the American Oriental Society*
JATBA *Journal d'Agriculture tropicale et de Botanique appliquée*
JBASA *Journal of the British Astronomical Association*
JCS *Journal of the Chemical Society*
JDMV *Jahresber. d. deutschen Math. Vereins*
JEA *Journal of Egyptian Archaeology*
JEP *Journal de l'Ecole (royale) Polytechnique* (Paris)
JESL *Journal of the Ethnological Society* (London)
JGPR *Journal of Geophysical Research* (formerly *Terrestrial Magnetism*)
JGSC *Ti Li Hsüeh Pao* (*Journal of the Geographical Society of China*)
JHI *Journal of the History of Ideas*
JHM *Journal of the History of Mathematics* (Moscow)
JHOAI *Jahreshefte d. österreichischen archäol. Instituts* (Vienna)
JHS *Journal of Hellenic Studies*
JIN *Journal of the Institute of Navigation*
JJAG *Japanese Journal of Astronomy and Geophysics*
JJHS *Kagakūshi Kenkyū* (*Japanese Journal of the History of Science*)
JMHP *Jen Min Hua Pao* (*People's Illustrated*)
JMJ *Japanese Meteorological Journal*
JMJP *Jen Min Jih Pao* (*People's Daily*)
JNES *Journal of Near Eastern Studies*
JOSHK *Journal of Oriental Studies* (Hong Kong University)
JPH *Journal de Physique*
JPOS *Journal of the Peking Oriental Society*
JRAI *Journal of the Royal Anthropological Institute*
JRAM *Journal f. reine u. angewandte Mathematik* (Crelle's)
JRAS *Journal of the Royal Asiatic Society*
JRAS/B *Journal of the (Royal) Asiatic Society of Bengal*
JRAS/KB *Journal (Transactions) of the Korea Branch of the Royal Asiatic Society*
JRAS/M *Journal of the Malayan Branch of the Royal Asiatic Society*
JRAS/NCB *Journal of the North China Branch of the Royal Asiatic Society*

JRGS	*Journal of the Royal Geographical Society*
JRSA	*Journal of the Royal Society of Arts*
JRSS	*Journal of the Royal Statistical Society*
JS	*Journal des Savants*
JSHB	*Journal suisse d'Horlogerie et Bijouterie*
JUB	*Journal of the University of Bombay*
JWCBRS	*Journal of the West China Border Research Society*
JWCI	*Journal of the Warburg and Courtauld Institutes*
JWH	*Journal of World History* (Unesco)
KDVS/HFM	*Kgl. Danske Videnskabernes Selskab* (Hist.-filol. Medd.)
KDVS/MFM	*Kgl. Danske Videnskabernes Selskab* (Math.-fysiske Medd.)
KHCK	*Kuo-Hsüeh Chhi Khan* (*Chinese Classical Quarterly*)
KHLT	*Kuo-Hsüeh Lun Tsung* (*Chinese Classical Review*)
KHS	*Kho-Hsüeh* (*Science*)
KHSSC	*Kho-Hsüeh Shih-Chieh* (*Scientific World*)
KHTP	*Kho-Hsüeh Thung Pao* (*Scientific Correspondent*)
KKHTC	*Kōkogaku Zasshi* (*Archaeological Miscellany*)
KMO/SM	*Scientific Memoirs of the Korean Meteorological Observatory, Chemulpo*
KNVSF/T	*Forhandlinger d. kgl. Norske Videnskabers Selskabs* (Trondheim)
KR	*Korean Repository*
KSHP	*Kuo Sui Hsüeh Pao* (*Chinese Classical Journal*)
KSP	*Ku Shih Pien* (Essays on Ancient History)
KVRS	*Kleine Veröffentlichungen d. Remeis Sternwarte*
LN	*La Nature*
LP	*La Pensée*
MA	*Man*
MAI/NEM	*Mémoires de l'Académie des Inscriptions et Belles Lettres* (Paris), *Notices et Extraits des MSS.*
MAL/FMN	*Memorie d. Accademia* (*r.* now *naz.*) *dei Lincei* (Cl. Sci. Fis. Mat. e Nat.)
MAL/MSF	*Memorie d. Accademia* (*r.* now *naz.*) *dei Lincei* (Cl. Sci. Mor. Stor. e Filol.)
MAN	*Mathematische Annalen*
MBIS	*Miscellanea Berolinensia ad Incrementum Scientiarum*
MCB	*Mélanges chinois et bouddhiques*
MCM	*Macmillan's Magazine*
MC/TC	*Techniques et Civilisations* (formerly *Métaux et Civilisations*)
MDGNVO	*Mitteilungen d. deutsch. Gesellschaft f. Natur- u. Volkskunde Ostasiens*
MEM	*Meteorological Magazine*
MG	*Mathematical Gazette*
MGSC	*Ti Chih Chuan Pao* (*Memoirs of the Chinese Geological Survey*)
MJ	*Mining Journal, Railway and Commercial Gazette*
MM	*Mining and Metallurgy*
MMI	*Mariner's Mirror*
MMT	*Messenger of Mathematics*
MN	*Monumenta Nipponica*
MPAW	*Monatsber. d. preuss. Akad. Wiss. Berlin*
MQ	*Modern Quarterly*
MRAI/DS	*Mémoires présentés par divers savants à l'Académie royale des Inscriptions et Belles Lettres* (Paris)
MRAS/B	*Memoirs of the Asiatic Society of Bengal*
MRAS/DS	*Mémoires présentés par divers savants à l'Académie royale des Sciences* (Paris)
MRASP	*Mémoires de l'Acad. royale des Sciences* (Paris)
MRDTB	*Memoirs of the Research Department of the Tōyō Bunko* (Tokyo)
MS	*Monumenta Serica*
MS/M	*Monumenta Serica* Monograph Series
MSAF	*Mémoires de la Société* (*nationale*) *des Antiquaires de France*
MSKGK	*Meiji Seitoku Kinen Gakkai Kiyō* (*Reports of the Meiji Memorial Society*)
MSOS	*Mitteilungen d. Seminars f. orientalische Sprachen* (Berlin)
MSRLSB/S	*Mém. de la Soc. Roy. des Lettres et des Sci. de Bohême* (Cl. Sci.)
MSRM	*Mémoires de la Soc. Russe de Mineralogie*
MUSEON	*Le Muséon* (Louvain)
MWR	*Monthly Weather Review*
MZ	*Meteorologische Zeitschrift*
N	*Nature*
NA	*Nautical Almanac*
NALC	*Nova Acta; Abhdl. d. kaiserl. Leop.-Carol. deutsch. Akad. Naturf. Halle*
NAP	*Nova Acta Acad. Petropol.*
NAT	*Nordisk Astronom. Tidskrift*
NAW	*Nieuw Archief voor Wiskunde*
NC	*Numismatic Chronicle*
NCH	*North China Herald*
NCM	*North China Mail*
NCR	*New China Review*
NGM	*National Geographic Magazine*

NGWG/PH — *Nachrichten v. d. k. Gesellsch.* (*Akademie*) *d. Wiss. z. Göttingen* (Phil.-hist. Klasse)
NQ — *Notes and Queries*
NTBB — *Nordisk Tidskrift för Bok- och Biblioteksväsen*
NW — *Naturwissenschaften*

O — *Observatory*
OAA — *Orientalia Antiqua*
OAZ — *Ostasiatische Zeitschrift*
OE — *Oriens Extremus* (Hamburg)
OLZ — *Orientalische Literatur-Zeitung*
OMO — *Österreichische Monatschrift f. d. Orient*
OR — *Oriens*
ORA — *Oriental Art*
ORR — *Orientalia* (Rome)
ORT — *Orient*
OSIS — *Osiris*

PA — *Pacific Affairs*
PAAAS — *Proceedings of the American Academy of Arts and Sciences*
PAI — *Paideuma*
PAPS — *Proceedings of the American Philosophical Society*
PASP — *Publications of the Astronomical Society of the Pacific*
PBA — *Proceedings of the British Academy*
PC — *People's China*
PDM — *Periodico di Matematiche*
PGA — *Proceedings of the Geologists' Association* (London)
PGS — *Proceedings of the Geological Society* (London)
PHR — *Philosophical Review*
PIAJ — *Proceedings of the Imperial Academy, Japan*
PJ — *Pharmaceutical Journal*
PMASAL — *Papers of the Michigan Academy of Science, Arts and Letters*
PMG — *Philosophical Magazine*
PMGP — *Papers in Meteorology and Geophysics* (Japan)
PMJP — *Mitteilungen aus Justus Perthes' geographischer Anstalt* (Petermann's)
PNHB — *Peking Natural History Bulletin*
POPA — *Popular Astronomy*
PPMST — *Proceedings of the Physico-Mathematical Society of Japan* (Tokyo) (continuation of *PTMS*)
PQ — *Philological Quarterly*
PRDA — *Priroda* (Moscow)
PRIA — *Proceedings of the Royal Irish Academy*
PRSA — *Proceedings of the Royal Society* (Ser. A)
PRSB — *Proceedings of the Royal Society* (Ser. B)
PRSG — *Publicaciones de la Real Sociedad Geográfica* (Spain)
PSA — *Proceedings of the Society of Antiquaries*
PTMS — *Proceedings of the Tokyo Mathematico-Physical Society* (cont. as *Phys.-Math. Soc. Japan*, *PPMST*)
PTRS — *Philosophical Transactions of the Royal Society*

QBCB/C — *Quarterly Bulletin of Chinese Bibliography* (Chinese ed.; *Thu Shu Chhi Khan*)
QBCB/E — *Quarterly Bulletin of Chinese Bibliography* (English ed.)
QJGS — *Quarterly Journal of the Geological Society of London*
QJRMS — *Quarterly Journal of the Royal Meteorological Society of London*
QMSF — *Quaderni del Museo di Storia delle Scienze, Firenze*
QSGM/A — *Quellen u. Studien z. Geschichte d. Mathematik* (Abt. A, Mathematik)
QSGM/B — *Quellen u. Studien z. Geschichte d. Mathematik* (Abt. B, Astronomie u. Physik)
QSGNM — *Quellen u. Studien z. Geschichte d. Naturwiss. u. d. Medizin* (contin. of *Archiv f. Gesch. d. Math., d. Naturwiss. u. d. Technik* (*AGMNT*) formerly *Archiv f. d. Gesch. d. Naturwiss. u. d. Technik* (*AGNT*))

RA — *Revue archéologique*
RAA/AMG — *Revue des Arts asiatiques* (*Annales du Musée Guimet*)
RAL/FMN — *Rendiconti d. Accademia* (*r.* now *naz.*) *dei Lincei* (Cl. Sci. Fis. Mat. e Nat.)
RAL/MSF — *Rendiconti d. Accademia* (*r.* now *naz.*) *dei Lincei* (Cl. Sci. Mor. Stor. e Filol.)
RASC/J — *Journal of the Royal Astronomical Society of Canada*
RAS/M — *Memoirs of the Royal Astronomical Society*
RAS/MN — *Monthly Notices of the Royal Astronomical Society*
RAS/ON — *Occasional Notes of the Royal Astronomical Society*
RCB — *Revue coloniale Belge*
REI — *Revue des Etudes islamiques*
RGI — *Rivista Geografica Italiana*
RGM — *Revista General de Marina* (Spain)
RGS — *Revue générale des Sciences pures et appliquées*
RHM — *Revue d'Histoire des Missions*
RM — *Reflets du Monde* (Brussels)
RQS — *Revue des Questions scientifiques* (Brussels)

RSISI *Rivista Scientifico-Industriale delle principali Scoperte ed Invenzioni fatti nelle Scienze e nelle Industrie*
RSO *Rivista di Studi Orientali*

SA *Sinica* (originally *Chinesische Blätter f. Wissenschaft u. Kunst*)
SAM *Scientific American*
SBE *Sacred Books of the East* Series
SC *Science*
SCI *Scientia*
SCSML *Smith College Studies in Modern Languages*
SG *Shinagaku* (*Sinology*)
SGZ *Shigaku Zasshi* (*Historical Journal*)
SHAW/PH *Sitzungsber. d. Heidelberg Akad. d. Wissenschaften* (Phil.-hist. Klasse)
SHR *Scottish Historical Review*
SHTC *Shih-Hsüeh Tsa Chih* (*Historical Journal*)
SHST *Ssu-Hsiang yü Shih-Tai* (*Thought and the Age; Journal of Chekiang University*)
SIR *Sirius*
SJJ *Seismological Journal of Japan*
SK *Sanjutsu Kyōiku* (*Mathematical Education*)
SKY *Sky and Telescope* (formerly *Sky*)
SM *Scientific Monthly* (formerly *Popular Science Monthly*)
SMA *Scripta Mathematica*
SPAW/PH *Sitzungsber. d. preuss. Akad. Wiss. Berlin* (Phil.-hist. Klasse)
SPCK Society for the Promotion of Christian Knowledge (London)
SPMSE *Sitzungsber. d. physik. med. Soc. Erlangen*
SPR *Science Progress*
SPRDS *Scientific Proceedings of the Royal Dublin Society*
SR *Shirin* (*Journal of History*)
SS *Science and Society* (New York)
SSA *Scripta Serica, Bulletin bibliographique* (Centre franco-chinois d'Etudes sinologiques, Peking)
SSE *Hua-Hsi Ta-Hsüeh Wên Shih Chi Khan* (*Studia Serica; West China Union University Literary and Historical Journal*)
SSE/M *Studia Serica* Monograph Series
ST *Die Sterne*
STTC *Shih Ti Tsa Chih* (*Chekiang University Journal of History and Geography*)
SUHVS *Skrifter utgifna af K. Humanistiska Vetenskaps Samfundet i Upsala*
SWAW/MN *Sitzungsber. d. (österreichischen) Akad. Wiss. Wien* (Vienna) (Math.-nat. Klasse)
SWAW/PH *Sitzungsber. d. (österreichischen) Akad. Wiss. Wien* (Vienna) (Phil.-hist. Klasse)
SWYK *Shuo Wên Yüeh Khan* (*Philological Monthly*)
SYR *Syria*

TAIMME *Transactions of the American Institute of Mining and Metallurgical Engineers* (formerly *TAIME*)
TAPS *Transactions of the American Philosophical Society*
TAS/B *Transactions of the Asiatic Society of Bengal* (Asiatick Researches)
TASJ *Transactions of the Asiatic Society of Japan*
TBGZ *Tōkyō Butsuri Gakko Zasshi* (*Journal of the Tokyo College of Physics*)
TCAAS *Transactions of the Connecticut Academy of Arts and Sciences*
TCLP *Ti Chih Lun Phing* (*Geological Review*)
TFTC *Tung Fang Tsa Chih* (*Eastern Miscellany*)
TG/K *Tōhō Gakuhō, Kyōto* (*Kyoto Journal of Oriental Studies*)
TG/T *Tōhō Gakuhō, Tōkyō* (*Tokyo Journal of Oriental Studies*)
TH *Thien Hsia Monthly* (Shanghai)
THTC *Ti Hsüeh Tsa Chih* (*Geographical Miscellany*)
TIAU *Transactions of the International Astronomical Union*
TIYT *Trudy Instituta Istorii Yestestvoznania i Tekhniki* (Moscow)
TM *Terrestial Magnetism and Atmospheric Electricity* (continued as *Journal of Geophysical Research*)
TMB *Bulletin du Musée ethnogr. du Trocadéro*
TNH *Tōa no Hikari* (*Light of East Asia*)
TNS *Transactions of the Newcomen Society*
TNZI *Transactions of the New Zealand Institute*
TOCS *Transactions of the Oriental Ceramic Society*
TP *T'oung Pao* (*Archives concernant l'Histoire, les Langues, la Géographie, l'Ethnographie et les Arts de l'Asie Orientale*, Leiden)
TSBA *Transactions of the Society of Biblical Archaeology*
TSSC *Transactions of the Science Society of China*
TSYK *Thu Shu Yüeh Khan* (*Library Monthly*)
TTB *Tekn. Tidskr. Bergsvetenskap*

TTWT	*Tokyo Astron. Observ. Journal*
TWHP	*Thien Wên Hsüeh Pao* (*Chinese Journal of Astronomy*)
TYG	*Tōyō Gakuhō* (*Reports of the Oriental Society of Tokyo*)
TYKK	*Thien Yeh Khao Ku Pao Kao* (*Reports on the History of Agriculture*)
UBHJ	*University of Birmingham Historical Journal*
UMN	*Unterrichtsblätter f. Math. u. Naturwiss.*
UNESC	*Unesco Courier*
VAG	*Vierteljahrsschrift d. astronomischen Gesellschaft*
VBW	*Vorträge d. Bibliothek Warburg*
VKAWA/L	*Verhandelingen d. Koninklijke* (*Nederl.*) *Akad. v. Wetenschappen te Amsterdam* (Afd. Letterkunde)
VS	*Variétés Sinologiques* Series
WNZ	*Wiener numismatische Zeitschrift*
YC	*Yü Chou* (*The Universe; Journ. of the Chinese Astron. Soc.*)
YCHP	*Yenching Hsüeh Pao* (*Yenching University Journal of Chinese Studies*)
Z	*Zalmoxis: Revue des Etudes religieuses*
ZA	*Zeitschrift f. Astronomie*
ZAE	*Zeitschrift f. angew. Entomologie*
ZASS	*Zeitschr. f. Assyriologie*
ZDMG	*Zeitschrift d. deutsch. morgenländischen Gesellschaft*
ZDPV	*Zeitschrift d. deutsch. Palästina-Vereins*
ZGEB	*Zeitschrift d. Gesellschaft f. Erdkunde* (Berlin)
ZMNWU	*Zeitschrift f. Math. u. Naturwiss. Unterricht*
ZMP	*Zeitschrift f. Math. u. Physik*
ZSF	*Zeitschrift f. Sozialforschung*

A. CHINESE BOOKS BEFORE +1800

Each entry gives particulars in the following order:

(*a*) title, alphabetically arranged, with characters;

(*b*) alternative title, if any;

(*c*) translation of title;

(*d*) cross-reference to closely related book, if any;

(*e*) dynasty;

(*f*) date as accurate as possible;

(*g*) name of author or editor, with characters;

(*h*) title of other book, if the text of the work now exists only incorporated therein; or, in special cases, references to sinological studies of it;

(*i*) references to translations, if any, given by the name of the translator in Bibliography C;

(*j*) notice of any index or concordance to the book if such a work exists;

(*k*) reference to the number of the book in the *Tao Tsang* catalogue of Wieger (6), if applicable;

(*l*) reference to the number of the book in the *San Tsang* (Tripiṭaka) catalogues of Nanjio (1) and Takakusu & Watanabe, if applicable.

Words which assist in the translation of titles are added in round brackets.

Alternative titles or explanatory additions to the titles are added in square brackets.

It will be remembered (p. 685 above) that in Chinese indexes words beginning *Chh-* are all listed together after *Ch-*, and *Hs-* after *H-*, but that this applies to initial words of titles only.

Where there are any differences between the entries in these bibliographies and those in vol. 1, the information here given is to be taken as more correct.

References to the editions used in the present work, and to the *tshung-shu* collections in which books are available, will be given in the final volume.

Abbreviations

C/Han	Former Han.
H/Han	Later Han.
H/Shu	Later Shu (Wu Tai).
H/Thang	Later Thang (Wu Tai).
J/Chin	Jurchen Chin.
L/Sung	Liu Sung.
N/Chou	Northern Chou.
N/Chhi	Northern Chhi.
N/Sung	Northern Sung (before the removal of the capital to Hangchow).
N/Wei	Northern Wei.
S/Chhi	Southern Chhi.
S/Sung	Southern Sung (after the removal of the capital to Hangchow).

An Thien Lun 安天論.
Discourse on the Conformation of the Heavens.
Chin, *c.* +320 (prob. +336).
Yü Hsi 虞喜.
(Fragment in *YHSF*, ch. 77.)

Ao Yü Tzu Hsü Hsi So Wei Lun 聱隅子歔欷瑣微論.
Whispered Trifles by the Tree-stump Master.
Sung, *c.* +1040.
Huang Hsi 黄晞.

Chan Kuo Tshê 戰國策.
Records of the Warring States.
Chhin.
Writer unknown.

Chan Yuan Ching Yü 湛淵靜語.
Tranquil Conversations by the Limpid Deep.
Yuan, +14th century.
Pai Thing 白珽.

Chang Chhiu-Chien Suan Ching 張邱建算經.
Chang Chhiu-Chien's Mathematical Manual.
N/Wei, or L/Sung or S/Chhi, between +468 and +486
Chang Chhiu-Chien 張邱建.

Chang Chhiu-Chien Suan Ching Hsi Tshao 張邱建算經細草.
Detailed solutions of the Problems in *Chang Chhiu-Chien's Mathematical Manual.*
Sui, late +6th.
Liu Hsiao-Sun 劉孝孫.

Chang Tzu Chhüan Shu 張子全書.
Complete Works of Master Chang (Tsai) (d. +1077) with commentary by Chu Hsi.
Sung (ed. Chhing), *editio princeps* +1719.
Chang Tsai.
Ed. Chu Shih 朱軾 & Tuan Chih-Hsi 段志熙.

Chen-La Fêng Thu Chi 眞臘風土記.
Description of Cambodia.
Yuan, +1297.
Chou Ta-Kuan 周達觀.

Chêng Khai-Yang Tsa Chu 鄭開陽雜著.
Collected (Geographical) Writings of Chêng Jo-Tsêng.
Ming, *c.* +1570 (first pr. 1932).
Chêng Jo-Tsêng 鄭若曾.

Chêng Lei Pên Tshao 證類本草.
Reorganised Pharmacopoeia.
N/Sung, +1108, enlarged +1116; and J/Chin, +1204.
Thang Shen-Wei 唐愼微.
Cf. Hummel (13), Lung Po-Chien (*1*).

Chêng Mêng 正蒙.
Right Teaching for Youth [or, Intellectual Discipline for Beginners].
Sung, *c.* +1060.
Chang Tsai 張載.

Chi Chung Chou Shu 汲冢周書.
The Books of (the) Chou (Dynasty) found in the Tomb at Chi.
See *I Chou Shu*.

Chi Ho Yuan Pên 幾何原本.
Elements of Geometry (Euclid's) [first six chapters].
Ming, +1607.
Tr. begun by Li Ma-Tou (Matteo Ricci) 利瑪竇 & Hsü Kuang-Chhi 徐光啓.
Completed by Alexander Wylie & Li Shan-Lan (1857).

Chi Jan
See *Chi Ni Tzu*.

Chi Ni Tzu 計倪子.
[= *Fan Tzu Chi Jan* 范子計然.]
The Book of Master Chi Ni.
Chou, −4th century.
Attrib. Fan Li (Chi Jan) 范蠡 (計然).

Chiang-Yin Hsien Chih 江陰縣志.
Topography of Chiang-yin and its District [in Chiangsu Province].
Sung, +1194 (continued in Ming and Chhing).
Original writer unknown.

Chiao Li Chiao Phing Shih Lüeh 交黎剿平事略.
Materials on the Pacificatory Expedition to Southern Annam (Cochin China) during the Li (Dynasty).
Ming, +1551.
Chang Ao 張鏊.

Chieh An Chi 戒菴集.
Notes from the Fasting Pavilion.
Ming, *c.* +1500.
Chin Kuei 靳貴.

Chieh Tzu Yuan Hua Chuan 芥子園畫傳.
The Mustard-Seed Garden Guide to Painting.
Chhing, +1679.
Li Li-Ong (preface) 李笠翁.
Wang Kai (text and illustrations) 王槪.

Chien Phing I Shuo 簡平儀說.
Description of a simple Altazimuth Quadrant [astronomical instrument].
Ming, +1611.
Hsiung San-Pa (Sabbathin de Ursis) 熊三拔.

Chien Wên Chi 見聞記.
Records of Things Seen and Heard.
Thang, +800.
Fêng Yen 封演.

Chih Fang Wai Chi 職方外紀.
On World Geography.
Ming, +1623.
Ai Ju-Lüeh (Giulio Aleni) 艾儒略.

Chih Lin Hsin Shu 志林新書.
New Book of Miscellaneous Records
Chin, +4th century.
Yü Hsi 虞喜.

Chin Chhi Chü Chu 晉起居注.
Daily Court Records of the Emperors of the Chin Dynasty.
Pre-Sui.
Liu Tao-Hui 劉道會.

Chin Lou Tzu 金樓子.
Book of the Golden Hall Master.
Liang, *c.* +550.
Hsiao I 蕭繹.
(Liang Yuan Ti 梁元帝.)

Chin Nang Chhi Mêng 錦囊啓蒙.
Brocaded Bag of (Mathematical Books) for the Relief of Ignorance.
Yuan or Ming, +14th century.
Writer unknown.
(In *Yung Lo Ta Tien*, chs. 16343–4.)

Chin Shih 金史.
History of the Chin (Jurchen) Dynasty [+1115 to +1234].
Yuan, *c.* +1345.
Tho-Tho (Toktaga) 脫脫 & Ouyang Hsüan 歐陽玄.
Yin-Tê Index, no. 35.

Chin Shu 晉書.
History of the Chin Dynasty [+265 to +419].
Thang, +635.
Fang Hsüan-Ling 房玄齡.
A few chs. tr. Pfizmaier (54–57).

Ching-Chai Ku Chin Chu 敬齋古今黈.
The Commentary of (Li) Ching-Chai (Li Yeh) on Things Old and New.
Sung, +13th century.
Li Yeh 李冶.

Ching Chhu Sui Shih Chi 荊楚歲時記.
Annual Folk Customs of the States of Ching and Chhu [i.e. of the districts corresponding to those ancient States: Hupei, Hunan and Chiangsi].
Prob. Liang, *c.* +550, but perhaps partly Sui, *c.* +610.
Tsung Lin 宗懍.
See des Rotours (1), p. cii.

Ching Shih Ta Tien.
See *Yuan Ching Shih Ta Tien*.

Ching Shu Suan Hsüeh Thien Wên Khao 經書算學天文考.
Investigations on the Astronomy and Mathematics of the Classics.
Chhing, +1797.
Chhen Mou-Ling 陳懋齡.

Ching Thien Kai 經天該.
Comprehensive Rhymed Catalogue of Stars.
Ming, prob. +1601.
Li Ma-Tou (Matteo Ricci) 利瑪竇 & Li Wo-Tshun 李我存.

Ching Tien Chi Lin 經典集林.
Collected Classical Fragments.
Chhing, *c.* +1790.
Hung I-Hsüan (ed.) 洪頤煊.

Chiu Chang Suan Ching.
See *Chiu Chang Suan Shu*.

Chiu Chang Suan Shu 九章算術.
Nine Chapters on the Mathematical Art.
H/Han, +1st century (containing much material from C/Han and perhaps Chhin).
Writer unknown.

Chiu Chang Suan Shu Hsi Tshao Thu Shuo 九章算術細草圖說.
Careful Explanation of the *Nine Chapters on the Mathematical Art*, with diagrams.
Chhing, *c.* +1795.
Li Huang 李潢.

Chiu Chang Suan Shu Yin I 九章算術音義.
Explanations of Meanings and Sounds of Words occurring in the *Nine Chapters on the Mathematical Art*.
Sung.
Li Chi 李籍.

Chiu Chia Chin Shu Chi Pên 九家晋書輯本.
Collected Texts of Nine Versions of the *History of the Chin Dynasty*.
Thang.
Ed. Thang Chhiu 湯球 (Chhing).

Chiu Huang Pên Tshao 救荒本草.
Famine Herbal.
Ming, (+1395), +1406.
Chou Ting Wang (prince of the Ming) 周定王.
Preserved in chs. 46–59 of *Nung Chêng Chhüan Shu* (q.v.).

Chiu Jih Lu 就日錄.
Daily Journal.
Sung.
Mr Chao 趙氏.
(Kuan Pu Nai Tê Ong; the Old Gentleman of the Water-Garden who attained Success through Forbearance 灌圃耐得翁.)

Chiu Kung Hsing Chhi Li-Chhêng 九宮行碁立成.
Divination Table arranged according to the Nine Palaces in which the Chess Pieces Move.
Thang or Wu Tai.
Writer unknown.
(Tunhuang MS. S/6164.)

Chiu Thang Shu 舊唐書.
Old History of the Thang Dynasty [+618 to +906].
Wu Tai, +945.
Liu Hsü 劉昫.

Cho Kêng Lu 輟耕錄
[sometimes *Nan Tshun Cho Kêng Lu*].
Talks (at South Village) while the Plough is Resting.
Yuan and Ming, +1366.
Thao Tsung-I 陶宗儀.

Chou I.
See *I Ching*.

Chou I Tshan Thung Chhi.
See *Tshan Thung Chhi*.

Chou Li 周禮.
Record of the Rites of (the) Chou (Dynasty) [descriptions of all government official posts and their duties].
C/Han, perhaps containing some material from late Chou.
Compilers unknown.
Tr. E. Biot (1).

Chou Li I I Chü Yao 周禮疑義舉要.
Discussion of the Most Important Doubtful Matters in the *Record of Rites of the Chou Dynasty*.
Chhing, +1791.
Chiang Yung 江永.

Chou Pei Suan Ching 周髀算經.
The Arithmetical Classic of the Gnomon and the Circular Paths (of Heaven).
Chou, Chhin and Han. Text stabilised about the −1st century, but including parts which must be of the late Warring States period (*c.* −4th century) and some even pre-Confucian (−6th century).
Writers unknown.

Chou Pei Suan Ching Yin I 周髀算經音義.
Explanation of Meanings and Sounds of Words occurring in the *Arithmetical Classic of the Gnomon and the Circular Paths of Heaven*.
Sung.
Li Chi 李籍.

Chou Shu 周書.
History of the (Northern) Chou Dynasty [+557 to +581].
Thang, +625.
Linghu Tê-Fen 令狐德棻.

Chu Fan Chih 諸蕃志.
Records of Foreign Peoples.
Sung, *c.* +1225. (This is Pelliot's dating; Hirth & Rockhill favoured between +1242 and +1258.)
Chao Ju-Kua 趙汝适.
Tr. Hirth & Rockhill (1).

Chü Mi Wei Shan Fu 聚米爲山賦.
Essay on the Art of Constructing Mountains with Rice [relief-maps].
Thang, *c.* +845.
Chiang Fang 蔣防.

Chu Shu Chi Nien 竹書紀年.
The Bamboo Books [annals].
Chou, −295 and before, such parts as are genuine. (Found in the tomb of An Li Wang, a prince of the Wei State, r. −276 to −245; in +281).
Writers unknown.
Tr. E. Biot (3).

Chu Tzu Chhüan Shu 朱子全書.
Collected Works of Master Chu [Hsi].
Sung (ed. Ming), *editio princeps* +1713.
Chu Hsi 朱熹.
Ed. Li Kuang-Ti 李光地 (Chhing).
Partial trs. Bruce (1); le Gall (1).

Chu Tzu Wên Chi 朱子文集.
Selected Writings of Master Chu [Hsi].
Sung.
Chu Hsi 朱熹.
Ed. Chu Yü 朱玉 (Chhing).

Chu Tzu Yü Lei 朱子語類.
Classified Conversations of Master Chu [Hsi].
Sung, *c.* +1270.
Chu Hsi 朱熹.

Chu Tzu Yü Lei (*cont.*)
Ed. Li Ching-Tê 黎靖德 (Sung).
Chu Yeh Thing Tsa Chi 竹葉亭雜記.
Miscellaneous Records of the Bamboo Leaf Pavilion.
Chhing, +17th century.
Yao Yuan-Chih 姚元之.
Chuang shih Suan Hsüeh 莊氏算學.
Mr Chuang's Treatise on Mathematics [geometry].
Chhing, *c.* +1720.
Chuang Hêng-Yang 莊亨陽.
Chuang Tzu 莊子.
[= *Nan Hua Chen Ching*.]
The Book of Master Chuang.
Chou, *c.* −290.
Chuang Chou 莊周.
Tr. Legge (5); Fêng Yu-Lan (5); Lin Yü-Thang (1).
Yin-Tê Index no. (suppl.) 20.
Chung Hsi Ching Hsing Thung I Khao 中西經星同異考.
Investigation of the Similarities and Differences between Chinese and Western Star-Names.
Chhing, +1723.
Mei Wên-Mi 梅文鼏.
Chhang-an Chih Thu 長安志圖.
Maps to illustrate the History of the City of Eternal Peace [Chhang-an (Sian), ancient capital of China].
Sung, *c.* +1075.
Li Hao-Wên 李好文.
Chhang-Chhun Chen Jen Hsi Yu Chi 長春眞人西遊記.
The Western Journey of the Taoist (Chhiu) Chhang-Chhun.
Yuan, +1228.
Li Chih-Chhang 李志常.
Chhao Shuo 潮說.
Discussion of the (Sea) Tides.
Sung, early +11th century.
Chang Chün-Fang 張君房.
Chhao Yeh Chhien Tsai 朝野僉載.
Stories of Court Life and Rustic Life.
Thang, +8th century, but much remodelled in the Sung.
Chang Tso 張鷟.
Chhêng Chhu Thung Pien Suan Pao 乘除通變算寶.
Precious Reckoner for Mutually Varying Quantities.
Sung, end +13th century.
Yang Hui 楊輝.
(In *Yang Hui Suan Fa*, q.v.)
Chhi Chêng Thui Pu 七政推步.
On the Motions of the Seven Governors [planetary ephemerides prepared according to the methods of the Muslim astronomers in China].
Ming, +1482.
Pei Lin 貝琳.
Chhi Chhiao Hsin Phu 七巧新譜.
New Tangram Puzzles.
Chhing.
See *Sang Hsia-Kho* (1).
Chhi Hsiu Lei Kao 七修類稿.
Seven Compilations of Classified Manuscripts.
Ming, *c.* +1530.
Lang Ying 郎瑛.
Chhi Ku Suan Ching 緝古算經.
Continuation of Ancient Mathematics.
Thang, *c.* +625.
Wang Hsiao-Thung 王孝通.
Chhi Kuo Khao 七國考.
Investigations of the Seven (Warring) States.
Ming.
Tung Yüeh 董說.
Chhi-tan Kuo Chih 契丹國志.
Memoirs of the Liao [Chhi-tan Tartar Kingdom].
Sung and Yuan, mid. +13th century.
Yeh Lung-Li 葉隆禮.
Chhi Tung Yeh Yu 齊東野語.
Rustic Talks in Eastern Chhi.
Yuan, *c.* +1290.
Chou Mi 周密.
Chhi Yao Hsing Chhen Pieh Hsing Fa 七曜星辰別行法.
The Different Influences of the Seven Luminaries and the Constellations [lists the *hsiu* and the stars in them; also gives their gods, diseases and remedies].
Thang, *c.* +710 (perhaps later in the +8th century).
Ascr. I-Hsing 一行.
TW/1309.
Chhi Yao Jang Tsai Chüeh 七曜攘災決 (訣).
Formulae for avoiding Calamities according to the Seven Luminaries [astronomical text, giving planetary ephemerides from +794, *nakshatra* (*hsiu*) extensions with ecliptic data, planetary names of the days of the Sogdian 7-day week, and associations of planets with the five Chinese elements, with the seven stars of the Great Bear, etc.].
Thang, +9th century, after +806.
(Probably partly tr. from Sogdian.)
Chin Chü-Chha 金俱吒.
TW/1308.
Chhiao Hsiang Hsiao Chi 樵香小記.
Woodsmoke Perfume Jottings.
Chhing.
Ho Hsiu 何琇.
Chhieh Yün 切韻.
Dictionary of the Sounds of Characters [rhyming dictionary].
Sui, +601.
Lu Fa-Yen 陸法言.
See *Kuang Yün*.
Chhien Han Shu 前漢書.
History of the Former Han Dynasty [−206 to +24].

Chhien Han Shu (*cont.*)
H/Han (begun about +65), *c.* +100.
Pan Ku 班固 and (after his death in +92) his sister Pan Chao 班昭.
Partial trs. Dubs (2), Pfizmaier (32–34, 37–51), Wylie (2, 3, 10), Swann (1).
Yin-Tê Index, no. 36.

Chhien Hsiang Li Shu 乾象曆術.
Calendrical Science based on the Celestial Appearances.
H/Han, *c.* +206.
Tshai Yung 蔡邕 & Liu Hung 劉洪.
(Preserved in later compilations.)

Chhien-Lung Fu Thing Chou Hsien Chih 乾隆府廳州縣志.
Chhien-Lung reign-period Geography of the (Chinese) Empire.
Chhing, +1787.
Hung Liang-Chi 洪亮吉.

Chhih Hsiu Liang Chê Hai Thang Thung Chih 勅修兩浙海塘通志.
Historical Account of the Coastal Protection Works of Chekiang Province, prepared by Imperial Command.
Chhing, +1751.
Fang Kuan-Chhêng 方觀承.

Chhih Pei Ou Than 池北偶談.
Chance Conversations North of Chhih (-chow).
Chhing, +1691.
Wang Shih-Chên 王士禎.

Chhih Shui I Chen 赤水遺珍.
Pearls Recovered from the Red River [the recognition of the value of ancient and medieval Chinese mathematics].
Chhing, +1761.
Mei Ku-Chhêng 梅榖成.

Chhih Ya 赤雅.
Information about the Naked Ones [Miao and other tribespeople].
Ming, +16th or +17th century.
Kuang Lu 鄺露.

Chhin-Ting Ho Yuan Chi Lüeh 欽定河源紀略.
See *Ho Yuan Chi Lüeh*.

Chhin-Ting I Hsiang Khao Chhêng 欽定儀象考成.
See *I Hsiang Khao Chheng*.

Chhin-Ting Ku Chin Thu Shu Chi Chhêng 欽定古今圖書集成.
See *Thu Shu Chi Chhêng*.

Chhin-Ting Shou Shih Thung Khao 欽定授時通考.
See *Shou Shih Thung Khao*.

Chhing Hsiang Tsa Chi 青箱雜記.
Miscellaneous Records on Green Bamboo Tablets.
Sung, early +11th century.
Wu Chhu-Hou 吳處厚.

Chhing I Lu 清異錄.
Records of the Unworldly and the Strange.
Wu Tai, *c.* +950.
Thao Ku 陶穀.

Chhing Po Pieh Chih 清波別志.
Additional Green-Waves Memoirs.
Sung, +1194.
Chou Hui 周煇.

Chhing Yü Chih Yü Shu 請雨止雨書.
Prayers for and against Rain.
Han.
Writer unknown.

Chhiung Thien Lun 穹天論.
Discourse on Heaven's Vault.
Chin, *c.* +265.
Yü Sung 虞聳.
(Fragment in *YHSF*, ch. 77.)

Chhou Hai Thu Pien 籌海圖編.
Illustrated Seaboard Strategy.
Ming, +1562.
Chêng Jo-Tsêng 鄭若曾.
Also attributed to Hu Tsung-Hsien 胡宗憲.

Chhou Jen Chuan 疇人傳.
Biographies of (Chinese) Mathematicians and Astronomers.
Chhing, +1799.
Juan Yuan 阮元.
With continuations by Lo Shih-Lin 羅士琳, Chu Kho-Pao 諸可寶 and Huang Chung-Chün 黃鍾駿.
(In *Huang Chhing Ching Chieh*, chs. 159ff.)

Chhu Hsüeh Chi 初學記.
Entry into Learning [encyclopaedia].
Thang, +700.
Hsü Chien 徐堅.

Chhü I Shuo Tsuan 祛疑說纂.
Discussions on the Dispersal of Doubts.
Sung, *c.* +1230.
Chhu Yung 儲泳.

Chhu Tzhu 楚辭.
Elegies of Chhu (State).
Chou, *c.* −300 (with Han additions).
Chhü Yuan 屈原
(& Chia I 賈誼,
Yen Chi 嚴忌,
Sung Yü 宋玉,
Huainan Hsiao-Shan 淮南小山 *et al.*).
Partial tr. Waley (23).

Chhü Wei Chiu Wên 曲洧舊聞.
Old Matters Discussed beside the Wei River.
Sung, *c.* +1130.
Chu Pien 朱弁.

Chhun Chhiu 春秋.
Spring and Autumn Annals [i.e. Records of Springs and Autumns].
Chou; a chronicle of the State of Lu kept between −722 and −481.
Writers unknown.
Cf. *Tso Chuan*; *Kungyang Chuan*; *Kuliang Chuan*.
See Wu Khang (1); Wu Shih-Chhang (1).
Tr. Couvreur (1); Legge (11).

Chhun Chhiu Wei Khao I Yu 春秋緯考異郵.
Apocryphal Treatise on the *Spring and Autumn Annals;* Investigation of Discrepancies and Mistakes.

Chhun Chhiu Wei Khao I Yu (*cont.*)
C/Han, −1st century.
Writer unknown.

Chhun Chhiu Wei Yuan Ming Pao 春秋緯元命苞.
Apocryphal Treatise on the *Spring and Autumn Annals*; the Mystical Diagrams of Cosmic Destiny [astrological-astronomical]
C/Han, *c.* −1st century.
Writer unknown.

Chhung-Chên Li Shu 崇禎曆書.
Chhung-Chên reign-period Treatise on (Astronomy and) Calendrical Science.
Ming, +1635 (first form of the Jesuit astronomical encyclopaedia).
See *Hsin Fa Suan Shu*.

Chhung Hsiu Ko Hsiang Hsin Shu 重修革象新書.
Revision of the *New Elucidations of the Heavenly Bodies* (by Chao Yu-Chhin).
Ming.
Wang Wei 王禕.

Chhung Hsü Chen Ching 冲虛眞經.
See *Lieh Tzu*.

Erh Chhêng Chhüan Shu 二程全書.
Complete Works of the Two Chhêng Brothers [Neo-Confucian philosophers]. Contains: *Honan Chhêng Shih I Shu, Honan Chhêng Shih Wai Shu, I-Chhuan I Chuan, Erh Chhêng Sui Yen*, etc.
Sung, *c.* +1110, collected +1323.
Chhêng I 程頤 & Chhêng Hao 程顥.
Coll. Than Shan-Hsin 譚善心 (Yuan).
Ed. Yen Yü-Hsi 閻禹錫 (Ming, +1461).

Erh Ya 爾雅.
Literary Expositor [dictionary].
Chou material, stabilised in Chhin or C/Han.
Compiler unknown.
Enlarged and commented on *c.* +300 by Kuo Pho 郭璞.
Yin-Tê Index no. (suppl.) 18.

Fa-Hsien Chuan 法顯傳.
See *Fo Kuo Chi*.

Fa-Hsien Hsing Chuan 法顯行傳.
See *Fo Kuo Chi*.

Fa Mêng Chi 發蒙記.
Tutorial Record.
Chin, late +3rd century.
Shu Hsi 束晳.
(In *YHSF*, ch. 62.)

Fa Wei Lun 發微論.
Effects of Minute Causes [geomantic].
Sung, *c.* +1170.
Tshai Yuan-Ting 蔡元定.

Fa Yen 法言.
Admonitory sayings [in admiration, and imitation, of the *Lun Yü*].
Hsin, +5.
Yang Hsiung 楊雄.
Tr. von Zach (5).

Fa Yuan Chu Lin 法苑珠林.
Forest of Pearls in the Garden of the Law [Buddhist encyclopaedia].
Thang, +668.
Tao-Shih (monk) 道世.

Fan Thien Huo-Lo Chiu Yao 梵天火羅九曜.
The *Horā* of Brahma and the Seven Luminaries.
Thang, +874.
Attrib. to I-Hsing, but not by him.
TW/1311.

Fang-Chou Tsa Yen 方洲雜言.
Reminiscences of (Chang) Fang-Chou.
Ming, +1430 to +1470.
Chang Ning 張寧.

Fang Yü Shêng Lan 方輿勝覽.
Triumphant Vision of the Great World [geography].
Sung, +1240.
Chu Mu 祝穆.

Fêng Chhuang Hsiao Tu 楓窻小牘.
Maple-Tree Window Memories.
Sung, early +13th century (after +1202).
Yuan Chhiung 袁褧.

Fêng Su Thung I 風俗通義.
The Meaning of Popular Traditions and Customs.
H/Han, +175.
Ying Shao 應劭.
Chung-Fa Index no. 3.

Fêng Thu Chi 風土記.
Record of Airs and Places.
Chin, +3rd century.
Chou Chhu 周處.

Fo Kuo Chi 佛國記
[=*Fa-Hsien Chuan* or *Fa-Hsien Hsing Chuan*].
Records of Buddhist Countries [also called Travels of Fa-Hsien].
Chin, *c.* +420.
Fa-Hsien (monk) 法顯.
Tr. Rémusat (1), Beal (1), Legge (4), H. A. Giles (3).

Fo Shuo Pei Tou Chhi Hsing Yen Ming Ching 佛說北斗七星延命經.
Sūtra spoken by a Bodhisattva on the Delaying of Destiny according to the Seven Stars of the Great Bear [contains Western zodiacal animal cycles].
Thang.
Tr. unknown.
TW/1307.

Fo Tsu Thung Chi 佛祖統紀.
Records of the Lineage of Buddha and the Patriarchs.
Sung, +1270.
Chih-Phan (monk) 志盤.
N/1661.

Fu Hou Ku Chin Chu 伏侯古今注.
Commentary of the Lord Fu on Things New and Old.

Fu Hou Ku Chin Chu (*cont.*)
H/Han, *c.* +140.
Fu Wu-Chi 伏無忌.
(Only fragments, as in *YHSF*, ch. 73.)

Fu Tzu 傅子.
Book of Master Fu.
Chin, +3rd century.
Fu Hsüan 傅玄.

Hai Chhao Chi Shuo 海潮輯說.
Collected Writings on the Sea Tides.
Chhing, +1781.
Yü Ssu-Chhien (ed.) 俞思謙.

Hai Chhao Fu 海潮賦.
Essay on the Tides.
Thang, +850.
Lu Chao 盧肇.

Hai Chhao Lun 海潮論.
Discourse on the Tides.
Wu Tai, *c.* +900.
Chhiu Kuang-Thing 邱光庭.

Hai Chhao Thu Hsü 海潮圖序.
Preface to Diagrams of the Tides.
Sung, +1025.
Yü Ching 余靖.

Hai Chhao Thu Lun 海潮圖論.
Illustrated Discourse on the Tides.
Sung, +1026.
Yen Su 燕肅.

Hai Han Wan Hsiang Lu 海涵萬象錄.
The Multiplicity of Phenomena.
Ming.
Huang Jun-Yü 黃潤玉.

Hai Kuo Wên Chien Lu 海國聞見錄.
Record of Things Seen and Heard about the Coastal Regions.
Chhing, +1744.
Chhen Lun-Chhiung 陳倫炯.

Hai Nei Shih Chou Chi 海內十洲記.
Record of the Ten Sea Islands.
Alleged Han; prob. +4th or +5th century.
Attrib. Tungfang Shuo 東方朔.

Hai Tao Suan Ching 海島算經.
Sea Island Mathematical Manual.
[Originally a supplement to the *Chiu Chang Suan Shu* and known before the Thang as *Chhung Chha*. The Sui bibliography also has a *Chiu Chang Chhung Chha Thu* 九章重差圖.]
San Kuo, +263.
Liu Hui 劉徽.

Hai Thao Chih 海濤志
[or *Hai Chhiao Chih* 海嶠志].
On the Tides.
Thang, *c.* +770.
Tou Shu-Mêng 竇叔蒙.

Hai Yao Pên Tshao 海藥本草.
Drugs of the Southern Countries beyond the Seas [or Pharmaceutical Codex of Marine Products].
Thang, *c.* +775 (or early +10th century).
Li Hsün (acc. to Li Shih-Chen) 李珣.
Li Hsien (acc. to Huang Hsiu-Fu) 李玹.
Preserved in *Pên Tshao Kang Mu*, etc.

Han Wu Ti Nei Chuan 漢武帝內傳.
The Inside Story of Emperor Wu of the Han.
Chin, +4th century, or at least pre-Sui.
Anon. (ascr. Pan Ku), perhaps Ko Hung 葛洪.
TT/289.

Han Yü Thung 函宇通.
General Survey of the Universe [incl. Ko Chih Tshao 格致草 on astronomy and Ti Wei 地緯 on geography].
Chhing, +1648.
Hsiung Ming-Yü 熊明遇 & Hsiung Jen-Lin 熊人霖.

Ho Kuan Tzu 鶡冠子.
Book of the Pheasant-Cap Master.
A very composite text, stabilised by +629, as is shown by one of the MSS. found at Tunhuang. Much of it must be Chou (−4th century) and most is not later than Han (+2nd century), but there are later interpolations, including a +4th- or +5th-century commentary which has become part of the text and accounts for about a seventh of it (Haloun (5), p. 88). It contains also a lost 'Book of the Art of War'.
Attrib. Ho Kuan Tzu 鶡冠子.
TT/1161.

Ho Lin Yü Lu 鶴林玉露.
Jade Dew from the Forest of Cranes.
Sung.
Lo Ta-Ching 羅大經.

Ho Shuo Fang Ku Chi 河朔訪古記.
Archaeological Topography of the Regions North of the Yellow River.
Yuan.
Na-Hsin 納新.

Ho Thu Wei Chi Yao Kou 河圖緯稽耀鈎.
Apocryphal Treatise on the *River Diagram*; Investigation of the Full Circle of the (Celestial) Brightnesses.
C/Han, *c.* −50.
Writer unknown.

Ho Thu Wei Kua Ti Hsiang 河圖緯括地象.
Apocryphal Treatise on the *River Diagram*; Examination of the Signs of the Earth.
C/Han, *c.* −50.
Writer unknown.

Ho Yuan Chi 河源記.
Records of the Source of the (Yellow) River.
Yuan, *c.* +1300.
Phan Ang-Hsiao 潘昂霄.

Ho Yuan Chi Lüeh 河源紀略.
The Sources of the (Yellow) River, and a History of our Knowledge of them [published by Imperial Authority].
Chhing, +1784.
Ed. by a number of scholars including A-Mi-Ta 阿彌達 and Wang Nien-Sun 王念孫.

Honan Chhêng shih I Shu 河南程氏遺書.
Remaining Records of Discourses of the Chhêng brothers of Honan [Chhêng I and Chhêng Hao, +11th-century Neo-Confucian philosophers].
Sung, +1168.
Chu Hsi (ed.) 朱熹.
In *Erh Chhêng Chhüan Shu*, q.v.

Honan Chhêng shih Wai Shu 河南程氏外書.
Supplementary Records of Discourses of the Chhêng brothers of Honan [Chhêng I and Chhêng Hao, +11th-century Neo-Confucian philosophers].
Sung.
Chu Hsi (ed.) 朱熹.
In *Erh Chhêng Chhüan Shu*, q.v

Honzō-Wamyō 本草和名.
Synonymic Materia Medica with Japanese Equivalents.
Japan, +918.
Fukane no Sukehito 深根輔仁.

Hou Ching Lu 候鯖錄.
Waiting for the Mackerel.
Sung.
Chao Tê-Lin 趙德麟.

Hou Chou Shu 後周書.
See *Chou Shu*.

Hou Han Shu 後漢書.
History of the Later Han Dynasty [+25 to +220].
L/Sung, +450.
Fan Yeh 范曄.
The monograph chapters by Ssuma Piao 司馬彪.
A few chs. tr. Chavannes (6, 16), Pfizmaier (52, 53).
Yin-Tê Index, no. 41.

Hu Shih Lun 弧矢論.
Discussion on Arcs and Sagittae.
Ming, *c.* +1550.
Thang Shun-Chih 唐順之.

Hu Shih Suan Shu 弧矢算術.
Calculations of Arcs and Segments.
Ming, +1552.
Ku Ying-Hsiang 顧應祥.

Hua Ching.
See *Pi Chhuan Hua Ching*.

Hua Man Chi 畫墁集.
Painted Walls.
Sung, +1110.
Chang Shun-Min 張舜民.

Hua Yang Kuo Chih 華陽國志.
Record of the Country South of Mount Hua [historical geography of Szechuan down to +138].
Chin, +347.
Chhang Chhü 常璩.

Huai Nan Hung Lieh Chieh 淮南鴻烈解.
See *Huai Nan Tzu*.

Huai Nan Thien Wên Hsün Pu Chu 淮南天文訓補注.
Commentary on the Astronomical Chapter in *Huai Nan Tzu*.
Chhing, +1788.
Chhien Thang 錢塘.
Reprinted as Appendix (Pt. 6) to Liu Wên-Tien (*2*), Com. Pr. Shanghai, 1923, 1926.

Huai Nan Tzu 淮南子.
[=*Huai Nan Hung Lieh Chieh*.]
The Book of (the Prince of) Huai Nan [compendium of natural philosophy].
C/Han, *c.* −120.
Written by the group of scholars gathered by Liu An (prince of Huai Nan) 劉安.
Partial trs. Morgan (1); Erkes (1); Hughes (1); Chatley (1); Wieger (2).
Chung-Fa Index no. 5.
TT/1170.

Huang Chhao Ching Shih Wên Pien 皇朝經世文編.
Collected Essays by Chhing Dynasty officials on Social, Political and Economic Problems. [Contains many +18th-century writings.]
See Ho Chhang-Ling (*1*).

Huang Chhao Li Chhi Thu Shih 皇朝禮器圖式.
Illustrated Description of Sacrificial Vessels, Official Robes and Insignia, Musical Instruments and Astronomical Apparatus used during the Chhing Dynasty.
Chhing, +1759 and +1766.
Ed. Tung Kao 董誥.

Huang Chhing Chih Kung Thu 皇清職貢圖.
Illustrated Records of Tributary Peoples in the time of the Chhing Dynasty.
Chhing, +1763.
Fu Hêng 傅恆.

Huang Chhing Ching Chieh 皇清經解.
Collection of (more than 180) Monographs on Classical Subjects written during the Chhing Dynasty.
See Yen Chieh (*1*).

Huang Tao Tsung Hsing Thu 黃道總星圖.
Star-Maps arranged according to Ecliptic Co-ordinates [celestial latitudes and longitudes].
Chhing, +1746.
Tai Chin-Hsien (Ignatius Kögler) 戴進賢.

Huang Yü Khao 皇輿考.
Geographical Investigations.
Ming, +1588.
Chang Thien-Fu 張天復.

Hun I 渾儀.
On the Armillary Sphere.
H/Han, *c.* +117.
Chang Hêng 張衡.
(Fragment in *YHSF*, ch. 76.)

Hun Kai Thung Hsien Thu Shuo 渾蓋通憲圖說.
On Plotting the Co-ordinates of the Celestial Sphere and Vault [stereographic projections for the astrolabe].

Hun Kai Thung Hsien Thu Shuo (cont.)
Ming, +1607.
Li Chih-Tsao 李之藻.

Hun Thien Fu 渾天賦.
Ode on the Celestial Sphere.
Thang, +676.
Yang Chiung 楊炯.

Hun Thien Hsiang Shuo or *Chu* 渾天象說(注).
Discourse on Uranographic Models.
San Kuo, *c.* +260.
Wang Fan 王蕃.
(*CSHK* (San Kuo sect.), ch. 72, pp. 1*a* ff.)

Hung Fan Wu Hsing Chuan 洪範五行傳.
Discourse on the Hung Fan chapter of the *Shu Ching* in relation to the Five Elements.
C/Han, *c.* −10.
Liu Hsiang 劉向.

Huo Kung Chhieh Yao 火功挈要.
Essentials of Gunnery.
Ming, +1643.
Chiao Chu 焦勗.
With the collaboration of Thang Jo-Wang (J. A. Schall von Bell) 湯若望.

Huo Lung Ching 火龍經.
The Fire-Drake (Artillery) Manual.
Ming, +1412.
Chiao Yü 焦玉.
The first part of this book, in three sections, is attributed fancifully to Chuko Wu Hou (i.e. Chuko Liang), and Liu Chi 劉基 (+1311 to +1375) appears as co-editor, really perhaps co-author.
The second part, also in three sections, is attributed to Liu Chi alone, but edited, probably written, by Mao Hsi-Ping 毛希秉 in +1632.
The third part, in two sections, is by Mao Yuan-I 毛元儀 (*fl.* +1628) and edited by Chuko Kuang-Jung 諸葛光榮 whose preface is of +1644, Fang Yuan-Chuang 方元壯 & Chung Fu-Wu 鍾伏武.

Hsi Chêng Kêng-Wu Yuan Li 西征庚午元曆.
Yuan Dynasty Calendar of the Kêng-Wu Year (+1210) made during the Western Expedition (of Chingiz Khan).
Yuan (before the conquest of China), *c.* +1220.
Yehlü Chhu-Tshai 耶律楚材.
(Now only in chs. 56 and 57 of the *Yuan Shih.*)

Hsi Chhi Tshung Yü 西溪叢語.
Western Pool Collected Remarks.
Sung, +11th century.
Yao Khuan 姚寬.

Hsi Chhing Ku Chien 西清古鑑.
Hsi Chhing Catalogue of Ancient Mirrors (and Bronzes).
Chhing, +1751.
Liang Shih-Chêng 梁詩正.

Hsi Ching Tsa Chi 西京雜記.
Miscellaneous Records of the Western capital.
Liang or Chhen, mid +6th century.
Attrib. to Liu Hsin 劉歆 (C/Han) and to Ko Hung 葛洪 (Chin), but probably Wu Chün 吳均.

Hsi-Chou Yen Phu 歙州硯譜.
Hsichow Inkstone Record.
Sung, +1066.
Hung Ching-Po 洪景伯.

Hsi-Hsia Chi Shih Pên Mo 西夏紀事本末.
Rise and Fall of the Hsi-Hsia State.
See Chang Chien (*1*).

Hsi Pu Thien Ko 西步天歌.
The Western *Song of the March of the Heavens.*
Chhing, +18th century.
Writer unknown, but presumably a Jesuit or a Chinese friend of the Jesuits.

Hsi Shih Chi 西使記.
Notes on an Embassy to the West.
Yuan, +1263.
Chhang Tê 常德. (Writer, Liu Yü 劉郁.)

Hsi-Yang Fan Kuo Chih 西洋番國志.
Record of the Barbarian Countries in the Western Ocean [relative to the voyages of Chêng Ho].
Ming, +1434.
Kung Chen 鞏珍.

Hsi-Yang Hsin Fa Li Shu 西洋新法曆書.
Treatise on (Astronomy and) Calendrical Science according to the New Western Methods.
Chhing, +1645 (second form of the Jesuit astronomical encyclopaedia).
See *Hsin Fa Suan Shu.*
A part of this title was conserved in the +1726 reprint in *TSCC*, *Li fa tien*, chs. 51–72 and 85–88, as *Hsin Fa Li Shu.*

Hsi Yu Chi.
See *Chhang-Chhun Chen Jen Hsi Yu Chi.*

Hsi Yü Chi.
See *Sêng Hui-Sêng Shih Hsi Yü Chi*; and *Ta Thang Hsi Yü Chi.*

Hsi Yu Lu 西遊錄.
Record of a Journey to the West.
Yuan, +1225.
Yehlü Chhu-Tshai 耶律楚材.

Hsi Yü Thu Chi 西域圖記.
Illustrated Record of Western Countries.
Sui, +610.
Phei Chü 裴矩.

Hsi Yü Wên Chien Lu 西域聞見錄.
Things seen and Heard in the Western Countries.
Chhing, +1777.
Chhun Yuan Chhi-shih-i Lao-jen 椿園七十一老人. [The 71-year-old Gentleman of the Cedar Garden.]

Hsia Hsiao Chêng 夏小正.
Lesser Annuary of the Hsia Dynasty.
Chou, between −7th and −4th centuries.
Writers unknown.

Hsia Hsiao Chêng (*cont.*)
Incorporated in *Ta Tai Li Chi*, q.v.
Tr. R. Wilhelm (6), Soothill (5).

Hsiahou Yang Suan Ching 夏侯陽算經.
Hsiahou Yang's Mathematical Manual.
L/Sung or Chhi, +450 to +500, or perhaps early +6th century.
Hsiahou Yang 夏侯陽.

Hsiang Chieh Chiu Chang Suan Fa Tsuan Lei 詳解九章算法纂類.
Detailed Analysis of the Mathematical Rules in the *Nine Chapters* and their Reclassification.
Sung, +1261.
Yang Hui 楊輝.

Hsiang Chung Chi 湘中記.
Records of Hunan.
Chin, *c.* +375.
Lo Han 羅含.

Hsiang Ming Suan Fa 詳明算法.
Explanations of Arithmetic.
Yuan and Ming, +14th century.
Perhaps Chia Hêng 賈亨; also attrib. An Chih-Chai 安止齋 and Ho Phing-Tzu 何平子.
(In *Yung-Lo Ta Tien*, chs. 16343–4.)

Hsiang Wei Hsin Phien 象緯新篇.
New Account of the Web of Stars.
Ming, early +16th century.
Wang Kho-Ta 王可大.

Hsiao-An Hsin Fa 曉菴新法.
(Wang) Hsiao-An's New (Astronomical) Methods.
Ming, +1643.
Wang Hsi-Shan 王錫闡.

Hsiao Hsiang Thing Yü Lu 瀟湘聽雨錄.
Listening to the Rain at Hsiao-hsiang.
Chhing.
Chiang Yü 江昱.

Hsiao Hsüeh Kan Chu 小學紺珠.
Valuable Observations on Elementary Knowledge.
Sung, +13th century.
Wang Ying-Lin 王應麟.

Hsiao Tai Li Chi.
See *Li Chi*.

Hsiao Wei Tzhu Hsiung Thu 孝緯雌雄圖.
Apocryphal Treatise on the *Filial Piety Classic*; Diagrams of Male and Female (Influences).
C/Han, −1st century.
Writer unknown.

Hsieh Chhuan Chi 斜川集.
Hsieh River Record.
Sung.
Su Kuo 蘇過.

Hsieh Lüeh 蟹略.
Monograph on the Varieties of Crabs.
Sung, *c.* +1185.
Kao Ssu-Sun 高似孫.

Hsieh Phu 蟹譜.
Discourse on Crustacea (cf. *Hsü Hsieh Phu*).
Sung.
Fu Kung 傅肱.

Hsien Chhuang Kua I Chih 閑窗括異志.
Strange Things seen through the Barred Window.
Sung.
Lu Ying-Lung 魯應龍.

Hsien Pin Lu 咸賓錄.
Record of All the Guests.
Ming, +1590.
Lo Yüeh-Chhiung 羅曰褧.

Hsien-Shun Lin An Chih 咸淳臨安志.
Hsien-Shun reign-period Topographical Records of the Hangchow District.
Sung, +1274.
Chhien Yüeh-Yu 潛說友.

Hsin-Chai Tsa Tsu 心齋雜俎.
Miscellanea of (Chang) Hsin-Chai.
Chhing, *c.* +1670.
Chang Chhao 張潮.

Hsin Fa Li Shu 新法曆書.
See *Hsi-Yang Hsin Fa Li Shu*.

Hsin Fa Piao I 新法表異.
Differences between (the Old and) the New (Astronomical and Calendrical) Systems.
Chhing, +1645.
Eventually incorporated in *Hsin Fa Suan Shu*, q.v.
Thang Jo-Wang (J. A. Schall von Bell) 湯若望.

Hsin Fa Suan Shu 新法算書.
Treatise on Mathematics (Astronomy and Calendrical Science) according to the New Methods.
Chhing, +1669, +1674. This encyclopaedia was first issued in the Ming (+1635) as *Chhung-Chên Li Shu* (q.v.), then in the Chhing (+1645) as *Hsi-Yang Hsin Fa Li Shu* (q.v.).
Thang Jo-Wang (J. A. Schall von Bell) 湯若望, Têng Yü-Han (Johann Schreck (Terrentius)) 鄧玉函, Lo Ya-Ku (James Rho) 羅雅谷, Lung Hua-Min (Nicholas Longobardi) 龍華民, Hsü Kuang-Chhi 徐光啓, Li Chih-Tsao 李之藻, Li Thien-Ching 李天經 and others.

Hsin I Hsiang Fa Yao 新儀象法要.
New Design for an Armillary Clock.
Sung, +1094.
Su Sung 蘇頌.

Hsin Kho-Lou Ming 新刻漏銘.
Inscription for a New Clepsydra.
Liang, +507.
Lu Chhui 陸倕.

Hsin Lun 新論.
New Discussions.
H/Han, *c.* +20.
Huan Than 桓譚.

Hsin Pien Tui Hsiang Ssu Yen 新篇對象四言.
Newly Revised Reader with Four Characters to the Line and Pictures to match. [The

Hsin Pien Tui Hsiang Ssu Yen (*cont.*)
oldest printed illustrated children's primer in any civilisation.]
Ming, +1436.
Compiler unknown.

Hsin Shu 新書.
New Book.
C/Han, −2nd century, but the present text may be partly Thang or pre-Thang.
Chia I 賈誼.

Hsin Thang Shu 新唐書.
New History of the Thang Dynasty [+618 to +906].
Sung, +1061.
Ouyang Hsiu 歐陽修 & Sung Chhi 宋祁.
Partial trs. des Rotours (1); Pfizmaier (66–74).
Yin-Tê Index, no. 16.

Hsin Thien Lun 昕天論.
Discourse on the Diurnal Revolution (of the Heavens).
San Kuo, *c.* +250.
Yao Hsin 姚信.
(In *YHSF*, ch. 76.)

Hsing Chha Shêng Lan 星槎勝覽.
Triumphant Visions of the Starry Raft. [Account of the voyages of Chêng Ho, whose ship, as carrying an ambassador, is thus styled.]
Ming, +1436.
Fei Hsin 費信.

Hsing Ching 星經.
The Star Manual.
A pre-Thang compilation of Chou and Han star-catalogues, probably with later additions (now incomplete).
Shih Shen 石申, Kan Tê 甘德, Wu Hsien 巫咸 *et al.*
TT/284.

Hsing Shui Chin Chien 行水金鑑.
Golden Mirror of the Flowing Waters (cf. *Hsü Hsing Shui Chin Chien*).
Chhing, +1725.
Fu Tsê-Hung 傅澤洪.

Hsiu Chen Thai Chi Hun Yuan Thu 修眞太極混元圖.
Veritable Restored Chart of the Supreme Pole and the Original Chaos.
Sung.
Hsiao Tao-Tshun 蕭道存.
TT/146.

Hsiu Chung Chi 袖中記.
Sleeve Records.
Liang, *c.* +500.
Shen Yo 沈約.

Hsiu Fang So Chih 修防瑣志.
Brief Memoir on Dyke Repairs.
Chhing.
Li Shih-Lu 李世祿.

Hsiu Yao Ching 宿曜經.
Hsiu and Planet Sūtra.
See *Wên-Shu-Shih-Li Phu-Sa...*, etc.

Hsiu Yao I Kuei 宿曜儀軌.
The Tracks of the *Hsiu* and Planets.
Thang, *c.* +710.
I-Hsing 一行.
TW/1304.

Hsü Hsi So Wei Lun.
See *Ao Yü Tzu Hsü Hsi So Wei Lun.*

Hsü Hsia-Kho Yu Chi 徐霞客遊記.
Diary of the Travels of Hsü Hsia-Kho.
Chhing, +1776 (written +1641).
Hsü Hsia-Kho 徐霞客.

Hsü Hsieh Phu 續蟹譜.
Continuation of the *Discourse on Crustacea* (cf. *Hsieh Phu*).
Chhing.
Chhu Jen-Huo 褚人獲.

Hsü Hsing Shui Chin Chien.
See Lei Shih-Hsü & Yü Chêng-Hsieh (*1*) in Bibliography B.

Hsü Hung Chien Lu 續弘簡錄.
The *Mass of Records* (General History of the Middle Ages, from the beginning of the Thang dynasty) continued.
Chhing, +1699.
Sun Yuan-Ping continuing the work of Shao Ching-Pang 邵經邦.

Hsü Ku Chai Chhi Suan Fa 續古摘奇算法.
Continuation of Ancient Mathematical Methods for Elucidating the Strange (Properties of Numbers).
Sung, +1275.
Yang Hui 楊輝.

Hsü Po Wu Chih 續博物志.
Supplement to the *Record of the Investigation of Things* (cf. *Po Wu Chih*).
Sung, mid +12th century.
Li Shih 李石.

Hsü Shih Shuo 續世說.
Continuation of the *Discourses on the Talk of the Times.*
See *Shih Shuo Hsin Yü.*
Sung, *c.* +1157.
Khung Phing-Chung 孔平仲.

Hsüan-Chi I Shu 璇璣遺述.
Records of Ancient Arts and Techniques [lit. of the Circumpolar Constellation Template].
Chhing.
Chieh Hsüan 揭暄.

Hsüan Chung Chi 玄中記.
Mysterious Matters.
Date uncertain, pre-Sung, perhaps +6th century.
Mr Kuo 郭氏.

Hsüan-Ho Fêng Shih Kao-Li Thu Ching 宣和奉使高麗圖經.
Illustrated Record of an Embassy to Korea in the Hsüan-Ho reign-period.
Sung, +1124 (+1167).
Hsü Ching 徐兢.

Hsüan-Ho Po Ku Thu 宣和博古圖.
[=*Po Ku Thu Lu*, q.v.]

Hsüan-Ho Po Ku Thu (*cont.*)
Hsüan-Ho reign-period Illustrated Record of Ancient Objects. [Catalogue of the archaeological museum of the emperor Hui Tsung.]
Sung, +1111 to +1125.
Wang Fu 王黼 or 黻, *et al.*

Hsüan-Ho Shih Phu 宣和石譜.
Hsüan-Ho reign-period Treatise on Stones.
Sung, *c.* +1122.
Tsu Khao 祖考.

Hsüeh Chai Chan Pi 學齋佔畢.
Glancing into Books in a Learned Studio.
Sung, +13th century.
Shih Shêng-Tsu 史繩祖.

Hsüeh Li Hsiao Pien 學曆小辯.
Minor Disputation on Calendrical Science [an answer to the *Li Tshê* and *Li Yuan*, q.v.].
Ming, +1631.
Thang Jo-Wang (J. A. Schall von Bell) 湯若望.

Hsün Tzu 荀子.
The Book of Master Hsün.
Chou, *c.* −240.
Hsün Chhing 荀卿.
Tr. Dubs (7).

I Chao Liao Tsa Chi 猗覺寮雜記.
Miscellaneous Records from the I-Chao Cottage.
Sung, +12th century.
Chu I 朱翌.

I Ching 易經.
The Classic of Changes [Book of Changes].
Chou with C/Han additions.
Compilers unknown.
See Li Ching-Chhih (*1*, *2*); Wu Shih-Chhang (1).
Tr. R. Wilhelm (2); Legge (9); de Harlez (1).
Yin-Tê Index no. (Suppl.) 10.

I Ching 藝經.
Treatise on Arts and Games.
San Kuo (Wei), +3rd century.
Hantan Shun 邯鄲淳.

I Chou Shu 逸周書.
[=*Chi Chung Chou Shu*.]
Lost Books of the Chou (Dynasty).
Chou, −245 and before, such parts as are genuine. (Found in the tomb of An Li Wang, a prince of the Wei State, r. −276 to −245; in +281).
Writers unknown.

I Chuan 易傳.
Explanations of the (*Book of*) *Changes*.
N/Wei, *c.* +490.
Kuan Lang 關朗.

I Hsiang Khao Chhêng 儀象考成.
The Imperial (Astronomical) Instruments [official description].
Chhing, +1744; enlarged +1757 to contain the Chhien-Lung star-catalogue.
Tai Chin-Hsien (Ignatius Kögler) 戴進賢; Pao Yu-Kuan (Anton Gogeisl) 鮑友管; Liu Sung-Ling (Augustin von Hallerstein) 劉松齡 and Fu Tso-Lin (Felix da Rocha) 傅作霖.

I Ku Yen Tuan 益古演段.
New Steps in Computation.
J/Chin (Yuan), +1259.
Li Yeh 李治.

I Lung Thu 易龍圖.
The Dragon Diagrams of the (*Book of*) *Changes*.
Wu Tai, *c.* +950.
Chhen Thuan 陳摶.

I Shu Kou Yin Thu 易數鉤隱圖.
The Hidden Number-Diagrams in the (*Book of*) *Changes* Hooked Out.
Sung, early +10th century.
Liu Mu 劉牧.

I Shu Pien 蛾術編.
The Antheap of Knowledge [Miscellaneous Essays].
(The first character is usually pronounced *ê*, moth; but here should be read *i*, ant.)
Chhing, *c.* +1770, but not published till +19th century.
Wang Ming-Shêng 王鳴盛.

I Thu Ming Pien 易圖明辨.
Clarification of the Diagrams in the (*Book of*) *Changes* [historical analysis].
Chhing, +1706.
Hu Wei 胡渭.

I Tsung Chin Chien 醫宗金鑑.
Golden Mirror of Medicine (compiled by Imperial Order).
Chhing, +1743.
Ed. O-Erh-Thai (Ortai) 鄂爾泰.

I Wei Chhien Tso Tu 易緯乾鑿度.
Apocryphal Treatise on the (*Book of*) *Changes*; a Penetration of the Regularities of Chhien (the first Kua).
C/Han, −1st century.
Writer unknown.

I Wei Ho Thu Shu 易緯河圖數.
Apocryphal Treatise on the (*Book of*) *Changes*; The Numbers of the *River Diagram*.
H/Han.
Writer unknown.

I Wei Thung Kua Yen 易緯通卦驗.
Apocryphal Treatise on the (*Book of*) *Changes*; Verifications of the Powers of the Kua.
C/Han, −1st century.
Writer unknown.

I Wu Chih 異物志.
Memoirs of Marvellous Things.
San Kuo.
Hsüeh Yü 薛珝.

I Yin 易音.
Dictionary of the Original Sounds of Words in the (*Book of*) *Changes*.

I Yin (*cont.*)
Chhing, +1667 (in *Yin Hsüeh Wu Shu*).
Ku Yen-Wu 顧炎武.

I Yü Thu Chih 異域圖志.
Illustrated Record of Strange Countries.
Ming, *c.* +1420 (written between +1392 and +1430); pr. +1489.
Compiler unknown.
Cf. Moule (4); Sarton (1), vol. 3, p. 1627.
(A copy is in the Cambridge University Library.)

I Yuan 異苑.
Garden of Strange Things.
Pre-Sui, prob. +5th century.
Liu Ching-Shu 劉敬叔.

Ichidai Yōki 一代要記.
Essentials of the History of an Age [a brief history of the emperors, from Ingyō (+412/+453) to Hanazono (+1309/+1318)].
Japan, +14th century.
Writer unknown.

Jih Hua (*Chu Chia*) *Pên Tshao* 日華諸家本草.
Master Jih-Hua's Pharmacopoeia (of All the Schools).
Sung, *c.* +970.
Ta Ming [Jih-Hua Tzu] 大明.

Jih Wên Lu 日聞錄.
Daily Notes.
Yuan, *c.* +1380.
Li Chhung 李翀.

Kaisan-ki Kōmoku 改算記綱目.
Comprehensive Summary of Integration [early calculus].
Japan, +1687.
Mochinaga Toyotsugu 持永豐次 & Ōhashi Takusei 大橋宅清.

Kan Ying Ching 感應經.
On Stimulus and Response (the Resonance of Phenomena in Nature).
Thang, *c.* +640.
Li Shun-Fêng 李淳風.

Kan Ying Lei Tshung Chih 感應類從志.
Record of the Mutual Resonances of Things.
Chin, *c.* +295.
Chang Hua 張華.

Kao Hou Mêng Chhiu 高厚蒙求.
Investigation of the Dimensions of the Universe. [Celestial and terrestrial cartography.]
Chhing, *c.* +1799.
Hsü Chhao-Chün 徐朝俊.

Kao-Li Thu Ching.
See *Hsüan-Ho Fêng Shih Kao-Li Thu Ching.*

Kao Sêng Chuan 高僧傳.
Biographies of Outstanding (Buddhist) Monks [especially those noted for learning and philosophical eminence].
Liang, between +519 and +554.
Hui-Chiao (monk) 慧皎.
TW/2059.

Kêng Hsin Yü Tshê 庚辛玉册.
Precious Secrets of the Realm of Kêng and Hsin [i.e. all things connected with metals and minerals, symbolised by these two cyclical characters. On alchemy and pharmaceutics. Kêng-Hsin is also an alchemical synonym for gold].
Ming, +1421.
Ning Hsien Wang (prince of the Ming) 寧獻王.

Khai Fang Shuo 開方說.
Theory of Equations of Higher Degrees.
Chhing, end +18th century.
Li Jui 李銳.

Khai-Pao Pên Tshao 開寶本草.
Khai-Pao reign-period Pharmacopoeia.
Sung, *c.* +970.
Liu Han 劉翰 & Ma Chih 馬志.

Khai-Yuan Chan Ching 開元占經.
The Khai-Yuan reign-period Treatise on Astrology (and Astronomy).
Thang, +729.
(Some parts, such as the *Chiu Chih* (*Navagrāha*) calendar, had been written as early as +718.)
Chhüthan Hsi-Ta 瞿曇悉達.

Khao Ku Thu 考古圖.
Illustrations of Ancient Objects.
Sung, +1092.
Lü Ta-Lin 呂大臨.

Khao Kung Chi 考工記.
The Artificers' Record [a section of the *Chou Li*, q.v.].
Chou and Han, perhaps originally an official document of Chhi State, incorporated *c.* −140.
Compiler unknown.
Tr. E. Biot (1).
Cf. Kuo Mo-Jo (*1*); Yang Lien-Shêng (7).

Khung Tshung Tzu 孔叢子.
The Book of Master Khung Tshung.
Ascr. H/Han, prob. later.
Attrib. Khung Fu 孔鮒.

Khung Tzu Chia Yü 孔子家語.
Table Talk of Confucius.
H/Han or more probably San Kuo, early +3rd century (but compiled from earlier sources).
Ed. Wang Su 王肅.
Partial trs. Kramers (1); A. B. Hutchinson (1); de Harlez (2).

Ko Chih Ching Yuan 格致鏡原.
Mirror of Scientific and Technological Origins.
Chhing, +1735.
Chhen Yuan-Lung 陳元龍.

Ko Chih Ku Wei.
See Wang Jen-Chün (*1*).

Ko Chih Tshao 格致草.
Scientific Sketches.

Ko Chih Tshao (cont.)
Ming, +1620 [+1648].
Hsiung Ming-Yü 熊明遇.

Ko Hsiang Hsin Shu 革象新書.
New Elucidation of the Heavenly Bodies.
Yuan.
Chao Yu-Chhin 趙友欽.
Revised by Wang Wei of the Ming as the *Chhung Hsiu Ko Hsiang Hsin Shu*, q.v.

Ko Ku Yao Lun 格古要論.
Handbook of Archaeology.
Ming, +1387.
Tshao Chao 曹昭.

Ko Yuan Lien Pi Li Thu Chieh 割圓連比例圖解.
Explanation of the Determination of Segment Areas (using infinite series).
Chhing, before 1800, but pub. 1819.
Tung Yu-Chhêng 董祐誠.

Ko Yuan Mi Lü Chieh Fa 割圓密率捷法.
Quick Method for Determining Segment Areas (using infinite series).
Chhing, +1774.
Ming An-Thu 明安圖.

Ku Chin Chu 古今注.
See *Fu Hou Ku Chin Chu.*

Ku Chin Chu 古今黈.
See *Ching-Chai Ku Chin Chu.*

Ku Chin Lü Li Khao 古今律厤考.
Investigation of the (Chinese) Calendars, New and Old.
Ming, *c.* +1600.
Hsing Yün-Lu 邢雲路.

Ku Chin Suan-Hsueh Tshung Shu 古今算學叢書.
See Liu To (*1*).

Ku Kho Tshung Chhao 古刻叢鈔.
Collection of Ancient (Thang and Sung) Inscriptions.
Ming, +14th century.
Thao Tsung-I 陶宗儀.

Ku Suan Chhi Khao 古算器考.
Enquiry into the History of Mechanical Computing Aids.
Chhing, *c.* +1700 (in *Li Suan Chhüan Shu*, q.v.).
Mei Wên-Ting 梅文鼎.

Ku Wei Shu 古微書.
Old Mysterious Books.
[A collection of the apocryphal Chhan-Wei treatises.]
Date uncertain, in part C/Han.
Ed. Sun Chio 孫瑴 (Ming).

Kua Ti Chih 括地志.
Comprehensive Geography.
Thang, +7th century.
Wei Wang-Thai 魏王泰.
(Fragments reconstituted by Sun Hsing-Yen in +1797.)

Kua Ti Hsiang.
See *Ho Thu Wei Kua Ti Hsiang.*

Kuai Shih Tsan 怪石贊.
Strange Rocks.
Chhing, +1665.
Sung Lo 宋犖.

Kuan Khuei Chi Yao 管窺輯要.
See *Thien Wên Ta Chhêng Kuan Khuei Chi Yao.*

Kuan Shih Lu 觀石錄.
On Looking at Stones.
Chhing, +1668.
Kao Chao 高兆.

Kuan Tzu 管子.
The Book of Master Kuan.
Chou and C/Han. Perhaps mainly compiled in the Chi-Hsia Academy (late −4th century) in part from older materials.
Attrib. Kuan Chung 管仲.
Partial trs. Haloun (2, 5); Than Po-Fu *et al.*

Kuang Ya 廣雅.
Enlargement of the *Erh Ya*; *Literary Expositor* [dictionary].
San Kuo (Wei), +230.
Chang I 張揖.

Kuang-Yang Tsa Chi 廣陽雜記.
Collected Miscellanea of (Master) Kuang-Yang (Liu Hsien-Thing).
Chhing, *c.* +1695.
Liu Hsien-Thing 劉獻廷.

Kuang Yü Thu 廣輿圖.
Enlarged Terrestrial Atlas.
Yuan, +1320.
Chu Ssu-Pên 朱思本.
First printed, and the word *Kuang* added, by Lo Hung-Hsien 羅洪先 (Ming, *c.* +1555).

Kuang Yün 廣韻.
Enlargement of the *Chhieh Yün*; *Dictionary of the Sounds of Characters.*
Sung.
(A completion by later Thang and Sung scholars, given its present name in +1011).
Lu Fa-Yen *et al.* 陸法言.

Kuei Chhien Chih 歸潛志.
On Returning to a Life of Obscurity.
J/Chin, +1235.
Liu Chhi 劉祁.

Kuei Hai Yü Hêng Chih 桂海虞衡志.
Topography and Products of the Southern Provinces.
Sung, +1175.
Fan Chhêng-Ta 范成大.

Kuei-Hsin Tsa Chih 癸辛雜識.
Miscellaneous Information from Kuei-Hsin Street (in Hangchow).
Sung, late +13th century, perhaps not finished before +1308.
Chou Mi 周密.
See des Rotours (1), p. cxii.

Kuei-Hsin Tsa Chih Hsü Chi 癸辛雜識續集.
Miscellaneous Information from Kuei-Hsin Street (in Hangchow); First Addendum.

Kuei-Hsin Tsa Chih Hsü Chi (*cont.*)
Yuan, +1308.
Chou Mi 周密.
See des Rotours (1), p. cxii.

Kuei Thien Lu 歸田錄.
On Returning Home.
Sung, +1067.
Ouyang Hsiu 歐陽修.

Kung Chhi Shih Hua 䂬溪詩話.
River-Boulder Pool Essays [literary criticism].
Sung, +1168.
Huang Chhê 黃徹.

Kungsun Lung Tzu 公孫龍子.
The Book of Master Kungsun Lung (cf. *Shou Pai Lun*).
Chou, −4th century.
Kungsun Lung 公孫龍.
Tr. Ku Pao-Ku (1); Perleberg (1); Mei Yi-Pao (3).

Kungyang Chuan 公羊傳.
Master Kungyang's Commentary on the *Spring and Autumn Annals*.
Chou (with Chhin and Han additions), late −3rd and early −2nd centuries.
Attrib. Kungyang Kao 公羊高, but more probably Kungyang Shou 公羊壽.
See Wu Khang (1).

Kuo Yü 國語.
Discourses on the (ancient feudal) States.
Late Chou, Chhin and C/Han, containing early material from ancient written records.
Writers unknown.

Lan Chen Tzu 嬾眞子.
Book of the Truth-through-Indolence Master.
Sung, between +1111 and +1117.
Ma Yung-Chhing 馬永卿.

Lao Hsüeh An Pi Chi 老學庵筆記.
Notes from the Hall of Learned Old Age.
Sung.
Lu Yu 陸游.

Lei Kung Yao Tui 雷公藥對.
Answers of the Venerable (Master) Lei concerning Drugs.
Perhaps L/Sung, at any rate before N/Chhi.
Prob. by Lei Hsiao 雷斆.
Later attrib. Lei Kung, legendary minister of Huang Ti.
Comm. by Hsü Chih-Tshai 徐之才.

Lei Phien 類篇.
Dictionary of Character Sounds.
Sung, +1067.
Ssuma Kuang 司馬光.

Li Chi 禮記.
[=*Hsiao Tai Li Chi.*]
Record of Rites [compiled by Tai the Younger].
(Cf. *Ta Tai Li Chi.*)
Ascr. C/Han; *c.* −70/−50, but really H/Han, between +80 and +105, though the earliest pieces included may date from the time of the *Analects* (*c.* −465 to −450).
Attrib. ed. Tai Shêng 戴聖.
Actual ed. Tshao Pao 曹褒.
Trs. Legge (7); Couvreur (3); R. Wilhelm (6).
Yin-Tê Index no. 27.

Li Fa Hsi Chhuan 曆法西傳.
History of Western Calendrical Science (and Astronomy).
Chhing, +1656.
Eventually incorporated in *Hsin Fa Suan Shu*, q.v.
Thang Jo-Wang (J. A. Schall von Bell) 湯若望.

Li Hai Chi 蠡海集.
The Beetle and the Sea [title taken from the proverb that the beetle's eye view cannot encompass the wide sea—a biological book].
Ming, late +14th century.
Wang Khuei 王逵.

Li Hsiang Khao Chhêng 曆象考成.
Compendium of Calendrical Science and Astronomy (compiled by Imperial Order) [based on the work of Cassini and Flamsteed; part of the *Lü Li Yuan Yuan*, q.v.].
Chhing, +1713 (+1723).
Ed. Mei Ku-Chhêng 梅瑴成 & Ho Kuo-Tsung 何國宗.
Cf. Hummel (2), p. 285; Pfister (1), pp. 647, 648, 653.

Li Sao 離騷.
Elegy on Encountering Sorrow [ode].
Chou, *c.* 295.
Chhü Yuan 屈原.

Li Shih A-Pi-Than Lun 立世阿毗曇論.
Lokasthiti Abhidharma Śāstra; Philosophical Treatise on the Preservation of the World [astronomical].
India, tr. into Chinese +558.
Writer unknown.

Li shih I Shu 李氏遺書.
Mathematical Remains of Mr Li (Jui).
Chhing, +1765 to +1814; printed +1823.
Li Jui 李銳.

Li shih Yao Lu 李氏藥錄.
Mr Li's Record of Drugs.
San Kuo (Wei), *c.* +225.
Li Tang-Chih 李當之.

Li Suan Chhüan Shu 曆算全書.
Complete Works on Calendar and Mathematics.
Chhing, +1723.
Mei Wên-Ting 梅文鼎.

Li Suan Shu Mu 曆算書目.
Mathematical Bibliography.
Chhing, +1723.
Mei Wên-Ting 梅文鼎.

Li Tai Chung Ting I Chhi Khuan Chih Fa Thieh 歷代鍾鼎彝器款識法帖.
Description of the Various Sorts of Bronze Bells, Cauldrons, Temple Vessels, etc. of all Ages.
Sung, +11th century.
Hsüeh Shang-Kung 薛尙功.

Li Tai Lun Thien 歷代論天.
Discussions on the Heavens in Different Ages.
Chhing, +1790.
Yang Chhao-Ko 楊超格.

Li Tai Thung Chien Chi Lan 歷代通鑑輯覽.
Essentials of History.
Chhing, +1767.
Lu Hsi-Hsiung (ed.) 陸錫熊.

Li Thi Lüeh 曆體略.
An Account of Astronomy and Calendrical Science.
Ming.
Wang Ying-Ming 王英明.

Li Tshê 曆測.
Calendrical Measurements.
Ming, +1631.
Wei Wên-Khuei 魏文魁 & Wei Hsiang-Chhien 魏象乾.

Li Yuan 曆元.
Origins of Calendrical Science.
Ming, +1631.
Wei Wên-Khuei 魏文魁 & Wei Hsiang-Chhien 魏象乾.

Liang Chhi Man Chih 梁溪漫志.
Bridge Pool Essays.
Sung, +1192.
Fei Kun 費衮.

Liang Shan Mo Than 兩山墨談.
Jottings from Two Mountains.
Ming.
Chhen Thing 陳霆.

Liang Shu 梁書.
History of the Liang Dynasty [+502 to +556].
Thang, +629.
Yao Chha 姚察 and his son Yao Ssu-Lien 姚思廉.

Liang Ssu Kung Chi 梁四公記.
Tales of the Four Lords of Liang.
Thang, *c.* +695.
Chang Yüeh 張說.

Lieh Tzu 列子.
[=*Chhung Hsü Chen Ching.*]
The Book of Master Lieh.
Chou and C/Han, −5th to −1st centuries. Ancient fragments of miscellaneous origin finally cemented together with much new material about +380.
Attrib. Lieh Yü-Khou 列禦寇.
Tr. R. Wilhelm (4); L. Giles (4); Wieger (7).
TT/663.

Lin Chio Chi 麟角集.
The Unicorn Horn Collection (of Examination Essays).
Thang, +9th century.
Wang Chhi 王棨.

Ling Hsien 靈憲.
The Spiritual Constitution (or Mysterious Organisation) of the Universe [cosmological and astronomical].
H/Han, *c.* +118.
Chang Hêng 張衡.
(In *YHSF*, ch. 76.)

Ling Nan I Wu Chih 嶺南異物志.
Record of Strange Things South of the Passes.
Thang, early +9th century.
Mêng Kuan 孟琯.

Ling Thai Pi Yuan 靈臺秘苑.
The Secret Garden of the Observatory [astronomy, including a star list, and State astrology].
N/Chou, *c.* +580.
Revised in Sung and much probably added in Ming; two different books circulate under this title and authorship.
Yü Chi-Tshai 庾季才.

Ling Wai Tai Ta 嶺外代答.
Information on What is Beyond the Passes (lit. a book in lieu of individual replies to questions from friends).
Sung, +1178.
Chou Chhü-Fei 周去非.

Liu-Chhiu Kuo Chih Lüeh 琉球國志略.
Account of the Liu-Chhiu Islands.
Chhing, +1757.
Chou Huang 周煌.

Liu Ching Thien Wên Pien 六經天文編.
Treatise on Astronomy (and Calendrical Science) in the Six Classics.
Sung, *c.* +1275.
Wang Ying-Lin 王應麟.

Liu Ching Thu 六經圖.
Illustrations of Objects mentioned in the Six Classics.
Many editions, the first in Sung, *c.* +1155.
Yang Chia 楊甲.

Lo-Fou Shan Chih 羅浮山志.
History and Topography of the Lo Fou Mountains (north of Canton).
Chhing, +1716 (but based on older histories).
Thao Ching-I 陶敬益.

Lo Shu Wei Chen Yao Tu 洛書緯甄曜度.
Apocryphal Treatise on the *Lo Shu* Diagram; Examination of the Measured (Movements) of the (Celestial) Brilliances.
Han.
Writer unknown.

Lo-Yang Chhieh Lan Chi 洛陽伽藍記.
Description of the Buddhist Temples of Loyang.
N/Wei, *c.* +530.
Yang Hsüan-Chih 楊衒之.

Lou-Kho Fa 漏刻法.
Treatise on the Clepsydra.
N/Wei, *c.* +440.
Li Lan 李蘭.

Lou-Kho Fu 漏刻賦.
Ode on the Clepsydra.
Chin, *c.* +295.
Lu Chi 陸機.

Lou Shui Chuan Hun Thien I Chih 漏水轉渾天儀制.
Method for making an Armillary Sphere revolve by means of water from a Clepsydra [perhaps only a part of the *Hun I*, to which it is attached in *YHSF*, ch. 76].
H/Han, *c.* +117.
Chang Hêng 張衡.

Lü-Chai Shih Erh Pien 履齋示兒編.
Instructions and Miscellaneous Information for the Use of Children of his own Family by (Sun) Lü-Chai.
Sung, +1205.
Sun I 孫奕.

Lü Ho-Shu Wên Chi 呂和叔文集.
Collected Writings of Lü Wên (Lü Ho-Shu).
Thang, *c.* +850.
Lü Wên 呂溫.

Lu I Chi 錄異記.
Strange Matters.
Sung.
Tu Kuang-Thing 杜光庭.

Lü Li Chih 律曆志.
Memoir on the Calendar.
H/Han, +178.
Liu Hung 劉洪 & Tshai Yung 蔡邕.
(In *Hou Han Shu*, ch. 13.)

Lü Li Yuan Yuan 律曆淵源.
Ocean of Calendrical and Acoustic Calculations (compiled by Imperial Order) [includes *Li Hsiang Khao Chhêng*, *Shu Li Ching Yün*, *Lü Lü Chêng I*, q.v.].
Chhing, +1723; printing probably not finished before +1730.
Ed. Mei Ku-Chhêng 梅瑴成 & Ho Kuo-Tsung 何國宗.
Cf. Hummel (2), p. 285; Wylie (1), pp. 96ff.

Lu-Ling I Wu Chih 廬陵異物志.
Strange Things of Lu-Ling.
Sung.
Writer unknown.

Lü Lü Chêng I 律呂正義.
Collected Basic Principles of Music (compiled by Imperial Order) [part of *Lü Li Yuan Yuan*, q.v.].
Chhing, +1723.
Ed. Mei Ku-Chhêng 梅瑴成 & Ho Kuo-Tsung 何國宗.
Cf. Hummel (2), p. 285.

Lü Lü Hsin Lun 律呂新論.
New Discourse on Music and Acoustics.
Chhing, *c.* +1740.
Chiang Yung 江永.

Lu Shan Chi 廬山記.
Description of Mount Lu (near the Poyang Lake).
Chhen Shun-Yü 陳舜俞.

Lu Shih 路史.
The Peripatetic History [a collection of fabulous and legendary material put together in the style of the dynastic histories, but containing much curious information on techniques].
Sung.
Lo Pi 羅泌.

Lü Shih Chhun Chhiu 呂氏春秋.
Master Lü's Spring and Autumn Annals [compendium of natural philosophy].
Chou (Chhin), −239.
Written by the group of scholars gathered by Lü Pu-Wei 呂不韋.
Tr. R. Wilhelm (3).
Chung-Fa Index, no. 2.

Lu Thang Shih Hua 麓堂詩話.
Foothill Hall Essays [literary criticism].
Ming, +1513.
Li Tung-Yang 李東陽.

Lun Hêng 論衡.
Discourses Weighed in the Balance.
H/Han, +82 or +83.
Wang Chhung 王充.
Tr. Forke (4).
Chung-Fa Index no. 1.

Lun Yü 論語.
Conversations and Discourses (of Confucius) [perhaps Discussed Sayings, Normative Sayings, or Selected Sayings]; Analects.
Chou (Lu), *c.* −465 to −450.
Compiled by disciples of Confucius (chs. 16, 17, 18 and 20 are later interpolations).
Tr. Legge (2); Lyall (2); Waley (5); Ku Hung-Ming (1).
Yin-Tê Index no. (Suppl.) 16.

Lun Yüeh Shih 論月蝕.
On Lunar Eclipses.
H/Han, +200.
Liu Hung 劉洪.
(In *Hou Han Shu*, ch. 12, p. 17*a*.)

Lung-Chhüan Hsien Chih 龍泉縣志.
Local Topography of Lung-chhüan (Chekiang).
Chhing, +1762.
Ku Kuo-Chao (ed.) 顧國詔.

Ma-Ku Shan Hsien Than Chi 麻姑山仙壇記.
Notes on the Altars to the Immortals on Ma-Ku Mountain.
Thang, *c.* +760.
Yen Chen-Chhing 顏眞卿.

Man Shu 蠻書.
Book of the Barbarians [itineraries].
Thang, *c.* +862.
Fan Chho 樊綽.

Mê Chi.
See *Mo Chi.*
Meigetsuki 明月記.
Moonlight Diary [from +1192 onwards].
Japan, *c.* +1200.
Fujiwara no Sadaiye 藤原定家 commonly called Teika (+1162/+1241).
Mêng Chhi Pi Than 夢溪筆談.
Dream Pool Essays.
Sung, +1086; last supplement dated +1091.
Shen Kua 沈括.
Mêng Liang Lu 夢粱錄.
The Past seems a Dream [description of the capital, Hangchow].
Sung, +1275.
Wu Tzu-Mu 吳自牧.
Mêng Tzu 孟子.
The Book of Master Mêng (Mencius).
Chou, *c.* −290.
Mêng Kho 孟軻.
Tr. Legge (3); Lyall (1).
Yin-Tê Index no. (Suppl.) 17.
Min Shui Yen Than Lu.
See *Shêng Shui Yen Than Lu.*
Ming Huang Tsa Lu 明皇雜錄.
Miscellaneous Records of the Brightness of the Imperial Court (of Thang Hsüan Tsung).
Thang, +855.
Chêng Chhu-Hui 鄭處誨.
Ming I Pieh Lu 名醫別錄.
Informal Records of Famous Physicians.
Before Liang, perhaps San Kuo or Chin.
Comm. by Thao Hung-Ching (*c.* +500) 陶宏景.
Ming I Thien Wên Shu 明譯天文書.
Astronomical Books translated (by Imperial Order of the) Ming.
Ming, +1382.
Tr. Hai-Ta-Erh 海達兒 (alternatively Hei-Ti-Erh 黑的兒).
Ming Shih 明史.
History of the Ming Dynasty [+1368 to +1643].
Chhing, +1739.
Chang Thing-Yü 張廷玉 *et al.*
Ming Thang Ta Tao Lu 明堂大道錄.
Studies on the Great Tradition of the Ming Thang [the cosmological temple, astronomical observatory, and emperor's ritual house, of antiquity].
Chhing, *c.* +1736.
Hui Shih-Chhi 惠士奇.
Mo Chi 默記.
Things Silently Recorded [affairs of the capital city].
Sung, +11th century.
Wang Chih 王銍.
Mo Ching 墨經.
See *Mo Tzu.*
Mo Kho Hui Hsi 墨客揮犀.
Fly-Whisk Conversations of the Scholarly Guest.
Sung, *c.* +1080.
Phêng Chhêng 彭乘.
Mo-Têng-Chia Ching 摩登伽經.
Mātangī Sūtra(?); The Book of Mātangī [contains astrological details and a list of *hsiu* with the number of stars in each].
Ascr. San Kuo, *c.* +225, but actually must be approximately +8th century.
Supposedly tr. from Sanskrit, Chu Lü-Yen 竺律炎.
N/645, TW/1300.
Mo Tzu (incl. *Mo Ching*) 墨子.
The Book of Master Mo.
Chou, −4th century.
Mo Ti (and disciples) 墨翟.
Tr. Mei Yi-Pao (1); Forke (3).
Yin-Tê Index no. (Suppl.) 21.
TT/1162.
Mu Thien Tzu Chuan 穆天子傳.
Account of the Travels of the Emperor Mu.
Chou, before −245. (Found in the tomb of An Li Wang, a prince of the Wei State, r. −276 to −245; in +281.)
Writer unknown.
Tr. Eitel (1); Chêng Tê-Khun (2).

Nan Chhi Shu 南齊書.
History of the Southern Chhi Dynasty [+479 to +501].
Liang, +520.
Hsiao Tzu-Hsien 蕭子顯.
Nan Chou I Wu Chih 南州異物志.
Strange Things of the South.
Chin, +3rd or +4th century.
Wan Chen 萬震.
Nan Fang Tshao Mu Chuang 南方草木狀.
Records of the Plants and Trees of the Southern Regions.
Chin, +3rd century.
Chi Han 稽含.
Nan Hao Shih Hua 南濠詩話.
Essays of the Retired Scholar Dwelling by the Southern Moat [literary criticism].
Ming, +1513.
Tu Mu 都穆.
Nan Hu Chi 南湖集.
Southern Lake Collection (of Poems).
Sung, +1210.
Chang Tzu 張鎡.
Nan Hua Chen Ching 南華眞經.
See *Chuang Tzu.*
Nan I I Wu Chih 南裔異物志.
Strange Things from the Southern Borders.
H/Han, end +2nd century.
Yang Fu 楊孚.
Nan-Ni-Chi-Shih-Fu-Lo-Thien Shuo Chih Lun Ching 難儞計濕嚩囉天說支輪經.
Manual of the Chih Cycle, spoken by Nan-Ni-Chi-Shih-Fu-Lo-Thien [astrological

Nan-Ni-Chi-Shih-Fu-Lo-Thien Shuo Chih Lun Ching (*cont.*)
and geomantic text, which includes the European zodiacal animal cycle].
Sung, *c.* +985.
Tr. (from Sanskrit?), Fa-Hsien 法賢.
TW/1312.

Nan Shih 南史.
History of the Southern Dynasties [Nan Pei Chhao period, +420 to +589].
Thang, *c.* +670.
Li Yen-Shou 李延壽.

Nan Tshun Cho Kêng Lu 南村輟耕錄.
See *Cho Kêng Lu*.

Nung Chêng Chhüan Shu 農政全書.
Complete Treatise on Agriculture.
Ming. Composed, +1625 to +1628; printed, +1639.
Hsü Kuang-Chhi 徐光啓.
Ed. Chhen Tzu-Lung 陳子龍.

Nung Shu 農書.
Treatise on Agriculture.
Sung, +1149; printed +1154.
Chhen Fu (Taoist) 陳旉.

Nung Shu 農書.
Treatise on Agriculture.
Yuan, +1313.
Wang Chên 王禎.

Pao Phu Tzu 抱樸(朴)子.
Book of the Preservation-of-Solidarity Master.
Chin, early +4th century.
Ko Hung 葛洪.
Partial trs. Feifel (1, 2); Wu & Davis (2); etc.
TT/1171–1173.

Pei Chhi Shu 北齊書.
History of the Northern Chhi Dynasty [+550 to +577].
Thang, +640.
Li Tê-Lin 李德林 and his son Li Pai-Yao 李百藥.
A few chs. tr. Pfizmaier (60).

Pei Chou Shu 北周書.
See *Chou Shu*.

Pei Shih 北史.
History of the Northern Dynasties [Nan Pei Chhao period, +386 to +581].
Thang, *c.* +670.
Li Yen-Shou 李延壽.

Pei Shih Chi 北使記.
Notes on an Embassy to the North.
J/Chin and Yuan, +1223.
Wukusun Chung-Tuan 烏古孫仲端.

Pei Thang Shu Chhao 北堂書鈔.
Book Records of the Northern Hall [encyclopaedia].
Thang, *c.* +630.
Yü Shih-Nan 虞世南.

Pei Tou Chhi Hsing Nien Sung I Kuei 北斗七星念誦儀軌.
Mnemonic Rhyme of the Seven Stars of the Great Bear and their Tracks.
Thang, *c.* +710.
I-Hsing 一行.
TW/1305.

Pên Tshao Kang Mu 本草綱目.
The Great Pharmacopoeia.
Ming, +1596.
Li Shih-Chen 李時珍.
Paraphrased and abridged tr. Read & collaborators (1–7) and Read & Pak (1) with indexes.

Pên Tshao Pieh Shuo 本草別説.
Informal Remarks on the Pharmacopoeia.
Sung, +1090.
Chhen Chhêng 陳承.

Pên Tshao Shih I 本草拾遺.
Omissions from Previous Pharmacopoeias.
Thang, *c.* +725.
Chhen Tshang-Chhi 陳藏器.

Pên Tshao Yen I 本草衍義.
The Meaning of the Pharmacopoeia Elucidated.
Sung, +1116.
Khou Tsung-Shih 寇宗奭.
TT/761.

Phei Wên Yün Fu 佩文韻府.
Encyclopaedia of Phrases and Allusions arranged according to Rhyme.
Chhing, +1711.
Ed. Chang Yü-Shu 張玉書 *et al.*

Phien Tzu Lei Pien 駢字類編.
Collection of Phrases and Literary Allusions.
Chhing, +1728.
Ed. Ho Chhuo 何焯 *et al.*

Phu Chi Fang 普濟方.
(Simple) Prescriptions for Everyman.
Ming, *c.* +1410.
Chu Hsiao (Chou Ting Wang) 朱橚, assisted by Chao Yen-Chhêng 趙俺誠 and others.

Pi Chhuan Hua Ching 秘傳花鏡.
The Mirror of Flowers [horticultural and zootechnic manual].
Chhing, +1688.
Chhen Hao-Tzu 陳淏子.

Pi Li Tui Shu Piao 比例對數表.
Logarithm Tables with Explanations.
Chhing, +1650.
Hsüeh Fêng-Tsu 薛鳳祚.

Pi Nai Shan Fang Chi.
See Pao Chhi-Shou (*1*).

Pi Shu Chien Chih.
See *Yuan Pi Shu Chien Chih*.

Pi Shu Lu Hua 避暑錄話.
Conversations while Avoiding the Heat of Summer.
Sung, *c.* +1150.
Yeh Mêng-Tê 葉夢得.

Piao I Lu 表異錄.
Notices of Strange Things.

Piao I Lu (*cont.*)
Ming.
Wang Chih-Chien 王志堅.

Pien Chêng Lun 辨正論.
Discourse on Proper Distinctions.
Thang, *c.* +630.
Fa-Lin 法琳.

Po Ku Thu Lu 博古圖錄.
See *Hsüan-Ho Po Ku Thu.*

Po Wu Chi 博物記.
Notes on the Investigation of Things.
H/Han, *c.* +190.
Thang Mêng (*b*) 唐蒙.

Po Wu Chih 博物志.
Record of the Investigation of Things (cf. *Hsü Po Wu Chih*).
Chin, *c.* +290 (begun about +270).
Chang Hua 張華.

Po Ya Chhin 伯牙琴.
The Lute of Po Ya [legendary lutanist, G/1662].
Sung, +13th century.
Têng Mu 鄧牧.

Pu-Li Kho Than 步里客談.
Discussions with Guests at Pu-li.
Sung, *c.* +1110.
Chhen Chhang-Fang 陳長方.

Pu Thien Ko 步天歌.
Song of the March of the Heavens [astronomical].
Sui, end +6th century.
Wang Hsi-Ming 王希明.
Tr. Soothill (5).

Samguk Sagi 三國史記.
History of the Three Kingdoms (of Korea) [Silla (Hsin-Lo), Kokuryo (Kao-Chü-Li) and Pakche (Pai-Chhi) −57 to +936].
Korea, +1145 (imperially commissioned by King Injong), repr. +1394, +1512.
Kim Pu-Sik (Chin Fu-Shih) 金富軾.

San Fu Huang Thu 三輔皇圖.
Description of the Three Districts in the Capital (Chhang-an; Sian).
Chin, late +3rd century, or perhaps H/Han.
Attrib. to Miao Chhang-Yen 苗昌言.

San Kuo Chih 三國志.
History of the Three Kingdoms [+220 to +280].
Chin, *c.* +290.
Chhen Shou 陳壽.
Yin-Tê Index, no. 33.

San Tshai Thu Hui 三才圖會.
Universal Encyclopaedia.
Ming, +1609.
Wang Chhi 王圻.

San-Yü Chui Pi 三餘贅筆.
Long-Winded Discussions at San-yü.
Ming, late +16th century.
Tu Ang 都卬.

Sêng Hui-Sêng Shih Hsi Yü Chi 僧惠生使西域記.
Record of Western Countries, by the monk Hui-Sêng.
N/Wei, *c.* +530.
Hui-Sêng (monk) 惠生.
Preserved in ch. 5 of *Lo-Yang Chhieh Lan Chi*, q.v.
Tr. Beal (1); Chavannes (3).

Sha-Chou Thu Ching 沙州圖經.
Topography of Shachow [Tunhuang].
Thang.
Writer unknown.
Tr. L. Giles (8, 9).

Shan Chü Hsin Hua 山居新話.
Conversations on Recent Events in the Mountain Retreat.
Yuan, +1360.
Yang Yü 楊瑀.
Tr. H. Franke (2).

Shan Hai Ching 山海經.
Classic of the Mountains & Rivers.
Chou and C/Han.
Writers unknown.
Partial tr. de Rosny (1).
Chung-Fa Index, no. 9.

Shan Ho Liang Chieh Khao.
See *Thien Hsia Shan Ho Liang Chieh Khao.*

Shang Shu Shih Thien 尚書釋天.
Discussion of the Astronomy in the *Shang Shu* section of the *Historical Classic.*
Chhing, +1749 to +1753.
Shêng Pai-Erh 盛百二.

Shang Shu Wei Khao Ling Yao 尚書緯考靈曜.
Apocryphal Treatise on the *Shang Shu* Section of the *Historical Classic*; Investigation of the Mysterious Brightnesses.
C/Han, −1st century.
Writer unknown.
(Now contained in *Ku Wei Shu*, chs. 1 and 2.)

Shê-Thou-Chien Thai-Tzu Erh-shih-pa Hsiu Ching 舍頭諫太子二十八宿經.
Śārdūlakarṇā-vadāna sūtra(?) [contains list of *hsiu* with number of stars in each.]
San Kuo or Chin, *c.* +300.
Tr. from Sanskrit, Chu Fa-Hu 竺法護 (Dharmarakṣa).
N/646, TW/1301.

Shen Hsien Chuan 神仙傳.
Lives of the Divine Hsien. (Cf. *Lieh Hsien Chuan* and *Hsü Shen Hsien Chuan.*)
Chin, early +4th century.
Attrib. Ko Hung 葛洪.

Shen I Ching 神異經 (or *Chi* 記).
Book of the Spiritual and the Strange.
Ascr. Han, but prob. +4th or +5th century.
Attrib. Tungfang Shuo 東方朔.

Shen Nung Pên Tshao Ching 神農本草經.
Pharmacopoeia of the Heavenly Husbandman [no longer exists as a separate work but forms the basis of all subsequent pharmaceutical compendia].
H/Han, *c.* +2nd century.

Shen Nung Pên Tshao Ching (*cont.*)
Reconstituted and commented by Miu Hsi-Yung (+1625); see *Shen Nung Pên Tshao Ching Su.*

Shen Nung Pên Tshao Ching Su 神農本草經疏.
Commentary on the Text of the *Pharmacopoeia of the Heavenly Husbandman.*
Ming, +1625.
Miu Hsi-Yung 繆希雍.

Shen Tao Ta Pien Li Tsung Suan Hui 神道大編曆宗算會.
Assembly of Computing Methods connected with the Calendar.
Ming, +1558.
Chou Shu-Hsüeh 周述學.

Shêng Chi Tsung Lu 聖濟總錄.
Imperial Medical Encyclopaedia (issued by authority).
Sung, *c.* +1111.
Ed. twelve physicians.

Shêng Hui Hsüan Fang 聖惠選方.
Glory-of-the-Sages Choice Prescriptions.
Sung, +1048.
Ho Hsi-Ying 何希影.

Shêng Shou Wan Nien Li 聖壽萬年曆.
The Imperial Longevity Permanent Calendar.
(One of the books in the Calendrical Opus, *Li Shu* 曆書.)
Ming, +1595.
Chu Tsai-Yü (prince of the Ming) 朱載堉.

Shêng Shui Yen Than Lu 澠水燕談錄.
Fleeting Gossip by the River Shêng.
Sung.
Wang Phi-Chih 王闢之.

Shih Chi 史記.
Historical Records [or perhaps better: Memoirs of the Historiographer(-Royal); down to −99].
C/Han, *c.* −90 [first pr. *c.* +1000].
Ssuma Chhien 司馬遷, and his father Ssuma Than 司馬談.
Partial trs. Chavannes (1); Pfizmaier (13–36); Hirth (2); Wu Khang (1); Swann (1), etc.
Yin-Tê Index, no. 40.

Shih Chia Chai Yang Hsin Lu 十駕齋養新錄.
Interpretations of the New by the Old; Miscellaneous Notes from the Shih-Chia Study.
Chhing, *c.* +1790.
Chhien Ta-Hsin 錢大昕.

Shih Ching 詩經.
Book of Odes [ancient folksongs].
Chou, −9th to −5th centuries.
Writers and compilers unknown.
Tr. Legge (8); Waley (1); Karlgren (14).

Shih I Chi 拾遺記.
Memoirs on Neglected Matters.
Chin, late +3rd or +4th century.
Wang Chia 王嘉.

Shih Liao Pên Tshao 食療本草.
The Nutritional Medicine Pharmacopoeia.
Thang, *c.* +670.
Mêng Shen 孟詵.

Shih Lin Kuang Chi 事林廣記.
Guide through the Forest of Affairs [encyclopaedia].
Sung, between +1100 and +1250; first pr. +1325.
Chhen Yuan-Ching 陳元靚.
(A Ming edition of +1478 is in the Cambridge University Library.)

Shih-Lin Yen Yu 石林燕語.
Informal Conversations of (Yeh) Shih-Lin (Yeh Mêng-Tê).
Sung, +1136.
Yeh Mêng-Tê 葉夢得.
See des Rotours (1), p. cix.

Shih Phin 石品.
Hierarchy of Stones.
Ming, +1617.
Yu Chün 郁濬.

Shih Shuo Hsin Yü 世說新語.
New Discourse on the Talk of the Times [notes of minor incidents from Han to Chin]. Cf. *Hsü Shih Shuo.*
L/Sung, +5th century.
Liu I-Chhing. 劉義慶
Commentary by Liu Hsün 劉峻 (Liang).

Shih Su 詩疏.
Studies on the *Book of Odes.*
Thang, *c.* +640.
Khung Ying-Ta 孔穎達.

Shih Tzu 尸子.
The Book of Master Shih.
Ascr. Chou, −4th century; probably +3rd or 4th century.
Attrib. Shih Chiao 尸佼.

Shih Wu Chi Yuan 事物紀原.
Records of the Origins of Affairs and Things.
Sung, *c.* +1085.
Kao Chhêng 高承.

Shih Wu Lun 時務論.
Discourse on Time [calendrical].
San Kuo, *c.* +237.
Yang Wei 楊偉.

Shih Yao Erh Ya 石藥爾雅.
Synonymic Dictionary of Minerals and Drugs.
Thang, +818.
Mei Piao 梅彪.
TT/894.

Shou Shih Li I Ching 授時曆議經.
Explanations and Manual of the *Shou Shih* (Works and Days) Calendar.
Yuan, +1280.
Kuo Shou-Ching 郭守敬 (with many colleagues and assistants).
(Now only in chs. 52, 53, 54, 55 of the *Yuan Shih.*)

Shou Shih Thung Khao 授時通考.
Complete Investigation of the Works and Days (published by Imperial Order)

Shou Shih Thung Khao (*cont.*)
[treatise on agriculture, horticulture and all related technologies].
Chhing, +1742.
Ed. O-Erh-Thai (Ortai) 鄂爾泰.
Shou Shou Wu Yo Yuan Fa 授受五嶽圓法.
The Received Method of Drawing the Circles of the Five (Sacred) Mountains.
Thang or Wu Tai.
Writer unknown.
(Tunhuang MS. S/3750.)
Shu Chêng Chi 述征記.
Records of Military Expeditions.
Chin, +3rd or +4th century.
Kuo Yuan-Sêng 郭緣生.
Shu Ching 書經.
Historical Classic [Book of Documents].
The 29 'Chin Wên' chapters mainly Chou (a few pieces possibly Shang); the 21 'Ku Wên' chapters a 'forgery' by Mei Tsê 梅賾, *c.* +320, using fragments of genuine antiquity. Of the former, 13 are considered to go back to the −10th century, 10 to the −8th, and 6 not before the −5th. Some scholars accept only 16 or 17 as pre-Confucian.
Writers unknown.
See Wu Shih-Chhang (1); Creel (4).
Tr. Medhurst (1); Legge (1, 10); Karlgren (12).
Shu Hsüeh 數學.
Mathematical Astronomy.
Chhing, *c.* +1750.
Chiang Yung 江永.
Shu Hsüeh Thung Kuei 數學通軌.
Rulês of Mathematics.
Ming, +1578.
Ko Shang-Chhien 柯尙遷.
Shu I Chi 述異記.
Records of Strange Things.
Liang, early +6th century.
Jen Fang 任昉.
See des Rotours (1), p. ci.
Shu Li Ching Yün 數理精蘊.
Collected Basic Principles of Mathematics (compiled by Imperial Order) [Part of *Lü Li Yuan Yuan*, q.v.].
Chhing, +1723.
Ed. Mei Ku-Chhêng 梅瑴成 & Ho Kuo-Tsung 何國宗.
Cf. Hummel (2), p. 285.
Shu Shu Chi I 數術記遺.
Memoir on some Traditions of Mathematical Art.
H/Han, +190 (?).
Hsü Yo 徐岳.
Shu Shu Chiu Chang 數書九章.
Mathematical Treatise in Nine Sections.
Sung, +1247.
Chhin Chiu-Shao 秦九韶.
Shu Ssu Shuo Ling 書肆說鈴.
Discussions of a Bibliographer.
Ming.
Yeh Ping-Ching 葉秉敬.
Shu Tu Yen 數度衍.
Generalisations on Numbers.
Chhing, +1661 and +1721.
Fang Chung-Thung 方中通.
Shu Wu I Ming Su 庶物異名疏.
Disquisition on Strange Names for Common Things.
Ming.
Chhen Mou-Jen 陳懋仁.
Shu Yü Chou Tzu Lu 殊域周咨錄.
Complete Description of Foreign Parts.
Ming, +1520.
Yen Tshung-Chien 嚴從簡.
Shui Ching 水經.
The Waterways Classic [geographical account of rivers and canals].
Ascr. C/Han, prob. San Kuo.
Attrib. Sang Chhin 桑欽.
Shui Ching Chu 水經注.
Commentary on the *Waterways Classic* [geographical account greatly extended].
N/Wei, late +5th or early +6th century.
Li Tao-Yuan 酈道元.
Shui Tao Thi Kang 水道提綱.
Complete Description of Waterways.
Chhing, +1776.
Chhi Shao-Nan 齊召南.
Shun-Yu Lin-An Chih 淳祐臨安志.
Shun-Yu reign-period Topographical Records of the Hangchow District.
Sung, *c.* +1245.
Shih Ê 施諤.
Shuo Fu 說郛.
Florilegium of (Unofficial) Literature.
Yuan, *c.* +1368.
Thao Tsung-I 陶宗儀.
See Ching Phei-Yuan (1).
Shuo Wên.
See *Shuo Wên Chieh Tzu*.
Shuo Wên Chieh Tzu 說文解字.
Analytical Dictionary of Characters.
H/Han, +121.
Hsü Shen 許愼.
So-Nan Wên Chi 所南文集.
Collected Writings of (Chêng) So-Nan (Chêng Ssu-Hsiao).
Yuan, *c.* +1340.
Chêng Ssu-Hsiao 鄭思肖.
Sou Shen Chi 搜神記.
Reports on Spiritual Manifestations.
Chin, *c.* +348.
Kan Pao 干寶.
Partial tr. Bodde (9).
Sou Tshai I Wên Lu 搜采異聞錄.
Collection of Strange Things Heard.
Sung.
Yung Hêng 永亨.
Ssu Chhao Wên Chien Lu 四朝聞見錄.
Record of Things Seen and Heard at Four Imperial Courts.

Ssu Chhao Wên Chien Lu (*cont.*)
Sung, early +13th century.
Yeh Shao-Ong 葉紹翁.
Ssu Mên Ching 四門經.
(Astrological) Manual of the Four Gates [distribution of the *hsiu* among the four palaces].
Thang, *c.* +780.
Tr. (probably from Sogdian) Ching-Ching 景淨 (Adam the Nestorian).
Ssu Min Yüeh Ling 四民月令.
Monthly Ordinances for the Four Sorts of People (Scholars, Farmers, Artisans and Merchants).
H/Han.
Tshui Shih 崔寔.
Ssu Yuan Yü Chien 四元玉鑑.
Precious Mirror of the Four Elements [algebra].
Yuan, +1303.
Chu Shih-Chieh 朱世傑.
Suan Ching Shih Shu 算經十書.
The Ten Mathematical Manuals.
Collected and issued, Sung, +1084.
Ed. Tai Chen 戴震 (Chhing).
Printed in the Palace Collection, Wu Ying Tien Chü Chen Pan Tshung Shu 武英殿聚珍版叢書 +1794.
Also pub. separately by Khung Chi-Han 孔繼涵 about +1760.
Cf. Hummel (2), p. 697.
Suan Fa Chhü Yung Pên Mo 算法取用本末.
Alpha and Omega of Simple Calculations.
Sung, late +13th century.
Yang Hui 楊輝.
Suan Fa Chhüan Nêng Chi 算法全能集.
Record of 'Do-Everything' Mathematical Methods.
Yuan or Ming, +14th century.
Chia Hêng 賈亨.
(In *Yung-Lo Ta Tien*, ch. 16343/4.)
Suan Fa Thung Pien Pên Mo 算法通變本末.
Alpha and Omega of the Mathematics of Mutually Varying Quantities.
Sung, end +13th century.
Yang Hui 楊輝.
(In *Yang Hui Suan Fa*.)
Suan Fa Thung Tsung 算法統宗.
Systematic Treatise on Arithmetic.
Ming, +1593.
Chhêng Ta-Wei 程大位.
Suan Hsüeh Chhi Mêng 算學啓蒙.
Introduction to Mathematical Studies.
Yuan, +1299.
Chu Shih-Chieh 朱世傑.
Suan Hsüeh Hsin Shuo 算學新說.
A New Account of the Science of Calculation (in Acoustics and Music).
Ming, +1603.
Chu Tsai-Yü (prince of the Ming) 朱載堉.
Sui-Chhu Thang Shu Mu 遂初堂書目.
Catalogue of the Books in the Sui-Chhu Hall.
Sung.
Yü Mou 尤袤.
Sui Chhü Yü Thu Chih 隋區宇圖志.
Records and Maps of the Districts of the Empire of the Sui (Dynasty).
Sui, *c.* +600.
Ed. Yü Shih-Chi 虞世基 & Hsü Shan-Hsin 許善心.
Sui Chu Chou Thu Ching Chi 隋諸州圖經集.
Collection of Local Topographies and Maps of the Sui Dynasty.
Sui, *c.* +600.
Ed. Lang Wei-Chih 郎蔚之.
Sui Hsi Yü Thu.
See *Hsi Yü Thu Chi.*
Sui Shu 隋書.
History of the Sui Dynasty [+581 to +617].
Thang, +636 (annals and biographies); +656 (monographs and bibliography).
Wei Chêng 魏徵 *et al.*
Partial trs. Pfizmaier (61–65); Balazs (7, 8); Ware (1).
Sui-Yin Man Lu 隨隱漫錄.
(Chhen) Sui-Yin's Random Observations.
Sung, late +13th century.
Chhen Sui-Yin 陳隨隱.
Sun Tzu Suan Ching 孫子算經.
Master Sun's Mathematical Manual.
San Kuo, Chin or L/Sung.
Master Sun (full name unknown) 孫子.
Sung Chhuang Pai Shuo 松窗百說.
A Hundred Observations from the Pine-tree Window.
Sung, +1157.
Li Chi-Kho 李季可.
Sung Hui Yao 宋會要.
History of the Administrative Statutes of the Sung Dynasty.
Sung.
Chang Tê-Hsiang 章得象.
Sung I Min Lu 宋遺民錄.
Sung Officials who refused to serve the Yuan Dynasty.
Ming, +1479.
Chhêng Min-Chêng 程敏政.
Sung Shih 宋史.
History of the Sung Dynasty [+960 to +1279].
Yuan, *c.* +1345.
Tho-Tho (Toktaga) 脫脫 & Ouyang Hsüan 歐陽玄.
Yin-Tê Index, no. 34.
Sung Shu 宋書.
History of the (Liu) Sung Dynasty [+420 to +478].
S/Chhi, +500.
Shen Yo 沈約.
A few chs. tr. Pfizmaier (58).
Sung Ssu-Hsing Tzu-Wei Shu 宋司星子韋書.
Book of the Astrologer (Shih) Tzu-Wei of the State of Sung.

Sung Ssu-Hsing Tzu-Wei Shu (*cont.*)
Chou, early −5th century.
Shih Tzu-Wei 史子韋.

Sung Ssu Tzu Chhao Shih 宋四子抄釋.
Selections from the Writings of the Four Sung (Neo-Confucian) Philosophers [excl. Chu Hsi].
Sung (ed. Ming, +1536).
Ed. Lü Jan 呂柟.

Sung Yuan Hsüeh An 宋元學案.
Schools of Philosophers in the Sung and Yuan Dynasties.
Chhing, *c.* +1750.
Huang Tsung-Hsi 黃宗羲 & Chhüan Tsu-Wang 全祖望.

Ta Chhing Hui Tien 大清會典.
History of the Administrative Statutes of the Chhing Dynasty.
Chhing: 1st ed. +1690; 2nd +1733; 3rd +1767; 4th +1818; 5th +1899.
Ed. Wang An-Kuo 王安國 and many others.

Ta Chhing I Thung Chih 大清一統志.
Comprehensive Geography of the (Chinese) Empire (under the Chhing dynasty).
Chhing, *c.* +1730.
Ed. Hsü Chhien-Hsüeh 徐乾學.

Ta Fang Têng Ta Chi Ching 大方等大集經.
Mahā-vaipulya-mahā-saṃnipāta Sūtra [includes calendrical and Western zodiacal animal cycle material, association of planets with *hsiu*, etc.].
N/Chou or Sui, between +566 and +585.
Tr. from Sanskrit, Na-Lien-Thi-Yeh-Shê 那連提耶舍 (Narendrayaśas).
N/61; TW/397.

Ta-Kuan Ching Shih Chêng Lei Pên Tshao 大觀經史證類本草.
Ta-Kuan reign-period Reorganised Pharmacopoeia.
See *Chêng Lei Pên Tshao*.

Ta Ming Hui Tien 大明會典.
History of the Administrative Statutes of the Ming Dynasty.
Ming, 1st ed. +1509, 2nd ed. +1587.
Ed. Shen Shih-Hsing 申時行 *et al.*

Ta Ming I Thung Chih 大明一統志.
Comprehensive Geography of the (Chinese) Empire (under the Ming dynasty).
Ming, *c.* +1450 (+1461?).
Ed. Li Hsien 李賢.

Ta Pao Chi Ching 大寶積經.
Mahāratnakūta Sūtra.
Tr. +541.
N/23; TW/310.

Ta Tai Li Chi 大戴禮記.
Record of Rites (compiled by Tai the Elder). (Cf. *Hsiao Tai Li Chi*; *Li Chi*.)
Ascr. C/Han, *c.* −70/−50 but really H/Han, between +80 and +105.
Attrib. ed. Tai Tê 戴德; in fact probably ed. Tshao Pao 曹褒.
See Legge (7).
Tr. Douglas (1); R. Wilhelm (6).

Ta Thang Hsi Yü Chi 大唐西域記.
Records of the Western Countries in the time of the Thang.
Thang, +646.
Hsüan-Chuang (monk) 玄奘.
Ed. Pien Chi 辯機.
Tr. Julien (1); Beal (2).

Ta Yen Hsiang Shuo 大衍詳說.
Explanation of Indeterminate Analysis [algebra]
Sung, *c.* +1180.
Tshai Yuan-Ting 蔡元定.

Ta Yuan I Thung Chih 大元一統志.
Comprehensive Geography of the (Chinese) Empire (under the Yuan Dynasty).
Yuan, *c.* +1310.
Editor not known.

Tan Chhien Tsung Lu 丹鉛總錄.
Red Lead Record.
Ming, +1542.
Yang Shen 楊愼.

Tan Fang Chien Yuan 丹方鑑源.
Original Mirror of Alchemical Preparations.
Sung, +11th century.
Tuku Thao 獨孤滔.
TT/918.

Tao Tê Ching 道德經.
Canon of the Tao and its Virtue.
Chou, before −300.
Attrib. Li Erh (Lao Tzu) 李耳 (老子).
Tr. Waley (4); Chhu Ta-Kao (2); Lin Yü-Thang (1); Wieger (7); Duyvendak (18); and very many others.

Tao Tsang 道藏.
The Taoist Patrology [containing 1464 Taoist works].
All periods, but first collected and printed in the Sung. Also printed in J/Chin (+1186 to +1191), Yuan, and Ming (+1445, +1598 and +1607).
Index by Wieger (6), on which see Pelliot's review.
Yin-Tê Index no. 25.

Thai Hsüan Ching 太玄經.
Canon of the Great Mystery.
C/Han, *c.* +10.
Yang Hsiung 揚雄.

Thai I Chin Ching Shih Ching 太乙金鏡式經.
The Golden Mirror of the Thai-I (Star); a Divining Board Manual.
Thang.
Wang Hsi-Ming 王希明.

Thai Pai Yin Ching 太白陰經.
Canon of the White (and Gloomy) Planet (of War; Venus) [treatise on military affairs].
Thang, +759.
Li Chhüan 李筌.

Thai-Phing Huan Yü Chi 太平寰宇記.
Thai-Phing reign-period General Description of the world [geographical record].
Sung, +976 to +983.
Yüeh Shih 樂史.

Thai-Phing Yü Lan 太平御覽.
Thai-Phing reign-period Imperial Encyclopaedia [lit. the Emperor's Daily Readings].
Sung, +983.
Ed. Li Fang 李昉.
Some chs. tr. Pfizmaier (84–106).
Yin-Tê Index no. 23.

Thai-Tshang Chou Chih 太倉州志.
Topography of Thai-tshang (Chiangsu).
Ming, *c.* +1500.
Ed. Chhien Su-Lo 錢肅樂.

Than Chai Thung Pien 坦齋通編.
Miscellaneous Records of the Candid Studio.
Sung, *c.* +1220.
Hsing Khai 邢凱.

Thang Chhüeh Shih 唐闕史.
Thang Memorabilia.
Wu Tai, +10th century.
Kao Yen-Hsiu 高彥休.

Thang Hui Yao 唐會要.
History of the Administrative Statutes of the Thang Dynasty.
Sung, +961.
Wang Phu 王溥.

Thang Pên Tshao 唐本草.
Pharmacopoeia of the Thang Dynasty. (Cf. the *Hsin Hsiu Pên Tshao*, q.v.)
Thang, +660.
Ed. Su Kung 蘇恭.

Thang Shu.
See *Chiu Thang Shu* and *Hsin Thang Shu.*

Thang Yü Lin 唐語林.
Miscellanea of the Thang dynasty.
Sung, collected *c.* +1107.
Wang Tang 王讜.

Thieh Wei Shan Tshung Than 鐵圍山叢談.
Collected Conversations at Iron-Fence Mountain.
Sung, *c.* +1115.
Tshai Thao 蔡絛.

Thien Ching 天鏡.
Mirror of the Heavens.
Liang, +6th century.
Writer unknown.

Thien Hsia Chün Kuo Li Ping Shu 天下郡國利病書.
Merits and Drawbacks of all the Countries in the World [geography].
Chhing, +1662.
Ku Yen-Wu 顧炎武.

Thien Hsia Shan Ho Liang Chieh Khao 天下山河兩戒考.
An Investigation of the Two Regions of the Earth [a study of the *fên yeh* system].
Chhing, +1723.
Hsü Wên-Ching 徐文靖.

Thien Hsüeh Hui Thung 天學會通.
Towards a Thorough Understanding of Astronomical Science.
Chhing, +1650.
Hsüeh Fêng-Tsu 薛鳳祚.

Thien Kung Khai Wu 天工開物.
The Exploitation of the Works of Nature.
Ming, +1637.
Sung Ying-Hsing 宋應星.

Thien Mou Pi Lei Chhêng Chhu Chieh Fa 田畝比類乘除捷法.
Practical Rules of Arithmetic for Surveying.
Sung, end +13th century.
Yang Hui 楊輝.
(In *Yang Hui Suan Fa.*)

Thien Pu Chen Yuan 天步眞原.
True Course of Celestial Motions.
Chhing, *c.* +1646.
Mu Ni-Ko (Nicholas Smogołecki) 穆尼閣 & Hsüeh Fêng-Tsu 薛鳳祚.

Thien Ti Jui Hsiang Chih 天地瑞祥志.
Record of Auspicious Phenomena in the Heavens and the Earth.
Thang, *c.* +666.
Shou-Chen (monk) 守眞.

Thien Tui 天對.
Answers about Heaven.
Thang, *c.* +800.
Liu Tsung-Yuan 柳宗元.

Thien Wên 天問.
Questions about Heaven [ode].
Chou, *c.* −300.
Chhü Yuan 屈原.
Tr. Erkes (8).

Thien Wên Chih 天文志.
On the Heavens. [A memorial preserved in the *Chin Shu, History of the Chin Dynasty.*]
H/Han, *c.* +180.
Tshai Yung 蔡邕.

Thien Wên Lei 天文類.
Collection of (Old) Astronomical Writings.
Chhing, *c.* +1799.
See Wylie (1), p. 101.

Thien Wên Lu 天文錄.
Résumé of Astronomy.
S/Chhi or Liang, *c.* +500.
Tsu Kêng-Chih 祖暅之.
(Fragments in *Khai-Yuan Chan Ching.*)

Thien Wên Lüeh 天問畧.
Explicatio Sphaerae Coelestis.
Ming, +1615.
Yang Ma-No (Emanuel Diaz) 陽瑪諾.

Thien Wên Ta Chhêng Kuan Khuei Chi Yao 天文大成管窺輯要.
Essentials of Observations of the Celestial Bodies through the Sighting Tube.
Chhing (incorporating Sung and Yuan material), +1653.
Huang Ting 黃鼎.

Thien Wên Ta Hsiang Fu 天文大象賦.
Essay on the Great Constellations in the Heavens.
Sui or Thang, early +7th century.
Li Po 李播.

Thien Wên Yao Lu 天文要錄.
Record of the Most Important Astronomical Matters.
Thang, *c.* +664.
Li Fêng (prince of the Thang) 李鳳.

Thien-Yuan Li Li Chhüan Shu 天元曆理全書.
Complete Treatise on the Thien-Yuan Calendar.
Chhing, +1682.
Hsü Fa 徐發.

Thien Yuan Yü Li Hsiang I Fu 天元玉曆祥異賦.
Essay on (Astronomical and Meteorological) Presages.
Ming, +1425 (never printed).
Chu Kao-Chih 朱高熾 (emperor of the Ming).
(MS. in Camb. Univ. Library.)

Thou Lien Hsi Tshao 透簾細草.
The Mathematical Curtain Pulled Aside.
Yuan, *c.* +1355.
Writer unknown.
(In *Yung-Lo Ta Tien*, ch. 16343/4.)

Thu Ching Pên Tshao 圖經本草.
See *Pên Tshao Thu Ching* and *Thu Ching Yen I Pên Tshao* in the Addenda to this Bibliography, p. 723.

Thu Hsiu Chen Chün Tsao-Hua Chih Nan 土宿眞君造化指南.
Guide to the Creation, by the Earth's Mansions Immortal.
See *Tsao-Hua Chih Nan*.

Thu Hsiu Pên Tshao 土宿本草.
The Earth's Mansions Pharmacopoeia.
See *Tsao-Hua Chih Nan*.

Thu Shu Chi Chhêng 圖書集成.
Imperial Encyclopaedia.
Chhing, +1726.
Ed. Chhen Mêng-Lei 陳夢雷 *et al.*
Index by L. Giles (2).

Thu Shu Pien 圖書編.
On Maps and Books [encyclopaedia].
Ming, +1562, +1577, +1585.
Chang Huang 章潢.

Thui Pu Fa Chieh 推步法解.
Analysis of Celestial Motions.
Chhing, +1750.
Chiang Yung 江永.

Thung-Chhi Hsien Chih 峒谿纖志.
Brief Notes from Thung-chhi [ethnography of tribal peoples].
Chhing, +17th century.
Chhen Ting 陳鼎.

Thung Chih 通志.
Historical Collections.
Sung, *c.* +1150.
Chêng Chhiao 鄭樵.

Thung Chih Lüeh 通志略.
Compendium of Information [part of *Thung Chih*, q.v.].

Thung Hsüan Chen Ching 通玄眞經.
See *Wên Tzu*.

Thung Su Wên 通俗文.
Commonly Used Synonyms.
H/Han or Chin.
Fu Chhien 服虔.

Thung Tien 通典.
Comprehensive Institutes [reservoir of source material on political and social history].
Thang, *c.* +812.
Tu Yu 杜佑.

Thung Wên Suan Chih 同文算指.
Treatise on European Arithmetic [lit. Combined Languages Mathematical Indicator].
Ming, +1614.
Li Ma-Tou (Matteo Ricci) 利瑪竇 & Li Chih-Tsao 李之藻.

Thung Yuan Suan Fa 通原算法.
Origins of Mathematics.
Ming, +1372.
Yen Kung 嚴恭.
(In *Yung-Lo Ta Tien*, ch. 16343/4.)

Ti Ching 地鏡.
Mirror of the Earth.
Liang, +6th century.
Writer unknown.

Ti Ching Thu 地鏡圖.
Illustrated Mirror of the Earth.
Liang, +6th century.
Writer unknown.

Ti Wei 地緯.
Outlines of Geography.
Ming, +1624 (pr. +1638 and +1648).
Hsiung Jen-Lin 熊人霖.

Ting Chü Suan Fa 丁巨算法.
Ting Chü's Arithmetical Methods.
Yuan, +1355.
Ting Chü 丁巨.
(In *Yung-Lo Ta Tien*, ch. 16343/4.)

Tongguk Yeji Seungnam 東國輿地勝覽.
Geographical Vista of the Eastern Kingdom (Korea).
Korea (Choson) begun *c.* +1470, finished +1530.
Sŏ Kŏ-Jong 徐居正.

Tsao-Hua Chhien Chhui 造化鉗鎚.
The Hammer and Tongs of Creation [i.e. Nature].
Ming, *c.* +1430.
Ning Hsien Wang (prince of the Ming) 寧獻王.

Tsao-Hua Chih Nan 造化指南 [= *Thu Hsiu Pên Tshao*].
Guide to the Creation [i.e. Nature].
Thang, Sung or Ming.
Thu Hsiu Chen Chün (the Earth's Mansions Immortal) 土宿眞君.

Tsao-Hua Chih Nan (*cont.*)
(Preserved only in quotations as in *PTKM*.)

Tsao-Hua Ching.
See *Tung-Hsüan Ling-Pao Chu Thien Shih-Chieh Tsao-Hua Ching*.

Tsao-Hua Ching Lun Thu 造化經論圖.
Illustrated Discourse on the *Creation Canon*.
Ming.
Chao Chhien 趙謙.

Tshan Thung Chhi 參同契.
The Kinship of the Three; or, The Accordance (of the *Book of Changes*) with the Phenomena of Composite Things.
H/Han, +142.
Wei Po-Yang 魏伯陽.
Tr. Wu & Davis (1).

Tshan Thung Chhi Khao I 參同契考異.
A Study of the *Kinship of the Three*.
Sung, +1197.
Chu Hsi (originally using pseudonym Tsou Hsin) 朱熹 (鄒訢).

Tshê Liang Fa I 測量法義.
Essentials of Surveying [trigonometry].
Ming, +1607.
Li Ma-Tou (Matteo Ricci) 利瑪竇 & Hsü Kuang-Chhi 徐光啓.

Tshê Liang I Thung 測量異同.
Similarities and Differences (between Chinese and European) Surveying Technique [trigonometry].
Ming, +1631.
Hsü Kuang-Chhi 徐光啓.

Tshê Suan 策算.
On the Use of the Calculating-Rods.
Chhing, +1744.
Tai Chen 戴震.

Tshê Thien Yo Shuo 測天約說.
Brief Description of the Measurement of the Heavens.
Ming, +1628.
Têng Yü-Han (Johannes Terrentius, i.e. Johann Schreck) 鄧玉函.

Tshê Yuan Hai Ching 測圓海鏡.
Sea Mirror of Circle Measurements.
J/Chin (Yuan), +1248.
Li Yeh 李冶.

Tshê Yuan Hai Ching Fên Lei Shih Shu 測圓海鏡分類釋術.
Classified Methods of the *Sea Mirror of Circle Measurements*.
Ming, +1550.
Ku Ying-Hsiang 顧應祥.

Tso Chuan 左傳.
Master Tsochhiu's Enlargement of the *Chhun Chhiu* (*Spring and Autumn Annals*) [dealing with the period −722 to −453].
Late Chou, compiled between −430 and −250, but with additions and changes by Confucian Scholars of the Chhin and Han, especially Liu Hsin. Greatest of the three commentaries on the *Chhun Chhiu*, the others being the *Kungyang Chuan* and the *Kuliang Chuan* but, unlike them, probably originally itself an independent book of history.
Attrib. Tsochhiu Ming 左邱明.
See Karlgren (8); Maspero (1); Chhi Ssu-Ho (1); Wu Khang (1); Wu Shih-Chhang (1); Eberhard, Müller & Henseling.
Tr. Couvreur (1); Legge (11); Pfizmaier (1–12).

Tso Chuan Pu Chu 左傳補注.
Commentary on *Master Tsochhiu's Enlargement of the Chhun Chhiu*.
Chhing, +1718.
Hui Tung 惠棟.

Tu Chhêng Chi Shêng 都城紀勝.
The Wonder of the Capital [Hangchow].
Sung, +1235.
Mr Chao 趙氏.
(Kuan Pu Nai Tê Ong, The Old Gentleman of the Water-Garden who achieved Success through Forbearance 灌圃耐得翁).

Tu Hsing Tsa Chih 獨醒雜志.
Miscellaneous Records of the Lone Watcher.
Sung, +1176.
Tsêng Min-Hsing 曾敏行.

Tu I Chih 獨異志.
Things Uniquely Strange.
Thang.
Li Jung 李冗.

Tu-Li-Yü-Ssu Ching 都利聿斯經.
The Tu-Li-Yü-Ssu (Astrological) Manual [probably two separate works combined].
Thang, *c.* +800.
Tr. from the Sogdian, Chhü-Kung 璩公.

Tu Shih Fang Yü Chi Yao 讀史方輿紀要.
Essentials of Historical Geography.
Chhing, +1667.
Ku Tsu-Yü 顧祖禹.

Tu-Yang Tsa Pien 杜陽雜編.
The Tu-yang Miscellany.
Thang, end +9th century.
Su Ê 蘇鶚.

Tuan-Chhi Yen Phu 端溪硯譜.
Tuan-chhi Inkstone Record.
Sung, *c.* +1135.
Writer unknown.

Tun-Huang Lu 燉煌錄.
Record of Tunhuang [local topography].
Thang.
Writer unknown.
Tr. L. Giles (8, 9).

Tung Hsi Yang Khao 東西洋考.
Studies on the Oceans East and West.
Ming, +1618.
Chang Hsieh 張燮.

Tung-Hsiao Shih Chi 洞霄詩集.
Poems by Visitors to the Tung-Hsiao [Taoist Temple at Hangchow].
Yuan, +1302.
Mêng Tsung-Pao 孟宗寶.

Tung-Hsüan Ling-Pao Chu Thien Shih-Chieh Tsao-Hua Ching 洞玄靈寶諸天世界造化經.
Creation Canon; The Origin of all Heavens and all Worlds: a work of the Tung-Hsüan Scriptures of Ling-Pao Chün [second person of the Taoist Trinity].
Perhaps Thang.
Writer unknown.
TT/318.

Tung Kung Ku Shih 東宮故事.
Stories of the Eastern Palace.
C/Han, before −48.
Chang Chhang 張敞.

Tung Ming Chi 洞冥記.
Light on Mysterious Things.
Ascr. Han; prob. +5th or +6th century.
Attrib. Kuo Hsien 郭憲.

Tung-Pho Chhüan Chi or *Chhi Chi* 東坡全集(七集).
The Seven Collections of (Su) Tung-Pho [i.e. Collected Works].
Sung, down to +1101, but put together later.
Su Tung-Pho 蘇東坡.

Tung Thien Chhing Lu (Chi) 洞天清錄(集).
Clarification of Strange Things [Taoist].
Sung, *c.* +1240
Chao Hsi-Ku 趙希鵠.

Wamyō-Honzō.
See *Honzō-Wamyō.*

Wan Chhing Lou Thu Pien 萬青樓圖編.
Study of (Star-) Maps from the Myriad Bamboo Tablet Studio.
Chhing, *c.* +1740.
Shao Ang-Hsiao 邵昂霄.

Wan Yung Chêng Tsung Pu-Chhiu-Jen Chhüan Pien 萬用正宗不求人全編.
The 'Ask No Questions' Complete Handbook for General Use.
Ming, +1609.
Yü Wên-Thai 余文台.

Wei Lüeh 魏略.
Memorable Things of the Wei State (San Kuo).
San Kuo (Wei) or Chin, +3rd or +4th century.
Yü Huan 魚豢.

Wei Lüeh 緯略.
Compendium of Non-Classical Matters.
Sung, +12th century (end).
Kao Ssu-Sun 高似孫.

Wei Shu 魏書.
History of the (Northern) Wei Dynasty [+386 to +550, including the Eastern Wei successor State].
N/Chhi, +554, revised +572.
Wei Shou 魏收.
See Ware (3).
One ch. tr. Ware (1, 4).

Wên Hsien Thung Khao 文獻通考.
Comprehensive Study of (the History of) Civilisation.
Sung, begun *c.* +1254, finished *c.* +1280, but not published until +1319.
Ma Tuan-Lin 馬端臨.
A few chs. tr. Julien (2); St Denys (1).

Wên Hsin Tiao Lung 文心雕龍.
On the Carving of the Dragon of the Literary Mind [literary criticism, earliest book]; or, Anatomy of the Literary Mind.
Liang.
Liu Hsieh 劉勰.

Wên Hsüan 文選.
General Anthology of Prose and Verse.
Liang, +530.
Ed. Hsiao Thung (prince of the Liang) 蕭統.

Wên-Shu-Shih-Li Phu-Sa chi Chu Hsien so Shuo Chi-Hsiung Shih Jih Shan-O Hsiu Yao Ching 文殊師利菩薩及諸仙所說吉凶時日善惡宿曜經.
Sūtra of the Discourses of the Bodhisattva Mānjusrī and the Sages on Auspicious and Inauspicious Times and Days, and on the Good and Evil *Hsiu* and Planets [includes the planetary names of the days of the Sogdian 7-day week, the Western zodiacal animal cycle, etc.]. Short title: *Hsiu* and Planet Sūtra.
Tr. Thang, +759.
Tr. Pu-Khung (Amoghavajra) 不空.
With commentary by Yang Ching-Fêng 揚景風, +764.
N/1356; TW/1299.

Wên Tzu 文子.
[=*Thung Hsüan Chen Ching.*]
The Book of Master Wên.
Han and later, but must contain pre-Chhin material; probably took its present form about +380.
Attrib. Hsin Yen 辛研 or 銒.

Wu-An Li Suan Shu Mu 勿菴曆算書目.
Bibliography of Mei Wên-Ting's (Wu-An's) Mathematical Writings.
Chhing, +1702.
Mei Wên-Ting 梅文鼎.

Wu Chhuan Lu 吳船錄.
Account of a Journey by boat to Wu [from Chhêngtu in Szechuan to Chiangsu].
Sung, +1177.
Fan Chhêng-Ta 范成大.

Wu Ching Lei Pien 五經類編.
Classified Information on the Five Classics.
Chhing, +1673.
Chou Shih-Chang 周世樟.

Wu Ching Suan Shu 五經算術.
Arithmetic in the Five Classics.
N/Chhi, +6th century.
Chen Luan 甄鸞.

Wu Ching Tsung Yao 武經總要.
Collection of the most important Military Techniques (compiled by Imperial Order).

Wu Ching Tsung Yao (*cont.*)
Sung +1040 (+1044).
Ed. Tsêng Kung-Liang 曾公亮.
Wu Chün Chih 吳郡志.
Topography of the Suchow Region (Chiangsu).
Sung, +1192; addenda, +1229.
Fan Chhêng-Ta 范成大.
Wu Chung Shui Li Shu 吳中水利書.
The Water-Conservancy of the Wu District.
Sung, +1059.
Shan O 單鍔.
Wu Hsing Hsing Tu Chieh 五星行度解.
Analysis of the Motions of the Five Planets.
Ming, +1640.
Wang Hsi-Shan 王錫闡.
Wu Li Hsiao Shih 物理小識.
Small Encyclopaedia of the Principles of Things.
Chhing, +1664.
Fang I-Chih 方以智.
Wu Li Lun 物理論.
Discourse on the Principles of Things [astronomical].
San Kuo, end +3rd century.
Yang Chhüan 楊泉.
Wu Li-Pu Shih Hua 吳禮部詩話.
On the Principles of Poetry, by Mr Wu, of the Ministry of Rites.
Yuan, early +14th century.
Wu Shih-Tao 吳師道.
Wu Li Thung Khao 五禮通考.
Comprehensive Investigation of the Five Kinds of Rites.
Chhing, +1761.
Chhin Hui-Thien 秦蕙田.
Wu Lu 吳錄.
Record of the Kingdom of Wu.
San Kuo, +3rd century.
Chang Pho 張勃.
Wu Pei Chih 武備志.
Treatise on Armament Technology.
Ming, +1628.
Mao Yuan-I 茅元儀.
Wu Pei Pi Shu 武備秘書.
Confidential Treatise on Armament Technology [a compilation of selections from earlier works on the same subject].
Chhing, late +17th century (repr. 1800).
Shih Yung-Thu 施永圖.
Wu shih Pên Tshao 吳氏本草.
Mr Wu's Pharmaceutical Codex.
San Kuo (Wei), *c.* +225.
Wu Phu 吳普.
Wu Tshao Suan Ching 五曹算經.
Mathematical Manual of the Five Government Departments.
Chin, +4th century.
Compilers unknown.
Wu Yo Chen Hsing Thu 五嶽眞形圖.
Map of the True Topography of the Five Sacred Mountains.
Chin?
Anonymous.

Yang Hui Suan Fa 楊輝算法.
Yang Hui's Methods of Computation.
Sung, +1275.
Yang Hui 楊輝.
Yao Tien 堯典.
The Canon of Yao.
(A chapter of the *Shu Ching*, q.v.)
Yeh Huo Pien 野獲編.
Memoirs of a Mission Achieved in the Wilds [an embassy to Samarqand +1368 to +1398].
Ming, *c.* +1400.
Shen Tê-Fu 沈德符.
Yeh Kho Tshung Shu 野客叢書.
Collected Notes of the Rustic Guest.
Sung, +1201.
Wang Mou 王楙.
Yen Fan Lu 演繁露.
Extension of the *String of Pearls on the Spring and Autumn Annals* [on the meaning of many Thang and Sung expressions].
Sung, +1180.
Chhêng Ta-Chhang 程大昌.
See des Rotours (1), p. cix.
Yen Phu 硯譜.
Inkstone Record.
Sung, late +10th century.
Shen Shih 沈仕 or Su I-Chien 蘇易簡.
Yen Shih 硯史.
On Inkstones.
Sung, *c.* +1085.
Mi Fu 米芾.
Tr. v. Gulik (2).
Yen Tan Tzu 燕丹子.
(Life of) Prince Tan of Yen (d. −226) [a more extensive version of the biography of Ching Kho (q.v.) in *Shih Chi*, ch. 86, apparently containing some authentic details not therein].
Probably H/Han, +2nd century.
Writer unknown.
Tr. Chêng Lin (1).
Yen Thieh Lun 鹽鐵論.
Discourses on Salt and Iron [record of the debate of −81 on State control of commerce and industry].
C/Han, *c.* −80.
Huan Khuan 桓寬.
Partial tr. Gale (1); Gale, Boodberg & Lin.
Yen Tzu Chhun Chhiu 晏子春秋.
Master Yen's Spring and Autumn Annals.
Chou, prob. −4th century.
Attrib. Yen Ying 晏嬰.
Ying Hai Lun.
See Chang Tzu-Mu (*1*).
Ying Tsao Fa Shih 營造法式.
Treatise on Architectural Methods.
Sung, +1097; printed +1103; reprinted +1145.

Ying Tsao Fa Shih (*cont.*)
Li Chieh 李誡.

Ying Yai Shêng Lan 瀛涯勝覽.
Triumphant Visions of the Boundless Ocean [relative to the voyages of Chêng Ho].
Ming, +1451.
Ma Huan 馬歡.
Re-written (before +1500) by Chang Shêng (*b*) 張昇.

Yo-Yang Fêng Thu Chi 岳陽風土記.
Customs and Notable Things of Yo-yang [in N. Hunan].
Sung.
Fan Chih-Ming 范致明.

Yü Chien 寓簡.
Allegorical Essays.
Sung.
Shen Tso-Chê 沈作喆.

Yü-Chih Lü Lü Chêng I 御製律呂正義.
See *Lü Lü Chêng I.*

Yü-Chih Shu Li Ching Yün 御製數理精蘊.
See *Shu Li Ching Yün.*

Yü Hai 玉海.
Ocean of Jade [encyclopaedia].
Sung, +1267 (first pr. Yuan, +1351).
Wang Ying-Lin 王應麟.

Yü Kung 禹貢.
The Tribute of Yü.
(A chapter of the *Shu Ching*, q.v.)

Yü Kung Chui Chih 禹貢錐指.
A Few Points in the Vast Subject of the *Tribute of Yü* [geographical chapter in the *Shu Ching*].
Chhing, +1697 and +1705.
Hu Wei 胡渭.

Yü Kung Shuo Tuan 禹貢說斷.
Discussions and Conclusions regarding the Geography of the *Tribute of Yü.*
Sung, *c.* +1160.
Fu Yin 傅寅.

Yü Thang Chia Hua 玉堂嘉話.
Noteworthy Talks in the Academy.
Yuan, +1288.
Wang Yün 王惲.

Yü Thu 輿圖.
Terrestrial Map.
Yuan, +1320.
Chu Ssu-Pên 朱思本.
See *Kuang Yü Thu.*

Yü Ti Chi Shêng 輿地紀勝.
The Wonders of the World [geography].
Sung, +1221.
Wang Hsiang-Chih 王象之.

Yü Ti Tsung Thu 輿地總圖.
General World Atlas.
Ming, +1564.
Shih Ho-Chi 史霍冀.

Yü-Ting I Hsiang Khao Chhêng 御定儀象考成; later re-issued as *Chhin-* 欽 *Ting I Hsiang Khao Chhêng.*
See *I Hsiang Khao Chhêng.*

Yü-Ting Li Hsiang Khao Chhêng 御定曆象考成.
See *Li Hsiang Khao Chhêng.*

Yu-Yang Tsa Tsu 酉陽雜俎.
Miscellany of the Yu-yang Mountain (Cave [in S.E. Szechuan].
Thang, +863.
Tuan Chhêng-Shih 段成式.
See des Rotours (1), p. civ.

Yü Yin Wên Ta 玉音問答.
Jade-Sound Questions and Answers.
Sung.
Hu Chhüan 胡銓.

Yuan-Chhun Thang Pi Chi.
See *Yuan-Shun Thang Pi Chi.*

Yuan Chien Lei Han 淵鑑類函.
The Deep Mirror of Classified Knowledge [literary encyclopaedia; a conflation of Thang encyclopaedias].
Chhing, +1710.
Ed. Chang Ying 張英 *et al.*

Yuan Ching Shih Ta Tien 元經世大典.
Institutions of the Yuan Dynasty.
Yuan, +1329 to +1331.
Partly reconstructed and ed. Wên Thing-Shih (1916) 文廷式.
Cf. Hummel (2), p. 855.

Yuan Ching Shuo 遠鏡說.
The Far-Seeing Optick Glass [account of the telescope].
Ming, +1626.
Thang Jo-Wang (J. A. Schall von Bell) 湯若望.

Yuan-Ho Chün Hsien Thu Chih 元和郡縣圖志.
Yuan-Ho reign-period General Geography.
Thang, +814.
Li Chi-Fu 李吉甫.

Yuan Pi Shu Chien Chih 元秘書監志.
Collection of Official Records of the Yuan Dynasty.
Yuan, *c.* +1350.
Wang Shih-Tien 王士點 & Shang Chhi-Ong 商企翁.

Yuan Shih 元史.
History of the Yuan (Mongol) Dynasty [+1206 to +1367].
Ming, *c.* +1370.
Sung Lien 宋濂 *et al.*
Yin-Tê Index, no. 35.

Yuan-Shun Thang Pi Chi 援鶉堂筆記.
Pen Jottings from the Yuan-Shun Hall [notes on the classics].
Chhing, late +18th century, but not printed until 1838.
Yao Fan 姚範.

Yüeh Chhiao Shu 越嶠書.
The Peaks of Annam [geography].
Ming, +1540.
Li Wên-Fêng 李文鳳.

Yüeh Chüeh Shu 越絕書.
Book of the Former State of Yüeh.

Yüeh Chüeh Shu (*cont.*)
H/Han, *c.* +52.
Writer unknown.
Yüeh Ling 月令.
Monthly Ordinances (of the Chou Dynasty).
Chou, between −7th and −3rd centuries.
Writers unknown.
Incorporated in the *Hsiao Tai Li Chi* and the *Lü Shih Chhun Chhiu*, q.v.
Tr. Legge (7), R. Wilhelm (3).
Yün Chhi Yu I 雲溪友議.
Discussions with Friends at Cloudy Pool.
Thang, *c.* +875.
Fan Lü 范攄.
Yün Chi Chhi Chhien 雲笈七籤.
Seven Bamboo Tablets of the Cloudy Satchel [a great Taoist collection].
Sung, +1025.
Chang Chün-Fang 張君房.
TT/1020.
Yün Lin Shih Phu 雲林石譜.
Cloud Forest Lapidary.
Sung, +1133.
Tu Wan 杜綰.
Yün Lu Man Chhao 雲麓漫抄.
Random Jottings at Yün-lu.
Sung, +1206 (referring to events of about +1170 onwards).
Chao Yen-Wei 趙彥衛.
Yün Shih Chai Pi Than 韻石齋筆談.
Jottings from the Sounding-Stone Studio.
Chhing, early +17th century.
Chiang Shao-Shu 姜紹書.
Yung Chhêng Shih Hua 榕城詩話.
Plantain City (Fuchow) Essays [literary criticism].
Chhing, +1732.
Hang Shih-Chün 杭世駿.
Yung-Lo Ta Tien 永樂大典.
Great Encyclopaedia of the Yung-Lo reign-period [only in manuscript].
Amounting to 22,877 chapters in 11,095 volumes, only about 370 being still extant.
Ming, +1407.
Ed. Hsieh Chin 解縉.
See Yuan Thung-Li (*1*).
Yünnan Thung Chih 雲南通志.
Local Topography of Yunnan.
Chhing, +1730.
Ed. O-Erh-Thai (Ortai) 鄂爾泰.
Yüyang Kung Shih Phu 漁陽公石譜.
Treatise on Stones by the Venerable Yüyang.
Sung, end +11th century.
Yüyang Kung 漁陽公.

ADDENDA TO BIBLIOGRAPHY A

Pên Tshao Thu Ching 本草圖經.
The Illustrated Pharmacopoeia.
Sung, *c.* +1070 (presented +1062).
Su Sung 蘇頌.
Now contained only as quotations in the *Thu Ching Yen I Pên Tshao* (*TT*/761) and later pharmacopoeias.
Thu Ching Pên Tshao 圖經本草.
See *Pên Tshao Thu Ching*.
The name belonged originally to a work prepared in the Thang (*c.* +658) which by the 11th century had become lost. Su Sung's *Pên Tshao Thu Ching* was prepared as a replacement. The name *Thu Ching Pên Tshao* has often been applied to Su Sung's work, but wrongly.
Thu Ching Yen I Pên Tshao 圖經衍義本草.
The Illustrated and Elucidated Pharmacopoeia.
Largely a conflation of the *Pên Tshao Yen I* and the *Pên Tshao Thu Ching* but with many additional quotations.
Sung, *c.* +1120.
Khou Tsung-Shih 寇宗奭.
TT/761.

B. CHINESE AND JAPANESE BOOKS AND JOURNAL ARTICLES SINCE +1800

Anon. (*1*).
Hsia Chiang Thu Khao 峽江圖考.
Illustrations of the (Yangtze) River Gorges.
In *Hsia Chiang Chiu Sêng Chhuan Chih*, q.v.
Late 19th century.

Anon. (*2*).
Hsia Chiang Chiu Sêng Chhuan Chih 峽江救生船志.
Record of the (Yangtze) River Gorges Lifeboat Service.
Late 19th century.

Anon. (*5*).
Tsui Ku Lao ti Thien Wên Thai 最古老的天文台
The Oldest Astronomical Observatory.
HCH, 1956 (no. 10), 39.

Anon. (*6*).
Seiyō Kohan Nihon Chizuchiki 西洋古版日本地圖集.
Early Printed Maps and Atlases of Japan made in Western Countries, 1552 to 1840.
Tenri (University) Central Library.
Photo Series, No. 4.
Tenri, Nara, 1954.

Aoyama, Sadao (*1*) 青山定雄.
Kochishi Chizu Nado no Chōsa 古地誌地圖等の調査.
In Search of Old Geographical Works and Maps.
TG/T, 1935, **5** (Suppl. vol.), 123.

Aoyama, Sadao (*2*) 青山定雄.
Gendai no Chizu ni tsuite 元代の地圖について.
On Maps of the Yuan Dynasty.
TG/T, 1938, **8**, 103.

Aoyama, Sadao (*3*) 青山定雄.
Ri-chō ni okeru nisan no Chōsen Zenzu ni tsuite 李朝に於ける二三の朝鮮全圖について.
On a Few General Maps from Korea, of the Yi (Li) Dynasty.
TG/T, 1939, **9**, 143.

Aoyama, Sadao (*4*) 青山定雄.
Sōdai no Chizu to sono Tokushoku 宋代の地圖とその特色.
The Maps of the Sung Dynasty and their Characteristics.
TG/T, 1940, **11**, 415.

Aoyama, Sadao (*5*) 青山定雄.
Nansō Junyū no Sekkoku Chirizu ni tsuite 南宋淳祐の石刻墜理圖について.
Concerning the General Map of China carved on stone in the Shun-Yu reign-period of the Southern Sung Dynasty. (One of four presented by Huang Shang in +1190/+1194.)
TG/T, 1940, **11**, 39.
Abstr. *MS*, 1942, **7**, 366.

Aoyama, Sadao (*8*) 青山定雄.
Rikkyoku-an Shozō no Yochizu ni tsuite 栗棘庵所藏の輿地圖について.
On an Atlas (entitled *Yü Ti Thu*) possessed by the Rikkyoku-an Cloister (one of those made by Huang Shang in +1193).
TYG, 1955, **37**, 471.

Chang Chhi-Yün (*1*) (ed.) 張其昀.
Hsü Hsia-Kho hsien-sêng Shih Shih San Pei Chou Nien Chi Nien Khan 徐霞客先生逝世三百週年紀念刊.
Essays in Commemoration of the 300th Anniversary of the Death of Hsü Hsia-Kho (+1586 to +1641) [11 contributors].
Chekiang Univ. Res. Inst. Hist. and Geogr. Publications No. 4: Tsunyi, 1942.

Chang Chien (*1*) 張鑑.
Hsi-Hsia Chi Shih Pên Mo 西夏紀事本末.
Rise and Fall of the Hsi-Hsia State.
c. 1830.

Chang Hui-Chien (*1*) 張慧劍.
Li Shih-Chen 李時珍.
Biography of Li Shih-Chen (+1518/+1593; the great pharmaceutical naturalist).
Jen Min Pub. House, Shanghai, 1954, repr. 1955.

Chang Hung-Chao (*1*) 章鴻釗.
Shih Ya 石雅.
Lapidarium Sinicum; a Study of the Rocks, Fossils and Minerals as known in Chinese Literature.
Chinese Geol. Survey, Peiping: 1st ed. 1921, 2nd ed. 1927.
MGSC (ser. B), no. 2.
Crit. P. Demiéville, *BEFEO*, 1924, **24**, 276.

Chang Li-Chhien (*1*) 張禮千.
'*Tung Hsi Yang Khao*' *chung chih Chen Lu* 東西洋考中之針路.
The Compass-Bearings in *Studies on the Oceans, East and West*.
TFTC, 1945, **41** (no. 1), 49.

Chang Tun-Jen (*1*) 張敦仁.
'*Chhi Ku Suan Ching*' *Hsi Tshao* 緝古算經細草.
Analysis of the Methods of the *Continuation of Ancient Mathematics*.
1801.

Chang Tun-Jen (*2*) 張敦仁.
Chhiu I Suan Shu 求一算術.
On Indeterminate Analysis [mathematics].
c. 1801.

Chang Tzu-Mu (*1*) 張自牧.
Ying Hai Lun 瀛海論.
Discourse on the Boundless Sea (i.e. Nature). [An attempt to show that many post-Renaissance scientific discoveries had been anticipated in ancient and medieval China.]
c. 1885.

Chang Yin-Lin (*3*) 張蔭麟.
'*Chiu Chang*' *chi Liang Han chih Shu Hsüeh* 九章及兩漢之數學.
The *Nine Chapters on the Mathematical Art*, and Mathematics in the two Han dynasties.
YCHP, 1927, **2**, 301.

Chang Yin-Lin (*4*) 張蔭麟.
Chi Yuan Hou Erh Shih-Chi Chien Wo Kuo ti-I Wei Ta Kho-Hsüeh Chia; *Chang Hêng* 紀元後二世紀間我國第一位大科學家; 張衡.
Chang Hêng; Our first Great Scientist (+2nd century).
TFTC, 1925, **21** (no. 23), 89.

Chang Yü-Chê (*1*) 張鈺哲.
Chung-Kuo Ku Tai Thien-Wên Niao Khan 中國古代天文鳥瞰.
Bird's-eye view of Ancient Chinese Astronomy.
YC, 1946, **16**, 17.

Chang Yung-Li (*1*) 張永立.
Shêng Ching chung chih Yuan Chou Li 聖經中之圓周率.
On the Determinations of π in Ancient Chinese Literature.
CLTC, 1943, **1**, 267.

Chao Jan-Ning (*1*) 趙然凝.
Yen Li Ming Than 衍率冥探.
A View of the Secrets of Indeterminate Analysis.
TFTC, 1944, **41**, 55.

Chao Yuan-Jen (*1*) 趙元任.
Chung Hsi Hsing Ming Khao 中西星名考.
A Study of the Names of Asterisms in China and the West.
KHS, 1917, **3**, 42.

Chêng Hao-Shêng (*1*) 鄭鶴聲.
Chêng Ho 鄭和.
Biography of Chêng Ho [great eunuch admiral of the +15th century].
Victory Pub., Chungking, 1945.

Chiang Shao-Yuan (*1*) 江紹原.
Chung-Kuo Ku Tai Lü-Hsing chih Yen-Chiu 中國古代旅行之研究.
Le Voyage dans la Chine Ancienne.
Fr. transl. by Fan Jen.
Shanghai, 1934.

Chou Chhing-Chu (*1*) 周清澍.
Wo Kuo Ku Tai Wei Ta ti Kho-Hsüeh Chia; Tsu Chhung-Chih 我國古代偉大的科學家;祖冲之.
A Great Chinese Scientist; Tsu Chhung-Chih (mathematician, engineer, etc.).
Essay in Li Kuang-Pi & Chhien Chün-Hua (q.v.), p. 270.
Peking, 1955.

Chou Kuang-Ti (*1*) 周光地.
Hsin Hsing 新星.
Novae.
YC, 1945, **15**, 35.

Chu Fang-Pu (*1*) 朱芳圃.
Chia Ku Hsüeh (*Wên Tzu Pien*) 甲骨學文字編.
Oracle-Bone Script and Characters (Collected Identifications).
Com. Press, Shanghai, 1933.

Chu Kho-Chen (*1*) 竺可楨.
Erh-shih-pa Hsiu Chhi-Yuan chih Ti-Tien yü Shih-Chien 二十八宿起源之地點與時間.
On the Place and Time of Origin of the twenty-eight *Hsiu*.
CHHP, 1944, **18**, 1.

Chu Kho-Chen (*2*) 竺可楨.
Erh-shih-pa Hsiu Chhi-Yuan chih Shih-Tai yü Ti-Tien 二十八宿起源之時代與地點.
On the Date and Place of Origin of the twenty-eight *Hsiu*.
SHST, 1944, no. 34, 1.

Chu Kho-Chen (*3*) 竺可楨.
Lun i Sui-Chha ting '*Shang Shu* (*Yao Tien*)' *Ssu Chung Chung Hsing chih Nien Tai* 論以歲差定尙書堯典四仲中星之年代.
Calculation of the Date of the *Yao Tien* chapter of the *Historical Classic* from the four quadrantal *hsiu* and the precession of the equinoxes.
KHS, 1926, **11** (no. 2), 1637; reprinted in Hsü Ping-Chhang (*1*), p. 333.

Chu Kho-Chen (*4*) 竺可楨.
Pei Sung Shen Kua tui-yü Ti-Hsüeh chih Kung-Hsien yü Chi-Shu 北宋沈括對於地學之貢獻與紀述.
The Contributions to the Earth Sciences made by Shen Kua in the Northern Sung Period.
KHS, 1926, **11**.

Chu Kho-Chen (*5*) 竺可楨.
Chung-Kuo Ku Tai tsai Thien-Wên Hsüeh shang ti Wei Ta Kung-Hsien 中國古代在天文學上的偉大貢獻.
The Most Important Contributions of Ancient Chinese Astronomy.
KHTP, 1951, **2**, 215.
Repr. in *Womên Wei Ta ti Tsu Kuo*, Peking, 1951, p. 62 我們偉大的祖國

Chu Kho-Chen (*6*) 竺可楨.
Wei shen-mo Yao Yen-Chiu Wo Kuo Ku Tai Kho-Hsüeh Shih 爲什麼要研究我國古代科學史.
Why we need Researches on the History of Science in our Country in Ancient and Medieval Times.
JMJP, 1954, 27 Sept.

Chu Shih-Chia (*1*) 朱士嘉.
Sung Yuan Fang-Chih Khao 宋元方志考.
Study of Local Topographies in the Sung and Yuan Periods.
THTC.

Chu Shih-Chia (*2*) 朱士嘉.
Lin-an San Chih Khao 臨安三志考.
A Study of the three (Sung) Topographies of the Hangchow District.
YCHP, 1936, **20**, 421.

Chu Shih-Chia (*3*) 朱士嘉.
Chung-Kuo Ti Fang-Chih Tsung-Lu 中國地方志綜錄.
Union Catalogue of the Local Topographies [preserved in thirty-five Public and fifteen Private Collections].
Com. Press, Shanghai, 1935.

Chu Shih-Chia (*4*) 朱士嘉.
Kuo-Hui Thu-Shu-Kuan Tsang Chung-Kuo Fang-chih Mu-Lu 國會圖書館藏中國方志目錄.
Catalogue of the Local Topographies preserved in the Library of Congress [Washington].
U.S. Govt. Printing Office, Wash. D.C., 1942.

Chu Wên-Hsin (*1*) 朱文鑫.
Li Fa Thung Chih 曆法通志.
History of Chinese Calendrical Science.
Com. Press, Shanghai, 1934.

Chu Wên-Hsin (*2*) 朱文鑫.
Li Tai Jih Shih Khao 歷代日食考.
A Study of the Eclipses recorded in Chinese History.
Com. Press, Shanghai, 1934.

Chu Wên-Hsin (*3*) 朱文鑫.
Thien-Wên Hsüeh Hsiao Shih 天文學小史.
Brief History of Astronomy.
Com. Press, Shanghai, 1935.

Chu Wên-Hsin (*4*) 朱文鑫.
Thien-Wên Khao Ku Lu 天文考古錄.
A Study of the Chinese Contribution to Astronomy.
Com. Press, Shanghai, 1933.

Chu Wên-Hsin (*5*) 朱文鑫.
'*Shih Chi (Thien Kuan)' Shu Hêng Hsing Thu Khao* 史記天官書恆星圖考.
A Study of the Fixed Stars recorded in the *Thien Kuan* chapter of the *Historical Records*.
Com. Press, Shanghai, 1927; repr. 1934.

Chhang Fu-Yuan (*1*) 常福元.
Thien-Wên I Chhi Chih-Lüeh 天文儀器志略.
On (Chinese) Astronomical Instruments.
Peiping, *c.* 1930.

Chhang Fu-Yuan (*2*) 常福元.
Hsü Po-Chêng, Than Yün, Liang Shu chih Pi-Chiao 許伯政譚澐兩書之比較.
On the (eighteenth-century) Mathematicians and Astrologers Hsü Po-Chêng and Than Yün.
FJHC, 1930, **2**, 111.

Chhen Chieh (*1*) 陳杰.
Suan Fa Ta Chhêng 算法大成.
Complete Survey of Mathematics.
1843.

Chhen Kao-Yung *et al.* (*1*) 陳高傭.
Chung-Kuo Li Tai Thien Tsai Jen Huo Piao 中國歷代天災人禍表.
Register of Natural Calamities and Man-made Misfortunes through the Centuries in China (−246 to +1911). 2 vols.
Chinan Univ. Press, Shanghai, 1940; note by A. W. Hummel, *QJCA*, 1949, **6**, 27.

Chhen Phan (*1*) 陳槃.
Ku Chhan-Wei Shu Lu Chieh Thi 古讖緯書錄解題.
Remarks on some Works of the Occult Science of Prognostication in Ancient China (the Chhan-Wei or Weft Classics).
AS/BIHP, 1945, **10**, 371; 1947, **12**, 35.

Chhen Phan (*2*) 陳槃.
Ku Chhan-Wei Shu Lu Chieh Thi 古讖緯書錄解題.
Further Remarks on Some Works of the Occult Science of Prognostication in Ancient China.
AS/BIHP, 1950, **22**, 85.

Chhen Phan (*3*) 陳槃.
Chhan-Wei Shih Ming 讖緯釋名.
The Origin of the Name Chhan-Wei (Weft Classics).
AS/BIHP, 1943, **11**, 297.

Chhen Phan (*4*) 陳槃.
Chhan-Wei Su Yuan 讖緯溯源.
The Origin of the (content of the) Chhan-Wei (Weft Classics) [attempted reconstruction of a text of Tsou Yen].
AS/BIHP, 1943, **11**, 317.
Ref. W. Eberhard, *OR*, 1949, **2**, 193.

Chhen Shu-Phêng (*1*) 陳述彭.
Yünnan Thang-Lang Chhuan Liu Yü chih Ti-Wên 雲南螳螂川流域之地文.
Geomorphology of the Thang-Lang River Valley, Yunnan (Kunming and its lake).
JGSC, 1948, **15** (nos. 2, 3, 4), 1.

Chhen Tsun-Kuei (*1*) 陳遵媯.
Ku Chin Hsing Ming Tui Chao 古今星名對照.
Identifications of Chinese and Western Names of Asterisms.
Acad. Sin., Nanking.

Chhen Tsun-Kuei (*2*) 陳遵嬀.
Chung-Kuo Thien-Wên Hsüeh Shih Chhu Lun 中國天文學史初論.
Introduction to the History of Chinese Astronomy.
YC, 1945, **15**, 9.

Chhen Tsun-Kuei (*3*) 陳遵嬀.
Hêng Hsing Thu Piao 恆星圖表.
Atlas of the Fixed Stars, with identifications of Chinese and Western Names.
Com. Press, Shanghai, 1937, for Academia Sinica.

Chhen Tsun-Kuei (*4*) 陳遵嬀.
Chhien Han Liu Hui Chi Shih 前漢流彗紀事.
Records of Comets in the Han Dynasty.
YC, 1945, **15**, 43.

Chhen Tsun-Kuei (*5*) 陳遵嬀.
Chung-Kuo Ku-Tai Thien-Wên-Hsüeh Chien Shih 中國古代天文學簡史.
Outline History of Ancient (and Medieval) Chinese Astronomy.
Jen-Min, Shanghai, 1955.

Chhen Tsun-Kuei (*6*) 陳遵嬀.
Chhing Chhao Thien-Wên I Chhi Chieh Shuo 清朝天文儀器解說.
The Astronomical Instruments of the Chhing Dynasty (and those surviving from Yuan and Ming also).
Peking Observatory and Chinese Assoc. for the Advancement of Science.
Peking, 1956.

Chhen Wei-Chhi 陳維祺; with Yeh Yao-Yuan 葉耀元, Sun Pin-I 孫䋋翼 *et al.* (*1*).
Chung-Hsi Suan-Hsüeh Ta Chhêng 中西算學大成.
Compendium of Chinese and Western Mathematics [and Physics].
1889.

Chhen Wên-Thao (*1*) 陳文濤.
Hsien Chhin Tzu-Jan Hsüeh Kai Lun 先秦自然學概論.
History of Science in China during the Chou and Chhin periods.
Com. Press, Shanghai, 1934.

Chhen Yin-Kho (*2*) 陳寅恪.
'*Chi Ho Yuan Pên' Man Wên I Pên Pa* 幾何原本滿文譯本跋.
Short Account of the Manchu Version of (Euclid's) *Geometry*.
AS/BIHP, 1931, **2** (no. 3), 281.

Chhêng Hung-Chao (*1*) 程鴻詔.
'*Hsia Hsiao Chêng' Chi Shuo* 夏小正集說.
Collected Commentaries on the *Lesser Annuary of the Hsia Dynasty*.
Before 1865.

Chhien Chün-Hua (*1*) 錢君曄.
Sung Tai Cho-Yüeh ti Kho-Hsüeh Chia; Shen Kua 宋代卓越的科學家；沈括.
A Chekiang Scientist of the Sung Dynasty; Shen Kua (mathematician, astronomer, cartographer, etc.).
Essay in Li Kuang Pi & Chhien Chün-Hua (q.v.), p. 288.
Peking, 1955.

Chhien Pao-Tsung (*1*) 錢寶琮.
Chung-Kuo Suan-Hsüeh Shih 中國算學史.
A History of Chinese Mathematics.
National Research Institute of History and Philol. Monographs, Ser. A, No. 6.
Academia Sinica, Peiping, 1932.
Pt. I (Pt. II never published).

Chhien Pao-Tsung (*2*) 錢寶琮.
Ku Suan Khao Yuan 古算考源.
Über den Ursprung der chinesischen Mathematik.
Com. Press, Shanghai, 1930.

Chhien Pao-Tsung (*3*) 錢寶琮.
'*Ying Pu Tsu' Shu Liu-Chhuan Ou-Chou Khao* 盈不足術流傳歐洲考.
On the Transmission of the Rule of 'Too Much and Not Enough' (algebraic Rule of False Position) from China to Europe.
KHS, 1927, **12** (no. 6), 707.

Chhien Pao-Tsung (*4*) 錢寶琮.
'*Thai I' Khao* 太一考.
Investigation of the Meaning of the Term 'Great Unique' or 'Heavenly Unity' (including the pole-star).
YCHP, 1932, no. 12, 2449; *CIB*, 1937, **1**, 60.

Chhien Pao-Tsung (*5*) 錢寶琮.
Han Jen Yüeh Hsing Yen-Chiu 漢人月行研究.
On the Motions of the Moon as Understood by the Han astronomers.
YCHP, 1935, **17**, 39.

Chhien Wei-Yüeh (*1*) 錢維樾.
Hêng Hsing Thu 恆星圖.
Atlas of Fixed Stars.
1839.

Endō, Toshisada (*1*) 遠藤利貞.
Nihon Sūgaku-shi 日本數學史.
History of Mathematics in Japan.
Iwanami Shoten, Tokyo, 1896; revised and enlarged ed. 1918.

Entsū (*1*) 圓通.
Bukkoku Rekishō-hen 佛國曆象編.
On the Astronomy and Calendrical Science of Buddha's Country (actually India and China).
Kyoto, 1810.

Fan Wên-Thao (*1*) 范文濤.
Chêng Ho Hang Hai Thu Khao 鄭和航海圖考.
Study of the Maps connected with Chêng Ho's Voyages.
Com. Press, Chungking, 1943.

Fêng Chhêng-Chün (*1*) 馮承鈞.
Chung-Kuo Nan-Yang Chiao-Thung Shih 中國南洋交通史.

Fêng-Chhêng Chün (*1*) (*cont.*)
History of the Contacts of China with the South Sea Regions.
Com. Press, Shanghai, 1937.

Fêng Kuei-Fên (*1*) 馮桂芬.
Hsien-Fêng Yuan Nien Chung Hsing Piao 咸豐元年中星表.
Table of Meridian Passages of 100 Stars in Right Ascension and Declination for 1851.
1851.

Fêng Yu-Lan (*1*) 馮友蘭.
Chung-Kuo Chê-Hsüeh Shih, 中國哲學史. 2 vols.
History of Chinese Philosophy.
Com. Press, Chhangsha, 1934; 2nd ed. 1941.
Tr. Bodde (See Fêng Yu-Lan, 1).

Fêng Yün-Phêng & Fêng Yün-Yuan 馮雲鵬, 馮雲鵷.
Chin Shih So 金石索.
Collection of Carvings, Reliefs and Inscriptions.
(This was the first modern publication of Han tomb-shrine reliefs.)
1821.

Fujiwara, M. (*1*) 藤原松三郎.
Shina Sugaku no Kenkyu 支那數學史の研究.
Miscellaneous Notes on the History of Chinese [and Korean] Mathematics.
TMJ, 1939, **46**, 284; 1940, **47**, 35, 309; 1941, **48**, 78.

Fujiwara, M. (*2*) 藤原松三郎.
Wasan no Kenkyu 和算史の研究.
Miscellaneous Notes on the History of Japanese Mathematics.
TMJ, 1939, **46**, 123, 135, 295; 1940, **47**, 49, 322; 1941, **48**, 201; 1943, **49**, 90.

Hashimoto, Masukichi (*1*) 橋本增吉.
'*Shokyō*' *no Kenkyū* 書經の研究.
Researches on the *Historical Classic*.
TYG, 1912, **2**, 283; 1913, **3**, 331; 1914, **4**, 49, 369.

Hashimoto, Masukichi (*2*) 橋本增吉.
Shina Kodai Rekihōshi Kenkyū 支那古代曆法史研究.
Über die astronomische Zeiteinteilung im alten China.
(Tōyō Bunkō Ronsō, no. 29 東洋文庫論叢)
Tokyo, 1943.

Hashimoto, Masukichi (*4*) 橋本增吉.
'*Shokyō (Gyōten)*' *no Shichūsei ni tsuite* 書經堯典の四中星に就いて.
On the Four Stars Culminating at Dusk at the Equinoxes and Solstices, recorded in the *Yao Tien* chapter of the *Historical Classic*.
TYG, 1928, **17** (no. 3), 303.

Ho Chhang-Ling (*1*) 賀長齡.
Huang Chhao Ching Shih Wên Pien 皇朝經世文編.
Collected Essays by Chhing Officials on Social, Political and Economic Problems.
1826, with continuations in 1882, 1888, 1897, etc.

Ho Kuan-Chou & Chêng Tê-Khun (*1*) 何觀洲, 鄭德坤.
'*Shan Hai Ching*' *tsai Kho-Hsüeh shang chih Phi-Phan chi Tso-Chê chih Shih-Tai Khao* 山海經在科學上之批判及作者之時代考.
A Critical Examination of the Scientific Value of the *Classic of Mountains and Rivers*, with an investigation of the Date of its Authorship.
YCHP, 1930, **7**, 1347.

Hosoi, Sō (*1*) 細井淙.
Nihon Kagaku no Tokushitsu (*Sūgaku*) 日本科學之特質(數學).
Special Characteristics of Japanese Science (Mathematics).
In *Tōyō Shisō* collection 東洋思潮, Trends of Oriental Thought, vol. 12.
Iwanami, Tokyo, 1935.

Hou Hsüeh-Yü (*1*) 侯學煜.
Chih Shih Chih Wu 指示植物.
Indicator Plants [geobotanical prospecting].
Acad. Sinica, Peking, 1952.

Hu Hou-Hsüan (*1*) 胡厚宣.
Chhi-Hou Pien Chhien yü Yin Tai Chhi-Hou chih Chien-Thao 氣候變遷與殷代氣候之檢討.
Climatic Changes and an Enquiry into the Climatic Conditions of the Yin (Shang) Period.
BCS, 1944, **4**, 1.

Hu Tao-Ching (*1*) 胡道靜.
'*Mêng Chhi Pi Than*' *Chiao Chêng* 夢溪筆談校證.
Complete Annotated and Collated Edition of the *Dream Pool Essays* (of Shen Kua, +1086).
2 vols.
Shanghai Pub. Co., Shanghai, 1956.

Huang Chieh (*1*) 黃節.
Chêng Ssu-Hsiao Chuan 鄭思肖傳.
Life of Chêng Ssu-Hsiao.
KSHP, 1904, **1** (no. 3), pt. 3, p. 8*a*.

Huang Mu (*1*) 黃模.
'*Hsia Hsiao Chêng*' *Fên-Chien* 夏小正分箋.
Analysis of the *Lesser Annuary of the Hsia Dynasty*.
c. 1802.

Huang Ping-Wei (*1*) 黃秉維.
Ti-Li Hsüeh chih Li-Shih Yen-Pien 地理學之歷史演變.
The Historical Development of Geographical Science (in China).
CLTC, 1943, **1**, 237.

Hung Chen-Hsüan (*1*) 洪震煊.
'Hsia Hsiao Chêng' Su-I 夏小正疏義.
Rectification of the Text of the *Lesser Annuary of the Hsia Dynasty*.
In *HCCC*, chs. 1318–1321. *c.* 1810.

Hsi Tsê-Tsung (*1*) 席澤宗.
Ku Hsin Hsing Hsin Piao 古新星新表.
A New Catalogue of Ancient Novae.
TWHP, 1955, 3 (no. 2), 183.

Hsi Tsê-Tsung (*2*) 席澤宗.
Tshung Chung-Kuo Li-Shih Wên Hsien ti Chi Lu Lai Thao Lun Chhao Hsin-Hsing ti Pao-Fa yü Shê Tien-Yuan ti Kuan-Hsi.
從中國歷史文獻的紀錄來討論超新星的爆發與射電源的關係.
On the Identification of Strong Discrete Sources of Radio-Emission with Novae and Supernovae Recorded in the Chinese Annals.
TWHP, 1954, *2* (no. 2), 177.

Hsiang Ta (*1*) 向達.
Han Thang Chien Hsi-Yü chi Hai Nan Chu Kuo Ku Ti-Li Shu Hsü Lu 漢唐間西域及海南諸國古地理書敍錄.
On the Geographical Books written between the Han and the Thang Dynasties on the Western Regions, and the Foreign Countries of the South Seas.
BNLP, 1930, **4** (no. 6), 23.

Hsiang Tsung-Lu (*1*) 向宗魯.
'Yüeh Ling' Chang Chü Su Chêng Hsü Lu 月令章句疏證敍錄.
Explanations and Rectifications of certain passages in the *Monthly Ordinances*.
Com. Press, Chungking, 1945.

Hsieh Chhing-Kao (*1*) 謝清高.
Hai Lu 海錄.
Ocean Memories (an account of the world, by an illiterate sailor who between 1783 and 1797 visited all the principal trading ports in Europe, Asia, Africa and America).
Canton, 1820.

Hsieh Chia-Jung (*1*) 謝家榮.
Chung-Kuo Yün Shih chih Yen-chiu Fu Piao 中國隕石之研究附表.
Additional List of Chinese Records of Meteorites.
KHS, 1923, **8**, 920.

Hsü Chhun-Fang (*1*) 許蒓舫.
Ku Suan Fa chih Hsin Yen-Chiu 古算法之新研究.
New Researches on Old Chinese Mathematics.
Shanghai, 1935; Supplementary Volume 1945.

Hsü Chhun-Fang (*2*) 許蒓舫.
Chung Suan Chia ti Tai Shu-Hsüeh Yen-Chiu 中算家的代數學研究.
A Study of Chinese Mathematics in all Ages.
Chung-Kuo Chhing-Nien Pub., Peking, 1954.

Hsü Chhun-Fang (*3*) 許蒓舫.
Chung Suan Chia ti Chi-Ho Hsüeh Yen-Chiu 中算家的幾何學研究.
A Study of Geometry in (Ancient and Medieval) Chinese Mathematics.
Chung-Kuo Chhing-Nien Pub., Peking, 1954.

Hsü Chhun-Fang (*4*) 許蒓舫.
Ku Suan Chhü Wei 古算趣味.
Mathematical Recreations in Old China.
Chhing-Nien, Peking, 1956.

Hsü Chung Shu (*2*) 徐中舒.
Yin Jen Fu Hsiang chi Hsiang chih Nan Chhien 殷人服象及象之南遷.
The Domestication of the Elephant by the Shang people, with Notes on its Southward Migration.
AS/BIHP, 1930, **2**, 60.

Hsü Sung (*1*) 徐松.
Hsi Yü Shui Tao Chi 西域水道記.
Account of the River Systems of the Western Regions (Sinkiang).
1823.
Anal. Himly (8).

Iijima, Tadao (*1*) 飯島忠夫.
Shina Kodaishi to Temmongaku 支那古代史と天文學.
Ancient Chinese History and Astronomical Science.
Tokyo, 1925.

Iijima, Tadao (*2*) 飯島忠夫.
Temmon Rekihō to In-Yō Gogyō-Setsu 天文曆法と陰陽五行說.
Astronomy and Calendrical Science in relation to the theories of the Yin and Yang, and the Five Elements.
Koseisha, Tokyo, 1943.

Iijima, Tadao (*3*) 飯島忠夫.
Shina Korekihō yoron 支那古曆法餘論.
Further Notes on Chinese Astronomy in Ancient Times.
TYG, 1922, **12**, 46.

Iijima, Tadao (*4*) 飯島忠夫.
Shina no Jōdai ni okeru Girisha Bunka no Eikyō to Jukyō Keiten no Kansei 支那の上代に於ける希臘文化の影響と儒敎經典の完成.
Greek Influence on the Ancient Civilisation of China and the Compilation of the Confucian Classics (with reference to calendrical science and astronomy).
TYG, 1921, **11**, 1, 183, 354.

Iijima, Tadao (*5*) 飯島忠夫.
Kandai no Rekihō yori mitaru 'Saden' no Gisaku 漢代の曆法より見たる左傳の僞作.
The Problem of the Falsification of the *Tso Chuan* in relation to the Han Calendar System (2 parts); with, A Further Dis-

Iijima, Tadao (*5*) (*cont.*)
cussion on the Date of Composition of the *Tso Chuan*.
TYG, 1912, **2**, 28, 181; 1919, **9**, 155.

Iijima, Tadao (*6*) 飯島忠夫.
'*Shu Ching*', '*Shih Ching*' *chih Thien-Wên Li Fa* 書經詩經之天文曆法.
Astronomy and Calendrical Science in the *Historical Classic* and the *Book of Odes*.
KHS, 1928, **13**, 18.

Iijima, Tadao (*7*) 飯島忠夫.
Shina no Koreki to Rekijitsu Kiji 支那の古曆と曆日記事.
Calendars and Calendrical Expressions in Ancient China.
TYG, 1929, **17** (no. 4), 449; **18** (no. 1), 58.

Imamura, Akitsune (*1*) 今村明恆.
1800-Nen mae no Jishinkei 千八百年前の地震計.
A Seismograph of Eighteen Hundred Years Ago.
In *Shina Bunka Dansō* 支那文化談叢.
Tokyo, 1942.

Ishida Mikinosuke (*1*) 石田幹之助.
'*Tu-Li-Yü-Ssu Ching*' *to sono Itsubun* 都利聿斯經とその佚文.
The *Tu-Li-Yü-Ssu Ching* and its Fragments [astrological manual of the Thang showing Iranian influence].
In Tōyō-shi Ronsō 東洋史論叢.
Commemoration volume for Haneda Tōru, p. 49 羽田亨.
Kyotō, 1950.

Jao Tsung-I (*1*) 饒宗頤.
Chhang-sha Chhu Mo Shih Chan Shen Wu Thu Chüan Khao-Shih 長沙楚墓時占神物圖卷考釋.
A Study of an Astrological (and Calendrical) Diagram [on silk, with pictures of strange beings], from a tomb of the Warring States period (Chhu State) at Chhangsha.
JOSHK, 1954, **1**, 69.

Jen Chao-Lin (*1*) 任兆麟.
'*Hsia Hsiao Chêng*' *Chu* 夏小正註.
Commentary on the *Lesser Annuary of the Hsia Dynasty*.
Before 1820.

Jung Kêng (*2*) 容庚.
Han Chin Wên Lu 漢金文錄.
Han Inscriptions on Bronzes.
Historical Institute Monographs, no. 5.
Acad. Sin. Peiping, 1931.

Kaetsu, Denichirō (*1*) 加悅傳一郎.
Suan Fa Yuan Li Kua Nang 算法圓理括囊.
(Tr. of *Sampō Yenri* (*Enri*) *Katsunō*.)
Comprehensive Discussion of Circle Theory Computations.
1851; preface, 1853.
Tr. 1874.

Kamada, Shigeo (*1*) 鎌田重夫.
Waga Kuni ni okeru Shina Temmongakushi Kenkyū no Kinjō 我が國に於ける支那天文學史研究の近狀.
Die heutige Situation der Studien zur chinesischen Astronomiengeschichte in Japan.
SGZ, 1945, **56**, 96.

Kanda, Kiichiro (*1*) 神田喜一郎.
'*Sangai-Kyō*' *yori mitaru Shina Kodai no Sangaku Sūhai* 山海經ヨリ観タル支那古代ノ山嶽崇拜.
Mountain Worship in Ancient China as seen in the *Classic of Mountains and Rivers*.
SG, 1922, **2**, 332.

Kanda, Shigeru (*1*) 神田茂.
Nihon Temmon Shiryō 日本天文史料.
Japanese Astronomical Records (of Eclipses, Comets, etc.).
Tokyo, 1935.

Kanda, Shigeru (*2*) 神田茂.
Taiyō Kokuten no Tōyō ni okeru Kiroku 太陽黑點の東洋に於ける記錄.
Catalogue of Sun-spots observed throughout the Ages in China and Japan.
TTWT, 1933, **1** (1), 37.

Kanda Shigeru (*3*) 神田茂.
Hoki oyobi sono Ruisho ni tsuite 簠簋及びその類について
The Oldest Japanese Almanacs and Similar Books.
JJHS, 1952 (no. 23), 21.

Kao Lu (*1*) 高魯.
Chung-Kuo Li Shih shang ti Jih Shih 中國歷史上的日蝕.
Eclipses Recorded in Chinese History.
KHSSC, 1941, **10**, 327.

Katō Heizaemon (*1*) 加藤平左衛門.
Wasan no Gyōreshiki Tenkai ni tsuite no Kentō 和算の行列式展開ニ就テの檢討.
On the Expansion of Determinants in Japanese Mathematics.
TMJ, 1939, **45**, 338.

Ko Li-Phu (A. W. Grabau) (*1*) 葛利普.
Chung-Kuo Ti-Chih Shih 中國地質史.
History of Geology in China.
Shanghai, 1924.

Ku Chieh-Kang (*8*) 顧頡剛.
Chhan-Jang Chhuan-Shuo Chhi Yü Mo Chia Khao 禪讓傳說起於墨家考.
A study of the Mohist Origin of the Legendary Abdications (of the ancient emperors). [Contains the view that the *Yao Tien* chapter of the *Shu Ching* was written as late as the Han.]
KSP, vol. 7 (pt. 3), 30–109.

Kuo Mo-Jo (*3*) 郭沫若.
Chia Ku Wên Tzu Yen-Chiu 甲骨文字研究.
Researches on the Characters of the Oracle-

Kuo Mo-Jo (*3*) (*cont.*)
Bones [including astronomical and calendrical data].
2 vols. Peiping, 1931.
German abstract by W. Eberhard, *OAZ*, 1932, **8**, 225.

Kuo Pao-Chün (*1*) 郭寶鈞.
Hsün-hsien Hsin-tshun Ku Tshan Mu chih Chhing Li 濬縣辛村古殘墓之清理.
Preliminary Report on the Excavations at the Ancient Cemetery of Hsin-tshun village, Hsün-hsien, Honan.
TYKK, 1936, **1**, 167.

Lai Chia-Tu (*1*) 賴家度.
'*Thien Kung Khai Wu' chi chhi Chu chê; Sung Ying-Hsing* 天工開物及其著者; 宋應星.
The *Exploitation of the Works of Nature* and its Author; Sung Ying-Hsing.
Essay in Li Kuang-Pi & Chhien Chün-Hua (q.v.), p. 338.
Peking, 1955.

Lao Nai-Hsüan (*1*) 勞乃宣.
Ku Chhou Suan Khao Shih (continued as *Hsü Pien*) 古籌算考釋(續編).
Investigation into the History of the Ancient Computing-Rods.
1886.

Lei Hsüeh-Chhi (*1*) 雷學淇.
Ku Ching Thien Hsiang Khao 古經天象考.
Investigation of Celestial Phenomena as recorded in the Ancient Classics.
1825.

Li Chao-Lo (*1*) 李兆洛.
Li Tai Yü Ti Yen Ko Hsien Yao Thu 歷代輿地沿革險要圖.
Historical Atlas of China.
1838, repr. 1879.

Li Chao-Lo (*2*) 李兆洛.
Hêng Hsing Chhih-Tao Ching-Wei Tu Thu 恆星赤道經緯度圖.
Map of the Fixed Stars according to Equatorial Co-ordinates (Right Ascension and Declination).
1855.

Li Chao-Lo (*3*) 李兆洛.
Huang Chhao I Thung Ti Yü Chhüan Thu 皇朝一統地輿全圖.
Comprehensive Atlas of the Chhing Dynasty.
Peking, 1832.

Li Ching-Chhih (*1*) 李鏡池.
'*Chou I' Kua Ming Khao Shih* 周易卦名考釋.
A Study of the Names of the sixty-four Hexagrams in the *Book of Changes*.
LHP, 1948, **9** (no. 1), 197 and 303.

Li Ching-Chhih (*2*) 李鏡池.
'*Chou I' Shih Tzhu Hsü Khao* 周易筮辭考續.
A Further Study of the Explicative Texts in the *Book of Changes*.
LHP, 1947, **8** (no. 1), 1 and 169.

Li Jui (*1*) 李銳.
Li Shih Suan-Hsüeh I Shu 李氏算學遺書.
Collected Works of Li Jui (mathematician, +1768 to +1817).
Contains seven treatises on as many ancient calendars from early Chou to the middle of the +5th century.
Juan Yuan ed. & pr. 1823.

Li Kuang-Pi 李光璧 & Chhien Chün-Hua 錢君曄 (*1*) (ed.).
Chung-Kuo Kho-Hsüeh Chi-Shu Fa-Ming ho Kho-Hsüeh Chi-Shu Jen Wu Lun Chi 中國科學技術發明和科學技術人物論集.
Essays on Chinese Discoveries and Inventions in Science and Technology, and on the Men who made them.
San Lien Shu Tien, Peking, 1955.

Li Kuang-Pi 李光璧 & Lai Chia-Tu 賴家度 (*1*).
Han Tai ti Wei Ta Kho-Hsüeh Chia; Chang Hêng 漢代的偉大科學家; 張衡.
A Great Scientist of the Han Dynasty; Chang Hêng (astronomer, mathematician, seismologist, etc.).
Essay in Li Kuang-Pi & Chhien Chün-Hua (q.v.), p. 249.
Peking, 1955.

Li Ming-Chhê 李明徹.
Yuan Thien Thu Shuo 圜天圖說.
Illustrated Discussion of the Fields of Heaven.
1819, with Supplement, 1821.

Li Nien (*1*) 李儼.
Chung-Kuo Shu-Hsüeh Ta Kang 中國數學大綱.
Outline of Chinese Mathematics in History (vol. 1 only appeared).
Com. Press, Shanghai, 1931.

Li Nien (*2*) 李儼.
Chung-Kuo Suan-Hsüeh Shih 中國算學史.
A History of Chinese Mathematics.
Com. Press, Shanghai, 1937.

Li Nien (*3*) 李儼.
Chung-Kuo Suan-Hsüeh Hsiao Shih 中國算學小史.
Brief History of Chinese Mathematics.
Com. Press, Shanghai, 1930.

Li Nien (*4*) 李儼.
Chung Suan Shih Lun Tshung 中算史論叢.
Gesammelte Abhandlungen ü. die Geschichte d. chinesischen Mathematik.
3 vols. 1933–5; 4th vol. (in 2 parts), 1947.
Com. Press, Shanghai.

Li Nien (*5*) 李儼.
Tsui Chin Shih Nien Lai Chung Suan Shih Lun Wên Mu-Lu 最近十年來中算史論文目錄.
Bibliography of Papers on the History of Chinese Mathematics during the last ten years.
HITH, 1948, **15**, 16.

Li Nien (*6*) 李儼.
Li Nien so Tshang Chung-Kuo Suan-Hsüeh Shu Mu-Lu (*Hsü*) *Pien* 李儼所藏中國算學書目錄(續)編.
Catalogue of [448] Chinese Mathematical Books in the Library of Li Nien.
KHS, 1920, **5**, 418, 525; 1925, **10** (no. 4); 1926, **11**, 817; 1933, **17**, 1005; 1934, **18**, 1547.

Li Nien (*7*) 李儼.
Tunhuang Shih Shih Li-Chhêng Suan Ching 敦煌石室立成算經.
The MS. Mathematical Table recovered from the Tunhuang caves [S/930 BM].
BNLP, 1935, **9**, 39; *QBCB/C*, 1939 (n.s.), **1**, 386.

Li Nien (*8*) 李儼.
Shang Ku Chung Suan Shih 上古中算史.
Ancient History of Chinese Mathematics.
KHS, 1944, **27**, 16.

Li Nien (*9*) 李儼.
Chu Suan Chih Tu Khao 珠算制度考.
History of the Abacus.
YCHP, 1931, **1** (no. 10), 2123.
Reprinted in Li Nien (*4*), vol. 3, p. 38.

Li Nien (*10*) 李儼.
Chung-Kuo Suan-Hsüeh Lüeh Shuo 中國算學略說.
Aspects of Chinese Mathematics.
KHS, 1934, **18**, 1135.

Li Nien (*11*) 李儼.
Chung Suan Chia chih Chi Shu Lun 中算家之級數論.
On the Treatment of Series by Chinese Mathematicians.
KHS, 1929, **13**, 1139, 1349.
Reprinted in Li Nien (*4*), vol. 3, p. 197.

Li Nien (*12*) 李儼.
Chung-Kuo Suan-Hsüeh Shih Yü Lu 中國算學史餘錄.
Additional Remarks on the History of Mathematics in China.
KHS, 1917, **3**, 238.

Li Nien (*13*) 李儼.
Tui Shu chih Fa-Ming chi chhi Tung Lai 對數之發明及其東來.
The Discovery of Logarithms and their Transmission to China and Japan.
KHS, 1927, **12**, 109; and two later papers in same vol.

Li Nien (*14*) 李儼.
Chung Suan Shih chih Kung Tso 中算史之工作.
The Work of Chinese Historians of (Chinese) Mathematics [with a register of them].
KHS, 1928, **13**, 785.

Li Nien (*15*) 李儼.
San-shih Nien Lai chih Chung-Kuo Suan-Hsüeh Shih 三十年來之中國算學史.
Thirty Years' Progress in Researches into the History of Chinese Mathematics.
KHS, 1947, **29**, 101.

Li Nien (*16*) 李儼.
Chung Suan Chia chih Yuan Chui Chhü Hsien Shuo 中算家之圓錐曲綫說.
History of the Development of Conic Sections in China.
KHS, 1947, **29**, 115.

Li Nien (*17*) 李儼.
Jih Suan Thuo-Yuan Chou Shu 日算橢圓周術.
(Eighteenth and Nineteenth Century Chinese and) Japanese Work on the Perimeter of Ellipses.
KHS, 1949, **31**, 297.

Li Nien (*18*) 李儼.
I-Ssu-Lan Chiao yü Chung-Kuo Li Suan chih Kuan-Hsi 伊斯蘭教與中國曆算之關係.
Islam in relation to Chinese Mathematics and Calendrical Science.
HCLT, 1941, **5** (nos. 3 and 4).

Li Nien (*19*) 李儼.
Erh-shih-pa Nien Lai Chung Suan Shih Lun-Wên Mu-Lu 二十八年來中算史論文目錄.
Bibliography of Papers on Chinese Mathematics during the past 28 years (i.e. since 1912).
QBCB/C, 1940 (n.s.), **2**, 372.
Cf. Li Nien (*5*); Li Nien & Yen Tun-Chieh (*1*).

Li Nien (*20*) 李儼.
Chung-Kuo Ku Tai Shu-Hsüeh Shih-Liao 中國古代數學史料.
Materials for the Study of the History of Ancient Chinese Mathematics.
Chung-Kuo Kho-Hsüeh Thu Shu I Chhi Kung-Ssu (Chinese Scientific Book and Instrument Co.), for Science Society of China, Shanghai, 1954.

Li Nien (*21*) 李儼.
Chung Suan Shih Lun Tshung 中算史論叢 (second series).
Collected Essays on the History of Chinese Mathematics, vol. 1, 1954; vol. 2, 1954; vol. 3, 1955; vol. 4, 1955; vol. 5, 1955.
Sci. Pub. House, Peking.

Li Nien & Yen Tun-Chieh (*1*) 李儼, 嚴敦傑.
Khang-chan i lai Chung Suan Shih Lun Wên Mu-lu 抗戰以來中算史論文目錄.
Bibliography of Papers on the History of Chinese Mathematics during the period of the Second World War.

Li Nien & Yen Tun-Chieh (*1*) (*cont.*)
QBCB/C, 1944, **5** (no. 4), 51.

Li Shan-Pang (*1*) 李善邦.
Ni Shih Ti Chen I Yuan Li chi Shê Chi Chih Tsao Ching Kuo 霓式地震儀原理及設計製造經過.
The Principle and Plan of a (War-Time) Horizontal Seismograph, and how it was constructed.
GM, 1945, no. 3.

Li Shih-Chhüan (*1*) 李石泉.
Yü Tso Thu 玉作圖.
Illustrations of Jade-Working Techniques.
In Bishop, Bushell, Kunz *et al.* (1), q.v. (with transl. by Bushell), New York, 1906.

Li Shih-Hsü & Yü Chêng-Hsieh (*1*) 黎世序, 俞正燮.
Hsü Hsing Shui Chin Chien 續行水金鑑.
Continuation of the *Golden Mirror of the Flowing Waters*.
1832.

Li Ssu-Kuang (*1*) 李四光.
'*Tshang Sang Pien Hua' ti Chieh-Shih* 「滄桑變化」的解釋.
An Elucidation of the Phrase 'Changing from Blue Sea to Mulberry Groves'.
Tzhu-hsing Local Government, Tzhu-hsing. 1942.

Li Thao (*1*) 李濤.
Wei-Ta ti Yao-Hsüeh Chia Li Shih-Chen 偉大的藥學家李時珍.
The Great Pharmacologist Li Shih-Chen (+1518/1593) (biography), 20 pp.
Peking, 1955.

Li Tso-Hsien (*1*) 李佐賢.
Ku Chhüan Hui 古泉滙.
Treatise on (Chinese) Numismatics.
1859.

Lin Chhing (*1*) 麟慶.
Hung Hsüeh Yin-Yuan Thu Chi 鴻雪因緣圖記.
Illustrated Record of Memories of the Events which had to happen in My Life.
1849.
See Hummel (2), p. 507.

Liu Chao-Yang (*1*) 劉朝陽.
Chia Ku Wên chih Jih Erh Kuan Tshê Chi Lu 甲骨文之日珥觀測記錄.
Mention of the Solar Corona on the Oracle-Bone Inscriptions (Shang dynasty).
YC, 1945, **15**, 15.

Liu Chao-Yang (*2*) 劉朝陽.
Oppolzer chi Schlegel yü Kühnert so Thui Suan chih Hsia Tai Jih Shih Oppolzer 及 Schlegel 與 Kühnert 所推算之夏代日食.
On the Calculations of Oppolzer, Schlegel and Kühnert concerning dates of eclipses in the Hsia dynasty.
YC, 1945, **15**, 29.

Liu Chao-Yang (*3*) 劉朝陽.
Thien-Wên Hsüeh Shih Chuan Hao 天文學史專号.
Current Studies in the History of (Chinese) Astronomy [exposition and criticism of the views of T. Iijima].
CST/HLJ, 1929, nos. 94–6, 1–69.
Cf. Eberhard (10).

Liu Chao-Yang (*4*) 劉朝陽.
'*Shih Chi (Thien Kuan)' Chih Khao* 史記天官志考.
Investigations on the Authenticity of the *Thien Kuan* (astronomical) Chapter of the *Historical Records*.
CST/HLJ, 1929, nos. 73–4, 1–60.

Liu Chao-Yang (*5*) 劉朝陽.
Yin Li Yü Lun 殷曆餘論.
Further Studies in the Yin (Shang) dynasty Calendar.
YC, 1946, **16**, 5.

Liu Chao-Yang (*6*) 劉朝陽.
Chou Chhu Li Fa Khao 周初曆法考.
The Calendar of the Early Chou Period.
SSE/M (ser. B), no. 2, 1944.
(rev. W. Eberhard, *OR*, 1949, **2**, 179.)

Liu Chao-Yang (*7*) 劉朝陽.
Wan Yin Chhang Li 晚殷長曆.
Chronology of the late Yin (Shang) Period.
SSE/M (ser. B), no. 3, 1945.
(rev. W. Eberhard, *OR*, 1949, **2**, 183.)

Liu Chao-Yang (*8*) 劉朝陽.
Yin Mo Chou Chhu Jih Yüeh Shih Chhu Khao 殷末周初日月食初考.
On the Eclipse Records of the Late Yin (Shang) and Early Chou Periods.
BCS, 1944, **4**, 85; 1945, **5**, 1.
(rev. A. Rygalov, *HH*, 1949, **2**, 432.)

Liu Chao-Yang (*9*) 劉朝陽.
'*Hsia Shu' Jih Shih Khao* 夏書日食考.
On the Solar Eclipse in the *Hsia Shu* (Section of the *Historical Classic*).
BCS, 1945, **5**, 1.

Liu Chao-Yang (*10*) 劉朝陽.
Tshung Thien-Wên Li Fa Thui Tshê 'Yao Tien' chih Pien Chhêng Nien Tai 從天文曆法推測堯典之編成年代.
The Use of Modern Astronomical Methods in determining the Date of the *Yao Tien* chapter in the *Historical Classic*.
YCHP, 1930 (no. 7).

Liu Fêng-Wu (*1*) 劉風五.
Hui Chiao Thu tui-yü Chung-Kuo Li Fa ti Kung-Hsien 回敎徒對於中國曆法的貢獻.
Contributions of Islam to Chinese Calendrical Science.
CNCK, 1944, **1**, 240.

Liu Fu (*1*) 劉復.
Hsi Han Shih-Tai ti Jih Kuei 西漢時代的日晷.
Sun-Dials of the Western Han Period.
KHCK, 1932, **3**, 573.

Liu Hsien-Chou (*1*) 劉仙洲.
Chung-Kuo Chi-Hsieh Kung-Chhêng Shih-Liao 中國機械工程史料.
Materials for the History of Engineering in China.
CHESJ, 1935, **3**, and **4** (no. 2), 27.
Reprinted Chhinghua Univ. Press, Peiping, 1935. With supplement.
CHER, 1948, **3**, 135.

Liu Ming-Shu (*2*) 劉名恕.
Chêng Ho Hang Hai Shih Chi chih Tsai Than 鄭和航海事蹟之再探.
Further Investigations on the Sea Voyages of Chêng Ho.
BCS, 1943, **3**, 131.
Crit. ref. by A. Rygalov, *HH*, 1949, **2**, 425.

Liu Ping-Hsüan (*1*) 劉冰弦.
Chung-Kuo Tai Shu Ming Chu 'I Ku Yen Tuan' Phing Chieh 中國代數名著益古演段評介.
Discussion and Explanation of the noted Chinese algebraical Work *New Steps in Computation* (Li Yeh, +1259).
TFTC, 1943, **39**, 33.

Liu Than (*1*) 劉坦.
Lun Hsing Sui Chi Nien 論星歲紀年.
On the Jupiter Cycle and the Calendar.
Acad. Sinica History Institute Pub.
Peking, 1955.

Liu To (*1*) (ed.) 劉鐸.
Ku Chin Suan-Hsüeh Tshung Shu 古今算學叢書.
Collection of Ancient and Modern Mathematical Works.
Suan-Hsüeh Publishers, Shanghai (?), 1898.

Liu To (*2*) 劉鐸.
Jo-Shui Chai Ku Chin Suan-Hsüeh Shu Lu 若水齋古今算學書錄.
Jo-Shui Studio Bibliography of Mathematical Books.
Math. Bookshop and Pub. Co., Shanghai, 1898.

Liu Tshao-Nan (*1*) 劉操南.
'*Chou Li "Chiu Shu"' Chieh* 周禮「九數」解.
What were the 'Nine Calculations' in the *Record of Rites of the Chou* (*Dynasty*)?
ISP/WSFK, 1944, no. 19.

Liu Wên-Tien (*2*) 劉文典.
Huai Nan Hung Lieh Chi Chieh 淮南鴻烈集解.
Collected Commentaries on the *Huai Nan Tzu* Book.
Com. Press, Shanghai, 1923, 1926.

Liu Yen (*1*) 六嚴.
Hêng Hsing Chhih-Tao Ching-Wei Tu Thu 恆星赤道經緯度圖.
Map of the Fixed Stars according to Equatorial Co-ordinates (Right Ascension and Declination).
1851.

Lo Fu-I (*1*) 羅福頤.
Chhuan Shih Ku Chhih Thu Lu 傳世古尺圖錄.
Illustrated Record of Known Foot-Measures of Ancient Times.
Privately printed, Manchuria, 1936.

Lo Shih-Lin (*1*) 羅士琳.
Suan-Hsüeh Pi Li Hui Thung 算學比例匯通.
Rules of Proportion and Exchange.
1818.

Lo Thêng-Fêng (*1*) 駱騰鳳.
I Yu Lu 藝游錄.
The Pleasant Game of Mathematic Art.
c. 1820.

Ma Chien (*1*) 馬堅.
Hui Li Kang Yao 回曆綱要.
Brief Survey of the Muslim Calendar.
Chung-Hua, Peking, 1950, 2nd ed. 1955.
Contains details of traditional Muslim calendrical science, and of the mission of Jamāl al-Dīn, and reproduces (with corrections and an extension of 47 years) Haig's *Comparative Tables of Muhammedan and Christian Dates* (London, 1932).

Ma Hêng (*1*) 馬衡.
'*Sui Shu* (*Lü Li Chih*)' *Shih-wu Têng Chhih* 隋書律曆志十五等尺.
The Fifteen different Classes of Measures as given in the Memoir on Acoustics and Calendar (by Li Shun-Fêng) in the *History of the Sui Dynasty*.
Pr. pr. Peiping, 1932, with translation by J. C. Ferguson (see Ma Hêng, 1).

Ma Kuo-Han (*1*) (ed.) 馬國翰.
Yü Han Shan Fang Chi I Shu 玉函山房輯佚書.
The Jade-Box Mountain Studio Collection of (Reconstituted) Lost Books.
1853.

Mao I-Shêng (*1*) 茅以昇.
Chung-Kuo Yuan Chou Li Lüeh Shih 中國圓周率略史.
History of the Determinations of π in China.
KHS, 1917, **3**, 411.

Masuda, Tadao (*1*) 增田忠雄.
Sōdai no Chizu to Minzoku Undō 宋代の地圖と民族運動.
On Sung Maps in relation to the Political Situation of the Times.
SR, 1942, **27**, 65.

Mêng Wên-Thung (*1*) 蒙文通.
Chung-Kuo Ku Tai Pei Fang Chhi-Hou Khao Lüeh 中國古代北方氣候考略.
On the Ancient Climate of North China.
SHTC, 1930, **2**.

Mikami, Yoshio (*1*) 三上義夫.
Chung-Kuo Suan-Hsüeh chih Thê-Sê 中國算學之特色.
Special Characteristics of Chinese Mathematics.

Mikami, Yoshio (*1*) (*cont.*)
Orig. pub. in Japanese, *TYG*, 1926, **15** (no. 4); **16** (no. 1).
Tr. into Chinese by Lin Kho-Thang 林科棠.
Com. Press, Shanghai, 1934.

Mikami, Yoshio (*2*) 三上義夫.
Shina Shisō Kagaku; Sūgaku 支那思想；科學 (數學).
Science in China; Mathematics. Iwanami Koza Lectures 岩波講座.
In *Tōyō Shisō* collection 東洋思潮.
Trends of Oriental Thought, vol. 2.
Tokyo, 1934.

Mikami, Yoshio (*3*) 三上義夫.
Wasan-shi Kenkyū no Seika 和算史研究の成果.
My Studies in Japanese Mathematics.
JJHS, 1951 (no. 19), 33.

Mikami, Yoshio (*4*) 三上義夫.
Loria Hakushi no Shina Sūgaku Ron
Loria 博士の支那數學論.
Dr Loria on Chinese Mathematics.
TYG, 1923, **12** (no. 4), 500.

Mikami, Yoshio (*5*) 三上義夫.
'*Chūjin den'-ron—Awasete van Hée-shi no Shosetsu o Hyōsu* 疇人傳論一併せて *van Hée* 氏の所説を評す.
Criticism of van Hée's Account of (Juan Yuan's) *Biographies of* (*Chinese*) *Mathematicians and Astronomers.*
TYG, 1927, **16** (no. 2), 185; (no. 3), 287.

Mikami, Yoshio (*6*) 三上義夫.
Yenri (*Enri*) *no Hatsumei ni Kansuru Ronshō* 圓理の發明に關する論證.
A Study of the Invention of the 'Circle-Principle'.
SGZ, 1930, **41** (nos. 7–10); *TBGZ*, 1931, nos. 472–5.

Mikami, Yoshio (*7*) 三上義夫.
Seki Kōwa to Bibungaku 關孝和と微分學.
Seki Kōwa and the Differential Calculus.
TBGZ, 1931, no. 480; 1934, no. 510.

Mikami, Yoshio (*8*) 三上義夫.
Seki Kōwa no Gyōseki to Kei-Han no Sanka narabini Shina no Sampō to no Kankei Oyobi Hikaku 關孝和の業績と京坂の算家並に支那の算法との關係及比較.
The Achievements of Seki Kōwa and his Relations with the Mathematicians of Osaka and Kyoto and with the Chinese Mathematicians.
TYG, 1932, **20** (no. 2), 217, 543, **21** (no. 1), 45, (no. 3), 352, (no. 4), 557, **22** (no. 1), 54.

Mikami, Yoshio (*9*) 三上義夫.
Shin-chō Jidai no Katsuenjutsu no Hattatsu ni Kansuru Kōsatsu 清朝時代の割圓術の發達に關する考察.
An Investigation of the Development of Rules for the Measurement of Circle Segments during the Chhing Dynasty.
TYG, 1930, **18**, (no. 3), 301, (no. 4), 439.

Mikami, Yoshio (*10*) 三上義夫.
Kijo-ka Kakkō 歸除歌括考.
A Study of the History of the Division Table in Verse for the Abacus.
SK, 1937, no. 177.

Mikami, Yoshio (*11*) 三上義夫.
Sō Gen Sūgakujō ni okeru Endan oyobi Shakusa no Igi 宋元數學上に於ける演段及び釋鎖の意義.
The Significance of the *Yen-Tuan* and *Shih-So* Methods of solving Equations in the Sung and Yuan Dynasties.
SK, 1937, no. 179.

Mikami, Yoshio (*12*) 三上義夫.
Shina Kodai no Sūgaku 支那古代の數學.
Ancient Chinese Mathematics.
MSKGK, 1932, no. 37.

Mikami, Yoshio (*13*) 三上義夫.
'*Shūhi Sankyō*' *no Temmonsetsu* 周髀算經の天文說.
On the Astronomical Theory of the *Chou Pei Suan-Ching* (*Arithmetical Classic of the Gnomon and the Circular Paths of Heaven*).
TBGZ, 1911 (no. 235), 241.

Mikami, Yoshio (*14*) 三上義夫.
Ancient Mathematics of the Liu-Chhiu Islands.
TNH, 1916, **11** (no. 9).

Mikami, Yoshio (*15*) 三上義夫.
Indo no sūgaku to Shina to no Kankei 印度の數學と支那との關系.
The Relations of Indian and Chinese Mathematics.
TNH, 1917, **12** (no. 6), 15.

Mikami, Yoshio (*17*) 三上義夫.
Nihon Bōenkyō-shi 日本望遠鏡史.
L'Histoire du Téléscope au Japon.
TBGZ, 1936, nos. 534, 535.

Mikami, Yoshio (*18*) 三上義夫.
Wakan Sūgaku-shijō ni okeru Senran oyobi Gunji no Kankei 和漢數學史上に於ける戰亂及び軍事の關係.
Influences of War and Military Affairs on the History of Mathematics in China and Japan.
SK, 1937, no. 182.

Mikami, Yoshio (*21*) 三上義夫.
Wasanshi Kenkyū no Seika
和算史研究の成果.
My Studies on Japanese Mathematics (posthumously published).
JJHS, 1951 (no. 19), 33.

Miu Yüeh (*1*) 繆鉞.
Li Yeh Li Chih Shih I 李冶李治釋疑.
The Uncertainty about the correct Name of Li Yeh [great +13th-century mathematician].
TFTC, 1943, **39**, 41.

Mori, Shikazō (*1*) 森鹿三.
Rikkyoku-an Shōzo Chizu Kaisetsu 栗棘庵所藏地圖解說.
On a (Sung) map (of about +1270) preserved in the Li-Chi Hall (of the Tung Fu Ssu [Tō Fuku Ji] Temple at Kyoto).
TG/K, 1941, **11**, 545.

Naba, Toshisada (*1*) 那波利貞.
Enanshi ni Mietaru Megane ni tsuite 淮南子に見えたる金目に就いて.
On the 'Metal Eye' mentioned in the *Huai Nan Tzu* Book.
SG, 1928, **3**, 606.

Naito, Torajiro (*1*) 內藤虎次郎.
Ti Li Hsüeh Chia Chu Ssu-Pên 地理學家朱思本.
The Geographer Chu Ssu-Pên [+14th century].
GB, 1920, **11**, 1.
Reprinted in *Dokushi Sōroku* 讀史叢錄, Kyoto, 1929, p. 391.
Chinese tr. by Wu Han 吳晗 in *QBCB/C*, 1933, **7**, no. 2, p. 11.

Nakayama, Heijirō (*1*) 中山平次郎.
Koshiki Shina Kyōkan Enkaku 古式支那鏡鑑沿革.
On Ancient Chinese Mirrors. [Includes recognition of the Wu Liang tomb-shrine TLV board scene as one of magical operations.]
KKHTC, 1919, **9**, 85, 145, 189, 280, 323, 381, 465.

Nōda, Chūryō (*1*) 能田忠亮.
'*Shūhi Sankyō' no Kenkyū* 周髀算經の研究.
An Enquiry concerning the *Chou Pei Suan Ching* (Arithmetical Classic of the Gnomon and the Circular Paths of Heaven).
Acad. of Oriental Culture, Kyoto Institute, Monograph Ser. no. 3, Kyoto, 1933.

Nōda, Chūryō (*2*) 能田忠亮.
'*Raiki* (*Getsurei*)' *Temmon-kō* 禮記月令天文考.
An Enquiry concerning the Astronomical Content of the *Yüeh Ling* (Monthly Ordinances) in the *Li Chi* (Record of Rites).
Acad. of Oriental Culture, Kyoto Institute, Monograph Ser. no. 12, Kyoto, 1938.

Nōda, Chūryō (*3*) 能田忠亮.
Temmon Rekihō 天文曆法.
(Ancient Chinese) Astronomy and Calendrical Science.
In *A History of Chinese Science and Economics* 支那科學經濟史, vol. 8 of *Chinese History and Geography* 支那地理歷史大系.
Tokyo, 1942.

Nōda, Chūryō (*4*) 能田忠亮.
Tōyō Temmongaku-shi Ronsō 東洋天文學史論叢.
Discussions on the History of Astronomy in East Asia.
Tokyo, 1944.

Nōda, Chūryō (*5*) 能田忠亮.
Kan-Seki '*Seikyō*'-*kō* 甘石星經考.
A Study of the *Hsing Ching* (the Star Catalogue attributed to) Kan (Tê) and Shih (Shen) [−4th century].
TG/K, 1931, **1**, 1.

Nōda, Chūryō (*6*) 能田忠亮.
'*Gyō-ten' ni mietaru Temmon* 堯典に見えたる天文.
On the Four Culminating Stars of the *Yao Tien* (chapter in the *Historical Classic*).
TG/K, 1937, **8**, 118.

Nōda, Chūryō (*7*) 能田忠亮.
'*Kashōsei' Seishōron* 夏小正星象論.
On the Astrometry in the *Hsia Hsiao Chêng* (*Lesser Annuary of the Hsia Dynasty*).
TG/K, 1941, **12**, 209.

Nōda, Chūryō (*8*) 能田忠亮.
'*Shikyō' no nisshoku ni tsuite* 詩經の日蝕に就て.
The Solar Eclipse recorded in the *Shih Ching* (*Book of Odes*).
TG/K, 1936, **6**, 204.

Nōda, Chūryō (*9*) 能田忠亮.
Rekigaku Shiron 曆學史論.
A History of the Calendar.
Tokyo, 1948.

Ogawa, Takuji (*1*) 小川琢治.
Kinsei Seiyō Kōtsū Izen no Shina Chizu ni tsuite 近世西洋交通以前の支那地圖に就て.
A Historical Sketch of Cartography in China before the modern Intercourse with the Occident.
CZ, 1910, **22**, 407, 512, 599.
Reprinted in Ogawa (*2*).

Ogawa Takuji (*2*) 小川琢治.
Shina Rekishi Chiri Kenkyū 支那歷史地理研究.
Studies in Chinese Historical Geography, 2 vols.
Kyoto, 1928.

Ogura, Kinnosuke & Ōya, Shinichi (*1*) 小倉金之助, 大矢眞一.
Mikami Yoshio Hakushi to sono Gyōseki Mikami Yoshio Sensei Chosaku Rombun Mokuroku 三上義夫博士 (*1875–1950*) とその業績三上義夫先生著作論文目錄.
Life and Work of Dr Yoshio Mikami [historian of mathematics], 1875–1950, with bibliography of his publications.
JJHS, 1951, (No. 18), 1, 9.

Ong Wên-Hao (*1*) 翁文灝.
Chung-Kuo Shan Mo Khao 中國山脈考.
Investigation of the Mountain Ranges in Chinese History (from the *Yü Kung* chapter of the *Historical Classic* onwards).

Ong Wên-Hao (*1*) (*cont.*)
KHS.
Ong Wên-Hao (*2*) 翁文灝.
Kansu Ti Chen Khao (and *Piao*) 甘肅地震考(表).
A Study of Earthquakes in Kansu Province (with a Register of their Recorded Occurrences from −780 to +1909).
KHS, 1922, **6**, 1079, 1197, 1201; 1923, **7**, 105.

Pao Chhi-Shou (*1*) 保其壽.
Pi-Nai Shan Fang Chi 碧奈山房集.
Pi-Nai Mountain Hut Records [on three-dimensional magic squares].
c. 1880.
Phan Chung-Hsiang (*1*) 潘鍾祥.
Tsên Yang Chhü Hsün Chao Chin Shu Kuang Mo 怎樣去尋找金屬礦脈.
How Modern Prospecting for Gold Deposits is carried out.
KHTP, 1951, **2**, 135.

Sang Hsia-Kho (*1*) 桑下客.
Chhi Chhiao Hsin Phu 七巧新譜.
New Tangram Puzzles.
1813, 1815, many editions.
Sekino, T., Yatsui, S., Kuriyama, S., Oba, T., Ogawa, K. & Nomori, T. (*1*) 關野貞; 谷井濟一; 栗山俊一; 小場恆吉; 小川敬吉; 野守健.
Rakuryo-gun Jidai no Iseki 樂浪郡時代の遺蹟.
Archaeological Researches on the Ancient Lo-Lang District (Korea); 1 vol. text, 1 vol. plates.
Sp. Rep. Serv. Antiq. Govt. Gen, Chosen, 1925, no. 4, and 1927, no. 8.
Shen Wên-Hou (*1*) 沈文侯.
Hsing Thu 星圖.
Celestial Atlas (with markings in both Chinese and Western identifications).
Preface by Shih Yen-Han 石延漢.
Fukien Meteorological Survey 福建省氣象局.
Yung-an, 1940.
Shih Chang-Ju (*1*) 石璋如.
Yin-Hsü tsui chin chih Chung-Yao Fa-Hsien Fu Lun Hsiao Thun Ti-Tshêng 殷墟最近之重要發現附論小屯地層.
The Most Important Recent Discoveries at Yin-Hsü (Anyang), with a Note on the Stratification of the Site.
AS/CJA, 1947 (no. 2).
Shih Chang-Ju (*2*) 石璋如.
Honan Anyang Hou-Kang ti Yin Mo 河南安陽後岡的殷墓.
Burials of the Yin (Shang) Dynasty at Hou-Kang, Anyang.
AS/BIHP, 1948, **13**, 21.
Shih Chang-Ju (*3*) 石璋如.
Chhuan-Shuo chung Chou Tu ti Shih-Ti Khao-Chha 傳說中周都的實地考察.
A Field Investigation of the Traditional Sites of Settlements of the Chou People.
AS/BIHP, 1949, **20B**, 91.
Shinjō, Shinzō (*1*) 新城, 新藏.
Chung-Kuo Shang Ku Thien-Wên 中國上古天文.
Chinesische Astronomie (Antike).
Chinese translation by Shen Hsüan 沈璿.
Com. Press, Shanghai, 1936.
Shinjō, Shinzō (*2*) 新城, 新藏.
Kanshi Gogyō-setsu to Senkyoku Reki 干支五行說と顓頊曆.
The (ancient) Chuan-Hsü Calendar and the Systems of the Cyclical Characters and the Five Elements.
SG, 1922, **2**, 387, 495.
Shinjō, Shinzō (*3*) 新城, 新藏.
Tung Yang Thien-Wên Hsüeh Shih Yen-Chiu 東洋天文學史研究.
Researches on the History of Astronomy in East Asia.
Tokyo, 1929.
Chinese translation by Chhen Tsun-Kuei 陳遵嬀.
Nat. Sci. Soc., Shanghai, 1933.
Shinjō, Shinzō (*4*) 新城, 新藏.
Shina Jōdai no Rekihō 支那上代の曆法.
Calendrical Science in Ancient China.
G, 1913, **4**, 666, 743; (no. 7 onwards), 16, 183.
Shinjō, Shinzō (*5*) 新城, 新藏.
Futatabi 'Saden', 'Kokugo' no Seisaku Nendai o ronzu 再び左傳國語の製作年代を論ず.
Further Remarks on the Date of Compilation of the *Tso Chuan* and the *Kuo Yü* [astronomical and calendrical evidence].
G, 1920, **11**, 619.
Shinjō, Shinzō (*6*) 新城, 新藏.
Kandai ni mietaru Shoshu no Rekihō o ronzu 漢代に見えたる諸種の曆法を論ず.
On the various forms of Calendar System found in the Han Period.
G, 1920, **11**, 701, 785, 966.
Sun Hai-Po (*1*) 孫海波.
Chia Ku Wên Pien 甲骨文編.
On the Oracle-Bone Writing.
Peiping, 1934.
Sun Wên-Chhing (*1*) 孫文青.
'Chiu Chang Suan Shu' Phien Mu Khao 九章算術篇目考.
A Study of the Chapter Headings in the *Nine Chapters on the Mathematical Art.*
CLHP, 1932, **2**, 321.
Sun Wên-Chhing (*2*) 孫文青.
Chang Hêng Chu Shu Nien Piao 張衡著述年表.
Chronological List of the Writings of Chang Hêng [eminent mathematician and astronomer of the +2nd century].
CLHP, 1932, **2**, 105.

Sun Wên-Chhing (*3*) 孫文青.
Chang Hêng Nien Phu 張衡年譜.
Chronological Biography of Chang Hêng.
CLHP, 1933, **3**, 331.

Sun Wên-Chhing (*4*) 孫文青.
Chang Hêng Nien Phu 張衡年譜.
Life of Chang Hêng [enlargement of (*3*)].
Com. Press, Shanghai, 1935; 2nd ed. Chungking, 1944.

Sung Ching-Chhang (*1*) 宋景昌.
Shu Shu Chiu Chang Cha Chi 數書九章札記.
Notes on the *Mathematical Treatise in Nine Sections* (of Chhin Chiu-Shao, +1247).
1842.

Takeda, Kusuo (*1*) 武田楠雄.
Mindai ni okeru Sansho Keishiki no Hensen 明代における算書形の變遷.
Change in the Form of Chinese Mathematical Books in the Ming Dynasty.
JJHS, 1953 (no. 26), 13.

Takeda, Kusuo (*2*) 武田楠雄.
Mindai Sūgaku no Toku-shitsu; 'Sampō Tōsō' Seiritsu no Katei 明代數學の特質; 算法統宗成立の過程.
The Character of Chinese Mathematics in the Ming Dynasty; the Process of Formulation in the *Suan Fa Thung Tsung*.
JJHS, 1954 (no. 28), 1, (no. 29), 8, (no. 34), 12.

Tasaka, Kōdō (*1*) 田坂興道.
Tōzen Seru Isuramu Bunka no Issokumen ni tsuite 東漸せるイスラム文化の一側面に就いて.
About an Aspect of Islamic Culture moving Eastwards [on the astronomical instruments of Jamāl al-Dīn and their identification].
SGZ, 1942, **53**, 401.

Tasaka, Kōdō (*2*) 田坂興道.
Seiyō Rekihō no Tōzen to fuifui Rekihō no Ummei 西洋曆法の東漸と回回曆法の運命.
The Introduction of European Astronomy to China, and the Muslim Calendar-makers.
TYG, 1947, **31** (no. 2), 141.

Têng Yen-Lin 鄧衍林 & Li Nien 李儼 (*1*)
Peiping Ko Thu-Shu-Kuan so Tshang Chung-Kuo Suan-Hsüeh Shu Lien-Ho Mu-Lu 北平各圖書館所藏中國算學書聯合目錄.
Union Catalogue of Chinese Mathematical Books in the libraries of Peiping.
Library Association of China, Peiping, 1936.
Ref. A. W. Hummel, *ARLC/DO*, 1937, 179.

Thai-Hsü (abbot) 太虛 & Chhen-Khung (monk) 塵空.
Chin-Yün Shan Chih 縉雲山志.
Record of the Cloud-Girdled Mountain (Temple) [in Szechuan north of Chungking].
Han-Tsang College of Buddhism, Chungking, 1942.

Than Chhi-Hsiang *et al.* (*1*) 譚其驤.
Kuo-Li Peiping Thu-Shu-Kuan Fang-chih Mu-Lu 國立北平圖書館方志目錄.
Catalogue of the Local Topographies in the National Library at Peiping.
Peiping, 1934 (Suppl. 1936).

Thang Jung-Tso (*1*) 唐榮作.
Yü Shuo 玉說.
Discourse on Jade.
In Bishop, Bushell, Kunz *et al.* (1), *q.v.* (with transl. by Bushell), New York, 1906.

Thung Shih-Hêng (*1*) 童世亨.
Li Tai Chiang Yü Hsing Shih I Lan Thu 歷代疆域形勢一覽圖.
Historical Atlas of China.
Com. Press, Shanghai, 1922.

Ting Chhien (*1*) 丁謙.
'*Yuan Ching Shih Ta Tien' Thu Ti-Li Khao Chêng* 元經世大典圖地理考證.
Investigation of the Maps and Geographical Information contained in the *Institutions of the Yuan Dynasty*.
In *Phêng-Lai Hsüan Ti-Li Hsüeh Tshung-Shu* 蓬淶軒地理學叢書.

Ting Chhü-Chung (ed.) (*1*) 丁取忠.
Pai Fu Thang Suan-Hsüeh Tshung-Shu 白芙堂算學叢書.
White Hibiscus Hall Collection of Ancient Mathematical Works.
1875.
See van Hée (5).

Ting Chhü-Chung (*2*) 丁取忠.
Ssu Hsiang Chia Ling Hsi Tshao 四象假令細草.
Explanation of the Detailed Steps in Sample Calculations (according to the method of) the Four Elements.
c. 1870.

Ting Fu-Pao 丁福保 & Chou Yün-Chhing 周雲青 (*1*).
Ssu Pu Tsung Lu Thien Wên Pien 四部總錄天文編.
Bibliography of Astronomical Books to supplement the *Ssu Khu Chhüan Shu* encyclopaedia.
Com. Press, Shanghai, 1956.

Ting Wên-Chiang (*1*) 丁文江.
Biography of Sung Ying-Hsing 宋應星 (author of the *Thien Kung Khai Wu*, Exploitation of the Works of Nature).
In the *Hsi Yung Hsüan Tshung-Shu* ed. Thao Hsiang 喜詠軒叢書, 陶湘.
1929.

Tuan Fang 端方 and collaborators.
Thao Chai Tshang Shih Chi 陶齋藏石記.
Description of Inscribed Stones in the Studio (Museum) of the Thao Family.
Peking, 1909.

Tung Thung-Ho (*1*) 董同龢.
'*Chhieh Yün Chih Chang Thu' Chung Chi-Ko Wên-thi* 切韻指掌圖中幾個問題.

Tung Thung-Ho (*1*) (*cont.*)
The Problem of the Authorship of the *Tabular Key to the Dictionary of the Sounds of Characters*.
AS/BIHP, 1946, **17**, 193.

Tung Tso-Pin (*1*) 董作賓.
Yin Li Phu 殷曆譜.
On the Calendar of the Yin (Shang) Period.
Academia Sinica, Lichuang, 1945.
(Prelim. notes *AS/BIHP*, 1936, **7**, 45; *SSE*, 1941, **2**, 1.)
See also Tung Tso-Pin (*6*).
Crit. K. Yabuuchi (*13*).

Tung Tso-Pin (*2*) 董作賓.
Tunhuang Hsieh-Pên Thang Ta-Shun Yuan Nien Tshan Li Khao 敦煌寫本唐大順元年殘曆考.
A Study of the Tunhuang MS. Calendrical Fragment of the Ta-Shun reign-period of the Thang Dynasty (+890).
QBCB/C, 1942, **3** (no. 3), 7.

Tung Tso-Pin (*3*) 董作賓.
Yin Tai Yüeh Shih Khao 殷代月食考.
On the Lunar Eclipses of the Yin (Shang) Period.
AS/BIHP, 1950, **22**, 139.

Tung Tso-Pin (*4*) 董作賓.
Yin Tai chih Li-Fa Nung-Yeh yü Chhi-Hsiang 殷代之曆法農業與氣象.
The Calendar in relation to Agriculture and Climate in the Yin (Shang) Period.
SSE, 1946, **5**, 1.
Germ. résumé, W. Eberhard, *OR*, 1949, **2**, 185.

Tung Tso-Pin (*5*) 董作賓.
Yin Wên Wu Ting Shih Pu Tzhu chung I Hsün Chien chih Chhi-Hsiang Chi-Lu 殷文武丁時卜辭中一旬間之氣象記錄.
Meteorological Records of Ten-Day Periods contained in Oracle-Bone Inscriptions from the time of the Yin (Shang) King, Wên Wu Ting.
CHHP, 1943, **17**, no. 17.

Tung Tso-Pin (*6*) 董作賓.
Yin Li Phu Hou Chi 殷曆譜後記.
Supplementary Notes to the *Yin Li Phu* (*1*).
AS/BIHP, 1948, **13**, 183.

Tung Tso-Pin (*7*) 董作賓.
Chou Chin Wên chung 'Sêng-Pa', 'Ssu-Pa' Khao 周金文中生霸死霸考.
On the Terms 'Sêng-Pa' and 'Ssu-Pa' (Phases of the Moon) in Bronze Inscriptions of the Chou Period.
Fu Ssu-Nien Commemoration Volume, p. 139 傅故校長斯年先生紀念文集.
Nat. Taiwan University, Taipei, Formosa, 1952.

Tung Tso-Pin (*8*) 董作賓.
Chi San Pai yu Liu Hsün yu Liu Jih' Hsin Khao 稘三百有六旬有六日新考.
A New Interpretation of the phrase 'the year has 366 days' in the *Yao Tien* (chapter of the *Historical Classic*).
SSE, 1940 (repr. 1949), **1**, 24.

Tung Tso-Pin, Lin Tun-Chen & Kao Phing-Tzu (*1*) 董作賓, 劉敦楨, 高平子.
Chou Kung Tshê Ching Thai Thiao Chha Pao Kao 周公測景臺調查報告.
Report of an Investigation of the Tower of Chou Kung for the Measurement of the Sun's (Solstitial) Shadow.
Com. Press (for Academia Sinica), Chhangsha, 1939.

Ueta, J. (*1*) 上田穰.
Seki-shi 'Seikyō' no Kenkyū 石氏星經の研究.
Investigations on the *Star Manual* of Shih (Shen).
Tōyō Bunkō Ronsō, no. 12 東洋文庫論叢.
Tokyo, 1929.

Wada, Y. (*1*) 和田雄治.
Seisō Ei-so Ryōchō no Soku-u-ki 世宗英祖兩朝之測雨器.
Korean Rain-gauges from the Reigns of Sejong (+1419 to +1450) and Yŏngjo (+1725 to +1776).
JMJ, 1911, no. 3.
Chinese résumé *KHS*, 1916, **2**, 582.

Wang Chen-To (*1*) 王振鐸.
Han Chang Hêng Hou-Fêng Ti-Tung I Tsao-Fa chih Thui-Tshê 漢張衡候風地動儀造法之推測.
A Conjecture as to the Construction of the Seismograph of Chang Hêng in the Han Dynasty.
YCHP, 1936, **20**, 577.

Wang Chen-To (*6*) 王振鐸.
Ti-Tung I 地動儀.
(Chang Hêng's) Seismograph.
JMHP, April 1952.

Wang Chih-Chhun (*1*) 王之春.
Kuo Chhao Jou Yuan Chi 國朝柔遠記.
Record of the Pacification of a Far Country [his ambassadorship to Russia].
c. 1890.

Wang Ching-Hsi (*1*) 汪敬熙.
Kho-Hsüeh Fang-Fa Man Than 科學方法漫談.
Random Discussions on Scientific Method.
Com. Press, Chungking, 1940, repr. 1944.

Wang Hsien-Chhien (*1*) (ed.) 王先謙.
Huang Chhing Ching Chieh Hsü Pien 皇清經解續編.
Continuation of the *Collection of Monographs on Classical Subjects written during the Chhing Dynasty*.
1888.
See Yen Chieh (*1*).

Wang I-Chung (*1*) 王以中.
'*Shan Hai Ching' Thu yü Wai-Kuo Thu* 山海經圖與外國圖.
The Pictures of the *Classic of Mountains and Rivers* in relation to analogous Pictures in the Literatures of other Countries.
STTC, 1937, **1**, 23.

Wang Jen-Chün (*1*) 王仁俊.
Ko Chih Ku Wei 格致古微.
Scientific Traces in Olden Times.
1896.

Wang Kuo-Wei (*2*) 王國維.
Yin Pu Tzhu so chien Hsien Kung Hsien Wang Khao. 殷卜辭所見先公先王考.
Information about the ancient kings and rulers on the Oracle-Bones.
In *Kuan Thang Chi Lin* 觀堂集林.
In *Hai-ning Wang Ching-An hsien-sêng I Shu* 海寧王靜安先生遺書.
48 vols.
Com. Press, Chhangsha, 1940.

Wang Kuo-Wei (*3*) 王國維.
'*Sêng Pa, Ssu Pa' Khao* 生霸死霸考.
Investigation of the terms 'Sêng Pa' and 'Ssu Pa' (phases of the moon), and the Seven-Day Week System (in the Chou Dynasty).
In *Kuan Thang Lin Chi* 觀堂集林.

Wang Yung (*1*) 王庸.
Chung-Kuo Ti-Li Hsüeh Shih 中國地理學史.
History of Geography in China.
Com. Press, Chhangsha, 1938.

Wang Yung 王庸 & Mao Nai-Wên (*1*), 茅乃文.
Kuo-Li Peiping Thu-Shu-Kuan Chung-Wên Yü Thu Mu-Lu 國立北平圖書舘中文輿圖目錄.
Catalogue of the Chinese Maps in the National Library at Peiping.
Peiping, 1933.

Wei Chü-Hsien (*2*) 衛聚賢.
Shu Mu Tzu 數目字.
Finger-Reckoning in Ancient China.
SWYK, 1943, **3**, 93.

Wei Chü-Hsien (*3*) 衛聚賢.
Ku Shih Yen-Chiu 古史研究.
Studies in Ancient History [researches and discussions on the interpenetration of Indian and Chinese civilisations before the Chhin Dynasty].
Shanghai, 1934.

Wên I-To (*2*) 聞一多.
'*Thien Wên' Shih Thien* 天問釋天.
On the Cosmology of the *Thien Wên* (*Questions about the Heavens*, by Chhü Yuan, *c.* −300).
In *Wên I-To Chhüan Chi* 聞一多全集.
Khaiming, Shanghai, 1948, vol. 2, p. 313.
Repr. from *CHJ*, 1936, **9** (no. 4).

Wu Chhêng-Lo (*2*) 吳承洛.
Chung-Kuo Tu Liang Hêng Shih 中國度量衡史.
History of Chinese Metrology [weights and measures].
Com. Press, Shanghai, 1937.

Wu Chhi-Chhang (*1*) 吳其昌.
Han i-chhien Hêng Hsing Fa-Hsien Tzhu Ti Khao 漢以前恆星發現次第考.
A Study of the Identification of the Fixed Stars known before the Han.
CLTC, 1943, **1** (no. 3), 273.

Wu Chhi-Chhang (*2*) 吳其昌.
Sung Tai chih Ti-Li Hsüeh Shih 宋代之地理學史.
History of Geography in the Sung Period.
KHLT, 1927, **1**, 37.

Wu Ta-Chhêng (*1*) 吳大澂.
Chhüan Hêng Tu Liang Shih Yen Khao 權衡度量實驗考.
An Investigation of (Ancient Chinese) Weights and Measures.
1894; repr. 1915.

Wu Ta-Chhêng (*2*) 吳大澂.
Ku Yü Thu Khao 古玉圖考.
Investigation of Ancient Jade Objects.
1889; repr. 1919.

Yabuuchi, Kiyoshi (*1*) 藪內清.
Zui Tō Rekihōshi no Kenkyū 隋唐曆法史の研究.
Researches on the Calendrical Science of the Sui and Thang Periods.
Tokyo, 1944.

Yabuuchi Kiyoshi (*2*) 藪內清.
Chūgoku no Temmongaku 中國の天文學.
(Introduction to the History of) Chinese Astronomy.
Koseisha, Tokyo, 1949.
Crit. Yang Lien-Shêng, *HJAS*, 1951, **14**, 671.

Yabuuchi, Kiyoshi (*3*) 藪內清.
Sūgaku 數學.
History of Chinese Mathematics (to the Thang).
In *A History of Chinese Science and Economics* 支那科學經濟史.
Vol. 8 of *Chinese History and Geography* 支那地理歷史大系.
Tokyo, 1942.

Yabuuchi Kiyoshi (*4*) 藪內清.
Chūgoku no Tokei 中國の時計.
Ancient Chinese Time-Keepers.
JJHS, 1951, (no. 19), 19.

Yabuuchi, Kiyoshi (*5*) 藪內清.
Chūgoku ni okeru Isuramu Temmongaku 中國に於けるイスラム天文學.
The Introduction of Islamic Astronomy to China (in the Yuan Dynasty).
TG/K, 1950, **19**, 65.

Yabuuchi, Kiyoshi (*6*) 藪內清.
Tō 'Kaigen Senkyō' chū no Seikyō 唐開元占經中の星經.

Yabuuchi, Kiyoshi (*6*) (*cont.*)
The Star Catalogue in the *Khai-Yuan Chan Ching* (*Treatise on Astrology and Astronomy*) of the Thang.
TG/K, 1937, **8**, 56.

Yabuuchi, Kiyoshi (*7*) 藪內清.
Ryō Kan Rekihō-Kō 兩漢曆法考.
On the Calendar Reforms of the Former and Later Han Dynasties (−104 and +85).
TG/K, 1940, **11**, 327.

Yabuuchi, Kiyoshi (*8*) 藪內清.
In Shū yori Zui ni itaru Shina Rekihō-shi 殷周より隋に至る支那曆法史.
A History of Chinese (Astronomy and) Calendrical Science from the Yin (Shang) and Chou Periods to the Sui Dynasty.
TG/K, 1941, **12**, 99.

Yabuuchi, Kiyoshi (*9*) 藪內清.
Tō-dai Rekihō ni okeru Fu Jitten Getsuri Jutsu 唐代曆法に於ける步日躔月離術.
The Calendar in Thang Times and the making of Ephemerides for Sun and Moon Motions.
TG/K, 1943, **13**, 189.

Yabuuchi, Kiyoshi (*10*) 藪內清.
Sōdai no Seishuku 宋代の星宿.
Descriptions of the Constellations in the Sung Dynasty.
TG/K, 1936, **7**, 42.

Yabuuchi, Kiyoshi (*11*) (ed.) 藪內清.
'*Tenkō Kaibutsu*' *no Kenkyū* 天工開物の研究.
A Study of the *Thien Kung Khai Wu* (The Exploitation of the Works of Nature, +1637).
Tokyo, 1953.
Chinese tr. of the eleven critical essays, by Su Hsiang-Yü *et al.*, Chung-Hua Tshung-Shu Wei Yuan Hui, Taipei, 1956.

Yabuuchi, Kiyoshi (*12*) 藪內清.
Shina Sūgakushi Gaisetsu 支那數學史概說.
Outline of Chinese Mathematics.
Kyoto, 1944.

Yabuuchi, Kiyoshi (*13*) 藪內清.
Indai no Rekihō 殷代の曆法.
The Calendar of the Shang (Yin) Period (an essay review of Tung Tso-Pin's *Yin Li Phu*).
TG/K, **21**, 217.

Yabuuchi, Kiyoshi (*14*) 藪內清.
Tō Sō Rekihō-shi 唐宋曆法史.
Calendrical Science in the Thang and Sung Dynasties.
TG/K, 1943, **13**, 491.

Yabuuchi, Kiyoshi (*15*) 藪內清.
Gen Min Rekihō-shi 元明曆法史.
Calendrical Science in the Yuan and Ming Dynasties.
TG/K, 1944, **14**, 264.

Yabuuchi, Kiyoshi (*16*) 藪內清.
Asuka & Nara Jidai no Shizengaku 飛鳥奈良時代の自然科學飛鳥奈良時代の文化.
The Natural Sciences in the Asuka Period (+593/+628) and the Nara Period (+710/+784).
Art. in *Asuka & Nara Jidai no Bunko*, p. 101.
Tokyo, 1950 (?).

Yabuuchi, Kiyoshi (*17*) 藪內清.
Kinsei Chugoku ni tsuta-erareta Seiyō Temmongaku 近世中國に傳之られた西洋天文學.
The Introduction of Western Astronomy into (Seventeenth-century) China.
JJHS, 1955, (no. 32), 15.

Yabuuchi, Kiyoshi 藪內清
& Nōda, Chūryo (*1*) 能田忠亮.
'*Kansho (Ritsurekishi)*' *no Kenkyū* 漢書律曆志の研究.
Researches on the Calendrical chapters of the *History of the Former Han Dynasty*.
Kyoto, 1947.

Yang Shou-Ching (*1*) 楊守敬.
Li Tai Yü Ti Thu 歷代輿地圖.
Historical Atlas of China.
1911.

Yao Chen-Tsung (*1*) 姚振宗.
'*Hou Han Shu*' *I Wên Chih* 後漢書藝文志.
The Bibliography in the *History of the Later Han Dynasty*.
1895.

Yao Ming-Hui (*1*) 姚明輝.
Chung-Kuo Fa-Ming Ti Yuan Shuo 中國發明地圓說.
On the Chinese discovery of the Roundness of the Earth.
TLHTC, 1915, **6** (no. 5).

Yao Pao-Hsien (*1*) 姚寶獻.
Chung-Kuo Li Shih shang Chhi-Hou Pien Chhien chih i Hsin Yen-Chiu 中國歷史上氣候變遷之一新研究.
A New Study of Climatic Changes in Chinese History.
CST/HLJ, 1935, **1**, 1.
English résumé, *CIB*, 1937, **1**, 39.

Yeh Tê-Lu (*1*) 葉德祿.
Chhi Yao Li Ju Chung-Kuo Piao 七曜曆入中國表.
On the coming to China of the 'Seven Luminary' Calendars [Persian and Manichaean elements in Chinese calendrical science during the Thang period].
FJHC, 1942, **11**, 137.

Yeh Thang (*1*) 葉棠.
Hêng Hsing Chhih-Tao Chhüan Thu 恆星赤道全圖.
Complete Map of the Fixed Stars, based on Equatorial Co-ordinates.
1847.

Yeh Yao-Yuan (*1*) 葉耀元.
Chung Hsüeh 重學 (chapters).
Mechanics and Dynamics.
In *Chung-Hsi Suan-Hsüeh Ta Chhêng* 中西算學大成.
Complete Textbook of Chinese and Western Mathematics (and Physics).
Thung-Wên, Shanghai, 1889.

Yen Chieh (*1*) (ed.) 嚴杰.
Huang Chhing Ching Chieh 皇清經解.
Collection of (more than 180) Monographs on Classical Subjects written during the Chhing Dynasty.
1829; 2nd ed. Kêng Shen Pu Khan, 1860.
Cf. Wang Hsien Chien (*1*).

Yen Kho-Chün (*1*) (ed.) 嚴可均.
Chhüan Shang-Ku San-Tai Chhin Han San-Kuo Liu Chhao Wên 全上古三代秦漢三國六朝文.
Complete Collection of Prose Literature (including Fragments) from Remote Antiquity through the Chhin and Han Dynasties, the Three Kingdoms and the Six Dynasties.
Finished 1836; published 1887–93.

Yen Tun-Chieh (*1*) 嚴敦傑.
'*Suan Hsüeh Chhi Mêng' Liu Chhuan Khao* 算學啓蒙流傳考.
History of the Fortunes of the *Introduction to Mathematical Studies* (by Chu Shih-Chieh, +1299).
TFTC, 1945, **41**, 31.

Yen Tun-Chieh (*2*) 嚴敦傑.
Chü-Yen Han Chien Suan Shu 居延漢簡算書.
The Mathematical Writings on the Han bamboo tablets discovered at Chü-yen (Edsin Gol).
CLTC, 1943, **1**, 315.

Yen Tun-Chieh (*3*) 嚴敦傑.
Chhou Suan Suan-Phan Lun 籌算算盤論.
A Discussion on the History of the Counting-Rods and the Abacus.
TFTC, 1944, **41**, 33.

Yen Tun-Chieh (*4*) 嚴敦傑.
Shanghai Suan-Hsüeh Wên Hsien Shu Lüeh 上海算學文獻述略.
(Old Chinese) Mathematical Books in Shanghai.
KHS, 1939, **23**, 72.

Yen Tun-Chieh (*5*) 嚴敦傑.
Tsu Kêng Pieh Chuan 祖暅別傳.
Biography of Tsu Kêng-Chih (+6th-century mathematician).
KHS, 1941, **25**, 460.

Yen Tun-Chieh (*6*) 嚴敦傑.
Sung Yuan Suan-Hsüeh Tshung Khao 宋元算學叢考.
Investigations into the Origins of the Mathematics of the Sung and Yuan Dynasties.
KHS, 1947, **29**, 109.

Yen Tun-Chieh (*7*) 嚴敦傑.
Ou-Chi-Li-Tê 'Chi Ho Yuan Pên' Yuan Tai Shu Ju Chung Kuo Shuo 歐幾里得幾何原本元代輸入中國說.
On the Coming of Euclid's Geometry to China in the Yuan Dynasty.
TFTC, 1943, **39** (no. 13), 35.

Yen Tun-Chieh (*8*) 嚴敦傑.
Sui Shu Lü Li Chih Tsu Chhung-Chih chih Yuan Lü Chi Shih Shih 隋書律曆志祖沖之之圓率記事釋.
What the *History of the Sui Dynasty* records about Tsu Chhung-Chih's calculation of the Value of π.
HITC, 1933, **15** (no. 10).

Yen Tun-Chieh (*9*) 嚴敦傑.
Sung Yuan Suan Shu yü Hsin Yung Huo-Pi Shih Liao 宋元算書與信用貨幣史料.
The Relation between the use of Paper Money in the Sung and Yuan Dynasties and the Mathematical Books of that time.
ISP/WSFK, 1943, no. 38.

Yen Tun-Chieh (*10*) 嚴敦傑.
Hui Li Chia Tzu Khao 回曆甲子考.
Discussion of the Sexagenary Calendar System among the Muslims.
KHS, 1949, **31**, 291.

Yen Tun-Chieh (*11*) 嚴敦傑.
Chin I-Wei-Yuan Li Tou Fên Khao 金乙未元曆斗分考.
Investigation of the I-Wei-Yuan Calendar of the (Jurchen) Chin Dynasty.
TFTC, 1945, **41**, 30.

Yen Tun-Chieh (*12*) 嚴敦傑.
With comments by Lu Shih-Hsien 魯實先.
Pei Chhi Tung Hsün Chêng Yuan-Wei Chia Yin Yuan Li Chi Nien Khao 北齊董峻鄭元偉甲寅元曆積年考.
Study of the Chia-Yin-Yuan Calendar of the Northern Chhi Dynasty prepared by Tung Hsün and Chêng Yuan-Wei (+575).
CH, 1945, **33**, 14.

Yen Tun-Chieh (*13*) 嚴敦傑.
Kuei Chü Chuan 規矩磚.
Han Bricks with Designs of Compasses and Carpenters' Squares.
SWYK, 1941, **3** (no. 4), 63.

Yen Tun-Chieh (*14*) 嚴敦傑.
Chung-Kuo Ku Tai Shu-Hsüeh ti Chhêng-Chiu 中國古代數學的成就.
Contributions of Ancient and Medieval Chinese Mathematics.
Chung-hua Chhüan Kuo Kho-Hsüeh Chi-Shu Phu-Chi Hsieh-Hai.
Peking, 1956.

Yen Yü (*3*) 燕羽.
Chung-Kuo Ku-Tai kuan-yü Chih-Wu Chih

Yen Yü (3) (cont.)
shih Kuang Tshang ti Chi Tsai 中國古代關於植物指示礦藏的記載.
A Note on the Use of Plants as Ore-Indicators in Ancient China.
Essay in Li Kuang-Pi & Chhien Chün-Hua (q.v.), p. 163.
Peking, 1955.

Yen Yü (5) 燕羽.
Shih-liu Shih-Chi ti Wei Ta Kho-Hsüeh Chia; Li Shih-Chen 十六世紀的偉大科學家; 李時珍.
A Great Scientist of the +16th century; Li Shih-Chen (pharmaceutical naturalist).
Essay in Li Kuang-Pi & Chhien Chün-Hua (q.v.), p. 324.
Peking, 1955.

Yin Tsan-Hsün (1) 尹贊勳.
Chung-Kuo Ku-Sêng-Wu Hsüeh chih Kên-Miao 中國古生物學之根苗.
The Beginnings of Palaeontology in China.
TCLP, **12**, 63.

Yü Phing-Po (1) 俞平伯.
Chhin Han Kai Yüeh Lun 秦漢改月論.
On the Calendar Reforms of the Chhin and Han Dynasties.
CHJ, 1937, **12**, 435.

ADDENDA TO BIBLIOGRAPHY B

Anon. (8) (ed.).
Chung-Kuo Ti-Chen Tzu-Liao Nien Piao 中國地震資料年表.
Register of Earthquakes in Chinese Recorded History (−1189 to +1955).
2 vols., Kho-Hsüeh, Peking, 1956.

Chhien Pao-Tsung (7) 錢寶琮.
Shou Shih Li Fa Lüeh Lun 授時曆法略論.
On the Shou Shih Calendar of Kuo Shou-Ching (+1281).
TWHP, 1956, *4* (no. 2), 193.

Hung I-Pyŏn (1) 洪以燮.
Chōsen Kagakushi 朝鮮科學史.
A History of the Natural Sciences in Korea.
Tokyo, 1944.

Hsi Tsê-Tsung (3) 席澤宗.
Sêng I-Hsing Kuan Tshê Hêng-Hsing Wei Chih ti Kung-Tso 僧一行觀測恆星位置的工作.
On the Observations of Star Positions [and Proper Motions] by the monk I-Hsing (+683 to +729) [and on the date of the *Hsing Ching*].
TWHP, 1956, *4* (no. 2), 212.

Lung Po-Chien (1) 龍伯堅.
Hsien Tshun Pên Tshao Shu Lu 現存本草書錄.
Bibliographical Study of Extant Pharmacopoeias (from all periods).
Jen Min Wei Sêng, Peking, 1957.

Ting Fu-Pao & Chou Yün-Chhing (2) 丁福保, 周雲青.
Ssu Pu Tsung Lu Suan Fa Pien 四部總錄算法編.
Bibliography of Mathematical Books to supplement the *Ssu Khu Chhüan Shu* encyclopaedia.
Com. Press, Shanghai, 1957.

Yang Khuan (4) 楊寬.
Chung-Kuo Li-Tai Chhih Tu Khao 中國歷代尺度考.
A Study of the Chinese Foot-Measure through the Ages.
Com. Press, Shanghai, 1938; revised and amplified ed. 1955.

C. BOOKS AND JOURNAL ARTICLES IN WESTERN LANGUAGES

ABETTI, G. (1). *The History of Astronomy.* Sidgwick & Jackson, London, 1954, tr. by B. B. Abetti from the Italian original *Storia dell'Astronomia* (rev. W. M. H. Greaves, *N*, 1955, **176**, 323).

ADAMS, F. D. (1). *The Birth and Development of the Geological Sciences.* Baillière, Tindall & Cox, London, 1938.

ADLER, E. N. (1) (ed.). *Jewish Travellers.* Routledge, London, 1930. (Broadway Travellers Series.)

ADNAN ADIVAR, A. (2). *La Science chez les Turcs Ottomans.* Maisonneuve, Paris, 1939.

AHLENIUS, K. (1). 'En Kinesisk Varlskarta från 17 Århundradet' (Verbiest's map *ca.* +1680). *SUHVS*, 1904, no. 8 (4).

AHMAD, NAFIS (1). *The Muslim Contribution to Geography.* Muhammad Ashraf, Kashmiri Bazar, Lahore, 1947.

AHRENS, W. E. M. G. (1). *Mathematische Spiele.* In *Encyklopädie d. math. Wiss.* vol. 1 (2), pp. 1081 ff. Teubner, Leipzig, 1900–4. Separate, enlarged, edition, Teubner, Leipzig, 1911. 3rd ed. 1916.

AIRY, G. B. (1). 'Comparison of the Chinese Record of Solar Eclipses in the Chun-Tsew [*Chhun Chhiu*] with the Computations of Modern Theory.' *RAS/MN*, 1884, **24**, 167.

ALLEN, C. W. (1). *Astrophysical Quantities.* Univ. of London Press, London, 1955.

ALLEN, J. (1). *Scale Models in Hydraulic Engineering.* Longmans, London, 1947.

ALLEY, R. (4). 'The Tai-kou; some Notes on the Clothing Hook of Old China.' *EW*, 1956, **7**, 147.

ALMAGIÀ, R. (1). 'La Dottrina della Marea nell'Antichità classica e nel Medio Evo.' *MAL/MN*, 1905, 1947.

AMBRONN, L. (1). 'Die Beziehungen der Astronomie zu Kunst und Technik.' Art. in *Astronomie*, ed. J. Hartmann. Pt. III, Sect. 3, vol. 3 of *Kultur d. Gegenwart*, p. 566. Teubner, Berlin & Leipzig, 1921.

AL-ANDALUSĪ, YAḤYĀ AL-MAGHRIBĪ (1). *Risālat al-Khiṭa wa'l-Īghūr* (On the Calendar of the Chinese and Uighurs). MS.

ANDRADE, E. N. DA C. (1). 'Robert Hooke' (Wilkins Lecture). *PRSA*, 1950, **201**, 439. (The quotation concerning fossils is taken from the advance notice, Dec. 1949.) Also *N*, 1953, **171**, 365.

ANDREWS, M. C. (1). 'The Study and Classification of Mediaeval Mappae Mundi.' *AAA*, 1926, **75**, 61.

ANDREWS, W. S. (1). *Magic Squares and Cubes.* Chicago, 1908.

ANON. (3). 'Alum Works at Peh-Kwan.' *NCM*, Shanghai, 1917.

ANON. (4). 'Significance of Trace Elements in Plants and Animals.' *N*, 1947, **159**, 206.

ANON. (25). *Les Instruments de Mathématiques de la Famille Strozzi faits en 1585–1586 par Erasmus Habermehl de Prague.* [Sale catalogue.] Frederik Muller, Amsterdam, 1911.

ANON. (26). 'The Days of our Years; has the time come to change our present Calendar?' *UNESC*, 1954, **7** (no. 1), 28.

ANON. (27) (ed.). *The World Encompasséd; an Exhibition of the History of Maps, held at the Baltimore Museum of Art....* (Catalogue.) Walters Art Gallery, Baltimore, 1952.

ANON. (28). 'Report on the past twenty-five years' work in Japan on the History of Chinese and Japanese Astronomy' (to the Commission for the History of Astronomy of the International Astronomical Union). *TIAU*, 1954, **8**, 626.

ANON. (33) (ed.). *Miscellanea Curiosa, being a Collection of some of the Principal Phaenomena in Nature Accounted for by the Greatest Philosophers of this Age, together with Several Discourses read before the Royal Society for the Advancement of Physical and Mathematical Knowledge.* Vol. 1, Senex, London, 1705. Vol. 2, Senex & Price, London, 1706. Continued as: *Miscellanea Curiosa, containing a Collection of Curious Travels, Voyages and Natural Histories of Countries, as they have been Delivered in to the Royal Society.* Vol. 3, Senex & Price, London, 1707.

ANQUETIL-DUPERRON, A. H. (1) (tr.). (*a*) *Zendavesta, Ouvrage de Zoroastre.* Paris, 1771. (*b*) *Zendavesta, Zoroaster's lebendiges Wort, worin die Lehren und Meinungen dieses Gesetzgebers, ingleichen die Ceremonien des heiligen Dienstes der Parsen...aufbehalten sind.* Riga, 1776.

ANQUETIL-DUPERRON, A. H. (2) (tr.) *Boun-dehesch.* Paris, 1771. (A French translation of the Hindi version of the *Bundahišn*, encyclopaedia of +1178).

AL-ANSARĪ. See al-Dimashqī.

APIANUS, PETRUS [PETER BIENEWITZ] (1). *Astronomicon Caesareum.* Ingolstadt, 1540.

APIANUS, PETRUS [PETER BIENEWITZ] & FRISIUS, GEMMA [VAN DER STEEN] (1). *Cosmographia, sive Descriptio Universi Orbis Petri Apiani et Gemmae Frisii, jam demum integritati suae restituta,....* Birckmann, Köln, 1574; Arnold Bellerus, Antwerp, 1584.

ARAKAWA, H. (1). 'On the Relation between the Cyclic Variation of Sun-spots and the Historical Rice Crop Famines in Japan' (+1750 onwards). *PMGP*, 1953, **4**, 151.
ARATUS OF SOLI. See Mair, G. R.
ARCHER-HIND, R. D. (1) (tr.). *The 'Timaeus' of Plato.* London, 1888.
ARCHIBALD, R. C. (1). 'Outline of the History of Mathematics' (and mathematical Astronomy). *AMM*, 1949, **56** (Supplement), 1.
ARCHIBALD, R. C. (2). 'The Cattle Problem [attributed to Archimedes; an indeterminate equation]. *AMM*, 1918, **25**, 411.
ARKELL, W. J. & TOMKEIEV, S. I. (1). *English Rock Terms, chiefly as used by Miners and Quarrymen.* Oxford, 1953.
ARLINGTON, L. C. & LEWISOHN, W. (1). *In Search of Old Peking.* Vetch, Peiping, 1935.
ASHLEY-MONTAGU, M. F. (1) (ed.). *Studies and Essays in the History of Science and Learning* (Sarton Presentation Volume). Schuman, New York, 1944.
ASHMOLE, ELIAS (1). *Theatrum Chemicum Britannicum.* London, 1652.
AUSTIN, R. G. (1). 'Greek Board-Games.' *AQ*, 1940, **14**, 257.

BAADE, W. (1). *ASTRO*, 1943, **97**, 125.
BAAS-BECKING, L. G. M. (1). 'Historical Notes on Salt and Salt-Manufacture.' *SM*, 1931, **32**, 434.
BABBAGE, C. (1). 'Logarithms in Chinese Form.' *RAS/MN*, 1827, **1**, 9.
BACHMANN, P. G. H. (1). *Die Elemente der Zahlentheorie.* Teubner, Leipzig & Berlin, 1921.
BADDELEY, J. F. (1). 'Father Matteo Ricci's Chinese World Maps.' *GJ*, 1917, **50**, 254.
BAGCHI, P. C. (1). *India and China; a thousand years of Sino-Indian Cultural Relations.* Hind Kitab, Bombay, 1944. 2nd ed. 1950.
BAGROW, L. (1). 'Ortelii Catalogus Cartographorum.' *PMJP*, 1929, Ergänzungsband, **43**, no. 199; 1930, Ergänzungsband, **45**, no. 210.
BAGROW, L. (2). *Die Geschichte der Kartographie.* Safari, Berlin, 1951.
BAGROW, L., MAZAHERI, A. & YAJIMA, S. (1). 'The Origin of Medieval Portulans.' In *Proc. VIIIth International Congress of the History of Science, Florence*, 1956 (Sect. 3, History of Geography and Geology).
BAILEY, K. C. (1). *The Elder Pliny's Chapters on Chemical Subjects.* 2 vols. Arnold, London, 1929 and 1932.
BAILLIE, G. H. (1). *Clocks and Watches; an historical Bibliography.* NAG Press, London, 1951.
BAILLY, J. S. (1). *Traité de l'Astronomie Indienne et Orientale; Ouvrage qui peut servir de Suite à l'Histoire de l'Astronomie Ancienne.* Debure, Paris, 1787.
BAILLY, J. S. (2) [& VOLTAIRE]. *Lettres sur l'Origine des Sciences, et sur celle des Peuples de l'Asie, précédées de quelques lettres de M. de Voltaire à l'auteur.* 2 vols. Debure, Paris, 1777.
BALDINI, P. G. (1). *Saggi di Dissertazioni Accademiche pubblicamente lette nella nobile Accademia Etrusca dell'antichissima città di Cortona*, vol. 3, p. 185. Rome, 1741.
BALL, W. W. ROUSE (1). *A Short Account of the History of Mathematics.* Macmillan, London, 1888.
BALL, W. W. ROUSE (2). *Mathematical Recreations and Essays.* 11th ed., ed. H. S. M. Coxeter. Macmillan, London, 1939.
BALSS, H. (1). *Antike Astronomie; aus griechischen und lateinischen Quellen mit Text, Übersetzung und Erlaüterungen Geschichtlich dargestellt.* Heimeran, München, 1949.
BAND, W. & BAND, C. (1). *Dragon Fangs.* London, 1947.
BANFI, A. (1). *Galileo Galilei.* Ambrosiana, Milan, 1949 (rev. P. Labérenne, *LP*, 1950 (No. 31), 155).
BARANOVSKAIA, L. S. (1). 'Pervaia Rabota po Matematike na Mongolskom Yazyke (On a Mathematical Work in the Mongol Language; the *Sochinenie o Koordinatach* of +1712).' *TIYT*, 1954, **1**, 53.
BARANOVSKAIA, L. S. (2). 'Iz Istorii Mongol'skoi Astronomii (On the History of Astronomy among the Mongols).' *TIYT*, 1955, **5**, 321.
BARGELLINI, P. (1). *The Cloisters of Santa Maria Novella and the Spanish Chapel.* Tr. H. M. R. Cox. Arnaud, Florence, n.d. (1954).
BARLOW, C. W. C. & BRYAN, G. H. (1). *Elementary Mathematical Astronomy.* UTP, London, 1946.
BARNARD, F. P. (1). *The Casting-Counter and the Counting-Board.* Oxford, 1916.
DE BARROS, JOÃO (1). *Terceira Decada da Asia.* Lisbon, 1563. Later ed. 1628.
BARTHOLD, W. (1). *Zwölf Vorlesungen ü. d. Geschichte d. Türken Mittelasiens.* Germ. tr. T. Menzel. Collignon, Berlin, 1935.
BARTON, W. H. (1). 'Sky Clocks and Calendars.' *SKY*, 1940, **4** (no. 11), 7.
DE BARY, W. T. (1). 'A Re-appraisal of Neo-Confucianism.' *AAN*, 1953, **55** (no. 5), pt. 2, 81. (American Anthropological Association Memoir, no. 75.)
BASTIAN, A. (1). *Reisen in China.* Jena, 1871.

Bate, John (1). *The Mysteryes of Nature and Art: conteined in foure severall Tretises, the first of Water Workes, the second of Fyer Workes, the third of Drawing, Colouring, Painting and Engraving; the fourth of Divers Experiments, as wel serviceable as delightful; partly collected, and partly of the author's peculiar practice and invention.* Harper & Mab, London, 1634, 1635.

Bateman, A. M. (1). *Economic Mineral Deposits.* Wiley, New York, 1950; Chapman & Hall, London, 1950.

Bauer, M. (1). *Edelsteinkunde.* 1896. (Eng. tr. by L. J. Spencer, with additions 'Precious Stones', London, 1904.)

Baxendall, D. (1). *Catalogue of the Collections in the Science Museum, South Kensington, with Descriptive and Historical Notes and Illustrations; Mathematics. I. Calculating Machines and Instruments.* HMSO, London, 1926.

Baxter, W. (1) (tr.). 'Pleasure not Attainable according to Epicurus.' In *The Works of Plutarch.* Ed. Morgan. London, 1694.

Bayley, E. C. (1). 'On the Genealogy of Modern Numerals.' *JRAS*, 1883, **14**, 335; **15**, 1.

Beal, S. (1) (tr.). *Travels of Fah-Hian [Fa-Hsien] and Sung-Yün, Buddhist Pilgrims from China to India (+400 and +518).* Trübner, London, 1869.

Beal, S. (2) (tr.). *Si Yu Ki [Hsi Yü Chi], Buddhist Records of the Western World, transl. from the Chinese of Hiuen Tsiang [Hsüan-Chuang].* 2 vols. Trübner, London, 1884. 2nd ed. 1906.

Beazley, C. R. (1). *The Dawn of Modern Geography.* 3 vols. (vol. 1, +300 to +900; vol. 2, +900 to +1260; vol. 3, +1260 to +1420.) Vols. 1 and 2, Murray, London, 1897 and 1901. Vol. 3, Oxford, 1906.

Beck, T. (1). *Beiträge z. Geschichte d. Maschinenbaues.* Springer, Berlin, 1900.

Becker, O. & Hofmann, J. E. (1). *Geschichte der Mathematik.* Athenäum, Bonn, 1951 (rev. O. Neugebauer, *CEN*, 1953, **2**, 364).

Becker, W. (1). *Sterne und Sternsysteme.* Steinkopf, Dresden & Leipzig, 1950.

van Beek, A. (1). *Beschrijving van een Chineeschen Zonne- en Maanwijzer, met Kompas.* Amsterdam, 1851.

Beer, A. (1). 'The Astronomical Significance of the Zodiac of the Quṣayr 'Amra.' In K. A. C. Cresswell, *Early Muslim Architecture*, p. 296. Oxford, 1932.

Belpaire, B. (2). 'Le Folklore de la Foudre en Chine sous la Dynastie des Thang (un document nouveau)' (the *Lei Min Chuan* of Shen Chhi-Chi, +779). *MUSEON*, 1939, **52**, 163.

Bendig, H. (1) (tr.). 'The *Yün Lin Shih Phu* [Cloud Forest Lapidary]'. Unpub.

Benndorf, O., Weiss, E. & Rehm, A. 'Zur Salzburger Bronzescheibe mit Sternbildern.' *JHOAI*, 1903, **6**, 32.

Berger, H. (1). *Geschichte d. wissenschaftlichen Erdkunde d. Griechen.* Leipzig, 1903.

Berlage, H. P. (1). Art. on seismological instruments. In *Handbuch d. Geophysik*, ed. B. Gutenberg, vol. 4, pp. 299 ff. Bornträger, Berlin, 1932.

Bernal, J. D. (1). *Science in History.* Watts, London, 1954 (Beard Lectures at Ruskin College, Oxford).

Bernard-Maître, H. (1). *Matteo Ricci's Scientific Contribution to China*, tr. by E. T. C. Werner. Vetch, Peiping, 1935. Orig. pub. as 'L'Apport Scientifique du Père Matthieu Ricci à la Chine', Hsienhsien, Tientsin, 1935; rev. Chang Yü-Chê, *TH*, 1936, **3**, 583.

Bernard-Maître, H. (5). *Le Père Matthieu Ricci et la Société Chinoise de son Temps* (1552 to 1610). 2 vols. Hsienhsien, Tientsin, 1937.

Bernard-Maître, H. (7). 'L'Encyclopédie Astronomique du Père Schall, *Chhung-Chên Li Shu* (+1629) et *Hsi-Yang Hsin Fa Li Shu* (+1645); La Réforme du Calendrier Chinois sous l'Influence de Clavius, Galilée et Kepler.' *MS*, 1937, **3**, 35, 441.

Bernard-Maître, H. (8). 'Les Étapes de la Cartographie Scientifique pour la Chine et les Pays Voisins dépuis le 16e jusqu'à la fin du 18e siècle.' *MS*, 1935, **1**, 428.

Bernard-Maître, H. (11). 'Ferdinand Verbiest, Continuateur de l'œuvre Scientifique d'Adam Schall.' *MS*, 1940, **5**, 103.

Bernard-Maître, H. (12). 'Galilée et les Jésuites des Missions d'Orient.' *RQS*, 1935, 356.

Bernard-Maître, H. (13). *La Mappemonde Ricci du Musée historique de Pékin.* Politique de Pékin, Peiping, 1928.

Bernier, François (1). *Bernier's Voyage to the East Indies; containing The History of the Late Revolution of the Empire of the Great Mogul; together with the most considerable passages for five years following in that Empire; to which is added A Letter to the Lord Colbert, touching the extent of Hindostan, the Circulation of the Gold and Silver of the world, to discharge itself there, as also the Riches Forces and Justice of the Same, and the principal Cause of the Decay of the States of Asia—with an Exact Description of Delhi and Agra; together with* (1) *Some Particulars making known the Court and Genius of the Moguls and Indians; as also the Doctrine and Extravagant Superstitions and Customs of the Heathens of Hindustan,* (2) *The Emperor of Mogul's Voyage to the Kingdom of Kashmere, in 1664, called the Paradise of the Indies....* Dass (for SPCK), Calcutta, 1909. (Substantially the same title-page as the editions of 1671 and 1672.)

BERNSTEIN, JOSEPH (1). 'Tsunamis' (tidal waves in the Sea of Japan). *SAM*, 1954, **191** (no. 2), 60.
BERRY, A. (1). *A Short History of Astronomy.* Murray, London, 1898.
BERTHELOT, ANDRÉ (1). *L'Asie Centrale et Sud-Orientale d'après Ptolemée.* Payot, Paris, 1930.
BERTHELOT, M. (2). *Introduction à l'Etude de la Chimie des Anciens et du Moyen-Age.* Lib. Sci. et Arts, Paris, 1938 (repr.). 1st ed. 1888.
BERTHELOT, M. (3). Review of de Mély (1), *Lapidaires Chinois. JS*, 1896, 573.
BERTHELOT, M. & RUELLE, C. E. (1). *Collection des Alchimistes Grecs.* Steinheil, Paris, 1888.
BERTHOUD, F. (1). *Histoire de la Mésure du Temps par les Horloges.* Paris, 1802.
BERTRAND, G. & MOKRAGNATZ, H. (1). 'Sur la Présence du Nickel et du Cobalt chez les Végétaux et dans la Terre Arable.' *CRAS*, 1922, **175**, 112, 458; **179**, 1566; 1930, **190**, 21.
BETTINI, MARIO (1). *Apiaria Universae Philosophiae Mathematicae, in quibus Paradoxa, et nova pleraque Machinamenta ad Usus eximios traducta et facillimis Demonstrationibus confirmata opus....* Ferronius, Bologna, 1645.
BETTINI, MARIO (2). *Recreationum Mathematicarum Apiaria Novissima Duodecim—quae continent Militaria, Stereometrica, Conica, et novas alias jucundas Praxes ac Theorias, in omni Mathematicarum Scientiarum Genere....* Ferronius, Bologna, 1660.
BEVAN, E. R. (1). 'India in Early Greek and Latin Literature.' In *CHI*, vol. 1, ch. 16, p. 391. Cambridge, 1935.
BEZOLD, C. (1). 'Sze-ma Ts'ien [Ssuma Chhien] und die babylonische Astrologie.' *OAZ*, 1919, **8**, 42.
BEZOLD, C., KOPFF, A. & BOLL, F. (1). 'Zenit- und Aequatorialgestirne am babylonischen Fixsternhimmel.' *SHAW/PH*, 1913, **4**, no. 11.
BIELENSTEIN, H. (1). 'An Interpretation of the Portents in the *Ts'ien Han Shu* [*Chhien Han Shu*].' *BMFEA*, 1950, **22**, 127.
BIELENSTEIN, H. (2). 'The Restoration of the Han Dynasty.' *BMFEA*, 1954, **26**, 1–20 and sep. Göteborg, 1953.
BIERNATZKI, K. L. (1). 'Die Arithmetik d. Chinesen.' *JRAM*, 1856, **52**, 59. (Translation of Wylie, 4.)
BIGOURDAN, G. (1). *L'Astronomie; Evolution des Idées et des Méthodes.* Paris, 1911.
BILFINGER, G. (1). *Zeitmesser d. antiken Völker.* Stuttgart, 1886.
BIOT, E. (1) (tr.). *Le Tcheou-Li ou Rites des Tcheou* [*Chou*]. 3 vols. Imp. Nat., Paris, 1851. (Photographically reproduced, Wêntienko, Peiping, 1930.)
BIOT, E. (3) (tr.). *Chu Shu Chi Nien* [Bamboo Books]. *JA*, 1841 (3e sér.), **12**, 537; 1842, **13**, 381.
BIOT, E. (4) (tr.). 'Traduction et Examen d'un ancien Ouvrage intitulé *Tcheou-Pei*, littéralement "Style ou signal dans une circonférence".' *JA*, 1841 (3e sér.), **11**, 593; 1842 (3e sér.), **13**, 198 (emendations). (Commentary by J. B. Biot, *JS*, 1842, 449.)
BIOT, E. (5) (tr.). 'Table Générale d'un Ouvrage chinois intitulé...*Souan-Fa Tong-Tsong*, ou Traité Complet de l'Art de Compter....' *JA*, 1839 (3e sér.), **7**, 193.
BIOT, E. (6). (On Pascal's Triangle in the *Suan Fa Thung Tsung*.) *JS*, 1835 (May).
BIOT, E. (7). 'Note sur la connaissance que les chinois ont eue de la Valeur de Position des Chiffres.' *JA*, 1839 (3e sér.), **8**, 497.
BIOT, E. (8). 'Catalogue des Étoiles Extraordinaires observées en Chine depuis les Temps Anciens jusqu'à l'An 1203 de notre Ère.' *CT*, 1846 (Additions), p. 60.
BIOT, E. (9). 'Catalogue des Comètes observées en Chine depuis l'an 1230 jusqu'à l'an 1640 de notre Ère.' *CT*, 1846 (Additions), p. 44. (With 'Supplément pour les Etoiles Extraordinaires observées sous la dynastie Ming, qui peuvent se rapporter a des Apparitions de Comètes' immediately following.)
BIOT, E. (10). 'Recherches faites dans la Grande Collection des Historiens de la Chine sur les Anciennes Apparitions de la Comète de Halley.' *CT*, 1846 (Additions), p. 69.
BIOT, E. (11). 'Catalogue Général des Etoiles Filantes et des autres Météores observées en Chine pendant 24 siècles depuis le 7ème Siècle av. JC jusqu'au milieu du 17ème de notre Ere.' *MRAS/DS*, 1848, **10**, 129, 415. (Also *CRAS*, 1841, **12**, 986; 1841, **13**, 203; 1842, **14**, 699; 1846, **23**, 1151.)
BIOT, E. (12). 'Sur la Direction de la Queue des Comètes.' *CRAS*, 1843, **16**, 751.
BIOT, E. (13). 'Recherches sur la Température Ancienne de la Chine.' *JA*, 1840 (3e sér.), **10**, 530.
BIOT, E. (15). 'Études sur les Montagnes et les Cavernes de la Chine d'après les Géographies Chinoises.' *JA*, 1840 (3e sér.), **10**, 273.
BIOT, E. (16). 'Sur la Hauteur de Quelques Points Remarquables des Territoires Chinois.' *JA*, 1840 (3e sér.), **9**, 81.
BIOT, E. (17). 'Notice sur Quelques Procédés Industriels connus en Chine au XVIe Siècle.' *JA*, 1835 (2e sér.), **16**, 130.
BIOT, J. B. (1). *Études sur l'Astronomie Indienne et sur l'Astronomie Chinoise.* Lévy, Paris, 1862.
BIOT, J. B. (2). Review of *The Oriental Astronomer*, by H. R. Hoisington. Batticotta, Ceylon, 1848. *JS*, 1859, 197, 271, 369, 401, 475. (Reprinted in J. B. Biot (1), p. 7.)

BIOT, J. B. (3). Review of E. B. Burgess' translation of the *Sūrya Siddhānta*. New Haven, Conn., 1860. *JS*, 1860. (Reprinted in J. B. Biot (1), p. 155.)

BIOT, J. B. (4). Review of L. Ideler's *Über die Zeitrechnung d. Chinesen*. Berlin, 1839. 'Ueber die Zeitrechnung d. Chinesen von Ludwig Ideler, [Sur la Chronologie des Chinois]; dissertation lue à l'Académie des Sciences de Berlin, le 16 Fév., 1837, et depuis considérablement augmentée, Berlin, 1839, in 4°.' *JS*, 1839, p. 721; 1840, pp. 27, 73, 142, 227, 264; addenda, pp. 309, 372. This was the memoir afterwards cited as 'Exposé Méthodique de l'Astronomie Chinoise'. The table of circumpolar correlations is on p. 246. Part of this had the collaboration of Abel Rémusat as well as E. Biot. It should have been reprinted in J. B. Biot (1).

BIOT, J. B. (5). 'Précis de l'Histoire de l'Astronomie Chinoise.' *JS*, 1861, pp. 284, 325, 420, 468, 573, 604. (Reprinted in J. B. Biot (1), p. 249.)

BIOT, J. B. (6). 'Sur les Nacshatras, ou Mansions de la Lune, selon les Hindoux; extrait d'une description de l'Inde redigée par un voyageur Arabe du XIe Siècle.' (al-Bīrūnī's *nakshatra* list, based on a tr. by Munk.) *JS*, 1845, 39.

BIOT, J. B. (7). 'Résumé de la Chronologie Astronomique.' *MRASP*, 1849, **22**, 380.

AL-BĪRŪNĪ, ABŪ-AL-RAIḤĀN MUḤAMMAD IBN-AḤMAD. *Ta'rīkh al-Hind* (History of India). See Sachau (1).

BISHOP, H. R., BUSHELL, S. W., KUNZ, G. F., LI SHIH-CHHÜAN, LILLEY, R., THANG JUNG-TSO *et al.* (1). *Investigations and Studies in Jade; The Heber R. Bishop Collection.* 2 vols. Elephant folio, privately printed, New York, 1906 (rev. and crit. E. Chavannes, *TP*, 1906, **7**, 396).

VON BISSING, F. W. (1). 'Steingefässe.' In *Catalogue Générale du Musée du Caire.* Cairo.

BLACK, D. M. (1). 'The Brocken Spectre of the Desert View Watch Tower, Grand Canyon, Arizona.' *SC*, 1954, **119**, 164.

BLACK, J. DAVIDSON (1) (ed.). *Fossil Man in China.* Peiping, 1933.

BLAGDEN, C. O. (1). 'Notes on Malay History.' *JRAS/M*, 1909, no. 53.

BLOCHET, E. (3). *Catalogue des MSS. Arabes.* Bibliothèque Nationale, Paris, 1925.

BLUNDEVILLE, T. (1). *Mr Blundevil, his Exercises, contayning Eight Treatises the Titles whereof are set down in the next printed page; Which Treatises are very necessary to be read and learned of all young Gentlemen, that have not been exercised in such Disciplines, and yet are desirous to have Knowledge as well in Cosmographie, Astronomie and Geographie, as also in the Art of Navigation, in which Art it is impossible to profit without the help of these or such like Instructions*....7th ed., corr. and enlarged, R. Hartwell. Bishop, London, 1636.

I Very easie Arithmetick.
II First principles of cosmography, especially a plaine Treatise of the Spheare.
III Plaine and full description of the Globes, as well Terrestriall as Celestiall.
IV Peter Plancius his universall Map (+1592).
V Mr Blagrave his Astrolabe.
VI First and chiefest Principles of Navigation.
VII Briefe description of universall Maps and Cards, also the true use of Ptolomie his Tables.
VIII The true order of making of Ptolomie his Tables.

BOBROVNIKOV, N. T. (1). 'The Discovery of Variable Stars.' *ISIS*, 1942, **33**, 687.

BODDE, D. (5). 'Types of Chinese Categorical Thinking.' *JAOS*, 1939, **59**, 200.

BODDE, D. (6). 'The Attitude towards Science and Scientific Method in Ancient China.' *TH*, 1936, **2**, 139, 160.

BODDE, D. (15). *Statesman, General and Patriot in Ancient China.* Amer. Oriental Soc. New Haven, Conn., 1940.

BOFFITO, G. (1). *Gli Strumenti della Scienza e la Scienza degli Strumenti, con l'Illustrazione della Tribuna di Galileo.* Seeber, Florence, 1929.

BÖKER, R. (1). *Die Entstehung d. Sternsphäre Arats.* Leipzig, 1948.

BOLL, F. (1). *Sphaera.* Teubner, Leipzig, 1904.

BOLL, F. (2). 'Die Sternkataloge des Hipparch und des Ptolemaios.' In *Bibliotheca Mathematica*, 1901, 3rd ser. **2**, 185.

BOLL, F. (3). 'Die Entwicklung des astronomischen Weltbildes im Zusammenhang mit Religion u. Philosophie.' Art. in *Astronomie*, ed. J. Hartmann. Pt. III, Sect. 3, vol. 3 of *Kultur d. Gegenwart*, p. 1. Teubner, Leipzig & Berlin, 1921.

BOLL, F. (4). 'Der ostasiatische Tierzyklus im Hellenismus.' *TP*, 1912, **13**, 699. (Reprinted in Boll (5), p. 99.)

BOLL, F. (5). *Kleine Schriften zur Sternkunde des Altertums.* Koehler & Ameland, Leipzig, 1950.

BOLL, F. & BEZOLD, C. (1). 'Antike Beobachtung färbiger Sterne.' *ABAW/PH*, 1918, **89**, (30), no. 1.

BOLL, F., BEZOLD, C. & GUNDEL, W. (1). (*a*) *Sternglaube, Sternreligion und Sternorakel.* Teubner, Leipzig, 1923. (*b*) *Sternglaube und Sterndeutung; die Gesch. ü. d. Wesen d. Astrologie.* Teubner, Leipzig, 1926.

BOLTON, L. (1). *Time Measurement; an Introduction to Means and Ways of reckoning Physical and Civil Time.* Bell, London, 1924.
BONELLI, M. L. (1). 'Globi Terrestri e Celesti.' *QMSF*, 1950, 8 pp.
BOODBERG, P. A. (1). 'Chinese Zoographic Names as Chronograms.' *HJAS*, 1940, **5**, 128.
BORCHARDT, L. (1). 'Die altägyptische Zeitmessung.' In E. v. Bassermann-Jordan's *Die Geschichte d. Zeitmessung u. d. Uhren.* de Gruyter, Berlin, 1920.
BORKENAU, F. (1). *Der Übergang vom feudalen zum bürgerlichen Weltbild.* Paris, 1934.
BOSANQUET, R. H. M. & SAYCE, A. H. (1). 'Babylonian Astronomy.' *RAS/MN*, 1879, **39**, 454; 1880, **40**, 105, 565.
BOSCH, F. D. K. (1). 'Le Motif de l'Arc-à-Biche à Java et au Champa.' *BEFEO*, 1931, **31**, 485.
BOSMANS, H. (1). 'Une Particularité de l'Astronomie Chinoise au 17ème Siècle' (the substitution by Verbiest of sexagesimal for the Chinese decimal graduation of degrees and minutes). *ASSB*, 1903 (1), **27**, 122. (Contains also bibliographical details on Verbiest's *Astronomia Europaea.*)
BOSMANS, H. (2). 'Ferdinand Verbiest, Directeur de l'Observatoire de Péking.' *RQS*, 1912, **71**, 196, 375.
BOSON, G. (1). 'Alcuni Nomi di Pietre nelle Inscrizioni Assiro-Babilonesi.' *RSO*, 1914, **6**, 969.
BOSON, G. (2). 'I Metalli e le Pietri nelle Inscrizioni Assiro-Babilonesi.' *RSO*, 1917, **7**, 379.
BOUCHÉ-LECLERCQ, A. (1). *L'Astrologie Grecque.* Leroux, Paris, 1899.
BOUCHÉ-LECLERCQ, A. (2). *Histoire de Divination dans l'Antiquité.* 4 vols. Leroux, Paris, 1879–82.
BOULLIAU, I. (1) (ed. & Latin tr.). *Works of Theon of Smyrna.* Paris, 1644. French tr. by J. Dupuis, Paris, 1892: 'Ce qui est utile en Mathématiques pour comprendre Platon'.
BOUSSET, W. (1). *Hauptprobleme der Gnosis.* Vandenhoek & Ruprecht, Göttingen, 1907. (*Forsch. z. Rel. u. Lit. d. alt.- u. neuen Testaments*, no. 10.)
BOXER (1) (ed.). *South China in the Sixteenth Century; being the Narratives of Galeote Pereira, Fr. Gaspar de Cruz, O.P., and Fr. Martin de Rada, O.E.S.A. (1550–1575).* Hakluyt Society, London, 1953 (Hakluyt Society Pubs., 2nd series, no. 106).
BOYER, C. B. (1). *The Concepts of the Calculus.* Columbia Univ. Press, New York, 1939.
BOYER, C. B. (2). 'Fundamental Steps in the Development of Numeration.' *ISIS*, 1944, **35**, 153.
BOYER, C. B. (3). 'Refraction and the Rainbow in Antiquity.' *ISIS*, 1956, **47**, 383.
BRAHE, TYCHO (1). *Astronomiae Instauratae Mechanica.* Wandesburg, 1598. [See Raeder, Strömgren & Strömgren.]
BRAHE, TYCHO (2). *De Nova et Nullius Aevi Memoria Prius Visa Stella, jampridem Anno a nato Christo 1572 mense Novembri primum Conspecta, Contemplatio Mathematica.* L. Benedictus, Hafniae, 1573. (Facsimile edition issued by the Royal Danish Science Society, Hafniae (Copenhagen), 1901.) English translation by V.V.S., Nealand, London, 1632.
BRAMLEY-MOORE, S. (1). *Map Reading in a Nutshell.* Pearson, London, 1941.
BRAUNMÜHL, A. (1). *Geschichte d. Trigonometrie.* 2 vols. Leipzig, 1900, 1903.
BRAVAIS, A. (1). 'Mémoire sur les Halos et les Phénomènes optiques qui les accompagnent.' *JEP*, 1847, **18** (no. 31), 1–270; cf. 1845, **18** (no. 30), 77, 97.
BRÉHIER, L. (2). *Vie et Mort de Byzance.* Albin Michel, Paris, 1947. [Evol. de l'Hum. series, no. 32.]
BRELICH, H. (1). 'Chinese Methods of Mining Quicksilver.' *MJ*, 1905, **77**, 578, 595.
BRENNAND, W. (1). *Hindu Astronomy.* Straker, London, 1896.
BRETSCHNEIDER, E. (1). *Botanicon Sinicum; Notes on Chinese Botany from Native and Western Sources.* 3 vols. Trübner, London, 1882 (printed in Japan). (Reprinted from *JRAS/NCB*, 1881, **16**.)
BRETSCHNEIDER, E. (2). *Medieval Researches from Eastern Asiatic Sources; Fragments towards the Knowledge of the Geography and History of Central and Western Asia from the thirteenth to the seventeenth century.* 2 vols. Trübner, London, 1888.
BRETSCHNEIDER, E. (4). 'Chinese Intercourse with the countries of Central and Western Asia during the fifteenth century. II. A Chinese Itinerary of the Ming Period from the Chinese Northwest Frontier to the Mediterranean Sea.' *CR*, 1876, **5**, 227. (Reprinted (abridged) in Bretschneider (2), vol. 2, p. 329.)
BRETSCHNEIDER, E. (7). 'Chinese Intercourse, etc. I. Accounts of Foreign Countries.' *CR*, **4**, 312, 385.
BREWER, H. C. (1). *Bibliography of the Literature on the Minor Elements and their Relation to Plant and Animal Nutrition.* Chilean Nitrate Educ. Bureau, New York, 1948.
BRITTEN, F. J. (1). *Old Clocks and Watches, and their Makers* (6th ed.). Spon, London, 1932.
BROCKELMANN, C. (2). *Geschichte d. arabischen Literatur.* Felber, Weimar, 1898. Supplementary volumes, Brill, Leiden, 1937.
BROMEHEAD, C. E. N. (1). Geology in Embryo (up to A.D. 1600). *PGA*, 1945, **56**, 89.
BROMEHEAD, C. E. N. (2). 'Aetites, or the Eagle-Stone.' *AQ*, 1947, **21**, 16.
BROMEHEAD, C. E. N. (3). 'A Geological Museum of the Early Seventeenth Century.' *QJGS*, 1947, **103**, 65.
BROOKS, C. E. (1). 'The Climatic Changes of the Past 1000 Years.' *EXP*, 1954, **10**, 153.

BROWN, B. (1). *Astronomical Atlases, Maps and Charts, an historical and general Guide.* Search, London, 1932.
BROWN, LLOYD A. (1). *The Story of Maps.* Little Brown, Boston, 1949.
BRUCE, J. P. (1) (tr.). *The Philosophy of Human Nature, translated from the Chinese, with notes.* Probsthain, London, 1922. (Chs. 42–8, inclusive, of *Chu Tzu Chhüan Shu.*)
BRUNET, P. & MIELI, A. (1). *L'Histoire des Sciences (Antiquité).* Payot, Paris, 1935.
BUDGE, E. A. WALLIS (3) (ed.). *Cuneiform Texts from Babylonian Tablets, etc. in the British Museum.* 41 vols. 1896–1931.
BULLING, A. (1). 'Descriptive Representations in the Art of the Chhin and Han Period.' Inaug. Diss., Cambridge, 1949.
BULLING, A. (5). 'Die Kunst der Totenspiele in der östlichen Han-zeit.' *ORE*, 1956, **3**, 28.
BUNBURY, E. H. (1). *History of Ancient Geography among the Greeks and Romans from the earliest Ages till the Fall of the Roman Empire.* 2 vols. Murray, London, 1879 and 1883.
DE BURBURE, A. (1). 'Quelques Précédents Expansionnistes Belges dans l'Hémisphère Chinois' (on Verbiest). *RCB*, 1951, **6**, 305.
BURGESS, E. (1) (tr.). *Sūrya Siddhānta; translation of a Textbook of Hindu Astronomy, with notes and an Appendix.* Ed. P. Gangooly, introd. by P. Sengupta. Calcutta, 1860. (Reprinted 1935.)
BURGESS, J. (1). 'On Hindu Astronomy.' *JRAS*, 1893, p. 758.
BURLAND, C. A. (1). 'A 360-Day Count in a Mexican Codex.' *MA*, 1947, **47**, 106.
BURTT, E. A. (1). *The Metaphysical Foundations of Modern Science.* New York and London, 1925.
BUSHELL, S. W. (3). 'The Early History of Tibet.' *JRAS*, 1880, N.S., **12**, 435.

CABLE, M. & FRENCH, F. (1). *The Gobi Desert.* Hodder & Stoughton, London, 1942.
CABLE, M. & FRENCH, F. (2). *China, her Life and her People.* Univ. of London Press, London, 1946.
CAHEN, C. (2). 'Quelques Problèmes économiques et fiscaux de l'Iraq Buyide [Buwayhid], d'après un Traité de Mathématiques [the *Kitāb al-Ḥāwī*...written between +1025 and +1050]'. *AIEO/UA*, 1952, **10**, 326.
CAJORI, F. (2). *A History of Mathematics.* 2nd ed. Macmillan, New York, 1919. (Repr. 1924.) (The 1st ed. does not contain the section on Chinese Mathematics.)
CAJORI, F. (3). *A History of Mathematical Notations.* 2 vols. Open Court, Chicago, 1928, 1929.
CAJORI, F. (4). *A History of the Logarithmic Slide-Rule.* New York, 1909.
CAJORI, F. (5). *A History of Physics, in its elementary branches, including the evolution of Physical Laboratories.* Macmillan, New York, 1899.
CALDER, I. R. F. (1). 'A Note on Magic Squares in the Philosophy of Agrippa of Nettesheim.' *JWCI*, 1949, **12**, 196.
CAMMANN, S. (2). 'The TLV Pattern on the Cosmic Mirrors of the Han Dynasty.' *JAOS*, 1948, **68**, 159.
CANTOR, M. (1). *Vorlesungen ü. d. Gesch. d. Mathematik.* 4 vols. Teubner, Leipzig, 1880–1908. Crit. Mikami (5).
CAPELLE, W. (1). *Berges- und Wolkenhöhen bei griechischen Physikern. Stoicheia*, no. 5, 1916.
CARDAN, JEROME (1). *De Subtilitate.* Nürnberg, 1550 (ed. Sponius); Basel, 1560. (Account in Beck (1), ch. 9.)
CARDAN, JEROME (3). *Artis Magnae sive de Regulis Algebraicis*, Nuremberg, 1545.
CARTON, C. (1). (*a*) *Notice Biographique sur le P. Verbiest, Missionnaire de la Chine.* (From *Annales de la Soc. de l'Emulation pour l'Histoire et les Antiquités de la Flandre Occidentale*, 1839, **1**, 83.) Vandecasteele-Werbrouck, Bruges, 1839. (*b*) *Biographie du R.P. Verbiest, Missionnaire en Chine.* Chabannes, Bruxelles, 1845. (From *Album Bibliographique des Belges Célèbres.*)
CARUS, P. (2). *Chinese Thought.* Open Court, Chicago, 1907.
CARY, G. (1). *The Medieval Alexander.* Ed. D. J. A. Ross, C.U.P., Cambridge, 1956. (A study of the origins and versions of the Alexander-Romance; important for medieval ideas on the flying-machine and the diving-bell or bathyscaphe.)
CARY, M. (1). 'Maës, qui et Titianus.' *CQ*, 1956, **6** (n.s.), 130.
CASANOVA, P. (1). 'La Montre du Sultan Noūr al-Dīn.' *SYR*, 1923, **4**, 282.
CASSINI, J. D. (1). 'Réflexions sur la Chronologie Chinoise.' *MRASP*, 1730 (1690), **8**, 300.
CAYLEY, H. (1). 'The Jade Quarries of the Kuen Lun.' *MCM*, 1871, **24**, 452.
DE CELLES, F. BEDOS (1). *La Gnomonique Pratique*.... Briasson *et al.* Paris, 1760.
CHALMERS, J. (1). 'Appendix on the Astronomy of the Ancient Chinese.' In Legge's transl. of the *Shu Ching*, p. [90] (Chinese Classics, vol. 3), Hongkong, 1865. Chinese tr. by Hsiang Ta, *KHS*, 1926, **11** (no. 12).
CHAMBERLAIN, B. H. (1). *Things Japanese.* Murray, London. 2nd ed. 1891; 3rd ed. 1898.
CHANG CHUN-MING (1). 'The Genesis and Meaning of Huan Khuan's "Discourses on Salt and Iron".' *CSPSR*, 1934, **18**, 1.

CHANG HUNG-CHAO (1). 'Lapidarium Sinicum; A Study of the Rocks, Fossils and Minerals as Known in Chinese Literature' (in Chinese with English summary). Chinese Geological Survey, Peiping, 1927. *MGSC* (ser. B), no. 2.

CHANG YÜ-CHÊ (1). 'Chang Hêng, a Chinese Contemporary of Ptolemy.' *POPA*, 1945, **53**, 1.

CHANG YÜ-CHÊ (2). 'Chang Hêng, Astronomer.' *PC*, 1956 (no. 1), 31.

CHAPMAN, F. R. (1). 'On the Working of Greenstone or Nephrite by the Maoris.' *TNZI*, 1892, **24**, 479.

CHATLEY, H. (1). MS. translation of the astronomical chapter (ch. 3, Thien Wên) of *Huai Nan Tzu*. Unpublished. (Cf. note in *O*, 1952, **72**, 84.)

CHATLEY, H. (5). 'Chinese Natural Philosophy and Magic.' *JRSA*, 1911, **59**, 557.

CHATLEY, H. (8). 'Ancient Chinese Astronomy.' *ASR*, 1938, Jan. (Reprinted by the China Society.)

CHATLEY, H. (9). 'Ancient Chinese Astronomy.' *RAS/ON*, 1939, no. 5, 65.

CHATLEY, H. (10). 'The Date of the Hsia Calendar *Hsia Hsiao Chêng*.' *JRAS*, 1938, 523.

CHATLEY, H. (10*a*). 'The Riddle of the *Yao Tien* Calendar.' (Appendix to Chatley 10), *JRAS*, 1938, p. 530.

CHATLEY, H. (11). '"The Heavenly Cover", a Study in Ancient Chinese Astronomy.' *O*, 1938, **61**, 10.

CHATLEY, H. (12). 'The Lunar Mansions in Egypt.' *ISIS*, 1940, **31**, 394.

CHATLEY, H. (13). 'Ancient Egyptian Star Tables and the Decans.' *O*, 1943, **65**, 121.

CHATLEY, H. (14). 'The Egyptian Celestial Diagram.' *O*, 1940, **63**, 68.

CHATLEY, H. (15). 'The Cycles of Cathay' (planetary cycles). *JRAS/NCB*, 1934, **65**, 36. (Addendum, *O*, 1937, **60**, 171.)

CHATLEY, H. (16). 'Thai Chi Shang Yuan; the Chinese Astrological Theory of Creation.' *JRAS/NCB*, 1926, **67**, 7; *TH*, 1937, **4**, 49. (Addendum, *O*, 1937, **60**, 171.)

CHATLEY, H. (17). 'Planetary Periods from the Han Histories.' *CJ*, 1936, **24**, 172.

CHATLEY, H. (18). 'The Chinese Calendar' [the 24 *chhi*]. *CJ*, 1933, **19**.

CHATLEY, H. (19). 'The True Era of the Chinese Sixty-Year Cycle.' *TP*, 19, **34**, 138.

CHATLEY, H. (20). 'The Sixty-Year and other Cycles.' *CJ*, 1934, **20**.

CHATLEY, H. (21). 'Sinological Notes' (on de Saussure (25), concerning planetary conjunctions, and on the Yao Tien chapter of the *Shu Ching*). *JRAS/NCB*, 1934, **65**, 187.

CHATLEY, H. (22). 'The Hangchow Bore.' *ASR*, 1950; *DHA*, 1950.

CHAUVENET, W. (1). *Manual of Spherical and Practical Astronomy*. Lippincott, Philadelphia, 1900.

CHAVANNES, E. (1). *Les Mémoires Historiques de Se-Ma Ts'ien* [Ssuma Chhien]. 5 vols. Leroux, Paris, 1895–1905. (Photographically reproduced, in China, without imprint and undated.)
1895 vol. 1 tr. *Shih Chi*, chs. 1, 2, 3, 4.
1897 vol. 2 tr. *Shih Chi*, chs. 5, 6, 7, 8, 9, 10, 11, 12.
1898 vol. 3 (i) tr. *Shih Chi*, chs. 13, 14, 15, 16, 17, 18, 19, 20, 21, 22.
vol. 3 (ii) tr. *Shih Chi*, chs. 23, 24, 25, 26, 27, 28, 29, 30.
1901 vol. 4 tr. *Shih Chi*, chs. 31, 32, 33, 34, 35, 36, 37, 38, 39, 40, 41, 42.
1905 vol. 5 tr. *Shih Chi*, chs. 43, 44, 45, 46, 47.

CHAVANNES, E. (3) (tr.). 'Le Voyage de Song [Sung] Yün dans l'Udyāna et le Gandhāra.' *BEFEO*, 1903, **3**, 379.

CHAVANNES, E. (6) (tr.). 'Les Pays d'Occident d'après le *Heou Han Chou*.' *TP*, 1907, **8**, 149. (Ch. 118, on the Western Countries, from *Hou Han Shu*.)

CHAVANNES, E. (7). 'Le Cycle Turc des Douze Animaux.' *TP*, 1906, **7**, 51.

CHAVANNES, E. (8). 'L'Instruction d'un Futur Empereur de Chine en l'an 1193' (on the astronomical, geographical, and historical charts inscribed on stone steles in the Confucian temple at Suchow, Chiangsu). In *Mémoires concernant l'Asie Orientale* (publ. Acad. des Inscriptions et Belles Lettres), vol. 1, p. 19. Leroux, Paris, 1913.

CHAVANNES, E. (9). *Mission Archéologique dans la Chine Septentrionale*. 2 vols. and portfolio of plates. Leroux, Paris, 1909–15. (Publ. de l'Ecole Franç. d'Extr. Orient, no. 13.)

CHAVANNES, E. (10). 'Les Deux Plus Anciens Spécimens de la Cartographie Chinoise.' *BEFEO*, 1903, **3**, 214. (With addendum by P. Pelliot.) (Rectifications in Chavannes (8), p. 20.)

CHAVANNES, E. (13). 'Le Calendrier des Yin [Shang].' *JA*, 1890 (8e sér.), **16**, 463.

CHAVANNES, E. (14). *Documents sur les Tou-Kiue* (*Turcs*) [*Thu-Chüeh*] *Occidentaux, recueillis et commentés par E.C.* .... Imp. Acad. Sci., St Petersburg, 1903.

CHAVANNES, E. (16). 'Trois Généraux Chinois de la Dynastie des Han Orientaux.' *TP*, 1906, **7**, 210. (Tr. ch. 77 of the *Hou Han Shu* on Pan Chhao, Pan Yung and Liang Chhin.)

CHAVANNES, E. & PELLIOT, P. (1). 'Un Traité Manichéen retrouvé en Chine, traduit et annoté.' *JA*, 1911 (10e sér.), **18**, 499; 1913 (11e sér.), **1**, 99, 261.

CHEN. See Chhen.

CHÊNG CHIN-TÊ (DAVID) (1). 'On the Mathematical Significance of the Ho Thu and Lo Shu.' *AMM*, 1925, **32**, 499.

CHÊNG CHIN-TÊ (DAVID) (2). 'The Use of Computing Rods in China.' *AMM*, 1925, **32**, 492.

CHÊNG, K. Y. (1). 'The Floods and Droughts of the Lower Yangtze Valley and their Predictions.' *Memoirs Nat. Inst. Meteorol. Nanking*, 1937, no. 9.

CHÊNG LIN (1) (tr.). *Prince Dan of Yann* [*Yen Tan Tzu*]. World Encyclopaedia Institute, Chungking, 1945.

CHÊNG TÊ-KHUN (2) (tr.). 'Travels of the Emperor Mu.' *JRAS/NCB*, 1933, **64**, 124; 1934, **65**, 128.

CHÊNG TÊ-KHUN (3). *Chinese Jade*. Chhêngtu, Sze., 1945. (West China Union University Museum, Guidebook Series, no. 1.)

CHÊNG TÊ-KHUN (4). 'An Introduction to Chinese Civilisation' (mainly prehistory). *ORT*, 1950, Aug. p. 28, 'Early Inhabitants'; Sept. p. 28, 'The Beginnings of Culture'; Oct. p. 29, 'The Building of Culture'.

CHESNEAUX, J. & NEEDHAM, J. (1). *Les Sciences en Extrême-Orient du 16ème au 18ème Siècle*. In *Histoire Génerale des Sciences*, vol. 2, ed. R. Taton, Presses Universitaires de France, Paris (in the press).

CHHEN, KUAN-SHÊNG (1). 'Matteo Ricci's Contribution to, and Influence on, Geographical Knowledge in China.' *JAOS*, 1939, **59**, 325, 509.

CHHEN, KUAN-SHÊNG (2). 'A Possible Source for Ricci's Notices on Regions near China.' *TP*, 1939, **34**, 179.

CHHEN KUAN-SHÊNG (3). '*Hai Lu*; Forerunner of Chinese Travel Accounts of Western Countries.' *MS*, 1942, **7**, 208.

CHHEN TA (1). 'Chinese Migrations with special reference to Labour Conditions.' *Bull. U.S. Bureau Labour Statistics*, no. 340, Washington, 1923.

CHHIBBER, H. L. (1). *The Mineral Resources of Burma*. London, 1934.

CHHIEN CHUNG-SHU (2). 'China in the English Literature of the Eighteenth Century.' *QBCB/E*, 1941 (N.S.), **2**, 7.

CHHU TA-KAO (2) (tr.). *Tao Tê Ching, a new translation*. Buddhist Lodge, London, 1937.

CHIANG SHAO-YUAN (1). *Le Voyage dans la Chine Ancienne, considéré principalement sous son Aspect Magique et Religieux*. Commission Mixte des Œuvres Franco-Chinoises (Office de Publications), Shanghai, 1937. Transl. from the Chinese by Fan Jen.

CHILDE, V. GORDON (8). 'The Oriental Background of European Science.' *MQ*, 1938, **1**, 105.

CHING PHEI-YUAN (1). 'Etude Comparative des diverses éditions du *Chouo Fou* [*Shuo Fu*].' *SSA*, 1946, no. 1.

CHOU HUNG-SHIH (1). 'We found the Source of the Yellow River.' *CREC*, 1954, **3** (no. 2), 2.

CHRISTIE, E. R. (1). *Through Khiva to Golden Samarkand*. London, 1925.

CHU KHO-CHEN (1) (CHU COCHING) = *1, 2*. 'The Origin of the Twenty-eight Mansions in Astronomy.' *POPA*, 1947, **55**, 62.

CHU KHO-CHEN (2). 'Some Chinese Contributions to Meteorology.' *GR*, 1918, **5**, 136.

CHU KHO-CHEN (3). 'Climatic Pulsations in China.' *GR*, 1926, **16**, 274.

CHU KHO-CHEN (4). 'Climatic Changes during Historic Time in China.' *TSSC*, 1932, **7**, 127; *JRAS/NCB*, 1931, **62**, 32.

CHU KHO-CHEN (5). *The Climatic Provinces of China*. Memoir No. 1, Academia Sinica Nat. Inst. of Meteorology, Nanking, 1930.

CHU KHO-CHEN (6). 'A Preliminary Study on the Weather Types of Eastern China.' *TSSC*, 1926, **4**, 33.

CHU KHO-CHEN (7). 'The Contribution of Chinese Scientists to Astronomy in the Early and Middle Ages' (in Russian). *PRDA*, 1953, **42** (no. 10), 66.

CHU KHO-CHEN (8). 'The Origin of the Twenty-eight Lunar Mansions.' Paper presented to the VIIIth International Congress of the History of Science, Florence, 1956.

CIAMPINI, J. (1). 'An Abstract of a Letter wrote some time since by Signior John Ciampini of Rome to Fr. Bernard Joseph a Jesu-Maria, etc., concerning the Asbestus, and manner of spinning and weaving an Incombustible Cloath thereof.' *PTRS*, 1701, **22**, 911.

CLARK, G. N. (1). *Science and Social Welfare in the Age of Newton*. Oxford. 2nd ed. 1949.

CLARK, R. T. RUNDLE (1). 'The Legend of the Phoenix, a Study in Egyptian Religious Symbolism.' *UBHJ*, 1949, **2**, 1; 1950, **3**, 114.

CLARK, W. E. (1). 'On the zero and place-value in Indian mathematics.' In *Indian Studies in Honour of C. R. Lanman*, p. 217.

CLARK, W. E. (2). '[History of] Science [in India].' Art. in *The Legacy of India*. Ed. G. T. Garrett, p. 335. Oxford, 1937.

CLARK, W. E. (3) (tr.). *The Āryabhaṭīya of Āryabhaṭa; an ancient Indian Work on Mathematics and Astronomy*. Univ. of Chicago Press, Chicago, 1930.

CLARKE, J. & GEIKIE, A. (1). *Physical Science in the Time of Nero, being a Translation of the 'Quaestiones Naturales' of Seneca, with notes by Sir Archibald Geikie*. Macmillan, London, 1910.

CLERKE, A. M. (1). *The System of the Stars*. Black, London, 1905.

COEDÈS, G. (2). 'A Propos de l'Origine des Chiffres Arabes.' *BLSOAS*, 1931, **6**, 323.

COHN, W. (1). 'The Deities of the Four Cardinal Points.' *TOCS*, 1940, **18**, 61.

COLEBROOKE, H. T. (1). 'On the Indian and Arabian Divisions of the Zodiack.' *TAS/B* (Asiatick Researches), 1807, **9**, 323.

COLEBROOKE, H. T. (2) (tr.). *Algebra, with Arithmetic and Mensuration, from the Sanskrit of Brahmagupta and Bhāskara.* Murray, London, 1817.

COLSON, F. H. (1). *The Week; an essay on the Origin and Development of the Seven-Day Cycle.* Cambridge, 1926.

COLTON, A. L. (1). (On apparent size of the sun at low altitudes.) In *Meteors and Sunsets observed by the Astronomers of the Lick Observatory in 1893, 1894 and 1895.* Lick Observatory Contributions, no. 5, p. 71, Sacramento, 1895.

CONRADY, A. & ERKES, E. (1). *Das älteste Dokument zur chinesische Kunstgeschichte, Tien-Wên, die 'Himmelsfragen' d. K'uh Yuan, abgeschl. u. herausgeg. v. E. Erkes.* Leipzig, 1931. Critiques: B. Karlgren, *OLZ*, 1931, **34**, 815; H. Maspero, *JA*, 1933, **222**, 59; Hsü Tao-Lin, *SA*, 1932, **7**, 204.

CONZE, E. (5). *Der Satz vom Widerspruch; zur Theorie des dialektischen Materialismus.* Beltz (pr. pr.), Hamburg, 1932.

COOLIDGE, J. L. (1). *A History of Geometrical Methods.* Oxford, 1940.

COOPER, E. R. (1). 'The Davis Backstaff or English Quadrant.' *MMI*, 1944, **30**, 59.

COPERNICUS, NICHOLAS (1). *De Revolutionibus Orbium Coelestium.* Nürnberg, 1543.

CORDIER, H. (1). *Histoire Générale de la Chine.* 4 vols. Geuthner, Paris, 1920.

CORDIER, H. (3). 'Description d'un Atlas Sino-Coréen Manuscrit du Musée Britannique'. In *Recueil de Voyages et de Documents pour servir à l'Histoire de la Géographie depuis le 13e siècle jusqu'à la fin du 16e; Section cartographique.* Leroux, Paris, 1896.

CORDIER, H. (4). *Les Monstres dans la légende et dans la Nature (Etudes sur les Traditions Tératologiques).* Pr. pub., Paris, 1890.

CORDIER, H. (7). 'The Life and Labours of Alexander Wylie.' *JRAS*, 1887, **19**, 351.

CORDIER, H. (8). *Essai d'une Bibliographie des Ouvrages publiés en Chine par les Européens au 17e et au 18e Siècle.* Leroux, Paris, 1883.

CORDIER, H. (11). *Mélanges d'Histoire et de Géographie Orientales.* Paris, Vol. 1, 1914; vol. 2, 1920.

CORROZET, G. & CHAMPIER, C. (1). *Le Catalogue des antiques erections des Villes et Cités, Fleuves et Fontaines, assisses es troys Gaules, cestassavoir Celticque Belgicque et Aquitaine, contenant deulx Livres....Avec ung petit traicté des Fleuves et Fontaines admirables, estans esdictes Gaules....*Juste, Lyon, *c.* 1535.

CORTESÃO, A. (1). *Cartografia e Cartografos Portugueses dos Seculos...15e e 16e.* 2 vols. Lisbon, 1935.

COSTARD, G. (1). 'A Letter...concerning the Chinese Chronology and Astronomy.' *PTRS*, 1747, **44**, 476.

COULING, S. (1). *Encyclopaedia Sinica.* Kelly & Walsh, Shanghai; O.U.P., Oxford & London, 1917.

COURANT, M. (1). *Bibliographie Coréenne.* Paris, 1895. (Pub. Ecole Langues Or. Viv. (3e sér.), no. 19.)

COUVREUR, F. S. (1) (tr.). *Tch'ouen Ts'iou [Chhun Chhiu] et Tso Tchouan [Tso Chuan]; Texte Chinois avec Traduction Française.* 3 vols. Mission Press, Hochienfu, 1914.

COUVREUR, F. S. (2). *Dictionnaire Classique de la Langue Chinoise.* Mission Press, Hsienhsien, 1890 (photographically reproduced, Vetch, Peiping, 1947).

COUVREUR, F. S. (3) (tr.). *'Li Ki' [Li Chi], ou Mémoires sur les Bienséances et les Cérémonies.* 2 vols. Hochienfu, 1913.

COWELL, P. H. & CROMMELIN, A. C. D. (1). See Olivier (1), p. 102. 1908.

CRANMER, G. E. (1). 'Denver's Chinese Sundial.' *POPA*, 1950, **58**, 119.

CRAWLEY, R. (1) (tr.). *Thucydides' 'History of the Peloponnesian War'.* Everyman edition; Dent, London, 1910.

CREEL, H. G. (1). *Studies in Early Chinese Culture* (1st series). Waverly, Baltimore, 1937.

CREEL, H. G. (4). *Confucius; the Man and the Myth.* Day, New York, 1949; Kegan Paul, London, 1951. Reviewed D. Bodde, *JAOS*, 1950, **70**, 199.

CRESSEY, G. B. (1). *China's Geographic Foundations; A Survey of the Land and its People.* McGraw-Hill, New York, 1934.

CRESSEY, G. B. (2). 'The Evolution of Chinese Cartography.' *GR*, 1934, **24**, 497.

CREW, H. & DE SALVIO, A. (1) (tr.). *Dialogues concerning Two New Sciences of Galileo.* New York, 1914.

CROMBIE, A. C. (1). *Robert Grosseteste and the Origins of Experimental Science.* Oxford, 1953.

CROMBIE, A. C. (2). *Augustine to Galileo; the History of Science, +400 to +1650.* Falcon, London, 1952.

CRONIN, V. (1). *The Wise Man from the West* (biography of Matteo Ricci). Hart-Davies, London, 1955.

DA CRUZ, GASPAR (1). *Tractado em que se cõtam muito por estẽco as cousas da China.* Evora, 1569, 1570 (the first book on China printed in Europe), tr. Boxer (1), and originally in *Purchas his Pilgrimes*, vol. 3, p. 81. London, 1625.

CULIN, S. (1). 'Chess and Playing-Cards; Catalogue of Games and Implements for Divination exhibited by the U.S. National Museum in connection with the Dept. of Archaeology and Palaeontology of the University of Pennsylvania at the Cotton States and International Exposition, Atlanta, Georgia, 1895.' *ARUSNM*, 1896, 671 (1898).

CULIN, S. (2). 'Chinese Games with Dice and Dominoes.' *ARUSNM*, 1893, 491.

CUMONT, F. (3). *Recherches sur le Manichéisme. I. La Cosmogonie Manichéenne.* Lamertin, Bruxelles, 1908.

CUNNINGHAM, R. (1). 'Nangsal Obum' (The story of,). *JWCBRS*, 1940, A, **12**, 35.

CURTIS, H. D. (1). 'Visibility of Faint Stars through Sighting-tubes and Holes.' *Lick Observatory Bulletin*, Sacramento, 1901 (no. 38), **2**, 67.

DAHLGREN, E. W. (1). 'Les Débuts de la Cartographie du Japon.' *AEO*, 1911, **4**, 1.

DALTON, O. M. (1). 'The Byzantine Astrolabe at Brescia' [of +1062]. *PBA*, 1926, **12**, 133.

DAMRY, A. (1). 'Le Père Verbiest et l'Astronomie Sino-Européenne.' *CET*, 1913, **34** (no. 7), 215.

DANJON, A. (1). *Astronomie Générale; Astronomie Sphérique et Eléments de Mécanique Céleste.* Sennac, Paris, 1952.

DANJON, A. & COUDERC, P. (1). *Lunettes et Téléscopes.* Paris, 1935.

DARESTE, C. (1). *Recherches sur la Production Artificielle des Monstruosités; ou, Essais de Tératogénie Expérimentale.* Reinwald, Paris, 1877.

DAS, S. R. (1). 'Astronomical Instruments of the Hindus.' *IHQ*, 1928, **4**, 256.

DAS, S. R. (2). 'The Alleged Greek Influence on Hindu Astronomy.' *IHQ*, 1928, **4**, 68.

DATTA, B. (1). 'On the Origin and Development of the Idea of Per Cent.' *AMM*, 1927, **34**, 530.

DATTA, B. (2). *The Science of the Śulba, A Study in Early Hindu Geometry.* Univ. Press, Calcutta, 1932.

DATTA, B. (3). 'Vedic Mathematics.' In *Cultural Heritage of India*, vol. 3, p. 382.

DATTA, B. & SINGH, A. N. (1). *History of Hindu Mathematics.* 2 vols. Motilal Banarsi Das, Lahore, 1935 (vol. 1), 1938 (vol. 2) (rev. O. Neugebauer, *QSGM/A*, 1936, **3**, 263).

DAUDIN, P. (1). '*L'Unité de Longueur dans l'Antiquité Chinoise.* Saigon, 1939.

DAUMAS, M. (1). *Les Instruments Scientifiques aux 17e et 18e Siècles.* Presses Univ. de France, Paris, 1953.

DAVIS, J. F. (2). 'On the Chinese Year.' *PTRS*, 1823, **113**, 91.

DAVIDSON, T. (1). 'On some Fossil Brachiopods of the Devonian Age from China.' *PGS*, 1853, **9**, 353.

DAVIS, T. L. & CHHEN KUO-FU (1). 'The Inner Chapters of *Pao Phu Tzu*.' *PAAAS*, 1941, **74**, 297. (Transl. of chs. 8 and 11; précis of the remainder.)

DAVISON, C. (1). *Founders of Seismology.* Cambridge, 1927.

DAWSON, W. R. (1). 'Ancient Egyptian Mathematics.' *SPR*, 1924, **19**, 50.

DEFOE, DANIEL (1). *The Consolidator; or, Memoirs of Sundry Transactions from the World in the Moon.* London, 1705. In Talboys' ed. vol. 9, p. 211.

DEFRÉMERY, C. & SANGUINETTI, B. R. (1) (tr.). *Voyages d'ibn Batoutah.* 5 vols. Soc. Asiat., Paris, 1853–9. (Many reprints.)

DEHERGUE, J. (2). (*a*) 'Gaubil, Historien de l'Astronomie Chinoise.' *BUA*, 1945 (3e sér.), **6**, 168. (*b*) 'Le Père Gaubil et ses correspondants (1639–1759).' *BUA*, 1944 (3e sér.), **5**, 354. (The 1945 paper has a very full index (with Chinese characters) to persons and other proper names, and subjects, in Gaubil, 1–5.)

DELAMBRE, M. (1). *Histoire de l'Astronomie.* 6 vols. 1817–27. (Vol. 1 includes *Ancienne Astronomie des Chinois*, pp. 347–99.) Courcier, Paris, 1817–27.

DELPORTE, E. (1). *Atlas Céleste.* (International Astronomical Union, Report of Commission 3.) Cambridge, 1930.

DELPORTE, E. (2). *Délimitation Scientifique des Constellations* (*Tables et Cartes*). (International Astronomical Union, Report of Commission 3.) Cambridge, 1930.

DEMBER, H. & UIBE, M. (1). (*a*) 'Versuch einer physikalischen Lösung des Problems der sichtbaren Grössenänderung von Sonne und Mond in verschiedenen Höhen über dem Horizont.' *ANP*, 1920 (4th ser.), **61**, 353. (*b*) 'Über die Gestalt des sichtbaren Himmelsgewölbes.' *ANP*, 1920 (4th ser.), **61**, 313.

DEMIÉVILLE, P. (2). Review of Chang Hung-Chao (*1*). *BEFEO*, 1924, **24**, 276.

DERHAM, W. (1). *Philosophical Experiments and Observations of Dr Robert Hooke.* London, 1726.

DESTOMBES, M. (1). 'Les Observations Astronomiques des Chinois au Lin-Yi.' In Stein, R. A. (1), p. 315.

DESTOMBES, M. (2). 'Globes Célestes et Catalogues d'Etoiles orientaux [Arabes] du Moyen-Age.' *Proc. VIIIth International Congress of the History of Sciences, Florence, 1956*, 313.

DESTOMBES, M. (3). 'Note sur le Catalogue d'Etoiles du Calife al-Ma'mūn.' *Proc. VIIIth International Congress of the History of Sciences, Florence, 1956*, 309.

DESTOMBES, M. (4). 'L'Orient et les Catalogues d'Etoiles au Moyen Age.' *A/AIHS*, 1956, **9**, 339.

DIAZ DEL CASTILLO, BERNAL (1). *The True Story of the Conquest of New Spain*, 1520. Ed. A. P. Maudsley. Routledge, London, 1928.

DICKS, D. R. (1). 'Ancient Astronomical Instruments.' Inaug. Diss. London, 1953. Abridged in *JBASA*, 1954, **64**, 77.

DICKSON, L. E. (1). *History of the Theory of Numbers*. 1st ed. Carnegie Institution, Washington, 1919–20; Carn. Inst. Publ. no. 256. 3 vols. 2nd ed. Stechert, New York, 1934. (Crit. Y. Mikami, *TBGZ*, 1921, nos. 354, 356, 362.)

DIELS-FREEMAN: FREEMAN, K. (1). *Ancilla to the Pre-Socratic Philosophers; a complete translation of the Fragments in Diels' 'Fragmente der Vorsokratiker'*. Blackwell, Oxford, 1948.

DIELS, H. (1). *Antike Technik*. Teubner, Leipzig & Berlin, 1914. 2nd ed. 1920 (rev. B. Laufer, *AAN*, 1917, **19**, 71).

DILLER, A. (1). 'The Parallels on the Ptolemaic Maps.' *ISIS*, 1941, **33**, 4.

AL-DIMASHQĪ, MU'AYYAD AL-DĪN AL-'URDĪ (1). *Risālat fī Kaifīya al-arṣād wa mā yuḥtāja ilā 'ilmihi wa 'amalihi min ṭuruq al-muwaddīya ilā ma'rifa 'audāt al-Kawākib*. (The Art of Astronomical Observations and the theoretical and practical knowledge needed to make them; and the methods leading to the understanding of the regularities of the stars.) See Jourdain (1); Seemann (1).

DINGLE, H. (1). 'The Essential Elements in the Scientific Revolution of the +17th century.' In *Actes du VIIe Congrès Internat. d'Histoire des Sciences*, p. 272. Jerusalem, 1953.

DIRINGER, D. (1). *The Alphabet; a Key to the History of Mankind*. Philos. Library, New York, 1948 (with foreword by Sir Ellis Minns).

DITTRICH, A. (1). 'Solstitium Hi-Kungi quod a. a. Chr. n. 656 evenit.' *MSRLSB/S*, 1952 (1953), no. 7.

DOBELL, CLIFFORD (2). 'Dr Uplavici.' *ISIS*, 1930, **30**, 268.

DOODSON & WARBURG (1). *Admiralty Manual of Tides*. 1941.

DORÉ, H. (1). *Recherches sur les Superstitions en Chine*. 15 vols. T'u-Se-Wei Press, Shanghai, 1914–29.
Pt. I, vol. 1, pp. 1–146: 'superstitious' practices, birth, marriage and death customs (*VS*, no. 32).
Pt. I, vol. 2, pp. 147–216: talismans, exorcisms and charms (*VS*, no. 33).
Pt. I, vol. 3, pp. 217–322: divination methods (*VS*, no. 34).
Pt. I, vol 4, pp. 323–488: seasonal festivals and miscellaneous magic (*VS*, no. 35).
Pt. I, vol. 5, sep. pagination: analysis of Taoist talismans (*VS*, no. 36).
Pt. II, vol. 6, pp. 1–196; pantheon (*VS*, no. 39).
Pt. II, vol. 7, pp. 197–298: pantheon (*VS*, no. 41).
Pt. II, vol. 8, pp. 299–462: pantheon (*VS*, no. 42).
Pt. II, vol. 9, pp. 463–480: pantheon, Taoist (*VS*, no. 44).
Pt. II, vol. 10, pp. 681–859: Taoist celestial bureaucracy (*VS*, no. 45).
Pt. II, vol. 11, pp. 860–1052: city-gods, field-gods, trade-gods (*VS*, no. 46).
Pt. II, vol. 12, pp. 1053–1286: miscellaneous spirits, stellar deities (*VS*, no. 48).
Pt. III, vol. 13, pp. 1–263: popular Confucianism, sages of the Wên Miao (*VS*, no. 49).
Pt. III, vol. 14, pp. 264–606: popular Confucianism, historical figures (*VS*, no. 51).
Pt. III, vol. 15, sep. pagination: popular Buddhism, life of Gautama (*VS*, no. 57).

DOUGLAS, R. K. (1). Miscellaneous translations. *OAA*, 1882, **1**, 1.

DOUTHWAITE, A. W. (1). 'Analyses of Chinese Inorganic Drugs.' *CMJ*, 1890, **3**, 53.

DRACHMANN, A. G. (1). 'Hero's and Pseudo-Hero's Adjustable Siphons.' *JHS*, 1932, **52**, 116.

DRACHMANN, A. G. (2). 'Ktesibios, Philon and Heron; a Study in Ancient Pneumatics.' *AHSNM*, 1948, **4**, 1–197.

DRACHMANN, A. G. (3). 'Heron and Ptolemaios.' *CEN*, 1950, **1**, 117.

DRACHMANN, A. G. (6). 'The Plane Astrolabe and the Anaphoric Clock.' *CEN*, 1954, **3**, 183.

DRECKER, J. (1). *Zeitmessung und Sterndeutung in geschichtlicher Darstellung*. Berlin, 1925.

DRECKER, J. (2). 'Gnomon und Sonnenuhren.' Inaug. Diss., Aachen, 1909.

DRECKER, J. (3). *Die Theorie der Sonnenuhren*. de Gruyter, Berlin, 1925. (Gesch. d. Zeitmessung u.d. Uhren, Liefg. E.)

DREYER, J. L. E. (1). 'On the Origin of Ptolemy's Catalogue of Stars.' *RAS/MN*, 1917, **77**, 528; 1918, **78**, 343.

DREYER, J. L. E. (2). 'The Instruments in the Old Observatory in Peking.' *PRIA*, 1883 (2nd ser.), **3**, 468; *C*, 1881, **1**, 134.

DREYER, J. L. E. (3). *Tycho Brahe; a Picture of Scientific Life and Work in the +16th Century*. Black, Edinburgh, 1890.

DREYER, J. L. E. (4). *History of the Planetary Systems from Thales to Kepler*. Cambridge, 1906. Reissued, in photolitho form with paper covers, as *A History of Astronomy from Thales to Kepler*, Dover, New York, 1953.

DROVER, C. B. (1). 'A Mediaeval Monastic Water-Clock.' *AHOR*, 1954.

DUBOIS-REYMOND, C. (1). 'A Chinese Sun-Dial.' *JRAS/NCB*, 1914, **45**, 85.

DUBS, H. H. (2) (tr., with assistance of Phan Lo-Chi and Jen Thai). *History of the Former Han Dynasty, by Pan Ku; a Critical Translation with Annotations.* 3 vols. Waverly, Baltimore, 1938.

DUBS, H. H. (5). 'The Beginnings of Alchemy.' *ISIS*, 1947, **38**, 62.

DUBS, H. H. (6). 'A Military Contact between Chinese and Romans in −36.' *TP*, 1940, **36**, 64.

DUBS, H. H. (7). *Hsün Tzu; the Moulder of Ancient Confucianism.* Probsthain, London, 1927.

DUBS, H. H. (20). 'The Growth of a Sinological Legend; A Correction to Yule's "Cathay".' (The supposed bringing of Greek astronomical books to China in +164.) *JAOS*, 1946, **66**, 182.

DUBS, H. H. (21). 'The Conjunction of May −205.' *JAOS*, 1935, **55**, 310. Reprinted in Dubs (2), vol. 1, p. 151.

DUBS, H. H. (22). 'Canon of Lunar Eclipses for Anyang and China, −1400 to −1000.' *HJAS*, 1947, **10**, 162.

DUBS, H. H. (23). 'Eclipses during the first 50 years of the Han Dynasty.' *JRAS/NCB*, 1935, **66**, 73. Reprinted, in altered form, in Dubs (2), vol. 1, p. 288.

DUBS, H. H. (24). 'Solar Eclipses during the Former Han Dynasty.' *OSIS*, 1938, **5**, 499. Reprinted, emended, in Dubs (2), vol. 3, pp. 546 ff.

DUBS, H. H. (26). 'The Date of the Shang Period.' *TP*, 1951, **40**, 322. Postscript, *TP*, 1953, **42**, 101.

DUBS, H. H. (29). 'An Ancient Military Contact between Romans and Chinese.' *AJP*, 1941, **62**, 322.

DUBS, H. H. (30). 'A Roman Influence on Chinese Painting.' *CP*, 1943, **38**, 13.

DUDENEY, H. E. (1). 'Magic Squares.' *EB* (14th ed.), vol. 14, p. 627.

D[UDGEON], J[OHN] (1). Letter commenting on Wylie (15), and identifying Fukienese *mi* (Sunday) as the first Syllable of Mithras and *mitra*. *CRR*, 1871, **4**, 195.

DUHEM, P. (1). *Etudes sur Léonard de Vinci.* 3 vols. Hermann, Paris.
Vols. 1, 2: 'Ceux qu'il a lus et ceux qui l'ont lu.' 1906, 1909.
Vol. 3: 'Les précurseurs Parisiens de Galilée.' 1913.
(Vol. 1: Albert of Saxony, Bernardino Baldi, Themon, Cardan, Palissy, etc.
Vol. 2: Nicholas of Cusa, Albertus Magnus, Vincent of Beauvais, Ristoro d'Arezzo, etc.
Vol. 3: Buridan, Soto, Nicholas d'Oresme, etc.)

DUHEM, P. (3). *Le Système du Monde; Histoire des Doctrines Cosmologiques de Platon à Copernic.* 5 vols. Paris, 1913–17.

DUMONT, M. (1). 'Un Professeur de Mathématiques au 9e siècle; Mohammed ibn Mousa al-Khowarizmi.' *RGS*, 1947, **54**, 7.

DUMOUTIER, G. (1). 'Etude sur un Portulan annamite du 15e Siècle.' *BGHD*, 1896, 141.

DUNLOP, D. M. (1). *History of the Jewish Khazars.* Princeton Univ. Press, Princeton, N.J., 1954.

DUNLOP, D. M. (2). 'Scotland according to al-Idrīsī, *c.* +1154.' *SHR*, 1947, **26**, 114; 1955, **34**, 95.

DUPUIS, J. (1) (tr.). *Works of Theon of Smyrna.* Paris, 1892.

DURAND, V. & DE LA NOË, G. (1). 'Cadran Solaire Portatif trouvé au Crêt-Châtelard, commune de St Marcel-de-Felines (Loire).' *MSAF*, 1896 (6e sér.), **7** [**57**], 1.

DURET, CLAUDE (1). *Histoire de l'Origine des Langues de Cest Univers.* Paris, 1613.

DUYVENDAK, J. J. L. (1). 'Sailing Directions of Chinese Voyages' (a Bodleian Library MS.). *TP*, 1938, **34**, 230.

DUYVENDAK, J. J. L. (6). Comments on Pasquale d'Elia's *Galileo in Cina*. *TP*, 1948, **38**, 321.

DUYVENDAK, J. J. L. (8). *China's Discovery of Africa.* Probsthain, London, 1949. (Lectures given at London University, Jan. 1947; rev. P. Paris, *TP*, 1951, **40**, 366.)

DUYVENDAK, J. J. L. (9). 'The True Dates of the Chinese Maritime Expeditions in the Early Fifteenth Century.' *TP*, 1939, **34**, 341.

DUYVENDAK, J. J. L. (10). 'Ma Huan Re-examined.' *VKAWA/L*, 1933 (n.s.), **32**, no. 3.

DUYVENDAK, J. J. L. (11). 'Voyages de Tchêng Houo [Chêng Ho].' In Yusuf Kamal, *Monumenta Cartographica*, 1939, vol. 4, pp. 1411 ff.

DUYVENDAK, J. J. L. (12). 'The Last Dutch Embassy to the Chinese Court' (+1794 to +1795). *TP*, 1938, **34**, 1, 223; 1939, **35**, 329.

DUYVENDAK, J. J. L. (16). 'An Illustrated Battle-Account in the History of the Former Han Dynasty.' *TP*, 1938, **34**, 249.

DUYVENDAK, J. J. L. (17). 'The "Guest-Star" of +1054.' *TP*, 1942, **36**, 174.

DUYVENDAK, J. J. L. (20). 'A Chinese *Divina Commedia*.' *TP*, 1952, **41**, 255. (Also sep. pub., Brill, Leiden, 1952.)

DUYVENDAK, J. J. L., MAYALL, N. V. & OORDT, J. H. (1). 'Further Data bearing on the Identification of the Crab Nebula with the Supernova of +1054.' *PASP*, 1942, **54**, 91.

DYE, D. S. (1). *A Grammar of Chinese Lattice.* 2 vols. Harvard-Yenching Institute, Cambridge, Mass., 1937. (Harvard-Yenching Monograph Series, nos. 5, 6.)

EARLE, A. M. (1). *Sundials and Roses of Yesterday*. Macmillan, New York & London, 1902.

EASTLAKE, F. W. (1). 'Finger-Reckoning in China.' *CR*, 1880, **9**, 249, 319.

EBERHARD, W. (6). 'Beiträge zur kosmologischen Spekulation Chinas in der Han-Zeit.' *BA*, 1933, **16**, 1.

EBERHARD, W. (9). *A History of China from the Earliest Times to the Present Day*. Routledge & Kegan Paul, London, 1950. Tr. from the Germ. ed. (Swiss pub.) of 1948 by E. W. Dickes. Turkish ed. *Čin Tarihi*, Istanbul, 1946. (Crit. K. Wittfogel, *AA*, 1950, **13**, 103; J. J. L. Duyvendak, *TP*, 1949, **39**, 369; A. F. Wright, *FEQ*, 1951, **10**, 380.)

EBERHARD, W. (10). 'Neuere chinesische und japanische Arbeiten zur altchinesischen Astronomie.' *AM*, 1933, **9**, 597.

EBERHARD, W. (11). 'Frühchinesische Astronomie.' *FF*, 1933, **9**, 252. (Summary of Eberhard & Henseling and of Eberhard, Müller & Henseling, q.v.)

EBERHARD, W. (12). 'Untersuchungen an astronomischen Texten des Chinesischen Tripiṭaka.' *MS*, 1940, **5**, 208.

EBERHARD, W. (13). 'Das astronomische Weltbild im alten China.' *NW*, 1936, **24**, 517.

EBERHARD, W. (14). Review and critique of Liu Chao-Yang (*6*) on the ancient calendars. *OR*, 1949, **2**, 179.

EBERHARD, W. (15). 'Chinesische Volkskalender und buddhistisches Tripiṭaka.' *OLZ*, 1937, p. 346.

EBERHARD, W. (16). 'Index of Words in the Papers of W. Eberhard and his collaborators on Chinese Astronomy.' *MS*, 1942, **7**, 242.

EBERHARD, W. & HENSELING, R. (1). 'Beiträge z. Astronomie d. Han-Zeit, I. Inhalt des Kapitels ü. Zeiteinteilung d. Han-Annalen.' *SPAW/PH*, 1933, **23**, 209.

EBERHARD, W. & MÜLLER, R. (1). 'Contributions to the Astronomy of the Han Period. III. The Astronomy of the Later Han.' *HJAS*, 1936, **1**, 194.

EBERHARD, W. & MÜLLER, R. (2). 'Contributions to the Astronomy of the San Kuo Period.' (Includes translation of Wang Fan's *Hun Thien Hsiang Shuo*.) *MS*, 1936, **2**, 149. (Crit. H. Maspero, *JA*, 1939, **231**, 459.)

EBERHARD, W., MÜLLER, R. & HENSELING, R. (1). 'Beiträge z. Astronomie d. Han-Zeit. II.' *SPAW/PH*, 1933, **23**, 937.

EDGAR, J. H. (1). 'Tibetan Numbers.' *JWCBRS*, 1936, **8**, 170.

EDKINS, J. (6). 'Local Value in Chinese Arithmetical Notation.' *JPOS*, 1886, **1**, 161.

EDKINS, J. (7). 'Star-Names among the Ancient Chinese.' *CR*, 1887, **16**, 257, 357. (Summary in *N*, 1888, **39**, 309.)

EDKINS, J. (8). 'The Babylonian Origin of Chinese Astronomy and Astrology.' *CR*, 1885, **14**, 90.

EDKINS, J. (10). 'On the Poets of China during the Period of the Contending States and of the Han Dynasty' (Chhü Yuan, etc.). *JPOS*, 1889, **2**, 201.

EDWARDS, R. (1). 'The Cave Reliefs at Ma-hao [near Chiating, Szechuan].' *AA*, 1954, **17**, 5, 103.

EICHELBERGER, W. S. (1). '[History of our Knowledge of] the Distances of the Heavenly Bodies.' *ARSI*, 1916, 169.

EISLER, ROBERT (1). *The Royal Art of Astrology*. Joseph, London, 1946. (Crit. H. Chatley, *O*, 1947, **67**, 187.)

EISLER, ROBERT (2). 'The Polar Sighting-Tube.' *A/AIHS*, 1949, **2**, 312.

EISLER, ROBERT (3). 'Early References to Fossil Fishes.' *ISIS*, 1943, **34**, 363.

EITEL, E. J. (1) (tr.). 'Travels of the Emperor Mu.' *CR*, 1888, **17**, 233, 247.

EKMAN, W. V. (1). *Instructions for the Use of the Ekman Current Meter (1932 Pattern)*. Pr. pr. Henry Hughes & Son, Ltd., London, n.d.

D'ELIA, PASQUALE (1). 'Echi delle Scoperte Galileiane in Cina vivente ancora Galileo (1612–1640).' *AAL/RSM*, 1946, (8a ser.), **1**, 125. Republished in enlarged form as 'Galileo in Cina. Relazioni attraverso il Collegio Romano tra Galileo e i gesuiti scienzati missionari in Cina (1610–1640).' *Analecta Gregoriana*, **37** (Series Facultatis Missiologicae A (N/1)), Rome, 1947. (Reviews: G. Loria, *A/AIHS*, 1949, **2**, 513; J. J. L. Duyvendak, *TP*, 1948, **38**, 321; G. Sarton, *ISIS*, 1950, **41**, 220.)

D'ELIA, PASQUALE (2) (ed.). *Fonti Ricciani; Storia dell'Introduzione del Cristianismo in Cina*, 3 vols. Libreria dello Stato, Rome, 1942–9. Cf. Trigault (1); Ricci (1).

D'ELIA, PASQUALE (3). 'The Spread of Galileo's Discoveries in the Far East.' *EW*, 1950, **1**, 156. (English résumé of (1).)

D'ELIA, PASQUALE (4). 'Presentazione della prima Traduzione Cinese di Euclide' (of Matteo Ricci & Hsü Kuang-Chhi) [with notes on the problem of the Yuan version]. *MS*, 1956, **15**, 161.

D'ELIA, PASQUALE (5). *Il Mappamondo Cinese del P. Matteo Ricci S.J. (3a edizione, 1602) conservato presso la Biblioteca Vaticana*. Vatican Library, Rome, 1938.

ELIADE, MIRCEA (1). *Le Mythe de l'Eternel Retour; Archétypes et Répétition*. Gallimard, Paris, 1949.

ELIADE, MIRCEA (2). *Traité d'Histoire des Religions*. Payot, Paris, 1949.

ELIADE, MIRCEA (4). 'Metallurgy, Magic and Alchemy.' *Z*, 1938, **1**, 85.

ELIADE, M. (5). *Forgerons et Alchimistes*. Flammarion, Paris, 1956.

ELLIOTT-SMITH, SIR GRAFTON (1). (*a*) *The Ancient Egyptians and the Origin of Civilisation.* London, 1923. (*b*) *The Diffusion of Culture.* London, 1933. (*c*) *Human History.* London, 1934 (2nd ed.).

ERKES, E. (1) (tr.). 'Das Weltbild d. *Huai Nan Tzu*' (transl. of ch. 4). *OAZ*, 1918, **5**, 27.

ERKES, E. (8). 'Chhü Yüan's *Thien Wên.*' *MS*, 1941, **6**, 273.

ERKES, E. (11). Observations on Karlgren's 'Fecundity Symbols in Ancient China' (9). *BMFEA*, 1931, **3**, 63.

ERKES, E. (17). 'Chinesische-Amerikanische Mythenparallelen.' *TP*, 1925, **24**, 32.

VAN ESBROECK, G. (1). 'Commentaires Etymographiques sur les Jades Astronomiques.' *MCB*, 1951, **9**, 161.

VAN ESBROECK, G. (2). 'Les Sept Etoiles Directrices.' *MCB*, 1951, **9**, 171.

EUCLID. See Heath (1).

EVANS, JOAN (1). *Magical Jewels of the Middle Ages and the Renaissance, particularly in England.* Oxford, 1922.

EVELYN, JOHN (1). 'Of a Method of making more lively Representations of Nature in Wax than are extant in Painting, and of a New Kind of Maps in Bas-Relief; both practised in France.' *PTRS*, 1665, **1**, 99.

FABRE, M. (1). *Pékin, ses Palais, ses Temples, et ses Environs.* Librairie Française, Tientsin, 1937.

FABRICIUS, J. A. (1). *Bibliotheca Graeca....* Edition of G. C. Harles, 12 vols. Bohn, Hamburg, 1808.

FARAUT, F. G. (1). *L'Astronomie Cambodgienne.* Schneider, Saigon; 1910.

FARRINGTON, B. (4). *Greek Science (Thales to Aristotle); its meaning for us.* Penguin Books, London, 1944.

FARRINGTON, B. (8). 'The Rise of Abstract Science among the Greeks.' *CEN*, 1953, **3**, 32.

FARRINGTON, B. (15). 'The Greeks and the Experimental Method.' *D*, 1957, **18**, 68.

FARRINGTON, O. C. (1). 'Amber; its Physical Properties and Geological Occurrence.' *FMNHP/GLS*, no. 3, Chicago, 1923.

FARRINGTON, O. C. & LAUFER, B. (1). 'Agate, Physical Properties and Origin; Archaeology and Folklore.' *FMNHP/GLS*, no. 8, Chicago, 1927.

FEDCHINA, V. N. (1). 'The Chinese 13th-century Traveller [Chhiu] Chhang-Chhun' (in Russian). In *Iz Istorii Nauki i Tekhniki Kitaya* (Essays in the History of Science and Technology in China), p. 172. Acad. Sci. Moscow, 1955.

FEER, L. (1). 'Fragments extraits du Kandjour.' App. 5. 'Les Etages Célestes et la Transmigration traduit du livre Chinois *Lou-tao-tsi* [*Liu Tao Chi*].' *AMG*, 1883, **5**, 529.

FEIFEL, E. (1) (tr.). *Pao Phu Tzu, Nei Phien*, chs. 1–3. *MS*, 1941, **6**, 113.

FEIFEL, E. (2) (tr.). *Pao Phu Tzu, Nei Phien*, ch. 4. *MS*, 1944, **9**, 1.

FEIFEL, E. (3) (tr.). *Pao Phu Tzu, Nei Phien*, ch. 11. *MS*, 1946, **11**, 1.

FELDHAUS, F. M. (1). *Die Technik der Vorzeit, der Geschichtlichen Zeit, und der Naturvölker* (encyclopaedia). Engelmann, Leipzig and Berlin, 1914.

FELDHAUS, F. M. (4). *Die geschichtlichen Entwicklung d. Zahnrades.* Stolzenberg, Berlin-Reinickendorf, 1911.

FELDHAUS, F. M. (22). 'Die Uhren des Königs Alfons X von Spanien.' *DUZ*, 1930, **54**, 608.

F[ELDHAUS], F. M. (23). 'Chinesische Logarithmen.' *GTIG*, 1917, **4**, 240.

FÊNG, S. H. (FONG, S. H.) (1). 'The *Hua Yang Kuo Chih.*' *JWCBRS*, 1940, A, **12**, 225.

FÊNG YU-LAN (5) (tr.). *Chuang Tzu; a new selected translation with an exposition of the philosophy of Kuo Hsiang.* Commercial Press, Shanghai, 1933.

FÊNG YU-LAN (6). 'Mao Tsê-Tung's "On Practice", and Chinese Philosophy.' *PC*, 1951, **4** (no. 10), 5.

FERGUSON, J. C. (3). (*a*) 'The Chinese Foot Measure.' *MS*, 1941, **6**, 357. (*b*) *Chou Dynasty Foot Measure.* Privately printed, Peiping, 1933. (See also a note on a graduated rule of *c.* +1117, *TH*, 1937, **4**, 391.)

FERGUSON, J. C. (7). 'Political Parties of the Northern Sung Dynasty.' *JRAS/NCB*, 1927, **58**, 35.

FERGUSON, T. (1). *Chinese Researches, Pt. I. Chinese Chronology and Cycles.* London, 1881.

FERRAND, G. (2) (tr.). *Voyage du marchand Sulaymān en Inde et en Chine rédigé en +851; suivi de remarques par Abū Zayd Haṣan (vers +916).* Bossard, Paris, 1922.

FESTUGIÈRE, A. G. (1). '*La Révélation d'Hermès Trismégiste, I. L'Astrologie et les Sciences Occultes.* Gabalda, Paris, 1944; rev. J. Filliozat, *JA*, 1944, **234**, 349.

FILLIOZAT, J. (7). 'L'Inde et les Echanges Scientifiques dans l'Antiquité.' *JWH*, 1953, **1**, 353.

FILLIOZAT, J. (8). 'La Pensée Scientifique en Asie Ancienne.' *BSEIC*, 1953, **28**, 5. (On transmissions from Mesopotamia to China and India, on pneumatic medicine, the lunar mansions and the mathematical zero.)

FILLIOZAT, J. (9). 'Le Symbolisme du Monument du Phnom Bǎkhèṅ.' *BEFEO*, 1953, **49**, 527.

FINOT, L. (1). *Les Lapidaires Indiens.* Bouillon, Paris, 1896. (Biblioth. de l'École des Hautes Etudes, no. 111.)

FINSTERBUSCH, K. (1). 'Das Verhältnis des *Schan-hai-djing* [*Shan Hai Ching*] zur bildenden Kunst.' *ASAW/PH*, 1952, **46**, 1–136. Crit. L. Lanciotti, *EW*, 1954, **4**, 1.

FISCHER, JOSEPH (1). *Claudius Ptolemaeus; Geographiae Codex Urbinas Graecus 82 phototypice depictus.* 4 vols. (MS. of *c.* +1200.) Brill, Leiden, 1932.

FISCHER, OTTO (2). *Chinesische Malerei der Han-Dynastie.* Neff, Berlin, 1931.

FISCHER, T. (1). *Sammlung mittelälterlicher Welt- und See-Karten italienischen Ursprungs.* Venice, 1877. (Beitr. z. Gesch. d. Erdkunde und d. Kartographie in Italien im Mittelalter.)

FISHER, R. A. (1). 'Reconstruction of the Sieve of Eratosthenes.' *MG*, 1929, **14**, 565.

FISHER, R. A. & YATES, F. (1). *Statistical Tables for Biological, Agricultural and Medical Research.* Oliver & Boyd, Edinburgh, 1938, 1953.

FONG. See Fêng.

DE FONTENELLE, B. (1). *Entretiens sur la Pluralité des Mondes.* Brunet, Paris, 1698. (1st ed. 1686; Eng. trs. 1688, 1702.)

[FORBES, R. J.] (4*a*). *Histoire des Bitumes, des Epoques les plus Reculées jusqu'à l'an 1800.* Shell, Leiden, n.d.

FORBES, R. J. (4*b*). *Bitumen and Petroleum in Antiquity.* Brill, Leiden, 1936.

FORBES, R. J. (5). 'The Ancients and the Machine.' *A/AIHS*, 1949, **2**, 919.

FORBES, R. J. (8). *Metallurgy* [*in the Mediterranean Civilisations and the Middle Ages*] in *A History of Technology*, ed. C. Singer *et al.* O.U.P., Oxford, 1956, vol. 2, p. 41.

FORKE, A. (3) (tr.). *Me Ti* [*Mo Ti*] *des Sozialethikers und seiner Schüler philosophische Werke.* Berlin, 1922. (*MSOS*, Beibände, **23–5**.)

FORKE, A. (4) (tr.). *Lun Hêng, Philosophical Essays of Wang Chhung*: Vol. 1, 1907. Kelly & Walsh, Shanghai; Luzac, London; Harrassowitz, Leipzig. Vol. 2, 1911 (with the addition of Reimer, Berlin). (*MSOS*, Beibände, **10** and **14**.) (Crit. P. Pelliot, *JA*, 1912 (10e sér.), **20**, 156.)

FORKE, A. (6). *The World-Conception of the Chinese; their Astronomical, Cosmological and Physico-philosophical Speculations* (Pt. 4 of this, on the Five Elements, is reprinted from Forke (4), vol. 2, App. I). Probsthain, London, 1925. German tr. *Gedankenwelt des chinesischen Kulturkreis.* München, 1927. Chinese tr. *Chhi-Na Tzu-Jan Kho-Hsüeh Ssu-Hsiang Shih.* Crit. B. Schindler, *AM*, 1925, **2**, 368.

FORKE, A. (9). *Geschichte d. neueren chinesischen Philosophie* (i.e. from beg. of Sung to modern times). de Gruyter, Hamburg, 1938. (Hansische Univ. Abhdl. a. d. Geb. d. Auslandskunde, no. 46 (ser. B, no. 25).)

FORKE, A. (13). *Geschichte d. alten chinesischen Philosophie* (i.e. from antiquity to beg. of Former Han). de Gruyter, Hamburg, 1927. (Hamburg. Univ. Abhdl. a. d. Geb. d. Auslandskunde, no. 25 (ser. B, no. 14).)

FOTHERINGHAM, J. K. (1). 'The Calendar.' *NA*, 1929 (1931), p. 734.

FOTHERINGHAM, J. K. (2). *Historical Eclipses.* Halley Lecture. Oxford, 1921. (Abstr. *JBASA*, 1921, **32**, 197.)

FOTHERINGHAM, J. K. (3). 'The Story of Hi and Ho.' *JBASA*, 1932, **43**, 248.

FOTHERINGHAM, J. K. (4). 'The New Star of Hipparchus and the Dates of Birth and Accession of Mithridates.' *RAS/MN*, 1919, **79**, 162.

FRANCK, H. A. (1). *Roving through Southern China.* Century, New York, 1925.

FRANK, J. (1). *Die Verwendung des Astrolabs nach al-Chwarizmi* [*al-Khwārizmī*]. Mencke, Erlangen, 1922. (Abhdl. z. Gesch. d. Naturwiss u. d. Med. no. 3.)

FRANKE, H. (1). 'Some Remarks on the Interpretation of Chinese Dynastic Histories' (with special reference to the Yuan). *OR*, 1950, **3**, 113.

FRANKE, H. (2). 'Beiträge z. Kulturgeschichte Chinas unter der Mongolenherrschaft.' (Complete translation and annotation of the *Shan Chü Hsin Hua* by Yang Yü, +1360.) *AKML*, 1956, **32**, 1–160.

FRANKE, H. (8). 'Some Remarks on Yang Yü and his *Shan Chü Hsin Hua*.' *JOSHK*, 1955, **2**, 302.

FRANKE, O. (5). 'Zur Frage der Einführung des Buddhismus in China.' *MSOS*, 1910, **13**, 295.

FRANKE, O. (6). 'Der kosmische Gedanke in Philosophie und Staat d. Chinesen.' *VBW* (1925/1926), 1928, p. 1. (Reprinted in Franke (8), p. 271.)

FRÉCHET, M. (1). *Les Mathématiques et le Concret.* Presses Univ. de France, Paris, 1955; rev. P. Labérenne, *LP*, 1956 (no. 69), 140.

FREEMAN, K. (1). *The Pre-Socratic Philosophers, a companion to Diels' 'Fragmente der Vorsokratiker'.* Blackwell, Oxford, 1946.

FRÉRET, N. (1). 'De l'Antiquité et de la Certitude de la Chronologie Chinoise.' *Mémoires de Littérature tirés des Registres de l'Acad. Roy. des Inscriptions et Belles-Lettres*, 1736, **10**, 377; 1743, **15**, 495; 1753, **18**, 178.

FRESCURA, B. & MORI, A. (1). 'Cartografia dell'Estremo Oriente; Un Atlante Cinese della Magliabecchiana di Firenze.' *RGI*, 1894, **1**, 417, 475.

FRITSCHE, H. (1). *On Chronology and the Construction of the Calendar with special regard to the Chinese Computation of Time compared to the European.* (Lithographed handwriting.) Laverentz, St Petersburg, 1886.

FRITZ, H. (1). *Die Periode d. solaren und erdlichen Phaenomäne.* Zürich, 1896. Eng. tr. by W. W. Reed, *MWR*, 1928, **56**, 401.

FROST, A. H. & FENNELL, C. A. M. (1). 'Magic Squares.' *EB* (13th ed.), vol. 17, p. 310.

FU, T. (1). 'Chinese Astronomy.' *ERE*, vol. 12, p. 74.

FUCHS, W. (1). *The 'Mongol Atlas' of China by Chu Ssu-Pên, and the 'Kuang Yü Thu'.* Fu-Jen Univ. Press, Peiping, 1946. (*MS/M*, no. 8) (rev. J. J. L. Duyvendak, *TP*, 1949, **39**, 197).

FUCHS, W. (2). *Der Jesuiten-Atlas der Khang-Hsi Zeit, seine Entstehungsgeschichte nebst Namenindices für die Karten der Mandjurei, Mongolei, Osttürkestan und Tibet, mit Wiedergabe der Jesuiten-Karten in Originalgrösse.* Fu-Jen Univ. Press, Peiping, 1943. 1 vol. + 1 box of plates. (*MS/M*, nos. 3 and 4.)

FUCHS, W. (3). 'Materialen zur Kartographie d. Mandju-Zeit.' *MS*, 1936, **1**, 386; 1938, **3**, 189.

FUCHS, W. (5). 'A Note on Fr. M. Boym's Atlas of China.' *IM*, 1953, **9**, 71.

FUNG YU-LAN. See Fêng Yu-Lan.

FUNKHOUSER, H. G. (1). 'A Note on a Tenth Century Graph.' *OSIS*, 1936, **1**, 260.

GABRIELI, G. (1). 'Giovanni Schreck Linceo, gesuita e missionario in Cina e le sue Lettere dell'Asia.' *RAL/MSF*, 1936 (sér. 6), **12**, 462.

GABRIELI, G. (2). 'I Lincei e la Cina.' *RAL/MSF*, 1936 (ser. 6), **12**, 242.

GAFFAREL, L. (1). *Curiositez Inouyes, sur la Sculpture Talismanique des Persans, horoscope des Patriarches, et Lecture des Estoiles.* [Paris], 1650 (1st edition, 1637).

GALEN. See Kühn.

GALILEO, GALILEI (1). *Opera.* Florence, 1842.

LE GALL, S. (1). *Le Philosophe Tchou Hi, Sa Doctrine, son Influence.* T'ou-se-wei, Shanghai, 1894 (*VS*, no. 6). (Incl. tr. of part of ch. 49 of *Chu Tzu Chhüan Shu.*)

GALLAGHER, L. J. (1) (tr.). *China in the 16th Century; the Journals of Matthew Ricci, 1583–1610.* Random House, New York, 1953. (A complete translation, preceded by inadequate bibliographical details, of Nicholas Trigault's *De Christiana Expeditione apud Sinas* (1615). Based on an earlier publication: *The China that Was; China as discovered by the Jesuits at the close of the 16th Century; from the Latin of Nicholas Trigault.* Milwaukee, 1942.) Identifications of Chinese names in Yang Lien-Shêng (4). Crit. J. R. Ware, *ISIS*, 1954, **45**, 395.

GAMOW, G. (1). 'Supernovae.' *SAM*, 1949, **181** (no. 12), 19.

GANDZ, S. (1). 'On the Origin of the term Root.' *AMM*, 1926, **33**, 261; 1928, **35**, 67.

GANDZ, S. (2). 'The Origin of the Ghubar Numerals, or Arabian Abacus and the Articuli.' *ISIS*, 1931, **16**, 393.

GANDZ, S. (3). 'Die Harpendonapten oder Seilspanner und Seilknüpfer.' *QSGM/B*, 1930, **1**, 255.

GANDZ, S. (4). 'Notes on Egyptian and Babylonian Mathematics.' In *Sarton Presentation Volume*, ed. Ashley-Montagu, M. F., p. 453. Schuman, New York, 1944.

GANDZ, S. (5). 'The *Mishnāh ha Middot*, the first Hebrew Geometry, of about +150.' *QSGM/A*, 1932, **2**.

GARDNER, C. T. (1). 'On Chinese Time.' *JESL*, 1870, **2**, 26. (Not seen.)

GARDNER, C. T. (2). 'The Tablet of Yü' (at the Yu-Lin Temple near Ningpo). *CR*, 1873, **2**, 293.

GASPARDONE, E. (1). 'Matériaux pour servir à l'Histoire d'Annam.' *BEFEO*, 1929, **29**, 63.

GATTY, A., EDEN, H. K. F. & LLOYD, E. (1). *A Book of Sun-Dials.* London, 1900.

GAUBIL, A. (1). Numerous contributions to *Observations Mathématiques, Astronomiques, Géographiques, Chronologiques et Physiques tirées des Anciens Livres Chinois ou faites nouvellement aux Indes et à la Chine par les Pères de la Compagnie de Jésus*, ed. E. Souciet. Rollin, Paris, 1729, vol. 1.

(*a*) Remarques sur l'Astronomie des Anciens Chinois en général, p. 1.

(*b*) Eclipses ⊙ Sexdecim in Historia aliisque veteribus Sinarum libris notatae et a Patre Ant. Gaubil e Soc. Jesu computatae, p. 18. (The first is the *Shu Ching* eclipse attributed to −2155; then follows the *Shih Ching* eclipse attributed to −776; then five *Tso Chuan* eclipses (−720 to −495), then one of −382 and finally 3 Han ones.)

(*c*) Observations des Taches du Soleil, p. 33.

(*d*) Observation de l'Eclipse de ☾ du 22 Déc. 1722 à Canton, p. 44.

(*e*) Observatio Eclipsis Lunae totalis Pekini 22 Oct. 1725, p. 47.

(*f*) Occultations ou Eclipses des Etoiles Fixes par la Lune, observées à Péking en 1725 & 1726, p. 59.

(*g*) Observations de Saturne, p. 69.

(*h*) Observations de Jupiter, p. 71.

(*i*) Observations de ♃ et de ses Satellites; Conjonctions ou Approximations de ♃ à des Étoiles Fixes, tirées des anciens livres d'Astronomie Chinoise (+73 to +1367), p. 72.
(*j*) Observations des Satellites de ♃, faites à Péking en 1724, p. 80.
(*k*) Observations de Mars, p. 95.
(*l*) Observations de Vénus, p. 98.
(*m*) Observations de Mercure, p. 101.
(*n*) Observations de la Comète de 1723 faites à Péking d'abord par des Chinois et ensuite par les PP. Gaubil & Jacques, p. 105.
(*o*) Observations géographiques (à) l'Ile de Poulo-Condor, p. 107.
(*p*) Plan de Canton, sa longitude et sa latitude, p. 123.
(*q*) Extrait du Journal du Voyage du P. Gaubil et du P. Jacques de Canton à Péking, etc., p. 127.
(*r*) Plan (& Description) de Péking, p. 136.
(*s*) Situation de Poutala, demeure du grand Lama, des sources du Gange et des pays circonvoisins, le tout tiré des Cartes Chinoises et Tartares, p. 138.
(*t*) Mémoire Géographique sur les Sources de l'Irtis et de l'Oby, sur le pays des Eleuthes et sur les Contrées qui sont au Nord et à l'Est de la Mer Caspienne, p. 141.
(*u*) Relation Chinoise contenant un itinéraire de Péking à Tobol, et de Tobol au Pays des Tourgouts, p. 148.
(*v*) Remarques sur le commencement de l'Année Chinoise, p. 182.
(*w*) Abrégé Chronologique de l'Histoire des Cinq Premiers Empereurs Mogols, p. 185.
(*x*) Observations Physiques (Lézard Volant à Poulo-Condor, Melon de Hami), p. 204.
(*y*) Observations sur la Variation de l'Aiman, p. 210.
(*z*) Observations Diverses, p. 223.

GAUBIL, A. (2). *Histoire Abrégée de l'Astronomie Chinoise.* (With Appendices 1, Des Cycles des Chinois; 2, Dissertation sur l'Eclipse Solaire rapportée dans le *Chou-King* [*Shu Ching*]; 3, Dissertation sur l'Eclipse du Soleil rapportée dans le *Chi-King* [*Shih Ching*]; 4, Dissertation sur la première Eclipse du Soleil rapportée dans le *Tchun-Tsieou* [Chhun Chhiu]; 5, Dissertation sur l'Eclipse du Soleil, observée en Chine l'an trente-et-unième de Jésus-Christ; 6, Pour l'Intelligence de la Table du *Yue-Ling* [*Yüeh Ling*]; 7, Sur les Koua; 8, Sur le Lo-Chou (recognition of Lo Shu as magic square).) In *Observations Mathématiques, Astronomiques, Géographiques, Chronologiques et Physiques, tirées des anciens livres Chinois ou faites nouvellement aux Indes, à la Chine, et ailleurs, par les Pères de la Compagnie de Jésus*, ed. E. Souciet. Rollin, Paris, 1732, vol. 2.

GAUBIL, A. (3). *Traité de l'Astronomie Chinoise.* In *Observations Mathématiques, etc.*, ed. E. Souciet. Rollin, Paris, 1732, vol. 3.

GAUBIL, A. (4). *Histoire de l'Astronomie Chinoise.* In *Lettres Edifiantes et Curieuses, écrites des Missions Étrangères; Nouvelle Edition—Mémoires des Indes et de la Chine*, vol. 26, pp. 65–295. Mérigot, Paris, 1783. (Reprinted in vol. 14 of the 1819 edition.)

GAUBIL, A. (5). *Traité de la Chronologie Chinoise; divisé en 3 parties, composé par le P. Gaubil, Missionaire à la Chine, et publié pour servir de suite aux Mémoires Concernant les Chinois*, ed. S. de Sacy. Treuttel & Wurtz, Paris, 1814.

GAUBIL, A. (6). 'Recherches Astronomiques sur les Constellations et les Catalogues Chinois des Etoiles Fixes, sur le Cycle des Jours, sur les Solstices et sur les Ombres Méridiennes du Gnomon observés à la Chine.' 1734 MS. at the Observatory, Paris, partly published by Laplace, see Gaubil (7, 8, 9). See also J. B. Biot, 'Notice sur des Manuscrits Inédits du Père Gaubil et du Père Amiot, par feu E. Biot.' *JS*, 1850, 302.

GAUBIL, A. (7). 'Des Solstices et des Ombres Méridiennes du Gnomon, observés à la Chine; extrait d'un Manuscrit envoyé en 1734 à M. Delisle, Astronome, par le P. Gaubil, missionaire jésuite.' *CT*, 1809, 382.

GAUBIL, A. (8). 'Observations Chinoises, depuis l'an 147 avant J.C., envoyées par le P. Gaubil en Nov. 1749' (on planetary conjunctions, and eclipses). *CT*, 1810, 300.

GAUBIL, A. (9). See Laplace. 'Mémoire sur la Diminution de l'Obliquité de l'Ecliptique, qui résulte des Observations Anciennes.' *CT*, 1811, 429.

GAUBIL, A. (10). 'Catalogue des Comètes observées en Chine.' MS. at the Observatory, Paris.

GAUBIL, A. (12). *Histoire de Gentchiscan* [*Chingiz Khan*] *et de toute la Dinastie des Mongous ses Successeurs, Conquérans de la Chine.*' Briasson & Piget, Paris, 1739.

GAUCHET, L. (6). 'Note sur la Généralisation de l'Extraction de la Racine Carrée chez les anciens Auteurs Chinois, et quelques Problèmes du *Chiu Chang Suan Shu*.' *TP*, 1914, **15**, 531.

GAUCHET, L. (7). 'Note sur la Trigonométrie sphérique de Kouo Cheou-King' (Kuo Shou-Ching). *TP*, 1917, **18**, 151.

GAUTHIER, H. (1). *La Température en Chine.* Shanghai, 1918.

GEERTS, A. J. C. (1). *Les Produits de la Nature Japonaise et Chinoise, comprenant la Dénomination, l'Histoire et les Applications aux Arts, à l'Industrie, à l'Economie, à la Médecine, etc. des Substances qui dérivent des Trois Règnes de la Nature et qui sont employées par les Japonais et les Chinois: Partie Inorganique et Minéralogique*...(only part published). 2 vols. Lévy, Yokohama, 1878; Nijhoff, 's Gravenhage, 1883. (A paraphrase and commentary on the mineralogical chapters of the *Pên Tshao Kang Mu*, based on Ono Ranzan's commentary in Japanese.)

GEIKIE, SIR A. (1). *The Founders of Geology*. Macmillan, London, 1905.

GELASIUS, A. (1). *De Terraemotu Liber*. Bologna, 1571.

GELDNER, K. F. (1) (tr.). *Der 'Ṛg Veda'*. 3 vols. Harvard Univ. Press, Cambridge (Mass.), 1951. (Harvard Oriental Series, nos. 33, 34, 35.)

GERINI, G. E. (1). *Researches on Ptolemy's Geography of Eastern Asia (Further India and Indo-Malay Peninsula)*. Royal Asiatic Society and Royal Geographic Society, London, 1909. (Asiatic Society Monographs, no. 1.)

GIBSON, G. E. (1). 'The Vedic Nakshatras and the Zodiac.' In *Semitic and Oriental Studies presented to Wm. Popper*, p. 149. Univ. of Calif. Press, Berkeley, 1951.

GILBERT, O. (1). *Die meteorologischen Theorien d. griechischen Altertums*. Berlin, 1909.

GILBERT, WILLIAM (1). *De Magnete*. Short, London, 1600. Ed. and tr. S. P. Thompson, Chiswick, London, 1900.

GILES, H. A. (3) (tr.). *The Travels of Fa-Hsien*. Cambridge, 1923.

GILES, H. A. (5). *Adversaria Sinica*:
1st series, no. 1, pp. 1–25. Kelly & Walsh, Shanghai, 1905.
no. 2, pp. 27–54. Kelly & Walsh, Shanghai, 1906.
no. 3, pp. 55–86. Kelly & Walsh, Shanghai, 1906.
no. 4, pp. 87–118. Kelly & Walsh, Shanghai, 1906.
no. 5, pp. 119–144. Kelly & Walsh, Shanghai, 1906.
no. 6, pp. 145–188. Kelly & Walsh, Shanghai, 1908.
no. 7, pp. 189–228. Kelly & Walsh, Shanghai, 1909.
no. 8, pp. 229–276. Kelly & Walsh, Shanghai, 1910.
no. 9, pp. 277–324. Kelly & Walsh, Shanghai, 1911.
no. 10, pp. 326–396. Kelly & Walsh, Shanghai, 1913.
no. 11, pp. 397–438 (with index). Kelly & Walsh, Shanghai, 1914.
2nd series, no. 1, pp. 1–60. Kelly & Walsh, Shanghai, 1915.

GILES, H. A. (8). 'The Character *hsi*' [evening tide]. *NCR*, 1921, **3**, 423.

GILES, L. (2). *An Alphabetical Index to the Chinese Encyclopaedia [Chhin-Ting Ku Chin Thu Shu Chi Chhêng]*. British Museum, London, 1911.

GILES, L. (4) (tr.). *Taoist Teachings from the Book of Lieh Tzu*. Murray, London, 1912; 2nd ed. 1947.

GILES, L. (5). *Six Centuries at Tunhuang*. China Society, London, 1944.

GILES, L. (6). *A Gallery of Chinese Immortals ['hsien']; selected biographies translated from Chinese sources [Lieh Hsien Chuan, Shen Hsien Chuan, etc.]*. Murray, London, 1948.

GILES, L. (8) (tr.). 'A Chinese Geographical Text of the Ninth Century' (S/367). *BLSOAS*, 1931, **6**, 825. (About the Tunhuang district.)

GILES, L. (9) (tr.). 'The *Tunhuang Lu*.' *JRAS*, 1914, 703.

GILES, L. (10). 'Translations from the Chinese World Map of Father Ricci.' *GJ*, 1918, **52**, 367; 1919, **53**, 19.

GILES, L. (12). *Glossary of Chinese Topographical Terms*. Geogr. Sect., General Staff, War Office, London, 1943.

GILLE, B. (3). 'Léonard de Vinci et son Temps.' *MC/TC*, 1952, **2**, 69.

GILLIS, I. V. (1). *The Characters 'chhao' and 'hsi'* [the tides]. Gest Chinese Research Library (McGill Univ. Montreal) Publication, Standard Press, Peiping, 1931.

GINZEL, F. K. (1). *Handbuch d. mathematischen und technischen Chronologie, das Zeitrechnungswesen d. Völker*. 3 vols. Hinrichs, Leipzig, 1906.

GINZEL, F. K. (2). *Spezieller Kanon der Sonnen- und Mond-Finsternisse f. d. Ländergebiet d. klassischer Altertumswissenschaften u. d. Zeitraum v. 900 v. C. bis 600 n. C.* Berlin, 1899.

GINZEL, F. K. (3). 'Die Zeitrechnung.' Art. in *Astronomie*, ed. J. Hartmann. Pt. III, Sect. 3, vol. 3 of *Kultur d. Gegenwart*, p. 57. Teubner, Leipzig and Berlin, 1921.

VON GLASENAPP, H. (1). *La Philosophie Indienne, Initiation à son Histoire et à ses Doctrines*. Payot, Paris, 1951 (no index).

GLATHE, A. (1). 'Die chinesische Zahlen.' *MDGNVO*, 1932, **26**, B, 1.

GODARD, A. (1). *Les Monuments de Marāghah*. Paris, 1934.

GODFRAY, H. (1). 'Dials and Dialling.' *EB*, vol. 8, p. 149.

GODWIN, FRANCIS, BP. (1). *The Man in the Moone; or, A Discourse of a Voyage Thither, by Domingo Gonsales, the Speedy Messenger*. London, 1638, 1657 and 1768, ed. G. McColley (2).

DE GOEJE, J. (1) (ed.). *Bibliotheca Geographorum Arabicorum.* 8 vols. Brill, Leiden.
GOETZ, W. (1). 'Die Entwicklung des Wirklichkeitssinnes vom 12. bis 14. Jahrhundert.' *AKG*, 1937, **27**, 33.
GOLDSCHMIDT, V. M. (1). 'The Principles of Distribution of Chemical Elements in Minerals and Rocks.' *JCS*, 1937, 655.
GOODRICH, L. CARRINGTON (4). 'Measurements of the Circle in Ancient China.' *ISIS*, 1948, **39**, 64.
GOODRICH, L. CARRINGTON (5). 'The Abacus in China.' *ISIS*, 1948, **39**, 239.
GOODRICH, L. CARRINGTON (6). 'Early Mentions of Fossil Fishes.' *ISIS*, 1942, **34**, 25.
GOODRICH, L. CARRINGTON (8). 'China's First Knowledge of the Americas.' *GR*, 1938, **28**, 400.
GOODRICH, L. CARRINGTON (9). *Introduction to Chinese History, and Scientific Developments in China.* Sino-Indian Cultural Soc., Santiniketan, 1954. (Sino-Indian Pamphlets, no. 21.)
GOODRICH, L. CARRINGTON (11). 'Geographical Additions of the 14th and 15th Centuries' (further MSS. of *Ta Yuan I Thung Chih* and *Huan Yü Thung Chih*). *MS*, 1956, **15**, 203.
GOODYER, J. (1) (tr.). *The Greek Herbal of Dioscorides, illustrated by a Byzantine, A.D. 512, englished by John Goodyer, A.D. 1664*, ed. R. T. Gunther. Oxford, 1934.
GOSCHKEVITCH, J. (1). 'Über das chinesische Rechnenbrett.' In *Arbeiten der kaiserlichen Russischen Gesandtschaft zu Peking*, vol. 1, p. 293. Berlin, 1858.
GOULD, R. T. (1). (*a*) *The Marine Chronometer; its History and Development.* Potter, London, 1923. (rev. F. D[yer], *MMI*, 1923, **9**, 191). (*b*) *The Restoration of John Harrison's Third Timekeeper.* Lecture to the British Horological Institute, 1931. Reprint or pamphlet, n.d.
GOW, J. (1). *A Short History of Greek Mathematics.* Cambridge, 1884.
GRABAU, A. W. (1). 'Palaeontology.' Art. in *Symposium on Chinese Culture*, ed. Sophia Zen. IPR, Shanghai, 1931, pp. 152 ff.
GRAHAM, A. C. (1). *The Philosophy of Chhêng I-Chhuan (+1033 to +1107) and Chhêng Ming-Tao (+1032 to +1085).* Inaug. Diss. London, 1953.
GRANET, M. (1). *Danses et Légendes de la Chine Ancienne.* 2 vols. Alcan, Paris, 1926.
GRANET, M. (5). *La Pensée Chinoise.* Albin Michel, Paris, 1934. (Evol. de l'Hum. series, no. 25 *bis*.)
GRANT, R. (1). *History of Physical Astronomy, from the Earliest Ages to the middle of the Nineteenth Century.* Baldwin, London, 1852.
GRAVEROL, F. (1). Letter which includes mention of 17th-century water-clock. *JS*, 1691, 75.
GRIFFITH, W. (1). *Journal of Travels in Assam, Burma, etc.* Calcutta, 1847.
GRIMALDI, P. (1). *Explicatio Planisphaerii* (Atlas Céleste Chinois). Peking, 1711 (printed in China). (Taken from I. G. Pardies' *Globi Coelestis in Tabulas Planas Redacti Descriptio.* Paris, 1674. The six charts form the sides of a cube circumscribing the sphere. Cf. Houzeau, p. 46.)
GROENEVELDT, W. P. (1). *Notes on the Malay Archipelago and Malacca.* 1876. In *Miscellaneous Papers relating to Indo-China*, 2nd series, 1887, vol. 1, p. 126.
DE GROOT, J. J. M. (2). *The Religious System of China.* Brill, Leiden, 1892.
Vol. 1, Funeral rites and ideas of resurrection.
Vols. 2, 3, Graves, tombs, and *fêng-shui.*
Vol. 4, The soul, and nature-spirits.
Vol. 5, Demonology and sorcery.
Vol. 6, The animistic priesthood (*wu*).
GROS, L. (1). *La Théorie du Baguenodier.* Lyons, 1872.
GROSSMANN, H. (1). 'Die gesellschaftlichen Grundlagen der mechanistischen Philosophie und die Manufaktur.' *ZSF*, 1935, **4**.
GROTH, P. H. (1). *Entwicklungsgeschichte d. mineralogischen Wissenschaften.* Berlin, 1926.
DE GUIGNES, C. L. J. (2). 'Planisphère Céleste Chinois, avec des Explications, le Catalogue Alphabétique des Etoiles, et la Suite de tous les Comètes observées à la Chine, depuis l'an 613 avant J.C. jusqu'à l'an 1222 de l'Ere Chrétienne, tirées des Livres Chinois.' *Mémoires des Savants Etrangers* (*Académie Royale des Sciences*), 1782, **10**, 1.
VAN GULIK, R. H. (2). *Mi Fu on Inkstones; a Study of the 'Yen Shih', with Introduction and Notes.* Vetch, Peking, 1938.
GUNDEL, W. (1). 'Astronomie, Astralreligion, Astralmythologie und Astrologie; Darstellung u. Literaturbericht 1907–1933.' In *Jahresber. ü. d. Fortschritte d. klassischen Altertumswissenschaft*, ed. K. Münscher, 1934, **243**, 1.
GUNDEL, W. (2). *Dekane und Dekansternbilder.* Augustin, Glückstadt & Hamburg, 1936. (Stud. d. Bibl. Warburg, no. 19.)
GUNDEL, W. (3). 'Neue astrologische Texte des Hermes Trismegistos.' *ABAW/PH*, 1936, n.F. **12**. (Crit. Neugebauer (9), p. 68.)
GUNTHER, R. T. (1). *Early Science in Oxford.* 14 vols. Oxford, 1923–45. (The first pub. Oxford Historical Soc.; the rest privately printed for subscribers.)
Vol. 1 1923 Chemistry, Mathematics, Physics and Surveying.

Vol. 2 1923 Astronomy.
Vol. 3 1925 Biological Sciences and Biological Collections.
Vol. 4 1925 The [Oxford] Philosophical Society.
Vol. 5 1929 Chaucer and Messahalla on the Astrolabe.
Vol. 6 1930 Life and Work of Robert Hooke.
Vol. 7 1930 Life and Work of Robert Hooke (contd.).
Vol. 8 1931 Cutler Lectures of Robert Hooke (facsimile).
Vol. 9 1932 The *De Corde* of Richard Lower (facsimile) with introd. and tr. by K. J. Franklin.
Vol. 10 1935 Life and Work of Robert Hooke (contd.).
Vol. 11 1937 Oxford Colleges and their men of science.
Vol. 12 1939 Dr Plot and the correspondence of the [Oxford] Philosophical Society.
Vol. 13 1938 Robert Hooke's *Micrographia* (facsimile).
Vol. 14 1945 Life and Letters of Edward Lhwyd.

GUNTHER, R. T. (2). *The Astrolabes of the World.* 2 vols. Oxford, 1932.

GÜNTHER, S. (1). 'Die Anfänge u. Entwicklungsstadien des Coordinaten-princips.' *ANHGN*, 1877, **6**, 3.

GURJAR, L. V. (1). *Ancient Indian Mathematics and Vedha.* Ideal Book Service, Poona, 1947. (Crit. A. H. Neville, *N*, 1948, **161**, 580.)

GUSTAVS, A. & DALMAN, D. G. (1). 'Der Saatrichter zur Zeit d. Kassiten.' *ZDPV*, 1913, **36**, 310.

HAENISCH, E. (1). 'Ein chinesischer Baedeker aus dem 13. Jahrhundert' (on Wang Hsiang-Chih's *Yü Ti Chi Shêng*). *OAZ*, 1919, **7**, 201.

HAGERTY, M. J. (1) (tr.). 'Han Yen-Chih's *Chü Lu*' (Monograph on the Oranges of Wên-Chou, Chekiang), with introduction by P. Pelliot. *TP*, 1923, **22**, 63.

DU HALDE, J. B. (1). *Description Géographique, Historique, Chronologique, Politique et Physique de l'Empire de la Chine et de la Tartarie Chinoise.* 4 vols. Paris, 1735; The Hague, 1736. (Eng. tr. R. Brookes, London, 1736, 1741.)

HALL, A. R. (1). *Ballistics in the Seventeenth Century; a Study in the Relations of Science and War, with reference principally to England.* Cambridge, 1951.

HALLBERG, I. (1). *L'Extrême-Orient dans la Littérature et la Cartographie de l'Occident des 13e, 14e et 15e siècles; Etudes sur l'Histoire de la Géographie.* Inaug. Diss., Upsala. Zachrisson, Göteborg, 1907.

HALLEY, EDMUND (1). 'Considerations on the Change of the Latitudes of the principal Fixt Stars.' *PTRS*, 1718, **30** (no. 355), 736.

HALOUN, G. (4). 'Zur Üe-Tsï [Yüeh-Chih]-Frage.' *ZDMG*, 1937, **91**, 243.

HALOUN, G. (5). 'Legalist Fragments, I; *Kuan Tzu* ch. 55, and related texts.' *AM*, 1951 (n.s.), **2**, 85.

AL-HAMDĀNĪ. See Rashīd al-Dīn al-Hamdānī.

HANBURY, DANIEL (1). *Science Papers, chiefly Pharmacological and Botanical.* Macmillan, London, 1876.

HANSFORD, S. H. (1). *Chinese Jade Carving.* Lund Humphries, London, 1950.

DE HARLEZ, C. (1). *Le Yih-King [I Ching], Texte primitif, Rétabli, Traduit et Commenté.* Hayez, Bruxelles, 1889.

DE HARLEZ, C. (2) (tr.). *Kong-Tze-Kia-Yu [Khung Tzu Chia Yü]; Les Entretiens Familiers de Confucius.* Leroux, Paris, 1899; and *BOR*, 1893, **6**; 1894, **7**.

HART, I. B. (3). 'The Scientific Basis of Leonardo da Vinci's work in Technology—an Appreciation.' *TNS*, 1956, **28**, 105.

HARTMANN, J. (1). 'Die astronomischen Instrumente des Kardinals Nikolaus Cusanus.' *AGWG/MP*, 1919 (n.F.), **10**, no. 6.

HARTNER, W. (1). 'Die astronomischen Angaben des Hia Siau Dscheng' [*Hsia Hsiao Chêng*]. Appendix to R. Wilhelm (6), p. 413. (And *SA*, 1930, **5**, 237.)

HARTNER, W. (2). 'Einige astronomische Bemerkungen' [to *Lü Shih Chhun Chhiu*]. Appendix to R. Wilhelm (3), p. 507.

HARTNER, W. (3). 'The Astronomical Instruments of Cha-Ma-Lu-Ting, their Identification, and their Relations to the Instruments of the Observatory of Maragha.' *ISIS*, 1950, **41**, 184.

HARTNER, W. (4). 'Chinesische Kalendarwissenschaft.' *SA*, 1930, **5**, 237.

HARTNER, W. (5). 'Das Datum der *Shih-Ching* Finsternis.' *TP*, 1935, **31**, 188.

HARTNER, W. (6). 'The Pseudoplanetary Nodes of the Moon's Orbit in Hindu and Islamic Iconographies; a Contribution to the History of Ancient and Mediaeval Astrology.' *AI*, 1938, **5**, 113.

HARTNER, W. (8). 'The Obliquity of the Ecliptic according to the *Hon Han Shu* and Ptolemy.' Communication at the 23rd International Congress of Orientalists, Cambridge, 1954. Also pub. in Silver Jubilee volume of Zinbun Kagaku Kenkyu-Syo, Kyoto University, 1954, p. 177.

HARTNER, W. (9). *Le Problème de la Planète Kaïd.* Lecture at the Palais de la Découverte, Paris (No. D, 36). Univ. of Paris, Paris, 1955.

HARTNER, W. (10). 'Zahlen und Zahlensysteme bei Primitiv- und Hochkultur-Völkern.' *PAI*, 1943, **2**, 268.

HARZER, P. (1). *Die exakten Wissenschaften im alten Japan.* Rede z. Feier d. Geburtstages S. Maj. des d. Kaisers Königs v. Preussen Wilhelm II. Lipsius & Tischer, Kiel, 1905. (Crit. Y. Mikami, *JDMV*, 1906, **15**, 253; *TBGZ*, 1906, nos. 174, 175.)

HASKINS, C. H. (1). *Studies in the History of Mediaeval Science.* Harvard Univ. Press, Cambridge, Mass., 1927.

HATT, G. (1). (*a*) 'Asiatic Motifs in American [Amerindian] Folklore.' In Singer Presentation Volume, *Science, Medicine and History*, vol. 2, p. 389, ed. E. A. Underwood. Oxford, 1954, (*b*) 'Asiatic Influences in American Folklore.' *KDVS/HFM*, 1949, **31**, no. 6.

HAUDRICOURT, A. & NEEDHAM, J. (1). *La Science Chinoise au Moyen-Age*, in *Histoire Générale des Sciences*. Vol. 1, ed. R. Taton. Presses Universitaires de France, Paris, 1957.

HAUSER, F. (1). *Über d. 'Kitāb fi al-Hijal' (das Werk ü. d. sinnreichen Anordnungen) d. Banū Mūsa* [*+803 to +873*]. Mencke, Erlangen, 1922. (Abhdl. z. Gesch. d. Naturwiss. u. Med. no. 1.)

HAYASHI, TSURUICHI (1). 'The *Fukudai* and Determinants in Japanese Mathematics.' *PTMS*, 1910, **5**, 254.

HAYASHI, TSURUICHI (2). 'Brief History of Japanese Mathematics.' *NAW*, 1905, **6**, 296 (65 pp.); 1907, **7**, 105 (58 pp.). (Crit. Y. Mikami, *NAW*, 1911, **9**, 373.)

HEATH, R. S. (1). *A Treatise on Geometrical Optics.* C.U.P., Cambridge, 1887.

HEATH, Sir THOMAS (1) (tr.). *The Thirteen Books of Euclid's Elements.* 3 vols. Cambridge, 1926.

HEATH, SIR THOMAS (2) (tr.). *Apollonius Pergaeus; Treatise on Conic Sections, edited in modern notation, with introduction, including an essay on the earlier history of the subject.* Cambridge, 1896.

HEATH, SIR THOMAS (3). 'Greek Mathematics and Astronomy.' *SMA*, 1938, **5**, 215.

HEATH, SIR THOMAS (4). *Greek Astronomy* (anthology of translations with introduction). Dent, London, 1932.

HEATH, SIR THOMAS (5). *Diophantos of Alexandria; a Study in the History of Greek Algebra.* Cambridge, 1885, 1910.

HEATH, SIR THOMAS (6). *A History of Greek Mathematics.* 2 vols. Oxford, 1921.

HEATH, SIR THOMAS (7). *Aristarchus of Samos and the Ancient Copernicans.* Oxford, 1913.

HEATH, SIR THOMAS (8) (tr.). *The Works of Archimedes.* Cambridge, 1897; with supplement, 1912. (Reissued, Dover, New York, n.d. (1953).)

HEAWOOD, E. (1). 'The Relationships of the Ricci Maps.' *GJ*, 1917, **50**, 271.

VAN HÉE, L. (1). 'Problèmes Chinois du Second Degré.' *TP*, 1911, **12**, 559.

VAN HÉE, L. (2). 'Algèbre Chinoise.' *TP*, 1912, **13**, 291. (Comment by H. Bosmans, *RQS*, 1912, **72**, 654.)

VAN HÉE, L. (3). 'Les Cent Volailles, ou l'Analyse Indéterminée en Chine.' *TP*, 1913, **14**, 203, 435.

VAN HÉE, L. (4). 'Li Yeh, Mathématicien Chinois du XIIIe siècle.' *TP*, 1913, **14**, 537.

VAN HÉE, L. (5). (*a*) 'Bibliotheca Mathematica Sinensis Pé-Fou.' *TP*, 1914, **15**, 111. (*b*) 'Le Grand Trésor des Mathematiques Chinoises.' *A*, 1926, 18. (*c*) 'The Great Treasure-House of Chinese Mathematics.' *AMM*, 1926, 117.

VAN HÉE, L. (6). 'Première Mention des Logarithmes en Chine.' *TP*, 1914, **15**, 454.

VAN HÉE, L. (7). 'Le *Hai Tao Suan Ching* de Lieou.' *TP*, 1920, **20**, 51.

VAN HÉE, L. (8) (tr.). 'Le Classique de l'Ile Maritime, ouvrage chinois du IIIe Siècle.' *QSGM/B*, 1932, **2**, 255.

VAN HÉE, L. (9). 'La Notation Algébrique en Chine au XIIIe Siècle.' *RQS*, 1913 (3e sér.), **24**, 574.

VAN HÉE, L. (10). 'The *Chhou Jen Chuan* of Juan Yuan (+1764 to +1849).' *ISIS*, 1926, **8**, 103. (Crit. Y. Mikami (4), (5).)

VAN HÉE, L. (11). 'The Arithmetical Classic of Hsiahou Yang.' *AMM*, 1924, **31**, 235.

VAN HÉE, L. (12). 'Le Précieux Miroir des Quatre Eléments.' *AM*, 1932, **7**, 242.

VAN HÉE, L. (13). 'Algèbre Chinoise' (note on Shen Kua and Kuo Shou-Ching). *ISIS*, 1937, **27**, 321.

VAN HÉE, L. (14). 'Euclide en Chinois et Mandchou.' *ISIS*, 1939, **30**, 84.

VAN HÉE, L. (15). 'Le Zéro en Chine.' *TP*, 1914, **15**, 182.

VAN HÉE, L. (16). 'Les Séries en Extrême-Orient.' *A*, 1930, **12**, 18.

HEIBERG, J. L. (1). 'Geschichte d. Mathematik u. Naturwissenschaften im Altertum.' Art. in *Handbuch d. Altertumswissenschaft*, vol. 5 (1), 2, Beck, München, 1925.

HEIDEL, W. A. (1). *The Frame of the Ancient Greek Maps.* Amer. Geogr. Soc., New York, 1937. (Amer. Geogr. Soc. Res. Ser. no. 20.)

VON HEIDENSTAM, H. (1). *Report on the Hydrology of the Hangchow Bay and the Chhien-Thang Estuary.* Whangpoo Conservancy Board, Shanghai Harbour Investigation, 1921 (series I, no. 5).

HELLMANN, G. (1). *Beiträge z. Geschichte der Meteorologie.* (Veröffentl. d. Kgl. Preuss. Meteorol. Inst. no. 273.) Behrend, Berlin, 1914.

HELLMANN, G. (2). 'Beiträge z. Erfindungsgeschichte meteorologischer Instrumente.' *APAW/MN*, 1920, no. 1. (Thermometer, barometer, rain gauge, weathercock, windrose.)

HELLMANN, G. (3). 'Die Meteorologie in den deutschen Flugschriften und Flugblättern des XVI. Jahrhunderts; ein Beitrag z. Gesch. d. Meteorologie.' *APAW/MN*, 1921, no. 1.

HELLMANN, G. (4). 'Versuch einer Geschichte d. Wettervorhersage im XVI. Jahrhundert.' *APAW/MN*, 1924, no. 1.

VAN HELMONT, J. B. (1). *Oriatrike, or Physick Refined....* Tr. J. C[handler], London, 1662.

HENCKEL, J. F. (1). 'Flora Saturnisans.' In *Pyritologie, ou Histoire Naturelle de la Pyrite; avec le Flora Saturnisans, démontrant l'Alliance entre les Végétaux et les Minéraux; et les Opuscules Minéralogiques.* Tr. from the German by the Baron d'Holbach and A. H. Charas. Hérissant, Paris, 1760.

HENNIG, R. (4). *Terrae Incognitae; eine Zusammenstellung und Kritische Bewertung der wichtigsten vorcolumbischen Entdeckungsreisen an Hand der darüber vorliegenden Originalberichte.* 2nd ed. 4 vols. Brill, Leiden, 1944. (Includes most of the Chinese voyages of exploration, Chang Chhien, Kan Ying, etc.)

HENNING, W. B. (1). 'An Astronomical Chapter of the *Bundahišn*.' *JRAS*, 1942, 229.

HENTZE, C. (1). *Mythes et Symboles Lunaires (Chine ancienne, Civilisations anciennes de l'Asie, Peuples limitrophes du Pacifique)*, with appendix by H. Kühn. de Sikkel, Antwerp, 1932. (Crit. *OAZ*, 1933, **9 (19)**, 33.)

HERRMANN, A. (1). *Historical and Commercial Atlas of China.* Harvard-Yenching Institute, Cambridge, Mass., 1935.

HERRMANN, A. (8). 'Die Westländer in d. chinesischen Kartographie.' In Sven Hedin's *Southern Tibet; Discoveries in Former Times compared with my own Researches in 1906–1908*, vol. 3, pp. 91–406. Swedish Army General Staff Lithographic Institute, Stockholm, 1922. (Add. P. Pelliot, *TP*, 1928, **25**, 98.)

HERRMANN, A. (9). 'Die älteste chinesischen Weltkarten' (Phei Hsiu's work). *OAZ*, 1924, **11**, 97.

HERRMANN, A. (10). 'Die älteste Reichsgeographie Chinas und ihre kulturgeschichtliche Bedeutung.' *SA*, 1930, **5**, 232.

HERRMANN, A. (11). 'Die ältesten chinesischen Karten von Zentral- und West-Asien.' *OAZ*, 1919, **8**, 185.

D'HERVEY DE ST DENYS. See Saint-Denys.

HESSEN, B. (1). 'The Social and Economic Roots of Newton's *Principia*.' In *Science at the Cross-Roads.* Papers read to the 2nd International Congress of the History of Science and Technology. Kniga, London, 1931.

HEYD, W. (1). *Histoire du Commerce du Levant au Moyen Age* (Fr. tr.). Harrassowitz, Leipzig, 1886 and 1923. 2nd ed. 2 vols. 1936.

HIGGINS, K. (1). 'The Classification of Sundials.' *ANS*, 1953, **9**, 342.

HIGGINS, W. H. (1). *The Names of the Stars and Constellations compiled from the Latin, Greek and Arabic, with their derivations and meanings; together with the 28 Moon-Stations of the Zodiac [sic] known to the Arabs.* Clarke, Leicester, 1882.

HILDBURGH, W. L. (1). 'Chinese Methods of Cutting Hard Stones.' *JRAI*, 1907, **37**, 189.

HILL, SIR JOHN (1) (tr.). *Theophrastus [of Eresus] History of Stones, with an English version and Critical and Philosophical Notes.* London, 1744, 1746.

HILLER, J. E. (1). 'Die Minerale d. Antike.' *AGMNT*, 1930, **13**, 358.

HILLER, J. E. (2). 'Die Mineralogie Anselmus Boetius de Boodts [*Gemmarum et Lapidum Historia*].' *QSGNM*, 1941, **8**, 1.

HIMLY, K. (1). 'Über zwei chinesische Kartenwerke. I. Einiges ü. das *Kuang Yü Thu*. II. *Han Kiang I Pei Sz' Shöng Pien Yü Thu*: Karte der Gränzen der vier nördlich vom Han-Strome belegenen Provinzen.' *ZGEB*, 1879, **14**, 181.

HIMLY, K. (8). 'Ein chinesisches Werk ü. d. westliche Inner-Asien.' (On the *Hsi Yü Shui Tao Chi* by Hsü Sung, 1823.) *ENB*, 1902, **3**, 1–77.

HIND, A. M. (1). *Introduction to the History of Woodcuts.* 2 vols. London, 1935.

HIND, J. R. (1). 'On Two Ancient Occultations of Planets by the Moon observed by the Chinese.' *RAS/MN*, 1877, **37**, 243.

HIND, J. R. (2). 'On the Past History of the Comet of Halley.' *RAS/MN*, 1850, **10**, 51.

HIND, J. R. (3). *The Comets; a descriptive Treatise upon those Bodies....* Parker, London, 1852.

HINKS, A. R. (1). *The Portolan Chart of Angellino de Dalorto; 1325 A.D., with a note on surviving Charts and Atlases of the 14th century.* Royal Geogr. Soc. London, 1929.

HINKS, A. R. (2). *Map Projections.* Cambridge, 1921.

HIPPOCRATES. See Littré.

HIRAYAMA, K. (1). 'On the Comets of +373 and +374.' *O*, 1911, **34**, 193.

HIRAYAMA, K. & OGURA, S. (1). 'On the Eclipses recorded in the *Shu Ching* and *Shih Ching*.' *PPMST*, 1915 (2nd ser.), **8**, 2.

HIRSCHBERG, W. (1). 'Lunar calendars and matriarchy.' *AN*, 1931, **26**, 461.

HIRTH, F. (1). *China and the Roman Orient.* Kelly & Walsh, Shanghai; G. Hirth, Leipzig and Munich, 1885. (Photographically reproduced in China with no imprint, 1939.)

HIRTH, F. (2) (tr.). 'The Story of Chang Chhien, China's Pioneer in West Asia.' *JAOS*, 1917, **37**, 89. (Translation of ch. 123 of the *Shih Chi*, containing Chang Chhien's Report; from § 18–52 inclusive and 101 to 103. § 98 runs on to § 104, 99 and 100 being a separate interpolation. Also tr. of ch. 111 containing the biogr. of Chang Chhien.)

HIRTH, F. & ROCKHILL, W. W. (1) (tr.). *Chau Ju-Kua; His work on the Chinese and Arab Trade in the 12th and 13th centuries, entitled 'Chu-Fan-Chi'.* Imp. Acad. Sci., St Petersburg, 1911. (Crit. by G. Vacca, *RSO*, 1913, **6**, 209; P. Pelliot, *TP*, 1912, **13**, 446; E. Schaer, *AGNT*, 1913, **6**, 329; O. Franke, *OAZ*, 1913, **2**, 98; A. Vissière, *JA*, 1914 (11e sér.), **3**, 196.)

HITTI, P. K. (1). *History of the Arabs.* 4th ed. Macmillan, London, 1949.

HO PENG-YOKE. See Ho Ping-Yü.

HO PING-YÜ (1). 'Astronomy in the *Chin Shu* and the *Sui Shu*.' Inaug. Diss. Singapore, 1955.

HO PING-YÜ & NEEDHAM, JOSEPH (1). *Ancient Chinese Observations of Solar Haloes and Parhelia* (in the press).

HOANG, P. See Huang, P.

HOBSON, R. L. (1). *Catalogue of the George Eumorfopoulos Collection of Chinese Pottery.* London, 1925.

HOCHE, R. (1) (ed.). *The 'Eisagoge Arithmetike'* (εἰσαγωγὴ ἀριθμητική) *of Nicomachus of Gerasa.* Teubner, Leipzig, 1866.

HODOUS, L. (1). *Folkways in China.* Probsthain, London, 1929.

HOGBEN, L. (1). *Mathematics for the Million.* Allen & Unwin, London, 1936. 2nd ed. 1937.

VON HOHENHEIM, THEOPHRASTUS PARACELSUS. See Paracelsus.

HOLMES, U. T. (1). 'The Position of the North Star about +1250.' *ISIS*, 1940 (1947), **32**, 14.

HOLMYARD, E. J. & MANDEVILLE, D. C. (1). *Avicennae 'De Congelatione et Conglutinatione Lapidum', being sections of the 'Kitāb al-Shifā'; the Latin and Arabic texts edited with an English translation of the latter and with critical notes.* Geuthner, Paris, 1927.

HOLTZAPFEL, C. & HOLTZAPFEL, J. J. (1). *Turning and Mechanical Manipulations.* Holtzapfel, London, 1852–94.
I Materials (C.H.), 1852.
II Cutting Tools (C.H.), 1856.
III Abrasive and Miscellaneous Processes (C.H. rev. J.J.H.), 1894.
IV Hand, or Simple, Turning (J.J.H.), 1881.
V Ornamental, or Complex, Turning (J.J.H.), 1884.

HOMMEL, F. (1). 'Über d. Ursprung und d. Alter d. arabischen Sternnamen und insbesondere d. Mondstationen.' *ZDMG*, 1891, **45**, 616.

H[OOKE], R[OBERT] (1). 'The Preface; An Account of a Voyage made by the Emperor of China into Corea and the Eastern Tartary in the year 1682...; A Relation of a second Voyage of the said Emperor...; An Explanation necessary to justifie the Geography supposed in these Accounts; Some Observations and Conjectures concerning the Character and Language of the Chinese....' *PTRS*, 1686, **16**, 35. (Ref. to printing on p. 65; abacus, p. 66.) Abridged in *Some Observations and Conjectures concerning the Chinese Characters, made by R.H., R.S.S.* In *Miscellanea Curiosa*, London, 1707, vol. III, pp. 212–32. (For full title see Anon., 33.)

HOOVER, H. C. & HOOVER, L. H. (1) (tr.). *Georgius Agricola 'De Re Metallica', translated from the first Latin edition of 1556, with biographical introduction, annotations and appendices upon the development of mining methods, metallurgical processes, geology, mineralogy and mining law from the earliest times to the 16th century.* 1st ed. *Mining Magazine*, London, 1912; 2nd ed. Dover, New York, 1950.

HOPKINS, L. C. See Yetts (12).

HOPKINS, L. C. (6). 'Pictographic Reconnaissances. II.' *JRAS*, 1918, 387.

HOPKINS, L. C. (11). 'Pictographic Reconnaissances. VII.' *JRAS*, 1926, 461.

HOPKINS, L. C. (12). 'Pictographic Reconnaissances. VIII.' *JRAS*, 1927, 769.

HOPKINS, L. C. (14). 'Archaic Chinese Characters. I.' *JRAS*, 1937, 27.

HOPKINS, L. C. (19). 'The Human Figure in Archaic Chinese Writing; a Study in Attitudes.' *JRAS*, 1929, 557.

HOPKINS, L. C. (26). 'Where the Rainbow Ends.' *JRAS*, 1931, 603.

HOPKINS, L. C. (27). 'Archaic Sons and Grandsons; a Study of a Chinese Complication Complex.' *JRAS*, 1934, 57.

HOPKINS, L. C. (30). 'Sunlight and Moonshine.' *JRAS*, 1942, 102.

HOPKINS, L. C. (35). 'The Chinese Numerals and their Notational Systems.' *JRAS*, 1916, 35, 737.

HORNBLOWER, G. D. (2). 'Early Dragon Forms.' *MA*, 1933, **33**, 79 (rev. and crit. A. de C. Sowerby, *CJ*, 1933, **19**, 64).

HORNER, W. G. (1). 'A New Method of Solving Numerical Equations of all Orders by Continuous Approximation.' *PTRS*, 1819, **109**, 308.

HORWITZ, H. T. (2). 'Ü. ein neueres deutsches Reichspatent (1912) und eine Konstruction v. Heron v. Alexandrien.' *AGNT*, 1917, **8**, 134.

HORWITZ, H. T. (7). 'Beiträge z. Geschichte d. aussereuropäischen Technik.' *BGTI*, 1926, **16**, 290.

HORWITZ, H. T. (8). 'Zur Geschichte d. Wetterschiessens.' *GTIG*, 1915, **2**, 122.

HOSE, C. & MCDOUGALL, W. (1). *Pagan Tribes of Borneo*. Macmillan, London, 1912.

HOSIE, A. (1). 'Sun-Spots and Sun-Shadows observed in China, −28 to +1617.' *JRAS/NCB*, 1878, **12**, 91. (Summary in *N*, 1879, **20**, 131.)

HOSIE, A. (2). 'Droughts in China, +620 to +1643.' *JRAS/NCB*, 1878, **12**, 51.

HOSIE, A. (3). 'Floods in China, +630 to +1630.' *CR*, 1879, **7**, 371.

HOUGH, W. (2). 'Collection of Heating and Lighting Utensils in the United States National Museum.' *BUSNM*, 1928, no. 141.

HOUTSMA, M. T. (1) (ed.). *Arabic text of 'Kitāb al-Buldān' [Book of the Countries] by Aḥmad ibn Abī Ya'qūb al-'Abbāsī (+892)*. 2 vols. Leiden, 1883.

HOUZEAU, J. C. (1). *Vade Mecum de l'Astronomie*. Hayez, Brussels, 1882.

HOUZEAU, J. C. & LANCASTER, A. (1). *Bibliographie Générale de l'Astronomie*. 2 vols. Hayez, Brussels, 1887.

HOWORTH, SIR HENRY H. (1). *History of the Mongols*. 3 vols., Longmans Green, London, 1876–1927.

HU SHIH (5). 'A Note on Chhüan Tsu-Wang, Chao I-Chhing and Tai Chen; a Study of Independent Convergence in Research as illustrated in their works on the *Shui Ching Chu*.' In Hummel (2), p. 970.

HUANG, P. (2). *Catalogue des Tremblements de Terre signalés en Chine d'après les Sources chinoises* (−1767 to +1895). 2 vols. Shanghai, 1909–13. (*VS* no. 28.)

HUANG, P. (3). *Catalogue des Eclipses de Soleil et de Lune relatées dans les Documents chinois et collationnées avec le Canon de Th. Ritter v. Oppolzer*. Shanghai, 1925. (*VS* no. 56.)

HUARD, P. (1). 'La Science et l'Extrême-Orient' (mimeographed). Ecole Française d'Extr. Or., Hanoi, n.d. (1950). (Cours et Conférences de l'Ec. Fr. d'Extr. Or. 1948–9.) This paper, though admirable in choice of subjects and intention, was written in difficult circumstances; it contains many serious mistakes and must be used with circumspection (rev. Gauchet, *A/AIHS*, 1951, **4**, 487).

HUARD, P. (2). 'Sciences et Techniques de l'Eurasie.' *BSEIC*, 1950, **25** (no. 2), 1. This paper, though correcting a number of errors in Huard (1), still contains many mistakes and should be used only with care; nevertheless it is valuable on account of several original points.

HUBAUX, J. & LEROY, M. (1). *Le Mythe du Phénix dans les Littératures Grecque et Latine*. Liège, 1939.

HUBBLE, E. (1). 'Supernovae.' *PASP*, 1941, **53**, 141.

HUBER, E. (2). 'Termes Persans dans l'Astrologie Bouddhique Chinoise.' (Part 7 of *Etudes de Littérature Bouddhique*.) *BEFEO*, 1906, **6**, 39.

HUC, R. E. (1). *Souvenirs d'un Voyage dans la Tartarie et le Thibet pendant les Années 1844, 1845 et 1846* [with J. Gabet]. Revised ed. 2 vols. Lazaristes, Peking, 1924. Abridged ed., *Souvenirs d'un Voyage dans la Tartarie, le Thibet et la Chine...*, ed. H. d'Ardenne de Tizac, 2 vols., Plon, Paris, 1925. Eng. trs. by W. Hazlitt, *Travels in Tartary, Thibet and China during the years 1844 to 1846*, Nat. Ill. Lib. London, n.d.; also ed. P. Pelliot, 2 vols., Kegan Paul, London, 1928.

HUDSON, G. F. (1). *Europe and China; A Survey of their Relations from the Earliest Times to 1800*. Arnold, London, 1931.

HUGGINS, M. L. (1). Art. 'Armilla.' *EB* (11th ed.), vol. 2, p. 575.

HUGHES, A. J. (1). *History of Air Navigation*. Allen & Unwin, London, 1946.

HUGHES, E. R. (1). *Chinese Philosophy in Classical Times*. Dent, London, 1942. (Everyman Library, no. 973.)

HUGHES, E. R. (7) (tr.). *The Art of Letters, Lu Chi's 'Wên Fu', A.D. 302; a Translation and Comparative Study*. Pantheon, New York, 1951. (Bollingen Series, no. 29.)

HUGHES, E. R. (8). 'The Ideational Psychology of Chang Hêng's *Ssu Hsüan Fu*.' Lecture to the Far Eastern Association, Philadelphia, 1951. Unpub. typescript.

HULBERT, H. B. (1). 'An Ancient Map of the World.' *AGS/B*, 1904, 600.

HUMMEL, A. W. (1). 'Phonetics and the Scientific Method.' *ARLC/DO*, 1940, 169.

HUMMEL, A. W. (2) (ed.). *Eminent Chinese of the Chhing Period*. 2 vols. Library of Congress, Washington, 1944.

HUMMEL, A. W. (6). 'Astronomy and Geography in the Seventeenth Century [in China].' On Hsiung Ming-Yü's work. *ARLC/DO*, 1938, 226.

HUMMEL, A. W. (7). 'A Ming Encyclopaedia [*Wan Yung Chêng Tsung Pu Chhiu Jen Chhüan Pien*] with Pictures on Tilling and Weaving [*Kêng Chih Thu*] and on Strange Countries [*I Yü Thu Chih*].' *ARLC/DO*, 1940, 165.
HUMMEL, A. W. (8). 'A View of Foreign Countries in the Ming Period.' *ARLC/DO*, 1940, 167.
HUMMEL, A. W. (9). 'Gazetteers.' *ARLC/DO*, 1931–2, 193.
HUMMEL, A. W. (10). 'The *Kuang Yü Thu*.' *ARLC/DO*, 1937, 174.
HUMMEL, A. W. (11). 'The Beginnings of World Geography in China.' *ARLC/DO*, 1938, 224.
HUMMEL, A. W. (12). 'Sixteenth-Century Geography.' *ARLC/DO*, 1933–4, 7.
HUMMEL, A. W. (13). 'The Printed Herbal of +1249.' *ISIS*, 1941, **33**, 439; *ARLC/DO*, 1940, 155.
HUNTER, W. (1). 'Some Account of the Astronomical Labours of Jayasinha, Rajah of Ambhere or Jaynagar.' *TAS/B* (Asiatick Researches), 1799, **5**, 177, 424.
HUTCHINSON, A. B. (1) (tr.). 'The Family Sayings of Confucius.' *CRR*, **9**, 445; **10**, 17, 96, 175, 253, 329, 428.
HUTCHINSON, G. EVELYN & WOLLACK, A. (1). 'Biological Accumulators of Aluminium.' *TCAAS*, 1943, **35**, 73.
HUTTMANN, W. (1). 'On Chinese and European Maps of China.' *JRGS*, 1844, **14**, 117.
HUYGENS, CHRISTIAAN (1). *The Celestial Worlds Discover'd; or, Conjectures Concerning the Inhabitants, Plants and Productions of the Worlds in the Planets....* Childe, London, 1698.

IBA, YASUAKI (1). 'Fragmentary Notes on Astronomy in Japan' (and China, Korea, etc.). *POPA*, 1934, **42**, 243; 1937, **45**, 301; 1938, **46**, 89, 141, 263.
IDELER, L. (1). *Über die Zeitrechnung d. Chinesen.* Berlin, 1839.
IDELER, L. (2). *Untersuchungen ü. den Ursprung und die Bedeutung der Stern-namen; ein Beytrag z. Gesch. des gestirnten Himmels.* Weiss, Berlin, 1809.
IMAMURA, A. (1). 'Tyōkō and his Seismoscope.' *JJAG*, 1939, **16**, 37.
INWARDS, R. (1). *Weather Lore.* Royal Meteorol. Soc., Rider, London, 1950 (rev. L. Dufour, *AIHS*, 1951, **4**, 225).
IONIDES, S. A. & IONIDES, M. L. (1). *One day telleth Another.* Arnold, London, 1939.

JACOB, K. G. (1). 'Neue Studien den Bernstein im Orient betreffend.' *ZDMG*, 1889, **43**, 353.
JÄGER, F. (1). 'Leben und Werk des P'ei Kü' [Phei Chü]. *OAZ*, 1920, **9**, 81, 216.
JÄGER, F. (3). 'Über chinesische Miao-Tse Albums.' *OAZ*, 1914, **4**, 81; 1915, **5**, 266.
JALABERT, D. (1). 'La Flore Gothique, ses Origines, son Evolution du 12e au 15e siècle.' *BMON*, 1932, **91**, 181.
JASTROW, M. (2). *Religion of Babylonia and Assyria.* Boston, 1898. *Die Religion Babyloniens und Assyriens.* Giessen, 1905.
JAVILLIER, M. (1). 'L'Oeuvre Biochimique et Agronomique de Jules Raulin et ses Développements en France.' *AUL* (Fascicule Spécial: 'l'Université de Lyon en 1948 et 1949'), 1950, p. 69.
JEANS, J. H. (1). 'The Converse of Fermat's Theorem.' *MMT*, 1897, **27**, 174.
JENSEN, P. C. A. (1). *Der Kosmologie der Babylonier.* Trübner, Strassburg, 1890.
JEREMIAS, A. (1). *Handbuch d. altorientalischen Geisteskultur.* Leipzig, 1913.
JOHNSON, M. C. (1). 'Greek, Muslim and Chinese Instrument Design in the Surviving Mongol Equatorials of +1279.' *ISIS*, 1940 (1947), **32**, 27.
JOHNSON, M. C. (2). *Art and Scientific Thought; Historical Studies towards a Modern Revision of their Antagonism.* Faber & Faber, London, 1944. (Reprints Johnson (1), pp. 95–109.)
JOHNSON, S. J. (1). 'Remarks on Ancient Chinese Eclipses' (those in the *Chhun Chhiu*). *RAS/MN*, 1875, **35**, 13.
JONES, G. H. (1). 'Life and Times of Ch'oe Ch'i-Wun' [Tshui Chih-Yuan]. *JRAS/KB*, 1903, **3**, 1.
JONES, W. R. (1). *Minerals in Industry.* Penguin, London, 1950.
JOURDAIN, A. (1). 'Mémoire sur l'Observatoire de Méragha et les Instruments employés pour y observer.' *Magasin Encyclopédique*, 1809 (6), **84**, 43; and sep. Paris, 1810.
JOYCE, T. A. (1). *Mexican Archaeology.* London, 1914.
JULIEN, STANISLAS (1) (tr.). *Voyages des Pèlerins Bouddhistes.* 3 vols. Impr. Imp., Paris, 1853–8. (Vol. 1 contains Hui Li's Life of Hsüan Chuang; vols. 2 and 3 contain Hsüan Chuang's *Hsi Yu Chi.*)

KALYANOV, V. (1). 'Dating the *Arthaśāstra*.' Papers presented by the Soviet Delegation at the 23rd International Congress of Orientalists, Cambridge, 1954. (Indian Studies, pp. 25, 40, Russian with English abridgement.)
KAMAL, YUSSUF (PRINCE) (1) (ed.). *Monumenta Cartographica Africae et Aegypti.* 14 vols. Privately published, 1935–9.
KAMIENSKI, M. (1). 'Halley's Comet and Early Chronology.' *JBASA*, 1956, **66**, 127.

KANDA, SHIGERU (1). 'Ancient Records of Sun-spots and Aurorae in the Far East, and the Variation of the Period of Solar Activity.' *PIAJ*, 1933, **9**, 293.

KAO LU (1). *The Peking Observatory*. Peiping, 1922.

KAPLAN, S. M. (1). 'On the Origin of the TLV-Mirror.' *RAA/AMG*, 1937, **11**, 21.

KARI-NIYAZOV, T. N. (1) (Member of the Uzbek Academy of Sciences). *Astronomicheskaia Shkola Ulugbeka* (The Astronomical School of Ulūgh Beg). Acad. Sci. Moscow, 1950 (in Russian).

KARLBECK, O. (1). *Catalogue of the Collection of Chinese and Korean Bronzes at Hollwyl House, Stockholm*. Stockholm, 1938.

KARLGREN, B. (1). *Grammata Serica; Script and Phonetics in Chinese and Sino-Japanese*. *BMFEA*, 1940, **12**, 1. (Photographically reproduced as separate volume, Peking, 1941.)

KARLGREN, B. (8). 'On the Authenticity and Nature of the *Tso Chuan*.' *GHA*, 1926, **32**, no. 3. (Crit. H. Maspero, *JA*, 1928, **212**, 159.)

KARLGREN, B. (9). 'Some Fecundity Symbols in Ancient China.' *BMFEA*, 1930, **2**, 1.

KARLGREN, B. (11). 'Glosses on the Book of Documents' [*Shu Ching*]. *BMFEA*, 1948, **20**, 39.

KARLGREN, B. (12) (tr.). 'The Book of Documents' (*Shu Ching*). *BMFEA*, 1950, **22**, 1.

KARLGREN, B. (14). (tr.). *The Book of Odes; Chinese Text, Transcription and Translation*. Museum of Far Eastern Antiquities, Stockholm, 1950. (A reprint of the translation only from his papers in *BMFEA*, **16** and **17**.)

KARLGREN, B. (16). *Analytical Dictionary of Chinese and Sino-Japanese*. Geuthner, Paris, 1923.

KAROW, O. (1). 'Der Wörterbücher der Heian-zeit und ihre Bedeutung für das japanische Sprachgeschichte; I, Das *Wamyōruijushō* des Minamoto no Shitagao.' (Contains, p. 185, particulars of the *Wamyō-honzō* (Synonymic Materia Medica with Japanese Equivalents) by Fukane no Sukehito, +918.) *MN*, 1951, **7**, 156.

KARPINSKI, L. C. (1) (tr.). *Robert of Chester's Latin Translation of the Algebra of al-Khwārizmī, with an Introduction, Critical Notes, and an English Version*. Univ. of Michigan Studies, New York, 1915.

KARPINSKI, L. C. (2). *History of Arithmetic*. Rand & McNally, New York, 1925.

KARPINSKI, L. C. (3). 'The Unity of Hindu Contributions to Mathematics.' *SCI*, 1928, 381.

KAYE, G. R. (1). 'Notes on Indian Mathematics—Arithmetical Notation.' *JRAS/B*, 1907 (n.s.), **3**, 475.

KAYE, G. R. (2). 'The Use of the Abacus in Ancient India.' *JRAS/B*, **4**, 293.

KAYE, G. R. (3). *Indian Mathematics*. Thacker & Spink, Calcutta, 1915.

KAYE, G. R. (4). *The Astronomical Observatories of Jai Singh*. Government Printing Office, Calcutta, 1918. (Archaeological Survey of India, New Imperial Series, vol. 40.) (Summaries by H. v. Kluber, *NAT*, 1932, **13**; 1933, **14**; *ST*, 1932, **12**, 81.)

KAYE, G. R. (5). *A Guide to the Old Observatories at Delhi, Jaipur, Ujjain and Benares*. Government Printing Office, Calcutta, 1920.

KEITH, A. BERRIEDALE (4). 'The Period of the Later Saṃhitās, the Brahmaṇas, the Āraṇyakas and the Upanishads.' *CHI*, vol. 1, ch. 5.

KEITH, A. BERRIEDALE (5). *The Religion and Philosophy of the Vedas*. 2 vols. Harvard Univ. Press, Cambridge (Mass.), 1925. (Harvard Oriental Series, nos. 31, 32.)

KEITH, A. BERRIEDALE (6) (tr.). *The Veda of the Black Yajus School entitled 'Taittirīya Saṃhitā'*. 2 vols. Harvard Univ. Press, Cambridge (Mass.), 1914. (Harvard Oriental Series, nos. 18, 19.)

KENDREW, W. G. (1). *Climate*. Oxford, 1930.

KENNEDY, J. (2). 'The Gospels of the Infancy, the *Lalita Vistara*, and the *Vishnu Purana* or the Transmission of Religious Ideas between India and the West.' *JRAS*, 1917, 209, 469.

KEYES, C. W. (2) (tr.). *Cicero's 'De Re Publica'* (contains in Bk. VI the *Somnium Scipionis*). Loeb Cl. Library, New York, 1928.

KHANIKOV, N. (1). 'Analysis and Extracts of *al-Kitāb Mīzān al-Ḥikma* (Book of the Balance of Wisdom), an Arabic work on the Water-Balance, written by al-Khāzinī in the +12th century.' *JAOS*, 1860, **6**, 1.

AL-KHWĀRIZMĪ, ABŪ ABDALLĀH MUḤAMMAD IBN MŪSĀ (+9th century). *Ḥisāb al-Jabr wa-l-Muqābalah* [Calculation of Integration and Equation]. See Karpinski (1).

AL-KHWĀRIZMĪ, MUḤAMMAD IBN-AḤMAD (+10th century). *Mufātiḥ al-'Ulūm* [Keys of the Sciences]. See van Vloten (1).

KIANG CHAO-YUAN. See Chiang Shao-Yuan.

KIELY, E. R. (1). *Surveying Instruments; their History and Classroom Use*. Bur. of Publications, Teachers' Coll., Columbia Univ., New York, 1953.

KIMBLE, G. H. T. (1). *Geography in the Middle Ages*. Methuen, London, 1938.

KING, C. W. (1). *The Natural History of Precious Stones and of the Precious Metals*. Bell & Daldy, London, 1867.

KING, C. W. (2). *The Natural History of Gems or Decorative Stones*. Bell & Daldy, London, 1867.

KING, H. C. (1). *The History of the Telescope*. Griffin, London, 1955; rev. D. W. Dewhirst, *JBASA*, 1956, **66**, 148.

KINGSMILL, T. W. (1). 'Comparative Table of the Ancient Lunar Asterisms.' *JRAS/NCB*, 1891, **26**, 44.
KINGSMILL, T. W. (2). 'The Chinese Calendar, its Origin, History and Connections.' *JRAS/NCB*, 1897, **32**, 1.
KIRCH, C. (1). 'Brevis Disquisitio de Eclipsi Solis, quae a Sinensibus anno 7 Quangvuti sive anno 31 aerae christianae vulgaris, notata est.' *MBIS*, 1723, **2**, 133.
KIRCH, C. (2). 'Annotationes Breves in antiquissimam Observationem Astronomicam, scilicet notabilem illam Conjunctionem Planetarum quae sub Chuen-Hio, Sinarum Imperatore, facta perhibetur' (on the legendary conjunction of −2448). *MBIS*, 1727, **3**, 165.
KIRCHER, ATHANASIUS (2). *Ars Magna Lucis et Umbrae.* Rome, 1646.
KIRCHNER, G. (1). 'Amber Inclusions.' *END*, 1950, **9**, 70.
KIRFEL, W. (1). *Die Kosmographie der Inder nach d. Quellen dargestellt.* Bonn & Leipzig, 1920.
KIRFEL, W. (2). *Der Rosenkranz* (on the history of the Rosary). Walldorf, Hessen, 1949.
KIRKWOOD, D. (1). 'Sun-spots.' *PAPS*, 1869, **11**, 94. (Summary in *N*, 1869, **1**, 284.)
KLAPROTH, J. (1). *Lettre à M. le Baron A. de Humboldt, sur l'Invention de la Boussole.* Dondey-Dupré, Paris, 1834. Germ. tr. A. Wittstein, Leipzig, 1884; résumés P. de Larenaudière, *BSG*, 1834, Oct.; anon. *AJ*, 1834 (2nd ser.), **15**, 105.
KLAPROTH, J. [?] (2). 'On the geographical and statistical Atlas of China, entitled "Kwang Yu Thoo" [*Kuang Yu Thu*], and on Chinese Maps in General.' *AJ*, 1832, **9**, 161.
KLAPROTH, J. (3). 'Mémoire sur les Khazars.' *JA*, 1823 (1e sér.), **3**, 153.
KLEBS, A. C. (1). 'Incunabula Scientifica et Medica; Short Title List.' *OSIS*, 1937 (1938), **4**, 1–359.
KLEBS, L. (1). 'Die Reliefs des alten Reiches (2980–2475 v. Chr.); Material zur ägyptischen Kulturgeschichte.' *AHAW/PH*, 1915, no. 3.
KLEBS, L. (2). 'Die Reliefs und Malereien des mittleren Reiches (7.–17. Dynastie, *c.* 2475–1580 v. Chr.); Material zur ägyptischen Kulturgeschichte.' *AHAW/PH*, 1922, no. 6.
KLEBS, L. (3). 'Die Reliefs und Malereien des neuen Reiches (18.–20. Dynastie, *c.* 1580–1100 v. Chr.); Material zur ägyptischen Kulturgeschichte.' Pt. I. 'Szenen aus dem Leben des Volkes.' *AHAW/PH*, 1934, no. 9.
KLEINWACHTER, G. (1). 'Origin of the "Hindu" Numerals.' *CR*, 1883, **11**, 379; **12**, 25.
KNOBEL, E. B. (1). 'On a Chinese Planisphere.' *RAS/MN*, 1909, **69**, 436.
KNOBEL, E. B. (2). *Notes on an Ancient Chinese Calendar* (*Hsia Hsiao Chêng*). Alabaster, London, 1882.
KNOBEL, E. B. (3). *Ulūgh Beg's Catalogue of Stars, revised from all the Persian MSS. existing in Gt. Britain, with a Vocabulary of Persian and Arabic [astronomical] words.* Carnegie Inst. Pubs. no. 250, Washington, 1917.
KNOTT, C. G. (1). 'The Abacus in its Historic and Scientific Aspects.' *TASJ*, 1886, **14**, 18.
KNOTT, C. G. (2). *Physics of Earthquake Phenomena.* Oxford, 1908.
VON KOBELL, F. (1). *Geschichte der Mineralogie, 1650–1860* München, 1864.
KOMROFF, M. (1). *Contemporaries of Marco Polo.* Boni & Liveright, New York, 1928. (Wm. Rubruck, John of Plano Carpini, Odoric of Pordenone, Benjamin of Tudela.)
KONANTZ, E. L. (1). 'The Precious Mirror of the Four Elements.' *CJ*, 1924, **2**, 304.
KOOP, A. J. (1). *Early Chinese Bronzes.* Benn, London, 1924.
KOPPERS, W. (1). 'Lunar calendars and matriarchy.' *AN*, 1930, **25**, 981.
KÖPRÜLÜ, M. F. (1). 'Maraga Rasathanesi hakkinda Bazi Nottar' (in Turkish). *BAU*, 1942, nos. 23, 24.
KOSIBOWICZ, E. (1). 'Un Missionaire Polonais Oublié, le Père Jean Nicolas Smogułęcki S.J., missionaire en Chine au XVIIe Siècle.' *RHM*, 1929, **6**, 1.
KOYRÉ, A. (1). 'The Significance of the Newtonian Synthesis.' *A/AIHS*, 1950, **29**, 291.
KOYRÉ, A. (2). *Etudes Galiléennes.* 3 vols. Vol. 1. *A l'Aube de la Science Classique* (i.e. Newtonian). Vol. 2. *La Loi de la Chute des Corps; Descartes et Galilée.* Vol. 3. *Galilée et la Loi d'Inertie.* Herrmann, Paris, 1939. (*ASI*, nos. 852–4.)
KOYRÉ, A. (3). 'Galileo and the Scientific Revolution of the Seventeenth Century.' *PHR*, 1943, **52**, 333.
KOYRÉ, A. (4). 'Galileo and Plato.' *JHI*, 1943, **4**, 424.
KRAFT, J. L. (1). *Adventure in Jade.* Holt, New York, 1947.
KRAUSE, M. (1). 'Die Sphärik von Menelaus aus Alexandria.' *AGWG/PH*, 1936, **3**, no. 17. (Also Berlin, 1936, sep. enlarged.)
KREITNER, G. (1). *Im fernen Osten.* 2 vols. Vienna, 1881.
KRETSCHMER, K. (1). 'Marino Sanuto der ältere und die Karten des Petrus Vesconte.' *ZGEB*, 1891, **26**, 352.
KRETSCHMER, K. (2). 'Die physische Erdkunde im christlichen Mittelalter.' In Penk's *Geographische Abhandlungen*, 1899, **4**, no. 1.

KROEBER, A. L. (1). *Anthropology.* Harcourt Brace, New York, 1948.

KU HUNG-MING (1) (tr.). *The Discourses and Sayings of Confucius.* Kelly & Walsh, Shanghai, 1898.

KU PAO-KU (1). *Deux Sophistes Chinois; Houei Che [Hui Shih] et Kong-souen Long [Kungsun Lung].* Presses Univ. de France (Imp. Nat.), Paris, 1953. (Biblioth. de l'Institut des hautes Etudes Chinoises, no. 8). Crit. P. Demiéville, *TP*, 1954, **43**, 108.

KUBITSCHEK, W. (1). 'On the Salamis abacus.' *WNZ*, 1899, **31**, 393.

KUGLER, F. X. (1). *Die Babylonische Mondrechnung; Zwei Systeme der Chaldäer ü. d. Lauf des Mondes und der Sonne.* Herder, Freiburg i/B, 1900.

KUGLER, F. X. (2). *Sternkunde und Sterndienst in Babel.* 2 vols. Aschendorff, Münster, 1907–24. Ergänzungen (in 3 parts, Part 3 by J. Schaumberger), Aschendorff, Münster, 1913–35.

KÜHN, K. G. (1) (tr.). *Galen 'Opera'.* 20 vols. Leipzig, 1821–33. (Medicorum Graecorum Opera quae exstant, nos. 1–20.)

KÜHNERT, F. (1). 'Der chinesische Kalender nach Yao's Grundlagen und die wahrscheinlich allmähliche Entwicklung und Vervollkommung desselben.' *TP*, 1891, **2**, 49.

KÜHNERT, F. (2). 'Das Kalenderwesen bei d. Chinesen.' *OMO*, 1888, **14**, 111.

KÜHNERT, F. (3). 'Über die Bedeutung d. drei Perioden Tschang, Pu, Ki; sowie ü. d. Elementen u. d. sogenannten Wahlzyklus d. Chinesen.' *SWAW/PH*, 1892, **125**, no. 4.

KÜHNERT, F. (4). 'Heisst bei d. Chinesen jeder einzelne Solar-term *Tsiet-K'i* [*chieh-chhi*] und ist ihr unsichtbarer Wandelstern *Ki* [*Chi*] thatsächlich unser Sonnencyclus von 28 julianischen Jahren?' *ZDMG*, 1890, **44**, 256.

KÜHNERT, F. (5). 'Über die von den Chinesen *Tê-Sing* oder Tugendgestirn genannte Himmelserscheinung' [earth-shine]. *SWAW/MN*, 1901, **110** (2).

KUNZ, G. F. (1). *The Curious Lore of Precious Stones.* Lippincott, Philadelphia, 1913.

KUNZ, G. F. (2). *The Magic of Jewels and Charms.* Lippincott, Philadelphia, 1915.

KUO MAI-YING (1). 'How to use the Chinese Abacus or *Suan-Phan*.' *NCR*, 1921, **3**, 127, 309.

KWAUK, MAIYING YOMING. See Kuo Mai-Ying.

KYESER, KONRAD (1). *Bellifortis* [Handbook of Military Engineering]. MSS. Cod. Phil. 63 Göttingen Univ. 1405; Donaueschingen, 1410. (See Sarton (1), vol. 3, p. 1550.)

DE LACOUPERIE, TERRIEN (2). 'The Old Numerals, the Counting-Rods, and the *Swan-Pan* in China.' *NC*, 1883 (3rd ser.), **3**, 297.

DE LACOUPERIE, TERRIEN (4). *Catalogue of Chinese Coins from the 7th cent. B.C. to A.D. 621 including the series in the British Museum*, ed. R. S. Poole. British Museum, London, 1892.

LALOY, L. (1). *Aristoxène de Tarente.* Paris, 1904.

LANDERGREN, S. (1). 'Om Spektralanalytiska Metoder och deras Användning vid Malm undersökningar och Malmprospektering (in Swedish). *TTB*, 1939, 65.

LANGHORNE, J. & W. (1) (tr.). *Plutarch's 'Lives'.* 6 vols. London, 1823.

LANGLOIS, V. (1). 'Géographie de Ptolémée: Reproduction Photolithographique du manuscrit Grec du Monastère de Vatopédi au Mont Athos, exécutée d'après les clichés obtenus sous la direction de M. Pierre de Sévastianov et précedée d'une introduction historique...' [MS. of *c.* +1250]. Didot, Paris, 1867.

LAPLACE, P. S. (1). 'Mémoire sur la Diminution de l'Obliquité de l'Ecliptique, qui résulte des observations anciennes.' *CT*, 1811, 429.

LAPLACE, P. S. (2). *Exposition du Système du Monde*, 6th ed. Paris.

v. LASAULX, E. (1). 'Die Geologie der Griechen und Römer.' *ABAW*, 1851.

LAST, H. (1). 'Empedocles and his Klepsydra again.' *CQ*, 1924, **18**, 169.

LATTIN, H. P. (1). 'The Eleventh Century MS. Munich 14436: its Contribution to the History of Coordinates, of Logic, and of German Studies in France.' *ISIS*, 1948, **38**, 205.

LATTIN, H. P. (2). 'The Origin of our Present System of Notation according to the theories of Nicholas Bubnov.' *ISIS*, 1933, **19**, 181.

LAUFER, B. (1). *Sino-Iranica; Chinese Contributions to the History of Civilisation in Ancient Iran. FMNHP/AS*, 1919, **15**, no. 3 (Pub. no. 201) (rev. and crit. Chang Hung-Chao, *MGSC*, 1925 (ser. B), no. 5).

LAUFER, B. (3). *Chinese Pottery of the Han Dynasty* (Pub. of the East Asiatic Cttee. of the Amer. Mus. Nat. Hist.). Brill, Leiden, 1909. (Reprinted Tientsin, 1940.)

LAUFER, B. (7). *Chinese Grave-Sculptures of the Han Period.* London & New York, 1911.

LAUFER, B. (8). *Jade; a Study in Chinese Archaeology and Religion. FMNHP/AS*, 1912. Repub. in book form, Perkins, Westwood and Hawley, South Pasadena, 1946 (rev. P. Pelliot, *TP*, 1912, **13**, 434).

LAUFER, B. (9). 'Ethnographische Sagen der Chinesen.' In *Aufsätze z. Kultur u. Sprachgeschichte vornehmlich des Orients Ernst Kuhn gewidmet* (Kuhn Festschrift). Marcus, Breslau (München), 1916, p. 199.

LAUFER, B. (11). 'Asbestos and Salamander.' *TP*, 1915, **16**, 299.

LAUFER, B. (12). 'The Diamond; a Study in Chinese and Hellenistic Folk-Lore.' *FMNHP/AS*, 1915, **15**, no. 1. (Publ. no. 184.)

LAUFER, B. (13). 'Notes on Turquoise in the East.' *FMNHP/AS*, 1913, **13**, no. 1. (Publ. no. 169.)

LAYARD, H. (1). *Discoveries among the Ruins of Nineveh and Babylon*. London, 1845.

LEAVENS, D. H. (1). 'The Chinese Suan-Phan.' *AMM*, 1920, **27**, 180.

LECAT, M. (1). *Histoire de la Théorie des Déterminants à plusieurs Dimensions*. Ghent, 1911.

LECOMTE, LOUIS (1). *Nouveaux Mémoires sur l'Etat présent de la Chine*. Anisson, Paris, 1696. (Eng. tr. *Memoirs and Observations Topographical, Physical, Mathematical, Mechanical, Natural, Civil and Ecclesiastical, made in a late journey through the Empire of China, and published in several letters, particularly upon the Chinese Pottery and Varnishing, the Silk and other Manufactures, the Pearl Fishing, the History of Plants and Animals, etc. translated from the Paris edition, etc.*, 2nd ed. London, 1698. Germ. tr. Frankfurt, 1699–1700.)

LEE, EDWARD BING-SHUEY. See Li Ping-Shui.

LEGGE, J. (1) (tr.). *The Texts of Confucianism, translated*: Pt. I. *The 'Shu Ching', the religious portions of the 'Shih Ching', the 'Hsiao Ching'*. Oxford, 1879. (*SBE*, no. 3; reprinted in various eds. Com. Press, Shanghai.) For the full version of the *Shu Ching* see Legge (10).

LEGGE, J. (2) (tr.). *The Chinese Classics, etc.*: Vol. 1. *Confucian Analects, The Great Learning, and the Doctrine of the Mean*. Legge, Hongkong, 1861; Trübner, London, 1861.

LEGGE, J. (3) (tr.). *The Chinese Classics, etc.*: Vol. 2. *The Works of Mencius*. Legge, Hongkong, 1861; Trübner, London, 1861.

LEGGE, J. (4) (tr.). *A Record of Buddhistic Kingdoms; an account by the Chinese monk Fa-Hsien of his travels in India and Ceylon (+399 to +414) in search of the Buddhist books of discipline*. Oxford, 1886.

LEGGE, J. (5) (tr.). *The Texts of Taoism*. (Contains (*a*) *Tao Tê Ching*, (*b*) *Chuang Tzu*, (*c*) *Thai Shang Kan Ying Phien*, (*d*) *Chhing Ching Ching*, (*e*) *Yin Fu Ching*, (*f*) *Jih Yung Ching*.) 2 vols. Oxford, 1891; photolitho reprint, 1927. (*SBE*, nos. 39 and 40.)

LEGGE, J. (7) (tr.). *The Texts of Confucianism*: Pt. III. *The 'Li Chi'*. 2 vols. Oxford, 1885; repr. 1926. (*SBE*, nos. 27 and 28.)

LEGGE, J. (8) (tr.). *The Chinese Classics, etc.*: Vol. 4, Pts. 1 and 2. *'Shih Ching'; The Book of Poetry*. 1. The First Part of the *Shih Ching*; or, the Lessons from the States; and the Prolegomena. 2. The Second, Third and Fourth Parts of the *Shih Ching*; or the Minor Odes of the Kingdom, the Greater Odes of the Kingdom, the Sacrificial Odes and Praise-Songs; and the Indexes. Lane Crawford, Hongkong, 1871; Trübner, London, 1871. Repr., without notes, Com. Press, Shanghai, n.d.

LEGGE, J. (9) (tr.). *The Texts of Confucianism*: Pt. II. *The 'Yi King'* [*I Ching*]. Oxford, 1882, 1899. (*SBE*, no. 16.)

LEGGE, J. (10) (tr.). *The Chinese Classics, etc.*: Vol. 3, Pts. 1 and 2. *The 'Shoo King'* [*Shu Ching*]. Legge, Hongkong, 1865; Trübner, London, 1865.

LEGGE, J. (11) (tr.). *The Chinese Classics, etc.*: Vol. 5, Pts. 1 and 2. *The 'Ch'un Ts'eu' with the 'Tso Chuen'* (*Chhun Chhiu* and *Tso Chuan*). Lane Crawford, Hongkong, 1872; Trübner, London, 1872.

LEMAÎTRE, S. (1). *Les Agrafes Chinoises jusqu'à la fin de l'Epoque Han*. Art et Hist., Paris, 1939.

LEMOINE, J. G. (1). 'Les Anciens Procédés de Calcul sur les Doigts en Orient et en Occident.' *REI*, 1932, 1.

LENORMANT, F. (1). *La Divination et la Science des Présages chez les Chaldéens*. Maisonneuve, Paris, 1875.

LENORMANT, F. (2). *La Magie chez les Chaldéens et les Origines Accadiennes*. Maisonneuve, Paris, 1874. Eng. tr. *Chaldean Magic* (enlarged), Bagster, London, 1877.

LENZ, H. O. (1). *Mineralogie der alten Griechen und Römer....* Thienemann, Gotha, 1861.

LEONARDO DA VINCI. See MCCURDY, E.

LEROI-GOURHAN, ANDRÉ (1). *Evolution et Techniques*. Vol. 1. *L'Homme et la Matière*, 1943; vol. 2. *Milieu et Techniques*, 1945. Albin Michel, Paris.

LEUPOLD, J. (2). *Theatrum Arithmetico-Geometricum*. Breitkopf, Leipzig, 1774.

LÉVI, S. (1). 'Les Missions de Wang Hiuen-Ts'e [Wang Hsüan-Tshê] dans l'Inde.' *JA*, 1900 (9e sér.), **15**, 297, 401.

LEVI DELLA VIDA, G. (1). 'Appunti e Quesiti di Storia Letteraria Arabe (4. Due Nuove Opere del Matematico al-Karajī).' *RSO*, 1934, **14**, 249.

LEVY, H. (1). *Modern Science; a Study of Physical Science in the World Today*. Hamilton, London, 1939.

LEYBOURN, W. (1). *The Art of Numbring by Speaking-Rods; Vulgarly Termed Nepeir's Bones*. London, 1667.

LI CHI (1). 'Chinese Archaeology.' Art. in *Symposium on Chinese Culture*, ed. Sophia Zen. IPR, Shanghai, 1931, pp. 184 ff.

LI CHI (2). *The Formation of the Chinese People; an Anthropological Enquiry.* Harvard Univ. Press, Cambridge, Mass., 1928.

LI NIEN (1). 'The Interpolation Formulae of Early Chinese Mathematicians.' *Proc. VIIIth International Congress of the History of Science, Florence*, 1956.

LI NIEN (2). 'Tsu Chhung-Chih, great mathematician of ancient China.' *PC*, 1956 (no. 24), 34.

LI PING-SHUI (1). *Modern Canton.* Mercury Press, Shanghai, 1936.

LIBRI-CARRUCCI, G. B. I. T. (1). *Histoire des Sciences Mathématiques en Italie depuis la Renaissance des Lettres jusqu'à la Fin du 17ème Siècle.* 4 vols. Renouard, Paris, 1838–40.

LILJEQUIST, G. (1). *Halo Phenomena and Ice-Crystals.* Scientific Results of the Norwegian-British-Swedish Antarctic Expedition, 1949–1952, vol. 2, pt. 2, Norskpolarinstitutt, Oslo, 1956.

LILLEY, S. (1). 'Mathematical Machines.' *N*, 1942, **149**, 462; *D*, 1945, **6**, 150, 182; 1947, **8**, 24.

LILLEY, S. (4). 'Cause and Effect in the History of Science.' *CEN*, 1953, **3**, 58.

LIM BOON-KÊNG. See Lin Wên-Chhing.

LIN CHI-KAI (1). 'L'Origine et le Développement de la Méthode Expérimentale.' Inaug. Diss., Paris, 1931.

LIN WÊN-CHHING (1) (tr.). *The 'Li Sao'; an Elegy on Encountering Sorrows, by Chhü Yüan of the State of Chhu* (ca. *338 to 288 B.C.*)....' Com. Press, Shanghai, 1935.

LIN YÜ-THANG (1) (tr.). *The Wisdom of Lao Tzu* [and Chuang Tzu] *translated, edited and with an introduction and notes.* Random House, New York, 1948.

LIU CHAO-YANG (1). 'On the Observabilities of α Scorpii in the Three Dynasties' (heliacal rising of Antares). *SSE*, 1942, **3**, 21. (Crit. W. Eberhard, *OR*, 1949, **2**, 184; A. Rygalov, *HH*, 1949, **2**, 416.)

LIU CHAO-YANG (2). 'Fundamental Questions about the Yin [Shang] and Chou Calendars.' *SSE*, 1945, **4**, 1. Also separately, enlarged, as *SSE/M* (ser. B), no. 2 (rev. Eberhard, *OR*, 1949, **2**, 179).

LOCKE, L. L. (1). *The Quipu.* Amer. Mus. Nat. Hist., New York, 1923.

LOEWENSTEIN, P. J. (1). 'Swastika and Yin-Yang.' *China Society Occasional Papers* (n.s.), no. 1. China Society, London, 1942.

VAN LOHUIZEN DE LEEUW, J. E. (1). *The 'Scythian' Period; an Approach to the History, Art, Epigraphy and Palaeography of North India from the 1st century B.C. to the 3rd century A.D.* Brill, Leiden, 1949.

LONES, T. E. (1). *Aristotle's Researches in Natural Science.* London, 1912.

LONG, G. W. (1). 'Indochina faces the Dragon.' *NGM*, 1952, **102**, 287 (302).

LORD, L. E. (1). 'The Touchstone.' *CLJ*, 1937, **32**, 428.

LORIA, G. (1). *Storia delle Matematiche dall'Alba della Civiltà al Secolo XIX.* 3 vols. ('L'Enigma Cinese' is in vol. 1, p. 261.) Sten, Torino, 1929. (New edition Hoepli, Milano, 1950.)

LORIA, G. (2). (*a*) 'Che cosa debbono le Matematiche ai Cinesi.' *Bollettino della Mathesis*, 1920, **12**, 63. (*b*) 'Documenti Relativi all'Antica Matematica dei Cinesi.' *A*, 1922, **3**, 141.

LORIA, G. (3). 'Chinese Mathematics.' *SM*, 1921, **12**, 517. (Crit. Y. Mikami, *TYG*, 1923, **12**, no. 4.)

DE LA LOUBÈRE, S. (1). *A New Historical Relation of the Kingdom of Siam, by Monsieur de la Loubère, Envoy-Extraordinary from the French King to the King of Siam, in the years 1687 and 1688, wherein a full and curious Account is given of the Chinese Way of Arithmetick and Mathematick Learning.* Tr. A.P., Gen[t?] R.S.S. [i.e. F.R.S.]. Horne Saunders & Bennet, London, 1693 (from the Fr. ed. Paris, 1691).

LOVELL, A. C. B. (1). 'The New Science of Radio-Astronomy.' *N*, 1951, **167**, 94.

LOVISATO, (1). 'Antiche Osservazioni Cinese delle Macchie Solari.' *RSISI*, 1875, **7**, 1.

LOWITZ, TOBIAS (1). 'Description d'un Météore remarquable.' *NAP*, 1794, **8**, 384.

LUCKEY, P. (1). *Die Rechenkunst bei Ǧamšīd b. Mas'ūd al-Kāšī* [*Jamshid ibn Mas'ūd al-Kāshī*] *mit Rückblicken auf die ältere Geschichte des Rechnens.* Steiner, Wiesbaden, 1951 (Abhdl. f. die Kunde des Morgenlandes, no. 31, i). A study of the *Miftāḥ al-Ḥisāb* (Key of Computation), *c.* +1427. Cf. Rosenfeld & Yushkevitch (1).

LUCKEY, P. (2). *Der Lehrbrief ü. d. Kreisumfang von Ǧamšīd b. Mas'ud al-Kāšī* [*Jamshid ibn Mas'ūd al-Kāshī*]. Berlin, 1953. A study of the *Risālat al-Moḥīṭīje* (Treatise on the Circumference), *c.* +1427. Cf. Rosenfeld & Yushkevitch (1).

LUCKEY, P. (3). 'Zur islamischen Rechenkunst und Algebra des Mittelalters.' *FF*, 1948, **24** (nos. 17–18), 199.

LUCKEY, P. (4). 'Ausziehung des *n*-ten Wurzel und der binomische Lehrsatz in der islamischen Mathematik.' *MAN*, 1948, **120**, 217.

LUCKEY, P. (5). 'Beiträge z. Erforschung d. islamischen Mathematik.' *ORR*, 1948, **17**.

LUNDMARK, K. (1). 'Suspected New Stars recorded in Old [Chinese] Chronicles, and among Recent Meridian Observations.' *PASP*, 1921, **33**, 219, 225.

LUNDMARK, K. (2). 'The Messianic Ideas and their Astronomical Background.' In *Actes du VII Congrès Internat. d'Histoire des Sciences*, p. 436. Jerusalem, 1953.

LURIA, S. (1). 'Die Infinitesimal-Theorie der antiken Atomisten.' *QSGM/A*, 1932, **2**, 106.

LYONS, H. G. (1). 'An Early Korean Rain-Gauge.' *QJRMS*, 1924, **50**, 26.
LYONS, H. G. (2). 'Ancient Surveying Instruments.' *GJ*, 1927, **69**, 132.

McCARTNEY, E. S. (1). 'Fossil Lore in Greek and Latin Literature.' *PMASAL*, 1924, **3**, 23.
McCOLLEY, G. (1). 'The 17th century Doctrine of a Plurality of Worlds.' *ANS*, 1936, **1**, 385.
McCOLLEY, G. (2). '*The Man in the Moone* and *Nuncius Inanimatus* for the first time edited, with introduction and notes, from unique copies of the first editions of London, 1629 and London 1638.' *SCSML*, 1937, **19**, 1.
McCRINDLE, J. W. (7). *The Christian Topography of Cosmas (Indicopleustes), an Egyptian monk.* London, 1897. (Hakluyt Society Publications (ser. 1), no. 98.)
McCURDY, E. (1). *The Notebooks of Leonardo da Vinci, arranged, rendered into English, and introduced by....* 2 vols. Cape, London, 1938.
McGOVERN, W. M. (2). *Manual of Buddhist Philosophy; I, Cosmology* (no more published). Kegan Paul, London, 1923.
McGOWAN, D. J. (1). 'Methods of Keeping Time known among the Chinese.' *CRRR*, 1891, **20**, 426. (Reprinted *ARSI*, 1891 (1893), 607.)
McGOWAN, D. J. (3). 'The Bore on the Chhien-thang River.' *JRAS* (Trans.)/*NCB*, 1854, **1** (no. 4), 33.
McKEON, R. (1). 'Aristotle's Conception of the Development and Nature of Scientific Method.' *JHI*, 1947, **8**, 3.
McLINTOCK, W. F. P. & SABINE, P. A. (1). *A Guide to the Collection of Gemstones in the Geological Museum (Museum of Practical Geology).* HMSO, London, 1951 (2nd ed.).
MA, C. C. (1). 'On the Origin of the Term Root in Chinese Mathematics.' *AMM*, 1928, **35**, 29.
MA HÊNG (1). *The Fifteen Different Classes of Measures as given in the 'Lü Li Chih' of the 'Sui Shu'*, tr. J. C. Ferguson. Privately printed, Peiping, 1932. (Ref. W. Eberhard, *OAZ*, 1933, **9 (19)**, 189.)
MAASS, E. (1). 'Salzburger Bronzetafel mit Sternbildern.' *JHOAI*, 1902, **5**, 196.
v. MÄDLER, J. H. (1). *Geschichte der Himmelskunde von der ältesten bis auf die neueste Zeit.* Westermann, Braunschweig, 1873.
MAENCHEN-HELFEN, O. (1). 'The Later Books of the *Shan Hai Ching*' (with a translation of chs. VI–IX). *AM*, 1924, **1**, 550.
MAENCHEN-HELFEN, O. (2). 'Two Notes on the Diamond in China.' *JAOS*, 1950, **70**, 187.
MAGALHAENS, GABRIEL (1). *A New History of China, containing a Description of the Most Considerable Particulars of that Vast Empire.* Newborough, London, 1688.
MAHDIHASSAN, S. (1). 'Cultural Words of Chinese Origin; Monsoon.' *CS*, 1949, **18**, 347.
MAHDIHASSAN, S. (2). 'Cultural Words of Chinese Origin' [*firoza* (Pers.) = turquoise, *yashb* (Ar.) = jade, *chamcha* (Pers.) = spoon, *top* (Pers., Tk., Hind.) = cannon, *silafchi* (Tk.) = metal basin]. *BV*, 1950, **11**, 31.
MAHDIHASSAN, S. (3). 'Ten Cultural Words of Chinese Origin' [*huqqa* (Tk.), *qaliyan* (Tk.) = tobacco-pipe, *sunduq* (Ar.) = box, *piali* (Pers.), *findjan* (Ar.) = cup, *jaushan* (Ar.) = armlet, *safa* (Ar.) = turban, *qasai, quasab* (Hind.) = butcher, *Kah-Kashan* (Pers.) = Milky Way, *tugra* (Tk.) = seal]. *JUB*, 1949, **18**, 110.
MAHDIHASSAN, S. (4). 'The Chinese Names of Ceylon and their Derivatives.' *JUB*, 1950, **19**, 80.
MAIR, G. R. (1) (tr.). *The 'Phaenomena' of Aratus* (Loeb Classics). Heinemann, London, 1921.
MALYNES, G. (1). *Consuetudo, vel Lex Mercatoria; or, The Antient Law-Merchant.* London, 1662.
MAO TSÊ-TUNG (1). *On Practice* (Orig. Supplement to *People's China*). Peking, 1951.
AL-MAQQARÎ, IBN-MUḤAMMAD AL-TILIMSANĪ (1). *Nafḥ al-Ṭīb*, etc. [Breath of Perfumes from the Boughs of Andalusia]. (History of the scholars of Muslim Spain.) Ed. Dozy *et al.*, Leiden, 1855 to 1861.
MARAKUEV, A. V. (1). *Weights and Measures in China* (in Russian). Vladivostok, 1930. (Summary by P. Pelliot, *TP*, 1932, **29**, 219.)
MARAKUEV, A. V. (2). *The Development of Mathematics in China and Japan* (in Russian). Vladivostok, 1930. (Summary by E. Gaspardone, *BEFEO*, 1932, **32**, 552.)
MARCH, B. (3). *Some Technical Terms of Chinese Painting.* Amer. Council of Learned Societies, Waverly, Baltimore, 1935. (ACLS Studies in Chinese and Related Civilisations, no. 2.)
MARGOULIÈS, G. (3). *Anthologie raisonnée de la Littérature Chinoise.* Payot, Paris, 1948.
MARTINI, M. (1). *Sinicae Historiae Decas Prima.* Munich, 1658; Amsterdam, 1659. (French tr. Paris, 1667.)
MARTINI, M. (2). *Novus Atlas Sinensis.* 1655. (See Schrameier (1) and Szczesniak (4).)
MARYON, H. & PLENDERLEITH, H. J. (1). *Fine Metal-Work [in Early Times before the Fall of the Ancient Empires].* In *A History of Technology*, ed. C. Singer *et al.* O.U.P., Oxford, 1954, vol. 1, p. 623.
MASCART, E. (1). *Traité d'Optique*, 3 vols. and atlas. Gauthier-Villars, Paris, 1893.
MASPERO, H. (1). 'La Composition et la date du *Tso Chuan*.' *MCB*, 1931, **1**, 137.
MASPERO, H. (2). *La Chine Antique.* Boccard, Paris, 1927 (Histoire du Monde, ed. E. Cavaignac, vol. 4) (rev. B. Laufer, *AHR*, 1928, **33**, 903). Revised ed., with characters, Imp. Nat. Paris, 1955.

MASPERO, H. (3). 'L'Astronomie Chinoise avant les Han.' *TP*, 1929, **26**, 267. (Abstract by Vacca, 5.)
MASPERO, H. (4). 'Les Instruments Astronomiques des Chinois au temps des Han.' *MCB*, 1939, **6**, 183.
MASPERO, H. (5). 'Le Songe et l'Ambassade de l'Empereur Ming.' *BEFEO*, 1910, **10**, 95, 629.
MASPERO, H. (8). 'Légendes Mythologiques dans le *Chou King* [*Shu Ching*].' *JA*, 1924, **204**, 1.
MASPERO, H. (14). *Etudes Historiques; Mélanges Posthumes sur les Religions et l'Histoire de la Chine*, vol. III, ed. P. Demiéville. Civilisations du Sud, Paris, 1950. (Publ. du Mus. Guimet, Biblioth. de Diffusion, no. 59), rev. J. J. L. Duyvendak, *TP*, 1951, **40**, 366.
MASPERO, H. (15). 'L'Astronomie dans la Chine Ancienne; Histoire des Instruments et des Découvertes.' Paper prepared for *SCI* in 1932 but not printed till 1950 in Maspero (14), p. 15.
MASPERO, H. (16). Review of Giles' *Adversaria Sinica*. *BEFEO*, 1909, **9**, 595.
MASPERO, H. (25). 'Le *Ming-Thang* et la Crise Religieuse Chinoise avant les Han.' *MCB*, 1951, **9**, 1.
MASSON-OURSEL, P., DE WILLMAN-GRABOWSKA, H. & STERN, P. (1). *L'Inde Antique et la Civilisation Indienne*. Albin Michel, Paris, 1933. (Evol. de l'Hum. Series, Préhist, no. 26.)
MATHER, K. F. & MASON, S. L. (1). *A Source-Book in Geology*. McGraw Hill, New York & London, 1939.
MATTHIESSEN, L. (1). 'Zur Algebra der Chinesen' (extract from a letter to Cantor correcting a mistake made by Biernatzki in translating Wylie; concerning the indeterminate analysis of Sun Tzu). *ZMP*, 1876, **19**, 270; *ZMNWU*, 1876, **7**, 73.
MATTHIESSEN, L. (2) 'Über das sogenannte Restproblem in den chinesischen Werken *Swan-King* von Sun-Tsze u. *Tayen Lei Schu* von Yih-Hing.' *JRAM*, 1881, **91**, 254.
MATTHIESSEN, L. (3). 'Vergleichung der indischen Cuttaca und der chinesischen Tayen-Regel, unbestimmte Gleichungen und Congruenzen ersten Grades aufzulösen.' Verhandlungen d. 30sten Versammlung deutscher Philologen u. Schulmänner, in Rostock, 1875, p. 125. Teubner, Leipzig, 1876.
MATTHIESSEN, L. (4). 'Die Methode Tá jàn (Ta Yen) im *Suán-King* von Sun-tsè und ihre Verallgemeinerung durch Yih-Hing im I. Abschnitte des *Tá jàn li schū*.' *ZMP*, 1881, **26** (Hist. Lit. Abt.), 33.
MATTHIESSEN, L. (5). 'Le Problème des restes dans l'Ouvrage chinois *Swan-King* de Sun-tsze et dans l'Ouvrage *Ta-yen-lei-schu* de Yih-hing.' *CRAS*, 1881, **92**, 291.
MAXE-WERLY, L. (1). 'Notes sur des Objets antiques.' *MSAF*, 1887, **48**, 170.
MAYERS, W. F. (3). 'Chinese Explorations of the Indian Ocean during the 15th century.' *CR*, 1875, **3**, 219, 331; 1875, **4**, 61.
MAZAHERI, A. (2). Review of Moucharrafa & Ahmad's edition of al-Khwārizmī's *Kitāb al-Jabr w'al-Muqabala*. *A/AIHS*, 1951, **4**, 504.
MEDHURST, W. H. (1) (tr.). *The 'Shoo King'* [*Shu Ching*], *or Historical Classic* (Ch. and Eng.). Mission Press, Shanghai, 1846.
MEEK, T. J. (1). 'Early Babylonian Wheel-Maps.' *AQ*, 1936, **10**, 223.
MEI YI-PAO (1) (tr.). *The Ethical and Political Works of Motse*. Probsthain, London, 1929.
MEISSNER, B. (1). *Babylonien und Assyrien*. 2 vols. Winter, Heidelberg, 1920 to 1925.
MEISTER, P. W. (1). 'Buddhistische Planetendarstellungen in China.' *OE*, 1954, **1**, 1.
MELLOR, J. W. (1). *Modern Inorganic Chemistry*. Longmans Green, London, 1916. (Often reprinted.)
DE MÉLY, F. (1). *Les Lapidaires Chinois*. Vol. 1 of *Les Lapidaires de l'Antiquité et du Moyen Age*. Leroux, Paris, 1896. Contains facsimile reproduction of the mineralogical section of (*Ho Han*) *San Tshai Thu Hui*, chs. 59 and 60, from a Japanese edition (rev. M. Berthelot, *JS*, 1896, 573).
DE MÉLY, F. (2). 'Le "De Monstris" Chinois et les Bestiaires Occidentaux.' *RA*, 1897 (3e sér.), **31**, 353.
DE MÉLY, F. (3). *Les Lapidaires Grecs*. Paris, 1898, 1902.
DE MÉLY, F. (5). 'Les Pierres de Foudre chez les Chinois et les Japonais.' *RA*, 1895 (3e sér.), **27**, 326.
DE MENDOZA, JUAN GONZALES (1). *Historia de las Cosas mas notables, Ritos y Costumbres del Gran Reyno de la China, sabidas assi por los libros de los mesmos Chinas, como por relacion de religiosos y oltras personas que an estado en el dicho Reyno*. Rome, 1585 (in Spanish). Eng. tr. Robert Parke, 1588 (1589), *The Historie of the Great & Mightie Kingdome of China and the Situation thereof; Togither with the Great Riches, Huge Citties, Politike Gouvernement and Rare Inventions in the same* [undertaken 'at the earnest request and encouragement of my worshipfull friend Master Richard Hakluyt, late of Oxforde']. Reprinted in Spanish, Medina del Campo, 1595; Antwerp, 1596 and 1655; Ital. tr. Venice (3 editions), 1586; Fr. tr. Paris, 1588, and 1589; Germ. and Latin tr. Frankfurt, 1589. Ed. G. T. Staunton, Hakluyt Soc. Pub. 1853.
MENNINGER, K. (1). *Zahlwort und Ziffer; aus der Kulturgeschichte unserer Zahlsprache, unserer Zahlschrift und des Rechenbretts*. Hirt, Breslau, 1934.
MENON, C. P. S. (1). *Early Astronomy and Cosmology*. Allen & Unwin, London, 1932.

MERTON, R. K. (1). 'Science, Technology and Society in Seventeenth Century England.' *OSIS*, 1938, **4**, 360.

M[ERTON], R. K. (2). 'The Comet of Hipparchus, and Pliny.' *SKY*, 1940, **4** (no. 11), 4.

MEYER, F. (1). 'Fraunhofer als Mechaniker und Konstrukteur.' *NW*, 1926, **14**, 533.

DE MEYNARD, C. BARBIER & DE COURTEILLE, P. (1) (tr.). *Les Prairies d'Or* (the *Murūj al-Dhabab* of al-Mas'ūdī, +947). 9 vols. Paris, 1861–77.

MEZ, A. (1). *Die Renaissance des Islams.* Winter, Heidelberg, 1922. (Part tr. Margoliouth & Khuda al-Buksh, *IC*, **2**, 92.)

MICHEL, H. (1). 'Les Jades Astronomiques Chinois; une Hypothèse sur leur Usage.' *BMRAH*, 1947, 31 (crit. Chang Yü Chê, *TWHP*, 1956, **4**, 257).

MICHEL, H. (2). (*a*) 'Les Jades Astronomiques Chinois.' *CAM*, 1949, **4**, 111. (*b*) 'Chinese Astronomical Jades.' *POPA*, 1950, **58**, 222. (*c*) 'Astronomical Jades.' *ORA*, 1950, **2**, 156.

MICHEL, H. (3). *Traité de l'Astrolabe.* Gauthier-Villars, Paris, 1947 (rev. F. Sherwood Taylor, *N*, 1948, **162**, 46).

MICHEL, H. (4). 'Du Prisme Méridien au *Siun-Ki* [*Hsüan-Chi*].' *CET*, 1950, **66**, 23.

MICHEL, H. (5). 'Le Calcul Mécanique; à propos d'une Exposition récente.' *JSHB*, 1947 (no. 7), 307.

MICHEL, H. (7). 'Sur les Jades Astronomiques Chinois.' *MCB*, 1951, **9**, 153.

MICHEL, H. (8). 'A Propos de Terminologie.' *CET*, 1951, **67**.

MICHEL, H. (9). 'Un Service de l'Heure Millénaire.' *CET*, 1952, **68**, 1.

MICHEL, H. (10). *Montres Solaires* [Portable Sun-Dials]. Catalogue des Cadrans Solaires du Musée de la Vie Wallonne. Musée Wallon, Liège, 1953. (Reprinted from *CET*, 1952, **68**, 253.)

MICHEL, H. (11). 'La Mesure du Temps.' *RM*, 1952, no. 3.

MICHEL, H. (12). *Introduction à l'Etude d'une Collection d'Instruments anciens de Mathématiques.* de Sikkel, Antwerp, 1939.

MICHEL, H. (13). 'Le Rectangulus de Wallingford, précédé d'une Note sur le Torquetum.' *CET*, 1944, **60** (nos. 11, 12), 1.

MICHEL, H. (14). 'Les Tubes Optiques avant le Télescope.' *CET*, 1954, **70** (nos. 5, 6), 3.

MICHEL, H. (17). 'Sur l'Origine de la Théorie de la Trépidation.' *CET*, 1950 (nos. 9 and 10), 52.

MICHELL, J. (1). 'Conjectures concerning the Cause, and Observations upon the Phenomena, of Earthquakes....' *PTRS*, 1761, **51**.

DES MICHELS, A. (1) (tr.). *Histoire Géographique des Seize Royaumes* [*Shih-liu Kuo Chiang Yü Chih*, +1798, by Hung Liang-Chi]. Leroux, Paris, 1891. (Pub. Ecole Langues Orient. Viv. 3e sér. no. 11.)

MIELEITNER, K. (1). 'Geschichte d. Mineralogie im Altertum und Mittelalter.' *FMKP*, 1922, **7**, 427.

MIELI, ALDO (1). *La Science Arabe, et son Rôle dans l'Evolution Scientifique Mondiale.* Brill, Leiden, 1938.

MIELI, A. (2). *Panorama General de Historia de la Ciencia.* Vol. I, *El Mundo Antiguo; griegos y romanos.* Vol. II, *El Mundo Islámico e el Occidente Medieval Cristiano.* Espasa-Calpe, Buenos Aires, 1946. (Nos. 1 and 5 respectively of Colección Historia y Filosofía de la Ciencia, ed. J. Rey Pastor.)

MIKAMI, Y. (1). *The Development of Mathematics in China and Japan.* Teubner, Leipzig, 1913 (Abhdl. z. Gesch. d. math. Wissenschaften mit Einschluss ihrer Anwendungen, no. 30) (rev. H. Bosmans, *RQS*, 1913, **74**, 641).

MIKAMI, Y. (2). 'Notes on Native Japanese Mathematics. I, The Pythagorean Theorem.' *AMP*, 1913 (3rd ser.), **20**, 1; 1914, **22**, 183.

MIKAMI, Y. (3). 'Arithmetic with Fractions in Old China.' *AMN*, 1911, **32**, no. 3.

MIKAMI, Y. (4). 'Chinese Mathematics' (reply to van Hée (10) on the *Chhou Jen Chuan*). *ISIS*, 1928, **11**, 123. (Japanese version, Mikami (5).)

MIKAMI, Y. (5). 'A Remark on the Chinese Mathematics in Cantor's *Geschichte d. Mathematik.*' *AMP*, 1909, **15**, 68; 1911, **18**, 209 ('Further Remarks').

MIKAMI, Y. (6). 'Mathematics in China and Japan.' In *Scientific Japan, Past and Present*, 3rd Pan-Pacific Science Congress Volume, p. 177. Tokyo, 1926.

MIKAMI, Y. (7). 'Chronological Table of the History of Science in China and Japan, +16th century.' *A*, 1941, **23**, 211.

MIKAMI, Y. (8). 'On a Japanese Astronomical Treatise [*Kwanshō Zusetsu*, 1823] based on Dutch Works [esp. J. F. Martinet's *Katechismus d. Natuur*, 1779].' *NAW*, 1911, **9**, 231.

MIKAMI, Y. (9). 'Hatono Sōha and the Mathematics of Seki [Takakusn]' (identity of Petrus Hartsingius Japonensis). *NAW*, 1911, **9**, 158.

MIKAMI, Y. (10). 'On an Astronomical Treatise [*Kenkon Bensetsu*] composed [in +1650] by a Portuguese in Japan' (Sawano Chūan = Giuseppe Chiara). *NAW*, 1912, **10**, 61.

MIKAMI, Y. (11). 'On a Japanese MS. of the Seventeenth Century concerning European Astronomy' (the *Namban Tenchi-ron* of *c.* +1670). *NAW*, 1912, **10**, 71.

MIKAMI, Y. (12). 'A Japanese Buddhist View of European Astronomy' (the *Bukkoku Rekishō-hen* of Entsū, +1810). *NAW*, 1912, **10**, 233.

MIKAMI, Y. (13). 'On Maeno [Ryōtaku's] Description of the Parallelogram of Forces' in the MS. *Hon-yaku Undō-hō* (*c.* +1780). *NAW*, 1913, **11**, 76.

MIKAMI, Y. (14). 'On Shizuki [Tadao's] Translation [*Rekishō Shinsho*, +1798 to +1802] of Keill's Astronomical Treatise [*Introductio ad veram Physicam et veram Astronomiam*, by J. Keill, Oxford, 1705; Dutch tr. by J. Lulofs, 1740].' *NAW*, 1913, **11**, 1.

MIKAMI, Y. (15). 'On the Dutch Art of Surveying as Studied in Japan' (a MS. entitled *Kiku-jutsu Denrai no Maki* (Book of the Successions of the Art of Surveying) by Shimizu Taiemon, 1717). *NAW*, 1911, **9**, 301, 370.

MIKAMI, Y. (16). 'A Chinese Theorem on Geometry.' *AMP*, 1905.

MIKAMI, Y. (17). 'Zur Frage abendländischer Einflüsse auf die japanische Mathematik am Ende des siebzehnten Jahrhunderts.' *BM*, 1907, **7** (no. 3).

MIKAMI, Y. (18). 'A Question on Seki's Invention of the "Circle-Principle".' *PTMS*, 1909 (2nd ser.), **4**, 442.

MIKAMI, Y. (19). 'The Circle-Squaring of the Chinese.' *BM*, 1910, **10**.

MIKAMI, Y. (20). 'The Influence of the Abacus on Chinese and Japanese Mathematics.' *JDMV*, 1911, **20**, 380.

MIKAMI, Y. (21). 'On the Establishment of the *Yenri* Theory in Old Japanese Mathematics.' *PTMS*, 1932 (3rd ser.), **12**, 43.

MILHAM, W. I. (1). *Time and Timekeepers; History, Construction, Care, and Accuracy, of Clocks and Watches.* Macmillan, New York, 1923.

MILLER, K. (1). *Mappaemundi, die ältesten Weltkarten.* 6 vols. Stuttgart, 1895–8.

MILLER, K. (2). *Die Peutingersche Tafel oder Weltkarte des Castorius.* Stuttgart, 1916.

MILLER, K. (3). *Die Erdmessung im Altertum und ihr Schicksal.* Stuttgart, 1919.

MILLER, K. (4). *Mappae Arabicae; arabische Welt- u. Länderkarten des 9.–13. Jahrhunderts....* Priv. published, Stuttgart.

MILLS, J. V. (1). 'Malaya in the *Wu Pei Chih* Charts.' *JRAS/M*, 1937, **15** (no. 3), 1.

[MILLS, J. V.] (2). 'The Sanbiasi Chinese World-Map; a printed "Ricci-type" Map of the World (Canton, *c.* 1648).' In *A Selection of Precious Manuscripts, Historic Documents and Rare Books, the majority from the renowned collection of Sir Thomas Phillipps (1792–1872), offered for sale by W. H. Robinson Ltd.* Robinson, London, 1950 (Catalogue no. 81).

MILLS, J. V. (3). 'Notes on Early Chinese Voyages.' *JRAS*, 1951, 3.

MILLS, J. V. (4). MS. Translation of ch. 9 of the *Tung Hsi Yang Khao.* (Studies on the Oceans East and West.) Unpub.

MILLS, J. V. (5). MS. Translation of *Shun Fêng Hsiang Sung* (MS.) (Fair Winds for Escort). Bodleian Library, Laud Orient. MS. no. 145. Unpub.

MILLS, J. V. (7). 'Three Chinese Maps.' (Two Coastal Charts (*c.* 1840) and a copy of the Chhien-Lung map of China, +1775.) *BMQ*, 1953, 65.

MILLS, J. V. (8). 'Chinese Coastal Maps.' *IM*, 1954, **11**, 151.

MILNE, J. (1). *Earthquakes and other Earth Movements.* Kegan Paul, London, 1886. (Later editions do not contain the account of the seismograph of Chang Hêng.)

MINAKATA, K. (1). 'The Constellations of the Far East.' *N*, 1893, **48**, 542.

MINAKATA, K. (2). 'Chinese Theories of the Origin of Amber.' *N*, 1895, **51**, 294.

MINGUZZI, C. & VERGNANO, O. (1). 'Il Contenuto di Nichel nelle ceneri di *Alyssum bertolonii* Desv.' *ASTSN*, 1948, ser. A, **55**, 3.

MITCHELL, W. M. (1). 'History of the Discovery of the Solar Spots.' *POPA*, 1916, **24**, 22, 82, 149, 206, 290, 341, 428, 488, 562.

DE MOIDREY, J. [& HUANG, P.] (1). 'Observations Anciennes de Taches Solaires en Chine.' *BAS*, 1904, **21**, 1.

MOIR, A. L. & LETTS, M. (1). *The World Map in Hereford Cathedral and its Pictures.* De Cantilupe (for the Cathedral Chapter), Hereford, 1955. The second part reprinted from *NQ*, 1955.

MONTUCLA, J. E. (1). *Histoire des Mathématiques....* 4 vols, 1758. 2nd edn. Agasse, Paris, An 7 de la République (1799–1802). (Account of Chinese astronomy in vol. 1, pp. 448–80.)

MOODY, E. A. (1). 'Galileo and Avempace [Ibn Bājjah]; the Dynamics of the Leaning Tower Experiment.' *JHI*, 1951, **12**, 163, 375.

MOORE, W. U. (1). 'The Bore of the Tsien-Tang Kiang [Chhien-thang R.].' *JRAS/NCB*, 1889, **23**, 185.

MORAN, H. A. (1). *The Alphabet and the Ancient Calendar Signs; Astrological Elements in the Origin of the Alphabet* (with foreword by D. Diringer). Pacific Books, Palo Alto, Calif. 1953 (lithoprinted).

MOREAU, F. (1). *Eléments d'Astronomie.* Bruxelles, 1942.

DE MORGAN, A. (1). *Budget of Paradoxes*, vol. 2, p. 66. Open Court, Chicago, 1915.

MORGAN, E. (1) (tr.). *Tao the Great Luminant; Essays from Huai Nan Tzu, with introductory articles, notes and analyses.* Kelly & Walsh, Shanghai, n.d. (1933?).
MORLEY, S. G. (1). *The Ancient Maya.* Stanford Univ. Press, Palo Alto, California, 1946.
MOUCHEZ, ADMIRAL (1). 'L'Observatoire de Pékin.' *LN*, 1888, **16** (no. 808), 406.
MOULE, A. C. (1). *Christians in China before the year 1550.* SPCK, London, 1930.
MOULE, A. C. (3). 'The Bore on the Chhien-Thang River in China.' *TP*, 1923, **22**, 135 (includes much material on tides and tidal theory).
MOULE, A. C. (4). 'An Introduction to the *I Yü Thu Chih.*' *TP*, 1930, **27**, 179.
MOULE, A. C. (5). 'The Wonder of the Capital' (the Sung books *Tu Chhêng Chi Shêng* and *Mêng Liang Lu* about Hangchow). *NCR*, 1921, **3**, 12, 356.
MOULE, A. C. (6). 'From Hangchow to Shangtu, +1276.' *TP*, 1915, **16**, 393.
MOULE, A. C. & PELLIOT, P. (1) (tr. and annot.). *Marco Polo (+1254 to +1325); The Description of the World.* Routledge, London, 1938.
MOULE, G. E. (1) (tr.). 'The Obligations of China to Europe in the Matter of Physical Science acknowledged by Eminent Chinese; being Extracts from the Preface to Tsêng Kuo-Fan's edition of Euclid, with brief introductory observations.' *JRAS/NCB*, 1873, **7**, 147.
MOULE, G. T. (1). 'The Hangchow Bore.' *NCR*, 1921, **3**, 289.
MUGLER, O. (1). 'Edelsteinhandel im Mittelalter und im 16-Jahrhundert, mit Excursen ü. den Levante- und Asiatischen-Handel überhaupt.' Inaug. Diss., München, 1928.
MUIR, SIR T. (1). *The Theory of Determinants in the Historical Order of its Development.* London, 1890. (Many subsequent editions.)
MUIRHEAD, W. (1). '[Glossary of Chinese] Mineralogical and Geological Terms.' In Doolittle, J. (1), vol. 2, p. 256.
MULDER, W. Z. (1). 'The *Wu Pei Chih* Charts.' *TP*, 1944, **37**, 1.
MÜLLER, C. (1). *Fragmenta Historicorum Graecorum*, 5 vols. Didot, Paris, 1841–51.
MÜLLER, F. W. K. (1). 'Der Weltberg Meru nach einem japanischen Bilde.' *ENB*, 1895, **1** (no. 2), 12.
MÜLLER, F. W. K. (3). 'Die "persischen" Kalenderausdrücke im Chinesischen Tripiṭaka.' *SPAW/PH*, 1907, **25**, 458.
MÜLLER, R. (1). 'Die astronomischen Instrumente des Kaisers von China in Potsdam.' *AT*, 1931 (no. 2), 120.
MULLET, C. (1). 'Essai sur la Minéralogie Arabe.' *JA*, 1868 (6e sér.), **11**, 5, 109, 502.
MUNTNER, S. (1). *Rabbi Shabtai Donnolo (+913 to +985)*; First Section, Medical Works, on the occasion of the millennium of the earliest Hebrew book in Christian Europe, edited, with a commentary, by S.M.; Second Section, Contributions to the History of Jewish Medicine [including a Biography of R. Shabtai Donnolo, a cosmographical Introduction to the *Sefer Yesirah*, and a discussion of the countries of origin of Drug Plants and the Trade in them, together with notes on the land routes, etc.] (in Hebrew). Mosad Haraw Kook, Jerusalem, 1949.
VAN MUSSCHENBROEK, P. (1). *Introductio ad Philosophiam Naturalem.* 2 vols., Luchtmans, Leiden, 1762; Padua, 1768.

AL-NADĪM, ABU'L-FARAJ IBN ABŪ YA'QŪB (1). *Fihrist al-'ulūm* [Index of the Sciences], ed. G. Flügel. Leipzig, 1871–2.
NAFIS AHMAD. See Ahmad, Nafis.
NAGASAWA, K. (1). *Geschichte der Chinesischen Literatur, und ihrer gedanklichen Grundlage.* Transl. from the Japanese by E. Feifel. Fu-jen Univ. Press, Peiping, 1945.
NAKAMURA, H. (1). 'Old Chinese World-Maps preserved by the Koreans.' *IM*, 1947, **4**, 3.
NALLINO, C. A. (2) (ed. and Latin tr.). *Al-Battānī sive Albatenii opus Astronomicum, Arabice editum, Latine versum, adnotationibus instructum a....* 3 vols. Milan, 1899–1907.
VAN NAME, A. (1). 'On the Abacus of China and Japan.' *JAOS*, 1875, **10**, cx.
NAPIER, JOHN (of Merchistoun) (1). *Rabdologiae, seu Numerationis per Virgulas Libri Duo: cum Appendice de Expeditissimo Multiplicationis Promptuario; quibus accessit Arithmeticae Localis Liber unus.* Edinburgh, 1617; Leiden, 1626 (trs. Verona, 1623; Berlin, 1623).
NARRIEN, J. (1). *An Historical Account of the Origin and Progress of Astronomy, with plates illustrating, chiefly, the ancient systems.* Baldwin & Cradock, London, 1833.
NAWRATH, A. (1). *Indien und China; Meisterwerke der Baukunst und Plastik* (album of photographs). Schroll, Vienna, 1938.
NEAL, J. B. (1). 'Analyses of Chinese Inorganic Drugs.' *CMJ*, 1889, **2**, 116; 1891, **5**, 193.
NEEDHAM, JOSEPH (2). *A History of Embryology.* Cambridge, 1934.
NEEDHAM, JOSEPH (29). 'Mathematics and Science in China and the West.' *SS*, 1956, **20**, 320.
NEEDHAM, JOSEPH (30). 'Prospection Géobotanique en Chine Médiévale.' *JATBA*, 1954, **1**, 143.
NEEDHAM, JOSEPH & WANG LING. See Wang & Needham.

NEEDHAM, MARCHAMONT (1). *Medela Medicinae; A Plea for the Free Profession, and a Renovation of the Art of Physick....* Lownds, London, 1665.

NESSELMANN, G. H. F. (1). *Die Algebra der Griechen.* Berlin, 1842.

NEUBURGER, A. (1). *The Technical Arts and Sciences of the Ancients.* Methuen, London, 1930. Tr. H. L. Brose from *Die Technik d. Altertums.* Voigtländer, Leipzig, 1919. (The English version inexcusably omits all the references to the literature.)

NEUGEBAUER, O. (1). 'The History of Ancient Astronomy; Problems and Methods.' *JNES*, 1945, **4**, 1; reprinted in enlarged version, *PASP*, 1946, **58**, 17, 104.

NEUGEBAUER, O. (2). 'The Water-Clock in Babylonian Astronomy.' *ISIS*, 1947, **37**, 37.

NEUGEBAUER, O. (3). 'The Astronomical Origin of the Theory of Conic Sections.' *PAPS*, 1948, **92**, 136.

NEUGEBAUER, O. (4). 'The Study of Wretched Subjects' (a defence of the study of ancient and medieval pseudo-sciences for the unravelling of the threads of the growth of true science, and for the understanding of the mental climate of the early discoveries). *ISIS*, 1951, **42**, 111.

NEUGEBAUER, O. (5). 'A Greek Table for the Motion of the Sun' (date unknown but with Indian connections). *CEN*, 1951, **1**, 266.

NEUGEBAUER, O. (7). 'The Early History of the Astrolabe.' *ISIS*, 1949, **40**, 240.

NEUGEBAUER, O. (8). 'Babylonian Planetary Theory.' *PAPS*, 1954, **98**, 60.

NEUGEBAUER, O. (9). *The Exact Sciences in Antiquity.* Princeton Univ. Press, Princeton, N.J. 1952 (Messenger Lectures at Cornell University on mathematics and astronomy in Babylonia, Egypt and Greece) (rev. *ISIS*, 1952, **43**, 69).

NEUGEBAUER, O. (10) (ed.). *Babylonian Ephemerides of the Seleucid Period for the Motion of the Sun, the Moon, and the Planets.* 3 vols. Vol. I, Introduction; the Moon. Vol. II, The Planets; Indexes. Vol. 3, Plates. Lund Humphries, London, 1955. (Pub. of the Institute for Advanced Study, Princeton, N.J.)

NEUGEBAUER, O. (11). 'Tamil Astronomy.' *OSIS*, 1952, **10**, 252.

NEUGEBAUER, O. (12). Essay review of the first two volumes of Renou & Filliozat (1), giving a summary of our knowledge of the passage of Greek geometrical astronomy to Northern India and of Babylonian algebraical astronomy to Southern India. *A/AIHS*, 1955, **8**, 166.

NEUGEBAUER, P. V. (1). *Astronomische Chronologie.* Berlin, 1929.

NEUGEBAUER, P. V. (2). *Tafeln zur astronomischen Chronologie.* I. Sterntafeln von 4000 vor Chr. bis zur Gegenwart nebst Hilfsmitteln zur Berechnung von Sternpositionen zw. 4000 vor Chr. und 3000 nach Chr.... Hinrichs, Leipzig, 1912. II. Tafeln für Sonne, Planeten und Mond, nebst Tafeln der Mondphasen für die Zeit 4000 vor Chr. bis 3000 nach Chr.... Hinrichs, Leipzig, 1914.

NEUMANN, K. F. (1). *Asiatische Studien.* Leipzig, 1837.

NEUMANN, K. F. (2). 'Catalogue des Latitudes et des Longitudes de plusieurs Places de l'Empire Chinois.' *JA*, 1834 (2e sér.), **13**, 87.

NICOLSON, M. H. (1). *Voyages to the Moon.* Macmillan, New York, 1948.

NIEUHOFF, J. (1). *L'Ambassade* [1655–1657] *de la Compagnie Orientale des Provinces Unies vers l'Empereur de la Chine, ou Grand Cam de Tartarie, faite par les Sieurs Pierre de Goyer & Jacob de Keyser; Illustrée d'une tres-exacte Description des Villes, Bourgs, Villages, Ports de Mers, et autres Lieux plus considerables de la Chine; Enrichie d'un grand nombre de Tailles douces, le tout recueilli par Mr Jean Nieuhoff*...(title of Pt. II: *Description Generale de l'Empire de la Chine, où il est traité succinctement du Gouvernement, de la Religion, des Mœurs, des Sciences et Arts des Chinois, comme aussi des Animaux, des Poissons, des Arbres et Plantes, qui ornent leurs Campagnes et leurs Rivieres; y joint un court Recit des dernieres Guerres qu'ils ont eu contre les Tartares*). de Meurs, Leiden, 1665.

NIEUWHOFF. See Nieuhoff.

NIVISON, D. S. (1). 'The Problem of "Knowledge" and "Action" in Chinese Thought since Wang Yang-Ming.' In *Studies in Chinese Thought*, ed. A. F. Wright, *AAN*, 1953, **55** (no. 5), 112 (Amer. Anthropol. Assoc. Memoirs, No. 75).

NŌDA, C. (1) = (*1*). *An Enquiry concerning the 'Chou Pei Suan Ching'.* Academy of Oriental Culture, Kyoto Institute, Kyoto, 1933. (Toho Bunka Gakuin Kyoto Kenkyusho Memoirs, no. 3.)

NŌDA, C. (2) = (*2*). *An Enquiry concerning the Astronomical Writings contained in the 'Li Chi, Yüeh Ling'.* Academy of Oriental Culture, Kyoto Institute, Kyoto, 1938. (Toho Bunka Gakuin Kyoto Kenkyusho Memoirs, no. 12.)

NOEL, FRANCIS [FRANTIŠEK], S.J. (1). *Observationes Mathematicae et Physicae in India et China factae a Patre Francisco Noel SJ ab anno 1684, usque ad annum 1708.* University Press, Prague, 1710. Cf. Slouka, pp. 161 ff. Pp. 56 ff. 'Varia ad Astronomiam Sinicam Spectantia': (*a*) the Stems and Branches, (*b*) diagram of a sexagenary cycle from +1684 to +1744, (*c*) list of 28 *hsiu*, (*d*) list of 24 *chieh chhi*, (*e*) rough correlation of star-catalogue arranged by zodiacal signs with Chinese star-group names, (*f*) discussion of Chinese metrology. The Royal Astronomical Society copy contains copious annotations, especially of Chinese characters, inserted by John Williams.

NOEL, FRANCIS (2). *Philosophia Sinica; Tribus Tractatibus primo cognitionem primi Entis Secundo Ceremonias erga Defunctos tertio Ethicam juxta Sinarum mentem complectens.* Univ. Press, Prague, 1711. (Cf. Pinot (2), p. 116.)

NOEL, FRANCIS (3). *Sinensis Imperii Libri Classici Sex, nimirum Adultorum Schola* [Ta Hsüeh], *Immutabile Medium* [Chung Yung], *liber Sententiarum* [Lun Yü], *Mencius, Filialis Observantia* [Hsiao Ching], *Parvulorum Schola* [San Tzu Ching?] *e Sinico Idiomate in Latinum traducti....* Univ. Press, Prague, 1711.

NOEL, FRANCIS (4). *Historica Notitia Rituum et Ceremoniarum Sinicarum in Colendis Parentibus ac Benefactoribus defunctis, ex ipsis sinensium auctorum libris desumpta....* Univ. Press, Prague, 1711. (Censored and withdrawn shortly after publication, therefore very rare.)

NOLTE, F. (1). *Die Armillarsphäre.* Mencke, Erlangen, 1922. (Abhdl. z. Gesch. d. Naturwiss. u. d. Med. no. 2.)

NORDENSKIÖLD, A. E. (1). *Periplus; an Essay on the Early History of Charts and Sailing Directions,* tr. F. A. Bather. Stockholm, 1897.

NORDENSKIÖLD, E. (1). 'Le Quipu Péruvien du Musée du Trocadéro.' *TMB,* 1931 (no. 1), **16**.

NOWOTNY, K. A. (1). 'The Construction of Certain Seals and Characters in the Work of Agrippa of Nettesheim.' *JWCI,* 1949, **12**, 46.

AL-NUWAIRĪ, AHMAD IBN-ABD AL-WAHHAB (1). *Nihayat al-arab fi funun al-adab* (Aim of the Intelligent in the Arts of Letters), ed. Ahmad Zaki pasha. Cairo, 1923–.

D'OHSSON, MOURADJA (1). *Histoire des Mongols depuis Tchinguiz Khan jusqu'à Timour Bey ou Tamerlan.* 4 vols. van Cleef, The Hague and Amsterdam, 1834–52.

OLDENBERG, H. (2). 'Nakshatra und Sieou.' *NGWG/PH,* 1909, 544.

OLIVIER, C. P. (1). *Comets.* Baillière, Tindal & Cox, London, 1930.

OLMSTED, J. W. (1). 'The "Application" of Telescopes to Astronomical Instruments [with graduated arcs for measuring angles].' *ISIS,* 1949, **40**, 213.

OLSCHKI, L. (2). *Galilei und seine Zeit.* Halle, 1927.

OLSCHKI, L. (3). 'Galileo's Philosophy of Science.' *PHR,* 1943, **52**, 349.

OLSCHKI, L. (4). *Guillaume Boucher; a French Artist at the Court of the Khans.* Johns Hopkins Univ. Press, Baltimore, 1946 (rev. H. Franke, *OR,* 1950, **3**, 135).

OMORI, F. (1). 'A Note on Old Chinese Earthquakes.' *SJJ,* 1893, **1**, 119.

D'OOGE, M. L., ROBBINS, F. E. & KARPINSKI, L. C. (1) (tr.). *Nicomachus of Gerasa; 'Introduction to Arithmetic', translated into English, with Studies in Greek Arithmetic.* Macmillan, New York, 1926. (Univ. of Michigan Studies, Humanistic Series, no. 16.)

OORDT, J. H. (1). 'Note on the Supernova of +1054.' *TP,* 1942, **36**, 179.

OPPENHEIM, S. (1). 'Über d. Perioden d. Sonnenflecken.' *FF,* 1928, **4**, 128.

OPPERT, M. (1). 'Tablettes Assyriennes...Prédictions tirées des Monstruosités....' *JA,* 1871 (6e sér.), **18**, 449.

VON OPPOLZER, T. (1). 'Ü. d. Sonnenfinsternis d. *Schu King* [*Shu Ching*].' *MPAW,* 1880, 166.

VON OPPOLZER, T. (2). *Canon der Finsternisse.* Vienna, 1887.

ORE, OYSTEIN (1). *Cardano, the Gambling Scholar.* Princeton Univ. Press, Princeton, N.J., 1953.

OUSELEY, SIR WILLIAM (1). *The Oriental Geography of Ebn Haukal.* London, 1800.

OZANAM, M. (1). *Recreations in Mathematics and Natural Philosophy....* Enlarged by M. Montucla. Eng. tr. C. Hutton. 4 vols. Kearsley, London, 1803. From *Récréations Mathématiques et Physiques.* Paris, 1694. (For bibliogr. see Rouse Ball, 2.)

PALMQVIST, S. & BRUNDIN, N. (1). On geobotanical and biogeochemical prospecting. See *CA,* 1939, **33**, 6762; 1942, **36**, 3402. See also Landergren (1).

PANCRITIUS, M. (1). 'Lunar calendars and matriarchy.' *AN,* 1930, **25**, 879, 889.

PANETH, F. A. (1). *The Frequency of Meteorite Falls throughout the Ages.* Art. in *Vistas in Astronomy,* ed. A. Beer. Pergamon Press, London & New York, 1956, vol. II, p. 1681.

PANNEKOEK, A. (1). 'Planetary Theories.' *POPA,* 1947, **55** (Kidinnu and Ptolemy); 1948, **56** (Copernicus, Kepler, Newton and Laplace). (Also issued separately, repaginated, for private circulation.)

PANNIER, L. (1). *Les Lapidaires Français du Moyen Âge....* Paris, 1882.

PAPINOT, E. (1). *Historical and Geographical Dictionary of Japan.* Overbeck, Ann Arbor, Mich. 1948. Lithoprinted from original ed. Kelly & Walsh, Yokohama, 1910. Eng. tr. of *Dictionnaire d'Histoire et de Géographie du Japon.* Sanseido, Tokyo; Kelly & Walsh, Yokohama, 1906.

DE PARAVEY, C. H. (1). *Illustrations de l'Astronomie hiéroglyphique et des Planisphères et Zodiaques retrouvés en Egypte, en Chaldée, dans l'Inde et au Japon, etc.* Paris, 1869.

PARKER, E. H. (4). 'Notes on Chinese astronomy.' *CR,* 1887, **15**, 182.

PARTINGTON, J. R. (1). *Origins and Development of Applied Chemistry.* Longmans Green, London, 1935.

PASCAL, BLAISE (1). *Traité du Triangle Arithmétique* (1654). Pub. posthumously, Paris, 1665.

PASQUALE D'ELIA. See d'Elia, Pasquale.

PEASE, A. S. (1). 'Fossil Fishes Again.' *ISIS*, 1942, **33**, 689.

PECK, A. L. (1) (tr.). *Aristotle; The Generation of Animals*. Loeb Classics series, Heinemann, London, 1943.

PEEK, A. P. (1). 'Notes on the so-called "Black Lime" (*chhing hui* or *mo hui*) of China.' *CIMC/MR*, 1885, no. 29, 40.

PELLAT, C. (2). 'Le Traité d'Astronomie pratique et de Météorologie populaire d'Ibn Qutayba.' *ARAB*, 1954, **1**, 84.

PELLAT, C. (3). 'Dictons rimés, *anwā* [proverbs about heliacal risings and settings in relation to weather], et Mansions Lunaires chez les Arabes.' *ARAB*, 1955, **2**, 17.

PELLIOT, P. (2). 'Les Grands Voyages Maritimes Chinois au début du 15e Siècle.' *TP*, 1933, **30**, 237; 1935, **31**, 274.

PELLIOT, P. (9). 'Mémoire sur les Coutumes de Cambodge' [a translation of Chou Ta-Kuan's *Chen-La Fêng Thu Chi*]. *BEFEO*, 1902, **1**, 123. Revised version: Paris, 1951, see Pelliot (33).

PELLIOT, P. (16). 'Le Fou-Nan' [Cambodia]. *BEFEO*, **3**, 57.

PELLIOT, P. (17). 'Deux Itinéraires de Chine à l'Inde à la Fin du 8e Siècle.' *BEFEO*, 1904, **4**, 131.

PELLIOT, P. (18). Review of G. Ferrand's *Voyage du Marchand Arabe Sulayman*...(on Sung and Arab maps). *TP*, 1922, **21**, 405.

PELLIOT, P. (19). 'Note sur la Carte des Pays du Nord-Ouest dans le *King Che Ta Tien*' [*Yuan Ching Shih Ta Tien*]. *TP*, 1927, **25**, 98.

PELLIOT, P. (20). Introduction to *Jades Archaïques de Chine appartenant à Mons. C. T. Loo*. van Oest, Paris & Brussels, 1925.

PELLIOT, P. (21). 'Les Prétendus Jades de Sou-Tcheou [Suchow, Kansu].' *TP*, 1913, **14**, 258.

PELLIOT, P. (25). *Les Grottes de Touen-Hoang* [*Tunhuang*]; *Peintures et Sculptures Bouddhiques des Epoques des Wei, des Thang et des Song* [*Sung*]. Mission Pelliot en Asie Centrale, 6 portfolios of plates. Paris, 1920–4.

PELLIOT, P. (33). *Mémoires sur les Coutumes de Cambodge de Tcheou Ta-Kouan* [Chou Ta-Kuan]; version nouvelle, suivie d'un Commentaire inachevé, Maisonneuve, Paris, 1951. (Œuvres Posthumes, no. 3.)

PELLIOT, P. (41). *Les Débuts de l'Imprimerie en Chine*. Impr. Nat. & Maisonneuve, Paris, 1953. (Œuvres Posthumes, no. 4.)

PELLIOT, P. (42). 'Neuf Notes sur des Questions d'Asie Centrale.' *TP*, 1928, **26**, 201.

PELLIOT, P. & MOULE, A. C. (1). 'An Ancient Seismometer' (Chang Hêng's). *TP*, 1924, **23**, 36.

PELSENEER, J. (2). 'Les Influences dans l'Histoire des Sciences.' *A/AIHS*, 1948, **1**, 347.

PERLEBERG, M. (1) (tr.). *The Works of Kungsun Lung Tzu, with a Translation from the parallel Chinese original text, critical and exegetical notes, punctuation and literal translation, the Chinese commentary, prolegomena and Index*. Privately printed. Hongkong, 1952. (Crit. J. J. L. Duyvendak, *TP*, 1954, **42**, 383.)

PERNICE, E. (1). *Galeni de ponderibus et mensuris Testimonia*. Bonn, 1888.

PERNTER, J. M. & EXNER, F. M. (1). *Meteorologische Optik*. Braumüller, Vienna & Leipzig, 1922.

PERRY, W. J. (1). *The Children of the Sun*. Methuen, London, 1923.

PETECH, L. (1). *Northern India according to the 'Shui Ching Chu'*. Ist. Ital. per il Medio ed Estremo Oriente, Rome, 1950. (Rome Oriental Series, no. 2.)

PETECH, L. (3). 'Una Carta Cinese del Secolo 18.' *AUON*, 1954, **5** (n.s.), 1.

PETERS, C. H. F. & KNOBEL, E. B. (1). *Ptolemy's Catalogue of Stars, a Revision of the Almagest*. Carnegie Institution, Washington, publ. no. 86, 1915.

PETRUCCI, R. (1). 'Sur l'Algèbre Chinoise.' *TP*, 1912, **13**, 559.

PETRUCCI, R. (3) (tr.). *Encyclopédie de la Peinture Chinoise* [the *Chieh Tzu Yuan Hua Chuan*]. Laurens, Paris, 1918.

PFISTER, L. (1). *Notices Biographiques et Bibliographiques sur les Jésuites de l'Ancienne Mission de Chine* (*+1552 to +1773*). 2 vols. Mission Press, Shanghai, 1932 (*VS* no. 59).

PFIZMAIER, A. (34) (tr.). 'Die Feldherren Han Sin, Pêng Yue, und King Pu' (Han Hsin, Phêng Yüeh and Ching Pu). *SWAW/PH*, 1860, **34**, 371, 411, 418. Tr. *Shih Chi*, chs. 90 (in part), 91, 92, *Chhien Han Shu*, ch. 34; not in Chavannes (1).

PFIZMAIER, A. (39) (tr.). 'Die Könige von Hoai Nan aus dem Hause Han' (Huai Nan Tzu). *SWAW/PH*, 1862, **39**, 575. Tr. *Chhien Han Shu*, ch. 44.

PFIZMAIER, A. (43) (tr.). 'Die Geschichte einer Gesandtschaft bei den Hiung-Nu's' (Su Wu). *SWAW/PH*, 1863, **44**, 581. Tr. *Chhien Han Shu*, ch. 54 (second part).

PFIZMAIER, A. (64) (tr.). 'Die fremdländischen Reiche zu den Zeiten d. Sui.' *SWAW/PH*, 1881, **97**, 411, 418, 422, 429, 444, 477, 483. Tr. *Sui Shu*, chs. 64, 81, 82, 83, 84.

PFIZMAIER, A. (67) (tr.). 'Seltsamkeiten aus den Zeiten d. Thang' I and II. I, *SWAW/PH*, 1879, **94**, 7, 11, 19. II, *SWAW/PH*, 1881, **96**, 293. Tr. *Hsin Thang Shu*, chs. 34–6 (Wu Hsing Chih), 88, 89.

PFIZMAIER, A. (70) (tr.). 'Über einige chinesische Schriftwerke des siebenten und achten Jahrhunderts n. Chr.' *SWAW/PH*, 1879, **93**, 127, 159. Tr. *Hsin Thang Shu*, chs. 57, 59 (in part: I Wên Chih including agriculture, astronomy, mathematics, war, five-element theory).

PFIZMAIER, A. (83) (tr.). 'Das *Li Sao* und die Neun Gesänge.' *DWAW/PH*, 1851, **3**, 159, 175.

PFIZMAIER, A. (92) (tr.). 'Kunstfertigkeiten u. Künste d. alten Chinesen.' *SWAW/PH*, 1871, **69**, 147, 164, 178, 202, 208. Tr. *Thai-Phing Yü Lan*, chs. 736, 737 (magic), 750, 751 (painting) and 752 (inventions and automata).

PFIZMAIER, A. (94) (tr.). 'Beiträge z. Geschichte d. Perlen.' *SWAW/PH*, 1867, **57**, 617, 629. Tr. *Thai-Phing Yü Lan*, chs. 802 (in part), 803.

PFIZMAIER, A. (95) (tr.). 'Beiträge z. Geschichte d. Edelsteine u. des Goldes.' *SWAW/PH*, 1867, **58**, 181, 194, 211, 217, 218, 223, 237. Tr. *Thai-Phing Yü Lan*, chs. 807 (coral), 808 (amber), 809, (gems), 810, 811 (gold), 813 (in part).

PHILIP, A. (1). *The Calendar; its History, Structure, and Improvement.* Cambridge, 1921.

PHILLIPS, G. (1). 'The Seaports of India and Ceylon, described by Chinese Voyagers of the Fifteenth Century, together with an account of Chinese Navigation....' *JRAS/NCB*, 1885, **20**, 209; 1886, **21**, 30 (both with large folding maps).

PHILLIPS, G. (2). 'Précis translations of the *Ying Yai Shêng Lan*.' *JRAS*, 1895, 529; 1896, 341.

PI, H. T. (1). 'The History of Spectacles in China.' *CMJ*, 1928, **42**, 742.

PIGNATARO, D. (1). Earthquake intensity scale. In G. Vivenzio's *Istoria di Tremuoti avvenuti nella Provincia della Calabria ulteriore e nelle Città di Messina nell'anno 1783.* Naples, 1788.

PINOT, V. (1). *La Chine et la Formation de l'Esprit Philosophique en France (1640–1740).* Geuthner, Paris, 1932.

PINOT, V. (2). *Documents inédits relatifs à la Connaissance de la Chine en France de 1685 à 1740.* Geuthner, Paris, 1932.

PLANCHET, J. M. (1). 'La Mission de Pékin.' *BCP*, 1914, **1** (no. 6), 211.

PLANCHON, M. (1). *L'Horloge; son Histoire rétrospective, pittoresque et artistique.* Laurens, Paris, 1899; 2nd ed. 1912.

PLEDGE, H. T. (1). *Science since 1500.* HMSO, London, 1939.

PLUMMER, H. C. (1). 'Halley's Comet and its Importance.' *N*, 1942, **150**, 249.

POGO, A. (1). 'Egyptian Water-Clocks.' *ISIS*, 1936, **25**, 403.

POLIAK, A. N. (1). *Khazaria* (in Hebrew). Tel Aviv, 1944.

POLIAK, A. N. (2). 'The Jewish Khazar Kingdom in Mediaeval Geographical Science.' In *Actes du VIIe Congrès Internat. d'Histoire des Sciences*, p. 488. Jerusalem, 1953.

POLLARD, A. W. (1). *Early Illustrated Books; a History of the Decoration and Illustration of Books in the 15th and 16th Centuries.* Kegan Paul, London, 1917.

PONDER, E. (1). 'The Reputed Medicinal Properties of Precious Stones.' *PJ*, 1925, **61**, 686, 750.

POPE, A. U. (1). *A Survey of Persian Art.* 6 vols. Oxford, 1939.

POPE-HENNESSEY, U. (1). *Early Chinese Jades.* London, 1923.

POPE-HENNESSEY, U. (2). *Jade Miscellany.* London, 1946.

POSEPNY, F. (1). 'The Genesis of Ore Deposits.' *TAIME*, 1893, **23**, 197.

POTTS, R. (1). *Euclid's Elements of Geometry, chiefly from the text of Dr Simson, with explanatory Notes; together with a Selection of Geometrical Exercises from the Senate-House and College Examination Papers; to which is prefixed, An Introduction, containing a Brief Outline of the History of Geometry.* Cambridge, 1845.

POWELL, J. U. (1). 'The Simile of the Clepsydra in Empedocles.' *CQ*, 1923, **17**, 172.

PRATT, J. H. (1). 'On Chinese Astronomical Epochs.' *PMG*, 1862 (4th ser.), **23**, 1.

PRICE, D. J. (1). 'Clockwork before the Clock.' *HORJ*, 1955, **97**, 810; 1956, **98**, 31.

PRICE, D. J. (2) (ed.). *The Equatorie of the Planetis* (probably written by Geoffrey Chaucer), with a linguistic analysis by R. M. Wilson. C.U.P., Cambridge, 1955.

PRICE, D. J. (3). 'A Collection of Armillary Spheres and other Antique Scientific Instruments.' *ANS*, 1954, **10**, 172.

PRICE, D. J. (4). 'The Prehistory of the Clock.' *D*, 1956, **17**, 153.

PRICE, D. J. (7). 'Medieval Land Surveying and Topographical Maps.' *GJ*, 1955, **121**, 1.

PROSTOV, E. V. (1). 'Early Mentions of Fossil Fishes.' *ISIS*, 1942, **34**, 24.

PRUETT, J. H. (1). 'Motion of Circumpolar Stars.' *SKY*, 1951, **10**, 98.

PRZYŁUSKI, J. (6). 'L'Or, son Origine et ses Pouvoirs Magiques.' *BEFEO*, 1914, **14**, 1.

PUINI, C. (1). 'I Muraglione della Cina.' *RGI*, 1915, **22**, 481.

PUINI, C. (2). 'Idee Cosmologiche della Cina Antica.' *RGI*, 1894, **1**, 618; 1895, **2**, 1.

PUINI, C. (3). 'Qualche appunto circa l'Opera Geografico del Padre Matteo Ricci.' *RGI*, 1912, **19**, 679.

PULLEYBLANK, E. G. (3). 'A Sogdian Colony in Inner Mongolia.' *TP*, 1952, **41**, 317.

PULLEYBLANK, E. G. (4). *Chinese History and World History.* Inaugural Lecture at the University of Cambridge. C.U.P. 1955.

PURCHAS, S. (1). *Hakluytus Posthumus, or Purchas his Pilgrimes, contayning a History of the World in Sea Voyages and Lande Travells.* 4 vols. London, 1625. 2nd ed. *Purchas his pilgrimage, Or Relations of the world and the religions observed in all ages and places discovered.* London, 1626.

RAEDER, H., STRÖMGREN, E. & STRÖMGREN, B. (1) (tr.). *Tycho Brahe's Description of his Instruments and Scientific Work, as given in his 'Astronomiae Instauratae Mechanica'* (Wandesburgi, 1598). Munksgaard, Copenhagen, 1946. (Pub. of K. Danske Videnskab. Selskab.)

RAMMING, M. (1). 'The Evolution of Cartography in Japan.' *IM*, 1937, **2**, 17.

RANDALL, J. H. (1). 'The Development of Scientific Method in the School of Padua.' *JHI*, 1940, **1**, 177.

RANDALL, J. H. (2). 'The Place of Leonardo da Vinci in the Emergence of Modern Science.' *JHI*, 1953, **14**, 191.

RANGABÉ, A. R. (1). 'Lettre de Mons. Rangabé à Mons. Letronne sur une Inscription Grecque du Parthénon, etc. etc.' (including description of the Salamis abacus), with an appended 'Note sur l'Echelle Numérique d'un Abacus Athénien, etc.' by Letronne. *RA*, 1846, **3**, 295.

RANKAMA, K. (1). 'Some Recent Trends in Prospecting; Chemical, Biogeochemical and Geobotanical Methods.' *MM*, 1947, **28**, 282.

RANKAMA, K. (2). 'On the Use of Trace Elements in some Problems of Practical Geology.' *BCGF*, 1941, **22**, no. 126, 90.

RANKAMA, K. & SAHAMA, T. G. (1). *Geochemistry.* Univ. Chicago Press, Chicago, 1950.

RASHĪD AL-DĪN AL-HAMDĀNĪ. *Jāmi' al-Tawārīkh* (Collection of Histories). See Quatremère.

RAVEN, C. E. (1). *Natural Religion and Christian Theology.* (1st series of the Gifford Lectures, *Science and Religion*, for 1952.) Cambridge, 1953.

RAVENSTEIN, E. G. (1). *Martin Behaim; his Life and his [terrestrial] Globe.* London, 1908.

RAY, J. (1). *Miscellaneous Discourses concerning the Dissolution and Changes of the World, wherein the Primitive Chaos and Creation, the General Deluge, Fountains, Formed Stones, Sea-shells found in the Earth, Subterranean Trees, Mountains, Earthquakes, Volcanoes... are largely examined.* London, 1692.

RĀY, P. C. (1) (tr.). *The Mahābhārata.* 22 vols. Bhārata Press, Calcutta, 1889.

RĀY, P. C. (1). *A History of Hindu Chemistry, from the Earliest Times to the middle of the 16th cent. A.D., with Sanskrit Texts, Variants, Translation and Illustrations.* 2 vols. Chuckervarty & Chatterjee, Calcutta, 1904, 1925. New and revised ed. in one volume, ed. P. Ray, Indian Chemical Society, Calcutta, 1956. Re-titled *History of Chemistry in Ancient and Medieval India.*

READ, BERNARD E. (with LIU JU-CHHIANG) (1). *Chinese Medicinal Plants from the 'Pên Tsh'ao Kang Mu' A.D. 1596... a Botanical Chemical and Pharmacological Reference List.* (Publication of the Peking Nat. Hist. Bull.). French Bookstore, Peiping, 1936 (chs. 12–37 of *Pên Tshao Kang Mu*) (rev. W. T. Swingle, *ARLC/DO*, 1937, 191).

READ, BERNARD E. (2) [with LI YÜ-THIEN]. *Chinese Materia Medica; Animal Drugs.*

		Serial nos.	Corresp. with chaps. of *Pên Tshao Kang Mu*
Pt. I	Domestic Animals	322–349	50
II	Wild Animals	350–387	51 *A* and *B*
III	Rodentia	388–399	51 *B*
IV	Monkeys and Supernatural Beings	400–407	51 *B*
V	Man as a Medicine	408–444	52

PNHB, 1931, **5** (no. 4), 37–80; **6** (no. 1), 1–102. (Sep. issued, French Bookstore, Peiping, 1931.)

READ, BERNARD E. (3) [with LI YÜ-THIEN]. *Chinese Materia Medica; Avian Drugs.*

Pt. VI	Birds	245–321	47, 48, 49

PNHB, 1932, **6** (no. 4), 1–101. (Sep. issued, French Bookstore, Peiping, 1932.)

READ, BERNARD E. (4) [with LI YÜ-THIEN]. *Chinese Materia Medica; Dragon and Snake Drugs.*

Pt. VII	Reptiles	102–127	43

PNHB, 1934, **8** (no. 4), 297–357. (Sep. issued, French Bookstore, Peiping, 1934.)

READ, BERNARD E. (5) [with YU CHING-MEI]. *Chinese Materia Medica; Turtle and Shellfish Drugs.*

	Serial nos.	Corresp. with chaps. of *Pên Tshao Kang Mu*
Pt. VIII Reptiles and Invertebrates	199–244	45, 46

PNHB (Suppl.), 1939, 1–136. (Sep. issued, French Bookstore, Peiping, 1937.)

READ, BERNARD E. (6) [with YU CHING-MEI]. *Chinese Materia Medica; Fish Drugs.*

Pt. IX Fishes (incl. some amphibia, octopoda and crustacea)	128–198	44

PNHB (Suppl.), 1939. (Sep. issued, French Bookstore, Peiping, n.d. prob. 1939.)

READ, BERNARD E. (7) [with YU CHING-MEI]. *Chinese Materia Medica; Insect Drugs.*

Pt. X Insects (incl. arachnidae etc.)	1–101	39, 40, 41, 42

PNHB (Suppl.), 1941. (Sep. issued, Lynn, Peiping, 1941.)

READ, BERNARD E. (8). *Famine Foods listed in the 'Chiu Huang Pên Tshao'.* Lester Institute, Shanghai, 1946.

READ, BERNARD E. & PAK, C. (PHU CHU-PING) (1). *A Compendium of Minerals and Stones used in Chinese Medicine, from the 'Pên Tshao Kang Mu'. PNHB*, 1928, **3** (no. 2), i–vii, 1–120. (Revised and enlarged, issued separately, French Bookstore, Peiping, 1936 (2nd ed.).) Serial nos. 1–135, corresp. with chs. of *Pên Tshao Kang Mu*, 8, 9, 10, 11.

RECORDE, ROBERT (1). *Whetstone of Witte.* London, 1557.

REDGRAVE, S. R. (1). *Erhard Ratdolt and his Work at Venice* (printer of astronomical books with colour-blocks). Bibliographical Soc. London, 1899.

REDIADIS, P. (1). Account of the Anti-Kythera machine (+2nd century). In J. Svoronos, *Das Athener Nationalmuseum*, Textband 1.

REEVES, J. (1). *Chinese Names of Stars and Constellations, collected at the request of Dr Morrison for his Chinese Dictionary.* Canton, 1819. (Morrison's dictionary appeared at Macao in 1815.) Not seen.

REGIOMONTANUS (JOHANNES MÜLLER of Königsberg) (1). 'De Torqueto, Astrolabio, Regula, Baculo', etc. In *Scripta.* Nürnburg, 1543.

REHM, A. (1). 'Parapegmastudien.' *ABAW/PH*, 1941 (n.F.), **19**, 22.

REICH, S. & WIET, G. (1). 'Un Astrolabe Syrien du 14ᵉ Siècle' (+1366). Portable equatorial sundial oriented by observation of Sun's altitude and used to determine Qiblah direction. *BIFAO*, 1939, **38**, 195.

REINACH, T. (1). *Mithridate Eupator, Roi de Pont.* Paris, 1890.

REINAUD, J. T. (1) (tr.). *Relation des Voyages faits par les Arabes et les Persans dans l'Inde et la Chine dans le 9ᵉ siècle de l'ère Chrétienne.* 2 vols. Paris, 1845. Re-translation of, and commentary on, the MSS. translated more than a century earlier by E. Renaudot, q.v.

REINAUD, J. T. & FAVÉ, I. (1). *Du Feu Grégeois, des Feux de Guerre, et des Origines de la Poudre à Canon, d'après des Textes Nouveaux.* Dumaine, Paris, 1845. (Crit. rev. by D[efrémer]y, *JA*, 1846 (4ᵉ sér.), **7**, 572; E. Chevreul, *JS*, 1847, 87, 140, 209.)

REISCHAUER, E. O. (2) (tr.). *Ennin's Diary; the Record of a Pilgrimage to China in Search of the Law* (the *Nittō-Guhō Junrei Gyōki*). Ronald Press, New York, 1955.

RÉMUSAT, J. P. A. (1) (tr.). *Fa Hian, 'Foe Koue Ki', traduit par Rémusat, etc.* Paris, 1836. Eng. tr. *The Pilgrimage of Fa Hian; from the French edition of the 'Foe Koue Ki' of Rémusat, Klaproth and Landresse, with additional notes and illustrations.* Calcutta, 1848. (Fa-Hsien's *Fo Kuo Chi.*)

RÉMUSAT, J. P. A. (3). 'Antoine Gaubil.' *BU*, 1856, **16**, 1.

RÉMUSAT, J. P. A. (4). 'Catalogue des Bolides et des Aérolithes observées à la Chine et dans les Pays Voisins, tiré des Ouvrages Chinois.' *JPH*, 1819, **88**, 348.

RÉMUSAT, J. P. A. (5). 'Observations Chinoises sur la Chute des Corps Météoriques.' In *Mélanges Asiatiques.* Dondey, Paris, 1825.

RÉMUSAT, J. P. A. (6). Translation of the *Chen La Fêng Thu Chi. Nouvelles Mélanges Asiatiques*, vol. 1, p. 134.

RÉMUSAT, J. P. A. (7) (tr.). *Histoire de la Ville de Khotan, tirée des Annales de la Chine et traduite du Chinois; suivie de Recherches sur la Substance Minérale appelée par les Chinois Pierre de Iu [Jade] et sur le Jaspe des Anciens.* (Tr. of *TSCC, Pien i tien*, ch. 55.) Paris, 1820.

[RENAUDOT, EUSEBIUS] (1) (tr.). *Anciennes Relations des Indes et de la Chine de deux Voyageurs Mahometans, qui y allèrent dans le Neuvième Siècle, traduites d'Arabe, avec des Remarques sur les principaux Endroits de ces Relations.* (With four Appendices, as follows: (i) Eclaircissement touchant la Prédication de la Religion Chrestienne à la Chine; (ii) Eclaircissement touchantl 'Entrée des Mahometans dans la Chine; (iii) Eclaircissement touchant les Juifs qui ont esté trouvez à la Chine; (iv) Eclaircissement sur les Sciences des Chinois.) Coignard, Paris, 1718. Eng. tr. London, 1733. The title of Renaudot's book, which was presented partly to counter the claims of the pro-Chinese party in religious and learned circles (the Jesuits, Golius, Vossius etc., see Pinot (1), pp. 109, 160, 229, 237), was misleading. The two documents translated were: (*a*) The account of Sulaimān al-Tājir (Sulaiman the Merchant), written by an anonymous author in +851. (*b*) The completion *Silsilat al-Tawārīkh* of +920 by Abū Zayd al-Ḥasan al-Shīrāfī, based on the account of Ibn Wahb al-Baṣrī, who was in China in +876 (see Mieli (1), pp. 13, 79, 81, 115, 302; al-Jalīl (1), p. 138; Hitti (1), pp. 343, 383; Yule (2), vol. 1, pp. 125–33). Cf. Reinaud (1); Sauvaget (2).

RENOU, L. & FILLIOZAT, J. (1). *L'Inde Classique; Manuel des Etudes Indiennes*, Vol. 1, with the collaboration of P. Meile, A. M. Esnoul and L. Silburn, Payot, Paris, 1947. Vol. 2, with the collaboration of P. Demiéville, O. Lacombe, & P. Meile, Ecole Française d'Extrême Orient, Hanoi; Impr. Nationale, Paris, 1953.

DE REPARAZ-RUIZ, G. (1). 'Historia de la Geografía de España.' In *España, la Tierra, el Hombre, el Arte*, vol. 1. Martin, Barcelona, 1937.

DE REPARAZ-RUIZ, G. (2). 'Les Précurseurs de la Cartographie Terrestre.' *A/AIHS*, 1951, **4**, 73.

REPSOLD, J. A. (1). *Zur Geschichte d. astronomischer Messwerkzeuge*, 2 vols. Leipzig. 1908–1914.

RESCHER, O. (1). *Eš-Šaqā'iq en-No'mānijje von Taškõprüzade, enthaltened die Biographien der türkischen und im osmanischen Reiche wirkenden Gelehrten....* Phoenix, Constantinople-Galata, 1927.

REY, ABEL (1). *La Science dans l'Antiquité.* Vol. 1: *La Science Orientale avant les Grecs*, 1930, 2nd ed. 1942; Vol. 2: *La Jeunesse de la Science Grecque*, 1933; Vol. 3: *La Maturité de la Pensée Scientifique en Grèce*, 1939; Vol. 4: *L'Apogée de la Science Technique Grecque* (*Les Sciences de la Nature et de l'Homme, les Mathématiques, d'Hippocrate à Platon*), 1946. Albin Michel, Paris. (Evol. de l'Hum. sér. complémentaire.)

RICCI, FRANCESCO (1). *Nuova Pratica Mercantile, nella quale con modo facile s'esprimento tutti sorte di conti, che possono occorrere nella mercantila, con la radice quadrata e cuba...e sue approssimationi.* Macerata, 1659.

RICCI, MATTEO (1). *I Commentarj della Cina*, 1610. MS. unpub. till 1911 when it was edited by Venturi (1); since then it has been edited and commented on more fully by d'Elia (2).

RICCI, MATTEO (2). World Map of +1602. Photographically reproduced. Chinese Soc. Histor. Geography, Peiping, 1936.

RICHARDSON, L. J. 'Digital Reckoning among the Ancients.' *AMM*, **23**, 7.

RICO Y SINOBAS, M. (1). '*Libros del Saber de Astronomia*' *del Rey D. Alfonso X de Castilla.* Aguado, Madrid, 1864.

RIGGE, W. F. (1). 'A Chinese Star-Map Two Centuries Old.' *POPA*, 1915, **23**, 29.

RIPA, MATTEO (1). *Memoirs of Father* [*Matteo*] *Ripa during thirteen years' Residence at the Court of Peking in the service of the Emperor of China; with an account of the foundation of the College for the Education of Young Chinese at Naples.* Selected and transl. from Italian by F. Prandi, Murray, London, 1844.

RITTER, C. (1). *Die Erdkunde im Verhältnis z. Natur und z. Gesch. d. Menschen.* Reimer, Berlin, 1837.

RITTER, C. (2). *Die Erdkunde von Asien.* 5 vols. Berlin, 1837.

ROBIN, P. A. (1). *Animal Lore in English Literature.* Murray, London, 1932.

ROBINSON, F. B. (1). 'The Astronomical Observatory in Peking.' *AAR*, 1930, 37.

ROBINSON, W. O., LAKIN, H. W. & REICHEN, L. E. (1). 'The Zinc Content of Plants on the Friedensville Zinc slime ponds in relation to Biogeochemical Prospecting.' *EG*, 1947, **42**, 572.

ROCKHILL, W. W. (1). 'Notes on the Relations and Trade of China with the Eastern Archipelago and the Coast of the Indian Ocean during the 15th Century.' *TP*, 1914, **15**, 419; 1915, **16**, 61.

RODET, L. (1). 'Le Souan-Pan et la Banque des Argentiers.' *BSMF*, 1880, **8**, 158. (Contains a translation by A. Vissière of part of the section of the *Suan Fa Thung Tsung* dealing with abacus computations.)

ROHDE, A. (1). *Die Geschichte d. wissenschaftlichen Instrumente vom Beginn der Renaissance bis zum Ausgang des 18. Jahrh.* Klinkhardt & Biermann, Leipzig, 1923. (Monographien d. Kunstgewerbes, no. 16.)

VON ROHR, M. (1). *Joseph Fraunhofers Leben, Leistungen und Wirksamkeit.* Akad. Verlagsgesellsch. Leipzig, 1929.

ROHRBERG, A. (1). 'Das Rechnen auf dem chinesischen Rechenbrett.' *UMN*, 1936, **42**, 34.

ROME, A. (1). 'Les Observations d'Equinoxes et de Solstices dans le ch. 1 du livre 3 du Commentaire sur l'*Almagest* par Théon d'Alexandrie.' *ASSB*, 1937, **57**, 213; 1938, **58**, 6.

ROSEN, F. (1) (tr.). *The Algebra of Mohammed ben Musa, edited and translated* [*from the Arabic*] (with preface and notes). Royal Asiatic Society, London, 1831. (Oriental Translation Fund.)

ROSENFELD, B. & YUSHKEVITCH, A. P. (1) (tr. and ed.). 'The Mathematical Tractates of Jamshīd Ghiyāth al-Dīn al-Kāshī (d. +1436)' (in Russian). The *Miftāḥ al-Ḥisāb* (Key of Computation) and the *Risālat al-Moḥīṭīje* (Treatise on the Circumference). *JHM*, 1954, **7**, 11–449. Notes by Yushkevitch & Rosenfeld from pp. 380 ff.

DE ROSNY, L. (1) (tr.). *Chan-Hai-King* (*Shan Hai Ching*); *Antique Géographie Chinoise*. Maisonneuve, Paris, 1891.

DE ROSNY, L. (4). *Tchoung-Hoa Kou-Kin Tsaï; Textes Chinois Anciens et Modernes, traduits pour la première fois dans une langue européenne* (a chrestomathy). Maisonneuve, Paris, 1874. Contains excerpts of texts and translations from *Chuang Tzu, Chu Fan Chih, San Tshai Thu Hui*, etc.

ROSS, W. D. (1). *Aristotle*. Methuen, London, 1930.

ROTHMAN, R. W. (1). 'On an Ancient Solar Eclipse observed in China.' *RAS/M*, 1840, **11**, 47.

DES ROTOURS, R. (1). *Traité des Fonctionnaires et Traité de l'Armée, traduits de la Nouvelle Histoire des Thang* (chs. 46–50). 2 vols. Brill, Leiden, 1948 (Bibl. de l'Inst. des Hautes Etudes Chinoises, no. 6) (rev. P. Demiéville, *JA*, 1950, **238**, 395).

DES ROTOURS, R. (2) (tr.). *Traité des Examens* (*Hsin Thang Shu*, chs. 44, 45). Leroux, Paris, 1932. (Bib. de l'Inst. des Hautes Etudes Chinoises, no. 2.)

ROXBY, P. M. (2). 'The Major Regions of China.' *G*, 1938, **23**, 9.

ROXBY, P. M. (3). 'China as an Entity; the Comparison with Europe.' *G*, 1934, 1.

ROXBY, P. M. (4). *The Far Eastern Question in its Geographical Setting*. Geogr. Assoc. Aberystwyth, 1920.

RUDOLPH, R. C. (1). 'Han Tomb Reliefs from Szechuan.' *ACASA*, 1950, **4**, 29.

RUDOLPH, R. C. (2). 'Early Chinese References to Fossil Fish.' *ISIS*, 1946, **36**, 155.

RUFUS, W. C. (1). 'The Celestial Planisphere of King Yi Tai-Jo' [of Korea]. *JRAS/KB*, 1913, **4**, 23; *POPA*, 1915, **23**, 6.

RUFUS, W. C. (2). 'Astronomy in Korea.' *JRAS/KB*, 1936, **26**, 1.

RUFUS, W. C. (3). 'A Political Star Chart of the Twelfth Century.' *RASC/J*, 1945, **39**, 33. Correspondence with H. Chatley, 280; comment H. Chatley, *O*, 1947, **67**, 33.

RUFUS, W. C. & CHAO, CELIA (1). 'A Korean Star-Map.' *ISIS*, 1944, **35**, 316.

RUFUS, W. C. & LEE WON-CHUL (1). 'Marking Time in Korea.' *POPA*, 1936, **44**, 252.

RUFUS, W. C. & TIEN HSING-CHIH (1). *The Soochow Astronomical Chart*. Univ. of Michigan Press, Ann Arbor, 1945 (rev. H. Chatley, *O*, 1947, **67**, 33).

RUSKA, J. (1). 'Die Mineralogie in d. arabischen Litteratur.' *ISIS*, 1913, **1**, 341.

RUSSELL, S. M. (1). 'Discussion of Astronomical Records in Ancient Chinese Books.' *JPOS*, 1888, **2**, 187.

RYLE, M. & RATCLIFFE, J. A. (1). 'Radio-Astronomy.' *END*, 1952, **11**, 117.

SACHAU, E. (1) (tr.). *Alberuni's India*. 2 vols. London, 1888; reprint, 1910.

SAGUI, C. L. (1). 'Economic Geology and Allied Sciences in Ancient Times.' *EG*, 1930, **25**, 65.

SAID, RUSHDI (1). 'Geology in Tenth Century Arabic Literature.' *AJSC*, 1950, **248**, 63.

SAINT-DENYS, D'HERVEY, M. J. L. (1) (tr.). *Ethnographie des Peuples Etrangers à la Chine; ouvrage composé au 13e siècle de notre ère par Ma Touan-Lin...avec un commentaire perpétuel*. Georg & Mueller, Geneva, 1876–83. 4 vols. (Translation of chs. 324–48 of the *Wên Hsien Thung Khao* of Ma Tuan-Lin.) Vol. 1. Eastern Peoples; Korea, Japan, Kamchatka, Thaiwan, Pacific Islands (chs. 324–7). Vol. 2. Southern Peoples; Hainan, Tongking, Siam, Cambodia, Burma, Sumatra, Borneo, Philippines, Moluccas, New Guinea (chs. 328–32). Vol. 3. Western Peoples (chs. 333–9). Vol. 4. Northern Peoples (chs. 340–8).

SALMON, M. (1). *L'Art du Potier d'Etain*. (Descriptions des Arts et Métiers, vol. 27, Acad. Roy. des Sciences.) Montard, Paris, 1788.

SALMONY, A. (1). *Carved Jade of Ancient China*. Berkeley, Calif., 1938.

SALUSBURY, T. (1) (ed.). *Mathematical Collections and Translations of Galileo*. London, 1661.

SAMBURSKY, S. (1). *The Physical World of the Greeks*, tr. from the Hebrew edition by M. Dagut; Routledge & Kegan Paul, London, 1956.

SAMMADAR, J. N. (1). 'Rain Measurement in Ancient India.' *QJRMS*, 1912, **38**, 65.

SANTAREM, M. VICOMTE (1). *Essai sur l'Histoire de la Cosmographie et de la Cartographie pendant le Moyen Age, et sur les Progrès de la Géographie après les Grandes Découvertes du XVe siècle; pour servir d'introduction et explication à l'Atlas composé de Mappemondes et de Portulans et d'autres Monuments Géographiques depuis le VIe siècle de notre Ere jusqu'au XVIIe*. 3 vols. Maulde & Renou, Paris, 1849–52.

SANTAREM, M. VICOMTE (2). *Atlas composé de Mappemondes et de Cartes Hydrographiques et Historiques*. Maulde & Renou, Paris, 1845.

DE SANTILLANA, G. (1). *The Crime of Galileo*. Univ. of Chicago Press, Chicago, 1955; rev. P. Labérenne, *LP*, 1956 (no. 69), 133.

SARGENT, C. B. (2). 'Index to the Monograph on Geography in the History of the Former Han Dynasty.' *JWCBRS*, 1940, A, **12**, 173.

SARTON, G. (1). *Introduction to the History of Science.* Vol. 1, 1927; Vol. 2, 1931 (2 parts); Vol. 3, 1947 (2 parts). Williams & Wilkins, Baltimore (Carnegie Institution Pub. no. 376).

SARTON, G. (2). 'Simon Stevin of Bruges; the first explanation of Decimal Fractions and Measures (+1585); together with a history of the decimal idea, and a facsimile of Stevin's *Disme*.' *ISIS*, 1934, **21**, 241; 1935, **23**, 153.

SARTON, G. (3). 'Early Observations of Sun-Spots.' *ISIS*, 1947, **37**, 69.

SARTON, G. (4). 'The Earliest Reference to Fossil Fishes.' *ISIS*, 1941, **33**, 56.

SARTON, G. (5). 'Decimal Systems Early and Late.' *OSIS*, 1950, **9**, 581.

DE SAUSSURE, L. (1). *Les Origines de l'Astronomie Chinoise.* Maissoneuve, Paris, 1930. Commentaries by E. Zinner, *VAG*, 1931, **66**, 21; A. Pogo, *ISIS*, 1932, **17**, 267. This book (posthumously issued) contains eleven of the most important original papers of de Saussure on Chinese astronomy (3, 6, 7, 8, 9, 10, 11, 12, 13, 14). It omits, however, the important addendum to (3), 3*a*, as well as the valuable series (16). Unfortunately the editing was slovenly. Although the reprinted papers were re-paged, the cross-references in the footnotes were unaltered; Pogo, however (*loc. cit.*), has provided a table of corrections by the use of which de Saussure's cross-references can be readily utilised.

DE SAUSSURE, L. (2). 'Prolégomènes d'Astronomie Primitive Comparée.' *ASPN*, 1907 (4e sér. **23**), **112**, 537.

DE SAUSSURE, L. (2*a*). 'Notes sur les Étoiles Fondamentales des Chinois.' *ASPN*, 1907 (4e sér. **24**), **112**, 19, 96.

DE SAUSSURE, L. (3). 'Le Texte Astronomique du Yao Tien.' *TP*, 1907, **8**, 301. (Reprinted as introduction to (1).)

DE SAUSSURE, L. (3*a*). 'Le Texte Astronomique du Yao Tien; Note Rectificative et Complémentaire.' *TP*, 1907, **8**, 559.

DE SAUSSURE, L. (4). 'L'Astronomie Chinoise dans l'Antiquité.' *RGS*, 1907, **18**, 135. (Commentary on de Saussure's work up to this time by P. Puiseux, *JS*, 1908, 512; reprinted *TP*, 1908, **9**, 708.)

DE SAUSSURE, L. (5). 'Le Cycle de Jupiter.' *TP*, 1908, **9**, 455. (This paper contains many mistakes, in the opinion of the author, who desired that it should be considered as cancelled and replaced by (12). See (1), p. 421 fn.)

DE SAUSSURE, L. (6). 'Les Origines de l'Astronomie Chinoise: l'Origine des *Sieou* [*hsiu*].' *TP*, 1909, **10**, 121. (Reprinted as [A] in (1).)

DE SAUSSURE, L. (7). 'Les Origines de l'Astronomie Chinoise; les Cinq Palais Célestes.' *TP*, 1909, **10**, 255. (Reprinted as [B] in (1).)

DE SAUSSURE, L. (8). 'Les Origines de l'Astronomie Chinoise; La Série Quinaire et ses Dérivés.' *TP*, 1910, **11**, 221. (Reprinted as [C] in (1).)

DE SAUSSURE, L. (9). 'Les Origines de l'Astronomie Chinoise: La Série des douze *tche*.' *TP*, 1910, **11**, 457. (Reprinted as [D] in (1).)

DE SAUSSURE, L. (10). 'Les Origines de l'Astronomie Chinoise; Le Cycle des Douze Animaux.' *TP*, 1910, **11**, 583. (Reprinted as [E] in (1).)

DE SAUSSURE, L. (11). 'Les Origines de l'Astronomie Chinoise: La Règle des *cho-ti* [*shê-thi*].' *TP*, 1911, **12**, 347. (Reprinted as [F] in (1).)

DE SAUSSURE, L. (12). 'Les Origines de l'Astronomie Chinoise: Le Cycle de Jupiter.' *TP*, 1913, **14**, 387; 1914, **15**, 645. (Reprinted as [G] and [G*bis*] in (1).)

DE SAUSSURE, L. (13). 'Les Origines de l'Astronomie Chinoise; Les Anciennes Etoiles Polaires.' *TP*, 1921, **20**, 86. (Reprinted as [H] in (1).)

DE SAUSSURE, L. (14). 'Les Origines de l'Astronomie Chinoise; Le Zodiaque Lunaire.' *TP*, 1922, **21**, 251. (Reprinted as [I] in (1).)

DE SAUSSURE, L. (15). 'Le Zodiaque Lunaire Asiatique.' *ASPN*, 1919 (5e sér. **1**), **124**, 105.

DE SAUSSURE, L. (16*a*, *b*, *c*, *d*). 'Le Système Astronomique des Chinois.' *ASPN*, 1919 (5e sér. **1**), **124**, 186, 561; 1920 (5e sér. **2**), **125**, 214, 325. (*a*) Introduction: (i) Description du Système, (ii) Preuves de l'Antiquité du Système; (*b*) (iii) Rôle Fondamental de l'Étoile Polaire, (iv) La Théorie des Cinq Eléments; (v) Changements Dynastiques et Réformes de la Doctrine; (*c*) (vi) Le Symbolisme Zoaire, (vii) Les Anciens Mois Turcs; (*d*) (viii) Le Calendrier, (ix) Le Cycle Sexagésimal et la Chronologie, (x) Les Erreurs de la Critique. Conclusion.

DE SAUSSURE, L. (16*e*). 'Le Système Cosmologique des Chinois.' *RGS*, 1921, **32**, 729.

DE SAUSSURE, L. (17). 'Origine babylonienne de l'Astronomie Chinoise.' *ASPN*, 1923 (5e sér. **5**), **128**, 5.

DE SAUSSURE, L. (18). 'Origine Chinoise du Dualisme Iranien.' *JA*, 1922 (11e sér. **20**), **201**, 302. (Abstract only, refers to (19).)

DE SAUSSURE, L. (19). 'Le Système Cosmologique Sino-Iranien.' *JA*, 1923 (12e sér. **1**), **202**, 235.

DE SAUSSURE, L. (20). 'La Cosmologie Religieuse en Chine, etc.' Congrès Internat. de l'Hist. des Religions, 1923, p. 79.

DE SAUSSURE, L. (21). 'La Série Septénaire, Cosmologique et Planétaire.' *JA*, 1923 (12e sér. **3**), **204**, 333; *NCR*, 1922, **4**, 461.

DE SAUSSURE, L. (22). 'Une Interpolation du *Che Ki* (*Shih Chi*); Le Tableau Calendarique de 76 Années.' *JA*, 1922 (11e sér. **20**), **201**, 105.

DE SAUSSURE, L. (22*a*). 'Une Interpolation du *Che Ki*; Note complémentaire.' *JA*, 1924 (12e sér. **5**), **206**, 265.

DE SAUSSURE, L. (23). 'Sur l'Inanité de la Chronologie Chinoise officielle.' *JA*, 1923 (12e sér. **2**), **203**, 360.

DE SAUSSURE, L. (24). 'Note sur l'Origine Iranienne des Mansions Lunaires Arabes.' *JA*, 1925 (12e sér. **6**), **207**, 166.

DE SAUSSURE, L. (25). 'La Chronologie Chinoise et l'Avènement des Tcheou.' *TP*, 1924, **23**, 287; 1932, **29**, 276 (posthumous). (Crit. H. Chatley, *JRAS/NCB*, 1934, **65**, 187.)

DE SAUSSURE, L. (26). (*a*) 'La Relation des Voyages du Roi Mou.' *JA*, 1921 (11e sér. **16**), **197**, 151; (11e sér. **17**), **198**, 247. (*b*) 'The Calendar of the *Muh T'ien Tsz Chuen*.' *NCR*, 1920, **2**, 513. (Comments by P. Pelliot, *TP*, 1922, **21**, 98.)

DE SAUSSURE, L. (26*a*). 'Le Voyage de Mou Wang et l'Hypothèse d'Ed. Chavannes.' *TP*, 1921, **20**, 19.

DE SAUSSURE, L. (27). 'Le Cycle des Douze Animaux et le Symbolisme Cosmologique des Chinois.' *JA*, 1920 (11e sér. **15**), **196**, 55. (Fig. 9 of this paper needs correction according to the note on p. 278 of (26).)

DE SAUSSURE, L. (28). 'La Symétrie du Zodiaque Lunaire Asiatique.' *JA*, 1919 (11e sér. **14**), **195**, 141.

DE SAUSSURE, L. (28*a*). 'The Lunar Zodiac.' *NCR*, 1921, **3**, 453.

DE SAUSSURE, L. (29). 'L'Horométrie et le Système Cosmologique des Chinois.' Introduction to A. Chapuis' *Relations de l'Horlogerie Suisse avec la Chine; la Montre 'Chinoise'*. Attinger, Neuchâtel, 1919.

DE SAUSSURE, L. (30). 'Astronomie et Mythologie dans le *Chou King* [*Shu Ching*].' *TP*, 1932, **29**, 359. (Appendix to (25); the last statement of de Saussure's views on Chinese Astronomy.)

DE SAUSSURE, L. (31). 'L'Etymologie du nom des monts K'ouen-Louen.' *TP*, 1921, **20**, 370.

DE SAUSSURE, L. (32). 'On the Antiquity of the Yin-Yang Theory.' *NCR*, 1922, **4**, 457.

DE SAUSSURE, L. (33). 'Note on a difficulty of Pelliot's concerning a calendar of −63 found among the Tunhuang documents.' *TP*, 1914, **15**, 463. (The inequality of the seasons not recognised till about +550.)

DE SAUSSURE, L. (34). 'La Tortue et le Serpent' (an amulet showing the Great Bear, and the Black Tortoise of the Northern Palace, with one of its constellations). *TP*, 1920, **19**, 247.

DE SAUSSURE, R. (1). 'Léopold de Saussure (1866–1925).' *ISIS*, 1937, **27**, 286.

SAUVAIRE, M. H. (1). 'On a Treatise on Weights and Measures by Eliya' (Elias bar Shinaya, +975 to +1049; Syriac). *JRAS*, 1877, **9**, 291.

SAYCE, A. H. (1). 'Astronomy and Astrology of the Babylonians.' *TSBA*, 1879, **3**, 145.

SAYCE, A. H. (2). *Babylonian Literature*. Bagster, London, n.d. (1877).

SAYILI, AYDIN (1). 'The "Observation Well".' In *Actes du VIIe Congrès Internat. d'Histoire des Sciences*. Jerusalem, 1953, p. 542.

SCHAFER, E. H. (1). 'Ritual Exposure [Nudity, etc.] in Ancient China.' *HJAS*, 1951, **14**, 130.

SCHAFER, E. H. (4). *The History of the Empire of Southern Han according to chapter 65 of the 'Wu Tai Shih' of Ouyang Hsiu*. Art. in Silver Jubilee Volume of the Zinbun Kagaku Kenkyusyo, Kyoto University, Kyoto, 1954, p. 339.

SCHAFER, E. H. (5). 'Notes on Mica in Medieval China.' *TP*, 1955, **43**, 265.

SCHAFER, E. H. (6). 'Orpiment and Realgar in Chinese Technology and Tradition.' *JAOS*, 1955, **75**, 73.

SCHEFER, C. (2). 'Notice sur les Relations des Peuples Mussulmans avec les Chinois depuis l'Extension de l'Islamisme jusqu'à la fin du 15e Siècle.' In *Volume Centénaire de l'Ecole des Langues Orientales Vivantes, 1795–1895*. Leroux, Paris, 1895, pp. 1–43.

SCHEINER, CHRISTOPHER (1). *Rosa Ursina sive Sol, ex admirando Facularum et Macularum suarum Phenomeno Varius....* Phaeus, Bracciani, 1630.

SCHIAPARELLI, G. (1). *Scritti sulla Storia della Astronomia Antica*. Zanichelli, Bologna, 1925.

SCHJELLERUP, H. C. F. C. (1). 'Recherches sur l'Astronomie des Anciens; II, On the Total Solar Eclipses Observed in China in the Years B.C. 708, 600 and 548.' *C*, 1881, **1**, 41.

SCHJÖTH, F. (1). *The Currency of the Far East; the Schjöth Collection at the Numismatic Cabinet of the University of Oslo, Norway*. Aschehong, Oslo, 1929; Luzac, London, 1929. (Pubs. of the Numismatic Cabinet of the University of Oslo, no. 1.)

SCHLACHTER, A. & GISINGER, F. (1). *Der Globus, seine Entstehung und Verwendung in der Antike*. Teubner, Leipzig & Berlin, 1927. (*ΣΤΟΙΧΕΙΑ*, Stud. z. Gesch. d. antik. Weltbildes u. d. griechischen Wiss., no. 8.)

SCHLEGEL, G. (5). *Uranographie Chinoise, etc.* 2 vols. with star-maps in separate folder. Brill, Leyden, 1875. (Crit. J. Bertrand, *JS*, 1875, 557; S. Günther, *VAG*, 1877, **12**, 28. Reply by G. Schlegel, *BNI*, 1880 (4e volg.), **4**, 350.)

SCHLEGEL, G. (7). *Problèmes Géographiques; les Peuples Étrangers chez les Historiens Chinois.*
(*a*) Fu-Sang Kuo (ident. Sakhalin and the Ainu). *TP*, 1892, **3**, 101.
(*b*) Wên-Shen Kuo (ident. Kuriles). *Ibid.* p. 490.
(*c*) Nü Kuo (ident. Kuriles). *Ibid.* p. 495.
(*d*) Hsiao-Jen Kuo (ident. Kuriles and the Ainu). *TP*, 1893, **4**, 323.
(*e*) Ta-Han Kuo (ident. Kamchatka and the Chukchi) and Liu-Kuei Kuo. *Ibid.* p. 334.
(*f*) Ta-Jen Kuo (ident. islands between Korea and Japan) and Chhang-Jen Kuo. *Ibid.* p. 343.
(*g*) Chün-Tzu Kuo (ident. Korea, Silla). *Ibid.* p. 348.
(*h*) Pai-Min Kuo (ident. Korean Ainu). *Ibid.* p. 355.
(*i*) Chhing-Chhiu Kuo (ident. Korea). *Ibid.* p. 402.
(*j*) Hei-Chih Kuo (ident. Amur Tungus). *Ibid.* p. 405.
(*k*) Hsüan-Ku Kuo (ident. Siberian Giliak). *Ibid.* p. 410.
(*l*) Lo-Min Kuo and Chiao-Min Kuo (ident. Okhotsk coast peoples). *Ibid.* p. 413.
(*m*) Ni-Li Kuo (ident. Kamchatka and the Chukchi). *TP*, 1894, **5**, 179.
(*n*) Pei-Ming Kuo (ident. Behring straits islands). *Ibid.* p. 201.
(*o*) Yu-I Kuo (ident. Kamchatka tribes). *Ibid.* p. 213.
(*p*) Han-Ming Kuo (ident. Kuriles). *Ibid.* p. 218.
(*q*) Wu-Ming Kuo (ident. Okhotsk coast peoples). *Ibid.* p. 224.
(*r*) San Hsien Shan (the magical islands in the Eastern Sea, perhaps partly Japan). *TP*, 1895, **6**, 1.
(*s*) Liu-Chu Kuo (the Liu-Chu islands, partly confused with Thaiwan, Formosa). *Ibid.* p. 165.
(*t*) Nü-Jen Kuo (legendary, also in Japanese fable). *Ibid.* p. 247.
A volume of these reprints, collected, but lacking the original pagination, is in the Library of the Royal Geographical Society. Chinese transl. under name Hsi Lo-Ko. (rev. F. de Mély, *JS*, 1904.)

SCHLEGEL, G. & KÜHNERT, F. (1). 'Die *Schu-King* [*Shu Ching*] Finsterniss.' *VKAWA/L*, 1890, **19** (no. 3).

SCHLOSSER, M. (1). 'Die fossilen Säugethiere Chinas.' *ABAW/MN*, 1903, **22** (incl. trade in fossils).

VON SCHLÖZER, K. (1). *Abu Dolef Misaris ben Mohalhel de Itinere Asiatico Commentarius.* Berlin, 1845.

SCHMELLER, H. (1). *Beiträge z. Geschichte d. Technik in der Antike und bei den Arabern.* Mencke, Erlangen, 1922. (Abhdl. z. Gesch. d. Naturwiss. u. d. Med. no. 6.)

SCHMIDT, M. C. P. (1). *Kulturhistorische Beiträge. II. Die Antike Wasseruhr.* Leipzig, 1912.

SCHNABEL, P. (1). *Berossos und die babylonisch-hellenistische Literatur.* Teubner, Leipzig, 1923.

SCHNABEL, P. (2). 'Recognition of Babylonian planetary ephemerides material in later Indian texts.' *ZASS*, 1924, **35**, 112; 1927, **37**, 60.

SCHOFF, W. H. (3). '*The Periplus of the Erythraean Sea*'; *Travel and Trade in the Indian Ocean by a Merchant of the First Century, translated from the Greek and annotated, etc.* Longmans Green, New York, 1912.

SCHÖPFER, W. H. (1). (*a*) 'Les Répercussions hors de France de l'Oeuvre de Jules Raulin (1836–1896) relative au Zinc, Oligo-Elément.' *AUL* (Fascicule Spécial: 'L'Université de Lyon en 1948 et 1949'), 1950. (*b*) 'La Culture des Plantes en Milieu Synthétique: Les Précurseurs. *A/AIHS*, 1951, **4**, 681.

SCHOTT, A. (1). 'Das Werden der babylonisch-assyrischen Positions-astronomie und einige seiner Bedingungen.' *ZDMG*, 1934, **88**, 302.

SCHOTT, W. (2). 'Ueber ein chinesisches Mengwerk, nebst einem Anhang linguistischer Verbesserungen zu zwei Bänden der Erdkunde Ritters' [the *Yeh Huo Pien* of Shen Tê-Fu (Ming)]. *APAW/PH*, 1880, no. 3.

SCHOVE, D. J. (1). 'Sun-spots and Aurorae.' *JBASA*, 1948, **58**, 178.

SCHOVE, D. J. (2). 'The Sun-spot Cycle before +1750.' *TM*, 1947, **52**, 233.

SCHOVE, D. J. (3). 'Sun-spot Epochs, −188 to +1610.' *POPA*, 1948, **56**, 247. Table superseded by that in Schove (7).

SCHOVE, D. J. (4). 'Chinese Raininess through the Centuries.' *MEM*, 1949, **78**, 11.

SCHOVE, D. J. (5). 'The Earliest Dated Sun-spot.' *JBASA*, 1950, **61**, 22, 126.

SCHOVE, D. J. (6). 'Sun-spots, Aurorae and Blood Rain: the Spectrum of Time.' *ISIS*, 1951, **42**, 133.

SCHOVE, D. J. (7). 'Sun-spot Maxima since −649.' *JBASA*, 1956, **66**, 59.

SCHOVE, D. J. (8). 'Halley's Comet; I, −1930 to +1986.' *JBASA*, 1955, **65**, 285.

SCHOVE, D. J. (9). 'The Comet of David and Halley's Comet.' *JBASA*, 1955, **65**, 289.

SCHOVE, D. J. (10). 'Halley's Comet and Kamienski's Formula.' *JBASA*, 1956, **66**, 131.

SCHOVE, D. J. (11). 'The Sun-spot Cycle, −649 to +2000.' *JGPR*, 1955, **60**, 127.

SCHOY, K. (1). 'Gnomonik d. Araber.' In E. v. Bassermann-Jordan's *Die Geschichte d. Zeitmessung u. d. Uhren*, vol. 1. de Gruyter, Berlin, 1923.

SCHRAMEIER, D. (1). 'On Martin Martini' [and his *Novus Atlas Sinensis* of 1655]. *JPOS*, 1888, **2**, 99.

SCHRÖDINGER, E. (1). *Science and Humanism.* Cambridge, 1951.

SÉBILLOT, P. (1). *Les Travaux Publics et les Mines dans les Traditions et les Superstitions de tous les Peuples.* Paris, 1894.

SÉDILLOT, J. J. E. (1). *Traité des Instruments Astronomiques des Arabes composé au 13e siècle par Aboul Hassan Ali de Maroc* (Abū 'Alī al-Ḥasan ibn 'Ali ibn 'Umar al-Marrākushī). Pub. with introduction by L. P. E. A. Sédillot. 2 vols. Imp. Royale, Paris, 1834–5.

SÉDILLOT, L. P. E. A. (1). 'Mémoire sur les Instruments Astronomiques des Arabes, pour servir de complément au Traité d'Aboul Hassan.' *MRAI/DS*, 1844, **1**, 1. (Also sep. Imp. Royale, Paris, 1841–5.)

SÉDILLOT, L. P. E. A. (2). *Matériaux pour servir à l'Histoire comparée des Sciences Mathématiques chez les Grecs et chez les Orientaux.* 2 vols. Didot, Paris, 1845–9.

SÉDILLOT, L. P. E. A. (3). *Prolégomènes des Tables Astronomiques d'Oloug Beg* [Ulūgh Beg ibn Shahrukh]. (*a*) Notes, Variantes et Introduction. Didot, Paris, 1847 (first printed Ducrocq, Paris, 1839). (*b*) Traduction et Commentaire. Didot, Paris, 1853.

SÉDILLOT, L. P. E. A. (4). 'De l'Astronomie et des Mathématiques chez les Chinois.' *BBSSMF*, 1868, **1**, 161.

SÉDILLOT, L. P. E. A. (5). *Courtes Observations sur quelques points de l'Histoire de l'Astronomie et des Mathématiques chez les Orientaux.* Lainé & Havard, Paris, 1863.

SÉDILLOT, L. P. E. A. (6). *Lettre sur quelques points de l'Astronomie Orientale.* Paris, 1834. (Probably = 'Lettre au Bureau des Longitudes.' *Moniteur*, 28 July 1834.)

SÉDILLOT, L. P. E. A. (7). 'Nouvelles Recherches pour servir à l'Histoire de l'Astronomie chez les Arabes; Découverte de la Variation [third inequality in lunar motion] par Aboul-Wefā, astronome du 10e siècle' [Abū'l Wafā al-Buzjānī, +940 to +997]. *JA*, 1835 (2e sér.), **16**, 420. (Also pub. sep. Impr. Roy. 1836.)

SÉDILLOT, L. P. E. A. (8). *Recherches Nouvelles pour servir à l'Histoire des Sciences Mathématiques chez les Orientaux.* Paris, 1837. Repr. from 'Notices de plusieurs Opuscules Mathématiques qui composent le MS. Arabe no. 1104 de la Bibliothèque Royale.' *MAI/NEM*, 1838, **13** (no. 1), 126.

SEEMANN, H. J. (1). 'Die Instrumente der Sternwarte zu Marāghah nach den Mitteilungen von al-'Urdī.' *SPMSE*, 1928, **60**, 15.

SEEMANN, H. J. (2). *Das Kugelförmige Astrolab.* Mencke, Erlangen, 1925. (Abhdl. z. Gesch. d. Naturwiss. u. d. Med. no. 8.)

SELIGMANN, K. (1). *The History of Magic.* Pantheon, New York, 1948.

SENGUPTA, P. C. (1). 'History of the Infinitesimal Calculus in Ancient and Medieval India.' *JDMV*, 1931, **41**, 223.

SENGUPTA, P. C. (2). 'The Age of the Brahmanas.' *IHQ*, 1934, **10**.

SENGUPTA, P. C. (3). 'Hindu Astronomy.' Art. in *Cultural Heritage of India*, vol. 3, pp. 341–78, Calcutta, 1940.

SERGESCU, P. (1). *Les Recherches sur l'Infini Mathématique jusqu'à l'Établissement de l'Analyse Infinitésimale.* Herrmann, Paris, 1949. (*ASI*, no. 1083.)

SHAKESHAFT, J. R. (1). 'Radio-Astronomy.' *ADVS*, 1953, 294.

SHAMASASTRY, R. (1) (tr.). *Kautilya's 'Arthaśāstra'.* With introd. by J. F. Fleet. Wesleyan Mission Press, Mysore, 1929.

SHAW, H. (1). *Applied Geophysics.* HMSO (Science Museum), London, 1938.

SHAW, W. NAPIER (1). *The Drama of Weather.* Cambridge, 1933.

SHAW, W. NAPIER & AUSTIN, E. (1). *Manual of Meteorology.* 4 vols. Cambridge, 1926; 2nd ed. 1932.

SHINJŌ, S. (1). 'On the Development of the Astronomical Sciences in the Ancient Orient.' In *Scientific Japan, Past and Present.* 3rd Pan-Pacific Science Congress Volume, Tokyo, 1926. (Chinese translation by Chhen Hsiao-Hsien in *CST/HLJ*, 1929, nos. 94–6; see Eberhard, 10.)

SHIROKOGOROV, S. M. (2). 'Lunar Calendars and Matriarchy.' *AN*, 1931, **26**, 217.

SHKLOVSKY, I. S. (1). 'Novae and Radio Stars' (in Russian). *ANSSR/AC*, 1953 (no. 143), 1; *CRAS/USSR*, 1954, **94**, 417.

SHKLOVSKY, I. S. & PARENAGO, P. P. (1). 'Identification of the Supernova of +369 with a powerful Radio Star in Cassiopeia' (in Russian). *ANSSR/AC*, 1952 (no. 131), 1. (Partial Eng. tr. *HANL*, 1953 (no. 70), 6.)

SHORT, JAMES (1). 'Description and Uses of an Equatorial Telescope.' *PTRS*, 1749, **46**, 241.

SIEBERG, A. (1). *Handbuch d. Erdbebenkunde.* Wieweg, Braunschweig, 1904.

SILCOCK, A. (1). *Introduction to Chinese Art.* London, 1935.

SIMON, E. (1). 'Über Knotenschriften und ähnliche Knotenschnüre d. Riukiuinseln.' *AM*, 1924, **1**, 657.

SIMONELLI, J. P., KÖGLER, I. & DELLA BRIGA, M. (1). *Scientiae Eclipsium ex Imperio et Commercio Sinarum Illustratae.* Pt. I (J. P. Simonelli). Rubeis, Rome, 1744. Pt. II (I. Kögler). Marescandoli, Lucca, 1745. Pts. III, IV (M. della Briga). Marescandoli, Lucca, 1747.

SINGER, C. (2). *A Short History of Science, to the Nineteenth Century.* Oxford, 1941.

SINGER, C. (8). *The Earliest Chemical Industry; an Essay in the Historical Relations of Economics and Technology, illustrated from the Alum Trade.* Folio Society, London, 1948.

SINGER, C. & SINGER, D. W. (1). 'The Jewish Factor in Mediaeval Thought.' In *Legacy of Israel*, ed. E. R. Bevan and C. Singer. Oxford, 1928.

SINGER, D. W. (1). *Giordano Bruno; His Life and Thought, with an annotated Translation of his Work 'On the Infinite Universe and Worlds'.* Schuman, New York, 1950.

SINGH, A. N. (1). 'A Review of Hindu Mathematics up to the +12th Century.' *A*, 1936, **18**, 43.

SINGH, A. N. (2). 'On the Use of Series in Hindu Mathematics.' *OSIS*, 1936, **1**, 606.

SINOR, D. (1). 'Autour d'une Migration de Peuples au 5e siècle.' *JA*, 1947, **235**, 1.

SINYAKOVA, A. (1). 'Lead Accumulation in Plants.' *CRAS/USSR*, 1945, **48**, 414; *CA*, 1946, **40**, 4113.

SION, J. (1). *Asie des Moussons.* Vol. 9 of *Géographie Universelle.* Colin, Paris, 1928.

SIREN, O. (1). (*a*) *Histoire des Arts Anciens de la Chine.* 3 vols. van Oest, Brussels, 1930. (*b*) *A History of Early Chinese Art.* 4 vols. Benn, London, 1929. Vol. 1, Prehistoric and Pre-Han; Vol. 2, Han; Vol. 3, Sculpture; Vol. 4, Architecture.

SISCO, A. G. & SMITH, C. S. (1) (tr.). *'Bergwerk und Probierbüchlein', a translation from the German of the 'Bergbüchlein', a sixteenth century book on mining geology...and of the 'Probierbüchlein', a sixteenth century work on assaying...with technical annotations and historical notes.* Amer. Inst. Mining & Metall. Engineers, New York, 1949.

SISCO, A. G. & SMITH, C. S. (2) (tr.). *Lazarus Ercker's Treatise on Ores and Assaying (Prague, 1574), translated from the German edition of 1580.* Univ. Chicago Press, Chicago, 1951.

DE SITTER, W. (1). 'On the System of Astronomical Constants.' *BAN*, 1938, **8** (no. 307), 216, 230.

DE SLANE, BARON MCGUCKIN (1). *Description de l'Afrique Septentrionale.* Algiers, 1857, 1858; new ed. Paris, 1910, 1913. (Tr. of al-Bakrī's *Kitāb al-masālik w'al-mamālik.*)

SLOLEY, R. W. (1). (*a*) 'Primitive Methods of Measuring Time, with special reference to Egypt.' *JEA*, 1931, **17**, 166. (*b*) 'Ancient Clepsydrae.' *AE*, 1924, 43.

SLOUKA, HUBERT *et al.* (1) (ed.). *Astronomie v Československu od dob Nejstarších do Dneška.* State Publishing House, Prague, 1952.

SMART, W. M. (1). *A Textbook on Spherical Astronomy.* Cambridge, 1936.

SMETHURST, GAMALIEL (1). 'An Account of a new invented arithmetical Instrument called a *Shwan-pan*, or Chinese Accompt-Table.' *PTRS*, 1749, **46**, 22. (With note by C. M[ortimer], Sec. R.S.)

SMITH, C. A. MIDDLETON (1). 'Chinese Creative Genius.' *CTE*, 1946, **1**, 920, 1007.

SMITH, D. E. (1). *History of Mathematics.* Vol. 1. *General Survey of the History of Elementary Mathematics*, 1923. Vol. 2. *Special Topics of Elementary Mathematics*, 1925. Ginn, New York.

SMITH, D. E. (2). 'Chinese Mathematics.' *SM*, 1912, **80**, 597.

SMITH, D. E. (3). 'Unsettled Questions concerning the Mathematics of China.' *SM*, 1931, **33**, 244.

SMITH, D. E. (4). 'The History and Transcendence of π.' In J. W. A. Young (ed.), *Monographs on Topics of Modern Mathematics relevant to the elementary field*, p. 396. Longmans Green, New York, 1911.

SMITH, D. E. & KARPINSKI, L. C. (1). *The Hindu-Arabic Numerals.* Ginn, Boston, 1911.

SMITH, D. E. & MIKAMI, Y. (1). *A History of Japanese Mathematics.* Open Court, Chicago, 1914. (rev. H. Bosmans, *RQS*, 1914, **76**, 251.)

SMITH, R. A. (1). 'On sinking water-pot clepsydras in Britain.' *PSA*, 1907, **21**, 319; 1915, **27**, 76.

SNELLEN, J. B. (1). '*Shoku Nihongi*; Chronicles of Japan.' *TAS/J*, 1934 (2e ser.), **11**, 151.

SOLGER, F. (1). 'Astronomische Anmerkungen zu chinesischen Märchen.' *MDGNVO*, 1922, **17**, 133.

SOLOMON, B. S. (1). '"One is No Number" in China and the West.' *HJAS*, 1954, **17**, 253.

SOONAWALA, M. F. (1). *Maharaja Sawai Jai Singh II of Jaipur and his Observatories.* Jaipur Astronomical Society, Jaipur, n.d. (1953) (rev. H. Spencer Jones, *N*, 1953, **172**, 645).

SOOTHILL, W. E. (4). 'The Two Oldest Maps of China Extant.' *GJ*, 1927, **69**, 532. (Incorporates an otherwise unpublished contribution of A. Hosie, 1924. Followed by discussion including E. Heawood.)

SOOTHILL, W. E. (5) (posthumous). *The Hall of Light; a Study of Early Chinese Kingship.* Lutterworth, London, 1951. (On the Ming Thang; also contains discussion of the *Pu Thien Ko* and transl. of *Hsia Hsiao Chêng.*)

SOPER, A. C. (1). 'Hsiang Kuo Ssu, an Imperial Temple of the Northern Sung.' *JAOS*, 1948, **68**, 19.

SOUCIET, E. See Gaubil, A.

SOUSTELLE, J. (1). *La Pensée Cosmologique des anciens Mexicains; Représentation du Monde et de l'Espace.* Hermann, Paris, 1940.

SOYMIÉ, M. (1). 'L'Entrevue de Confucius et de Hiang T'o [Hsiang Tho].' *JA*, 1954, **242**, 311.

SPASSKY, I. G. (1). 'The Origin and History of the Russian "schioty" (abacus).' Art. in *Historical-Mathematical Researches*, 4th ed. Moscow, 1952.

SPENCER, L. J. (1). *A Key to Precious Stones.* London & Glasgow, 1936.

SPENCER-JONES, SIR HAROLD (1). *General Astronomy.* Arnold, London, 1946 (2nd ed. reprinted).

SPENCER-JONES, SIR HAROLD (2). 'The Royal Greenwich Observatory.' *PRSB*, 1949, **136**, 349.

SPEYER, W. (1). 'Beitrag z. Wirkung von Arsenverbindungen auf Lepidopteren.' *ZAE*, 1925, **11**, 395.

SPINDEN, H. J. (1). *Ancient Civilisations of Mexico and Central America.* Amer. Mus. Nat. Hist., New York, 1946.

SPIZEL, G. (1). *De Re Litteraria Sinensium Commentarius.* Leiden, 1660.

SPRAT, THOMAS (1). *The History of the Royal Society of London, for the Improving of Natural Knowledge.* 3rd ed. Knapton *et al.* London, 1722.

VAN DER SPRENKEL, O. (1). *Chronology, Dynastic Legitimacy, and Chinese Historiography.* Contribution to the Far East Seminar in the Conference on Asian History, London School of Oriental Studies, July, 1956.

STANLEY, C. A. (1). '[Glossary of Chinese] Geographical Terms.' In Doolittle, J. (1), vol. 2, p. 268.

STAPLETON, H. E. (1). 'Sal-Ammoniac; a Study in Primitive Chemistry.' *MRAS/B*, 1905, **1**, 25.

STAUNTON, SIR GEORGE T. (1) (tr.). '*Ta Tsing Leu Lee*' [*Ta Chhing Lü Li*]; *being the fundamental Laws, and a selection from the supplementary Statutes, of the Penal Code of China.* Davies, London, 1810.

STAUNTON, SIR GEORGE T. (2). *An Authentic Account of an Embassy from the King of Great Britain to the Emperor of China....* Nicol, London, 1797. 2nd ed. 2 vols. 1798.

STEERS, J. A. (1). *An Introduction to the Study of Map-Projections.* Univ. Press, London, 1946.

STEIN, R. A. (1). 'Le Lin-Yi; sa localisation, sa contribution à la formation du Champa, et ses liens avec la Chine.' *HH*, 1947, **2** (nos. 1–3), 1–300.

STEIN, SIR AUREL (6). 'Notes on Ancient Chinese Documents, discovered along the Han Frontier Wall in the Desert of Tunhuang.' *NCR*, 1921, **3**, 243. (Reprinted with Stein (7), Chavannes 12), and Wright, H. K. (1) in brochure form, Peiping, 1940.)

STEIN, SIR AUREL (7). 'A Chinese Expedition across the Pamirs and Hindukush, A.D. 747.' *NCR*, 1922, **4**, 161. (Reprinted, with Stein (6), Chavannes (12) and Wright, H. K., in brochure form, Peiping, 1940.)

STEINSCHNEIDER, M. (1). 'Die Europäischen Übersetzungen aus dem Arabischen bis mitte d. 17. Jahrhunderts.' *SWAW/PH*, 1904, **149**, 1; 1905, **151**, 1; *ZDMG*, 1871, **25**, 384.

STEINSCHNEIDER, M. (2). (*a*) Über die Mondstationen (Naxatra) und das Buch Arcandam.' *ZDMG*, 1864, **18**, 118. (*b*) 'Zur Geschichte d. Übersetzungen aus dem Indischen in Arabische und ihres Einflusses auf die Arabische Literatur, insbesondere über die Mondstationen (Naxatra) und daraufbezügliche Loosbücher.' *ZDMG*, 1870, **24**, 325; 1871, **25**, 378. (The last of the three papers has an index for all three.)

STEINSCHNEIDER, M. (3). 'Euklid bei den Arabern.' *ZMP*, 1886, **31** (Hist.-Lit. Abt.), 82.

STEINSCHNEIDER, M. (4). *Gesammelte Schriften*, ed. H. Malter & A. Marx. Poppelauer, Berlin, 1925.

STENSEN, NICHOLAS (1). *The Prodromus to a Dissertation concerning Solids naturally contained within Solids, laying a Foundation for the Rendering of a Rational Accompt both of the Frame and the several Changes of the Masse of the Earth, as also of the various Productions in the same.* Tr. H[enry] O[ldenburg]. Winter, London, 1671.

STEVENSON, E. L. (1). *Terrestrial and Celestial Globes; their History and Construction....* 2 vols. Hispanic Soc. Amer. (Yale Univ. Press), New Haven, 1921.

STEVENSON, E. L. (2). *Portolan Charts; their Origin and Characteristics....* Hispanic Soc. Amer., New York, 1911.

STILES, W. (1). *Trace Elements in Plants and Animals.* Cambridge, 1948.

STOLICZKA, F. (1). 'Note regarding the Occurrence of Jade in the Upper Karakash Valley on the Southern Borders of Turkistan.' In Forsyth, T. D., *Report of a Mission to Yarkand in 1873 under command of Sir T.D.F.* Calcutta, 1875.

STØRMER, C. (1). *The Polar Aurora.* O.U.P., Oxford, 1955.

STRATTON, F. J. M. (1). 'Novae.' Art. in *Handbuch d. Astrophysik*, vol. 6, p. 251. Springer, Berlin, 1928.

STRONG, E. W. (1). *Procedures and Metaphysics.* Univ. Calif. Press, Berkeley, 1936.

STRUIK, D. J. (1). 'Outline of a History of Differential Geometry.' *ISIS*, 1933, **19**, 92.

STRUIK, D. J. (2). *A Concise History of Mathematics.* 2 vols. (pagination continuous). Dover, New York, 1948.

STUHLMANN, C. C. (1). 'Chinese Soda.' *JPOS*, 1895, **3**, 566.

STÜHR, P. F. (1). *Untersuchungen ü. d. Ursprünglichkeit u. Altertümlichkeit d. Sternkunde unter den Chinesen u. Indern u. ü. d. Einfluss d. Griechen auf den Gang ihrer Ausbildung.* Berlin, 1831.

SÜHEYL ÜNVER A. (3). *Türk Pozitif Ilimler tarihinden bir bahis Ali Kuşci Hayati ve eserleri.* (New Material on Natural Science during the Reign of Muḥammad the Conqueror especially concerning the coming of 'Ali Ibn Muḥammad al-Qūshchī from Samarqand to Constantinople.) In Turkish, illustr. Istanbul, 1948. (İstanbul Universitesi Fen Fakultesi Monografileri (Ilim Tarihi Kismi), no. 1.)

SULAIMĀN AL-TĀJIR (1) (attrib.). *Akhbār al-Ṣīn wa'l-Hind* (Information on China and India), +851. See Renaudot (1), Reinaud (1), Ferrand (2), Sauvaget (2).

SUTER, H. (1). *Die Mathematiker und Astronomen der Araber und ihre Werke*. Teubner, Leipzig, 1900. (Abhdl. z. Gesch. d. Math. Wiss. mit Einschluss ihrer Anwendungen, no. 10; supplement to *ZMP*, **45**.) Additions and corrections in *AGMW*, 1902, no. 14.

SUTER, H. (2). 'Das Buch der Seltenheiten der Rechenkunst von Abū Kamil al-Misrī.' *BM*, 1910 (3ᵉ sér.), **11**, 100.

SWALLOW, R. W. (1). *Ancient Chinese Bronze Mirrors*. Vetch, Peking, 1937.

SWANN, NANCY L. (1) (tr.). *Food and Money in Ancient China; the Earliest Economic History of China to +25* (with tr. of [*Chhien*] *Han Shu*, ch. 24 and related texts, [*Chhien*] *Han Shu*, ch. 91 and *Shih Chi*, ch. 129). Princeton Univ. Press, Princeton, N.J., 1950. (rev. J. J. L. Duyvendak, *TP*, 1951, **40**, 210; C. M. Wilbur, *FEQ*, 1951, **10**, 320; Yang Lien-Shêng, *HJAS*, 1950, **13**, 524.)

SYKES, SIR PERCY (1). *The Quest for Cathay*. Black, London, 1936.

SYLVESTER, J. J. (1). *Collected Mathematical Papers*. Cambridge, 1904.

SZCZESNIAK, B. (1). 'The Penetration of the Copernican Theory into Feudal Japan.' *JRAS*, 1944, 52.

SZCZESNIAK, B. (2). 'Notes on the Penetration of the Copernican Theory into China from the 17th to the 19th Centuries.' *JRAS*, 1945, 30.

SZCZESNIAK, B. (3). 'Notes on Kepler's *Tabulae Rudolphinae* in the Library of the Pei Thang in Peking.' *ISIS*, 1949, **40**, 344.

SZCZESNIAK, B. (4). 'Athanasius Kircher's *China Illustrata*.' *OSIS*, 1952, **10**, 385. (This paper contains many misprints, and all transcriptions of titles, etc., should be checked.)

SZCZESNIAK, B. (6). 'The Writings of Michael Boym.' *MS*, 1955, **14**.

SZCZESNIAK, B. (7). 'Matteo Ricci's Maps of China.' *IM*, 1955, **11**.

AL-TĀJIR, SULAIMĀN. See Sulaimān al-Tājir.

TAKI, SEI-ICHI (1). *Three Essays on Oriental Painting*. Quaritch, London, 1910.

TALBOYS, D. A. (1) (ed.). *The Novels and Miscellaneous Works of Daniel Defoe*. 20 vols. Oxford, 1840–1.

TANNERY, P. (2) (tr.). 'The *Παράδοσις εἰς τὴν εὕρεσιν τῶν τετραγώνων ἀριθμῶν* of Manuel Moschopoulos' (on magic squares). *AAEEG*, 1886, 88. (Reprinted in Tannery (3), 1920, vol. 4, p. 27.)

TANNERY, P. (3). *Mémoires Scientifiques*. 17 vols. Paris, 1912–46.

TAQIZADEH, S. H. (1). On the horoscope of the coronation of Khosrov Anosharvan (+531) in an astrological work of Qasram (+889). *BLSOAS*, 1938, **9**, 128.

TARDIN, J. (1). *Histoire naturelle de la Fontaine qui brusle près de Grenoble; Avec la recherche de ses Causes et principes, et ample Traicté des feux sousterrains*. Linocier, Tournon, 1618.

TARN, W. W. (1). *The Greeks in Bactria and India*. Cambridge, 1951.

TATON, R. (1). *Le Calcul Mécanique*. Presses Univ. de Fr., Paris, 1949.

TAYLOR, E. G. R. (1). (*a*) 'Ideas on the Shape and Habitability of the Earth prior to the Great Age of Discovery.' *H*, 1937, **22**, 54. (*b*) *Ideas on the Shape, Size and Movements of the Earth*. Historical Association, London, 1943. (Hist. Ass. Pamphlets, no. 126.)

TAYLOR, E. G. R. (2). 'Some Notes on Early Ideas of the Form and Shape of the Earth.' *GJ*, 1935, **85**, 65.

TAYLOR, E. G. R. (3). *Tudor Geography, +1485 to +1583*. Methuen, London, 1930.

TAYLOR, E. G. R. (4). *Late Tudor and Early Stuart Geography, +1583 to +1650*. Methuen, London, 1934.

TAYLOR, E. G. R. (5). 'Position Fixing in Relation to Early Maps and Charts.' *BBSHS*, 1949, **1**, 25.

TAYLOR, E. G. R. (7). *The Mathematical Practitioners of Tudor and Stuart England*. C.U.P., Cambridge, 1954; rev. D. J. Price, *JIN*, 1955, **8**, 12.

TAYLOR, F. SHERWOOD (2). 'A Survey of Greek Alchemy.' *JHS*, 1930, **50**, 109.

TAYLOR, F. SHERWOOD (3). *The Alchemists*. Heinemann, London, 1951.

TAYLOR, JOHN, M.D., H.E.I. Co's Bombay Medical Establishment. (1) (tr.). *Lilawati, or a Treatise on Arithmetic and Geometry, by Bhascara Acharya, translated from the Original Sanskrit*. Rans, Bombay, 1816.

THIBAUT, G. (1). 'On the Hypothesis of the Babylonian Origin of the so-called Lunar Zodiac.' *JRAS/B*, 1894, **63**, 144.

THIBAUT, G. (2). *Astronomie, Astrologie und Mathematik* [*der Inder*] in *Grundriss der Indo-Arischen Philologie und Altertumskunde* (Encyclopaedia of Indo-Aryan Research), ed. G. Bühler & F. Kielhorn, Bd. 3, Heft 9.

THIEL, R. (1). *Und es ward Licht; Roman der Weltallforschung*. Rowohlt, Hamburg, 1956.

THIELE, G. (1). *Antike Himmelsbilder*. Weidmann, Berlin, 1898.

THOM, A. (1). 'The Solar Observatories of Megalithic Man.' *JBASA*, 1954, **64**, 396.

THOM, A. (2). 'A Statistical Examination of the Megalithic Sites in Britain.' *JRSS*, 1955, **118**, 275.

THOMAS, F. W. (1). 'Notes on the "Scythian Period" [Śaka Era].' *JRAS*, 1952, 108.

THOMAS, O. (1). *Astronomie; Tatsachen und Probleme.* Bergland-Buch, Graz, Vienna, Leipzig & Berlin, 1934. 7th ed. Bergland-Buch, Salzburg, 1956 (rev. A. Beer, *O*, 1957, **77**, 161).

THOMPSON, D'ARCY W. (1). 'Excess and Defect; or the Little More and the Little Less.' *M*, 1929, **38**, 43.

THOMPSON, C. J. S. (1). *The Mystery and Lore of Monsters; with accounts of some Giants, Dwarfs and Prodigies.* Williams & Norgate, London, 1930.

THOMPSON, J. E. S. (1). *Maya Hieroglyphic Writing; an Introduction.* Carnegie Institute Pubs. no. 589. Washington, D.C., 1950; rev. D. H. Kelley, *AJA*, 1952, **56**, 240.

THOMPSON, R. C. (1). *Reports of the Magicians and Astrologers of Nineveh and Babylon* [in the British Museum on cuneiform tablets]. 2 vols. Luzac, London, 1900.

THOMSON, JOHN (1). *Illustrations of China and its People; a Series of 200 Photographs with letterpress descriptive of the Places and the People represented.* 4 vols. Sampson Low, London, 1873–4. French tr. by A. Talandier & H. Vattemare, 1 vol. Hachette, Paris, 1877.

THOMSON, J. O. (1). *History of Ancient Geography.* Cambridge, 1948.

THORNDIKE, L. (1). *A History of Magic and Experimental Science.* 6 vols. Columbia Univ. Press, New York: vols. 1 and 2, 1923; 3 and 4, 1934; 5 and 6, 1941.

THORNDIKE, L. (2). *The Sphere of Sacrobosco and its Commentators* (Text, Commentaries and Translation). Chicago, 1949.

THORNDIKE, L. (3). 'Franco de Polonia and the Turquet.' *ISIS*, 1945, **36**, 6.

THORNDIKE, L. (4). 'Thomas Werkworth on the Motion of the Eighth Sphere.' *ISIS*, 1948, **39**, 212.

THORNDIKE, L. (5). 'A Weather Record for +1399 to +1406.' *ISIS*, **32**, 304.

THORNDIKE, L. (6). 'The Cursus Philosophicus before Descartes.' *A/AIHS*, 1951, **4**, 16.

THORNDIKE, L. & SARTON, G. (1). 'Tacuinum and Taqwīm.' *ISIS*, 1928, **10**, 489.

THUREAU-DANGIN, F. (1). (*a*) 'Sketch of a History of the Sexagesimal System.' *OSIS*, 1939, **7**, 95. (*b*) *Esquisse d'une Histoire du Système sexagésimal.* Geuthner, Paris, 1932.

THUREAU-DANGIN, F. (2). 'L'Origine de l'Algèbre.' *CRAIBL*, 1940, 292.

THUROT, C. (1). 'Recherches historiques sur le Principe d'Archimède.' *RA*, 1868.

TING, V. K. See Ting Wên-Chiang.

TING WÊN-CHIANG (2). 'Notes on Records of Droughts and Floods in Shensi, and the supposed Desiccation of Northwest China.' Hyllningskrift tillägnad Sven Hedin (Hedin Festschrift). *GA*, special no. 1925, p. 453.

TING WÊN-CHIANG (3). 'On Hsü Hsia-Kho (+1586 to +1641), Explorer and Geographer.' *NCR*, 1921, **3**, 325.

TOOLEY, R. V. (1). *Maps and Map-Makers.* Batsford, London, 1949.

TORGASHEV, B. P. (1). *The Mineral Industry of the Far East.* Chali, Shanghai, 1930.

TRIGAULT, NICHOLAS (1). *De Christiana Expeditione apud Sinas.* Vienna, 1615; Augsburg, 1615. Fr. tr.: *Histoire de l'Expédition Chrétienne au Royaume de la Chine, entrepris par les PP. de la Compagnie de Jésus, comprise en cinq livres...tirée des Commentaires du Matthieu Riccius*, etc. Lyon, 1616; Lille, 1617; Paris, 1618. Eng. tr. (partial): *A Discourse of the Kingdome of China, taken out of Ricius and Trigautius.* In *Purchas his Pilgrimes.* London, 1625, vol. 3, p. 380. Trigault's book was based on Ricci's *I Commentarj della Cina* which it follows very closely, even verbally, by chapter and paragraph, introducing some changes and amplifications, however. Ricci's book remained unprinted until 1911, when it was edited by Venturi (1) with Ricci's letters; it has since been more elaborately and sumptuously edited alone by d'Elia (2). Eng. tr. (full), see Gallagher (1).

TROLLOPE, M. N., BP. (1). 'Korean Books and their Authors' with 'A Catalogue of Some Korean Books in the Chosen Christian College Library.' *JRAS/KB*, 1932, **21**, 1 and 59.

TROPFKE, J. (1). *Geschichte d. Elementar-Mathematik in systematischer Darstellung mit besonderer Berücksichtigung d. Fachwörter.* de Gruyter, Berlin & Leipzig, 1921–4.

Vol. 1 (3rd ed.). 1930. Rechnen.
Vol. 2 (3rd ed.). 1933. Allgemeine Arithmetik.
Vol. 3 (3rd ed.). 1937. Proportionen, Gleichungen.
Vol. 4 (3rd ed.). 1940. Ebene Geometrie.
Vol. 5 (2nd ed.). 1923. Ebene Trigonometrie, Sphärik und sphärische Trigonometrie.
Vol. 6 (2nd ed.). 1924. Analyse, analytische Geometrie.
Vol. 7 (2nd ed.). 1924. Stereometrie. Verzeichnisse.

TSU WÊN-HSIEN. See Wên Hsien-Tzu.

TSUCHIHASHI, P. & CHEVALIER, S. (1). 'Catalogue d'Etoiles observées à Pékin sous l'Empereur Kien-Long [Chhien-Lung], XVIIIe siècle.' *Annales de l'Observatoire Astronomique de Zô-sè* [Zikkawei], 1914 (1911), **7**, no. 4. (Translation of part of the Star Catalogue prepared by the Astronomical Bureau under the directorship of Fr. I. Kögler (Tai Chin-Hsien) between +1744 and +1757—*Chhin-Ting I Hsiang Khao Chhêng*.) Comments on the planispheres by W. F. Rigge, *POPA*, 1915, **23**, 29.

TSUDA, S. (1). 'On the Dates when the *Li Chi* and the *Ta Tai Li Chi* were edited.' *MRDTB*, 1932, **6**, 77.

TYTLER, J. (1). 'Essays on the Binomial Theorem as known to the Arabs.' *TAS/B* (Asiatick Researches), 1820, **13**, 456.

UCCELLI, A. (1) [with the collaboration of G. SOMIGLI, G. STROBINO, E. CLAUSETTI, G. ALBENGA, I. GISMONDI, G. CANESTRINI, E. GIANNI & R. GIACOMELLI]. *Storia della Tecnica dal Medio Evo ai nostri Giorni*. Hoeppli, Milan, 1945.

UCCELLI, A. (3). *Enciclopedia Storica delle Scienze e delle loro Applicazioni*. Hoeppli, Milan, n.d. (1941). Vol. 1, *Le Scienze Fisiche e Matematiche*.

VON UEBERWEG, F. & HEINZE, M. (1). *Grundriss d. Geschichte d. Philosophie*. 4 vols. Mittler, Berlin, 1898 (but many editions).

UETA, J. (1)=(*1*). *Shih Shen's Catalogue of Stars, the oldest Star Catalogue in the Orient*. Publications of the Kwasan Observatory (of Kyoto Imperial University), 1930, **1** (no. 2), 17. (A portrait of Dr and Mrs Ueta will be found in *POPA*, 1936, **44**, 121.)

UHDEN, R. (1). 'Die antiken Grundlagen d. mittelälterlichen Seekarten.' *IM*, 1935, **1**, 1.

UNGER, E. (1). 'Early Babylonian Wheel-maps.' *AQ*, 1935, **9**, 311.

UNGER, E. (2). 'From the Cosmos Picture to the World Map.' *IM*, 1937, **2**, 1.

UNGERER, A. *Les Horloges Astronomiques et Monumentales les plus remarquables de l'Antiquité jusqu'à nos Jours* (preface by A. Esclangon). Ungerer, Strasbourg, 1931.

AL-'URDĪ. See al-Dimashqī.

USHER, A. P. (1). *A History of Mechanical Inventions*. McGraw-Hill, New York, 1929. 2nd ed. revised, Harvard Univ. Press, Cambridge, Mass., 1954 (rev. Lynn White, *ISIS*, 1955, **46**, 290).

VACCA, G. (1). 'Note Cinesi.' *RSO*, 1915, **6**, 131. [(*a*) A silkworm legend from the *Sou Shen Chi*. (*b*) The fall of a meteorite described in *Mêng Chhi Pi Than*. (*c*) Invention of movable type printing (*Mêng Chhi Pi Than*). (*d*) A problem of the mathematician I-Hsing (chess permutations and combinations) in *Mêng Chhi Pi Than*. (*e*) An alchemist of the +11th century (*Mêng Chhi Pi Than*).]

VACCA, G. (3). 'Sulla Matematica degli antichi Cinesi. *BBSSMF*, 1905, **8**, 1.

VACCA, G. (4) (tr.). Translation of *Chou Pei Suan Ching*. *BBSSMF*, 1904, **7**.

VACCA, G. (5). 'Due Astronomi Cinesi del IV Sec. AC e i loro cataloghi stellari.' Zanichelli, Bologna, 1934. (Offprint from *Calendario del r. Osservatorio Astronomico di Roma*, n.s., 1934, **10**.) A review of Maspero (3).

VACCA, G. (6). 'Note sulla Storia d. Cartografia Cinese ' *RGI*, 1911, **18**, 113.

VACCA, G. (7). 'Della Piegatura della Carta applicata alla Geometria.' *PDM*, 1930 (ser. 4), **10**, 43.

VACCA, G. (10). 'Sull'Opera geografica del P. Matteo Ricci.' *RGI*, 1941, **48**, 1.

DE LA VALLÉE POUSSIN, L. (7) (tr.). *Troisième Chapitre de 'l'Abhidharmakoṣa', Kārikā, bhāṣya et vyākhyā... Versions et textes établis* (*Bouddhisme; Etudes et Matériaux; Cosmologie; Le Monde des Etres et le Monde-Réceptacle*). Kegan Paul, London, 1918.

VANHÉE, P. L. See van Hée, L.

VÄTH, A. (1) (with the collaboration of L. van Hée). *Johann Adam Schall von Bell, S.J., Missionär in China, Kaiserlicher Astronom und Ratgeber am Hofe von Peking; ein Lebens- und Zeitbild*. Bachem, Köln, 1933. (Veröffentlichungen des Rheinischen Museums in Köln, no. 2.) Crit. P. Pelliot, *TP*, 1934, 178.

VENTURI, P. T. (1) (ed.). *Opere Storiche del P. Matteo Ricci*. 2 vols. Giorgetti, Macerata, 1911.

VERBIEST, F. (1). *Astronomia Europaea sub Imperatore Tartaro-Sinico Cam-Hy* [*Khang-Hsi*] *appellato, ex Umbra in Lucem Revocata...*. Bencard, Dillingen, 1687. This is a quarto volume of 126 pp., edited by P. Couplet. A folio volume with approximately the same title (Verbiest, 2) had appeared in 1668, consisting of 18 pp. Latin text and 250 plates of apparatus on Chinese paper, only one of which, the general view of the Peking Observatory, was re-engraved in small format for the 1687 edition. The large version is rare and I have only seen some loose plates from it. Cf. Houzeau, p. 44; Bosmans.

VERBIEST, F. (2). *Liber Organicus Astronomiae Europaeae apud Sinas Restitutae, sub Imperatore Sino-Tartarico Cam-Hy* [*Khang-Hsi*] *appellato...*. Peking, 1668.

VERNET, J. (1). 'Influencias Musulmanas en el Origen de la Cartografía Nautica.' *PRSG*, 1953, Ser. B, no. 289.

VERNET, J. (2). 'Los Conocimientos Náuticos de los Habitantes del Occidente Islámico.' *RGM*, 1953, 3.

DES VIGNOLES, A. (1). 'De Cyclis Sinensium Sexagenariis.' *MBIS*, 1734, **4**, 24, 245; 1737, **5**, 1.

DES VIGNOLES, A. (2). 'De Conjunctione Planetarum in China Observata' [−2448]. *MBIS*, 1737, **5**, 193.

DA VINCI, LEONARDO. See McCurdy, E.
VIOLLE, B. (1). *Traité Complet des Carrés Magiques.* 3 vols. Paris, 1837.
DE VISSER, M. W. (2). *The Dragon in China and Japan.* Müller, Amsterdam, 1913. Orig. in *VKAWA/L*, 1912, **13** (no. 2).
VISSIÈRE, A. (1). 'Recherches sur l'Origine de l'Abaque Chinois et sur sa dérivation des anciennes Fiches à Calcul.' *BG*, 1892, 28.
VAN VLOTEN, G. (1) (ed.). *Arabic text of 'Mufātiḥ al-'Ulūm' (The Keys of the Sciences) by Muḥammad ibn-Aḥmad al-Khwārizmī* (+976). Leiden, 1895.
VOGT, H. (1). 'Versuch einer Wiederherstellung von Hipparchs Fixsternverzeichnis.' *ASTNR*, 1925, **224**, cols. 17 ff.
VOGT, T. & BERGH, H. (1). 'Geochemical and Geobotanical Methods for Ore Prospecting.' *KNVSF/T*, 1946, **19** (no. 21), 76.
VOLPICELLI, Z. (1). 'Chinese Chess' [*wei chhi*]. *JRAS/NCB*, 1894, **26**, 80.
VOSS, ISAAC (1). *Variarum Observationum Liber.* Scott, London, 1685.

WADA, S. (1). 'The Philippine Islands as known to the Chinese before the Ming Dynasty.' *MRDTB*, 1929, **4**, 121.
WADA, T. (1). 'Schmuck und Edelsteine bei den Chinesen.' *MDGNVO*, 1904, **10**, 1.
WADA, Y. (1). 'A Korean Rain-Gauge of the +15th Century.' *QJRMS*, 1911, **37**, 83 (translation of Wada, *1*); *KMO/SM*, 1910, **1**; *MZ*, 1911, 232. Figure reproduced in Feldhaus (1), col. 865.
VAN DER WAERDEN, B. L. (1). *Ontwakende Wetenschap; Egyptische, Babylonische en Griekse Wiskunde.* Noordhoff, Groningen, 1950. (Histor. Bibl. voor de exacte Wet. no. 7.)
VAN DER WAERDEN, B. L. (2). 'Babylonian Astronomy. II. The Thirty-six Stars.' *JNES*, 1949, **8**, 6. (I, in *EOL*, 1948, **10**, 424, deals with the Old Babylonian Venus computation tablets; and III, in *JNES*, 1951, **10**, 20, with other astronomical computations. The 36 stars are the month-stars of the Three Roads of Anu, Ea, and Enlil.)
VAN DER WAERDEN, B. L. (3). *Science Awakening.* Engl. tr. of (1) by A. Dresden with additions of the author. Noordhoff, Groningen, 1954.
WAGNER, A. (1). 'Über ein altes Manuscript der Pulkowaer Sternwarte' (with additional note by J. L. E. Dreyer). Chinese and Persian MS. believed to be from the time of Kuo Shou-Ching and Jamāl al-Dīn. *C*, 1882, **2**, 123.
WALES, H. G. QUARITCH (1). *The Making of Greater India; a Study in Southeast Asian Culture Change.* Quaritch, London, 1951.
WALES, H. G. QUARITCH (2). 'The Sacred Mountain in Old Asiatic Religion.' *JRAS*, 1953, 23.
WALEY, A. (1) (tr.). *The Book of Songs.* Allen & Unwin, London, 1937.
WALEY, A. (4) (tr.). *The Way and its Power; a study of the 'Tao Tê Ching' and its Place in Chinese Thought.* Allen & Unwin, London, 1934. (Crit. Wu Ching-Hsiung, *TH*, 1935, **1**, 225.)
WALEY, A. (5) (tr.). *The Analects of Confucius.* Allen & Unwin, London, 1938.
WALEY, A. (10). *The Travels of an Alchemist* [Chhiu Chhang-Chhun's journey to the court of Chingiz Khan]. Routledge, London, 1931. (Broadway Travellers Series.)
WALEY, A. (11). *The Temple, and other Poems.* Allen & Unwin, London, 1923.
WALEY, A. (13). *The Poetry and Career of Li Po (+701 to +762).* Allen & Unwin, London, 1950.
WALEY, A. (17) (tr.). *Monkey, by Wu Chhêng-Ên.* Allen & Unwin, London, 1942.
WALEY, A. (21). 'The Eclipse Poem [in the *Shih Ching*] and its Group.' *TH*, 1936, **3**, 245.
WALEY, A. (23). *The Nine Songs; a study of Shamanism in Ancient China [the 'Chiu Ko' attributed traditionally to Chhü Yuan].* Allen & Unwin, London, 1955.
WALLACE, T. (1). *The Diagnosis of Mineral Deficiencies in Plants.* HMSO, London, 1943. (ARC Monograph.)
WALLIS, JOHN (1). *De Algebra Tractatus.* 1685. In *Opera*. Oxford, 1693.
WALTERS, R. C. S. (1). 'Greek and Roman Engineering Instruments.' *TNS*, 1922, **2**, 45.
WANG KUO-WEI (2). 'Chinese Foot-Measures of the Past Nineteen Centuries.' *JRAS/NCB*, 1928, **59**, 112. (Tr. A. W. Hummel & Fêng Yu-Lan.)
WANG LING (2). *The 'Chiu Chang Suan Shu' and the History of Chinese Mathematics during the Han Dynasty.* Inaug. Diss. Cambridge, 1956.
WANG LING (3). 'The Development of Decimal Fractions in China.' *Proc. VIIIth Internat. Congress of the History of Science, Florence, 1956*, p. 13.
WANG LING (4). 'The Decimal Place-Value System in the Notation of Numbers in China.' Communication to the XXIIIrd International Congress of Orientalists, Cambridge, 1954.
WANG LING (5). On the Indeterminate Analysis of I-Hsing and Chhin Chiu-Shao (in the press).
WANG LING & NEEDHAM, JOSEPH (1). 'Horner's Method in Chinese Mathematics; its Origins in the Root-Extraction Procedures of the Han Dynasty.' *TP*, 1955, **43**, 345.

WANG, W. E. (1). *The Mineral Wealth of China.* Com. Press, Shanghai, 1927.

WANG YÜ-CHHÜAN (1). *Early Chinese Coinage.* Amer. Numismatic Soc., New York, 1951. (Numismatic Notes and Monographs, no. 122.)

WARD, F. A. B. (1). *Time Measurement.* Pt. I. *Historical Review.* (Handbook of the Collections at the Science Museum, South Kensington.) HMSO, London, 1937.

WARREN, SIR CHARLES (1). *The Ancient Cubit.* London, 1903.

WARREN, H. V. & DELAVAULT, R. E. (1). (*a*) 'Biogeochemical Investigations in British Columbia.' *GP*, 1948, **13**, 609. (*b*) 'Further Studies in Biogeochemistry.' *BGSA*, 1949, **60**, 531.

WARREN, H. V. & HOWATSON, C. H. (1). 'Biogeochemical Prospecting for Copper and Zinc.' *BGSA*, 1947, **58**, 803.

WATTENBERG, D. (1). 'Die Supernovae des Milchstrassensystems.' *ZNF*, 1949, A, **4**, 228.

WEBER, A. (1). 'Die Vedische Nachrichten von den Naxatra (Mondstationen).' *APAW*, 1860, 283; 1861, 349.

WEBSTER, E. W. (1) (tr.). *Aristotle's 'Meteorologica'.* Oxford, 1923.

WEIDNER, E. F. (1). *Handbuch d. babylonischen Astronomie. I. Der babylonische Fixsternhimmel.* Leipzig, 1915. (Assyriologische Bibliothek, no. 23.)

WEIDNER, E. F. (2). 'Enuma-Anu-Enlil.' *AOF*, 1942, **14**, 172, 308.

WEIDNER, E. F. (3). 'Ein babylonisches Kompendium der Himmelskunde.' *AJSLL*, 1924, **40**, 186.

WEINBERGER, W. M. (1). 'An Early Chinese Bronze Foot Measure.' *ORA*, 1949, **2**, 35.

WEINSTOCK, S. (1). 'Lunar Mansions and Early Calendars.' *JHS*, 1949, **69**, 48. (Lunar mansions in a +4th-century Greek papyrus and a +15th-century Byzantine MS.)

WELLMANN, M. (1). 'Die Stein- u. Gemmen-Bücher d. Antike.' *QSGNM*, 1935, **4**, 86.

WÊN HSIEN-TZU (1). 'Observations of Halley's Comet in Chinese History.' *POPA*, 1934, **42**, 191.

WÊN HSIEN-TZU (2). 'A Statistical Survey of Eclipses in Chinese History.' *POPA*, 1934, **42**, 136.

WÊN SHION TSU. See Wên Hsien-Tzu.

WENLEY, A. G. (1). 'The Question of the Po Hsiang Shan Lu.' *ACASA*, 1948, **3**, 5.

WERNER, E. T. C. (1). *Myths and Legends of China.* Harrap, London, 1922.

WESTPHAL, A. (1). (*a*) Über die chinesisch-japanische Rechenmaschine.' *MDGNVO*, 1873, **8**, 27. (*b*) 'Über das Wahrsagen auf der Rechenmaschine.' *MDGNVO*, 1873, **8**, 48. (*c*) 'Über die chinesische *Swan-Pan.*' *MDGNVO*, 1876, **9**, 43.

WEYL, HERMANN (1). *Symmetry.* Princeton Univ. Press, Princeton, N.J., 1952.

WHEELER, R. E. M. (4). *Rome beyond the Imperial Frontiers.* Bell, London, 1954.

WHEWELL, WILLIAM (1). *History of the Inductive Sciences.* Parker, London, 1847. 3 vols. (Crit. G. Sarton, *A/AIHS*, 1950, **3**, 11.)

WHITE, LYNN (2). 'Natural Science and Naturalistic Art in the Middle Ages.' *AHR*, 1946, **52**, 421.

WHITE, W. C. & MILLMAN, P. M. (1). 'An Ancient Chinese Sun-Dial.' *RASC/J*, 1938, **32**, 417.

WHITE, W. C. & WILLIAMS, R. J. (1). *Chinese Jews; a Compilation of Matters relating to Khaifêng-fu.* Univ. Press, Toronto, 1942. 3 vols. Vol. 1, Historical. Vol. 2, Inscriptional. Vol. 3 (with R. J. Williams), Genealogical.

WHITEHEAD, A. N. (1). *Science and the Modern World.* Cambridge, 1926.

WHITEHEAD, A. N. (7). *Essays in Science and Philosophy.* Rider, London, 1948.

WHITNEY, W. D. (1). *On the Lunar Zodiac of India, Arabia and China.* Art. no. 13 in *Oriental and Linguistic Studies*, 2nd series, p. 341. Scribner, New York, 1874. 2nd ed. 1893. Sep. pub. Riverside, Cambridge (Mass.), 1874.

WHITNEY, W. D. (2). 'On the Views of [J. B.] Biot and [A.] Weber respecting the Relations of the Hindu and Chinese systems of Asterisms—with an addition on [Max] Müller's views respecting the same subject.' *JAOS*, 1864, **8**, 1–94.

WHITNEY, W. D. & LANMAN, C. R. (1) (tr.). *Atharvaveda Saṃhitā.* 2 vols. Harvard Univ. Press, Cambridge (Mass.), 1905. (Harvard Oriental Series, nos. 7, 8.)

WIEDEMANN, E. (1). 'Zur Mineralogie bei den Muslimen.' *AGNT*, 1909, **1**, 208.

WIEDEMANN, E. (2). 'Arabische Studien ü. d. Regenbogen.' *AGNT*, 1913, **4**, 453.

WIEDEMANN, E. (9). 'Über eine astronomische Schrift von al-Kindī' [on an instrument similar to Ptolemy's triquetrum]. *SPMSE*, 1910, **42**, 294. (Beiträge z. Gesch. d. Naturwiss. no. 21 A.)

WIEDEMANN, E. (10). 'Über das Schachspiel und dabei vorkommende Zahlenprobleme.' *SPMSE*, 1908, **40**, 41. (Beiträge z. Gesch. d. Naturwiss. no. 14 (4).)

WIEDEMANN, E. & HAUSER, F. (4). 'Über die Uhren im Bereich der Islamischen Kultur.' *NALC*, 1915, **100**, no. 5. Incl. transls. of the *Kitāb fī Ma'rifat al-Ḥiyal al-Handasīya* (Treatise on the Knowledge of Geometrical (i.e. Mechanical) Contrivances) by Ibn al-Razzāz al-Jazarī (*fl.* +1180 to +1206) written in +1206; and of the *Book on the Construction and Use of* (*Striking Water-*) *Clocks* by Riḍwān al-Khurāsānī al-Sa'ātī (*fl. c.* +1160 to +1230) written in +1203.

WIEGER, L. (1). *Textes Historiques.* 2 vols. (Ch. and Fr.). Mission Press, Hsienhsien, 1929.

WIEGER, L. (2). *Textes Philosophiques* (Ch. and Fr.). Mission Press, Hsienhsien, 1930.

WIEGER, L. (3). *La Chine à travers les Ages; Précis, Index Biographique et Index Bibliographique.* Mission Press, Hsienhsien, 1924. Eng. tr. E. T. C. Werner.

WIEGER, L. (4). *Histoire des Croyances Religieuses et des Opinions Philosophiques en Chine depuis l'origine jusqu'à nos jours.* Mission Press, Hsienhsien, 1917.

WIEGER, L. (6). *Taoisme.* Vol. 1. *Bibliographie Générale*: (1) Le Canon (Patrologie); (2) Les Index Officiels et Privés. Mission Press, Hsienhsien, 1911. (Crit. P. Pelliot, *JA*, 1912 (10e sér.), **20**, 141.)

WIEGER, L. (7). *Taoisme.* Vol. 2. *Les Pères du Système Taoiste* (tr. selections of Lao Tzu, Chuang Tzu, Lieh Tzu). Mission Press, Hsienhsien, 1913.

WIENER, P. P. (2). 'The Tradition behind Galileo's Methodology.' *OSIS*, 1936, **1**, 733.

WILHELM, HELLMUT (1). *Chinas Geschichte; zehn einführende Vorträge.* Vetch, Peiping, 1942.

WILHELM, HELLMUT (7). 'Der *Thien Wên* Frage.' *MS*, 1945, **10**, 427.

WILHELM, RICHARD (2) (tr.). *'I Ging' [I Ching]; Das Buch der Wandlungen.* 2 vols. (3 books, pagination of 1 and 2 continuous in first volume). Diederichs, Jena, 1924. Eng. tr. C. F. Baynes (2 vols.). Bollingen-Pantheon, New York, 1950.

WILHELM, RICHARD (3) (tr.). *Frühling u. Herbst d. Lü Bu-We* (the *Lü Shih Chhun Chhiu*). Diederichs, Jena, 1928.

WILHELM, RICHARD (4) (tr.). *'Liä Dsi'; Das Wahre Buch vom Quellenden Urgrund; [Lieh Tzu] 'Tschung Hü Dschen Ging'; Die Lehren der Philosophen Liä Yü-Kou und Yang Dschu.* Diederichs, Jena, 1921.

WILHELM, RICHARD (6) (tr.). *'Li Gi', das Buch der Sitte des älteren und jüngeren Dai* [i.e. both *Li Chi* and *Ta Tai Li Chi*]. Diederichs, Jena, 1930.

WILHELM, RICHARD (7). *Chinesische Volksmärchen.* Diederichs, Jena, 1914.

WILKINS, JOHN, BP. (1). *Discovery of a World in the Moon, tending to prove that 'tis probable that there may be another habitable World in that Planet.* London, 1638.

WILLEY, B. (1). *The Seventeenth Century Background.* Chatto & Windus, London, 1934.

WILLIAMS, J. (1). 'Notes on Chinese Astronomy' [presentation of planispheres]. *RAS/MN*, 1855, **15**, 19.

WILLIAMS, J. (2). 'On an Eclipse of the Sun recorded in the Chinese Annals as having occurred at a very early period of their History' [the *Shu Ching* eclipse]. *RAS/MN*, 1863, **23**, 238.

WILLIAMS, J. (3). 'On the Eclipses recorded in *Chun Tsew*' [*Chhun Chhiu*]. *JRAS/MN*, 1864, **24**, 39.

WILLIAMS, J. (4). 'Solar Eclipses observed in China from −481 to the Christian Era.' *RAS/MN*, 1864, **24**, 185.

WILLIAMS, J. (5). 'Chinese Observations of Solar Spots.' *JRAS/MN*, 1873, **33**, 370.

WILLIAMS, J. (6). *Observations of Comets from −611 to +1640, extracted from the Chinese Annals,... with an appendix comprising the tables necessary for reducing Chinese time to European reckoning; and a Chinese Celestial Atlas.* Strangeways & Walden, London, 1871.

WILLIAMS, S. WELLS (1). *The Middle Kingdom; A Survey of the Geography, Government, Education, Social Life, Arts, Religion, etc. of the Chinese Empire and its Inhabitants.* 2 vols. Wiley, New York, 1848; later eds. 1861, 1900; London, 1883.

WILSON, MILES (1). *The History of Israel Jobson, the Wandering Jew. Giving a Description of his Pedigree, Travels in this Lower World, and his Assumption thro' the Starry Regions, conducted by a Guardian Angel, exhibiting in a curious Manner the Shapes, Lives, and Customs of the Inhabitants of the Moon and Planets; touching upon the great and memorable Comet in 1758, and interwoven all along with the Solution of the Phaenomena of the true Solar System, and Principles of Natural Philosophy, concording with the latest Discoveries of the most able Astronomers. Translated from the Original Chinese by M.W.* London, 1757. (Cf. G. K. Anderson, *PQ*, 1946, **25**, 303.)

WINTER, H. J. J. (4). 'The Optical Researches of Ibn al-Haitham.' *CEN*, 1954, **3**, 190.

WINTER, H. J. J. (5). 'Muslim Mechanics and mechanical Appliances.' *END*, 1956, **15** (no. 57), 25.

WINTER, H. J. J. & ARAFAT, W. (1). 'The Algebra of 'Umar Khayyāmī.' *JRAS/B*, 1950, **16**, 27.

WITTFOGEL, K. A. (2). 'Die Theorie der orientalischen Gesellschaft.' *ZSF*, 1938, **7**, 90.

WITTFOGEL, K. A. (3). 'Meteorological Records from the Shang [dynasty] Divination Inscriptions.' *GR*, 1940, **30**, 110. (Crit. Tung Tso-Pin, *SSE*, 1942, **3**.)

WITTFOGEL, K. A. (4). *Wirtschaft und Gesellschaft Chinas; Versuch der wissenschaftlichen Analyse einer grossen asiatischen Agrargesellschaft—Erster Teil, Produktivkräfte, Produktions- und Zirkulationsprozess.* Hirchfeld, Leipzig, 1931. (Schriften d. Instit. f. Sozialforschung a. d. Univ. Frankfurt a. M. III, 1.)

WITTFOGEL, K. A., FÊNG CHIA-SHÊNG *et al.* (1). *History of Chinese Society (Liao), +907 to +1125.* *TAPS*, 1948, **36**, 1–650 (rev. P. Demiéville, *TP*, 1950, **39**, 347; E. Balazs, *PA*, 1950, **23**, 318).

WITTKOWER, R. (1). 'Marvels of the East; a Study in the History of Monsters.' *JWCI*, 1942, **5**, 159.

WOEPCKE, F. (1) (tr.). *L'Algèbre d'Omar Alkhayyāmi.* Duprat, Paris, 1851.

WOEPCKE, F. (2). 'Recherches sur l'Histoire des Sciences Mathématiques chez les Orientaux d'après des Traités inédits Arabes et Persans.' *JA*, 1855 (5e sér.), **5**, 218.

WOEPCKE, F. (3) (tr.). *Extrait du Fakhri [of Abū Bakr al-Ḥasan al-Ḥāsib al-Karajī, =al-Karkhī, c. +1025], précedé d'un Mémoire sur l'Algèbre indéterminée chez les Arabes.* Paris, 1853.
WOEPCKE, L. (1). *Disquisitiones Archaeologico-Mathematicae circa Solaria Veterum.* Inaug. Diss. Berlin, 1847.
WOHLWILL, E. (1). 'Zur Geschichte d. Entdeckung der Sonnenflecken.' *AGNT*, 1909, **1**, 443.
WOLF, A. (1). *A History of Science, Technology and Philosophy in the 16th and 17th Centuries.* Allen & Unwin, London, 1935.
WOLF, A. (2). *A History of Science, Technology and Philosophy in the 18th Century.* Allen & Unwin, London, 1938.
WOLF, R. (1). *Handbuch d. Astronomie, ihrer Geschichte und Litteratur.* 2 vols. Schulthess, Zürich, 1890.
WOLF, R. (2). *Geschichte d. Astronomie.* Oldenbourg, München, 1877.
WOLF, R. (3). *Handbuch d. Mathematik, Physik, Geodäsie und Astronomie.* 2 vols. Schulthess, Zürich, 1869 to 1872.
WOOD, R. W. (1). *Physical Optics.* Macmillan, New York, 2nd ed. 1911, 3rd ed. 1934.
WOOLARD, E. W. (1). 'The Historical Development of Celestial Coordinate Systems.' *PASP*, 1942, **54**, 77.
WOOTTON, A. C. (1). *Chronicles of Pharmacy.* 2 vols. Macmillan, London, 1910.
WORCESTER, G. R. G. (1). *Junks and Sampans of the Upper Yangtze.* Inspectorate-General of Customs, Shanghai, 1940. (China Maritime Customs Pub., Ser. III. Miscellaneous, no. 51.)
WRIGHT, J. K. (1). *Geographical Lore of the Time of the Crusades.* New York, 1925. (Amer. Geogr. Soc. Res. Series, no. 15.)
WU KHANG (1). *Les Trois Politiques du 'Tchounn Tsieou' [Chhun Chhiu] interprétées par Tong Tchong-Chou [Tung Chung-Shu] d'après les principes de l'école de Kong-Yang [Kungyang].* Leroux, Paris, 1932. (Includes tr. of ch. 121 of *Shih Chi*, the biography of Tung Chung-Shu.)
WU LU-CHHIANG & DAVIS, T. L. (2) (tr.). 'An Ancient Chinese Alchemical Classic; Ko Hung on the Gold Medicine, and on the Yellow and the White; being the 4th and 16th chapters of Pao Phu Tzu', etc. *PAAAS*, 1935, **70**, 221.
WYLIE, A. (1). *Notes on Chinese Literature.* 1st ed. Shanghai, 1867. Ed. here used, Vetch, Peiping, 1939 (photographed from the Shanghai 1922 ed.).
WYLIE, A. (2). 'History of the Hsiung-Nu' (tr. of the chapter on the Huns in the *Chhien Han Shu*, ch. 94). *JRAI*, 1874, **3**, 401; 1875, **5**, 41.
WYLIE, A. (3). 'The History of the South-western Barbarians and Chao Sëen' [Chao-Hsien, Korea] (tr. of ch. 95 of the *Chhien Han Shu*). *JRAI*, 1880, **9**, 53.
WYLIE, A. (4). 'Jottings on the Science of the Chinese; Arithmetic.' *North China Herald*, 1852 (Aug.–Nov.), nos. 108, 111, 112, 113, 116, 117, 119, 120, 121. Repr. *Shanghai Almanac and Miscellany*, 1853. Repr. *Chinese and Japanese Repository*, 1864, **1**, 411, 448, 494; **2**, 22, 69. Repr. *Copernicus*, 1882, **2**, 169, 183. Incorporated in Wylie (5), Sci. Sect., p. 159. Germ. tr. K. L. Biernatzki, q.v., 1856. Review and brief abridgement, J. Bertrand, *JS*, 1869, 317, 464; French tr. O. Terquem, *Nouv. Ann. Math.* 1862 (2e sér.), **1** (pt. 2), 35, 1863 (2e sér.), **2**, 529, *Bull. Bibl. Hist.* 1863, **2**, 529.
WYLIE, A. (5). *Chinese Researches.* Shanghai, 1897. (Photographically reproduced, Wêntienko, Peiping, 1936.)
WYLIE, A. (6). *List of Fixed Stars.* Incorporated in Wylie (5), p. 346 (Sci. Sect. p. 110). Also reprinted in Doolittle (1), p. 617.
WYLIE, A. (7). *The Mongol Astronomical Instruments in Peking.* In *Travaux de la 3e Session, Congrès Internat. des Orientalistes*, 1876. Incorporated in Wylie (5), p. 237 (Sci. Sect. p. 1).
WYLIE, A. (8). *Eclipses Recorded in Chinese Works.* Reprinted in Wylie (5) (Sci. Sect. p. 29).
WYLIE, A. (9). *Asbestos in China.* Reprinted in Wylie (5) (Sci. Sect. p. 141).
WYLIE, A. (10) (tr.). 'Notes on the Western Regions, translated from the *Ts'een Han Shoo* [*Chhien Han Shu*], Bk. 96.' *JRAI*, 1881, **10**, 20; 1882, **11**, 83. (Chs. 96 A and B, as also the biography of Chang Chhien in ch. 61, pp. 1–6, and the biography of Chhen Thang in ch. 70.)
WYLIE, A. (11). 'The Magnetic Compass in China.' *NCH*, 1859, 15 March. Reprinted in Wylie (5) (Sci. Sect. p. 155).
WYLIE, A. (13). [*Glossary of Chinese*] *Mathematical and Astronomical Terms.* In Doolittle, J. (1), vol. 2, p. 354.
WYLIE, A. (14). 'Notes of the Opinions of the Chinese with regard to Eclipses.' *JRAS/NCB*, 1866, **3**, 71.
WYLIE, A. (15). 'On the Knowledge of a Weekly Sabbath in China.' *CRR*, 1871, **4**, 4, 40. Reprinted in Wylie (5) (Hist. Sect. p. 86.)

YABUUCHI, KIYOSHI (1). 'Indian and Arabian Astronomy in China.' Art. in Silver Jubilee Volume of the Zinbun Kagaku Kenkyusyo, Kyoto University, Kyoto, 1954, p. 585.

YAJIMA, S. (1). 'Bibliographie du Dr Mikami Yoshio; Notice Biographique.' In *Actes du VIIe Congrès Internat. d'Histoire des Sciences*, Jerusalem, 1953, p. 646.
YAMPOLSKY, P. (1). 'The Origin of the Twenty-Eight Lunar Mansions.' *OSIS*, 1950, **9**, 62. Mainly a translation from the Japanese of the essential views of Iijima Tadao and Shinjō Shinzō.
YANG HSIEN-YI & YANG, GLADYS (1) (tr.). *The 'Li Sao' and other Poems of Chu* [*Chhü*] *Yuan*. Foreign Languages Press, Peking, 1953.
YANG LIEN-SHÊNG (1). 'A Note on the so-called TLV-Mirrors and the Game *Liu-Po*.' *HJAS*, 1945, **9**, 202.
YANG LIEN-SHÊNG (2). 'An Additional Note on the Ancient Game *Liu-Po*.' *HJAS*, 1952, **15**, 124.
YANG LIEN-SHÊNG (4). *Topics in Chinese History*. Harvard Univ. Press, Cambridge, Mass. 1950. (Harvard-Yenching Institute Studies, no. 4.)
YANG LIEN-SHÊNG (5). 'Notes on the Economic History of the Chin Dynasty.' *HJAS*, 1945, **9**, 107. [With tr. of *Chin Shu*, ch. 26.]
YANG LIEN-SHÊNG (6). Review of Yabuuchi Kiyoshi's edition of the *Thien Kung Khai Wu* (*Tenkō Kaibutsu no Kenkyū*). Tokyo, 1953. *HJAS*, 1954, **17**, 307.
YAO SHAN-YU (1). 'The Chronological and Seasonal Distribution of Floods and Droughts in Chinese History (−206 to +1911).' *HJAS*, 1942, **6**, 273.
YAO SHAN-YU (2). 'The Geographical Distribution of Floods and Droughts in Chinese History (−206 to +1911).' *FEQ*, 1943, **2**, 357.
YAO SHAN-YU (3). 'Flood and Drought Data in the *Thu Shu Chi Chhêng* and the *Chhing Shih Kao*.' *HJAS*, 1944, **8**, 214.
AL-YA'QŪBĪ, i.e. AḤMAD IBN ABĪ YA'QŪB AL-'ABBĀSĪ. *Kitāb al-Buldān* (Book of the Countries). See Houtsma.
YETTS, W. P. (5). *The Cull Chinese Bronzes*. Courtauld Institute, London, 1939.
YETTS, W. P. (7). 'Glass in Ancient China.' *ILN*, 1934, 732.
YETTS, W. P. (16). *Catalogue of the Collection of Ancient Chinese Bronzes from the collection of Mr A. E. K. Cull lent to the School of Oriental Studies of the University of Durham to mark the Coronation of H.M. Queen Elizabeth II*. School of Oriental Studies, Durham, 1953.
YI IK-SEUP (1). 'A Map of the World.' *KR*, 1892, **1**, 336.
YODER, H. S. (1). 'The Problem of Jadeite.' *AJSC*, 1950, **248**, 227, 312.
YOSHINO, Y. (1). *The Japanese Abacus Explained*. Tokyo, 1938.
YULE, SIR HENRY (1) (ed.). *The Book of Ser Marco Polo the Venetian, concerning the Kingdoms and Marvels of the East, translated and edited, with Notes, by H.Y....*, ed. H. Cordier. Murray, London, 1903 (reprinted 1921). 3rd ed. also issued, Scribners, New York, 1929. With a third volume, *Notes and Addenda to Sir Henry Yule's Edition of Ser Marco Polo*, by H. Cordier. Murray, London, 1920.
YULE, SIR HENRY (2). *Cathay and the Way Thither; being a Collection of Mediaeval Notices of China*. Hakluyt Society Pubs. (2nd ser.), London, 1913–15 (1st ed. 1866). Revised by H. Cordier. 4 vols. Vol. 1 (no. 38), *Introduction; Preliminary Essay on the Intercourse between China and the Western Nations previous to the Discovery of the Cape Route*. Vol. 2 (no. 33), *Odoric of Pordenone*. Vol. 3 (no. 37), *John of Monte Corvino and others*. Vol. 4 (no. 41), *Ibn Baṭṭuṭah and Benedict of Goes*. (Photographically reproduced, Peiping, 1942.)
YULE & CORDIER. See Yule (1).
YUSHKEVITCH, A. P. (1). *On the Achievements of Chinese Scholars in the field of Mathematics* (in Russian), in *Iz Istorii Nauki i Tekhniki Kitaya* (Essays in the History of Science and Technology in China), p. 130. Acad. Sci. Moscow, 1955.

VON ZACH, F. X. (1). 'Über ältere Chinesische Beobachtungen.' *ZA*, 1816, **2**, 299 (302).
ZEUTHEN, H. G. (1). 'Sur l'Origine de l'Algèbre.' *KDVS/MFM*, 1919, **2**, no. 4.
ZEUTHEN, H. G. (2). *Die Geschichte der Mathematik im XVI. und XVII. Jahrhundert*. Leipzig, 1903.
ZEUTHEN, H. G. (3). *Die Geschichte der Mathematik im Altertum und Mittelalter*. Copenhagen, 1896. (French tr. Paris, 1902.)
ZILSEL, E. (2). 'The Sociological Roots of Science.' *AJS*, 1942, **47**, 544.
ZILSEL, E. (3). 'The Origin of William Gilbert's Scientific Method.' *JHI*, 1941, **2**, 1.
ZILSEL, E. (4). 'The Genesis of the Concept of Scientific Progress.' *JHI*, 1945, **6**, 325.
ZILSEL, E. (5). 'Copernicus and Mechanics.' *JHI*, 1940, **1**, 113.
ZINNER, E. (1). *Geschichte d. Sternkunde, von den ersten Anfängen bis zur Gegenwart*. Springer, Berlin, 1931.
ZINNER, E. (2). 'Entstehung und Ausbreitung d. Copernikanischen Lehre.' *SPMSE*, 1943, **74**, 1.
ZINNER, E. (3). *SIR*, 1919, **52** (nos. 2–8).

ZINNER, E. (6). 'Gerbert und das See-rohr.' *BNGBB*, 1952, **33**, 39; *KVRS*, 1952, no. 7.

ZINNER, E. (7). 'Die ältesten Rädeuhren und modernen Sonnenuhren; Forschungen über den Ursprung der modernen Wissenschaft.' *BNGBB*, 1939, **28**, 1–148.

ZINNER, E. (8). *Deutsche und Niederländische astronomische Instrumente des 11–18 Jahrhunderts*. Beck, München, 1956.

VON ZITTEL, K. A. (1). *Geschichte d. Geologie u. Paläontologie bis Ende des 19. Jahrhunderts*. München & Leipzig, 1899. (Gesch. d. Wissenschaft in Deutschland, no. 23.) Eng. tr. M. M. Ogilvie-Gordon, *History of Geology and Palaeontology to the End of the 19th Century*. London, 1901.

ADDENDA TO BIBLIOGRAPHY C

BEREZKINA, E. I. (1) (tr.). 'Drevnekitaïskii Traktat *Matematika v deviati Knigach*.' (The ancient Chinese Work *Chiu Chang Suan Shu*, 'Nine Chapters on the Mathematical Art'). *JHM*, 1957, **10**, 423–584.

BERNARD-MAÎTRE, H. (14). 'Note complémentaire sur l'Atlas de Khang-Hsi.' *MS*, 1946, **11**, 191.

BERNARD-MAÎTRE, H. (15). 'Les Sources Mongoles et Chinoises de l'Atlas Martini (1655).' *MS*, 1947, **12**, 127.

BERNARD-MAÎTRE, H. (16). 'La Science Européene au Tribunal Astronomique de Pékin (17e–19e siècles).' Palais de la Découverte, Paris, 1952 (*Conférences*, Sér. D, no. 9).

DEMIÉVILLE, P. (2). Review of Chang Hung-Chao (*1*), *Lapidarium Sinicum*. *BEFEO*, 1924, **24**, 276.

DINGLE, H. (2). 'Astronomy in the +16th and +17th centuries.' art. in *Science, Medicine and History*, Singer Presentation Volume, ed. E. A. Underwood, vol. 1, p. 455. Oxford, 1953.

FÊNG YU-LAN (1). *A History of Chinese Philosophy*, vol. 1, *The period of the Philosophers (from the beginnings to c. B.C. 100)*, tr. D. Bodde; Vetch, Peiping, 1937; Allen & Unwin, London, 1937. Vol. 2, *The Period of Classical Learning (from the 2nd. century B.C. to the 20th century A.D.)*, tr. D. Bodde; Princeton Univ. Press, Princeton, N.J., 1953. At the same time, Vol. 1 was re-issued in uniform style by this publisher. Translations by Bodde of parts of Vol. 2 had appeared earlier in *HJAS*. (See Fêng Yu-Lan, *1*).

GROUSSET, R. (1). *Histoire de l'Extrême-Orient*. 2 vols. Geuthner, Paris, 1929. (Also appeared in *BE/AMG*, nos. 39, 40).

KEPLER, JOHANNES (1). *Tabulae Rudolphinae, quibus Astronomicae Scientiae, Temporum longinquitate collapsae Restauratio continetur a Tychone Brahe...concepta et destinata anno 1564...post mortem auctoris...Johannes Keplerus in lucem extulit*. (With large world-map). Ulm and Prague, 1627–30.

KLEINSCHNITZOVÁ, FLORA (1). 'Ex Bibliotheca Tychoniana Collegii Soc. Jesu Pragae ad S. Clementem.' *NTBB*, 1933, **20**, 73.

KOJIMA TAKASHI (1). *The Japanese Abacus, its Use and Theory*. Tuttle, Tokyo, 1954.

KOYRÉ, A. (5). *From the Closed World to the Infinite Universe*. (Noguchi Lectures). Johns Hopkins Univ. Press, Baltimore, 1957.

MOULE, A. C. (14). 'A Note on the Chinese Atlas of the Magliabecchian Library (at Florence).' *JRAS*, 1919, 393. (The copy of the *Kuang Yü Khao* brought back in 1598 by Carletti and translated with a Chinese collaborator in 1701).

MOULE, A. C. (15). *Quinsai, with other Notes on Marco Polo*. Cambridge, 1957.

NEEDHAM, JOSEPH, BEER, ARTHUR & HO PING-YÜ (1). '"Spiked" Comets in Ancient China.' *O*, 1957, **77**, 137.

NEEDHAM, JOSEPH, WANG LING & PRICE, DEREK J. (1). *Heavenly Clockwork; the Great Astronomical Clocks of Medieval China*, Cambridge (in the press). (Antiquarian Horological Society Monographs, no. 1).

SZCZESNIAK, B. (8). 'The 17th-Century Maps of China; an Inquiry into the compilations of European Cartographers'. *IM*, 1956, **13**, 116.

TOKAREV, V. A. (1). 'The most Ancient Chinese Book on Minerals and Mining' (identifications of seventeen minerals mentioned in the *Shan Hai Ching*) (in Russian). *MSRM*, 1956, **85**, 393.

WINTER, H. J. J. (6). 'Notes on *al-Kitāb Suwar al-Kawakib* (Book of the Fixed Stars) *al-Thamaniya al-Arba'in* of Abū'l-Husain 'abd al-Rāhman ibn 'Umar al-Sufī al-Rāzī (commonly known as al-Sufī), +903 to +986'. *A/AIHS*, 1955, **8**, 126.

YANG LIEN-SHÊNG (7). 'Notes on N. L. Swann's "Food and Money in Ancient China".' *HJAS*, 1950, **13**, 524.

GENERAL INDEX

by Muriel Moyle

NOTES

(1) Articles (such as 'the', 'al-', etc.) occurring at the beginning of an entry, and prefixes (such as 'de', 'van', etc.) are ignored in the alphabetical sequence. Saints appear among all letters of the alphabet according to their proper names. Styles such as Mr, Dr, if occurring in book titles or phrases, are ignored; if with proper names, printed following them.

(2) The various parts of hyphenated words are treated as separate words in the alphabetical sequence. It should be remembered that, in accordance with the conventions adopted, some Chinese proper names are written as separate syllables while others are written as one word.

(3) In the arrangement of Chinese words, Chh- and Hs- follow normal alphabetical sequence, and *ü* is treated as equivalent to *u*.

(4) References to footnotes are not given except for certain special subjects with which the text does not deal. They are indicated by brackets containing the superscript letter of the footnote.

(5) Explanatory words in brackets indicating fields of work are added for Chinese scientific and technological persons (and occasionally for some of other cultures), but not for political or military figures (except kings and princes).

TABLE OF CHINESE DYNASTIES

	夏	HSIA kingdom (legendary?)		c. −2000 to c. −1520
	商	SHANG (YIN) kingdom		c. −1520 to c. −1030
	周	CHOU dynasty (Feudal Age): Early Chou period		c. −1030 to −722
	春秋	CHOU dynasty (Feudal Age): Chhun Chhiu period		−722 to −480
	戰國	CHOU dynasty (Feudal Age): Warring States (Chan Kuo) period		−480 to −221
First Unification	秦	CHHIN dynasty		−221 to −207
	漢	HAN dynasty: Chhien Han (Earlier or Western)		−202 to +9
		HAN dynasty: Hsin interregnum		+9 to +23
		HAN dynasty: Hou Han (Later or Eastern)		+25 to +220
	三國	SAN KUO (Three Kingdoms period)		+221 to +265
First Partition	蜀	SHU (HAN)	+221 to +264	
	魏	WEI	+220 to +264	
	吳	WU	+222 to +280	
Second Unification	晉	CHIN dynasty: Western		+265 to +317
		Eastern		+317 to +420
	劉宋	(Liu) SUNG dynasty		+420 to +479
Second Partition		Northern and Southern Dynasties (Nan Pei chhao)		
	齊	CHHI dynasty		+479 to +502
	梁	LIANG dynasty		+502 to +557
	陳	CHHEN dynasty		+557 to +587
	魏	Northern (Thopa) WEI dynasty		+386 to +535
		Western (Thopa) WEI dynasty		+535 to +554
		Eastern (Thopa) WEI dynasty		+534 to +543
	北齊	Northern CHHI dynasty		+550 to +577
	北周	Northern CHOU (Hsienpi) dynasty		+557 to +581
Third Unification	隋	SUI dynasty		+581 to +618
	唐	THANG dynasty		+618 to +906
Third Partition	五代	WU TAI (Five Dynasty period) (Later Liang, Later Thang (Turkic), Later Chin (Turkic), Later Han (Turkic) and Later Chou)		+907 to +960
	遼	LIAO (Chhitan Tartar) dynasty		+937 to +1125
		West LIAO dynasty (Qarā-Khiṭāi)		+1125 to +1211
	西夏	Hsi Hsia (Tangut Tibetan) state		+990 to +1227
Fourth Unification	宋	Northern SUNG dynasty		+960 to +1126
	宋	Southern SUNG dynasty		+1127 to +1279
	金	CHIN (Jurchen Tartar) dynasty		+1115 to +1234
	元	YUAN (Mongol) dynasty		+1260 to +1368
	明	MING dynasty		+1368 to +1644
	清	CHHING (Manchu) dynasty		+1644 to +1911
	民國	Republic		+1912

N.B. When no modifying term in brackets is given, the dynasty was purely Chinese. Where the overlapping of dynasties and independent states becomes particularly confused, the tables of Wieger (1) will be found useful. For such periods, especially the Second and Third Partitions, the best guide is Eberhard (9). During the Eastern Chin period there were no less than eighteen independent States (Hunnish, Tibetan, Hsienpi, Turkic, etc.) in the north. The term 'Liu chhao' (Six Dynasties) is often used by historians of literature. It refers to the south and covers the period from the beginning of the +3rd to the end of the +6th centuries, including (San Kuo) Wu, Chin, (Liu) Sung, Chhi, Liang and Chhen.

ERRATA TO VOLUME I

p. 31, 4th paragraph, 2nd line. *For* minister *read* premier.
p. 36, 13th line. *For* 1280 *read* 1648.
p. 37, *substitute an asterisk for the dot at the junction of* S- *and* -êng.
p. 49, 5th paragraph, 9th and 12th lines. *For* Su *read* So.
p. 56, 'Provinces', after Chekiang. *For* 淅 *read* 浙.
'Place-names', after Amoy (city). *For* 夏 *read* 厦.
p. 57, line 31. *For* Liuchu *read* Liuchhiu. *For* 琉珠 *read* 琉球.
p. 67, 2nd line. *For* Black *read* Caspian.
p. 77, note d, 4th line. Huang chi tien *should be in italics.*
p. 78, 32nd line. *For* +937 *read* +907.
33rd line. *For* +1125 *read* +1144.
p. 79, 3rd paragraph, 5th and 6th lines. *For* first President *read* Secretary-General.
p. 93, footnote d. *For* pa *read* po.
p. 108, last paragraph, 2nd line. *For* Hsü Shih *read* Hsü Fu.
p. 122, 9th line. *For* last *read* second.
p. 125, 6th line from end. *For* Weichhi *read* Yüchhih.
p. 128, 6th line. *For* +610 *read* +636.
p. 137, 7th line. *For* a mud-boat *read* mud-boots.
p. 138, footnote b, line 1. *For* second *read* third.
p. 142, footnote 1. *For* 章 *read* 璋.
p. 145, 6th line. *For* volumes *read* chapters.
p. 154, 18th line. *For* late −5th century *read* mid −4th.
note g. *Read* Cf. p. 227 below.
p. 165, 16th line. *For* −296 *read* −245.
23rd line. *For* Su Tan-Chi *read* Su Ta-Chi.
p. 166, 9th line after illustration. *For* +2nd century *read* −3rd century.
p. 174, note f, 5 lines from end. *For* p. 185 *read* p. 186.
p. 180, 2nd paragraph, 10th line. *For* Fan Chhang *read* Fan fang.
p. 181, 6th and 13th lines. *For* Ganchow *read* Kanchow.
p. 182, note c. *For* p. 195 *read* p. 196.
p. 186, last paragraph, 5th line from end. *For* p. 204 *read* p. 205.
p. 192, fourth column from the right. *For* before +429 *read c.* +264.
p. 199, section (2), 7th line. *For* Yuan *read* Liang.
p. 200, 1st line. *For* Kallenberg (1) *read* Kallenberg, 1.
p. 202, note b, 3rd line. For *Hsiu* read *Hsin*.
p. 207, 6th line. *For* Chiu-Mo-Lo-Shih-Pho *read* Chiu-Mo-Lo-Shih.
p. 209, note b, 2nd line. For +663 read +647.
p. 212, 2nd line. For *Yo-Yang Tsa Tsu* read *Yu-Yang Tsa Tsu*.
p. 219, 3rd line after quotation. For *tshao-hsieh* read *tshao-shu*. For 草寫 read 草書.
last line but one. For *chih* read *chiu*.
note 8. *For* 炙 *read* 灸.
p. 220, 2nd paragraph, 8th line. *For* claims from them *read* claims for them.
p. 233, note g. For *Elegies*, III, 12 read *Elegies*, IV, 3.
p. 258, 11th line, first column. *For* +510 *read* +520.
18th line from bottom, first column. *For* +670 *read* +629.

14th line, second column. *For* +640 *read* +636.
27th line, second column. *For* +670 *read* +629.
p. 262, 4th line, first column. For *Yo-Yang Tsa Tsu* read *Yu-Yang Tsa Tsu.*
5th line, first column. *For* Yo-Yang *read* Yu-Yang.
p. 264, 14th line. For *Pieh Hao Su Yin* read *Pieh Hao So Yin.*
p. 269, under BIOT. *For* 1930 *read* 1940.
p. 277, under DE GROOT. *For* (1) *Die Hunnen* etc. *read* Vol. 1, *Die Hunnen* etc., *and for* (2) *Die Westlände* etc. *read* Vol. 2, *Die Westlände* etc.
under *Chhun Chhiu*. *For* 118 *read* 117.
p. 302, 10th line, second column. *For* Chiu-Mo-Lo-Shih-Pho *read* Chiu-Mo-Lo-Shih.
p. 317, 7th line from bottom, second column. *For* Weichhi *read* Yüchhih.
p. 318, 23rd line, second column. For *Yo-Yang Tsa Tsu* (The Yo-Yang Miscellany) read *Yu-Yang Tsa Tsu* (The Yu-Yang Miscellany).

ERRATA TO VOLUME II

p. 19, 2nd paragraph, 1st line. *For* (*c.* −298 to −238) *read* (*c.* −305 to −235).
p. 74, note b. *For* Dr E. Galazs and Dr Lu Bwei-Djen *read* Dr E. Balazs and Dr Lu Gwei-Djen.
p. 133, note c, last line. *For* +6th century *read* +7th century.
p. 164, note c. *For* Lin Tung-Chi *read* Lin Thung-Chi.
p. 305, 10th line. *For* −1150 *read* −1050.
13th line. *For* −1100 *read* −1020.
p. 360, 20th line. *For* Ming *read* Yuan.
p. 378, note e, 1st line. *For* Yao Shan Yün *read* Yao Shan-Yu.
p. 394, note b. *For* Laufer (6) *read* Laufer (8).
p. 486, note e, 2nd line. For *chen* read *chêng.*
p. 493, 9th line. For *miao wan wu erh yen chih yeh* read *miao wan wu erh yen chê yeh.*
p. 516, 3rd paragraph, 3rd line. *For* Ling *read* Lin.
p. 625, last line. *For* 1925 *read* 1935.
p. 633, 18th line. *For* HUANG HSIN-CHI *read* HUANG HSIU-CHI.
p. 634, under KARLGREN. *For* Shanghai *read* Peiping.
p. 637, 30th line. *For* LIN TUNG-CHI *read* LIN THUNG-CHI.

39079

ST. MARY'S COLLEGE OF MARYLAND
ST. MARY'S CITY, MARYLAND